FALLEN

NEVER FORGOTTEN

Vietnam Memorials In The USA

To Susan,

Ronny Ymbras
67·68
RVN

Ronny Ymbras, Matt Ymbras & Eric Rovelto

RU Airborne Inc.

RUAIRBORNE.COM

PUBLISHED BY

RU AIRBORNE INC.
57 SLEIGHT PLASS RD,
POUGHKEEPSIE, NY 12603

ISBN 978-0-692-60531-8
LCCN 2016900005

TEXT: ERIC ROVELTO
PRODUCTION: MATT YMBRAS
LAYOUT: MATT YMBRAS
CONCEPT: RONNY YMBRAS
INFO GATHERING/FACT CHECKER: RONNY YMBRAS
PHOTOGRAPHY: VARIOUS, MENTIONED IN CHAPTER

FOR INFORMATION ON HOW TO OBTAIN BULK COPIES OF THIS BOOK OR ONE OF OUR OTHER PRODUCTS OR SERVICES PLEASE CONTACT US ON OUR WEBSITE: RUAIRBORNE.COM OR BY TELEPHONE AT 914 497 7137.

PRINTED IN THE USA

Table of Contents

Foreword

by Ronny Ymbras

It was the mid-1960s. I wasn't thinking about the Vietnam War. I was playing ball and dating women. A football injury prevented me from running track in college so I started thinking about joining the service. I wasn't sure which branch, the Air Force, the Marines, or the Army's Green Berets. It was a very confused time and so was I. My brother Tom was a paratrooper with the 509th PIR in Germany in the early 1960s. It was inspiring to see his picture in those shiny jump boots and in his Class A uniform in Paris on leave. At that time there where a lot of movies that pumped up patriotism in the theaters, such as "The Longest Day", "A Bridge To Far" and "Battle Ground". Who knew that

seven days after my high school graduation I would join the Army to go airborne, never realizing how that decision would affect me for the rest of my life...

This book is not about the war, it's about the aftermath. When we returned home from the war we were treated like crap at the airports. People called us baby killers and threw dog shit at us. That really hurt deep inside. I thought that we as a unit, an army and a country did what we could to help another country from being bound to communism. I disconnected publicly for about 13 years even though I quietly stayed in touch with the TV News about what was going on in the war. But I didn't talk to vets about any of it. I was a little depressed and was separating from my friends. I was suffering from PTSD and did not know it.

Fourteen years after coming home President Ronald Reagan dedicated the Vietnam Memorial in Washington DC. From that moment on a lot of healing began, for not only myself but for millions of other veterans. I met up with some guys at The Wall and went back to a hospitality suite at the Hilton Hotel where I met a bunch of other guys from the 101st Airborne Division. It was a wonderful experience. We couldn't get enough of talking to each other about what had taken place back then and in the time since leaving Vietnam.

After that I started to go to Washington at least once a year. I joined the 101st Airborne Division Association. I later became chapter secretary of the Anthony C McAuliffe's NY NJ Chapter in 1988 and 1989. Being involved with this kind of organization taught me about publically honoring the guys we left behind and whose names are on The Wall. With participation in additional organizations such as The Disabled Veterans of America, Veterans of Foreign Wars, the American Legion, the Vietnam Veterans of America and the 82nd Airborne Division Association with whom I jumped, I found the worthiness of our service. I enjoyed talking to other vets. We spoke the same language, sometimes without literally saying a word. I spoke to high schools groups and church groups about the war. I got in touch with the national reunion group from the 101st and started attending those reunions and conventions. The healing was continual, seeing so many guys from so many outfits and so many eras. The Vietnam War has the largest number of surviving veterans. It was one of our saddest and longest wars. We served in good faith and received no gratitude or recognition upon returning home. That had a serious impact on our lives psychologically, emotionally and physically. There were no parades, marching bands or hanger receptions. Each soldier, sailor, marine or flyboy came home in solitude and faced disgrace. We did not lose the war, we lost the will of the country. It felt like a thankless task.

But never the less it always seems to come back to The Wall, the names on The Wall, the faces of honor in the memorial of the three soldiers standing across from the monument, the gifts left there by friends and family members, and the battle brothers who always show up rain or shine. The lines get longer and the

crowds get larger. The ceremonies become more meaningful and detailed, honoring particular battles and particular groups, placing recognition on these veterans who deserve and need it deep in their soul. If you go to The Wall on Memo-

rial Day or Veterans Day, you are sure to be moved by the somber celebrations and gatherings that take place. Many tears are shed and much pain exposed, only comforted by the fellow veterans themselves. And don't forget "Rolling Thunder" and the thousands of motorcyclists who pay their respects every year from all over the country. The Wall is an instrument of healing.

In our book "Fallen not Forgotten", we try to pay tribute to many of the local monuments around the country. It's impossible to get every one in every state in every county in every city. We sought to choose the state memorial, a memorial closer to people's hearts or a new memorial. Most memorials were put up 20 or 30 years ago and since then many facets have been added to the memorial sites. Helicopters have been put in place, some with mannequins and lights (Colorado's western slope and Delaware) and that sound of a HUEY that comes on when you walk under the chopper in New Jersey. Names have been added to the walls, some missing in action have been found and

have become KIA and changed on the walls and monuments around the country. Other monuments and statues have been added such as that of nurses who served in Vietnam, like the site in Somers, NY. Gold Star Mother Monuments, as well as war dog statues as in North Carolina and numerous other states. Many others have added lights for 24 hour visitation and placed memorial benches like Roswell, Georgia for a time of reflection. Most sites have been kept up very well, but some have not and become worn looking . For instance Philadelphia had to open up there site so vandals could be better seen, this in turn made it become one of the nation's finest and well-kept monuments after a large amount of private funding. Some memorials are at highway rest areas in remote places in the country such as the nation's first Vietnam State Memorial is in Sharon, Vermont. We try to represent each state memorial, The Wall in Washington and the Traveling Wall as best we could. By reaching out to many people as possible. Gathering this info was not easy, in many places we got a lot of information in others almost none. But we learned about about what has transpired since the first monument in Vermont to the latest one that was put up in Arlington, Texas, with the help of the South Vietnamese population. The inclusion of the Westchester County site in Somers

NY is personal for me. I was there for its dedication, carried the 101st chapter flag in the parade and honored three guys I went to school and played ball with. May they rest in peace, Pete Mitchell, Pete Bushey and Jeff Dodge.

When I meet a brother in arms at a Vietnam War Memorial and we spend some time together it is amazing how you gain a sense of peace, camaraderie and brotherhood without even knowing each other. As veterans in different outfits from the same era in a faraway place we focus on the troops who never made it home and whose names are written on The Wall.

Enjoy the read. Look up a name. It's all there for you, the veteran and the American public who now honors these incredible veterans from a time that a country was fighting against itself with these warriors in the cross fire of politics. Visit the memorial near you and honor those Veterans who made the ultimate sacrifice... AMEN

PHILIP I KRI K Jr

EDWIN L ARMSTRONG

JOHN H BARNES

PHILIP C BENN

JACK E DERRICO

ALVIN R GIBBLE

JOHN R HORTON

GLEN D HUBBARD

WAYNE D KRUEGER

ROGER M LINK

JOE W SMITH

HOYLE TERRY Jr

DON J YELVERTON

"Greater love has no one than this, that one lay down his life for his friends."

John 15:13

Photos by Nate Stirrat

The Lower Alabama Vietnam Memorial

2703 Battleship Pkwy, Mobile, Alabama, 36602

The Lower Alabama Vietnam Memorial is located near the USS Alabama and the infamous B-52 named Calamity Jane. The memorial was made to remember the service members from Alabama who were lost in the Vietnam War. Other war memorials are located close by giving visitors the opportunity to pay their respects to the veterans of all foreign wars. The memorial is open to the public, but for those interested, the community holds events at the location on Memorial Day, the Fourth of July, and Veterans Day.

The memorial consists of multiple elements. The bronze monument was made by Richard Arnold, while the sculpture of a POW/MIA Bracelet was made by Dr. Barry Booth. As a whole, the memorial was dedicated on November 7, 1998. The Lower Alabama Vietnam War Memorial also features a brick memorial wall with engraved bricks that are funded by donations people make in order to sponsor a brick for a loved one.

The Lower Alabama Vietnam War Memorial consists of two granite memorial walls, one with the names of the lost service members from Mobile and Baldwin counties and another with the names of those who passed or went missing from the whole state of Alabama. Standing guard over the two walls is a recently replaced Huey helicopter that was used during the war.

A brick wall honoring the fallen was also added to the memorial. The bricks feature the names of veterans as well as the names of people who donated money to support the memorial. This portion of the memorial was done originally as a fundraising project for the maintenance of the memorial. Those who'd like to help can obtain an order form for an engraved brick at the memorial's gift shop

In addition, there is a bronze sculpture of a Vietnam veteran holding a dog tag in his hand while he appears to be searching the memorial wall for the names of his lost friends. There is also a POW/MIA Bracelet statue to honor the soldiers who went missing in action or who were captured by the enemy.

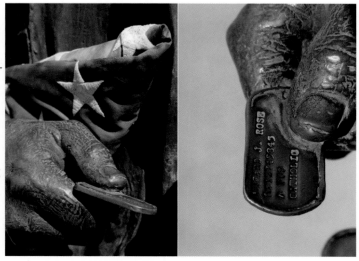

The Names of those from the state of Alabama
Who Made the Ultimate Sacrifice

ABRAMS TIMOTHY C JR
ABSTON JAMES ESTUS JR
ACTON MARION FRANKLIN
ADAMS JAMES CONRAD
ADAMS SPENCER
ADAMS WALTER LEE
ADAMSON LARRY ONEAL
ADDISON O'NEAL
ADKISON CARL ELMUS
AKINS ADRIAN ALAN
ALEXANDER BOBBY RAY
ALEXANDER DAVID J JR
ALLEN GRANVILLE JOEL JR
ALLEN JOHNNY JR
ALLEN ROBERT WARREN
ALLEN WILLIAM TERRY
ALLUMS ALLEN WAYNE
ALLUMS FREDERICK LARRY
AMBROSE EDWARD
ANDERSON ROY JR
ANDERSON WILLIAM ALLISON
ANDREWS CLIFTON BISHOP
ANDREWS COLEY L
ANDREWS HOWARD RIVERS JR
ARD HENRY
ARNOLD HAROLD
ARRINGTON SAMUEL W JR
ASH JOHN SILVY
ASKEW THOMAS EARL
AUSBORN DONALD EUGENE
AUSTIN WILLIE JR
AVERY JOHN MARK
AVERY RONNIE G
AYERS LESLEY STEVEN
BADGETT LEAGRANT
BAILEY JOHN HOWARD
BAKER ERNEST AUSTIN JR
BAKER JERRY SCRUGGS
BAKER MELVIN
BAKER RAYMOND DELMAR
BALDWIN LARRY GLENN
BALL JIMMY REX
BALLEW ARTHUR CLAY
BANNON PAUL WEDLAKE
BARBER CHADWICK MC FALL
BARBER ERNEST MC DONALD
BAREFIELD JAMES ARTHER
BARGE FREDERICK DOUGLAS
BARKSDALE WILLIAM HOW-
ARD
BARNARD LEWIS CECIL
BARNES RICHARD LOUIS
BARNETT DONALD EUGENE
BARRON DANNY LANCE
BARSOM GEORGE KASPER III
BARTLETT DONNIE STEPHEN
BARTON DAVID ALLEN
BASON WILLIAM ALFRED II
BASS ROY LEE
BATTLE HAROLD JAMES
BEARD ALEXANDER
BEARDEN LEE V

BEARDEN RICHARD DEWAYNE
BEASLEY GEORGE HUTCHIN-
SON
BEATON ROBERT LOUIS
BEAVERS JAMES DAVID
BECK JOHN THERON
BELL DAVID TOMIE
BELL JEROME
BELLOMY WILLARD GORDON
BELT ARTHUR LAVINE
BENJAMIN RICHARD
BENNETT DANIEL MURPHY
BENNETT JACOB
BENNETT MELVIN LESLIE
BENNETT WILLIAM GEORGE
BENOSKI JOSEPH JR
BENTFORD ANANIAS
BENTLEY COBBIE JAMES
BERRYMAN WILLIAM ERNEST
BEXLEY ROBERT EDWARD
BICE JIMMIE RAY
BILLINGS WILL DANNY
BINION THOMAS
BISHOP JAMES ARTHUR
BISHOP WOODROW WILSON
JR
BLACK JIMMY P
BLACK LARRY PAUL
BLACKMON DENNIS GLENN
BLACKSTON DONALD LAMAR
BLAKELY JOSSLYN F JR
BLALOCK JAMES TERRELL
BLANKENSHIP LARRY J
BOBE RAYMOND EDWARD
BOBO CHARLES GLEN
BOLES FLETCHER W II
BOOKER THOMAS ARTHUR
BOONE WILLIAM EDWARD IV
BOSTON DONALD EARL
BOUYER JAMES EARL
BOWLEN HAROLD DAVID
BOYD ANANIAS
BOYER LARRY EUGENE
BRACKIN RANDY CARROLL
BRADBERRY ARTHUR MILTON
BRADLEY RUBIN FLETCHER
BRADLEY TYRONE CARLOS
BRAMLETT HOWARD WAYNE
JR
BRANNON PAUL DEWITT
BREWER BOBBIE HERALD
BREWSTER OLLIS
BREWTON JOHN COOKE
BRIGHT BILLIE WAYNE
BROADHEAD JACK PHILLIP
BROCK EDWARD LEE
BROCK JAMES WALTER III
BROOKS JAMES FOSTER
BROOKS JESSIE MICHAEL
BROOKS WILLIAM LEE
BROWN BOBBY JAMES
BROWN CARL

BROWN CARL LEE
BROWN COLBURN
BROWN HUGH BERNARD III
BROWN JAMES HOMER
BROWN JAMES PHILLIP
BROWN JOHN HENRY
BROWN LARRY ALLEN
BROWN PAUL O NEAL
BROWN WALTER EVANS JR
BRUNSON ROBERT WADE
BRYAN FRANKLIN DELANO
BRYANT ROGER JERREL
BUCKLEY ROBERT EARL
BULLARD KENNY WAYNE
BUNCH CLAUDE MARVIN
BURCH KENNETH EDWARD
RAY
BURNETT DONALD FRED-
ERICK
BURNHAM DONALD DAWSON
BURT JAMES HOWARD
BUSBY MONTE REX
BUSBY SAM WILLIAM
BUSH JAMES
BUTTS LONNIE R
BYRD GUY ALBERT
BYRD LONNIE VERNON
CAHELA GERALD ALAN
CALDWELL HENRY JR
CALENDER MARSHALL LEE
CALHOUN FRANCHOT TONE
CAMERON BOBBY WAITS
CAMPBELL THOMAS ALLEN
CAMPBELL THOMAS JOHN D
CANADA GEORGE
CANIDATE JAMES ELLIS
CANNION WILLIAM
CANNON LARRY GEORGE
CANTRELL LEWIS EDWARD
CARDWELL HENRY WATERS
CARGILE CLAUDE HARMON
CARLTON JAMES EDMUND JR
CARMICHAEL ALFRED JR
CARPENTER THOMAS JR
CARSON CHARLES N JR
CARSTARPHEN HAROLD JR
CARTER HAMP JR
CARTER HARRY GIBSON
CARTER JERALD WAYNE
CARTER NATHANIEL EARL III
CARTHAGE OTIS JR
CARVER JERRY DEWAYNE
CARVER JERRY LEON
CARY WILLIE B
CASH BENNY DALE
CAULEY AUBREY
CAUSEY BEN ELMORE JR
CHAFFIN ALLAN RAY
CHAMBERS OSCAR EDWARD
CHAMBERS PAUL RICHARD
CHAMBERS ROBERT O
CHAMBLEE JIMMIE LADON

CHANDLER LARRY DELYNN
CHANDLER LEONARD ONEAL
CHANEY LARRY WILLIAM
CHAPMAN ANDEE JR
CHAPMAN WILLIE JAMES
CHASTANG JOHNELL LA-
VERNE
CHASTANT RODNEY RENE
CHILDERS PHILLIP DON
CHILDERS VIRGIL EUGENE
CHRISTIAN LYTELL B
CHURCHWELL DONALD
WALTER
CLANTON CHARLES BENJA-
MIN
CLANTON LOUIS LAMAR
CLARK BOBBY DEAN
CLARK DORIS WAYNE
CLARK FRANCIS EVERETTE
CLARK J C JR
CLARK LARRY GENE
CLARK RICHARD
CLARK ROBERT LEE
CLEMENTS RANDALL KELVIN
CLEMMONS JACK ELLIOTT
CLEVELAND ALBERT FRANK-
LIN
CLEVELAND BRENT PHILLIP
CLINE DONALD LEO
COATS DOUGLAS
COCHRAN AARON WASH-
INGTON
COERS BARRY BRYANT
COKER SAMUEL EARL
COLEMAN GEORGE
COLEMAN JIMMY LEE
COLLEY MICHAEL IRA
COLLIER JERRY LAMAYNE
COLLIER WILLIE LESTER
COLLINS JEROME LISTON
COLSTON LOUIS JR
COMBS JOHN BEECHLY
COMPTON JOHNNIE RAY
CONNELL OSCAR ALLEN
COOK LARRY DAVIDSON
COOK MARLIN CURTIS
COOPER HERMAN LEE
COOPER JEFFREY LANCE
COOPER WILLIAM MORRIS
COPELAND SAMUEL CHAM-
PION
COTNEY ELMER EUGENE
COTTON THOMAS WAYNE
COTTRELL WILLIE JAMES
COUGHLIN ARTHUR RAY-
MOND
COUSETTE JOSEPH
COX CHARLES EDWARD
CRAFT JAMES DAVID
CRAIG CLAYTON GEROME
CRAIN ROBERT VICTOR
CRAVEY JOHN JAMES
CREAGHEAD CLARENCE
CREAR WILLIS CALVIN
CRENSHAW JOE EDWARD
CRENSHAW WILLIAM AN-
DERSON
CREWS THOMAS FRANKLIN
CROCKETT JAMES LARRY
CRODY RONALD ISAAC
CROFFORD CLINTON E
CROW RODGER PINKNEY
CROWE RONALD GARY
CROWELL SAMUEL GERALD
CRUITT MICHAEL DOUGLAS
CRUMP JACK VANN
CULVER ALFONZIE
CUNNINGHAM CAREY ALLEN
CUPP ERNEST BRYAN
DAILEY FRANCIS EDWIN
DALHOUSE JOHN DUDLEY
DANIEL ELIJAH JR
DANIEL ROBERT G
DANIELS WALTER EUGENE
DARBY JIMMY EARL
DARWIN JAMES DAVID

DAVENPORT JAMES DONALD
DAVENPORT JAMES HUEY
DAVIE BOOKER T JR
DAVIES TIMOTHY SCOTT
DAVIS ALBERT
DAVIS CECIL LEROY
DAVIS CHARLES WILLIAM
DAVIS CURRY BARRY
DAVIS EMMETT LEE
DAVIS MICHAEL EDWARD
DAVIS WILLIAM FRANCIS JR
DAVIS WILLIE LOUIS
DAWES WILLIAM LE GRAND
DAY CHARLES TYRONE
DE PRIEST JOHN THOMAS
DEAS CHARLES MILTON
DEDMAN LESLIE PAUL
DEES EDGAR ALLEN JR
DEICHELMANN SAMUEL
MACKAL
DENNEY JIMMIE BRYSON
DENNIS JAMES WALTER JR
DENNIS WILLIAM EARL
DICKENS DAVID RUDOLPH
DILBECK LONNIE ADKEN
DILLARD THOMAS MANUEL
DILLWORTH EARL JR
DISMUKES RAYMOND KYLE
DIXON LEE ARTICE
DIXON LELAND FRANCIS
DIXON LEO CHESTER
DIXON LOUIS KRIMMIT
DOBYNES JOSEPH JAMES
DORAN THOMAS E
DORFMAN WILLIAM DAVID
DOROUGH JERRY EUGENE
DOWNS JAMES LARRY
DOWNS VERNON LEROY JR
DRYSDALE CHARLES DOUG-
LAS
DUBOSE FRED CLINTON III
DUCKWORTH JAMES EDWARD
DUFFY PATRICK EDWARD
DUKE BILLY WAYNE
DUNAWAY GORDON HERBERT
DUNCAN THOMAS DAVID
DUNN RALPH GERALD
DURALL ROBERT MICHAEL
EARNEST CHARLES M
EARP BILLY WAYNE
EATMAN EARNEST JR
EDWARDS FREDDIE LEE JR
EDWARDS JOSEPH WILLIAM
EGGLESTON ROBERT
EIDSON SAMUEL ARLEN
EILAND GRADY LOUIS
ELENBURG ALVIN ROBERT
ELENBURG JAMES WALTER
ELLARD CLAUDE ERNEST JR
ELLIOTT ERNEST LEE
ELLIS WILLIAM RICHARD
ELROD JIMMY CHARLES
ENFINGER KENNETH EARL
ENGLISH DENNIS LAVERNE
ERVIN CLIFFORD LEON
ERWIN EARL JR
ESTES DONALD CARTHEL
EVANS ANDREW C
EVANS DOUGLAS MC ARTHUR
EVANS HERMAN
EVANS JAMES LARRY
EVANS JERRY THOMAS
EVANS JOHNNIE LEE
EVANS RODNEY JOHN
EVANS THOMAS C
FAULKS WILLIE JAMES
FERGUSON WILLIAM EDWIN
FIELDS JAMES LEWIS
FIELDS JAMES RONALD
FIELDS WILLIAM MICHAEL
FINCH LAMONT WILKERSON
FLOYD JOHN DOUGLAS
FORD CHARLES WALKER
FORD CLIFFORD EUGENE JR
FORD EDWARD
FORD GLENN EDWARD

Alabama

FORD ROBERT
FOREMAN AUBURN WOOD JR
FORRESTER JOEL WAYNE
FOSTER JOE ALBERT JR
FOWLER ROBERT ALLEN
FOX AMOS OLIVER
FOX CARL JAMES
FRANKLIN CLARENCE RICHARD
FRANKLIN IRA MELTON JR
FRANKLIN JAMES ANTHONY
FREEMAN DAVID HAROLD
FREEMAN GARRY DON
FREEMAN JIMMY GRANT
FROWNER EDWARD
FRYE BOBBY SAM
FULGHUM JACKIE JUNIOR
GAINES ALLAN JOSEPH
GAINES WORDELL
GAMBLE JAMES HENRY
GANTT SAMUEL LEE
GARDNER FRED MICHAEL
GARDNER ROBERT EUGENE
GARDNER ROY EDWARD
GARDNER WILLIAM HUGH JR
GARDNER WILLIE JR
GARNER JACKIE WAYNE
GARNER WILLIE FRANK
GARRISON CARL FRANKLIN
GARTH RAYMOND
GASTON ROSS ALLEN
GAUSE BERNARD JR
GAUTNEY EARL
GENTLE CLYDE GLENN
GENTRY OSCAR JR
GIDDENS HORACE GILBERT JR
GILDER LEWIS C
GILES LEONARD EARL
GILES WILLIE JR
GILL ROBERT EARL
GILMORE RONALD
GIPSON CHARLES DONNETTE
GLASS ARTHUR
GLENN RICHARD J
GLOVER FREDDIE BEE
GLOVER LAWRENCE WALTER
GLOVER ROBERT BRANCH
GODWIN JOHNIE REESE JR
GODWIN WILLIAM RILEY
GOHAGIN JAMES RAYFORD
GOLDEN GEORGE KENNETH
GONZALEZ LARRY EUGENE
GOODWIN PAUL VENON
GORDON ERNEST LEE
GORDON THOMAS LESLIE
GOREE CARLTON TRAVIS
GRACE LARRY EDWARD
GRAHAM ALLEN UPTON
GRAHAM ROGER LEE
GRAY DELACY
GRAYSON RAMON LEE
GRAYSON RONNIE PAUL
GREENE JAMES ETHERIDGE JR
GREER LARRY WAYNE
GREGORY WILLIAM ROBERT
GRIMSLEY LEE ELDRIGE
GROVE RICHARD CRAIG
GULLEY PERCY LEE JR
GUNN TERRY SIDNEY
GUNNELS MICHAEL DAVID
GUNTER MELVIN WISTER
GURLEY THOMAS
GUY BENNY ROSS
GUYER RONALD LYNN

HADLEY VERLON
HALL ADOLPHUS JR
HALL BYRON ROYCE
HALL JEFFERSON DAVIS
HALL LAVLE JIMMY
HALL RONALD HUGH
HAMILTON EUGENE DAVID
HAMILTON ULYS FORD
HAMM DONALD CURTIS
HAMMAC JOSEPH EARL
HAMMER BILLY GENE
HAMMONDS JAMES ROBERT
HAMNER CHARLES
HAMNER JOHN ALBERT
HAMNER THEODORE S III
HANDLEY HOWARD BROWN
HANKINS JOEL RICHARD
HANSEN LOWELL C
HANSON CLYDE WENDELL
HARDY WARREN JR
HARGROVE JAMES MABRON
HARGROVE OLIN JR
HARPER RICHARD EARL
HARRELL RONNIE
HARRIS BENJAMIN
HARRIS CARL COLEMAN
HARRIS CLEVELAND SCOTT
HARRIS EDWARD LEON
HARRIS FRANK CAY
HARRIS GARY BLUITT
HARRIS JERRY LEE
HARRIS NATHANIEL
HARRISON CLEOPHIS
HARWELL GARY CURTIS
HASTY WILLIAM DONALD
HATFIELD BILLY T
HAVEARD DAVID MARSHALL
HAWK JAMES RICHARD
HAWKINS DANNIE LEE
HAWTHORNE ANDREW GEORGE
HAYES HARRY ELLIS
HAYES JOHNNY VANCE
HAYES LAWRENCE ALLEN
HEAD MARVIN JR
HEARD JAMES ROBERT JR
HEARD ROBERT LOUIS
HEATH CHARLES EDWARD
HEATH KENNETH EDWARD
HEGLER MOSE JR
HENDON JOHN LEWIS
HENDRIX PAUL GEORGE
HERMAN LAWRENCE JOHN III
HERRING DAVID BOUNDS
HERVAS AARON KAMALA
HEYER EDWARD ELIAS
HICKS PRENTICE WAYNE
HICKS WOODIE LEE
HIGGINBOTHAM RICHARD LEE
HIGGINS JERRY WAYNE
HILL EDDIE LEE JR
HILL JERRY DWAIN
HILL THOMAS MARVIN JR
HILL WILLIAM B JR
HILLEY ROBERT LEE
HILLMAN JOSEPH III
HILYER BROADUS DALE
HIMES MICHAEL BRUCE
HITT ROY MARVIN JR
HOCUTT LARRY KEITH
HODGES BENNIE EDWARD
HODGES FERMAN BOBBY
HODGES JAMES DALE
HOGAN JERRY FRANKS
HOGANS WALTER JIM
HOLBROOK HORACE ALVIE
HOLBROOK VERNON GLEN
HOLDEN ALFRED JEFFERSON
HOLKEM JIMMY RAY
HOLLAND JAMES LARRY
HOLLAND ROBERT JOSEPH
HOLLAWAY PHILIP STEPHEN
HOLLIMON BILLY MICHAEL
HOLLIS JAMES AUGUSTUS
HOLLOWELL WILLIAM BYARD
HOLMES EARNEST PAUL JR
HOLMES LEONARD HUGH
HOOD CHARLES EARNEST
HOOD EUGENE
HOOVER WILLIE JR
HORSLEY LARRY FRANK
HOSE JOHN WALLACE JR

HOULDITCH JULIUS C JR
HOWARD CLARENCE WILLIAM
HOWARD EDWARD EMANUEL
HOWARD JAMES J
HOWARD THEODORE
HOWELL PRESTON LEE
HUBBARD ALFRED WILLIE
HUBBARD ROBERT WALKER
HUDGENS JOHN WAYNE
HUDSON JIMMY DALE
HUDSON JOHNNY
HUFF JAMES EDMOND
HUGGINS BOBBY GENE
HUGHES ERROL ARTHUR
HUGHES MACKLIN OTIS
HUGHEY EDWARD WENDELL
HUIE ROBERT ANDREW
HULLETT NATHAN EARL
HUNT LARRY FRANK
HUNT WILLIAM DICKSON
HUNTER MILTON CHARLES
HURD LAWRENCE ADAMS
HURST ROOSEVELT JR
HURST WILLIAM JOSEPH
HYATT WAYNE REUBEN
INGRAM CHARLIE BERNARD JR
ISAAC JAMES EDWARD JR
ISAAC WILL JR
JACKSON ADAM
JACKSON BILLY LEE
JACKSON CRAWFORD JR
JACKSON DAVID ROLLAND
JACKSON GEORGE EMMETT
JACKSON THOMAS CLAYTON
JACKSON THOMAS WINFORD
JACKSON WILLIAMS OTIS
JACOBS PERRY OWEN
JAMES GERALD
JAMES RAYMON HORACE JR
JAMES WILLIE JR
JENKINS FRANK PAUL JR
JENKINS JOYFUL J
JENKINS WILLIAM CLARENCE
JOHNS CAREY LEE
JOHNS MICHAEL WAYNE
JOHNSON ALLEN LOUIS
JOHNSON ARMSTEAD
JOHNSON CURTIS
JOHNSON FREDDIE LEE
JOHNSON HARRY J
JOHNSON HARVEY DOUGLAS
JOHNSON JAMES EARL III
JOHNSON JAMES LARRY
JOHNSON JEROME
JOHNSON JERRY REED
JOHNSON JIMMY EARL
JOHNSON JOE EDWARD
JOHNSON JOSEPH WALLACE
JOHNSON LILE LAMAR JR
JOHNSON OBBIE

JOHNSON RALPH EDWARD
JOHNSON RICHARD S JR
JOHNSON SANFORD STEVEN
JOHNSON THOMAS ALLEN
JOHNSON WILLIAM HORACE JR
JONES ALBERT JUNIOR
JONES BRUCE DALE
JONES JACK MARION
JONES JAMES GRADEY
JONES JIMMIE LEE
JONES JOE LOUIS
JONES JOHN HENRY
JONES JOHN HENRY JR
JONES JOHNNY MACK
JONES LARRY NEAL
JONES LOUIS HENDERSON
JONES NEIL WADE
JONES RALPH WAYNE
JONES THOMAS DEWITT
JOSHUA JAMES EDWARD JR
JUNE JEREMIAH
KEEFE FLOYD MILTON
KEENE GLEN CAMERON JR
KELLEY LARRY DEAN
KELLEY WILLIAM ROBERT
KELLY DONALD LYNN
KELLY JAMES MATHEW
KENDRICK JAMES CALVIN
KENNEDY JAMES
KENNEDY ROBERT JR
KENNEY JOSEPH HAYDEN
KERSEY MAX DUANE
KETCHIE SCOTT DOUGLAS
KEY ROGER EUGENE
KIGER JAMES ANTHONY
KING ARGESTLAR JR
KING DOYLE GAYLON
KING FELIX DELOACH JR
KING ROBERT HENRY
KINNEY RANDLE
KIRBY RANCE A
KIRKSEY ROBERT LOUIS
KISTLER RUSSELL WILFORD
KITCHENS FRANK MCCRARY JR
KIZZIAH JERRY WAYNE
KNIGHT MACK ARTHUR
KNIGHT RALPH MAX
KNIGHTEN JACKEY VAN
KUGLER TERRY GUS
KUHSE MICHAEL DARRELL
LAFFERTY DAVID NELSON
LAGRAND ROBERT HENRY
LAIS ROBERT WALLACE
LAMB HOWARD SIDNEY
LANE GERALD BRUCE
LANEY BILLY RAY
LANGLEY FRANCIS LEE
LARRY JOHN DAVIS JR
LASSETER KENNETH RAY
LASSITTER JOHN IRVING
LATTA CHARLES R

LAVENDER ROBERT EDWARD
LAWRENCE GARRY FRANK
LAWRENCE GREGORY PAUL
LAY WILLIE RAY
LAYTON JAMES RICHARD
LEATHERWOOD JAMES
LEATHERWOOD WILLIAM E JR
LECATES ROBERT BURTON
LEDBETTER DAVID WAYNE
LEDGER GILBERT
LEE CHARLIE FRANK
LEE GEORGE BLUE
LEE JAMES FRANKLIN
LEGG JOHN DUANE
LELAND LEROY JR
LEONARD MATTHEW
LEONARD SIDNEY LAMAR
LESLIE ROGER LAMAR
LEWIS GRADY LEONARD
LEWTER STANLEY REED
LIKELY JAMES THOMAS
LILLEY JOSEPH EMMETT
LINDSAY JAMES RICKEY
LISENBY DONALD EUGENE
LITTLE HENRY LEON
LITTLE JOHN EDGAR
LITTLE WALLACE SYLVESTER
LITTLEFIELD ROBERT HENRY
LLOYD RODNEY DALE
LLOYD RONALD EDWARD
LOCKETT CLEO
LOCKHART CLARENCE
LOCKRIDGE JACK RAY
LONG CHARLES EDWARD
LONG MICHAEL DAVID
LONG RAY FRANK
LOONEY MILFORD JR
LOTT JUNIOR EDWARD
LOVE J C
LOVELADY RONALD DAVID
LOVELL JAMES RICHARD
LOVETT TERRY WAYNE
LOWE LOUIS CARDELL
LOWERY DALTON BUSTER
LUCY ARCHIE THOMAS
LUEALLEN EDGAR BOWIE
LUNDY LONNIE EUGENE
LYLE JOHN BRUCE
LYLE LARRY VANN
LYNN STEPHEN DAVID
MADDEN JAMES FLOYD
MAGNUSSON FRED WAYNE
MAHER CHRISTOPHER LORING
MALEC PAUL WILLIAM
MALLORY DAVID ALLEN
MANESS JAMES EMORY
MANGRUM GEORGE THOMAS
MANN CARL WILLIAM
MANNING WILLIAM TERRY
MARSH BOBBY JOE
MARSHALL JAMES CONRAD
MARSHALL JAMES HENRY

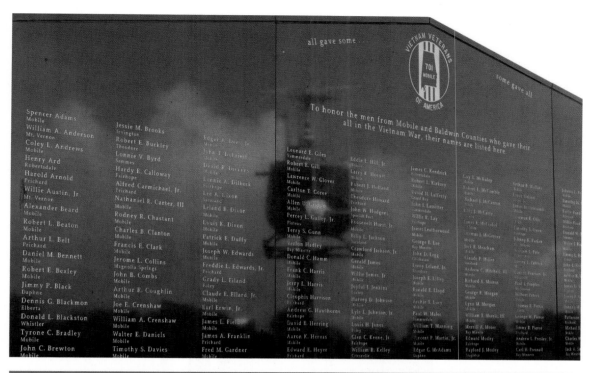

NAILEN JAMES PATRICK
NALL JOHN TRUMAN
NARAMORE DAVID ZOHLEEH JR
NATHAN RALPH EUGENE
NEELY DAN LEE
NELSON CHARLES
NELSON DARROL OREN
NELSON LEROY
NELSON ROBERT THOMAS
NELSON ROGER TILTON
NEWCOMB JAMES DWIGHT
NICHOLAS TOMMY L
NICHOLS LARRY J
NISEWONGER EDWARD EARL
NIX EDWARD LEWIS
NOBLE ALLEN EARL
NORRIS VAN ALLEN
NORSWORTHY JIMMY LAYNE
NORTHINGTON WILLIAM CLYDE
O NEAL VICTOR HUBERT
OAKES CHRISTOPHER COLUMBU
ODOM JOHN THOMAS
ODOM WILLIAM CLINTON JR
OGLETREE YOUNG DAVID
OLIVER HENRY MC CARTY
OLIVER ROGER LEE
OTIS SHERMAN ELDRIDGE
OVERTON WILLIAM HILLIARD
OWENS DAVID RAY
OWENS DEWEY RAY
OWENS THOMAS EARL
PACE DANNY WAYNE
PAGE ROY DONALD
PALMER GILBERT SWAIN JR
PALMER WILLIAM HERSCHELL
PARKER CARTER JR
PARKER JOHNNY KENDRICK
PARKER LONNIE RONALD
PARKER UDON
PARKER WILLIAM AVALON
PARR RONALD EUGENE
PATE WILLIAM LAWRENCE
PATRICK DANNY LEON
PATTERSON BOOKER T JR
PATTERSON SAMUEL LEE
PATTILLO RALPH NATHAN
PATTY DUDLEY RANDOLPH
PAYNE LAWRENCE EDWARD
PEARSON CARL OSCAR JR
PENCE JAMES THOMAS
PENDERGRASS VERNON FRANKL
PENDLEY WILLIAM GRANT
PENLAND MARVIN KENNY
PENN ROOSEVELT FRANKLIN
PEOPLES EDDIE DONALD
PEOPLES HOWARD GREGORY
PEOPLES PAUL JOSEPH
PERKINS JAMES BARNEY
PERKINS WARDELL
PERRY GEORGE EDWARD
PERRY JAMES EARL
PERRY ROBERT LEWIS
PETERS WILBERT
PETERSON JULIUS LEE
PETTIS BILLY WAYNE
PETTIS THOMAS EDWIN
PHILLIPS ELBERT AUSTIN
PHILLIPS HOWARD EDWARD
PHILLIPS JAMES LESTER
PHILLIPS LEONARD
PHILLIPS ORMAN DORR
PHILLIPS WILLIAM RUSSEL
PIERCE EDWARD DAVIS
PIERCE GEORGE WASHINGTON
PIERCE JIMMY RAY
PIKE EDWARD MORRIS
PIPER EDWARD ROGER
POLK KENNETH ERBIE
POOL HAROLD LAVEROL
POOLE CONRAD EARL
POOLE THOMAS DEWITT
POPE CHARLES DEAN
POWELL ABRAHAM
POWELL ALBERT CHARLES
PRESLEY ANDREW LEE JR
PRESLEY MELTON HOWARD
PRESNALL CARL HAMBY
PRESSLEY CORNELIUS
PRICE MARLIN LADON

MARTIN CHARLES EDWARD
MARTIN HUBERT WILLIAM
MARTIN JOHNNY COCHRAN
MARTIN RUFUS MICHAEL
MARTIN VINCENT PATRICK JR
MARTINDALE PAUL VAUGHAN
MARVIN JOSEPH
MARZENELL EDWARD JR
MASON EARNEST LEE JR
MATHIS DAVID LINWOOD
MATTHEWS CHARLES TONEY
MATTHEWS ROBERT L
MAXWELL WILLIAM ELBERT
MC ADAMS EDGAR GREGORY
MC AULEY GUY THOMAS
MC BRIDE GRADY E III
MC BRIDE HERMAN ALVIN
MC CAIG ROBERT LEE
MC CAIN MICHAEL CLINTON
MC CALL CLIFFORD
MC CAMBLE ROBERT LEE
MC CARTY BILLY JOE
MC CARY CHARLES WAYMAN
MC CLENDON JOHN NEWT JR
MC CORKEL JAMES EDWARD
MC CUTCHEN GEORGE
MC DONALD DAVID LETCHER
MC DONALD JOSEPH WAYNE

MC DUFFIE LARRY RAY
MC GEE DANNY ALBERT
MC GEE ROBERT LEWIS JR
MC GEEVER THOMAS JOSEPH
MC GINTY CALVIN A JR

MC GOWAN IRA EUGENE
MC HANEY CARL JAMERSON
MC KELVEY JAMES DANIEL
MC LEMORE TAYLOR HENRY
MC LEOD CHARLES WILLIAM
MC LEOD LAMAR
MC LESTER SHERMAN DOUGLAS
MC LIN LOUIS WILLIAM III
MC MANUS CHARLES VERNE
MC MILLIAN SOLOMON LEON
MC MURRAY JOHNNIE RAY
MC MURTREY WILLIAM NEWTON
MC NABB JERRY WAYNE
MC VAY JOHN EARL
MCAPHEE SAMUEL LEE
MCCAIN MARVIN RAYMOND JR
MEACHAM JACK BENNIE
MEADS HERBERT LYNN
MENEFEE GENE ALLEN

MERRITT ALLEN TWIGGS IV
MICHAEL DON LESLIE
MICKENS EDDIE JAMES
MIDDLEBROOKS ROBERT NEAL
MILAM LEWIS EDWARD
MILES ELIJAH JR
MILLER CLAUDE PAUL
MILLER FRANK LEONARD III
MILLER GREEN EDWARD JR
MILLER JOHNNIE ROBERT
MILLER ORMOND MITCHELL
MILLER RICHARD CHARLES
MILLIGAN JOHNSON MARCUS
MILLS ROBBIE RAY
MIMS KENNETH EDWARD
MINOR MATTHEW JR
MINOR RANDY MICKEL
MITCHELL ANDREW C III
MITCHELL EUGENE EMMETT
MITCHELL HOMER JR
MITCHELL JOSEPH ROBERT JR
MITCHELL PERRY ADKINS
MOIREN RICHARD ALLEN
MOLTON KENNETH WAYNE
MONCRIEF JAMES RAY
MONCUS BENNIE RAY
MONROE WILBER DEAN

MONTGOMERY DONALD LEE
MOON JERRY RUDOLPH
MOONEY JAMES
MOORE DALLAS HENRY
MOORE JOSEPH M
MOORE LEONARD DAVID
MOORE ROBERT LOUIS
MOORE ROY LEE
MORGAN CARL EUGENE
MORGAN GEORGE ROBERT
MORGAN JESSE FRANK
MORGAN LYNN MARTIN
MORRIS WILLIAM T III
MORRISON BILLY JOE
MOSER MERRILL ANDREW
MOSIER ROBERT KEAL
MOSLEY EDWARD
MOSLEY RAYFORD JUNIOR
MOSS JACK JR
MOTLEY JOHN LARRY JR
MOTON EDDIE LEE JR
MULLINS ARTHUR BRENT
MUNDY ROBERT HAL
MURFF WILLIAM EDWARD
MURPHREE IRA JEROME
MURRAY DARNELL PATRICK
MURRELL ERVIN JEROME
MURRY EUGENE

Alabama

PRICE MICHAEL KEATON
PRIDGEN GARY MORGAN
PRINCE GARRY GARNETT
PRINCE HARRY GORDON JR
PRYEAR JOHNNIE LEE
PUCKETT JEAN WAYNE
PURCELL LARRY JOE
PURSER CHARLES EDWARD
PYLE TIMOTHY HOWARD
RAGSDALE JOSEPH MICHAEL
RAIFORD MARK PHILLIP
RAINWATER JAMES ALVIN JR
RAND EARLIE
RANDALL JAMES ARTHUR
RANDALL SIMON
RASPBERRY LAWRENCE
RATCLIFF JACKIE LEE
RAWLINS JAMES PATRICK
RAYNOR JAMES DANIEL
REAID ROLLIE KEITH
REED WILLIE
REYNOLDS JOHN HENRY
RHODES RAY ANTHONY
RICE ROBERT IVAN
RICH RONALD DUDLEY
RICHARD JERRY GORDON
RICHARDS ROBERT
RICHARDSON DONALD
WILLIAM
RIDDLE BOBBY
RIDGEWAY WILLIE JAMES
RITCH JOHN GWIN
ROBERSON JOSEPH THOMAS
ROBERTSON BENJAMIN F JR
ROBERTSON JOHN HARTLEY
ROBINSON CHARLIE JR
ROBINSON HERMAN RAY
ROBINSON JIMMIE LEE
ROBINSON JOHN LEO
ROBINSON WILLIE JAMES
ROBISON LARRY WAYNE
ROCKETT ALTON CRAIG JR
RODGERS BOBBY RAY
ROGERS CLAYTON GEORGE JR
ROGERS RONNIE VAGO
ROGERS WILLIAM T IV
ROSS LUTHER JULIAN JR
RUFF RONALD CALVIN
RUFFIN JAMES THOMAS
RUNNELS GLYN LINAL JR
RUSH THEODORE MARSHALL
RUSHING MICHAEL GEAN
RUSSELL CHARLES TERRY
RUSSELL FLOYD H JR
RUTHERFORD MICHAEL
TOXEY
SABLAN FRANK AGUAN
SALTER CHARLES LOWELL
SALTER DWAYNE LAMONT
SALTER FRANK DEMON
SALTER ROBERT WAYNE
SAMPLES LARRY JUNIOR
SANDERS GLENN EDWARD
SANDERS JESSIE FRANKLIN
SANDERS RODNEY RAYFORD
SAPP WILLIAM DANIEL
SARGENT GEORGE THOMAS
JR
SAWYER PAUL LEWIS JR
SCARBROUGH ARTHUR
BENJAMI
SCARBROUGH ENNIS RALPH
SCHMALE WILLIAM OTTO
SCHOFIELD CECIL CLAYTON

SCHOOLEY JAMES DANIEL
SCOTT JAMES FRANK
SCOTT JIMMIE L
SCOTT JOHNNY MAJOR JR
SCOTT PATTERSON JR
SCOTT TRAVIS HENRY JR
SCROGGINS DOUGLAS SIDNEY
SEABORN WILLIAM HERMAN
JR
SEAWRIGHT WILLIAM J JR
SELLERS CHARLES RAYFORD
SELLERS MELVIN LOUIS
SELLERS PHILIP DOYLE
SENN THOMAS LARRY
SEWELL JOHNNIE BRUCE
SEWELL LORENZO
SHAFER GLENN WESLEY
SHARPE RONNIE
SHAW JAMES DOUGLAS
SHAW WILLIAM MARSHALL JR
SHEDD ALTON
SHEFFIELD ANTHONY D
SHELTON CHARLES HOWARD
SHELTON JOSEPH HENRY III
SIMMONS OBIE CLYDE
SIMPKINS WILMER FRANKLIN
SIMS CLINT JOSEPH
SIMS MICHAEL EUGENE
SIMS THOMAS JAMES
SISK HARRY DUNCAN
SKINNER JAMES ALLEN
SMILEY GEORGE ROBERT
SMITH AUTHOR C
SMITH CHARLES WARREN
SMITH CLIFTON BRADLEY
SMITH CLINTON DANIEL
SMITH DAVID WILLARD
SMITH DONALD WAYNE
SMITH GARY
SMITH HENRY BEALL JR
SMITH HURLEY ALVIN
SMITH JACK A
SMITH JAMES BUFORD
SMITH JAMES DAVID
SMITH JEFFERY W
SMITH JIM L
SMITH JOE WILKINS
SMITH JOHN LEE
SMITH LOUGHTON
SMITH MALCOLM CARLIS
SMITH MOSE JR
SMITH NORRIS RAY
SMITH RICKY GENE
SMITH RONNIE WAYNE
SMITH ROY
SMITH SAMUEL DAVID
SMITH SAMUEL THOMAS JR
SMITH THOMAS TIMOTHY
SMITH WILLIAM CARY
SMITH WILLIAM HOYT
SPEAKS MAC WAYNE
SPENCER CORDELL
SPIVEY HARLEY EDWIN
SPRADLIN GERALD DOUGLAS
STABLER JOHN LESLIE
STALLINGS JOHN LARRY
STAMEY JIMMY EDWARD
STAMPS JOHNNY GREEN
STANDRIDGE PAUL RICHARD
STANLEY JAMES MITCHELL
STANLEY JAMES STEVEN
STANLEY JOE HARRY
STEELE TOWNSER JR
STEPHENS GERALD WAYNE
STEPHENS JAMES ROWE
STEPHENS LARRY EUGENE
STEPHENSON WAYMOND
NELSON
STERNS RANDOLPH JOEL
STEWART CHARLIE ACES JR
STEWART SAM WILLIAM
STINSON WILLIAM SHERRIL
STOFFREGEN ROY DIXON
STOKES KENNETH LARRY
STONE ROGER ALLEN
STOREY CHARLES WILLIAM
STORY J C
STOVALL CHARLES ALLEN
STOVES MERRITT III
STRACNER WILLIAM ELLIS
STRIBBLING GWYMAN
STUDDARD FINIS RONEY
STURMA CHARLES FRANK
SUGGS JAMES DAVID

SULLIVAN ARNOLD HOSEA
SUMMERLIN J C
SUTTLE WILLIAM EARL
SUTTON JAMES KENNETH
SUTTON TRAVIS ROBERT
SWAIN LEE WESLEY JR
TANTON CHARLIE THOMAS
TAYLOR CHARLES STOCKTON
TAYLOR CLARENCE
TAYLOR CLIFTON THOMAS
TAYLOR DE WAYNE
TAYLOR ELMER JACK
TAYLOR JIMMY B
TAYLOR ROBERT HILDRETH
TAYLOR STEVIE
TERRY ARIE
TERRY BILL HENRY JR
TERRY WILLIAM JAMES
THACKERSON WALTER A JR
THOMAS HOWARD RAY JR
THOMAS JIMMY RAY
THOMAS LARRY BENJAMIN
THOMAS ROY EDWARD
THOMAS TENNYSON AARON
THOMAS WILTON HERMAN
THOMPSON BENJAMIN A JR
THOMPSON FARLEY DEE
THRIFT FRED LEWIS
TILLER ROBERT
TINDALL BRUCE GARLAND
TISDALE HENRY CARLOS
TODD CARLOS FRANKLIN
TOLBERT REGINALD GAY
TOLBERT RODERICK KEN-
NETH
TOLSMA RAYMOND EARL

TOMLIN BARRY COLEY
TOMLINSON GARY PRESTON
TOSH JAMES CHRISTOPHER III
TOUART JOHN ELLIOTT
TOWNES MORTON ELMER JR
TOWNSEND ROOSEVELT
TOYER LEE ARTHUR
TRAINHAM JOHNNY WILLIAM
TRAYLOR FRED EDWARD
TRAYLOR WAYNE MCKEN-
NELY
TROUPE HERMAN LEE
TRUELOVE JAMES MELVIN
TURBERVILLE CHARLES
WAYNE
TURNER ANDERSON
TURNER CLAUDE TYLER
TURNER DAVID LEE
TURNER GEORGE ALLEN
TURNER LOUIS G
TURNER WILLIAM OLIVER
UNDERWOOD DANIEL
LEDARE
UPNER EDWARD CHARLES
UPTAIN DAVIS
VINSON HENRY MITCHELL
VINSON WALTER WAYNE
VIX STEPHEN AUGUST JR
VOYLES FLOYD
WADDLE SAMMIE WAYNE
WADE STEVEN MICHAEL
WADSWORTH HARRY MAR-
SHALL
WALBRIDGE GEORGE WILCOX
WALDREP JIMMY RAY
WALDROP RAYMOND CLAR-

ENCE
WALKER CHARLES CLARENCE
WALKER CHARLIE LEWIS
WALKER CLIFFORD C
WALKER ROBERT LEE JR
WALKER WILLIE TERRY JR
WALLACE FRANKIE LEE
WALLACE GARY FRANK
WALLACE WILLIE LEWIS
WARD CARL GENE
WARD WAYNE LEVOYER
WARE DON HAMUEL
WARE MACK ARTHUR
WARE MATTHEW
WASHINGTON WILLIAM F JR
WATKINS HAROLD EUGENE
WATKINS JOEL KEITH
WATSON JOHNNY MACK
WATTS JOHN RAYMOND
WATTS ROY DELANO
WAXTON WILBERT EUGENE
WEED MORGAN WILLIAM
WEIMORTS ROBERT FRANK-
LIN
WELBORN MELVIN O NEAL
WELLS BENJAMIN GARETH
WELLS BILLY
WESLEY MARVIN JR
WESTBROOK DENNIS FRANK-
LIN
WESTBROOK ROY THOMAS
WHAN VORIN EDWIN JR
WHITE CHARLES EDWARD
WHITE DONALD RICHARD
WHITE JAMES DAVIS
WHITE JOHN OLIVER

WHITE LEAMUEL ARTIS
WHITE MICHAEL EUGENE
WHITE RAYMOND
WHITE ROBERT WAYNE
WHITE TED ARNOLD
WHITESIDE JOHN CURTIS
WIGGINS DAVID ROGER
WIGINTON GARY RAY
WILDER STEVE CLIFTON
WILKINSON JOSEPH E III
WILLIAMS BOBBIE LEE
WILLIAMS DONALD LEE
WILLIAMS DONALD
WINSLOW
WILLIAMS GENE WILLIAM
WILLIAMS JIMMY LAVERNE
WILLIAMS JOHNNY JR
WILLIAMS LARRY DOUGLAS
WILLIAMS MELVIN JAMES
WILLIAMS MELVIN JOE
WILLIAMS PAUL EDWARD
WILLIAMS ROBERT CLEVEN
WILLIAMS ROBERT JOHN
WILLIAMS SHERMAN ELLIOT
WILLIAMS THADDEUS E JR
WILLIAMS TOMMIE LEE
WILLIAMSON HOWARD
LANIER
WILLIAMSON JAMES CALVIN
WILLIS LARRY WAYNE
WILLIS RAYMOND CONLUIS
WILSON DALE KEITH
WILSON FRED
WILSON GERALD W
WILSON LEVI JAMES
WILSON ROBERT THOMAS

WILSON WILLIE GENE
WINCHESTER LARRY ALDEN
WINSTON JAMES CLENNON
WINSTON WILLIAM CURTIS
WINTER JOHN WESLEY
WISE RICHARD MARVIN
WOOD DAVID MITCHELL
WOOD LARRY DAVID
WOOD LEWIS EDWARD
WOODALL CHARLES MINOR
JR
WOODARD HARRY DONALD
WOODS ABRAHAM
WOODS JAMES ARLIE
WOODS JERRY OTIS
WOOLEY DONALD
WOOLSEY HILTON EDWARD
WORRELL HURSTON EDWARD
WRIGHT JAMES EARL
WRIGHT JERDY ALBERT JR
WYROSDIC WILLIAM EVER-
ETT
YEEND RICHARD C JR
YERION JEFFERY ALLEN
YOUNG CLAUDE
YOUNG WILLARD FRANK
YOUNGBLOOD JIMMY DEAN
ZEIGLER EUGENE

Photos by Daryl Pederson

Anchorage Veteran's Memorial
Delaney Park, 10th ave and L St, downtown Anchorage

The Anchorage Veteran's Memorial, originally dedicated in 1952, is located in Anchorage, Alaska between 9th and 10th Avenues on L Street. It was created in order to honor the local veterans from all wars including those lost and missing due to the Vietnam War. In 1987, the memorial as it stands now was completed thanks to the Veterans' Action Committee. After Vietnam Veteran and Anchorage mayor Tony Knowles pushed to have a better veterans' memorial, a committee was formed and an updated memorial to be located on I Street between the same avenues began to come to life.

In order to keep costs low, the veterans' memorial was complete mainly through volunteer labor and materials including services from the Alaska National Guard and the veterans' community. Fundraising also helped pay for the memorial ($115,000 was raised). To properly maintain the memorial and revamp it for the current generation, $1.5 million was said to be needed. Thanks to joint efforts from a Blue Ribbon Task Force formed by Mayor Mark Begich in 2007, the necessary funds were found through multiple allotments from the Municipality of Anchorage as well as the fundraising efforts of volunteers. The renovation was finished in October of 2013 with the rededication ceremony taking place in May of 2014.

The original memorial was made for quiet reflection and consisted of a flagpole situated in a flower bed, bronze plates honoring the Alaskan soldiers who were lost in WWII, the Korean War, the Vietnam Wars, Grenada, the Gulf War, and the wars in Iraq and Afghanistan. There is also a statue of a soldier honoring WWII veterans as well as a memorial dedicated to the Purple Heart awardees.

In the renovation, a new monument known as the Fallen Warrior statue was added to the memorial. The well-known image of a soldier's boots, rifle, and helmet are featured in the monument. Two screen panels were also added to the memorial.

14

The Names of those from the state of Alaska Who Made the Ultimate Sacrifice

ANDERSON THOMAS EDWARD
BANTA MICHAEL DEAN
BARR EDWARD NASUESAK
BARR THOMAS M
BAUER RICHARD GENE
BETTS LARRY LE ROY
BROWN CHARLES EDWARD
BROWN DAVID DEE JR
BULLOCK GARY EDWIN
CHILDERS WILLIAM STEVEN
CHMIEL DONALD GEORGE
COOK CLINTON ARTHUR
COX DANIEL FRANKLIN
DUNCAN WILLIAM BRADLEY
EATON WILLIAM ALBERT
ELISOVSKY DAVID HENRY
ELLIOTT ROBERT THOMAS III
FERRY DAVID LYNN
FLEENER NICK ULYSSES
GAMBLE CHARLES F JR
GRIFFIN DALE ANTHONY
GULLIKSEN HOWARD WAYNE
HARMON DANIEL LEE
HIBPSHMAN WILLIAM EARL
HOKE MICHAEL THOMAS
HORN JERRY VERNE
INT-HOUT KURT
IVEY SAM
JONES WAYNE ELMER
KETZLER GILBERT JR
KILBUCK GEORGE GREGORY
KITO DONALD HARRY
KOSLOSKY HOWARD MARK
LA BELLE KERMIT HAROLD JR
LANG MICKEY DANIEL
LAPE DAVID ALEN
LEE ROBERT EDWARD
LINGLEY NORMAN LEWIS
MC INERNEY PATRICK M
MILLHOUSE KENNETH BRUCE
PAULSEN WARREN
PIASKOWSKI WILLIAM
FRANCI
PLETT LARRY JOE
PRENTICE DAVID SHELTON
RAINEY LLOYD STEVEN
RICE ANDREW WILLIAM JR
RICHARDSON FLOYD WHIT-
LEY
RIDLEY NORMAN FRANKLIN
ROBISON DONALD ROBERT
SANDERS DONALD RAY
SIMEONOFF FREDERICK M
SPERL DONALD WALTER
THOMPSON WILLIAM AR-
THUR
VANDERPOOL JOE WAYNE
WALTERS WILLIAM FRANCIS
WHITNEY ARTHUR JOSEPH JR
YOUNG DAVID REESE JR

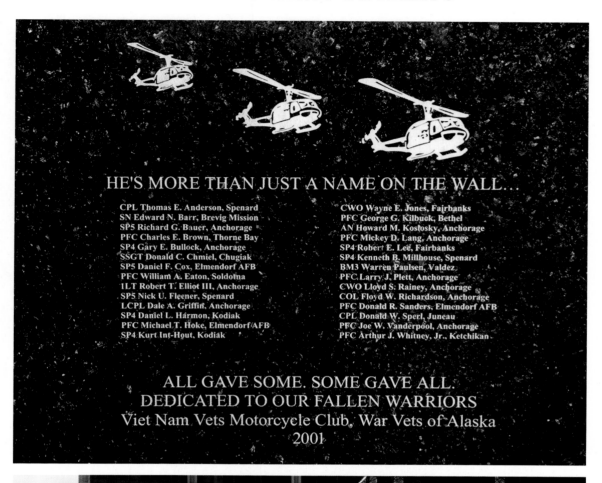

HE'S MORE THAN JUST A NAME ON THE WALL...

CPL Thomas E. Anderson, Spenard
SN Edward N. Barr, Brevig Mission
SP5 Richard G. Bauer, Anchorage
PFC Charles E. Brown, Thorne Bay
SP4 Gary E. Bullock, Anchorage
SSGT Donald C. Chmiel, Chugiak
SP5 Daniel F. Cox, Elmendorf AFB
PFC William A. Eaton, Soldotna
1LT Robert T. Elliot III, Anchorage
SP5 Nick U. Fleener, Spenard
LCPL Dale A. Griffin, Anchorage
SP4 Daniel L. Harmon, Kodiak
PFC Michael T. Hoke, Elmendorf AFB
SP4 Kurt Int-Hout, Kodiak

CWO Wayne E. Jones, Fairbanks
PFC George G. Kilbuck, Bethel
AN Howard M. Koslosky, Anchorage
PFC Mickey D. Lang, Anchorage
SP4 Robert E. Lee, Fairbanks
SP4 Kenneth B. Millhouse, Spenard
BM3 Warren Paulsen, Valdez
PFC Larry J. Plett, Anchorage
CWO Lloyd S. Rainey, Anchorage
COL Floyd W. Richardson, Anchorage
PFC Donald R. Sanders, Elmendorf AFB
CPL Donald W. Sperl, Juneau
PFC Joe W. Vanderpool, Anchorage
PFC Arthur J. Whitney, Jr., Ketchikan

ALL GAVE SOME. SOME GAVE ALL.
DEDICATED TO OUR FALLEN WARRIORS
Viet Nam Vets Motorcycle Club, War Vets of Alaska
2001

Photos by Dave Moody

Vietnam Memorial in Wesley Bolin Memorial Park

205 S 17th Ave, Phoenix, Arizona, 85007

The Wesley Bolin Memorial Plaza is a park located in Phoenix, Arizona at the entrance of the State Capitol. It houses multiple memorials including the Vietnam Memorial. The Plaza itself was created in 1978 but the Vietnam Memorial, which honors the Vietnam War veterans from Arizona, wasn't dedicated until 1985. As a whole, the memorial is made up of a bronze sculpture along with an honor roll monument for the fallen and missing servicemen.

The bronze sculpture known as the "The Fallen Warrior" was created by bronze sculptor Jasper D'Ambrosi before his 1986 death. It was his aim to present the soldiers in his sculpture as young men who were not only heroes but victims of the Vietnam War. While D'Ambrosi's statue was cast in bronze, the memorial wall consists of 10 black granite columns that were engraved with the names of the deceased and missing Vietnam veterans. The whole memorial was administered by the city of Phoenix in 1985 before the actual dedication occurred in November of the same year.

The black granite slabs that make up the honor roll monument of the Vietnam War Veterans Memorial feature 623 Arizonians who never made it back from their tour in the South Pacific. There is also a timeline that outlines the Vietnam War from the original Ho Chi Minh-led uprising in 1941 to the 1975 fall of Saigon.

The most noteworthy feature, however, is the D'Ambrosi bronze statue. "The Fallen Warrior" is made up of three life-size Vietnam soldiers clad in combat gear. One soldier is wounded and outstretched on the ground as a second soldier props his head up as he kneels beside his fallen brother. A third soldier stands in front of the other two and pulls at the outstretched hand of the kneeling soldier.

The memorial is complete with dedication plaques at the standing flagpoles.

The Names of those from the state of Arizona Who Made the Ultimate Sacrifice

ADAIR DALLAS TYLER JR
ADIKAI ALVIN JR
AGUAYO OSCAR JR
AGUILAR JAMES DANIEL
AGUIRRE FILBERTO JR
ALBRIGHT BUCK EDWARD
ALDAY FRANK TISNERO
ALLEN HENRY GERHARDT
ALLENBERG JAMES PATTEE
ALLRED ORIN LARRY
ALVAREZ ESTEBAN MORALES
ALVAREZ JOSE RICARDO L
ALYEA WALTER JOHN
ANDERSON GARY JOHN
ANDERSON LEE E
ANDERSON ROBERT KEITH
ANDRADE RICHARD
ANDREWS ROBERT LEE JR
ANDREYKA THEODORE E JR
APOLINAR FORTINO JAMES
APPLEGATE PAUL ORBEN
ARAGON JOSEPH MANUEL
ARKIE VALLANCE GALEN
ARLENTINO DUDNEY NELSON
ARMSTRONG WALTER LEE
ARNN JOHN OLIVER
ARNOLD ROY LEE
ARRINGTON JOSEPH PHILLIP
ASPLUND MARCUS RAY
ATWELL WILLIAM ALBERT
AUSTIN ALBERT DELGADO JR
AUSTIN OSCAR PALMER
BABCOCK RONALD LESTER
BAILEY LARRY EUGENE
BAKER VINCENT B
BANKS HENRY DUANE
BAREFIELD BOBBY JOE
BARNES ALLEN ROY
BARNES WILLIAM ACKER
BARNETT BENJAMIN FRANK-
LIN
BARRIGA ARTURO
BARTON JAMES EUGENE
BATEMAN MARK ANDREW
BATES BRIAN WILLIAM
BATES PAUL JENNINGS JR
BAUTISTA JESUS ESTRADA
BAY RONALD STEPHEN
BAYNE MICHAEL JOHN
BEACH SAM FESTIS JR
BECKER JAMES FRANCIS
BEGODY HAROLD L

BELL OSCAR CHARLIE JR
BENNETT ANTHONY LEE
BENNETT WAYNE
BENNETT WILLIAM RAY-
MOND
BENTON ROBERT DANIEL
BIA MICHAEL HOWARD
BILDUCIA CONRADO FRAN-
CISC
BILES MICHAEL LYNN
BILLIE LARRY ROGERS
BIRCH JOEL RAY
BISJAK HOWARD ROBERT
BLACKWATER DWIGHT
THOMAS
BLACKWELL KENNETH G
BLANCHETTE MICHAEL R
BLUDWORTH MICHAEL
VERNON
BOEHM BRADLEY WAIN-
WRIGHT
BOHANNON EDWARD JEAN
BOIS CLAIRE RONALD ALAN
BOJORQUEZ SISTO BO-
JORQUEZ
BOND DAVID ARTHUR
BONILLAS GUILLERMO TRUJIL
BORG MICHAEL ROYCE
BORIEO RICHARD DAVID
BOYCE JOHN FRANKLIN
BOYKINS RANDY RONELL
BRAXTON JAMES HAROLD
BRENNER LARRY RAY
BRINKOETTER JAMES ALBERT
BROADSTON SCOTTY RAY
BROWN ANTHONY BARTOW
BROWN DONALD ALAN
BROWN JAMES RONALD
BROWN JAMES SCOTT
BROWN RICK SAMUEL
BROWNLEE KENNETH DUANE
BRYANT WILLIAM J JR
BURNS JOHN PATRICK
BURNSIDE DERRILL LEE
BYASSEE NORMAN KELLY
CAAMANO LEONARD OLGUIN
CALDERON RICHARD TORRES
CALDWELL JOE
CAMPBELL STEVE DANIEL
CANALES DAVID JOSEPH
CANNON FRANCIS EUGENE
CARABEO LEONARD

CARDENAS JOE CANDELAR-
IA R
CARLBORG ALAN GEORGE
CARLSON RICHARD THEO-
DORE
CARRASCO RALPH
CARRILLO JOE JR
CARTER JACK DAVID
CASEY TOM GAYLE
CASSELL ROBIN BERN
CATES WILLIAM LLOYD
CHAIRA FRANCISCO PERAZA
CHAVES ALLEN FRED
CHAVEZ ROBERT L
CHESTER ALVIN
CHIAGO GREGORY BUR-
KHART
CHILDERS MELVIN RONALD
CHRISTMAN JERRY NOLAN
CHRISTMAN LAWRENCE PAUL
CHRISTOPHER ADOLPHUS
CINTINEO GIACOMO JAMES
CLARK JESSE LEWIS II
CLARK RICHARD GARLAND
CLASSEN EARL THOMAS
CLAW PETER YAZZIE
CLIFFORD WILLIAM HENRY
COBB THERON WALLACE
COCHRAN LARRY ALAN
COFFEY STEVEN LYNN
COFFIN JEFFREY ALAN
COKER JAMES LEE
COLE JON
COLLINS BILLY G
COLLUMS BOBBY G
CONLEY GREEN
CONRY JOHN TIMOTHY
CONTRERAS RICHARD
AGUIRRE
COOK JIMMY LEE
COOK RONALD JOHN
COONS CLIFFORD KENT
CORNELIUS JOHNNIE CLAY-
TON
CORNWELL LEROY JASON III
CORPUS DAVID JOSEPH
COURTRIGHT MICHAEL
EUGENE
COX JOHN DAVIES JR
CRIBB EDWARD BERNARD
CRICK DALE EUGENE
CRIDER RUSSELL DUANE

CRONIN JAMES RUSSELL
CROOK ELLIOTT
CROW JAMES DENNIS
CRUZ TONY
CURRAN JAMES R
CURRAN JOHN DEHAAS
CZECHOWSKI JOHN LOUIS JR
DALE BENNIE
DALE CHARLES ALVA
DANIEL FRED JACOBO
DANIELS RUSSELL GLEN
DARLING DENNIS THOMAS
DAVIDSON CHARLES ALLEN
DAVIS DONALD ALLEN
DAVIS FLOYD ROBERT
DAVIS GARY LYNN
DAVIS HARRY K
DAVIS JAMES MARK
DAVIS LEONARD DOUGLAS
DAVIS LEONARD RAY
DAVIS WESLEY WAYNE
DAW JERRY LORENZO
DE MASI MICHAEL ARMOND
DEAL FLOYD ANDREW
DEES JERRY RICHARD
DELANEY JAMES PERRY
DELLECKER HENRY FLOYD
DELOZIER JOHN ADRIAN
DENIPAH DANIEL DEE
DENNEY ALAN WAYNE
DENT BRUCE JAMES
DESCHAMPS RAMON
DIANDA CASIMIRO
DICKERSON WILLIAM CLINT
DIMMER MICHAEL PHILLIP
DOMINGUEZ FRANK L
DOMINGUEZ MICHAEL J
DORSETT ROY GEREAD
DOWNING JOHN LESLIE
DRANE JOHN WILBUR
DRAPER ROBERT DALE
DRAPER WILFRED
DRYE JACK LEE
DUNAGAN MICHAEL DENNIS
DUNDAS STEVEN WILLIAM
DUNLAP WILLIAM CHARLES
DUSCH GEORGE EDWARD
ECKLUND ARTHUR GENE
ELIAS JUAN ANGEL
ELLIS DONALD RAY
ELMORE DONALD ROBERT
EMBREY RICHARD LYNN
ENGELMAN RICHARD
GEORGE
ENOS LEONARD ARVIN
ENRIGHT ROBERT EARL
EPPERSON ROY ALLEN
ERNSBERGER RANDALL
WAYNE
FELKINS WILBURN DANIEL
FELTS DAN OWEN
FENTER CHARLES FREDERICK
FENTON JAMES WILLIARD
FIESLER ROBERT NATHAN
FIGUEROA ANTHONY H JR
FINCHER DONALD B
FLORES MANUEL SOLARES
FLORES ROBERT LEE
FOOTE WALTER BRUCE
FOURMENTIN GREGG R
FRANCISCO PATRICK PHILLIP
FRAZER FREDRICK HARRY
FRENCH DENNIS
FRITZ LEONARD EUGENE
FULTS LAWRENCE ARTHUR JR
FURCH JOE HENRY
GALLEGO MICHAEL
GALVEZ TOM
GARCIA ARTHUR MARTINEZ
JR
GARCIA CLIVE JR
GARCIA JUAN MANUEL
GARCIA LARRY ROBERT
GARCIA STEVEN VARGAS
GAULT CLINTON MONROE JR
GAYNE JEFFREY LEE

GERAGHTY MERRILL THOMAS
GIFFORD WILLIAM GARY
GILLILAND DENNIS ELBERT
GODFREY JOHNNY HOWARD
GOFF STANLEY ARTHUR
GOMEZ JESSIE YUTZE
GONZALES GERARDO
HOLQUIN
GONZALEZ JOSE LUIS
GOSNEY DURWARD DEAN
GOSS CLARENCE EUGENE
GOSSETT WILLIAM O
GRAVES GARY EVERETT
GRAY DALE ALAN
GRAY JAMES
GREEN JERRY L
GREEN KENNETH LEON
GRIJALVA GERONIMO LOPEZ
GRISBY GARY BERNARD
HAIN ROBERT PAUL
HALL GARY C
HAMBLIN RONALD B
HAMILTON WILLIAM EUGENE
HANKINS BRUCE LYNN
HANNA DONALD RAY
HARBOTTLE JAMES LAVERN
HARDY CHARLES MC RAE
HARDY LINCOLN
HARMON CAREY DEAN
HARRINGTON JOHN DEE
HARRISON CHIP RUSSELL
HASKINS JOHN MERLE
HATHAWAY STEPHEN WORTH
HAWK RANDALL LEE
HAWKINS ROBERT LEWIS
HAWKINS THOMAS G
HAWTHORNE GENE
HAYNES JOHN ONA
HEISSER KENNETH HAROLD
HENLING RICHARD RAY
HENRY DANIEL LEE
HENRY GEORGE D JR
HEPNER STEPHEN THOMAS
HERNANDEZ ALEX JAMES
HERNANDEZ HUMBERTO
ROBLES
HERRERA FRANK G
HICKS DONALD
HICKSON LEONARD MARTIN
HIGGINS EDWIN RAY
HIGGINS ROY JOHN JR
HILL JOHN WALTER III
HILL MICHAEL ALAN
HILLER MICHAEL JAMES
HILLS CLARENCE S
HIMES JACK LANDEN
HOAGLAND GEORGE A III
HOBBS LARRY L
HOGAN GARY LEE
HOLCK PAUL ALAN
HOLLISTER JOHN FREDERICK
HOLLMEN WILLIAM HARRY
HOLMES ALLAN WILLIAM
HOLTON STANLEY GENE
HOOD TERRANCE LEE
HOPKINS LEROY JR
HOPPER EARL PEARSON JR
HOWELL JAMES RILEY
HUELSKAMP RONALD JAMES
HUFF JACKIE EUGENE
HUGHES JOHN HOWARD
HULSE RICHARD DAVID
HUNT BOB CLARENCE JR
HUNT LEIGH WALLACE
HURD JERRY ALAN
HUSKON BENNY LEO
INGRAM ALLEN WADE
JACKSON RALFORD JOHN
JAMISON JAN DWAIN
JAUREGUI DAVID CRUZ
JEFFORDS DERRELL BLACK-
BUR
JENSEN FRANK ALFRED
JIMENEZ JOSE FRANCISCO
JOHNSON DARRELL LEE
JOHNSON DAVID ALVIN

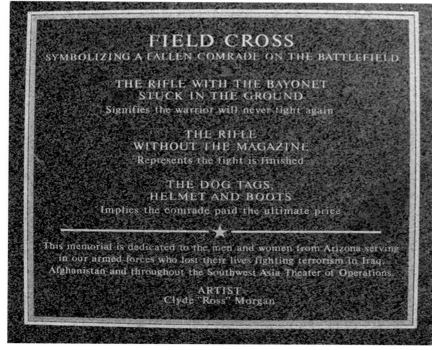

FIELD CROSS
SYMBOLIZING A FALLEN COMRADE ON THE BATTLEFIELD

THE RIFLE WITH THE BAYONET
STUCK IN THE GROUND
Signifies the warrior will never fight again

THE RIFLE
WITHOUT THE MAGAZINE
Represents the fight is finished

THE DOG TAGS,
HELMET AND BOOTS
Implies the comrade paid the ultimate price

This memorial is dedicated to the men and women from Arizona serving
in our armed forces who lost their lives fighting terrorism in Iraq,
Afghanistan and throughout the Southwest Asia Theater of Operations.

ARTIST
Clyde "Ross" Morgan

Arizona

JOHNSON DAVID ARTHUR
JOHNSON ROG
JOHNSTON DAVID WILLIAM
JONES JERRELL RAY
JONES MICHAEL BRUCE
JORDAN LAWRENCE WILLIA
JORDAN WILLIAM ARLIN
KALINA EDWARD CHARLES
KEE WILSON BEGAY
KELLEY VICTOR BRUCE
KERSEY J D WILLIAM
KING BRADFORD STANLEY
KINSEY JOE EDWARD
KIRKSEY JAMES WALTER
KISER WILLIAM BROOKS
KNEPPER WARREN ORISON JR
KNIGHT BILLY MELTON
KOESTER JOEL FREDERICK
KONIGSFELD PHILIP LORNE
KONOPA CARL RAYMOND
KOSIK JOSEPH III
KRISELL JAMES LEE
KROEHLER KENNETH RICH-
ARD
KUVIK GENE LAWRENCE
LACKEY VERNON HARVIC
LADENSACK ROBERT JOSEPH
LAMBERT STEVE NATHANIEL
LAMPRECHT MARK AUGUST
LASER JAMES DALE
LASZLO JOSEPH
LATHON JAMES
LAUFFER BILLY LANE
LAURENCE JOE ROBERT
LAWSON WILLIAM ROY
LEE BILL GREGORY
LEE DENNIS VARIS
LEE JAMES ANDREW
LEE NED
LEE ROBERT
LEGLEU SAMUEL
LEHMAN MILLARD WESLEY
LENTZ EDWARD MARTIN
LERMA GERONIMO
LEYVA FRANK MONTANO
LITHERLAND THOMAS
EDWARD
LITTLETON JOHN WAYNE
LITZLER JAMES WILLIAM
LONG ROBERT ORRIE
LONGDAIL DENNIS LEE
LOPEZ EDDIE CESARIO
LOPEZ PERFECTO NUNEZ

LOPEZ ROBERT DIAS
LOPEZ ROBERT FRANCISCO
LUBBEHUSEN GERALD
MARTIN
LUCAS MICHAEL RICHARD
LUKENBACH MAX DUANE
LUMPKINS LARRY RICHARD
LUND ARNOLD ATWOOD
LUTRICK DARRELL LEROY
LYONS JOSEPH WALTER
MACHADO FRANCISCO JR
MADRID ERNEST
MAKIN ALLEN THEODORE II
MALONEY OSCAR
MANSFIELD PATRICK LEROY
MARQUEZ GERALDO
MARRIETTA HAROLD JOSEPH
MARTIN CLIFFORD B JR
MARTIN EDWIN WOODS JR
MARTIN WILLIAM EVERETT
MASON DANIEL
MC CHESNEY JOHN T III
MC CULLOUGH RONALD
JAMES
MC GINNIS CHRISTOPHER
MAR
MC INTOSH RANDALL LEE
MEDEGUARI RENE
MELIUS JOHN STERLING
MENDENHALL THOMAS DEAL
MENDEZ SALVADOR JOE
MENDOZA ALBERT MANUEL
MERRETT JAMES ALLEN
MESQUITA FERNANDO
OLIVAS
MESSER DARRYL
MEYER ARTHUR WILLIAM
MILLER MARSHALL GREGORY
MILLER MICHAEL ANDREW
MOLINA SIMON ROSALINO
MONCAVAGE DAVID JOHN
MONCAYO JOSE ROBERTO
MONROE CHARLES CALEB
MONTANO FRANCISCO
ANDREW
MONTGOMERY WILLIAM
JOHN
MONTIJO MICHAEL
MONTOYA MANUEL TOMAS
MOORE ELGAN LEROY
MORALES ANTONIO RUIZ
MORENO ALFRED JR
MORENO JOSE LUIS
MORENO MIGUEL ORTEGA
MORTON DOUGLAS GEORGE
MORTON MATTHEW EDWARD
JR
MUIR JAMES
MUNOZ ROJELIO OLIVAN II
MURRIETTA FRANK A
NEAD ELWOOD FRANKLIN JR
NEWTON MELVIN DEW
NEWVILLE VAN HAROLD
NIETO JESUS DIEZ JR
NORVELL RAYMOND FRANK
NORVELLE CLYDE L JR
NORZAGARAY SALVADOR
LOPEZ

O BRIEN ARTHUR ALEN
O BRIEN WILLARD DONALD
O CONNOR MORTIMER
LELANE
OCHOA JESUS
OCHOA RALPH RICHARD
OGDEN RUSSELL KEVIN
OGILVIE GORDON WILSON
OLEA FRANCISCO HERRERA
OLSON ERICK OWEN
OROZCO TONY SALAZAR JR
ORTIZ ANTONIO OLIVAREZ
OVIEDO MICHAEL LESLIE
OWEN JOHN WILSON
OXLEY JAMES EDWARD
PAHISSA WILLIAM ANTHONY
PARKER MICHAEL LEE
PARKS CALVIN ALAN
PARTON JOHN EDWARD
PASHANO JACK POOLA
PATTEN JIMMIE
PATTERSON DANIEL CHARLES
PATTERSON WILLIAM WESLEY
PAULSEN MICHAEL
PEARSON BRUCE FULLER
PENA JOHN L
PENNINGTON FRED MELVIN
PERKINS DAVID DRAKE
PERRY ELMER REID
PETE FRANKLIN DANNY JR
PICKETT DARREL MONROE
PIKE DENNIS STANLEY
POLANCO JOSE YBARRA JR
POLESETSKY BRUCE
POROVICH STEVE
POWELL LARRY DEAN
POWERS LOWELL STEPHEN
PULS ROBERT LAWRENCE
QUESNEY JOSE MANUEL
QUINTERO FERNANDO
MENDOZA
RAMIREZ ARMANDO
RAMIREZ RICHARD JR
RAMIREZ ROBERTO MAN-
DOZA
RANDALL JOHN MICHAEL
RAPER ALVIN LOUIS
RASMUSSEN JOHN WILLIAM
RAZO FRANK AMBROSE
REED GUY RICHARD
REID JON ERIC
REINHOLD MICHAEL J
RENDON GUADALUPE
REYES GILBERT
REYNOLDS JAMES STEPHEN JR
RHINE RICHARD ALLEN
RHOADES FREDERICK PAUL
RICE DENNIS KELLY
RICH JOHN ALLAN
RIDENOUR EDWIN MICHAEL
RITER JAMES LEE
RIZO ALBERT MARTINEZ
ROBERTS JAMES AARON F JR
ROBERTS JOHN J
ROBINETTE CHARLES ED-
WARD
ROBINSON CHARLES DAVID
ROBINSON JERRY LYNN

RODRIGUEZ JOE STELO
RODRIGUEZ PAUL M JR
ROMERO MICHAEL ANDREW
ROMERO ROBERT ANTHONY
ROMERO ROBERT LUIS
ROMO JOHN ROGER
ROSENSTOCK MARK LAMONT
ROUSH SAMUEL EMMERSON
ROWLAND RICHARD LEE JR
ROYBAL THOMAS MICHAEL JR
RUIZ PETER GEORGE
SAAVEDRA ROBERT
SALCIDO GEORGE ARTHUR
SALINAS ANTONIO MONTANO
SANCHEZ PAUL FRANK
SANCHEZ RUDOLPHO
SANDOVAL RANDALL JACK
SANTA CRUZ JOSE ANGEL
SANTOR ROBERT PAUL
SCARBOROUGH JAMES
ARTHUR
SCHAFFER DAVID THOMAS
SCHIBI JAMES LEE
SCHOENEWALD DAVID
CHARLES
SCHRADER RONALD BRUCE
SCHWARZ DONALD EDWIN
SCORSONE GEORGE ANTHO-
NY
SCOTT GREG BRADFORD
SEDILLO JUAN NATIVIDAD
SEGAR CALVIN RUSSELL
SERNA HERMAN
SERSHON LAURENCE G
SEXTON JEFFREY ROSS
SHARPE WILLIAM A JR
SHAVER CLINTON WILLIAM JR
SHEVLIN HUGH JOHN
SHOCKEY ROBERT L
SHRUM WILLIAM LAWRENCE
SILVAS JORGE ALVARADO
SIMMONS BILLY JOE
SIMMONS JAMES CHARLES
DAN
SINN BRADLEY LOUIS
SIQUEIROS MANUEL MEN-
DOZA
SKAGGS HAROLD ALONZO
SLIM JIMMIE FARRELL
SLOCUM WILLIAM SCOTT
SMITH ANDREW WILLIAM
SMITH TERRENCE GLEN
SMOLIK VERNON KENNETH JR
SOLANO MIKE ANTHONY
SOLIS DAVID TOBIAS
SOTO BRAVIE
SOUTHARD CHARLES A III
SOWERS CHARLES HENRY II
STADDON PETER BRUCE
STALEY ROBERT E
STANDS DANIEL GILBERT JR
STANLEY RICHARD ALLEN
STEVENS JOHN WARNER JR
STEWART GREGORY WILLIAM
STOCKETT RICHARD LEE
STOWE JEFFREY CHARLES
STRAHL RICHARD WILLIAM
STUBBS BILLY RAY

SUDDUTH ROBERT THOMAS
SUTER JERRY TIMOTHY
SUTTON FRANK
TARKINGTON CURTIS RAY
TEDRICK WARREN GAMBIEL
JR
TEN HUSKIE YAZZIE B
TERRELL ALVA RAY
TERSTEEGE PAUL FRANCIS
THOMAS JAMES CALVEN
THOMPSON RICHARD LEWIS
JR
THRASHER JOHN DOUGLAS
THURSBY RICHARD ALLEN
TILLOU JOHN FREDERICK JR
TODD GEORGE ALBERT
TOMPKINS PHILLIP WARREN
TORRES MANUEL ROMERO
TOSCHIK MARK JOSEPH
TOTH WILLIAM CHARLES
TRUJILLO JACOB ROMO
TSOSIE ALBERT
TSOSIE LEE DINO
TUOHY JACKIE ALLEN
TURNER JOHN MICHAEL
URIAS DAVID SOQUI
VALDEZ MODESTO
VALENCIA ROSALIO
VALENZUELA PEDRO
VALENZUELA RODOLFO
VALLE MANUEL BURROLA
VALO HENRY LOUIS JR
VAN FREDENBERG ALLEN
JOHN
VAN LOON FRANK C JR
VANCE KERRY LAVERNE
VARNER HARRY KAY
VASQUEZ JOSE MARIA
VASQUEZ MARTIN MENDOZA
VERNO JOHN ARTHUR
VILLALOBOS HENRY ESTREL-
LA
WAHL JOHNNIE MITCHELL
WALKER BRADLEY A
WALLACE DIXIE DE
WALLING CHARLES MILTON
WALTERS CRAIG COLLINS
WEBSTER DAVID O NEIL
WEEKS CURTIS MILLER JR
WEITZEL GEORGE MARTIN
WEST LARRY JOE
WHEELER JAMES ATLEE
WHITE DONALD MERLE JR
WHITE FRED DONALD
WHITE SAMUEL MARLAR JR
WHITEHEAD WILLIAM J
WHITMER ALFRED VAN
WILBANKS LESLIE JOE
WILKES EULIS NEIL JR
WILLETT FRANKLIN DAVID
WILLIAMS ROBERT JR
WILSON JOHN THOMAS
WILSON JOHN WILLIAM
WILSON ROBERT LAURENCE
WILSON WILLIAM NEIL
WISE JOSEPH ROBERT
WOOD DELBERT ROY
WOODLAND DOUGLAS MEAD

WOOLRIDGE THOMAS AL-
PHONSE
WORLEY DON FRANKLIN
WRIGHT JEFFERY LYNN
YAZZIE LEONARD LEE
YBARRA MANUEL GUTIERREZ
YCOCO GEORGE ROJAS
YESCAS ANTONIO GILBERTO
YOAKUM DAVID LEWIS

YOUNG ROBERT EARNEST
ZELESKI PHILIP EDWARD
ZIGALLA LEONARD JAMES
ZODY RICHARD LEE

Arkansas Vietnam Veteran's Memorial

1500 W. 7th st, Little Rock, AR 72201

The Arkansas Veterans Memorial is located near the Arkansas capitol building on the southeast corner of the ground (at the intersection of 6th and Woodlane) and was created in order to honor the 662 Arkansas-born fallen service members who were lost in the Vietnam War. Events regarding the memorial usual occur annually around Veteran's Day and Memorial Day and usually consist of a ceremony and a parade honoring the fallen and their families.

This memorial was officially dedicated in November of 1987. The original memorial was made by Stephen Gartmann while the addition known as "The Grunt" was created by John Deerling, an Arkansas Democrat and political cartoonist for a local Little Rock newspaper. The Grunt was placed on the location six months after the rest of the memorial was placed, but the memorial was dedicated as a whole. General Westmoreland attended the dedication and actually pushed a fellow veteran in a wheelchair during the parade.

The Arkansas Veterans Memorial was funded through private funds as well as state funds. The project was originally approved in 1983. At that time, $150,000 was allocated by the government for the project. Private funds were collected to match that amount to complete the project. After the money was collected, then-governor Bill Clinton created a committee of veterans to manage the completion of the memorial.

The Arkansas Veterans Memorial consists of a memorial wall made of white and black granite slabs that features the names of the 662 Arkansas service members who were either killed in action or went missing in action during the Vietnam War. The memorial also features a statue of an infantryman which has the US military branches names inscribed on its base. The monument is made complete by the American flag and the Arkansas state flag. The entire memorial is lit by flood lights for convenient night viewing.

The Names of those from the state of Arkansas
Who Made the Ultimate Sacrifice

ACOSTA JOHN WAYNE
ALLEN JAMES HARLEN
ANDREWS FRED EUGENE
ANTHONY CAREY C
ARMSTRONG BILLY STANLEY
AVERY HARVEY CHARLES
AVEY EVERAL FLOYD
AYLOR GERALD LEON
BAGGETT CHARLES RICHARD
BAILEY MICHAEL WILSON
BAILEY THOMAS HAROLD
BAILEY WILSON PAUL
BAKER AQUILA
BAKER CURTIS RICHARD
BAKER DANNY RAY
BALLARD RONNIE EDSEL
BARNETT STUART LEE
BARNWELL RAY MAX
BARR JOHN FREDERICK
BATEMAN JESSIE RAYMOND
BATES CARL CALVIN JR
BATES TERRY HOYTE
BEARE CHARLES HAWKINS
BEASLEY ODELL DANIEL
BELL MARVIN EARL
BENNETT HAROLD GEORGE
BEST ANDREW THOMAS
BILES CALVIN WEBB
BIRMINGHAM TERRY WAYNE
BLEVINS HIRIS WAYNE
BLISARD REX WAYNE
BOATRIGHT WILLIAM ARVEL
BOATWRIGHT JACKLIN
MEGGS
BOBO EDWARD LEE
BOLING LESLEY JR
BONNER DON W
BOSTON CHARLES EDWARD
BOUDRA KILBERN DEAN
BOWDEN RICKY LYNN
BOYCE ALTON
BOYCE SAMUEL MINOR
BOZEMAN CHARLES LEE
BRADFORD WILLIE B
BRADLEY RICKY CURTIS
BRADSHAW THEODORE
JACKSON
BRANSCUM ARLIS RAY
BRATTON DARRELL DWANE
BRIDGES FRED J JR
BRIGHT ROY EVERETT
BROOKS WILLIAM ROGER
BROWN JAMES LEE JR
BROWN LOUIS
BROWN VERNON JR
BROWN WILLIAM HENRY
BURGETT BOYCE DALE
BURKETT CURTIS EARL
BURNLEY JOHN MOORE

BURRIS JOHN CHARLES
BURTON HORACE LEE
CALHOUN LARRY GENE
CANTRELL JERRY DALE
CARPENTER DOUGLAS JOE
CARTWRIGHT JIMMY
CECIL ROGER DALE
CHAMBERS JOSEPH LEE
CHARLES EDWARD WILLIAM
CHARLTON JERRY DEAN
CHESNUTT CHAMBLESS M
CHILDRESS J M
CLARK MAURICE
CLARK ROBERT LEE
CLARK ROOSEVELT
CLARK TIMOTHY EUGENE
CLAYBORN BILLY JOE
CLEMENT NEWTON STEVE
CLEMENTS GARY MAXWELL
CLEVELAND HARDY EDWARD
CLIFTON WILLIAM A
CLOSSER HENRY VERNON
COCHRAN JAMES CLIFFORD
CODY WESLEY OTERIA
COKER DOUGLAS CARROLL
COLDREN THOMAS L
COLE MARVIN RAY
COLWYE JAMES LEON
CONRAD CARLOS WADE
COOK JOHN EDWARD
COOK MARVIN JR
COTNER MORRISON AUTHER
COUCH FREDDIE LEE
COUCH ROY EVERETT
COVER BOBBY CECIL
CRADDOCK FRED BURKETT JR
CRAIN RONALD EDWARD
CRAWFORD JAMES EUGENE
CRELIA BILLY DUANE
CRISMON LONNIE JOE
CROCKETT DELMAR LEE JR
CROCKETT JOEL
CROOK WILLIAM FELTON JR
CROW KENNETH LELAND
CROWDER HAROLD EDWARD
CROWDER MICHAEL
CUMBIE HAROLD ERVIN
CUMMINGS NATHANIEL
CUNNINGHAM BILLY
CUNNINGHAM JESSE J JR
CUPPLES GARY CURTIS
CURBOW BILLY JOE
CURTIS JERRY JAMES
DACUS WILLIAM FLOYD
DANIELS JOHNIE NATHANIEL
DANSBY EMMIT CHARLES
DARR CHARLES EDWARD
DAVIS CLAUD ALBERT
DAVIS DON EDWARD

DAVIS JOHNNY F
DAVIS RONALD EUGENE
DE VASIER BILLY KIETH
DEAN LAWRENCE CHARLES
DECKER TEE WALLACE
DENSON FLOYD CORNELIUS
DICKASON CLYDE LEROY
DIETZ DONALD WILLIAM
DILL GARVIN WAYNE
DILLON RICHARD HALL JR
DIXON TOMMY JOE
DOBBS DONEL JOE
DODGE JEWELL FLETCHER
DOKES CHARLES WILLIE
DOWNS CARL LESTER
DUNCAN HERMAN DERL
DUNHAM BOBBY JOE
DUNLAP JERRY
EAKIN SHELTON LEE
EARNEST JUNIOR BARNETT
EASON JOSHUA WAY
EDNEY DONALD WAYNE
EDWARDS CHARLES M
EFIRD FRANKLIN D ROOSEVEL
ELLEDGE MICHAEL STEWART
ELLIOTT ROBERT THOMAS
ELLIS CHARLES WESTLEY JR
ELLIS CONEY
ELLISON WILLIE JR
ENGLISH ROBERT PRESTON
EPPERSON STEVEN GILL
ETHERIDGE HAMPTON A III
EVANS CLEVELAND JR
EVANS GERALD LEE
EVANS SAMMY GRAY
EVERETT JERRY DON
FARRIER GERALD WYATTE
FINCHER CECIL FRANKLIN JR
FISHER DENNIS WAYNE
FITTS GERALD LAMPLEY
FLAGG JAMES EDWARD
FLORENCE DEXTER BUSH
FORD ALVIN WALLACE
FORD HAROLD JOSEPH
FORT RAYMOND JR
FORTE GERALD WAYNE
FREELAND GUY THOMAS
FREEMAN CHESTER LEON
FREEMAN FLEMMON PAUL
FRIAR FREDDIE LYNN
FROST JAMES ALLEN
FUDGE JOHN T
FULFORD JIMMY DON
FULLER CARIO
FURR FREDERICK EDWARD
FUTRELL GARY THOMAS
GARNER ERNEST LEROY
GENSEMER DAVID DANIEL III
GENTRY JIMMIE FERREL

GETTER JAMES LEE
GILKER JOHN VICTOR
GILLIAM TERRY LYNN
GLENN CHARLES PHILLIP
GOACHER CARL FRANKLIN
GODWIN HARRY M
GODWIN SOLOMON HUGHEY
GOLDEN JIMMY LEE
GOODMAN JAMES DONALD
GORDON ALVIN JR
GORDON HUBERT HASKEL JR
GOREE WILLIE VANN
GOSS DANNY LEON
GOULD EDWARD DEAN
GRAF BARRY WADE
GRAHAM CHARLES HERBERT
GRANT ROBERT EARL
GRAY DANNY MICHEAL
GRAY GREGORY VAUGHAN
GRAY SEVIER JR
GREEN JAMES DAVID
GREEN TIMOTHY JOSEPH
GREEN WILLIAM HERSCHELL
GREEN WILLIE JR
GREENE JOHN WAYNE
GRIFFITH EDWARD WILSON
GRISSOM JOHNNY PAUL
GUIRE JOHN CHARLES
HACKWORTH CHARLES
LEHMAN
HAGGARD DARRELL LYNN
HALE WILLIAM ROBERT
HALL JOSEPH LINDSEY
HALSELL JOHN EDMOND
HALSTEAD MICHAEL CLAY
HAMBY JACKIE DWAYNE
HAMILTON DONALD PAUL
HAMM FRANKLIN ALVIN
HAMM GERALD EUGENE
BOOTH
HAMPTON MICHAEL DE-
WAYNE
HANDLEY ANTHONY WIL-
LIAM
HANNA WILLIAM ANTHONY
HARDIN PHILLIP RALPH
HARDING JOHN H
HARMON EDEWIN CLEO
HARMON JERRY WILSON
HARPER LARRY NEIL
HARRELL DONALD AUGUSTUS
HARRINGTON HUGH LEE
HARRIS JAMES RONALD
HARRIS NOEL AUSTIN JR
HARRIS RANDALL LYNN
HARRIS ROBERT TAYLOR
HARRISON PAUL ALVIN
HARRISON SAMMY RAY
HARSSON JERRY DON

HARTWICK BILLY WAYNE
HARVEY RANDALL LLOYD
HASTINGS BOBBY GENE
HAZELWOOD THOMAS
GERALD
HEBERT SYRIAC JR
HEDGES DANIEL MACOM
HEGLER FLOYD JR
HELTON JAMES EDWARD
HENDRIX EARNEST L
HENRY GEARLD ALBERT
HIBBLER JOE JUNIOR
HICKS JAMES BEN
HILDEBRAND HERBERT S
HILL CARL WAYNE
HILL CHESTER EUGENE
HILL JAMES EDWARD
HINES WILBURT NATHAN
HINTON OVERTIS JR
HIVELY BENNIE RAY
HIX RICHARD LAWSON
HOLLINGSWORTH JOHN
ANDREW
HOLLOWAY FREDDY LEE
HOLMAN DONALD WOODS
HOLMES KEITH DANIEL
HOLT JAMES RICHARD
HOLT JAMES WILLIAM
HOLZER BOBBY LEE
HOMSLEY VICTOR JORY
HONEYCUTT JAMES EARL
HOOPER JULIAN R
HOOVER THOMAS LEE
HOUSE JOHN CHARLES
HOUSLEY JAMES DAVID
HOUSTON JOHN WESLEY
HOUSTON MARVIN LYNN
HOWE SIDNEY A
HUBBARD ROGER LEE
HUFF JAMES A
HUFFINE DENNIS WILLARD
HUGHES JAMES ALVIN
HUIE ROBERT DOTSON JR
HUMBLE CHARLES RAY
HUNTER HAROLD CLAYTON II
HUNTER JOHN ROBERT
HUTCHINS MARION RAY
HUTCHINS TOLER LEE JR
HUTCHISON CHARLES
RANDEL
HUTSON CARL RICKIE
HUTSON RICKS ARBRA
INGRUM JOHN DANIEL
ISBELL OTIS EDWARD
JACKS MARK DOUGLAS
JEFFERS ODES WINSTON
JEFFRIES JAMES HERBERT
JENKINS PAUL LAVERNE
JERMANY RILEY

Arkansas

JOHNSON ALEXANDER JR
JOHNSON DAVID EARL
JOHNSON DAVID HAROLD
JOHNSON HENRY
JOHNSON LEDELL JR
JOHNSON ROBERT THOMAS
JONES OTIS CECIL JR
JONES ROY MITCHELL
KASIAH CLAUDE CHARLES
KEETER MARVIN ROSS
KEIFER JOE HAROLD
KEIM JAMES ROBERT
KELEHER KEVIN REYNOLDS
KELLEY LARRY MILTON
KEMP ROBERT VICTOR
KENNEDY CHARLES F
KERTIS HENRY LEE JR
KEVER DWAYNE ELBERT
KIMBROUGH HAROLD BRUCE
KING ELI J B
KING GARLAND BRYAN JR
KING JOHNNY RAY
KINGERY DONALD LEE
KLASSEN FRANCIS JAMES
KNEBEL THOMAS EDWARD
KNIPPERS WILLARD RUSSELL
KOLB LEROY JR
LANGFORD RICHARD HENRY
LANGSTON EVERETT EUGENE
LANIER JERRY DON
LATIMER WILBUR DALE
LAVENDER RICHARD ALLEN
LEDBETTER SANFORD JAMES
LEE CHESTER LLOYD
LEE HUBERT LEON JR
LEMON JOE LEE
LEWIS ROY ROBERT
LEWIS WILLIAM EWING
LINDLER JESSIE RAY
LIVINGSTON BILLY DALE
LLOYD JAMES VERNON
LONG RAYMOND LEON JR
LUCAS LARRY JACK
LYNCH STEPHEN WILLIAM
LYONS MARION WAYNE
LYONS WALTER JOHN
MAILHES LAWRENCE SCOTT
MANN CHARLES CLIFTON JR
MANN GARLAND RAY
MARCUSSEN GLENNON
MAROON JAMES WILLMER
MARSH CLARK LYNWOOD
MARTIN JEAN D
MASK JOE JUNIOR
MASON WILLIAM HENDERSON
MASSEY JAMES
MATHEWS HENRY DON
MAXWELL JAMES RICKEY
MAY ERNEST
MAY FARRIS ELDON
MAYER PAUL EVANS
MAYHUE DON N
MAYS E G JR
MAYS MC ELREE JR
MC CRYSTAL JAMES LARRY
MC DANIEL CHESTER
MC DONALD JERRY VERNON
MC ELHANEY BOBBY GENE
MC ELROY GRADY EDWARD
MC FADDEN FLOYD
MC FALLS BILLY CESAR
MC GAUGHEY WILLIE LEE
MC GEE BOLEN PONDEXDER
MC GINNIS ROBERT RAY
MC HENRY JAMES CARTHELL

MC KINLEY LEVERNE WILLIAM
MC MILLIN DONNELL DEAN
MCCOY RALPH LINDSEY JR
MELODY EDWARD BRUCE
MERONEY VIRGIL KERSH III
METCALF CHARLES EUGENE
METZ DANNY RAY
MICHLES EARL R
MILLER ARNEZ FRANKLIN JR
MILLER CARL JEROME
MILLS ANDREW LEE
MITCHELL LARRY LEVERN
MITCHELL ROBERT STEVENS
MIZE CLIFFORD N
MODESITT SAMUEL LEE
MOODY LARRY GENE
MOON WALTER HUGH
MOORE CHARLES RAY
MORAN RICHARD ALLAN
MORDEN BOBBY LEON
MOREAU THOMAS MICHAEL
MORGAN JAMES EDWARD

MORGAN JAMES SHEPPARD
MORGAN WILLIE LORENZO JR
MOSELEY HAROLD EUGENE
MOSLEY BERNIE JACK
MYERS JOHN EARL
NIXON SAMUEL RAY
NIXON WILLIAM DALE
NOE FRANK RAY
NOLEN BOBBIE ELDON
NORMAN JAY ROY
NOTHERN JAMES WILLIAM JR
OAKLEY LINUS LABIN
OWEN ROBERT DANEL JR
OWENS JERRY LYNN
PALMER JERRY ALLEN
PALMER LAYMON
PAMPLIN JOHN MAC
PARKER BENNY BRUCE
PARKER LONNIE EDWARD
PARRISH CONNIE WAYNE
PATTISON RONALD ALAN
PAYNE JAMES CARL
PEARCE CHARLES HUBERT JR

PENDERGIST RONALD LYNN
PENDERGRASS WILLIE CLEBER
PERRY RICHARD WILLIAM
PETERS STEPHEN FREDERICK
PHILLIPS OTIS LAMONT
PILLOW RONALD EDWARD
PIPPINS WILLIE SR
PLANTS OTIS EUGENE
POE JOSEPH BYRON
POLK GARY DON
POOL LARRY GAY
POOLE JOHN EDWARD
POOLE PERRY LEE
PORCHIA BOBBY RAY
PRICE ARTHUR HOUSTON
PUGH ROBERT EARL
PURIFOY HUBERT J
PURTELL ROBERT BUCK
RAINWATER JEWEL LEE
RAMSEY ROBERT LEE JR
RAMSEY VIRGUS FREDRICK JR
RASBERRY MIKE RAYMOND
RASH RONALD WAYNE

RAY MICHAEL WAYNE
REATHER WALLACE LEE JR
REED FLOYD LARDINO JR
REEVES GREGORY KEITH
REINECCIUS KARL LEWIS
REMMEL HARMON L III
RICE JOHN EDWARD
RICHARDS MICHEAL EDWARD
RILES JAMES CALVIN
RILEY EUGENE LEE
RIVERA RAYMOND NITO
ROAR WILLIAM ARTHUR
ROBBINS DENNIS TRUMAN
ROBBINS LEROY BRIAN
ROBERTS EDDIE LEROY
ROBERTS NOEL WAYNE
ROBERTSON GERALD WILLIAM
ROBINSON CLARENCE
ROBINSON HORACE VALLEY JR
ROBINSON WINSTON TERRY
ROBISON EDWARD KEITH

ROEHRICH RONALD L
ROGERS CLAUDE BENTON
ROGERS GUINN JUNELL
RONE JAMES ROBERT
ROSE JESSE BEA
ROSS JERRY WAYNE
ROSSON PHILLIP ENOS
RUCKS OTIS JAMES
RUSSELL ARTHUR JAMES
RUSSELL KENNETH MUREL
RUST HENRY WILSON JR
SABATINI ROBERT JOSEPH
SANDERS EDWARD LEON
SANDERS JAMES ALBERT
SANDERS WAYNE JACKSON
SATTERFIELD JOHN STEPHEN
SAUCIER ROBERT ARTHUR
SAULS ELBERT JAMES
SCHAFFER JOHN FERDINAND
SCHARNBERG RONALD
OLIVER
SCHLUTERMAN DAVID FRANK
SCHMIDT JOHN JOSEPH
SCRIVNER BROOKS MICHAEL
SCUCCHI JOHN GLENN
SEAVERS STANLEY JOSEPH
SHADDON ROY GENE
SHARPE JAYE ARTHUR
SHEPPARD BEN OGILVIE JR
SHERRILL RICHARD WAYNE
SHEWMAKE JOHN DANIEL SR
SHIREMAN PAUL JR
SHORES MALTON GENE
SILER GARY HUBERT
SIMMONS EDGAR LEE
SIMS PONDER RAY
SIMS WILLIAM JESS
SINGLETON JAMES ARNOLD
SISCO BILLY JOE
SMITH GEORGE FREETH
SMITH JAMES CHRISTOPHER
SMITH JIMMY DON
SMITH JIMMY JOE
SMITH WILLIAM TAFT
SMITH WILLIAM WARD
SMOCK WILLIAM HASKELL
SNIDOW STEPHEN ALLEN
SOLOMON DOUGLAS ED-
WARD
SPRATLIN MICHAEL STEPHEN
STAGGS LARRY DEAN
STANTON EMMETT CHARLES
STERLING RICHARD JOE
STEVENSON GEORGE MARK
STOKES PAUL AMOS
STONE JAMES LAWRENCE
STOUT JOHN HENRY
STRIBLING JESSE B
STROBLE COY EDWARD
STROUD EDWARD EUGENE
STURGEON WALTER
SUBLETT MICHEL KENT
SULLIVAN NEIL BRIAN
SUMPTER EDDY GALE
SWARTZLANDER ELIE ED-
WARD
SWINNEY GEORGE EDWARD
TABLER ROY TOM
TALLEY BILLY J
TALLEY TEDDY GENE
TATUM IVRA ALLEN
TAYLOR CHARLES MINOR III
TAYLOR DAVID F III
TAYLOR EDD DAVID
TEAGUE JAMES ERLAN
TEAS CLARENCE A
TEETER NORMAN WADE
TEFTELLER GORDON RAY
TELL BRITT JR
TETTLETON DAVID DEWAYNE
THACKER FREDRICK AN-
THONY
THOMAS HOUSTON FRANK-
LIN
THOMAS MICHAEL HERMAN
THOMPSON SOLOMON
EUGENE
THROWER FREDRICK LAMAR
TORRENCE WILLIE CHARLES
TRAVIS LYNN MICHAEL
TRUELOVE JERRY ALLEN
TURNAGE THOMAS ALFRED
TURNER ARTHUR TRAVIS
TURNER LARRY THOMAS

UPSHAW OLEN LEE
VARNADO CLARENCE
VAUGHAN JAMES LLOYD
VINSON AARON
VORIES JOHN LLOYD
WALKER CHARLES EDWARD
WALKER CLARENCE
WALL JIMMIE PAUL
WALLER THERMAN MORRIS
WALTERS ROBERT LOUIS
WARD GARRY WALLACE
WARREN SAMMIE LEE
WASHINGTON JOHN
WATERS TROY LEE

WATSON JOE NATHAN
WAYMIRE BILLY JOE
WELLS RICHARD FOY
WELLS ROY VON
WEST DANNY RAY
WEST MOUNCE EDWARD
WHALER ARCHIE LEON
WHEAT PRYOR L
WHITE CHARLES
WHITE DONALD NISLER
WHITE JERRY DEAN
WHITTON TEDDY GENE
WICKLIFFE JOHN NORMAN
WILDMAN MELVIN ALVIN

WILES JOHNNY
WILLIAMS ARTHUR C JR
WILLIAMS BURNELL JR
WILLIAMS HARVEY LEE
WILLIAMS JOEL JR
WILLIAMS LEE ARTHUR
WILLIAMS NATHANIEL
MEARLO
WILLIAMS PLUMMER
WILLIAMS TERRY LUTHER
WILLISON FRANKLIN JOE
WILSON DAVID
WILSON RICHARD JR
WILSON WILLIAM ROBERT

WITHEY HOWARD HUGH
WOLFE WILLIAM EDWARD JR
WOMACK ROY ARNOLD
WOOLEY HENRY EUGENE
WORST KARL EDWARD
YOUNG JIMMY RANDLE
YOUNG LE ROY JR
YOUNG ROBERT ALLEN JR
YOUNG THOMAS FRANKLIN

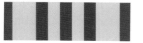

Photos by Michael English

California Vietnam Veterans Memorial

Capitol Park at 15th and L in Sacramento, CA 95814

Behind the State Capitol building in Sacramento, California there is a park that hosts the California Vietnam Veterans Memorial. This memorial is meant to pay homage to the Vietnam War veterans, especially those that were lost in combat or were never found. According to records, it's one of the most frequented memorials of all 15 that are located in Capitol Park. The veterans' memorial is the site of numerous ceremonies across the calendar year.

The project was set into motion in 1983 when a group of veterans located in Sonoma and Marin counties, led by Herman Woods, talked state Assemblyman Richard Floyt into introducing a bill to erect a Vietnam War memorial. Assembly Bill 650 was signed into law by then-governor George Deukmejian in September of that same year. A design contest was held a year later with Michael Larson and Thomas Chytrowski taking the prize. The California Vietnam Veterans Memorial's final design was revealed in May 1985. Rolf Nord Kriken was responsible for the bronze statues that are a large part of the memorial including the bronze map of Vietnam that sits at the memorial's entrance.

The memorial was complete in 1988 and officially dedicated in December of the same year. The whole project cost $2.5 million which was made up entirely of donations instead of state appropriations.

The California Vietnam Veterans Memorial features 22 black granite panels that boast the names of the over 5,000 dead or missing California veterans. There are also life-size bronze statues that represent the varying realities of life during the Vietnam War. The memorial is set in concentric circles that are surrounded by cherry trees. The outermost ring is made up of precast concrete planters that are broken up to create three entrances. Pylons with lamps symbolizing eternal flames flank the three entrances. Inside the inner circle is a sculpture of a young combat soldier holding his M-16and a letter. The memorial also features bronze reliefs depicting different scenes like combat scenes as well as dedication plaques. Every year on the day before Memorial day an event including the reading of all names on the memorial takes place. This event usually lasts about twelve hours.

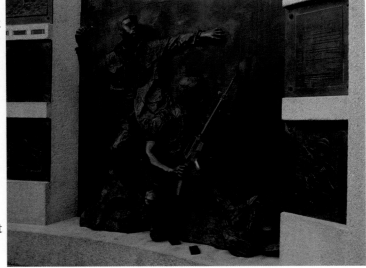

The Names of those from the state of California Who Made the Ultimate Sacrifice

ABBIE DONALD PAUL
ABBOTT EDWARD DONALD
ABBOTT JOHN
ABEYTA ERNEST
ABINA ROBERT THOMAS
ABLES ELMER ROBERT LEE JR
ABNER CARL EDWARD
ABRAHAM PAUL LEONARD
ACEVEDO RICHARD JOSEPH
ACHICA EDDIE
ACHOR TERRENCE WILLIAM
ACKLEY GERALD LEVIE
ACOSTA DANIEL
ACOSTA JOHN MICHAEL
ACOSTA LOYD DEAN
ACUNIA EDGAR
ADACHI THOMAS YUJI
ADAIR HARVEY GENE
ADAME GILBERT JIMMIE
ADAMS GILLES DAVID
ADAMS JAMES EDWARD
ADAMS JAMES LINDELL
ADAMS JOHN ROBERT
ADAMS JOHN WILBURN
ADAMS KENNETH STANLEY
ADAMS LEE AARON
ADAMS LEON HENRY
ADAMS PAUL EDWIN
ADAMS RICKY FAY
ADAMS WILLIAM ERNEST
ADAY ROBERT LEE
ADCOCK RICHARD LYNN
ADDISON HARVEY CHARLES
AGUIAR JUAN DANIEL
AGUILAR MIKE JOHN
AGUILAR OSCAR
AGUILAR RUDOLPH RENE
AGUILERA DANIEL
AGUILLON FELIZARDO
CUENCA

AGUIRRE GEORGE
AGUIRRE JOSEPH ANTHONY
AHERN JOHN BERNARD
AKAMU ALBERT KAIWI
AKINS SAMUEL LEROY
AKSTIN JAMES MICHAEL
ALAGNA PETER LEONARD
ALANIZ FEDERICO JR
ALARCON ARTURO FRAGOSO
ALBASIO JOHN ANTHONY
ALBERT RAYMOND HOWARD
JR
ALBERTINI JOSEPH ALFRED
ALBERTON BOBBY JOE
ALCANTAR FRANK COSME
ALCORN DALE ROBERT JR
ALDERSON BENJAMIN
ROBERT
ALEGRE DANIEL ALBERT
ALENCASTRE ANTHONY
ALBERT
ALESHIRE RONALD LEE
ALEX CHARLES RAY
ALEXANDER GEORGE WOOD
ALEXANDER ROBERT EMMET
ALFORD MARK CARL
ALFORD MICHAEL LYNN
ALFRED THOMAS SAMUEL
ALLEN DAVID ANDREW
ALLEN GARY CHARLES
ALLEN JERRY JOE
ALLEN JOSEPH HAROLD
ALLEN LARRY HUGH
ALLEN REX THOMAS
ALLEN RICHARD GRAHAM
ALLEN TERRY JAMES
ALLEN WAYNE ANDERSON
ALLENDER FRANK ROSS JR
ALLMAN JONATHAN WAYNE
ALLRED FRANK LEROY JR

ALLRED REX CHARLES
ALMANZA JOHN JERALD
ALMARAZ RONALD PAUL
ALMEIDA JOE JR
ALONZO LUIS
ALTIERI ALLAN JOSEPH
ALVARADO ALFRED FRED-
ERICK
ALVARADO LEONARD LOUIS
ALVAREZ ALEX JIM
ALVAREZ CHARLES ALLEN
ALVAREZ GEORGE CALDERON
ALVAREZ JIMMIE MARRON
ALVAREZ JOSE CARMEN
ALVAREZ MICHAEL BYRON
ALVERAZ CYRIL ANTHONY
AMATO RICHARD C
AMESCUA STEVEN EPEFANIO
AMEY SAMUEL ALLEN
AMISON ROOSEVELT JR
AMSPACHER WILLIAM HARRY
JR
ANANIAN JOHN MOSES
ANDERS EDWARD JAMES
ANDERSEN MICHAEL NILE
ANDERSEN WILLIAM T JR
ANDERSON CLINTON H JR
ANDERSON CLINTON RUSSELL
ANDERSON DONALD LEROY
JR
ANDERSON EDWIN P JR
ANDERSON JAMES BOYD
ANDERSON JAMES EDWARD
ANDERSON JAMES HOWARD
ANDERSON JAMES JR
ANDERSON KENNETH TERRY
ANDERSON KENT STUART
ANDERSON ROBERT DOUG-
LAS
ANDERSON ROBERT EUGENE

ANDERSON TERRANCE
WESLEY
ANDERSON VINCENT CRAIG
ANDERSON WARREN LESTER
ANDERSON WILLIAM OLIN
ANDREWS RONALD L
ANDREWS VAUN
ANELLA JAMES DAVID
ANGEL MICHAEL EUGENE
ANGELLEY GERALD DWAIN
ANGLIM PATRICK EMMETT
ANNIS CHARLES DOUGALS
ANNIS ROBIN RICHARD
ANNOS GEORGE RICHARD
ANTHONY JOHN EDWARD
ANTHONY LIONEL S
ANTOGNINI JOSEPH III
ANTONE FRANK GEORGE
ANTUNANO GREGORY
ALFRED
APELLIDO RAYMOND HUGH
APODACA JACK MICHAEL
APODACA PETER MICHAEL
APPLEBY IVAN DALE
APPLEBY RICKEY EUGENE
APPLEGATE KENNETH
CHARLES
APPLETON DANNY ELBERT
ARAGON ALONSO JR
ARAMBULA PAUL TEJEDA
ARELLANO LE ROY FRED
ARENT KENNETH JACOB
ARIAS RICHARD
ARIAS WILLIAM CIP JR
ARMENDAREZ MIKE
ARMENTA HERIBERTO
ARMENTA RUBEN MAXIMO
ARMITSTEAD STEVEN RAY
ARMOND ROBERT LAURENCE
ARMSTRONG DONALD GLENN

ARMSTRONG JOHN HENRY
ARMSTRONG KENNETH
DANIEL
ARNALL ROBERT D
ARNOLD ODIS DANIEL
ARRAIZ JAMES PAUL
ARREDONDO JOSE MARIA R
ARREDONDO THOMAS
ALFRED
ARREGUIN JOE
ARREY FRANK JR
ARTAVIA JOSEPH GREGORY
ARVIZU XAVIER AMADO
ASADA RONALD KAZUO
ASCHENBRENER ERVIN G JR
ASHFORD GREGORY MICHAEL
ASIRE DONALD HENRY
ASUNCION HENRY FRANCE
ATKINS DOYLE
ATKINSON JERRY DOYLE
ATUATASI SA JR
AUBERT THOMAS CLIFFORD
AULD ROGER MARTIN JR
AUSTIN EDDIE PAUL
AUSTIN LARRY D
AUSTIN ROBERT
AUSTIN STEPHEN EDWARD
AUSTON KENNETH JOE
AVILA JESUS V
AVILA JOHN MANUEL
AVILA MANUEL JR
AXTON EDWIN EVERETTE
AYALA GILBERT JR
AYERS DAVID WILLIAM
AYLOR CHARLES VINCENT
AZBILL ROY GORDON
AZNOE KENNETH EUGENE
BAADE ROBERT RICHARD II
BABCOCK JOHN RICHARDS
BABYAK LAWRENCE JOSEPH

California

BACA FRANK MARTIN
BACA RICHARD DAVID
BACCUS JIMMY DEVER
BACHELOR DON RAY
BACHER MARK WARREN
BACKEN DENNIS DONALD
BACKMAN ROBERT EUGENE
BADOSTAIN TIMOTHY ERNEST
BAFILE JOHN ANTHONY
BAGGETT FRANK ALLEN
BAGLIO RICHARD ANTHONY
BAILEY ALFRED LEON JR
BAILEY BERNARD PHILLIP
BAILEY ROBERT BENTON
BAILEY TERRY JOE
BAINTER NEAL VINCENT
BAIRD ROBERT STANLEY
BAIRD RONALD EUGENE
BAKER ALTON EUGENE
BAKER BARRY JAY
BAKER DONALD LEE
BAKER JOHN THOMAS
BAKER ROBERT NELSON
BAKER ROBERT OLIVER JR
BAKER RONALD RAY
BAKER STANLEY WELLINGTON
BAKER STEVEN DEWITT
BAKER TONY ANDERSON
BAKKIE DONALD KEITH
BALADES DAVID ZAVALA
BALBIRNIE JAMES FREDERICK
BALCH JAMES IVERSON
BALDIZON-IZQUIERDO CARLOS
BALDWIN LARRY DEAN
BALL JAMES MARVIN
BALLARD MEL ROY
BALLARD ROBERT LEE
BALLAUF CHARLES ALAN
BALLIN JOE MAGDALENO JR
BALTERS STEPHEN A JR
BAMFORD THOMAS CAMPBELL
BAMFORD THOMAS CAMPBELL
BANAGA SALVADOR M L JR
BANGLOS GARY ALAN
BANGS CHRISTOPHER DELBERT
BANKS ROBERT ALAN
BANNER STEVE ARTHUR
BANUELOS ALBERT A JR
BAPTISTA PAUL ALIPIO
BARBA PHILLIP JOSE
BARBEE RICHARD LORDY
BARBER DAVID EDWIN
BARBER MELVIN
BARBER ROGER LEE
BARDEN ARNOLD WINFIELD JR
BARGER FERDINAND ORA JR
BARKER DANA RANDOLPH
BARKER GARY LEE
BARKER ROBERT LEE JR
BARKFELT DAVID WILLIAM
BARKLEY STEPHEN RICHARD
BARLEEN THOMAS LYLE
BARNARD GARY MICHAEL
BARNES FRANCIS ARCHER
BARNES MARVIN DONALD
BARNES RICHARD LEIGH
BARNES ROY DWIGHT

BARNES WALTER FRASIER
BARNETT JOHN FRANK
BARNETT STEVEN PAUL
BARNHOLDT TERRY JOE
BARNHOUSE DARREL EMERSON
BARON DOUGLAS KEN
BAROVETTO JOHN LAWRENCE
BARRAGAN REYNALDO LEON JR
BARRERA JOSE GILBERT
BARRERA MANUEL
BARRERA RAUL ROY JR
BARRERAS FRANK III
BARRETT CHARLES WESLEY
BARRETT FREDERICK HARRY
BARRETT MICHAEL BARRY
BARRETT ROBERT LEE JR
BARRIOS JAMES PATRICK
BARRIOS MARCELLO NUNEZ
BARRON JEFFREY MICHAEL
BARTALOTTI ALFONSO PAUL
BARTELS STEPHEN DONALD
BARTH THOMAS FREDRICK
BARTH WAYNE ROBERT
BARTHOLOMEW MICHAEL M
BARTLEBAUGH DENNIS LEE
BARTLETT JOHN REX
BARTLOW GARY WILLIAM
BARTMAN STEVEN DOUGLAS
BARTON HAROLD BRUCE
BARTON JERE ALAN
BARTON VAL E
BARTON VIRGIL WAYNE
BASILE PATRICK LYNN
BASS GARY NOLAN
BASTYR DOUGLAS BRUCE
BATES JAMES EDWARD
BATES WAYNE SHERWOOD
BATH ELDRIDGE JACK
BATTAGLIA CHRISTOPHER PAU
BATTIEST ANDREW
BATTY DENNY ALBERT
BAUCHIERO HAROLD
BAUDER JAMES REGINALD
BAUER CURTIS DEAN
BAUER JAMES NEIL
BAUM DOUGLAS BRUCE
BAUM MICHAEL LEE
BAUMANN RENE GEORGES
BAUMGARDNER DAVID LEON
BAUMGARDNER DUANE ROY
BAXLEY DENNIS WAYNE
BAXTER IVERY LEE
BAXTER JERRY
BAYLES GERALD WILLIAM
BAYLES STEPHEN ERNEST
BAYS LEE R
BAZA JOSEPH CRUZ
BAZELL FRANK DAVID
BAZULTO SALVADOR
BEACH ARTHUR JAMES
BEALL TYSON VANCE
BEALS RONNIE HERBERT
BEAMAN RONALD RALPH
BEAMON THOMAS KEITH
BEARD BILLIE LESTER
BEARDSLEY JEFFREY THOMAS
BEARDSLEY RONALD ALLEN
BEAUDETTE LARRY MICHAEL
BEAUMONT ROBERT EUGENE
BECK GREGORY GEORGE
BECK STEVEN LEE
BECKERMANN FRED B JR
BEDAL ARTHUR EUGENE
BEDARD BARRY JOSEPH
BEDOLLA JOSEPH LOPEZ
BEDRA THEODORE FRANK
BEDSWORTH BILLIE MICHAEL
BEECHE RAFAEL EDUARDO
BEEK JOHN LAWRENCE
BEENE JAMES ALVIN
BEHM STANLEY WILLIAM
BELKNAP RONALD LEE
BELL EDWARD ALLEN
BELL GERALD DEAN
BELL HENRY DANIEL JR
BELL HOWARD CLAYTON
BELL JAMES EVERRETT
BELL JOHN MARTIN
BELL MARK WAYNE
BELL SAMUEL WAYNE

BELLAMY JOHN MICHAEL
BELLOMO TERRENCE JOHN
BELLRICHARD LESLIE ALLEN
BELNAP GLEN DEAN
BELON MARC BRADLEY
BELTRAN ROBERT JOSEPH
BELTRAN ROBERT LEON
BELVEAL JAMES ALLEN
BENADUM RICHARD DENNIS
BENDOR JOHN LEE
BENEGAS VINCENT JOSEPH
BENES WAYNE JOSEPH
BENNETT BRIAN JOHN
BENNETT CLIFTON E
BENNETT DANIEL JOSEPH
BENNETT DONALD LUCIAN
BENNETT PHILIP MARK
BENNETT ROBERT LLOYD
BENNETT ROBERT M
BENSON KEITH LLOYD
BENSON LEE DAVID
BENSON ROBERT WILLIAM
BENTLEY BORIS ROMAN BENJA
BENTON GREGORY REA JR
BENTON THOMAS HOWARD
BERARD JAMES EUGENE
BERBERT KARL ROBERT
BERG MYRON WALDO
BERG RAY WILLIAM JR
BERGREN THOMAS HOWARD
BERMUDEZ JESUS ROJAS
BERNAL RAYMOND JR

BERNARD RANDALL BRUCE
BERNSTEIN BRUCE BRYANT
BERRY DAVID JOE
BERRY JAMES EDWARD
BERRY MICHAEL GEORGE
BERRY RALPH THOMAS
BERRY ROY VERNON JR
BERRY WILLIAM AARON
BERTAGNA LAWRENCE JOSEPH
BERTOMEN NARCISO JR
BERTSCH BRENT JOHN
BERUMEN JUAN BOSCO
BETANCOURT GABRIEL
BETCHEL DAVID BROOKS
BETHARDS EDWARD WAYNE
BETTENCOURT DANIEL F JR
BETTGER GENE LYLE
BEVARD BOBBY LEE
BEVERFORD TIMOTHY WAYNE
BIANCHINI MICHAEL LINN
BIBER JOSEPH FRANK
BIBLER WILSON E JR
BICKLE JIMBOB
BIEMERET ARTHUR THOMAS
BIERMAN CARROLL MONROE JR
BIERNACKI JAMES RICHARD
BIESER KARL ROY
BILLHIMER GARY ARTHUR
BILLINGSLEY RICHARD WAYNE

BINGMAN RONALD HOWARD
BINNS DAVID RICHARD
BIRCH DANIEL PATRICK
BIRCH JOHN MACY
BIRCHIM JAMES DOUGLAS
BIRD HAROLD ALVIN
BIRD KENNETH ROBERT
BIRDEN LEE ROY
BIRDSELL GEORGE DAVID
BIRDWELL MICHAEL DEL
BIRKET SCOTT LEE
BISCAILUZ ROBERT LYNN
BISHOP DANIEL EDWARD
BISHOP MICHAEL RICHARD
BISHOP RICHARD LAVERN
BISHOP RUSSELL LAVERNE
BLACK PAUL VERNON
BLACKBURN ELBERT FRANK
BLACKWOOD GORDON BYRON
BLAGDON EDWIN ELLIS
BLAIR ROCKY LEE
BLAKELY MARTIN GEORGE
BLAKELY WILLIAM
BLAKEY MICHAEL ARCHIE
BLANEY THOMAS ARTHUR
BLANKENSHIP DONALD LEE
BLANKENSHIP JAMES ARLIA
BLASINGAME NORMAN LEE
BLESSMAN WILLIAM DAVID
BLEVINS ANTHONY JAMES
BLINDER RICHARD BART
BLINER JOHN EDWARD

BLOODWORTH DONALD BRUCE
BLOOM RICHARD MCAULIFFE
BLOSKY GENE ORVILLE
BLOWERS RICHARD LYLE
BLUME DALE L
BOBBITT ARTHUR
BOBBITT JERRY KEITH
BOCANEGRA FELIX RAMON
BOCANEGRA HUGO ARTHUR
BOCK JIMMIE VAN
BODA JAMES ALBERT
BODENSCHATZ JOHN EUGEN JR
BOELZNER ROBERT CRAIG
BOETCHER HAROLD EDWARD
BOETS PETER QUIRINUS JR
BOETTCHER WALTER R JR
BOGGS CLIFFORD ALLEN
BOHLER ROBERT RONALD
BOHLIG JAMES RICHARD
BOHNER LEONARD ALLEN
BOLES JOEY LEE
BOLSTER CHRISTOPHER ORAN
BOLTON DAVID JOSEPH
BOLTON DENNIS LEWIS
BONJOUR KEVIN EARL
BONNER WILLIAM ROBERT
BOOMSMA ROGER ALLEN
BOOTH EMMETT LEE
BOOTH JAMES ERVIN
BOOTS JAMES ALLEN
BORDERS JOHN WILLIAM JR
BORGEN CARL LEE
BORGES MICHAEL EDWARD
BORJA DOMINGO R S
BORQUEZ LAWRENCE GABE
BORREGO EDWARD LEE
BORRUSO JOSEPH JR
BOSBERY DONALD CHARLES
BOSCH ERIC ALAN
BOSH ANTHONY ROBERT
BOSWORTH TERRY LEE
BOTTAN DANIEL JACQUES
BOUDREAUX ALLEN JOHN
BOWDEN BYRON BILL
BOWEN DUANE CURTIS
BOWEN JOHN LEWIS
BOWEN LARRY WILLIAM
BOWEN THOMAS RAY
BOWERS RICHARD ALAN
BOWLES THEOPHILUS
BOWLING ROY HOWARD
BOWMAN PAUL JR
BOWMAN STEPHEN WESLEY
BOX HOUSTON CLIFFORD JR
BOX JOHN ROBERT
BOYCE LAWRENCE STEVEN
BOYD BILLY RAY
BOYD BRADLEY MONROE
BOYD JOHN LEE
BOYD ROBERT CARL
BOYD WILLIAM DEMARR
BOYDSTON OSCAR DAN
BOYER BARNEY EVANS
BOYER DONALD WILLIAM
BOYER ROBERT LEE
BOYLE DAVID JOSEPH
BOYLE HARRY LEWIS
BOZARTH ALVIN RAY
BOZEMAN PERRY LEONARD
BRACE BRUCE WAYNE
BRACKINS VERNON EDWARD
BRADFORD SHERMAN DUANE
BRADFORD WILLIAM JONATHAN
BRADLEY MICHAEL LEE
BRADLEY RAY EUGENE
BRADSHAW HENRY LEE
BRADY JAMES GREGORY
BRADY THOMAS GERARD
BRADY THOMAS GERARD
BRADY THOMAS PAUL
BRANCATO MICHAEL GEORGE
BRANTLEY MARK CURTIS
BRASHEAR WILLIAM JAMES
BRATTON FREDDY LAMAR
BRECK GARY ANTHONY
BREDA DENNIS JOHN
BREITENBACH BERNARD PAUL
BRENTON MICHAEL JOSEPH
BRESHEARS KENNETH LESTER
BRESHEARS RONALD CHRIS

BRETCHES RAYMOND DEAN
BREWER GEORGE HENRY
BREWER JOHN NEWTON
BRIC WILLIAM HENRY III
BRICE RONNIE
BRICMONT FRANCIS PETER JR
BRIDGES BERRY JOE
BRIDGES PHILLIP WAYNE
BRIERLY JAMES KENNETH
BRIGGS LARRY ISHMUEL
BRIGGS THOMAS HAROLD C
BRIGHT RICHARD
BRINK JAMES RICHARD
BRITT AQUILLA FRIEND
BRITTEN LAWRENCE ALAN
BRITTON MURRY LAWRENCE
BRITTON SHERRICK CAMDEN
BROADBECK JOHN GILBERT
BROCKER THOMAS GEORGE
BROCKINGTON CURTIS
BROCKMAN FRANCIS CARL III
BRODNIK FRANKLIN VINCENT
BRODRICK STEVEN PARKER
BROGOITTI BRUCE CLAYTON
BROOKENS WILLARD JR
BROOKS BENJIMAN
BROOKS LAWRENCE ARTHUR
BROOKSHIRE GEORGE DEWEY
BROPHY DANIEL RALPH
BROSTROM DAVID CHARLES
BROTHERS GERALD JOHN
BROTZMAN MICHAEL RAY
BROWN ALVIN RAY
BROWN BARRETT CHAMBERLAND
BROWN BARRY EDWARD
BROWN BRUCE EDWARD
BROWN CHARLES LYNN
BROWN CLEMMIE JR
BROWN DON CHARLES
BROWN DONALD HUBERT JR
BROWN DONALD LYNN
BROWN GALE LEE
BROWN GARY WAYNE
BROWN GEORGE MICHAEL
BROWN GERALD RAY
BROWN JAMES EDWARD
BROWN JOE MAC
BROWN LAURENCE GORDON
BROWN MARC ALAN
BROWN MARSHALL JASON
BROWN MICHAEL GREGORY
BROWN MICHAEL R
BROWN RANDOLPH JR
BROWN RICHARD ALBERT
BROWN RICHARD ALLEN
BROWN RICHARD LEE
BROWN RICHARD TYRONE
BROWN ROBERT ALVA II
BROWN ROGER LOUIS
BROWN RONALD A
BROWN RONALD HOWARD
BROWN RUSSELL LEE
BROWN TANNER MARTIN JR
BROWN WILLIAM THEODORE
BROWNE EDWARD RAYMOND
BROWNING MICHAEL LOUIS
BROWNING WILLIAM FRANK
BROYLES IVAN JOSEPH
BROYLES LANHAM ODELL
BRUBAKER THOMAS GEORGE
BRUCE DENNIS RAY
BRUCE WILLIAM JACK
BRUCKART DONALD LEE
BRUCKNER PATRICK LOUIS
BRUHN GARY WILLIAM
BRUMLEY BOB GENE
BRUMMET PAUL DOUGLAS
BRUNAT MICHAEL F
BRUNER MARK LEROY
BRUNNER O D
BRUNTON STEPHEN CORNELL
BUCHECK ROBERT MARTIN
BUCKHOLDT LEO BUDDY
BUCKLES RICHARD LEE
BUCKLEY FRANCIS RICHARD
BUEHLER ROBERT HENRY
BUERK WILLIAM CARL
BUGARIN BENJAMIN
BULLARD VICTOR WALKER JR
BULLARD WILLIAM HARRY
BULLINGTON JAMES ALLEN
BULLIS KRAG COLT SR

BUMGARNER BRUCE HOWARD
BUMGARNER THOMAS EDWARD
BUNCH RAYMOND LEE JR
BUNCH WILLIAM LLOYD
BUNDAGE CECIL ODELL
BUNDY LINCOLN E
BURBAGE RAYMOND DOUGLAS
BURCHARD MARK WAYNE
BURCIAGA ROBERT
BURFOOT PHILLIP DUANE
BURGE THOMAS GUY
BURGESS ROBERT HOWARD
BURICH JOHN ANTHONY JR
BURKE JOHN PATRICK
BURKE WILLIAM DAVIDSON JR
BURKE WILLIAM JAMES JR
BURKEY KERMIT EDWARD
BURKHART EUGENE WAYNE
BURLESON CLARENCE PAUL
BURLINGAME STEPHEN FRANK
BURNETT DAVID LEIGH
BURNETT JOSEPH DARRYL
BURNS DEAN HARRY
BURNS JAMES DAVID
BURNS KEN DWIGHT
BURNS MICHAEL EDWARD
BURNS MICHAEL THOMAS
BURNS RICHARD ALLEN
BURNS WILLIAM CARL JR
BURRI MIGUEL RAMON
BURRIS LEONARD CHARLES
BURRIS VICTOR ANTONIEO
BURROLA SAMMY JR
BURSON DAVID RICHARD
BUSH MARK JOEL
BUSHAY BYRON HALEY
BUSS RONALD FRANK
BUSTAMANTE ARTHUR
BUSTAMANTE MICHAEL ANDREW
BUSTAMANTE STANLEY R JR
BUTCHER GALE W JR
BUTCHER JOHN HENRY JR
BUTLER DAVID LEROY
BUTLER EDWARD WAYNE
BUTLER GARY WILLIAM
BUTLER JOSEPH MILTON
BUTLER KENNETH ALLAN JR
BUTLER LARRY DON
BUTOROVIC STEVE
BUTTERFIELD DOUGLAS HOLMA
BUTTERFIELD MARVIN JEAN
BUTTRY RICHARD RUSSELL
BUTTS JERRY EUGENE
BUTTS JOHN MICHAEL
BUYS KENNETH ALLEN
BUZZARD LARRY B
BYAM MICHAEL LEROY
BYERS KENNETH EDWARD
BYRD BOBBY JOHN
BYRD CHARLES
BYRD JAMES CARMEN
BYRNE JAMES RONALD
BYRNES ROBERT SCOTT
CABANO GEORGE ANGELO JR
CAGNACCI JOSEPH MARIO
CAGUIMBAL PEPITO
CAIN ROBERT KEITH II
CAIQUEP JOSE
CAIRNS ROBERT ALEXANDER
CALDERON FELIX ANTONIO
CALDERON JULIO ALFREDO
CALDERON LOUIS OSCAR
CALDWELL EVERETTE BRENT
CALHOUN ROBERT DARRELL
CALL RICHARD JOSEPH
CALLAGHAN THOMAS LEONARD
CALLAHAM JOHN MARSHALL JR
CALLAN PHILIP MICHAEL
CALLANAN RICHARD JOSEPH
CALLEN RICHARD JAMES
CALLE-ZULUAGA FERNANDO
CALLIS DAVID GEORGE
CALPH GENE ELWOOD
CALVILLO ROBERT JESS
CALZIA FRANK VINCENT

CAMARENA-SALAZAR EDUARDO
CAMERO SANTOS
CAMERON KENNETH ROBBINS
CAMERON ROBERT JOHN
CAMP WILLIAM GORDON
CAMPAIGNE JERRY ALAN
CAMPBELL CARLIN MARTIN JR
CAMPBELL EUGENE CHARLES
CAMPBELL JACK DONALD
CAMPBELL JAMES LEE
CAMPBELL ROBERT DEAN
CAMPBELL ROBERT JOHN
CAMPBELL THOMAS EDWARDS
CAMPOS RICHARD FREDERICK
CANADA CLYDE LEE ROY
CANAVAN MARTIN JOSEPH JR
CANDELAS JOHN FRANK
CANDLER GREGORY JAMES
CANRIGHT STEVEN CRAIG
CANTRELL PHILLIP GENE
CAPUANO GEORGE ANTHONY
CAPUTO MICHAEL JOHN
CARAMELLA PAUL DOANE
CARAWAY JOHNNIE J
CARBAJAL CARLOS GUZMAN
CARD WAYNE NORMAN
CARDENAS DANIEL JR
CARDIN WILLIS GLEN
CARDINALE JAMES ANTHONY
CARDWELL JAMES MELVIN
CAREY JOHN JR
CAREY MICHAEL WILLIAM
CARIVEAU WILLIAM JOSEPH
CARL ARTHUR JACK
CARLETON RONALD DEE
CARLEY RAYMOND MONTELL
CARLI DAVID ARTHUR
CARLIN DAVID ALLEN
CARLOCK JOHN RONALD
CARLSON JAMES CLARK
CARLSON RICHARD ALLAN
CARMONA EFREN
CARNEVALE DAVID JAMES
CARNEY TYRONE EDWARD
CARPENTER CHARLES
CARPENTER GARY RALPH
CARPENTER ROGER NELVIN
CARPENTER SCOTT MARSHALL
CARR GREGORY VERNON
CARR JOHN PARM III
CARR ROGER JAMES
CARRANO JACKIE ANDREW
CARRANZA MARTIN
CARRASCO ARTHURO
CARRASCO DANIEL
CARRIER DANIEL LEWIS
CARRIERE OSCAR ROLAND
CARRILLO GEORGE J JR
CARRILLO JIMMY
CARRILLO JOSE CASTANEDA
CARRILLO RICHARD
CARROLA EDWARD
CARROLL JOHN THOMAS
CARROLL PETER RICHARD
CARROLL TIMOTHY MICHAEL
CARSON CLARENCE JASPER JR
CARTER DENNIS RAY
CARTER GREG ROY
CARTER JACKIE CHARLES
CARTER JAMES LOUIS
CARTER LESLIE DEAN JR
CARTER MICHAEL BOYD
CARTER RODNEY BALAAM
CARTER THOMAS JAMES
CARTER THOMAS LEE
CARTER THURL GUY III
CARTER WENDELL LOUIL
CARUSO THOMAS EDWARD
CARVAJAL JOSEPH CARLOS
CARVER BOBBY DON
CARVER RICHARD ALAN
CASARES MANUEL
CASE GLENN EDWARD
CASE ROBERT DON
CASE THURLE EUGENE JR
CASEY DANIEL GENE
CASEY LIAM SOUEPH
CASEY RICHARD WILLIAM

CASH DAVID MANFRED
CASIAS CHRISTOPHER
CASON DAVID ALLAN
CASPER RONALD JEROME DENT
CASSELMAN RODNEY WILLARD
CASTANEDA BENJAMIN BELTRA
CASTANEDA BENJAMIN FRANK
CASTANEDA HUGO CARLOS
CASTILLO ARTHUR JOHN
CASTILLO CHARLES MIKE
CASTILLO DAVID RIVAS
CASTILLO GREGORIO PEDRO
CASTILLO JOSE
CASTILLO MANUEL ANGEL
CASTILLO MANUEL GRIJALVA
CASTILLO THOMAS
CASTLE ROBERT EDWARD
CASTLEBERRY JIMMIE LYNN
CASTRO ALFONSO ROQUE
CASTRO JESSE ROMERO
CASTRO JOE
CASTRO JORGE ARTURO
CASTRO JUAN JOSE
CASTRO LOUIS
CASTRO REINALDO ANTONIO
CATELLI CHARLES JOHN
CATINO STEVEN LYNN
CATRON GARRY WAYNE
CATT JOSEPH FRANCIS JR
CAUDILLO JOSEPH
CAUSEY DAVID LOUIS
CAVARZAN DUANE EARL
CAVENDER JIM RAY
CECIL JACK WILSON
CELANO FRANK ANTHONY
CENTENO CHARLES MANUEL
CENTENO EDWARD LOUIS
CENTER ROBERT LEE
CERVANTES GEORGE ANDREW
CERVANTES GERALD
CHABOT DON WILLIAM
CHACON RIGOBERTO COTO
CHACON ROBERT REINHARD
CHAMBERLAIN DALE STEWART
CHAMBERS STEVEN DOYLE
CHAN PETER
CHANDLER ROBERT HUGHESTON
CHANEY DAVID MICHAEL
CHANEY DONALD LEE
CHANEY DOUGLAS DALE
CHANEY NORMAN J
CHANEY THOMAS CLIFFORD
CHAP DE LAINE ARNOLD A JR
CHAPA ARMANDO JR
CHAPIN CHARLES CLARK
CHAPP ROBERT ANTHONY
CHARLES RONNIE JOE
CHARLESWORTH CHAD ALLEN
CHASE CHARLES JOSEPH
CHASE MARK RICHARDSON
CHATFIELD WENDELL OLIVER
CHAUDOIN ROBERT CONN
CHAVEZ CARLOS JR
CHAVEZ EDUARDO
CHAVEZ JESUS ERNEST JR
CHAVEZ RUDOLFO
CHAVIRA STEPHEN
CHERNEY PETER FREDERICK
CHEROFF MICHAEL
CHERRSTROM RONALD PAUL
CHERRY DANIEL PARKS
CHERRY WILLIAM TEMEN JR
CHERVONY EDDIE EDWIN
CHESTER DENNIS EDWARD
CHIACCHIO JOSEPH S JR
CHILCOTT RONALD HARRY
CHILDERS JAMES STANLEY BE
CHISLOCK LEONARD JAMES
CHISUM DAVID
CHO HERBERT POK DONG
CHOATE RANDALL BINGHAM
CHOI WILLIAM DAVID
CHRISMAN REX GORDON
CHRISTENSEN HAROLD ROY
CHRISTER EUGENE MERL
CHRISTIANSEN EUGENE F

California

CHRISTIANSON DAVID B
CHRISTIANSON RONALD F
CHRISTIE DENNIS RAY
CHRISTOPHER ANTHONY PHILL
CHRISTY DONALD RAY
CHRISTY RICHARD THOMAS
CHRYSLER MEDFORD ADARINE
CHUBB JOHN JACOBSEN
CHUNKO GEORGE DAVID
CHURCH JOHN LEONARD
CHURCH LEVAN ARLIN
CHURCHILL LAWRENCE JEFFRE
CHUTE STEPHEN FORREST
CIARFEO GLENN THOMAS
CIGAR FREDDIE JOE
CINCOTTA THOMAS ANTONE
CINTRON JIMMIE DUAYNE
CISNEROS JOSE B
CISNEROS MARIO ALVAREZ
CLAEYS EDWARD ORAN
CLAIRE KENNETH WILLIAM
CLANTON LARRY JACK
CLARK DONALD EUGENE
CLARK DOYLE WAYNE
CLARK ERNEST LEE
CLARK GARY RICHARD
CLARK JAMES LEE
CLARK LONNIE WARREN
CLARK MICHAEL BURRISS
CLARK PAUL LESLIE
CLARK RAYMOND CHARLES
CLARK ROBERT ALAN
CLARK STANLEY SCOTT
CLARK STEPHEN WILLIAM
CLARK THORNE M III
CLARK WILLIE C
CLARKE WALTER KIRT
CLASEN MICHAEL ROY
CLAUSEN HARLAND GENE JR
CLAY RUSSELL LELAND
CLAYTON TOMMY MAKIN
CLEAVES MICHAEL DAVID
CLEEM LARRY LLOYD
CLEMENTS WILLIAM RICHARD
CLENDENEN CHARLES CURTIS
CLEVELAND JAMES
CLEVELAND LARRY MICHAEL
CLIFFORD HAROLD JOHN
CLIFTON MANCOL RAYMOND
CLIREHUGH ROBERT W JR
CLOSE DONALD EDWARD
CLOUD JOSEPH JR
CLUTTER CARL NORMAN
COALSON STEPHEN EDWARD
COATS CHARLES ALEX
COBARRUBIAS ROBERTO
COBARRUBIAS ROBERTO
COBB JOHN WESLEY
COBB TYLER WILLIAM JR
COBOS ALFRED
COCHRAN MICHAEL DALE
COCHRANE GREGG LAWRENCE
COCKERHAM JOHN WILLIE JR
COE PAUL THOMAS
COFFEY RICHARD ARTHUR
COHEN ROBERT BRUCE
COLE ROBERT OWEN
COLE SAM JR
COLE WAYNE MICHAEL

COLEMAN DONALD HUSTON
COLEMAN GARY TERRENCE
COLLAZO RAPHAEL LORENZO
COLLIER TIMOTHY LYNN
COLLINS BRUCE WAYNE
COLLINS CLINT
COLLINS JONATHAN III
COLLINS MICHAEL RAYMOND
COLLINS RICHARD FRANK
COLLINS RICHARD GLEN
COLOMBERO JAMES STEPHEN
COLYEAR CURTIS CRAIG
COMACHO PETER FRANK JR
COMBS ALFRED HENRY JR
COMBS ALLAN EUGENE
COMBS JAMES MILES
COMBS KENNETH DALE
COMPTON ROBERT WILLIAM
COMPTON WILLIAM EDGAR III
CONLEY ROBERT FRANK
CONLEY TERRY LEWIS
CONLIN JEFFREY FRANCIS
CONN DAVID BRUCE
CONNELLY RICHARD JOHN
CONNER JACK WILLIAM
CONNER MELVIN HUBBARD JR
CONNIFF THOMAS JOSEPH
CONNOLLY GEORGE THOMAS
CONRAD ROY EUGENE
CONTRERAS BENITO JR
CONTRERAS JOHN JENARO
CONTRERAS MIGUEL ZARAGOZA
CONWAY TERRY MIKEL
COOK AUSTIN BRUCE
COOK CHARLES JOSEPH
COOK CHARLES JR
COOK CHRISTOPHER CORWIN
COOK DENNIS PHILIP
COOK GEORGE KENNETH
COOK JOHN PHILLIP
COOK JOHN W
COOK MICHAEL DEAN
COOK MICHAEL FRANK
COOK ROBERT PAUL
COOK WILLIAM DONALD JR
COOKE HAROLD THOMAS
COOKE LARRY HOUSTON
COOKE ROBERT ALLEN
COOLEY OCIE DANIEL
COOMER RICHARD ROSS
COONEY JAMES HENRY
COONROD ROBERT LEE
COONS PETER MICHAEL
COOPER DONALD NATHANIEL
COOPER DONALD RAY
COOPER EDWIN EARL
COOPER JAMES RICHARD
COOPER ROBERT GEAN
COOPER ROY ELDON
COPLEY WILLIAM MICHAEL
COPP THOMAS ELLIOTT
COPPERNOLL DAVID WILLIAM
CORBIN RONALD JAMES
CORDERO WILLIAM EDWARD
CORDOVA SAM GARY
CORNEJO ALFRED JOSEPH
CORNELL EDWARD MICHAEL
CORNELL RICKY LYNN
CORONA DOMINIC ANTHONY
CORONA FRANK RODRIQUEZ
CORONA RUDOLPH RALPH III
CORR PAUL JR
CORRIE GARY ALLEN
CORRIE MARK LANE
CORRIGAN MICHAEL JOSEPH
CORTEZ ALBERT ROMERO
CORTEZ RICHARD
CORYELL MICHAEL NOBLE
COSSA WILLIAM EDWARD JR
COSTA WILLIAM CARL
COSTANTINI FRANK JOSEPH JR
COSTON RICHARD JAMES
COTA ERNEST KENO
COTE ROBERT PAUL
COTTERELL JACK PATRICK
COTTRELL THOMAS LEE
COULSON THOMAS EUGENE
COUNCILL ARTHUR COBY III
COURTEAU EDWARD GERARD
COURTEMANCHE CALLEN

JAMES
COURTNEY JOE RAY JR
COUTRAKIS GEORGE
COVER LAWRENCE LEROY
COVERT RICHARD DEAN JR
COWAN DANNY ALLEN
COWAN PAUL ALLEN
COWELL RICHARD JOHN
COWELL ROBERT BLANCO
COX FREDDIE JAMES JR
COX GARY ALLEN
COX GARY WAYNE
COX GREGORY ELLIS
COX LARRY JAMES
COX MICHAEL MILTON
COX RICHARD PAUL
CRABB WINFORD R
CRAIG GARY RAYMOND
CRAIG MICHAEL DENNIS
CRAIG THOMAS EDWARD
CRAIN CHARLES ERNEST
CRAMBLET HOWARD EARL
CRANDALL TIMOTHY ALLEN
CRANE ROBERT IRVING
CRANFORD THOMAS WILLIAM
CRANMER FOSTER
CRAPO RONALD CARL
CRAW DONALD DWIGHT JR
CRAWFORD CLAUDE LEE
CRAWFORD JAMES PATRICK
CRAWFORD STANLEY WENDEL
CREAL CARL MARTIN
CREASON W K UTAH
CRIPE MERL L
CRISWELL ROBERT REED
CRITES ROBERT LINCOLN JR
CROKE ROBERT STANLEY
CRONE DONALD EVERETT
CROOK JAMES PEYTON
CROPPER CURTIS HENRY
CROSBY JAMES ALLEN
CROSS HUGH W
CROSS MONROE WARD
CROSS MONROE WARD
CROSSEN MICHAEL O
CROUSE JEFFREY CHARLES
CROW CHARLES CURTIS
CROWE CARL WAYNE
CROWE RICHARD EYRE
CROXEN RICHARD LYNN
CRUISE KENNETH T JR
CRUM DARYL WAYNE
CRUMP ERSKINE LOGAN
CRUZ JOHNNY MANUEL
CRUZ OSCAR
CRUZ PETE FRANK
CRUZ RAPHAEL
CRYAN KENNETH MICHAEL
CRYAR MICHAEL GEORGE
CUELLAR PILAR JOSEPH
CULBERTSON SAMUEL KENT
CULLEN DENNIS JOHN
CULVER ROBERT WAYNE
CULVERHOUSE LEON THOMAS
CUMMINGS DAVID GUY
CUMMINGS DONALD LOUIS JR
CUMMINGS JAMES BARTON JR
CUMMINGS RONALD EUGENE
CUMMINS LANNY DEE
CUMMINS THOMAS WAYNE
CUNEO ANDREA JR
CUNNINGHAM DAVID CARSON
CUNNINGHAM GEORGE MICHAEL
CUNNINGHAM JOSEPH W JR
CUNNINGHAM NORMAN NORTHRO
CUOZZO FRANK XAVIER
CURCI ANTHONY BOY
CURLEY RAYMOND NELSON
CURRENCE WILLIAM ALLEN
CURRY JIMMY DOUGLAS
CURRY ROBERT ERVEN
CURTIS GARY STILLMAN
CURTIS RICHARD
CURTIS ROGER DALE
CURTTRIGHT LARRY BRENT
CUSSON THOMAS LEE
CUTHBERT LOWRY TAYLOR
CUTHBERT STEPHEN HOWARD

CZARNECKI STEVEN CHARLES
D AIELLO MICHAEL DENNIS
D EMANUELE ROBERT PAUL
DAFFER JOSEPH JOHN
DAHL KENNETH ALAN
DAHLIN DAVID COURTNEY
DAILEY KEVIN MELBOURNE
DAILY THOMAS BLAKE
DAL POZZO ANTHONY JR
DALENTA ZBIGNIEW JOSEF
DALRYMPLE ROGER EARL
DALTON ROBERT LE ROY
DAMON MICHAEL PATRICK
DANCE ROBERT LYNN
DANCE ROBERT LYNN
DANIEL JOHNNIE LLOYD
DANIELS BRUCE WILLIAM
DANIELS HARLAN EUGENE
DANIELS LARRY PHILLIP
DANIELS THOMAS JAMES
DANIELS TOMMY LEON
DANIELS WILLIAM JR
DANNA SAMUEL DON
DANSER GARY RICHARD
DARCY MICHAEL CHAVEZ
DARDEN OTIS JAMES
DARRAH GARY KOYLE
DAUGHERTY DONALD D
DAUGHTON JOSEPH D JR
DAVENPORT ROBERT MALCUM
DAVID MICHAEL DENNIS
DAVIDOVE ERNEST FREDERIC
DAVIS ALAN EUNICE
DAVIS ALBERT JACKSON
DAVIS ALFRED LEE
DAVIS BRENT EDEN
DAVIS CARLOS RAY
DAVIS CHARLES EDWARD
DAVIS DALE LEROY
DAVIS DANNY CRAIG
DAVIS DUANE ROSS
DAVIS EDWARD THOMSON JR
DAVIS FREDRIC BRUCE
DAVIS GARY JAMES
DAVIS JAMES ALLEN
DAVIS JAMES MIKE
DAVIS JEFFREY LYNN
DAVIS JOHN EDWIN
DAVIS LARRY KENT
DAVIS RAY RENE
DAVIS ROBERT SCOTT
DAVIS ROY HENRY
DAVIS SAMUEL VERNELL
DAVIS THOMAS JOEL
DAVIS WALTER SCOTT
DAVIS WILLIAM STANLEY
DAVISON DAVID MICHAEL
DAWES JOHN JAMES
DAWSON DANIEL GEORGE
DAWSON FRANK ARTHUR
DAWSON PAUL GLEN
DAWSON THOMAS JOE JR
DAWSON WILLIAM JOHN
DAY DOUGLAS WAYNE
DAY KEVIN LLOYD
DAY PETER EVAN
DE ABRE JAMES MICHAEL
DE AMARAL CHARLES F JR
DE ARO STEPHEN WAYNE
DE BOLT MICHAEL LLOYD M
DE GRAW CHARLES IVAN
DE GROOT MAARTEN
DE LA PAZ ABEL JOSEPH
DE LA PENA GILBERT
DE LA ROSA LARRY A JR
DE LA TORRE JOSE MANUEL
DE LA TORRE LUIS
DE LACY MICHAEL CHARLES
DE LAPP WILLIAM C III
DE MARA JUAN JOSEPH
DE MARCO FRANK JOHN
DE MARCUS JERRY DENNIS
DE MELLO BRYAN JOE
DE MELLO CLYDE LAWRENCE
DE MELLO ROBERT BRUCE
DE NARDO FRANK MICHAEL JR
DE ROO JOHN ALBERT
DE ROO LANCE AARON
DE SOTO ERNEST LEO
DE VEGTER PAUL ANTHONY
DE VORE EDWARD ALLEN JR

DEAL OLIVER EVANS JR
DEAN MICHAEL FRANK
DEARDORFF HEROLD TROY
DEARING JERRY WAYNE
DEAVER JACK
DECELLE ROBERT EUGENE II
DEDMAN JULIAN DEAN
DEEBLE JAMES FREDERICK
DEHERRERA RAYMUNDO F
DEHNKE DALE WILLARD
DEITRICK GEORGE DOUGLAS
DEKKER GEORGE WILLIAM
DEL CASTILLO MARCO OSCAR
DELANEY THOMAS ALAN
DELANO THOMAS FRANCIS
DELEIDI RICHARD AGUSTINE
DELGADO FRANCISCO PENA
DELGADO JOHN PEDRO
DELGADO JOSE ALEJANDRO
DELGADO RAY
DELGADO RAYMOND RODRIGUEZ
DELGADO RICHARD FALCON
DELLAMANDOLA GREGORY JOHN
DELP KENNETH HARVEY
DEMMON DAVID STANLEY
DENLINGER DAVID WOOD
DENMAN WILLIAM LUTHER
DENMARK ROBERT LEE
DENNA DAVID RAMIRO
DENNIS BLAIR EDWARD
DENNIS HAYVARD JR
DENNY JACKIE LEE
DENT GARY LYNN
DENTON DAVID ANDREW
DEOCAMPO GREGORIO MANESE
DERIG PATRICK MARTIN
DERRICKSON THOMAS G II
DERVISHIAN SARKIS
DESILETS WILLIAM JAMES
DESMOND RAY GLEN
DETRICK ROBERT LLOYD
DEVERS LESLIE ALLEN JR
DEVORE KENNETH ROY
DEWEY ERIC MELVIN
DI BARI LOUIS SCOTT
DI FIGLIA FRANK ANTHONY
DI NAPOLI MICHAEL JOSEPH
DIAZ DANIEL
DIAZ GARY MICHAEL
DIBBLE GORDON JOHN
DICK MANUEL LEVI
DICKSON THOMAS GEORGE
DICKSON WILLIAM DOUGLAS
DIEBALL DENNIS RAY
DIEFENDERFER THOMAS EDWAR
DIEHL HARRY G
DIETZ LAWRENCE ALFRED II
DIETZ WOLF-DIETER
DIGGS JOHN FRANCIS
DILLON DAVID ANDREW
DILLON DENNIS JAMES
DILLON GEORGE ALFRED CHED
DINEEN TIMOTHY JOHN
DIONNE DONALD THOMAS SR
DIORIO MARK STEVEN
DISMAYA EDDIE JR
DISMAYA EDDIE JR
DISPENSIERO DOUGLAS LOUIS
DIVES THOMAS LAMONTE JR
DIXON LEE CHRIS
DIXON MICHAEL KENNETH L
DOANE GEORGE ALFRED
DOCK RAYMOND LEE JR
DODD BILLY FRANCIS
DODD LAWRENCE RUDIN
DODGE GREGORY ALEXIS
DODGE RONALD WAYNE
DODSON DAVID LEE
DOLAN DAVID PATRICK
DOLIM STEVEN FRANCIS JR
DOMINGUEZ ERNESTO
DOMINGUEZ MICHAEL CHARLES
DONAHE WARREN LEE
DONNELLY VERNE GEORGE
DONOVAN PATRICK JOHN
DONOVAN TOMMY CLAYTON II
DOOLEY MICHAEL BANION

DOOLITTLE RONALD LOUIS
DORAN PATRICK MICHAEL
DORAN SEAN TIMOTHY
DORMAN CHARLES DUDLEY
DORMAN DANIEL GENE
DORNBERGH WILLIAM L JR
DORSEY CARLITO LADORES
DOTSON DENNIS WILLIAM
DOTSON EUGENE LEWIS
DOUCET WILLIAM BRADLEY
DOUGLAS CHARLES MAC
DOUGLAS JOHNNIE LEE
DOUGLASS GERALD TYLER JR
DOWDELL STEPHEN
DOWNING JOHN FREDERICK
DOWNS JACK DENNIS
DRAKE EARLE AVON
DRAKE JOHN DE WITT
DRAKE RICHARD GUY
DRISCOLL JOHN RAYMOND III
DRYER RICHARD EUGENE
DUARTE GERALD MICHAEL
DUARTE JOHN FRANK JR
DUCE ROGER L
DUCK CURTIS LAMAR
DUCOMMUN RONALD LLOYD
DUEMAN MERLE L
DUEMLING RALPH NELSON
DUENAS JUAN LEON GUER-
RERE
DUENAS ROBERTO CERVANT-
ES
DUENSING JAMES ALLYN
DUESSENT CHARLES PAUL
DUFFEY GERALD THOMAS
DUFFIELD JOHN DAVID
LOCKW
DUFFY FRANCIS JOSEPH
DUFFY VINCENT EDWARD
DUGAS MICHAEL JEAN
DUGGAN THOMAS PATRICK
DUGGAN WILLIAM JOSEPH
DUHE BYRON RANDALL
DUKE ALAN RAY
DUKEHART STEPHEN ERNEST
DULAY SALVADOR REDILLA
DULLEY KENNETH LAW-
RENCE
DUNAGAN JIMMY LYN
DUNCAN JAMES ROBERT
DUNCAN RICHARD WINER-
FRED
DUNCAN ROY WILLIAM
DUNCAN TIMOTHY JOSEPH
DUNGEY RIM MICHAEL
DUNKIN JAMES EDGAR JR
DUNN CREIGHTON ROBERT
DUNN DONALD LEORY
DUNN GREGORY LYNN
DUNNING TIMOTHY CHARLES
DUNSING DENNIS PAUL
DUPERRY PETER ALFRED
DURAN JUAN CHAIRES JR
DURAN PABLO
DURBIN THOMAS FREDERICK
DURELL ALGER EDGAR JR
DURHAM SAMUEL RAY
DUROY ALLEN JACQUES
DURR LAVALL
DUTRA ROBERT LEONARD
DUVAL MICHAEL EUGENE
DWYER PATRICK WILLIAM
DYCUS RICKEY DALE
DYE DANIEL GROVER
DYE DANNY DAVID
DYE JAMES CLETUS
DYER DAVID WAYNE
DYER DENNIS EARL
DYER FREDERICK LEE
DYER JAMES RICHARD
DYER TERRY BROOKS
DYKE ROBERT LOUIS
DYKES RICHARD MONROE
EADE RAYMOND FREDRICK
EARLES FRED THOMAS
EARLEY JOHN RICHARD
EASON EDWIN RAYMOND
EASTON JOHN WILLIAM
EASTON ROBERT GLENN
EATON BRUCE HORACE
ECKENRODE DANIEL EDNEY
ECKENRODE MARCUS RICH-
ARD
ECKL THOMAS ANTHONY

ECKOFF DALE ARNOLD
EDGAR TERRECE EUGENE
EDGERTON ARTHUR DONALD
JR
EDIE KURT CHARLES
EDWARDS BOBBY BRANCE
EDWARDS DONALD MAC
EDWARDS JERRALD LEROY
EDWARDS JOHN LEONARD
EDWARDS JOHNNY LAW-
RENCE
EDWARDS ROBERT THEO-
DORE
EGGLESTON ROBERT RICH-
ARD
EGOLF KLAUS DIETER
EISENACHER CHARLES JOHN
EISENBRAUN WILLIAM
FORBES
EISMAN JAMES FREDRICK
ELBRACHT WILLIAM MI-
CHAEL
ELGIN ROBERT GERALD
ELIA REESE CURRENTI JR
ELIA REESE CURRENTI JR
ELIAS PORFIRIO ELIAS
ELIASON WENDELL THEO
ELKINS JEROME
ELKINTON MICHAEL
ELLIOTT ANDREW JOHN
ELLIOTT BROCK DENNIS
ELLIOTT DAVID RAY
ELLIS DENNIS FLOYD
ELLIS GENE HOWARD JR
ELLIS GEORGE WALTER
ELLIS JAMES ALVIN
ELLIS MICHAEL LE ROY
ELLSWORTH JAMES OLIVER
ELLSWORTH LAWRENCE
ELMORE ALLAN LADD
ELSON JEFFREY CHARLES
EMERINE JERRY OWEN
EMERSON WAYNE HERSCHEL
ENARI MARK NIGGOL
ENBODY MICHAEL WILLIAM
ENCINAS ESEQUIEL MARTI-
NEZ
ENDERIZ VICTOR ANTHONY
ENDSLEY KENNETH RICHARD
ENEDY ROBERT JOHN
ENGBERSON ROBERT LEWIS
ENGEL GREGORY CHARLES
ENGELMAN THOMAS ALMET
ENGEN ROBERT JOSEPH
ENGLAND ROBERT BLAIR II
ENGLAND RONALD LEE
ENGLANDER LAWRENCE JESSE
ENGLE DARRELL LEROY
ENGSTROM LOREN EUGENE
ENRIQUEZ LUCAS RODRIQUEZ
ENRIQUEZ NICHOLAS BEN
ENRIQUEZ TERRY MICHAEL
ENYEART RAYMOND R JR
EPHRIAM DAVID BURNELL
EPLEY ROGER LEE
ERB PATRICK DOUGLAS
ERICKSON DONALD THEO-
DORE
ERICKSON LEONARD DANIEL
ERICKSON PHILIP CHARLES
ERNST EDWARD JOSEPH
ERSCHOEN ARTHUR RAY-
MOND
ERVIN GLEN OTIS
ERVIN GREGORY ALLEN
ERVIN JAMES WILBUR
ESCAGEDA JESUS
ESCALANTE DOMINGO JR
ESCALERA RICHARD MEDINA
ESCAMILLA JOSEPH
ESCOBAR EDWARD ANGIANO
ESCOBAR JESUS GUTIERREZ
ESCOBAR JOSEPH SANCHEZ
ESCOBAR SANTIAGO HER-
RERA
ESCOBEDO DANIEL
ESCOTT KENNETH ROBERT
ESKEW RONNIE JOE
ESPARZA JOSEPH DAVID
ESPARZA MALCOLM MAR-
CELLIN
ESPARZA MALCOLM MAR-
CELLIN
ESPINOZA ALFONZO LOUIS JR

ESPINOZA MIKE PATRICIO
ESQUEDA ARTHUR DIAZ
ESQUIVEL JAIME
ESTES BRIAN ROBERT
ESTES DENNIS REX
ESTOK MICHAEL DAVID
ESTRADA MARIO PEREDA
ESTRADA MAXIMINO
ESTRADA RANDOLPH PHILLIP
ESTRADA ROY LEE
EVANS CHRIS STEVEN
EVANS DONALD RAY
EVANS DONALD WARD JR
EVANS GARFIELD
EVANS MICHAEL JOHN
EVELAND MICKEY EUGENE
EVERSULL ANTHONY PATRICK
EVERTS DENNIS LEE
EWALD ROBERT CLARENCE
EWART JOHN ANDREW
EWING JERRY LEW
EWING LON BARRY
EWING TIMOTHY DAVID
EXNER FRED ANTONY III
EYNON JOHN PATRICK
FACCHINI STEPHEN DALE
FACKRELL CLINTON BLAIR
FAGE ROBERT FREDERICK JR
FALES PHILIPPE B
FANFA ANTHONY JOHN
FANNIN CLAYTON ALLEN
FANUA FIAPAI JR
FARAN DANIEL EDWARD

FARDEN KENNETH ROY
FARLEY MARSHALL COLIN
FARR DAVID EARL
FARRELL BRUCE CHARLES
FAULKNER MICHAEL AN-
THONY
FAULKS DANIEL CLYDE JR
FAVATA SAM JOSEPH
FAY MICHAEL ANDREW
FAY PATRICK DENNIS
FAZZAH GEORGE RICHARD
FEATHERSTON CLIO C JR
FEDOROFF ALEXANDER
FEIGENBUTZ TERRENCE R
FELAND THEODORE GLEN
FELKNER DAVID WILLIAM
FELT DAVID LEVANT
FELTON RUBY EDWARD III
FENN MELVIN B
FENTON ROBERT ALLEN
FERGUSON BLAINE M
FERGUSON EDWARD KEN-
NETH
FERGUSON MARK ANDREW
FERGUSON MERL WAYNE
FERGUSON MICHAUEL DON
FERGUSON RICHARD EUGENE
FERGUSON RONALD DENNIS
FERGUSON WARREN JOHN JR
FERGUSON WILLIAM BOYD
FERGUSSON ROBERT C L
FERNANDEZ EARL WILLIAM
FERNANDEZ MANUEL AN-

SELMO
FERNANDEZ ROBERT SAN-
CHEZ
FERNANDEZ XAVIER
FEROUGE RONALD WALTER
FERRALEZ RICHARD
FERRANTE GILBERT
FERRARI ARNOLD JAY
FERRO PHILIP ANTHONY
FERRULLA ROBERT SAMUEL
FEWELL TIMOTHY FLOYD
FIDEL HONORIO MORAN JR
FIELD JAMES ROLAND
FIELDING DAVID ANDREW
FIERRO ALEJANDRO FRAN-
CISC
FIGUEROA FRANK NUNEZ
FIGUEROA JAVIER PUENTES
FILES ALBERT CLIFTON JR
FILIPIAK PETER JAN
FILIPPELLI JOHN MARIO
FILIPPI GERALD FRANCIS
FILLMORE RONALD RICHARD
FINK ROBERT ALTON
FINLEY LELAND PATRICK
FIORENTIN JOHN VELCO
FISCHER GREGORY JAMES
FISH WILLIAM ARRON
FISHER DENNIS FRANKLIN
FISHER EDWARD STEPHAN
FISHER JIMMY LEE
FISHER JOHN WILLIAM
FISHER LA MARR

FISHER RICKIE DAVIS
FISHER ROBERT GENE
FITCHETT REGINALD WILLIAM
FITZGERALD TERENCE PATRIC
FITZSIMMONS JAMES PATRICK
FITZSIMMONS PATRICK G
FIX WILLIAM LEROY
FLADGER RALPH SAMUEL
FLANNERY MICHAEL EDWARD
FLANNERY ROBERT EDWARD JR
FLANSAAS DANIEL ROBERT
FLEER ROBERT DEAN
FLEISCHMANN DALE FRANK JR
FLEMING BERNARD JOHN
FLETCHER KIM WILLIAM
FLICKINGER JAMES EDWARD
FLICKINGER JAMES HERBERT
FLINN JOHN LEROY
FLOHR GEORGE JR
FLORES DANIEL
FLORES EDWARDO
FLORES FELIX FRANK
FLORES JOSE DEJESUS
FLORES JOSE LUIS
FLORES MONICO JR
FLORES RICHARD JAVIER
FLOREZ REYNALDO B
FLOWER CARL DAVID
FLOWERS WILSON NATHENIAL
FLOYD GARLAND DALE
FLOYD GEORGE ALLEN
FOBAIR ROSCOE HENRY
FOGG ALBERT RANDOLPH III
FOLEY LONNIE DEE
FOLMAR MASON OPHELIA
FOLSOM ROBERT ELMER
FOLSOM TERENCE J
FORBES PAUL GLENN JR
FORBES RICHARD ALLAN
FORCE DAVID LEE
FORD DAVID TODD
FORD EARL EUGENE
FORD ERNEST DOW
FORD PATRICK OSBORNE
FORD RICHARD WAYNE
FORREST STEPHEN CALEB
FORRESTER LAWRENCE BRADFO
FORSHEY ROBERT ERNEST
FORSTER RONALD EDWARD
FORTNER FREDERICK JOHN
FORTUNE RODGER LEE
FOSTER EDDIE DALE
FOSTER JOHN MICHAEL
FOSTER LAWRENCE EUGENE
FOSTER PAUL HELLSTROM
FOSTER ROBERT LEE
FOULKE JEFFREY HOWARD
FOULKS RALPH EUGENE JR
FOUST MICHAEL THORNTON
FOWLER DANIEL CHARLES
FOWLER LEWIS LOREN
FOWLER WILLIAM HOLT III
FOX JAMES CARL
FOX THOMAS JOSEPH JR
FRAKES DWIGHT GLENN
FRAKES KENNETH DEAN
FRANCIS JAMES PATRICK
FRANCIS STEVEN DAVID
FRANCKOWIAK JOSEPH

RALPH
FRANCO FRANCISCO
FRANKEL JOHN PAUL
FRANKLIN GEORGE STEVE
FRANKLIN JAMMIE VAN
FRANKLIN JEFF LEE JR
FRANKLIN JOHN HENRY
FRANKS BARRY RICHARD
FRATTALI MICHAEL ANGELO
FRAZIER REX LEONARD
FREDSTI STEFFAN MICHAEL
FREEDLE FRANK LOUIE
FREEMAN DARELL GOODWIN
FREEMAN RICHARD BARTON
FREGOSO MARCO AURELIO
FREILING JOHN RICHARD JR
FREITAS ROBERT EDWIN
FRENCH FRED
FRENYEA EDMUND HENRY
FRENZELL HERBERT ERNEST
FREY DANIEL ALAN
FREY DEAN LEE
FREY JESSE CLIFFORD
FRICK EDSALL A
FRICK JOHN ALAN
FRIEL BRUCE GARY
FRIEND RICHARD ALLEN
FRISBEE DENNIS WAYNE
FRITS ORVILLE BILL
FRITSCHE ROBERT JR
FROMME FREDERICK W JR
FRONTELLA MELVIN LAWRENCE
FROST HERBERT CORNELIUS
FRY BILLY G
FRY JAMES RAY JR
FRY STEPHEN MICHAEL
FRYE DONALD PATRICK
FRYE RICHARD ALAN
FRYE TERRANCE DONALD
FRYER BENNIE LAMAR
FRYER ROBERT RISLEY
FRYMAN ROY ALLEN
FUENTES ANTONIO MORILLIA
FUENTES HECTOR
FUJIMOTO DONALD SHUICHI
FULLER CARROLL BRUCE
FULLER DENNIS EARL
FULLER STANLEY CARL
FULLERTON JAMES PRICE
GABALDON TONY EIDDIE
GABBERT DENNIS ERWIN
GABRIEL MEREDITH ALTON
GAFTUNIK ROBERT ERNEST
GAFTUNIK STEVEN JOHN
GAGE MICHAEL ARTHUR
GAINES MELVIN CLYDE
GALABIZ JOHN ROSALEZ
GALAMBOS JOSEPH GARY
GALAN DAVID LUIS
GALE DAVID LEE
GALIANA RUDOLPH STEVEN
GALLAGHER ARTHUR TERRY
GALLARDO JOHNNY JOE
GALLEGOS GABRIEL
GALLION GAYLEN RAY
GALLO PETER JOSEPH
GALVEZ JOE ANGEL
GALVIN JAMES PATRICK
GALVIN RALPH FORRESTER
GAMBLE HARRY PAUL
GARCIA ANDREW PEREZ
GARCIA ANGEL ANTONIO
GARCIA ANTONIO MENDEZ
GARCIA ARNOLD FALCON
GARCIA CHRISTOPHER
GARCIA DOMINGO YABBARA
GARCIA EDILBERTO
GARCIA FRANK JOSEPH
GARCIA GREGORIO M
GARCIA HENRY JR
GARCIA ISAAC RAMIREZ JR
GARCIA JEROME
GARCIA JOHN
GARCIA JOHNNY PHILLIP
GARCIA MANUEL MENDEZ JR
GARCIA MARCAS JOSE
GARCIA MARCIAL BONDOC
GARCIA MELESSO
GARCIA NICKOLAS
GASTELUM
GARCIA PEDRO CORDOVA
GARCIA RAYMOND IGNACIO
GARCIA RAYMOND JR

GARCIA RICHARD
GARCIA RICHARD CALUDE
GARCIA WILLIAM
GARCIA-GARAY JUAN
GARDENHIRE JIMMY MARYLAND
GARDINER ROBERT PAUL
GARDNER GLENN VIRGIL
GARDNER GORDON DWIGHT
GARDNER JACK ELROY
GARIBAY GUADALUPE B L
GARNICA ANDY
GARRETT LAWRENCE CASEY
GARRETT MICHAEL SHERIDAN
GARRINGER DAVID FRANK
GARTON TOMMY RAY
GARVER PHILLIP EUGENE
GARZA ARNOLD GARZA
GARZA JOHN ANGEL
GARZA RICHARD JR
GASSER DONALD LEROY
GASTELUM EUGENE
GATES ALFRED ALAN
GATEWOOD GERALD PETER
GATEWOOD GERALD PETER
GATTI GARY FRANCIS
GAU LOUIS ELLIE
GAULT BILL EDGER
GAVIA JOSEPH JESS
GAY LONNIE JAMES
GAYER KENNETH EUGENE
GAYMON STEPHEN H
GAYNOR JAMES THOMAS
GAYOSSO JOE FRANK
GAYTAN HUGO ARAUX
GAZDAGH JAMES ALEX
GEAR GARY WAYNE
GEBHART CARL MERLIN JR
GEE ALAN TIMOTHY
GEE GREGORY JOSEPH
GEIGER ROBERT CHARLES
GEISER DAVID JEROME
GELIEN WALTER JOHN
GELLERMAN KENNETH GILBERT
GEMAS TERRY DALE
GENDRON ROBERT MICHAEL
GEORGE RAYMOND JAMES

GEORGE RICHARD EUGENE
GEORGES JERRY HAROLD
GERARD LAWSON DOUGLAS
GERHARDT ERNEST KAY
GERO JOHN ANTHONY
GEROME MICHAEL ANTHONY
GERSHEN HOWARD DEXTER
GEYER LEROY CLYDE
GIANNELLI GIUSEPPE
GIBBINS ROBERT WAYNE
GIBSON WALTER LEWIS
GIFFORD HOWARD M
GILBERT THOMAS ORVAL
GILBERTSON CARL LOUIS
GILBERTSON VERLAND ANSEL
GILBRETH DON RILEY
GILES GARY LYN
GILES JAMES
GILLASPY EDWARD ANDREW
GILLES ALAN CLARK
GILLIAM JOHN PERRY
GILLIAM MARTIN GALE
GILLIAM ROBERT WENDELL
GILLILAND JOHN HENRY III
GILMOR DAVID LEE
GILMORE PETER WARREN
GILSON MICHAEL ANTHONY
GINES MANUEL LOUIS
GINGERICH GREGORY RAY
GIOVANACCI RICHARD ALLEN
GIOVANNELLI GARY LEE
GIPSON RICKY DUANE
GIRARD CHARLES PIERRE
GLANCY LEE O DAY III
GLASPIE WILLIAM HAROLD
GLASS DONALD ROBERT
GLASSON WILLIAM ALBERT JR
GLEIM ARTHUR FREDERICK JR
GOAR LARRY L
GODINES MIKE MORA
GODINEZ ALEJANDRO RAY
GODOY PETER JR
GODWIN JAMES R
GODWIN JOSEPH SAMUEL
GOELLER MICHAEL DENNIS
GOEREE WIM
GOETSCH WAYNE AUGUST
GOFF ALAN SHERMAN
GOIAS EVERETT WILLIAM

GOINS CHESTER LEON
GOLDBIN CHARLES HENRY
GOLDEN JOHN MICHAEL
GOLDMEYER CHARLES HENRY
GOLDSMITH DANIEL ERIC
GOLLIHER PATRICK CARL
GOLSH STEPHEN ARTHUR
GOMES MICHAEL CHARLES
GOMEZ ANDRES ARMANDO R
GOMEZ HENRY
GOMEZ LAMBERT ANSELMO
GOMEZ MANUEL JOSEPH
GOMEZ ROBERT
GOMEZ ROBERT RAZO
GOMEZ-DIAZ RIGOBERTO
GOMEZGUTIERREZ MANUEL
GOMOLICKE LEONARD MICHAEL
GONDERMAN FRANK LEE
GONSALEZ LUIS MARTINEZ
GONZALES CARLOS M
GONZALES DAVID
GONZALES JIM ROY
GONZALES JIMMY
GONZALES JOE JULIAN
GONZALES JUAN E
GONZALES MERCED HERMAN
GONZALES PAUL ALFRED
GONZALES RICHARD CASTILLO
GONZALES ROBERT
GONZALEZ ALBERT
GONZALEZ BERNARDINO JR
GONZALEZ CARLOS
GONZALEZ JOAQUIN CHRISTOP
GONZALEZ JOSE ALBERTO
GONZALEZ RICHARD
GONZALEZ ROBERT ESPINOZA
GOOD KENNETH NEWLON
GOODEN MICHAEL ANTHONY
GOODINE DAVID ROBERT
GOODLIN JERRY LEE
GOODMAN GREG FREDRIC
GOODMAN LAWRENCE RAY JR
GOODNIGHT PETER RAY
GOODWIN ROBBIN ADAIR
GORDON JOHN HEBER
GORDON JOHNNY LEE

Dedication

The people of the State of California dedicate this Monument to the memory of those Californians who died, or remain missing, in the

Vietnam War
1959-1975

and in doing so, Honor all the men and women who served during that war.

All Gave Some.
Some Gave All.

The names of all contributors to the California Vietnam Veterans Memorial are enclosed within this treasury.

GORE DAVID EDWARD
GORE DONALD EARL
GORE JERRY
GORSICH JAMES TONY
GORTON JACK BURT
GORVAD PETER LAWRENCE
GOSCH THOMAS CHARLES
GOSS RICHARD DEAN
GOSSE JOSE C
GOTCHER LARRIE JACK
GOTTI GALE EDWARD
GOUDELOCK WILLIAM ROGER
GOULD WILLIAM ANDREW
GRADY JERRY EDWARD
GRAF JOHN GEORGE
GRAFF ALLEN MICHAEL
GRAFF PAUL ARNOLD
GRAHAM ALBERT JR
GRAHAM DAVID BRUCE
GRAHAM GILBERT JAMES
GRAHAM JAMES HENRY
GRAHAM JOHN HARRY
GRAHAM ROY WAYNE
GRAHAM STEVEN LOUIS
GRALLA DONALD MICHEAL
GRAMLICK MICHAEL F
GRANADOS RICHARD
GRANELLE AMEDEE GEORGE JR
GRANEY DONALD CARYL
GRANILLO HENRY
GRANNAN MICHAEL STEPHEN
GRANSBURY GERALD ARLEN
GRANT HERBERT RAYMOND
GRANT STEPHEN LEE
GRANT WARREN HARVEY JR
GRANT WILLIAM RICHARD
GRANT WILLIE JR
GRANTHAM ROBERT EUGENE
GRAVEL BOBBY JOE
GRAVES HAROLD LEONARD
GRAVES THOMAS LAWRENCE
GRAVROCK STEPHEN HOWARD
GRAY BERNARD LEROY
GRAY KENNETH MERVIN
GRAY LEONARD CLARENCE JR
GRAY WARREN
GRAYSON WILLIAM RONALD
GRECO LEE ATTILIO
GRECU MICHAEL JOHN
GREEN BERNARD ALAN
GREEN BILLY MONROE
GREEN CHARLES VERNON
GREEN CLIFFORD NEWTON
GREEN DONALD EDWARD
GREEN DONALD GEORGE
GREEN KISH LEMONT
GREEN MICHAEL WAYNE
GREEN ROY COLYN
GREEN THOMAS FREDERICK
GREENE EDWARD LEONARD
GREENE ROBERT EARL
GREENLAW ALAN HEALD
GREENSTREET ROBERT LOWELL
GREENWAY ROGER KENNETH
GREER DENNIS DALE
GREER ROBERT LEE
GREGORIUS GEORGE
GREGORY MACK EDWARD
GREVILLE LEONARD GEORGE
GREY JAMES WILLIAM
GREY RODNEY CHARLES
GRIBBIN JAMES MICHAEL
GRIEGO CLARENCE
GRIFFIN GERALD LEE JR
GRIFFIN LOUIS FREDRICK
GRIFFITH MICKEY EUGENE
GRIFFITH WILLIAM CHAPIN
GRIFFITHS RAYMOND CARSON
GRIFFITHS ROBERT BRYNLEY
GRIGGS STEVEN THOMAS
GRIGSBY BARRY N
GRILLY DAVID A
GRIMES GARY WINGATE
GRIMES MICHAEL BRYAN
GRINNELL GEORGE ALLEN
GRISAFE MICHAEL F JR
GRISWOLD SCOTT CRAIG
GRITZ TOBY RICHARD
GROSS JAMES DAVID

GROSS MICHAEL ANTHONY
GROTHE LEWIS DANIEL
GROVE CORDELL
GROVE NORMAN DOYAL
GRUBBS GAREY LEE
GRUBER FREDERICK LOUIS
GRUBER JOHN HENRY
GRUNDY DALLAS GEORGE
GRUNDY JAMES LEROY JR
GRZESKOWIAK WALDEMAR S
GUARALDI THOMAS JOSEPH
GUARD MARTIN WILLIAM
GUARDADO DANIEL
GUARDADO RUDOLPH
GUARIENTI RALPH
GUENTHER BERT MARRION JR
GUENTHER JOHN CARL JR
GUERIN JOHN PETER
GUERRA JERRY EUGENE
GUERRA RAUL ANTONIO
GUERRA-HERNANDEZ RENE
GUERRERO FRANK ROBERT
GUERRERO JOSE F JR
GUERRERO RICHARD JOSEPH
GUERRERO VICTOR MANUEL
GUEST RAYMOND CALVIN
GUEVARA ERVELL MADRID
GUEVARA IRINEO
GUGLIELMONI TIMOTHY P
GUILLEN DAVID LAWRENCE
GUILLEN PHILLIP O
GUILLORY JAMES CLIFTON
GULIE JAMES PATRICK
GULLART SAMMY MANUEL
GUMMERE DAVID DEE
GUPTON RICHARD CHARLES
GURNIAS NICKALAS PEREZ
GURWITZ LEONARD ZACHARY
GUTHRIE STEVEN ALLEN
GUTIERREZ ERNEST LEMAS
GUTIERREZ RAYMOND RAMIREZ
GUY GEORGE ALLEN
GUYMON ALAN RUSSELL
GUZMAN JUAN ARAUJO
GUZMAN PETER DAVID
GUZMAN PHILLIP JR
GWINN MICHAEL JAMES
HAAKINSON WILLIAM H III
HAAS FREDERICK WILLIAM
HABERLEIN CRAIG
HADLEY JEROME CECIL
HADLEY JOSEPH AUSTIN
HADZEGA GEORGE STEPHEN
HAGEL RICHARD WILLIAM
HAGEN CRAIG LOUIS
HAGEN THOMAS FRANK
HAHN GARY GORDON
HAHN HARLAN LESLIE
HAHN PAUL EDWARD
HAIL WILLIAM WARREN
HAKES CLIFFORD EDWARD
HALBERT LEROY ERNEST JR
HALEY HARRISON LEROY
HALL JAMES KENNETH
HALL JAMES OSCAR JR
HALL JAMES WAYNE
HALL LEONARD JOHN
HALL LEWIS STEVEN
HALL RICHARD DAVID
HALL WALTER RAY
HALLBERG ROGER C
HALLIDAY GARY DEAN
HALPIN RICHARD CONROY
HALSTEAD WAYNE EDWIN
HALVORSON ERNEST JOSEPH
HAMBRICK HAROLD MICHAEL
HAMBY CLYDE RANDALL
HAMILTON EARLIE C JR
HAMILTON GLENN ANTHONY
HAMILTON ROLAND CHARLES
HAMMAN LEE THOMAS
HAMMAR JAMES LEROY
HAMMER ROBERT RALPH
HAMMETT RICHARD LEE
HAMMOND KENNETH JOE
HAMMOND LLOYD MARTIN JR
HAMPTON HENRY GARFIELD
HANBURY DAVID DELANY
HANCOCK CHARLES EDWARD
HANGER JACK DENNIS
HANKAMER GREGORY L
HANKS ERNEST BEAUEL III

HANN DAVID MICHEL
HANNIBAL JAMES EDWARD
HANNIGAN JOHN EDWARD III
HANSEN MARK JOHN
HANSEN WILLIAM JAMES
HANSON KENNETH GREGORY
HANSON STEPHEN PAUL
HARALDSON DAVID ALAN
HARAN RORY TIMOTHY T
HARANO ALLEN HIDEO
HARBORD ANTHONY GORDON
HARBOUR THOMAS JAMES
HARDEE JOSEPH ROBERT
HARDESTY ROBERT WARREN
HARDING DAVID LEE
HARDING TERRY ALAN
HARDING WARREN GUTHRIE JR
HARDMAN JAMES ALLEN
HARDY FRED DOUGLAS
HARDY JERRY RAY
HARDY JOHN KAY JR
HARE MICHAEL JAMES
HARE MICHAEL KENNETH
HARGRAVES MURVYN EUGENE
HARLEY JOHN LEWIS
HARMON NORMAN MARK
HARMON WAYNE CLARK
HARMS GARY LA MONTE
HARNDEN JIM LAWRENCE
HARPER MONTE RAY
HARRELL STEPHEN CARL
HARRINGTON TIMOTHY MICHAE
HARRIS ALLAN LYNN
HARRIS DEAN ALLEN
HARRIS JACK M
HARRIS JACK MARSTON
HARRIS MICHAEL LEO
HARRIS ROBERT EARL
HARRISON EDWARD TERRY JR
HARRISON FOSTER EARL
HARRISON JAMES RICHARD
HARRISON PAUL JAMES
HART ERNEST DWIGHT JR
HART JOSEPH LESTER
HARTLAND CHARLES LEE
HARTMAN JOHN WILLIAM
HARTMAN NICHOLAS MARK
HARTMAN ROBERT GLENN
HARTMAN WILLIAM TAYLOR
HARTNELL RICHARD MIGUEL
HARTY DAVID LEWIS
HARVEY JEFFERY ARNOLD
HARVEY MICHAEL GAIL
HARVEY NEIL EDWARD
HASKELL CHARLES WESLEY
HASKINS MICHAEL WAYNE
HASLET THOMAS EARL
HASSELMAN WILLIAM GEORGE
HASTINGS RICHARD WARREN
HASTINGS STEVEN MORRIS
HASUIKE SKYLER LANCE
HATA GLENN LEE
HATCH RICHARD LEE
HATLESTAD RICHARD L
HATLEY EDDIE LEE
HATTORI MASAKI
HATZELL MICHAEL MAXWELL
HAUG FRED GUNDER
HAUGEN ALAN ROBERT
HAUSER RAYMOND EDWARD
HAUSER VINCENT VANALSTYNE
HAUSRATH DONALD ARTHUR JR
HAVEL RICHARD THOMAS
HAWES ROBERT CARLBERN
HAWK JEFFREY ALLEN
HAWKING THOMAS HOWARD
HAWKINS ARTHUR LOREN JR
HAWKINS DON ALBERT
HAYDEN GLENN MILLER
HAYDEN JOHN LOREN
HAYDEN NEIL WILLIAM
HAYDEN WILLIAM LYLE
HAYEN EDWARD GARDNER II
HAYES DAN DAVID
HAYES DAVID ANTHONY
HAYES DENNIS LEO
HAYES FRED JOE

HAYES JOSEPH D
HAYES WAYNE NORMAN
HAYNER CLAIRE LOWELL
HAZEL LOUIS
HEADLEY FRANK EBERLY IV
HEAL MICHAEL JOSEPH
HEARNE MAURY WILLIAM
HEASTON DONALD LEROY
HEATH ISAAC EDWARD
HECKLER FREDRICK MERRIMAN
HEDERMAN PATRICK SHAWN
HEEKIN TERRY GENE
HEEP WILLIAM ARTHUR
HEEREN DARREL WAYNE
HEFLIN GLENN ELDEN
HEFNER FRANCIS JOE
HEIHN DENNIS RAY
HEIL RICHARD EDWARD
HEIM RICHARD WAYNE
HEIMARK DON RAY
HEIMBOLD JAMES REEVE
HEIN ROBERT CHARLES
HEINMILLER ROBERT LYNN
HEINTZ WAYNE DOUGLAS
HELGESON DALE GEORGE
HELLER ROBERT LEE
HELLMAN KENNETH RAYMOND
HEMPEL BARRY LEE
HENDERSON ARTHUR FRANKLIN
HENDERSON FRANK HAREL
HENDERSON GARLIN JERIS JR
HENDERSON HUGHLEN
HENDERSON JOHN MICHAEL
HENDERSON KAYLE DEAN
HENDERSON TOMMY RAY
HENDRICKS TERRY ALAN
HENDRICKSON ALAN EUGENE
HENNEBERG ROBERT JOSEPH
HENNINGER HOWARD WILLIAM
HENREY RICHARD DEE
HENRICKS CHARLES DRAYTON
HENRICKS FRED CARL
HENRICKSON COMBLY HANIBAL
HENRY DAVID ALAN
HENRY FRANCIS GILBERT
HENRY FREDERICK JOHN
HENRY GERALD RUSSELL
HENRY ROBERT GREGORY
HENRY RONALD JEROME
HENRY SCOTT D
HENRY WILLIAM RICHARD
HENSHAW THOMAS STOW
HENSLEY JOHN
HENSLEY LEROY
HENSON JOHN MICHAEL
HENSON LESLEY HERBERT
HERBST THOMAS WILLIAM
HERCHKORN RICHARD RONALD
HERD THOMAS LEON
HERDERICK CHRISTOPHER ERV
HEREDIA MIGUEL
HERMANSON STEPHEN MARK
HERMOSILLO JOSEPH REFUGIO
HERMS KLAUS JURGEN
HERNANDEZ ALEXANDER VERA
HERNANDEZ CLEMENTE DANIEL
HERNANDEZ DEMETRIO LOMELI
HERNANDEZ FELIPE
HERNANDEZ FRANK SANCHEZ
HERNANDEZ JULIO ALFONSO
HERNANDEZ LOUIS PETER
HERNANDEZ PEDRO ALEXANDRO
HERNANDEZ PHILLIP
HERNANDEZ RAMON ANTONIO
HERNANDEZ RAMON S JR
HERNANDEZ RENE ZAROGOZA
HERNANDEZ REYES C JR
HERNANDEZ REYNALDO

CASTIL
HERNANDEZ VICTOR REYES
HERNDON ROBERT EDWARD
HERRELL DANIEL BRUCE
HERRERA LARRY
HERRERA MICHAEL WARD
HERRERA PHILLIP ARNOLD JR
HERRGESELL OSCAR
HERRICK HENRY TORO
HERRIN DELMAR JOYCE JR
HERRING STEVEN WAYNE
HERRMANN WALTER EDWIN III
HERSHNER DENNIS L
HERTZLER RICHARD DALE
HESS ROBERT JAY
HESSMAN RONALD ANTHONY
HESTER LEO CLAUDE
HEWITSON PAUL CRAWFORD
HIATT BARRY CLINTON
HICKMAN DAVID ALAN
HICKOK ROGER ALAN
HICKS CHARLES DARNELL JR
HICKS DAVID LEE
HICKS EUGENE STANLEY
HICKS FRANK EDWARD
HICKS MICHAEL EUGENE
HICKS STEVEN GARY
HIGGINBOTHAM LARRY GENE
HIGGINS JOHN IGNATIUS JR
HIGGINS PATRICK ALBERT
HIGHFILL ROBERT RAY
HIGHSMITH GRANT HAMLY
HIGHT CHARLES BENNY
HIGUERA MANUEL
HILL ALLAN BRUCE
HILL ARTHUR SINCLAIR JR
HILL DALE EVAN
HILL JOHN ROBERT
HILL JOHN ROBERT
HILL MAURICE RICHARD
HILL MICHAEL WAYNE
HILL PAUL WAYNE
HILL PETER ALAN
HILL RANDALL STEVEN
HILL RODNEY DEAN
HIMMER LAWRENCE
HINES VAUGHN M
HINKSTON ROBERT FRANCIS
HINTON CHARLES COLEMAN JR
HIRES THOMAS MICHAEL
HIROKAWA ROCKY YUKIO
HIRSCHMANN FREDERICK III
HITSON FREDERICK ALTON
HOADLEY GARY ELLIS
HOBBS DOUGLAS ERNEST
HOBBS GARY LEE
HODEL MARK EDWARD
HODGE JOHN DAVID
HOEFFS JOHN HARVEY
HOFER RUSSELL GENE
HOFFMAN FREDRICK JEAN
HOGAN KRAIG SEWELL
HOLBROOK JAMES WENDELL
HOLCOMB DANIEL JENNINGS
HOLCOMB MELVIN DOUGLAS
HOLCOMBE THOMAS MARVIN
HOLDERMAN BRUCE EDWARD
HOLDREDGE DAVID LEE
HOLEMAN RONALD STEVEN
HOLEMAN WARREN DALE
HOLGUIN FRANK JOHN
HOLGUIN LUIS GALLEGOS
HOLLAND LAWRENCE THOMAS
HOLLAND LUEY VERNON
HOLLEY BOBBY ROY
HOLLIDAY CLYDE LEE
HOLLIS JAMES SHELTON
HOLLY RONNIE
HOLMES FREDERICK LEE
HOLMES HUGH BRYANT
HOLMES JERRY LEONARD
HOLMGREN ROY JAY
HOLOVITS LASZLO
HOLSCLAW GARY ARTHUR
HOLSTEIN JOHN L
HOLSTIUS MICHAEL JOHN
HOLT JAMES CHARLES
HOLT RAYMOND CLYDE
HOLTE BRENT ARTHUR
HOLTON GARY DENNIS
HOLTZ MICHAEL LEE

California

HOLZMAN MICHAEL WILLIAM
HOM CHARLES DAVID
HOMUTH RICHARD WENDAL
HONRATH JON ROY
HOOD ERNEST ERVIN
HOOPER BARRY WAYNE
HOOPER JOHN JOSEPH
HOOPER WARD LAWRENCE JR
HOOVER WILLIAM CLIFTON
HOPPE PATRICK BERT
HOPPER LARRY CHARLES
HORAL THOMAS GLEN
HORCAJO ROBERT ALBERT
HORN ALAN MURRAY
HORNE STANLEY HENRY
HORNER CARL NICHOLAS M
HORRELL GERALD ROBERT
HORRIDGE FREDERICK RAYMON
HORSLEY RICHARD WAYNE
HORTON ALBERT HUGH
HORTON MARSHAL LYNN
HOSAKA ISAAC YOSHIRO
HOTCHKISS MICHAEL JENNING
HOTTENROTH JAMES RANDALL
HOTTINGER FRED LEE
HOUGHTON JAMES CURTIS
HOURIGAN MICHAEL PATRICK
HOUSER DORIAN JAN
HOUSER JERRY LEE
HOUSTON JOHN DAVIS JR
HOWARD DAVID TERRELL
HOWARD DWANE GENE
HOWARD JAMES BYRON
HOWARD SYDNEY CLAUDE
HOWE LARRY WAYNE
HOWELL DONALD EDWARD
HOWELL GATLIN JERRYL
HOWELL MICHAEL WAYNE
HUBBARD GREGORY GEORGE
HUBBELL DAN ROBERT
HUBERTH ERIC JAMES
HUBNER DAVID ERVIN
HUBRINS EDDIE BARRY
HUCKABA THOMAS JAMES
HUDSON DENNIS NYE
HUDSON HENRY JR
HUDSON LEONARD PAUL
HUDSON ROBERT LARRY
HUEBNER HERMAN HENRY
HUERTA TOMMY
HUFF RICHARD ELLIOT
HUFF ROBERT RANDEL
HUFFMAN DAVID JAY
HUFFSTUTLER STEVEN RILEY
HUGGANS KENNETH RICHARD
HUGHES BRIAN GREGORY
HUGHES EDWARD COWART III
HUGHES JOHN CHARLES
HUGHES SAMUEL RUEBEN
HUGHES THOMAS GILBERT
HUICOCHEA-REYNA IGNACIO
HULSE GARY WAYNE
HULSE ROBERT MARK
HULTS PHILLIP FRANK
HUME CARL MICHAEL
HUMMINGBIRD FERRELL
HUMPHREY JAROLD EDWARD
HUMPHREY JOHN RICHARD
HUMPHREY WEDEN GARY

HUMPHRIES GARY DEAN
HUNNICUTT JASON DAVID
HUNSBARGER GERALD C III
HUNT BRUCE CHARLES
HUNT CALVIN GENE
HUNT HOOD HAL
HUNT JAMES ANTHONY
HUNT JOHN STUART
HUNT MARTIN MOSHER
HUNT PHILIP MICHAEL
HUNT RALPH WOMMACK
HUNT WILLIAM RAYMOND
HUNTER DELON
HUNTER DENNIS WAYNE
HUNTER RORY WILLIAM
HUNTLEY MICHAEL ALAN
HUNTLEY THOMAS MATTHEW
HUPE RUSSELL EDWARD
HURST HOWARD EUGENE
HURTADO ALBERT STEVEN
HURTADO JOHN BERNARD
HUSTEAD TERENCE MICHAEL
HUTCHINSON ROBERT S II
HUTSON MICHAEL GALE
HWANG GERALD RICHARD
HYLAND JOHN PETER
IANNICELLI RICHARD LEE
IBANEZ DI REYES
ICKE RALPH EDWARD II
IDE BEN HERVEY
IGGULDEN SCOTT WARREN
ILAOA FALEAGAFULA
ILLMAN WILLIAM STEVE
IMBACH JOHN III
INCROCCI RICHARD LAFAYETT
INGALLS GEORGE ALAN
INGRAM ROBERT HOWARD
INGRAM RONALD ERNEST
INGUILLO JOHN DEOGRACIAS
INIGUEZ DENNIS GLENN
INMAN PHILLIP LEE
INSCORE ROGER VERNON
INSEL LAURENCE ALAN
INSPRUCKER GLENN EDWARD
IRONSIDE STEVEN PAUL
IRVING JOHN WILLIAM JR
IRWIN RICHARD RAY JR
IRWIN VAN ALLEN
ISHMAN ALBERT JR
ISRAEL RALPH WALDO JR
ISSENMANN MICHAEL WILLIAM
IVES DAVID ALLEN
JACKSON ARTHUR JAMES
JACKSON BEN JR
JACKSON CHESTER LEE
JACKSON CHRISTOPHER A
JACKSON COLIN FRANK
JACKSON CURTIS DARRELL
JACKSON DAVID ANDREW
JACKSON DENNY MILBURN
JACKSON DONALD GENE
JACKSON DONNEY LYRCE
JACKSON G B JR
JACKSON JAMES ALBERT
JACKSON JAMES ARTHUR
JACKSON LITTLE JAY
JACKSON MICHAEL CHARLES
JACKSON NATHANIEL HARVEY
JACKSON OTIS E
JACKSON PAUL EDWARD
JACKSON ROBERT ELEE
JACOBS DENNIS WAYNE
JACOBS JOSEPH LEWIS
JACOBS WILLIAM JOHN
JACOBSEN TIMOTHY JOHN
JACOBSEN WALLACE RAY
JACOBSON WARNER CRAIG
JACQUES FELIX
JACQUES KENNEDY
JAGARD LARRY FRANK
JAGER ROLAND VINCENT JR
JAGIELO ALLEN DALE
JAKOBSEN PETER LAUST
JAMES CHARLES ROBERT
JAMES CLAUDE RAY
JAMES JOE NEAL
JAMES RONALD EUGENE
JAMISON ROCKWELL GRANT
JANEWAY JERRY LEE
JANOWICZ JOSEPH ANTHONY

JANS ROBERT ALLEN
JANSEN JEROME EDDIE WALLY
JARAMILLO JORGE M
JARICK RUSSELL WILLIAM
JAY ROBERT VERN
JEANTET FRANCIS LEON
JENEWEIN MARK ARDELL
JENKINS CHARLES WAYNE
JENKINS DENNIS ALAN
JENKINS STEVEN LEE
JENKS RICHARD DALE
JENNINGS RUDOLPH
JENSEN DOUGLAS GARY
JENSEN KENNETH VERN
JENSEN MICHAEL CHARLES
JERNBERG ROBERT STEVENS
JERNIGAN MARK THOMAS
JERO DAVID WAYNE
JERSE WILLIAM EDWARD
JESSIMAN THAD BAZILY
JETER DANNY WAYNE
JETT JIMMIE JOE
JETT MICHAEL STEVEN
JETT WILLIAM HOWARD
JEWITT BOB
JIMENEZ ISIDRO BRICENO
JOHNSEN JOHNNIE WAYNE
JOHNSON ANTHONY ERIC
JOHNSON BOBBY CAL
JOHNSON CHARLES AARON
JOHNSON CHARLES HOWARD
JOHNSON DANIEL COPE
JOHNSON DAVID JOSEPH
JOHNSON DENNIS CHARLES
JOHNSON DOUGLAS RAY
JOHNSON FLOYD RAY
JOHNSON FRANKLIN A
JOHNSON GARY LEE
JOHNSON GEORGE STEPHEN
JOHNSON GERALD
JOHNSON GERALD LYNN
JOHNSON HAROLD BENJAMIN
JOHNSON JAMES EDWARD
JOHNSON JAMES J L
JOHNSON JOHN HARRY
JOHNSON KENNETH MICHEAL
JOHNSON KENNETH PAUL
JOHNSON KIM WILLIAMS
JOHNSON LARRY LEE
JOHNSON LARRY RAY
JOHNSON LEONARD RICHARD
JOHNSON PHILIP HARRY
JOHNSON RALPH MARTIN
JOHNSON RAY ELDRIEGE
JOHNSON ROBERT ALLEN
JOHNSON ROBERT WILLIAM JR
JOHNSON RONALD JOE
JOHNSON STANLEY GARWOOD
JOHNSON STEPHEN AYER
JOHNSON STEVEN HOWARD
JOHNSON TIMOTHY ALAN
JOHNSON WILLIAM F
JOHNSON WILLIAM JOHN
JOHNSTON JERRY BERNIS
JOHNSTON RICHARD J
JOHNSTON ROBERT EARL JR
JOHNSTON TOMMY WAYNE
JOHNSTON WILLIAM E
JOJOLA HARRY DANIEL
JONES ABBOTT ROBERT
JONES CHARLES ALEXANDER
JONES CLIFFORD ALAN
JONES DAVID RUSSELL
JONES DONALD BYRON
JONES EARL TIMOTHY
JONES GEORGE WALLACE
JONES GRIFFITH ALFRED
JONES JAMES LESTER
JONES JAMES ROBERT
JONES JAMES WALTER
JONES JERRY DON
JONES JIMMIE WAYNE
JONES JOHN WALLACE
JONES LARRY ALLAN
JONES RANDOLPH ROBERT
JONES ROBERT EMMETT
JONES SCOTT WINFIELD
JONES TERRY EDWARD
JONES TOMMY
JORDAN JIMMY DALE
JORDAN LARRY MICHAEL

JORDAN LITAEL JR
JORDAN PAUL ROBERT
JORDAN STEPHEN ALAN
JORDAN STEVE EUGENE
JOSEPH JEFFREY JOEL
JOSEPH MICHAEL ARNOLD
JOSEPHSON HARTLEY MICHAEL
JOSLIN TERRY LEROY
JOUJON-ROCHE EDWARD
JOY DENNIS EARL
JOYNER STEPHEN DOUGLASS
JOYS JOHN WILLIAM
JUAREZ GEORGE ALBERT
JUAREZ JESSE GOMEZ
JUAREZ JOHN
JUDGE MARK WARREN
JUNGE JAMES CLARENCE
JURI ELGIN JOHN
JUSTUS MICHAEL EUGENE
KAJIWARA JAMES TOSHI
KALB MICHAEL DALE
KALFAS ALLAN GEORGE
KAMRATH JACK HARLAN
KANDLER TERRENCE ARTHUR
KANE MICHAEL
KANEKO JULIO
KANESKI ROBERT ADAM
KANNEL DONALD LEE
KAPPMEYER THEODORE C
KARASCH WOLFGANG WERNER
KARDELL DAVID ALLEN
KARGER BARRY EDWIN
KARGER RICHARD TILDON
KASCH FREDERICK MORRISON
KASKI DONALD ALBERT
KATZ ALLAN HARVEY
KAWACHIKA ARTHUR KAORU
KAWAMURA ROBERT KIYOSHI
KAYE MARK SAMUEL
KAZIKOWSKI JEFFREY G
KEAO JOHN K III
KEARNEY CHARLES DARYL
KECKLER ROBERT L
KEELER RALPH LEROY
KEELING LARRY DEWAYNE
KEEN DARYL LA VERNE
KEEN EDWIN THOMAS
KEENE THOMAS WILLIAM
KEENER RONALD FLOYD
KEIPER GEORGE FREDERICK
KEISTER LAWRENCE LEE
KEITH DANNY JOE
KEITH DENNIS MEVES
KELL JAMES STEWART
KELLER GREG
KELLER JAMES LOUIS
KELLER RICHARD LEON
KELLEY DANA RICHARD
KELLEY DAVID BRUCE
KELLEY JOE FRANKLIN
KELLEY ROGER VIRGIL
KELLISON DAVID GLENN
KELLY CHARLES PATRICK
KELLY DENNIS LEROY
KELLY ERIC MELVIN
KELLY GREGORY PAUL
KELLY JAMES RAYMOND III
KELLY LARRY LEE
KELLY ROBERT MICHAEL
KELLY TERRY LEON
KEMSKI GARY DOUGLAS
KENDALL JAMES D
KENNEDY ALLAN GORDON
KENNEDY BRUCE LEONARD
KENNEDY EDWARD HENRY
KENNEDY WILLIAM DANIEL III
KENNEDY WILLIAM EDWARD
KENNEY ELMER FREDERICK
KENOFFEL STEPHEN MICHAEL
KENT DOUGLAS BRIAN
KEPPLER JOHN MARTIN
KERBY MARTIN JOHN
KERR RONNIE ALBERT
KETCHUM WILLIAM ARNOLD JR
KETTER TERRY LEE
KEYES DANIEL DUANE
KIDD NORMAN RICHARD JR
KIEHL MICHAEL RAYMOND
KIESWETTER GERARD

MARTIN
KILEY MICHAEL JAMES
KILLGORE GENE DOUGLAS
KIM HARRY
KIMBLE CLEATUS PAUL
KIMMEL LEWIS ALBERT JR
KIMMEL STANLEY REGAN
KINCANNON RAYMOND OMER
KINDER BRADLEY ALLEN
KINDRED MICHAEL GEORGE
KINDRICK BRYCE LEROY
KING CHARLES MICHAEL JR
KING DAVID GLENN
KING DONALD GENE
KING GILBERT
KING JAMES ALLEN
KING JOHN CHESTER
KING NORTON ZIGMUND
KING ROBERT WAYNE
KING RONALD REED
KING RONALD RUNYAN
KINGMAN BARRY DEAN
KINNARD JAMES EDWARD
KINNARD WILLIAM LLOYD
KINSWORTHY LOYD EUGENE
KIPP DENNIS WALTER
KIPP DONALD LEE
KIRBY DONALD ROBERT III
KIRK MELVIN LYNN
KIRKPATRICK WILLIAM W
KIRSTEIN DANIEL LYNN
KISH ERNEST
KISTNER GUY DALE
KITNER RICHARD GRANVILLE
KITZMILLER JOHN LESTER
KLAUSING RONALD LAVERN
KLEIN MICHAEL KENNETH
KLENSKE HOWARD LEE
KLINCK HARRISON HOYT
KLINE HARVEY EDWARD II
KLINE ROBERT DANIEL
KLINE ROBERT FRANCIS JR
KLINGER HENRY CHESTER
KLINKE DONALD HERMAN
KLOESE WAYNE RICHARD
KLOSS THOMAS DONALD
KLOTZ JOHN ROBERT
KLUG RICHARD DUANE
KLUGE JAMES DONALD
KNAPP KENTON DON
KNEPP JACK DALE
KNEVELBAARD ANDY
KNIGHT CHESTER WILFORD
KNIGHT HENRY CLAY
KNOTT DAVID LLOYD
KNOTT DENNIS LEE
KNUDSEN HAROLD EUGENE JR
KNUTSON JAMES KEITH
KOEHLER JAMES KEVIN
KOEHLER WILSON COUCH
KOELL DICKIE DEAN JR
KOENIG JOHN MICHAEL
KOEPPE WALTER JR
KOFLER SIEGFRIED
KOHLER LUDWIG PETER
KOLAS ROBERT ALLEN
KOLEMAINEN MICHAEL WALTER
KOLLMANN GLENN EDWARD
KOMERS JOHN GEORGE
KOONCE ROBERT EDMUND
KOPFER JOHN JEROME
KORNOVICH FRANK DENNIS
KOSKY WALTER HENRY JR
KOSTICH ROBERT BOZO JR
KOTARSKI VINCENT R JR
KOTYLUK KENNETH EUGENE
KOVACS ZOLTAN ALAJOS
KOZEL PATRICK CHARLES
KRALIK WILLIAM JOHN
KRAMER DENNIS DALE
KRANER DAVID STANLEY
KRAUSE MANFRED WALTER
KRAUSMAN EDWARD L
KRAVITZ ARNOLD GARY
KRAVITZ JAMES STEPHEN
KRECH STEVEN DENNIS
KRILL RUSSELL WALTER
KRISSMAN RUDY PAUL
KROM MICHAEL LEE
KROTZER DONALD MORGAN
KROUSE JAMES CHARLES

KRUEGER LORNE COLEMAN
KRUG STEPHEN PAUL
KRUPKIN STEVEN HAROLD
KUBLER GARRY LEE
KUCHCINSKI RALPH WARREN
KUEBEL ANDREW MICHAEL
KUERSTEN JEFFERY DAVID
KUNKEL ALFRED HENRY JR
KUNKEL JOHN ROBERT
KUNTZ GENE RAY
KUNZLER DARYL ROY
KURILICH ROBERT VASO
KURTZ CHRISTOPHER LANDIS
KUYKENDALL HENRY JOSEPH
LA BARBER JAMES J
LA CHICA JOHN N
LA GROU RAYMOND LOUIS JR
LA MARR PHILLIPS
LA ROCHE JOEL MITCHELL
LA ROCHELLE MARCEL ADELAR
LA ROSA MARION DOMINIC
LABOWSKI LEONARD WILLIAM
LACKEY ROBERT EDGAR
LAFRAMBOISE PHILLIP DOUGL
LAGUNA MARIO MONTES
LAHNA GARY WILLIAM
LAINE WAYNE KEVIN
LAIRD PATRICK STEVEN
LAKE LARRY VERNON
LAKE RONALD LEE
LALICH DAVID HUGH
LAMB THOMAS ROBERT
LAMBERT JEFFREY EARL
LAMBERTSON PAUL BRUCE
LAMBORN KENNETH HOWARD
LANDERS EDMOND JOHN
LANDERS JACKY EUGENE
LANDERS RICHARD RAY
LANDON GARY JOSEPH
LANE DENNIS EUGENE
LANE RICHARD ARTHUR
LANELLI JACK DANIEL
LANGH THOMAS EARL
LANGHORNE LENNART G
LANGROCK DENNIS RAY
LANGSLOW ROBERT MALCOLM
LANZARIN LEONARD ALLAN
LAPORTE MICHAEL LOUIS
LAPP MELVIN CHARLES
LARA APIMENIO
LARA ARTURO MENDOZA
LARA CHEVO GARCIA
LARA LARRY CALIUSTUS
LARA SABINO JR
LARIMER KEITH WAYNE
LARMON TIMOTHY ELTON
LARRABEE STEVEN MICHAEL
LARSON LARRY JOSEPH
LARSON PETER SWINNERTON
LARSON RICHARD KEMP
LASTER ALVIN MACK JR
LATHROPE ROBERT MICHAEL
LAU CORNELIUS AFAI LAULII
LAU HOI TIN
LAUDERDALE RONALD GENE
LAUER CHARLES RUSSELL
LAUTERIO MANUEL ALONZO
LAVELLE PATRICK JAMES
LAVERY OWEN THOMAS
LAVIGNE GERARD ANDRE
LAW JERALD LEE
LAWRENCE BOBBY GENE
LAWRENCE ERNEST FREDERICK
LAWRENCE TORY DRAKE
LAWS RICHARD LEE
LAWSON DANIEL W
LAWSON DONALD VICTOR JR
LAWSON GARY DON
LAWSON LEO CHARLES
LAWSON WILLIAM CHARLES
LAWSON WILLIAM E
LAYTON RONALD DEAN
LAZAR GEORGE FEODRO
LE BARS STEVEN
LE BLANC ELOY FELIPE ESTE
LE DESMA LOUR
LE TOURNEAU JACK DATE
LEACH DICKIE LYNN

LEACH DICKIE LYNN
LEACH JAMES ANDREW
LEACH KENNETH RAYMOND
LEADBETTER ROGER GORDON
LEAKE RONALD JAMES JR
LEAL CHRISFINO DENNIS
LEAL FRANK DANIEL
LEAL JOHN BORGES
LEAL SALVADOR JR
LEASE WILLIAM FREDERICK
LEATUTUFU FAGALII LAITA
LEDBETTER LARRY DOUGLAS
LEDE ROY LEO
LEDESMA JOSEPH JR
LEE CHARLES RICHARD
LEE DONALD GERALD
LEE JAMES ALLEN
LEE JAMES GEORGE
LEE JOHN PATRICK
LEE LOREN VICTOR
LEE MICHAEL DURYEA
LEE PAUL EDGAR
LEE RALPH NORRIS
LEE RICHARD NORMAN
LEEMAN ROBERT ALLAN
LEFEVRE BERNARD LOUIS
LEGA JAMES GREGORY
LEGAT WILLIAM CHARLES
LEGLER STEVEN EDWARD
LEHR DAVID RICHARD
LEHRKE STANLEY LAWRENCE
LEIMBACH LARRY KENNETH
LEINEN GREGORY MICHAEL
LEMA ANTHONY LEROY
LEMUS CHARLES RUIZ JR
LEON GUERRERO KINNY SAN N
LEONARD ARNOLD LEE JR
LEONARD EDWARD N
LERCH EARL ROGER
LESTELLE JOHN ANDREW II
LESTER JAMES ROBERT
LETSON GARY WAYNE
LEVETT WILLIAM JAMES
LEVIN ROBERT PHILLIP
LEW SAI GIN
LEW VICTOR WALTER
LEW VINCENT GENE
LEWIS ADRON LEE
LEWIS ALLEN WAYNE
LEWIS CALVIN
LEWIS HAROLD ST CLAIR
LEWIS JAMES C RALPH
LEWIS JAMES ROBERT
LEWIS JOHN EDWIN
LEWIS JOHN STEPHEN
LEWIS JOSEPH ANTHONY
LEWIS RICHARD EUGENE
LEWIS RODGER DALE
LEWIS RONALD KEITH
LEWIS STEPHEN MIX
LEWTER DONALD EUGENE
LEYVA-PARRA-FRIAS FELIX F F
LICEA FRANCISCO XAVIER
LIDER FRED RODRIGUEZ
LIEBERMAN MAX
LIEBERNECHT VON MILES
LIEBESPECK JAMES WARREN
LIGGETT DURAND GARFIELD
LIGONS DARYL LEE
LILLUND WILLIAM ALLAN
LILLY DAVID ROSE
LILLY LAWRENCE EUGENE
LIND MORTEN ARVID JR
LINDBERG DAVID CARL
LINDEL JOHN RICHARD
LINDER GARRY HAROLD
LINDER GEORGE RICHARD
LINDSAY BRUCE STUART
LINDSEY REGINALD WALLACE
LINN DAVID WILLIAM
LIPSIUS MICHAEL GLENN
LIPTAK CHARLES LEWIS
LIRA ROBERT CHAGOYA
LISS LARRY WILLIAM
LITTLE RODNEY DWIGHT
LIZARRAGA MICHAEL WAYNE
LO FORTI PAUL ROSARIO
LOCATELLI VINCENT
LOCKETT EDWARD DEAN
LOCKHART ROY
LOCKHORST JOHN ELDON JR
LOGAN HALFORD

LOHMAN ROBERT THOMAS
LOHMEYER DOUGLAS EDWARD
LOISEL PATRICK MICHAEL
LONA GABRIEL
LONG DAN STEVEN
LONG FLOYD LESTER
LONG HAL RANDOLPH
LONG JOHN WADE JR
LONG LEONARD
LONG LEWIS BENTON
LONZO ANGELO ALBERT
LOOMIS BILLIE CLIFFORD
LOONEY JERRY WAYNE
LOPEZ ADRIAN SALOME
LOPEZ ARMANDO
LOPEZ EDWARD
LOPEZ EDWARD JOSEPH
LOPEZ FREDERICK GEORGE
LOPEZ GEORGE LEONARD
LOPEZ JOHN
LOPEZ JOHN EDWARD JR
LOPEZ JOSE ANGEL JR
LOPEZ JOSE ANTONIO
LOPEZ JOSE DE JESUS
LOPEZ LUPE PAUL
LOPEZ MAX ANDY
LOPEZ PAUL
LOPEZ PETER MITCHELL JR
LOPEZ RAYMOND
LOPEZ RICHARD HENRY
LOPEZ STEVE
LOPEZ VICTOR
LOPEZ-RAMOS LUIS ALFONSO
LOPRINO TERRY STEVEN
LORD NEAL ALEXANDER JR
LORENZO ROBERT J
LORTZ JOHN EDWARD III
LOSEL FRED GEORGE JR
LOSOYA ERNEST FELIPE
LOSOYA RAUL
LOSTUTTER GEORGE FRANCIS
LOTTA PHILLIP ANTHONY
LOVATO LAURIANO LAWRENCE
LOVE JOHN JR
LOVELAND RONALD RAY
LOVELL ERVIN
LOVENGUTH TERRANCE LEE
LOWE CLAYTON BENTLEY JR
LOWE EDWARD LEONARD
LOWERY MICHAEL AYR
LOWERY ROGER DALE
LOWRY JAMES EARL
LOZANO DONALD JAMES
LOZANO FERNANDO LEONARD
LOZENSKI RICHARD ORDELL
LUALLIN LEE ANDRES
LUBIN RICHARD MARC
LUC FRANK LEO
LUCAS ANDRE CAVARO
LUCAS CLYDE AUSTIN
LUCAS HERBERT GEORGE
LUCAS KARL
LUCAS MICHAEL ELSMERE
LUCAS MYRON DONALD
LUCAS PHILLIP WARREN
LUCE PAUL FRANKLIN
LUCERO JAMES CLIFFORD
LUCIA STEPHEN WAYNE
LUCIDO JOSEPH BERT
LUCKETT LARRY JOE
LUDWIG LEONARD R
LUGO ANTHONY SANTANA
LUHNOW GLENN EUGENE
LUKE ARNOLD WAYNE
LUKINS PAUL ROGER
LUNA HENRY THOMAS
LUNA JOE JR
LUNA ROBERT
LUNDBERG PETER THOMAS
LUNDY ALBRO LYNN JR
LUSCOMBE DOUGLAS EDWIN
LUSK DONNIE RAY
LUSSIER LARRY PAUL
LUTGE THOMAS ALBERT
LYLE TERRANCE RICHARD
LYMAN GERALD CLYDE
LYNCH JOHN WILLIAM III
LYNCH ROBERT LAWRENCE
LYNCH WILLIAM AFFLEY JR
LYON DONAVAN LOREN
LYONS CARL

LYONS CHESTER GEORGE
LYONS WILLIAM JOHN
LYTLE CLIFFORD JAMES
MABREY GARY MICHIEL
MAC GEARY FRED ERNEST
MAC INTOSH DONALD GORDON
MAC LEOD PHILLIP LESLEY
MAC MANUS JAMES FRANCIS
MACAGBA EDILBERTO DULA
MACHADO GARY ALLAEN
MACK DANNY RAY
MACK ROBERT LEE
MACKEY THOMAS EARL
MACKLIN RONALD WAYNE
MADDEN DAVID ALLEN
MADDOX RICHARD GREENE
MADDUX ROY RAYMOND JR
MADRID ADANO HERNENDEZ
MADRID MICHAEL PHILLIP
MADRIGAL-CORDERO RAFAEL A JR
MADRUGA MANUEL DOMINIC
MAES PEDRO MIGUEL
MAESTAS GILBERT MERILL
MAGEE JOHN EARL
MAGUIRE CHRISTOPHER J III
MAGUIRE ROBERT STANLEY
MAHONEY ALFRED RICHARD JR
MAHONEY RALPH MARTIN
MAHONEY THOMAS P III
MAIER DAVID ROY
MAILLOUX EARL ADELBERT
MAISEY REGINALD VICTOR JR
MAJOR ROBERT WARREN
MAKUCK MICHAEL PATRICK
MAKUH FRANK JOSEPH
MALDONADO ANTHONY GILBERT
MALLOBOX JESSE ARMANDO
MALLON THOMAS WINSTON
MALONE FELIX
MALONE JAMES EDGAR
MALONE JOHN EDWARD
MALONE SIDNEY JACK JR
MAMON CESAR JABONILLO
MANEMANN RICHARD RAYMOND
MANESS STEVEN WAYNE
MANGAN MICHAEL ROBERT
MANNING JERRY WAYNE
MANOWSKI EDWARD
MANSERGH WILLIAM A JR
MAPE JOHN CLEMENT
MARCELLO JOHN BERNARD
MARCRUM RONALD DEAN
MARCUS MICHAEL BOOTH
MARDIS JAMES ARNOLD JR
MAREK JOSEPH PENN
MARGOLIS ROBERT LYNN
MARIANI JOHN ROY
MARIN JULIAN
MARION HARRY LEWIS
MARKEN JOHN PAUL JR
MARKILLIE JOHN ROY
MARKS PHILLIP HADDON
MARKUSEN TOBIAS EARL
MARKWELL EUGENE LYNN
MARLAND INNES LEE
MARLAR OLIN DEWEY III
MARLOWE JACK WILLIAM
MARQUEZ FLORENCIO Q
MARQUEZ JOHN
MARQUEZ MARTIN JR
MARQUEZ PAUL JOSEPH
MARR JOHN AUSTIN
MARRUFO RODNEY ELMER JR
MARSH HERBERT LYNN
MARSHALL CHARLES RAY
MARSHALL MICHAEL ALLAN
MARSHALL RICHARD ALLAN
MARSHALL ROGER ROBERT
MARSHALL WILLARD DALE
MARTIN DAVID LEE JR
MARTIN ERNEST TYRONE
MARTIN FREDDIE KAY
MARTIN GEORGE PAUL
MARTIN GERALD
MARTIN GREGORY LAWRENCE
MARTIN JEFFREY LEA
MARTIN JIMMIE CARTER

MARTIN JOHN ANTHONY III
MARTIN JOHN CHARLES
MARTIN JOHN MAJOR
MARTIN JOSEPH CRAIG
MARTIN JOSEPH THOMAS
MARTIN KENNETH LEROY
MARTIN KENNETH WILLIAM
MARTIN LARRY ALLEN
MARTIN LARRY JOE
MARTIN MICHAEL EMMETT
MARTIN RAYMOND CHARLES
MARTIN RICHARD JODY
MARTIN STEPHAN JAMES
MARTIN STEVEN LARRY
MARTIN THOMAS CHARLES
MARTINEZ CHRIS RONALD
MARTINEZ ERNESTO
MARTINEZ ERNIE ROBLES
MARTINEZ ERNIE ROBLES
MARTINEZ EZEKIAL
MARTINEZ GEORGE VINCENT
MARTINEZ JOHN ANTHONY
MARTINEZ PAUL DINNES JR
MARTINEZ RODNEY DEAN
MARTINEZ STEVEN CATARINO
MARTINSEN LOREN DAUNE
MARTZ MELVIN LEE
MASADAS BEN OBSENIARES
MASSETH ROBERT EUGENE
MASSEY MICHAEL SEAN
MASSONE MICHAEL STACY
MASTELLER ALLAN DEAN
MASUDA ROBERT SUSUMU
MASUEN MICHAEL NICHOLAS
MATAYOSHI WALLACE KENJI
MATHERN EDWARD GERARD
MATHIAS ROBERT P
MATHIS BRENT EUGENE
MATSON GARY LEE
MATTA MICHAEL ERNEST
MATTERA FRANK JOHN JOE
MATTESON LYNN MICHAEL
MATTHEWS ALAN LEE
MATTHEWS BERNARD JULIAN
MATTHEWS FLOYD JOSEPH
MATTHEWS JAMES NEWTON
MATTIS WILLIAM CARROLL
MATTSON KENNETH EUGENE
MAULDIN EDDIE LEE
MAULDIN MICHAEL B
MAURER WALTER LAWRENCE
MAXAM LARRY LEONARD
MAXWELL ROBERT JAMES
MAY JAMES JR
MAY ROY EDWARD
MAYNARD THOMAS HARRY
MAYO DUDLEY WAYNE
MAYO GERALD FRANK
MAYS GEORGE M JR
MAZON THEODORE JR
MAZZA STEPHEN DARRELL
MC ADOO GLENN PAUL
MC AFEE CLYDE RICHARD
MC ALISTER DONALD LYNN
MC ALLISTER WILLIAM WALTE
MC ANDREW ROBERT CHARLES
MC ATEE DON JAY
MC BRIDE FITZ-RANDOLPH BU
MC BROOM LOYD LINDAL
MC CABE MARC WAYNE
MC CALL ALLAN LEE
MC CALL DOUGLAS HUDSON
MC CANN MICHAEL ROSS
MC CANTS ALFRED FRAZIER
MC CARTHY JOSEPH F JR
MC CARTHY TERRY ALAN
MC CARTY JOHN LEIGH
MC CARTY KENNETH LEON
MC CAULEY STEPHEN ARTHUR
MC CLAIN GARY THOMAS
MC CLELLAND MYRON
MC CLOUD GARY LEE
MC CLOYN JOSEPH
MC CLURE JACK DALE
MC COIG DONALD B
MC COLLUM DAVID VERNON
MC COMMONS MICHAEL RAY
MC CONAHAY MICHAEL PAUL
MC CONNAGHY WILLIAM P
MC CONNAUGHHAY DAN

33

California

DAILY
MC CONNELL DAVID WAYNE
MC CONNELL WILLIAM WALKER
MC CORD DAVID MICHAEL
MC CORKLE LESLIE LEROY
MC COSAR WINFORD
MC COY JOHN LOWERY
MC COY MERIL OLEN JR
MC CULLOUGH MICHAEL EUGEN
MC CURTAIN CHARLES RAY JR
MC DANIEL CRAIG ALLAN
MC DANIEL RICHARD BYERS
MC DAVID WILLIAM EARL
MC DAVID WILLIAM EARL
MC DONALD CLYDE D II
MC DONALD D LANCE
MC DONALD GEORGE COLUMBUS
MC DONALD JAMES HOWARD
MC DONALD KURT CASEY
MC DONALD LEWIS LEVI
MC DOUGAL BILLY DEAN
MC DOWELL MELVIN WARREN
MC DOWELL WILLIAM JOSEPH
MC ELROY DENNIS ARTHUR
MC ELYEA JAMES FRANK
MC ENTEE NEIL CHARLES
MC EUEN RONALD CURTIS
MC FALL GARY RICHARD
MC FARLAND KENNETH EARL
MC FARLAND LOUIS HENRY
MC FARLAND STEVEN LEE
MC GAR BRIAN KENT
MC GEHEE JOHN ALBERT
MC GERTY MICHAEL JOHN
MC GINLEY GERALD GREYDON
MC GIVERN WILLIAM DAVID
MC GLASSON JAMES CLARK
MC GLOCHLIN DAVID EARL
MC GLOTHIN RAYMOND DENNIS
MC GOVERN CHARLES MANLEY
MC GRIFF DANNY JAY
MC GUIRE JAMES WILLIAM
MC GUIRE RICHARD HAROLD
MC GUIRE TIMOTHY PAUL
MC HAM RICHARD HUGH
MC INNIS THEODORE VALENTI
MC INTYRE DUNCAN B
MC JIMSEY WILLIAM ROBERT
MC KAY DAVID GEORGE
MC KECHNIE DANIEL LEE
MC KEE JACK ROGER
MC KEE THOMAS EUGENE
MC KENZIE DOUGLAS N II
MC KENZIE JAMES ALLEN
MC KENZIE JAMES CALVIN
MC KENZIE RICHARD WAYNE
MC KIERNAN TIMOTHY JAMES
MC KINLEY STEPHEN WILLIAM
MC KINNEY ALBERT W JR
MC KINNEY RONALD EUGENE
MC KINNON JACK WILEY JR
MC KINNON LARRY DEE
MC KINSEY GERALD LEROY JR
MC KINSTRY JAMES J JR
MC LAUGHLIN JAMES PAUL

MC LAUGHLIN KIRK ALVIS
MC LAUGHLIN WILLIAM LAWRE
MC LEAN JAMES HENRY
MC LEAN RONALD WALSH
MC LELLAN JIMMY LEE
MC LELLAND MARVIN EDWARD
MC LEOD ROBERT LEE
MC LEOD ROBERT LEE
MC MAHON DOUGLAS DUANE
MC MAHON JOHN THOMAS
MC MAINS DONALD HENRY JR
MC MANUS MARK LAWRENCE
MC MASTER JOHN WILLIAM
MC MASTERS CHARLES ANTHON
MC MILLAN BRUCE FRANCIS
MC MINN DANNY LEE
MC MURDO JAMES ALFRED
MC MURPHY JAY DARRELL
MC NALLY HARRY MERLE
MC NALLY JAMES WILLIAM
MC NEARNEY PATRICK VICTOR
MC NEES RONALD HARVEY
MC NEILLY RONALD WILLIAM
MC NELLY WILLIAM ROBERT
MC NULTY MILTON KEITH
MC PHAIL FRANKLIN LLOYD
MC PHEARSON JAMES CARL
MC PHEE RANDY NEAL
MC PHERSON FRED LAWER
MC PHERSON LARRY RANDALL
MC ROBERTS CLIFFORD WAYNE
MC TAGGART WILLIAM JAMES
MCCORMICK THOMAS RANDOLPH
MCELVAIN JAMES RICHARD
MCGRATH WILLIAM DARRELL
MCIVER ALEXANDER
MCLEMORE JOHN WILSON JR
MCMULLEN GEORGE E III
MEADOR LARRY JOE
MEANS RONALD LEE
MEARNS GLENN RODNEY
MEARS CHARLES ROBERT
MEBUST OWEN EDWARD
MECKEL JOHN BLOCKER
MEDEIROS DENNIS JOSEPH
MEDINA DANIEL MICHAEL
MEDINA DAVID PHILLIP
MEDLIN PAUL CHARLES
MEEHAN DALE PATRICK
MEEK JOE LANELL
MEEK THOMAS OTIS
MEEKS CHARLES HENRY JR
MEISTER GEORGE FREDERICK
MELCHOR JOHN GLENN
MELENDEZ RUDOLPH
MELENDRES JOSEPH THOMAS
MELENDREZ ROBERT CHARLES
MELIM JON MICHAEL
MELVILLE TIMOTHY JAMES
MELVIN MICHAEL WAYNE
MENA JOSEPH ANGEL
MENANE JERRY BRUCE
MENDELL ALLAN
MENDENHALL THOMAS JAMES
MENDEZ MAURILIO
MENDIBLES RAYMOND G
MENDOZA ANTONIO
MENDOZA DAVID RAMIREZ
MENDOZA GILBERT
MENDOZA JOHN DEE
MENDOZA JOSE MEDEL
MENDOZA JOSEPH LOUIS
MENDOZA MARTIN ELBY
MENDOZA PETER ACOSTA
MENDOZA RONNIE ALLEN
MERCER WILLIAM IVAN
MERCKE TERRANCE LEE
MERENO MICHAEL
MERICANTANTE THOMAS LEE
MERIDITH GARY LEE
MERRILL DENNIS LEE
MESA JAMES GREGORY
MESA RICARDO
MESSER JON LAIRD
MESSER SIGURD MARTIN

MESSERLI STEVEN LOUIS
MESSING MITCHEL
METCALF RICHARD LOUIS
METOYER MICHAEL ESPY
METZGER RUSSELL EDWARD
METZLER CHARLES DAVID
MEYER GARY ANTONE
MEYER GARY PAUL
MEYER KENNETH ALLEN
MEYER OTTO PAUL III
MEZA JESUS JAMES
MICHAEL DENNIS STEVEN
MICHAEL LLOYD DONALD
MICHEHL THOMAS CHARLES
MIDDLEKAUFF DAVID
MIGUEL MICHAEL JOSEPH
MILDNER ROBERT MARC
MILES DALE ARTHUR
MILES MARK SCOTT
MILES RONALD DAVID
MILLARD KENNETH ARTHUR
MILLER CALVIN LEROY
MILLER CLEVE DAVIS
MILLER GARY AMES
MILLER GARY DEAN
MILLER GLENN EDWIN
MILLER JERRY ROBERT
MILLER JIMMY ALLEN
MILLER KENT FROEMMING
MILLER LOUIS CHARLES
MILLER MARK JEFFERY
MILLER MARVIN RAY
MILLER MICHAEL CLIFTON
MILLER MICHAEL JR
MILLER OREN KENNETH
MILLER PAUL WAYNE
MILLER RICHARD HERSHEL
MILLER ROBERT CHARLES
MILLER ROBERT GAIL
MILLER ROBERT LESTER
MILLER WALTER RAY JR
MILLIGAN RANDALL GALE
MILLNER MICHAEL
MILLS CARROLL RAY
MILLS JAMES BURTON
MILLS JOHN PAUL
MILLS PETER ROBERT
MILLS ROBERT THOMAS
MILLS TED DOUGLAS
MILOVICH ROBIN PATRICK
MINCEY ROBERT EARLE
MINER MICHAEL ROBERT
MINKO MICHAEL ANTHONY
MINNIEAR DWIGHT JED
MIRAMONTES ARTHUR FRED
MIRANDA JOE ALEMAN
MIRANDA MANUEL
MISA TULELE
MISA VIANE SOFENI
MISCHEAUX RENE CLARENCE
MITCHELL DAVID EUGENE
MITCHELL ERNEST DARRELL
MITCHELL GILBERT LOUIS
MITCHELL JAMES CARROLL JR
MITCHELL JAMES MCNALLY JR
MITCHELL MARK DAVID
MITCHELL MICHAEL JEFFREY
MITCHELL MICHAEL SIDNEY
MITCHELL MICHAEL THOMAS
MITCHELL ROGER C
MITTON WILLIAM JAMES
MOFFETT MELVIN GLEN
MOHNIKE PHILLIP SHERMAN
MOISE HERVE JEAN
MOLDENHAUER RUSSELL
MOLINA AGAPITO JR
MOLINA GEORGE GERONIMO
MOLINA MICHAEL JOSEPH
MOLOSSI ROBERT JOHN
MONAHAN MICHAEL JAMES
MONFILS DENNIS EUGENE
MONK SHERMAN DALTON
MONTAGUE STEPHEN GRIFFITH
MONTANA JIMMY CARLUS
MONTANEZ PARIS WILLIAM
MONTANO ANTHONY
MONTAPERT RONALD M
MONTELEONE GARY ROBERT
MONTELLANO MICHAEL A
MONTERROSO ALFONSO ALFRED
MONTERRUBIO ARMANDO
MONTES LEONARD DANIEL

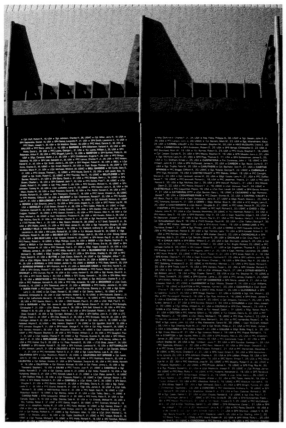

MONTES RAUL
MONTEZ FRANK JAMES
MONTGOMERY GEORGE WESLEY
MONTGOMERY MICHAEL MALLOR
MONTGOMERY STEVEN HUGH
MONTGOMERY WILLIAM EUGENE
MONTION ARTURO DANIEL
MONTOYA LOUIE GOOCH
MOODY STEWART ROBBINS
MOOERS WILLIAM MATHIAS
MOONEY JOHN HOWARD JR
MOONEY MICHAEL JAMES
MOORE BILLY EUGENE
MOORE CHARLES EDWARD JR
MOORE CHARLES SARGENT
MOORE DENNIS EUGENE
MOORE DERRYL LEE
MOORE ELLIOTT WAYNE
MOORE GALEN LEROY
MOORE GLENN DOUGLAS
MOORE JIMMY RAY
MOORE LARRY JAY
MOORE LESTER LEWIS
MOORE ROBERT NED
MOORE RONALD ALLAN
MOORE STEPHEN DOUGLAS
MOORE WALTER ZAMPIER JR
MOORE WILLIAM ROBERT
MOOTHART LARRY GRAYDON
MORA ERNEST LOPEZ
MORA GREGORIO MANUEL
MORA RAMIRO MICHAEL
MORALES ANGELO RAYMOND
MORALES TOMMY
MORAN ALBERTO HECTOR
MORAN RAY EDWARD JR
MORAND BRAD WILLIAM
MORELAND JAMES LESLIE
MORELAND STEPHEN CRAIG
MORELAND WILLIAM DAVID
MORELOS CATARINO JR
MORENO ADOLFO VALENZUELA
MORENO JOHN BOBBY
MORENO MARTIN WALTER
MORENO ROBERT
MORFORD LARRY HOWARD
MORGAN CHARLES ELZY

MORGAN DAVID ROBERT
MORGAN GLENDELL
MORGAN JOHN LOUIS JR
MORGAN MARK LAKE
MORGAN RONALD EDWARD
MORGAN STEPHEN EDWARD
MORGAN WILLIAM LESLIE
MORGENS CHRISTOPHER W
MORI BRUCE JUN
MORIARTY PATRICK DALE
MORITZ MICHAEL PERRY
MORLEY JEFFREY PAUL
MORRELL WILLIAM ALEXANDER
MORRILL DAVID WHITTIER
MORRILL MERWIN LAMPHREY
MORRIS DONALD WARREN
MORRIS GEORGE WILLIAM JR
MORRIS HARRY LEO JR
MORRIS HERMAN RAY
MORRIS JOHN FREDERICK
MORRIS KELLY STUART
MORRIS LYLE WAYNE
MORRIS THOMAS W
MORRISON GLEN MARK
MORRISON JOE HAROLD
MORROW DALE ARTHUR
MORROW RICHARD DAVID
MORTENSEN ALLAN DAVID
MORTON EDWARD EARL
MOS RONALD BRUCE
MOSBY STATUE JR
MOSCHETTI BILL ARTHUR
MOSES JAMES THOMAS
MOSES WILLIE LEE
MOSHIER JIM EDWIN
MOSS CHARLES LEE JR
MOTLEY LARRY KEITH
MOUSSEAU LLOYD FRANCIS
MOYNAHAN JOHN JAMES
MUCHA HENRY JR
MULGREW KEVIN SPEAR
MULHOLLAND ROBERT ALTON
MULICK MICHAEL WILLIAM
MULLAN CHARLES RICHARD JR
MULLEN JOSEPH WILLIAM JR
MULLEN LARRY DONALD
MULLER ALLEN DONALD
MULLER HAROLD BRADLEY
MULTHAUPT JAMES WAYNE

MUMMERT ROBERT STERLING
MUNATONES JOSE JR
MUNCY GILBERT HOWARD
MUNDEN DONALD MARTIN
MUNOZ CARLOS GARCIA
MUNOZ DAVID
MUNOZ DAVID LOUIE
MUNOZ JESUS ARTHUR
MUNOZ JOSE JR
MUNOZ LARRY
MUREN THOMAS RICHARD
MURPHY BILLY DAN
MURPHY JERRY RAY
MURPHY JOSEPH PATRICK
MURPHY LLOYD ALBERT
MURPHY MICHAEL PATRICK
MURPHY ROBERT EMMETT JR
MURPHY TIMOTHY XAVIER
MURPHY VINCENT PATRICK
JR
MURPHY WALTER EDWARD JR
MURRAY BRUCE ANDERSON
MURRAY DOUGLAS EARL
MUSGUIRE GLEN ALAN
MUSICK RAYMOND EARL JR
MUSSELMAN DONALD L
MYERS CHARLES DEAN JR
MYERS CHARLES LEE
MYERS EDWARD GEORGE
MYERS RICKY ALAN
MYRICK GEORGE FRANKLIN
NACCA CARL JR
NADAL BALDOMERO ARTURO
NAGATO YOSHIIWA
NAGENGAST CARL DELANE
NAJAR ADAM SERNA
NAJARIAN MICHAEL AN-
THONY
NAJERA MANUEL CHICK JR
NAKASHIMO MASASHI
NANCE DAVID EUGENE
NANCE KENNETH EDWIN
NAPIERSKIE DANIEL
NARANJO DAVID JESUS
NARCISSE ALVIN RAY
NARDELLI ROBERT JOSEPH
NASH DAVID EUGENE
NASH JAMES ROBERT
NASSER ROBERT BENJAMIN
NASTOR TONY VALDEZ
NATHAN JOHN ARTHUR
NAVA SALVADOR MARTINEZ
NAVONE VICTOR CHARLES JR
NAYLOR DENNIS EUGENE
NAYLOR LYNN PATTINSON
NAZABAL ARTURO ALBER-
TO JR
NEAL KENNETH LAWRENCE
NEAL STEPHEN BROWNING
NEGRANZA MARIANO R JR
NEGRINI WILLIAM LODI
NEISWENDER DANIEL LYNN
NELSON ALBERT OSCAR JR
NELSON ALLAN JOSEPH JR
NELSON BENJAMIN ROY JR

NELSON BOYD JEFFERY
NELSON DANIEL EUGENE JR
NELSON DONALD LAWRENCE
NELSON GARY NELS
NELSON JOHN EDWARD
NELSON LEON GROVER
NELSON MICHAEL ROY
NELSON TERRANCE WILLIAM
NELSON WILLARD EDWARD
NELSON WILLIAM DE WITT
NETHERLY ARTANZIE
NETTLE WILLIAM LEROY
NEUBACHER BRANDT STEELE
NEVAREZ ALEXANDRO
NEVES MANUEL CATANO
NEVILLE WILLIAM EDWARD
NEW GEORGE JR
NEWBERG ROBERT MARION
NEWBERRY JASPER NEWTON
JR
NEWBY GARY EUGENE
NEWMAN DENNIS EARL
NEWSOME JOHNNY
NEWSON LEROY JR
NEWTON DONALD STEPHEN
NEWTON LEONARD LEE
NEWTON WILLIAM J
NIBBELINK LEA EVERETT
NICHOLAS ROBERT GEORGE
NICHOLS DOUGLAS ELLS-
WORTH
NICHOLS JAMES ARTHUR
NICHOLS RANDE LEE
NICHOLSON DAVID LEONARD
NICHOLSON JAMES ALEX-
ANDER
NICHOLSON JAMES ARTHUR
NICKEL WARREN F JR
NIDEVER DAVID FRANK
NIEDERHAUSE STEPHEN
SCOTT
NIOUS ELVAIN ENNIS
NISHIMURA JOHN
NISHIZAWA GLENN NOBUYKI
NIX WARREN PAUL
NOBLE JOHN RODNEY
NOBLE MORRIS ALLAN
NOE TIM A
NOELKE RICHARD ALLEN
NOKES KENNETH CLIFFORD
NOOTZ GAYLORD EUGENE
NORA RAYMOND VERNON
NORDELL JOHN EDWARD JR
NORDQUIST GARY LEIGH
NORDSTROM VICTOR CARL
NORMAN CLAE TERRY
NORRIS JOHN ALEXANDER III
NORRIS WIELAND CLYDE
NORTH MICHAEL WALTER
NORTHERN JAMES ROBERT
ALL
NORTHUP DAVID WAYNE
NORWOOD RICHARD DALE
NOTT BYRON LEE JR
NOTTINGHAM RICHARD

LANCE
NOVAKOVICH JERRY A
NOYOLA RICHARD
NUESSE CHESTER KEITH
NUNEZ DAVID GUERRERO JR
NUNEZ FRED CONTRERAS
NUNEZ JESUS CARLOS
NUNEZ RUDOLPH ALGAR
NUNNALLY TIMOTHY CRAIG
NURISSO CHARLES WILLIAM
NUSSBAUMER STEVE OWEN
NYBERG LEONARD ERIC
NYHOF RICHARD E
NYSTROM THOMAS ALLEN
NYSTUL WILLIAM CRAIG
O BANNON ALBERT F JR
O BANNON ROBERT III
O BRIEN CHESTER LAVERN JR
O BRIEN RICHARD CONAWAY
O BRIEN ROBERT PHILLIP
O BRIEN STEPHEN
O CONNOR DENIS
O CONNOR DENNIS ALFRED
O CONNOR DENNIS KENNETH
O CONNOR JOHN VINSON JR
O CONNOR ROBERT LEE
O KEEFE ROBERT WILLIAM
O NEIL ROBERT WILLIAM
OATES ROBERT WAYNE
OBERDING FRED JR
OBERLE DAVID ALAN
OBERT RICHARD ROBERT
OBREGON RAUL ALBERT
OCAMPO ROBERT EGMEDIO
OCHOA LOUIE
OENS LAVERN OREN
OFFERDAHL WILLIAM BRUCE
OFSTEDAHL JERRY WAYNE
OGAMI TERRY Y
OGAS PHILLIP ARTHUR
OGDEN DAVID ELLIS
OGLESBY JOHN R
OGLETHORPE THOMAS JAY
OGREN JERRY LEWIS
OHARA STEVE MASAO
OHLSON GALEN ERICK
OKAMOTO DONALD RAY
OLAND DAVID MICHAEL
OLETA JESUS C JR
OLIVARES-MARTINEZ AR-
TURO
OLIVER ERSKINE JAY
OLIVER PAUL HAROLD
OLIVER RICK ALTON
OLIVERAS RUDY MICHAEL
OLLIVIER JOSE ANTONIO
OLMOS ALFONSO
OLMOS LUIS
OLSEN STEVEN WAYNE
OLSON ROBERT GARY
OLSZEWSKI JOSEPH VERNE
OMSTEAD DAVID KING
ONANA RALPH WHEELER
ONETO HARRY STEVEN JR
ONTIVEROS THOMAS J

ORNELAS-ARELLANO VICTOR
M
OROPEZA MANUEL GARCIA
OROSCO LARRY J
ORR RAYMOND FRANKLIN JR
ORR THOMAS JOSEPH
ORSUA CHARLES DAVID
ORTEGA FRANK DENNIS
ORTEGON RICARDO JOSEPH
ORTIZ ARTURO MARQUEZ
ORTIZ PEDRO
ORTIZ REINALDO SALVADOR
ORWIG MICHAEL JOHN
OSBORNE RICHARD GENE
OSBURN FRED HARRISON JR
OSHIRO WARREN SHIGEO
OSTER FRANK ALLEN
OSTROFF STEVEN LARRY
OSUNA ANTONIO RICHARD
OSUNA JOSE LUIS
OTSEN DAVID BRYAN
OTT PATRICK LOUIS
OTT WILLIAM AUGUST
OWEN ROBERT GARY
OWEN STEVEN CRAIG
OWENS ALBERT DANNY
OWENS BILLY RAY
OWENS JACK COLEMAN
OWENS REO
OWENS RICHARD LEE
OWNBY EDWARD ALLAN
JOSEPH
OZANNE JORDAN JAY
OZBUN JAMES D
PACE RONALD EARL
PACHECO JOSE ANTHONY
PACHECO ROBERT LEE
PACK FRED WALTER
PACO RICHARD MANUEL
PADDOCK MICHAEL JAMES
PADDOCK MICHAEL L
PADILLA ANTONIO DUARTE
PADILLA EDDIE JACK
PADILLA GARY TEOFILIO
PADILLA RALPH HENRY
PADILLA ROBERT LOUIS
PAEPKE DUANE CARL JR
PAGALING MICHAEL
PAGCALIUAGAN CEIZHAR
VALE
PAGE GEORGE MERRITT JR
PAGE GORDON LEE
PAGE JOHN ARTHUR
PAGE LUTHER JR
PAGE MICHAEL RANSOM
PAGE STEVE WILSON
PAGET MICHAEL GORDON
PAINE EDWARD ARTHUR
PAINE VICTOR LLEWELLYN
PAINTER JOHN RALPH JR
PAINTER ROBERT LEE
PAIZ JERRY
PALACIO JOE MAURICIO
PALACIO RAYMOND JESUS
PALACIOS CASIMIRO
PALACIOS LUIS FERNANDO
PALACIOS TONY
PALCZEWSKI EDMUND
LAWRENC
PALENSKE WILLIAM ALLEN
PALMA RAYMOND BARELA
PALMER DAVID LESLIE
PALMER RONNY LEROY
PALMQUIST STEVEN LEON-
ARD
PANAMAROFF WALTER JOHN
PAOPAO KEILA
PAPALAS ANTHONY STEVEN
PAPIN SAMUEL ALEXANDER
JR
PAPPAS DON LEE
PAPPAS RALPH BYRON
PARCELS REX LEWIS JR
PARHAM RICHARD LYNN
PARIS CRAWFORD BRIAN
PARISH FRANK BRENNAN
PARISH MICHAEL LAWRENCE
PARISI GUILLERMO
PARKEL GERALD PHILLIP
PARKER BENNIE FRANK
PARKER CHARLES LESLIE JR
PARKER JIMMIE EDWARD
PARKER KENNETH
PARKER LARRY THOMAS

PARKER MAXIM CHARLES
PARKER RONNIE EARL
PARKER ROY EUGENE
PARKER VICTOR RALPH
PARKER WILLIAM MONROE
PARKER WILLIAM THOMAS
PARKINSON GARY CONVERS
PARKS DAVID NORTON
PARKS STEPHEN EARL
PARMETER GERALD THOMAS
PARNELLA JOHN
PARRA LIONEL JR
PARRA MANUEL FRANCISCO
PARRA MANUEL FRANCISCO
PARSONS HENRY BENNETT III
PARTIDA CHARLIE LOPEZ
PARTRIDGE ALAN BRIAN
PARTRIDGE DOUGLAS EL-
WOOD
PASHMAN STEPHEN MARK
PASILLAS HENRY
PASTORES GEVIN PESCOZO
PATRICK RICHARD MICHAEL
PATTERSON BRUCE DIXON
PATTERSON DANIEL ARTHUR
PATTERSON DWAYNE MAX-
IFIEL
PATTERSON EDWARD LEON
PATTERSON GEORGE FRANCIS
PATTERSON JAMES KELLY
PATTERSON LARRY HART
PATTERSON RONALD OREN
PATTERSON STEVEN CRAIG
PATTON DORRIS EDWARD
PATTON JOHN PERRY
PAULK ROBERT MILTON
PAYNE ANDREW JAMES JR
PAYNE WILLARD FRANCIS
PAYTON VENUS DEWHIT JR
PEARCE MARVIN ROBERT
PEARCE ROBIN ANDREW
PEARCY ROBERT LESLIE
PEARLSTEIN JERROLD S
PEARSON ARNOLD C
PEARSON GEORGE WILLIAM
JR
PEARSON GREGORY JOHN
PEARSON JOHN HOWARD
PEARSON ROBERT LEON
PEARSON THOMAS RICKARD
JR
PEARSON VAN HARVEY
PEAVY THOMAS MICHAEL
PECORARO FRANK ANTHONY
PEDDY CHARLES LEE
PEDERSEN FRED LEWIS
PEDERSEN KENNETH RALPH
PEDERSEN WILLIAM A
PEDERSON JOE PALMER
PEDRICK CHARLES C II
PEEBLER CHRISTY ALBERT
PEGUERO RICHARD
PEHRSON DALE CHRISTO-
PHER
PELHAM LESTER LEON
PELLICANO JEAN PIERRE V
PENDERGRAFT RONNIE DEAN
PENE RONALD EDWARD
PENMAN RONALD STIRLING
PENN RONALD W
PEQUENO JUAN RODRIGUEZ
PERDOMO KRIS MITCHELL
PERDUE DON MELVIN
PEREZ ARTHUR CARLYLE
PEREZ ERNEST EUSTACE
PEREZ GUADALUPE
PEREZ JESUS ALBERT
PEREZ JOE FRANCISCO JR
PEREZ JOSEPH ESPINO
PEREZ JUAN J
PEREZ PETER
PEREZ RAYMOND LUNA
PEREZ RICHARD
PERICH JOHN WHILDEN
PERKINS RONALD JAMES
PERKINS WILLIAM THOMAS JR
PERRY ANTONE JR
PERRY CASEY CLAYTON
PERRY DONALD LEE
PERRY KENNETH RICHARD
PERRY STEVE JOSEPH LEONE
PETANOVICH NICHOLAS C
PETERMAN THOMAS HOW-
ARD

California

PETERS EDWARD THEODORE JR
PETERS JOHN DENIS
PETERS LAUVI PAUL PHILIP
PETERS WALTER JOHN
PETERSEN GALEN DEAN
PETERSEN GAYLORD DEAN
PETERSEN LAWRENCE LEE
PETERSEN RAYMOND ALLAN
PETERSEN WILLIAM ROBERT
PETERSON DAVID BRUCE
PETERSON DENNIE DONALD
PETERSON DENNIS WILLIAM
PETERSON DONALD MARTIN
PETERSON MICHAEL EUGENE
PETERSON RICHARD W
PETERSON STEPHEN RUSSELL
PETRICK RONALD PAUL
PETRIE JAMES ALLAN
PETTIGREW KENNETH DALE
PETTIT CRAIG STEVEN
PETTIT DENZIL DAL
PETTITT DONALD ACE
PETTY EUGENE
PFLASTER GARY LEWIS
PHELPS RONNIE LOUIS
PHILBIN RICHARD GRIFFITH
PHILLIPS ANTHONY BRUCE
PHILLIPS JOHN DAVID
PHILLIPS JOHN MICHAEL
PHILLIPS PAT ELLIS
PHILLIPS RONALD CHARLES
PHILLIPS THEODORE BERT
PHILLIPS THOMAS MILES
PHILLIPS WILLIAM JOSEPH
PHINN WILLIAM MARK
PHIPPS DONALD RAY
PHIPPS JAMES ALVIN
PHIPPS JIMMY WAYNE
PIATT RICHARD WEAVER
PICKETT WILTON RAY
PIERCE LARRY STANLEY
PIERCE LOY WENDELL
PIERCE SAMUEL HENRY JR
PIERCE WILLIAM EARVIN
PIGG THOMAS CHARLES
PIMENTEL RONNIE CARDOZA
PINA FRANK DAVID
PINALES LAWRENCE
PINAMONTI ERNEST ANTHONY
PINATELLI THOMAS MICHAEL
PINCHOT CRAIG D
PINER JOHN ROBERT
PING ROY MARTIN
PINK JOSEPH PATRICK
PINNEY JOHN SCOTT
PINOLE BABE
PIPER JAMES DENNIS
PIPKIN FRANK MEADOWS
PISHNER WILLIAM JR
PITT ROBERT LOUIS
PITT ROY SHARP
PITTENGER DONALD ALAN
PITTMAN ROBERT LOUIS
PITTS ROY EDWARD
PITTS TERRY DENNIS
PLACERES MOSES
PLAKE JAMES ROLAND
PLANCHON RANDALL T II
PLEASANT STEPHEN DONALD
PLOWMAN JAMES EDWIN
PLUMB GARY ANTHONEY
PODGORNY DENNIS RICHARD

POELSTRA DENNIS PATRICK
POFFENBARGER WILLIAM OSCA
POGRE BOB ELIA
POGUE JOSEPH DONALD
POGUE MICHAEL ALAN
POHL WILLIAM ANTHONY
POHLMAN JOHN HOWARD
POIRIER ROGER MILTON
PONCE PAUL
POOLE THOMAS LYNN
POPE DEREK BOYD
POPE SERVESTON DEVON
POPE THOMAS ROBERT
POPPA GERALD LELAND
PORT GARY CRAIG
PORTE ROBERT ANDREW
PORTELLO RODERICK CHARLES
PORTER MICHAEL GRANT
PORTER RONALD WILLIAM
PORTERFIELD DALE KYETTE
PORTUGAL IGNACIO JR
POSEY RALPH EDWARD
POSO JOHN RICHARD
POSTEN GERALD WAYNE
POTTER WILLIAM VERNON
POTTS WILMER
POULSON BRUCE WILLIAM
POWELL DANIEL LEE
POWELL ERWIN GILBERT JR
POWELL LESLIE ALLEN
POWELL REGINALD FOSTER
POWELL RUSSELL J
POWERS BRADLEY LELAND
POWERS RONALD LEE
POWERS STEVEN CHARLES
PRANGE JOSEPH WILDER
PRATER CALVIN RAY
PRATER DONALD HAROLD
PRATHER LAVON NEIL
PRATT JAMES IRVING
PRATT RICHARD CHESTER
PREDDY ROBERT LEE
PREDIGER FRANZ GERHARD
PREMENKO JOHN AL
PRENTICE DENNIS ALBERT
PRESBY THOMAS FRANK
PRESCOTT DENNIS LOUIS
PRESCOTT WILLARD SHERWIN
PRESTON ROSS MCCLLELAN
PREVOST KENNETH WAYNE
PRICE THOMAS GORDON
PRIEN DON
PRIESTHOFF JOHN HOWARD II
PRIETO TRINIDAD GUTIERREZ
PRINGLE EMMETT TERENCE
PRITCHARD CLARENCE R JR
PRITCHARD WILLIAM JOHN
PRITCHETT GREGORY GENE
PROCK DANIEL LEE
PROCTOR RICKEY ALLEN
PROCTOR WILLIAM C JR
PROSKY LEVERET ROSCOE
PROTTO ROBERT B JR
PROUDFOOT TIMOTHY COLE
PRYS ROBERT WILLIAM
PUENTES MIGUEL ANGEL
PUFF THOMAS JOE
PUFFENBARGER WILLIAM T
PUGH KENNETH WARD
PUGH RICHARD CARL
PULLAM JAMES LEE
PULLEN MELVIN LEWIS
PULSIFER NELSON F JR
PUMPHREY CORNEALUS JR
PURCELL DENNIS EDWARD
PURCELL GARY WILLIAM
PURDIE ROBERT DAVID
PURDIN PATRICK LAWRENCE
PURSELL CHARLES ALAN
PURSER JAMES LEAVELL
QUAGLIERI PAUL VINCENZO
QUEENER ULYSSES GRANT JR
QUERY ROBERT PETER
QUEZADA ARTHUR
QUIGLEY JAMES MICHAEL
QUIGLEY TIMOTHY ERNEST
QUILALANG ANASTACIO D JR
QUILL EDWARD BEEDING JR
QUINN ANTHONY LOUIS
QUINN DOUGLAS FRANK
QUINN MELVIN DARYL

QUINN PATRICK OWEN
QUINN TERRY LEE
QUINTANA JUAN CARLOS
QUINTANILLA JEFFERY IGLESIAS
QUINTERO JOSE HERNANDEZ
QUIROGA ALEX LEON
QUIROS CARLOS MANUEL
QUIROZ ALFRED MAURO
RABEY KENNETH TILDEN
RADER GARY PHILIP
RADTKE CARL LEONARD
RAGSDALE GARY WAYNE
RAGSDALE STEPHEN LEON
RAILLA JEAN ANTHONY
RAINBOLT JAMES EDWARD
RAINE DAVID SHELTON
RAINFORD EDWARD GEORGE
RAINS CLYDE EDWARD
RAINS VERNON BARTON
RAINVILLE RANDALL BRIAN
RALPH GARY RAY
RAMBUR MICHAEL JAMES
RAMEY JOE DON
RAMIREZ DAVID THOMAS
RAMIREZ EDUARDO CRUZ
RAMIREZ HILDEFONSO M
RAMIREZ JESUS P
RAMIREZ JOHN ARTHUR
RAMIREZ JOSEPH YBARRA JR
RAMIREZ JUAN
RAMIREZ LORENZO JR
RAMIREZ RALPH ALBERT JR
RAMIREZ VINCENT ALBERT
RAMOS BERNARDO KEALOHA
RAMOS GEORGE MICHAEL
RAMOS LUIS
RAMOS RICHARD
RAMSDEN GERALD LEE
RAMSEY MICHAEL WAYNE
RAMSEY RICHARD CHARLES
RANDALL LYNN MURRAY
RANDALL MICHAEL PAUL
RANDAZZO JOSEPH ANTHONY
RANDOLPH RICHARD ALAN
RANGEL RICHARD
RARIG ROBIN ARTHUR
RASCHEL THOMAS REGINALD
RASCHKE DEAN NELSON
RASCO KENNETH EDWARD
RASEY LARRY WAYNE
RASMUSSEN NEAL ARTHUR
RASMUSSON MICHAEL ALFRED
RASORI CARL RAYMOND
RATCLIFF TERRY WARD
RATHMANN EUGENE LE ROY
RATLEDGE DANIEL P JR
RATLIFF PAUL WAYNE
RATLIFF TERRY DIXON
RAUCH EDWARD HAROLD
RAUCH KIRK LESLIE
RAWLIN ROY VERNON
RAWSTHORNE EDGAR ARTHUR
RAY DARWIN ESKER
RAY ROBERT BRECKENRIDGE
RAY WILLIAM DAVID
RAYBURN STEPHEN LOUIS
RAYMOND EDWARD ROBERT III
READY ROBERT WILLIAM
REAGAN JOHN WALTER
REAUME PAUL EDMUND
REAVES HOMER LEE
REED DENNIS DALE
REED GARY WALTON
REED GREGG ERWIN
REED JAMES CLAYTON
REED MELVIN L JR
REED ROBERT WILLIAM
REED ROGER LEE
REEDER DAVID LEE
REEDY WILLIAM HENRY JR
REESE JOHN WILLIAM JR
REEVES GORDON MICHAEL
REGNOLDS JAMES RANDOLPH
REHE RICHARD RAYMOND
REHLING GUNTHER H
REICH WILLIAM GOODRO
REID DAVID STIRLING
REID EDWARD ROWAN JR
REID WILLIAM ALBERT

REILLY ALLAN VINCENT
REILLY ROBERT JOHN JR
REIS LUCIO JON
REMELTS WILLIAM HENRY II
REMILLARD GARRY EDWARD
RENFRO NORMAN A
RENNING RICHARD ANDREW
RENTERIA RUDOLPH SOTELO
RETZLOFF JAMES ROBERT JR
REVIER JOHN DAVID
REVIS RONALD JAMES
REYES EDWARD THOMAS
REYES HENRY R
REYES ROBERT ANTONIO
REYES RONALD
REYES RONALD DAVID
REYES RUBEN EVERARDO
REYNOLDS JOSEPH RAY
REYNOLDS MICHAEL MONROE
REYNOSO RENE
REZA LEONARD
REZENDE DANIEL DIAS
RHEA RANDOLPH VINCENT
RHOADS DANNY DAVID
RHODES WILLIAM BARTON
RICARDO SALVADOR ORTENCIO
RICE CAMERON A
RICE GREGORY LLOYD
RICE RONALD FRED
RICH DANNY KAYE
RICHARD DUANE LAWRENCE
RICHARDSON CHARLES A
RICHARDSON DAVID ALLEN
RICHARDSON DONALD HAROLD
RICHARDSON GARY LYLE
RICHARDSON ROBERT BROOKS
RICHEE JAMES BURNUS
RICHTER JAY DEE
RICK JOHN SCOTT
RICKER DARRELL BLANCHARD
RIDENHOUR DARWIN BRUCE
RIDEOUT DAVID JAMES
RIDGWAY CLYDE MOSES
RIDINGS LOUIS
RIGG WILLIAM CECIL
RIGGINS GARY RONALD
RIGGS WALTER RODERICK
RIGGS WILLARD WAYNE
RILEY ALDEN LAVERNE
RILEY JAMES CALVIN
RILEY RICKY VAUGHN
RILEY RONALD HOWARD
RINEHART FRED GEROLD
RINEHART JAMES DALE
RIST GARY MICHAEL
RITCHIE DOUGLAS REID
RIVENBURGH RICHARD WILLIA
RIVERA ERNEST ARBALLO JR
RIVERA JOSE A
RIVERA JULIAN CABRAL
RIVERA SILVESTRE MARTINEZ
RIVERA-BERMUDEZ JOSE ANTO
RIX DOUGLAS ALFRED
RIZZARDINI TIMOTHY JOSEPH
RIZZO ROBERT CHARLES
ROACH MARION LEE
ROARK ANUND C
ROBBINS CHARLES LESTER
ROBBINS JAY LEE JR
ROBBINS JERRY CLAYTON
ROBERSON ARTHUR PAUL
ROBERTS ARCHIE JAMES JR
ROBERTS CLIFFORD ALTON
ROBERTS CLIFFORD JOSEPH
ROBERTS KENNETH RAY
ROBERTS LOUIS WADE
ROBERTS MICHAEL ALLEN
ROBERTS STEPHEN LORD
ROBERTSON CLIFTON BOYD JR
ROBERTSON JOE CARROL
ROBERTSON JOHN ERNEST
ROBERTSON MERLE ELDON
ROBERTSON ROBERT ALLAN
ROBERTSON WILLIAM S III
ROBILLARD LARRY KENNETH

ROBIN DAVID ALAN
ROBINSON BRUCE ALLEN
ROBINSON FLOYD IRWIN
ROBINSON GUS BLAKELY
ROBINSON KENNETH JAMES
ROBINSON MARK EDWARD
ROBISON JIM BRUCE
ROBUSTELLINI DAVID W
ROCHA FELICIANO
ROCHA GEORGE XAVIER
ROCHA ROBERT SILAS
ROCHE JOHN
RODARTE ALEXANDER D
RODDAM RODDNEY ALLEN
RODDICK WILLIAM HENRY
RODGERS GARY GENE
RODGERS JOHN ARLINGTON
RODGERS JOHN THOMAS
RODREICK RONALD NELSON
RODRIGUES EUGENIO
RODRIGUES GARY WAYNE
RODRIGUES JOHN NETO
RODRIGUES JOSEPH MICHAEL
RODRIGUEZ BENITO BOBO
RODRIGUEZ ENCARNASION
RODRIGUEZ JESSE EMITERIO
RODRIGUEZ JOE IGNACIO
RODRIGUEZ LOUIS
RODRIGUEZ MANUEL JOE
RODRIGUEZ OSCAR FRANCISCO
RODRIGUEZ REGINALD JOSEPH
RODRIGUEZ ROGER ESPINOZA
RODRIGUEZ ROMIRO C
RODRIGUEZ RUDOLPH
RODRIQUEZ ARTURO CANTU
ROE LINUS ROBERT
ROE PHILLIP WILLARD
ROEDIGER CHRISS LESLIE
ROGERS HARVEY DAVID JR
ROGERS JAMES STEVEN
ROGERS MICHAEL FREDRICK
ROGERS ROBERT GENE
ROGERS RONALD LEE
ROGERS WAYNE JOHNATHEN
ROGOFF JAMES BILL
ROGONE JOHN PIO
ROLAND GEORGE RAY
ROLES JOHN WAYNE
ROLEY HERBERT WALLACE
ROLLASON WILLIAM DAVID
ROMERO JOSEPH MICHAEL
ROMERO PEDRO JR
ROMERO ROBERT WILLIAM
ROMERO ROBERTO ANDRESS
ROMERO VICTOR
ROMO FRANK GONZALES
ROMO ROBERT ALLEN
RORICK KENNETH ROY
ROSAR ROBERT JOHN
ROSE BARNES WARLAND JR
ROSE DANA GALE
ROSE DANIEL PATRICK
ROSE DAVID JON
ROSE DAVID LEE
ROSE FRANK JAMES JR
ROSE LEO JAMES
ROSE LEONARD DALE
ROSE MICHAEL ALLEN
ROSE NATHANIEL ROBERT
ROSE PAUL WARREN
ROSE ROGER CLARKE
ROSENBERGER DAVID ARTHUR
ROSENLUND NELS VERN II
ROSS DOUGLAS ALAN
ROSS GENE K
ROSS KENNETH EDWARD
ROSS LARRY EDWARD
ROSS LARRY THOMAS
ROSS PAUL R
ROSS RAYMOND JEFFERSON JR
ROSS ROBERT GARRY
ROSS ROBERT W
ROSS ROGER DALE
ROSSI THOMAS LOUIS
ROUNTREE RONALD CORBIN
ROWE BRUCE PHILLIP
ROWE JAMES GRAY JR
ROWE WILLIAM EDWIN
ROWEN GERALD LOYD II
ROWLAND WILLIAM MI-

CHAEL
ROWLES STEVEN ROBERT
ROY JAMES WILLIE III
ROYALL LESLIE WILLIAM III
RUBERG CHRISTOPHER EUGENE
RUBIO PETER PAUL
RUCHTI HEINZ
RUDDAN WILLIAM ANDREW
RUDOLPH ROBERT DAVID
RUEBEL JOSEPH PETER
RUFFNER RUSSELL MILES JR
RUGGE LLOYD TAYLOR
RUGGERI ANTONINO
RUIZ ANDREW ANDY
RUIZ GILBERT
RUIZ MANUEL
RUIZ RAMON RODRIGUEZ
RUIZ RAYMOND
RUIZ SALVADORE INIGUEZ
RUIZ WILLIAM JR
RUMLEY RICHARD ALLEN
RUMMERFIELD JAMES C JR
RUNDLE DANNY RAY
RUNGE ROBERT CARL
RUNNELS LLOYD CHISOLM JR
RUONAVAARA ROBERT EDWIN
RUSH DAVID CLYDE
RUSH ERVIN LEE
RUSH WILLIAM ARDIE
RUSHER ROBERT CHARLES
RUSS LEE HENDERSON
RUSSELL ALLEN BARTLEY
RUSSELL DONNIE HOWARD
RUSSELL GREGORY ALLEN
RUSSELL JAMES ROBERT
RUSSELL RICHARD LEE
RUSSELL RICHARD SHANNON
RUSSELL RONALD PATRICK
RUSSELL WAYNE
RUTTIMANN ALLAN
RUYFF RONALD PAUL
RYAN BERNARD STEVEN
RYAN DANIEL JOSEPH
RYAN FRANK D JR
RYAN JERRY VAN
RYAN RONALD ROYCE
RYE DILLARD GALE
RYGG CHARLES ALLEN
RYMOND NICHOLAS JAMES
RYSE ROY LOUIS JR
RYTTER PAUL E
SABEL JOEL MICHAEL
SABO LARRY MICHAEL
SADLER MITCHELL OLEN JR
SADLER THOMAS WAYNE
SAENZ FRANCISCO XAVIER
SAFFLE EDGAR JOE JR
SAIN DON RUE
SAITO SAMUEL RYOICHI
SALAZAR GILBERT SOLANO
SALDANA FERNANDO SAENZ
SALDANA RICHARD DAVID
SALDANO VINCENT
SALINAS JOE MANUEL
SALMON LARRY ANTHONY
SALZMAN LAVERN LEO
SAMPSELL JOEL WARREN
SAMSON FRANCISCO LEO JR
SAMSON MICHAEL ROMAN
SAN MARCOS EDMOND
SANCHEZ BENNY KUMIYAMA
SANCHEZ EDWARD JR
SANCHEZ JESSE
SANCHEZ JIMMY PINEDA
SANCHEZ JOSE ANGEL
SANCHEZ RALPH JR
SANCHEZ ROBERT PAUL JR
SANCHEZ SANTOS
SANCHEZ THOMAS JOSEPH
SAND RALPH THOMAS
SANDER JAMES KIETH
SANDER MICHAEL DENNIS
SANDERS FRANK BART
SANDERS HARVEY RICHARD
SANDERS RICHARD LEE
SANDERSON SANDER CHRIS
SANDLIN STEVEN RAY
SANDOVAL ALAN PAUL
SANDOVAL DANIEL FLORE
SANDOVAL GEORGE
SANDOVAL LOUIE JOE
SANDOVAL THOMAS FRED-

RICK
SANDS THOMAS MICHAEL
SANDSTROM ROBERT RICH-
ARD
SANDVIG VERNON DALE
SANTA-CRUZ DAVID FRANK
SANTOS JOHN F JR
SANTOS JOSEPH
SANTOS LAYNE MICHAEL
SANTOS MICHAEL EUGENE
SAPINOSA ALFRED ROBERT
SAPP CLARK EDWARD
SAPP WAYNE LEROY
SARAKOV HARRY DANIEL
SARSFIELD HARRY CARL
SATCHER CHARLES SHERLEE
SATHER RICHARD CHRISTIAN
SATO TAKESHI
SATTLER WILLIAM JOHN III
SAUER PHILIP HOWARD
SAUNDERS NICHOLAS GA-
BRIEL
SAUNDERS WILLIAM O JR
SAVAGE DOUGLAS PAUL
SAXTON GARY LEE
SCALES ASTOR JR
SCALISE EDWARD JOSEPH
SCANLON WILLIAM MANUEL
SCARPINATO JOHN ANDREW
SCHAEFFER FREDERICK WILLI
SCHARF CHARLES JOSEPH
SCHARFF LENNIE HAROLD
SCHAROSCH PATRICK FRAN-
CIS
SCHASRE DAVID M
SCHEIDT WILLIAM H
SCHELLING CHARLES HOW-
ARD
SCHIERMEYER WILLIAM D JR
SCHIMMELS EDDIE RAY
SCHLAMP GARY OLIN
SCHLOTE LOUIS CHRIS
SCHLOTTMAN JAMES ED-
WARD
SCHMIDT NORMAN
SCHMIDT RICKFORD RAY
SCHMIDT SCOTT LAWRENCE
SCHMIDT STEVEN WARREN
SCHMIDT WILLIAM JAMES
SCHMITZ RICHARD ALBERT
SCHNACK STEVEN SPENCER
SCHNELLER STEVEN OWEN
SCHOEL RENNY DEAN
SCHOENBAUM CRAIG RAY
SCHOEPKE ANTON JOHN
SCHRAMM BROCK ROWLAND
SCHROBILGEN WARREN H JR
SCHROEDER JERRY DEAN
SCHROEDER NICHOLAS LEE
SCHROEDER NICHOLAS LEE
SCHROEDER STANLEY A
SCHUBERT JOEL LUTHER
SCHULTZ DAVID ALAN
SCHULTZ EDWARD AUGUST
SCHULTZ GARY A
SCHULZE ROBERT EUGENE
SCHUYLER RONALD LEE
SCHWARTZ DAVID EARL
SCHWARTZ GARY STEVEN
SCHWARTZ KENNETH DALE
SCHWELLENBACH GARY
RALPH
SCOFIELD JOHN CHARLES
SCOGGINS JOHN PAUL
SCOTT BUSTER LEROY
SCOTT CHARLES F
SCOTT DARRYL KENNETH
SCOTT DAYNE YORK
SCOTT JAMES GEORGE
SCOTT O D
SCOTT THOMAS LEE
SCURR KENNETH WESLEY
SEAMAN JOHN CHARLES JR
SEARBY BARRY MARTIN
SEARFUS WILLIAM HENRY
SEARLES CHARLES PETER
SEARS GORDON BERT
SEASTROM WILLIAM LEONAR
SEAVEY DOUGLAS REX
SEBENS GAYLORD JAMES
SECANTI RICHARD MICHAEL
SEE RICHARD CHARLES
SEELEY JOHN STUART
SEEMAN JERI CALVIN

SEGUNDO PETE SPRULE
SEGURA STEVEN REY
SEHI GEORGE STEPHEN
SEIDEL WALTER JAMES
SEKVA ROBERT GLENN
SELF IRVING ALBERT
SELLERS RICHARD TAYLOR JR
SEMANS THOMAS EDWARD
SEMLER STANLEY KENTON
SEMMER PETER ANTHONY
SEMORE BOBBY ALLEN
SENGER MICHAEL MELVIN
SENNETT ROBERT RUSSELL
SERATTE JOHN STEVEN
SERNA LEOPOLDO PEREA
SERRANO GILBERT
SERRANO RODOLFO CAR-
RILLO
SERREM MARK MAC DONALD
SETTLE WILLIAM FOY
SETZER PAUL RAY
SEU MILTON J S
SEVENBERGEN JERRY L
SEVERLOH PAUL BRUCE
SEWELL DONALD MELVIN
SHAFFER JONATHAN PETER
SHAINA CONRAD WILLIAM
SHALHOOB TERRY WAYNE
SHAMEL JOHN CLARENCE
SHANKS DONALD WILFRED
SHANLEY MICHAEL HENRY JR
SHANNON GUY GENE JR
SHANNON JOHN PATRICK JR
SHANNON KENNETH MI-
CHAEL
SHANNON RANDELL FRANK
SHANNON STEPHEN CRAIG
SHAPIRO MILTON
SHARK EARL ERIC
SHARP DAVID JACKSON
SHARP LARRY DOUGLAS
SHARP PHILIP DEAN
SHARP SAMUEL ARTHUR JR
SHARPLESS ROBERT LEON
SHAUGHNESSY EDWARD
JEROME
SHAUGHNESSY JAMES J JR
SHAW JAMES ROBERT
SHAW THOMAS WILLIAM
SHEA JAMES PATRICK
SHEAHAN MICHAEL DAVID
SHELLEY MICHAEL OWEN
SHELTON ARTHUR ALEXAN-
DER
SHELTON CHARLES THOMAS
SHELTON EDWARD MINOT JR
SHELTON HENRY EARL
SHELTON RICHARD POWELL
SHELTON ROBERT SCOTT
SHEPARD LAWRENCE ROBERT
SHERIDAN ROBERT EDWARD
SHERLOCK JOSEPH V III
SHERMAN JOHN CALVIN
SHERMAN ROOSEVELT JR
SHERRELL DAVID FRANK
SHEWMAN RONALD JAMES
SHIELDS ALAN HARRY
SHIELDS DAVID THOMAS
SHIELDS JIMMY LEE
SHIELDS RICHARD DALE
SHINN WILLIAM CHARLES
SHOCKLEY RONALD DAVID
SHOEMAKER DAVID HOWARD
SHOMAKER JEROME CHARLES
SHOOK GEORGE LEONARD JR
SHORT CHARLES DUDLEY
SHORT LEWIS LEROY
SHORT MITCHELL CONRAD
SHORT RONALD LEE
SHOWERS DENNIS KARL
SHREVE JOSEPH LYNWOOD JR
SHRINER ROBERT LEE
SHRIVER JERRY MICHAEL
SHROUT SANFORD JR
SHUGART LYNN DOYLE
SHULTS WALTER GLENN
SHULTZ CHARLES EDGAR
SHUMATE BERLIN ROBERT
SHUSTER DARRYL WAYNE
SICILIA BRIGGS KINNEY
SIEDENTOPF MARK
SIEVERS DALE GLENN JR
SILAS THEODORE BUCHANAN
SILBAS ROSENDO FLORES

SILLER PETER LENHART
SILVA ANTONIO
SILVA FEDERICO
SILVA GEORGE LEE
SILVA THOMAS JOSEPH
SILVEIRA JOSE A C
SILVER GARETH MAC KENZIE
SILVERII LOUIS ZANE
SILVERS MITCHELL FRANK
SIMETH THOMAS JAMES SR
SIMMONDS JERRY LEE
SIMON PAUL JOSEPH
SIMONS ERNEST EUGENE
SIMPSON ALFRED FRANKLIN
SIMPSON LARRY DOUGLAS
SIMS JERRY G
SINGLETON GERALD BLAINE
SIOW GALE ROBERT
SIPES RICHARD EARL
SIRATT JACOB F III
SISCO JERRY DONALD JR
SISSON WINFIELD WADE
SITLER BARRY JAMES
SIX CHRISTOPHER JAMES ROY
SIZEMORE DONNIE RAY
SIZEMORE JAMES ELMO
SKAGGS RICHARD ALLAN
SKAKEL GEORGE WALTER
SKARPHOL ROBERT WAYNE
SKEEN RICHARD ROBERT
SKEINS RODRICK ALLAN
SKELLY STEVEN G
SKINNER HERBERT KIRK
SKINNER WALTER FRANCIS
SKIPPER HUGH G
SKIRVIN ORVAL L
SLEMSEK FRED ALBERT
SLOAN DOUGLAS DEAN
SLOAN LESLIE RAY
SLOAN ROBERT LELAND
SLOPPYE ROBERT ROYCE
SMART LESTER EDWARD JR
SMARTT MICHAEL CHRIS-
TOPHE
SMEAD CARL ROY
SMEVOLD EMIL HAROLD
SMILEY WILLIAM THOMAS
SMITH AMMONS EWING JR
SMITH ARCHIE D
SMITH BENNY JAMES
SMITH CHARLES CLARENCE JR
SMITH CHARLES LENET
SMITH CHRISTOPHER SCOTT
SMITH DANIEL JEFFREY
SMITH DAVID HUGH
SMITH DENNIS ALLEN
SMITH DONALD BRUCE
SMITH DONALD EUGENE
SMITH DONALD JAMES
SMITH DONALD RAY
SMITH DONALD RAY
SMITH FRANK JOHN
SMITH GARY KENNETH
SMITH GILBERT NOLAN
SMITH HARDING EUGENE SR
SMITH JACKIE LEE
SMITH JAMES DOUGLAS
SMITH JESSE LEE
SMITH JOHN CALVIN
SMITH JOHN CALVIN
SMITH LARRY MICHAEL
SMITH LARRY WAYNE
SMITH LAWRENCE CLAUDE
SMITH MARK EDWARD
SMITH MARLIN
SMITH MARSHALL R
SMITH MATTHEW EDWARD
SMITH MICHAEL BRUCE
SMITH MICHAEL FRANK
SMITH MICHAEL REX
SMITH PAUL LESLIE
SMITH PAUL WESLEY
SMITH PHILIP JEREMIAH
SMITH PHILIP JR
SMITH PHILLIP CHARLES
SMITH RICHARD JOHN
SMITH ROBERT CARROLL
SMITH ROBERT EUGENE JR
SMITH THOMAS LLOYD
SMITH TIMOTHY N JR
SMITH TULLIE ROSCOE JR
SMITH VENNIE LEE
SMITH VERNON PARR
SMITH WAYNE KEITH

SMITH WISELEE
SMOOT ROBERT GENE
SNEE FRANCIS JOSEPH JR
SNELL ESMOND EMERSON JR
SNELSON TERRIL WAYDE
SNYDER BOBBY CLYDE
SNYDER JOHN HERBERT
SNYDER PRESTON JOHN
SOKOLOF HARVEY GERALD
SOLANO RICHARD JOHN
SOLIS EUSEBIO
SOLIZ ENRIQUE LORENZO
SOLIZ THOMAS
SOLLEY JOHN JOSEPH
SOMMERS LARRY EUGENE
SOMMERS STEVEN ALLEN
SONES JOHN LESTER
SONSTEIN PAUL PHILLIP
SORCHINI ANDRES
SORENSEN DONALD ROBERT
SORENSEN ODIN EDGAR
SORENSON EUGENE A
SORICK STEVEN PAUL
SORROW CHARLES FINNEY JR
SOTELO LUIS ALONZO
SOTH MICHAEL JOSEPH
SOTO ARTHUR OLOGUE
SOTO MARTIN JESUS
SOTO RICARDO HINOJOSA
SOULE RONALD GLEN
SOURS BRUCE MICHAEL
SOUSA ROBERT PATRICK
SOUTAR WALTER JACK
SOUTHER JOHN MARTIN
SOUZA CHRIS ANTHONY
SOUZA FRANCIS LOUIS
SOUZA RAYMOND JOSEPH
SOWARD DOUGLAS
SPAFFORD GALON GENE
SPANN JAMES HALL
SPARKS GLENN LOUIS
SPARKS HENRY EUGENE
SPARKS JAMES EDWARD
SPAW JAMES ODIS
SPEAK ERIC B
SPEAR JOHN RANDALL
SPEER BYRON MORROW
SPENCER LEANDREW JR
SPENCER WARREN RICHARD
SPINALI DAVID JOHN
SPITTLER IRA JAMES III
SPITZER KENNETH LYLE
SPOTWOOD FRANK JR
SPRINGSTON THEODORE JR
SQUIRE BOYD EDWIN
SRSEN STEVE ALBERT
ST LOUIS BRUCE WAYNE
STAAB KURT CLARENCE
STACEY RALPH MCGUIN JR
STACK JOSEPH VINCENT
STADING GARY ALAN
STAFFORD JAMES HUBERT
STAFFORD PHILIP CLARK
STALLCUP ALVIN WAYNE
STALTER JOHN RAYMOND
STAMPFLI THEODORE AR-
THUR
STANDRIDGE JERRY WAYNE
STANDRING LAUREN WALTER
STANLEY DENNIS JOHN
STANLEY EURAL JR
STANSELL RICHARD NORRIS
STANTON SCOTT NEAL
STAPLETON OLLIE RAY
STARK COY FOSTER
STARKEY HENRY MORGAN
STARKEY JAMES WAYNE
STARR KIERAN JOHN
STATEN ROBERT JOSEPH
STAUD ROBERT NICOLAS
STEARNS MICHAEL FOR-
RESTER
STEELE STEVEN PATRICK
STEELE WALTER CHARLES
STEFANSKI STEVEN RUSSELL
STEIMER ROBERT FENTON
STEIMER THOMAS JACK
STELLE GERALD CAIN
STELZER CURTIS EDWIN
STEMAC STEPHEN JOSEPH
STEPHAN LARRY ROY
STEPHENS BOYD ADAM JR
STEPHENS HAYS CHARLES
STEPHENS LARRY ALLAN

California

STEPHENS ROGER DEAN
STEPHENSON BRUCE DONALD
STERLING JOHN CHARLES
STERUD MARTIN FREDERICK
STEVENS DENNIS LEE
STEVENS EDRICK KENNETH
STEVENS JOHN BRADFORD
STEVENS LARRY JAMES
STEVENS TAMADGE CECIL JR
STEVENS WALTER BRUCE
STEVENSON GARY GEORGE
STEVENSON JOHN RAYMOND
STEVER JAMES MITCHELL
STEWART DENNIS RAY
STEWART LELAND
STEWART LONNY LAWRENCE
STEWART PAUL CLARK
STEWART RICHARD JAMES
STEWART ROBERT LEE
STEWART SAMUEL KAY
STEWART WILLIAM WESLEY
STICKLER CLARK D
STIDHAM ERNEST JAMES
STILES CHARLES WALTER
STIRLING ELGIN LEROY
STIRLING JOHN F
STITH DARYL LA DON
STITT RICHARD WESLEY
STOGSDILL JACKIE DEAN
STOKER KENNETH GRANT
STOKES CHARLES EUGENE
STOKES JEFFREY RANDALL
STONE GREGORY MARTIN
STONE HARRY JAMES
STONE JOSEPH LAMAR
STONE RICHARD ARLAN
STONE WILLIAM EARL
STONESIFER DONALD LEE
STORBO RONALD LAWRENCE
STORK ROBERT JOHN JR
STORM RALPH DORMAN
STORZ GEORGE WILLIAM
STOTTS JAMES MARTIN
STOUT TERRY LEE
STOVER TOMMY GENE
STOWE LUTHER TONY
STRAIN KENNETH DALE
STRAND PHILIP STANLEY JR
STRANGE FLOYD WAYNE
STRANGE ROBERT GREER
STRAW BARRY MERCER
STREET BRENT ANTHONY
STREET TOBY WINDFIELD
STRIBLING VICTOR BERNARD
STRICKLAND DOUGLAS LEE
STRICKLIN ROBERT GUY
STRINGHAM WILLIAM
STERLIN
STROBBE DANIEL EDWIN
STROBLE JAMES JOHN
STROBRIDGE RODNEY LYNN
STROCK CHARLES FREDERICK
STRONG DANIEL LEROY
STRONG GRIDLEY BARSTOW
STRONG STANLEY GRANT
STROUSE LARRY DALE
STUCKEY WALTER
STUESSEL JAMES DAVID
STURDIVANT JASPER DEAN
STURDY ALAN MAC DONALD
STUTES WILLIAM BYRON
SUAREZ ENCARNACION
ALEGRE
SUAREZ JOSE WILFREDO

SUEDMYER LARRY DEAN III
SUGIMOTO LEONARD JAMES
SUGIURA TOM DENNIS
SUIAUNOA TUIOALELE T
SULLIVAN JOHN MICHAEL
SULLIVAN RAYMOND WALTER
SULLIVAN RICHARD ARTHUR
SULLIVAN RICHARD D JR
SULLIVAN THOMAS HOWARD
SUNDELL LARS PEDER
SUNIGA MICHAEL EDWARD
SUPNET EMILIO CABRERA JR
SUPNET RICHARD ARELLANO
SUTHERLAND HERBERT LEE
SUTHERLAND JAMES EDWARD
SUTTON BRYAN JAMES
SUTTON DENNIS LEE
SUYDAM JOHN HOWARD III
SUZUKI KENNY RYOSUKE
SWABBY BRENT LESLIE
SWAIM ALLAN GREGORY
SWANSON DONALD LLOYD
SWANSON KEITH LYLE
SWANSON TODD EARLE
SWARBRICK LAWRENCE
GORDON
SWAYZE RICHARD DAVID
SWICK ROBERT MICHAEL
SWIGART PAUL EUGENE JR
SWISHER CLIFFORD LEE
SWORDS SMITH III
SYSAK CRAIG ALAN
SZUTZ BRAD JOHN
TABET HENRY MARSIAL
TABOADA ADOLFO ANTO-
NIO JR
TACTAY EUGENE RICARDO JR
TAFAO FA'ASAVILIGA V
TAFOYA JOSEPH ERNEST
TAGATA LAAVALE FUATAU
TAGUE NICHOLAS ALLEN
TAKETA KEN HARRIS
TALKEN GEORGE FRANCIS
TALLEY GARY LEE
TALLEY HAROLD LEE
TALLMAN GEORGE
TALMON PETER GEORGE III
TAMAYO FRANCISCO MARIO
JR
TANDY MICHAEL GORDON
TAPIA MOISES
TAPIO HEINZ ARNOLD
TAPP MARSHALL LANDIS
TAPPAN FREDERICK HOWARD
TARANGO ERNESTO
TARRANCE WILLIAM BLAIR
TASCH JON
TASKER DAVID LEROY
TATE FENNELL
TATE FRED EUGENE
TAUAESE VALENTINO
TAUFI AOULIULITAU FAITUPE
TAYLOR ALBERT RUSSELL II
TAYLOR ALONZO HUGHES
TAYLOR DAVID THORNTON
TAYLOR DENNIS LEE
TAYLOR GARY DEAN
TAYLOR GEOFFREY RAYMOND
TAYLOR HERMAN L
TAYLOR JAMES HARRY
TAYLOR JERRY LEWIS
TAYLOR JESSE JUNIOR
TAYLOR JOHN RAYMOND
TAYLOR LARRY GENE
TAYLOR MARVIN JUSTIN
TAYLOR PAUL CLIVE O
TAYLOR RICHARD KENNETH
TAYLOR ROBERT DWIGHT
TAYLOR RUDY RONNIE
TAYLOR SHERMAN RAY
TAYLOR WAYNE OLIVER
TAYLOR WILLIAM DOUGLAS
TEAGUE BRUCE EDWARD
TEAL FRED THOMAS
TEBBETTS TERRY LEE
TEDDS MERVYN DONALD
TELA MOLIMAU ASOMALIU
TELLES PAUL GEORGE
TEMPLETON RAYMOND
WOODROW
TENORIO JIMMY JOE
TERRAZAS JUAN LUIS
TERRY MARVIN HALL
TESILLO ARMANDO

TESSMAN CLARENCE CLEM-
ENT
THARALDSON JEFFRY RAY
THEURKAUF HARRY LEE
THIELEN JOHN ROGER
THIERY JOHN
THIRKETTLE MICHAEL JOHN
THOMAN THEODORE VAIL
THOMAS ANDREW JACKSON
THOMAS CHARLES BLAKE
THOMAS DARWIN JOEL
THOMAS HARRY EUGENE
THOMAS JAMES EDWARD JR
THOMAS JERRY T
THOMAS JOHN RAYMOND
THOMAS KENNETH BEN
THOMAS MICHAEL EDWARD
THOMAS MONTE VERNON
THOMAS RICHARD ALAN
THOMAS RICHARD LYNN
THOMAS RUFUS ALFONZO JR
THOMAS TIM
THOMAS TIMOTHY ARMA
THOMAS TOMMY ROY
THOMAS WILLIAM DEWAYNE
THOMPSON BERNARD DAVID
JR
THOMPSON BRUCE WAYNE
THOMPSON CHARLES CLAIR
THOMPSON DENNIS EUGENE
THOMPSON EDGAR WAYNE
THOMPSON KENDALL WIL-
LIAM
THOMPSON LOUIS KENNETH
THOMPSON RICHARD MAR-
TIN
THOMPSON RICHARD VICK-
ERS
THOMPSON ROBERT ALAN
THOMPSON ROBERT EUGENE
THOMPSON ROBERT JR
THOMPSON ROBERT R
THOMPSON THOMAS DON-
ALD JR
THOMPSON WALTER LEE
THOMPSON WILLIAM DEWEY
JR
THONUES GUENTER ROBERT
THORNBURG VINCENT
ROBERT
THORNELL EDMUND FRANCIS
THORNLEY REX EDWIN
THORNTON ALAN WAYNE
THORNTON DAVID LESLIE
THORNTON WILLIAM A JR
THORPE DAVID LOUIS
THORPE DENNIS RAY
THORPE FRED ROBERT
TIBBETTS DAVID RAMSEY
TICE GARY DALE
TIDERENCEL JOHN WERNER
TIDWELL ERICH LINWOOD
TIFFANY DAVID L
TIFFIN RAINFORD
TIGHE CHARLES JOSEPH
TIGHE JOHN ROY
TIGHE RAYMOND HOWARD
TILLINGHAST BRADLEY OLEN
TILLITSON STANLEY SCOTT
TIMBOE ARTHUR RICHARD
TIMMONS DENNIS EDWARD
TIMMONS EDWARD HUGH
TIMOTHY WAYNE ELLIOTT
TINGLE KENNETH WAYNE
TIPTON TIMOTHY TAYLOR
TISCORNIA JOHN JOSEPH
TISDALL GARY DEAN
TITMAS JAMES III
TITSWORTH KENNETH CARL
TIVIS JOHNNY EARL
TODD ROBERT JAMES
TOGNAZZINI MILFORD
MARVIN
TOIA MATAU JR
TOLENTINO CLARENCE
TOLESON THOMAS NORMAN
TOLETTE RICHARD ROSS
TOLPAROFF ALEX ROBERT
TOMASINI RICHARD E JR
TOMKINS JOHN MICHAEL
TOMLINSON DAVID CULLEN
TOMLINSON DAVID MARLOW
TOMLINSON MICHAEL JAMES
TOMLINSON ROBERT DALE

TORESON ROBERT WAYNE
TORLIATT CHARLES PETER JR
TORRES ARCADIO JR
TORRES DAVID
TORRES FRANK CHICO JR
TORRES HIGINIO RODRIGUEZ
TORRES MANUEL PRIETO
TORRES MICHAEL ANGEL
TORRES RAYMOND
TORTORICE RICHARD JOHN
TORTORICI BRUCE
TOTH BERTALAN JAMES
TOTTY DELBERT CHAN
TOVAR ATILANO URIEGAS
TOWNSEND STEPHEN LANCE
TOWNSEND WILLIAM PAUL JR
TOY GERALD OSCAR
TOYIAS CHARLES LESLIE
TRAMELL DANIEL
TRANTHAM DONALD RAY
TREASURE ROBERT JOSEPH
TREJO JOHN MICHAEL
TREJO JOSEPH JR
TRESTER DAVID ALEXANDER
TREVINO JUAN RAMON
TRICKER CHARLES RUPERT
TRIDLE LEON PAUL
TRIMBLE JAMES MITCHELL
TRIPLETT GRADY THOMAS
TRISLER RICHARD LEE
TRITSCH PHILIP ALON
TRIVELPIECE STEVE MAURICE
TROELSTRUP THOMAS LEE
TROMBETTA TONY
TROYANO ROLAND DEAN
TRUESDALE STANLEY E
TRUJILLO FELIX MARCIAL
TRUJILLO FRANCISCO M
TRUJILLO RAYMOND AN-
THONY
TRUSSELL ROYCE WILLIAM JR
TRYON FRED ALBERT JR
TRYON LEE JR
TSCHUMI WILLIAM JOHN
TUBRE STEPHEN RENIER
TUCKER GREGORY CHARLES
TUFF MICHAEL STEPHEN
TULLER DENNIS J
TUNALL STANLEY WILLIAM
TUNNELL JOHN WALLACE
TUNSTILL FRANK JR
TURCOTTE RALPH JEAN
TURK JOHN GEOFFREY
TURNBULL GARY ALLEN
TURNER ALAN BRADFORD
TURNER DAVID ROBERT
TURNER EDDIE D
TURNER EUGENE
TURNER JEFFREY ARTHUR
TURNER KELTON RENA
TWITTY DANIEL RAY
TWYFORD THOMAS LIONEL
TYES ROBERT LEE
TYLER LARRY JOSH
UCKER DAVID JOHN
UDING STANLEY ROY
UELI PENI
UHL ROBERT DALE
UHLIG MICHAEL STEVEN
ULIBARRI EDWARD ANTHONY
ULLMER WILLIAM ARTHUR JR
ULLOA MANUEL GURROLA
UNDERWOOD WILLIAM
HENRY J
UNFRIED BARRY LON
UPCHURCH RODNEY CLEVE-

LAND
UPLINGER BARTON JOHN
UPRIGHT RUSSELL EDWARD
URIBE EDWARD ANTHONY
URNES JAMES LEE
URRABAZO HOMER
UTLEY RUSSEL KEITH
UYESAKA ROBERT JOSEPH
VAGNONE RICHARD BER-
NARD
VALADEZ RICHARD PAUL
VALADEZ TIMMY
VALDEZ ALFRED
VALDEZ DANIEL VIRAMON-
TES
VALDEZ DAVID MEDINA
VALDEZ ISMAEL JOSE JR
VALENCIA CLEMENT JR
VALENCIA FRANCISCO
MACEDO
VALENCIA RALPH MARIO
VALENZUELA CARLOS
VALENZUELA HENRY JR
VALENZUELA OSCAR JR
VALERO JOHN JUAN
VALLERAND LARKIN OSCAR
VALOV JAMES DAMION
VALSTAD CLYDE JULIUS
VALTIERRA JUAN BORJA
VAN CAMPEN THOMAS
CHARLES
VAN DERVORT EDWARD PAUL
VAN DUYNE ROBERT SCHUY-
LER
VAN FLEET DONALD WILLIAM
VAN GIESON ROBERT LESTER
VAN HOOK JAMES DOUGLAS
VAN HORN BARRY WILLIAM
VAN HORN CHARLES ALBERT
VAN LANT WAYNE G
VAN PATTEN ROBERT AN-
DREW
VAN VALKENBURG CLYDE
W JR
VAN WYK JOHN HENRY
VANDER DUSSEN GEORGE
VANDE-VEGTE DOUGLAS LEE
VANDEVENTER JAMES
CHARLES
VANDIVER FRED GERALD
VANOVER PAUL PHILLIP
VANTOL GARY LEON
VARGAS RAUL JOHN
VARNER DOUGLAS ALLEN
VARNER JERRY DANIEL
VARNI HOWARD STEVEN
VASQUES SELVESTER JOE
VASQUEZ CHARLES V JR
VASQUEZ DAVID
VASQUEZ JIMMY
VASQUEZ MARK ANTHONY
VASQUEZ MAX V III
VASQUEZ PATRICK JOHN
VATER DIETER RUDOLF
VAUGHAN DANIEL JOSEPH
VAUGHAN DONALD CHARLES
VAUGHAN ROBERT REDDING-
TON
VAUGHN DONALD WILBANKS
VAUGHN JOHN PATRICK
VAUGHN RICHARD WILLIAM
VAUGHT CRAIG STEPHEN
VAZQUEZ WILLIAM
VEACH RICHARD ELZIE
VEDROS RANDOLPH PAUL
VEGAS PAUL ISAAC

VELASCO MIKE RALPH
VELASQUEZ ANTONIO
VELASQUEZ CHARLES
VELASQUEZ ROBERT
VELOZ EDUARDO
VENDELIN THOMAS LESLIE
VENTERS ROGER LEE
VENUTI VINCENT JR
VERDUGO ADALBERTO R E
VERDUGO DANIEL ALEXAN-
DER
VERGANO ROBERT THOMAS
VERLINDEN CRAIG ALDEN
VERNES ROBERT FRANK
VESEY CHARLES HANSEN
VESEY JERROLD LOUIS
VIERAS JOSE LOUIS
VIERRA JOSEPH
VIGIL ALEXANDER
VIGIL ARTHUR VERNON
VILARDO RONALD ALLAN
VILLA ARMANDO
VILLALOBOS ARTHUR GARCIA
VILLALOBOS IGNACIO L
VILLALOBOS JUAN JESUS
VILLALOBOS PAUL RUBIO
VILLALPANDO RAYMOND JR
VILLANUEVA FRANCISCO JR
VILLANUEVA JOSE EDWARDO
VILLAROSA PAUL HERMAN
VILLARREAL RICARDO
VILLEGAS DANIEL JOHN
VILLEGAS RALPH PAUL
VINAL RICHARD ALDEN
VINASSA MICHAEL
VINCENT GEORGE
VINES RICHARD LARRY
VINGE TERRY LEE
VINTER STEVEN CHARLES
VIVILACQUA THEODORE R
VOGET DONALD GUSTAV
VOKES RICHARD HENRY
VOLHEIM MICHAEL CORY
VOLKE CLIFFORD JAMES II
VOLZ STEPHEN THOMAS
VOSSEN STANLEY JOSEPH
VRABLICK MICHAEL STEPHEN
VROOMAN NICHOLAS WHIT-
TIER
VULLO MICHAEL PHILLIP
VURLUMIS CHRIS C
WADE DONALD JAMES
WADSWORTH JOHN LANIER
WAFER DAVID EARL
WAFFORD RONALD LEMON
WAGNER DAVID FREDERICK
WAGNER JOHN LAWRENCE
WAGNER MARVIN LEROY
WAGNER RICHARD ALLEN
WAGNER ROBERT KAY
WAGNER RUSSELL MARK
WAGNER WALTER ADOLPH
WAGNER WAYNE DOUGLAS
WALBER RONALD JAMES
WALDRON DUANE EVERETT
WALDRON GEORGE ALLEN
WALDRON STEVEN
WALDROP RONALD TERRY
WALKER CLIFFORD WAYNE
WALKER GARY LAYNE
WALKER GEORGE THOMAS
LLOY
WALKER JAMES DANIEL
WALKER JAMES EDWARD JR
WALKER JERRY LEE
WALKER JOHN WESLEY
WALKER LARRY ALLEN
WALKER MICHAEL FREDER-
ICK
WALKER THOMAS JAMES
WALKER THOMAS MICHAEL
WALKER THOMAS RAY
WALKER WILLIAM JOHN
WALKINSHAW GEORGE
MYRON
WALL PAUL EVERETT
WALLACE ARNOLD BRIAN
WALLACE MICHAEL D
WALLER JOHN BUSSEY
WALROD RICHARD ARTHUR
WALTERS DONALD WESLEY
WALTERS JOHN BRADY
WALTERS WILLIAM E
WALTMAN JESSE LYLE

WALTON CRAIG LESLIE
WANAMAKER DANNY WAYNE
WANBAUGH RONALD NELSON
WANDRO JAMES MATTHEW
WANGLER LOSSY RAY
WARD DANNY EDWARD
WARD DENNIS CHARLES
WARD GEORGE WARREN
WARD JAMES CRAIG
WARD PATRICK MICHAEL
WARD ROBERT DAVID
WARD ROBERT WILLIAM
WARE KEITH LINCOLN
WARF LAWRENCE ROBERT
WARMBRODT JON FREDERICK
WARNER CHARLES WILLIAM F
WARNER HERBERT ALVIN JR
WARNOCK ROBERT MERRILL
WARRELMANN KLAUS
WARREN DONALD ALBERT
WARRINGTON CHARLES W JR
WASHBURN JOHNNY LEE
WASHBURN ROBERT GLEN
WATANABE JAMES RYOICHIRO
WATKINS DAVID EUGENE
WATKINS MARTIN LEE
WATKINS SHERDIN JAY
WATSON ALFONZA
WATSON ARTHUR JR
WATSON J V
WATSON JAMES EDWARD
WATSON JAMES OSMOND
WATSON LESLIE JAMES
WATSON RONALD R
WATSON STANLEY EUGENE
WATSON THOMAS EDWARD
WATSON WILMER
WATTERSON DENNIS RAY
WATTS THOMAS JAMES
WATTS THOMAS ROGER
WAUGH RANDALL MICHAEL
WAYSACK WILLIAM JOHN
WEATHERLY JACKIE DON
WEATHERSBEE ERNEST
MURRAL
WEAVER DOYLE WAYNE
WEAVER RICHARD ALLAN
WEAVER TIMOTHY PATRICK
WEBB DONALD RAY
WEBBER SCOTT STILLMAN
WEBER DAVID GERALD
WEBER PATRICK
WEBER PAUL FREDERICK
WEBER RAYMOND N
WEBORG JOHN CHARLES
WEBSTER FRANK ANTHONY
WEBSTER HOWARD GREGORY
WEBSTER RHENA CHARLES
WEBSTER WILBERT MICHAEL
WEDDENDORF ROBERT
GEORGE
WEDLOW KENNETH EDWIN
WEDMAN KENNETH ALBERT
WEED DONALD EDMOND
WEEDER RICHARD D
WEEKLEY CLIFFORD WAYNE
WEGER JOHN JR
WEHR JAMES LE ROY
WEIDEMIER PETER JOSEPH
WEIGT STEPHEN LENN
WEIKAL WILLIAM BYRON JR
WEIL RICHARD ANTHONY JR
WEIR JOHN RANDOLPH
WEISS RAYMOND DOUGLAS
WEITZEL KELLY WAYNE
WELCH MICHAEL ALLEN
WELCH RICHARD ERNEST
WELLER ROBERT ALLEN II
WELLS FRANK JR
WELLS JAMES EDWARD
WELLS RICHARD KENNETH
WENBAN BRUCE R
WENSEL NORMAN BYRON
WENSINGER RALPH ROBERT
WENTWORTH JOHN VESTER
WENTZ MITCHELL ALLEN
WENTZELL JEFFREY RAY-
MOND
WENZL RONALD ALBERT
WERNER NORBERT OTTO
WERNER ODELL JACK
WERNER STUART ARTHUR
WERNER WALLACE BRUCE
WESKE RICHARD ALWIN

WESOLOWSKI ALVIN JOHN JR
WESOLOWSKI JEFFREY SCOTT
WESSELMAN GARY LEROY
WEST BOBBY
WEST JAMES CLIFFORD JR
WEST JAMES OSCAR
WEST ROBERT WILKS
WEST RUSSELL UDELL
WESTBAY GAYLORD LEE
WESTBERRY VINCENT
DOUGLAS
WESTCOTT GARY PATRICK
WESTERN AARON HAROLD
WESTON ROBERT HUGH
WEYMOUTH THEODORE GAY
WHALEY JAMES GOODWIN
WHALEY LOY NEAL
WHEATLEY WILLIAM GEORGE
WHEELER DARRELL EUGENE
WHEELER NICOLAS
WHELAN MICHAEL PATRICK
WHIPKEY RICHARD ALLEN
WHIPPLE CLIFFORD LEROY
WHISENANT JOHN WILLIAM
WHISENHUNT JAMES HENRY
WHITAKER G W
WHITAKER JERE LEE
WHITAKER MICHAEL JOSEPH
WHITE CHARLES MOTT SR
WHITE DONALD HERBERT
WHITE FRANKLIN RALPH
WHITE JEFFREY MERLE
WHITE JOHN HERBERT JR
WHITE JOHNNY BRYAN
WHITE LEONARD RAY
WHITE MARVIN CHARLES
WHITE MONETTE VON
WHITE RALPH ERIC
WHITE RICHARD ALLEN
WHITE WILLIAM HENRY
WHITE WILLIAM JOSEPH JR
WHITEHEAD CHARLES F JR
WHITEMAN RICHARD LEE
WHITESIDES RICHARD
LEBROU
WHITING MALCOLM D III
WHITING RICHARD EDWARD
WHITLEY EMMANUEL DAVID
WHITLOW THOMAS JAMES JR
WHITMORE RICHARD ALLEN
WHITSON JIMMY ALAN
WHITT MARK ALAN
WHITTED ROBERT ALBIN
WHITTEN DAVID ELGA
WHITTEN PATRICK ANTHONY
WHITTIER MARK CROSBY
WHORTON DWAYNE JEFFER-
SON
WHYNAUGHT JEFFREY LYLE
WIDENER JOHN EDWARD
WIGGINS WALLACE LUTTRELL
WIGHTMAN DAVID LLOYD III
WIKLE RICKY LYNN
WILDAUER PAUL ARTHUR
WILDER ARLOS CLAYTON
WILES MARVIN BENJAMIN
CHR
WILKES ROBERT LEE
WILL FREDERICK REED
WILLARD FREDERICK R JR
WILLETT RICHARD JAMES
WILLEY JOHN JAMES
WILLIAMS BARRY HENNETT
WILLIAMS BRIAN JOHN
WILLIAMS CHARLES JAMES
WILLIAMS CHARLES ROBERT
WILLIAMS CHRISTIAN LARS
WILLIAMS CRAIG EMERY
WILLIAMS DAVID CHARLES
WILLIAMS DAVID GEORGE
WILLIAMS DAVID III
WILLIAMS DENNIS CRAIG
WILLIAMS DENNIS LEE
WILLIAMS FRANK DUVALL
WILLIAMS FREDERICK JOSEPH
WILLIAMS FREDRICK H JR
WILLIAMS GERALD PATRICK
WILLIAMS HAROLD DAVID
WILLIAMS JACK ELWIN
WILLIAMS JAMES ALEC
WILLIAMS JOHN DAVID
WILLIAMS JOHN VINSON JR
WILLIAMS JOHNNY
WILLIAMS LARRY ELLIS

WILLIAMS LARRY KEITH
WILLIAMS MICHAEL WALTER
WILLIAMS RICHARD D
WILLIAMS RICHARD FRANK
WILLIAMS ROBERT ALWYN
WILLIAMS THOMAS HOWARD
WILLIAMS THOMAS JOHN
WILLIAMS VINCENT RICHARD
WILLIAMSON DONALD RAY
WILLIAMSON JOHN CLAR-
ENCE
WILLINGHAM ELDON WAYNE
WILLIS DONALD CLYDE
WILLIS HAROLD EUGENE
WILLIS JOEL THOMAS
WILSON ADAM
WILSON CLAUDE DAVID JR
WILSON DANIEL L
WILSON DAVID ALLEN
WILSON DAVID HENRY II
WILSON DAVID LEWIS
WILSON DEAN CHARLES
WILSON FRANK LEONARD
WILSON GARY DUANE
WILSON JACK PYEATT JR
WILSON JAMES DAVID
WILSON JAMES WILLIE
WILSON KENNETH RICHARD
WILSON MICKEY ALLEN
WILSON PAUL
WILSON RONALD ALTON
WILSON RONALD KELLEY
WILSON ROYCE HAROLD JR
WILSON WILLIAM JEFFREY
WILTON STANLEY FRANK
WILTSE JAMES B JR
WILTSE RONALD ELLIS
WIMER FLOYD DANIEL
WIMER ROBERT ARNOLD
WINFREY RAYMOND MI-
CHAEL
WINKEMPLECK GEORGE
HAROLD
WINNINGHAM JOHN QUIT-
MAN
WINTERS DARRYL GORDON
WINTERS DAVID MARSHALL
WISHAM GEORGE MERRITT JR
WITHERELL GARY LEE
WITMER NOEL BRUCE
WITTLER LARRY ELDON
WITTMAN NARVIN OTTO JR
WOLFE BRIAN EDWARD
WOLFE JOSEPH KENT
WOLFE KENNETH WAYNE
WOLFE WILLIAM EDWARD
WONG EDWARD PUCK KOW
JR
WONNACOTT WALTER L
WOOD PATRICK LEE
WOOD PETER LORENZ
WOOD TODD LOUIS
WOOD WILLIAM MILTON JR
WOODALL GEORGE WALLY
WOODCOCK STEVEN JON
WOODRUFF EDWARD WAR-
REN
WOODS ALONZO DALE
WOODS ARTHUR LEE
WOODS CHARLES GORDON
WOODS JAMES CLARK
WOODS LARRY JAMES
WOODSIDE MICHAEL LEE
WOODSON JOHNNY
WOOLLEY MACK LEE JR
WOOLSEY WILLIAM JAY
WOOTTEN CARL DEE
WORD WILLIAM KENEITH
WORL LESLIE WAYNE
WORLEY KENNETH LEE
WORLEY ROBERT FRANKLIN
WORRELL GARY PAUL
WORTHINGTON LAURENCE D
WORTHINGTON ROBERT
LEROY
WRIGHT CHARLES FRED
WRIGHT DAVID DANIEL
WRIGHT DENNIS HAROLD
WRIGHT DENNIS PAUL
WRIGHT EDWARD TAYLOR
WRIGHT GARY GENE
WRIGHT GENE THOMAS
WRIGHT HENRY ARTHUR
WRIGHT JAMES JOSEPH

WRIGHT JAMES L
WRIGHT JAMES WILLIAM II
WRIGHT JAY L
WRIGHT JERRY DEAN
WRIGHT LARRY EDWARD
WRIGHT ROBERT EDWARD
WRIGHT SILAS CLIFFTON
WRIGHT STEVEN JAMES
WRIGHT THOMAS CLAY
WRIGHT TYRONE
WYATT ALVIE CLINTON
WYATT BILLY HERDON
WYATT JOHN WESLEY JR
WYKOFF THEODORE LEON-
ARD
WYMAN MURRAY JOHN
WYSEL MITCHELL BLAINE
XAVIER AUGUSTO MARIA
YABIKU TAKESHI
YAGER JOHN CORWIN
YAMANAKA ROGER KIMO
YAMANE BENJI
YAMASHITA AKIRA
YAMASHITA KENJI JERRY
YAMASHITA SHOJIRO
YANCHUK RICHARD PHILIP JR
YANEZ VICTOR MANUEL
YARNELL DANIEL DAVID
YATES BRUCE EDGAR
YATES ROBERT ALAN
YATES SAMUEL WALTER
YBANEZ JOSE
YBARRA FRANK RODRIQUEZ
YBARRA KENNETH FRANCIS
YEAKLEY JAMES D
YEE EDWARD
YELL GLEN HOWARD
YLLAN CHARLES DAVID
YOCHUM LAWRENCE WAYNE
YOHN WILLIAM LEON
YOUNG CARLOS AVILA
YOUNG DARYEL JOE
YOUNG ERNEST HAROLD III
YOUNG GARY LEE
YOUNG JAMES PAUL
YOUNG JOHN EDWARD
YOUNG JON MICHAEL
YOUNG LARRY CLAYTON
YOUNG LEWIS JOHN
YOUNG MICHAEL ROBERT
YOUNG RONALD LEE
YOUNG SAMUEL LEE
YOUNG THOMAS EVERETT
YOUNG WILLIAM GARY
YOUNGBLOOD WILLIAM
RONALD
YSGUERRA ROBERT MARTIN
YUKI DOUGLAS HARVARD
ZABALA SALVADOR JR
ZACHARZUK MICHAEL
PATRICK
ZAGER EDWARD ARTHUR
ZAMORA EDWARD
ZARAGOZA VICTOR
ZENICK ROBERT JAMES
ZERBA DOUGLAS PAUL
ZERBE MICHAEL RICHARD
ZIEGLER STANLEY BRUCE
ZILLGITT DONALD HENRY
ZIMBERLIN ROBERT E JR
ZIMMERLE RENE AUGUST
ZIMMERMAN DAVID ERVIN
ZIMMERMAN DAVID PAUL
ZIMMERMAN EDWARD AN-
THONY
ZIMMERMAN RAYMOND L
ZIRFAS EWALD
ZISS EMIL ROGER
ZOBOBLISH DONALD
ZOELLER LEE BENJAMIN
ZOLLER ERIC WARD
ZUCROFF STEVEN DALE
ZUFELT ROY GLENN
ZUMWALT EDWIN ALLEN
ZUNIGA CHARLES EDWARD
ZUNIGA EFRAIN JR
ZUNIGA JOSEPH ANTHONY
ZUNIGA LEON JR
ZUNIGA VICENTE

Photos by Bonnie McGuire

Colorado

Welcome Home Memorial

340 CO-340, Fruita, CO 81521

In order to make up for the "welcome home" that so many Vietnam veterans didn't get to enjoy, the Western Slope Vietnam War Memorial Park in Fruita, Colorado was created to give the veterans that welcoming feeling that they rightly deserved all those years ago. The memorial can be visited at any time, but special events are held there on Memorial Day and Veterans Day.

The park itself was founded by Jim Doody who wanted to create a Vietnam War memorial in honor of his brother and his brother's friend who served in the conflict. Doody's brother, Thomas, was a helicopter pilot who died in action in 1971. Most of the labor and materials were donated by the Associated Builders and Contractors of America while the bronze statue that is featured in the memorial was made by Richard Arnold, the same sculptor who created the memorial statue for the Mobile, Alabama monument. The memorial was originally dedicated in July of 2003, but the bronze "Welcome Home" memorial wasn't placed until November of 2007. Funds for the memorial as a whole came through donations and fundraising efforts while the city incorporated the land of the memorial grounds as part of the City Park's System. To ensure the proper maintenance of the memorial the city of Fruita asked that Doody raise $10,000 for the continued upkeep of the site, which he did.

The Western Slope Vietnam War Memorial has multiple features including the "Welcome Home" bronze sculpture by Arnold. The statue is a life-size depiction of parents greeting their soldier, who is still in his uniform, as he comes home from overseas. There is also a memorial wall consisting of black granite slabs featuring the names of the casualties of the Vietnam War as well as other soldiers who served in a branch of the armed forced between the years 1959 and 1975. A donation made to the memorial can have a qualified service member's name added.

A Huey helicopter mounted to a 15-foot perch is also part of the memorial. The platform the Huey rests on also features plaques with information about the replica Vietnam Service Ribbon as well as a list of 55 local, Western Slope casualties. A fun feature of this part of the memorial is the motion detector that turns on speakers playing helicopter sounds and music when a visitor walks passed. The entire memorial is enclosed by the Walk of Honor which features the names of veterans from all wars as well as community members who have donated money to support the monuments. The American flag and Colorado State flag fly over the memorial.

The Names of those from the state of Colorado Who Made the Ultimate Sacrifice

ABBOTT GUY FRANCIS
ABEYTA TONY GENEVEVO
ADAMS ARTHUR LLOYD JR
ADAMS RONALD WYATT
ADAMS WILLIAM EDWARD
ADKINS CHARLES LELAND
ADKISSON JAMES WILLIE
AGUIRRE RAYMOND
ALANIZ BENITO
ALGER GEORGE BERKLEY
ALLEN LARRY DEAN
ANDERSON EVERETT ROBERT
ANDERSON WILLIAM JR
ANSELMO WILLIAM FRANK
APODACA VICTOR JOE JR
ARAGON JOSE RUBEN
ARAGON RUEBEN THOMAS
ARANDA ISMAEL BENITO
ARCHULETTA RAY ADAM
ARIAS LUCIANO
ARNDT CRAIG ALAN
BAHL WALTER TIMOTHY
BAKER GERALD OTIS
BAKER RICHARD THOMAS
BAKER STANLEY LOYD
BALES SHAREL EDWARD
BALLARD ADAM DAVID
BANDY CURTIS ELBERT
BARBER THOMAS DAVID
BARELA BARTOLO AMADOR JR
BEALL WILLIAM EARNEST JR
BEARD CHARLES RAY
BEASLEY JAMES OTIS
BECK MICHAEL JAMES
BEESON ROBERT HENRY JR
BELARDE BENJAMIN JOSEPH
BENJAMIN JEFFERY JAMES
BENSBERG ROBERT TRAME
BERRY JOHN ALVIN
BIEKER CARL JOSEPH
BLAKE ARMIN JOCHAIM
BLATNICK RODGER ALAN
BLEA ROBERT DANIEL
BLESSING WILLIAM STANTON
BLOOM RONALD KEITH
BLUNN DAVID LEE
BOBIAN RALPH DANIEL
BORGENS JERRY LEE
BOSSER JOHNNY STEVE
BOSTON LEO SYDNEY
BOWELL TERRANCE LEE
BOWMAN BRUCE ALLEN
BRABO HENRY
BRACK DAVID ALLAN
BRANAUGH LARRY JAMES
BRANCIO DAVID MIKE
BRATRSOVSKY GERALD JOHN
BRIGHAM ROBERT GENE
BRISCOE LARRY

BROWN MICHAEL WADE
BROWN THOMAS LOUIS
BROWNLEE CHARLES RICH-
ARD
BURNELL SAM JR
BURTON JAMES EDWARD JR
BYARS RICHARD SCOTT
CAIN MICHAEL JOSEPH
CALKINS BYRON THOMAS
CAPRARO CLAUD WILLIAM
CARDENAS JOSEPH ARTHUR
CARDY BRUCE LEE
CARROLL FRANK JEROME
CARROLL GERALD FORD
CARSTENS GARY AMOS
CARTER JAMES WILLIAM
CASIAS HENRY ELOY
CASLER JOSEPH DUANE
CATALANO SAM JR
CATTERSON RONALD GENE
CHACON DAVID ANDREW
CHAVARRIA JOHN MAREZ
CHAVEZ GILBERT MICHAEL
CHAVEZ GREGORY ANTON
CHENOWETH AUSTIN RAY
CLARK BILLY FLOYD
CLARK FRANCIS WILLIAM
CLARK GEORGE ARTHUR
CLEVENGER DANIEL JOHN
CODDING RAY EDWIN
CONDREAY ERVIN LEE
CONDY LADD ROBERT
CONN DONALD WARREN JR
COOK DONALD ESTEL
COOPER KENNETH WILLIAM
CORBITT GILLAND WALES
CORDOVA RICHARD JOE
CORDOVA RUTILIO PROFIRIO
CORNELISON JOSEPH MI-
CHAEL
COX LARRY CHARLES
CRISMON FRANK SCOT
CROSSMAN MELVIN EUGENE
CUDE HERSHEL DUANE JR
CULLNAN LARRY LAWRENCE
CULVER RAYMOND WALTER
CUNNINGHAM BRUCE WAYNE
CUNNINGHAM STEPHEN RAE
DABNEY HAROLD THOMAS
DALTON CLARENCE ELMER JR
DANIELSON MARK GILES
DAUBENDIEK RAYMOND M
DAVIS DUDLEY
DAVIS HARRY LEE
DAVIS JOHN ALLEN
DAVIS JOHN CALVIN
DAVIS RICHARD LEE
DAVISON JACKIE LEE
DAY STEPHEN WAYNE
DE HERRERA BENJAMIN

DAVID
DE HERRERA PEDRO
DELGADO LEROY FRED JR
DENARDO JOSEPH FREDERICK
DENAVA JOHN JOSEPH
DEWITT JAMES PHILLIP
DICKINSON DAVID THOMAS
DICKSON ERIC VOUGHN
DICKSON ROBERT LEE
DITSON LYMAN RICHARD
DIX DONALD ANDREW
DOBY JOHN WILLIAM
DOMINGUEZ MICHAEL GENE
DONATHAN RICHARD PETE
DONOVAN LEROY MELVIN
DOODY THOMAS PATRICK
DORONZO PAUL FRANK
DOWNEY STEPHEN WOOD
DUNN RALPH ALLEN
DURAN ALFONSO MARQUEZ
DURAN AMIE JACOB
DURAN ELOY
DURAN ERNEST LOUIS
ECKVALL RICHARD ALLEN
ELDER JAMES BRYAN JR
ELLSWORTH ROBERT WAYNE
EMMONS THOMAS KENNETH
ETHERIDGE MICHAEL RAY-
MOND
ETHERTON STEVEN PAUL
EVANS RAYMOND E
FAHRENBRUCH RICHARD L
FARMER THOMAS LEONARD
FELDHAUS THOMAS VINCENT
FICKLIN GEORGE RAY
FIEBELKORN MARCUS GUY
FIELDS DANIEL LEE
FITZHUGH ROBERT PAUL
FLOREZ TONY MANUEL
FOGG DAVID EDWARD
FORD ALLEN D
FOSTER DONALD RAY
FRANZ BRUCE RONALD
FRAZIER CHARLIE JR
FREEMAN IVEL DOAN
FREGIA ROBERT RANDY
FULLER THOMAS LEE
FULLER TIMOTHY
GABBIN FRED LEE
GARCIA DAVID BENEDICTO
GARCIA EDWARD LEE
GARCIA JESUS MARIA
GARCIA WILLIE JR
GARDELL CLIFFORD MC
CARTH
GARNES JACK ALLEN
GEIGER LAWRENCE RAY-
MOND
GEORGE GERALD LEE JR
GEORGE GORDON MILTON JR

GIBSON WALTER CARL
GIDDINGS CLYDE ARTHUR
GILBERT JACK DEAN
GILCHRIST ROBERT MICHAEL
GIRARDO DAVID LAVERNE
GIRON FRANCISCO
GOBBO JERRY WAYNE
GODDARD ROCKFORD
WAYNE
GOMEZ ERNEST LAWRENCE
GOMEZ FELIDELPHIO BENJIMI
GOMEZ MARGARITO RO-
DRIQUEZ
GONZALES JESUS ANTONIO
GONZALES MICHAEL FILBERT
GORDON HUBERT ELTON
GRANT DENNIS HOWARD
GRAY GERALD DAN
GREEN GARRY GEORGE
GREEN GERALD
GREENE ELLIS DAVID
GREGORY ALFRED RAYMOND
GRIEGO ELOY SANTIAGO JR
GRIFFEE THOMAS LYNN
GRIFFITH PERRY WITT
GUTIERREZ JOE MARIA
HACKETT CHARLES K JR
HAGLUND VICTOR MILFORD
JR
HAINING PAUL LINN
HAKES JAMES DANIEL JR
HALCOMB CARLTON BARRY
HALE HAROLD LELAND
HALL ROBERT JAMES
HAMILTON GILBERT LEE
HAMM JAMES EDWARD
HANNON PATRICK JOSEPH
HANRATTY THOMAS MI-
CHAEL
HANSEN LESTER ALAN
HARMAN CHARLES DAVID
HARRINGTON GEORGE M
HAVENS ALAN DALE
HAWKINS MICKEY LEE
HECK DAVID WILLIAM
HEGGEN GREGORY LYNN
HEIL JACKIE PHILLIP
HELLER DAVID JUNIOR
HELMS JOHN RAY
HELRIEGEL DAVID
HELWIG ROGER DANNY
HENSLEY JACKIE VERNON
HERNANDEZ JOE JR
HERRERA LOUIS ANTHONY
HERRERA MANUEL
HERRERA MANUELITO
LEOPOLD
HERRITZ EVERETT AUGUST
HERSHISER DAVID OWEN
HERTZ ROBERT DALE

HINTON DENNIS EDWARD
HIRSCHLER RALPH DEAN JR
HOBSON CHRISTOPHER MARK
HOCKER NORMAN ROGER
HOFFMAN RODNEY LOUIS
HOFFMAN RONALD THOMAS
HOGAN RADFORD DOUGLAS
HOLLINGWORTH DAVID
MCLEAN
HOPKINS RAYMOND LEE
HOPKINS WILLIAM EDWARD
HORGAN DUANE FRANK
HORTON FRED HOWARD
HORVATH ROBERT JOHN
HOWARD ROBERT BAILEY
HOWELL DUANE GEORGE
HOWERTER EARL EVERETT JR
HRDLICKA DAVID LOUIS
HUFFSTUTLER KEITH VIN-
CENT
HUNT RONALD ALAN
HUNTER TROY HAZARD
HURLBERT ROY DOUGLAS
HURLEY KEVIN MICHAEL
HYDE ROBERT LEE
HYLAND DENNIS MICHAEL
ILES BRUCE ADRION
INCE JOHN DAVID
IVES TIMOTHY JAMES
JACKSON JOHN RAYMOND
JACKSON LLOYD WILLIAM JR
JACKSON ROBERT EUGENE
JACQUES JAMES JOSEPH
JEFFERSON PERRY HENRY
JENKINS ERIC DORAN
JENSEN JAMES MAYNARD
JINDRICH STEVEN FREDERICK
JOHNSON DAVID CHARLES
JOHNSON GREGORY RAN-
DOLPH
JOHNSON ROBERT WAYNE
JOHNSTON KENNETH DALE
JONES HAROLD LEE
JONES HARRY
JONES LAWRENCE EDWARD
JONES WILLIAM ARCHIE JR
KASTRINOS JEROLD LLOYD
KATZ RONALD CHRISTOPHER
KELLEY JERRY CONRAD
KELLY ERNEST CALVIN
KEMP CLAYTON CHARLES JR
KENNANN LARRY RUSSELL
KENTON STANLEY CHARLES
KERR EDWARD LEMOYNE
KETELS FLOYD DALE
KIBEL CHARLEY CHESTER
KIECKER PAUL FREDERICK
KINSER JACOB LEE
KINTON DONALD RAY
KOHL DANIEL KAYE

COLORADO WESTERN SLOPE VIETNAM CASUALTIES

CHARLES LELAND ADKINS
ADAM DAVID BALLARD
JOHN ALVIN BERRY
THOMAS LOUIS BROWN
SAM BURNELL JR
AUSTIN RAY CHENOWETH
BILLY FLOYD CLARK
GEORGE ARTHUR CLARK
MARK GILES DANIELSON
LEROY FRED DELGADO JR
JAMES PHILLIP DEWITT
ROBERT LEE DICKSON
JOHN WILLIAM DOBY
RICHARD PETE DONATHAN
JERRY ALAN HURD

LEROY MELVIN DONOVAN
THOMAS PATRICK DOODY
AMIE JACOB DURAN
GEORGE RAY FICKLIN
MARCUS GUY FIEBELKORN
ROBERT PAUL FITZHUGH
DAVID EDWARD FOGG
BRUCE RONALD FRANZ
EDWARD LEE GARCIA
JERRY WAYNE GOBBO
MICHAEL FILBERT GONZALES
THOMAS LYNN GRIFFEE
VICTOR MILFORD HAGLUND JR
PAUL LINN HAINING
WILLIAM ARTHUR KIMSEY JR
ROY GLENN ZUFELT

DAVID McLEAN HOLLINGWORTH
RAYMOND LEE HOPKINS
DAVID CHARLES JOHNSON
HAROLD LEE JONES
HARRY JONES
LARRY RUSSELL KENNANN
ROBERT MORGAN LIDDELL
JOSE ADORO MANZANARES
EDWARD KETTERING MARSH
LESLIE T MCMACKEN JR
PHILIP DENNIS MILLER
ROBERT EMILIO MONTOYA
ALAN EVERETT PILON
WILLIAM LEONARD SANDERSEN
DONALD CHARLES PATCH
DAVID LAVERNE GIRARDO

JOHN WILBUR SAUNDERS JR
RONNIE DEAN SCHULTZ
DANNY GILBERT SCHWARTZ
ALLEN DEAN SCOGGIN
PHILIP RAYMOND SHAFER
JOHN ROBERT SIEVERS
JAMES RICHARD SMITH
GARY LEON TRUJILLO
JOHN A VIALPANDO
DAVID RUSSELL WELCH
TROY DEAN WILHITE
JOHN PAUL WRIGHT
WALTER CLARENCE WRIGHT
EDWIN LLOYD YOUNGMAN
WILLIAM ROBERT ROMACK
LANE WAYNE WISEMAN

Colorado

KOHLER DELVIN LEE
KRIST MATTHEW JOHN
LA POINTE LARRY W
LA VOO JOHN ALLEN
LADEWIG MELVIN EARL
LAMB MICHAEL HUGH
LAMBERT DOUGLAS JOSEPH
LAMBERT TIMOTHY
LAND CHARLES DWAYNE
LANDES DREK ALLEN
LANMAN THOMAS DESMOND
LARGE GARY RAY
LARSON JEFFRY ARTHUR
LAWLOR ROBERT JAMES
LAYTON RONALD DEAN
LEAF JACK BRIAN
LEEPER WALLACE WILSON
LEMON JAMES RICHARD
LIDDELL ROBERT MORGAN
LOCKWOOD DONALD PAUL
LOPEZ JOSEPH PAUL
LOVING MARTIN EDWIN
LUCERO ALBERTO A
LUCERO PATRICK ARNOLD
LUCERO ROBERT FLORIENCIO
LUKOW MICHAEL EUGENE
LUND MICHAEL ORSON
LUNDY RANDY JOE
LUSHER THOMAS ROY
LYLE MICHAEL STEVEN
MAC KENNA JAMES JESSE
MACHATA RUDOLPH GEORGE
MACIEL PETER JR
MANLY FREDERICK LEE
MANNING DAVID LAWRENCE
MANZANARES CHARLES EDWARD
MANZANARES JOSE ADORO
MARCHESI JIMMY EUGENE
MARKLE WILLIAM CARL JR
MARQUEZ RONALD O
MARQUEZ VALENTINE
MARTIN CHARLES FREDERICK
MARTIN DUANE WHITNEY
MARTIN JOSEPH THOMAS
MARTIN RICHARD LE ROY
MARTINEZ ADOLPH ALFRED
MARTINEZ JUAN PATRICIO
MARTINEZ LE ROY FELIX
MARTINEZ MANUEL FLOYD
MARTINEZ MAURO
MARTINEZ PETE MICHAEL
MARTINEZ ROBERT LEE
MASCARENAS JOE LEO
MASCARENAS ROBERT RAY
MASON SVEN STERNING
MATTHEWS MICHAEL FRANKLIN
MC BRIDE ALBERT CAYRL
MC CLUNG LARRY EARL
MC CONNELL WILLIAM C IV
MC GOWEN CHARLES FRANK
MC MACKEN LESLIE T JR
MC VEY LAVOY DON
MC WETHY EDGAR LEE JR
MCCASLIN HAROLD JR
MEDINA JOHNNY
MENARD DOUGLAS FRED
MENDENHALL WILLIAM G
MERSHON DANIEL LEE
MESTAS PETER VINCENT
MIDCAP DAVID MICHAEL
MILLER DAVID HARVEY
MILLER JAMES OLEN
MILLER JOSEPH LLOYD
MILLER MANFRED BERTOLD

MILLER PETER RICHARD
MILLER PHILIP DENNIS
MILLER STANLEY GENE
MILLER WILLIAM ANGUS
MILLS RICKEY DUANE
MINCE LYNN ELDON
MINTON CHRISTOPHER ALAN H
MITCHELL LARRY GENE
MITCHELL THOMAS BARRY
MONDRAGON BENJAMIN ALLEN
MONTANO JOSE CLEMENTE
MONTOYA ALEXANDER
MONTOYA ROBERT EMILIO
MOORE JAMES LEE
MOORE ROBERT DELL
MORA JAMES J
MORENO RICARDO LEON
MORGAN BURKE HENDERSON
MUELLER KURT JR
MULLINS HAROLD EUGENE
MUNSON ALLEN ARTHUR
MUSICK MORRIS OLEN JR
MUSSMAN DENNIS ERVIN
MYHR BARRY BERNDT
MYRICK ALVA NORTEN II NAME
NARANJO MIGUEL ERNEST JR
NAVARRO NICHOLAS LEON
NAYLOR EDWARD REYNOLDS JR
NEEL CHARLES HERBERT JR
NEELEY DONALD LEE
NELSON DONALD EUGENE
NELSON SCOTT THOMAS
ODELL MICHAEL CHARLES
OGBOURN GAYMAN CRANDALL
ONTIS BILLIE JOE
ORSLAND GARY VARNER
ORTIZ JACOB
ORTIZ RANDALL ISAAC-JED
OSBORN DONALD KEITH
OTTO MICHAEL L
OWENS RANDY LEE
PACHECO EUGENE CARL
PACHECO GEORGE ARTHUR
PACKARD RONALD LYLE
PAPPENHEIM THOMAS HENRY
PARMENTIER ROGER DAVID
PARSONS DOUGLAS BLANCHARD
PATRICK JERRY KENT

PAWLISH GEORGE FRANCIS
PEERY NORMAN DOUGLAS
PENN EDWIN ALLAN
PEONIO STEPHEN JOSEPH
PESCHEL JAMES DOUGLAS
PETERS EMMETT JACK
PETERSDORF CHARLES H JR
PETERSON RUSSELL GEORGE
PETTIE FLOYD WILLIAM III
PFEIFER DENNIS WAYNE
PHILLIPS JERRY LEN
PIERCE LEON JOSEPH
PILON ALAN EVERETT
PIZER WESLEY IRWIN
PLATT GARY W
POLLOCK DOUGLAS RAY
POMEROY CARLYLE B JR
POTTER JERRY LEE
PRICE LARRY JUNIOR
PRIDEAUX JAMES EARL
PRITCHARD GALE STEWART
PURVIS PHILIP ALAN
QUINTANA FRANKLIN HARRY A
RACEY BRADFORD GREG
RALSTON FRANK DELZELL III
RATLIFF FRED ALEXANDER
RAUSCHKOLB JAN
RAY THOMAS PAUL
REEVES ROBERT LINTON
REICHERT STEVEN EDWARD
RENDON JOSEPH
RENTERIA LOUIS JESUS
REYES HAROLD
REYNOLDS MARTIN DANIEL
RINGENBERG JEROME JOSEPH
ROBBINS JOHN WILLIAM
ROBERTS CHARLES PRICE
ROCHA DANIEL ALBERT
RODRIGUEZ PAUL DAVID
ROESLER JOHN ONDERDONK
ROMACK WILLIAM ROBERT
ROMERO ARTHUR WILFRED
ROMERO BENNIE
ROMERO JOSEPH
ROMERO RICHARD
ROONEY TERRENCE MANN
ROSS ROBERT LEE
ROUSE GREGORY MICHAEL
ROYBAL ANTHONY WILFRED
RUPKE DARYL JAMES
RUYBAL DANNY GILBERT
RYNNING KENNETH DEAN
SAIZ RONALD JAMES

SALAS DANIEL STEPHEN
SALEH CHRISTOPHER RUBEN
SALIMAN NORMAN SHELDON
SANCHEZ NICK ENRIQUE
SANDERS WILLIAM LEROY
SANDERSEN WILLIAM LEONARD
SANDOVAL JOSE RAMON
SANDVIG LAMOINE LOWELL
SANTISTEVAN BENNY M JR
SANTORELLA ROBERT H
SARACINO FRANK DE PAUL JR
SAUNDERS JOHN WILBUR JR
SCHAEFFER ARLON GLENN
SCHAFER CHARLES EDWARD
SCHERF MICHAEL GREGORY
SCHMIDT PAUL EDWARD
SCHMITT GARY WALTER
SCHNEIDER ROBERT DEAN
SCHRAM FREDERICK LLOYD
SCHRENK DONALD GEORGE
SCHUKAR GENE LEROY
SCHULTZ RONNIE DEAN
SCHWARTZ DANNY GILBERT
SCOGGIN ALLEN DEAN
SCOTT JAMES HOWARD
SCRIBER LEON R
SEILER CLYDE
SELIX JAMES MICHAEL
SENA FRED JR
SENTI DONALD LEE
SERNA ERNEST
SEWARD KENNITH MARION
SHADE WILLIAM STEVE
SHAFER PHILIP RAYMOND
SHAPARD MICHAEL ROBERT
SHAPLEY ELDON LYLE
SHEEHAN DANIEL MORELAND
SHIELDS RUSSELL ALLEN
SHIMODA WESLEY
SHOWALTER WALDEMAR D
SIEVERS JOHN ROBERT
SILVA CLAUDE ARNOLD
SIMMONS JOHN WAYNE
SIMPSON JOSEPH LOUIS
SITTON DAVID THOMAS
SKEEN STEVEN JAMES
SKINNER BRIAN KAY
SMILIE BLAINE PATRICK
SMITH DENNIS MICHAEL
SMITH DUANE CHARLES
SMITH JAMES RICHARD
SNOVER DAVID DARRELL
SOLANO PORFIRIO SAM

SPARKS DAVID LEO
SPAULDING RICHARD LEE
SPELLMAN JOSEPH VICTOR
SPINUZZI JAMES CARL
STAAB RICHARD EUGENE
STARK DAVID PAUL
STARKWEATHER JEROME FRANK
STASKO THOMAS WILLIAM
STEADMAN JAMES EUGENE
STEARNS ROGER HORACE
STEELE THOMAS WILLIAM
STEPHENS NATHANIEL H JR
STETSON KENNETH EARL
STICE LARRY DOUGLAS
STOLPA RAYMOND VINCENT
STONE WILLIAM J B
STOTLER MICHAEL DEAN
STREAMER FRANK MARION
STRUNK WILLIAM LOCKE
STUART JOE BEN JR
SUBLETTE GARY LYNN
SULLIVAN WILLIAM LEE
SUNIGA JOHN ANTHONY JR
SWANSON JON EDWARD
TABOR BRUCE WAYNE
TAGGART LARRY JOEL
TALTY PATRICK ANTHONY
TAYLOR LARRY ROBERT
TERWILLIGER RODGER EDSON
TESSADRI JIMMY JOE
THADEN GARY DENNIS
THOMAS BARRY DON
THOMAS RONALD GENE
THOMPSON DONALD R
THOMPSON KARL LUDWIG
THOMSON ROBERT BRIAN
THORNE ROBERT WALTER
THOUVENELL ARMAND RENE
TOMSIC MICHAEL PATRICK
TOOTHAKER JAMES ALLAN
TOTH RONALD C
TREVARTON LARRY GEORGE
TRIPP DENNIS ROBERT
TRUJILLO GARY LEON
TRUJILLO VICTOR DAVID
TUCKER TIMOTHY MICHAEL
TURNER JAMES LOUIS
UHL RAYMOND RIEDE
VADEN ROBERT LEE
VALASQUEZ PETE ANTHONY
VALDEZ JOHN BEN
VALDEZ JUAN PEDRO

VAN BEBER ELDON CHRIST
VANWEY WILLIAM EARL
VARGAS FRANKIE LEYBA
VARGAS LAWRENCE JAMES
VELASQUEZ JOHN ROBERT
VELASQUEZ JOSE HILARIO
VESTAL STEVE ALAN
VIALPANDO JOHN A
VIGIL DAVID LORENZO
VIGIL FREDERICK ANTHONY
VIGIL LOUIS DAVID
VOSS MICHAEL ALAN
WALKER BRUCE CHARLES
WALKER LA VERNE
WALKER MICHAEL CLYDE
WALLACE RUSSELL LEWIS
WALLER JAMES LEONARD
WARNER DAN EDISON
WARREN MICHAEL WALTER
WATKINS TOBY JACK
WATSON KENNETH GARY
WEIMER JERRY ALAN
WELCH DAVID RUSSELL
WELLS LARRY DEAN
WENTZEL MERLYN LEE
WESKAMP ROBERT LARRY
WESTBROOK JIMMY WAYNE
WHALEN GARLAND GUY
WHALEN MICHAEL JAMES
WHALEN RODRICK PIUS
WHIPS FLETCHER DANNY
WHITE DONALD LEE
WHITE OSCAR LEE
WHITNEY PHILIP LEONARD
WILEY PHILLIP TONY
WILHITE TROY DEAN
WILLIAMS FLOYD LEE JR
WILLIAMS LAWRENCE DEAN
WILLIAMS LESLIE WAYNE
WILLIAMS THOMAS HANS-
FORD
WILLIAMS VERE LOYD JR
WILLIAMS VIRGIL LAWRENCE
WILSON WILLIAM BRUCE
WIMP ROBERT G
WINGET KENNETH WAYNE
WINSLOW JERRY G
WOOD CARL MITCHELL
WOOD JOHN CLIFFORD
WORRELL ROBERT LEE
WRIGHT JOHN PAUL
WRIGHT LEE ROY
YOUNG CHARLES HARRY
YOUNGMAN EDWIN LLOYD
YUGEL LOUIS ARTHUR
ZARINA DONN PETER

43

Photos by James Nalesky

Connecticut Vietnam Veteran's Memorial

Lake St, Coventry, CT 81521

The Connecticut Vietnam Veterans Memorial is situated in Coventry's Veterans' Memorial of the Green and was created by the Connecticut Vietnam Veterans Memorial Committee which was started by Jean Risley, the sister of a Vietnam veteran who never made it home.

The Connecticut Vietnam Veterans Memorial came about in a rather unusual fashion. Nearby middle school students were asked to do a research project regarding the 612 local soldiers who lost their lives in Vietnam. The students did so and compiled the information into a book they dubbed "612." Risley heard about this project and used as inspiration to put together the aforementioned committee. Her efforts were not the first to push for a state memorial for the veterans, but others failed to manifest for multiple reasons. After a lot of work Risley, along with the committee, raised the necessary funds for the memorial and ground was broken in August of 2007 with the dedication coming in 2008 on Armed Forces Day.

The Connecticut Vietnam Veterans Memorial was placed in the Veteran's Memorial of the Green in order to provide visitors with a peaceful spot to reflect on the sacrifice made by those lost in the war.

It's made of black granite and features the names of those 612 service members etched in white. Their names are paired with the infamous words "All gave some, some gave all" which is etched at the top of the monument. There are also murals etched into the stone flanking the names. Two stone benches are situated near the monument in order to give visitors a place to sit as they reflect on the veterans. The seals of each branch of the US armed forces appear on the base of the memorial while the American flag, the Connecticut state flag, and the POW/MIA flag fly above it all. Each year this location hosts Memorial day services and parade.

The Names of those from the state of Connecticut Who Made the Ultimate Sacrifice

AINSWORTH JOHN MATHEW
ALHO ANTONIO LOPEZ
AMBRUSO RICHARD DICK
ANDERSEN MARTIN WEIGNER
ANDERSON EVERETT LEE
ANDREWS STUART MERRILL
ARGENTA ALLEN CHARLES
ARTKOP ARTHUR JAMES
ASHE RONALD A
ATHERDEN LESTER ROBERT
AUSTIN CHARLES DAVID
AYERS DOUGLAS EDWARD
AZZARITO FRANK ANTHONY JR
BACHLEDA BERND
BACHMAN CHARLES W JR
BACZALSKI JOSEPH
BAGNALL ROBERT SALMON
BAILEY ARTHUR WILLIAM JR
BAILEY LORING M JR
BAKER VERNON HOWARD II
BANKOWSKI ALFONS ALOYZE
BANNING JAMES HENRY JR
BANNON WILLIAM JOHN JR
BARBER SIDNEY EMERY
BARKER FRANK AKELEY JR
BARMMER TIMOTHY MICHAEL
BARNES ERIC MARVIN
BARNES LEROY FRANCIS
BARREIROS SILVINO FERNAND
BATH JOHN MICHAEL
BATSON JAMES CHARLES
BAUCHMANN EARL JOHN
BAYNES ERNEST JOHN

BEAMAN ROBERT JON
BEARD WILLIAM ARTHUR
BEAVER HEARNE W
BEDWORTH GRIFFITH BRONSON
BELTON CALVIN
BENICEWICZ RICHARD C
BENTLEY WALTER EARL
BENTSON PETER MORGAN
BERGEVIN CHARLES LEE
BERGSTROM WALLACE CARL JR
BERNARD HENRY WILFRED JR
BERRY MALCOLM CRAYTON
BERRY ROBERT ERVA
BEVERIDGE DOUGLAS JAMES
BICKFORD RICHARD OLIVER
BIEBER EDWARD L
BIEHL OSCAR JR
BINGHAM DAVID RICHARD
BISHOP EDWARD JAMES JR
BITTNER ROBERT EDWARD JR
BLACK HARRY ELSWORTH
BLANCHARD THOMAS JOSEPH
BLANCHETTE RAYMOND
BLANTIN ERIC GEORGE
BLOUNT JOHNIE LEE JR
BOURDEAU GERALD LEE
BOWE THOMAS JOHN
BOYD GERALD DAVID
BRAMAN DONALD LEON
BRANSON JAMES ALLAN
BRAUN PAUL JOSEPH
BRAY CHARLES EDWIN JR
BRENNAN STEPHEN JOHN

BREWER SAMUEL WALTER
BRIDGES ROBERT EARL
BROADHURST RICHARD EDWARD
BROOKS WILLIAM FRANCIS
BROWN ALEXANDER CAMERON
BROWN JOSEPH ORVILLE
BROWN RICHARD CRAIG
BRUNO PAUL JOSEPH
BRYANT CREED LORENZIO
BURDICK WILLIAM F JR
BURGESS RAYMOND ARTHUR
BURKE DAVID MOY JR
BURKE ROGER VINCENT PAUL
BURNS JAMES ARTHUR
BUSH ROBERT EDWARD
BYRD CLIFFORD LAMONT
CABLES GORDON LEONARD
CANCEL PEDRO O
CANFIELD MATTHEW M JR
CAPUTO RICHARD P
CARINCI JOSEPH ANTHONY
CARLEY MICHAEL JOHN
CARLSON DAVID LAWRENCE
CARNEGIE THOMAS EDWARD
CARNEY JAMES PATRICK JR
CARRASQUILLO-SOLTERO REINALD
CARSON WILLIAM D
CARTER D C
CASSIDY DAVID ALEXANDER
CASSIN RICHARD ALBERT
CAUTHERN ROGER ROBERT
CAVANAUGH THOMAS JAMES
CHABOT RICHARD EARL
CHENIS MARK CONSTANT
CHIALASTRI THOMAS ANTHONY
CHMURA MICHAEL LOUIS
CHOQUETTE ROBERT G JR
CHOWKA ANDREW DANIEL
CHRISTENSEN EDWARD JOHN
CHRISTY ALBERT GEORGE JR
CIESIELSKI STANLEY M
CINOTTI RALPH SILVIO
CLARK FREDERICK RALPH
CLARK GEORGE WILLIAM
CLARKE JOHN KEARNAN
CLEARY PETER MCARTHUR
CLIFFORD MICHAEL WILLIAM
COHN WILLIAM PAUL JR
COLLETTE CURTIS DAVID
COLLIER STEVEN EDWARD
CONKLIN RICHARD DOUGLAS
CONNORS FERGUS FRANCIS JR
CONTINO RAYMOND FRANK
COOLEY ROBERT KARL
CORR JOHN GEYER
COSTANZO RALPH PAUL
COVEY WILLIAM F JR
CREAMER JAMES EDWARD JR
CRONIN BRIAN JOHN
CRONKHITE CHRISTOPHER
CROSBY ARTHUR ALLEN JR
CRUGNOLA MARIO CHARLES JR
CULLINAN JOHN PATRICK
CURRIE GEORGE CRAWFORD
CURRY JACK HENRY
CURTIN JOHN GERALD
CUSHEN KENNETH
DA COSTA JACK RICHARD
DANIELS JAMES AURTHUR
DAUTEN FREDERICK W JR
DAVI JOSEPH NICHOLAS
DAVIS EDWIN PHILLIP
DAVIS WILLIE CECIL
DAWSON NORMAN EDWARD JR
DE ANGELIS RICHARD NICHOL

DE BARBER JOHN THOMAS
DE CARLO JAMES ANTHONY
DE FORGE DAVID HENRY
DE ORIO WILLIAM JOSEPH JR
DE SANTIS STEPHEN ANTHONY
DEBERNARDO FRANK JR
DEL GRECO VICTOR JR
DEMORE MICHAEL GEORGE
DESILLIER RICHARD GILL
DIAZ ANGEL LUIS
DIEDRICKSEN ALAN LEE
DINDA MICHAEL JOSEPH
DIXON MARK HANNAY
DOLAN WILLIAM JOHN
DONESKI HENRY JOHN
DONNELLY RAYMOND PETER
DONOHUE FRANCIS CHARLES
DONOVAN THOMAS EDWARD
DORIA ALDO ANTHONY
DORSEY EDWARD ROBERT
DOUD NORMAN KENT
DOUGAN MICHAEL JAMES
DOWD FRANCIS JOSEPH JR
DOYLE JOHN FRANCIS
DU FAULT JAMES RAYMOND
DUBIEL PETER PHILIP
DUEL EDWARD KENNETH
DUMIN PAUL MICHAEL
DUNN MERL THOMAS JR
DUNN RICHARD EDWARD
DUNNING ALLAN LOMBARD JR
DUNNING WILLIAM MARTIN
DUPUIS CLEMENT ARTHUR
DURKIN JOSEPH WILLIAM JR
DWYER ROBERT KEEFE
DWYER THOMAS D
EADDY ISHMELL
EDDY EDMUND FRANCIS
EDMOND THOMAS ALLEN
EDWARDS GEORGE FREDERICK
EDWARDS PAUL WILLIAM
ELIA ROBERT A
EMOND DAVID BRUCE
ENGLISH DARYL LEE
EQUI RUSSELL LLOYD
ERICSON WILLIAM F II
ESTEN JOHN ERNEST
FABRISI PAUL EUGENE
FABRIZIO JAMES
FACONDINI RICHARD MICHAEL
FALCON ANIBAL
FALK FREDERICK JOHN JR
FANT RUSSELL THOMAS
FARMER MICHAEL MELVIN
FARNHAM ALLEN STEARNS
FECTEAU GENE EDWARD
FENTON WILLIAM CHARLES JR
FERGUSON DONALD PORTER
FERGUSON PETER CLARENCE
FERGUSON WHITNEY T III
FERRY RAY LEONARD III
FIALKO DAVID ANDREW JR
FIEDLER DREW
FIRMNECK ALLAN PAUL
FITTON CROSLEY JAMES JR
FITZGERALD GEORGE RICHARD
FLEMING DUNCAN HARTWELL
FLETCHER RANDALL SCOTT
FLYNN JAMES GERALD
FORGUE GERALD HENRY
FOSTER DANIEL WILLIAM
FOWLER CLAUDIE
FOY MICHAEL JOSEPH
FRANCE RONALD LYNN
FRANCOLINI JOSEPH DAVID
FRASER DOUGLAS PAUL
FRATTO MICHAEL JOHN
FRAZIER LEROY
FREEMAN DAVID MICHAEL

FREEMAN SAMUEL DIGGES III
FRITSCH THOMAS WILLIAM
FRITZER THOMAS ALBERT JR
FROST FRANK RUDOLPH III
GAGNE RENALD LUDGER
GAIDIS ALFRED JAMES
GALLAGHER JOHN THEODORE
GANDY KENT ELLSWORTH
GEER STEPHEN JAMES
GEISEN JOHN BENNETT JR
GERMAN BROMLEY HOWARD
GIANELLI ANTHONY
GIDVILAS MICHAEL
GILBREATH RICHARD ARTHUR
GILLY RONALD ALAN
GOETT JOHN KENNETH
GOOD CURTIS
GOODALE LEON RUSSELL JR
GOULET RONALD MARCEL
GRAHAM ALBERT F JR
GRANDPRE EDWARD FREDERICK
GRASSO JOHN M JR
GRAVELINE RICHARD PAUL
GRAVIL JOHN ALLEN
GREENE JAMES LEONARD JR
GREGORY FRANCISCO
GRILLO JOSEPH JOHN JR
GRILLO LAWRENCE HUGH
GRISWOLD GARY CLIFFORD
GUARINO RAYMOND BLAISE
GUILLET ANDRE ROLAND
GUROVICH JOHN EDWARD
GUTHRIE ARCHIE LEE
HABUREY EDWARD JAMES
HANSEN EDWARD ROBERT
HANSON JOHNNIE RAY
HARDWICK EDWARD MAHLON
HARDY ROOSEVELT JR
HARRELL LENWOOD THOMAS
HARRIS GEORGE WILLIAM
HARRIS ROY EDWARD
HEALEY JOHN JOSEPH JR
HEINZ ROGER WILLIAM
HENDRICKS JAMES THOMAS
HENSLEY RICHARD DAVID
HESFORD PETER DEAN
HIGHT DAVID KEITH
HILL THOMAS ARTHUR
HINES JONNY
HOLLAND JOSEPH PHILLIP
HOLM ARNOLD EDWARD JR
HOLMAN RAYMOND CLARK
HOPE RICHARD MICHAEL
HOPKINS ROBERT E
HORTON FLOYD MONROE
HOWARD ROBERT LOUIS
HUBBARD JOHN R
HUGHES PAUL ARNOLD
HUNTER RUSSELL PALMER JR
ILLINGWORTH JOHN JAMES
JACKSON BARRY
JACKSON CHARLES ARTHUR
JACOB ROBERT MICHAEL
JOBST KURT KARL JR
JOHNSON EUGENE RICHARD
JOHNSON JOHN LAFAYETTE
JOHNSON PETER WYETH
JOHNSON RICHARD ARNO JR
JONES GEORGE EDWARD JR
JUREK EDWARD JOSEPH II
KASKE RICHARD ALAN
KASPER ROBERT EDWARD
KATONA JOHN JAMES JR
KEITHLINE RICHARD WARD
KELLER TIMOTHY WAYNE
KELLY BARNEY JOE
KERR CHARLES DAVID
KIEWLEN FRANK JOSEPH JR
KIGHT MICHAEL AARON
KING PATRICK WILLMER
KIRK DAVID MICHAEL
KISER ROBERT JESSE

Connecticut

KOCIPER ANTONINE GEORGE
KOSOVICH GEORGE C JR
KRAJEWSKI DONALD JOSEPH
KUHLMANN CHARLES FREDERIC
KURTYKA GEORGE ALBERT
LABRECQUE WILLIAM F JR
LAFOND ROLAND ROBERT
LAMOREUX EDWARD DONALD
LANGSTON JIMMY LEE
LAVALLEE KARL JOSEPH
LAVIGNE JOSEPH EVERETT
LAVINE KENNETH ANTHONY JR
LAVOIE CLARENCE ROSAIRE
LAYMAN ROBERT EMMETT JR
LE BLANC FRANCIS JOSEPH
LE BLOND DONALD CHESTER
LE MAY RICHARD DRIGGS JR
LEONARD JAMES STEVEN
LERNER IRWIN STUART
LESCARBEAU GEORGE GERALD
LESKA ROBERT JOHN
LEVESQUE ROLAND PHILLIP
LEVY GERALD
LEWIS GARY LYNN
LILIENTHAL MARK ALLEN
LILLEY FRANK JOHN
LILLY WILLIAM JOSEPH
LIND FRED ANDREW
LINKS RICHARD FREDERICK
LONSDALE GEORGE EDWARD
LOUGHRAN JOSEPH M JR
LUBESKY GEORGE A JR
LUKASIEWSKI STEPHEN JAMES
MACDOUGAL JAMES HOWARD
MACHIE MICHAEL ALLEN
MACHOWSKI JOSEPH ANTHONY
MACIUSZEK PAUL JOSEPH
MALCOLM JOHN DANIEL
MALONEY JEFFERY RAYMOND

MANAREL CHARLES ROSS
MANN JAMES EDWARD
MANNING JAMES HOLDEN
MANNING JOHN WARREN
MANSELLE EUGENE L III
MANSFIELD JOHN MICHAEL
MARC AURELE LIONEL LUCIEN
MARCY WILLIAM LINCOLN
MARINELLI ELMO
MARINO CARL JOHN
MARKARIAN WILLIAM ARAM
MARKS JOHN
MARKUS JERRY
MARX ROBERT GARRY
MASSEY HARRY
MATIS WALTER FRANCIS E JR
MC ARTHUR JAMES STEPHEN
MC CULLOUGH ALBERT
MC DERMOTT TERRENCE M
MC DONALD WALTER RAYMOND
MC GILTON CALVIN EUGENE
MC GLOTHLIN ALEXANDER J
MC GOULDRICK FRANCIS J JR
MC KENZIE WAYNE ROBERT
MC KINLEY GERALD WAYNE
MC LELLAN ARTHUR CHARLES
MC MANUS TRUMAN JOSEPH
MC NEELY JAMES WILLIAM
MEDIATE ALAN WAYNE
MEEKER MARC JEFFERY
MEGLIO WILLIAM MICHAEL JR
MEISEL WILLIAM W JR
MELNICK STEPHEN JOHN
MENNONE MICHAEL GIOVANNI
MICLOSKEY KEN E
MIGLIERINA ROBERT LEO
MILLER KEITH ALLAN
MILLER WILLIAM EDWARD
MITCHELL ROBERT WALTER
MITCHELL WILLIAM BRUCE
MIXTER DAVID IVES
MLYNARSKI ROBERT LUCIAN
MONAHAN EDWARD JAMES JR
MOODY ALFRED JUDSON FORCE
MORASCINI JOHN V
MORIARTY PATRICK O NEAL
MORIARTY PETER GIBNEY
MOSER PAUL KIERSTEAD
MOTT JAMES FRANKLIN
MUISENER JACK ELLSWORTH
MULLER ERIC P
MULWEE ISAIAH JR
MYERS WALTER HARVEY JR
NAVARRO FRANK GEA
NEAL THOMAS MARTIN
NEALE CHRISTOPHER JONATHA
NEGER ROGER LEE
NELSON HENRY J
NELSON ROBERT CHARLES
NESTICO PHILIP FRANK
NEWSOME ROY C

NICKERSON WILLIAM BREWSTE
NIELSEN HAROLD RICHARD
NOBERT CRAIG ROLAND
NOLAN ROBERT FRANK
NORTON ROBERT LYON
NYMAN MICHAEL STUART
O BRIEN CLYDE HAROLD
OAKLEY WILLIAM JOSEPH
ORRICO JOHN THOMAS
OTT EDWARD LOUIS III
OUELLETTE DAVID DENNIS
OUELLETTE DONALD ROLAND
PACKARD CARL EDWARD JR
PALMENTA EDWARD VINCENT
PALMIERI JOHN JOSEPH
PALOWSKI RICHARD EDWIN
PAPARELLO JOSEPH JOHN
PAQUIN MICHAEL BRADLEY
PARKER RALPH JOHN JR
PARKER THOMAS EDWARD
PARSONS CLIFFORD E JR
PASCALE GEORGE JOHN
PASTORE JAMES JOSEPH JR
PATIENCE WILLIAM R JR
PEALER ELIAS BENSON JR
PECK STEPHEN GRADY
PECORA JOSEPH ANTHONY JR
PELLEGRINO JOHN PETER
PELLETIER LAWRENCE JOSEPH
PENDERGAST ROBERT LEE
PEPE GEORGE WILLIAM
PERRY STEPHEN TUCKER
PERRY THOMAS HEPBURN
PFAFFMANN CHARLES BROOKS
PHILLIPS PAUL HENRY
PIERCE BERNARD LAWRENCE
PLATOSZ WALTER
PLOURDE CLAYTON
POLGLASE WILLIAM RAULISON
POST JAMES HARVEY JR
PREIRA DOMINIC J JR
PREVOST ALBERT MICHAEL
PRICE ANTHONY ALOYSIUS
PRICE DWIGHT ANTHONY
PRINDLE ASHTON HAYWARD
PROVOST DAVID ARMAND
PUGH KENNETH LEE
PURDY LOUIS JAMES
PYLE WILTON STROUD
QUEY DAVID MICHAEL
RABER JOHN HAROLD
RADER FREDERICK M III
RAMOS ROBERTO
RANDALL RONALD MITCHELL
RAYMOND THEODORE PAUL
READ ROBERT BERTON
REGAN RICHARD JAMES
REITWIESNER JOHN CHARLES
REMER CHARLES BRADLEY JR
REMUTH LAWRENCE GUS-

TAVE
RENSHAW FRANKLIN MASON
REPACI DONALD SHELDON
REPOLE RICHARD GLENN
RHUDA ROBERT ARTHUR
RICH RICHARD
RICHARD ROLAND ARMAND
RICHARDS CHARLES EDWARD
RICHARDS GARY CHARLES
RICHARDS THOMAS STEPHEN
RIGGS ROBERT CHARLES
RINES EVERETT EDWARD
RIVERA-RUIZ ANDRES
ROBBINS WAYNE DUSTIN
ROBERTS GERALD JASON JR

ROBERTSON PIERCE IRVING
ROBINSON HENRY MILLARD JR
ROBINSON HOWARD CLINTON
ROGERS DAVID ALAN
ROLFS GERHARD M
ROMANIELLO ANTHONY JOSEPH
ROSENSTREICH AARON LIEB
ROTKO RUSSELL JOSEPH JR
ROWLEY CHARLES STODDARD
ROWSON GEOFFREY THOMAS
ROY ALLEN JAYSON
ROY RICHARD W

ROZZI WILLIAM ALLEN
RYAN MICHAEL JOHN
RYDER ALDO EUGENE
RYE BRYAN A
RYON JOHN W
SAEGAERT DONALD RUSSELL
SAMPT JOHN FRANCIS
SANDBERG JOEL ALEXIS
SANDERS MELVIN HILTON
SAPP ISAAC
SARMENTO HENRY MICHAEL
SAUER CHARLES EDWARD
SAUNDERS EARNEST ROLLIN
SAWYER BRADFORD PRESTON
SCANLON MICHAEL JOHN
SCHEIDEL ROBERT L JR
SCHMECKER JOHN LEONARD
SCHULTZ JOHN ROBERT
SEARS LEON
SEBASTIAN ALTON BROWN-
ING
SEDGWICK ROBERT CHARLES
SERVERA-BAEZ RAMON
AURELIO
SFERRAZZA ANGELO JOSEPH
SHAVIES GEORGE ARTHUR
SHAY THOMAS WILLIAM
SHEA DANIEL JOHN

SHEA JOHN FRANCIS
SHEEHY DAVID LAWRENCE
SHEEHY RONALD J
SHELTON JEROLD JEROME
SHERMAN JOHN BROOKS
SHONECK JOHN REGINALD
SHORTALL STEPHEN ADAMS
SILVER LAWRENCE JAY
SILVERMAN SHELDON
SINCERE JAMES WALTER
SIPP PETER ELMER
SLACK STEVEN GEORGE
SMERIGLIO ALBERT PETER
SMITH ALAN IVAN
SMITH ARTHUR ALBERT
SMITH CLINTON ARNOLD
SMITH HOWARD BRUCE
SMITH JAMES GORDON
SMITH MICHAEL DAVID
SPEATH DAVID PAUL
SPENARD NORMAN JOSEPH
GEO
SPIRITO ANTHONY JOSEPH JR
ST JOHN WILLIAM LUKE
STEVENS ROBERT FRANCIS
STEWART GERALD HILAND
STODDARD NORMAN R JR
STOLARUN RICHARD RAY-

MOND
STRANO JAMES CLINTON
STRYCHARZ STEPHEN S JR
STULL JAY WEBSTER
STURGEON DONALD FRED-
ERIC
SUAREZ RAYMOND JR
SULLIVAN FRANCIS JORDAN
SURPRENANT NORMAN
ROGER
SWAN DAVID MARTIN
TARSI WILLIAM JAMES
TAYLOR DENNIS GILBERT
TAYLOR JOHN STEWART
TCHAKIRIDES IRVING BURR
TERLECKI WALTER ALEXAN-
DER
TESSMAN RICHARD CARL
THIBEAULT GILBERT
THOMAS BRUCE MAYNARD
THOMPSON ROBERT JAMES
THORIK PAUL JR
THORNE LARRY ALAN
TIERNEY BRIAN EDWARD
TIGHE THOMAS DANIEL
TIGNER JEFFREY SANDERS
TILLQUIST ROBERT ARNOLD
TINGLEY THOMAS JAMES

TINO JOHN FRANCIS JR
TOWNE PETER CLARK
TRIPP PETER LEADBETTER
TYRELL THOMAS JUDD
TYSZKA EDWARD MICHAEL
URBAN RICHARD EDWARD
URSIN WILLIAM NORMAND
VAGNONE MICHAEL JOHN
VAN CEDARFIELD JAMES RAY
VARELA DANIEL
VAUGHAN RAYMOND WALTER
JR
VAUTOUR DAVID
VELEZ JUAN ANTONIO
VISCONTI LAWRENCE GUY
VOEGTLI JOHN SARGEANT
VOLLHARDT PHILIPP R
VOUGHT WARREN DEMAR-
EST JR
WALSH DONALD KEVIN
WALSH FRANCIS ANTHONY JR
WARNER BRUCE BYERLY
WAYMAN DONALD MICHAEL
WELCH JOHN HENRY III
WEST WILLIAM EDWARD
WESTWOOD NORMAN PHILIP
JR
WHALEY HENRY LEE

WHITE DOUGLAS EDWARD
WHITE ROBERT JAMES
WHORFF JOHN DENNIS
WILKERSON JUNIOR
WILKINSON STEPHEN DAVID
WILLEY DONALD MORRIS
WILLIAMS FRANKIE ROSS
WILLIAMS JAMES JOSEPH
WILLIAMS MALCOLM GEORGE
WISNIEWSKI CHARLES J JR
WOBLE JOHN B
WOLCHESKI RICHARD JOHN
WOODS LAWRENCE DANE
WORKS JONATHAN P
WRIGHT HERMAN W O JR
YOUNG DOUGLAS WHITING
YOUNG JOHN F
ZABOROWSKI WILLIAM JOHN
ZALE JOSEPH PAUL
ZAMBRANO BERNARD AN-
THONY
ZASTOWSKY DONALD JOHN

Photos by Steven LePage

Kent County Veterans Memorial Park

555 Bay Rd, Dover, Delaware, 19901

The Kent Veteran's Memorial Park is located near the Kent County Levy Court Building and plays host to a number of war memorials including the Vietnam War Memorial, a Korean War memorial, a Gold Star Mother and Families memorial, a war dog memorial, and a memorial for the conflict in the Middle East. The memorial park holds annual events on Memorial Day, Veterans Day, and Christmas Eve in order to give the community a collective time to reflect on the sacrifices made by local heroes.

The Kent County Veterans Memorial Park sits on an acre and a half of land and was commissioned by the Kent County Vietnam Veterans Chapter 850. The whole park was done through private donations and fundraisers. The Vietnam War memorial, along with the other memorials, were developed by Chapter 850. The monument for the Vietnam War was dedicated in 2009, but there are new editions that continue to pop up throughout the years. For the Vietnam War Memorial, that new edition came in the form of a Huey helicopter that flew during the war. It was dedicated in May of 2014. The Vietnam memorial was a collaborative effort of Paul Davis and other veterans who were looking to honor those who never returned from the war as well as those who did return, but didn't get the warm welcome they hoped to receive.

While the park itself hosts a number of different memorials, the Vietnam War Memorial features multiple components. The original memorial consists of a stone monument featuring the inscription "In honor of the men and women who served during Vietnam and have returned with broken bodies, minds and spirits. May God have a special place in His heart for you." The memorial also features the names of the fallen on a stone slab along with an etched map of Vietnam, Laos, and Cambodia.

The 2014 edition to the memorial was a Huey helicopter that was used during the war and is said to have saved a soldier from Dover, Delaware. It's situated on a high platform over the original monument. To complete this portion of the memorial, two mannequins dressed in military uniforms are seated inside of the Huey.

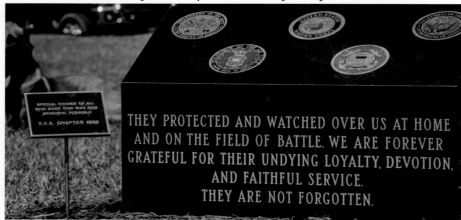

The Names of those from the state of Delaware Who Made the Ultimate Sacrifice

ADAMS THOMAS B
AIKIN GEORGE LEE
ALEXANDER ROBERT DAVID
ALLEY DOUGLAS DWIGHT
AMOROSO FRANCIS BRADFORD
ANDERSON CHARLES RICHARD
ANDERSON ROBERT CARL
ARNOLD DAVID MORGAN
AYRES GERALD FRANCIS
BAILEY DONALD RAY
BAKER WILLIAM S
BIRD LEONARD ADRIAN
BISCHOF WOLFRAM WALTHER
BLEACHER RONALD THOMAS
BOWMAN RICHARD ALAN
BOWMAN ROBERT MICHAEL
BOXLER CHARLES EVERETT
BRANYAN PAUL F JR
BRITTINGHAM ELMORE JR
BRITTINGHAM LINDEN WAYNE
BROWN WERNER CURT II
BUNTING WILLIAM JOSEPH
BURRIS REGINALD WAYNE
BUTLER WILBERT RUDOLPH
CASSIDY JEFFREY TYRONE
CHAMBERLAIN RICHARD MORRI
CHASON THEODORE JOSEPH
CLOUGH BRUCE EDWARD
COLLINS TOBY ERNEST
CONNELL JAMES JOSEPH
COPPAGE GEORGE HERMAN III
CRIPPS GEORGE WARREN
CROSBY ROLIN JAMES
CUBBAGE CLIFTON
CUFF DONALD MERRITT
DADISMAN MICHAEL RAYMOND
DAWSON DONALD EDWARD JR
DEMPSEY GARY LEE
DENNISON RICHARD SAMUEL
DI PASQUANTONIO MICHAEL
DIEFFENBACH ROBERT W JR
DOBRZYNSKI RAYMOND PAUL
DOLBOW BRUCE EDWARD
DONAWAY ROBERT HUGHES
DONNELLY JAMES VOELKEL
DOWNS LLOYD J
DOYLE ROBERT WALTER
FALKENAU ROBERT ARTHUR
FAULKNER ELMER LEE JR
FAULKNER RICHARD J
FLAHERTY KEVIN GREGORY
FONES PAUL MARK
FULLER ROBERT JOHN
GAWORSKI FRANCIS XAVIER
GEISSINGER ALAN GWINN
GIBSON DONALD LEE
GLOVER JAMES ALBERT
GOTT JOHN JOSEPH JR
GRANT GENE TYNDALL
HADDICK HAROLD WILLIAM
HALL VAUGHN O NEIL

HAMILTON DONALD PHILIP
HASTINGS DAVID LYNN
HAYDEN JON JAMES
HENRICKSON JAN VICTOR
HESS GENE KARL
HETZLER RAYMOND CURTIS
HILL ARTHUR STANLEY JR
HILL RICHARD KENNETH
HITCHENS LAWRENCE EDWARD
HOLDEN JAMES EDWARD
HOWARD SYLVESTER JOSEPH
JESTER WAYNE CLIFFORD
JOHNSON JAMES HAROLD JR
JOHNSON ROBERT ERNEST
JONES FRANK WARREN
JONES WILLIAM EDWARD
KAMINSKI JOSEPH M JR
KENTON DONALD E
LA SALLE LAWRENCE LEE
LEATHERBURY LOUIS ANTHONY
LOWDON GRAHAM NORRIS JR
LOWMAN WILLIAM LOUIS
LUDWIG RAYMOND JAMES
MC FALLS HARRY PRESTON
MEREIDER ROBERT JOHN
MILLER GLENN WILLARD
MILLER JAMES LEE
MOMCILOVICH MICHAEL JR
MORRIS ARTHUR CYRUS JR
MORRIS JAMES THURMAN JR
MOSES JESSE LEE
MURPHY WILLIAM JOSEPH
NELSON FRANK WILLIAM JR
PAOLETTI SAMUEL
PARTIN GEORGE EDWARD
PERRY GEORGE FRANCIS III
POLLARD WILLIAM ALFRED
PORTER CHARLES EDWARD
POTTS LARRY FLETCHER
PRESS VICTOR EUGENE
PROTACK THOMAS JOHN
QUICK PAUL WAYNE III
RAY NOLAN REED
RIPANTI JAMES LAWRENCE
ROBINSON LIONEL LARUE
RODOWICZ MICHAEL JOHN
SMITH FREDERICK E
SORNSON EDWIN HAROLD
STEVENSON RICHARD CHARLES
SUDLER EDMUND LAWRENCE
TAMS ROBERT NIELSEN
THOMPSON RALPH LAYTON JR
TIDWELL JOSEPH STANLEY
TRACY ROBERT LOUIS
TRESSLER DANIEL ARK JR
WEBB EARL RAY JR
WEBB HOWARD LEE
WELDIN JACOB ROBINSON
WILLIAMS GEORGE HARVEY
WILLING EDWARD ARLE
WILSON RODNEY WAYNE

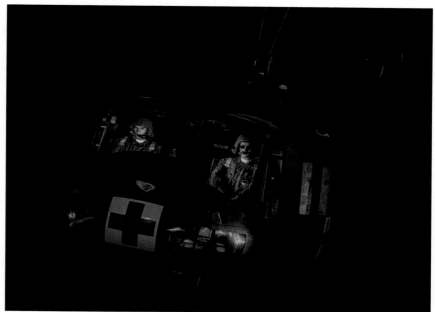

Photos by George Scapin

Wall South at the Veterans Memorial Park

200 S. 10th ave, Pensicola, FL 32502

The Wall South in Pensacola, Florida is a Vietnam War memorial that was made to be a half-scale replica of the famous memorial wall that was made in Washington DC. This version of the memorial is located in Veteran's Memorial Park which plays host to other war memorials including a WWI and WWII memorials as well as one dedicated to the children of 20th-century veterans.

While the memorial is a replica of the Washington DC memorial, the inspiration behind creating the Wall South memorial actually came about after a visit from the "Moving Wall," a mobile memorial wall that makes stops in towns across the nation to provide people with a chance to reflect on the lives lost in the Vietnam War. This wall made a stop in Pensacola back in 1987 and left a deep impact on the veterans who reflected and grieved for their lost friends during the wall's visit to their town. When the wall's time in Pensacola was over, the veterans decided that they wanted to have their own permanent memorial to those lost in the war.

The Vietnam Veterans of Northwest Florida raised funds through garage sales, sports tournaments, 5k and 10k runs, car washes, and more in order to collect the money to make their dream come to life. Finally, in 1992, the Wall South was unveiled in Veterans Memorial Park and is currently the only permanent memorial that is an exact replica of the National Vietnam Memorial.

While the memorial came along nicely, there were some hiccups over the years. Due to the location of Pensacola, the park is prone to hurricane damage and, in 2004, Hurricane Ivan hit the area hard and left a displayed Huey helicopter destroyed. The city of Pensacola maintains the site.

The memorial is situated on five and a half acres of parkland right across from the waterfront. The monument consists of a black granite wall of 64 panels. It stands 256 feet long and is over eight feet high at its highest and a mere two inches at its lowest points. Etched into the stone are the 58,219 names of the service members who were killed or missing in action. As mentioned before, there was a Huey helicopter on display near the monument before it was destroyed in a hurricane. In 2007, it was replaced with an AH-1JC Cobra that was repainted to make it resemble a Huey as it'd look during the war.

The Names of those from the state of Florida Who Made the Ultimate Sacrifice

AARON EUGENE ALLEN
AARONSON WILLIAM F IV
ABERNETHY WILLIAM FORMAN
ABNEY DANIEL THOMAS JR
ABRAMS SAMUEL JR
ADAMITZ IAN WILLIAM
ADAMS BOYED TIMOTHY
ADAMS JAMES ROBERT
ADAMS ROYCE HORACE
ADAMS SAMUEL
ADDAIR KYLE ASHCOM
AKEL RICHARD LOUIS
ALBRIGHT PETER HENRY
ALBRITTON GERALD WAYNE
ALBURY LELAND W JR
ALDAY DANNY WADE
ALEXANDER KERRY
ALLEN EDWIN CHARLES
ALLEN GARY LEE
ALLEN HENRY LEWIS
ALLEN HERBERT MARSHALL
ALLEN WILLIAM JR
ALLEN WILLIAM ORLANDO
ALLEY JAMES HAROLD
ALLWOOD JOSEPH WAYNE BRYA
ALSTON RUBEN CLEVELAND
AMICK TIMOTHY DAVID
AMODIAS OSVALDO
ANDERSON ALTO JR
ANDERSON ARTHUR JAMES
ANDERSON DAVID MICHAEL
ANDERSON HOWARD D
ANDERSON IVY THOMAS
ANDERSON LUCIUS JR
ANDERSON PHILLIP RUSSELL
ANDERSON RALPH TOMMY
ANDERSON ROBERT JAMES
ANDERSON WILLIAM MARK
ANDREWS ARTHUR LEE
ANDREWS CHRISTOPHER
ANNABLE JEFFREY DALE
ANSLOW WALTER HAROLD
ANTHONY CHARLIE C
ARD HOWARD CARLTON
ARD RANDOLPH JEFFERSON
ARNOLD DANIEL RAYMOND
ARNOLD DAVID L
ARNOLD GARY WAYNE

ARNOLD LOUIS BROWARD
ARNOLD REID CARLTON
ARNOLD ROBERT
ARONHALT LARRY DUANE
ARROYO-BRENES GILBERT D
ARTMAN TIMOTHY HAROLD
ASH FREDERIC NATHANIEL
ASKIN JAMES FREDERICK
ATWELL DONALD WILLIAM JR
ATWOOD CHARLES AARON JR
AUE OTTO WAYNE
AUVE CHARLES PAUL
AVELLA JOHN JOSEPH
AYERS CHARLES DAVID
AYERS DANNY R
BABERS HENRY DENNIS
BABULJAK STEPHEN
BACOTE MOSES JUNE
BAGGETT JOSEPH BRADSHAW
BAGGETT WAYNE CARLOS
BAILEY FRED EARL
BAIR CHARLES JACOB
BAKER HOWARD RANOLD
BAKER SAMUEL J
BALDWIN NORMAN EARL
BALL LUTHER EDWARD JR
BALLANCE EDMOND TELLO
BALLANCE NORMAN L III
BALLARD JOHN RICHARD
BALLARD NORMAN CASEY
BALLOU DAVID ALLAN
BAPTISTE MICHAEL BRADFORD
BARBER MORRIE CURTISS
BARFIELD JERRY
BARFIELD LARRY BRUCE
BARKER HOWARD CLEVELAND
BARKSDALE JAMES WILLIAM
BARNES CEPHAS JR
BARNES JOHN LUMSDEN
BARNES RODGER GLYNN
BARNETT KENNETH LEE
BARRINGTON PAUL V JR
BARROW ERIC B JR
BARTLE BARRY GEORGE
BARTLETT BRUCE EUGENE
BARTON ROBERT JAMESON
BASCO JOSEPH FLOYD JR
BASS DUNCAN EDWARD JR

BASSETT ROY DOUGLAS JR
BATCHER LARRY GENE
BATTAGLIA PHILIP J JR
BATTLE JOHN HENRY
BATTLE RONALD KENNETH
BATTS WILLIAM GEORGE
BAXTER JAMES COLON
BAXTER TERRY LEE
BAYONET THOMAS WYLIE
BEALL CHARLES RICHARD
BEAN JOHN ROBERT
BEASLEY JAMES TERRY
BEASLEY JOHNNIE HAROLD
BEASLEY MERRILL VAN
BEAUCHAMP KEVIN PATRICK
BEAVER JAMES HAROLD
BECK MARTIN ROBERT
BECK WINFIELD WESLEY
BECKWITH WALTER LEE JR
BEGGS LARKIN MCDONALD JR
BEHM CHRIS ROGER
BELL GARY JOSEPH
BELL JAMES B JR
BELL JOHN JR
BELL LEROY LEMUEL
BELL LESTER
BELL MALCOLM FRANK
BELL OLIVER JR
BELL RUBEN JR
BELLAMY LARRY RONALD
BELLAMY ROBERT LEE
BENAK JOSEPH FRANK
BENEDICT JOSEPH WAYNE
BENNETT DANIEL MORRIS
BENNETT HOWARD DUNCAN
BENNETT ROBERT LEWIS
BENTON HENRY
BENZING BRUCE MARTIN
BERRIOS JOHN RICHARD
BERRY PAUL L
BEYDA IRWIN
BIEHL GARY LADD
BINTLIFF RONNIE HANKINS
BIRDSELL GORDON DOUGLAS
BIRDWELL GEORGE ALFRED
BISHOP MARK RONALD
BISHOP ROSTEN WAYNE
BISZ RALPH CAMPION
BIXEL MICHAEL SARGENT
BLACK ROBERT DENNIS JR

BLACKBURN DAVID RAY
BLACKMON DAVID OTIS
BLACKSHEAR JAMES GUY
BLACKWELL KENNETH HORACE
BLAIR CHARLES DOUGLAS
BLANN STEPHEN
BLOOMFIELD HARRY GENE
BOATWRIGHT RAYMOND LAVOY
BOGER RHINE HART
BOGIAGES CHRISTOS C JR
BOLYARD LARRY CHARLES
BOMAR FRANK WILLIS
BONANNO FREDERICK MONOTT
BOOTH HERBERT W JR
BORDERS WARDELL
BORG JOHN MICHAEL
BOSTON JAMES JR
BOUTON JEFFERY DALE
BOUTWELL AMOS HAYES
BOWE ROBERT WILLIAM
BOWEN ARCHIE SHERROD
BOWEN HAMMETT LEE JR
BOWENS FRANK
BOWMAN ROBERT CARLOSS
BOYETT GARY RODNEY
BRACKER DAVID EUGENE
BRADEE GARY LEROY
BRADY JAMES HOMER
BRADY TERRY PHILIP
BRANCH DAVID WESLEY
BRANCH LOUIS WILLIAM
BRANNON CLAYTON CHARLES
BRANTLEY ALEXANDER BRYANT
BRASS BASIL PLANE
BRAY RICHARD LOYD III
BREGLER JOHN RAMSEY
BREITNITZ LAWRENCE W
BREMERMAN DALE VINCENT JR
BRIDGES WILLIE GENE
BRIGHAM JAMES WOODROW JR
BRIGHT RALPH NORTH
BRISTOW NORMAN KENNETH
BRITT HOWARD LINTON

BRITTLE ADRIAN COOGIE JR
BROCK MARVIN ZION JR
BROCKMAN ROBERT DAVID
BRODT JAMES HENRY
BROGDON DONALD RAY
BROMMANN HENRY RICHARD
BROOKER DANIEL SCANLON
BROOKS DAVID LEROY
BROOKS GREGORY PAUL
BROOKS JIMMIE LYNN
BROOKS STEVEN KARL
BROSHEAR SARGENT J
BROTHERS BENJAMIN M III
BROWN AUBREY SHAWN
BROWN BARRY LEE
BROWN BENJAMIN FREEMAN JR
BROWN BYRON LEA
BROWN DANIEL MARTIN
BROWN DAVIS FREEMAN
BROWN DONALD RAY
BROWN EUGENE
BROWN GARY WAYNE
BROWN GEORGE R
BROWN GERALD KEITH
BROWN HERMAN FRANK
BROWN JONATHAN
BROWN JOSEPH L JR
BROWN LARRY
BROWN LARRY WAYNE
BROWN MANCE
BROWN NORMAN DALE
BROWN STEVEN ALAN
BROWN WILLIAM B
BROWN WILLIAM LEO
BROWN WILLIE
BROWN WILLIE LEE JR
BROXTON ARTHUR JR
BRUCE SAMUEL JR
BRUNGARD GUY JOSEPH
BRUNSON LOUIS
BRYANT CEASAR
BRYANT CULLIE WILSON
BRYANT JOHN DARRALL
BRZEZINSKI BERNARD FRANCI
BUCHY JAMES LOUIS
BUCKLES WILLIAM THOMAS
BUFF CHARLES FREDERICK
BUHOLTZ TONY LEE

Florida

BULLARD KARL LEE
BULLARD THOMAS C
BULLOCK LEON DANIEL
BUNDY NORMAN LEE
BUNN JAMES ALBERT
BUNTING DENNIS LAMAR
BURCH JAMES ROBERT JR
BURKE JOHN ROLAND
BURKE MARION MCCLAIN
BURKHART WALTER GUY
BURNETT PAUL WAYNE
BURNEY DAVID FRANK
BURNS LEONARD WESLEY
BURNS MARVIN MELTON
BURRIS FRANKLIN IVAN JR
BURROWS MARVIN EUGENE
BURTON THOMAS JOHN
BUSH JOHN ROBERT
BUSH OTIS LEE
BUSTAMANTE GILBERTO
BUTLER EARLIE JAMES JR
BUTLER FRED III
BUTLER JAMES MICHAEL
BUTLER RANDOLPH TODD
BYARS STEVE EUGENE
BYE ROBERT ANTHONY
BYRD GEORGE BENJAMIN JR
BYRD NATHANIEL
BYRD REGINALD TYRONE
CADEAU ROBERT KENNETH
CAHALL EDWIN LEWIS
CAIN ALLEN
CALDWELL RICHARD BRUCE JR
CALFEE JACK WAYNE

CALLANAN JOHN V
CALLOWAY HARDY EUGENE
CAMPBELL DOUGLAS JOHN
CAMPBELL MICHAEL
CAMPBELL RONALD GATES
CAMPBELL WILLIAM ARTHUR
CANNON EDWARD EUGENE
CANNON EMORY STEPHEN
CANNON HENRY TUCKER
CANNON JOHN HENRY
CANTER WILLIAM LINDLEY
CANTRELL ROBERT OWEN
CAREY JAMES DOUGLAS JR
CARLISLE LARRY DEXTER
CARNLEY RUDY AVON
CARRICARTE LOUIS ANTHONY
CARROLL JAMES JOSEPH
CARROLL LARRY MARTIN
CARROLL WALTER JACKSON
CARTER BRUCE WAYNE
CARTER CLYDE WALTER
CARTER GEORGE WILLIAM
CARTER GREGORY
CARTER JAMES BASIL
CARTER JOHNNIE JR
CARTER JOSEPH JR
CARTER MARK JERALD
CARTER PAUL LAMAR
CARTER TERREL ELBERT
CARTER THOMAS ANTHONY
CARTER VERNON THOMAS JR
CASSUBE RICHARD HUGH
CASTLEMAN RICKEY DON
CATO ROBERT O NEAL
CATO WILLIE FRED
CAULEY EUGENE JR
CERES THOMAS ALLEN
CHAMBERLAIN ALLEN B
CHAMBERLIN GEORGE E JR
CHAMBERS CORNELIUS J B
CHAMBLEE WILLIAM DONALD
CHARLES TERRY LEE
CHARLTON JOHN WILLIAM
CHASE LEO CURTIS JR
CHASE VICTOR EDWARD
CHATMAN TYRONE
CHAVOUS SAMUEL CALHOUN JR
CHEANEY PRUITT HENRY
CHERRY CHARLES EDWARD

CHESSHER CHARLES MICHAEL
CHISHOLM RONALD LEE
CHRISS BRAD DONALD
CHRISTMAS LOYE THOMAS
CLARK ARTHUR BOYD
CLARK BRIAN JAMES
CLARK CHARLES EDWARD
CLARK CHARLES II
CLARK ISAAC NATHANIEL
CLARK JOHN HOWARD JR
CLARK PAUL FRANKLIN
CLARK PHILLIP LESLIE
CLARK WILLIAM MARSHALL
CLARK WILLIAM STEPHEN
CLARKE DAVID ERROL
CLARKSON GERALD JOSEPH
CLAUSEN LAWRENCE CHRISTIAN
CLAYTON BILLY JACK
CLAYTON JESSE NATHANIEL
CLEMMONS DOUGLAS FRANK
CLEMONS EDWARD
CLEMONS JOSEPH
CLEMONS LARRY RAYMOND
CLEVELAND RICHARD GROVER
CLIFTON TERRY W
CLINE PAUL HAROLD
CLOSE SANFORD JR
CLOTFELTER MARK DENNIS
CLOUD HARRY JAMES
COBB ROY WILLIAM
COCHRAN ROBERT MC LAIN JR
CODY CLYDE TERRY
COHAN STEPHEN
COLANGELO GEORGE PEYTON
COLBURN DENVER DEWEY JR
COLE LEGRANDE OGDEN JR
COLE RICHARD WILSON
COLEMAN JAMES JR
COLEMAN OLIVER JR
COLEMAN RALPH
COLEMAN RONALD DEAN
COLLINS GUY FLETCHER
COLLINS HORACE CLEVELAND
COLLINS MICHAEL HOWARD
COLLUM WILLIAM EDWARD
COMER HOWARD BRISBANE JR
CONE LEROY

CONELLY MITCHELL PAULLIS
CONNELL CHARLES ANTHONY
CONNER IDUS JAMES
CONNER PAUL ALLAN
CONNOLLY KEVIN THOMAS
CONRAD GEORGE DEWEY JR
CONWAY RAYMOND TERRENCE
COOK CALVIN LEON
COOK HAROLD CLARENCE
COOK JIMMIE DEE
COOPER GARY RAY
COOPER ROBERT WAYNE
COPAS ARDIE RAY
CORBIN THOMAS BERRY
CORBITT DEWAYNE
CORCORAN WILLIAM RICHARD
CORDELL TERRY DENVER
CORNELL ROBERT LESLIE
CORNS RONALD FREEMAN
CORTES-ROSA RAMON
CORWIN MICHAEL HARRY
COSSON WILBUR LYNN
COSTELLO RUSSELL RALPH
COTHRAN CURTIS EDGAR
COWAN ALPHONSO DEDRICK
COX GEORGE TOLLOVAR
COX JIMMIE DON
COX LESTER WAYNE
COX MACK CECIL
COY JESSIE EDDIE LEE
COYMAN PETER R
CRADDOCK CARY
CRAIG JAMES LEWIS JR
CRAIG WILLIAM HOVER JR
CRAWFORD BILLY MAX
CRAWFORD RICHARD
CREMER RONALD MARVIN
CREWS CHARLES RICHARD
CREWS ROBERT LOUIS
CROCE JOHN JOE
CROMWELL EARL LEE
CROMWELL ROBERT WALTER
CROSBY CHARLES DAVID
CROSBY FREDERICK PETER
CRUCE CLAYTON LEON
CRUICKSHANK WILLIAM ROY
CRUTCHER JOE ALBERT
CULBERTH ROBERT LEE JR
CULBREATH JOHNIE KING

CULLEN KENNETH ARTHUR
CULLINS ALVIN
CUMBIE WILLIAM THOMAS
CUMMINGS JAMES EDWARD
CUMMINGS PAUL JOSEPH JR
CUMMINGS WILLIAM LARRY
CUMMINS STEVEN TRAVIS
CURRIER RICHARD JAMES JR
CURTIS THOMAS GUY JR
CUTINHA NICHOLAS JOSEPH
DALTON DAVID JAMES
DAMRON WILLIAM THOMAS
DANCER WALTER JAMES
DANFORD JAMES ISAH E
DANIELS GARY LEONARD
DARBY JOHN FREDERICK III
DARDEN LARRY EUEL
DARVILLE EDWARD ROBINSO III
DAVENPORT CHARLES E II
DAVIDSON RONALD LARRY
DAVIS BENTLEY THOMAS
DAVIS BLAKELY IRVING JR
DAVIS EMMETT LARUE
DAVIS ERNEST PETTWAY
DAVIS JAMES ALBERT
DAVIS JOSEPH WILLIAM
DAVIS TOM JR
DAVIS WILLIAM FORREST
DAVIS WILLIAM TERRELL
DAVIS WILLIAM W JR
DE JESUS JOAQUIN
DEAN CHRISTOPHER J JR
DEAN JAMES ROBERT JR
DEANE WILLIAM LAWRENCE
DECAIRE JACK LEONARD
DECKER ALLAN GEORGE
DEEN DAVID KEITH
DEESE DANNY EUGENE
DEESON MICHAEL DANIEL
DEITSCH CHARLES EDWARD
DELANEY KENNETH LEON
DELARA FRANKLIN VICTORY
DELONG EVERETT EUGENE JR
DELPHIN BARRY RONAL
DEMALINE JOHN THOMAS
DEMPS HENRY VAN
DEMPSEY THERON SPENCER
DENHAM GAIL JR
DENHOFF THOMAS EDWARD
DE-NICOLA ALLEN
DENNARD MACK JR

DENTON GREGORY JOHN D
DEROSIER MICHAEL DOUG-
LAS
DETMER DONALD GARY
DEUERLING WILLIAM JOSEPH
DEUTSCH HENRY ALBERT
DICK ALAN JAY
DICKERSON CHARLES C JR
DICKEY CHARLES C JR
DIEHL STANLEY GENE
DIERS RICHARD WALTER
DIETZ GARY PHILIP
DIGSBY LEROY
DILLARD JAMES L III
DILLENDER WILLIAM ED-
WARD
DILLON JAMES DALE
DINKINS MICHAEL GARY
DIPOLO ROLAND FORREST
DIXON CHARLES ALVIN
DIXON DAVID ERNEST
DIXON MORRIS FRANKLIN JR
DIXON ROBERT DALE
DIXON TERRENCE GLADE
DODD JAMES WILLIAM
DOELGER-LANDIVAR HER-
MANN
DOMINIQUE GARY MARK
DONKER LEO MICHAEL
DORNELLAS RICHARD AL-
LISON
DOUGLAS JOHNNIE LAMAR
DOXEY JAN DEAN
DRAKE MICHEAL JOHN
DRAWDY RYLAND WHITNEY
DRUMMOND EMANUEL
FRANK JR
DUBBELD ORIE JOHN JR
DUBOSE DOUGLAS SCOTT
DUCK WILLIAM WHITBY
DUCKETT ARLEN JACKSON JR
DUFFY THOMAS KNOWLES
DUKES ARTHUR ROGER JR
DUKES THOMAS LESTER
DUNCAN JAMES HENRY
DUNCAN ROBERT RAY
DUNDAS MICHAEL C
DUNLOP THOMAS EARL
DURDEN TROY
DUSCHEK RUDI HERMANN
DUVALL GARY LEE
DYER WILLIE GENE
EALUM CARREL GORUM
EBERLE RONALD EARL
EDENFIELD RONALD DAVID
EDGAR ROBERT JOHN
EDWARDS DENNETTE A III
EDWARDS RICHARD JR
EDWARDS THOMAS CLIFFORD
ELDER WILLARD FRANCIS
ELLIS ALTON LEE
ELLIS ALTON STARLING
ELLIS PRESTON HENRY
ELLIS WALTER GENE MERVIN
ELLIS WILLIAM WALTER III
EMMONS JUDSON WAYNE
EMRICH ROGER GENE
ENGRAM RANDAL CLYDE
EPLEY KENNETH KEITH
EPPERSON CHARLES WILLIAM
ESTERS CHARLES JR
EUBANKS RANDOLPH
EVANS RICHARD WAYNE
EVANS ROBERT DILLON
EVERETT LEROY
EXUM EZEKEIL THEODORE
FARMER HARRY EARL
FARTO CARLOS ANGEL
FENNELL WILLIAM ERVIN
FERGUSON LOWELL VERNON
JR
FERNANDEZ-LESTON EN-
RIQUE
FERRA-FLORES PEDRO
FERRELL TENNIS CRISPIAN
FIELDS CHARLIE
FIELDS FREDERICK LEE
FIELDS KENNETH WAYNE
FIELDS ROBERT WAYNE
FIKE ARTHUR HARRY
FIKE RUSSELL LARRY
FILLINGIM THURMAN ELBY
FINNEY ARTHUR THOMAS
FISCHER KENNETH EDWARD

FISCHER ROY SCOTT
FISHBECK JAY JOHN
FLANIGAN JOHN NORLEE
FLEMING HORACE HIGLEY III
FLEMISTER HUGH ROBERT
FLETCHER DONNITH HOW-
ARD
FLINT WILLIAM NEIL
FLOURNOY PAUL DOUGLAS
FLOYD ROBERT GENE
FLYNN WILLIAM VINCENT
FOARD WALLACE BILLANY JR
FOLDEN THOMAS
FONSECA-VARGAS HORA-
CIO A
FORD LEONARD DAVID
FORD RALPH LEE
FORD RANDOLPH WRIGHT
FORD RUSSELL THOMAS
FOREHAND JERRY
FORSHEY JOHN DANIEL
FORTE FREDERICK C JR
FORTNER JOHN LYNWOOD
FOSTER ISIAH
FOSTER TOMPKINS GRIFFEN
FOWLER JOEL CAROL
FOX REINIS
FOX ROBERT CHARLES
FRANGELLA FRANK A
FRANKLIN CHARLES ROBERT
FRANKLIN EUGENE
FRANKS JOHN HOWDEN
FRANKS MONROE
FREDA ROBERT
FREDERICK DOUGLAS LLOYD
FREDERICK JAMES CARL
FREDERICK STEVEN EDWARD
FREDRICK BRIAN RANDALL
FREEMAN ARDENIA
FREEMAN EUGENE LARRY JR
FREEMAN MOULTON LAMAR
FRIDDLE GLENN MARK
FROSIO ROBERT CLARENCE
FROST RICHARD HAMMOND
FRYE KEVIN MARK
FRYER WILLIE JAMES
FULFORD JOHN THOMAS
FULFORD VARL ESTON
FULGHAM EDWARD BRAX-
TON JR
FUQUA JAMES F
FURNISH THOMAS HAROLD
FUSSELL FELTON ROGER
GAFFNEY RONALD SEFTON
GAINES BYRON ADAMS JR
GAINES CHARLES A
GAINES DOUGLAS JR
GAINOUS JOHN CHARLES
GALAN RICHARD
GALLANT HENRY JOSEPH
GALLION DAVID ANDREW
GAMBLE HENRY HWEY
GANDY CLAUDDELL
GANOE BERMAN JR
GANZY CLYDE WAYNE
GARBETT JIMMY RAY
GARCIA MIGUEL RAMOS
GARDNER JAMES DALE
GARNER RONALD RAY
GARNET OWEN NIEL
GASKIN DAVID WILLIAM
GASSMAN FRED ALLEN
GAST WILLIAM RAYMOND
GATES ROBERT ALFRED
GAY GERALD GILBERT
GAYLOR GERALD H
GEDDIS HENRY LEO JR
GEIGER CHARLES RICHARD
GEIGER ISADORE SAMUEL JR
GENERAL CARL LEWIS
GENTRY ROBERT BARRY
GEOGHAGEN GEORGE EDDIE
GERDON ROY CLINTON
GERONIMO CHARLES AN-
THONY
GIBSON JOHN BROWN JR
GIBSON PETER
GIERMAN WALTER EDWARD II
GILES CLEM CALVIN
GILLIS GARY WAYNE
GILLMAN DONALD STANDLEY
GINDLESBERGER GERALD
THOM
GINN MICHAEL PATRICK

GIPSON GEORGE WESLEY
GLAZAR AARON ZANE
HOWARD
GLENN MICHAEL O ROY
GLOVER FRED RICHARD
GODBOLDT WILLIE FRANK
GODFREY JOHNNY LEE
GODWIN STANLEY MAURICE
GOI LOUIS CHARLES
GOLD EDWARD FRANK
GOLDBERG WILLIAM JACOB
GOLDEN KENNETH D JR
GOLDEN WILLIAM JOSEPH
GOLDING JAMES RICHARD
GOLDMAN HAROLD
GOLIGHTLY ROLLIN EUGENE
GOMEZ RICARDO JOSE
GOMEZ ROBERT ARTHUR
GOMEZ STEVE
GONZALEZ HECTOR
GOODHUE MARLIN JAMES
GOODMAN JACK LANCE
GOODMAN MARVIN FOY JR
GOODSON THOMAS HENRY
GOODWIN ALVIN MAYNARD
GOODWINE ISOM JUNIOR
GOOLSBY JAMES RUEL
GORDON CHARLES A
GORDON OTIS JR
GORMICAN DAVID C
GORSKE ROBERT EDWARD
GOSWICK WESLEY IRA
GOTT RODNEY HERSCHEL
GOULD JOHNNY WAYNE
GRAHAM TERRY DURAND
GRANBERRY JOHNIE FRANK-
LIN
GRANT JAMES WOOD
GRANT JOHNNIE
GRANT STEPHEN MITCHELL
GRAY BOBBY ELMER
GRAY WILLIAM GEORGE
GRECO STEVEN JAMES
GREEN JEREMIAH
GREEN JIMMIE WAYNE
GREEN WILLIE FRANK
GREENE ARTHUR WILLIE
GREENE BRADFORD BARTON
GREENE DONALD BRICE
GREER MATTHEW ERNEST
GREESON DAVID CURTIS
GREESON JOHN EGBERT
GREGORY EULAS FAY
GREGORY HERBERT LEE III
GREGORY JOHN HENRY JR
GRIENER JAMES G
GRIFFIN DAVID SCOTT
GRIFFIN EUGENE
GRIFFIN JAMES ROGER
GRIFFIS JAMES LARRIAN
GRONQUIST CARL EUGENE JR
GROOMS RONALD KEITH
GROOVER RICHARD ANTHO-
NY
GUCOFSKI STEPHEN DOUG-
LAS
GUDE MARVIN JOSEPH
GUNTER WILLIAM CLAYTON
GUNTHER JOHN JACOB
GURR HERMAN LEROY
GWINN RICHARD ALFRED
HAAS CHARLES GEORGE
HACKETT DANIEL HAROLD
HACKETT JAMES EDWARD
HADDEN HERBERT MICHAEL
HADLEY SHERRY JOE
HADSOCK WILLIAM ALFRED
HAGA JOSEPH CLAYTON
HAGERICH WILLIAM CLYDE
HAILE RICHARD GUSTAVE JR
HAIR ROBERT LEE
HALL CHARLES EDWARD
HALL JAMES HAYES
HALL JERRY RAY
HALL JOHN DEAN
HALL MICHAEL ROBERT
HALL MILTON LEE
HALL PATRICK LINDSEY
HALL WILLIS ROZELLE
HALLOCK WILBUR LEWIS J JR
HAMILTON EDWARD
HAMILTON GEORGE KIRT-
LAND
HAMILTON VIRGIL VERN

HAMMAN THOMAS RALPH
HAMPTON ENOCH
HAMPTON FREDERICK
JORDAN
HANCOCK EUGENE SCOTT
HANKINS GREGORY EUGENE
HANKISON TOMMY LEE
HANSON THOMAS PATTER-
SON
HARBOT FREDERIC RICHARD
HARDWICK TOMMY
HARE ANGUS LAYAFETTE
HARLESS CARL CLARENCE
HARP WILLIAM
HARPER JOSEPH JAMES
HARPER TONY
HARRELL ROGER PARRY
HARRELL SAMUEL
HARRELL STANLEY MOORE
HARRINGTON PATRICK JAMES
HARRIS GRADY HERSHALL
HARRIS HARLIN JR
HARRIS JOSEPH RICHARD
HARRIS RODNEY CARSWELL
HARRIS THOMAS WYATT
HART THOMAS TRAMMELL III
HARTER ROBERT LOUIS
HARTNEY JAMES CUTHBERT
HARTSFIELD BILLY JACOB
HARTSON STANLEY GERALD
HARVEY ALAN DARYL
HARVEY JACK ROCKWOOD
HARVIN JIMMIE LEE
HASKETT EDWARD O DAY
HATCH KENNETH NEAL
HATCHER ROBERT LEE
HATFIELD CHARLES DAVID
HATHAWAY JOHN HOOPER V
HATHAWAY STEVE
HATHAWAY WALTER SAMUEL
HAWKINS KENNETH JEROME
HAWLEY LAWRENCE CHESTER

HAYDEN RALPH PARKER
HAYES ALBERT JUDSON
HAYES JAMES EDWARD
HAYES QUENTIN
HAYS JOHN HULSEY
HAYWARD JOHN KENT
HEATH BRIAN CHARLES
HEATH MICHAEL FREDERICK
HEIDE HENRY NICHOLAS II
HEINSELMAN THEODORE E
HEIZER GARY PAUL
HELVESTON ROBERT FULTON
HENDERSON EDWARD E JR
HENDERSON JAMES D
HENDERSON LEON
HENESY HAROLD THOMAS
HENRY LINDY EDWARD
HENSLEY JOHN THOMAS
HERALD ROBERT
HERNANDEZ NOEL BARBARO
HERRING HAROLD JOERENZO
HERRING JOHN WILLIE
HERSHBERGER GARY PATRICK
HESSION PATRICK B
HESTER DONALD VOL JR
HESTER LEO CLAUDE JR
HESTER VANESTER LAMAR
HESTLE ROOSEVELT JR
HETRICK CARL POST
HETTERLY JOHN DONALD SR
HETTICH DONALD LEE
HEWITT BRIAN CHARLES
HEWITT CHARLES GLEN
HICKEY JOHN PATRICK
HICKMAN DAVID ALLEN
HICKS BENNY JOE
HICKS WILBUR LEE
HIEMENZ JAMES BORLAND
HIGGINBOTHAM HAROLD S
HIGGINS EDWARD HUBERT
HILERIO ALBERT JR
HILL ALVIN GENE

53

Florida

HILL CLARENCE MITCHELL
HILL EDWIN CHARLES
HILL JIMMY LEE
HILL WILLIAM ERNEST
HILLIARD JAMES GILBERT
HINES JOE RAYMOND
HINOJOSA RUDOLPH JR
HINSON JAMES HARVEY
HITCHCOCK RALPH JOHN
HIX WILLIAM COLQUETH JR
HODGE ANDREW HERMAN
HODGE WILLIAM REUBEN
HODGSKIN JAMES G JR
HOFFMANN CHARLES J III
HOGBIN RONNIE ELLIS
HOHMAN JOHN MICHAEL
HOLDER RANDOLPH CHESTER
HOLLINGSWORTH MICHAEL
DEN
HOLLINGSWORTH VERNICE
HOLLOWAY CHARLES ED-
WARD
HOLLOWAY PAUL DAVID
HOLMES CLEVELAND
HOLMES SAM JR
HOLMES SAMMY LEE
HOLTON JOHN THOMAS JR
HOOD RICHARD E JR
HOOPS FRANKLIN WERNER JR
HOPEWELL DONALD CLEM-
ENT
HOPPER DANIEL EUGENE
HOPPS GARY DOUGLAS
HORNADAY RALPH J
HORNE KENNETH RAY
HOUDASHELT FRANCIS
GERALD
HOUSTON JOHN LUCIUS
HOWARD BILLY
HOWARD GARY EDWARD
HOWARD JAMES T
HOWELL JAMES LAURENCE
HUBBARD THOMAS LEE
HUBBARD TONY GENE
HUBER LEON FAIRDEN
HUBER STEPHEN LEE
HUDSON JOSEPH JR
HUGGINS FRAZIER DANIEL
HUGHART HAROLD GRAN-
VILLE
HUGHES JEFFREY REX
HUGHES JERRY NELSON
HUGHES SAM ZEB
HUGHEY MICHAEL ALLEN
HULSLANDER ROSS THOMAS
HUNT ARTHUR WALTON III
HUNT MARSHALL WIMBERLY
HUNT WILLIAM HOWARD
HUNTER WILLIE HAYWARD
HURD ERNEST LEON
HURLOCK CURTIS WOOD-
ROW
HURLOCK PETER CLIFTON
HURST JAMES RANDOLPH
HUSTON DALE MARTIN
HUTCHESON GEORGE DEWEY
HUTCHINS LUCIOUS
HUTCHINSON WAYNE ALLEN
HUTCHISON CHESTER K
HUTTING ROY DONALD
HUYLER CECIL
HYSMITH HAROLD FRANKLIN
ILGENFRITZ HERBERT E JR
INGRAM ELIJAH
INGRAM JERRY GRANT
IRVIN PAUL EDWARD
IRWIN ROBERT JOSEPH

ISRAEL JOHN WALLACE
IVEY DORRIS ALBERT
IVEY TOMMY HUBERT
JACKSON ARELINN LEWIS
JACKSON CLARENCE JAMES
JACKSON EDDIE LEE
JACKSON FREDDIE
JACKSON GERALD ARTHUR
JACKSON HERMAN
JACKSON JAMES TERRY
JACKSON MARK
JACKSON MURRAY JUNIOR
JACKSON NATHANIEL JR
JACKSON RICHARD CURTIS
JACKSON THORNTON ISHAM
JACOBS GARY ORLAND
JAMES KENNETH BRADLEY
JAMES PAUL JOSEPH
JANSENIUS RAYMOND LEE
JAY HARVEY LEON
JEFFERSON JAMES MILTON
JENKINS ANDREW EARL
JENKINS FRED CARLTON
JENKINS JAMES LUCKY
JENKINS LARRY
JENKINS LARRY BARNEY
JENKINS ROBERT HENRY JR
JENNINGS JASPER LEWIS
JERKINS WILLIAM EDGAR
JEWETT GUY LEONARD
JOHNS DONALD CECIL
JOHNS ERNEST LEE
JOHNS LAMARR LEE
JOHNS RONALD ELMER
JOHNSON ARTIE EUGENE
JOHNSON BUFORD GERALD
JOHNSON CARLTON JERRY
JOHNSON ERIC WAYNE
JOHNSON FRED ARTHUR
JOHNSON GEORGE DENNIS
JOHNSON GORDON MICHAEL
JOHNSON JAMES ALVIN
JOHNSON JAMES BRUCE SR
JOHNSON JAMES JR
JOHNSON JOHN WAYNE
JOHNSON LARRY
JOHNSON LEO FRED
JOHNSON NORRIS FELTON
JOHNSON PERRY DAVID
JOHNSON RAYMOND EUGENE
JOHNSON SANFORD LEE
JOHNSON WALLACE B III
JOHNSON WILLIAM EDWARD
JOHNSON WILLIAM THEO-
DORE
JONES ALDON CECIL
JONES ANTHONY BERNARD
JONES CLARENCE LLOYD
JONES DWIGHT HUBERT
JONES EDWIN
JONES ERVIN
JONES FREDERICK OLEN
JONES JAMES LYNN
JONES JAMES RANDALL
JONES JIMMIE LEE
JONES JOSEPH RICHARD
JONES JULIUS FRAZIER
JONES MARCUS CLAUDE
JONES MILFORD
JONES PAUL
JONES SHERMAN LAWRENCE
JONES WALTER HOLT II
JONES WILLIAM STANLEY
JONES WILLIE DONALD
JONES WILLIE GERALD
JOYCE DERRELL WALTER
JOYNER CARL HENRY
JUCKETT ELMER L III
KAMINSKY JOHN PERRY
KEATHLEY CHARLES BRIAN
KECK JAY LYNN
KEE JULIAN STANLEY JR
KEENE GRAT ALBERT
KEHRLI HERBERT ALBERT
KEITH JIMMIE EUGENE
KELLEY KARL ELTON JR
KELLEY KENDRICK KING III
KELLEY MAHLON LEWIS
KELLEY WILLIAM FRANCIS
KELLY SEEBER J
KELSEY STRAUGHAN D JR
KEMP MARWICK LEROY
KENNEDY MICHAEL JOSEPH
KEPCZYK TADEUSZ MARIAN

KERN BRUCE ALAN
KERR JOHN CREIGHTON
GILLE
KERSEY WILLIAM RUSSELL JR
KETT RANDOLPH CHARLES
KEY LESTER
KICKLITER JAMES THOMAS
KIDD GEORGE R
KIDD KENNETH EDWARD
KIEME BRUCE DOUGLAS
KIESER CHARLES DAVID
KIMBLER LAWRENCE RUTH-
ERFO
KINDER LARRY WADE
KING EARL HUGO
KING JAMES HENRY
KING LAURENCE MICHAEL
KING LEWIS
KIRBY MICHAEL CHARLES
KIRIK MARTIN EUGENE
KIRKLAND WILLIE LEE
KIRKPATRICK MICHAEL WARD
KIRKPATRICK RONALD RENE
KITTLE STEPHEN RANDALL
KLEBER HARRY WILLIAM
KLECKLEY FREDDIE LEE
KLINE BRUCE EUGENE
KNIGHT CARLOS LARUE
KNIGHT HUBERT CHARLES
KNIGHT PETER STANLEY
KNIGHT RICHARD VINCENT
JR
KNIGHTON ELI WHITNEY JR
KNOWLES JAMES D
KNOWLES WILLIE JR
KNUTH LAWRENCE DOUGLAS
KOCKRITZ JEFFRY LETSON
KOHN ROBERT A
KOPETSKI MICHAEL BENJA-
MIN
KRAFT NOAH MORRIS
KREGELOH DONALD RICH-
ARD
KRUPSKI WALTER BENJAMIN
KUBIK KENNETH ARTHUR
KUHNS KURT LLOYD
KURLIN WAYNE CARLTON
KUSHNER DANIEL KENT
KYLE DONALD CHARLES
LABRECQUE ROBERT WIL-
LIAM
LACAGNINA RALPH VINCENT
LADNER JAY WESLY
LADSON LAFON WINSTON
LAFRENIERE PAUL JOSEPH JR
LAKE JOHN ROACH JR
LAKER CARL JOHN
LAKIN JOHN HAYES
LAMB BILLY WAYNE
LAMB THEODORE
LAMBERT LEE MATHEWS
LAMBERT WILLIAM GLENN
LAMN JAMES FRANKLIN
LAND DAVID ALFRED
LAND FRED EMERY
LANDERS KENNETH JEFFER-
SON
LANDERSHEIM LARRIE JOHN
LANE MICHAEL D
LANG ERNEST ALPHONSO
LANGFORD ROGER LEO
LANGLEY WASHINGTON
MORRIS
LANGSTON ROBERT EBERT
LANTZ PETER J
LAURIE MICHAEL J
LAVEZZOLI PAUL RICHARD
LAWLER THOMAS FREDERICK
LAWRENCE BILLY EVERETT
LAWRENCE FRANCIS M JR
LAWSON WARREN STEPHEN
LE FEBURE RONALD DEAN
LEDBETTER THOMAS ISAAC
LEDGERWOOD DAVID GAIL
LEE DONALD LAMAR
LEE EDGAR
LEE GEORGE JR
LEE JOE LEWIS
LEE ROBERT LIST JR
LEER JOHN EDWARD
LEFFLER RICHARD JOHN
LEFFLER RUSSELL ALAN
LEGATE RICHARD EDWARD
LENNARD BENJAMIN EDWIN

JR
LENTZ DOUGLAS ALAN
LEONARD BILLY
LEVINS FREDERICK RICHARD
LEWIS DON ROBERT
LEWIS JAMES ROBBINS JR
LEWIS JOE
LEWIS JOHNNY ELMER
LEWIS STEVEN
LIESER ROBERT DARYL
LILLIE JOE HENRY
LINDER JAMES JR
LINDSAY GREGORY THAYER
LINDSEY DANIEL HINSON
LINTHICUM DON WILLIAM
LISENBY JAMES ARNOLD
LISLE JACK MCBRIDE
LIVENGOOD STEVE ALLEN
LLAMAZALES HUMBERTO
LOCKWOOD KENNETH
CHARLES
LOFMAN LANCE MICHAEL
LOFTHEIM DENNIS DEAN
LOGES JOHN EARL
LOISELLE BRUCE WAYNE
LONG BILLIE MONROE
LONG CHARLES EDWARD
LONG WARREN LARUE
LOTT HARVEY EUGENE
LOUDERMILK JAMES ELLIS
LOVE CHARLES WILLIAM JR
LOVELL PATRICK DARREN
LOWE ROBERT BREWSTER
LOWE WILLIAM EARL
LOWES RICHARD SMITH
LOWRY JIMMY CLINT
LUCCI CHRISTOPHER DUTCH-
ER
LUCE RICHMOND ROSS
LUCKEY JAMES ALFRED
LUKE RONALD HAROLD
LUTES MARK STANTON
LUTZ JOSEPH PATRICK
MACKEY ROBERT EUGENE
MACON SAMUEL CORNELIUS
MAGRUDER DOUGLAS GRA-
HAM
MAIN WILLIAM TERRY
MALESZEWSKI PAUL EDWARD
MALOY GARY LEE
MANGRUM RICHARD GALE
MANIGO EUGENE
MANINGER RAYMOND
MARCINE
MANNING DENNIS DEWAIN
MANOR JAMES
MANOS ARTHUR
MARCH FRANK JR
MARINELLI ANTHONY JOHN
MARKLAND DONALD P III
MARSHALL FREDDIE JR
MARTIN DAVID EARL
MARTIN JOHN SANFORD
MARTIN LARRY EUGENE
MARTIN STEVE LAIL
MARTIN WAYNE OSCAR
MARTIN WILLIAM GEORGE
MARTINEZ THOMAS MICHAEL
MARTINO THOMAS JOSEPH
MASCIALE VINCENT TOMMY
MASEDA GERALD LEE
MASLINSKI DWIGHT ANDREW
MASLYN EDWARD JAMES
MASON BOBBY G
MASON ROBERT ERNEST
MASTERS EDWARD ULYSES
MASTERS JAMES MADISON JR
MATEJECK WALTER LAW-
RENCE
MATHENY RUSSELL LEE
MATHIAS ROBERT
MATHIS SAMUEL JUDSON
MATTHEWS AITKEN L JR
MATTHEWS HOLLEY DEWITT
MATTHEWS RONNIE EUGENE
MATTHEWS SETH HAYDEN III
MAULDEN LORENZO CO-
LUMBUS
MAY RICHARD GEORGE
MAYHAIR WILLIAM HERBERT
MAYS RAYMOND RALIFORD
MAZARIEGOS FRANCISCO
ALBE
MC ALUM ERNEST E

MC BRIDE ELLIS A JR
MC CALL BILLIE RAY
MC CAREY GUY HECTOR
MC CARTY JOHN DAVIS
MC CLAIN FRED JULOUS
MC CLENDON WILLIE JAMES
MC CLENTON HENRY
MC CLOSKEY SCOTT SIMONS
MC COMB AUBURN DALE
MC CONNICO DONALD
MC CORMICK BRUCE ALLEN
MC CORVEY EDWARD JR
MC COY BOOKER TEE JR
MC COY LARRY WILLIAM
MC CRAE JAMES HENRY
MC CRAY EUGENE
MC CRAY FRANK JR
MC CURLEY TIMOTHY LEWIS
MC DONALD DAVID HAROLD
MC ENANY KEITH ALLEN
MC FALL KENNETH LEWIS
MC FARLAND SYLVESTER
WARR
MC GEE THOMAS LEE
MC GILL DAVID LOREN
MC GRATH EDWARD CHARLES
MC GUCKIN JOSEPH
MC INTOSH ROBERT A
MC INTOSH WILLIE EDWARD
MC KEE WALTER ROY
MC KENDRICK GARY RAY-
MOND
MC KINLEY JAMES MARION
MC KINNEY IVORY LEE
MC KINNEY WESLEY JUNIOR
MC KINNIE CHARLES W JR
MC KINNON CLARENCE LEE
MC KINNON TITUS JR
MC KNIGHT MATTHEW
OWEN
MC LAUGHLIN OLEN BURKE
MC LELLAN JOHN MALGER
MC LEMORE TILGHMAN
RICHAR
MC NABB RICHARD DALE
MC NAC DONALD CHARLES
MC NUTT FRANK ELLIOTT
MC PHEE DOUGLAS WAYNE
MC PHERSON WILLIAM
JOSEPH
MC PHETERS CHET EUGENE
MC WATERS DALTON HUBERT
MCANDREWS MICHAEL
WILLIAM
MCKAY EUGENE HENRY III
MCKELLIPS RANDOLPH
BURNS

MCLEOD DAVID VANCE JR
MCSWINEY CHARLES A JR
MEACHAM RICHARD W JR
MEAD DALE WALTER
MEANS JOHNNY
MEDLIN JACKIE MONROE
MEEK JAMES BRANNON
MEISHEID ALAN JAMES
MELTON JACKIE LEE
MEMORY AL DEWITT
MERCER JACOB EDWARD
MERRICKS ALVIN
MERRIHEW GLEN FREDERICK
MERRY DONALD LEWIS
MESA MANUEL E JR
MESSER EARLEY JOSEPH
MIHALEK ADELBERT F IV
MILES JOHN ELMER
MILIKIN RICHARD M III
MILLENDER ROBERT CLIF-
FORD
MILLER CHARLES
MILLER CHARLES EDWARD
MILLER CLARENCE STEPHEN
MILLER EARNEST LEE
MILLER EDWARD MARTIN
MILLER HOLLIS GREGORY
MILLER J C THEODORE
MILLER JAMES BERNARD
MILLER MALCOLM THOMAS
MILLER MICHAEL MERLIN
MILLER RICHARD DANCY
MILLER RICHARD W
MILLER ROBERT EARL
MILLER WILBUR JAMES JR
MILLICAN MALCOLM ED-
WARD
MINCEY JOHN H
MINEAR MARK WENDELL
MISIUTA EDWARD MICHAEL
MITCHELL CURTIS
MITCHELL JULIUS AUGUSTA
MITCHELL THOMAS PETER
MIZE JAMES WESLEY JR
MIZE MELVIN LAMAR
MOBLEY WARREN HERBERT
MOLE MALCOLM GEOFFREY
MONSEWICZ LLOYD JOEL
MONTANA HAROLD LLOYD
MOODY ARTHUR R III
MOODY JERRY MARCUS
MOONEY GENE ALLEN JR
MOORE DAVID T
MOORE HERCULES LEE
MOORE HERMAN A
MOORE JOHNNY LEE
MOORE WILLIE JAMES

MOORMAN CECIL ROY
MORALES HAROLD WAYNE
MORGAN DENNIS LACO
MORGAN JOSEPH JR
MORGAN SHELTON
MORNINGSTAR GEORGE
AARON
MORRIS KENNETH BRYAN
MORTON JERREL CARL
MOSS RICHARD LEE
MOTES CARL GILBERT
MOULTRIE CALVIN
MOUTARDIER ODES HERMAN
MUCKLEROY JAMES RICHARD
MUIR JOHN DAVID
MULCAHY JOHN MARTIN
MULLEN DANIEL JERRY
MULLINS JAMES EDWARD
MUNGIN LAWRENCE DAVID
III
MURPHY BARRY DANIEL
MURPHY MICHAEL THOMAS
MURRAY GORDON CHESTER
MUSA HENRY ALFRED JR
MYERS HOMER JULIUS
MYERS WILLIAM LATHEM JR
NAIL ROBERT MELVIN
NAILS EDDIE LEE JR
NAIMO JOSEPH PETER JR
NALEY RICHARD HERBERT
NEAL DENNIS PAUL
NELSON ARCHIE LEE JR
NELSON JAN HOUSTON
NELSON LAWSON DWIGHT
NELSON ROBERT MELVIN
NELSON TONEY JR
NELSON WILLIAM IRVIN II
NERECK LAWRENCE THOMAS
NEUBAUER RONALD GEORGE
NEURA TED PETER JR
NEWKIRK JAMES EDWARD
NEWKIRK TERRY CURTIS
NEWLAND LONNIE PITTS
NEWMAN RONALD ELLIS
NEWMAN STANLEY HAROLD
NICHOLAS DENIS
NICHOLS HUBERT CAMPBEL
JR
NICHOLSON GERALD W JR
NICKERSON WILLIAM WAL-
TER
NIEDERMEYER JOHN GARY
NIELSEN MAGNUS CARL
NIXON JESSE ERNEST
NOBLE LEWIS RAULERSON
NOLES GARY EDWIN
NORMAN CALVIN JR

NORMAN MICHAEL WARREN
NORMAN W H
NORRIS BILLY RAYVON
NORRIS JERRY A
NORRIS JOSEPH ROBERT
NORTH DENNIS COLE
NORTON KENNETH DEAN
O BRIEN GARY MALCOLM
O CONNOR MICHAEL BARRY
O ROURKE RONALD PATRICK
O STEEN CHARLES ROBERT
ODUM MICHAEL RALPH R
OLIVER ROMMIE
OLSON GERALD EVERETT
OLSON JAMES ROBERT
OSBORN GEOFFREY HOLMES
OSBORN RICHARD D
OSBORNE EDWIN NELMS JR
OSBORNE RONALD CHARLES
OSTEEN MICHAEL STEVEN
OSTEEN RICHARD ARNOLD JR
OSTERMEYER WILLIAM
HENRY
OTT RAYMOND EARL
OUTLAW CHARLES REUBEN JR
OVERTON DANNY WAYNE
OWENS BENNETT HOWELL JR
OWENS HAROLD EUGENE
OWENS KENNETH GRANT
PACE JAMES ALVIN
PACKARD DAN BRUEN
PADDLEFORD FRED HAROLD
PADRON IRENARDO FELIX
PAGE JIMMY EDWARD
PALMER HUBERT
PALMER JOHNNY LEE
PARKER DALE WARREN
PARKER JOHN JACKSON
PARKER OTIS
PARKER WOODROW WILSON
II
PAROUNAGIAN GEORGE JR
PARSELLS JOHN WILLIAM
PARTIN DANIEL ROSS
PATE WILLIAM
PATRICK ALBERT EARL
PATRICK JERRY LEE
PATTERSON JAMES ROBERT
PAULK ELIAS JOHNSON
PEACOCK JACK ALLAN
PEACOCK NATHAN EDDLOW
JR
PEARSON JAMES ROY
PEDINGS BILLY DEAN
PEEK RUSSELL JAMES
PEEKS LEEROY ELDRED
PEELER GLOVER AUSTIN III

Florida

PENDLETON GEORGE JR
PENDLEY ROBERT GLENN
PENNY WILLIAM VICTOR
PEREZ CELSO A
PEREZ-VERDEJA RAFAEL
PERRY HAL EDWARD
PETERS JOSEPH CRAIG
PETERSON JAMES WILLIAM
PETERSON ROBERT WALKER
PETERSON THOMAS WAYNE
PETRIE RICHARD JEFFREY
PETTIJOHN JAMES EARL
PETTIS LORENZO RICHARD
PHARIS RONALD WASHINGTON
PHILLIPS DAVID JOSEPH JR
PHILLIPS JERRY NEWTON
PHILLIPS MARK JOHN
PHILLIPS ROY FRANKLIN
PHILLIPS THOMAS FRANK
PICKETT WILLIE CLARENCE
PIERCE JOSEPH HOWARD JR
PILK ROBERT HARRISON
PINA GERALD MARTIN
PIPPIN HENRY LEE
PIRKLE WILLIAM ITHEL
PITTMAN ROBERT EDWARD
PITTS CHARLES R
PITTS CLEVELAND
PITTS FREDDIE RICHARD
PITTS JAMES ELSWORTH
PITTS WAYNE MONROE
PLATA MARVIN JAMES
PLUMMER RICHARD EUGENE
POCHER WILLIAM THORNTON
POFF JERRY WAYNE
POLLARD SIDNEY GERALD
PONCURAK RAYMOND JOSEPH
PONDER DERRELL LOIAL
PORTER SANDY HILLY
POWELL JOHNNIE EARL
POWELL RICHARD WARREN JR
POWERS MARK FREDERICK
PRATT DAVID ALVIN
PRESIDENT ERNEST
PRICE JOSEPH MICHAEL
PRIDDY RICHARD THOMAS
PRINE ROBERT WAYNE
PRITCHARD ROBERT BRUCE
PROCTOR JAMES PATRICK
PROCTOR JOHNNY LEE
PRUDEN FREDERICK WILLIAM
PUGH ROBERT EARL
PULLARA ANGELO
PURCELL CHARLES KENT II
PURNELL ADRIAN FLOYD
PUTNAM CHARLES LANCASTER
QUIGLEY HENRY LEROY
QUINTANA-SOTO LUIS E
RABREN LARRY WAYNE
RADTKE LE ROY CARL JR
RAGANS HERBERT RANDOLPH
RAGIN WILLIAM DAVID HOWSA
RAINEY CHARLIE
RAINEY THOMAS BALLARD
RALYA WARREN HENRY JR
RAMBERG MICHAEL JOHN
RANDALL ROBERT JOHN JR
RATLIFF OSCAR E
RATLIFF THOMAS HENRY
RAVELO-TORIBIO ELPIDIO J
READ CHARLES HAROLD W JR
REAGAN ROBERT WILLIAM

REALI GUIDO SILVESTRO JR
RECUPERO RICHARD ANTHONY
REDDING WALTER LEE
REED MARION EUGENE
REEVES DOYLE WELLS
REGISTER BILLY ELWOOD
REHBERG JAMES HERBERT
REID LEROY JR
REILLY RAYMOND PATRICK
REIN CHARLES FREDERICK
REMBERT LESLIE EUGENE
REYNOLDS DONALD J
RHODEN TALMADGE
RHODES CLIFFORD G
RHODES DAVID FREDERICK
RHODES JOHN OWEN
RHODES RONALD JAMES
RHODES WILLIE MICHAEL
RHUE CHARLES RUSSELL
RHYNES GLOUSTER
RICE GEORGE WARREN
RICE MCKINLEY JR
RICE MICHAEL RAY
RICHARD NORMAN LEO
RICHARDSON ARPHALIA L JR
RICHARDSON SCOTT DOUGLAS
RICHARDSON WILLIAM L JR
RICKEL DAVID J
RICKELS FREDERICK DALE
RIGDON RONALD MICHAEL
RIGGS DORSE
RIGSBY RANDY MARVIN
RILEY BOBBY LEE
RILEY DON ROBERT
ROBERSON JOHN TARRY
ROBERTS ALBERT FRED
ROBERTS CLAUDE
ROBERTS DAVID JOHN
ROBERTS FRANK JAMES
ROBERTS KERMIT BRUCE
ROBERTS LESTER LEE
ROBERTS PAUL MICHAEL
ROBERTS WAYNE LEROY
ROBERTSON JOHN CHESTER
ROBINSON SHEPPARD JR
ROBINSON TOMMY LEE
ROGERS GARY HENDERSON
ROGERS ROY JAMES
ROGERSON CHARLES ROLAND
ROHN HERSHEL HILLARY JR
ROHWELLER ROBERT TIM
ROKER JONATHAN CECIL
ROLAND CHARLES EDWARD
ROLLE JOHN BERKLEY
ROLLE MELVIN
ROLLER WILLIAM EUGENE
ROLLINGS JIMMY
ROOP FRANK
ROQUE FLORENTINO R
ROSENWASSER LEE EDWARD
ROSS LEWIS DEWAYNE
ROTHAR PHILLIP EDWARD JR
ROUSSEAU RICHARD LEE
ROWLAND JOHN WILLIAM JR
ROYAL JAMES NORMAN
RUGGS RANDALL
RUIS DEWEY DOLEN JR
RUIZ PASTOR FRANCISCO
RUNNELS JAMES MIKEL
RUSHING JAMES MONROE
RUSS JAMES ALVIN
RUSS RICHARD JR
RUSSELL JOE TRAVIS
SAINT CLAIR CLARENCE H JR
SALMELA ROBERT EARL
SAMPLER LEWIS EUGENE
SAMPSON RANDOLPH
SAMS MICHAEL DOUGLAS
SAMUELS ELZIE EUGENE
SANDERS JACK ALAN JR
SANDERS JON HUBBARD
SANDNER ROBERT LOUIS
SANKS JERRY WILLIE RAY
SANSING JERRY RUSSELL
SAPP RONALD ALLEN
SARGENT KENNETH PAGE
SATER REGINALD MARK
SAULS GEORGE HAROLD JR
SAUNDERS FREDERIC C JR
SAUNDERS GEOFFREY D R
SAUNDERS MICHAEL JOSEPH
SAVAS SAM MICHAEL JR

SAWYER WILLIAM LELON JR
SAXTON JAMES HERSHEL JR
SCARBOROUGH RUSSELL WILLI
SCHALTENBRAND WAYNE KEITH
SCHICKEL MICHAEL JOSEPH
SCHMIDT KARL ALBERT JR
SCHMITT FRANCIS BARON
SCHMITT RICHIE HUMES
SCHNEIDER JOHN MILLARD
SCHOFIELD ROBERT LOUIS
SCHOLLARD JOHN ANDREW
SCHRADER RUDOLF AUGUST
SCHRYVER PETER EDWARD
SCHULTZ ERNEST M III
SCHWANBECK KARL WILLIAM
SCOFIELD HARVEY DREW
SCOTT DON RUSSELL
SCOTT HERBERT WILLIAM III
SCOTT JAMES ELVIN
SCOTT JAMES RAYMOND
SCOTT JEREMIAH
SCOTT JERRY NIXON
SCOTT JOSEPH ROBERT
SCOTT PRESTON ROOSEVELT
SCOTT ROOSEVELT
SCOTT WILLIAM NORMAN
SCOVILLE HOWARD JAMES
SCREEN MARVIN EDMUND
SCRIMSHAW WAYNE GREGORY
SCRIVEN SAMUEL T
SCRIVENER STEPHEN RUSSELL
SCRUGGS ALBERT JOSEPH
SEARCY ELTON LLOYD
SEAY WILLIAM WAYNE
SEGERS ROGER DALE
SEITZ CHRISTOPHER RICHARD
SELLERS FLOYD EUGENE
SELLERS JERRY ALAN
SETTLE FRANK LEROY
SEXTON JIMMY CLYDE
SEXTON LUTHER MANLEY JR
SEXTON PHILLIP EDWARD
SHAFER GARY CHRISTOPHER
SHAMP PAUL DAVID JR
SHAPPEE JAMES MONFRE
SHARAR RONALD LEE
SHAW ROBERT FLOYD
SHAW ROY EDWARD JR
SHEFFIELD EARNEST EARL
SHEFFIELD FREDRICK WAYNE
SHEFFIELD JAMES TIMOTHY
SHEFFIELD ROY HENRY
SHELDON KIMBALL HAYES
SHELLEY STEPHEN ANDREW
SHEPARD MORRIS RAYMONE JR
SHEPHERD HARRY JOSEPH JR
SHEPPARD THOMAS EDWARD
SHERMAN KENNETH LARMAR
SHIRLEY HAROLD GENE
SHIVER HENRY ARNOLD
SHIVER RICHARD WAYNE

SHORTER ROBERT LEE
SHUBERT EDWIN LENARD JR
SHUBERT JACKIE ECHOLS
SIKES CHARLES MICHAEL
SIKES THOMAS GARY
SILVER WILLIAM ROBERT
SIMKAITIS ERICH
SIMMONS ROSEVELT JR
SIMPSON BRUCE LAMAR
SIMS CLIFFORD CHESTER
SIMS HENRY JAMES
SIMS JEROME
SINGLETARY ALTON LAMER
SINGLETARY JAMES SAMUEL
SIROIS MAURICE LEO
SISK ROBERT DONALD
SIZEMORE DONALD EUGENE
SIZEMORE ROBERT RALPH JR
SLOAN ARTHUR JR
SMALL EUGENE
SMALLS BERNARD AUGUSTUS
SMALLWOOD THOMAS J JR
SMEDLEY LARRY EUGENE
SMILEY FRANKIE LEE
SMITH ADRIAN JAMES
SMITH ARTHUR WHORLOW
SMITH BARNEY MC COY
SMITH BOBBY LEE
SMITH BRIAN FREDERICK
SMITH CHARLES MARCELLEUS
SMITH DENNIS JR
SMITH DONALD P
SMITH GEORGE HENRY
SMITH GILBERT JR
SMITH HERBERT EUGENE
SMITH JACK STEPHEN
SMITH JAMES HENRY
SMITH JAMES PRATT
SMITH JIMMY HERMAN
SMITH JOHN ARCHER
SMITH JOSEPH FREDERICK JR
SMITH LARRY ELDON
SMITH LEON BOYD II
SMITH LLOYD HENRY
SMITH LUTHER AUGUSTUS
SMITH MILTON FRANCIS
SMITH NEAL ARTHUR
SMITH OTIS THOMAS
SMITH RICHARD ALAN SR
SMITH RICHARD JR
SMITH ROBERT JOE
SMITH WILLIAM
SNITCHLER HOWARD WILLIAM
SOLES DONALD RAYMOND
SOLIS ISMAEL
SOLOMON ECKWOOD HAROLD JR
SOSA-CAMEJO FELIX
SPAINHOWER CLAYTON MARQUI
SPARKS CLIFFORD EDWARD
SPEER RICHARD MICHAEL
SPENCER LEROY JR
SPICER JONATHAN NATHAN-

IEL
SPIVEY JAMES WILLIAM
SPRADLIN ROGER WAYNE
SPRAY VICTOR GENE
SPRINGFIELD THOMAS EARL
SPRINKLE MICHAEL DUANE
SPROTT ARTHUR ROY JR
SQUAIRE JAMES EDWARD
STAFFORD RONALD WADE
STAHL PHILLIP THOMAS
STANLEY MARION HENRY
STARCHER DAVID WAYNE
STARLING WALTER LEO
STAUDT RUSSELL MARVIN
STAVLAS PANORMITIS
STEDMAN PAUL FRANCIS
STEELE ROBERT HUGH
STEIMBACH JOSEPH JOHN
STEPHENS DAVID ALLEN
STEPHENS JOHNNIE PERRY JR
STEPHENS WILLIE DOUGLAS
STEVENS CHARLES WAYNE
STEVENSON THOMAS G JR
STEWART DONNY RAY
STEWART EDWARD SAMUEL
STEWART PETER JOSEPH
STEWART RALPH CARSON
STEWART SAMUEL R III
STODDARD JAN MARTINE
STOKES BARTLEY THOMAS III
STOKES DONNIE MONROE
STOKES JAMES MICHAEL
STONE JAMES MARVIN
STONE JERRY MICHAEL
STOVALL CARL ROGERS
STOVALL WILLIAM DALE
STRAHAN WALTER SPERRING
STRICKLAND BILLIE GRANVILLE
STRICKLAND JAMES S JR
STRINGER ISAAC JR
STROUSE PAUL EDWIN
SULLIVAN GLENN
SUMMERSILL EARL PHILLIP
SUPRENANT CHARLES E JR
SURBER HERBERT DONALD
SUTHERLAND JOHN ALVIN
SUTTON BEN FREDERICK
SWAIN WALTER LEE
SWEAT HERBERT HOOVER JR
SWEET JOHN HARLAN
SWEET ROGER WILLIAM
SWINDLE ROBERT EARL
SZIDOR JOSEPH DANIEL
SZPONDER ROBERT ALLAN
TABER MARTIN LESTER
TANNER KENNETH PAUL
TARRANCE JAMES CURTIS
TAYLOR DAVID STUART JR
TAYLOR GERALD K
TAYLOR JAMES
TAYLOR JAMES EDWARD
TAYLOR JAMES ROBERT
TAYLOR ROBERT LYMAN
TAYLOR WILLIAM EDWARD
TEAL RAYMOND WILSON

"IN MEMORY OF THOSE WHO SERVED"

TEMPLES KENNETH RAY
TERRELL ROBERT EARL
TERRY ROBERT LOUIS
THAIN HARRY LINDSAY
THAMES JAMES FRANKLIN
THEISEN GEORGE DANIEL
THIELE JOHN ARTHUR JR
THIELEN MICHAEL JOSEPH
THIGPEN WILLIE JUNIOR
THOMAS ARTHUR WAYNE
THOMAS CHARLES ELLIS
THOMAS CHARLES F IV
THOMAS FRANK HERBERT JR
THOMAS FREDDIE LEE JR
THOMAS ISAAC JR
THOMAS JAMES RICHARD
THOMAS JESS
THOMAS JIMMIE LEE
THOMAS JOHN DAVID
THOMAS OSCAR LEE
THOMAS OSCAR LOW JR
THOMAS ROBERT JAMES
THOMAS WAYNE EARL
THOMAS WILLIAM ARTHUR JR
THOMPSON FREDERICK C JR
THOMPSON HERBERT LEON
THOMPSON JOHN LEE
THOMPSON LESLIE DALE
THOMPSON MORGAN
THOMPSON WILLIAM DARRELL
THOMPSON WILLIAM JOSEPH
THORNTON ROBERT EDWARD
TIFFANY RAYMOND ELLIS
TILLER WALTER LEON
TIMMONS BRUCE ALLAN
TINDELL JAMES FRANKLIN
TIPTON LYNWOOD AUSTIN
TITTLE WILLIAM EDWARD
TITUS CHARLES M
TOMLINSON JAMES HOWARD SR
TOSH BRENT JOHN
TOWNSEND BRUCE
TOWNSEND CHARLES ROLAND
TRAVER JOHN GROVE III
TRAVERS LOUIS WESLEY
TRIEST LEON BUTLER
TRIPLETT JAMES MICHAEL
TRUE MALCOLM ROSCOE JR

TRUETT QUINCY HIGHTOWER
TRUSHAW JAMES EDWARD
TUCKER CARL WESLEY
TUCKER JAMES ERIC
TUCKER JERRY JAMES
TUGGLE JACK DE WAYNE JR
TURNAGE EARNEST LEE
TURNBULL JOSEPH PARKHILL
TURNER EARL RALPH JR
TURNER LAWRENCE FRANK
TURNER RAYMOND RIVERS
TURNER STEPHEN FREDRICK
TWOMEY RAYMOND LEE
TYNER ROBERT EMMETT
TYNES EARL KENNETH
TYNES GREGORY ALLAN
TYPE WALTER JOHN JR
TYSON CHARLES FLOYD III
UNDERWOOD RONALD EUGENE
UNGARO DOMINIC JR
UNGER DON LEE
VACENOVSKY DENNIS EDWARD
VALDEZ LEROY EDWARD
VALE CHARLES
VALRIE DWIGHT THEODORE
VAN NORMAN JOHN R III
VAN VLECK JOHN JOSEPH
VANDERZICHT JOHN ROBERT
VAUGHN JOHN CARL
VELASQUEZ ROBERT JEROME
VELTMAN TIMOTHY ANDREW
VERNON DONALD GENE
VICKERS DAVID ERWIN
VICKERY FREDERICK M III
VICKREY CHARLES CRAIG
VITCH RALPH ALLAN
VOLLMERHAUSEN JOHN M JR
WALDEN JAMES LARRY
WALDEN JAMES ROBERT
WALDEN MARION FRANK JR
WALDRON JERRY MONROE
WALKER ISUM MERRILL
WALKER JAMES ALFRED JR
WALKER MICHAEL DWAYNE
WALKER ROBERT LEE
WALKER STEPHEN CHRISTIAN
WALKER THOMAS
WALKER WILLIAM WAYMAN

WALL JAMES NEIL
WALLACE CHARLES JR
WALLACE DANIEL LEON
WALLACE JACKIE ELMORE
WALLACE LEMON JR
WALLACE LEROY
WALLER JAMES HARRELL
WALLER ROBERT WILLIE
WALTERS FREDERICK F
WARD JAMES LARRY
WARD JOHNNY NEWTON JR
WARDROBE RICHARD ALFRED
WARNER CHARLES WILLIAM JR
WARNER DOUGLAS LEROY
WASHENIK GARY LEE
WASHINGTON BOOKER THOMAS
WATERBURY RICHARD MEAD
WATSON COLUMBUS JR
WATSON DONNIE EDWARD
WATTS FRANK TAYLOR
WATTS RICHARD JOE
WEBB DANIEL DAVID
WEBER DELBERT ELLIS
WEEKS HOWARD DANIEL
WEHRHEIM LOUIS JOSEPH
WEIMAN EDWARD OTTO
WEINMAN DONALD FREDERICK
WEISS DOUGLAS JOHN
WEISSMUELLER COURTNEY EDW
WELCH DAVID
WELCH JOSHUA JR
WELCH TERRY
WELLS EVERETT EARL JR
WENCKER CLIFFORD L
WENDEROTH GERALD F P
WERNET DAVID PAUL
WEST EDGAR LEO JR
WEST KENNETH WADE
WHALEN ROBERT JAMES
WHIDDON TOMMY LEON
WHISENANT PERRY SHELTON
WHITE BOBBY BLAKE
WHITE CALVIN PERRY
WHITE GEORGE PRESTON
WHITE HARRY RAY JR
WHITE JAMES BLAIR
WHITE JAMES E

WHITE JOHN ARTHUR
WHITE LARIS JR
WHITE ULYSSES
WHITE WILLIAM EDMOND III
WHITMAN JERRY RONALD
WHITMORE JAMES ROBERT
WHITTEN ROBERT EUGENE
WIEBEN OTTO TOM
WIELAND DAVID ERIC
WIGGINS AUBREY ALAN
WIGGINS JOSEPH
WIGGINS RONALD HOWARD
WIGGINS STEPHAN MAX
WIGGINS VERNON MIKELL
WILDES MICHAEL LAYTEN
WILDY SHIRLEY JR
WILHOIT ROBERT STEVE
WILKERSON DAVID LEE
WILKINSON DALE SLOAN
WILKINSON DENNIS EDWARD
WILLIAMS ALEXANDER
WILLIAMS ALFONZIA
WILLIAMS ALLAN JAMES
WILLIAMS BEN HAROLD
WILLIAMS CURTIS F JR
WILLIAMS DEREX JR
WILLIAMS EDDIE LEE
WILLIAMS EDWARD WAYNE
WILLIAMS FRANK CURTIS
WILLIAMS FRANK EMANUEL
WILLIAMS FRANK WAYNE JR
WILLIAMS GENRETT
WILLIAMS JAMES GORDON
WILLIAMS JOHNNIE LEE JR
WILLIAMS JOSEPH PIERCE
WILLIAMS LAMAR LONGO
WILLIAMS LAURENCE E
WILLIAMS LEROY C
WILLIAMS MAURICE THERON
WILLIAMS PHILLIP W
WILLIAMS PONDEXTUER E
WILLIAMS RAY FRANCIS
WILLIAMS ROBERT CURTIS
WILLIAMS ROBERT EARL
WILLIAMS ROBERT EARL
WILLIAMS ROBERT L
WILLIAMS SAMUEL
WILLIAMS THOMAS EARL JR
WILLIAMS WARREN
WILSHER EVERETT NELSON
WILSON HAROLD STANLEY

WILSON JEROME I
WILSON JOHN ROBERT
WILSON RAYMOND WESLEY
WINSLOW WILLIAM DAVID
WINSTON ERNEST GREGORY
WINTERS JEROME CORDELL
WINTERS WALTER RAY
WITHAM KENNETH LEROY
WITHERSPOON JOHNELL
WITTMAN ROBERT KEITH
WOLFE JACK LEE
WOODALL JERRY RUSS
WOODARD JOSEPH
WOODRUFF DONALD COLES
WOODS ROBERT EDWIN
WOOTEN DAVID DARYL
WORLDS JAMES ALLEN
WORRELL JAMES R JR
WRIGHT CLIFFORD DEVON
WRIGHT HENRY BERTRAM
WRIGHT JAMES DAVID JR
WRIGHT JAMES WALTER
WRIGHT JERRY GORDON
WRIGHT LARRY
WRIGHT WYLEY JR
WYATT PHILLIP EDGAR
WYATT ROBERT PAUL
WYNN LEONARD ANDREA
WYNNE PATRICK EDWARD
WYRICK DAVID KEITH
YARBER VERNON LEE
YEAGER LARRY GENE
YORK JOEL CRAIG
YOUMANS DAN RANDEL
YOUNG BARCLAY BINGHAM
YOUNG CHARLIE M
YOUNG DAVID
YOUNG JAMES HOWARD
YOUNG LAURENCE ATWOOD
YOUNG ROBERT EARL
YOUNG ROGER LEE
YOUNG WILLIAM
ZAMORA WILFREDO PANTALEON
ZANE TILDEN BRUCE
ZAPPINI JOSEPH VINCENT JR

Photos by Sue Buxton

Faces of War Memorial

38 Hill St, Roswell, GA, 30075

The haunting and beautiful Faces of War Memorial is a monument to recognize the service members of the Vietnam War. It was dedicated in 1998 and features 50 faces cast in bronze that exhibit ranging emotions representing all of those in the war. It can be found in a municipal park in Roswell, Georgia, just near the town's City Hall in the Memorial Garden located at 38 Hill Street.

The Faces of War Memorial was created by architect Zachary Henderson and sculptors Don and Teena Haugen. The memorial is made of bronze, colonial bricks and Georgia marble capstones while the surrounding plaza and walkways are made of memorial bricks that were available for supporters to purchase.

Besides the 50 faces on the memorial that are cast in bronze, the monument also features a full-size bronze standing soldier extending his arm to hold the outstretched hand of a young girl (also cast in bronze) who is actually positioned outside of the standing wall monument creating a three-dimensional effect that is rather stunning. There is also a waterfall that flows down the backdrop of the 50 faces which essentially gives off a mirror effect where visitors can see their reflections in the memorial. The memorial also features an inscription describing the meaning behind the symbols in the monument. As a whole, the monument stands at 14-feet tall and is 20-feet wide.

The Faces of War Memorial came about when pilot Wes Mc-Cann saw a Vietnam memorial on a military base in Savannah that he thought was subpar. He knew the service members who served in Vietnam deserved better so he gathered support from fellow pilot Al Leland. The two pilots and the small group of volunteers who were out to help the cause raised $200,000 over four years.

Since its dedication in 1998, the Faces of War Memorial is the centerpiece of the town's Memorial Day Ceremony. People from all walks of life gather there annually to take their turn and share their feelings and stories with all present. The memorial is open for visitors daily from 8 a.m. to 8 p.m.

The Names of those from the state of Georgia Who Made the Ultimate Sacrifice

ADAIR SAMUEL YOUNG JR
ADAMS CHARLES HENRY
ADAMS HERBERT NORMAN
ADAMS NEIL JR
ADAMS RUSSELL BYRD
ADDINGTON ZACK TAYLOR
ADKINS WAYNE LAWRENCE
AHOUSE WILLIAM C
ALFRED BRUCE CROCKLIN
ALLEN DAVID MARTIN
ALLEN GUS
ALLEN LARRY MICHAEL
ALLEN RAYMOND
ALLEN TERRY JR
ALSTON KENNITH
ALTMAN DAVID BRANTLEY
AMERSON CARLTON
AMOS FLOYD LEHMAN
AMOS WILLIE FRANK
ANDERSON ARTIS WESLEY
ANDERSON FRANKLIN
EMMETT
ANDERSON GARY
ANDERSON JOHN ERNEST
ANDERSON LARRY JAMES
ANDERSON OLIVER
ANDERSON RONALD CARLIS
ANDREWS HORACE
ANTHONY DAVID MARSHALL
ARLINE SOLOMAN DAVID JR
ARMSTRONG ATWELL ASBELL
ARNOLD MOSES ANTHONY
ARNOLD PHILLIP FRED
ARNOLD WILLARD DAVID
ARTHUR JESSE JAMES III
ARTHUR WILLIAM PRESCOTT
ASHLEY CHARLES R JR
ATHEY BURDER SMITH III
ATKINS DON LARRY
AVERY WILLIAM CALVIN
AYERS WILLIAM HERSCHEL
BACON BARNARD
BAGLEY JERRY
BAILEY GENE THOMAS
BAILEY JAMES ALVIN
BAILEY LARRY WILLIAM
BAKER ELBERT JAMES JR
BAKER GEORGE WAYNE
BAKER JOHN WESLEY JR
BALLANGER ENOCH ANDREW
BALLARD GILBERT FLOYD
BALLEW HENRY JR
BARBER MANNIE ALFRED
BARBRE SAMUEL DAVID
BARKER JACK LAMAR
BARKER OSCAR JR
BARLOW JESSIE LEE
BARNABY ROLAND NATHAN-
IEL
BARNES BERNARD
BARNES ROBERT LEE
BARNES TOMMY LEE
BARRETT DONALD
BARRETT JOHN HAROLD
BARRETT MICHAEL OWEN
BARROW THOMAS MELVIN JR
BASS JACKIE DENNIS
BEAM RAYMOND
BEAUFORD SAMUEL PORTER
BEAVERS CHARLES EVAN
BECK DAVID MICHAEL
BECKWORTH HARLEY DANIEL
BEDGOOD JIMMY
BELCHER HERBERT EUGENE
BELFLOWER JAMES H
BELL JOHN DARVIN
BELLEW GUY LESTER
BEMBRY SNYDER PATTISHALL
BEMIS EARLE JOHN
BENDER GERNOT
BENFIELD DON CURTIS
BENFORD HOWARD GILBERT
BENNETT CHARLES DUANE
BENNETT GEORGE ROGERS
BENNETT JAMES STEPHEN
BENNETT JOHN ARTHUR
BENSON DAVID EUGENE

BERNARD WILLIAM EDWIN
BERRY CHARLIE E
BERRY JOE CLEVELAND
BERRY WILLIAM ARTHUR
BETLEYOUN GOLA CALVIN
BIDDULPH THOMAS ARTHUR
BIGGERS LEWIS LAMAR
BIRD GEORGE ALLISON III
BISHOP EDGAR LEE
BISHOP JOSEPH ADRIAN
BLACK HARVEY
BLAIR KENNETH RAY
BLANKS THOMAS LEE
BLOUNT JAMES CURTIS
BLOUNT JOHN WILLIAM
BOGART CHARLES ROBERT
BOGGS PASCHAL GLENN
BOLES ROBERT MADISON
BOLING CHARLES GEORGE
BOLTON MELVIN
BONNER ROGER LEE JR
BONNER WILLIAM EDWARD
BOREN TOM EDWIN
BOSWELL JOHNNIE LEE
BOWDEN JAMES RONALD
BOWEN HARVEY LEWIS JR
BOWEN ROLAND MICHAEL
BOWMAN DANIEL RAYMOND
BOYNTON FRANK
BOZEMAN DWIGHT ERVYN
BRACKINS ALLEN
BRADLEY STANLEY THOMAS
BRADSHAW FLOYD LEE III
BRANNEN JAMES ROBERT
BRANT RICHARD F JR
BRANTLEY DAVID WATSON
BRANTLEY TROY ELLIS JR
BRASWELL DONNY JOE
BREWER GRADY LEE
BRICKER WILLIAM EDWARD
BRIDGES R B JR
BRIGHAM ALBERT
BRINSON HUBERT F
BRITT BILLY WINFORD
BRITT KENNETH JOHN
BRITT TED DENNIS
BRITTIAN CHARLES HENRY JR
BROCK DANIEL LEE
BROCK JAMES ALBERT

BROCK RANDY HOFFMAN
BROOKS DONALD RAY
BROOKS JOHN WOODRUFF
BROOKS LEON RAY
BROOKS THOMAS JR
BROOKS WILLIE LEWIS
BROWN ALBERT LEE
BROWN BENTON
BROWN CURTIS LEE
BROWN DENNIS ADRAIN
BROWN DONALD WILLIAM
BROWN DOUGLAS
BROWN EDDIE JR
BROWN JAMES ANDERSON II
BROWN JAMES THARPE JR
BROWN JOE DAVID
BROWN ROBERT LEWIS
BROWN ROGER RAY
BROWN TERRY LEE
BROWN TOMMY LEE
BROWN WILLMATT
BROWNING BILL GWINN
BROWNING CLEVELAND
BRUNT ARTHUR LEE
BRYAN DAVID GRADEY
BRYANT DAVID EUGENE
BRYANT FREDDIE JAMES
BRYANT GEORGE EDWARD
BRYSON JOHNNY RAY
BUFFINGTON FRED
BUFFINGTON SAMMY
BULLOCH SAMUEL VIEL JR
BURCH KENNETH EUGENE
BURDETT EDWARD BURKE
BURGAMY ERNIE LEE
BURGESS CLEATIS LYNN
BURGESS WILLIAM C JR
BURNETT RICHARD JAMES
BURNEY CHARLIE LEE
BURNHAM RICHARD FLOYD
JR
BURROUGHS EMANUEL FERO
BURROUGHS JAMES MICHAEL
BUSBY WILLIAM RUSSELL
BUSH JAMES HOWARD JR
BUSH MILTON JACKSON
BUSSEY JIMMY LEE
BYNUM FRANKLIN DELANO
BYRD WALTER FRANK JR

BYRD WILLIAM LARRY
CABE DENNIS STEWART
CAGLE ALLEN JAMES
CAGLE RANDY GRAHAM
CAIL JOHN EDWARD JR
CALDWELL ALLEN HAYES
CALHOUN JOHNNY C
CALHOUN PATRICK PALMER
CALHOUN RODERICK WESLEY
CALL JOHN GRANVILLE
CALLAHAN WELBORN A JR
CALLAWAY ALLAN BROOKS
CALLAWAY LEWIS ANDRES III
CAMACHO GREGORIO MENO
CAMERON JAMES FREDERICK
CAMERON THOMAS STEWART
CAMP ANTHONY LORIN
CAMP JOHN WAYNE
CAMPBELL GEORGE ALLEN
CAMPBELL GORDON ALLAN
CAMPBELL STEPHEN MAN-
TON
CAMPBELL THOMAS DAVID
CAMPFIELD MELVIN
CANNADY WOODROW
MICHAEL
CANNON BRUCE ALTON
CANNON RONALD LAMAR
CAPPAERT JON MICHAEL
CAREY CHARLES B
CARLAN JACK MORRIS
CARPENTER CHARLES ED-
WARD
CARPENTER WILLIAM
JOHNNY
CARROLL JOE DAVID
CARROLL JOHN LEONARD
CARROLL LARRY DAVID
CARROLL ROY ARNOLD
CARTER HAROLD E
CARTER LARRY REAUMAINE
CASE THOMAS FRANKLIN
CASEY JOHNNY DALE
CASTLEBERRY ROY LEE
CAWTHORNE WILLIAM
BAYLES
CHAMBLEE DANIEL LEE
CHAMBLISS LUTHER
CHAMPION JOSEPH

CHANDLER ANTHONY
GORDON
CHAPMAN BILLY
CHARLES BILLY
CHASIN STEPHEN C
CHATMON CHARLES LOUIS
CHESHIRE ALLEN DONIHUE
CHESSER HARRY EDWARD
CHISHOLM JAMES
CHRISTIE JAMES MILLER
CLARK ARTHUR
CLARK BILLY EARL
CLARK CLYDE JR
CLARK HARLOW GARY JR
CLARK JAMES LESTER
CLARK JOHN MICHAEL
CLARK LEM
CLARK RONALD CLEVELAND
CLARK THOMAS
CLARK WILLIAM JR
CLARKSON JAMES LA FAYETTE
CLEMENTS DAWSON
CLEMENTS LONNIE EDWARD
CLEMENTS WALTER LEE
CLEVELAND CLARK EDWARD
CLEVELAND RONALD
CLINE CHARLES WILLIAM
CLINE WILLIAM LOUIS
CLOUD MILAM EDWARD
CLOVIS FRANKLIN
COCHRAN WILLIAM SHER-
WOOD
COFER JAMES TERRELL
COFIELD JESSIE CLIFFORD
COGGINS JAMES TERRY
COKER DENNIS SANDERS
COKLEY TROY WESLEY
COLBAUGH HOWARD LEBRON
COLE TIMOTHY JR
COLEMAN DONNIE
COLEMAN JOHN THOMAS
COLEMAN LENARD
COLLIER ALONZO CARLTON
COLLIER LARRY EUGENE
COLLIER NOAH CHANDLER JR
COLLINS ALBERT
COLLINS DOUGLAS WOOD-
ROW
COLLINS FLOYD EUGENE JR

Georgia

COLLINS FRANKLIN THOMAS
COLON-MOTAS ESTEBAN
CONAWAY LONDON
CONDON FRANK ALLOYSIUS
CONDREY GEORGE THOMAS III
CONE RALPH ASBERRY
CONKLE JOE THOMAS
CONLEY JAMES GRADY
CONNER JESSIE WENDELL
CONNER LORENZA
COOK CHARLES
COOK JERRY ROBERT
COOK JOHN WILLIAM JR
COOK PATRICK HENRY JR
COOPER CURTIS
COOPER JAMES ARTHUR
COOPER JAMES ENNIS
COOPER JOHN RANDOLPH JR
COOPER ULYSSES CORNELIUS
COOPER WILLIAM EARL
COOPER WILLIE A
COPELAND DAVID LEE
COPELAND JAMES RANDALL
CORBETT DONALD JUNE
CORBETT ISAAC JOSEPH
CORDELL RALPH DURWARD
CORDLE DONALD CALVIN
CORN JACK ALVIN
CORRY CHARLES MICHAEL
COUCH JULIAN WAYNE
COUCH ROBERT EDWARD
COUSAR WILLIAM JAMES
COUSIN ROBERT LEE
COVINGTON WILLIAM LEE
COWAN JAMES ALTON JR
COX ALLAN LAMAR
COX JACKSON ELLIOTT
COY JAMES ANTHONY
CRAIG THOMAS RICHARD JR
CRAWFORD HAROLD JEROME
CRAWFORD WALTER NORMAN
CREECH BILLY GENE
CREWS ARTHUR BLAKE
CRISP JOHN HAROLD
CROCKETT JAMES BRANNAH
CROOK JIMMY RAY
CROOM RUFUS RAY
CROW EDWARD DAVID
CROY WILLARD WINSTON
CRUMPTON EUGENE HAYWARD
CRUSE GARY ROBERT
CUMBEE JIMMY DEAN
CURRY JIMMY LEE
DANIEL CANTRELL MONRO III
DANIEL JAMES ROBERT
DANIELL THOMAS FRED
DANIELS EARL JR
DANIELS JAMES DENNIS
DARRISAW CURTIS
DAUGHERTY RALPH OLEN JR
DAVIS ANDREW JAMES JR
DAVIS BILLY CHARLES
DAVIS DANIEL RICHARD
DAVIS ELIGAH LAMAR
DAVIS HERBERT CARSON
DAVIS JAMES LEE
DAVIS JAMES LEROY
DAVIS JERRY LLOYD
DAVIS JERRY VANOID
DAVIS JOHN GAYLEALON
DAVIS LARRY FRANKLIN
DAVIS MARVIN ROYCE
DAVIS RODNEY MAXWELL
DAVIS WILLIAM DEWITT
DAWSON THOMAS PHILLIP

DE LOACH DAVID LLOYD
DEACON JAMES DALA
DEESE JACK DEMPSEY
DELAIGLE THEUS EVERETTE
DELANEY WARREN C
DENNIS WALTER KENON
DEVINE JOHN WILLIE
DICKENS DELMA ERNEST
DICKERSON THOMAS GERALD
DISMUKE ALBERT ROYCE
DIXON CARLTON LEO
DIXON DONALD WAYNE
DOBYNS RUSSELL MARTIN JR
DODD CHARLES DAVID
DODSON DAVID PAUL
DOMINICK CHARLIE JUNIOR
DORSEY ROBERT LEE
DORSEY ROGER
DOTSON MICHAEL LEE
DOUBERLY JAMES ODEN
DOWD CARTER WAYNE
DOWDELL MARVIN
DOWDY MITCHEL ANTHONY
DOWNS ARTHUR MITCHELL
DOZIER JOHN TILLMAN II
DRURY JACKY LEE
DUCKETT THOMAS ALLEN
DUFF PHILLIP RANDALL
DUKE LARRY WADE
DUMAS OLIVER DEWITT
DUNCAN BENJAMIN WAYNE
DUNCAN GLENN CHRISTIE
DUNCAN ONNIE DAVID
DUPREE BENNY RAY
DUPREE BILL JAKE
DURHAM HAROLD BASCOM JR
DWIGHT WILLIAM LAMAR
DYKES ROBERT LEE JR
EARLES ARTHUR JAMES
EAVES FRANK GEORGE
EBERHART SAMUEL HOUSTON
ECHOLS ROBERT EDWIN
EDWARDS HARRY SANFORD JR
EDWARDS JOHN H JR
ELDER GRADY LEE
ELDRED ROBERT EDWARD
ELLIOTT ANTHONY EDWIN
ELLIS JAMES LEE JR
ELLIS JOE HENRY
ELLIS JOHN PATRICK
ELLISON ALTON LEON
ELROD DAVID LAMAR
ELROD JAMES THOMAS
ELROD WILLIAM CARROLL JR
ELY DANIEL GERARD
EMBREY GRADY KEITH
ENGLAND MICHAEL
ENGLISH WILLIAM WELTON JR
ESTES NEDWARD CLYDE JR
ETHERIDGE JAMES RALPH
EUKEL DAVID DEAN
EVANS ALBERT
EVANS CHARLES JAMES
EVANS FREEMON
EVANS MICHAEL EUGENE
EVANS PAUL RAYMOND
EVANS RUSSELL
EVEREST ROBERT K III
EVERETT CLARENCE E
EVITT WAYNE LEE
FAIRCLOTH ARTHUR CRAIG
FAIRCLOTH ELLIS LOVINE
FAIRCLOTH HENRY FLOYD
FAIRCLOTH JOHNNIE WILLIAM
FALKNER RUFUS PERRY JR
FANN DANNY WAYNE
FARMER BOBBY GENE
FARMER CHARLIE WILL JR
FAUST TIMOTHY RAY
FELTS EUGENE JR
FENNELL ALTON JIMMY
FERGUSON EARL
FETHEROLF LARRY STEVEN
FICKLING ROY EDWARD
FIELDS BOBBY JENE
FIELDS HERMAN THURSTON
FIELDS ROBERT JR
FIELDS WILLIE STEPHEN
FINKEL KENNETH IAN
FINLEY WILLIAM EDWARD
FITZGERALD PAUL L JR

FLOWERS LAWRENCE BUFORD
FLOWERS WILLIAM EDWARD TH
FLOYD ALAN GREGORY
FLOYD ALVIN WINSLOW
FOLMAR HARRIS ALAN
FORBES THOMAS LEROY
FORD CHARLES EDWARD
FORDHAM JERRY LEE
FORDHAM KENNETH CHARLES
FORDHAM RUSSELL CARRELL
FOSTER WILLIE JAMES
FOUCHE PAUL JERRY
FOUNTAIN KENNETH LOREN
FOWLER DONALD RANDALL
FOWLER JAMES ROBERT
FOWLER MICHAEL EDWARD
FOWLER WILLIAM RAY
FOX JAMES DARRYL
FOX RICHARD HERBERT
FOY THOMAS LAMAR
FRADY HARVIE RENA
FRALEY CHARLES ALBERT
FRANCIS OSCAR THOMAS
FRANCIS PAUL JAMES
FRANK HARRY BERNARD JR
FRANKLIN JOHN ALVIN
FRAZIER ULYSSES VAN
FRAZIER WILLIE JAMES
FREEMAN BOBBY
FREEMAN RUBE ALFRED
FREEMAN WILLIE LEE
FRIDDLE KENNETH CLAYTON
FULLER GEORGE RONALD
FULLER JOHN LUTHER JR
FULLER JOHNNY THOMAS
FULLERTON FRANK EUGENE
FURNEY WILLIS LEE
GADDIS FRED AUSBUN
GADDIS JONATHAN ROYAL
GADDY WILLIE GENE
GAINES GREGORY RANDALL
GAINES THOMAS GALE
GALLAGHER JOHN HENRY
GALLOWAY SAM HARRIS
GAMBRELL JOHN LAWRENCE
GANTT JOHNNY EDWARD
GARRETT DONALD WAYNE
GARRETT ERNEST WILLIAM
GARRETT ROBERT EUGENE
GARRETT TOMMIE
GARTH ROBERT WILTON JR
GATES JAMES WALTER
GAY JOHN BEN
GAY WILLIAM ELLIS JR
GAYMAN JOHN DUFF
GAZAWAY CHARLIE TIDWELL
GENTRY DENNIS WAYNE
GENTRY JERRY WAYNE
GIBBS KENNETH SAMUEL
GIBSON JOHN
GIBSON WALTER MURRAH
GILES CLAUDE VERNOR
GIPSON ROBERT PAUL
GITTENS ERIC EUGENE

GLENN LIVINGSTON
GLENN MICHAEL ROBERT
GLISSON JOHNNY WILLIAM
GLOVER DONALD LAWRENCE
GLOVER WILLIE EDWARD
GNANN HENRY MOUZON
GOBER BILLY WILLIE
GODFREY BILLY LAMAR
GODFROY RALPH DONALD
GODOWNS ROY WILLARD
GOEN RICHARD DEAN
GOINGS JOSEPH HUBERT JR
GOLLAHON JOHN DAVID
GOODALE JOSEPH DANIEL JR
GORE CALVIN THOMAS
GOSWICK LARRY EUGENE
GOUDELOCK FORREST
GRAHAM LARRY ALONZA
GRAHAM THOMAS JR
GRANT ED NATHAN LOUIS
GRANT HOUSTON JR
GRANT PHILO DERRICK III
GRAVES LARRY
GRAVES RANDOLPH EDWIN
GRAY CLIFFORD
GRAY HERBERT HOOVER
GREEN DONALD ALBIE
GREEN EUGENE
GREEN HAROLD ALFRED
GREEN JEWELL ROBERT
GREEN RONALD FRANK
GREEN STANLEY NORRIS
GREEN THOMAS HENRY
GREENE CHARLES EDWARD
GREENE JOHN EDWARD
GREENE WILLIE
GRIFFIN BOBBY RONALD
GRIFFIN CARLTON
GRIFFIN GARLAND ALEX JR
GRIFFIN JAMES DONALD
GRIFFIN OSCAR LEE
GRIFFIS JOSEPH E
GRIFFIS WILLIAM ARLAND
GRIFFITH ROBERT SMITH
GRIMES LLOYD HAROLD II
GROOMS RICHARD JAMES
GROSS BILLY MONROE
GROSS STEPHEN RUSSELL
GROVNER ALLEN JEROME
GUTHRIE CHARLES LARRY
GUTHRIE DANNY EUGENE
GUTTILLA CHARLES RICHARD
HADLEY WILLIAM YANCEY
HAGAN JOHN ROBERT
HAIGLER CECIL MORRIS
HAIRE BENJAMIN WAYNE
HALE HOLLIS RAY
HALL ARVEL HUGH
HALL CHARLES WAYNE
HALL ELMORE LAWRENCE
HALL JACKIE WAYNE
HALL PERRY WOODROW JAMES
HALLMAN PAUL TRUVILLE
HALSEY JOHN CALVIN
HALSTEAD STEPHEN LLOYD

HAMBRICK JAMES JR
HAMBY KIRBY LYNN
HAMBY LANNY MAYES
HAMILTON JAMES EDWARD
HAMILTON MARK LELAND
HAMLIN ROBERT WAYNE
HAMMOND BILLY JOE
HAMMOND CAREY JR
HAMMOND HERBERT LEE
HAMPTON HORACE ALVESTER
HANCOCK WILLIAM TYLER
HARDEN ROBERT WESLEY JR
HARDING STEVE ALDMAN
HARDISON ALLEN CARSON
HARDWICK ROBERT CUSPERT
HARMON JAMES EDWARD
HARMON LEWIS ANDREW
HARPER ALLAN G
HARPER CLARENCE EUGENE JR
HARPER JAMES CECIL JR
HARPER JOHN DAVID JR
HARRELL WILLIAM FRANKLIN
HARRIS CLINTON EUGENE JR
HARRIS EDGAR
HARRIS ISAAC
HARRIS JAMES LARRY
HARRIS JAMES THOMAS
HARRIS MARVIN
HARRIS PERRY LEE
HARRISON DONALD LEE
HARRISON JOSEPH
HARRISON LARRY THOMAS
HARRISON LONNIE HUGHLEN
HARRISON RICKY GENE
HART FRED D
HART VERNON
HARTENHOFF DUANE LELAND
HARWELL RONALD EUGENE
HATCHER CARLOS RANDALL
HATCHER RICHARD ANTHONY
HATTEN GEORGE EDWARD
HAUGABOOK WILLIE CLARENCE
HAWES JAMES DALE
HAWKINS RALLS
HAYES DAVID BARTOW
HAYES IVEY JACKSON
HAYSLIP BOBBY VERNON
HEAD BILLY RAY
HEARD EARNEST JR
HEATON TOMMY CALVIN
HELTON JOHN KENNETH
HEMBREE JAMES THOMAS JR
HENDERSON BILLY HUGH
HENRY WILLIE LEE
HERENDON DAVID MACK
HERREN EVERETT DELOY
HERREN RONALD WAYNE
HERRING ELIJAH JR
HERRINGTON JEROME
HERTLEIN GEORGE B III
HESTER MARTIN DAVID

HESTER STEVEN LEWIS
HESTERLEE RAY WESLEY
HEWITT AUBREY LAURENCE
HICKS EARLIE HENRY JR
HIGDON LEONARD THOMAS
HIGGINBOTHAM OSCAR K JR
HIGH THEODORE W IV
HILL LARRY EDWIN
HINES ARTHUR
HINSON RALPH LEON
HIPP JOSEPH EARNEST
HISCOCK STEPHEN MAYO
HOAG JAMES DEAN
HODGES JOSEPH
HODGES RICHARD PRESTON
HOGAN DAVID CLEVELAND
HOLBROOK CARL EUGENE
HOLCOMB DANIEL
HOLLAND EDDIE HERMAN
HOLMES HAROLD ANTHONY
HOLMES JOSEPH
HOLMES LARRY LAMAR
HOLMES NATHAN
HOLT JACK ENGLISH
HOLT RICHARD ANCIL
HOLTON LEON G
HOLTZCLAW PHILIP BRUCE
HOLTZCLAW THOMAS J III
HONAKER RAYMOND KERMIT
JR
HONOUR CHARLES M JR
HOOD CARLTON HARVEY
HOOKS WILEY DEAN
HOPKINS JAMES HARRISON
HOPKINS PERRY BERNARD
HORNE LAMAR
HOUNSHELL JEFFREY DAVID
HOWARD HAZE III
HOWARD JIMMIE
HOWARD LEWIS JR
HOWARD RONALD HERBERT
HOWARD ROY LEE
HOWELL BEN WILLIS
HOWELL CALVIN LAMAR JR
HOWELL DWIGHT BRINSON
HUBBARD LAMAR
HUDDLESTON THOMAS PATE

HUDSON CALVIN CLIFFORD
HUDSON JOE DAVID
HUDSON ROY
HUDSON SAMUEL BERNARD
HUFSTETLER JAMES THOMAS
HUGHES CARL PATRICK
HUGHES CARL WAYNE
HUGHES JESSE HOWARD
HUGHIE WARNER PRATER
HULSEY LARRY BRYSON
HULSEY ROGER
HUMPHREY JOHNNY WIL-
LIAM
HUMPHRIES BENNIE FRANK
HUNTER KENNETH RONALD
HUNTER ROBERT GERALD
HUNTER WASHINGTON
HURSTON HUGH LARRY
HUTTO CURTIS WOODROW
HYATT JOSEPH LAMAR
HYDE LLOYD PATTERSON
HYMAN WALLACE
IDLETT JAMES
INGLETT GERALD WAYNE
INMAN JAMES WESLEY
IRELAND LEROY
ISBELL MARSHALL HOWARD
IVEY HERMAN FRED
JACKSON ALLEN VERONE
JACKSON CALVIN OTIS
JACKSON CLARENCE
JACKSON FRED ORR JR
JACKSON JAMES WESLEY JR
JACKSON TERRY KENT
JACKSON WILLIE
JACOBS WILLIE BREWSTER
JAEGER JULIUS PATRICK
JAMES BOBBY
JAMES DAN NINKEY
JAMES MARK EVERETT
JAMES SAMUEL REESE II
JANCA LOUIS EMIL
JANES NICKLOS BYRON
JARRARD JERRY EDWIN
JARRELL SIDNEY WADE
JARVIS WILLIAM THOMAS
JAY BRIAN EDWARD

JEFFERSON NELSON JR
JENKINS ANTHONY LEROY
JENKINS BARNETTE GAR-
TRELL
JENKINS ROBERT WILEY
JENNINGS ELTON LEE JR
JINKS RAYMOND ARTHUR
JIVENS JERRY
JOHN WILLIAM THOMAS
JOHNSON CHARLES BUFORD
JR
JOHNSON CLIFFORD THOMAS
JOHNSON DAVID LEE
JOHNSON EMORY FRANKLIN
JOHNSON FOREST DENVER JR
JOHNSON FREDDIE LEE
JOHNSON GEORGE MILTON
JOHNSON HORACE JR
JOHNSON JERRY
JOHNSON JIMMY ALVIN
JOHNSON MARION EDWARD
JOHNSON MELVIN
JOHNSON MILO PRESTON
JOHNSON MILTON
JOHNSON ROBERT MILTON
JOHNSON RUSSELL LEE
JOHNSON TOMMY
JOHNSON WALTER BILLY M
JOHNSON WILLIAM C JR
JOHNSON WILLIE C JR
JOHNSTON DONALD RAY
JOINER WILLIAM FRANKLIN
JONES ALLEN WINFRED
JONES BOBBY JOE
JONES BOBBY MARVIN
JONES BOYD EUGENE
JONES CLEVELAND DAVID
JONES CURTIS
JONES DANIEL CLEVELAND
JONES GARY C
JONES GEORGE EDWARD JR
JONES HALCOTT PRIDE JR
JONES JAMES E
JONES JAMES WALTER
JONES JOSEPH BARRY
JONES JOSEPH MELVIN
JONES LARRY PAUL

JONES LEROY ELTON
JONES LUTHER M
JONES LYNN
JONES MITCHELL JR
JONES ROBERT LEE
JONES RONNIE JOE
JONES RUSSELL JR
JONES STERLING M JR
JONES THOMAS HOWARD
JONES THOMAS STEVEN
JONES WILLIAM JR
JORDAN JERRY KENNETH
JORDAN JOSEPH LAMAR
JOSLYN JAMES EUGENE
JUNKINS JOHNNY JUERGEN
JUSTICE RICHARD LEE
KAYS JAMES G
KELLEY NATHANIEL
KELLY CHARLES L
KELLY JOEL RAY
KELLY PAUL EDWARD JR
KELLY STEPHEN ALLEN
KELSALL BILLY ALLEN
KENDRICK HOMER PHILLIP
KENERLY WARREN EUGENE
KENT WAYNE LEE
KERR JAMES CLAYTON
KESLER LAWRENCE DAVID
KEY HULUS EDGAR JR
KILE JOHN TERRENCE
KILLINGSWORTH SCOTT E
KILPATRICK LARRY RONALD
KIMBLE EDDIE CLAUDE
KIMBRELL GORDON T JR
KINASZ MONTE CLIFFORD
KINES EDWARD WRAY
KING ALEXANDER
KING CHARLES RAY
KING DE WAYNE
KING EDWARD EARL
KING JOHNNY
KING LARRY EUGENE
KING MICHAEL ELI
KING RAYFORD HENRY
KIRBY BOBBY ALEXANDER
KISER LEON EMMANUEL
KITCHENS HARRY MOSS

KITCHENS PERRY CASTEL-
LION
KNIGHT BILLY
KNIGHT TROY LEE
KNOWLES NATHANIEL
KOSTER KENNETH LEROY
KRETSCHMANN WOLFRAM J
KROBOTH STANLEY NEAL
KROEGER JOHN CURTIS
LAMB BRICEY ELROD
LAMB DONALD CAROL JR
LAMB LARRY NESBIT
LAMBERTON GEORGE MAGEE
II
LANCE SAMUEL STEPHEN
LANE THOMAS
LANG MAINOR DAVID JR
LANGFORD ROBERT CANDLER
LANGLEY BILLY GUINN
LANIER JAMES ARTHUR
LASLIE JOSEPH TAYLOR JR
LASSITER DAVID STEVEN
LATINI GERALD LEOPOLD
LAWRENCE DENNIS ROLAND
LAWRENCE LARRY EUGENE
LAY GENE WENDELL
LEACH DOUGLAS HORACE
LEACH ROY HAMILTON
LEAGUE ROY BARRY
LEARY SOLOMON
LEAVELL MELVIN RANDOLPH
LEBOFF JOHNNY HANS
LEE ANTHONY IRVIN
LEE HOMER VIRGIL
LEE MOSES CALVIN
LEE TRAVIS BERTRAND JR
LEIGH NEWELL FERRELL JR
LESTER JAMES THOMAS
LEWIS NATHANIEL
LEWIS ROBERT ALAN
LIGHTSEY DANNY LEE
LIGHTSEY JOHN HENRY
LINDBLOOM CHARLES DAVID
LINDSEY EDWARD BYRON
LINDSEY WILLIAM JEFFERSON
LINSE KENNETH DAVID
LIPSCOMB MELVIN

Georgia

LITTLE DONNIE HUGH
LLOYD ORLAND THOMAS
LOCKLAR TED T
LONG DOUGLAS LEONARD JR
LONG HERLIHY TOWNSEND
LONG JOHNNY F
LORD ARTHUR JAMES
LOTT DOUGLAS HUGH JR
LOWE JERE RONE
LOWERY DONALD STEVEN
LUMPKIN HARRY JAMES
LUSE MICHAEL JOHN
LYNN WELDON GEORGE
MACK JOSEPH BINGHAM JR
MACK WILLIE EDWARD
MACKEN CHARLES DAVIS
MACKEY VERTIS L
MALLARD MORRIS A JR
MALLOY JOHN PERRY
MANAC DON
MANN EDDIE MORRIS
MANN JOHN HAROLD
MANNING RALPH E
MANNING RALPH EDWARD
MARBUTT GARY THOMAS
MARION CURTIS
MARKS TOMMY LEE
MARTIN BENNIE LOUIS
MARTIN CHARLES THOMAS
MARTIN HENRY OLIN III
MARTIN HENRY RONALD
MARTIN JOHN WARREN
MARTIN WILLIAM HAROLD
MASSEY MICHAEL JAY
MATHIS ROGER EDWARD
MATHIS RONNIE THOMAS
MATHIS RUBIN II
MATTOX JOHN RICHARD
MATTOX WILBUR FLORENCE
MAXIE NORMAN
MAXWELL CHARLES D
MAYBURY THOMAS VINCENT
MAYER JUERGEN AUGUST
MAYO MARVIN LACY
MAYS EMMITT JR
MC BURROWS WENDELL D
MC CADEN JAMES LEE
MC CAN CLAUDE JR
MC CARLEY CHARLES D JR
MC CASKILL WILLIAM
MC CLURE THURLO MERIDA
MC COLLUM ROBERT HENRY
MC COOK GREGORY MADISON
MC CRANIE DAVID CARROLL
MC CRARY CLIFFORD PAUL
MC CUAIG GLENN RICHARD
MC DANIEL JERRY JACKSON
MC DANIEL MORRIS LAROSCO
JR
MC DONALD JOHN ETHRIDGE
MC DONALD THOMAS R JR
MC DOWALL FRANCIS JR
MC DOWELL GERALD LEE
MC DURMON CALVIN LAVON
MC ELRATH WINSTON JR
MC GEE DANNY DEAN
MC GEE SAMUEL RUSSELL III
MC GILL ROBERT ANDREW
MC GRUDER EDWARD
MC GUIRE WILLIAM EDGAR
MC HELLON GEORGE S
MC KENZIE JACKIE RAY
MC KENZIE JOHNNY RAY
MC KENZIE WILLIE JAMES
MC KIBBEN RAY
MC KINLEY WAYNE HOUSTON
MC KINNIE HERMAN

MC LAMB HARRY LAWRENCE
MC LENDON RALPH WERNER
MC MURRAY ODIE C
MC WILLIAMS ROY M
MCDOWELL WILLIAM CLAYTON
MCRAE DAVID EDWARD
MCWHORTER HENRY STERLING
MEADOWS ARTIS WILBUR JR
MEADOWS ROY LESTER
MEANS VERNON
MEARS GUY LAMAR JR
MEGAR HERBERT LEONARD JR
MELTON CLIFFORD DEAN
MELTON DENNIS CAROL
MELVIN STANLEY TRACY
MERCER JIMMY HENRY
MERCHANT LONNIE VESTER
MERRITT JOHN CLINTON
MEWBORN WESLIE DAVID
MICHAEL JAMES ALBERT
MILAM WILBUR LAWRENCE
III
MILES PALMER BEACH
MILLER CHARLES RUSSELL
MILLER HERBERT
MILLER J D
MILLER ROY WALDO JR
MILLIRONS JAMES EUGENE
MIMBS BILLIE
MINCEY DORSEY III
MINER WILLIAM DAVID
MINNITEE JOHNSON JR
MITCHELL FRED W JR
MITCHELL GARY HENTON
MITCHELL LARRY LEON
MITCHELL LEROY
MOBLEY WILLIE ROY
MOLDAVAN EDWARD A
MONFORT BENNIE FRANK
MONSON JOSEPH
MOOG PHILLIP JACOB
MOORE CLYDE VERNON
MOORE EDWARD LAMAR JR
MOORE JAMES ROBERT JR
MOORE KENNETH DEE
MOORE LAWRENCE MICHAEL
MORGAN GREGORY SCOTT
MORGAN JACKIE MARCELL
MORGAN LUTHER JR
MORGAN MAJOR BOONE JR
MORRIS JAMES LOGAN
MORRIS JIMMY TONY
MORRIS JOHN NATHAN
MORRIS LEON LOPEZ

MORRIS RAYMOND MURPHY
MORRIS ROBERT DAVIS
MORRIS ROBERT WESLEY
MORRIS THOMAS HALL
MOSLEY ROBERT LEE
MOSS CHARLES NATHAN
MOTES JAMES JACKSON
MULKEY TERRY LEE
MULLIS CHARLES EDWARD
MULLIS MARVIN BURNETT JR
MURPHY DAVID
MURPHY LARRON DAVID
MURPHY PAUL WILLIAM JR
MURPHY ROY LYNWOOD
MURRAY LARRY DONNELL
MURRAY MICHAEL VAN
NASH THOMAS STEVEN
NASWORTHY MALVIN LOWE
JR
NATIONS MICHAEL CLAY
NAVARRO JAMES LEE
NEAL CHARLIE THOMAS
NEELEY EDDIE JOE
NEELEY LOWRENZO
NELSON MILES HENRY
NELSON RAYFORD
NESBITT JOHNNY B
NESMITH NEWMAN RAY
NEVILLE PATRICK MICHAEL
NEWBERRY GRAYSON HENRY
NIPPER DAVID
NORRID HOLLIS RONNEY
NORTON DAN BAKER
NORTON JOHN EMORY
NORTON MITCHELL EARL
O KEEFE TIMOTHY JOHN
O NEAL TONY LEE
ODEA THOMAS PATRICK
ODOM STEVEN CRAIG
ODUM JOSEPH BRITTON
OGLES KENNETH WAYNE
OLIVER DENNIS JERROD
OLSON CARL ANDREW
OLSON STEVEN ALLAN
OVERSTREET ROGER WAYNE
OWENBY CLYDE
OWENBY EUGENE OLIVER
OWENS THOMAS RUDOLPH
PADGETT ROBERT JERRY
PAINTER GARY WILLIAM
PALMER CARL LEE
PALMER LEON ALTON
PALMER MILLARD LAMAR
PANNELL WALTER THAXTON
PARHAM JOHN HOLT III
PARKER DAVID WAYNE

PARKER DONALD FREDRICK
PARKER PERLUM M JR
PARKER RICHARD ANTHONY
PARKER WAYMON M
PASLEY HENRY
PATE GARY
PATRICK JIMMIE LEE
PATRICK JIMMY RALPH
PATRICK MARION ELIJAH
PATTERSON JOHNNIE
PATTERSON MAXIE
PATTERSON THOMAS
PATTERSON WAYNE O NEAL
PATTILLO MINOR WESLEY
PAYNE GARY LEE
PAYNE HOWARD DAVID III
PAYNE LLOYD ADRIAN
PAYNE RONALD HARRY
PAYNE TROY DAVID JR
PEACOCK LEONARD EARL
PEARSON JOHN RUDOLPH
PELHAM EARL TIMOTHY JR
PENMAN JOHN RICHARD
PERRY CLYDE RANDOLPH JR
PERRY DENNIS MITCHELL
PERRY EARNEST
PETERSON ALBERT EUGENE JR
PETERSON JOE LEE
PFLASTERER GEORGE ROBERT
PHELPS HERBERT LEE
PHILLIPS RICHARD BRUCE
PHILLIPS ROBERT LITTLETON
PHILLIPS WILLIAM LEROY
PICKARD RICHARD JAMES
PICKETT MORRIS CALVIN
PIERCE PHILLIP MALCOLM JR
PIERCE ROBERT JAMES
PINSON LARRY GUNNELL
PIPER ROBERT ANTHONY
PITMAN PETER POTTER
PITTMAN EDGAR STEVAN
PITTMAN WILLIAM T
PLUMLEY JIMMIE LEE
PLUMMER SAMUEL RUDOLPH
POE JERRY LYNN
POOLE EARL LEROY
POOLE MELVIN
POOLE ORIS LAMAR
POPE EMMETT FELTON JR
PORTER OSCAR KILPATRIC JR
PORTER ROBERT LEE
POTTS BARTOW WESLEY JR
POUGH EDDIE LEE
POWELL CHARLES THOMAS
POWELL LARRY GENE
POWELL MICHAEL ANTHONY

POWELL MORRIS JAMES
POWELL RAYMOND ALAN
POWELL WILLIAM
PRATT FRED OMAR
PRATT RODNEY TERRANCE
PRESLEY JAMES HENRY
PRESSLEY JAMES EDWARD
PRESTON JAMES ARTHUR
PRESTON JOHNNY CALVIN
PRICE RONALD BRUCE
PRINCE JOSEPH STEPHEN
PRITCHETT CARL WAYNE
PROCTOR SAMUEL JR
PULLINS ROGERS JR
PULLUM HENRY JR
PURSER DAVID ARTHUR
PUTNAM CHARLES RICHARD
QUEEN DONALD WAYNE
RABB ROBERT IRA
RAINEY WILLIAM GEORGE
RAINWATER WILBUR DEAN
RAMPLEY CHARLES HOWARD
RAMSEY ROCKE DARRELL
RANDALL JAMES GARY
RANGE THOMAS RONNIE JR
RAULERSON CLIFFORD H JR
RAY DURWARD FRANK
RAY JAMES FLOYD
RAY WILLIAM COTTER
REACH WILLIAM THOMAS
REECE RONNEY DEAN
REED JERRY DONNIE
REESE JAMES ROBERT
REESE JAMES WILLIAM
REGISTER MAXIE DEAN
REGISTER ROY CARROLL
REICH MERRILL DALE JR
REID BENJAMIN HERSCHELL
REID DANNY ELIE
REID DAVID DONALD
REID JOHNNIE GENE
REID ROBERT WOODSON
REINEL RUSSELL EDWARD
RELEFORD ISIEAH JR
RESTREPO JAIME
RHODES JAMES ROBERT
RICE CLAUDE
RICE DONALD JEROME
RICHARDS DON JUNE
RICHARDS RONALD
RICHARDSON DANNY JOE
RICHARDSON EUGENE III
RICHARDSON ROBERT
RICHARDSON WILLIE LEE
RICHEY THOMAS EARL
RICKERSON ALBERT LEONARD

RICKERSON JAMES EDWARD
RIDDLE LARRY RAY
RIDER ARNOLD TILMAN
RIMES TERRY MARTIN
RISHER CLARENCE T III
RITCHIE EARNEST DEE
RIVERS JOHN WILSON
ROBBINS WILLIAM D
ROBERTS BEN
ROBERTS CHARLES W JR
ROBERTS LONNIE BARRY
ROBERTSON JOHNNY BILL JR
ROBINSON HERMAN DAVID
ROBINSON JOHN CALVIN II
ROBINSON WILLIE CLYDE JR
RODGERS JAMES HAMILTON
ROGERS CLARENCE W JR
ROGERS EDWARD LEROY
ROGERS JERRY EUGENE
ROGERS ROBERT RUSSELL
ROLAND JAMES CURTISS
ROPER CLAUDE TILLMAN
ROPER JOEL CLYDE
ROSS SANDY LEE
ROSS WILLIAM ALLEN
ROTTON JOHNNY STEVE
ROWE MICHAEL THOMAS
ROWELL ROGER JAMES
ROWLAND THOMAS W
ROYAL WILLIAM EARL
RUCKER JOHN MARSHALL
RUDOLPH RICHARD JOSEPH
RUDOLPH ROBERT GEORGE
RUIS FRANKLIN DWIGHT
RUMSEY MELVIN DARRYL
RUSH ROLAND EDWARD
RUSHIN LESTER
RUSHING KENNETH ROGER
RUSSELL CHARLES M III
RUSSELL PATRICK ANTHONY
SADBERRY BENJAMIN
SAGE REX RUSSELL
SANDS WILLIAM D III
SANFORD JAMES IRA
SANGSTER ROBERT LEONARD
SAPP WILLIAM EDWARD
SAULS ROBERT NED
SAVACOOL PAUL ROSS JR
SAVAGE JAMES TERRY
SAXON CLYDE EDWARD
SCALES DOUGLAS
SCARBOROUGH GEORGE
THOMAS
SCHOEPFLIN CHARLES
DUAINE
SCOTT ARTHUR EDWARD
SCOTT CLARENCE WALTER
SCOTT EDDIE JAMES
SCOTT JOHNNY FRED
SCOTT LARRY
SCOTT LEONARD STANLEY JR
SCOTT WILLIE CHARLES
SCRUGGS STUART JACKSON JR
SEGARS TOMMY QUINN JR
SEPULVEDA LAWRENCE
KENNET
SEWARD WILLIAM HENRY
SEXTON WESLEY ROBERT
SHAFFER EARL THOMAS SR
SHANNON EARL EDWIN
SHANNON JESSIE EDWIN
SHANNON LEROY JR
SHARMAN CHARLES W III
SHARP LUFKIN SCOTT
SHARPE MACK DONALD
SHARPLESS JOHN PAUL
SHAW JOHN ANDY
SHELTON CHARLES MURRY
SHERFIELD BRUCE JR
SHERIFF JAMES CHARLES JR
SHOOK ROBERT LYNN
SHULER HAROLD WILLIAM
SHUMAN ERNEST MAXWELL
JR
SHUMATE WILLIAM CLAYTON
SILLS TOMMIE LEE
SILVER LONNIE LEE
SIMMONS BURNELL
SIMMONS CHARLIE JR
SIMMONS EDWARD LAMAR
SIMMONS FRANK RUDOLPH
SIMMONS NATHAN BEDFORD
SIMPSON JOHN WILLIAM JR

SIMPSON JOHNNY CLEVE-
LAND
SIMPSON OTIS RAYMOND
SIMS CHARLES WAYNE
SIMS EDWARD CLEO
SIMS JAMES LARRY
SINGLETARY ROY LEE
SINGLETON JESSE W JR
SIRMANS ALBERT WILSON JR
SIRMANS RUFUS
SITTEN JOHNNY WAYNE
SIZEMORE WILLIAM D
SKIPPER JAMES EARL
SKRINE WILLIE B JR
SLATON ALVIN MAYNARD
SLOAN LEWIS LEONARD
SLOAN MAX EUGENE
SMALL BURT CHAUNCY JR
SMALLWOOD ERRAL DALE
SMALLWOOD JOHN JACKIE
SMARR KENNETH WAYNE
SMILEY JIMMIE TAVY
SMITH BILLIE HAYWOOD
SMITH BILLY EUGENE
SMITH CAREY WAYNE
SMITH CHARLES DANIEL
SMITH CHARLES EUGENE
SMITH CHARLES HERBERT
SMITH CHARLES LEE
SMITH DEAN JR
SMITH DONALD EUGENE
SMITH DONALD LAMAR
SMITH EDDIE LEE
SMITH EDGAR LARUE
SMITH EMORY MOREL
SMITH FORREST LLOYD
SMITH FRED WINSTON
SMITH GALEN MINOR
SMITH HENRY FLOYD
SMITH HERBERT JR
SMITH JERRY LYNN
SMITH JESSE E
SMITH JOE CLARENCE
SMITH JOHN RAYMOND
SMITH KENNETH DOUGLAS
SMITH LARRY DEAN
SMITH LEONARD HOWARD
SMITH MICHAEL ANTHONY
SMITH ROBERT JOSEPH
SMITH RONALD LARRY
SMITH RUSSELL LAMAR
SMITH THOMAS CLINTON JR
SMITH TOMMY LEE
SOGNIER JOHN WOODWARD
JR
SOLIVAN LOUIS
SOLOMON LEAVY CARLTON
SORRELLS BOBBY HORACE
SOUTHERLAND VERNON
DAVEY
SOWELL DONALD BRITTON
SPEARS BENJAMIN GEORGE
SPEIGHT WILLIAM ROBERT
SPENCE JAMES MAYNARD
SPERRY WILLIAM FORSYTH
SPILLERS GEORGE THOMAS
SPIRES ROBERT LEE
SPIVEY EDDIE LEE
SPIVEY JOHNNY WAYNE
SPOTANSKI SERGE WALTER
SPREWELL JOHN SPURGEON
SPROUSE LONNIE DAVID
SPURLIN DANIEL RAYMOND
ST GERMAINE RONALD
HUBERT
STAINES ERNEST MICHAEL
STALEY RONALD ALEX
STALEY THOMAS W JR
STALNAKER LAWRENCE
ARNOLD
STANFORD ERNEST LEE
STANLEY VIRGIL JR
STANSBURY RAYMOND L II
STAPLES THOMAS TRAMMEL
II
STARLEY JAMES ARTHUR
STARNES CULLEN GEORGE JR
STEGALL ALLAN JR
STEGALL ALTON LESKER
STEWART HOWARD WARREN
STEWART JAMES JR
STEWART WAYNE YEARWOOD
STINSON WILLIAM CLYDE JR

STOKES GUY LYNN JR
STOKES WAYNE JOSEPH
STONE DAVID
STONER JAMES CONLEY
STOWERS BENNY THOMAS
STREVEL WILLIAM COLUM-
BUS JR
STRICKLAND RANDY ALBERT
STRICKLAND WAYNE THAD
STRINGER ANTHONY ODELL
STROBO HENRY RONALD
STUCK LAWRENCE MILTON
SUIT GROVER LYNN
SULLIVAN EDDIE LEE
SUTERA LOUIS JR
SUTHERLAND BOBBY COL-
LINS
SUTTER RICHARD FURLONG
SWAIN TOMMY HERMAN
SWANCEY RANDALL FILL-
MORE
SWANN JOHNNY DELBERT
SWEAT NORMAN ROGER
SWIM PAUL EUGENE
SWORDS JOHN ARTHUR
SWYMER GEORGE T
SYKES DON CARLOS
TABB PHIL
TABOR CLIFFORD JR
TALLEY JAMES LANE
TANKSLEY ROBERT WILLIE
TANNER WILLIAM LA MARR
TATNALL CLYDE BENJAMIN
TATUM DORSEY L
TATUM HAROLD DEAN
TAYLOR CALVIN LEROY
TAYLOR DAVID ADOLPHUS
TAYLOR DEANE ARTHUR JR
TAYLOR JAMES ALTON
TAYLOR JERRY LEE
TAYLOR JIMMIE ELLIS
TAYLOR LANDUS S JR
TAYLOR NORMAN ALFRED
TAYLOR ROBERT ALLEN
TAYLOR ROBERT L
TAYLOR ROYNALD EDWARD
TEBOW WILLIAM JENNINGS
TEMPLETON BILLY
TENON JOHNNIE MERRITT
TERRELL WILLIAM LEE
TERRY EDDIE THOMAS
TERRY PATRICK WAYNE
THACKER GRADY
THAXTON JOHNNY R JR
THOMAS ALLISON LEWIS JR
THOMAS CHARLES FRANKLIN
THOMAS DAVID EUGENE
THOMAS FRED L
THOMAS JAMES RICHARD
THOMAS NATHAN
THOMAS OTHEL
THOMAS WILLIAM HENRY JR
THOMPSON MELVIN CARL
THOMPSON ONNIE JR
THOMPSON ROBERT AC-
QUINN
THOMPSON ROGER DARRIEL
THORNTON CARL LEE
THORNTON DWIGHT JACK-
SON
THORNTON FRANK JR
THORNTON LYNWOOD
KEETON
THURMOND EDWARD SCOTT
TIDWELL ROBERT PAUL
TIGNER JOHN HENRY
TILLMON WILLIE SANDFORD
TODD JOHN ANDREW
TODD LARRY RICHARD
TOLER ROBERT WILBER JR
TOMLINSON CLEMMIE JAMES
TOMPKINS HAROLD
TOOMBS WILLIAM HAYWARD
TRAIN WILLIAM FREW III
TREADWELL MILLARD LEON
JR
TRIBBLE PRESTON JR
TROUP RODRICK
TUCKER EARNEST ALFRED JR
TUCKER OLLIE
TUCKER RICHARD EUGENE
TUGGLE LORENZO
TURNER JAMES MACK

TURNER MICHAEL BARRY
TURNER MILAN ELLIOT
TURNER PHILIP GERALD
TURNER STANLEY
TUTEN RICHARD BAILEY
TYSON HAROLD RAY
TYUS JAMES DREWERY JR
ULMER HOWARD D JR
UR STANLEY EUGENE
USHER FREDDIE
VALLE FRANCISCO LOUIS
VAUGHAN EGBERT R
VAUGHAN HARRY KENNETH
VAUGHN CLAUDE FRANKLIN
VAUGHN JOSEPH DOUGLAS
VAVRIN FRANK NEAL
VESSELL WAYNE JACKSON
VICKERS ROBERT LEE
VOLLRATH JOHNNY DEWEY
WADE JAMES LEE JR
WADE NATHANIEL
WADLEY JOE ALTON
WAGES JERRY LEON
WAGNER DAVID LEE
WALDEN DAVID
WALDING JARED BRUCE
WALKER CHARLIE C
WALKER CLIFTON
WALKER HARDEN BERT
WALKER NEELY CLARENCE
WALKER WALTER LEWIS
WALKER WILLIE B JR
WALL GEORGE ROBERT
WALLACE GLEN EVERETT
WALLACE ROOSEVELT
WALLACE WILLIAM THOMAS
JR
WALLER JERRY GORDON
WALLIN DENNIS RAY
WALLS JOHN THOMAS
WALTERS FREDRICK STEPHEN
WALTHALL CHARLES EDWARD
WALTON HAROLD LEE
WARD GARY EDWIN
WARD TONY ROBERT
WARE GENE LAMAR
WARREN JOHN OWEN
WARREN ROBERT GLENN
WASHINGTON JAMES C JR
WASHINGTON JOHN WILLIE
WATERS MICHAEL ROY
WATKINS BRUCE LAMAR
WATKINS EARL WELDON JR
WATSON JOSEPH MICHAEL
WATSON THOMAS ARTHUR
WATSON TOMMIE
WATSON ULMER JOE
WATTS HENRY LINCOLN SR
WEAVER CLINTON JAMES
WEAVER JACK
WEAVER JOHN HERBERT
WEAVER JOSEPH ROBERT JR
WEAVER PHILIP WARREN
WEBB LEROY BOYD
WEBB OLIVER KENNETH
WEBB ROBERT MITCHELL JR
WEHUNT BILLY DEAN
WEISNER FRANKLIN LEE
WELLS BOBBY GENE
WELLS CONNIE VERGEL
WELLS JAMES RANDALL
WENRICK PHILIP BRUCE
WESLEY ERNEST LAMAR
WESSON LANNY LAMAR
WEST EUGENE EDWARD
WEST JAMES LARRY
WESTER WILBURN EDWARD
WESTON THOMAS JR
WHEELER JOHN MELVIN
WHEELER JOHNNY CECIL
WHEELER LARRY KENNETH
WHITE ALBERT DEWELL
WHITE HAROLD LEE
WHITE LEROY JR
WHITE NATHAN MONROE
WHITE RANDALL RAY
WHITE ROBERT RANDOLPH JR
WHITE THOMAS MITCHELL
WHITEHEAD CLARENCE
ALBERT
WHITEHEAD JEFF
WHITFIELD JAMES LEMAR
WHITFIELD LAWRENCE

ALLEN
WHITWORTH SAMMY HOW-
ARD
WILBANKS HILLIARD AL-
MOND
WILCOX JOHN ARTHUR JR
WILDER CHARLIE LARRY
WILEY JAMES JOSEPH
WILLBANKS CHARLES ED-
WARDS
WILLIAMS ARTHUR JR
WILLIAMS CAL WILLIS
WILLIAMS CHARLES CLINTON
WILLIAMS CHARLES ROSS
WILLIAMS DAVID JR
WILLIAMS DONALD
WILLIAMS FREDDY ROOS-
EVELT
WILLIAMS GERALD MARK
WILLIAMS HOWARD
WILLIAMS HOWARD EUGENE
JR
WILLIAMS IRA WINARD
WILLIAMS JAMES BERNARD JR
WILLIAMS JAMES THOMAS
WILLIAMS JIMMIE
WILLIAMS JOE JR
WILLIAMS KENNETH JERRY
WILLIAMS MILLIGAN RU-
DOLPH
WILLIAMS MOSES
WILLIAMS RAY
WILLIAMS ROOSEVELT
WILLIAMS TERRY DOUGLAS
WILLIAMS WILLIS WHITE
WILLIAMSON MILLARD
LEROY
WILLIS GLENN LEE
WILLS DAVID COLLIER JR
WILSON GEORGE A
WILSON GERALD LEE
WILSON LEON
WILSON RUDOLPH
WINDELER CHARLES CARL JR
WINDSOR DAVID WARREN JR
WINFREY DOUGLAS NELSON
WINKLES JAMES WILLIAM
WINSTON WILLIAM OVERTON
WISE JAMES CARL JR
WOLF DEWITT JOSEPH
WOOD CHARLES
WOOD DAVID BEAVERS
WOOD DONALD ROY
WOOD LARRY LESTER
WOOD ROBERT ABBOTT
WOOD ROBERT HELM
WOOD ROBERT WAYNE
WOOD THOMAS DANIEL JR
WOOD WILLIS LEROY
WOODALL RALPH TRAYLOR JR
WOODEN CHARLIE K
WOODRUFF ALTON DARNELL
WOODS RAY HOUSTON
WOODS WILLARD PAUL
WOODY JOHN HENRY
WORRELL MILTON JERRY
WORTHEN LARRY EUGENE
WRIGHT JAMES
WYATT CHARLES REMBERT
WYNN HARVEY EUGENE
WYNN JOSEPH RAY JR
YARBROUGH JAMES LAMAR
YARBROUGH LESTER GAR-
NELL
YARBROUGH LEVERETT E
YAWN TERRY LYNN
YEOMANS ALLEN CALVIN III
YORK GARY WILSON
YOUMANS JAMES NELSON
YOUNG BOBBY ARTHUR
YOUNG FRANKIE JR
YOUNG GEORGE LAMAR
YOUNG LARRY JOHN
YOUNGBLOOD BOYD JAMES
ZORN THOMAS ONEAL JR

Photos by Mitchel Viernes

Vietnam Veterans Memorial - National Cemetery of the Pacific

2177 Puowaina Dr, Honolulu, HI 96813

The National Cemetery of Pacific in Honolulu, Hawaii features 39 memorials that honor the country's veterans of most of the known wars, including the surprise attack on the military base at Pearl Harbor. The Vietnam Veterans Memorial is just part of a network of intertwined war memorials.

The Vietnam Veterans Memorial is a part of the larger Honolulu Memorial in the cemetery which was built in 1964 under the order of the American Battle Monuments Commission. The point of the memorial was to honor the achievements and trials of all members of American armed forces in the Pacific. The memorial originally was meant for WWII and Korean War veterans but grew in 1980 to include the veterans of the Vietnam War who were declared missing in action. In 2012, it expanded to include two more pavilions which play host to an orientation map and two mosaic maps that depict the war itself. The memorial was created with various marbles imported from Italy while the surrounding landscape features most of the original foliage.

On the whole, the Honolulu Memorial is made up of a flight of stairs that feature the "Courts of the Missing" along the sides of it. The names of over 20,000 service members who were declared missing, lost, or were buried at sea are featured on this memorial in alphabetical order by armed forces branch in order to make finding a name simpler. The top of the staircase is the "Court of Honor" which is dedicated to all branches of the US military. Here, a center tower holds the statue of a female figure holding laurel branches known as "Columbia." The memorial's chapel is also found at this spot. In order to honor the veterans of the Vietnam War, two more half courts were added to the original memorial at the base of the stairs. Here, visitors can find the 2,504 names of the missing veterans.

64

The Names of those from the state of Hawaii Who Made the Ultimate Sacrifice

ACERET PEPITO RIVERA
AHUNA ABRAHAM KAALELE
AKANA FRANKLIN RANDOLPH
AKI FRANCIS CLAYBURN JR
ALAMEDA WILLIAM KAPENA
ALCOS LARRY MELVIN
ALIPIO LESTER WARREN
ALLAGONEZ RODOLFO P
AMISONE FUIFUITAUA
ANDRADA WILFREDO BALAGOT
ANDRADE KENNETH SOARES
ANDRADE ROBERT D
ANDRADE ROBERT SOARES
ANDUHA HOWARD J
ANTONIO CATALINO B JR
ARAGON HENRY T
ARAKAKI WAYNE ALLEN
ARNADO FREDRICO
ARQUERO ELPIDIO ALLEN
ARRIAGA TONY R
AU HOY EARL CHUI MUN
BALAI ANDRES
BAN HERMAN HALEMANU
BARCENA BOBBY JOHN
BAYRON BENEDICTO PIOSALAN
BELL DAVID LEROY
BENJAMIN ROBERT WILLIAM
BILONTA LARRY KILITO
BINGHAM KLAUS YRURGEN
BLEVINS THOMAS A K
BODE ROBERT RUSSELL
BONGO ANTHONY
BRIGHTER JERRY KAOPUA
BROWNE WALTER D
BUELL NORMAN JOSEPH
CABANAYAN ALBERT
CABRERA JOHN WAIKANE
CAIN FREDERICK CHARLES
CAIRES CLYDE JOSEPH
CALIBOSO ROBERT MALUENDA
CAMPOS MAGNO
CANTOHOS RODNEY SALVADOR
CAPODANNO VINCENT ROBERT
CARLOS STEPHEN G
CASH MORRIS ELTON
CASTANEDA EUGENE
CASTILLO LEONARD BALDOMIR
CHING STEVEN SAM CHOY
CHOCK LINUS GERARD K
CHOW CALVIN KEALOHAOKALAN
CHUN REGINALD WUNG YETT
CHUNG DOUGLAS KAMKEE
CONCEPCION FRANCISCO JR
COSTELLO GEORGE SIMONDS
CRANE CHARLES HENRY
DEFRIES GAYLORD KILA
DEL ROSARIO JOSEPH JESUS
DICK SAMUEL EUGENE
DICKENS ELMER WILLIAM
DOANE JAMES ABRAHAM
DOIKE JOHN TOSHIO
DRYDEN RALPH MARION JR
ELLIS RICHARD LEIALOHA K
ENGLISH PHILIP DOMINIC K
FACTORA DOUGLAS GEORGE
FALEAFINE SISIFO
FLORES DOUGLAS
FOSTER TIMOTHY K
FUJIMOTO MASAICHI
FUJITA MELVIN SHOICHI
FUKUNAGA RODNEY TAMOTSU
FUNN GARY FRANCIS
GAA JOSEPH WILLIAM JR
GABRIEL JAMES JR
GALLEGO LAWRENCE
GIER DAVID JULIAN

GIFFARD SAMUEL POOKEAOKAL
GOMES ALLEN EDWARD
GONZALES ALEXANDER
GOROSPE LEONARD GORDON
GREENWOOD GEORGE R W K
HAMBLETON MARK EVAN
HAO JOSEPH N
HATADA FRED KAWAILANI MAS
HEDEMANN WAYNE HOWARD
HIRANO OWEN TETSUMI
HO ALVIN JOCK
HONDA KAORU
HOOPII BERNARD PALENPA JR
HORN JOHN ELIA
HOSE HERMAN BATER JR
HOWARD JAMES EDWARD JR
HU PATRICK HOP SUNG
HUGO FELICISIMO ARELIANO
IGARTA BENITO JR
IGNACIO ROY
IMAE HACHIRO
IOPA CHARLES MAHELONA
ISHIHARA JAMES HIROSHI
JACOBS RICHARD CHESTER
JARRETT FREDERICK
JEREMIA ALEKI
JOHNSON GIDEON PICHA
JOHNSON MICHAEL ELLIOTT
JOHNSON STEVE FREDDIE
KAAIHUE KENNETH R
KAAKIMAKA ALGERNON P JR
KAAWA JOHN RICHARD
KAHANA SAMUEL KAULUHAIMAI
KALANI CHARLES MANUWAHL
KALILI MELVYN HAMANA
KALUA SOLOMON JR
KAMA FRED KAIMI NAAUAO
KAMALOLO JOEL KAHALEALOHA
KANE MOMI NUHI
KANESHIRO EDWARD NOBORU
KAOPUIKI ALEXANDER A JR
KAPALU GEORGE KUAMOO
KAUHAIHAO JOHN KUULEI
KAUHANE ELIAS MAULILI
KAWAMURA GARY NOBORU
KAWAMURA TERRY TERUO
KEA EDWARD KIKAU
KEAHI GENE LUTHER
KEKAHUNA WILLIAM ANTONE
KIAHA RODNEY SIU
KIM EDWARD Y C
KINARD DIXON TALMADGE
KOBAYASHI ROY SHIGERU
KONG BRIAN WALLACE
KOVALOFF JOSEPH THOMAS
KUAHIWINUI MOSES IWANE J
LACEY PETER JOSEPH III
LAZARUS ROBERT LANI
LEANIO HILARIO B JR
LEE GLENN HUNG NIN
LEOPOLDINO LARRY GENE
LESLIE WENDELL WAYNE
LEVINTHOL JOHN JR
LEWIS ALLEN LANUI
LIMA KENNETH KAWIKA
LINDBERGH ROBERT RAYMOND
LINDSEY JAMES KAHILILAUIN
LITTLER JAMES L M III
LLANES HAROLD LEROY
LOO EDWARD LUKANA JR
LUIS GEORGE GREGORIO
LUM DAVID ANTHONY
MACHADO ROBERTS
MALABEY BENJAMIN KEALII
MAMIYA JOHN MICHIO
MAQUILING LEONARD GIDEON
MARSHALL DONALD FISHER II

MARTIN RICHARD D
MATSUURA ALAN YUKIO
MAUNAKEA RODNEY H
MC KILLIP MERRIL ANDREW
MCCORMICK MICHAEL TIMOTHY
MENO ROY FLORES
MEYERS RAYMOND EILERT
MILAR ALBERTO JR
MIYAKE GARY NOBUO
MIYAZAKI RONALD KAZUO
MOHL WOLFGANG TONY OTTO
NAKASHIMA MICHAEL SEIJI
NAKI WILLIAM III
NATARTE ROBERT ORTOGERO
NERVEZA DELMORE BYRON
NEWMAN CLIFFORD AUHUNA
NISHIYAMA MELVIN TETSUO
NOGUCHI ROCKNE MASAYOSHI
NUEKU ROBERT LANI
OGATA TERRANCE AKI
OKUMURA EARL AKIO
OLIVAR GILBERT
OPPERMAN HUGH DANIEL
OTAKE JOHN SADAO
PACHECO MICHAEL JEROME
PACOLBA ALFREDO
PADILLA RONALD MATTHEW
PAELE PETER JAMES
PAIALII PASIA
PAKELE FRED DALE
PANGAN ROGER ROGELIO
PANGANORAN ABRAHAM
PARESA EDWARD KENNETH
PARO RANDY CHARLEY
PASCAL IVAN KIMOKEO
PASCUA DALMACIO P JR
PASCUAL FLORENDO B
PAVAO RODNEY WAYNE
PEACOCK JOHN ROBERT II
PELEIHOLANI HAYWARD K H
PERPETUA ROQUE JR
PERREIRA ERROL WAYNE
PIMENTEL TEOFILO CASTILLO
PUHI DANIEL KIMOKEO
PUHI KEITH JON
QUINONES JULIO JR
RABACAL PATRICK WILLIAM
RAMOS ROLAND ROLANDO
RAMOS STEPHEN KEALOHA
REGO JOHN H
REMULAR RUDOLPH
RIBILLIA MARIANO JR
RIBUCAN VAN V
RICHARDSON RICKY WAYNE
RICKARD WALTER L
RIVEIRA ROBERT CHARLES
RODRIGUEZ ARTURO SERNA
RODRIGUEZ FRANK LOUIS
ROOT THOMAS RICHARD
SAGON RUDY MANTIAD
SAGON STANLEY INCILLO
SAKAI ERNEST SEICHI
SALAZAR JOHN
SALEAUMUA UINIFARETI
SALEMA GEORGE STANLEY
SALVATORE THOMAS ANTHONY
SANUT ALFREDO
SAROCAM JOSEPH
SASAKI ALLYSON YUKIO
SERAIN CALVIN ERNESTO
SHIBATA GLENN TEUGIO
SHIMABUKURO KENYU
SHINKAWA ROY YASUSHI
SIBAYAN FRANKLIN DANIEL
SILVERI DENNIS MICHAEL
SINCLAIR LEE ELDEN
SMITH ELMELINDO RODRIGUES
SNOWDEN THOMAS EDWARD
SODERSTROM MICHAEL DENNIS
SOLOMON SAMUEL K JR

SORIANO JAMES GABRIEL
SPILLNER ROBERT K
SPITZ GEORGE ROSS
STEVENS PHILIP HUGH
STORCH WILLIAM FRANK JR
STROMBECK EDWARD EARL
SUAPAIA DAVID KEALOHA
SUETOS GILBERT BRIAN
SUMIDA JERALD KATSUJI
TADENA ESTEBAN WALLACE
TADIOS LEONARD MASAYON
TAIRA CLIFFORD KAZUMI
TAKEHARA YOSHIO
TAKEMOTO KENNETH JAMES
TALAN ARISTON R JR
TANAKA MINORU
TANIMOTO MILES T
TATSUNO ALBERT HIROSHI
TAUANUU PELESASA SOLOMONA
TAYLOR WILLIAM ROBERT
TESORO RICHARD RAMIREZ
TOMA RICHARD HISAO
TOMLINSON JONES EUGENE
ULI SASA
VALDEZ FERNANDO MARCELO
VICTOR GEORGE M

VILLANUEVA FLORENCIO G
VILLON CASIMIRO
WARK DANIEL EDWARD
WEBB GEORGE GRANT KING JR
WELLS ORVILLE D
WHITE ROBERT HENRY
WIRE EUGENE CHARLES
WISCHEMANN DAVID EDWARDS
WOLTER STEVEN ROSS
WOODWARD STANLEY KAMAKI
YAMASHIRO EDWARD SATORU
YAMASHIRO NAOTO
YAMASHITA MELVIN MASAICHI
YANO RODNEY JAMES TAKASHI
YOSHIDA ELLIOT MATSUOH
YOSHINO KANJI
ZAMORA EUGENE CONSTANTINO

Photos by Brian Brown

Idaho State Vietnam Veterans Memorial

Freeman Park, Science Center Drive, Idaho Falls, ID, 83402

The Idaho State Vietnam Veterans Memorial is a place for quiet reflection in Idaho Falls, Idaho. The monument was placed in Freeman Park on Riverside Drive. The Vietnam memorial is meant to honor all of the men and women from Idaho that served during the South Pacific conflict; especially those who never made it home.

The Idaho State Vietnam Veterans Memorial was designed by fellow Vietnam veteran Tom Chriswell. The frame of the structure is of carbon steel and features a copper inlay of a map of Vietnam while the accompanying relief, sculpted by Chriswell is made bronze. It was dedicated after its completion in August of 1990 thanks to the Idaho Falls Vietnam Veterans known as Freedom Bird Inc.

The dedication of the Idaho State Vietnam Veterans Memorial states, "This memorial was built to convey our gratitude and appreciation of the men and women who served during this difficult time." The memorial consists of a large 24-foot stainless steel inverted "V" structure which is supposed to represent controversy surrounding the Vietnam War. The structure spans across a walkway allowing visitors to pass underneath the memorial. The steel structure is complete with a copper inlay depicting a map of Vietnam, while two copper dedication plaques boast the names of the 251 fallen Idaho service members, as well as a message reading "Requiescat in Pace Brothers – For as long as your names are remembered so too are you. Thank you for serving for us and your country." The memorial's bronze relief sculpture features depictions of a combat soldier, an American prisoner of war, and a woman wrapped in the American Flag gathered under the outstretched wings of an eagle. This is a representation of the freedom the country enjoys thanks to the sacrifices of the soldiers.

The Names of those from the state of Idaho Who Made the Ultimate Sacrifice

ADAKAI LEO JOE
AHLBERG THOMAS OLIVER
AKKERMAN DUANE CHARLES
ALLEY GERALD WILLIAM
ALLRED JAMES HERBERT
ANDERSEN REESE MARK
ANDERSON JAMES BARTON
ANDERSON VICTOR EDWARD
ANDERSON WILLIAM ED-
WARD
ASLETT ALLAN THEO
ATKINSON GLEN LAWRENCE
BALDWIN GERALD LEE
BAUMERT BRENT JOHN
BEASLEY PHILIP ARTHUR
BEASLEY WILLIAM RONALD
BEE ROSS MICHAEL
BELLAMY ANTHONY RODNEY
BENNETT BRUCE ROLLA
BENTON JOHNNY WILLIAM
BITTON GARY W
BLENKINSOP WILLIAM
DARWIN
BODAHL JON KEITH
BOGGESS EDWARD JAMES
BOHLSCHEID CURTIS RICH-
ARD
BOICOURT JESS BURTON JR
BOUSHELE GARY RAY
BOWLES BRUCE GREGORY
BOYLE JOHN ALEX
BRAUBURGER EVERETT W
BRENN HARRY MILTON
BROENNEKE LEONARD LEE
BRONSON RANDY K
BROWN ARLO FRANK

BRUMET ROBERT NEWTON
CARLSON RICHARD BUCK
CARRICO CLYDE ROBERT
CARSON CHAD LEONARD
CHAMBERS ROBERT STANLEY
CHAPMAN JOHNNY HOWARD
CHATTERTON DAVID ROGER
CINKOSKY DAVID EDWARD
CLARK CONN KAY
CLARK GRANT LEROY
CLAYBAUGH JAMES BRADLEY
COATS LARRY DALE
COBURN CLYDE RALPH
CORDON RALPH BRENT
CREASON JESS WILLIAM JR
CROUSON MICHAEL LEE
CROW CLYDE ARTHUR
CURTIS DAVID LEE
CURTIS JAMES MARVIN
DAMIANO LE ROY EDWARD
DE FORD ELMO LEE
DEFILIPPIS LARRY DALE
DENNIS RICHARD LESTER
DODSON JACK LEROY
DROWN LYLE EUGENE
EARP MICHAEL LEE
ELDRIDGE DONALD LEE
EMERY LOUIS CRAIG
ENDICOTT FRANKLIN DAVID
ENGLAND STEVEN GLENN
ESTES JERRY DUANE
EVANS GARY GENE
FAIRCHILD DAVID ACEL
FINLEY RAYMOND PATRICK
FLINT WINFIELD SCOTT
FLORES FRANCISCO JOHN

FOREMAN THOMAS ALLEN
FOSTER GARY JACK
FRAZIER GARY VIRGIL
FUNKE THOMAS GEORGE
GABRIEL GARRY LEE
GARCIA ALBARO QUEZADA
GOLDEN MERVIN DENNIS
GOODSELL BRUCE LYNN
GORDON ROBERT JERRY
GORTON RALPH SHOUP III
GREEN MICHAEL FRANK
GREEN ROBERT CARRELL
GREENHALGH LARRY DEE
GREGORY CHARLES CLARK
GRIFFIN WILLIAM JAMES
HAILE DONALD JACK
HANSEN CRAIG HAYES
HANSEN ROBERT WARREN
HARSHBARGER ERIC THOMAS
HENDRICKSON LONNIE
HILTON
HEPBURN WILLIAM BARTON
HERBERT DAVID EDWIN
HIRSCHI CRAIG W
HODGES TEDDY MERLIN JR
HOLLINGER GREGG NEYMAN
HOLLINGSWORTH HAL T
HOSKINS SHELDON DALE
HUNT WILLIAM BALT
HURIANEK JERRY ANTONE
HURST JOHN ALLEN
HUSTON TERRY FLOYD
IRELAND ELMER GLENN
JENSEN LLOYD BRUCE
JONES BRENT R
JONES DAVID SAMUEL

JONES HOWARD LEMUEL JR
KEARSLEY TOMMY L
KIMURA KAY KAZU
KOEFOD RODGER MAGNUS
LAMB COLIN EDWARD
LARISON ROBERT WILBUR
LARSON DALE K
LARSON JOHN GILBERT
LEMMONS WILLIAM E
LISH GILBERT RAY
LISTER JAMES JOHN
LOCKWOOD JAMES ALTON
LOHMAN HERMAN AUGUS-
TA JR
MACKAY NEILE COOPER
MAGGARD DANNY JOE
MAPES EDDIE D
MARTIN CLAYTON ARTHUR
MASSINE RICHARD PETER
MAYER RODERICK LEWIS
MC ARTHUR STEVEN MI-
CHAEL
MC CASLIN RAYMOND LOUIS
MC GINLEY DONALD SMITH
MC LAIN JAY DARWIN
MC MASTER MICHAEL LEE
MC NAMAR JIM CARL
MC NARY FRANKLIN DELANO
MCDONALD STEVEN JAMES
MEECHAN RICHARD JOSEPH
MERRELL STEVEN DEE
MILLSPAUGH CECIL RAY
MITCHELL JOHN E S JR
MITCHELL LONNIE RAY
MOON DEAN LEROY
MOORE DAN ROSS

MORLEDGE WILLIAM RALPH
MOTTISHAW RONALD GRANT
MOULTON LESTER NEAL
NAILLON DANNY L
NAKAYAMA JIMMY D
NELSON ROBERT WILLIAM
NIPP STEVEN HAROLD
O BRIEN MARK JAMES
OLIVER TROY ROBERT JR
OLSEN WILLIAM WHITBY
OLSON BENNETT WALFRED
PAINTER MICHAEL HARRIS
PETERSEN WILLIAM DONN
PETERSON BOBBY GENE
PETERSON JON DALE
PHELPS JESSE DONALD
PHILLIPS SAMUEL C III
PIVA JAMES EDWARD
POLETTI MICHAEL LEE
POWELL ROBERT ALLAN
POWERS JOHN LYNN
PRIEST MICHAEL LLOYD
PROBART LEWIS DEVERN
PUGMIRE MAX WELKER
REASONER FRANK STANLEY
REED CHRISTOPHER RAY
RICHARDS MICHAEL HUGH
RIOS ARTURO RECIO
RODRIQUEZ SAMUEL HENRI
ROTTER RALPH LEE
ROWE DOUGLAS NOEL
RUEPPEL RONALD BENTON
SANDOVAL VICENTE DIAZ
SAVELL FLOYD GWEN
SCHAFFNER MARSHALL GUST
SHAFF RONALD DEAN
SHIEFER JOHN FREDERICK
SHROPSHIRE GLEN EMERY
SKIDMORE VERLE JENNINGS
SMALL KENNETH LLOYD
SMART FRED STEVEN
SMITH ARIEL JAMES
SMITH BILLY GENE
SMITH GARY CLARENCE
SMITH JAMES ANDERSEN
SNYDER MICHAEL ALLAN
SPARKS JON MICHAEL
STAPELMAN RONALD LEE
STEELE GARY LYN
STODDARD CLARENCE W JR
TACKE RAYMOND LEROY
TANNER DAVID ARLINGTON
TEDROW DANIEL CLINE
TEWS HENRY JAMES WILLIAM
THOMAS TERENCE PIERCE
THORNTON LARRY C
TOOLEY JAMES EDWARD
TURNER KENNETH EUGENE
TURNER RODNEY CARL
VERMEESCH WESLEY WIL-
LIAM
VOLK GEORGE FRANCIS
WADE DOUGLAS JOHN
WALDRON HOWARD BERT
WALKER JAMES LLOYD
WARD JIMMY LEE
WARD JOHNNY LEE
WASSERMAN MICHAEL LEON
WATSON RUSSEL LEE
WEIDENBACH EDWARD
JOSEPH
WHEELER MICHEL T
WILLEY ROBERT LEON
WILLIAMS BILL GENE
WILLIAMS DANIEL EUGENE
WILLIAMS MORRIS EDWARD
WILLIS THOMAS MURTEN
WILSON JERRY BARBER
WOLTERS EUGENE EBEN
YAGUES ROBERT GENE
YOUNG KENNETH WILSON

Photos by Joseph Coplen

Illinois Vietnam Veterans Memorial

1500 Monument Ave, Springfield, IL 62702

_____

Oak Ridge Cemetery is not only the home of Abraham Lincoln's tomb, it also plays host to the Illinois Vietnam Veterans Memorial which is in the southwest corner of the cemetery. It was created to honor the men and women who didn't make it home from the Vietnam War. The memorial is open to the public, but there are special memorial services held on the first weekend in May, on Memorial Day, and on the first Saturday of December which is called the "Christmas Remembrance." On the first full weekend of May, there is also a 24-hour vigil at the memorial.

The memorial, a granite monument, was designed by Jerome Lager who was only 20-years-old at the time. It was funded by a combination of individual donations, corporate sponsorships, and public fundraisers. It was dedicated in May of 1988 and is currently maintained by the Illinois Historic Preservation Agency thanks to its status as a state historic site.

The memorial consists of a 58-foot wide granite slabs that features the names of the 3,000+ men and women who never made it home from Vietnam. There are five triangular-shaped walls to represent the five branches of the US military with the coinciding veterans' names for each. The slabs are laid out in a circular shape with the highest points at the innermost part. Positioned at the end of each slab are five upright granite pillars that have the insignias of the branches of the military. These granite walls are connected in the middle by a platform on which sits an eternal flame that has been holding strong since the dedication. There is also an inscription on the outside of the walls that reads, "To those who died honor and eternal rest, to those still in bondage remembrance and hope, and to those who returned gratitude and peace."

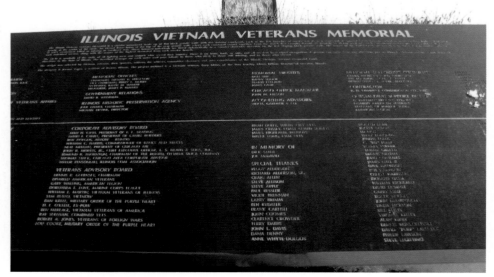

The Names of those from the state of Illinois Who Made the Ultimate Sacrifice

ABBATE RICHARD CLARK
ABRAMS JOHN ALAN
ACHAS ROBERT JOHN
ADAMS MICHAEL EDWARD
ADAMS WILLIAM RAYMOND
ADDUCI JOHN JOSEPH
ADKINS TERRY LEE
AGUADO ROBERT CHARLES
AHINZOW TONY
AHLFIELD ALAN PAUL
AJSTER JOSEPH ROBERT
ALBERT WILLIAM DAVID
ALBERTS JOHN CHARLES
ALDRIDGE NEIL WAYNE
ALEXANDER JAMES PATRICK
ALEXANDER ROY M
ALLEN BOBBY KENNETH
ALLEN HARRISON CHARLES JR
ALLEN HOWARD LLOYD
ALLEN MELVIN LEE
ALLES JAMES KENNETH
ALLMAN HENRY HAYDEN
ALLMEYER FREDERICK ALLEN
ALMAGUER BENJAMIN FRANCIS
ALMANZA PABLO
ALMANZA RICKY JEROME
ALONGI MICHAEL PETER JR
ALVIS ROY GENE
ALWAN HAROLD JOSEPH
AMADOR DIEGO
AMBROSE JAMES WILLIAM III
AMBROSINI JOHN STEVEN
AMESBURY HARRY ARLO JR
ANDERSON FRANCIS ALAN
ANDERSON GREGORY LEE
ANDERSON JUSTIN KENNETH
ANDERSON LEWIS CARL
ANDERSON MARK STEVEN
ANDERSON MEREDITH GLENN
ANDERSON MICHAEL FRANCIS
ANDERSON RANDALL BRUCE
ANDERSON RICHARD GUNNAR
ANDERSON ROBERT WILLIAM
ANDERSON ROGER CHARLES
ANDERSON STEPHEN ARTHUR
ANDRE CARL VAL
ANDRES KEITH JOHN
ANDREWS OTIS ELIZA
ANDREWS ROBERT P
AQUINO RAYMOND JOHN

ARCHIBALD DENNIS
ARIAZ EDWARD JOSEPH
ARMS JAMES WALTER
ARMSTRONG BARRY LEE
ARMSTRONG HERMAN ROBERT
ARNESON MARCUS EUGENE
ARNOLD DAVID BRUCE
ARNOLD DONALD EDWARD
ARNOLD JOE EDDY
ARNOLD JOHN CRAIG
ARNOLD KENNETH W
ARNOLD ROBIN LEE
ARSENEAU GALEN LEROY
ARTHUR ALLEN LEE
ARTMAN JAMES BOYD
ASBRIDGE LARRY GENE
ASCHER JAMES ALLAN
ASHENFELTER ALAN WAYNE
ATOR JOHNNIE WAYNE
AULTMAN GREGORY WAYNE
AVANT SHERMAN
AVERY DON WAYNE
AVGERINOS GEORGE RICHARD
AVINGTON LARASETT EARL JR
AYERS CARL BRACY JR
AZBELL JAMES ALLEN
BABB RICHARD CLARK JR
BABIARZ EDWARD MARTIN
BABICH JOHN MICHAEL
BADSING MICHAEL TERRANCE
BAETZEL ROBERT ALLEN
BAHL RICHARD HOWARD JR
BAILEY FLOYD CLARK
BAILEY JOSEPH JR
BAILEY RAYMOND
BAIRD JOHN ROBERT JR
BAIZE GARY CECIL
BAKER CURTIS EVERETT
BAKER DONALD LEE
BAKER EUGENE JR
BAKER JACK LESLIE
BAKER JON DOUGLAS
BAKER RAFTKEITH EROS
BAKER REGINALD
BAKER ROBERT JOHN
BALDWIN ROY LEE
BALDWIN WILLIAM CLARENCE
BALLARD MELVIN
BALLEW HENRY HERSCHEL
BALLEW ROLAND LEE

BANDY LARRY GENE
BANGERT STEPHEN RAY
BANISTER JOHN EDWARD
BANKS FLOYD JACKSON
BANNING TERRY LEE
BARAN BRUNO
BARBER ERNEST LEE
BARCELONA RALPH ANTHONY
BARDON BRUCE HAROLD
BARKER BOBBY LEE
BARKLEY JESSE LOUIS
BARLETT RALPH HARRY JR
BARNARD GARY ADRIAN
BARNES BRUCE MICHAEL
BARNES DONALD ALBON
BARNES ROBERT CROZIER JR
BARNES TOMMY LEE
BARNES WILLIE JAMES
BARNHART CARL RAY
BARRETT CLARE ARNOLD
BARTELS NORMAN WILLIAM
BARTLEY RALPH GILBERT JR
BARTON NORMAN LEE
BARTRAM GERALD EDWARD
BATEMAN JAMES AUSTIN
BATES LARRY LEE
BATES ROBERT ANTHONY
BATES VIRGIL JAY JR
BATT ROGER LEE
BATTAGLIA AUGUST THOMAS
BATTLE ULYSSES
BATTLES TROY CLEVELAND
BATTS LARRY
BAUER CARL TIMOTHY
BAUER CRAIG ARLEN
BAUM MICHAEL EDWARD
BAZEL MICHAEL GEORGE
BEALS ALLEN MACY
BEAM RAYMOND GLENN
BEAN GEORGE TYRUS
BEATTY LEONARD JR
BEAVER JAMES CLARKE
BEAVERS CHRISTOPHER WAYNE
BECK EDWARD CHARLES
BECK NORMAN ELMER
BECK PATRICK FRANCIS
BECK RICHARD JAMES JR
BECKER CHARLES WARNER
BECKER GARY EDWARD
BECKER HOWARD JOHN JR
BECKER LESTER ERWIN
BEDNARZ WILLIAM WALTER

BEELER RUSSELL RICHARD
BEHAR DANIEL SIMON
BEHRNS RICHARD JOHN
BEIERLE THOMAS LAWRENCE
BEIRNE MICHAEL JAMES
BELAND WILLIAM ANTHONY
BELCHAK PAUL JR
BELCHER VERNON EUGENE
BELL DONNELL
BELL JAMES WILLIAM
BELL JERRY W
BELL JOHN HENRY
BELL LARRY DEAN
BELL LEO JR
BELLES JOHN DAVID
BELSLY STEVEN DALE
BENEFIEL DUDLEY JAMES JR
BENFORD JONAS
BENNER FRED ALFRED
BENNETT DWIGHT FARWELL JR
BENNETT RICHARD CHARLES
BENOIST WILLIAM F III
BENSON ALLAN CAMERON
BERAN NICHOLAS MICHAEL JR
BERBLINGER KENNETH MICHAE
BEREK MICHAEL STANLEY
BERGER NICHOLAS ALLEN
BERGFIELD PHILLIP REX
BERKSON JOSEPH MIKE
BERNAL JOSE ROLANDO
BERNER EDGAR DAVIDSON
BERRY LARRY MICHAEL
BERTHOUX DALE PORTER
BERTOLINO FRED GORDON
BERTOLOZZI PAUL CHARLES
BESSENT SAMUEL ALONZO
BEST GARY ALLEN
BESZE GYORGY JANOS
BEUKE DENNIS ARTHUR
BEUSTER RONALD LEE
BEUTEL ROBERT DONALD
BEZECNY JOHN WILLIAM
BIBBS LEONARD JEROME
BIBBS WAYNE
BIEDRON ANDREW ALBERT JR
BIEGEL ROBERT CHARLES
BIENEMAN JOHN CHARLES
BIERBAUM LAWRENCE ANTHONY
BIGLEY GEORGE CARL
BILY WILLIAM CHARLES
BINEGAR BENJAMIN H JR
BISHOP DALE ALAN
BISHOP JAMES FRANKLIN
BISHOP JAMES WALTER
BISHOP WILLIAM WAYNE
BLACK CHARLES DUFFY
BLACK LEWIS DAVIS
BLACK RONALD LEE
BLAKE WAYNE VALGEEN
BLANCHFIELD MICHAEL R
BLANCO JOHN ALEXANDER JR
BLAND GARY PAUL
BLASEN RICHARD LEE
BLAUVELT RALPH LEIGH
BLEDSOE HOWARD TYRONE
BLOOMER TERRY LEE
BLUMER EDWARD EUGENE
BLUMER WILFORD LEE
BOAZ KENNETH WAYNE
BOCANEGRA ROJELIO
BOCK JERRY CHARLES
BODDEN TIMOTHY ROY
BODINE ROBERT LEE
BOEHM WILLIAM EUGENE
BOETJE WILLIAM WAYNE
BOGGUESS MAURICE
BOHAN PATRICK JOHN
BOICOURT ROBERT C
BOLAND WILLIAM JOSEPH JR
BOLLINGER ARTHUR RAY
BOLLMAN DONALD WARREN
BOLLMAN ROBERT NORMAN

BONEBRIGHT ROBERT ALLEN
BONERT RONALD JOSEPH
BONETTI PAUL JOSEPH
BONILLA HERMINIO AMELIO
BOOCHKO VICTOR
BOONE JAMES ARTHUR
BORCHART WILLIAM H
BORCZYK STEPHEN ZBIGNIEW
BORDES ANDREW MORLEY
BORNMAN DONALD WAYNE
BOROWSKI JOHN C
BORST LEROY J JR
BOSWELL WILLIAM HENRY
BOTES GEORGE
BOTTOM A J
BOTTOMS HAROLD GENE
BOUCHEZ DANNY PHILLIP
BOWER JIMMY CHARLES
BOWLIN ROBERT JOE
BOWMAN WILLIAM PARKER
BOYCE TERRY LAWRENCE
BOYD ERNEST JR
BOYD ROGER WILLIAM
BOYD WILBURN HUGH
BOYER JAMES ALAN
BOYER LARRY DEAN
BOYER RONALD EARNEST
BOYEV PETER KESTUTIS
BOYLE JAMES PATRICK
BRACKETT FRANK
BRADFORD ELLSWORTH SMITH
BRADFORD RODNEY
BRADLEY LOREN EUGENE
BRADSHAW JAMES THOMAS
BRAM AARON L
BRANCATO JOHN HARRISON JR
BRANCH JAMES ALVIN
BRAND JOSEPH WILLIAM
BRANDON JAMES BYRD
BRANHAM JAMES JEROME
BRANTLEY JOHN ARTHUR
BRASCHE GERALD WILLIAM
BRASS PAUL ROBERT
BRAUN HARRY WALKER JR
BRENCICH WILLIAM JOSEPH
BRENT DAVID ALLISON
BRENT GERALD ROBERT
BREWER DALE CONNARD
BRICKER CHARLES WILLIAM
BRIED ROBERT ALLAN
BRIGHTMAN MICHAEL DENNIS
BRIMM JOHN M
BRINKER KENNETH RAY
BRITT RONALD JEROME
BRITTENUM OSCAR LEE JR
BROCKMEYER DELBERT RAY
BRODERICK PATRICK EMMET
BROEFFLE IVAN CLIFFORD
BROMLEY THOMAS EDWARD
BROOKS ALLEN
BROQUIST STEVEN ANDRE
BROSE ALBERT C
BROWN BARRY LYNN
BROWN BILLY JAMES
BROWN BOBBY GENE
BROWN BRUCE GILBERT
BROWN CHARLES CHUCK
BROWN CLYDE ALVIN
BROWN DENNIS WILLIAM
BROWN ELYVIN LAVERNE
BROWN JAMES PATRICK
BROWN JOSEPH RAYMOND
BROWN RAYMOND
BROWN ROBERT EDWARD
BROWN STEVEN EUGENE
BROWN TERRANCE LEE
BROWN WALTER WILLIAM
BROWN WILLIAM HENRY JR
BROWNLEE ROBERT WALLACE JR
BROZICH ANTHONY GEORGE
BRUESKE HARRY DIETRICH

Illinois

BRUMMER MICHAEL LEE
BRUNNER DONALD RALPH
BRUNNER HANS WOLFGANG
BRUNSON GAZZETT BEN JR
BRYAN DAN E
BRYANT FRANCIS LEON
BRYANT MICHAEL STEVEN
BRYAR JOHN JOSEPH
BRYNELSEN THOMAS ALLEN
BUCHER BERNARD LUDWIG
BUCKLES RICHARD DEAN
BUCKLEY CARL DWAYNE
BUEHLER LEON CHRIST
BUFF WILLIAM REINHART
BUGAJSKY KERRY MICHAEL
BULLIS STANLEY ALLEN
BULTHUIS WILLIAM NELSON
BULTMAN ROY JAMES
BUNKER PARK GEORGE
BURCHETT GEORGE ELMER
BURGENER GERALD EUGENE
BURGER DIETER HANS
BURGETT JOSEPH SCOTT
BURGOYNE JAMES JOSEPH
BURKE MICHAEL JOHN
BURKE ROBERT CHARLES
BURKES DAVID E
BURKES DAVID RONALD JR
BURKHART MICHAEL JAMES
BURNS JAMES PATRICK
BURNS MARTIN JAMES
BURRIS JOSEPH SAMUEL III
BURTON JAMES ARTHUR
BURZAWA JOHN ANDREW JR
BUSCH ERIC PETER
BUTKUS ALAN PAUL
BUTLER BENNY LEE
BUTLER GEORGE RICHARD
BUTLER ROBERT LEE
BUTTERFIELD CALVIN
FRANKL
BYLON JOHN LOUIS
BYRD EATTERSON JR
BYRNE JOSEPH HENRY
BYRNES ROBERT HOWARD
CABRERA LOUIS XAVIER JR
CABY BILLY RAY
CACCIOTTOLO NEIL JOSEPH
CADIEUX THOMAS PAUL
CALANDRINO MICHAEL
THOMAS
CALIFF JAMES PATRICK
CALKINS CODY RAY
CAMPA JOHN JOSEPH
CAMPBELL ALLIE WILLIAM
CAMPBELL DONALD
CAMPBELL JACK
CAMPBELL JAMES HENRY JR
CAMPBELL JOHN DREW
CAMPBELL JOHN RUSSELL
CAMPBELL KENNETH
CAMPBELL WILLIAM EUGENE
CANELAKES PETER JOSEPH
CANNADA BRIAN JEFFREY
CAPEL JOHN BRUCE
CARAVETTA LARRY ANTHO-
NY
CARDENAS ARNOLDO J
CARDENAS RAMIRO
CAREY DANIEL EDWARD
CARLOCK RALPH LAURENCE
CARLONE JOHN JOSEPH II
CARLS TERRY ALAN
CARLSON JOHN EDWARD
CARLSON JOHN WERNER
CARMICHAEL HENRY ELLIS JR
CARMICHAEL SAMUEL LEE
CARNELL TALMADGE WAYNE
CAROLAN TIMOTHY JOHN
CARPENTER RALPH R JR

CARR MARTIN CODY
CARR ROBERT GEORGE
CARRELL LARRY DALE
CARRINGTON FRED EMERY
CARROLL WILLIAM EUGENE
CARSON MERVYN MAURICE
CARSON RICHARD RAY
CARTER DANIEL JR
CARTER LEONARD JAMES
CARTER RALPH WINFIELD
CARTLAND DONALD NOR-
MAN
CASSANO DANIEL
CASSATA ORRIN JOSEPH
CASSIDY PATRICK CHRISTIAN
CASTELLANOS SANTOS JR
CASTILLO LOUIS
CATES NORMAN GENE
CATLIN NORMAN RICHARD
CATTON JOHN LESLIE
CAULTON WILLIE RICHARD
CAUSEY JOHN BERNARD
CAVANAUGH JOHN CHARLES
CEDERLUND RONALD MI-
CHAEL
CELLETTI JERRY
CERIONE JAMES STANLEY III
CERVANTEZ EDWARD EDDY
CHALLBERG CURTIS PAUL
CHAMBERS JAMES DOUGLAS
CHAMBERS SAMUEL P III
CHAPMAN JERRY JUNIOR
CHAPMAN SIDNEY DAVID
CHAPPELL EDWARD LEWIS
CHASTINE KENNETH FOSTER
CHAVEZ ANTONIO GONZA-
LEZ
CHENOWETH IRVING S III
CHEPELY GENE E
CHILDERS STEPHEN ANDREW
CHILDRESS LEWIS CLAYTON
CHLEWA JOHN
CHRISTENSEN WILLIAM RAY
CHRONISTER JAMES VIRGIL
CHRYSTYNYCZ THEODORE
CHUNGES JERRY MICHAEL
CHURCH WILLIAM MALCOM
CHURCHILL STEVE JOHN
CICERO FEDELE ANTHONY
CISTARO RUDOLPH V JR
CIUPINSKI JAMES MICHAEL
CLAIR JAMES THOMAS
CLANCY CRAIG MICHAEL
CLANCY JOSEPH ALOYSIOUS
CLARBOUR DONALD ALAN
CLARE RICHARD STEVEN
CLARK HENRY PATRICK
CLARK NATHANIEL BUSTER
CLARK RICHARD CROSBY
CLARK WILLIAM JEROME III
CLATFELTER ROBERT DENNIS
CLAXTON CHARLES PETER
CLAY DOYLE GREGORY
CLAY KAROL
CLAYBROOK LARRY DEAN
CLAYPOOL RHONDAL GENE
CLAYTON JOHN W
CLEMENTS MARSHALL
EDWARD
CLENNON EDWARD FRANCIS
CLINCH JOSEPH RUSSLE
CLINTON DEAN EDDIE
CLOUGH DONNIE JOE
CLYMER DENNIS LEE
COATES EMORY THERON
COBB JOHNNY RAY
CODY WILLIAM DE BRECE
COFER ARTHUR WILLIAM
COFFEY JERRY EAIRD
COFFMAN CHARLES EUGENE
COFRAN WILLIAM EARL
COHEN HARRY
COLDREN ELDON DEAN JR
COLE THOMAS STEPHEN
COLEGATE WILLIAM KARL
COLEMAN DANIEL
COLEMAN JAMES IVORY
COLEMAN MICHAEL JOHN
COLEMAN RICHARD
COLFORD DARRELL LEE
COLLINS JAMES ALFRED
COLLINS MICHAEL LEE
COLLINS RAY
COLLINS VERNEL

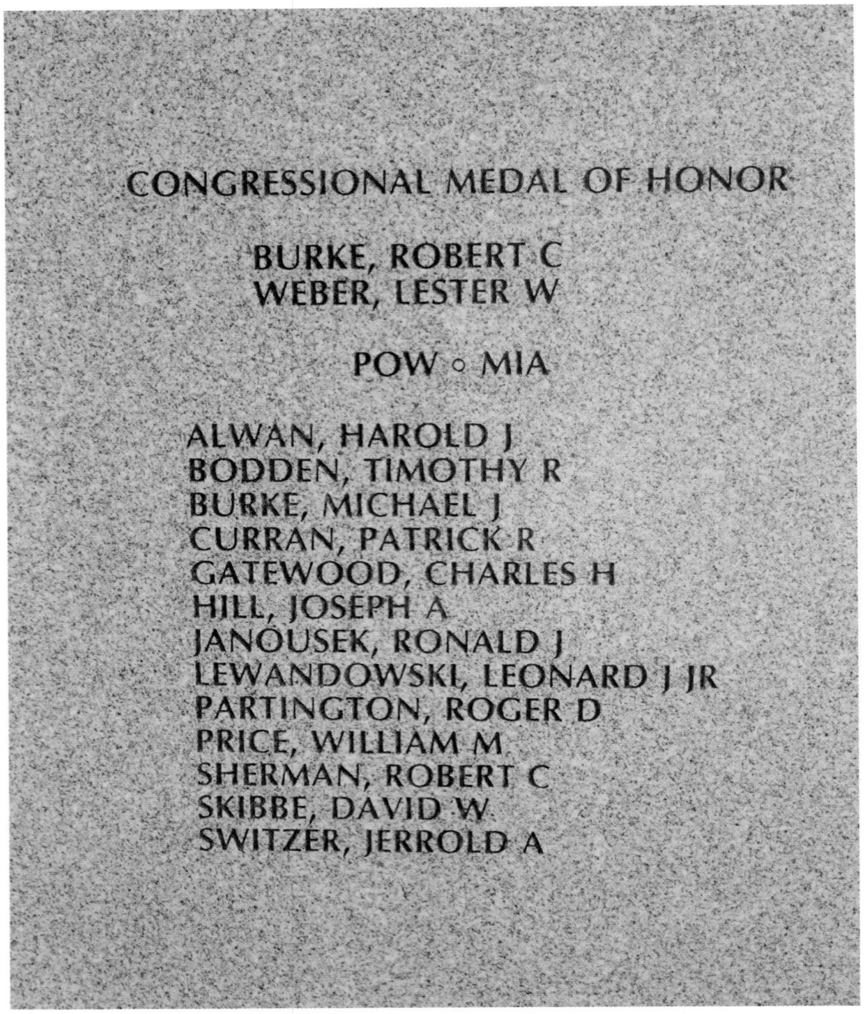

CONGRESSIONAL MEDAL OF HONOR

BURKE, ROBERT C
WEBER, LESTER W

POW ○ MIA

ALWAN, HAROLD J
BODDEN, TIMOTHY R
BURKE, MICHAEL J
CURRAN, PATRICK R
GATEWOOD, CHARLES H
HILL, JOSEPH A
JANOUSEK, RONALD J
LEWANDOWSKI, LEONARD J JR
PARTINGTON, ROGER D
PRICE, WILLIAM M
SHERMAN, ROBERT C
SKIBBE, DAVID W
SWITZER, JERROLD A

THE MARINES' HYMN

From the Halls of Montezuma
To the shores of Tripoli;
We fight our country's battles
In the air, on land and sea.

First to fight for right and freedom
And to keep our honor clean;
We are proud to claim the title
of United States Marine!

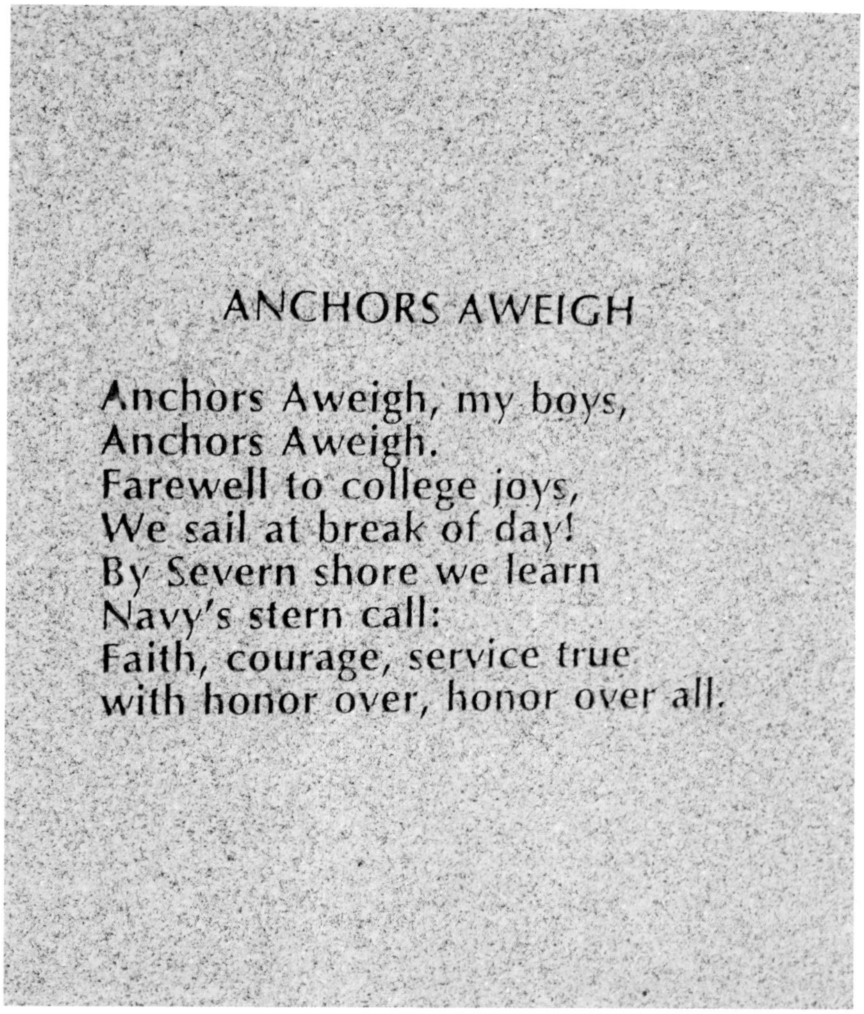

ANCHORS AWEIGH

Anchors Aweigh, my boys,
Anchors Aweigh.
Farewell to college joys,
We sail at break of day!
By Severn shore we learn
Navy's stern call:
Faith, courage, service true
with honor over, honor over all.

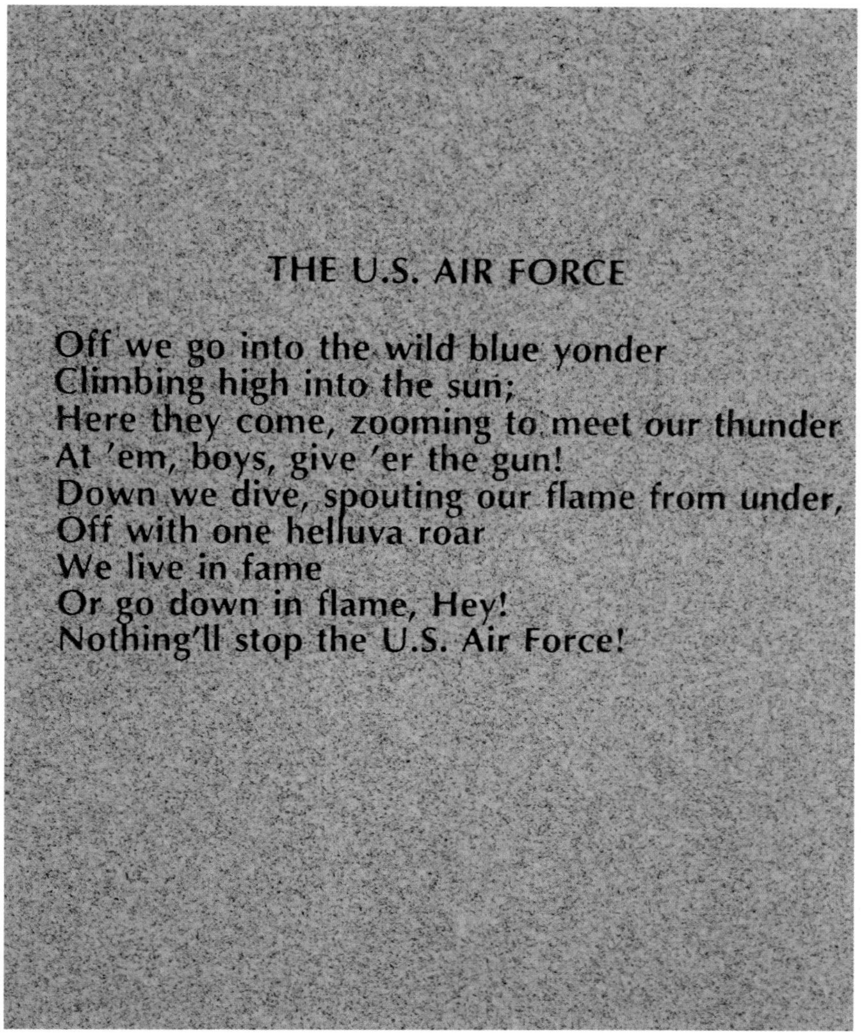

THE U.S. AIR FORCE

Off we go into the wild blue yonder
Climbing high into the sun;
Here they come, zooming to meet our thunder
At 'em, boys, give 'er the gun!
Down we dive, spouting our flame from under,
Off with one helluva roar
We live in fame
Or go down in flame, Hey!
Nothing'll stop the U.S. Air Force!

COLLINS WILLARD MARION
COLVINS RONALD EARL
COMBS FARRISH
CONDE-FALCON FELIX M
CONEY LAWRENCE NELSON
CONGIARDO THOMAS DEAN
CONNER PATRICK
CONNER THOMAS EARL
CONNOLLY THOMAS CHARLES
CONNOR PATRICK JAMES
COOK ALBERT ELMORE
COOK BERNARD JAMES
COOK BILLY LEE
COONEY PHILLIP BERNARD
COONEY THOMAS JOSEPH
COOPER AVERY LEE
COPACK JOSEPH BERNARD JR
COPLEY HENRY EUGENE JR
CORLEY ROBERT HAL
CORNELIUS MERLIN G JR
CORRIGAN DANNY JOSEPH
COSTANTINO RONALD JOSEPH
COURTNEY TERENCE FRANCIS
COUTURIAUX EUGENE JR
COVEY GENE TRACY
COWAN AARON DAVIS
COWAN HAROLD EUGENE
COWELL JAMES EDWARD
COX DANIEL RONEN
CRAGG GERALD
CRAIG CHARLES OWEN JR
CRAIG HARRY LEE
CRAIG ODELL
CRANDALL JOHN PAUL
CRAVENS ROBERT MILTON JR
CRAWFORD BOBBY DEAN
CRAWFORD LAWRENCE JOE
CRAWFORD RICHARD ALLEN
CREECH WILLIAM OWEN JR
CRICHTON CHARLES FREDERIC
CRITES RAYMOND
CRONIN DAVID MICHAEL
CRONIN JOHN EARL
CROON GALE WALTER
CROSS BENNIE LEE
CROUCH JACK EMANUEL JR
CRULL DALE ALTON
CRUM DUANE
CRUM EDWARD WALDREN
CRUTCHFIELD CHARLES ELLIS
CRUTHIRD GEORGE W
CUBIT BILLY RAY
CULLETON CARSON GREGORY
CUMMINGS JAMES THOMAS JR
CUNNINGHAM KENNETH LEROY
CUNNINGHAM ROBERT MAURICE
CURRAN PATRICK ROBERT
CURRAN ROBERT BRUCE
CURTIN JAMES CHRISTOPHER
CURTIN JOHN HENRY
CUTLER RICHARD ALLEN
CWIOK FRANK JOHN
CZERWIEC RAYMOND GEORGE
CZERWONKA AUGUST EMIL
DABBERT WILLIAM CARL
DABON NATHANIEL
DAFFRON THOMAS CARL
DAGNON MICHAEL ERWIN
DAHL ALBERT EUGENE
DAILEY BOBBY RAY
DAILEY WILLIAM GRANT JR
DALE DONALD MILTON
DALEY WILLIAM
DALIE LOUIE FRANK
DALLAPE TERRY LEE
DALTON JOHN MICHAEL
DALTON RANDALL DAVID
DANAY JERRY LEE
DANIEL MATHIS
DANIELS ELDRIDGE MAURICE
DANIELS LAWRENCE EDWARD
DARR AARON LEE
DASHO GEORGE ALBERT F JR
DAUGHERTY DENNIS MICHAEL
DAUGHERTY RICHARD LAWRENC
DAVIE RODNEY OWEN
DAVINO THOMAS ALPHONSE
DAVIS CLYDE
DAVIS EUGENE FESTER
DAVIS GARRY DON
DAVIS GLENN EDWIN
DAVIS GLENN PHILLIP
DAVIS JERALD C
DAVIS JOE MASON
DAVIS JOHN WILLIAM
DAVIS REX ALLEN
DAVIS ROBERT JOSEPH
DAVIS ROBERT NELSON
DAVIS RONALD CALEB
DAVIS RONNIE DEAN
DAVIS TERRY LEE
DAVIS WILLIE JR
DAVISON LARRY CHARLES
DAWSON ANDREW LEE
DAWSON DENNIS EUGENE
DAWSON MICHAEL DALE
DAWSON WAYNE EUGENE
DAYTON JAMES LESLIE
DE CARLO GENNARO JOSEPH
DE COSTE DAVID ANTHONY
DE CRAENE ALAN CHARLES
DE LEON RODOLFO
DE ROSA JOSEPH WILLIAM
DEAN ALAN JAMES
DEAN CARL EARLY JR
DEAN WILLIAM MEARL
DEBATES WILLIAM ARTHUR
DEBICKERO DENNIS RALPH
DECESARO JACK JR
DEDMAN TONY
DEDMON DONALD CLAY
DEIHL JOHN PERRY
DELANEY HERALD LEE
DELGADO MICHAEL JULIAN JR
DELGADO RUBEN
DEMATTEIS DAVID KELL
DEMERJIAN STEPHEN HAIG
DENCY KARL PETER
DENNY LAWRENCE EDWARD
DENTINO MERLE ALLEN
DERMONT DONALD EUGENE JR
DERRIG MICHAEL JAMES
DERRINGTON EARMON RAY
DES ROCHERS JAMES BRIAN
DETREMPE BARRY VICTOR
DEUEL WILLIAM TOWNSLEY
DEVEREAUX REESE
DEXTER HERBERT J
DI SANTIS WILLIAM RICHARD
DICKE DENNIS MICHAEL
DICKEN PERRY JR
DICKEY WILLIAM WALTER
DIDIER JOHN PAUL JR
DILLARD BERNARD
DILLON WILLIAM JERRY
DIMARZIO MARTIN JOHN
DIMOND ALVIN JAMES
DINE JAMES CHARLES
DINGMAN MILFRED HAROLD
DISCHERT JAMES RICHARD
DISSELKOEN DONALD GENE
DIVENS MELVIN
DIXON ALONZO LENORD
DIXON PATRICK MARTIN
DIXON RICHARD LEE
DOBBS JIMMIE LEE
DODD RICHARD WILLIAM
DODE FRED RICHARD
DODSON JERRY LEE
DOGGETT EDWARD JOSEPH
DOLIK PAUL EDWARD
DOMAN BENJAMIN VICTOR
DOMINIAK HOWARD STANLEY
DONAHUE JAMES ALLAN
DONAHUE JAMES T JR
DONNAL JOHN ANDREW
DONNELLY DAVID
DONOHO WILFORD LYNN
DONOVAN MICHAEL JOHN
DOOLEY DENNIS LYNN
DORSEY CECIL EVERETT
DORSEY DENNIS
DOSSETT JAMES EDWIN
DOUGLAS TERRY LEE
DOWDY JAMES RAY

Illinois

DOWJOTAS GERALD JAY
DOWNEY PATRICK H
DOYE RICKY LEE
DOYLE MICHAEL CHARLES
DRAKE RODNEY GEORGE
DRISKILL JERYL FRANKLIN
DROSZCZ DANIEL PATRICK
DRUSCHEL WILLIAM LENORD
DRYOEL DONALD L
DU BEAU GERALD EUGENE
DUCKETT CURTIS LEE
DUDEK JOSEPH WALTER
DUDLEY FOREST EDD
DUER THOMAS WADE
DUFFY THOMAS BENEDICT JR
DUGGAR KENNETH P
DUNBAR JOHN MICHAEL
DUNBAR ROBERT
DUNCAN WILLIAM JAY
DUNLAP LAWRENCE DAVID JR
DUNN LARRY
DUNN LAURENCE JOHN
DURFLINGER ROLLAND LEON
DVORATCHEK THOMAS
ANTHONY
DZIWISZ FRANK EDWARD JR
EADS DENNIS KEITH
EAKINS MARION TROY
EALEY WILLIS EDWARD
EALY WILLIAM DANIEL
EASLEY ODELL
EASTHAM MARTIN PHILLIP
EASTMAN EVERETT ALLAN
EATON EMMANUEL LLOYD
EAVES CARROLL WAYNE
EDGREN THOMAS GORDON
EDWARDS JAMES WALTER
EGAN TIMOTHY JAMES
EHLERS LARRY DEAN
EICKLEBERRY ROBERT DON-
ALD
EISENHOUR GLENN R
EL HONDAH DOVE
ELBEN MICHAEL WILLIAM
ELIZONDO FREDERICK H
ELLINGTON KENNETH JULIAN
ELLIOTT FRANK WILLIAM
ELLIOTT GERALD LEE
ELLIOTT TOMMY GENE
ELLIOTT WILLIAM KARL
ELLIS ALFRED
ELLIS BENNEL
ELSWICK LEX
ELYEA SIDNEY JOHN
ELZA RONALD LEE
EMBRY WILLIAM ESSIE
EMBRY WILLIAM ROBERT JR
EMERY ROBERT EDWARD
EMORY ILLINOIS JR
ENDERBY ROBERT FRANCIS
ENGESSER DANNY WRAY
ENGLAND RICHARD ALAN
ENGS RUSSELL LARNED III
EPPS HERSCHEL LEE JR
ERICKSON HOWARD W JR
ERICKSON RUSSELL MARTIN
ERICSON GARY WAYNE
ERLING WILLIAM NELS JR
ERVIN JOHN LEE
ESCH FRANK RYAN
ESKRIDGE JAMES EARL
ESPINOSA ELLIS CASIANO
ESSIG PHILLIP JOHN
EUSTACE ARTHUR BARNETT
JR
EVANOFF ALVIN LEE
EVANS CHARLES MICHAEL
EVANS JEFFERY WILLIAM
EVANS JOE
EVERSGERD MARLIN CHRIS

EVERSGERD NORMAN LEE
EVILSIZER DAVID NATHANIEL
FABRIS CHRIS FRANK
FAIRFIELD DENNIS HOWARD
FALLOON EDWIN JOSEPH
FANIS GEORGE NICHOLAS JR
FARNER JON MICHAEL
FARRELL DANIEL FRANCIS
FAULKNER MAURICE
FEDEROWSKI ROBERT ALLAN
FEEZEL HAROLD EUGENE
FEEZELL DAN GUINN
FELKAMP RONALD ALLEN
FERENCE MICHAEL WILLIAM
FERNANDEZ DENNIS
FERNANDEZ JAMES THOMAS
FERRELL JAMES LEE
FESSENDEN ROGER ALLEN
FIELDS ROBERT LOUIS III
FIESTER GLEN ALAN
FIGUEROA ANGELO
FIGUEROA FERNANDO
FIKE ROGER WESLEY
FILPI JOHN TAYLOR
FINCH PATRICK DALE
FINLEY MICHAEL PAUL
FINN MICHAEL BLAKE
FINZER BENJAMIN B
FIRAK ANTHONY MARIAN
FISCHER GEORGE ARTHUR
FISCHER NORMAN CHARLES
FISCHER WAYNE HENRY
FISHER DAVID LUTHER
FISHER DAVID WAYNE
FISHER OTIS SYLVESTER
FITCH GARY RAY
FITZGERALD DAVID EDWARD
FITZMAURICE TIMOTHY
GEORG
FITZPATRICK CURTIS L JR
FIVELSON BARRY FRANK
FLAMENT HOWARD L
FLANINGAM DAVID EUGENE
FLANNERY BRIAN MICHAEL
FLANNIGAN PHILLIP WAYNE
FLATLEY THOMAS MICHAEL
FLEISCHER DAVID ABRAM
FLEMING JERRY
FLEMING JOHN JAMES
FLEMING RICHARD ALAN
FLETCHER DAVID FOSTER
FLETCHER THOMAS THERON
FLOOD MICHAEL HAROLD
FLORES JOSE ANIBAL
FLOURNOY JEFFERY DONALD
FLOWERS RALPH EUGENE JR
FLYNN JOHN HENRY
FOAD MELVIN EUGENE
FOGLEMAN GEORGE EDWARD
FOHT STEPHEN CRAIG
FOLEY JAMES RICHARD
FONSECA JOHN
FORAN WILLIAM PATRICK
FORBES ARTHUR KIRKS
FORD BERNARD FRANCIS
FORD KENNETH LAVERNE
FORD MICHAEL EUGENE
FORD STEPHEN ROMO
FORDHAM JOHN LA VERNE
FOSTER MARK ANTHONY
FOSTER STEEN BRUCE
FOX BERNARD LYLE
FOX ROBERT ALAN
FOX RONALD LEE
FOZZARD ROBERT LEE
FRAKES ROBERT LEE
FRANCIS JOHN VINCENT
FRANKE BERNARD LEE
FRANKIEWICZ PHILIP ROBERT
FRANTA MICHAEL JOHN
FRAZER KENNETH CHARLES
FREDERICK JOHN WILLIAM JR
FREED ROBERT THOMAS
FREELAND TROIT DONOVAN
FREISE MELVIN JOHN
FRENCH ALLEN GEORGE
FRENCH JOY TRINT
FRENCL MICHAEL JAMES
FRENDLING EDWARD JOSEPH
FRERICKS LOUIS WAYNE
FRESE MICHAEL ALBERT
FREUND CARTER JOHN
FRIES DENNIS JEROME
FRIESE MICHAEL KEITH

FRIESON SAMUEL JEROME
FRISBY CHARLES LEE
FRIZZELL RONALD EUGENE
FROMM RONALD ALBERT
FROSSARD WILLIAM JOHN
FRYE JAMES KENNETH
FUGATE DALE LEROY
FULK MICHAEL RAYMOND
FULLER JAMES E
FULLER JAMES LARRY
FULLILOVE WILLIE KETCHERY
FULTZ MICHAEL KENT
FUNSTON JOSEPH ERNEST
GAFFNEY MICHAEL FRANCIS
GAGLIANO FRANK F
GAINES PHILLIP RAY
GAINES THOMAS LEE JR
GALARZA RUDOLPH JOSEPH
GALBAVY GEORGE RICHARD
GALLEGOS STEVEN
GALLOWAY CLARENCE
GAMBILL CHARLES RICHARD
GAMMON LARRY JAMES
GANION THOMAS FRANCIS
GARAPOLO FRANK WILLIAM
GARBER WAYNE ARTHUR
GARCIA ANTONIO
GARCIA JUAN REFUGIO
GARDNER ROBERT CHARLES
GARGUS FRANCIS BERTON
GARLICK RICHARD LEE
GARNER LARRY ARTHUR
GARRIGAN JOHN L
GARRINGER JAN DOUGLAS
GASKA LAWRENCE LEONARD
GASPERICH FRANK JOHN JR
GASSEN STEVEN CARL
GATEWOOD CHARLES HUE
GATLIN JERRY GENE
GATTIS CHARLES MANLEY JR
GAUS BRADLEY KENT
GAVARIA GEORGE LOUIS
GAVIN EZRA
GAYLORD GORDON MANSON
GAZZE JAMES ALBERT
GEDEON RUSSELL EUGENE
GEE LE ROY
GEE RAYMOND LEON JR
GEHL MICHAEL ARTHUR
GEIER WILLIAM MICHAEL
GEIS WILLIAM CHARLES
GELLER CHARLES GREGORY
GELONEK ROBERT EUGENE JR
GEMMATI ORONZO
GEORGE FRANK DANIEL
GERLACH STEVEN HENRY
GETLIN MICHAEL PETER
GETMAN CHARLES LESTER
GETZ ROBERT WILLIAM
GIBBS WILLIAM HARLEY JR
GIBSON DALE HENRY
GIBSON DONALD FREDERICK
GIBSON JOHN ARTHUR IV
GIEBE RICHARD JOHN
GILBERT JACK RICHCARDO
GILCHRIST FRANK R JR
GILGENBERG JOHN DANIEL
GILLEAN GARY LEWIS
GILLEN JOHN ALOYSIUS
GILMAN FREDERICK EUGENE
GILSON TIMOTHY LAWRENCE
GINTER DENNIS HARRY
GIPSON HOWARD WAYNE
GIRSCH ROBERT EDWARD
GISCHER GERALD MARION
GITHENS RICHARD EARL
GIUNTA MICHAEL ANTHONY
GIVENS ROGER LEE
GLASS WALTER LEWIS
GLASSFORD GARY BRUCE
GLAWE THOMAS DUANE
GLEATON MELVIN ROSCOE
GNIADEK ROBERT JOSEPH
GODDARD MYRON THOMAS
GODWIN RAYMOND WILLARD
GOEBEL THOMAS ANTHONY
GOECKNER EUGENE FRANK-
LIN
GOELZ EDWARD CHARLES
GOETHE SPENCER ALAN
GOLDEN CALVIN JR
GOLDSBERY JOHN ALLEN II
GOMEZ ARMONDO ABEL
GONCE RAY LONNIE

GONZALEZ DENNIS
GONZALEZ FRANCISCO JR
GONZALEZ VICTOR JR
GOODEN WILLIAM ELLIOTT
GOOSSENS MATTHEW RAY-
MOND
GORALSKI LEO STANLEY
GORDON DARWIN DALE
GORDON HENRY JOE
GORMAN KEVIN TERRENCE
GOSELIN ROBERT MARTIN
GOSNELL ODIS LEON
GOTTSCHALK WILLIAM
HENRY
GOUGH WILLIAM LYLE
GOWIN HARRY DALE
GOYNE ALLEN BENJAMIN JR
GRABLE MICKEY RAY
GRAFF JAMES HOWARD
GRAHAM GEORGE RICHARD
GRANATH JOHN EDWARD JR
GRANGE ARTHUR CHARLES
GRANT ARTHUR JOHN JR
GRANT JACKYA KEDERIS
GRANT ROBERT WILLIAM
GRASS LAWRENCE GEORGE
GRAY ALLEN RAY
GRAY DAVID ARTHUR
GRAY EDWIN MICHAEL
GRAY GERALD ALFRED
GRAY RONALD LEONARD
GRECO MICHEL JACK
GREEN ARTHUR WILLIAM
GREEN CHARLES JR
GREEN ERNEST
GREEN IVAN IVORY
GREEN KENNETH LESLIE
GREEN RICHARD AL
GREENE BEN JOHN
GREENE CARL MADASON
GREENE RICHARD HENRY
GREGG ROBERT STANLEY
GREGOIRE JOHN RICHARD
GREGORY GLENARD JAY
GRENSBACK THEODORE E JR
GRIFFEY JAMES RAY
GRIFFIN LEVESTER
GRINDOL PHILLIP WAYNE
GROENE DAVID
GROS RONNIE LEE
GROSS LARRY MICHAEL
GROSS RODGER THOMAS
GROVE WALTER BRENNEMAN
JR
GRUD THOMAS ANTHONY
GRUNEWALD BRUCE WALTER
GUBBINS EUGENE
GUENTHER TERRY ELMER
GUERIN WALTER THOMAS
GUIMOND PAUL GERALD
GUINN JIMMY HORACE
GULBRANTSON DAVID ARLIN
GULDAN JOHN ANTHONY
GULLEY RONALD WALTER
GURVITZ JEFFERY
GUSTAFSON EDWARD LEE
GUTOWSKI WALTER JOSEPH
GUTTMANN JOHN PETER JR
GUY ALLEN EDWARD
GUYETT GEORGE ERVIN
HAAKE DAVID OSCAR
HAAN DOUGLAS JOHN JR
HABADA TOM
HABERMAN NOLAN DONALD
HACEK JAMES DAVID
HACKER KURT ERIC
HACKETT WILLIAM RALPH JR
HADLEY JAMES STANTON JR
HAGERMAN ROBERT WARREN
HAGIE MICHAEL WADE
HAGSTROM RONALD EDWIN
HAHN DENNIS FRANCIS
HAHNER GEORGE LAWRENCE
JR
HAIN GEORGE ANTON
HAINES MICHAEL SCOTT
HALE PAUL EDWARD
HALEY PATRICK LAWRENCE
HALFORD CHARLES E
HALL DELBERT EUGENE
HALPIN WILLIAM FRANCIS
HAMILTON CHARLES RAY-
MOND
HAMILTON TIMOTHY MCKEE

HAMMETT DAVID A
HAMMOND PETE B
HAMPTON DAVID LEE
HANIK RAYMOND CONRAD
HANRAHAN JEROME M JR
HANSEN DONALD CHARLES JR
HANSEN LYLE WAYNE
HANSON MICHAEL LEROY
HARBIN GARY LEE
HARDEN JOHN MERRILL
HARDIG TERRY NEIL
HARDIMON EARNEST JR
HARGRAVE STEPHAN LEE
HARING KARL RICHARD
HARKER ROBERT DALE
HARMS FREDERICK WILLIAM
JR
HARRIS GLENN ALVIN
HARRIS HARRY JAMES
HARRIS HARVEY JR
HARRIS JESSE LEE
HARRIS JESSIE EARL
HARRIS MICHAEL PAUL
HARRIS PATRICK JAMES
HARRIS PRENTISS JR
HARRISON CELISTER JR
HARRISON DANA ALAN
HARRISON JIMMIE RAY
HARRISON THOMAS EDWARD
HARRY CLIFFORD ROBERT
HART ROBERT WILLIAM
HART SAMUEL NICHOLAS
HARTL JOSEPH MICHAEL
HARTMAN FRED ANDREW JR
HARTUNG THOMAS EDWARD
HARTZELL SAMMY LOWELL
HARVEY CARMEL BERNON JR
HARVEY CHARLES EDWARD
HARVEY RAYMOND
HARWOOD WILLIAM PHILLIP
HATTER JEROME GERALD
HAUSCHILDT JOHN CHARLES
HAVARD MICHAEL JOHN
HAVENS DANIEL LEE
HAWK JESSE VIRGINIUS III
HAWKEY LOUIE ELMER
HAWKINS JERRY PAVEY
HAWKS RICHARD FRANKLIN
HAWKS ROBERT JAMES
HAYES WILLIAM ALLEN
HAYS GEORGE BURNS
HAZLIP CHARLES EDWARD
HEALEY JAMES JAY
HEAPS JOHN WAYNE
HEARNS WILLIAM VAN
HEAVER BRIAN TRACY
HECIMOVICH ROBERT ALLEN
HECKWINE PETE GERALD
HEDGECOCK DONALD GENE
HEFFERNAN DANIEL JOSEPH
HEFT NORMAN ANTHONY
HEIDEMAN THOMAS EDWARD
HEILIG ROBERT FRANK JR
HEIMAN SHERLIN ANDREW
HEINEMEIER CHARLES
THOMAS
HEINRICH MICHAEL
HEISE THOMAS HOWARD
HELLYER WILLIAM EDWARD
HELM WILLIAM CARROLL
HEMMINGSON NELS IVAR
HENDEE LARRY KEITH
HENDERLIGHT BUDDY
EUGENE
HENDERSON JACK JR
HENDERSON JONATHAN
HENDERSON MONTE EUGENE
HENDERSON STEPHEN CARL
HENDRIX JOHN RUSSELL
HENGELS RAYMOND GEORGE
HENKE RICHARD ARTHUR
HENNINGSEN REID CHARLES
HENRICKS DONALD MERLE JR
HENRY EPHRIAM JR
HENRY GERALD EDWARD
HENSEY LAWRENCE LOUIS JR
HENSLEY JAMES CURTIS
HENSLEY MEDFORD S JR
HENSON CHARLES KENNETH
HENSON GWYN THAXTON
HERGERT THOMAS MALCOLM
HERIAUD FREDERICK
CHARLES
HERINGTON JOHN DONOVAN

HERMAN ALLAN JOSEPH
HERMANOWICZ JOHN JOSEPH
HERREID ROBERT DALE
HERRING THOMAS FOREST
HERSCHBACH DAVID EDWARD
HETLAND RONALD LEE
HICKS ROBERT LYLE
HIENSMAN STEVEN LANCE
HIGHSMITH JAMES ARTHUR
HILGART RICHARD PETER
HILL DAVID NOEL
HILL JOSEPH ARNOLD
HILL ROBERT WILSON
HILL RONALD ALLEN
HILLIARD DONALD RAY
HINES JAMES ROOSEVELT
HINKLE JAMES MELVIN
HINKLE WILLIAM CECIL
HIRSCH MARSHALL RAYMOND
HISSONG HARRY LEAVERN
HITCHINS LLOYD LYNN
HLAVACEK GLENN JOHN
HOBAN CHARLES JOHN III
HOCK ROBERT WILLIAM
HOCKETT JAMES RAYMOND
HODGE CHARLES EDWARD
HODGE THOMAS WAYNE
HOENIGES THOMAS LEO
HOFFMAN DANIEL ROBERT
HOFFMAN DAVID R
HOFFMAN EDWIN EARL
HOFFMAN ROBERT ALLEN
HOFFMANN RICHARD ALFONCE
HOFMANN EARL FREDERICK
HOGAN GORDON LEE
HOGAN JOHN WESLEY
HOJNACKI ROBERT FRANK
HOLBROOK CHARLES ALLEN
HOLCMAN MORRIS ELIOT
HOLDEN ROBERT FRANKLIN

HOLIAN GARY LEE
HOLLIDAY BERNARD
HOLLIE ROBERT LEE JR
HOLLINGSWORTH NICHOLAS LE
HOLLIS THOMAS WILLIAM
HOLLOMAN CLETUH JR
HOLLOWAY MICHAEL JAMES
HOLLWEDEL CHARLES WILLIAM
HOLMES ROBERT THOMAS
HOLTHOFF WILLIAM HENRY
HON JOHNNY JOE
HOOKER SANDY LEE
HOOKS DENNIS RAYE
HOOP ROBERT GENE
HOOTS DOUGLAS JAMES
HORCHEM NELSON LEPORT JR
HORN RAYMOND LEON
HOULE KIRK EDWARD
HOUSH ANTHONY FRANK
HOUSMAN ROBERT CHARLES
HOUSTON BENNIE LEE
HOWARD LAWRENCE EDWARD
HOWERTER BRUCE G
HOWES ROGER HAYDEN
HOWLAND JOHN CHARLES
HUART MARTIN REINHOLD JR
HUDSON RICHARD GREY
HUFF BRUCE NORMAN
HUFF CHARLES FRANK
HUGHES DENNIS FOX
HUGHES JAMES KENNETH
HUGHES ROBERT WAYNE
HUMES MAYNARD JEWEL
HUNDLEY MOSE CHILDS
HUNTER DAVID
HUNTLEY EDWARD GLENN
HURLEY NOEL
HURST ROBERT LEE
HUSKA MARTIN SAM
HUTCHISON STANLEY

ROBERT
HUTSON MICHAEL LOUIS
HUTTON JAMES EDWARD
INBODEN JAMES RAY
INBODEN STEVE LEE
INGRASSIA MICHAEL JOSEPH
IRBY CHARLES WILLIAM
IRELAND RONALD WAYNE
IRVING EARL ELESTER JR
IRWIN THOMAS EDWARD
ISAACSON MILFORD DON
IVY JESSE W JR
IWASKO EDWARD BERNARD
IYUA ARCHIE HUBERT JR
JABLONSKI MICHAEL JAMES
JACKSON BILLY DALE
JACKSON DAVID ERIC
JACKSON LEONARD JR
JACKSON TYRONE
JACKSON WILBUR DESMAR
JACOB RANDALL GORDON
JACOBS THOMAS CARLYLE
JACOBSGAARD DAVID KEITH
JACOBSON SCOTT NELSON
JAECKELS TOBY EDWARD
JAHN DENNIS EARL
JALLOWAY STEPHEN FRANK
JAMES CLIFFORD W
JAMES WILLIE LEE
JAMROCK PHILIP ROBERT
JANKA JAMES EDWARD
JANOUSEK RONALD JAMES
JANSEN LARRY WAYNE
JANSSEN ROBERT DEAN
JARANSON JAMES EDWARD
JAROLIMEK JAMES MICHAEL
JARRETT LEMOYNDUE
JARVIS LEE BRIAN
JARVIS ROGER ELLIS
JASINSKI RONALD NORMAN
JAVORCHIK JOHN CHARLES
JECMEN ANTON JAMES JR
JEFFERSON JIMMIE LEE
JEFFRIES CHARLES B JR
JELINEK FRANCIS PETER
JENKINS EARL DALE JR
JENNINGS BOBBY DALE
JENNINGS THOMAS ALVIN
JENNINGS WILLIAM CLARENCE
JENSEN JOHN JEFFREY
JENSEN RONALD CHARLES
JENSEN RONALD JOHN
JEPSON ARTHUR C JR
JERDE GERALD DEAN
JETERS DAROLD
JEWELL RONALD DEE
JOHANSEN RONALD
JOHNSON AARON GILBERT
JOHNSON ARNOLD EDWARD
JOHNSON CARROLL MARSHALL
JOHNSON CHARLES
JOHNSON CHARLES EDWARD
JOHNSON CLAYTON WINSLOW
JOHNSON DANIEL GENE
JOHNSON FRANK JR
JOHNSON GARY STEVEN
JOHNSON HENRY L
JOHNSON JAMES ALBERT
JOHNSON JERRY ALLEN
JOHNSON JERRY JACK
JOHNSON JESSE LEWIS
JOHNSON JOHN ANDRES
JOHNSON JOHN PETER
JOHNSON KENNETH CARL
JOHNSON LARRY ALLEN
JOHNSON LAWRENCE
JOHNSON MARLIN JAMES
JOHNSON MARSHALL D
JOHNSON MARTIN RAYMOND
JOHNSON MICHAEL ARTHUR
JOHNSON MICHAEL KIRK
JOHNSON NAPOLEON
JOHNSON OLIVER
JOHNSON ROBERT CHARLES
JOHNSON ROBERT EDWARD
JOHNSON RODNEY DEAN
JOHNSON TERRY MELVIN
JOHNSON VERNE LYLE JR
JOHNSON WYMAN TRUVOY
JOHNSTON TERRY RANDALL
JONES CLARENCE JR

JONES DAVID LEE
JONES DENNIS KEITH
JONES DONALD EUGENE
JONES GARY LEE
JONES JACK PAHL
JONES JAMES ANDREW
JONES LARRY WAYNE
JONES LOWEN LEON
JONES MILTON JOSEPH
JONES OMAR DAVID
JONES RICHARD WILLIAM
JONES ROBERT EUGENE
JONES RONNIE LEE
JONES WILLIAM JR
JONES WILLIAM THOMAS
JONSSON RONALD BRYNIEL
JORDAN ROBERT LEROY JR
JORDON ORVAL CLYDE III
JOSELANE HOWARD LEO
JOSEPH THOMAS EDWARD
JOSSENDAL RICHARD LYNN
JOWERS BEN JR
JOYCE THOMAS MICHAEL
JUAREZ MATEO
JUDY DAVID LYNN
JUSTICE RALPH ROGER
KABARA DENNIS FLOYD
KAGEBEIN DALE LEONARD
KALE MICHAEL ROBERT
KALTER JAMES MICHAEL
KARAS WALTER
KARSZNIA LESZEK STANLEY
KASPER GREGORY JOSEPH
KASZUBOWSKI DANIEL F
KAUGARS JOHN
KEAG ROBERT THOMAS
KEANE PATRICK BRENDAN
KEARNEY TIMOTHY WILLIAM
KEATS ROBERT GEORGE
KEDROSKI ALBERT ARTHUR JR
KEELING ARTHUR R
KEENE ROBERT MICHAEL
KEEP DONALD WAYNE
KEFER CHARLES HENRY JR
KEGLEWITSCH WILHELM LUDWI
KEHOE ROBERT ANTHONY
KELL LYLE FRANCIS
KELLER CHARLES LEE
KELLER LAWRENCE OSWALD JR
KELLER RONALD DALE
KELLERMANN ALLAN HOWARD
KELLETT JOHN EDWARD
KELLEY MICHAEL PATRICK
KELLUMS DENNIS ALLEN
KELLY STEPHEN GERE
KELLY STEPHEN JAMES
KELNHOFER JOSEPH ALLEN
KELTON RICHARD LANE
KENEIPP WARREN OWINGS JR
KENNEY TERRY JOE
KENTER MICHAEL WILLIAM
KERKSTRA HARRY WILLIAM
KERR CHARLES FRANKLIN
KERSTEN LESTER JOSEPH
KERWIN REVELRY LAWRENCE
KETELAAR ROBERT LEE
KETTMANN DANIEL RAY
KICK DANIEL LEE
KIESELBURG GARY ROBERT
KIESLER RAYMOND JOSEPH
KIESLING GERALD DENNIS
KIESTLER JAMES LARRY
KILVER PHILLIP HENRY
KIMBALL RICHARD NELSON JR
KIMBLE LESTER WILSON
KING GEORGE PAUL
KING GLEN EDWARDS
KING JAMES MICHEAL
KING JOHN TERRENCE
KING MONROE DEE
KINNETT GEORGE DELMER
KINNEY DELMER LANGLY
KINNEY MERLE ALLAN
KISALA WALTER
KISUCKY ANTHONY EDWARD
KITCHEN EDDIE JR
KITCHEN MICHAEL ROOSEVELT
KLAUS GEORGE PETER
KLAUSING THOMAS PATRICK

KLEINAU CARL EDWARD
KLOPMEYER JAMES MARTIN
KMIEC JOHN STANLEY
KNECHT PAUL HERBERT
KNECHTGES MICHAEL ALLEN
KNIGHTON HIRAM J JR
KNOEFERL KENNETH JOSEPH
KNOX DAVID
KNOX LEONARD WAYNE
KNOX MICHAEL JOSEPH
KNUDSEN JOHN HENRY
KNUTSON FELIX DELANO
KOBOR FRANK LOUIS
KOCH FRANKLIN LEROY
KOEHLER WALTER ALLEN
KOELPER DONALD EDWARD
KOLAR JERRY JOSEPH JR
KOLBECK FRANZ JOSEPH
KOLLMANN RICHARD LEON
KONOW MICHAEL JACOB
KOPEC EDWARD
KOPKA RICK EDWARD
KORANDO OLIVER KASPER
KOS JOHN JOSEPH
KOSAR RICHARD DENNIS
KOSKY RICHARD ALLEN
KOSOWSKI KENNETH JOSEPH
KOTULLA MICHAEL JERRARD
KOVANDA JOHN MARTIN
KOVAR JAMES RUSSELL
KOVARIK FRED GEORGE
KOWALK CHARLES NORBERT
KOWSKI EDWARD JOHN JR
KRAUHS CURTIS JOHN
KREIS SHERWOOD DAVID
KRELL ROBERT GAIL
KRETSINGER DONALD MAURICE
KRISAN DAVID ANTHONY
KROGER NEIL A
KRUEGER DAVID RUSSEL
KRUEGER GEORGE THOMAS
KRUPA RICHARD DIDACUS
KRYSTOSZEK GERALD MICHAEL
KUBIAK LEONARD
KUHN ROBERT WILLIAM
KUHNKE WILLIAM ANDREW
KUPPERSCHMIDT JEROME DEAN
KUROPAS MICHAEL VINCENT
KYSER JOHN THOMAS
LA BUNDY JOHN ARTHUR
LA FLEUR GERALD JOHN
LABAY JOSEPH STANLEY
LACKEY KEITH BERNELL
LADD LARRY ROBERT
LADD LEAMON RAY
LAKE JAMES LEE
LAKWA EDWARD JOHN
LAMA EDWARD BARTHOLOMEW
LAMBERT DONALD RAY
LANDERS CHARLES FRANCIS
LANDERS RONNIE RAY
LANDON WILLIAM GREGORY
LANE JAMES JOSEPH JR
LANG DAVID ROBERT
LANG WALTER ROBIN
LANGE CONRAD THOMAS
LANGE HANS DIETRICK
LANGENHORST HERBERT CYRIL
LANGHAM WILLIAM C
LANKFORD CHARLES BERNARD
LANKFORD JOHN WAYNE
LANTER RODGER PAUL
LARA HUMBERTO
LARSEN TERRY LEE
LARSON THOMAS LLOYD
LASCELLES DON HARRISON
LASSITER WILLIAM O III
LATIMER ROBERT NATHANIEL
LATIMER WILLIAM ROYCE
LATORIA DAVID JOSEPH
LAUBER ROBERT DEAN
LAVIN THOMAS PATRICK
LAVISH JOHN LARRY
LAWLER JOHN E JR
LAWLOR JAMES V
LAWRENCE JOHNNY HAROLD
LAYE EDGAR CARTHA JR
LAZZAROTTO ALBERT LOUIS

Illinois

LE FEVRE BRIAN FRANCIS
LEACH WILLIAM EDWARD
LEDBETTER ROGER DALE
LEE GARRETT FLORIS
LEE ROBERT MICHAEL
LEE STEPHEN MICHAEL
LEE STEVE DONALD
LEHMAN NELSON SAYLER JR
LEIF MICHAEL WAYNE
LEKOVISH DONALD F
LEMON JEFFREY CHARLES
LENDERMAN WAYNE MORRIS
LENOVER WILLIAM JOSEPH
LEOPOLD LESTER HAROLD
LEROY JEROME EDWARD
LESTER THOMAS LYNN
LESTON THOMAS JEROME
LETTO ROGER WILLIAM
LEUTENEGGER JOE CARL
LEVAN MELVIN VERNON
LEWANDOWSKI LEONARD J JR
LEWIS HARRY JR
LEWIS JOHN FREDERICK
LEWIS LESLIE ROSS
LEWIS RONALD EUGENE
LEWIS SINCLAIR BYRON JR
LIBERATI PETER JOSEPH
LIBERTY RONALD EDWARD
LINDER HERBERT III
LINDNER JOHN MICHAEL
LINDQUIST VIRGIL
LIS RICHARD JOHN
LISOWSKI ANDREW ZBIG-
NIEW
LITTLE WILLIAM WALTER III
LLOYD LOWELL RAY
LOCKHART CURTIS
LOGAN CHARLIE LEE
LOHENRY ROBERT RAYMOND
LOHREY JAMES WILLIAM
LONG CLYDE EDWARD JR
LONG DENNIS LANE
LONG SAMMIE JAMES
LONG THOMAS KENDRICK
LOPEZ HENRY ROBERT
LOPEZ LEOPOLDO AYALA
LOPEZ-VAZQUEZ LEONARDO
LOPP JAMES LEONARD
LOVELLETTE GEORGE RON-
ALD
LOWE RONALD BRUCE
LOWE STEVEN RAY
LOWERY CHARLES WILLIAM
LOWERY LARRY DEAN
LOZANO JOSEPH ALFRED
LUBAS JAMES ALEX
LUBERDA ANDREW PATRICK
LUCAS HOWARD LEWIS JR
LUCCHI AERIO JOSEPH JR
LUEBKE JOHN CHARLES JR
LUKENS DONALD GLEN
LUSTER DALE ALAN
LUTTRELL GARY ALLEN
LUTTRELL JOHN WALTER
LYDEN MICHAEL P
LYLES J L
LYLES MICHAEL ALLEN
LYNCH DANIEL MICHAEL
LYNN HOMER MORGAN JR
LYNN JOHN JOSEPH JR
LYNN ROBERT RAY
LYONS WILLIAM PERRY
MACDONALD GEORGE
DUNCAN
MACHACEK WILLIAM ALLEN
MACHALICA JOSEPH PAUL
MACHUT RICHARD RAY
MACK LARRY WESLEY
MACKEY LARRY ALLEN
MADDEN THOMAS ANDREW

II
MADDOX HAROLD WAYNE
MADDOX NOTLEY GWYNN
MADDY LARRY ROBERT
MADSEN MARK EUGENE
MADSEN WILLIAM JOSEPH
MADSON ROBERT WARREN
MAGGIO JOSEPH ANTHONY JR
MAGGIO RANDALL EUGENE
MAGNUSON DAVID JACK
MAGYAR BLAZE III
MAJKOWSKI DONALD HENRY
MAJOR STEVEN ROBERT
MAKSYMIW WALTER B
MALCZYNSKI MATTHEW
PAUL
MALECKI ROBERT RICHARD
MALNAR JOHN MARION
MALONE RICHARD CLAIR
MANDERFELD KENNETH JAY
MANHEIM VERNON ARTHUR
JR
MANN DAVID ROY
MANRIQUE RAMIRO JR
MANSTIS ANTHONY WAYNE
MANUEL LARRY GEORGE
MARA JOSEPH P
MARCHANT PAUL LAFON-
TAINE
MARROQUIN PEDRO JR
MARSCHALL ALAN FREDERIC
MARSHALL RICHARD CARL-
TON
MARTIN ARTHUR GLENN
MARTIN ASA JR
MARTIN JAMES C JR
MARTIN JOHN EUGENE
MARTIN LARRY
MARTIN ROBERT DENNIS
MARTIN ROBERT ELMER
MARTIN RONALD LEE
MARTIN WILLIAM DEAN
MARTINEZ PETER JOHN JR
MARTINEZ PETER STEVEN
MARTINEZ ROGELIO MANUEL
MARYFIELD WILLIAM RICH-
ARD
MASILLO JUAN
MASNY BERNARD JOSEPH
MASON HARRY STANLEY JR
MASON JAMES PHILLIP
MASON JOSEPH ANSON JR
MATE DONALD RICHARD
MATHENY BOBBY DANIEL
MATHENY LARRY DALE
MATHEWS CHARLES LEON
MATHEWS JAMES LEONARD
MATHIAS JOSEPH VERNON
MATHIESEN ERHARDT WIL-
LIAM
MATHISON MICHAEL K
MATSON ROBERT EDWIN
MATTHEWS JOSEPH
MATTHEWS KENT DOUGLAS
MATTSON BERNARD CHARLES
MATTSON PAUL EDWARD
MAUL RICHARD ALLEN
MAXSON JOHN ROBERT
MAY ALAN RICHARD
MAY DANIEL ARNOLD
MAY JOHN ALBERT
MAYBERRY LARRY EUGENE
MAYMON DAVID MARK
MC ALLISTER KENNETH
RALPH
MC ALLISTER ROBERT ALLEN
MC CALL DIMITRIOUS
CORTEZ
MC CALL PHILLIP GLEN
MC CANN JAMES KEVIN
MC CARTHY EDWARD JOSEPH
MC CLAIN MICHAEL DEE
MC CLANE MICHAEL JAMES
MC CLOUD STEVEN WILLIAM
MC CORD MICHAEL RAYE
MC CORMICK JAMES MILTON
MC CRANEY CLARENCE
MC CREIGHT TIMOTHY JOE
MC CUE WILLIAM JAMES
MC CULLOUGH PREZEL
MC DANIEL WILLIAM T
MC DANIELS WILLIAM
LAWREN
MC DONIAL WESLEY

MC ELROY GLENN DAVID
MC ELWEE JACKIE RAY
MC FARLAND LOUIE JUNNIE
MC FARLAND WILLIAM
LLOYD
MC GEATH RICHARD ALLEN
MC GEE CHARLES ADAM
MC GEE HERMAN
MC GHEE LARRY DALE
MC GLOTHLIN MICHAEL
JOHN
MC GOVERN TERRANCE
JAMES
MC GOWAN PAUL JOSEPH
MC GRATH THOMAS HOW-
ARD
MC GUIRE ANDY JR
MC GUIRE TIMOTHY PATRICK
MC HENRY EDWARD CURTIS
MC HUGH GARY ROBERT
MC INTIRE HERMAN LEROY
MC INTOSH ROBERT JAMES
MC INTOSH WALTER LESLI JR
MC KAY GERALD EUGENE
MC KEE DONALD WAYNE
MC KENNA JOHN MICHAEL
MC KENNA KENNETH R JR
MC KEON JOSEPH THOMAS JR
MC KINSON MICHAEL JAMES
MC LELLAN EMMETT DENEEN
MC MILLEN RONALD DEAN
MC NABB ALFRED LEE
MC NAMARA EDWARD
MICHAEL
MC NEIL SYLVESTER
MC PHILLIPS JAMES CRAIG
MC RIGHT ROGER LYNN
MC WHIRTER JAMES GILBERT
MCCLEER TOMMY MIKE
MCCORMICK CARL OTTIS
MCGLONE GERALD FIELD
MCMASTER JAMES THOMAS
MEAD THOMAS JOHN
MEADS KIM ELMER
MEEHAN DONALD LLOYD JR
MEEKER RAMON ARTHUR
MEIEROTTO EDWARD RALPH
MELOY LARRY JOHN
MELTZER EDWARD ALAN
MENSING STANLEY ALFRED
MERKER RAND RUSSELL
MERRELL DAVID RICHARD
MERSHON STEVEN VICTOR
MESSICK JAMES AVERY
METCALF JERRY EUGENE
METROS CARL DEANE
MEYER DAVID PAUL
MEYER EDWARD K
MEYER VAL GREGORY
MICHAEL JAMES RICHARD
MICHALOWICZ STANLEY
JULUI
MICHALOWSKI RAYMOND
JOHN
MIFFLIN JOHN RAY
MIKRUT JOHN THOMAS
MILAM CALVIN EDWARD
MILCO WILLIAM JOHN
MILES BLAINE STANLEY JR
MILES BRUCE EDWARD
MILES JOHN EMORY
MILLER CLARENCE ALVIE JR
MILLER DANIEL HAROLD
MILLER DAVID EDWARD
MILLER DENNIS CARL
MILLER GERALD CRAIG
MILLER GLENN RAY
MILLER JIMMIE
MILLER JOHN EDWARD
MILLER NORMAN ANTHONY
MILLER PAUL LYNN
MILLER ROBERT EDWIN III
MILLER ROBERT HENRY
MILLER ROBERT MICHAEL
MILLER RONALD YATES
MILLER TERRY DEAN
MILLER THOMAS LEONARD
MILLER TOMMY NEAL
MILLER WALTER CHARLES
MILLER WILLIAM FRANKLIN
MILLS DAVID LEE
MILLS GREG WENDELL
MILLS HANS LOTHAR
MINCEY JOHN ELLIS

MINDOCK RICHARD WILLIAM
MINK DONALD KENNETH
MIRANDA MICHAEL
MIRANDA PAUL ANDREW JR
MISHEIKIS THEODORE N JR
MITCHELL ALBERT JEAN
MITCHELL CHARLES IRVIN
MITCHELL CHRISTOPHER
MITCHELL JOSEPH WILLIAM
MITCHELL ROBERT LEE
MITCHELL STEPHEN PHILIP
MIZNER GARY LEE
MLODZINSKI BRUNO J JR
MOAKE CHARLES EDWARD JR
MOEHRING DEAN WARD
MOKE RUSSELL EUGENE
MOLL STEVEN WILLIAM
MOLLOY JOSEPH JAMES
MONKMAN DONALD EUGENE
MONROE GREGORY JAMES
MONROE JAMES HOWARD
MONTES MIGUEL ALEJANDRO
MONTGOMERY EDDIE JR
MONTGOMERY STANLEY
DYKUS
MONTOYA GUADALUPE
ESPARZA
MOODY CHARLES WILBURN
MOODY PAUL JAMES
MOODY RICHARD FINISA
MOODY ROBERT WILCOX
MOOMEY CHARLES RAY
MOON WILLIAM CHARLES
MOONEY ROBERT RAY
MOORE DALE WILLIAM
MOORE DENVER JR
MOORE DONALD EUGENE
MOORE HARRY TRUMAN
MOORE LEE ELMER JR
MOORE MAURICE
MOORE NELSON ROGER
MOORE RICHARD LYNN
MOORE ROBERT GENE
MOORE TERRY DWIGHT
MOORE WILLIAM JOHN
MORALES RAMON J
MORAN MICHAEL THOMAS
MORENO VICTOR AUREL-
LIANO
MOREY ALDEN FRANK JR
MORGAN GEORGE ALLEN
MORGAN JUNIOR RAY
MORGAN KENNETH DWIGHT
MORGAN MILLER EDWARD
MORGAN ROBERT FRANCIS
MORGAN ROGER WAYNE
MORRIS DONALD J
MORRIS WINSTON
MORRISON EDWARD JR
MORRISON GEORGE RAY
MORSE LEONARD ALAN
MORTENSEN TERRENCE JOHN
MOSELEY STEPHEN C
MOURGELAS DENNIS W
MOZDZEN DALE MICHAEL
MUCCI JOHN ROCCO
MUCHA LOUIS STEPHAN
MUEHE MARK RONALD
MUHR WARREN FRANCIS
MULLIN RICHARD ROCCO
MULLINEAUX STEVEN PAUL
MUMMERT ALLEN LAWRENCE
MUNSON RONALD LEE
MURDOCK STANLEY
MURRY WILLIE JAMES
MUSZALSKI GREGORY ALLAN
MYERS BILLY EUGENE
MYERS HAROLD EDWIN
MYERS LAWRENCE THOMAS
MYLES ANTON CAESAR
NAFFZIGER MARSHALL
EDWARD
NAGY JOHN PAUL
NANCE CHARLES THOMAS
NAUGHTON JOHN R JR
NAWROCKI ROBERT DENNIS
NEAL JOHNNY LEONARD
NEAL JONATHAN
NEAL NELSON DENFIELD
NEAL ROBERT JUNIOR
NEELEY WILLIAM MERRITT
NEILL JOE MELVIN
NELSON LEROY A
NELSON ROBERT ALLEN

NEMETH ANTHONY JOHN
NEVILLE WILLIAM ROY
NEWBERRY WAYNE ELLS-
WORTH
NEWBOULD WILLIAM
GEORGE
NEWMAN BOBBY JOE
NEWMAN FRANK CHARLES
NEWSOME DEAN OLIVER
NICHOLS PHILIP GWYN
NICHOLSON DAVID DONELL
NICHOLSON GLENN EDWARD
NICKELS LESLIE DAVID
NICOLA DENNIS GRANT
NICOLINI PETER JOSEPH
NIEMCZUK PETER RICHARD
NIESPODZIANY CASIMIR
NITZSCHE LEONARD ARTHUR
NODDIN WILLIAM DAVID
NOLAN JOSEPH PAUL JR
NORBUT GEORGE EDWARD
NORFLEET HENRY JR
NORRIS OTIS LESLIE JR
NORTON ROGER KAY
NOSEK WILLIAM ALLEN
NOVAK EDWARD JAMES
NOVAK GERALD FRANCIS
NOVOTNY JAMES ROBERT
NOWAK RONALD MICHAEL
NOWAKOWSKI WALTER JOHN
NULL HAROLD EDWARD
NUTT RICHARD E
O BRIEN EDWARD TERRY
O CONNOR JOHN FRANCIS
O CONNOR JOHN THOMAS
O CONNOR WILLIAM JAMES JR
O DEA THOMAS FRANCIS JR
O DONOVAN EDWARD
THOMAS
O GUINN MICHAEL EUGENE
O MALLEY FRED GILLESPIE
O MEARA LAWRENCE W
O REILLY TIMOTHY BOURKE
OCASIO VICTOR JR
ODENEAL JIMMY
ODLE JOHN CHARLES
ODOM VERN ERIC JR
O'DONNELL MICHAEL DAVIS
OEHLER GEORGE HERMAN
OLIVE MILTON LEE III
OLIVER FRED JR
OLMSTEAD JOHN PAUL
OLSEN DONALD WAYNE JR
OLSEN FLOYD WARREN
OLSEN JAMES DANIEL
OLSEN OLAF THOMAS
OLSON ALLEN EDWIN
OLSON CHARLES ANDREW
OLSON MARK ALLEN
OLSON MEADDOW JOHN
OLSON RANDALL ALAN
OLSON WILLIAM JAMES
ORDONEZ RAYMOND
ORLOWSKI JOSEPH M
ORMOND DENNIS ALAN
ORR GEORGE JAMES
ORR WARREN ROBERT JR
ORSZULAK KENNETH BEN-
EDICT
ORTIZ JOHN
ORTIZ JOHN MANUEL
ORTIZ MANUEL GALVAN
ORTMANN RALPH BERNARD
OSBORNE JOSEPH JR
OSKILANEC WILLIAM JOSEPH
OSTENSON RONALD LE ROY
OSTERTAG VERNON LEE
OTTE RICHARD LEE
OTTO WILLIAM FREDRICK
OVERMYER ROBERT JOE
OVERTON DOYLE WAYNE
OVERTURF PHILIP GENE
OWEN WILLIAM LEE JR
OWENS JAMES HOWARD JR
OWENS ROBERT ERNEST
PADILLA THOMAS
PAEZ JOSEPH FLAVIO
PAGAN EDWIN PEREZ
PAGE DAVID RONALD
PAGE JAMES ROBERT
PAGE JOHN MAC ARTHUR
PAGE LEWIS WAYNE
PAGE RONNIE
PAGE THELBERT G

PAHR WILLIAM JOHN JR
PAINTER DENNIS EARL
PALCOWSKI RICHARD WAYNE
PALENIK JAMES ANDRES JR
PALM DENNIS DU WAYNE
PALMER JAMES
PALMER LARRY RAY
PALMER LEROY JR
PAMPEL LOREN LEE
PANEK ROBERT JOSEPH SR
PANKIEWICZ JAMES MICHAEL
PANNELL JOSEPH
PAPE FRANK ALBERT
PAQUETTE RICHARD WALTER
PARKHURST VINCENT BER-
TRAM
PARKS GLENN ALLEN
PARR KEITH MASON
PARRIS JEROME JR
PARSONS DONALD EUGENE
PARTINGTON ROGER DALE
PARZYNSKI HERBERT JOSEPH
PASCARELLA FRANK MARIO
PASCHALL LES HOWARD
PASSAVANTI JOSEPH J III
PASTROVICH EUGENE AR-
THUR
PATERSON ROSS JAMES
PATRIZI ANTHONY
PATTERSON CLEVELAND
PATTERSON STANLEY F
PATTERSON TERRY ALLEN
PATTERSON WALLIS GILBERT
PAVEY DALE RUSSELL
PAYNE ROBERT PAUL
PAZDAN DENNIS SIGMUND
PEACH ROBERT ALAN
PEARSON EARL THOMAS JR
PEARSON ROBERT VERNER
PEARSON WAYNE EDWARD
PEDERSEN CLARK RUSSELL
PEDICONE JEROME JOHN
PEEK DENNIS LEE
PEELER WILLIAM GERALD
PEFFER GREGORY LEE
PEGGS ALBERT LEE
PEKNY CHARLES DENNIS
PELIKAN ROGER
PELZMANN GERALD F
PEMBERTON ALVIN LEWIS
PENDELL JERALD WAYNE
PENDOLA ANTHONY EUGENE
PENN CHARLES VARENCE
PENNAMON RICHARD STEVE
PENNINGTON JAMES E JR
PENNINGTON RICHARD W
PENSON HAROLD EUGENE
PENSONEAU TERRY
PEPPERS HAROLD DOUGLAS
PEREZ RAUL VICTOR
PEREZ VICTOR JR
PERILLO DONALD LEE
PERKINS CHARLIE JR
PERKINS DONALD DEAN JR
PERRIS FELIZ
PERRY ANDREW JR

PERRY KENNETH EDWARD
PERRY KENNETH MERLE
PETER LE ROY ALVIN
PETERS EDWARD KENT
PETERS GEORGE CHARLES
PETERS LARRY J
PETERS MICHAEL
PETERS ROBERT CHARLES
PETERSEN DONALD ROGER JR
PETERSON DONALD CARL
PETERSON HOWARD MATHIS
PETERSON JOHN KENNETH
PETERSON TED BARNETT
PETRAUSKAS KESTUTIS A
PETROSSI WILLIAM JR
PETTERSON CHARLES STAN-
LEY
PETTY WILLIE JR
PEYTON WILLIAM ALLEN
PHEIFFER MICHAEL LAVERNE
PHENEGAR WESLEY ROBERT
JR
PHILLIPS AQUILLA ANTHONY
PHILLIPS ERNEST
PHILLIPS JOHN ROBERT
PHILLIPS LLOYD FREEMAN
PHILLIPS ROY LEE
PHILLIPS TOMMIE
PHIPPS JAMES LARRY
PIANTKOWSKI EDWARD
JOSEPH
PIASECKI JOHN MICHAEL
PICKETT JOSEPH CHARLES JR
PICKETT MALCOLM JEROME
PIERCE CALVIN BOB
PIERCE JOHN ROBERT
PIERCE JOSEPH ROBERT CLIN
PIERCE LEO
PIERCE ROGER LEE
PIGG EDWARD WAYNE
PIKE DENNIS EUGENE
PILKINGTON CARL EDWARD
SR
PINKSTON ROBERT GENE
PINTO JOSEPH JOHN
PIPPIN DAVID WAYNE
PITTS BENJAMIN FREDERICK
PLACZEK PAUL GEORGE
PLOTE DALE EDWIN
PLUCINSKI JACK ALBERT
PLUMMER JAMES ARMAND
PLUMMER NEWTON RAY
POE JESSIE GERALD
POELING EUGENE FREDERICK
POHANCEK STEVE
POHLMAN CHARLES PAUL
POINTER DARRYL WARREN
ANT
POLLARD JAMES ROBERT JR
PONDOFF JOHN CHRISTO-
PHER
POOL JERRY LYNN
POOLER JOHN SHELBY
POPPENGA PATRICK EDWARD
PORTER DELBERT RAY
PORTER DONALD JOHN

POTE FREDDIE CHARLES JR
POTEET THOMAS JAMES
POTTS CLIFTON DENNIS
POUNDS ALVIN LEE
POWELL MORRIS
POWELL ROBERT
POWERS EDWARD DEAN
POWERS RONALD EUGENE
PRAIRIE LEROY PAUL
PRATT DONALD WILLIAM
PRATT WILLIAM TERRY
PRENGEL MICHAEL WAYNE
PRESCOTT MILTON EMMETT
JR
PRESS ROBERT M JR
PRESSER PAUL MICHAEL
PRICE GARY DONALD
PRICE GARY WAYNE
PRICE JAMES HENRY
PRICE WILLIAM MARSHALL
PRIDEMORE JAMES LESLIE
PRIP SOREN
PRITCHARD WILLIAM HENRY
PROFILET ROBERT C
PROIETTI ANTHONY AL-
PHONSE
PROMBO JOHN ANTHONY
PROSE THOMAS DEAN
PROTZ CLAUDE DOUGLAS
PROVENZANO ROBERT LEE
PRUETT DARREL EUGENE
PRUSKO PAUL STANLEY
PUETZ MICHAEL DUANE
PUZYREWSKI LESLIE
PYLE JOHN WILLIAM
QUALLS DAVID WAYNE
QUARTERMAN EARL QUIN-
NON
QUERRY HOWARD EMERSON
QUICK MICHAEL EDWARD
QUIN CULLEN WOOD
QUINN PATRICK THOMAS
QUIROZ JOSEPH ALBERT
RABURN WILLIAM FAY
RACHAL LIONEL THOMAS
RACINE FRANKLIN DOUGLAS
RACKHAUS JOHN PELL
RADER CHARLES WAYNE
RAINEY LARRY STEPHEN
RAJCEVAC HANS ANTHONY
RAMIREZ LOUIS JOSEPH
RAMM FERENC JOHN
RAMOS SAMUEL
RAMSDEN RANDALL EDWARD
RAMSEY ALAN RYAN
RANDALL EDDIE SAM JR
RANKINS SAMUAL KAYE
RANSON DAVID WILLIAM
RAPCZAK MARTIN JOSEPH
RAPP JOSEPH LOUIS
RASH HARRY DON
RASMUSSEN ROBERT MI-
CHAEL
RASSANO WILLIAM
RATHE PHILIP HENRY
RATTIN DENNIS MICHAEL

RAWSON WILLIAM ALLEN
RAYBORN DANNY KEITH
RAYCHEL JAMES DANIEL
REA PHILLIP KENNETH
REATHERFORD LARRY REX
REBER MICHAEL RICHARD
REDENIUS DAVID GARY
REDFEARN DON ALLAN
REDMOND DONALD MERLE
REDMOND JOSEPH VERN
REED EDWARD ROGER
REED GEORGE PARNELL
REED PHILIP PAUL
REED RICHARD LEON
REEL WILLIAM EDWARD
REILLY MICHAEL PATRICK
REKAU HAROLD EDWARD
RENDER CECIL LAVON
RENZ JAMES THOMAS
REVELLE GLENN
REXROAD RONALD REUEL
REYNOLDS FRANK EVERETT
REYNOLDS GARY EDWARD
RHODES JAMES LAWRENCE
RHODES RICHARD JAMES
RICE JEROME JAMES
RICE JESSE
RICHARD ANDREW GUS
RICHARDS DOUGLAS WAYNE
RICHARDS LEONARD JEFFREY
RICHARDSON ARTHUR GENE
RICHARDSON LEMOND
RICHARDSON OSSIE
RICHTMYRE CHARLES LAW-
RENC
RIDDLE WILLIAM MILLER
RIDGEWAY RICHARD
RIEPE EVERETT DALE
RIGGINS ROBERT PAUL
RILEY CHARLES JOHN
RILEY HARRY LEE JR
RILEY JAMES THOMAS
RILEY KIRK IRWIN
RIMMER JAMES EDWARD
RINGENBERGER ROBERT E
RIORDAN PATRICK CARLISLE
RIOS GERADO PEDRO
RISSI DONALD LOUIS
RIVES JOHN ARTHUR JR
ROACH CHARLES MICHAEL
ROBBLEY RICHARD PHILLIP
ROBERTS JOSEPH RAY
ROBERTS RONALD EUGENE
ROBERTSON JIMMY KARON
ROBERTSON THOMAS HARRY
ROBINSON CLIFFORD LEROY
ROBINSON HAROLD JACK JR
ROBINSON JAMES WILLIAM JR
ROBINSON ROBERT EDWARD
ROBINSON WALTER
ROCHACZ RICHARD JOHN
ROCHKES FRANCIS ALBERT
RODENBECK RODERICK
JAMES
RODRIGUEZ CESAR RODRIGO
RODRIGUEZ DENNIS JAMES

RODRIGUEZ JOSEPH
ROESLER ARTHUR CLEON
ROG EDWARD JOSEPH JR
ROGERS HERSHEL GALE
ROGERS ROBERT CHARLES
ROGERS ROBERT JAMES
ROGERS RONALD DEAN
ROGERS WILLIS JR
ROGIERS CHARLES JOSEPH
ROGOWSKI RONALD CHESTER
ROGUS ANDREW JOSEPH JR
ROHAN FREDERICK LEO
ROHRKASTE RONALD ED-
WARD
ROLLINS ARBAL JR
ROMANO WILLIAM ROSS
ROME ROOSEVELT SNOW
RONIGER JUNIOR FLOYD
RONZANI CHARLES KENNETH
ROSANOVA DANIEL FRANK
ROSE LARRY EMMETT
ROSE LAWRENCE CARROLL
ROSEN DANIEL ELMER
ROSS CHARLES GREGORY
ROSSI ALDO JR
ROST LEROY ALPHUS
ROTH BILLIE LEROY
ROTH BRUCE JONATHAN
ROTH IVAN DAVID
ROTHERY RICHARD ALAN
ROUSH ROBERT ROGERS
ROWELL RICHARD A
ROY JAMES DEAN
ROYALTY AMEL DOUGLAS
ROZANSKI EDWARD CHARLES
RUHLOFF GARY CARL
RUNSER ROBERT JOSEPH
RUSSELL CARL ERIC
RUSSELL CECIL LEE
RUSSELL LARRY GENE
RUSSELL RONNIE LEN
RYAN JOHN ROGER JR
RYAN WILLIAM DEAN
RYCKO RAYMOND ADAM
SABO ANDREW ROBERT
SACCO EDWARD STEVEN
SACHASCHIK JAMES HARRY
SACKS JAY CHARLES
SALA JAMES DONALD
SALAZAR ROY
SALONIES EDWARD JR
SAMPLE RONALD NEIL
SANDERS CHARLES WILLIAM
SANDERS HENRY CLYDE
SANDERS WILLIAM RAYMOND
SANDIDGE THEODORE
WILLIAM
SANDOVAL HECTOR MON-
TALVO
SANDS RICHARD EUGENE
SANSONE DONALD FRANK
SANTANA FLORENTINO JOHN
SANTELLANO LUIS ADRIAN
SANTUCCI VINICIO FREDEK
SAPP JON CHARLES
SAULSBERRY CLARENCE L JR
SAUNDERS DARRYL ELDRIDGE
SCALF DARYL GENE
SCARBROUGH ROGER ALLEN
SCHADDELEE WILLIAM D
SCHAEFER CHARLES HAROLD
SCHAEFER WILLIAM ERIC
SCHANEBERG LEROY CLYDE
SCHATZLEY MICHAEL DONN
SCHELL TERRY LEE
SCHERTZ JOHN EDWARD
SCHIRO GERALD ANTHONY
SCHLICK JOSEPH FRANCIS
SCHLOSSER STEVEN MICHAEL
SCHMELTZ JERRY E
SCHMIDT GERALD BERNARD
SCHMIDT RONALD EUGENE
SCHMIDT WILFRED F JR
SCHNEIDER JACK ARTHUR
SCHNELLER ANTHONY JOHN
JR
SCHOONVELD RICHARD JAY
SCHROEDER ALFRED M JR
SCHROEDER DONALD LEE
SCHROEDER WILLIAM RAY
SCHUEREN DANIEL RICHARD
SCHUKAR RONALD KEITH
SCHULMAN SHELDON BORIS
SCHULTZ DENNIS MELVIN

Illinois

SCHULTZ JAMES CHESTER
SCHULTZ KENNETH EUGENE
SCHULTZ WILLIAM JOHN
SCHULZE DAVID EDWARD
SCHUMACHER RONALD KENNETH
SCHURCH RONALD LEE
SCHUTZ RICHARD JAMES
SCHWALBACH GEORGE AUSTIN
SCHWARTZ MARTIN PETER
SCHWARTZ RANDALL FRANK
SCHWARZ FRANCIS ANTHONY
SCHWEBEL MICHAEL PHILIP
SCHWEIG VICTOR JOHN
SCHWICHOW RICHARD JOSEPH
SCHWICK MARTIN FRANK JR
SCHWIDERSKI RICHARD DEAN
SCHYSKA LEROY FLOYD
SCOTT DAVID LEE
SCOTT DORTY HINCHMAN JR
SCOTT GAYLAND OMER
SCOTT JOHN WALTER
SCOTT MICHAEL JON
SCOTT ROBERT MILLER
SCOTT THOMAS WILLIAM
SCROGGIN MICHAEL THOMAS
SCROGGINS JAMES LELAND
SCUDIERO PATRICK FRANK
SCULL JOHN FELLOWS JR
SCULLY PATRICK R JR
SEAGROVES MICHAEL ANTHONY
SEARGENT ROBERT LEE
SEARING JACK EDWARD
SEAY BOBBY DAREL
SEFRHANS JAMES
SEIBERT MICHAEL ROBERT
SEIDEL KENNETH WAYNE
SEISSER KENNETH ANTHONY
SEK MITCHELL FRANCIS
SELLETT STEPHEN CHARLES
SEPUT FREDERICK WILLIAM
SETTLEMIRE WILLIAM DAVID
SEVERSON PAUL ROY
SHAFER ROBERT LAURENCE
SHAFF MAURICE ALBERT JR
SHAFFER JACK LEON
SHAFFER LAWRENCE ALLEN
SHANER MICHAEL IRA
SHANKS JAMES EVERETT
SHARP KEITH FRANCIS
SHAW KRIS EDWARD
SHAW LEE ROY
SHEHORN TOMMY LOREN
SHELBY JAMES BENJAMIN
SHELBY JAY CLAYTON
SHELLEY DELMAR
SHELTON EARL S
SHELTON RONALD THOMAS
SHELTON WESLEY STEWART
SHEPARD RAYMOND ANDREW
SHEPLER ANTHONY GEORGE JR
SHERMAN ROBERT CARL
SHIANNA LOUIE JOHN
SHIELDS ELMER MATTHEW
SHIELDS GARY DON
SHILT RICHARD EUGENE
SHIMP ANDREW HARRY
SHINER JOHN ROBERT
SHIPLEY RONALD EUGENE
SHIPMAN MARVIN LEROY
SHIPMAN ROBERT DUANE
SHIRMANG RICHARD
SHOOT TERRY WILLIAM
SHOWMAKER RONALD

EUGENE
SHROBA THOMAS MICHAEL
SHUKAS JAMES CHRIS
SHUMBARGER DALE EARL
SIBLEY RALPH
SIDDONS JAMES GARLAND
SIEBEN EDWARD MICHAEL
SIEGER RAYMOND MARTIN
SIEKIERKA DONALD BERNARD
SIETSEMA DENNIS RAYMOND
SIMMERMON ROBERT JOHN
SIMMONS ELROY
SIMMONS RANDALL ROBERT
SIMPSON EDWARD MONROE
SIMPSON MICHAEL PAUL
SIMPSON WILLIAM JAMES
SIMS LARRY ROY
SINCAVAGE MICHAEL JOSEPH
SINK CHARLES ROBERT
SINKSEN ARTHUR DALE
SINTIC GREGORY JOHN
SIPKA RONALD WAYNE
SIPP RODGER WILLIAM
SIRCHER PAUL CHARLES
SIRIANNI PAUL JR
SIROUSA MICHAEL ANGELO
SKIBBE DAVID WILLIAM
SKINNER PHILLIP CRAIG
SLABINGER PETER WALTER
SLAGEL JAMES ALLAN
SLAGER CHARLES ALBERT
SLANE LYLE EDWARD
SLANE WILLIAM LLEWELLYN
SLAWEK JOSEPH DENNIS JR
SLIFKA JOHN JOSEPH
SLY RICHARD STEPHEN
SMILES WALTER LEROY
SMITH ALBERT EDWARD JR
SMITH BERNARD EDWARD
SMITH CARL GENE
SMITH CURTIS
SMITH DALE GENE
SMITH DONALD WOODROW
SMITH HAROLD VICTOR
SMITH HARVIE G
SMITH JAMES ALFRED
SMITH JAMES LEE
SMITH JEFFREY EARL
SMITH JOHN CHARLES
SMITH JOHN CURTIS JR
SMITH JOHN MARSHALL
SMITH JOHN RUSSELL
SMITH JOSEPH STANLEY
SMITH LUKE ANDREW JR
SMITH MARVIN GENE
SMITH PHILIP CORY
SMITH ROBERT LEE
SMITH ROBERT MICHAEL
SMITH SAMUEL JEROME
SMITH STANLEY BRUCE
SMITH STEPHEN JAY
SMOCZYNSKI THOMAS JOSEPH
SMRTNIK DONALD EUGENE
SNODGRASS DALLAS RAY
SNYDER LORA WILLIAM
SNYDER ROBERT LEE
SNYDER TERRY LEE
SOBACKI PETE WILLIAM
SOBOTA DANIEL JAMES
SOLCZYK RICHARD JOHN
SOLOMON JAMES VERDELL
SOLTOW NORMAN WILLIAM
SORIM ROLLEEN C
SOTO JOSEPH MARTINEZ
SOUHRADA TERRENCE LEE
SOUTHEY JAMES RUSSELL
SPAIN ERVIN
SPEAR MICHAEL JOHN
SPEARMAN WILLIAM T III
SPENCER HERBERT CHARLES
SPENCER JOHNNIE JR
SPENCER STEPHEN ALAN
SPENCER WENDELL
SPICER JERRY EUGENE
SPIEGEL ROBERT EUGENE
SPILKER KENNETH ALFRED
SPINO ANTHONY LAWRANCE
SPOHN JOHN SCOTT
SPRAGG HAROLD DEAN
SPRINGER TIMOTHY MICHAEL
SPROUSE JERRY WAYNE
SPURLOCK JOHN
ST PETER ROBERT EUGENE

ST PETERS JOHN DONALD
ST PIERRE DEAN PAUL
STACY MICHAEL LEIGH
STADEL CHUCK MICHAEL
STAGGS ROBERT DALE
STANFORD EARL MICHAEL
STANLEY FRANKIE
STANTON HAROLD E
STARK GORDON WILLIAM
STASSI JAMES STEPHEN
STATECZNY HARRY JOHN JR
STATON PAUL RAY
STAUDOHAR TERRENCE EDWARD
STEARNS FRANK EDWIN
STEC FRANK LOUIS
STEELMAN TEDDY WAYNE
STEIBEL FRANK DALE
STEIN PAUL ANDREW
STEIN PAUL HENRY JR
STEINHEBEL KENNETH ERWIN
STENDER PAUL ALAN
STEPHENS ALLEY OAKLEY
STEPHENSON FREDERICK DALE
STEPP DOW E
STEVENS ALLYN TROY
STEVENS MARVIN OWENS
STEVENSON JAMES DERRILL
STEVENSON MELVIN L
STEWART GEORGE CURTIS
STEWART RONALD WAYNE
STIEHLER GEORGE DENNIS
STIGEN WAYNE DOUGLAS
STITES JAMES JOHN
STOCKWELL BRYAN B
STONE EDWARD WILSON
STONE HARMON S JR
STONE HAROLD ALVIN
STORY JAMES CLELLON
STRANDE THOMAS ALVIN
STRATEGOS PETER STEPHEN
STREHLE ERNEST WILLIAM
STRNAD FRANK JAY
STROHL BILLIE RICHARD
STROM LARRY A
STROMBACK GLENN CHARLES
STROTHER CHATWIN ARNOLD
STUBSTAD GERALD EDWARD
STURTEVANT THEODORE JAMES
SUBERT GEORGE THOMAS
SUEDMEYER MERILL LAWRENCE
SULLIVAN BENJAMIN JOSEPH
SULLIVAN THOMAS EMERSON
SUPERCZYNSKI JOHN PAUL JR
SURWALD MICHAEL EDWARD
SUTTON JACK RICHARD
SWAN ROBERT RONALD
SWANSON JOHN WILLARD JR
SWANSON NELS WILLIAM
SWATSLEY MICKEY LYNN
SWEENEY ROBERT MICHAEL
SWEENEY TIMOTHY JAMES
SWITZER JERROLD ALLEN
SWOOPE RUDOLPH
SYKES DERRI
SYKES HAMP JUROME JR
SZCZUPAJ JAMES WALTER
TABOR CLAUDE EDWARD
TABOR EVERETT LEROY
TADEVIC RALPH DULANE
TAFT ROBERT EDMUND
TAGGART ISAAC
TAMMEN WILLIAM DWIGHT
TAPP NEWTON LEE
TARBERT CHARLES STANLEY
TARJANY RANDOLPH MICHAEL
TATARYN GEORGE LUBOMYR
TATE GARY DENTON
TATE KENNETH WAYNE
TATE TODD III
TATUM HERBERT ARTHUR
TAUBERMAN CHARLES G SR
TAYLOR DANIEL MORRIS
TAYLOR DONALD CLAUDE
TAYLOR GARY LYNN
TAYLOR LARRY DEAN
TAYLOR RONALD BURTON
TEBBE RONALD JOE
TEGTMEIER LESLIE JON

TELLIS ANDREW JESENEK
TEMPLETON JOHN ASHLEY
TENNIS THOMAS ROY
TERRELL EDDIE GEAN
TERRY JOHN FRANCIS JR
TERRY ORAL RAY
TESSARO MICHAEL JOHN
THARP GERALD LEROY
THAXTON DAVID EDWARD
THOELE NICHOLAS EUGENE
THOMAS JAMES LAWRENCE
THOMAS JAMES OLIVER
THOMAS KENNETH DEANE JR
THOMAS MARSHALL FLOYD
THOMAS NORMAN EUGENE
THOMAS REGINALD MICHAEL
THOMAS ROBERT ERVIN JR
THOMAS WALTER REED
THOMPSON THELBERT K JR
THOMPSON WILLIAM HOWARD
THORNE-THOMSEN CARL SPAUL
THORNTON TERRY LEE
THUNMAN RICHARD GWINN
TIEMAN EDWARD LEWIS
TIMMERMAN ALLAN DAVID
TIMSON DAVID OLIVER
TINKER JOHN GREGG
TOADVINE DENNIS ARRON
TOENNIES NORMAN GEORGE
TOEPRITZ RICHARD
TOMPKINS GLENN ALAN
TOMPKINS HARVEY JOSEPH
TOPPS RONNIE NEAL
TORRES ANGELO
TOTCOFF DENNIS STEVEN
TOTORA CHRIS ANTHONY
TOWER KENNETH KEITH JR
TOWLE JOHN CLINE
TRECINSKI LEON
TREEST NORMAN EUGENE
TREMAYNE JAMES RONALD
TRENT LESLIE ROLAND
TRENT WILLIAM DERRILL
TREVINO RUDOLPH ROBERT
TREZEK JERRY ALLEN
TRIPLETT MARK LEON
TRIPLETT A W
TROCK THEODORE ALLEN
TRODDEN PATRICK JOHN
TROIANELLO CLEMENT JOSEPH
TROLIA MICHAEL PATRICK
TROTTER THOMAS MICHAEL
TROYE DANIEL ROBERT
TRUCANO ALAN DALE
TRUMBLAY LEONARD JAMES
TRUSTY WILLIAM ROBERT JR
TRYGG STANLEY HERBERT JR
TUCKER ARTHUR L
TUCKER MELVIN EUGENE
TUCKER VALENTINE
TUNISON GEORGE ROBERT
TURBITT RICHARD JOHN JR
TURK EDWIN FRANCIS JR
TURK JON PETER
TURKSTRA ARTHUR JOHN
TURLEY RICHARD LYNN
TURNER KENNETH LEON
TURNER THOMAS GEORGE
TURONE NORMAN MICHAEL
TURSKEY HAROLD SEAN
TUTTLE CLETUS DALE
TWING ROBERT ANTHONY
TYLER ALLAN ROBERT
TYRKA PETER STEVEN
URBANIAK EDWARD
URDIALES ALFRED JR
URDIALES CHARLES A JR
UTECHT ROBERT STEPHEN
UTTER THOMAS DUANE
VADBUNKER JAMES PATRICK
VAICKUS ANTHONY JOSEPH JR
VAN DUYN JON FRANCIS
VAN GORDER WILLIAM JOSEPH
VAN GUNDY NELSON EARL
VAN HOOSIER JAMES D
VAN HORN JOHN RICHMOND
VAN WAMBEKE RONALD ARTHUR
VANCE SHERMAN DALE

VANDEN EYKEL MARTIN D II
VANDERKLOOT HARRY CORNELI
VANGELISTI MICHAEL J
VANGUNDY GEORGE JEFFERSON
VARGAS MARCELINO JR
VARICK ROBERT KITTRIDGE
VARNER RAYMOND ROBERT JR
VASILOPULOS JOHN WILLIAM
VAUGHN HOWARD GREGORY
VAULTZ JIMMY LEE
VAVROSKY PAUL PETER
VENEGAS VERNON BERNABE
VERHELST JAMES LAYMAN
VERSCHEURE JOHNNY DELBERT
VEVERA PHILIP JOHN
VICICH ALBERT LEE
VICTORY JOHN JOSEPH
VIEHWEG MICHAEL
VILKAS ALLEN RUDOLPH
VITALE WILLIAM MILLER
VLASAK WILLIAM JAMES JR
VOGEL DONALD FRANCIS
VOGLER GALE KURK
VOIGTS RICHARD SCOTT
VOSYLIUS VINCENT
VOTAVA JAMES JOSEPH JR
VOWLES JOHN WESLEY
VOYLES JOHN WALTER
WADE MICHAEL ALLEN
WAGMAN JEFFREY BURTON
WAGNER ROBERT JAMES
WAGSTAFF STEVEN RAY
WAINWRIGHT DAVID BARD
WAINWRIGHT MICHAEL ALBERT
WAJDA PHILIP JOHN
WALDEN DARRELL EDWARD
WALJESKI CHARLES
WALKER AARON
WALKER BARRY RONALD
WALKER EDDIE LEE
WALKER JULIUS LEMUEL JR
WALKER M B JR
WALKER RICHARD JR
WALLACE JAMES LEROY
WALLACE RICHARD FRANK
WALLEN GARY LEE
WALLS KENNETH MARION JR
WALSH CASPAR MARVIN III
WALSH WILLIAM THOMAS JR
WALTERS DONALD EDWARD JR
WALTHERS FRANK DANIEL
WALTON JOSEPH HERBERT
WALTRICH ROBERT JOHN
WARD JAMES HOWARD
WARD WALTER LEONARD
WARD WILLIAM
WARE JOE L
WARREN MICHAEL JAY
WARZECHA GERALD WALTER
WASHBURN WAYNE ARTHUR
WASHINGTON ANTHONY FELIX
WASHINGTON JAMES ERVIN
WASHINGTON LEONARD B JR
WASHINGTON MAURICE JOEL
WATERLOO MICHAEL JEFFERY
WATERS MELVIN LESTER
WATERS MICHAEL
WATKINS LARRY LANCE
WATSON SULLIVAN WALL
WATTS ROBERT LEE
WATTS RUSSELL DAVID
WEAVER JOHN SIMMONS
WEBB BRUCE DOUGLAS
WEBB NORVELL JOHNATHAN
WEBB STEVEN CHARLES
WEBER LESTER WILLIAM
WEBSTER ROBERT LEWIS
WEEKFALL EDDIE LEE
WEHRHEIM CHARLES GEORGE
WEIMER WILLIAM PATRICK
WEISS DAVID EARL
WEITZ DONALD EDWARD
WELCH JACK ALLEN
WELCH LARRY EUGENE
WELCH MICHAEL JOHN
WELGE BRUCE RICHARD

WELLINGHOFF RALPH ALVIN
WELLS BARRY SCOTT
WELLS KENNETH WAYNE
WELSH DANIEL
WENDOLOWSKI JAMES
FRANCIS
WENTE DANIEL LEWIS
WERBISKI PHILIP MICHAEL
WERDERMAN JAMES EDWARD
WERLE HAROLD FRANCIS
WERSCHING ADAM EDWARD
WEST DAVID EUGENE
WEST LARRY CHANDLER
WESTPHAL RONALD DALE
WHEELER JAMES
WHEELER MELVIN CARTER JR
WHITE GENE LEWIS
WHITE JAMES DARRELL JR
WHITE JAMES LEO
WHITE KENNETH LEROY
WHITE LARRY FREDERICK
WHITE LEE OWENS JR
WHITE LUCKY GAYLEN
WHITE OWEN JR
WHITE ROBERT LEE
WHITE STEPHEN O MEARA
WHITE TIMOTHY LEE
WHITE WILLIAM HENRY
WHITFIELD THOMAS MI-
CHAEL
WHITINGTON LARRY E
WHITLATCH GAIL LEE
WHITNEY BLAKE DOMINIC
WHITNEY HARLEY DAIRREL
WHITNEY ROBERT WALTER
WHITT JAMES EDWARD
WHITTINGTON RUSSEL
JOSEPH
WICKAM JERRY WAYNE
WIDDOWS JOHN WILLIAM
WIDERQUIST THOMAS CARL
WIEGAND ROY VICTOR
WIENEKE CARL JOSHUA
WIERZBA EDWIN RUDOLPH
WIEST DONALD RAY
WIGGINS TOMMY AUSTIN
WIKE JOHN MICHAEL
WILBERTON HAROLD JR
WILEY RICHARD DENNIS
WILFONG GIL STEVENS
WILKERSON GEORGE OLIVER
WILKERSON RICHARD LEE
WILKS WALTER ALBERT
WILLERT DIETER ERIC
WILLETT ROBERT LEE
WILLIAMS AUBREY
WILLIAMS EARNEST
WILLIAMS EUGENE VERNON
WILLIAMS GARY ROBERT
WILLIAMS JOHN DEWEY
WILLIAMS JOHNNY GLEN
WILLIAMS JOSEPH MICHAEL
WILLIAMS LONNIE
WILLIAMS LOUIS
WILLIAMS MICHAEL EARL
WILLIAMS NATHANIEL JR
WILLIAMS OTTAWAY LARSON
WILLIAMS PAUL EDWARD
WILLIAMS RALPH MAURICE
WILLIAMS RAYMOND
CHARLES
WILLIAMS RAYMOND LEWIS
WILLIAMS RICHARD ALLEN
WILLIAMS ROBERT CYRIL
WILLIAMS ROBERT EARL
WILLIAMS RUSSELL LOWELL
WILLIAMS STEVEN GARY
WILLIAMS THEODORE
ALFORD
WILLIAMS THEODORE JR
WILLIAMS VICTOR DEMOTT
WILLINGHAM PRESTON T JR
WILLIS DONNIS GLEN
WILLIS LARRY JOE
WILLOUGHBY JESSE LAVERN
WILSON JAMES HAROLD
WILSON MICHAEL
WILSON ROBERT LEE
WILSON ROGER LEE
WILSON RONALD LEE
WINSON JAMES JOSEPH
WISE SCOTT EDWARD
WISKUR JAMES CLYDE
WISNIEWSKI DENNIS EUGENE

WISNIOWICZ THOMAS LEO
WISSELL LAWRENCE JAMES
WITCHER LEONARD III
WITEK EDWARD JOSEPH
WITEK WILLIAM FRANK
WITHERS GARY WAYNE
WITHERSPOON JAMES
MARTIN
WITT JERRY PAUL
WITTE ROGER EARL
WITTEVRONGEL MICHAEL
CAMI
WITZIG RAYMOND GEORGE
WOEHRL MICHAEL JOHN
WOHRER JAMES FIELDING
WOJCIK LAWRENCE ADAM
WOLF DURWYN LEE
WOLFE JOEL DAVID
WOLTER RONALD ALAN
WOLTERS THEODORE AN-
THONY
WONDERLICH MICHAEL KAYE
WONER JOHN PERRY
WOOD JAMES LEONARD
WOOD JAMES WATSON
WOOD ROBERT DELUN
WOODALL JOHN BRAXTON
WOODARD JON ROBERT
WOODS CORDELL EMANUEL
WOODS EARL
WOODS GARY DORVIN
WOODS JOHN KEVIN
WOODS STEPHEN FORREST
WOODSON RICHARD EUGENE
WOOLBRIGHT JOHN WAYNE
WOOLDRIDGE PAUL M JR
WOOLFORD PAUL BURNELL
WOPINSKI BARRY MILTON
WORK GEORGE ALLEN
WORKMAN JOSEPH MYRON
WORMDAHL RICHARD GENE
WORTHEY DAVID ALLEN
WORTHEY DONNIE LEON
WRIGHT CLIFFORD IVAN
WRIGHT KENNETH MICHAEL
WRIGHT ROBERT LEROY
WRIGHT SCOTT ALAN
WRIGHT WILLIE ALFRED
WYANT WILLIAM DOUGLAS
WYLIE JOSEPH DUNN III
WYMAN MICHAEL JAMES
WYNN ROBERT LEE
WYNNE LARRY B
YARBER DAVID WAYNE
YARBER MICHAEL JEROME
YATES GLENDELL EUGENE
YBARRA DAVID
YOCUM GEORGE KENT
YONAN KENNETH JOSEPH
YOUMANS FREDERICK JOHN
YOUNG DALLAS CLYDE JR
YOUNG HORACE EARLE
YOUNG JAMES PAUL
YOUNG MARK DOUGLAS
YOUNG MICHAEL ALAN
YOUNG RAYMOND ALBERT
YOUNG WILLIAM RANDOLPH
YURGAITIS STANLEY GEORGE
ZACH RONALD LEE
ZACH WAYNE STEVE
ZAEHLER EARL HENRY
ZAGATA JOHN JOSEPH
ZALEWSKI STANLEY JR
ZAWISZA THEODORE LEO
ZELINKO GEORGE ALLEN
ZELLER GARY GENE
ZIEGLER EARL KAY
ZIEMANN RONALD JOHN
ZIMMERMAN KURT FREDRICK
ZIMMERMAN RICHARD
ELMER
ZIMMERMAN ROGER
ZIMMERMAN THOMAS ALLEN
ZINN RONALD LLOYD
ZIONTS CHARLES A
ZUCKER LOUIS CLAUDE
ZUKOWSKI ROBERT JOHN
ZUM MALLEN PHILIP OTTO JR

Photos by Hugh Sruillj

Community Veterans Memorial - Veterans Memorial Park

9710 Calumet Ave, Munster, IN 46321

The Community Veterans Memorial is located in Munster, Indiana's Veterans Park. The park is six and a half acres large and features multiple monuments and sculptures representing the multiple wars and conflicts that have happened across the 20th Century. It was created in order to remember all of the service members who participated in the wars, including the Vietnam War. The committee who put the park and memorials together also want to educate the visitors about the downsides of war instead of glorifying the heroics of war. This is done as a way to try and stop the cycle of war and make room for peace. The park is open from dawn to dusk and is free to the public, but there are special events held at the location. There is a Veterans Day ceremony every year as well as a Pearl Harbor Remembrance ceremony in the park.

Following the vision of Edward P. Robinson, the park memorials, including the Vietnam Memorial, was made by Julie Rotblatt-Amrany and her husband Omri and made possible through the efforts of Robinson and the Veteran's Committee. The park as a whole was dedicated officially in June of 2003 but was commissioned to the Rotblatt-Amrany pair in 1999.

The Vietnam Memorial is made up of multiple components and is made in a way to recreate the Vietnam War from the ground up. The artists recreated the rice paddies of Southeast Asia and set up their monuments within the recreation to make everything look more realistic. There are seven monuments set up throughout the Vietnam Memorial site which is linked to the other war memorial sites by a commemorative brick path that is made up of donated/sponsored bricks.

The memorial consists of a retired Huey helicopter sitting in a rice paddy, a map of Vietnam etched into granite with images of the war, three sculptures of M-16s with flowers sticking out of their barrels positioned on a stack of bricks, a wall sculpture depicting a nurse with a wounded soldier, a recreation of a Buddhist temple in ruins, and two bronze sculptures. One statue is a depiction of a creeping soldier with his weapon in his hand while the other, more somber statue, is that of a fallen combat soldier who is lying face down on red stones; his legs are missing.

78

The Names of those from the state of Indiana
Who Made the Ultimate Sacrifice

ABBOTT HAROLD WAYNE
ABBOTT JAMES TERRY
ABBOTT JOHN WILLIAM
ABEL ARNOLD GORDON
ACHER ROBERT PAUL JR
ACKERMAN JOHN ROBERT
ACKERMAN REX WILLIAM
ACREE BILLIE RAY
ACTON DAVID AUGUST
ADAMS JERRY DEAN
ADAMS ROBERT JAMES
AHART WILLIAM JUNIOR
ALBERTS DANIEL LOUIS
ALERT ROBERT JOSEPH JR
ALEXANDER CARL THEODORE
ALEXANDER JAMES HINES
ALEXANDER LAURIE LEON
ALL CARL KELLY
ALLEN EDDIE JAMES
ALLEN EDWARD JAMES
ALLEN JACK LEE
ALLEN JAMES OTIS
ALLISON WILLIAM EDWIN
ALLSTOTT MARK JOSEPH
ALSMAN WILLIAM FRANKLIN
ALVIS DONALD DEAN
AMEIGH JAMES KEITH
AMMON GLENDON LEE
ANDERSON RAL JEFRO JR
ANKRUM GLENN EUGENE
APPLEGATE JOSEPH CHARLES
ARCHIBALD GARY MICHAEL
ARIENS RICKY MICHAEL
ARMSTRONG DOUGLAS
WAYNE
ARMSTRONG WILLIAM L
ARNETT MAHLON RONNIE
ARNOLD JAMES EDWARD
ARNOLD MAJOR JR
ARRINGTON JOHN ROBERT
ASBURY BENTON FRANCIS
ASH PAUL MICHAEL
ASPER IVAN RICHARD JR
ATWOOD DAWSON JESSE
AUSTIN JAMES EARL
BAER MAX IRWIN
BAILEY JAMES ANTHONY
BAKER ELDON ALLEN
BAKER ROBERT LEE
BALLINGER JAMES ARTHUR
BANCROFT WILLIAM W JR
BANDELIER HOWARD WAYNE
BANEY CHARLES LYNN
BANKS LARRY CLAYTON
BAPP RONALD DALE
BARDACH ALAN JENSEN
BARKER JOHN WAYNE
BARKSDALE CULLEN JR
BARNETT CARL TAYLOR
BARNETT CLIFFORD C JR
BARNETT THOMAS MARTIN
BARRETT DONALD RICHARD
BARRETT LARRY WAYNE
BARTLEY DONALD RAY
BARTLEY RICHARD LOUIS
BASEY DWIGHT LEROY
BASHAM JAMES DARRYL
BATCHELOR MAX WAYNE
BAUGH LARRY MICHAEL
BAUGHMAN RONALD GENE
BAUM DAVID MICHALE
BEALS CHARLES ELBERT
BEASLEY PERCY JR
BEDELL JAMES WAYNE
BEECHER QUENTIN RIPPETOE
BEER MERLIN GAIL
BEESON WILLIAM DALE
BELL STEVEN ALLEN
BELL THOMAS LYNN
BENGE SAMUEL EDWARD
BENGE THOMAS CLAYTON
BENNETT EDWARD DALE
BENTON ARNOLD RAY
BERNOSKA WAYNE GARY
BERRY KURTIS AUREL

BERRY MICHAEL LEWIS
BERRY WILLIAM MC KINLEY
BERTA ROBERT DEWITT
BEYL DAVID ROBERT
BICKEL BARRY WAYNE
BIDDLE DANIEL ELLIS
BIEDRON MICHAEL PETER
BIGGER CALVIN HART
BINGHAM CHESTER ELMEARL
BIRD MICHAEL ALAN
BIRKHOLZ ROBERT EARL
BISHOP THOMAS WAYNE
BLACK DENNIS WALTER
BLACK MARK RYAN
BLACKERBY RALPH W
BLACKSTEN RONALD LEE
BLACKWELL JAMES LISMAN JR
BLANTON RICHARD PATRICK
BLASKO JAMES DEE
BLASKOVICH STEVE JR
BLEVINS JAMES ROBERT
BLOOM STEVEN GARY
BLOOMER JERRY ROBERT
BLOYER SHELDON EUGENE
BLYSTONE THOMAS MICHAEL
BOEHNE STEPHEN BRUCE
BOHNSACK JOHN EDWARD
BOICE LARRY LEE
BOLIN DANNY LEE
BOLTON DENNIS OPAL
BONDS BYRON DEAN
BONNELL LARRY GENE
BONNIE LEWIS ELI
BOOKER JERRY LABORN
BOOKER TERRY WAYNE
BOOTH STEPHEN FLOYD
BORGMAN NORRIS RAY
BORKHOLDER JERRY M
BORMAN JERALD ALLEN
BOROWSKI TADEUSZ JAN
BOWLING JAMES WISDOM
BOWMAN CLARENCE JR
BOWMAN JACK ALLEN

BOWMAN LESLIE VON
BOYCE GARY LEE
BRADLEY KENNETH ROBERT
BRAGG JOHN ROBERT
BRANAMAN KENNETH MERLE
BRANDENBURG STEVEN
KEITH
BRANDENBURG VERLIN
RICHAR
BRASWELL DANNIE GLENN
BREES WILLIAM MARION JR
BREINER STEPHEN EUGENE
BREWER JAMES EUGENE
BRIDGES ERNEST LARRY
BRINDLE WILLIAM VICTOR
BRINEGAR BARRY LYNN
BRISCOE CHESTER JR
BRITTAIN JOSEPH BRUCE
BRIX ROBERT CARL
BROADY TERRY LEE
BROOKINS DAVID EVERETT
BROOKS STEVEN RANDALL
BROUHARD MALCOLM KEITH
BROWN BOBBY JOE
BROWN BOBBY RAY
BROWN DAVID CLARENCE
BROWN DAVID GRANT
BROWN DEWITT WILCOX III
BROWN DONALD WAYNE
BROWN JOEL KENTON
BROWN JOHN STEPHEN
BROWN MARION C
BROWN RAYMOND
BROWN ROBERT NUGENT
BROWN TOM WILLIE
BROWN VAUGHN LEE
BRUBAKER MAX L
BRUCE DANIEL DEAN
BRYAN JOHN ALLEN
BRYANT DAVID ALTON
BRYANT WILLIAM JOHN
BUBALA RICHARD FRANCIS
BUCHANAN ELMER LEVERNE

BUCHANAN GILBERT EDWARD
BUCHANAN JEFFREY LYNN
BUCHANAN JOHN GARY
BUFFIN NICHOLAS JAY
BUHR THOMAS FREDERICK
BUNDY GLENN EDWARD
BUNDY MARK STEPHEN
BUNN DONALD WAYNE
BURCH JAMES EDWARD
BURKETT WILLIAM OMER
BURKHOLDER LARRY GENE
BURNS HOWARD FRANK
BURNS JUNIOR R
BURNS MORRIS EUGENE
BURTON DENNIS LEE
BURTON FRANK THOMAS
BURTON STEVEN DALE
BUSBY CHARLES FRANCIS
BUSTOS GREGORIO C
BUSTOS MIKE GARCIA
BUTLER ROBERT HERMAN JR
CABLE RICHARD ALLEN
CADENHEAD RANDALL JAMES
CAMPBELL TOMMIE JOE
CAMPFIELD ALBERT L
CANCEL RAMON PENA
CANUP WILLIAM DAVID
CARANASIOS EVANGELOS K
CAREY RONALD DUANE
CAREY THOMAS JOSEPH
CARLE GARY LEE
CARMICHAEL DALE EUGENE
CARNINE STEPHEN MICHAEL
CARPENTER FRED W
CARPENTER HOWARD R JR
CARPENTER JESSE DALE
CARPENTER TOMMY LEE
CARRICO DAVID AARON
CARROLL MANUEL LEROY
CARTER MICHAEL STEPHEN
CARTER WILLIAM ALLEN
CARVER HARRY FRANKLIN

CASTILLO PHILLIP
CAUDILL BILLY JOE
CAUGHEY JAMES EDWARD
CAWLEY WILLIAM BRACE JR
CHAFFIN DONALD ALAN
CHAMBERS RAYMOND EARL
CHANEY MICHAEL WAYNE
CHANEY ROY LEE
CHAPEL HOSY
CHAPMAN GARY WAYNE
CHAPMAN JOHN ROY
CHAPMAN LARRY LEE
CHAPPEY JOHN MICHAEL
CHARLES DAN EUGENE
CHEEK ROBERT MICHAEL
CHIFOS WILLIAM LEWIS
CHOMEL CHARLES DENNIS
CHURCHWARD STEVEN DEAN
CISAR THOMAS CHARLES
CLANTON HOWARD
CLARK GARY LEE
CLARK JERRY DOUGLAS
CLARK LAWRENCE
CLARK ROBERT LEWIS
CLARK ROBERT NELSON JR
CLARK STEVEN EUGENE
CLARK TERRY RICHARD
CLARK VINCENT ALLEN
CLAWSON WILLIAM K
CLAY EDWARD ROGER
CLELAND RONALD LOUIS
CLEM THOMAS DEAN
CLESTER DOUGLAS ARTHUR
CLEVENGER WILLIAM HENRY
CLEWLOW ROBERT LEE
CLIFFORD GARY ALAN
CLINGLER STANLEY MELVIN
CLODFELTER DARREL JAY
CLOVER WILLIAM FRANK JR
CODY ROBERT DEAN
COFFEY ROBERT WILLIAM
COGDELL WILLIAM KEITH
COLDEBERG DONALD RAY

Indiana

COLDIRON DALE WILLIAM
COLE ROBERT KENNETH
COLE ROBERT LEROY
COLEMAN CLARENCE LEROY
COLEMAN LARRY HAROLD
COLLINGSWORTH DELNO BILLY
COLLINS DAVID LEE
COLLINS ELTON BRADLEY
COLLINS HAROLD DUANE
COLLINS RONALD CHARLES
COLONE RONALD JAMES
COLVIN DAVID
CONLEY ROBERT L
CONNELL THOMAS MICHAEL
CONNELLY SAMUEL GERALD
CONOVER CHARLES RAYMOND
COOK DENNIS LYNN
COOK DONALD WARREN
COOK JOHN DALE
COOLEY RONALD MARVIN
COOPER DAVID ARTHUR
COOPER HOWARD KENNETH
COOPER MAURICE ALAN
CORWIN JOHN JAMES II
COTTRELL DARRELL WAYNE
COVEY CHARLES ALLEN
COVINGTON DARELL LEE
COX EVERETT FREDERICK
COX GARY WAYNE
COY BENJAMIN D JR
CRABTREE RANDALL LEWIS
CRAIG DICKEY
CRAIG WILLIAM ANDERSON
CRANDALL CHARLES EVERETT
CRANE DONALD LEONARD
CRAVEN LEONARD ISLER
CRAWFORD GORDON LEE
CRAWFORD JOHNNY RAY
CREAMER CHARLES FORAK III
CREASON RICHARD EARL
CREASY JERRY N
CRIPE DENNIS WRAY
CRIPE TOMMIE MAX
CRISWELL GEORGE DAVID
CRODY KENNETH LLOYD
CROSS LARRY EDWARD
CROY JOHN LEE
CROY WILLIAM MARK
CUEVAS FRANK OSCAR
CULL HERMAN RAY
CULLISON BARRY ANDREWS
CULP EVERETT T
CUMMINGS DALLAS DEWEY
CUMMINGS ROGER WAYNE
CUPP JOHN CHARLES
CUTLER DONALD EARL
CZARNY WILLIAM EUGENE
DALE TERRENCE MICHAEL
DANCE JACK RAY
DARLING LARRY WAYNE
DARNELL WILLIAM EUGENE
DAUGHERTY WILLIAM STANLEY
DAULTON WILLIAM MANSON
DAVIDSON JAMES RICHARD
DAVIS EVERETT
DAVIS GENE EDMOND
DAVIS JEFFREY ALAN
DAVIS JOHN POWERS
DAVIS RANDALL MARK
DAVIS ROBERT ALLEN
DAVIS ROBERT EUGENE
DAVIS RONALD
DAYTON JOHN EMERY
DE BUSK MICHAEL EUGENE
DE LA GARZA EMILIO A JR

DE WEESE WILLIAM CHARLES
DEAL LARRY KEITH
DEAN ANTHONY WILLIAM
DEATON JACK JOE
DEBOLT WILLARD CLINTON
DECKER WAYNE AUSTIN
DEITEMEYER THOMAS PAUL
DELAPLANE JAMES CHARLES
DELP RONALD MARVIN
DELPH SCOTT CLAYMON
DEMPSEY RONALD LEE
DENNING THOMAS GEORGE
DENNIS DANIEL MAURICE
DENNIS DAVID ALAN
DENNIS RONALD GENE
DENNY DAVID LESTER
DENT MICHAEL EARL
DEPP CHARLES WILLIAM
DEVANEY BRIAN JOHN
DEVORE RICHARD E
DICKERHOFF TERRY WAYNE
DICKERSON HAROLD
DICKERSON JOHN GREEN III
DICKERSON RICCARDO BURTON
DICKUS MICHAEL JOHN
DIEFENBACH LARRY ARTHUR
DILLS RONALD EUGENE
DINGUS CARL
DIPERT MARVIN LEE
DIXON DAVID LEE
DIXON GALE WILLIAM
DLUZAK DAVID MARTIN
DOADES FLOYD EUGENE
DONOHUE RONALD FRANCIS
DORSETT HARRY CLINTON
DORSHAK ROBERT JOSEPH
DOUGHERTY JOHN CHRISTIAN
DOUGLAS JAMES DALE
DOVER GEORGE RICHARD
DOWNING DAVID ALLEN
DOWNING MICHAEL WILLIAM
DOWNS JERRY WAYNE
DRAPER MARK GREGORY
DRAVES LARRY DANIEL
DRAZER THOMAS STEPHEN
DRINSKI DAREN LEE
DROHOSKY EDWARD DANIEL
DUCAT PHILLIP ALLEN
DUCKETT RONALD WARREN
DULIN ZETTIE J C
DUNCAN CHARLES EDWARD
DUVALL DEAN ARNOLD
EAMICK BRUCE ALLEN
EASTON DAVID EVERETT
EATON TOMMY RAY
EBERT MICHAEL LEROY
EDWARDS GARY STEPHEN
EGLY SHELLY
EGOLF RODGER LEE
EICHENAUER THOMAS LYNN
EILER LINDEN DALE JR
ELLENBERGER CAREY WAYNE
ELLINGER FRANKLIN MAX
ELLIS HERMAN JR
ELLIS ROBERT EARL
ELLIS ROBERT WAYNE
ELLIS RONALD LEE
ELLSWORTH RICHARD ALLEN
ELSTON ROBERT FRANKLIN
EMMART JAMES LEE
ENGELHARDT GARY WAYNE
ENGLE CHARLES EDWIN
ERB KARL FREDRICK
ERNHART STUART JAMES
ERWIN DONALD EDWARD
ERWIN LYAL HANCIL
ESTRADA GUILLERMO
ETTEL HENRY C JR
EULER MICHAEL DAN
EUTSLER JOHNNY NEIL
EVANS RUSSELL IRWIN
EYLER ALLAN DOUGLAS
FAIRCHILD DENNIS MELVIN
FALCONBURY EARL FERN
FANKBONER DANIEL ROSS
FARLEY MICHAEL LEE
FARMER NEIL PHILIP
FARRIS GEORGE K
FAULKNER EARL EUGENE
FEARS THOMAS JEFFERSON
FEGAN ROBERT MATHEW

FENNER MARK WILLIAM
FENNIMORE GREGORY SCOTT
FERGUSON WILLIAM GLEN
FEWELL JOHN PHILLIP JR
FIEGLE GERALD WILLIAM
FIELDS GARRISON DAVID
FINCHUM JACK WILLARD
FIRTH CHARLES VERNON
FISHER DAVID R
FISHER ROBERT LEROY
FLASKAMP JOHN EUGENE
FLECK GREGORY LAMAR
FLORES MANUEL SOTO
FLORY ROBERT LESTER JR
FOGLE LARY DALE
FOGLEMAN JOHNNY
FOLCK BENJAMIN THOMAS
FOLTZ PAUL RAYMOND
FORE JAMES EDWARD
FOREMAN JAMES LEE
FORNEY ALVIN CARVER
FOSTER BENNY EDWARD
FOSTER CURTIS LAMARR
FOUST DONALD CHARLES
FOWLER EUGENE RUSSELL
FOWLER KENNETH W
FOWLER LAWRENCE EUGENE
FOX CHARLES NATHAN
FOY STEVEN JOSEPH
FRAIN KENNETH MICHAEL
FRAKES WILLIAM DOUGLAS
FRALEY GARY THOMAS
FRANCIS TERRANCE DEAN J
FRAZER RONALD LLOYD
FREE LAWRENCE CAMERON
FREEMAN DONALD VERN
FREEMAN ROY ELDON JR
FREUND ERNEST ELWOOD JR
FRY WALTER ALLEN
FRYE MICHAEL BRUCE

FRYMAN JAMES OMER
FUHRMAN TERRY LEE
FULK DALE STEVEN
FULK JAMES WESLEY
FULKERSON ROBERT ALAN
FULWIDER DANIEL RAYMOND
FUNCK ALFRED
FUNKHOUSER CARL T
FURR WILLIAM RENARD
GALEY JAMES NORBERT
GALLAGHER FRANK R
GARD JAMES BARRY
GARRETSON RICHARD EUGENE
GARRETT DAVID FRANK
GARRISON NOEL KEITH
GARRITY WILLIAM KENNETH
GEISE MICHAEL DAVID
GENTH GARY ROY
GEORGE CLAUDE MARVIN
GEST DENNIS EUGENE
GETTELFINGER THOMAS J
GETTINGS GUY CLIFTON
GIBBS IRA EUGENE
GIBSON JOE BILL
GIBSON LAWRENCE EDWARD
GIBSON ROWLAND EDWARD
GIL CORNEL
GILBERT GERALD FREDERICK
GILBRECH RUSSELL EARL
GILL DAVID EUGENE
GILSINGER FREDERICK M JR
GIPSON BILLY EDWIN
GISH CALVIN ROBERT
GISH JAMES EDWARD
GLASPER JOHN JAMES
GLASSBURN JOE RICHARD
GLASSBURN MARVIN EDWARD
GLASSCOCK TERRY LEE

GLEGG JAMES EDWARD
GLOVER LARRY RAY
GODBEY SAMUEL EDWARD
GODFREY JOHN LARIMORE
GODSEY DANNY RAY
GOEN MARTIN DOUGLAS
GOFF CHARLES MITCHELL
GOFFREDO MICHAEL ANTHONY
GOHEEN RICHARD H
GOLC JOSEPH LOUIS JR
GOMEZ HAROLD
GOODE GEORGE SHERMAN
GOODNIGHT JACKIE LEE
GORDON LAWRENCE LEE
GORDON RICHARD DALE
GORDON WYATT CECIL
GORE FREDDY RAY
GOSE ELVIN WAYNE
GOSS LARRY JO
GOURLAY BRUCE ANDREW
GOWERS THOMAS ANTHONY
GRABBE JOHN ALBERT
GRAFT TERRY GENE
GRAHAM HARVEY GENE
GRANT CHARLES ROBERT
GRANT JAMES MICHAEL
GRAY ROBERT ALLEN
GRAY WILBUR LEWIS JR
GRCICH NICK JIM
GREEN GEORGE CURTIS JR
GREENWOOD FRANCIS DAVID
GRIBBLE RAY NEAL
GRIBLER DONALD ROSS
GRIMES THOMAS ALLEN
GROSS COLUMBUS VIRGLE
GROTHAUS DARYL ROBERT
GROVES RONALD LEE
GROW LA MOINE EUGENE
GRUBBS GEORGE EDWARD

GRUBE TERRY LEE
GUERRERO JOSEPH DONALD
GUTHRIE THOMAS LEON
HAASE DELBERT WAYNE
HACKETT ROBERT E
HACKNEY RONALD WAYNE
HADLEY STEPHEN WAYNE
HAGAN ROBERT ALBERT JR
HALE MICHAEL DAVID
HALE WILLIAM EARL
HALL DAVID EMERSON
HALL GEORGE MICHAEL
HALL MICHAEL JENNINGS
HALL RICKEY WAYNE
HALL ROBERT JOSEPH
HALL RONNIE ELMON
HALL STEPHEN THOMAS
HALLEY WILSON FITZGERALD
HAMEL TEDDY LEON
HAMILTON DICK DALE
HAMILTON MICHAEL GEORGE
HAMLET BERNARD JR
HAMMACK CAL THOMAS
HAMMACK LESLIE TOBIAS
HAMPTON STEVEN AARON
HANDLON JERRY LEE
HANDLY EDWARD CLARENCE
HANSELMAN ROBERT LOYD
HANSHEW FRED NEWTON JR
HANTZ HERMAN EUGENE
HAP EDWARD FRANK
HARDER STEPHEN
HARDESTY RICHARD LEE
HARDESTY ROBERT JOE
HARKINS EDISON AMOS III
HARLEY DONNIE RAY
HARP DOUGLAS RAY
HARPER BILLY NEAL
HARPER DONALD EUGENE JR
HARPER RALPH LEWIS
HARRELL JAMES ELMORE
HARRIS BRUCE RANDALL
HARRIS BURNIE
HARRIS JAMES CRAIG
HARRIS LYNN ARDEN
HARRIS MICHAEL LEROY
HARRISON DONALD LEWIS
HARTER FRANCIS WILLIAM
HARTER WILLIAM AARON
HARTWELL PATRICK ALAN
HARTWELL WILLIAM RAY-
MOND
HARVEY DARNELL
HASH JAMES RICHARD
HASH JONATHAN PAUL
HASTE RODGER DALE
HASTREITER RICHARD JAMES
HATCHETT EUREY LEE
HATTABAUGH PAUL RUSSEL
HATTON JAMES L
HAUGH JAMES CURTIS
HAWKINS JONATHON JEFFREY
HAWLEY JOHN HARRISON
HAWS HOMER HOWARD
HAYES ROBERT WAYNE
HAYES TIMOTHY LEE
HAYNES JAMES EDWARD
HAYWOOD MOSES JR
HAZEN RONALD L
HAZLETT ROBERT DALE
HEARN KENNETH LEE
HEATER DANIEL NEIL
HEATH LLOYD LAVERN
HEAVRIN MARK THOMAS
HEDDEN ROGER DALE
HEDGE BILLY WAYNE
HEFFNER DENNIS WAYNE
HEFTY JOHN ELLSWORTH
HEITGER MICHAEL LYNN
HEITMAN STEVEN WAYNE
HELD JOHN WAYNE
HEMBREE JAMES VERNON
HEMPHILL DAVID WAYNE
HENDERSON DERRICK
HENDRICKS STEPHEN ED-
WARD
HENNINGER JOHNIE MI-
CHAEL
HENRY DONALD RAY
HENTHORN HARRY THOMAS
HENZE RANDALL ALLEN
HEPPEN GEORGE HENRY JR
HEROY DEWEY WILLIAM
HEWITT SAMUEL EUGENE

HEYEN JOHN EUGENE
HICKMAN THOMAS STEVEN
HICKS CARMAN KEETON
HICKS DONALD GENE
HIDAY THOMAS MICHEAL
HIGGINBOTHAN ALLEN L
HIGGINS THOMAS ANDREW
HILL DENNIS EUGENE
HILLS JOHN RUSSELL
HIMES STEPHAN CARL
HINES TERRI LIEGH
HINKLE CARL RODMAN
HINSON REGGIE WESTEL
HINTON DENNIS RAY
HITCHCOCK LEE CHARL
HOCKETT DAVID ALLEN
HOECKELBERG THOMAS JOE
HOFER THOMAS EDWARD
HOFF RONALD ALVIN
HOFFMAN TERRY ALAN
HOLBROOK JEFFREY LYNN
HOLDEMAN ROBERT EUGENE
HOLLMAN DAVID LEE
HOLLOPETER RAYMOND
RICHAR
HOLLOWAY THOMAS EUGENE
HOLMES LONNIE MICHAEL
HOLMES RODGER DALE
HOLZKNECHT BERNARD LEE
HONAKER WILLIE ELSWORTH
H
HOOK WILLIAM WREN
HOOS WILLIAM ARTHUR JR
HOOVER REX MICHAEL
HORNBY DAVID EUGENE
HORTON CHARLES RONALD
HOSEA WILLIAM HADLEY
HOSKINS DONALD RUSSELL
HOUCHIN DARCY ALLEN
HOUCK EARL FRANKLIN JR
HOUSE GEORGE JONATHAN
HOUSTON MARK JOSEPH
HOVIS RONALD LEE
HOWARD DAVID LEROY
HOWARD SAMUEL HENRY
HOWE HARVEY GRANT JR
HOWELL CHARLES DENNIS
HOWELL HANCIL EVERT JR
HOWELL ROBERT LEE
HOWES GEORGE ANDREWS
HUBBARD DENNIS LEROY
HUDSON RONALD CHARLES
HUEBNER BURREL DALE
HUFFER KENNETH KIPLING
HUFFMAN DAVID KEITH
HUGHES CARL LEROY JR
HUGHES FURMAN DAVID
HUMPHREY CECIL HOWARD
JR
HUNT JAMES ROBERT
HUNT WILLIAM LARRIE
HURT DARRELL VON
HURT RONALD WAYNE
HUSK CLARENCE RAY
HUTCHINGS DAVID GEORGE
HUTH RALPH CHARLES
INGALLS BENJAMIN HARRI-
SON
IVY LEONARD CLARENCE JR
JACKSON ALLEN LEE
JACKSON CHARLES EDWARD
JACKSON DAVID CHARLES M
JACOBS JOHN CHARLES
JAMES ARTHUR LEROY
JAMES RICHARD DALE
JANKOWSKI LARRY FLOYD
JARBOE WILLIAM LEE
JARONIK ROBERT WALTER
JAROSCAK JOHN PAUL
JASPER DAVID CLAUD
JEFFERSON GARY DONALD
JENKINS GERALD THOMAS
JENNETT WILLIAM ROBERT
JILCOTT CHARLES B JR
JOHANNES URBAN HAROLD JR
JOHNS PAUL FREDERICK
JOHNSON CARL DAVID
JOHNSON DAVID ALLEN
JOHNSON GERALD LEE
JOHNSON JACK LEE
JOHNSON JAMES REED
JOHNSON JIMMIE LEE
JOHNSON RALPH EDWARD
JOHNSON TERRY ALAN

JOHNSON THOMAS EUGENE
JOHNSON THOMAS WAYNE
JOHNSON WILLIAM D JR
JOLLEY DAVID MARVIN JR
JONES CLARENCE EDWARD JR
JONES EDWARD CHARLES
JONES GARY BLAINE
JONES GRAYLAND
JONES GUY THOMAS
JONES JACKIE DALLAS
JONES JAMES EDWARD
JONES LARRY
JONES ROBERT MORRIS
JONES ROMAN LEE
JONES STEPHEN PERRY
JORDAN DANIEL WALTER
JORDAN ROY DOUGLAS
JOSEPH RONALD RAY
JUSTICE ROGER DALE
JUSTIS RONALD HENRY
KAHRE DONALD LEE
KAIL ROBERT MORTON
KALIL JAMES NOBLE
KAPPMEYER PAUL JOSEPH
KARAS WILLIAM JAMES
KASA KENNETH EUGENE
KASER RANDALL FRANK
KATRENICS JAMES NOEL

KATZENBERGER RAYMOND L
KAUFFMAN MICHAEL M II
KAUFMAN WAYNE ELDON
KAYS DAVID COLEMAN
KEELER JAMES EDMUND
KEESLING JOHN ARTHUR
KELLAMS GLENNIS RAY
KELLEMS RAYMOND EARL
KELLY JEROME RICHARD
KELLY MICHAEL EUGENE
KEMPF DENNIS JOSEPH
KENDALL KENNETH BRUCE
KENNEDY RAYMOND O
KEPPEN THOMAS ROGER
KERNEY JOHN OSCAR
KERR ROBERT WESLEY
KESSLER TIMOTHY ROBERT
KIGER JAMES ROBERT
KIKKERT ROBERT MERRILL
KINDT THOMAS PATRICK
KING FREDRICK BEN
KING JOHN EDWARD
KINSER ARTHUR WILLIAM
KIRKENDALL JOSEPH KEITH
KISTLER JAMES LEROY
KITTS MARIO CLAYTON
KLAPAK JOHN ROBERT JR
KLARIK STEVE

KLEIBER GEORGE L JR
KLEINT WILLIAM STANLEY
KLINKER MARY THERESE
KLUTE KARL EDWIN
KNEBEL DONALD JOSEPH
KNOCHEL CHARLES ALLEN
KOCH THOMAS MICHAEL
KOEHLER JOHN FRANCIS
KOLVEK MARK ANDREW
KRAFT MICHAEL EUGENE
KRAUS RONALD CALVIN
KRIDER JACK GALE
KROS ROGER ALLEN
KRYSKE LEO NEAL
KSIAZEK BENNIE
KUHLMAN ROBERT JOHN JR
KUHN CLIFFORD MARTIN
KUMMINGS JAMES ALBERT
KUPFERER JACK JOSEPH
KURDELSKI JAMES HOWARD
KURELLA MICHAEL J
KURTZ ROBERT WARNER
LABUDA ROBERT ALAN
LACKEY PHILLIP LANS
LACY TIMOTHY HOWARD
LAIER STEPHEN EUGENE
LAKE JOHN WILLIAM
LAKINS JAMES EARL

81

Indiana

LAMBDIN MARVIN DOUGLAS
LAMBTON BENNIE RICHARD
LANDIS DUANE GERALD
LANE LEONARD FRANCIS
LANIER JAMES PERRY
LANKASTER JOHN THOMAS JR
LANNING DAVID ALAN
LARAWAY WILLIAM DEAN
LARKINS CHARLES KENNETH
LARKINS VIRGIL LEE
LATHAM MICHAEL TERRY
LAUTZENHEISER MICHAEL
LAW JAMES DOUGLAS
LAWHON MICHAEL HOWARD
LAWSON KARL WADE
LAWSON ROGER W
LEACH TERRY VIRGIL
LEAP THOMAS EDWARD JR
LEASER ROGER RAY
LEAVELL RICHARD TYRONE
LEE BILLIE LEWIS
LEE DOM E
LEE JACK CHARLES
LEE MILAN LAVOY
LEE WILLIAM
LEMAY PAUL LOUIS
LEMING CHARLES R
LEMOND DOUGLAS ALAN
LENNON MARK STEVEN
LENZ GERALD FRANCIS
LEWELLEN WALTER EDWARD
LEWICKI STEVE WILLIAM
LEWIS MICHAEL
LEWIS ROBERT DEAN
LEZAMA JOSE JR
LIKENS BOBBY JOE
LINDEWALD CHARLES W JR
LIROT CHARLES PATRICK
LISMENTS VILIS
LITTLEPAGE THOMAS EARL
LOCHNER KEITH ALAN
LOCKE JACK ELSWORTH
LOFFER TERRY ALLEN
LOGAN RONNIE LEE
LOGSDON HERBERT JR
LOMAX MALCOLM EUGENE
LONG LORIN ELWOOD
LONG ROBERT LYNNE
LOPEZ ROBERT
LOTTES HERBERT JAMES
LOWERY DARYL LEE
LOY RANDELL HOOD
LOZIER KENNETH WAYNE
LUCAS STEVEN ERNEST
LUDBAN GLEN CHARLES
LUNDGREN LAWRENCE EMIL
LUSK BERNARD MERLE
LUTGEN CHESTER ARTHUR
LUTTEL KENNETH BERNARD
LUX ANTHONY RAYMOND
LYNCH PHILLIP EDMOND
LYNCH RICHARD E
LYNCH TIMOTHY JAMES
LYNN WILLIAM THOMAS
LYON JAMES MICHAEL
LYTLE RICHARD WALTER
MACY JOSEPH DEAN
MAGEE MITCHELL JR
MAHER LOUIS JOSEPH JR
MALLON JAMES JOSEPH JR
MALLORY WILLIAM EARL JR
MALOY TERRY LEE
MANGER JAMES ALLEN
MANN ROBERT LEE
MANNING THOMAS R
MANNS ROGER D
MAPLE HOLLIS GARNELL
MARCUM JERRY LEE
MARCUM KENNETH

MARKHAM RAYMOND PAUL
MARKOWSKI HENRY JOSEPH JR
MARKS DAVID ALAN
MARKS FRANK WILLIAM
MARQUARDT MERLIN EUGENE
MARTIN CHARLES JEFFREY
MARTIN DONALD EDWARD
MARTIN DONNIE JOE
MARTIN JERRY DEAN
MARTIN ROBERT ALAN
MARTZ DANIEL MORRIS JR
MARVIN ROBERT GERALD
MATCHETT LESLIE DAVID
MATHEWS WILLIAM JEROME
MATTINGLY LARRY FRANKLIN
MAUSEN STEPHEN GREGORY
MC ALLISTER WILLIAM DENNI
MC ANDREWS JOHN JOSEPH
MC BRIDE THOMAS LEO
MC CLELLAN FRANK EDWARD
MC CLELLAND CHESTER RAY
MC COY GEORGE FRANKLIN
MC CRAY GREGORY
MC DANIEL JOHN THOMAS
MC DANIEL KENNETH REED
MC DANIEL WAYNE IVAN
MC DANIELS CHARLES ALBERT
MC DONALD GENIE LEE
MC DOWELL LARRY JAMES
MC DUFFIE RONALD LEE
MC FETRIDGE GARRY CLAYTON
MC GEE RICHARD WAYNE
MC GEE STEPHEN DWAYNE
MC INTOSH DONALD RAY
MC KEE DAVID LEROY
MC KILLOP LESLIE WAYNE
MC KINLEY ALLEN
MC KINNEY LARRY ROBERT
MC KINNEY NEIL BERNARD
MCFARLAND RICK E
MCGARVEY JAMES MAURICE
MCKINNEY ROBERT DALE
MCLHERN MICHAEL SHEA
MEAD DENNIS MICHAEL
MEADOWS DAVID LEWIS
MEASLEY HENRY HERBERT JR
MEDJESKY VINCENT JOSEPH
MENEELY HERMAN RICHARD
MERKEL MICHAL ALVIN
MERRELL JAMES KEITH
MERRILL DAVID LOUIS
MESZAR FRANK III
METCALF HAROLD
METSKER THOMAS CURTIS
MEYER JOHN JOSEPH
MEYERS DAVID LEE
MEYERS GEORGE HENRY
MICHALAK WAYNE ALLEN
MICHALSKI JAMES
MIDDLETON JAMES EDWARD
MIDNIGHT FRANCIS BARNES
MIELKE BRUCE EDWARD
MILES GLENN EDWARD
MILLER BURRNON ELIHUE
MILLER CARY DUANE
MILLER DARYL L
MILLER EARL DAVID
MILLER EDWARD CLINTON
MILLER EVERETT GENE
MILLER GEORGE DANIEL
MILLER GERALD LEON
MILLER IVAN DEAN JR
MILLER JACK WAYNE
MILLER JERRY EUGENE
MILLER JOHN MICHAEL
MILLER MARVIN LEO
MILLER MICHAEL J
MILLER RICHARD L
MILLER RUSSELL PERRY
MILLER TOMMY ROGER
MILLS AUDLEY DUANE
MILLS DALE EDWARD
MILLS RICHARD THOMAS
MILLS ROGER DALE
MINCKS JIMMIE LEE
MINIX CLYDE
MINTON BOBBY
MIRANDA WILLIAM
MISTER DARNELL
MITCHELL HARRY E

MITCHELL HORACE GIBBS JR
MITCHELL MICHAEL LYNN
MOCK JOEL WILLIAM
MODGLIN JOHN LARRY
MONROE MARVIN EUGENE
MONTGOMERY LARRY
MONTGOMERY RONALD WAYNE
MOORE ALLAN JOHN
MOORE DAVID ALAN
MOORE FRED JR
MOORE GEORGE MONROE
MOORE JAMES LYNN
MOORE LEONARD IRVIN
MOORE LONZIA RAY
MOORE RALPH EDWARD
MOORE RANDY COIS
MOORE ROGER DEAN
MOORE WILLIAM JUNIOR JR
MORALES VICTOR
MORFORD LOREN LEE
MORGAN DENNIS EVERETT
MORGAN SAMUEL FLOYD
MORMAN WILLIAM EUGENE JR
MORRELL DENNIS RICHARD
MORRIS LARRY LEE
MORRIS WALTER JOSEPH
MORROW JAMES RALPH
MORTON GARY RAY
MOVCHAN DAVID EDWARD
MULLER STEPHEN PETER
MULLET STEVEN JAMES
MUNDELL GREGORY STAN
MUNDY HAROLD EUGENE
MURPHY RAY
MURRAY STEVEN EDWARD
MUSSELMAN HAROLD EARL
MUSSELMAN JAMES KEVIN
MUSSELMAN ROBERT EUGENE
MUVICH DENNIS ROBERT
MYERS JAMES ALEXANDER JR
MYERS PEARL WAYNE
MYERS WILLIAM HENRY
MYLES ROBERT RAY
NAGY STEVEN
NASH CALVIN CURTIS
NASH GEORGE ALFRED JR
NASH JOHN MICHEL
NEAL RONALD WAYNE
NEAL WILLIAM EDWARD
NEELEY DENNIS PAUL
NEELEY MARVIN EUGENE
NELLANS WILLIAM LEE
NEMETH JOSEPH STEVEN
NERINI THOMAS HAROLD
NEVIN PATRICK CHRISTOPHER
NEWBERRY LARRY GENE
NEWBURN LARRY STEPHEN
NICHOLS JOSEPH DAVID JR
NICKERSON MICHAEL KENT
NICODEMUS WILLIAM DEO
NIGGLE HARRY TILLMAN
NOBLE JAMES HERBERT
NOE FLOYD RUSSELL
NOLEN PAUL MICKLE
NORD DAVID LEE
NORMAN LANNY JOSEPH
NORTH CLAUDE EUGENE
NUNLEY JAMES E
NUNN JOSEPH LORAN
NURSE JOHN GORDON
NUZIARD RICHARD LEE
NYE AVERY MERRILL III
O BOYLE TERRENCE PATRICK
O CONNELL MICHAEL GRANT
O CONNOR DAVID CORNELIUS
O CONNOR TIMOTHY JOHN
O LEARY MICHAEL WILLIAM
ODIER STEVEN KENT
OGLESBY CHARLES DOYLE
OLDHAM KENNETH LINDLE
ONEAL STEVEN CLIFFORD
ONOHAN LOUIS GEORGE
ORCUTT LARRY LEE
ORN DEAN RUSSELL
ORT IVAN ALLEN JR
OSBORN BARRY LEE
OSBORN LYNN ARTHUR
OSBORN WILLIAM CHESTER
OSBORNE GEORGE JR
OVERPECK JAMES HARLEY

OWENS BEN
OWENS ROBERT FRANKLIN
PABEY JOSE ANTHONY
PACE GARY LYNN
PACKARD GEORGE RICHARD
PADGETT DAVID EUGENE
PADGETT JON LESLIE
PAHL RONALD G
PANULA REINO ARNE
PARCEL JOHN WILLIAM
PARISH EUGENE ALLEN
PARK IRVING GEON
PARK ROBERT LEE
PARKER THOMAS AQUINAS
PARLIAMENT KIM RANDLE
PARRETT JAMES RAY
PATE RICKY ALAN
PATTERSON LARRY GENE
PAUL JAMES RICHARD
PAULUS ROBERT DUANE
PAVEY CHESTER RAYMOND
PEDUE ROGER WILLIAM
PELL RANDALL LEE
PENNINGTON DALE ALLEN
PENRY MARVIN EUGENE
PERDUE WILLIAM CARMAN
PERKINS OFALEE
PERRY ROBERT LEE
PERRY THOMAS DAVID
PERSON ROBERT LEE
PERSONS HENRY HARVEY
PETERS RICHARD EUGENE
PETERSON JERRY LEE
PETTITT JOHN THOMAS SR
PHELPS HUGER LEE
PHILLIPS BOYCE DEAN
PHILLIPS CARL WAYNE
PHILLIPS KERRY WAYNE
PIERCE KENT DE WAYNE
PIERCE RICHARD A
PIERSON LARRY JAMES
PIERSON LYNN ALLEN
PINKERTON LLOYD D
PINTAR JAMES ALBERT
PLATT DAVID BORNE
PLATT LARRY DEAN
PODELL RICHARD W
POE STEVEN MELVIN
POLICH DAVID WILLIAM
POOR RUSSELL ARDEN
POOR VICTOR LYNN
PORTEOUS ROBERT RICHARD
PORTER DAVID RANDLE
POSEY GEORGE RAY
POWELL JAMES BENJAMIN JR
POWELL JAMES RICHARD
PRANGER GLENN AIREN
PRATT CAREY JAY
PRCHLIK WILLIAM CHARLES
PRIEST TERRENCE LEE
PRIESTHOFF THOMAS EUGENE
PRIETO ANTHONY RAYMOND
PRUSH MONTY DOUGLAS
PRY JERRY EARL
PUGH MICHEL LEE
PUTZ LAWRENCE JAMES JR
QUICK ROBERT EUGENE
RAGLE JAMES WILLIAM
RAINES ROBERT STEPHEN
RAINS MICHAEL EDWARD
RALPH JAMES TROY
RAMBERGER JERRY RAY
RANDALL MICHAEL ALLEN SR
RANDALL MICHAEL EUGENE SR
RANDOLPH RICHARD MANFORD
RANKIN DAVID GEOFFREY
RAUSCHER LARRY LEE
RAYMER CARROLL EDWARD JR
REA EMORY LEE
READ ALAN THOMAS
REASONER DAVID LEE
REASOR THOMAS W
RECTOR MICHAEL WILLIAM
REDIC TERRY PETE
REED KENNETH LEROY
REED PHILLIP EUGENE
REEDY GARY MARTIN
REFF CHARLES RICHARD
REHWALD ROYSE WAYNE
REIPLINGER ROBERT LEE

RENNER MATTHEW MARK
RENNER STEVEN RAY
RENO DENNIS KEITH
RETSECK JOHN DEAN JR
REYNOLDS JACKIE DEAN
REYNOLDS JAMES DEREK
REYNOLDS JAY WILLARD
REYNOLDS ROBERT LEE JR
RHODES DONALD RAY
RICE JOHN MICHAEL
RICHARDS RICKEY LEE
RICHARDSON CHARLES WAYNE
RICHARDSON WILLIAM H JR
RICHEY KENNETH ALAN
RICHTER MERVIN RALPH
RIEGEL TERRY LEE
RIGGLE MARK ANTHONY
RIGGS DONALD STEPHEN
RIGGS WILLIAM STEVEN
RIGNEY LARRY JAMES
RIGSBY BARRY LANE
RILEY DENNIS HARLEN
RIPPE LARRY ALLAN
ROACH RALPH EDWARD
ROARK ROY ROGERS
ROBBINS HUGH MILLER
ROBERTS DAVID OWEN
ROBERTS DENNIS RAY
ROBERTS JOHN LESLIE
ROBERTS PAUL MICHAEL
ROBERTSON ANDREW JAMES
ROBINSON CHARLES HARVEY
ROBINSON KENNETH DALE
RODRIGUEZ PEDRO ANGEL
ROE TIMOTHY ROY
ROGERS BILLY LEE
ROGERS CHARLES EDWARD
ROGERS RONALD EDWIN
ROMMEL MICHAEL RAY
ROSE GARY LEE
ROSE ROBERT JAMES
ROSE WESLEY HAROLD
ROSEBERRY ROGER DUANE
ROSEMAN MICHAEL DENNIS
ROSS CHARLES BRENT
ROSS CONRAD E
ROSS GERALD RAYDINE
ROTHENBUHLER LYNN HARLEY
ROTHRING HOWARD EARL JR
ROY LEONARD ALLAN
ROZOW JOHN
RUNKEL RONALD LEE
RUNKLE DANIEL C
RUNYON STEVEN THOMAS
RUSH LARRY ALLEN
RUSSELL HOLLIE BOYD
RUTH TERRY AUSTIN
RUTLEDGE GEORGE EDWARD
RYBOLT DONALD RAY
SABENS JERRY DEAN
SABIN RONALD
SAFRIT WILLIAM JULIUS
SAINT CLAIR BRADLEY ANDREW
SALLEE DOYLE EUGENE
SALTZ MARION NELSON
SALYER BILLY RAY
SAMPLE STEPHEN GEORGE
SANDERS CHARLES
SANDERS JACKIE LYNN
SAPPINGFIELD FRANKLIN A
SAUNDERS RALF IRVIN
SAYLOR CHARLES DUANE
SCHECK CLIFFORD HENRY
SCHEIBER RICHARD ALAN
SCHIMMEL STEVEN GEORGE
SCHMIDT DANNY RAY
SCHNEIDER THOMAS HERSCHAL
SCHOONOVER CHARLES DAVID
SCHROEDER DONALD RAY
SCHUCK DONALD PHILIP
SCHWUCHOW GERALD LEE
SCISNEY MICHAEL LYNN
SCOFIELD ROBERT LEE
SCOLLEY BENJAMIN ELMER
SCOTT DAVID AMOS
SCOTT KENNETH LEROY
SCOTT RICKEY LEROY
SCOTT WALTER MICHAEL
SECHREST JAMES RONALD

SEE WARD EUGENE
SEIDEL CRAIG LEE
SEMPSROTT BRUCE GORDON
SEVIER DAVID HOWARD
SFERRUZZI WILLIAM LEE
SHACKELFORD DON R
SHAFER ROYAL ROY
SHAFFER BRUCE WILLIAM
SHAFFER ROBERT EUGENE
SHANK EDWIN GERALD JR
SHAW GORDON ALLEN
SHAW STEPHEN WILLIAM
SHEELY ROBERT PAUL
SHEKELL STEVEN EDWARD
SHELTON ROBERT WAYNE
SHEPHERD RONALD STEVE
SHERELS CURLEY JR
SHERMAN HARLEY EDWARD
SHERMAN LARRY DEE
SHIRK STEVEN GLEN
SHOULDERS DONALD RAY
SHROYER ALAN CRAIG
SHUTTERS PATRICK ALAN
SIAMBONES GUS
SIDERS MARVIN ISAAC
SIGSBEE MICHAEL JAMES
SIMS JAMES WALTER
SINK MELVIN FRANCIS
SINNOTT DANIEL BERNARD
SIPPLE CONRAD ALAN
SIZELOVE EDWARD LEROY
SKAGGS LONNIE G
SKAGGS WILLARD JR
SKINNER DAVID LEE
SKLODOSKI LAWRENCE
SKORO JOHN PETER JR
SLACK DONALD FRANCIS JR
SLATER KENNETH EUGENE
SLAVENS WENDELL LEE
SLOAN GEORGE MICHAEL
SMALL ROBERT RAYMOND
SMENYAK MARK ANDREW
SMITH BOBBY DALE
SMITH CLIFFORD
SMITH DAVID WILLIAM
SMITH DONALD JOSEPH
SMITH FREDRICK JOE
SMITH IVAN RAY
SMITH JOHN BYRON
SMITH LARRY MICHAEL
SMITH LESLIE R
SMITH LYLE ELTON
SMITH LYNN LEROY
SMITH MICHAEL EUGENE
SMITH MICHAEL STEPHEN
SMITH PAUL ALLEN
SMITH ROBERT T
SMITH RONALD EUGENE
SMITH RONALD LEE
SMITH STEPHEN THOMAS
SMITH STEVEN LEE
SMITH WARDELL

SMITH WILLIAM GARY
SMOCK TERRY DANE
SNEAD WALTER MURRELL
SNIDER RONNIE M
SODEN ROBERT HARRY
SOMERVILLE WILLIAM
HAROLD
SOUCY RONALD PHILIP
SOUTH OSWALD CLAYTON JR
SOUTHERN RICKEY DALE
SOWDERS BARRY GENE
SOWERS ROBERT LEE
SPEER LOUIS LEON
SPENCER JERRY LEE
SPICER JERRY LOUIS
STAEHLI BRUCE WAYNE
STALEY ROBERT LEE JR
STALNAKER LEONARD ALLEN
STAMPER FRANK RAYMOND
STANLEY BUDDY ALFONZA
STANLEY CHARLES HUBERT
STANSBURY DAVID JOE
STATEN TYRONE JOSEPH
STATH ALLEN WAYNE
STEBBINS HARDY WESLEY JR
STEEN JAMES NELSON
STEFFUS GARY PAUL
STEGALL LORENZO
STEINBACHER STEVEN
MICHAE
STEUER FRED MARTIN
STEVENS GARY LYNN
STEVENS RICHARD CRAIG
STEVENS THOMAS ARTHUR JR
STEWART RICHARD
STICKLER ROBERT ALLEN
STILLIONS RONALD BRUCE
STOFKO STEVEN MICHAEL
STOKES HAROLD DEAN
STOLZ LAWRENCE GENE
STONE FOREST MICHAEL
STONEBRAKER KENNETH
ARNOL
STOOPS JONATHAN LYNN
STOPHER GALE JR
STOTLER LARRY PAUL
STRANGE PAUL ROBERT MACK
STRASZEWSKI GEORGE
STEPHE
STRATE BRUCE EDGAR
STUART JOHN FRANKLIN
STUCK RANSOM LEE
STUCKEY JOHN STEINER JR
STULTS EVEANS JERRY
SUCHKA BRADLEY EUGENE
SUMNER JAMES HOWELL
SUTT GEORGE STEVEN
SWAIM BRUCE ALAN
SWAIN ROBERT RAY
SWANGO JAMES RAY
SWEATT THEODORE ALFRED
SWISHER WILLIAM HENRY

TAFT THOMAS HAROLD
TALBOTT JAMES FRANKLIN
TAPPER FREDDIE LESLIE
TAYLOR JAMES EDWARD
TAYLOR MARK RANDALL
TAYLOR ROBERT EUGENE
TAYLOR STEVEN LESTER
TAYLOR TERRY DEAN
TEETER HILBERT WALTER
TERHUNE CHARLES PATRICK
TERRY DANIEL LEE
TERRY MICHAEL DEAN
THARP PAUL ARNOLD
THOMAS CHARLES WAYNE
THOMAS DAVID CARL
THOMAS FRED LOUIS JR
THOMAS JAMES LEON JR
THOMAS JERRY DENVER
THOMAS JOHN CLARENCE
THOMPSON DALE EUGENE
THOMPSON HARRY STEWART
THOMPSON JOHNNY WAYNE
THOMPSON RANDALL ALAN
TOBIAS JOHN CHILICOTT
TODD JOHN CALVIN
TOLBERT PAUL EDWARD
TOMASZEWSKI PHILIP PAUL
TOSH MICHAEL CLAY
TOTH ROBERT GENE
TOWNSEND JAMES LEE
TRAMPSKI DONALD JOSEPH
TRAUGHBER STEPHEN LEE
TRAVIS DALLAS RAY
TRAVIS EDMUND BURKE
TRAVIS MICHAEL WARREN
TREESH JAMES M
TRIBBETT LLOYD EUGENE
TRISSELL WOODROW N JR
TROUT BRADFORD LEE
TRUJILLO WILLIAM OWEN
TUHOLSKI GREGORY ALLEN
TUNGATE DAVID JESSEE
TUNNY NICHOLAS RANDLE
TUTTLE LAWRENCE KAY
TYSON DENNIS LEE
UNCAPHER VALENTINE
DANIEL
UPP JEFF HAROLD
URBELIS JOHN EDWARD
UTTER JAMES ROBERT
VALDEZ FRANCISCO NEVARES
VALENTINE DONALD LYNN
VAN ALSTINE MERLE O
VAN DRIESSCHE JOHN
VAN SESSEN RONALD A
VANCE DARRELL VERNON
VANCE JERRY DUANE
VANDIVIER JOHN DANIEL
VANNATTA JON DAVID
VAUGHT WILLIAM H III
VEACH ROBERT EUGENE
VEGA ANTONIO

VEHLING ROBERT WAYNE
VENEKAMP PHILLIP ROBERT
VERA VENANCIO
VERON MURRAY LEE
VINSON DONALD RAY
VOGELPOHL REX ALAN
VROOM JAMES LEE
WADE BARTON SCOTT
WADE ROBERT LE ROY
WAGES JESSE FLOYD
WAGNER RAYMOND ANTHO-
NY
WALDON TOMMY ANDREW
WALLACE GILBERT EARL
WALLACE JAMES RALPH
WALLACE WILLIAM ROBERT
WALLEN ERNIE LEE
WALTERS BOBBY JOE
WALTERS JOHN EDWARD
WALTERS KENNETH LEE
WALTERS MICHAEL
WALTERS TIM LEROY
WALTON LEWIS ALAN
WARBRITTON JERRY LEE
WARD ROGER LEE
WARD TERRY MICHAEL
WARDROP THOMAS III
WARNER WILFRED WESLEY JR
WARREN JAMES ROBERT JR
WARTHAN ALBERT WILLIAM
WASHINGTON CLARENCE
H JR
WASHINGTON ROBERT JAMES
WATSON HARRY ALLEN
WAYMAN BOBBY RAY
WAYMIRE MICHAEL KARL
WAYNE JAMES CLARK
WEAVER TERRY LEE
WEBB MARK JAMES
WEBER DANNY A
WEBER TERRY LEE
WEEDEN LARRY LEE
WEISHEIT LONNIE HAROLD
WEISMAN KURT FREDERICK
WEISNER GREGORY CHARLES
WEISS STEPHEN LEE
WEISSERT MICHAEL FRANCIS
WELLS JAMES EDWARD
WENZEL JAMES EDWARD
WESSEL RICHARD
WESTLAKE WILLIAM ARNOLD
WESTPHAL GLENN A
WHITAKER FRED DARREL
WHITE MARVIN RAY
WHITE STEPHEN MARK
WHITE STEVEN RUDOLPH
WHITIS LARRY MICHAEL
WHITTEN ROBERT FRANKLIN
WHITTLE JUNIOR LEE
WIEDEMANN ROBERT JOSEPH
WIESKUS WILLIAM CLEMENS
WILCOX WAYNE ALAN

WILES TERRY LEE
WILFONG ROBERT WESLEY
WILKERSON CHARLES
ROBERT
WILLARD KENNETH EUGENE
JR
WILLIAMS FRANKLIN DEAN
WILLIAMS GEORGE DAVIS JR
WILLIAMS J C JR
WILLIAMS JIMMIE KEITH
WILLIAMS JOHNNY
WILLIAMS LEROY WALTER
WILLIAMSON ROBERT JOE
WILLIS STEVEN CRAIG
WILSFORD MICHAEL STEPHEN
WILSON BOBBY
WILSON BOBBY JOE
WILSON DONALD CHARLES
WILSON DONALD MAURICE
WILSON EARL CLIFFORD
WILSON GORDON SCOTT
WINDBIGLER RICHARD
EDWARD
WINLAND PRESTON PAUL JR
WINTERS GENE TALBERT
WISEMAN RICHARD LEE
WITHEE JAMES MONTGOM-
ERY
WITHERSPOON JOE
WOLFE RICHARD EDWARD
WOLFINGTON RICHARD JR
WOOD RICHARD DALE
WOODS ALVIN RICHARD JR
WOODS MATTHEW
WOODSMALL MAX MARVIN
WOODWORTH JAMES LEROY
WORKMAN JAMES ARNOLD
WORKMAN LARRY E
WORREL THOMAS DUANE
WORTH ROY EDWARD
WOZNIAK RICHARD LOUIS
WRIGHT FRED YOUEL JR
WRIGHT STEPHEN LOUIS
WRIGHT TERRY TIM
WRIGHT THOMAS THAWSON
WYSONG JOSEPH WALTER
YEAKLEY ROBIN RAY
YERYAR DONALD FREDERICK
YODER BRUCE ALLEN
YORK IVOL MICHEAL
YOUNG JEFFREY JEROME
YOUNG JOHNNY LEON
YOUNG PAUL AARON
ZELLER LAWRENCE JOSEPH
ZIEL JOSEPH BERNARD
ZIMMERMAN EDWARD C JR

Photos by Mark Binson

A Reflection of Hope - Vietnam War Memorial

E Walnut St, Des Moines, IA 50319

The Vietnam War Memorial located in front of the Supreme Court building in Des Moines, Iowa is officially named "A Reflection of Hope" and was made to remember and honor the men and women who served in the Vietnam War. The memorial is open to the public continuously, but there is also a special ceremony annually on May 7th, Vietnam Veteran's Recognition Day. "A Reflection of Hope" was dedicated in 1984 upon completion and installation, but was rededicated in May of 2015 after another soldier's name was added to the wall. This soldier passed on in 1975 but was overlooked when the wall was first created.

The Vietnam War memorial was made by Mary Jane Fisher, Sam Grabarski, and Tim Salisbury and was made possible by The Citizens of Iowa, according to the inscription on the monument. It was commissioned by Jacqueline Day and the Iowa Vietnam Commission who raised more than $85,000 for the creation and installation of the memorial.

While the memorial is simple in its nature when compared to other elaborate multi-monument memorials, "A Reflection of Hope" is a constantly evolving memorial that promises more additions including a plaque to honor and remember the service members who are continuing to deal with the mental and/or physical health problems from their time in service in Vietnam.

The Vietnam War memorial is made up of slabs of black granite arranged in a curved fashion. The stones feature the names of 867 local service members who weren't lucky enough to make it home from battle. Black reflective granite, similar to the one used in the national memorial in Washington DC, was used to capture the light of the sun. The other side of the monument is decorated with etchings of the insignias of the five different branches of the US military. There is also a field cross to commemorate the fallen service members. There is also an inscription that reads, "A monument established by the citizens of Iowa to honor Iowans who served during the Vietnam War. These absent friends will never be forgotten."

The Names of those from the state of Iowa Who Made the Ultimate Sacrifice

ABLER JAMES LYNN
ABOLINS JANIS
ABRAHAMSON GARY LEE
ADAMS GLENN ARTHUR
ADAMS STEVEN HAROLD
ADAMS WILLIAM JAMES
AHRENDSEN DENNIS LYNN
ALDRIDGE HERBERT RAY
ALEXANDER NICHOLAS RICHAR
ALSTED STEPHEN PAUL
ANDERSON DAVID BRUCE
ANDERSON JOHN STEVEN
ANDERSON LARRY MICHAEL
ANDERSON WAYNE RICHARD
ANTILL MICHAEL EVAN
ANTRIM TOMMY EDWARD
ARNOLD ALLEN RAY
ASTLEY JOHN MICHAEL
ATKINSON GERALD THOMAS
AUSTIN LARRY DEAN
AVERY ALLEN JAMES
AYERS RICHARD LEE
BAKER PHILIP LOU
BALDWIN JOHN FRANK
BALDWIN KENNETH MAYNARD
BALFOUR WILLIAM JAY
BALL MICHAEL ROGER
BALLHEIM RICHARD ALAN
BARBER HARRY ADELBERT
BARKER WILLIAM GAYLAND
BARR MICHAEL MCKEE
BARRAGY WILLIAM JOSEPH
BARTA ROBERT CHARLES
BARTHOLOMEW CHARLES RICKY
BARTLOW RICHARD LEE
BARTMESS GARY WAYNE
BARTON DENNIS MICHAEL
BASS CHARLES WILLIAM
BATT DARYLE WAYNE
BAXTER KENNETH CARL
BEACH FLOYD IRVY
BEALL ROGER CLOYCE
BEAVER MICHAEL HUGH
BEECK RONALD MARVIN
BEHRENS THOMAS MARTIN
BENDER GARY DEAN
BENDER IVYL RAY
BENNETT MARTIN LEE
BENSON ALBERT DU WARD
BENSON DALE EARL
BERG HAROLD PETER
BERGANTZEL ALBION JOE
BERGQUIST VERNON GAIL
BERSTLER BILL LAVERN
BETTIS NORMAN RADEAN
BILDEN HARLAN TILPHER
BILLS LYLE PRESTON
BINDER CALVIN WILLIAM II
BINGHAM DENNIS WILLIAM
BIRKY HAROLD EDWIN
BISSEN HOWARD MATTHEW
BITTNER DARREL GENE
BIXBY VIRGIL MARTIN
BLACKMAN LARRY PAUL
BLEEKER LARRY DEAN
BLEWETT ROY ROGER
BLEYTHING LARRY DEAN
BLUBAUGH THOMAS EDWARD
BOAT MICHAEL TERRY
BOESHART RICHARD JOSEPH
BONESTROO KENNETH WAYNE
BOONE ALAN RANSOM
BOOTS STEPHEN ELDON
BORCHARD LEONARD E JR
BORNEMAN DEAN ALLEN
BORSCHEL LARRY DEAN
BOUSQUET JAMES ESTREM
BOWERS RICHARD SAULERS
BOYER WILLIAM KLINE
BRANDTS HARLAN RAY
BRENDEL LARRY WILLIAM

BRESNAHAN ALAN RAY
BRIESE STEPHEN CRAIG
BRINKMAN ROBERT JAY
BROCKWAY RANDALL LAWRENCE
BROOKHART GARY LEE
BROOKS WHEELER DAVID
BROWN DENNIS EDWARD
BROWN GENE WESLEY
BROWN WARREN KEITH
BRUCE DENNY LOWELL
BRUE EDWARD JAMES
BRUNS VERLYN CARL
BUCHANAN RONALD IVAN
BUCKLEY JIMMY LEE
BUDDE LARRY JOHN
BUDDI THOMAS LOUIS
BULL BILLY BRUCE
BUNN JERRY ARTHUR
BUNTING RONALD DELL
BURGESON THOMAS JON
BURKE KEVIN GAIL
BUTTZ HAROLD WARREN
CAIN DENNIS REED
CAIN DOUGLAS MICHAEL
CAMP JAMES DALE
CAMPBELL DAVID LAVERN
CARLSON DENNIS ALLEN
CARRA ANTHONY
CARRINGTON THOMAS WILLIAM
CARROLL ROGER EUGENE
CARSON PAUL DAVID
CARTER JOHN E JR
CASE CHARLES CECIL
CASON WILLIAM ARNOLD
CESAR RICHARD ALLEN
CHANDLER CONNIE LEROY
CHASE TERRY ALLEN
CHASE VERNON GLENN
CHATFIELD DANIEL H
CHEEK PHILIP DUANE
CHRISTIANSEN JOHN E JR
CLARK JERRY PROSPER
CLARK RONALD BLAIR
CLARK RONALD EMERY
CLARK WILLIAM MARTIN
CLAUSSEN HENRY ROBERT

CLAYTON CECIL ROGER
CLEFISCH DUANE ALAN
CLEMENS MICHAEL JOSEPH
CLENDENEN RICHARD DEAN
CLENDENIN CHARLES FISHER
COBB ROBERT JAN
COHRON JAMES DERWIN
COLLINS JACK LARELL
COLLISTER JERRY LEE
COMSTOCK ROBERT JAMES
CONCANNON RICHARD NEIL
CONE JOHN MILTON
CONNER EUGENE JOSEPH
COOK CHARLES WILLIAM
COOK DWIGHT WILLIAM
COOK JAMES BLACK
COOK KELLY FRANCIS
COOK PETER EVERETT
COOLEY LOUIS NEWTON JR
COONS GREGORY MAC
COOPER EDWARD THOMAS
COOPER LEONARD DEAN
CORNWELL HARRY JAY
COTTRELL DUANE ALLAN
COUSINS MERRITT THOMAS
COX HOWARD MAX
COZAD WILLIAM MORRIS
CRAVER DENNIS LINN
CRAVER DENNIS MARTIN
CRAWFORD LOWELL LAVAIN
CROOKS DOUGLAS EUGENE
CROSS ARIEL LINDLEY
CROUCH ALBERT B
CROUSE LESLIE DEWAYNE
CRUM STEVEN VINCENT
CRUMLEY HARRY RICHARD
CUFF DICK E
CULLEN RICHARD LEE
CUNNINGHAM JAMES LEON
CUTHBERT BRADLEY GENE
CUTTING JERRY WOODROW
DAHMS LARRY ALBERT
DART LAWRENCE MICHAEL
DAUTREMONT DENNIS DALE
DAVIS DAVID LEE
DAVIS FRANCIS JOHN
DAVIS ROBERT ROY
DAVIS ROLLIN DUANE

DE WITT DAVID CHARLES
DEAVER FREDERICK KENNETH
DEFENBAUGH KENNETH LEROY
DEHNER GEORGE EDWARD
DELEHANT THOMAS FRANCIS
DELL GEORGE DOUGLAS
DESPARD JEROLD VIRGIL
DEUTSCH BERNARD FRANCIS
DEYO ROBERT WILBUR JR
DICKERSON BERNARD W JR
DILLON PATRICK MAURICE
DINES JEFFERY THOMAS
DITCH DAVID KENNETH
DONAHUE RICHARD EARLE
DOSTAL THOMAS JEROME
DOWLING JOHN ROBERT
DOWNS WILLIAM GEORGE JR
DREIER MARK STEVEN
DREW EDWARD JOSEPH II
DRISKELL LARRY RAY
DUART BILLIE D
DUNEMAN ALLEN EUGENE
DUNTZ RONALD DE VERE
EARLYWINE GARY JAMES
EDWARDS STEVEN FRANK
EDWARDSON DAVID R
EGBERT DALE EDWARD
EILERS DENNIS LEE
EISCHEID THOMAS JOHN
EKART PAUL DAVID
EKLOFE SAMUEL ALVIN
ELLSWORTH MARK ALLEN
EMBREE RONALD EUGENE
ENGLISH MARK LEO
EVANS VANCE MARTIN
EVERETT LAUREN RAY
EWING MICHAEL LEE
EWOLDT ROBERT EDWIN
FAGERLIND MERLE KEITH JR
FANKHAUSER CARROLL EUGENE
FARLOW RANDALL LEE
FARNHAM ROBERT DALE
FARRELL WILLIAM DOUGLAS
FEDLER BRUCE JEROME
FELL DANIEL BOONE

FELTNER GERALD LEE
FERGUSON DENNIS DEAN
FERRIS DELMER LEE
FETTKETHER GERALD THOMAS
FISCH DAVID ALAN
FISHER RANDY LEE
FITZGERALD MICHAEL THOMAS
FLATTERY RICHARD T JR
FLEETWOOD DONALD LOUIS
FLESKES DAVID ALLEN
FOELL GERALD LLOYD
FOGARTY GEORGE ALLEN
FOGARTY JOHN JOSEPH III
FOLLON WILLIAM ELLYN
FORD MANZELLE ALAN
FOREMAN TERRY WILLIAM
FOSTER ALBERT DEAN
FOSTER CARL RICHARD
FOSTER JIMMIE LEE
FOWLER JOE LYNN
FRAKES JERRY ALLEN
FRANK THOMAS PAUL
FRANKEN ARLIN DALE
FRASHER GARY DEAN
FRAZIER JERRY RAY
FREEMAN RANDALL GAYLORD
FREESTONE WILLIAM FREDRIC
FRICKE PATRICK LOYAL
FRIEDHOFF DENNIS PATRICK
FRIES DANNY JOE
FRIESE WILLARD JOHN
FRITZE TIM LAVERN
FULLER MICHAEL DAVID
GARDNER GERALD LEE
GARRETT RICHARD B
GARRISON LARRY ALLEN
GAULOCHER FRANCIS LEROY
GEHRT MICHAEL DAVID
GERKEN RALPH BERNARD
GERONZIN ANSON THORNE
GERTSEN ROGER LEE
GIBERSON JERRY GUY
GILBERT HARVEY WOODFORD
GILBERT JAMES MICHAEL

Iowa

GLASPEY RICKY MAURICE
GLAZEBROOK FRANCIS E JR
GLENN DENNIS RAY
GOLL DAVID ROBERT
GOLLIDAY WILLIAM FRANK
GOODMAN RICHARD LEE
GORDON MARVIN EDWARD
GOSCH LARRY GENE
GOURLEY LAURENT LEE
GRADOVILLE CHARLES EDWARD
GRANT DALE EUGENE
GRAY NEWTON MORGAN JR
GRAY PAUL HOUSTON
GREEN JAMES EDWARD
GREEN TIMOTHY LEE
GREGO PHILLIP HARRY
GRETHEN GALEN DEAN
GRETTEN HENRY ARTHUR
GRIFFEY TERRENCE HASTINGS
GRONEWOLD LARRY MARSHALL
GROSS WAYNE WILLIAM
GUNDER DENNIS ANTHONY
GUNDERSON DAVID CRAIG
HAGEDORN LAWRENCE RAYMOND
HAHN MICHAEL DUANE
HAINES DENNIS ALLEN
HALBACH BRUCE CHARLES
HALL RICKY GENE
HALLEY RUSSELL LOUIS
HAMILTON DENNIS CLARK
HAMILTON PAUL GEORGE JR
HAMLIN DARRELL L
HAMMEL KENNETH DALE
HANEY ROBERT BRUCE JR
HANSON HOWARD EMERSON JR
HARGER DON R
HARMS LLOYD
HARRELL DON CLAIR
HARRIS ROBERT DUANE
HARRIS ROBERT ERNEST
HARRISON HARRY TODD
HARTOGH DAVID MICHAEL
HARTSOCK LONNIE DEAN
HARVEY ROBERT GEORGE
HASKINS DONALD DEAN
HASPER CHARLES MARTIN
HASS STEPHEN CRAIG
HATCHER JERRY DEAN
HATTING WILLIAM THEODORE
HAVLIK RICHARD ALLAN
HAYNES VERNON LEE
HEERMAN DENNIS RAY
HEGGEN KEITH RUSSELL
HEIN CHARLES JOHN JR
HEIN GARY LLOYD
HEISELMAN JOHN GERALD
HEITMANN KENNETH HARRY
HELLER MICHAEL LEO
HENDRICKS STEVEN WAYNE
HENRICH MYLLIN GERALD
HENSLEY RAYMOND ALBERT
HERIN JAMES EDWARD
HERRICK JAMES WAYNE JR
HESSE CHARLES THOMAS
HIBBS ROBERT JOHN
HICKERSON JERRY WARNER
HIEBER JACK JEAN
HIGGINS WILLIAM PATRICK
HILKIN LAWRENCE ARCHIE
HILTON DANIEL JEROME
HINDMAN TOMMY IVAN
HINMAN DWIGHT EARL
HIPKINS COLIN KEITH
HISE JAMES HAMILTON
HITES ROGER ALLEN
HODGSON DOUGLAS RALPH
HOFFMAN DOUGLAS ELMER

HOFFNER WAYNE HENRY
HOLLAND DOUGLAS C
HOLLAND DOUGLAS DEAN
HOLM DONALD HENRY
HOLMES LESTER EVAN
HOLST FREDERICK AUGUST
HOLT MARSHALL MYRON JR
HOLTORF DENNIS WAYNE
HORSKY ROBERT MILVOY
HOUG DOUGLAS DUANE
HOUSTON J H
HUBBARD WAYNE GENE
HUK PETER PAUL
HURLEY AILEY BERDEAN
HUTTON KENNETH KEITH
IVENER TERRY LEE
IVES WILLIAM ALLEN
JACKSON DEAN ALFRED
JACOBSEN DONALD LEROY
JAMES JACK LLEWELLYN
JAMESON LARRY DUANE
JANISH DAVID WILLIAM
JEDLICKA DONALD WILLIAM
JENSEN DENNIS RAY
JENSEN HAROLD NORGAARD
JEWELL STEVEN THURLOW
JOHNSON BEN JR
JOHNSON CHARLES A III
JOHNSON CHARLES EVERETT
JOHNSON DANNY WAYNE
JOHNSON DAVID HENRY
JOHNSON DENNIS OGDEN
JOHNSON ERIC BERNARD
JOHNSON GEORGE
JOHNSON JAMES DEAN
JOHNSON JAMES GORDON
JOHNSON LARRY HOWARD
JOHNSON RONALD EUGENE
JOHNSON RONALD JOHN
JOHNSON STEVEN CHARLES
JOHNSON WALTER BOYCE
JOHNSTON CHARLES W JR
JONES DAVID WILLIAMS
JONES LLOYD WESLEY
JONES PHILIP BOYD
JONES ROBERT TAYLOR JR
JOOSTEN CURTIS CHARLES
JUDGE DARWIN LEE
JUERGENS WILLIAM OWEN
JURGENS KENNETH WILLIAM
JURGENSEN DANIEL LEE
KADOUS DARYL LEE
KAHLSTORF KEITH ALAN
KAISER RONALD HARRY
KAPLAN DANIEL JAMES
KARR ROBERT EUGENE
KASTER JERRY LEE
KEEHNER CARROL GENE
KEELER BERT AUSTIN
KEENAN LAWRENCE JOHN
KERN WILLIAM FRANCIS
KILLEN JOHN DEWEY III
KILLIAN MELVIN JOSEPH
KING CHARLES DOUGLAS
KING ROBERT DOUGLAS
KINNY GERALD CARL
KIRCHOFF WILBUR GLEN
KLAAHSEN LAWRENCE JON
KLEFFMAN WILLIAM WALTER
KLEIN RUSSELL LEO
KLOOTWYK ROBERT IVAN
KNAPP TOMMY DUANE
KNAPPER EDWARD WILLIAM
KNIGHT RAYMOND HENRY
KOCK EUGENE JOHN GEORGE
KOEHLER NICKOLAS RAY
KOEPP DENNIS EDWARD
KOERNER RODNEY LEE
KOPRIVA JOHN GAYLORD
KOSANKE PAUL JON
KOUHNS DENNIS BEN
KREMER DONALD PAUL
KROMMENHOEK JEFFREY MARTI
KRUKOW ARDEN LEE
KRUSE JAMES ARTHUR
KRUSE KENDAL ROBERT
KUEHN DUANE JOSEPH
LA DAGE DENNIS ALLEN
LAIRD JAMES BYRON
LAKE RONALD ROY
LAKEY DONALD KAY
LAMMERS DONALD GARY

LAMPERT ARLYN LORANZ
LAMPHIER LARRY GENE
LARSON MARK ALLAN
LASCHE JAMES ALAN
LAWSON MICHAEL CARTER
LEAMON WILLIAM EUGENE
LEAZER TERRY FRANKLIN
LEDLIE DONALD RALPH
LEE JAMES MARVIN
LEFLER DAVID ALLEN
LENNON FREDERICK WILLIAM
LEONARD ROBERT BRUCE
LEONARDI JERRY LEE
LETSCH ROBERT DONALD JR
LEVIS DENNIS RICHARD
LEWIS GARY LEE
LEWIS MERRILL RAYMOND JR
LEWIS MICHAEL KEITH
LIEWER RICHARD GEORGE
LIMBACHER DURWARD ALLAN
LIMBERG DUANE EDWARD
LINK JOHN FRANCIS
LINT DONALD MICHAEL
LIZOTTE WARREN G H JR
LOCKWOOD JOHN LARRY
LOCKWOOD RICHARD JON
LOGAN WILLIAM LEON
LOHSE ARNOLD EDWIN HENRY
LONSDALE JOHN DAVID
LOVITT DAVID GLEN
LOWE JOHN CHRISTOPHER
LUCKSTEAD EDWIN JOSEF
LUERKENS MARVIN ALLEN
LUNDBY LORENCE MARION
LUSE KENNETH ALAN
LYTLE MICHAEL LINN
MAAG JOSEPH ANTHONY JR
MADISON JOHN B
MAJOR GERRY DEWAYNE
MALLONEE KENNETH A
MALONEY MICHAEL KEVIN
MANN DAVID LYLE
MANSON JOHN EDWARD
MANTERNACH MARVIN GEORGE
MARLIN EARL WILLIAM JR
MARTIN DENNIS KEITH
MARTIN RUSSELL DEAN
MAST RANDY LEE
MATHERS STEVEN ALLEN
MATTHEIS DENIS DUANE
MATTHEWS DAVID BRUCE
MATTHEWS GORDON BRUCE
MAURER JEFFREY ALAN
MAY DENNIS ARNOLD
MC BETH ROBERT STEVEN
MC CARL ROBERT JAMES
MC CLAIN JAMES LOUIS
MC CLAIN KENNETH ALLEN
MC CLAIN RICHARD AARON
MC CLATCHEY ROGER WAYNE
MC CLURG JOHN LLOYD
MC COMBS DAVID LEROY

MC CONAHAY BRIAN DUAINE
MC CORD ROGER CLAIR
MC COY EUGENE TAYLOR
MC DONALD ROBERT WILFRED
MC GRANE DONALD PAUL
MC GUIRE WAYNE THOMAS
MC KEEN GERALD CLAUDE
MC KIBBEN WILLIAM RUSSELL
MC MATH DAVID LEE
MC NAMARA DONALD WOODWARD
MC NETT JOE BILLY
MC QUINN BYRON DEAN
MC QUINN LEONARD LLOYD
MCBEAIN DUANE MARVIN
MCCONKEY WAYNE ALLEN
MCCUTCHEON FRANK STAN III
MCDOWELL STEVEN DOUGLAS
MEANS RONALD LEROY
MEIER CARROLL RODNEY
MEIER ROY ALAN
MEIGHAN RICHARD JAMES
MELOY JOHN PATRICK
MERRICK JAMES LEE JR
MERRITT CHARLES EVERETT
MEYER DAVID LEE
MEYER RONALD WILLIAM
MICHAEL DAVID WILLIAM
MIHALAKIS ELLAS LOUIS
MILES LYNN LEROY
MILIUS PAUL LLOYD
MILLER CHARLES CLAUDE
MILLER JAMES LEROY
MILLER JOEL LE ROY
MILLER MELVIN DALE
MILLER MERLIN EUGENE
MILLER TERRY VERNON
MILNER MICHAEL WAYNE
MITCHELL STEVEN MICHAEL
MOHRHAUSER WILLIAM RICHAR
MONEYSMITH HAROLD DEAN
MONFORE WILLIAM DAVID
MOORE GARY LEE
MOORE LARRY A
MORRIS MARSHALL KENNETH
MORRISON GLENN RAYMOND
MORSE CHARLES ALLEN
MORTICE THOMAS E III
MUELLER DAVID HAROLD
MULLEN MICHAEL EUGENE
MUNCH MICHAEL RAYMOND
MURPHY THOMAS JOSEPH
MURRAY RICHARD LEMOYNE
MUSCH DAVID IRA
MYCKA TONEY FRANCIS JR
MYERS CHARLES LOUIS JR
MYERS DAVID WENDELL
NEAVOR GARY ARNOLD
NEBEL THOMAS ALLEN
NEHRING LARRY JOSEPH
NELSON CLIFFORD DALE

NELSON DUANE MICHAEL
NEWELL TIM EDWIN
NEWENDORP JAMES VERNON
NIHSEN DALLAS LEE
NORRIS GEORGE CLYDE
NORRIS JAMES ALAN
NORTHUP EDWIN GILBERT
NORTON GERALD OWEN
NUTT WALTER LEE III
O CONNOR MICHAEL DONALD
O CONNOR MICHAEL MAURICE
O HARA ROBERT CHARLES
O NEILL CARROLL PAUL
O SHEA JAMES CHARLES
OAKES PAUL LAVERNE JR
ODELL DENNIS LYNN
OKLAND VERNON LEO
OLSEN CECIL CHANCEY
OLSON CARL JOHN
OLSON DENNIS GALE
OLSON DUANE ELMER
OLSON ROBERT FRANKLIN
OLSON ROGER LEWIS
OLTMAN DEAN WILLIAM
OSBORN EARL DOUGLAS
OSHEIM JON OWEN
PALEN CARL ANTHONY
PALMER ARNOLD RALPH
PARISH DAVID LEROY
PARK MARVIN EDWARD
PASSIG DUANE RINEHARDT
PAULSEN DAVID HENRY
PAXTON DONALD ELMER
PAYNE RICHARD JOSEPH
PEARSON DAVID L
PEASE WILLIAM HARRISON
PEDDICORD DONALD GLENN
PENA JESSE JOSEPH
PENCE JAMES HOWARD
PEREZ ERNESTO
PETERS WILLIAM LEE JR
PETERSEN MARK CARSON
PETERSON ANTHONY EARL
PETERSON DOUGLAS EUGENE
PETERSON KERMIT C JR
PHILLIPS WARREN EVERETT
PICKART DWAYNE ROBERT
PICKERING DONALD WILLIAM
PICKERING RUSSELL THOMAS
PICKING FRANKLIN WILLIAM
PIITTMANN ALAN DALE
PITZEN JOHN RUSSELL
PLATT JOHN HERBERT
PODHAJSKY NORBERT ALBERT
PODNAR ROBERT JOHN
POLLARD THOMAS LEROY
POOCK MYRON JEROME
POPPEMA LEROY WARREN
PORTER THOMAS ALAN
POTTER WILLIAM DON
POUNDSTONE THOMAS RICHARD
POWERS JAMES CONRAD

86

POWLES DONALD EUGENE
POWLISTHA GERALD STE-
PHEN
QUAM JOHN ELLSWORTH
QUINLAN FRANK JOSEPH JR
RAFFENSPERGER JAMES E JR
RAMSEY ERNEST LEROY
RANDALL TERRELL LYNN
RANDOLPH VERNON CHES-
TER
RANEY STEVEN LEON
RATH GARY KEITH
REAVIS BRETT GRANT
REECE HOWARD WAYNE
REED TERRY JOE
REED WAYNE FRANCIS
REES WILLIAM EDWARD
REEVES LOREN STEVEN
REID JOHN LEE
REIDY MARTIN JOHN
REILLY JOHN MICHEAL
REISTROFFER DANIEL PHILLI
RENDON RAPHAEL JOHNNY
REVLAND RICKEY DON
REX ROBERT F
REXROAT TERRY LYNN
RIAL JAMES ALPHONSE
RICH CRAIG ARTHUR
RICKELS JAMES BURNELL
RILEY JAMES FRANCIS
RINDONE MICHAEL GUSTAVE
RISH RICHARD LEE
RISTINEN ARMAND ERVIN
RITTER DENNIS LEE
ROBERTS TERRY
ROBINS JAMES MILTON
ROCHE KENNETH WAYNE
ROGERS CORDELL BRUCE
ROGERS CRAIG RAY
ROGERS GEORGE PATRICK
ROGGOW NORMAN LEE
ROHLFSEN LYLE ERVIN
ROSENBAUM GERALD
GEORGE
ROSS STANLEY DENNIS
ROTH LA ROY FREDERICH
RUDD RICHARD JOHN JR
RUDEN MATTHEW ALBERT
RULE TED JAMES
RUPE DONALD LEE
RUSHING STEPHEN ABRAM

RUTGERS DAVID LYNN
SAGERS RONALD RAY
SAMPERS JAMES WILLIAM
SAMS JOHN WILBUR JR
SANDERSON GAIL GENE
SCHARES ROBERT JOHN
SCHIMBERG JAMES PHILIP
SCHMIDT ALLAN LEE
SCHMIDT DALE HOWARD
SCHMITT JOHN KENNETH JR
SCHRADER FRANKLIN DANIEL
SCHROEDER MICHAEL ALLEN
SCHULTZ MICHAEL DOUGLAS
SCHULTZ ROBERT WILLIAM
SCHUMACHER DONALD
EUGENE
SCHUTT RANDALL KARL
SCHWARZ ROGER LEE
SCHWEBKE LARRY CHARLES
SCHWERDTFEGER JOSEPH
ALLE
SCOTT DONALD EUGENE
SCOTT KENNETT KEITH
SCOTT LARRY ROBERT
SCOTT STEVEN CLAYTON
SCULL GARY BERNARD
SCULLY RUSSELL CRAIG
SEARLES JEFFREY PAUL
SEARS STEVEN DWIGHT
SELLS ROBERT DEE JR
SETKA STANTON JAMES
SHACKELFORD RANDALL LEE
SHAIN ELWIN ROY
SHANK JOHN B
SHANNON ROBERT JOSEPH
SHERLOCK DAVID HENRY
SHONKA DARYL DAVID
SIEGEL DAVID DOUGLAS
SIMMONS NORBERT GENE
SISLEY RUSSELL JAY
SISSEL CHARLES EDWARD
SIVERLY DAVID LEE
SKIRVIN JOHN DARREL
SKOGERBOE DENNIS MI-
CHAEL
SLATER JOHN EDWARD
SLOAN TERRY PATRICK
SMALL BURTON EUGENE
SMALL VERNARD JAY
SMIDSTRA CHARLES RICHARD
SMITH CHARLES WENDLE

SMITH CRAIG LEWIS
SMITH FRANK GEORGE
SMITH GREGORY ALLAN
SMITH HAROLD LEE
SMITH JACK RAE
SMITH JEROME JOSEPH
SMITH LARRY EUGENE
SMITH ROBERT CARL
SMITH STANLEY RICHARD
SMITS HERMAN JR
SNETHEN ROBERT CARL
SNITKER CURTIS DEAN
SPARKS DONALD LEE
SPEER JAMES WALTER
SPOTSWOOD MICHAEL CARR
SQUIERS GARY LADD
STANSBARGER RICHARD
LAURE
STATON FRANK LYNN
STEELE ROBERT FRANKLIN
STEIN RONALD MARVIN
STEPHENSON KURT PATRICK
STERLING ROBERT ALLEN
STICKELS MARK GALEN
STILL JIMMIE DALE
STINN JOHN RICHARD
STOKES DAVID ALAN
STOLTENBERG REID WILLIAM
STOLTENBURG MARK ERNEST
STOLTZ STEVEN RAY
STONE OTTO JR
STRAIN EDWARD W
STRAUDOVSKIS JOHN
STRIEPE PAUL RAYMOND
STROSCHEIN RONALD
ROBERT
STRUBLE STANLEY DEAN
STUBBLEFIELD KENNETH R
STUDER LOREN FRANCIS
SUTHERLAND RICHARD
EUGENE
SWAIM RONALD GAIL
SWANEY RICKEY EUGENE
SWANSON JOHN EARNEST JR
SWEET RONALD STEVEN
SWIFT RICHARD C
TAFOLLA NABOR RICHARD
TEATSWORTH GARREL LEE
TEDESCO JAMES JOSEPH
TEW JERRY EUGENE
THOMASON KENNETH

ARTHUR
THOMPSON DONALD AR-
THUR
THOMPSON DONALD WAYNE
THOMPSON MELVIN EUGENE
THORPE WILLIAM DAVID
TIMMER AKKE JANS JR
TINDALL CORBIN CLARK
TORREY STEVEN MICHAEL
TRACY JOHN LEO
TROTTER PATRICK JOSEPH
TUCKER KENNETH WAYNE
ULFERS JOHN BURDETTE
UPTON DANIEL CARL
UTTER MICHAEL JOSEPH
VAN BALLEGOOYEN ROBERT
AR
VAN DALSEM MARC GREGORY
VAN EVERY EDWARD JR
VAN RIESEN ALVIN CHRIS
VANATTA RANDALL ALLEN
VERGAMINI DOUGLAS SILVIO
VILLARREAL MICHAEL
VOSS ROBERT J
WADE RICHARD ALLEN
WAGNER GREY H
WALKER LESTER TIMOTHY
WALL JOHN WALTER
WALTERS JIM JAMES
WALTHOUR SAMUEL W JR
WARD EUGENE AMBROSE
WARD TIMMIE JOE
WARREN GRAY DAWSON
WARTH WOODROE WARREN
WATERMAN CRAIG THOMAS
WATERMAN DENNIS WALTER
WEARMOUTH RONALD
VERNON
WEBB DONALD RAY
WEBER DENNIS LEE
WEBER WILLIAM EUGENE
WEHR MARVIN FRANCIS
WEHRHEIN RICHARD JOSEPH
WEIDNER FREDERICK WIL-
LIAM
WEIR GARY WAYNE
WELCH RICHARD M
WERNER JOHN FREDRICK
WEST GRAYSON JERALD
WEST JAMES WILLIAM
WESTBERG RICHARD

CHARLES
WESTERGARD TERRY MI-
CHAEL
WESTLY CYRIL JEFFREY
WETJEN GORDON JOHN
WEYKER DONALD DENNIS
WHITE ALLEN JOSEPH
WHITE STANLEY DEAN
WHITEMAN WAYNE FRANK
WHITFORD LAWRENCE W JR
WIDTFELDT PAUL FRANK JR
WILCOX THOMAS DEWEY
WILDMAN STEVEN EARL
WILKERSON DAVID HUNTER
WILLIAMS DENNIS NEIL
WILLIAMS GARY LYNN
WILLIAMS WALDO ALVA
WILSON BRYAN LEE
WILSON DONALD WAYNE
WILSON JEFFREY LYNN
WILSON JOHN JOSEPH
WILSON KEITH LESLIE
WILSON PAUL JOSIAH
WINGER JON RICHARD
WISELY DANIEL LEE
WISSINK STEVEN LEE
WOHLFORD LLOYD CYRUS JR
WOLF KENT CARTER
WOLFF WARREN KENNETH
WOLFORD CRAIG BENTON
WOOD DONALD FRED
WOOD REX STEWART
WOODARD STEPHEN LEE
WOODS CLAYTON LEON
WRIGHT BRUCE WILLIAM
WRIGHT VERNON ARTHUR
WRISBERG JOHN HOLGER III
WUTZKE WAYNE GARY
YASHACK RONALD ALLEN
YETMAR DENNIS JAMES
YORK LANNY ALLEN
YOUNGBEAR RICHARD CLIVE
ZAHN LELAND DALE
ZAPPIA MICHAEL LEE
ZIMMERMAN GORDON F
ZIMMERMAN TERRY RAY
ZINNEL HERBERT OWEN JR
ZITTERGRUEN LOUIS LLOYD

Photos by Rod Mikinsky

Vietnam War Memorial

Heritage Park, W 6th St & Washington St, Junction City, KS 66441

The Kansas Vietnam Memorial is located in Heritage Park in Junction City, Kansas. It is a beautifully maintained monument that was created to honor and remember the men and women who were either killed in action or went missing in action during the Vietnam War.

The Vietnam memorial was made by Dan Babcock with Bruce Memorials (now Wilbert Memorials) and was officially declared the state's designated official State Vietnam Veterans Monument by both the Kansas House of Representatives and the Kansas Senate in a resolution passed in May of 1991. The Kansas Vietnam Memorial is made of black granite and it was installed and dedicated on the Fourth of July in 1987.

The Vietnam War memorial stands at an impressive 13 feet by 46 feet and is made of seven black granite slabs. Four of the seven slabs feature the names of the 753 local service members who were killed during the Vietnam War. A fifth panel lists the full names of the 38 people who were still presumed missing at the time of the 1987 dedication.

The center panels feature etchings of two soldiers who are flanking depictions of the Purple Heart Medal and the Vietnam Campaign Medal. On the back of the memorial, visitors will find the etched insignias of the five branches of the US armed forces. The monument also features the inscription "History will remember the war. Will America remember her men?"

The memorial is well-maintained in order to be worthy of its official designation and, as such, is lit 24 hours a day, seven days a week for all visitors to enjoy at any time. Besides the Vietnam War memorial, Heritage Park also plays host to a memorial for the Purple Heart, the Civil War Arch, the 1st Infantry Memorial, and a POW/MIA memorial. It is a great place to attend Memorial day and veteran's day services.

The Names of those from the state of Kansas Who Made the Ultimate Sacrifice

ACHESON CHARLES RALPH
ACOSTA JOSE FRANCISCO
ADAM JOHN QUINCY
ADAMS THOMAS EDWARD
ADAMS WILLIAM CARL
ALDERMAN ANDREW ALBERT
ALLENDORF MICHAEL GEORGE
ALLGOOD FRANKIE EUGENE
ALVORD RONNIE EUGENE
AMERINE KENT L
AMMANN ALBERT FRANK
ANDERSON ALFRED EARL
ANDERSON DENIS LEON
ANDERSON JAMES RICHMOND
ANDERSON LANNIE RAY
APPLEGATE NEWELL F SR
ARB FRANCIS LOREN
ARBUTHNOT JAMES MALCOLM
ARNETT JAMES DOUGLAS
ARNOLD RICHARD EARL
AST STEVEN VINCENT
ATKINS JAMES
AVILA THOMAS ROBERT
BADWAY VICTOR WOLF JR
BAILEY CHARLES CLIFFORD
BALENTINE ROLAND JR
BALES CHARLES ROBERT
BALL MERLIN EUGENE
BANNON GARY CLIFFORD
BARKSDALE JERRY DEAN
BARNETT GARY JOE
BARR TERRY LEE
BARRETT GEORGE DWAYNE
BAUER LAWRENCE EDWARD
BAUGHMAN JOHN OLIVER
BELDEN LARRY GENE
BELL CHARLES MARTIN
BELLER WILLIAM RUSSELL JR
BENNEFELD STEVEN HENRY
BERGAN MERLIN HERMAN
BERGER GERALD DAVID
BERRIER TOMMY JOE
BERWERT PATRICK MICHAEL
BICKFORD RALPH NEVIN
BILLINGSLY LEE WAYNE
BINDER GARY LEE
BIRD DANNIE LEON
BIRD SAMUEL RICHARD
BISHOP JAMES LOUIS
BITTLE DOUGLAS ROBERT
BLAND ISAAC
BLANSCET MICHAEL JOHN
BLITCH BERNARD L
BOESE ROBERT LEE
BOOLIN CLARENCE HENRY
BORDERS DARELD NORVAL
BOWLES HARRY OWEN
BOWLING JOSEPH PERRY
BOWMAN DONALD ROBERT
BOYD RICHARD EUGENE
BRACK HARRY HUBERT JR
BRADBURY STEVEN WAYNE
BRADFORD ALLEN ROYAL
BRANDT GEAROLD LEE
BRAY RALPH OSCAR JR
BREEDING MICHAEL HUGH
BRENNER DAVID ALDEN
BRENNER DAVID GEORGE
BRENNER KENNETH JAMES
BRESHEARS ALAN WAYNE
BROOKS BARTON W
BROWN KENNETH LAVERN
BROWNING GARY LEE
BROWNING RAYMOND VENSON
BRULL MICHAEL JOSEPH
BRYANT BOBBY RAY
BUCHANAN JACK LYNN
BUCKRIDGE MARVIN DOUGLAS
BUNCH JAMES GEORGE JR
BURGESS LAWRENCE DEAN
BURKETT CLOYCE ORAL JR

BURNAM STEVEN WAYNE
BUSH EDWARD L
BUTTS DARRELL WAYNE
CALVIN STANLEY DEAN
CAMDEN JOHNNIE ROGER
CANADY TROY VERNAL
CANFIELD BOYD
CAREY BARTON WAINWRIGHT
CARLEY TIMOTHY LYNN
CARTER OTIS
CASE DANIEL CHARLES
CASTOR JAMES WILLIAM
CHADWICK BILLY RAYMOND
CHAPMAN KURTIS NOLAN
CHRISTESON LEONARD WAYNE
CHRISTIAN DAVID MARION
CLAFLIN RICHARD AMES

CLARK LARRY RAY
CLARK RONNIE LEE
CLARK TIMOTHY RICHARD
COE KENNETH EUGENE
COLLINS GARY DEAN
COMER WILLIAM MARVIN JR
CONRARDY RICHARD JOHN
CONROY RONALD LEE
COOMBS DAN L F III
COON KEITH DAVID ED WILL
COOPER CHARLES EDWARD
COOPER JAMES RAYMOND
COPELAND WILLIAM E II
CORCORAN BRUCE ANTHONY
CORR CLIFFORD WAYNE
COSTELLO LAWRENCE R
COURTNEY JAMES IRA
COWAN SAMUEL PAIGE JR
COX GARY DEAN

CRAIG EDWARD LEE
CRAIG JIMMY LEON
CRAIG REX LEE
CRAWSHAW STEEVE ALEXANDER
CRIST KENNETH ROY
CRUSE GEORGE LARRY
CUNNINGHAM CARL EDWIN
DANIELS JAMES MICHAEL
DARTY OMER GENE
DAVENPORT JOHN JUNIOR
DAVIS CHARLES EUGENE
DAVIS EMMETT RAY
DAVIS JAMES THOMAS
DAVIS JOHN LAWRENCE
DAVIS YALE REZIN JR
DECOW MELVIN DALE
DELAUGHDER DAVID LEE
DENNIS WILLIAM R III

DENTON DENNIS ALAN
DERRITT EDDIE RAY
DICKEY DERREL KEITH
DICKEY GARY LYNN
DICKINSON DANIEL ALBERT
DIEDERICH JOHN LEO
DILLON DONALD EUGENE
DILORENZO RAYMOND JOHN
DIMITT ROBERT VICTOR
DODGE WARD KENT
DODSON ERNEST DEAN
DONOVAN MICHAEL LEO
DOORNBOS DON MICHAEL
DORNON CHARLES WILLIAM
DORSEY LEWIS R G
DOTTER EDWIN EARL
DROUHARD PETER AUGUST
DUDLEY GARY WILLIAM
DUKELOW CORNELIOUS P II

Kansas

ECKELL JOHN W
ECKHART RUSS EUGENE
EISENHOUR JAMES DOYLE
EITEL JACK ORVAL
ELDRIDGE JAMES WILBUR
ELGAARD ROBERT JAMES
ELLIOTT ROBERT JOE
ELSENRATH JOHN JOE
EMBREY RALPH CURTIS II
EMERY JOE LOUIS
ENLOW PHILIP JAMES
EVANS DONALD ALLEN
EVANS GERALD BRUCE
EVANS JAMES JOSEPH
EVERHART WILLIAM JOSEPH
FEARNO JOSEPH BARNETT
FEW SAMUEL ARTHUR
FIFFE RICHARD LEE
FINCH JOHN WEBSTER
FISHER HENRY LEE
FONSECA MICHAEL JEROME
FORD JOSEPH AVERLL III
FORREST MONTE WAYNE
FOUNDS GERALD DEAN
FOWLER JAMES ISIAH JR
FRY ROLLAND KEITH
FUQUA HARRY IVAN JR
GADSON EDDIE DEAN
GANDY MICHAEL LEE
GARDNER SAMUEL RAY
GARRETT JAMES MICHAEL
GENTRY HENRY JR
GERBER MICHAEL EUGENE
GFELLER JOHN HERBERT
GIBSON DAVID PARKER
GILLASPIE PAUL THOMAS
GILLEN THOMAS ELDON
GIRDNER ROBERT ORAL
GLADDEN JAMES TAYLOR JR
GLAZE KENNETH LEE
GLYNN DENNIS WAYNE
GODSEY JAMES FREDERICK
GOOCH WESLEY LEE
GOODE JACK DEE
GOODNER ROBERT EUGENE
GOODWIN BOB JACK
GORDON JOHN SWAIN
GORMAN EDWARD THEODORE III
GOSSELIN PHILIP LYN
GOUDY RICHARD LEE
GRABER JOHN ALLEN JR
GRACE MARTIN JOSEPH JR
GRAHAM DENNIS LEE
GREATHOUSE ROBERT CHARLES
GREEN DENNIS BLAINE
GREEN LUTHER JR
GREER CHARLES ROBERT
GRETENCORD DEAN LEE
GRIFFIN KEITH D
GRIFFIN PATRICK JOSEPH
GROTHAUS ROBERT JOHN II
GUILLEN JOHN DAVID
GURTLER CHARLES RONALD
GUYER ALBERT MARSHALL
HADLEY LEO LARRY
HAGEMAN JOEL THOMAS
HALBOWER HARLOW KENNETH
HALE VICTOR
HANDSHUMAKER LLOYD E JR
HARBOUR DEXTER DUANE
HARRIS JOHN JAMES
HARRIS ROBERT LEE
HARRISON JIMMY KEN
HARROLD PATRICK KENDAL
HARVEY LARRY WAYNE
HASSETT JAMES PETER

HAUG RONALD LEE
HAWTHORNE WILLIAM ALLEN
HAYES GEORGE E
HAYS CLIFTON WALTER
HAYWOOD JAZREAL LEVITE
HAZELWOOD JOHN EDWARD
HEAD DAVID NEIL
HEFFNER STEVEN CLINTON
HEMMINGWAY CHARLES LYNN
HENDRICKSON GARY ARLAND
HENDRICKSON GAYLORD BLAIN
HENDRIX JERRY WAYNE
HERRIMAN RONNIE LEE
HESKETT JAMES EARL
HESS KERRY EUGENE
HEWITT THOMAS THEODORE
HIEBERT JOHN MICHAEL
HIGLEY LYNNFORD HARLOW
HILMES STEVEN LEE
HINES PHILIP BLAINE
HINES PHILLIP MASON
HINTON RODNEY GENE
HLADIK HAROLD HERBERT
HOEME FORREST DEAN
HOFFMAN WILLIAM DAVID
HOLCOMB REBEL LEE
HOLMES JACK EUGENE
HOLROYD JAMES LAWRENCE
HOLT ALLEN LEE
HOLTOM MARK RICHARD
HONEYCUTT BLAINE LEROY
HORINEK DONALD EDWARD
HOSKINS CHARLES LEE
HOUSEHOLTER TERRY AUGUST
HOWARD CHARLES EMORY
HOWARD STEVEN DALE
HUBERT MICHAEL NEIL
HUMPHREY JAMES GILBERT
HURST QUENTIN FOXX
HUTTIE FREDERICK E III
HUTTON EARL DEWITT
INGRUM JOSEPH HENRY
ISHMAN HUEY LEE
JACKSON GARY RAY
JAMESON RODGER LEE
JANKE THEODORE JR
JANTZ ROBERT WAYNE
JARVIS RONALD ALAN
JENKINS ROBERT EARL
JEWELL EUGENE MILLARD
JIM MARTIN JR
JIMENEZ EDUARDO
JOHNSON CHARLES FRANKLIN
JOHNSON FREDERICK PETER JR
JOHNSON GARY DALE
JOHNSON JIMMIE LE ROY
JOHNSON KENNETH DUANE
JOHNSON PHIL DAVID
JOHNSON ROBERT FRED JR
JONES MERLE ELDON
JONES RICHARD LEE
JORDAN FRANCIS EUGENE
JUSTICE WALTER EUGENE SR
KAEBERLE DANA JAMES
KARLIN DONALD DEAN
KARST CARL FREDERICK
KECK WARREN EDWARD
KENAGA GARY LYLE
KIER CHARLES RICHARD
KIMMEL ROBERT GENE
KINNAMON SAMMY EDWARD
KIRKENDOLL CLEE ANDREW
KIRKENDOLL JERRY WAYNE
KITCHEN DAVID LEE
KLAUS ARTHUR LEE
KLENDA DEAN ALBERT
KNOWLES KENNETH JOSEPH
KRAMER KEVIN CLINTON
KREHBIEL KENNETH DILLARD
KRIG DAVID LEE
LA FRANCE JON PATRICK
LA PLANT KURT ELTON
LACEY FRANKLIN D
LACKEY JACK VERNON JR
LAKE LLOYD DEAN
LAKIN ROGER ALAN
LAMON ROY ALLEN

LAND DAVID ALDEN
LARRABEE FLOYD MICHAEL
LARSON LOREN HENRY
LASKOWSKI ANTHONY JAMES
LE BOMBARB LONNIE GUY
LEFTWICH RAYMOND FRANCIS
LEHNHOFF EDWARD WILLIAM JR
LEIKAM NORMAN ALEXIUS
LEMLEY BILLY JOE
LEWALLEN JACKIE LEE
LEWIS CLARENCE PAUL
LEWIS DONALD ALLEN
LEWIS JESSIE ROY
LINDAHL JOHN CARL
LINDHOLM DAN VICTOR
LINDQUIST WILLIAM FRANCIS
LITTLE GARY DEAN
LLAMAS JOSE
LONG ELDON DALE
LONG GEORGE WENDELL
LONG JEROME ALBERT
LONG MELVIN RAY
LOVE DARRELL STEVE
LOVE JOHN ARTHUR
LOYD LONNY LEE
LUKERT EDWARD ROY
LUNDY MAURICE EDWARD
LYONS ROBERT PAUL
MAC DONALD ALLAN HERBERT
MALONE CHARLES WALTER
MALONE ROBERT GARY
MARTIN LARRY EUGENE
MARTIN STEVEN LOUIS
MARTIN WALTER WESLEY
MARTINEZ GEORGE FRANCIS
MARTINEZ PETER
MASLAK JOHN JOSEPH
MASON ROMAN GALE
MATHER ALVIN EUGENE
MATTHEWS WILLIAM CLAY
MC CUBBIN GLENN DEWAYNE
MC CULLOUGH PATRICK ELVIN
MC DERMOTT JOHN PATRICK
MC DONALD DANNY LEE
MC DONALD MICHAEL JAY
MC GINNIS LESTER CLEO II
MC GONIGLE WILLIAM DEE
MC GOVERN JEROME GEORGE
MC INTOSH DONALD WILLIAM
MC KAIN BOBBY LYN
MC KAY GERALD OTTO
MC KINNELL RICHARD LEE
MC TEER JEFFERY CLARK
MCLAREN ROBERT DALE
MEADE JAMES ROBERT
MEANS WILLIAM HARLEY JR
MEIS DONALD DAVID
MERYS ROBERT KEITH
METZGER GEORGE LEONARD
MILBRADT DALE LA VERNE
MILLER KENNETH WALTER
MILLER TIMMY LARRY
MISCHLER HAROLD LOUIS
MITCHELL CRAIG WESLEY
MIZER LENTON EUGENE
MOHLER TIMOTHY ALLEN
MONTEMAYOR JAMES MICHAEL
MOORE BILLY RAY
MOORE DONALD EUGENE
MOORE JAMES EZRA
MOORE WILLIAM RAY
MORA RAYMOND CASTILLO
MORGAN LELAND RAY
MORTIBOY WILLIAM SHELTON
MOTT JOHN ARTHUR
MOUNTS JERRY DUANE
MOWREY RICHARD LYNN
MUEHLBERG RONALD LEE
MUELLER STEVEN AL
MULLER DANIEL SCOTT
MUNDAY PHILLIP DEAN
MUNGER RONALD WILLIAM
MURRAY CARL EUGENE
MURRAY VIRGIL ARTHUR
MYERS GENE ALLEN
NAIL GARY DEAN
NEER GERALD KING

NELSON DAVID REID
NETH FRED ALBERT
NEUBURGER DANIEL LEO
NEVINS ELDON EUGENE
NEVINS FLOYD CHARLES
NEWMAN JERRY LEE
NICKS BENJAMIN ARNOLD III
NORTHROP JAMES LEEROY
NUFER JAMES LEO
O BRIEN PATRICK EDWARD
O CONNOR GERARD FRANCIS
O NEIL RILEY CHARLES JR
OATNEY ALLEN EUGENE
OGLE DAVID ROBERT
ORNELAS CARL JOHN
ORTEGA ERNEST
OSENBAUGH JAMES DALE
OWENS TIMOTHY EUGENE
PACHECO FELIX
PAPPAN BOBBY JACK
PARKER HARVEY R
PARRISH ROGER ALAN
PARSONS GARY REED
PARTRIDGE NORMAN WAYNE
PATRICK DARYL WAYNE
PATTERSON JEROME DEAN
PATTON WARD KARL
PAXSON STEVEN DUANE
PAYNE KENNETH RAY
PAYNE RICHARD NORMAN
PAYNE ROBERT ELGIN
PEEL LAWRENCE RAY
PERRYMAN RONALD GLEN
PETERSEN DANNY JOHN
PETETT LARRY WYNN
PETTY JOHN CABLE II
PHALP WILLIAM ANDERSON JR
PHILLIPS JACK WARREN
PHILLIPS MARLEN LE ROY
PHILLIPS MICHAEL GENE
PIERCE CLINTON DWIGHT
PIKE RAYMOND HORACE JR
POKE DONALD MAURICE
PONTING JOHN L
PORRAS JUAN
PORTER JAMES HOLLAND
POSEY CHARLES ALBERT
PRIETO RUBEN
PROBERTS WAYNE DOUGLAS
PROCINO NICHOLAS RALPH
PUGH DENNIS GERARD
PULLIAM DALE ALLAN
RAGER WILLIAM EARL
RAIMEY CHRISTOPHER LA G
RAMSEY MILTON HARDIN
RAY RANDY DAVID
REDD BOBBY EDWARD
REDMOND RALPH GEORGE
REGIER RAYMOND DEAN
REID JAMES EDWARD
REISSIG LARRY LEROY
RENFRO JACK DENNIS
REUKAUF LEE EDWARD
REYNOLDS WILLIAM LAWRENCE
RHODES LARRY WAYNE
RICHARDSON ARLEN DEL
RICHARDSON GARY WAYNE
RIEDEL ROBERT EUGENE
RIGBY OLIS RAY
RING HAROLD KENNETH
ROBERTS RICHARD STEPHEN
ROBINSON FLOYD HENRY
ROBINSON MICHAEL BERNARD
ROLFE GARY FAY
ROMINE ALBERT W
ROOT EDWARD CHARLES
ROSALES MARTIN ANGEL
ROSS DENNIS WAYNE
ROUCHON ALAN MICHAEL
ROYSTON LOUIS DON JR
RUCKLE CLINTON GEAN
RUDISILL DAYTON LUTHER
RUNKLE ROBERT LESLIE
RYAN DELBERT LEROY
SAENZ EDWARD LLOYD
SAGE TERENCE FAIRCHILD
SALINAS PHILLIP LOUIE
SAMUELS GEORGE LEROY
SANCHEZ FRANKIE
SASEK RICHARD JOHN
SCANLAN LAWRENCE

WALKER
SCATES CHARLES EDWARD JR
SCHALIPP MURVIN JR
SCHMELZLE JOHN JOSEPH
SCHOTH WILLIAM WESLEY II
SCHULTZ RONALD JAMES
SCHULTZ WILLIAM LEE
SCHULZ RONALD DOUGLAS
SCOTT DAVE RUSSELL
SCOTT LARRY EUGENE
SEGLEM RICHARD NOYCE
SEVICK JOHN FRANCIS
SHAMBAUGH DALE K
SHANK GARY LESLIE
SHANNON RICHARD DEAN JR
SHERMAN DANIEL L
SHERRILL AMOS CHESTER II
SHUE RUSSELL DALE
SIDENER WESLEY MELVIN
SIMMONS JOHN STEPHEN
SMALL NORMAN EUGENE
SMITH AARON BRUCE
SMITH KENNETH WAYNE
SMITH MAYNARD LEE
SMITH MICHAEL JOSEPH
SMITH RICHARD DEAN
SMITH STEPHEN LEE
SNOW LONNIE DALE
SONDERMAN THOMAS LEE
SORENSEN KENNETH LEE
SOUTHALL JOHN GEORGE
SOUTHERN EDWARD CHARLES
SPIESS JOHN CHARLES
SPRAGINS CARROLL WAYNE
SPRINGER GERALD WAYNE
SQUIER WILLIAM RUSSELL JR
STAHL EDWARD ARNOLD
STANDEFORD JAMES MICHAEL
STANDERWICK ROBERT L SR
STANFIELD GARY KELVIN
STEIMEL GREGG FRANCIS
STEPHENS LESTER AL
STEWART LEONARD KEITH
STEWART MICHAEL HENRY
STEWART WILLIAM STEVEN
STIGGINS DOUGLAS LEE
STONE GEORGE DAVIDSON
STRAUB CONRAD FRANCIS
STRUBE JAMES CLARENCE
STULTZ CHARLES GILBERT
SULLIVAN MICHAEL JAMES
SUTTER FREDERICK JOHN
SWAZICK DANNY GEORGE
SWENDER JACK SHIVELY
SWINK JACKIE LEE
SZIJJARTO STEPHEN JOSEPH
TALBURT RAYMOND THURL
TALIAFERRO GLEN JOHNSON
TAYLOR WALTER LEE JR
THOENNES MICHAEL WALTER
THOMAS PAUL EDWARD
THOMPSON CECIL TRUMAN
THOMPSON RICHARD W
THOMPSON ROBERT EUGENE
THOMPSON WILLIAM JOSEPH
TIDERMAN JOHN MARK
TINKUM ETHER ARNOLD
TODD JEROME DEAN
TODD JIMMIE LESTER
TOOMES WILLIS ALBERT
TRAIN STEVE WARREN
TRAVIS MICHAEL RICHARD
TREAS RICHARD LEE
TROWER GARY RAY
TROXEL CHARLES LEONARD
TRUBE DELBERT LEROY JR
TUCKER ROBERT EUGENE
TURNER JAMES EARL
TURNER JOHN HAROLD
TURNER ROBERT ELDON
URBAN ROBERT LEE
VAN WINKLE JESS H JR
VELASQUEZ DAVID ROBERT
VIEGRA LUZ
WALKER CHARLES BUTNER JR
WALLACE DONALD DEAN
WALLACE LANNY JOHN
WALTERS JOHN EDMOND
WARD RONALD RAY
WARDS DAN THOMAS
WARREN LARRY DEAN
WATTS LARRY DEAN

WEBB GARY ALAN
WEBB JOHNNY ROBERT
WEIS KENNETH D
WELCH JODIE VARNER JR
WELDON ROBERT P
WELLS BRIAN LEE
WELSCH CLARENCE LEON JR
WELSH LARRY DON
WEST JOHN MICHAEL
WEST PAUL EDWARD
WESTOVER DAVID EDWIN
WESTPHAL JERELD EUGENE
WHEELER MORRIS CRAIG
WHELCHEL RUSSELL DESMOND
WHINERY ROGER LEE
WHITE GLENN EARL
WHITE MICHAEL ALAN
WHITE MICHAEL LA VERN
WHITEHEAD THOMAS LEROY
WHITERS DONALD EMERY
WILLIAM RONALD WAYNE
WILLIAMS BILLIE JOE
WILLIAMS JOHN CHARLES
WILLIAMS ROGER RALPH
WILLIAMS THOMAS VERNON JR
WILSON BILLIE JOE
WILSON JOHN LEE
WILSON MICHAEL LUND
WILSON RODNEY DAVID
WINDLE PAUL RALPH
WINN DONALD DEAN
WINNINGHAM CLIFTON
WINTERS JOHN EDWARD
WISWELL SAMMY RAY
WOLFE RICHARD OGDEN
WOOD LLOYD JOSEPH SR
WOODY WILLARD EVERETT
WRIGHT RICHARD HUGH
WYRICK MICHAEL ALLEN
ZELLER MICHAEL CHARLES
ZUTTERMAN JOSEPH A JR

Photos by Estill Robinson

Kentucky Vietnam Veterans Memorial

365 Vernon Cooper Ln, Frankfort, KY 40601

The Kentucky Vietnam Veterans Memorial is located in Memorial Park overlooking the state Capitol in Frankfort, Kentucky. It was created to honor the men and women who served the nation during the Vietnam War. The memorial is always open to the public and is one of the most visited sites in the state, but there are certain days where visitors can check out special ceremonies. The Vietnam memorial plays host to these ceremonies on Armed Forces Day, Flag Day, Memorial Day, POW/MIA Recognition Day, and Veterans Day.

The Kentucky Vietnam Veterans Memorial was made by Helm Roberts who was not only an architect but was a veteran. Construction of the project began in November of 1987 and was completed by the summer of 1988 and ready for its dedication in November of the same year.

The memorial came about through the collective efforts of local Vietnam veterans, some members of the General Assembly, the National Guard, and a few members of the state's Executive Branch. The money needed for the project was raised through the Memorial Foundation known as KVVMF, a non-profit corporation that has been tapped to maintain the monument and grounds. While no tax dollars were used to create the memorial, the land was leased to the Foundation by the state "in perpetuity."

The Vietnam War memorial is made of 327 cut granite panels from a quarry in Georgia that weigh over 215 tons. The names featured on the stones were etched using the same lettering that appears on grave markers in Arlington National Cemetery. This memorial is set up to resemble a sundial and features the names of the 1,104 Kentucky citizens who passed in the Vietnam War. The names were placed in a way so the sundial's shadow highlights the veteran's name on the anniversary of their death. In addition to this, the sundial points to the inscription "Greater love hath no man than this, that a man lay down his life for his friends" at exactly 11:11 a.m. on November 11 each year. The memorial also features the POW/MIA flag, the American flag, and the Kentucky state flag flying over the monument.

The Names of those from the state of Kentucky Who Made the Ultimate Sacrifice

ABEL CHARLES SEABORN
ADAMS EMMITT COLON
ADAMS JAMES HENRY
ADAMSON FRANK LESLIE
AKER JEFFREY SCOTT
AKERS ARVEL DEWIT
AKERS DENNIS OWEN
ALDRIDGE WILLIE GENE
ALLEN CHARLES FRANKLIN II
ALLEN JAMES WILLIAM
ALLEN JOHN FRANKLIN
ALLEN KENNETH
ALVEY ALFRED ELI JR
ALVEY RONALD LOUIS
ANDERSON DELMER
ANDERSON JAMES DWIGHT
ANDERSON RONNIE COLEMAN
ANDERSON WALTER EVAN JR
ANDREW JOSEPH CARLISLE
ANTE JAMES LOUIS
APPLETON JOHN BURDETTE
ARMSTRONG JOSEPH LARRY
ARMSTRONG WARDELL LESTER
ASH PAUL ENGLISH JR
ASH ROBERT EVERETT
ASHBROOK DELMER VIRGIL
ASHBY CLAYBORN WILLIS JR
ASHCRAFT HARRY DANIEL
ATCHER HAROLD ALLEN
ATON PAUL DOUGLAS
AUSBROOKS RICHARD DAVID
BABBAGE EWING COTTRELL
BAGGARLY JIMMY RAY
BAILEY EVERETTE ROLAND
BAILEY JOHN SPENCER JR
BAILEY JOSEPH DANIEL
BALL CLYDE JAMES
BALL MICHAEL EDWARD
BALLENGER CARL AUGUSTUS
BANEY WILLIAM GERALD JR
BANTA LANNY WILSON
BARKER FLOYD JR
BARLOW EDWARD ARNOLD
BARNES THOMAS JACKSON JR
BARRETT WILLIAM KATHMAN
BARRICK BENJAMIN LUTHER
BASHAM EDWARD RAY
BATES HARRY E
BATES NORMAN WILLIAM
BATTERTON TROY HILLIS
BAUERLE FREDRICK E III
BEARD LEON
BEASLEY DONNIE RAY
BEATTYS LAWRENCE VICK
BECHTEL STEPHAN LEROY
BECKER JOHN BERTRAM
BEGLEY BURRISS NELSON
BELCHER JORDAN
BELL JAMES LYLE
BELLAMY WESLEY EARL
BENNETT CLYDE JAMES
BENNETT JAMES HARVEY
BERNING ROBERT RAYMOND
BERNING THOMAS JOSEPH
BERRY DONALD CARL
BERRYMAN LUTHER CLARK
BERTRAM DAVID MICHAEL
BIGGS DAVID OWEN
BINKLEY STEVEN RAY
BIRCH LARRY WAYNE
BISHOP ROGER EARL
BLACKBURN FREDDIE ANDRAY
BLACKBURN WILLIAM KENDALL
BLAGG PATRICK EARL
BLAIR KENNETH NEAL
BLANKENSHIP JEWELL C
BLATZ THOMAS LEE
BLAUT ROBERT J JR
BLEDSOE DONALD RAY
BLEVINS DANNY EUGENE
BOBB JOHN FRANKLIN
BOLT AUGUST FERREL

BOONE DANNY LEE
BOREN JIMMY FLOYD
BOSTICK BENJAMIN R IV
BOWLING BRADLEY
BOWMAN JESSE CARL
BOWMAN ROBERT EARL
BRADFORD CHARLES MARSHALL
BRADLEY GIVEN WEST
BRADLEY GLENN MARTIN
BRAGG JOE EDDY
BRAGHINI ROBERTO JR
BRAKE BOYD LAWERENCE
BRANK IRA CHARLES
BRANSON DANIEL ALEXANDER
BRATCHER CHARLIE ARUCE
BRAWNER FRANK EDWARD
BRAY ERVIL THOMAS
BREWER WILLIAM JACKSON JR
BRIDGEMAN BILLY WAYNE
BRIDGERS DOUGLAS STEPHEN
BRIGHTMAN HARRY PHILLIP
BROCK ARNOLD LEE
BROOKS CARL RAYMOND
BROWN DARIUS E
BROWN EDDIE WAYNE
BROWN HAROLD MILTON
BROWN THEODORE
BROWNING GEORGE ROBERT
BRUIN JOHN WILLIAM
BRUMAGEN ARTHUR
BRYANT KENNETH MARK
BUCKLEY JAMES ANDREW
BUESCHER JOHN FRANCIS
BULLOCK LARRY ALAN
BURDEN JOHN CURTIS
BURKHART WILLARD HARLEY
BURKHEAD DANNY DALE
BURNS ERVIN L
BURROWS ROGER THOMAS
BURTON ERNEST
BURTON HAROLD RAY
BUSBY WILLIAM LEON
BUSH PEARL
BUTCHER BRUCE EDWARD
BUTLER DONALD RAY
BUTLER THOMAS LYNN
BUTTRY DAVID EUGENE
BYRD HUGH MCNEIL JR
CALDWELL CHARLES WARREN E
CALLER MICHAEL JAY
CAMBRON JOSEPH TERRY
CAMPBELL ANDREW J
CAMPBELL LEONARD WAYNE
CAMPBELL ROBERT LEWIS
CAMPBELL RONALD EDWARD
CARPENTER EDDIE DEAN
CARR BENNY GILLIS
CARR BERTRAM ANTHONY
CARR ERNEST RAY
CARROLL DOUGLAS
CARSON OMER PRICE
CARTER CHARLES IRA
CARTER PAUL DEAN
CASE JAMES RUSSELL
CASWELL EDWIN DOUGLAS
CAUDILL ORVILLE
CAUDILL ROGER DALE
CAULEY ROGER DALE
CAYSON ALVIN LLOYD
CENTERS WILLIAM P JR
CHAFFINS ERNEST JR
CHAMBERS HARVEY ROBERT JA
CHAMPLIN JOHN ROBERT
CHANEY DAVID GLENN
CHAPPEL LUTHER MALCOLM
CHEATHAM JAMES MICHAEL
CLARK CHARLES RONALD
CLARK LAWRENCE EDWARD
CLAYBURN RONNIE LEE
CLEMENTS ROBERT ANDREW
CLEMONS WILLARD LEE
COATES JOHN WAYNE

COEN LOVELL FRANKLIN
COFFEY ROBERT ALLEN
COFFEY ROBERT DANIEL
COLE BILLY JOE
COLE JOHN MICHAEL
COLEGROVE ROBERT HOWARD
COLEMAN BILLIE LEE
COLEMAN CHARLES LOYD
COLEMAN PHILLIP RODNEY
COLEMAN RICHARD CLYDE
COLLETT ROBERT LEE JR
COLLINS CLYDE CECIL
COLLINS DAVID BURR
COLLINS ELZIE J JR
COLLINS WILLIAM DANIEL
COLSON RONALD SANDERS
COLWELL KEITH
COLWELL PAUL
COLYER WILLIAM WALTER
COMBS CHARLES
COMBS EDWARD ALTON
COMPTON DOUGLAS
CONKRIGHT JAMES EDWARD
CONLEY BILLY GENE
CONNER ROGER LEROY
COOMES JOSEPH ANTHONY
COOMES WILLIAM MICHAEL
COON JAMES THOMAS
COOPER CARL DALTON
COOPER ROGER DALE
COOTS JACKIE
COREY GEORGE EDWARD
CORNETT CARLOS WAYNE
CORNETT GREGORY DOUGLAS
COSSEY RICKY FAY
COTTEN JAMES L JR
COTTINGHAM JOHN EDWARD
COX CHARLES WILLIAM
COX CHESTER GARVIS
COX MITCHELL EDWARD
COX RUBE ARTHUR JR

COX WILLIAM GAYLE
CRAIN TRAVIS GLEN
CREECH PHILLIP GENE
CRISP THOMAS MIKELL
CROOKS EDWARD TAUL
CRUMP BUCKNER JR
CRUMP JESSIE LEE
CRUSE JAMES DALE
CRUSE MICHAEL LEE
CUNDIFF ROBERT EUGENE
CURRY WILLIAM RIEVES
DAILEY JAMES ALBERT
DAUGHERTY CECIL
DAULTON JAMES HAROLD
DAVIS CHARLIE BROWN JR
DAVIS MARCUS RAYMOND
DAVIS ROBERT LEWIS
DAWSON JAMES VERNON
DE MARCHES JOHN THOMAS
DEAN HOWARD HADDEN
DENHAM JAMES VIRL
DENNISON CORTLAND ELLIS
DICK BOYCE RAY
DICKERSON GEORGE EVERETT
DISHMAN WILLIAM ANDREW
DIXON WARREN MITCHELL
DOAN TERRY WAYNE
DOBSON CECIL LEE
DOBSON JAMES CARLINE
DOOM CHARLES LEONARD
DORE GARY AUSTIN
DORRIS CLAUDE HESSON
DOSS RAYMOND
DOTSON RICHARD WAYNE
DOWELL GARY LOUIS
DOWNARD CLYDE DAVID JR
DUNBAR CLARENCE WILSON
DUNCAN DONALD ROBERT
DUNCAN KENNETH EUGENE
DUNCAN WILLIAM M
DUNFORD FRANK BELLEW III
DUNIGAN JERRY WAYNE

DURHAM DAVID TERRELL
DURHAM RHONALD LEE
DUSCH PARIS DALE
DUTY ANTHONY
DUVALL RANDOLPH JR
DWYER MICHAEL ALLEN
DYE LARRY CLAY
EAKINS CHARLES ADRAIN
EDEN CHESTER WADE
EDWARDS JOHN NEWT
ELDRIDGE WILLIAM FRANKLIN
ELKINS GEORGE ANDREY
ELLIOTT LAVAUGHN
EMERSON ROBERT LOYD
EMRATH JOHN PHILLIP
ESTES MERLE EDWARD
ESTRIDGE CURTISS
ETHINGTON GLENN RAY
EVANS GARY LEE
EVANS WADDEL
FAIN JAMES LEONARD
FARMER JAMES GORDON
FAVORS BOBBY LEE
FAWBUSH STEVEN LEE
FEE PHILYAW
FELTY ROY LEE
FIECHTER JOHNNY PATTON
FIELDS BOBBY GEORGE
FIELDS JERRY
FIELDS JULIAN THOMAS
FIELDS KELLY
FIELDS LONNIE DALE
FITCH WILLIAM ANDREW
FITTS RICHARD LEE JR
FITZGERALD JOHN W JR
FLAMMER TIMOTHY MATTHEW
FLEEK CHARLES CLINTON
FLEISCHMANN MARTIN A
FLEMING WILLIAM GORDON JR
FLETCHER ROBERT MELVIN

Kentucky

FLOYD EDWIN ZEKE
FLYNN JIMMY EDWIN
FORD CHARLES WAYNE
FORD MELVIN
FORD RAYMOND LEE
FORD RAYMOND SYLVESTER
FOREMAN BOBBY LEE
FOSTER BILLY REX
FRALEY EZEKIEL JR
FRALEY WILLIAM CLIFFORD
FREDERICK CLIFTON JR
FREPPON JOHN CHARLES
FRUECHTENICHT CARL LEE
FRYE EARL WAYNE
FULKS CHARLES EUGENE
FURGERSON JAMES MURPHY
GABBARD THOMAS JEFFER-
SON
GALLAGHER WILLIAM JOSEPH
GALLOWAY ROBERT GLENN
GARRETT PAUL ELLIS
GAY DONALD COLEMAN
GAY KENNETH RAY
GAY MARVIN EDWARD
GAYLES LORENZA
GEER ROBERT SAMPSON
GEILEN DONATUS JOSEPH
GENTRY BOBBY LEE
GIBSON JAMES THURMAN
GIBSON STEVIE RAY
GILLESPIE ROY PAINTER
GILLIAM KENNETH RAY
GILLISPIE LARRY WAYNE
GIPSON RONNIE LEE
GIVENS BILLY R
GLASS PHILLIP SCOTT
GLISSON JAMES EDWIN
GODBEY EDGAR
GOFF AUBREY JR
GOODALL HERMAN GLEN-
DAEE
GOODE RODNEY MICHAEL
GOODRICH JEFFERY CAMON
GRAHAM CHARLES WAYNE
GRAHAM JOHN MEIGS
GRAHAM ROBERT OWEN
GRAVES JAMES EDDIE
GRAY ROBERT EDWARD
GRAY WALTER RAY
GREEN JEFFREY WALLACE
GREEN ROBERT CARL JR
GREENE LAWRENCE DOUG-
LASS
GREENWELL JOSEPH EDWARD
GRIFFITH DALE EUGENE
GRIMES JOHN R
GRISMER EDGAR JOSEPH
GROVES DENNIS MICHAEL
GROVES FERGUS COLEMAN II
GROVES JAMES DOUGLAS
GUILLAUME NORMAN EU-
GENE JR
GUM EDWARD SHERIDAN
HAFENDORFER CHARLES
THOMAS
HAGY JOSEPH ROBERT JR
HALE JOHN DOUGLAS
HALL BROWNIE
HALL CHESTER GENE
HALL CLARENCE
HALL DENNIS GAYLE
HALL DONALD
HALL GARY L
HALL JAMES MICHAEL
HALL THEODORE CROSSMAN
HAMILTON EDWARD SAMUEL

HAMILTON THOMAS SCOTT
HAMILTON WILLIE CHARLES L
HAMM JOHN WILLIAM
HAMMONS PHILIP
HAMPTON JOHN EDISON
HAMPTON ORVILLE
HANCOCK GERALD QUINN
HANNAH FREDDIE JARREL
HANSEN DONALD DELOY
HARDESTY EDWIN HOWARD
JR
HARDIN HAROLD MORRIS
HARDIN KENNETH ALLEN
HARDIN RICHARD ALLEN
HARDWICK JIMMY WAYNE
HARDY BUFORD
HARLOW REX DOUGLAS
HARMON RAY MELVIN
HARP THOMAS ALEXANDER
HARRIS JIMMY
HARRIS JIMMY LEO
HARRIS JOSEPH RANDAL
HARRIS KENNETH RAY
HARRIS KENNETH WARD
HARRIS MICHAEL R JR
HARRIS ROBERT EUGENE
HARRIS WILLIAM DEXTER
HART WILLIAM JOSEPH JR
HATCHER ANTHONY DWIGHT
HATHORNE JAMES COLEMAN
HAWKINS GARY WAYNE
HAWKINS WILLIAM EDWARD
HAYDON PAUL DEARING
HAYES BOBBY LEE
HAYES CHRISTOPHER LYNN
HAYES GEORGE FRANKLIN
HAYWOOD DONALD RAY
HELLMANN GERALD M JR
HELM CARL BENJAMIN
HELTON GLEASON CAY
HENDRIX CHARLES RODNEY
HENRY DENNIS LEE
HENSON CLIFTON
HERALD SHERRILL NEWTON
HEREAU DANNY EDWARD
HESTER JOE EDD
HETTICH ALAN JOSEPH
HICKS CHARLES LEE
HIGHLANDER MICKY RAY
HIGHTOWER JAMES LARRY
HILBERT CHARLES ALLEN
HILBERT JERRY LEE
HILL JOHN MICHAEL
HILL JOHN WESLEY JR
HINES GEORGE MCDONALD
HINKLE KENNETH DANIEL
HISLE GARY LEE
HOCKENSMITH DAVID BAKER
HODGE RUSSELL ADDISON
HOGAN DARRELL
HOGAN WILLIAM FRANCIS JR
HOGGE MICHAEL LEE
HOKENSON WAYNE ALLEN
HOLBROOK JERRY RAY
HOLLAND VERNON EDWARD
HOLLAND WILLIE J
HOLMES TERRY WAYNE
HOLTZCLAW GARY EARL
HONAKER RALPH
HOOKS RALPH MICHAEL
HOOPER STEVEN DALE
HOOSIER ROGER KEITH
HOPKINS WILLIAM ROBERT
HOPSON ROE JR
HORN DAYMON DONALD
HORN JACOB ANDREW
HORNBUCKLE CLARENCE E JR
HORNSBY JOHN R
HORSMAN JOSEPH BERNARD
HOSFORD LARRY DELANO
HOWARD LEON GAYE
HUBBLE WILLIAM BAKER
HUCKLEBERRY JAMES ROBERT
HUGHES FRANCIS ALLEN
HUGHES MITCHELL JR
HUGHES WILLIAM BURDICK
HUMPHREY CARL A
HUMPHREYS RUBERT GREG-
ORY
HUNT LEON ANDREW
HURRY SAMUEL GREEN
INMAN HARRY CHARLES III
ISOM DAVID
ISON ARNOLD E

JACKSON JAMES CHARLES
JACKSON JOHN RICHARD
JACKSON LARRY RICHARD
JACKSON MICHELE LEE
JACO ARNOLD NOEL
JAMES EDWARD LUCAS II
JAMES LEE ROY
JAMESON DAVID ALLEN
JAMROZY STANLEY MICHAEL
JARBOE LOWELL THOMAS
JEFFERSON JERRY WAYNE
JENKINS J CLIFFORD
JENNINGS BOBBY JOHN
JETT DANNY THOMAS
JEWELL DAVID PRESTON
JOHNS JOSEPH DARRYL
JOHNSON ANDY JR
JOHNSON DAVID LEE
JOHNSON DOHN WILLIAM
JOHNSON DOUGLAS ANDREW
JOHNSON JIMMY DONALD
JOHNSON LARRY PATRICK
JOHNSON LOWELL
JOHNSON NICHOLAS G SR
JOHNSON PAUL ALLEN
JOHNSON PAUL EDWARD
JONES BILLY JOE
JONES CHARLES SPENCER
JONES DAVID ALLEN
JONES DWIGHT DALE
JONES FREDDIE DAVID SR
JONES JERRY DEAN
JONES JOHN LEE
JONES LARRY HUGH
JONES MARSHALL KEENE
JONES OTIS ROBERT
JONES RONNIE CLYDE
JONES VERNON DOUGLAS
JORDAN GARY STEPHEN
JUETT WILLIAM LEE
JUSTICE DON MC CLELLAND
JUSTICE THOMAS LARRY
KAUFMAN THOMAS JAY
KAVICH ROBERT DALE
KAYS JERRY ALLAN
KEELER DICKIE GAYLE
KELLEY VIRGIL KINNAIRD JR
KELSO THOMAS JOSH JR
KEMP MITCHELL LYNN
KENDALL COLEY LEE
KIDD DONALD EUGENE
KIDD RHEA MARSHALL
KIHNLEY GEORGE MATTHEW
KING BILLY BROWN

KING HARRY CARLTON
KINGREY EDWARD LEO
KINNEY DONALD MACK
KIRBY GEORGE H JR
KIRCHNER HENRY JOSEPH JR
KIRN JAMES EDWARD
KNIGHT ALVIN COY
KNIGHT JOHNNIE DAVID
KRAMER ARTHUR THEODORE
JR
KUHLENHOELTER JIMMY
LANCASTER JOHN MANNING
LANDIS CHARLES DAVID
LANE ERNEST EDWARD JR
LANGNEHS MICHAEL WIL-
LIAM
LAW WILLIAM LARRY
LAWRENCE BOBBY JOE
LAWSON FREDDIE DON
LAWSON RONNIE
LAWSON STANLEY GARFIELD
LAWSON THOMAS JUNIOR
LEE EWELL JR
LEE JAMES RICHARD
LEFLER BERT DOUGLAS
LEONARD CHARLES RAY-
MOND
LEWIS CHARLES RATES
LEWIS DONALD RANDELL
LEWIS LAWRENCE EDWARD
LINDSAY ARTHUR DALE
LINDSEY JOHNNY WARNER
LITSEY MICHAEL LEWIS
LITTLETON DAVID ERNEST
LITTON GARY WAYNE
LIVELY PAUL JOHNSON
LIVINGSTON LARRY MONTEZ
LOCKARD DAVID LEE
LOCKHART RONALD JAY
LOPER MILES HILTON JR
LOPEZ RENE CERDA
LOSCHIAVO THOMAS LEE
LOUDENBACK DOUGLAS
FRANKL
LOVE RANDALL WAYNE
LOVELACE ROBERT KENNETH
LOVINS ARNOLD
LOWE WILLIE LEE
LOWERY CARL CECIL
LUCAS BILLY RAY
LUTTRELL JAMES LEE
LUTTRELL LLOYD IRVIN
LYNCH MICHAEL HENRY
LYON FRANK ELLIOT

LYONS MONTAGUE
MACKEY DAVID RANDELL
MADISON WILLIAM LOUIS
MAGGARD LARRY DWIGHT
MAGRUDER DAVID BYRON
MAHAN ROBERT CARY
MALAPELLI JOHN WAYNE
MALLORY CONNARD DAR-
RELL
MANTOOTH LEONARD HAYES
JR
MARCUM WALTER VERNON
MARSHALL CLIFFORD WAYNE
MARSHALL JAMES EDWARD
MARSHALL JIMMIE RAY
MARSHALL SAMUEL THOMAS
JR
MARTIN BILLY JOE RALPH
MARTIN HARRELD PIRTLE
MARTIN SAMUEL CALVIN
MASON LARRY JOE SR
MASTERSON ROBERT ALLEN
MATEJA ALAN PAUL
MATLOCK JOHN PHILLIP
MATLOCK MC KENLEY ODIS
MATTINGLY OSBORNE JR
MAYBERRY GERALD WAYNE
MAYNARD DARRELL WAYNE
MC COLLUM RONALD LEE
MC CORMICK CARL PHILIP
MC CUBBINS LARRY JAMES
MC FADDIN LARRY RONALD
MC GAUGHEY PAUL JR
MC GOWAN FRANCIS RUSSELL
MC GUIRE JEFFREY DURON
MC ILVOY JOSEPH RONALD
MC INTOSH ESTILL R
MC INTOSH RONALD
MC KEE MILFORD GERALD
MC KEE ROBERT EARL
MC KINNEY RAYMOND BRUCE
MC MACKINS REUBEN CARL
JR
MC MILLAN DONALD FRANK-
LIN
MC NAY GUY ECKMAN JR
MC NEES GEORGE WILLIAM
MC NEW RALPH DENNIS
MC STOOTS THOMAS HOW-
ARD
MCELRATH RALPH EDWARD
MEADOWS JERRY ROGER
MEDLEY CHARLES MICHAEL
MEE MARION EUGENE

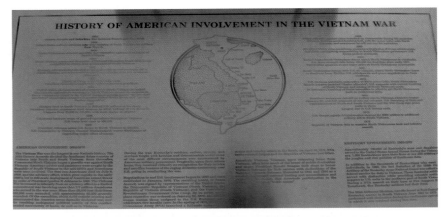

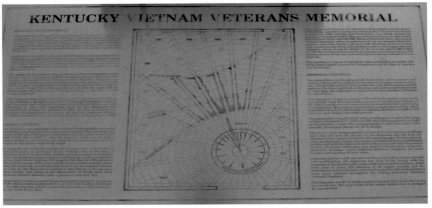

94

MEECE MAC HUGHLEN
MEFFORD BOBBY RAY
MEFFORD HARRELL SAMUEL
MELCZEK JOE ROGER
MESAROSH DONALD EARL
METCALF CARL JOSEPH JR
MIDKIFF O L
MIKESELL RONALD LEE
MILES DAVID LEE
MILEY JOSEPH WAYNE
MILLAY CHARLES FRANCIS
MILLER DONALD ROBERT
MILLER DONALD WAYNE
MILLER HUBERT WAYNE
MILLER JOHN WILLIAM
MILLER JOSEPH LORAN
MILLER LEON ABNER
MILLER MARSHALL
MILLER MICHAEL LEE
MILLIGAN JAMES ELDRIE
MILLINER WILLIAM PATRICK
MILLION RONALD LEE
MILLS RONNIE
MILLS VICTOR LANE
MINER JAMES ALLEN
MIRACLE JAMES JR
MITCHELL CARL BERG
MITCHELL DONALD WAYNE
MIZE FREDDIE D
MOLLETTE JAMES RONNIE
MOLNAR ISTVAN
MONTGOMERY CLIFFORD
MONTGOMERY OWEN RAYMOND
MOON LOWELL EDWIN
MOORE CARLOS DAVID
MOORE CHARLES EDWARD
MOORE EDWARD P
MOORE GARY LESLIE
MOORE JAMES THOMAS
MOORE JAMES WILLIAM JR
MOORE LARRY GENE
MOORE THOMAS PHILIP
MORGAN CHARLES VERNON
MORRIS WALTER KENNETH
MORRISON JOSEPH CASTLEMAN
MORTON GEORGE WINSTON
MOSIER ROBERT SHERMAN
MUELLER CARL WILLIAM
MUELLER MICHAEL DAVID
MULKEY JEFF
MULLINS EARNEST RANDALL
MUNRO IVAN HALL
MURRAY CECIL SCOTT
MUSS GLENN DAVID
NASH DAVID PAUL
NAYLOR EUGENE
NAYLOR RAYMOND LUKE
NEACE DENNIE
NEALIS TOMMY R
NESTER ROGER
NEWBERRY CLIFFORD LEE
NEWBY BOBBY GENE
NEWSOME KENNETH RAY
NIEWAHNER RONALD LEO
NIX JOHN DAVID
NOE GEORGE HOBERT
NOLEN KENNETH JOE
NORRENBROCK WILLIAM A
NORRIS JAMES RAPHAEL
NUNN CHARLES ROBERT
O BANION JAMES RUSSELL
O BRIEN ALDEN WALTON
OAKS WILLIE JAMES
OLDS JERRY DEAN
ONAN JERRY LANG
ORR EMMETT SOMERS
ORR PATRICK O'REILLY
OSBORNE CHARLES EDWARD
OSBORNE JAKE
OSBORNE ROBERT ALLEN
OVERSTREET WILLIAM LUTHER
OVERTON WINCE ISAAC JR
PALMER HENRY LEE
PANNELL HARRY CLAYBURN
PARCHER ROBERT HAROLD JR
PARKER BILLY RAY
PARKS FLOYD JUNIOR
PARSONS JOHN ROBERT
PATRICK BILLY RAY
PATRICK BOBBY GENE
PATTON GUY WESLEY

PAUL CLYDE EVERTTE JR
PAULIN JOHN THOMAS
PAULLEY OSCAR JR
PAYNE HERMAN GARFIELD
PEAK LAWRENCE JOSEPH
PEASE KENNETH WAYNE
PEAY HARVEY A
PEDIGO CHARLES DANIEL
PEMBLETON RONALD LEE
PENDYGRAFT GEORGE R
PENN CHARLES HUGHES
PERDUE DONALD M
PERKINS JOHNNIE KAY
PERRY CLAUDE
PETERS RALPH EDWARD
PFEISTER ROBERT
PHILBECK DONALD DEWAYNE
PHILLIPS ALTON RAY
PHIPPS ROBERT EARL
PICKETT KENNETH WALTER
PIERCE JAMES EVERETT
PIKE DONALD CLEAVER
PINKERTON BENJAMIN ROBERT
PIRRMAN RAYMOND LEE
PITTMAN RONNIE RAY
PITTS BILLY JAY
PLANCK EVERETT ALLEN
POLAND HARRY TURNER
POLSON EDWARD LEE
POOLE CHARLES BURTON
PORTWOOD JAMES JR
POTTER DON
POWELL BOBBY WAYNE
POWELL MARION DAVID
POWELL TROY EVERETT
POWERS EDWARD CLAUS
PRATER HARVEY WILLIAM
PRATHER MARTIN WILLIAM
PRESSON BILLIE TAYLOR
PRESTON LEONARD LEE JR
PREWITT WILLIAM EARL
PRICE BOBBY WAYNE
PRICE JAMES ERWIN
PRICE PAUL LEE
PRICE WILLIAM DAVID
PRICE WILLIAM JOSEPH
PRUITT JAMES ELMER
PRYOR ERNEST PAUL
PUCKETT ROGER DALE
PYLE JERRY WILLIAM
RADCLIFF DONALD GORDON
RAMAGE JAMES WAYNE
RAMEY THOMAS RANDELL
RANDALL LOUIS R
RANDOLPH RICHARD WAYNE
RANKIN DONALD IRVIN
RANKIN EDWARD LEE
RANSOM BRADLEY ROGERS
RATLIFF BILLY HARRISON
RATLIFF LARRY GENE
RAY DAVID L
RAY WILLIAM CLAYTON
REDMON LARRY RAY
REED BILLIE WAYNE
REED LARRY
REED OTTIS
REED RONALD LEE
RENFRO FRANKLIN JR
RETSCHULTE THOMAS HOWARD
REYNOLDS HARVEY CLAUDE
REYNOLDS LARRY LEE
RICE JOHNIE EDWARD JR
RICHIE CHARLES HOWARD
RICHMOND THOMAS GLEN
RIGGLE ROBERT FRANKLIN
RILEY ERNST
RISINGER GERALD LEE
ROBB RICHARD ALBERT
ROBERTS HERMAN DAVID
ROBERTS JERRY ARDELL
ROBERTS THEODORE IRWIN
ROBERTSON ROY ALLEN JR
ROBINSON JOSEPH BRUCE
ROBINSON NATHAN LYEN
RODEN GEORGE COLUMBUS JR
ROGERS CHARLES LEE
ROGERS HENRY LEWIS
ROSE JOHNNY JR
ROSE ROBERT LEE
ROSS JOSEPH SHAW
ROSSER ERNIE WAYNE

ROUNTREE GLEN EVERETT
ROWLAND GEORGE CLAYTON JR
ROWLAND GEORGE JR
ROWLETT JIMMIE HENRY
RUSH KENNETH
RUSSELL CHARLES GLENN
RUSSELL CHARLES PIERCE
RUTHERFORD DANNY LEWIS
RUTHERFORD ERNEST WAYNE
RUTHERFORD LARRY SCOTT
RUTHERFORD MELVIN NEAL
SALES CHARLES CARROLL
SALLEE RICHARD JR
SALLY HANK
SALYER FRED LAMARR
SAMS RICHARD BARRY
SANDERS ARTHUR JACKSON
SANDERS ELZIE JR
SANFORD ALBERT RUSSELL
SARGENT BILLY RAY
SAWYER FRANK W JR
SAYER ALBERT FRANCIS JR
SAYERS LARRY VENCIL
SCHAEFER WILLIAM HAYS
SCHAFER MARVIN ALBERT
SCHAFFNER JACK DOHN
SCHEIBER WILLIAM HENRY JR
SCHERLE WILLIAM JOSEPH JR
SCHNEIDER GARY LEE
SCHOBORG GARY ALLEN
SCHOLL CLIFFORD LEO
SCHULTE ALVIN CLAYTON
SCOTT SAMMY LEE
SCRUGGS JOSEPH ALLEN
SEARS EARNEST G
SEATON DAVID THOMAS
SEATON ROBERT WAYNE
SEBASTIAN BILLY JOE
SECRESS HARLAN
SENTERS BOBBY
SENTERS CHARLES DONALD
SESTER EUGENE
SEYMORE RICHARD MORRIS
SHAIN JERRY WAYNE
SHARP ALLEN MORRIS
SHARP STEPHEN ALLEN
SHEA LARRY
SHELLMAN VERNON LINDLY
SHELTON CHARLES ERVIN
SHELTON DAVID PRESTON
SHEPHERD MICHAEL ALLEN
SHERRILL JIMMY L
SHIFFLETT ALVIN MARION JR
SHIRLEY CARL EUGENE
SHOEMAKER KENNETH R JR
SHOLAR EDWIN FRANKLIN
SHOOPMAN PHILLIP RAY
SHORT JAMES EVERRTTE
SHRUM LEON JERRY
SHUFFITT KENNETH LEN
SICKELS ROBERT T
SIEGEL THEODORE FRANK
SILBERSACK RONALD VINCENT
SIMON JAMES MARTIN
SIMON MICHAEL WAYNE
SIMPSON CHESTER PAUL
SIMPSON DANNY ROY
SIMPSON RONALD EARL
SINGLETON ARTHUR DWIGHT
SIZEMORE CLARENCE
SKAGGS FREDERICK BRIAN
SLACK DENTON RAY
SMALL SAMUEL OLIVER
SMITH ALBERT DOUGLAS
SMITH AVERY GENE
SMITH BILLY
SMITH CHARLES ALLAN
SMITH DARRELL
SMITH DARRELL JACK
SMITH DONALD EMMETT
SMITH EDWARD ARTHUR
SMITH EUGENE IVAN
SMITH GARY HOLDEN
SMITH JAMES ALBERT
SMITH JAMES BRYAN
SMITH JAMES HOWARD
SMITH PATRICK LEROY
SMITH ROBERT LEWIS
SNIPES BILLY EUGENE
SOUTHERLAND CECIL WAYNE
SOWARD LOUIS RAY
SOWARDS DAVID MICHAEL

SPALDING AARON BERNARD
SPANGLER JAMES NELSON
SPENCER EUGENE
SPENCER JAMES PRICE
SPENCER NORMAN
SPILLMAN HAROLD RAY
SPIVEY WILLARD EARL
SQUIRES DAVID RAY
STAHL ALVIN THORNTON
STALLINGS RONALD CLARK
STAMM MONTE LEWIS
STATON ROBERT GARY
STCLAIR CHARLES DAVID
STEPHENS CLYDE J
STEPHENS MARVIN GENE
STEPP WILLIAM HOWARD
STEVENSON CHARLES ROYCE
STEVENSON WILLIAM LUTHER
STEWART DAVID WAYNE
STEWART JAMES LLOYD
STEWART MANFORD DALVIS
STILES MICHAEL PAUL
STINSON FLOYD ALBERT
STOLL WILLIAM KEEN JR
STONE ORMAN
STOUT JAMES ROBERT
STRADER CHARLES EDWARD
STRATTON EVERETT JR
STRAUGHN WILLIAM HERSCHEL
STRINGER JOHN CURTIS II
STRINGER ROY LEE
STURGEON IRA JACKIE
SULLIVAN HAROLD
SUMPTER BOBBY RECE
SUMPTER JOSEPH BOYD
SWANN ELLSWORTH
SWEATT GEORGE EDWARD
SWENCK ROBERT BENNETT
SWOPE CHARLES FREDERICK
TACKETT CLARENCE E
TACKETT GEORGE EDWARD
TACKETT RUBEN NOAH
TALLEY FLOYD G
TANNER STEVEN DALE
TAPP JOHN BETHEL
TARTER BOBBY LEE
TAULBEE DANNY JOE
TAYLOR HARRY EUGENE
TAYLOR LEE ROY
TAYLOR WILLIAM RUSSELL
TEAGUE ALONZO ALLEN
TERRELL CALVIN LEE
TERRY ANCEL JAMES
TERRY PHILIP ALLEN
TERRY RALPH PAUL
THARP CLAUDE WILLIAM
THAYER THOMAS EDWARD JR
THOMAS BRUCE EDWARD
THOMAS CHARLES EDWARD
THOMAS LEO TARLTON JR
THOMAS MICHAEL FRANCIS
THOMAS MURREL D
THOMAS STEPHEN EVANS
THOMPSON BARRY NEAL
THOMPSON GEORGE JR
THOMPSON MYRON
THOMPSON PHILIP BRUCE
THOMPSON WILLIAM P JR
TICHENOR QUINN WILLIAM
TIPTON JAY C
TODD KENNETH WAYNE
TODTENBIER JAMES LOUIS
TOLER DAVID BRUCE
TOLLIVER JIMMY ELLISON
TOMLINSON EDGAR LEE
TOON JERRY WAYNE
TORRES ESTEVAN
TOWATER JERALD RILEY
TRAIL RANDELL GENE
TRAINER DORRIS WAYNE
TRAVIS JAMES LEONARD JR
TRIGG ROBERT CARL
TROSPER JACKIE EDWARD
TUCKER WILLIE ROBERT
TULLY STEPHEN MEREDITH
TULLY WILLIAM BOYD
TURNER WILLIAM COY
TUTTLE ARLEN CLIFTON
UNDERWOOD JERRY DWAYNE
VAN HOOSE PAUL EDWIN
VAN WINKLE CURTIS GLENN
VANCE DENNIS LEE

VANCE WILLIAM JR
VANDERPOOL EDWARD LEE
VANVACTOR VICTOR HAROLD
VARNEY RONALD T
VEST DAVID WAYNE
VIBBERT CARLOS DYRAL
VICK WILLIAM LEON
VOELKER MARVIN ALDRID JR
WADLINGTON LOUIS WAYNE
WAFORD HUBERT EARL
WAGMAN NICHOLAS OWEN
WAGNER DAN JR
WALDROP KYLE
WALKER CECIL
WALKER GEORGE NELSON
WALLACE GARY ANTHONY
WALLS WALTER LEE
WALTERS DAVID MORGAN
WALTERS WILLIAM OWEN
WALTON HARRY THOMAS JR
WARD CHARLES EDWARD
WARD DANNY RUSSELL
WARNER LEWIS WILLIAM JR
WARREN WILLIAM THOMAS JR
WARTMAN CHESTER JAMES
WASHINGTON LAWRENCE O
WATKINS BILLY JOE
WATKINS MARION
WATSON WILLIAM L
WATTS ASTER
WATTS FLOYD
WATTS ROBERT WESLEY
WATTS SCHYLER
WEBB PAUL HENLEY
WEDDINGTON PHILLIP MURRY
WEILL JOHN BRUCE
WEISS RODERICK LEE
WELCH RANDALL EDWARD
WELLS GENE GORDON
WELLS RICHARD ARTHUR
WELLS TINSLEY JACK JR
WEST HOMER
WESTER DONALD LEE
WETMORE DOUGLAS MCARTHUR
WHALEN CHARLES ARTHUR
WHALEY WILLIAM ELDRED III
WHISMAN ERMIL LEE
WHITAKER STEVE RANDAL
WHITE MARCUS DELMAR
WHITE SYLVAIN LARRY
WHITEHEAD ALFRED EVARTS
WHITESIDE ROY RUDOLPH
WICKER HENRY RAY
WILHOIT HOWARD RAY JR
WILHOITE HENRY O NEIL
WILKINS BEN HENRY JR
WILKINSON GARY
WILLIAMS BILLY JOE
WILLIAMS CURTIS JOHN
WILLIAMS JAMES A
WILLIAMS JAMES S
WILLIAMS LARRY LEE
WILLIAMSON DON IRA
WILLIAMSON PETE ELLIS
WILSON EDMOND Q JR
WILSON ERNEST HOWARD
WILSON JAMES MURL
WILSON WENDELL LEWIS
WINCHESTER JIMMY DALE
WISDOM KERRY DEAN
WOOD HAROLD SHELBY JR
WOODRUFF DAVID GLENN
WOODS RONALD LEE
WOOLLEY KIRK ALLEN
WOOSLEY PERRY LEE
WORKMAN LIONEL
WRAY JIM ALLEN
WRIGHT KENNETH RAY
WRIGHT PHILLIP GERALD
WRIGHT RANDY BLAKE
WRIGHT WILLIAM ANDREW
WYNN FLOYD
YATES JAMES IRVINE
YATES MANNIFRED
YOUNG BOBBY
YOUNG GARY EDWARD
ZIPP MARION LOUIS

Photos by Epaul Julien

Vietnam War Memorial - Louisiana Superdome

Poydras St, New Orleans, LA, 70130

To commemorate the sacrifice that the men and women who served the US in the Vietnam War made, or were willing to make, a bronze statue seated on a marble pediment was erected in New Orleans, Louisiana, just outside of the Superdome on the Poydras street side of the arena. This specific street also plays host to other statues that were erected for various reasons.

The Vietnam War Memorial was commissioned by the Vietnam Veterans Leadership Program which is made up of Louisiana Vietnam veterans. Sculptor William Ludwig was selected to create the monument in February 1984. The bronze statue was completed and ready for dedication in November of that same year.

While the New Orleans Vietnam War Memorial was made to honor the veterans who served in the Vietnam War, the names, ranks, and towns that are usually featured on other such memorials aren't present here. Instead of listing the 882 names of the Louisiana Vietnam veterans, there is a timeline of major events behind the memorial.

The main portion of the memorial is the bronze sculpture which is made up of three troops clad in combat gear and their weapons as they carry a wounded brother off of the battlefield (one presumes based on their alert and aggressive stances and sense of urgency). There is a stone bench set to the side of the memorial which was placed in order to offer people a place to sit and reflect on the symbolism behind the Vietnam War Memorial as well as a dedication plaque featuring a quote from John F. Kennedy regarding the survival and success of liberty. Veterans and Memorial day services are held at the monument each year.

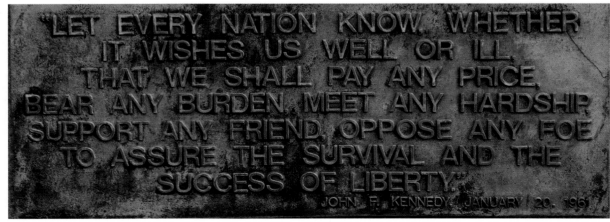

The Names of those from the state of Louisiana Who Made the Ultimate Sacrifice

ABBOTT JAMES EDWARD
ABRAM JAMES HENRY
ABSHIRE RICHARD FRANKLIN
ACHORD DAVID PAUL
ACKER HERMAN CAROL
ACKERMAN DENNIS CARLTON
ACOSTA JAMES ARTHUR JR
ADAM HOSEA DENNIS
ADAMS FRANK DAVID
ADAMS JOSEPH
AINSWORTH KENNETH JOHN
ALBRITTON JOHNNY BOYD
ALFORD THOMAS EARL III
ALLEN FREDDIE LEE
ALLEN ROBERT CLYDE
ALLEN RONALD JOSEPH
ANDERSON EDWARD EUGENE
ANDERSON VON STEVEN
ANDRY HILAIRE ALBERT JR
ANTHONY BENJAMIN JONES
ANTHONY CARL THOMAS
ANTHONY JOSEPH ROY
ARCENEAUX HERBERT JOHN JR
ARDENEAUX GARY JAMES
ARDOIN ROBERT GLEN
ARMSTRONG FRANK ALTON III
ARNAUD GARY WAYNE
AUGUST FRANK JOHN
AULL EARL DUBOIS
AUSTIN MICHAEL FRANCIS
BABIN CLARENCE JOSEPH JR
BABIN THOMAS DALTON JR
BAKER ISIAH III
BALDINI MICHAEL LOUIS
BALLARD ROBERT IRVING
BALLEW PATRICK DEWEY
BANDY MICHAEL J
BANKS LAVINE JOHN
BARBOSA ALVARO
BARNES DARRYL VERDUE
BARRETO LUIS JR
BARRETT THOMAS J JR
BARROWS IRVING DONALD
BARTHOLOMEW DAVE MARTIN
BARTHOLOMEW TILMEN VERGES
BASCO HARVEY LEE
BASS BUDROW JR
BATISTE CLEVELAND JR
BATISTE JOHN MILLIAN
BAUGH CHARLES LEE
BEALIN TROY
BEAN EUGENE JR
BECKWITH EDWARD COE
BEDGOOD JAMES DOUGLAS
BEESON MORRIS SAMPSON
BELCHER ROBERT ARTHUR
BENAIM GILBERT ALBERT
BENJAMIN FREDDIE JAMES
BENJAMIN ROBERT LEE
BENNETT BENJAMIN F JR
BENNETT ROBERT VERNON
BENNETT RONALD DAVID
BENNETT STEVEN LOGAN
BENOIT GARLAND DAVE
BERGERON ROY LOUIS
BERRY JACKIE WAYNE
BESSON LAWRENCE EUGENE
BILLEAUD WAYNE JAMES
BILLEAUD WILLIS J JR
BILLIOT RUDOLPH JOHN
BILLUPS EDDIE JR
BIRD KIM SOVEREEN
BLACK DENNIS BEDELLE
BLACKMON JAMES WILLIE
BLANCHARD ANDRUS JAMES
BLOCKER MURRIE LEE
BLOUNT GARY GEORGE
BLOUNT ROBERT LARRY
BOB CHESTER
BODIN ALLEN JAMES
BOGARD LONNIE PAT

BOOKER HARVEY WATKINS
BOOTY LARRY OVEID JULIAN
BORENSTEIN BORIS FRANZ M
BOUDREAUX KENNETH CHARLES
BOUDREAUX LEE JOSEPH JR
BOUINGTON JOHNNY WILLIAM
BOULLION ELLIAS
BOURG MILLER JOHN
BOURQUE BRADLEY JOHN
BRACKENS JOE JEFFERSON
BRADLEY LOUIS LLOYD JR
BRANCH GEORGE ALLEN
BRANNAN JAMES CURTIS
BRATTON JOHN LESLIE
BREAUX LEONARD
BREAUX SHELTON L
BRIDGES LONNIE
BRIDGES WILLIAM LEE
BRIGNAC JOSEPH PAUL
BROADTMAN HENRY ROBERT JR
BROOKS LEE MURRAY
BROOMFIELD TED DEWAINE
BROUSSARD ANDREW RICHARD
BROUSSARD GERALD GENE
BROUSSARD LEO JAMES JR
BROWN EARNEST CAESAR
BROWN HERMAN JR
BROWN IRVING JOHN JR
BROWN JAMES JR
BROWN KENNETH EARL
BROWN ROBERT
BROWN WARREN FRED
BROWN WILLIAM WESLEY
BROWNE RICHARD ALLAN
BRUMFIELD RICHARD LYNN
BRUNET ELDRIDGE MICHAEL
BRUPBACHER ROBERT MICHAEL
BRYANT EMMETT JOSEPH
BRYANT MELVIN GENE
BRYANT RUSSELL DAVID JR
BUFORD RALPH JOSEPH
BURGESS RUBEN ANTHONY
BURKE CHARLES MORRIS
BURNS ROBERT GEORGE
BURNS VICTOR LEE
BUSH CECIL FLOYD
BUTLER GORDON

CAIN CARL DENNIS
CALHOON DONALD EUGENE
CALHOUN DURL GENE
CALLOWAY PORTER EARL
CALP ALBERT FRANKLIN
CAMBAS VICTOR BYRON
CARBAJAL ADRIAN DAVID
CARMACK JOHN EDWARD
CARNLINE TROY MONROE
CARR CLINT EDWIN
CARTER GILL LESTER
CARTER JIMMY
CARTER MILFORD DONAVIN
CARTER SHELBY M
CATOIR JOSEPH GEORGE P JR
CHABERT GARY AUGUST
CHAPPELL KENNEY DEAN
CIRUTI JAMES DAVID
CLARK THOMAS JR
CLAUSE MICHAEL ALLEN
CLINE MARCUS EUGENE
COADY ROBERT FRANKLIN
COBB RAYMOND
COKER DAVID LANGSTON JR
COLE EMILE
COLEMAN JOSHUA
COLLETTE ROGER CHARLES
COLSON DONALD REGINALD
CONVERSON TYRONE
COOK DONALD JAMES
COOKE PAUL DONALD
COOPER IRA DAUNETTE
COOPER JOE
COOPER WILLIE JAMES
COPPLE RAMON ALLEN
CORDOVA JAMES THOMAS H
CORKERN JERRY WAYNE
CORLEY CLARENCE ALTON JR
CORMIER MELVIN GLENN
CORNETT DONALD C
COTTON MICHAEL
COX EDWARD ERLIN JR
COYLE LAVERNE DARTON
COZART ROBERT GORDON JR
CRADEUR DOUGLAS JOSEPH
CRAIG JAMES LARRY
CRAIG MERLIN JOSEPH
CROCKER DONALD JACK
CROOKS LESTER LOIS
CROSS HERBERT TERRELL
CROW DAVID LYNN
CROXDALE JACK LEE II

CULPEPPER ALLEN ROSS
CUTRER MARVIN EUGENE
DAIGLE BRADLEY TIMOTHY
DAIGLE JOSEPH DEWEY
DAIGREPONT ROBERT LYNN
DAIR ALBERT JOSEPH
DANIELS CARL STEPHEN
DANIELS RAYMOND LOUIS
DARBY JAMES MICHAEL
DARBY PAUL
DARDAR CLARENCE JR
DARDEAU OSCAR MOISE JR
DAUPHINE ROBERT JR
DAVID HAROLD LUCIEN JR
DAVIDSON DONALD FREDERICK
DAVIS BRUCE
DAVIS CLIFTON ANTHONY
DAVIS FRANK JR
DAVIS HARRIS VONZELL
DAVIS MELVIN ERNEST
DAVIS NEVITT DIEALL
DAVIS RAYMOND RANCE
DAVIS STEVEN FREDERICK
DAVIS THOMAS J III
DAW CECIL ERNEST
DE JEAN CHARLES ODEN III
DE LA HOUSSAYE ARTHUR J JR
DECAREAUX NORMAN E JR
DEDON CHARLES BERLIN
DELACROIX WILLIE JAMES
DELRIE JAMES EDWARD
DENNY ROGER EDWARD
DENTON SIDNEY EDWARD
DERRYBERRY ABRAHAM R III
DESSELLE RICHARD JUDE
DICKSON GROVER LEE
DIEZ ISAAC ANDREW JR
DILLARD JOHN ALBERT B JR
DIX STANLEY WESLEY
DOHERTY GUY WOODS
DOLAN HASKELL JUNIOR
DOSS HAROLD CONWAY JR
DOTY CHARLES
DOWNS CHARLES MILTON
DRAUGHN THOMAS EDWARD
DRINKHOUSE JOHN WATTS
DU BOIS RICHARD FRANCIS
DUGAS JOSEPH GERALD
DUKES PAUL DOUGLAS
DUNCAN MITCHELL JEROME
DUNN LESSELL JR

DUPLECHAIN ANDRUS FLOYD
DUPLESSIS GEORGE LLOYD
DUTHU ROY ANTHONY
EARLY HOWARD LEE
EASTERLING EARL K
EATON JACK
EDWARDS BILLY MARCUS
EDWARDS R V
ELDRIDGE THOMAS CHARLES
ELIE LEONARD WAYNE
ELLENDER TERRY LEE
EMANUEL WILLIAM FREDERICK
ENGLISH CARVER JOSEPH JR
ERWIN HUBERT AARON
EVANS DONALD LYNN
EVANS JOHNNY LEE
EVANS LONNIE DALE
EVERETT BOBBY JOE
EZELL WILLIAM BENJAMIN
FARRELL MICHAEL JAMES
FAUGHN ISSAC DAVID
FAULK THEODORE ALPHONSE
FEDELE JOHN ANTHONY
FEHRENBACH THERON CARL II
FELLER DAVID KENT
FENCEROY LOUIS EARL
FENCEROY WILLIAM CHARLES
FERBOS STANLEY
FEUCHT JAMES DONALD
FINN WILLIAM ROBERT
FIRMIN MITCHELL LAWRENCE
FITCH DANNIE
FLASHNER KENNETH MICHAEL
FLETCHER JOHN EARL
FLINT TROY LEE
FLOYD DAVID ALLEN
FLYNN GEORGE EDWARD III
FLYNN HAROLD BROWN
FLYNN RAYMOND PATRICK
FONTAINE MICHAEL ARTHUR
FONTENOT CHESTER JOSEPH C
FONTENOT GARY PAUL
FONTENOT HAROLD
FORD BOB JOE JR
FOREMAN ROBERT JR
FORET KENNETH JOHN
FOSTER BILLY RAY

Louisiana

FOURNET DOUGLAS BERNARD
FRANCEWAR JOHN EDWARD
FRANCIES DOLROY
FRANCIS WILLIE JR
FRANCIS WILLIE JR
FREITAG KENNETH LEE
FREY JOHN HARVEY
FURMAN EDMUND
GADDIS RALPH ARNOLD
GARCIA JAMES RONALD
GARCILLE DAVID LEE
GARNETT ISIAH CALVIN
GARRETT ALFRED DOUGLAS
GARRETT THOMAS C III
GASSAWAY AMBROSE
GATES JAMES WAYNE
GAUTHIER BRIAN JAMES
GAUTHIER GERALD PAUL
GEORGE D C
GERDES ALBERT BRUNO JR
GILLARD MICHAEL
GILLETT DONALD CHARLES
GILMORE WILLIAM ALAN
GINART DONALD FRANCIS
GIROD JOHN ALTON JR
GIROIR PAUL GERARD
GLASPER WILLIAM LAWRENCE
GLENN ROBERT LEE JR
GLOER DAVID LAWRENCE
GOLDMAN SAMMY WAYNE
GOLMON JIMMY DARWIN
GOODFELLOW CARL RAYMOND
GOODWIN RAYMOND RAY
GORDON JIMMIE LEE
GOSHORN EDWARD FRANCIS
GRACE JAMES WILLIAM
GRAHAM JOHNNIE LEE JR
GRAHAM JOSEPH HAROLD
GRAHAM MORRIS
GRANGER FLOYD IRA JR
GRANT JERRY
GRANT WESLEY ONEAL
GRANT WILLIE JR
GRAYSON HERMAN LEE
GRECO ERIC JOSEPH
GREER RALPH JERRY
GRIFFITH ERIC LAWRENCE
GRIMES ALVIN
GUIDRY MICHAEL JAMES
GUILLORY BUD AUGUSTINE
GUILLORY EARL J
GUILLORY EDWARD JOSEPH
GUILLORY GERALD JAMES
GUILLORY HUBIA JUDE
GUILLORY WENDELL
GUILLORY WILLIAM ALLEN
GULLUNG JOSEPH FRANK III
GUNTER WILLIAM ANTHONY JR
HAGGARD THOMAS EDWARD
HAINS ANTHONY JOSEPH JR
HAIRE CARSON EARL
HALBERT PATRICK HENRY
HALEY JERRY RANKIN
HALL DONALD WILFORD
HALL VINCENT JOSEPH
HAMILTON JAMES V
HAMILTON JAMES WILLIAM JR
HAMMONTREE BILLY LEON
HAMNER WALTER SCOTT
HAMPTON CHARLES VERNON JR

HAMPTON EDMOND
HAMPTON ISAAC DE VAND
HAMPTON RALPH LAMAR
HANKS DANNY DEAN
HARKLESS JAMES ANTHONY
HARRINGTON IRIS HILTON
HARRIS CALVIN
HARRIS WESLEY HOMER JR
HARRISON ALBERT LEWIS
HARRISON JOSEPH WAYNE
HART RANDOLPH GUY JR
HART TEDDY MYRLE
HARTWELL HAROLD JAMES
HAWSEY KENNETH
HAYES JAMES JR
HAYES KENNETH FRANCIS
HAYES PHILLIPS III
HAYS WILLIAM BRIAN
HAYWARD DAVID ROY
HAYWOOD GLENNON
HAYWOOD ROGERS LEMANDER
HAZELTON HERMAN
HEATH DOUGLAS RANDOLPH
HEBERT ALTON JOHN
HEBERT CALVIN RAYMOND JR
HEBERT CARROLL JAMES
HEBERT JAMES III
HELLBACH HAROLD JAMES
HENDERSON ANTHONY JOSEPH
HENDERSON CHARLES CLIFTON
HENRY CLARENCE IVORY
HENSLEY THOMAS TRUETT
HERRINGTON RONNIE JOE
HESTER CHARLES RICHARD
HETZLER LARRY GLEN
HILL IVORY JR
HILL JAMES LOUIS
HILL JAMES MARSHALL
HOGEMAN CARROLL GENE
HOLLENSHEAD WINSTON GEORG
HOLLOWAY LARRY DANIEL
HOLMES PHILLIP HEASE JR
HOLSOMBACK FRANK NOLAN
HONLEY JIMMIE CARROL
HORNBROOK RONALD RAY
HOSKINS JOHN THOMAS
HOSTETTER STUART GLEN
HOUGHTION ROBERT CHARLES
HOWARD JIMMY LEE
HOWARD ROY WILLIAM
HUBBARD MARVIN PETER
HUBER LEO JOHN III
HUGHENS FREDERICK EDWARD
HYATT TRACY ROY
HYMES JOHN LAMUEL
IMBORNONE DARRELL DAVID
IMPSON DOUGLAS GERALD
ISELY MICHAEL GENE
JACK WILSON JR
JACKSON ABRAHAM
JACKSON CARL EDWIN
JACKSON CHARLES
JACKSON DALTON EARL
JACKSON JOHN WENDELL
JACKSON LARRY ANTHONY
JACKSON RAY LEE
JACKSON WILLIE
JEFFERSON CARTER JR
JEFFERSON CLARENCE JR
JEFFERSON HERMAN LOUIS JR
JENKINS CHARLIE EDWARD
JENNINGS JAMES JR
JERSON JAMES RAY
JETT RUSSELL LANE
JOHNSON ANDREW
JOHNSON BOBBY GENE
JOHNSON CALVIN RAY
JOHNSON CHARLES EDWARD
JOHNSON CLAYTON HENRY
JOHNSON DENNY LAYTON
JOHNSON FREDDIE
JOHNSON GEORGE ALBIAN JR
JOHNSON GILTON WALTER
JOHNSON JESSE
JOHNSON MC ARTHUR
JOHNSON MILTON JAY
JOHNSON RAYMOND PAGE
JOHNSON VICTOR JR

JOHNSON WELLINGTON M
JOHNSTON BILLY NEAL JR
JOHNSTON PAUL KINARD
JOLES RICHARD WADE
JONES CHARLES WAYNE
JONES HENRY JR
JONES JAMES HARVEY
JONES MELVIN LEWIS
JONES REUBEN JR
JUNEAU MICHAEL JOSEPH
KARR JOHN PRESTON
KAUFMAN DAVID MITCHELL
KAY WALTER THOMAS JR
KELLOGG ALTON DELANEY
KELLY CHRISTOPHER
KEMP CHARLIE EDWARD
KEMPFF RONALD WARREN
KEY RICHARD JOHN
KING DONNIE LUSTER
KIRKWOOD DERYL RAMON
KLING LEROY JOHN OLIVER
KNIEPER PHILIP GEORGE JR
KNOWLTON WAYNE HOWARD
KOPFLER JOSEPH STARNS III
KUJAWA LARRY FRANK
LA COSTE THOMAS EMILE
LA FLEUR GREGORY L
LA POINT LARRY JOHN
LACEY EDWARD GENE
LACHNEY FLOYD CAMILLE
LAFLEUR ROBERT WAYNE
LANDRY EDDIE LEE
LANDRY HOWARD DENNIS
LANDRY JOSEPH RONALD
LANGLINAIS JACK PETE
LANNES SHERMAN DAVID JR
LASSITER JOHN ALFRED

LAVITE ANTHONY III
LE BLANC GERALD THOMAS
LE BOEUF MICHAEL JAMES
LE BOUEF ELTON JR
LE BOUEF WILTON PAUL
LE GRAND JOSEPH DALLAS JR
LE LEAUX MICHAEL JAMES
LEBLANC ALFRED LEROY
LEDFORD STEVE DENNIS
LEE ALAN JAMES
LEE ROBERT CHARLES
LEGAUX MERLIN PHILIP
LEGER MALCOLM FRANCIS
LEJEUNE HORACE JOSEPH JR
LEMOINE WILLIAM FRANCIS
LENOIR EUGENE
LEWIS CLARENCE HENRY
LEWIS DANIEL
LEWIS ELTON WILLIAM
LEWIS FREDDIE
LEWIS KENNETH JERNIGAN
LEWIS OTIS
LIGGINS CLAYTON
LILES ROBERT LEONEL JR
LINDSAY STEPHEN LEE
LINDSEY MARVIN NELSON
LINES RICHARD MICHAEL
LISTE DAVID ALLEN
LITTLE PETER
LOCKARD LEONARD WAYNE
LOCKETT LLOYD
LOMBAS DEXTER JOSEPH
LONCON LARRY JOSEPH
LONDON EARL
LONG JERRY ROY
LOUT BILLY BURKE
LOUVIERE MARVIN JOHN

LOWTHER HAROLD WAYNE
LUPO FRANCIS DAVID
LYONS MALCOLM JOSEPH
LYONS QUILLARD FRANK
MAC FETTERS DUNCAN ALEXAN
MACK CALVIN DAVID
MACK JAMES
MAGEE BOYD
MAGEE RALPH WAYNE
MAHL KENNETH ARTHUR
MALBROUGH CHARLES RAY
MALMAY THOMAS SIMON
MANNING GLENN ROBERT
MANNING RONALD
MARCANTEL ELBERT
MARCEAUX ERASTE JOHN
MARCOMBE STEVE GARY
MARSHALL BILLY RAY
MARTIN ANTHONY TONY
MARTIN ARTHUR JUSTILIEN
MARTIN DONALD LAWRENCE
MARTIN WILLIAM TORBERT
MATHIS HARRY JR
MAYE MICHAEL MC KENZIE
MAYES DAVE JR
MAZZANTI JOSEPH EDMUND
MC CARROLL IVY M JR
MC CARTER JAMES W JR
MC CLENDON WILLIAM W JR
MC DUFFY ROBERT LOUIS
MC KINNEY CHARLES ANTHONY
MC KINNEY JAMES ODAS
MC KNIGHT GEORGE PARKER
MC LAUREN CHARLES WILLIAM

MC MAHAN CHARLES DARNELL
MC MANUS JERRY DOYNE
MC MILLAN GERALD WAYNE
MC NEIL BRUCE ALAN
MC WILLIAMS FREDDIE
MCCLEARY GEORGE CARLTON
MCKINNEY MICHAEL GEORGE
MEAUX PAUL JAMES
MEDINE BERTRAND C JR
MEGEHEE JAMES WOOD
MENARD LOUIS UISVILLE JR
MENDOZA MILTON JOHN
MERCER POLLARD HUGH JR
MERRITT TERRY LEE
METOYER BRYFORD GLENN
METOYER JAMES EDWARD
MILLER ANTHONY
MILLER ARTHUR JR
MILLER CHARLIE REUBEN JR
MILLER LEON PETER
MILLER LEONARD CHARLES
MILLER ROBERT RICHARD
MIREMONT JAMES EDWARD
MISTRETTA ERIC PAUL
MITCHELL CHRIS ANTHONY JR
MOAK CLIFTON PEARCE
MOFFETT BILLY RAY
MOLAISON GORDON THOMAS
MONDAY ALVIN
MONTE SALVADOR LOUIS JR
MONTELEONE ERNEST J JR
MOORE ANTHONY LOUIS
MOORE KENNETH CHARLES
MOORE THOMAS
MOPPERT EUGENE MEYERS
MORELAND LARRY WAYNE
MORGAN CHARLES RICHARD
MORGAN WILLIAM JOOR
MORRIS EUGENE JR
MORRISON CARL PHILLIP
MOSES ABELL
MOUGIER JOHN EDGAR JR
MULLEN FRANK
MURPH SAMUEL ENNIS
MURPHY PATRICK RONALD
MYLES JAMES WALTER
MYLES PHILLIP MURRY
NAQUIN SIMON ADOLPH
NARCISSE PAUL
NATIONS JERRY LEE
NATIONS ROY LEE
NELSON DANIEL CARTER
NELSON GEORGE WASHINGTON
NELSON LARRY
NOBLE DANIEL JOSEPH
NUGENT MICHAEL RAY
O NEAL HAROLD JR
OGBURN GLENN ROY
OGEA WALLACE LEE
ONISHEA JESSE JAMES
ORGERON PHILIP JOSEPH
ORSO PRINIS WILSON
ORTEGO GERALD M
OWENS CLAUDE JAMES
PAGE RUSSELL ELWARD
PALM JOSH JR
PANQUERNE CHARLES PAUL
PARKER JOHN BOYD
PARKER RUDOLPH
PASSAFUME MICHAEL JAY
PATTISON JOHN JR
PAUL DANNY LEE
PAYNE ERNEST
PEA EDWARD EARL
PELLEGRIN O NEIL J JR
PENN HERMAN
PEOPLES JERRY WAYNE
PEOPLES PERRY LEE
PERALTA RAPHAEL ALEXANDER
PEREZ RICARDO JAMES
PERKINS LUTHER RIVES
PERRODIN CURTIS JOSEPH
PETERS ALBERT JAMES
PETTY MICHAEL HARRIS
PHARRIS WILLIAM VALRIE
PICHAUFFE CARL JOSEPH
PICHON HERMAN EDWARD
PICHON LOUIS ALPHONSE JR
PITRE FLOYD LEON

PITRE JORDY JOSEPH
PITRE KENNETH JOSEPH
PLUMMER REGGINALD WILLIAM
POLLARD RICHARD
POOL CHARLES LEO
POOLE CHARLIE SHERMAN
PORTER LEO
POTIER MILTON PHILLIP
POTTER ROBERT GLEN
POUNDS RONNIE LOUIS
POUSSON MICHAEL WAYNE
POWELL LIONELL
PRATHER HENRY LEE III
PREJEAN KENNETH ANDREW
PREWITT WILLIAM ROLAND
PRICE MICHAEL GLEN
PRIMM SEVERO JAMES III
PUGH PERCY ISAIAH
QUEBODEAUX WILLIAM C JR
RACHAL CHARLES WILLIAM
RAGLAND MASON ERWIN
RAHM ARNOLD JOHN
RAINEY VERNON EDWARD
RAMSEY ANTHONY LOUIS
RANDOLPH GEORGE
RANDOLPH LIONEL
RAO GLENN BURLEIGH
RATHBURN RICHARD ALLEN
RATLIFF JAMES LEE
REDDIX MISTER JR
REED ISREAL DALLAS
REED LEROY
REID JOSEPH CLARK
REID WINFIELD WALTER
REMEDIES RICHARD JARRELL
RENFROW BILLY JOE
RHODUS RAY WESLEY
RICHARD BYRON MATTHEW
RICHARD ROY JAMES
RICHARDSON JAMES DOUGLASS
RICHARDSON JESSIE
RICHARDSON MICHAEL WAYNE
RIVERE ALVIN PIERIE
RIVIERE FRANK IRA
ROBERSON DONALD RADFORD
ROBERTS CHARLES WADDELL
ROBERTS MARVIN JAMES
ROBERTSON JAMES WAYNE
ROBINSON FRANCIS JOSEPH
ROBLEDO RAUL
ROD RONALD FRANCIS
RODGERS CARROLL L
ROGERS ROBERT LEE
ROLLEN CLARENCE EDWARD
ROMAGOSA LAYNE JOSEPH
ROMERO GLENN WAYNE
ROMERO RONALD JAMES
ROSENTHAL MICHAEL D JR
ROSS ROBERT LEE
ROSS RONNIE ALLEN
ROUSSELL RALPH S JR
ROWLEY HARRY EMILUS
RUTH ALFRED DARNELL
SANCHEZ HERMAN PAUL
SANDERS JAMES EDGAR JR
SANDERS ROBERT HERNDON
SANTINAC LAWRENCE HAROLD
SCHMOLKE JOSEPH MICHAEL
SCHNITGER GERARD GEORGE
SCOTT ALVIN JOSEPH
SCOTT ROBERT LEE
SEAMSTER WILLIE PURFOY
SEEBODE JOHN CONRAD
SELDERS EUGENE
SEMON KENNETH RONALD
SENSAT MORRIS JOSEPH
SERCOVICH JOSEPH GEORGE
SEREX HENRY MUIR
SHARPLEY LEONARD COSBY
SHAW CLAIBORNE LAVELLE
SHUPTRINE ROBERT M
SIEBERT FREDERICK W JR
SIKES BOBBIE EARL
SILMAN GARY WILLIS
SIMMONS CHESTER JOHN
SIMMONS NATHANIEL
SIMON CURLEY JOHN
SIMON PAUL JOSEPH
SINCLAIR PATRICK EUGENE

SINEGAL HUBERT JR
SINGLETON CHARLIE JR
SINGLETON GEORGE JAMES
SISTRUNK DONALD WAYNE
SKINNER CLAIBORNE JOHN
SMITH AARON CHARLES
SMITH ALLAN EUGENE
SMITH DAVID LEON
SMITH HARRY WINFIELD
SMITH JOHN JR
SMITH LARRY F
SMITH MARCUS
SMITH MELVIN
SMITH ROBERT
SMITH ROBERT SR
SMITH WILLIAM DAVID
SMITH WINSTON JOHN
SMOOT CURTIS RICHARD
SNYDER WOODROW WILSON JR
SOLORZANO ROBERT ANGELO
SONGNE DARNELL JOSEPH
SONNIER FOSTER LEE
SORTER MICHAEL VINCENT
SOUTHER WALTER ALVIN III
SPELLMAN WAYNE JUDE
SPENCE EDGAR CLAY
STANFORD BOBBY GAYLE
STEIN ARMOND JOSEPH JR
STEIN CLAUDE JOSEPH
STEPHENS DANNY LYNN
STEVENS DONNY RAY
STEVENS RUDOLPH
STEVENSON DON EDDIE
STEWART EDWARD
STEWART JOHN EMANUEL
STEWART ULYSSES
STEWART VIRGIL GRANT
STIGALL ARTHUR DONALD
STOCKSTILL WALLACE A JR
STOCKWELL EDWARD E JR
STRICKLAND ROBERT CECIL
STUTES JAMES RONALD
STUTES KENNETH JOHN
TALTON BOBBY RAY
TARVER EDWARD
TATE TONY LARUE
TATNEY ERNEST JR
TAYLOR DUNCAN JR
TAYLOR JOHN LEWIS
TAYLOR STANLEY WADE
TERHUNE DARYL BERT JR
TERRELL DAVID WILLIS
TERRELL KEAVIN LEE
THERIOT PHILLIP FINNAN
THIBODEAUX EDWARD JOSEPH
THIBODEAUX MICHAEL L

THOMAS CHARLES JR
THOMAS CHARLIE BERNARD
THOMAS ISIAH
THOMAS JAMES RONALD
THOMAS KENNETH LEE
THOMPSON FREDDIE JR
THOMPSON JOHN H
THOMPSON JOHN ROY
THOMPSON PERRY EDDISON
THOMPSON ROBERT EUGENE
THORNTON EVANS JEROME
TIPTON CHARLES ROY
TRAINHAM THOMAS NEIL
TREGRE LARRY PETER
TULLIER LONNIE JOSEPH
TUNNEY MICHAEL JOSEPH
TURNER BERNARD EMERSON
TURNER FREDDIE
TUYES DONALD GLENN
TWEEDY VERNON RUBEN
UPTIGROVE JESSE
VALENTINE JOSEPH RONALD
VARNADO MICHAEL BANARD
VENABLE JOSEPH ALVIN
VERRETT DURWOOD WAYNE
VICE FARRELL JAMES
VIDRINE TERREL JAMES
VINCENT HALTON RAMSEY
VIOLA CARL DANIEL
VOLENTINE PHILIP ALVIN
WAGUESPACK GARY LOUIS
WALKER JAMES EDWARD
WALKER MICHAEL STEPHEN
WALKER WILLIE
WALLACE EPHRON JR
WALLACE VERNON MARTIN
WALLS ROBERT LEE
WALTER HERBERT
WALTON JIMMY RONALD
WANER LOUIS BERNARD II
WARREN BENJAMIN IRVIN
WARREN JEIDER JACKSON
WARREN RODIS JOHN
WARWICK DUNCAN ALBERT
WASHINGTON DAN THOMAS
WASHINGTON DOUGLAS MAC
WASHINGTON JAMES LEROY SR
WASHINGTON KING DAVID
WATERS CLARENCE EDGAR
WEAVER SAMMY LANE
WEBB EARL KENNON
WEEKLEY RUSSELL JOSEPH
WELCH ARTHUR NORMAN
WELCH THOMAS EDWARD
WELD JULIO CESAR
WELLS WALTER LOUIS
WEST PAUL BRADLEY

WESTCOTT RODNEY WAYNE
WHEAT GENE JOSEPH
WHELESS DOUGLASS TERRELL
WHITE AULDON KEITH
WHITE JAMES HARDY
WHITE KURNEY JOSEPH JR
WHITE TIMOTHY CHAMPREAUX
WHITLOCK IVAN PRESTON
WIESENDANGER LAWRENCE LOU
WIGGINS THOMAS WAYNE
WILCOX CHARLES EARL JR
WILEY THOMAS J
WILHELM LAWRENCE M JR
WILLIAMS CHARLES RAY
WILLIAMS DAVID BERYL
WILLIAMS ERIC
WILLIAMS ERNEST C JR
WILLIAMS FLOYD CHARLES
WILLIAMS HOWARD CLAYTON
WILLIAMS HUEY
WILLIAMS JIMMY
WILLIAMS JOHN WILLIAM
WILLIAMS LAWRENCE JR
WILLIAMS RAY MILTON
WILLIAMS WALTER
WILLIBER GERALD PAUL
WILLINGHAM WILMER JAY
WILLIS LEDELL
WILSON JAMES WILLIAM
WILSON VOMER OVID JR
WOLFE ABRA JOSEPH JR
WOLFE THURMAN WILLIAM
WOLTER ARTHUR GEORGE WILL
WOMACK ROBERT LEE
WOOD BOBBY CLYDE
WOODS DUREL STEVENS
WORLEY STEPHEN RAY
WRIGHT FRANK JR
WRIGHT SYLVESTER JR
WYATT TOMMY LLOYD
YELVERTON DON JUNIOR
YOUNG HERMAN DEAL
YOUNG LEROY JOSEPH
YOUNGBLOOD CHARLES EUGENE
YOUNGBLOOD DAVID WAYNE
ZERANGUE ALTON JOSEPH JR
ZERINGUE RALPH HENRY

Photos by Janet Robbins

Vietnam Veteran's Memorial - Cole Land Transportation Museum

405 Perry Rd, Bangor, ME 04401

The Vietnam War memorial in Bangor, Maine is located at the Galen Cole Land Transportation Museum. It contains several different monuments that were created to honor the veterans of the Vietnam War. This project was considered important because the Maine veterans who did return home from the war didn't get a proper welcome home.

The memorial is located outside of the museum and is accessible to the public 24 hours a day, seven days a week. There are, however, special events held at the memorial on Veterans Day and Memorial Day each year. At these ceremonies, qualified Vietnam War veterans receive commemorative walking sticks, as do the surviving veterans of other 20th century wars.

Galen Cole and his son Gary commissioned Glen and Diane Hines to make the three-person statue that is featured in the Vietnam Veterans Memorial Park. This particular sculpture is made of bronze and the left figure on the far left was actually modeled after Lance Corporal Eric Wardwell, Gary's childhood friend who died in Vietnam in September of 1967 as a member of the US Marine Corps. The memorial as a whole was dedicated on May 31, 2004, by Governor John Baldacci.

The Vietnam Veterans Memorial consists of multiple monuments. The first monument in the memorial is a Huey helicopter which sits off to the side of the memorial wall. The wall is made of granite and features the names of 339 Maine service members who weren't fortunate enough to make it home from the war. There is also the aforementioned bronze statue. This sculpture portrays three figures, a nurse helping a wounded soldier with the assistance of another serviceman. Rounding that out is a retired M60A1 tank that stands powerfully off to the side of the other monuments.

The Names of those from the state of Maine Who Made the Ultimate Sacrifice

ABDELLAH BRUCE ALLYN
ALBERT LOUIS BASIL JR
ALBERT RICHARD PATRICK
ALEXANDER DAVID HAROLD
ALLEN KEITH DOBSON JR
ALMON WILLIAM RUSSELL
ANDERSON CURTIS STEWART
ARMSTRONG HERBERT ELBRIDG
ARNOLD RICHARD W
ASSELIN LEO ROGER
AVORE MALCOLM ARTHUR
BAILEY GEORGE LEROY
BAILEY JON
BAKA JAMES ALEXANDER
BAKER CHARLES OAKES II
BAKER ROBERT LEE
BARKER PAUL LEROY
BARR ALLAN VAUGHN
BARTLETT DONALD HAROLD
BATCHELDER WILLIAM ROBERT
BAZEMORE THOMAS WAYNE
BEAN GUY ROBERT
BEAN STEPHEN LOUIS
BECHARD RAYMOND JOSEPH
BELANGER ALBERT LEE
BELANGER GEORGE
BELANGER JOSEPH KENNETH
BELANGER PAUL EDWARD
BELL GILBERT STEVENS JR
BERNARD THEODORE DANIEL
BERRY ROBERT LESTER
BERUBE RICHARD
BLAIN DENNIS KNUTE
BLAIR JOSEPH R L
BOIVIN EDWARD J
BOOBAR LARRY DANIEL
BORDUAS RAYMOND ARTHUR
BOSSIE KENNETH JAMES
BOSWORTH DAVID RUSSELL

BOYD ROBERT WHITE
BRETON HAROLD GEORGE
BROOKS JOHN HENRY RALPH
BROOKS LYLE GIBSON
BROWN CHARLES NORMAN
BROWN DONALD LEROY
BROWN GARDNER JOHN
BRYANT PHILIP SHERWOOD
BUBAR RICHARD PERLEY
BUCK HOLLIS WINFIELD
BUKER BRIAN LEROY
BURCHFIELD JOE STUART
BUXTON DALE RYAN
CALLINAN WILLIAM FRANCIS
CARTER ZANE AUBRY
CASH JOHN HAROLD JR
CAYFORD PHILLIP J JR
CHASE CLARENCE LAWRENCE
CHASE JAMES FRANCIS
CHAVARIE NORMAN JOSEPH
CHILDS CHRISTOPHER J III
CHURCHILL CARL RUSSELL
CLEARY JAMES WILLIAM
CLIFFORD JON IRVING
CLUKEY ROBERT LEOPOLD JR
COILEY CHARLES ROBERT
COLLIER RAYMOND LYN
COLOSANTI NORMAN EDWARD
CORO BERNARD LOUIS
CORRIVEAU RICHARD THOMAS
CORSON TERRY CHARLES
COX DAVID AUSTIN
CRANEY LESLIE LEE
CRANSON ROBERT DORIAN
CRESSEY JAMES DANIEL JR
CROCKER DENNIS OWEN
CROWELL ARTHUR ALBERT
CYR LAWRENCE JOSEPH
CYR PAUL LEO

CYR WAYNE CLIFTON
D ENTREMONT LARRY AIME
DAIGLE BENNETT JOSEPH
DALRYMPLE LESLIE AARON
DARCY EDWARD JOSEPH
DARLING GEORGE ROBERT
DAVAN BENEDICT MAHER
DAVENPORT BARD ELTON
DEAN DONALD BING
DECHENE ROBERT NORMAND
DEFORREST RONALD C
DELANO MERWIN A JR
DELISLE RODNEY JEROME
DERAGON MICHAEL HENRY
DEROSIER LAURIER DON
DESCHAINE NORMAND CAMILLE
DESCHENES JAMES GEORGE
DESCHENES MICHAEL HUBERT
DEW HENRY LOUIS
DICKINSON LESLIE A JR
DIPHILLIPO ROCCO
DORITY RICHARD CLAIR
DORR GERALD ANDREW
DOUGLAS HARVEY JAMES
DREW THEODORE GLENN
DROWN TERRY FRANCIS
DUBE ANDRE LOUIS
DUBE PETER LEE
DUFAULT JAMES RICHARD
DUFFY JOHN EVERETT
DULAC MALCOLM CYRIL
DUPREY DANNY LEE
DURANCEAU DAVID MARIUS
DYER BLENN COLBY
ELKINS WAYNE ROBERT
ELLINGSON JAMES EARL
ELLIOT ARTHUR JAMES II
ELLIOTT EDWIN ELLIS
ELLIS FRED MILTON

ELWELL DONOVAN KEITH
FARLEY DAVID LITTLEHALE
FAULKNER LARRY FREEMAN
FERGUSON THOMAS BERNARD
FERGUSON WILLIAM GLEN JR
FITCH DELLWYN ALLEN
FITCH RONALD RUSSELL
FLAHERTY ROGER ELLIS
FOGG DAVID BRUCE
FOSS DANIEL ARTHUR
FOSS JOSEPH RALPH
FOSTER DAVID DAN
FOSTER ROBERT ENOCH JR
FOURNIER JOSEPH DAVID
FREEMAN MARTIN LEE
FROST CARLTON ANDREW
GAGNE BERTRAND RONALD
GAGNON JOHN EDGAR
GAGNON JOSEPH DENNIS
GAGNON MORRIS DOMINIQUE
GAGNON PERCY CHARLES
GALLANT ROGER PAUL
GARRISON EARL STANLEY
GAUVIN ROGER EDWARD
GENESEO LOUIS J
GERALD DANA LEON
GETCHELL PAUL EVERETT
GLAUDE RICHARD PAUL
GODERRE JOHN ROGER
GODFREY JAMES WALTER
GODING ROBERT EARLE
GOGGIN RICHARD JAMES
GOOGINS DOUGLAS E JR
GRAUSTEIN ROBERT STEWART
GRAY STEPHEN FRANCIS
GRAY THOMAS EDWARD M JR
GREGOIRE MILES ROBERT
GRINNELL RICHARD RALPH
GUAY HERVE JOSEPH

GUERETTE ROLAND PHILIPPE
GUSTIN ANTHONY JOHN
HALL EDWARD SENIOR
HALL SAYWARD NEWTON JR
HALL WALTER LOUIS
HANLEY TERENCE HIGGINS
HARDY HERBERT FRANCIS JR
HARRINGTON WILLIAM FREDER
HASKELL LLOYD BURTON JR
HEAL HENRY ALBERT JR
HEATH JOSEPH EMERSON
HICKS LESLIE CLYDE
HICKS SHELDON WAYNE
HIGGINS KENNETH LEE
HODGKIN FOREST CLAYTON
HOPKINS CHESTER LEE
HOYT ARTHUR JAMES
HUNTLEY JOHN NORMAN
HURD COLIN PLUMMER
HUTCHINSON ALLEN MELVIN
JOHNDRO RODNEY GEORGE
JONES LEWIS CARLTON JR
JORDAN WILLIAM E III
KAHKONEN EDWIN MATTI JR
KIRKPATRICK RONALD IRVING
KNIGHT JOHN WALLACE
KNOWLTON BURNS WINSHIP JR
KRZYNOWEK PAUL S
LA BONTE ROGER EDWARD
LANE DAVID ALAN
LANGLEY WESTON JOSEPH
LANO LAWRENCE
LEIGHTON RAYMOND ELTON
LENTO STANLEY JOHN
LEPAGE REYNALD GERARD
LEVESQUE J B L
LIBBEY MALCOLM PIERCE
LIBBY JOHN H

Maine

THESE MAINE VIETNAM VETERANS GAVE THE ULTIMATE SACRIFICE

LOFSTROM LELAND EDDY
LOMBARD DURWOOD BERT
LOUGHRAN THOMAS WILLIAM
LOVLEY THOMAS GRANT
LUTTRELL BRUCE IRVING
MAC DONALD LESTER EARL
MAC KILLOP NEIL HOWARD
MACE DANA LEROY
MANCHESTER GARY ORAL C
MANCHESTER JOHN SMYTHE
MANN EDWARD LEONARD JR
MANSFIELD DONALD LEWIS
MATTHEWS GENE FLETCHER
MATTHEWS RICHARD LEE
MC AULIFFE EARLE EUGEN JR
MC DONOUGH JAMES M JR
MC EACHERN RANCE ALDEN JR
MC EWEN ROY CLIFFORD
MC GEE FREDDY ALFORD
MC GONAGLE MICHAEL JOHN
MC GUIRE HARRY JOHN III
MC HUGH FREDERICK WILLIAM
MC KECHNIE JAMES ALLEN
MC KENNEY NORMAN LAFOREST
MC MAHON THOMAS JOSEPH
MC MORROW JAMES JOSEPH
MCLAUGHLIN JAMES BRUCE
MEADE JOHNSON ASHLEY
MERRICK WALTER FORREST
MICHAUD BENTON

MICHAUD LEO EDWARD
MILLEDGE FREDERICK RAYMON
MILLETT LAURENCE ARTHUR
MILLS CHARLES HOMER
MOGAN JOHN EDWARD
MOODY THOMAS JOHN
MORGAN VAUGHAN SHAW
MORRILL BERNARD FRANCIS
MORRILL FRED WILLIAM
MUSETTI JOSEPH TONY JR

NADEAU LARRY JOSEPH
NADEAU ROBERT JOHN
NICHOLAS REGINALD
NICHOLSON JAMES PATON
NICKERSON BRADFORD SCOTT
NILE MAURICE J
NISKANEN MARTIN KEITH
O REILLY TARRY THOMAS
OLSEN WILLIAM FRANK
PARSONS RONALD ALLEN

PELKEY RAYMOND NELSON
PERKINS IRA HILTON JR
PERRON JOSEPH ADRIAN G
PETERS LAWRENCE VINCENT
PICKLES MICHAEL RICHARD
PLOURDE VICTOR M
POITRAS NORMAN GERALD JOS
POITROW EMERY NORMAN
POLAND LEON LOVELL JR
POLIQUIN MICHAEL EDWARD

POMERLEAU JAMES GERARD
PORTER STEVEN LINDSEY
PRESKENIS RICHARD JOSEPH
QUINN GREGORY CORNELIUS
QUIRION JOSEPH G L JR
REYNOLDS GEORGE R JR
REYNOLDS HAROLD W
RICHARDSON NELSON GRAFTON
RICKARDS LINWOOD PRESTON

ROBERTS JOHN WAYNE
RODERICK SCOTT JAMES
ROSSIGNOL RICHARD W
ROY ROBERT RICHARD
RUSSELL DONALD MYRICK
RUSSELL EDWARD T
RUSSELL LYNN JORDAN
RYDER EDWIN BYRON
SANDERS WILLIAM STEPHEN
SANGILLO WAYNE
SAVOY CLAYTON EDWARD
SAWYER JONATHAN ANSEL
SCHRIVER STEPHEN PAUL
SHANK RODNEY GEORGE
SHAY LAWRENCE WILLIAM JR
SHEA THOMAS COLLINS
SHEPARD HARRY CLIFTON JR
SHUMAN WILLIAM CONRAD
SKIDGEL DONALD SIDNEY
SKINNER JAMES CRAWFORD
SLOAT GREGORY ALEC
SMILEY RONALD OWEN
SMITH ELDON WAYNE
SMITH JAMES ALLEN
SMITH ROBERT JOHN
SOULE CHARLES HOWARD
SOULE WILLIAM D
SOUTHARD HAROLD ELLS-
WORTH
SOWERS JAMES RODNEY
SPENCER HAYWARD CARL
SPENCER WILLIAM EDWARD
SPROUL RAYMOND RONALD
ST JOHN RONALD GEORGE
STERRY RAYMOND EDWARD
STEVENS FRANCIS GEORGE
STEWART RICHARD JOHN
STIMPSON PAUL LEWIS
STROUT PHILIP WILLIAM
STUART CHARLES EDGAR
SUDSBURY PAUL EARL
SULLIVAN CHARLES E JR
SULLIVAN THOMAS MICHAEL
TAYLOR NEIL BROOKS
TEWKSBURY ROBERT W
THERIAULT HARRY EVERETT
THIBODEAU DAVID PAUL
THOMAS RONALD MEDFORD
II
THURSTON CLAIR HALL JR
TIBBETTS BRUCE HAROLD
TIBBETTS CLINTON EUGENE
TIBBETTS EDWARD W
TIBBETTS GORDON EDMUND
TOPPI CHRISTOPHER JOHN
TRACY GARY DALE
TROTT DONALD HERMON
TUELL DANIEL PAUL
TURNER GILBERT CRAIG JR
TURNER ROBERT ALLAN
UMEL MICHAEL PETER
VACHON WILBUR JOSEPH III
VESTER FREDRICK HAZER
VLAHAKOS PETER GEORGE
WALKER HAROLD EVERETT JR
WARD ALLEN LEA
WARDWELL ERIC MICHAEL
WATSON LORING WILLIAM
WATTS WAYNE ALAN
WEBSTER OWEN H
WEST JAMES RUSSELL
WEST PAUL ROBERT
WHIPPLE STEPHEN JOHN
WHITTEN MILAN ELMER
WIGGIN ROBERT JAMES
WIGHT ALONZO WILLIAM
WILLIAMSON ROBERT GREG-
ORY
WILLS ROBERT EMERY
WITHEE CLYDE WILLIAM
WITHEE EDWARD WILLIAM
WOODMAN STUART ALAN
YOUNG ROBERT B

MAINE VIETNAM
VETERANS MEMORIAL
1961 – 1975

DEDICATED MAY 31, 2004
BY GOVERNOR JOHN E. BALDACCI

103

Photos by John Shuler

Maryland Vietnam Veteran's Memorial

3001 East Dr, Baltimore, MD 21225

The Vietnam memorial at the Maryland State Veterans Park is located in Middle Branch Park in Baltimore, Maryland. The park offers visitors a serene location to reflect on the sacrifices made by the many fallen and surviving veterans while looking at the nearby Patapsco River.

The memorial came to fruition through the efforts of a group of veterans known as "The Last Patrol" along with other Marylanders. The names of all of the veterans to be honored were thoroughly researched and the next of kin were contacted. Once that was done, construction began on the memorial in October of 1988. It was funded through contributions by individuals, businesses, and the Maryland General Assembly. In total, $2,250,000 was raised.

In 1989, the Maryland Vietnam Veterans Memorial Commission transferred responsibility for the Maryland Vietnam Veterans Memorial to the Maryland Veterans Commission. The memorial was officially dedicated in May of 1989.

The Vietnam memorial is set up in a circular formation of stone and features a granite wall made up of 64 stones that feature the 1,046 names of the Maryland fallen. There are also 38 names of servicemen who were listed as missing in action.

The entrance of the memorial features two granite slabs with inscriptions. One inscription is about reflecting on the troops' sacrifices while the other inscription is a dedication to the service men and women.

There are also two 45-foot tall flagpoles that fly the American flag and the Maryland state flag. Under the state flag is the POW/MIA flag. The memorial also consists of 16 spires that symbolize each year of the Vietnam War. The memorial is made complete by beautifully landscaped greenery including decorative flower beds and manicured trees. The bayside location is wonderful for the Memorial day event it hosts each year.

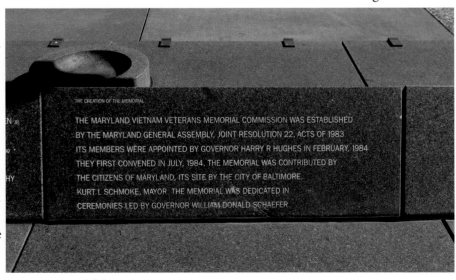

THE CREATION OF THE MEMORIAL

THE MARYLAND VIETNAM VETERANS MEMORIAL COMMISSION WAS ESTABLISHED BY THE MARYLAND GENERAL ASSEMBLY, JOINT RESOLUTION 22, ACTS OF 1983 ITS MEMBERS WERE APPOINTED BY GOVERNOR HARRY R HUGHES IN FEBRUARY, 1984 THEY FIRST CONVENED IN JULY, 1984. THE MEMORIAL WAS CONTRIBUTED BY THE CITIZENS OF MARYLAND, ITS SITE BY THE CITY OF BALTIMORE, KURT L SCHMOKE, MAYOR. THE MEMORIAL WAS DEDICATED IN CEREMONIES LED BY GOVERNOR WILLIAM DONALD SCHAEFER.

The Names of those from the state of Maryland Who Made the Ultimate Sacrifice

ABERNATHY ROBERT WILLIAM
ABSHEAR WILLIAM WALLACE
ADAMS RONALD M
ADAMS STEPHEN HAMILTON
AIAU HARVEY CHADWICK K
AIKEN DAVID ROSS
ALBI LOUIS VICTOR JR
ALLPORT JAMES SHERWOOD
AMOSS RUSSELL MONROE
ANASTASIO VINCENT JOHN
ANDERSON LARRY WAYNE
ANDREWS WILLIAM ALBERT
ANSELL JOHN ARTHUR JR
ANZELONE PAUL ROBERT
ARMENTROUT CHARLES F
ARMSTEAD GREGORY VAN
ARMSTRONG CHARLES JOSEPH
ARONHALT CHARLES E JR
ARROWOOD JAMES OSCAR
ATCHISON JAMES MITCHELL
AUD FRANCIS MATTHEW
AUGUSTUS DAVID RYAN
AVELLEYRA JOHN WILLIAM
AYD JACQUE JOSEPH
AYERS EDWARD FRANKLIN
BAILEY LELAND ALSTON
BAILEY TOLLIE
BAKER FREDERICK NORMAN
BAKER JAMES HOWARD JR
BAKER RONALD
BAKER VERNON R
BALDWIN ROBERT LLOYD
BARCLAY FREDERICK ALLEN
BARKLEY KENNETH PAUL JR
BARNARD HAROLD EDWARD
BARNES BARRIE VANE
BARNES JOHN HOWARD
BARNHART OTTO PHILIP
BARNHILL LARRY M
BARTHELME ALBERT LEWIS JR
BARTLEY KENNETH LEONARD
BATES JAMES LEON
BATES MELVIN CARROLL JR
BATES ROBERT JR
BAUER JOSEPH FREDERICK JR
BAYLOR ARTHUR JEROME
BAZEMORE EARL SHERMAN
BEARD JACK ALLEN
BEAUCHAMP JOHN HENRY JR
BEDSOLE CHARLES ARTHUR
BEGOSH MARTIN JOHN
BELCHER STEPHEN EDWIN
BELL CHRISTOPHER JAMES
BELL GEORGE ALBERT JR
BENJAMIN PHILLIP ERNEST
BENKE RONALD JOHN III
BERENDS JAMES
BERGER BARRY HOWARD
BERGMAN JACK STEPHEN JR
BEST BILLY HOWARD
BETHEA WILLIAM HENRY III
BETTS ALBERT LEON
BEYERLING JAMES LEROY
BIFARETI JOHN ANTHONY JR
BILLER HAROLD DOUGLAS
BINKO GEORGE
BIONDI JOHN MICHAEL
BISHOP ROGER WAYNE
BISSELL WILLIAM RONALD
BITTINGER ROBERT LEE JR
BIVENS FREDERICK WOOD JR
BLADES THOMAS NELSON
BLAKE DANNY LEE
BLAKE ROGER LEE
BLAKENEY GREGORY ALLEN
BLANCHFIELD RICHARD ALLEN
BLANDON GILBERT
BLANKENSHIP CHARLES HERMA
BLANTON JOHN JAMES
BLEND CLIFFORD CRAIG JR
BLOTTENBERGER MICHAEL J
BOARDMAN EDWARD ALLEN

BOBBITT GARLAND CLAUDE
BOCEK LEONARD JOSEPH
BOEHM WILLIAM JOSEPH
BOLAN ROBERT LOUIS
BOND WILLIAM ROSS
BONNETT GEORGE FABIAN
BORT HARRY JULIAN
BOUNDS GARY LEE
BOWERS JOHN MICHAEL
BOWLING DAVID BICKNEL
BOWMAN REGINALD ALONZO
BOWYER ARTHUR J
BOYCE MCDONALD E
BOYNTON CHARLES BENHAM JR
BRANDENBURG CHARLES FRANK
BRANDENBURG DALE
BRANOCK WILLIAM MICHAEL
BRASHEARS LARRY FRANKLIN
BRAZIER JOHN KENNETH
BRENT EDMUND DAVID JR
BREWER MICHAEL LEON
BRIDDELL CHARLES LILLETON
BRIDGETT PAUL EDWARD
BRISCOE JOHN ARNOLD
BRITT CHARLES JACKSON
BRITTAIN DANIEL SPENSER
BRITZ RONALD JOSEPH
BROOKE EARL THOMAS
BROWN GERALD BERNARD
BROWN JAMES TRULY
BROWN JOSEPH CLINTON
BROWN MARCUS JR
BROWN MICHAEL FRANCIS
BROWN ROBERT WILSON JR
BROWN THOMAS FRANCIS JR
BROWN WILLIAM JOSEPH
BRUCE HENRY MC DONALD
BRUCE LEE RAYMOND JR
BRUCE RICHARD BERT
BRUNSON LANCE DUNHAM
BRYANT ROSCOE EDWARD
BUCHANAN ROBERT BUTLER
BUDAHAZY JAMES DONALD II
BUDKA DAVID JOHN
BUGGS NATHANIEL JR
BUGOSH WILLIAM
BULLEN LAWRENCE RAN-

DOLPH
BURCH CLIFFORD GARLAND
BURDETTE LANNY JOE
BURLEY CLARENCE JOHN
BURNEY MARVIN
BURRIER PAUL THOMAS
BURROWS ROBERTS PATON
BURTON JOHN LEE
BUSH NATHANIEL
BUTLER CHARLES GILMAN JR
BUTTERWORTH DONALD H
BYERS JAMES NORMAN
BYRD NOLAN DARYL
BYRD NORMAN CECIL
CALDWELL DONALD PATRICK
CALL GERALD LEE
CALL JOHN HENRY III
CAMDEN FRANCIS EDWARD JR
CAMPANELLO DARRELL EDWARD
CAMPBELL JACK EDWIN
CAMPBELL WILLIAM LADD
CANAPP GARY EDWARD
CANIFORD JAMES KENNETH
CANNINGTON JAMES B JR
CANTLER DENNIS RICHARD
CAPASSO JOHN ALAN
CARBAUGH WOODROW FRANKLIN
CARLOZZI ROBERT MATTHEW
CARROLL JOSEPH KENNETH
CARTER ANDERSON JR
CARTER GEORGE ALBERT
CARTER HARVEY WILLIAM
CARTER WALTER CORBIN
CASE JAMES GILBERT
CASNER LEWIS EDGAR JR
CASSELL RONALD BRETT
CASSIDY WILLIAM EDWARD
CASTRO JOAQUIN
CHAPMAN MAURICE P JR
CHASE JOHN JOSEPH
CHASE JOHN LENWOOD
CHERRY ERVIN BENJAMIN
CHESLEY EUGENE NATHANIEL
CHILDRESS CALVIN JEFFERY
CHIN ALEXANDER SCHELEPH

CHINQUINA ROBERT NORRIS
CHMIEL LARRY VINCENT
CHRISTMAN WILLIAM J III
CHRISTOPHER WAYNE EDWARD
CLARK LARRY MONROE
CLAYTON DAVID NELSON
CLOVER LIONEL TIMOTHY
COIT LEON
COLE ERNEST WESLEY
COLE RAINER LOUIS
COLEMAN RONALD ALLEN
COLLIER GENE FRANCIS
CONCANNON FRANCIS BRYANT
COOK AUDREY JULIUS
COOK DAVID RICHARD
COOK ROBERT WILKINSON
COOK WILMER PAUL
COOK WILSON LEE
COOLEY JAMES EDWARD
COOPER OSCAR EDMOND
COOPER RICHARD WALLER JR
COPENHAVER GREGORY SCOTT
CORBIN WILLIAM JENNINGS
COSGRAVE GARY WAYNE
COSTELLO JEREMIAH FREDERI
COTTMAN ROBERT LEE
COX GEORGE MARION II
COX JEHU JUNIS JR
COYNE KEVIN MARK
CRAFT EZRA DELANO
CRAIG DEAN JOHN
CREW CARL JOSEPH
CRISSEY HARRY ELIAS JR
CRONKRITE WOODROW CHARLES
CROSBY GERALD LEE
CROZIER DAVID PAUL
CRUMP VICTOR LINCOLN
CRUTCHLEY DONALD CLAIR
CUMBERLAND PAUL ANTHONY
CUMBERPATCH JAMES R JR
CUNNINGHAM RICHARD SAVAGE
CURTIS BERNARD EUGENE
CURTIS JOSEPH PAUL

CUSTEN HENRY DAVID
CYMBALSKI KENNETH JULIAN
CZAJKOWSKI JOSEPH VERBERT
DAFFIN GARY ROBERT
DALE CHARLES RICHARD
DALE DENNIS HUMPHREY
DALTON TOUSSAINT O JR
DASTOLI JOSEPH PETER
DAUBERT WILLIAM JOHN
DAVIDSON DAVID ARTHUR
DAVIDSON ROBERT GRIFFIN
DAVIS DENNIS DEAN
DAVIS ELMER NEAL
DAVIS JAMES NORRIS
DAVIS JOHN ENGLISH
DAVIS LAWRENCE ARNOLD
DAVIS MELVIN GILMORE
DAVIS ROBERT JULIAN JR
DAVIS THOMAS RAYMOND
DAYTON WILLIAM CLARENCE
DE MARR JOHN CHARLES JR
DE MEY JOHN
DEAN CARL ANDREW
DEASEL JAMES JEROME JR
DEAVERS KENNETH LAMAR JR
DECK PATRICK A JR
DECKER STEVEN WILLIAM
DEFIBAUGH MICHAEL THOMAS
DEIBEL EDWARD PAUL III
DEITZ GORDON JAMES JR
DEITZ THOMAS MITCHELL
DEMBY GEORGE ALLEN
DERRILL CARROLL EDWARD
DEVINCENT EDWARD J JR
DICKENS RUSSELL W
DICKERSON JAMES EGBERT
DICKERSON ROBERT BOLT III
DICKERSON TOMMY EUGENE
DIEUDONNE CARROLL STEPHEN
DIGENNO MICHAEL
DIXON LINDEN BROOK
DIXON WILLIAM ALLEN
DIZE GEORGE HARLAND
DODD JAMES ERWIN
DOLAN THOMAS ALBERT
DONAHUE WELLINGTON

Maryland

MARTIN
DONALDSON EVERETTE LEROY
DONALDSON ROBERT D
DONNELL LAWRENCE HENRY
DORMAN DONALD RALPH
DORSEY GARDNER
DORSEY JAMES R JR
DOSS LUTHER JAMES JR
DRAKE TIMOTHY CALVIN
DROSD WALTER LLOYD
DUCAT BRUCE CHALMERS
DUFF BARRY WILLIAM
DUFFETT JAMES HENRY JR
DUNSMORE FRANK MELVIN JR
DURBIN RONALD WAYNE
DYDYNSKI STEPHEN MICHAEL
DYER LARRY EUGENE
DYSON LESLIE MILTON JR
EASON JOSEPH MILTON
ECKER TERRY LEE
ECTON HARRY LEON
EDER WILLIAM JOHN
EDWARDS JOHN PAUL
EGOLF CARL M
ELLIOTT RAYMOND LESTER
ELLIS GEORGE LEMUEL
EMERSON STEWART CHARLES
EPPS LAMONT GEORGE
EVANS CECIL VAUGHN
EVANS JEFFREY ALAN
FAISON EARL JR
FANNIN BRYANT D
FARMER CHARLES EDWARD
FEEZER JOHN HARVEY
FELL EDWARD WILLIAM JR
FICKUS JOHN ZANG
FIELDS CLINTON ANGELO
FIKE ROSS FRANCIS
FIKE THOMAS EUGENE
FISHER DONALD JAY
FITEZ HARRY SAMUEL JR
FLABBI GARY BERNARD
FLANAGAN SHERMAN EDWARD JR
FLETCHER GUY TALMADGE JR
FLINT RALPH PRESTON JR
FLONORY ORLANDO
FLOTT CHARLES LAWRENCE
FOGLER LEWIS JOHN
FOSSETT NORMAN ARCHIE
FOSTER WILLIAM HENRY
FOWLER WILLIAM EDWARD
FOY THOMAS WALTER
FRALEY CLAYTON EUGENE
FRANCE PHILLIP STANLEY
FRENG MARSHALL FRANKLIN
FULTON CLARENCE
FURLONG WILLIAM ROBERT JR
GAFFIGAN ROBERT MICHAEL
GAFFNEY EDWARD ALBERT
GARDNER ROBERT WAYNE
GAREISS KURT WILLIAM
GARLICK RODGER LYNN
GARRIS MICHAEL ANTHONY
GATES FRED HORATIO II
GEIST STEPHEN JONATHAN
GENTRY CHARLES EDWARD
GHEE JAMES FITZROY
GIBIS MICHAEL E
GIBSON JOHN CHRISTOPHER
GILBERT FRANKLIN BRADLEY
GILLIAM GEORGE HARVEY
GILLIAM RONALD STUART

GLENN EDWARD RALPH JR
GODEN RICHARD WALTER
GODMAN EARL ARTHUR
GOLDBERG BENJAMIN NILES
GOLDBERG STEWART B
GOLDER EDWARD ENOCH III
GOODMAN EDWARD LEON
GORDY MICHAEL EDWARD
GORRERA GEORGE MEDFORD JR
GORSCHBOTH ROLAND ALLEN
GOSNELL JACK MARTIN
GRAHAM JAMES ALBERT
GRANDEA AMBROSIO SALAZAR
GRASER JOHN WILLIAM
GRAVES WILLIAM D
GRAY CHARLES GONZIE
GRAY FRANCIS GARFIELD
GREEN EDDIE
GREEN LARRY
GREEN THOMAS OWEN
GREEN WALTER JR
GREENFIELD JOHN ARTHUR
GREGORY THOMAS JR
GRIGGS MICHAEL ALLEN
GROOM ROBERT ROXBURGH
GROOMES MURIEL STANLEY
GROVE KENNETH EDWARD
GUTRICK DONALD MAURICE
HACKNEY TATE TALMAGE III
HALL JAMES BUCKNER
HALL JOHN STERLING
HAMET DENNIS JOSEPH
HAMILTON ROGER DALE
HAMM HARRY DAVID
HAMMERSLA JAMES RUSSELL
HAMMOND LAWRENCE THEODORE
HAMRICK JAMES MADISON JR
HANNA GARY W
HARBERT CHARLES WALTER
HARDESTY MICHAEL OWEN
HARDING CHARLES CLIFFORD
HARE JOHN THOMAS
HARMAN CURTIS JOSEPH
HARMON ALPHONSO LEE
HARP MICHAEL LEE
HARRIS JAMES LOUIS
HARRIS JEFFREY LYNDOL
HARRIS JOHN LEE JR
HARRIS STEVE WESTLEL
HARRISON PAUL LEROY
HART DAVID MELDRUM
HARTMAN EUGENE WINFIELD
HARTSOCK ROBERT WILLARD
HATCHER CLAYBURN MC GEE
HAVAS STEPHEN LAWRENCE
HAYES LEROY ANTHONY
HAYES RICHARD EDWARD
HEARD JAMES BENEDICT
HENRY HOWARD BOYD
HENSON ALVAH WORRELL JR
HETTINGER ROBERT LANCE
HICKS NORMAN EDWARD JR
HICKS SILAS LUCAS JR
HICKS TERRIN DINSMORE
HIGGINBOTHAM ROBERT M
HILL JIMMY ARNOLD
HILL RICHARD GARFIELD
HILL THOMAS WAYNE
HILTON ROBERT LARIE
HIMES CLYDE STEVEN
HINES WILLIAM LINCOLN JR
HITCHCOCK RAYMOND R JR
HITZELBERGER GEORGE
HODGES DAVID LAWTON
HODGES WILLIAM JEFFREY
HOLCOMB JOHN PALMORE
HOLDWAY DAVID KEITH
HOLLEY ROBERT STANLEY III
HOLMAN SAMUEL L
HOLMES RONALD EUGENE
HOLTON LOUIS ALEXANDER JR
HOOK CHARLES WAYNE
HOOVER MICHAEL JAMES
HOPKINS AARON MILTON
HOPKINS DANNY LEE
HOPKINS WILLIAM KENIS
HORTON ROBERT BERNARD
HOUSE WILLIAM HANDSOME
HOUSE WILLIS FRANCIS

HUBARD THOMAS CARR JEFFER
HUGHES ERNEST JOSEPH
HUGHES MARION BENNETT JR
HULINGS WALTER VINCENT
HULLIHEN IRA HENRY
HUMPHREY RICHARD DAVID
HUSKEY STEPHEN JOSEPH
HUSSMANN KURT CHRISTOPHER
HUTZELL JOHN FRANKLIN
IMLER HAROLD EUGENE JR
IODICE TULLIO PATRICK JR
IRELAND PHILLIP EARL
ITALIANO HARRY RICHARD
ITZOE ROBERT ANTHONY
JACKSON ANGUS N
JACKSON CARROLL THOMAS
JAMACK CHARLES ANTHONY
JAMES FRANKLIN THEODORE
JAMES PERRY DEAN
JEFFERSON ROLAND IRA
JENKINS CECIL RAYMOND
JENKINS PHILIP PAUL
JENNINGS DONALD MELVIN
JENNINGS JAMES LEWIS
JOHNS FRANK HOWARD
JOHNS VERNON ZIGMAN
JOHNSON DAVID RUDOLPH
JOHNSON ENOCH
JOHNSON EUGENE MELVIN JR
JOHNSON JOHN HENRY JR
JOHNSON MICHAEL NEAL
JOHNSON ROBERT LEE
JOHNSON WILLIAM FRANK
JONES DONALD ERNEST
JONES DUBOIS ROBERT
JONES EVERETT SORRELL
JONES HOWARD LAWRENCE
JONES JAMES HOWARD
JONES KENNETH ROLAND
JONES LLOYD EDWARD
JONES MICHAEL EDWARD
JONES THOMAS EDWARD JR
JONES THOMAS WELDON
JOYCE JOHN MULLEN
JULIA JON ALBERT
JUSTICE EVERETT EUGENE JR
KALB LOUIS WILSON
KANE CHARLES WILLIAM
KEENEY JOSEPH FRANK
KELLER WAYNE ARNOLD
KELLEY JOSEPH HOWARD
KELLY JAMES MICHAEL
KENNEDY JAMES JR
KENNY RONALD MICHAEL
KERCOUDE ANTHONY KONSTANT

KESLING RONALD LEE
KESSING THOMAS EDWARD JR
KIDD MICHAEL LOU
KIEFFER WILLIAM LEWIS JR
KILCULLEN THOMAS MICHAEL
KILLMON FREDERICK RUSSELL
KIMMELL GEORGE SAMUEL
KINCAID BARRY EDWARD
KING HAROLD JUNIOR JR
KING ROBERT LEE
KING THOMAS RAY
KING WOODROW WILSON JR
KIRK ARNOLD DAVID
KIRK DAVID GRANT
KLINE HAROLD FRANKLIN
KLINE MARK LEE
KLIPPEN ARTHUR G
KNADLE ROBERT EDWARD
KNIGHT ORVILLE LEE
KOCH EDWARD STEPHEN
KOEHNE RODNEY HOWARD
KOON GEORGE KENNETH
KORPISZ ANTHONY JOSEPH JR
KOZLOWSKI JAMES MICHAEL
KRALICK KENNETH DONALD
KRAMER HOWARD MORRIS
KRAMER JOHN DAVID
KRANTZ FRANKLIN JOSHUA JR
KRAUSSER ALBERT OTTO
KROM KENNETH LIONEL
KRUG LINWOOD BROOKS
KUNSMAN LEONARD PAUL JR
LAMB EDWARD ALAN
LAMBIE JOHN ALOYSIUS JR
LAMM JONATHAN LEE
LAMORTE ARTHUR WILLIAM
LANCASTER KENNETH RAY
LARMAN CHARLES WILLARD
LAS HERMES PHILIPPE LUC
LAWRENCE GORDON LEE
LEACH JAMES KENNETH
LEAF JAMES WILLIAM
LEE CHARLES THOMAS
LEINO GLENN KARL
LEONARD LEROY EDWARD
LEONARD PAUL AUSTIN
LERNER ROBERT HENRY
LETMATE GEORGE CAROLL
LEWIS ANDREW LEON
LEWIS CHARLES HUGH JR
LIGONS RAYMOND
LINN JOHN HOLMES
LIPINSKI VERNON RAYMOND
LIVERMAN JOHN CLARENCE
LLOYD FREDDIE GEAN
LOCKARD ALAN CARROLL

LONG JAMES MCKINLEY
LONG JOE
LONG THOMAS CALVIN JR
LORBER DONN MICHAEL
LOSPINUSO JAMES
LOVE JAMES EDWARD
LOVE KENNETH HARLEN
LOWERY RICHARD HOMER
LUCAS ALLEN LEE
LUMSDEN WILLIAM WAYNE
LUNDELL WAYNE THOMAS
LUTZ WILMER THOMAS
LYTTON BALFOUR OLIVER JR
MAC IVER NEIL KIRK
MAC LAUGHLIN DONALD C JR
MAC VICKAR JAMES S JR
MACHEN ARTHUR WEBSTER III
MACK EARL
MACKLIN RAYMOND LOUIS
MACY JAMES ROBERT
MAGSAMEN FREDERICK JOHN
MANELLO FRANK RONALD
MANGUM THEODORE EDWARD JR
MANNION AUGUST GORDIAN JR
MARCH FREDERICK LUTHER
MARSH LEE ERNEST JR
MARTIN HAROLD DOUGLAS
MARTIN ROBERT HARRISON JR
MASON CHARLES GILBERT
MASON RAYMOND LEROY
MASON ROBERT DAVID
MASSIE GEORGE EDGAR
MATTARO DONALD JAMES JR
MATTINGLY GEORGE MICHAEL
MATTINGLY HARRY ALBERT JR
MAY DAVID MURRAY
MAZZA ROBERT WILLIAM
MC ALLISTER DONALD C JR
MC ARTHUR JEROME DANNIE
MC CARTHY PHILIP JAMES
MC CARTHY THOMAS WELLER
MC CLANAHAN DONALD LEE
MC CORKLE CHARLES THOMAS
MC CREARY STANLEY EUGENE
MC DANIELS JOHNNY ANDERSO
MC DERMOTT JOSEPH F III
MC DONALD ROY LAWRENCE
MC GINN JOHN ARTHUR
MC GOVERN JAMES GERALD
MC GOWAN WILLIAM LEWIS

MC LEAN ALEX LEON
MC MAHON JAMES EDWARD
MC MAHON THOMAS JOHN
MC ROBIE NORMAN WAYNE
MEEHAN RICHARD WOODS
MELLO ANTHONY JOSEPH
MELTON EARL JR
MENTZER ROBERT EDWIN JR
METZKER FRANKLIN H
MIDDLEBROOKS CARL MASON
MIGNINI WILLIAM DOUGLAS
MILBERRY RUSSELL E
MILLER CHRISTOPHER J J
MILLER CLARENCE DALE
MILLER JOSEPH JOHN JR
MILLER VERNON JOSEPH JR
MILLS ROGER BERTHA
MISKOWSKI EDWARD ANTHONY
MISSAR JOSEPH CYRIL JR
MITCHELL LONNIE WAYNE
MOFFITT THOMAS CARROLL D
MONNETT LEONARD ALLEN
MOORE JAMES JR
MOORE MAURICE HENRY
MOORE THOMAS WOODROW
MOORE WESLEY RICE JR
MORGAN LEONARD
MORRISON RICHARD KEITH
MORRISSEY JOHN DENNIS
MORSE HARRY MADISON
MORTIMER EDWARD LEWIS JR
MOSGROVE JAMES MAURICE JR
MOTE TERRY ALAN
MOULDEN JOHN
MOULTRIE OXLEY CARRINGTON
MUIR THOMAS WAYNE
MUIR THOMAS WILSON
MULCAHY JOHN EDWIN CHARLE
MULKEY HERBERT EUGENE JR
MURPHY FREDERICK WILLIAM
MURRAY HARRY WALTER
MUSE MARIO FOWLER JR
MUSTAIN JERRY WAYNE
MYERS GEORGE LEE
NEEF FREDERICK RICHARD
NEVILLE ANTHONY ANDREW
NEWCOMB CLARENCE MATTHEWS
NEWMAN GEORGE KENNARD
NEWMAN JOSEPH ERNEST
NICHOLS LARRY DONALD
NICKLOW DANNY EUGENE
NIELSON JOHN LEIF
NIEWENHOUS GERALD E JR
NITSCHE RICHARD EDMUND JR
NOETZEL WILLIAM WESLEY
NOHE JOSEPH EDWARD JR
NOMM TOIVO BERNHARD
NORRIS LINZA
NORTON GREGORY BERNARD
NOVELLO FRANCES F
NOVOSOD RAYMOND ORITIZ
NUTWELL JOHN SYLVESTER
O CONNELL ROBERT GENE
O KEEFE ROY TULANE
O SHELL DON THOMAS
OAKLEY WILLIAM LYNN
OGLESBY RONALD
OKANE JAMES BRUCE
OLDS ERNEST ARTHUR
OLIVER TONY SYLVESTER
ONEILL JAMES TIMOTHY
ORASH MICHAEL WILLIAM
ORETO JOSEPH ANTHONY
OSBORNE DAVID FRANKLIN
OSBORNE DAVID WILLIAM
OSBORNE DONALD GWYN
OTT RICHARD DEANE
OWEN DEAN GILMAN
OWENS GEORGE ADAM
OWENS THOMAS BREVARD
PAINTER ROBERT ALBERT JR
PARKER JAMES ALLEN
PARKER WILLIAM THOMAS III
PARKS DONALD JERALD
PARSONS CHARLES WALTER

PASQUALUCCI EMIDIO
PATZWALL JAMES GEORGE
PAYNE EUGENE JEROME III
PAYNE KYLIS THEROD
PAYNE LOUIS SR
PAYNE WALTER FRANKLIN
PEARSON NORMAN JAMES
PEARSON RUDOLPH
PEAY DOUGLASS FRANKLIN
PEGG DAVID BURTON
PENDER JOHN FRANCIS
PENNINGTON PAUL PATRICK
PENSON DANIEL L
PERKINS CALVIN MOORE
PERKINS WILLIAM ARTHUR JR
PERRON NORMAND PAUL
PHILLIPS BENJAMIN F JR
PHOEBUS FREDERICK ALLEN
PIACENTINO MICHAEL ALLEN
PINKNEY HARVEY TYRONE
PITTINGER CHARLES ROBERT
PLUNKARD JOHN FRANCIS
POE JOHN RAYMOND
POSEY ROBERT LEE JR
POSTORINO ERNEST
POTTS ROBERT JAMES
POWELL GEORGE THOMAS
POWELL RAYMOND JR
POWERS VERNIE HOMER
PRATHER JAMES W
PREIS MARK JOSEPH
PRENDERGAST ARTHUR ONEILL
PRICE MILLARD ERNEST JR
PRICE RUSSELL LEE
PRICE WILLIAM EUGENE
PRITT THOMAS EUGENE
PROCTOR FRANK MAURICE
PROPST WILLIAM EARL
PRUITT FRANCIS JOHN J
PRYOR THOMAS WILLIAM
PULLIAM CHARLES AUBREY
PULLIAM ERIC VINCENT
PUMPHREY EDWIN HOLLAND
PURKEY JAMES PAUL
PURWIN ANTONI BOGUSLAW
QUAITE DANNY JOE
QUESENBERRY BOBBY RAY
QUESENBERRY JOHN QUINCY
RAMP DAVID
RANDOLPH MICHAEL JAMES
RANSON MELVIN RENSELLAER
RASH WILLIAM GEORGE
RATCLIFFE CARL JR
RATLIFF EVERETT DUEL
RAWLINGS BENJAMIN JOSEPH
REDDING CHARLES V III
REED DAVID ALAN
REGAN PHILIP THOMAS JR
REHBERGER CHARLES GEORGE
REID JAMES ALFRED
REILLY WILLIAM F III
REYNOLDS LEVI RAY
REYNOLDS SHERWOOD
RHODES CURTIS ALLEN
RICE JERRY DAVID
RIETSCHY EDWARD CHARLES
RIFFLE JOSEPH HENRY
RINGGOLD LAWRENCE L JR
ROBERTS JOHN WILSON III
ROBERTSON GEORGE LORD
ROBINSON CLARENCE JR
ROBINSON CLINTON CURTIS
ROBINSON GEORGE
ROBINSON LARRY LEE
ROBINSON RALPH LEWIS
ROBINSON SAMUEL PERCELL
ROBINSON WILLIAM D JR
ROCKENBAUGH WAYNE M
RODENBERG JOHN FREDERICK
ROEHMER ROBERT PAUL
ROGERS BOBBY DALE
ROHLEDER DONALD WILLIAM
ROLFE MICHAEL DUANE
ROLLINS WADE HAMPTON
RONNEBERG HUGH JULIUS
ROSE CHARLES WILLIAM
ROSS WILLIE JAMES
ROWE RUSSELL ALLEN
ROWLEY JOSEPH PATRICK
ROY CLIFTON DOUGLAS

RUBY BLANE MARKWOOD
RUCKER RICKY LEE
RUDD CHARLES NIVEN
RUFF THOMAS VALENTINE JR
RUHL ROBERT WAYNE
RUPPERT FRANCIS GROVER
RUSSELL BERNARD
RUSSELL DAVID ALLEN
SAKELLARIS MICHAEL GEORGE
SANBOWER RONALD LEE
SANDERS STANLEY
SANDS JOSEPH GREGORY
SANSBURY RICHARD H
SAPP ALFRED GEORGE SR
SAUNDERS RONALD
SAUSE BERNARD JACOB JR
SAVAGE MORGAN ELBERT
SAVANUCK PAUL DAVID
SAVILLE JOHN DERWOOD JR
SAXON FRANK ROBERT
SAXON JAMES RUSSELL
SCARPULLA FRANK MARK JR
SCHAAF RICHARD ALLAN
SCHAAF WILLIAM JOHN
SCHAFER GARY RAY
SCHAP FRANK JOSEPH
SCHARON ROBERT E III
SCHETTLER HARRY ROBERT
SCHINDLER THOMAS JAMES
SCHLOTT DENNIS GUY
SCHROEDER GEORGE H JR
SCHROYER LAWSON J III
SCOBEL UWE-THORSTEN
SCOTT CHARLES LOUIS JR
SEEKFORD DANIEL LEONARD
SELAK JOHN RAYMOND
SELBY ROBERT B
SELDON DAVID SCOTT
SESSUMS KENNETH BRUCE
SETZER JERRY PHILIP
SEWELL JOHN FRANCIS JR
SEWELL WILLIAM JERRY
SHALLER WILLIAM HOWARD
SHANKLIN ROY EDWARD
SHARP JOHN DAVID
SHAY DONALD EMERSON JR
SHEGOGUE ROBERT STEPHEN
SHELTON HAROLD DAVID
SHERMAN STEVEN ROSS
SHIELDS ROBERT HAZEN II
SHIELDS STEPHEN EDWARD
SHOCKLEY DON LEE
SHORTER JOHN JOSEPH
SHUCK RICHARD LEE
SIGLER ADRIAN EDWARD
SIMMETH MAXIMILIAN HEINRI
SIMON RALPH
SINGLETON EDWARD JR
SIRBAUGH THOMAS EDWARD
SKINNER RICHARD AARON
SLATER FREDDIE LEON
SLAUGHTER WILLIAM A JR
SMALLWOOD JAMES FRANCIS
SMALLWOOD JIMMY ANDREW
SMAY ATLAS JASPER MORENE
SMITH ALFRED DOUGLAS JR
SMITH ALLEN THOMAS
SMITH BARRY WAYNE
SMITH CHARLES DANIEL
SMITH EUGENE WILLARD
SMITH JACK HOWARD
SMITH JOSEPH RAYMOND
SMITH RALPH JAMES
SMITH RICHARD CLIFTON
SMITH ROBERT EARL JR
SMITH RUSSELL FRANCIS
SMITH SAMUEL WALLACE
SMITH THOMAS ALEXANDER
SMITH VICTOR ARLON
SMITHSON PAUL WINTHROP
SNEAD BERNARD JAMES JR
SNODGRASS WILLIAM LEONARD
SNYDER RODGER CLAYBORN
SNYDER TERRANCE LEE
SOCHUREK FERDINAND J III
SOLTYS MICHAEL THADDEUS
SPARE WAYNE JOHN
SPARENBERG BENARD JOHN
SPATES WILLIAM RICHARD JR
SPRIGGS OTHA THOMAS JR
SPUDIS RONALD ANTHONY

SQUARRELL SAMUEL LUVENE
SROKA STEPHEN EUGENE
STACY WILLIAM ARTHUR JR
STAFFORD HAROLD RICHARD
STAMPS OLIVER CLIFTON
STATES JOHN WAYNE
STEFFE MICHAEL WILLIAM
STEGER DAVID NAYLOR
STEGMAN THOMAS
STEINER CHARLES THOMAS
STEINER LARRY ALLEN
STEINKIRCHNER KENNETH M
STEVENS HOWARD STANLEY
STEVENS JOSEPH NELSON
STEVENS WAYNE ALAN
STEVENSON LAWRENCE EDWARD
STEWART KENNETH ALAN
STEWART WILBERT JR
STEWART WILLIAM JAMES
STOCKMAN DAVID LYNN
STOCKS WILLIAM REED
STOKER RONALD EDWARD
STONE WILLIAM MARVIN JR
STONEKING DANNY MIRE
STONESIFER HARRY NELSON
STREEKS FRANK MORRIS JR
STROUD ALLEN RALPH
SUDLER DERRICK
SUMMERS JOHN THOMAS III
SUTTON EDMOND CEASAR
SWAB RICHARD EUGENE
SWAIN ROBERT HATCHER
SWIFT EUGENE EDWARD
SZABO ISTVAN
SZEKELY AKOS DEZSO
TAPP MARION NEAL
TASKER KENNETH EARL
TAYLOR JAMES RANDEL
TAYLOR MICHAEL PATRICK
TAYLOR RALPH LEE
TAYLOR RAYMOND NOVELL
TAYLOR ROBERT EMERSON
TAYLOR THEODORE F JR
TEARL MARK FRANCIS
TESAURO JOHN APOLLO
THOMAS BENJAMIN ANDREW
THOMAS JOHN CHARLES
THOMAS ROBERT JOHN
THOMPSON JOHN PATRICK
THOMPSON JOHN WALTER
THOMPSON STEPHEN MICHAEL
THORNE JOSEPH CLAYTON JR
THORNHILL WILLIAM JOHN
TIBBS EUGENE COSTELLA
TILGHMAN BENJAMIN
TINE JOHN RICHARD
TINSON PAUL DRAKE
TITUS DONALD ROBERT
TOLLIVER LARRY LEE
TOMIKEL DAVID HAROLD
TORRINGTON THOMAS JACOB
TRACY DOUGLAS LEE
TRAIL ROBERT HILL III
TRAYLOR MARTHELL JR
TREMBLAY JAMES ALLAN
TUCKER WILLIAM EUGENE JR
TURNER CHARLES HERBERT JR
TURNER THOMAS GAINES
TUROWSKI JOSEPH MARION JR
TWIGG JOSEPH RICHARD JR
TWIST ROBERT JAMES
TYLER ADOLPHUS NORWOOD
UMSTOT CLARENCE EDWARD
UNDERWOOD FRANKLIN W JR
VADEN ROBERT WILLIAM
VAN DANIKER JOSEPH MICHAEL
VARNEY WILLIE ROSS
VAUGHAN JAMES ODELL
VAUGHT HAROLD TIMOTHY
VON KLEIST AUSTIN RICHARD
WAGNER JAMES JOSEPH
WAIDMAN WILLIAM HERMAN JR
WALINSKI BERNARD GORDON
WALKER LINWOOD ALFERONIA
WALKER RICHARD LEE
WALSH ROBERT DALE
WARD ALEXANDER KEARNEY

WARD DONALD ROLAND
WARD JAMES PATRICK
WARD MARK HEYWOOD
WARING GEORGE BADEN
WARK CARLISLE OGDEN JR
WARNICK JERONE JAMES
WATERS ROBERT MITCHELL
WATKINS ALTON LAMOTTE JR
WATKINS HARRY LEE JR
WATKINS ROBERT JAMES JR
WATSON HARVEY RAYMOND
WATSON JAMES ANTHONY
WATSON PAUL EDWARD
WAXMAN TEDDY
WEBSTER FRANKLIN
WEHNER BRIAN CHARLES
WEISMAN DONALD EUGENE
WELCH DAVID ELMER
WERNSDORFER GERALD FRANCI
WERTMAN JOHN THOMAS
WEST DALLAS ARNOLD
WEST JOHN THOMAS
WHARTON HENRY MARVIN JR
WHITE THEODORE G JR
WHITEFIELD CHARLES ELMER
WHITLEY ARSELL
WHITTINGTON JOHN HEZEKIAH
WHOOLERY TRACY LEE
WIGFALL NEOPOLIS
WILEY MICHAEL RAY
WILFONG ROGER DALE
WILK WILLIAM ANTHONY
WILLARD ROBERT LEROY
WILLIAMS CHARLES
WILLIAMS RAYNER EDWARD
WILLIAMS SAMUEL LOUIS JR
WILLIAMS WALTER DOUGLAS
WILLINGHAM JOHN DAVIS
WILLS FRANCIS DESALES
WILSON IRVING MCKINLEY JR
WILSON JAMES EDWARD
WILSON JAN FRANKLIN
WILSON JOSEPH
WILSON LORNE JOHN
WILSON MONTY NORRIS
WILSON VIRGIL HENRY JR
WIMBROW NUTTER JEROME III
WIMMER WILLARD ALVON
WINGFIELD ALBERT GREEN JR
WINKLER DAVID DE SALES
WINSTON ALVESTER LEE
WINSTON THURMAN WILLIAM
WINTERMOYER TERRY
WISE EDWARD JOSEPH
WISNIEWSKI DAVID
WITHERSPOON CHARLES E
WOJCICKY JOHN LEO
WOOD BERTRAM JR
WOODBURN LARRY ALBERT
WOODLAND THOMAS S JR
WORTH JAMES FREDERICK
WRIGHT ALBERT FLOYD JR
WRIGHT DAVID IRVIN
WRIGHT DAVID LEO
WRIGHT DONALD LEE
WRIGHT HOWARD EUGENE
WRIGHT HOWARD OLIVER JR
YEAGER JOHN WILLIAM
YEAGER MICHAEL JOSEPH
YORKER ROBERT D
YOUNG DONALD EARL
YOUNGER HOWARD JAMES JR
YOUNGHAM JAMES DOMINIO
YOUNGKIN ANDREW WINTER JR
ZIEGLER DAVID BARTELS
ZIMMERMAN SANDY JR
ZIMMERMAN WILLIAM E JR
ZUMBRUN JAMES HENRY

Photo by William Roach & others

National Vietnam Veterans Memorial

5 Henry Bacon Dr NW, Washington, DC 20004

The National Vietnam Memorial Wall in Washington D.C. honors the US men and women who fought bravely for the US during the Vietnam War. It is situated on three acres of land in the US capitol and features several different monuments besides the memorial wall.

In April of 1979, the Vietnam Veterans Memorial Fund Inc. became a nonprofit organization and one of its first actions was to set up a memorial for the veterans of the Vietnam War. This project was pushed forward thanks to a wounded Vietnam War veteran named Jan Scruggs. Scruggs gathered support from other veterans including Navy chaplain Arnold Resnicoff who helped raise over $8 million in private donations.

It took another year for a site for the memorial to be found and approved by Congress. The site that was finally selected was the ground where a former World War I munitions building stood. At this time, a design competition was also announced in order to find the best design for the memorial. Over 1,000 people submitted designs for a chance to create their masterpiece and win $50,000 and the selection committee narrowed the list down to 232 and then down again to 39 before selecting the 1,026 entry submitted by artist Maya Lin.

Lin's design stirred up some controversy due to its dark-colored design, but after some concessions were made on both sides of the table, the groundbreaking for the memorial finally took place in March 1982. The highly reflective stone was imported from Bangalore, Karnataka, India and the cutting was done in Barre, Vermont. The typesetting for the names was done in Atlanta, Georgia by Datalantic before being sent to Memphis, Tennessee where they were etched into the stone.

In October of 1982, a flagpole was positioned off to the side of the wall near bronze sculptures. The whole memorial was dedicated a month later following a march by surviving Vietnam War veterans. The new total of dead and missing on the wall is 58,307. The missing have been denoted with a plus sign, if they are found to be dead this is changed to a star.

The Names of those from Washington DC Who Made the Ultimate Sacrifice

AARON RICHARD ALAN
ADAMS RUSSELL LEE
ADLER TOM ROBERT
ALEXANDER ROBERT
ALLEN DOUGLAS MELVIN
ANDERSON WENDELL WARREN
ANDREWS GEORGE ROBERT
ARTIS VERNON DARYLE
ASMUSSEN GLENN EDWARD
ATKINS JOSHUA ABRAHAM III
AUSTIN PAUL JASPER JR
AVERY JAMES LINWOOD
BARNES ALFRED JR
BATTLE WILLIAM ALFRED
BEAMON THEODORE M
BOUKNIGHT CALVIN
BRADSHER ROBERT JR
BRETT ROBERT RAYMOND
BROOKS RAYMOND AUGUSTA
BROWN CHARLES WILLIS E
BROWN DARIUS LLEWLYN DEMA
BROWN JACK MONTGOMERY JR
BROWN MELVIN BERNARD
BROWN NICHOLSON
BROWN RICHARD ALLEN
BROWN ROBERT MAURICE
BROWNING LEROY JACK
BRYAN BLACKSHEAR M JR
BUNN BENJAMIN JR
BURGE FREDERICK
BUTLER DEWEY RENEE
BUTLER WINSTON JR
CABNESS DERRICK CLIFFORD
CALE JAMES MARTIN
CAMP JOHN HOLMES JR
CARTER DONALD ODELL
CARTER REGINALD F JR
CHAMBERS ERNEST L JR
CHANEY ELWOOD DAVID JR
CHARITY EDWARD JR
CLEGG LESTER HOWARD
CLEMENCIA JEAN ROGER JR
COFIELD EDWARD CHARLES J
CONNERS RALPH WILSON JR
COOKE CALVIN COOLIDGE JR
COOPER JOHN OLIN III
CRITTENBERGER DALE J
CROOM MARION JR
CUNNINGHAM LARRY ALFONSO

DARDEN CLAUDE JR
DARGAN JEFFREY LYNN
DAVIS ERLE FLETCHER
DAY CALVIN SYLVESTER
DEAN THOMAS NELSON
DENNIS THADDEUS
DICKERSON DAVID DOWNING
DICKINSON ROBERT CHARLES
DODSON PAUL ALONZO SR
DORSEY JAMES VERNON JR
DOUGANS EMMETT ARTHUR
DOWNS EDWARD JOSEPH
DU CHARM PAUL MEDORE
DUCKETT JOSEPH L JR
DUNAWAY JAMES ROBERT
DURHAM JAMES WILLIAM JR
FARLEY ANDREW SIMMONS JR
FELLS WILLIAM HENRY
FIELDS ANTHONY THOMAS
FLOYD RONALD JAMES
FOREMAN DWIGHT GARY
FORMEY JERRY BERNARD
FOSTER GERALD ANTHONY
FRALEY LANDER RAY
FULLER JOHNNIE CHESTER
GARDNER RICHARD GEORGE
GARNETT LEON JR
GASTON JOHN RUFUS JR
GELLER ROBERT EARL
GEORGE JOHN WESLEY
GOODEN JOHNNIE
GORDON GERALD ELLIOTT
GORDON JAMES BERNARD L JR
GORDON WILLIAM SAMUEL
GOVAN ROBERT ALLEN
GRAHAM HENRY HERNDON
GRAVES FRANK
GRAY RAYMOND HENRY
GRAY ROSCOE CONKLIN JR
GREEN MELVIN LOUIS
GREEN NORMAN MORGAN
HAIRSTON JIMMY LEE
HAMILTON LEON GONZA JR
HARDIN RAYMOND HOWARD
HARE DAYTON LEO JR
HARGRAVES MANCE KENNETH
HENDERSON RALPH LEE
HENRY BISMARK WASHINGTON
HILL ANDREW LAMAR
HILL LEROY

HILL SID
HOBBS RONALD ROBERT
HOPKINS JOHN EDWARD
HOSKINSON HARRY RONALD
HURDLE PAUL EDWARD
IDE DONALD WILLIAM
IZZARD SAMUEL JULIUS
JACKSON ALEXANDER
JACKSON ANDREW
JACKSON FREDERICK LEROY
JACKSON JEROME ELLIS
JOHNSON ADOCK VEISO
JOHNSON ROOSEVELT JR
JOHNSON STERLING HENRY
JONES JEROME MICHAEL
JONES PHILIP ALFONSO
JONES RONALD
KEMP SAMUEL LEE
KLUEVER LARRY JOHN
KOLB RONALD VICTOR
KOWALEWSKI ZYGMUNT
KURI JACK
KURTZ LLOYD NELSON
LANGLEY DAVID FRANCIS
LASKIN FRANK HOWARD
LEWIS RONALD WILLIAM
LINK ROBERT CHARLES
LOFTON CHARLES EDWARD
LUCAS JOSEPH JR
MAHONEY HARRY CURTIS JR
MASON ALVIN PERNELL
MASON CHARLES JOSEPH L
MATTHEWS CALVIN BERNARD
MAYES JAMES RUSSELL
MAZAK STEFAN
MC CLAIN HARRY
MC CORMICK RICHARD ALAN
MC GONIGAL ALOYSIUS PAUL
MC INTYRE HOMER CLEO JR
MC QUEEN CLAUDE EDWARD
MELTON DAVID LAWRENCE
MILLS KYNARD
MITCHELL KENNETH
MITCHELL WILLIE JR
MOLDENHAUER PETER JAMES
MOORE JAMES RUSSELL
MOORE JOHN OTIS
MORGAN JAMES HENRY
MURPHY BOBBY LOUIS
NALLS JOHN LAURENCE
NEELY PAUL JAMESON
NORCIA JAMES JOSEPH
O DONNELL DOUGLAS

WILLIAM
ONLEY CLAUDE ALOYSIUS
PAGE WINGFIELD JR
PALMER DOUGLAS TERRY
PEARSON RICHARD ELLSWORTH
PHILLIPS MICHAEL LEON
PIERCE HARRY W JR
PLUNKETT RAYMOND LOUIS
PONDER WILLIE LE EARL
PORTER LARRY MICHAEL
POWELL GARRY REGINALD
POWELL JAMES WILLIAM JR
PRESTON ALVIN LEWIS
PRICE JAY ANTHONY
PRICE WILLIAM SIDWAY
PRINCE RAYMOND LOUIS
PROCTOR WILLIAM AMBROSE
PUGH EVERETT CHARLES
RACCA WILLIAM
RAMOS BRINSLEY BERNARD
REED HAROLD B
REESE JAMES HARRISON
REYNOLDS KENNETH ALDERSON
RICHARDSON BERNARD MCKINL
RIDOUT CHARLES SAMUEL
RINEHART JOSEPH LESTER
RISHER DAVID HORACE
ROBINSON JOHN JACKLON
RUFFIN CHARLES NATHANIEL
RUSSELL DAVID ADAMS
SANDERS LEO MELVIN
SAUNDERS EMANUEL LAWRENCE
SCOTT GRADY
SCOTT ROBERT EUGENE
SHELTON EDWARD LEE
SHERROD LOUIS
SHOUFF JOHNNY EDWARD
SIMONS AINSLEY CUDIE
SMALLWOOD EUGENE FENTON
SMITH ALBERT MERRIMAN
SMITH ALBERT PRESLEY
SMITH CECIL RAY JR
SMITH EDWARD SPENCER
SMITH FRED DOUGLAS JR
SMITH GEORGE JOHN JR
SMITH JOHN ROBERT JR
SMITH ROBERT WILBUR
SNIPES EDDIE WENDELL

SQUARE GREGORY
ST ONGE MICHAEL JOSEPH
STAFFORD HENRY LEE
STALLINGS FRANKLIN DELANO
STANLEY EARL
STEPHENS LLOYD ISAAC
STEWART JACK THOMAS
STEWART ROBERT ALLAN
STEWART TOMMY LANE
STRICKLER JOHN CLINE JR
TAYLOR DARRYL WADE
TAYLOR RANDY LEE
TEGLAS GEZA
THOMPSON JEROME
TIGNER LEE MORROW
TOLER JOSEPH BERNARD
TOMPKINS JAMES ERVIN
TOOGOOD MANSFIELD M JR
TURNER RICHARD
TYLER SYLVESTER GEORGE
WALKER JOHNNIE
WALL THOMAS JR
WALSH DAVID WILLIAM
WARNER HENRY LUKE III
WARREN MANASSEH BROCK
WASHINGTON ARTHUR DOUGLAS
WASHINGTON CHARLES L
WATKINS GERALD EDWARD
WATKINS KENNETH MAURICE
WHITE WILLIAM ERNEST JR
WHITTINGTON PERRY LEE
WILLIAMS COLERIDGE JR
WILLIAMS LAWRENCE
WILLIAMS LAWRENCE H JR
WILLIAMS RAYMOND LEON
WILLIAMS ROBERT LEE
WILLIAMS TIMOTHY LEROY
WILSON CARL RICHARD JR
WINDER WALTER RIDGEWAY J
WORTHINGTON JAMES AUTHOR
WRIGHT ANDREW SAMUEL
WRIGHT JAMES GREER SR
ZEIGLER WILLIAM HENRY

109

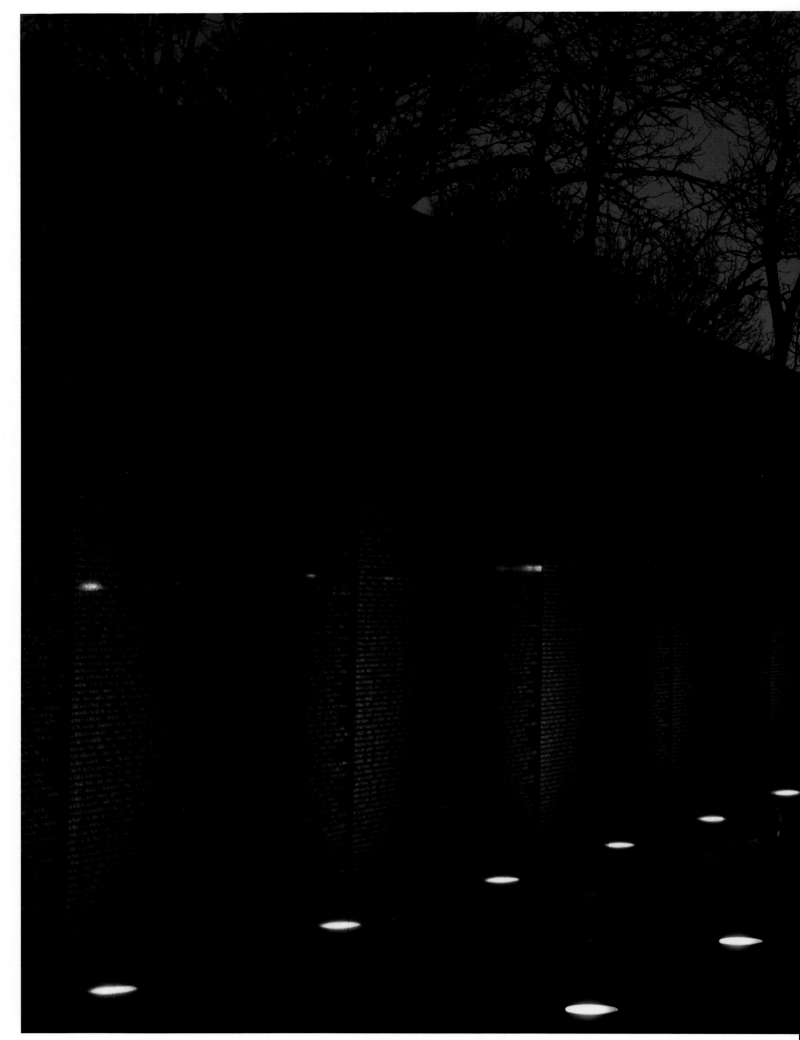

Photos by Debbie Helbing

Vietnam Veteran's Memorial

Biotech Park Area, Worcester, MA 01605

The Vietnam Veterans Memorial is located in Worcester, Massachusetts and was created to honor the men and women who sacrificed their lives in the Vietnam War. The memorial sits on four acres of land that includes the memorial itself, walking paths, and a serene pond. The site offers visitors a quiet place in nature to reflect on the troops. While it honors those specifically who were lost in the war, at its core it's dedicated to all of the service members who served from 1955 to 1975.

The Vietnam Veterans Memorial started in the 1980's when the Massachusetts Vietnam Veterans Memorial Fund was formed in Boston. The fund looked to build a veterans memorial at the Charlestown Navy Yard by September 1988, but it never manifested. When that effort failed, the Vietnam Veterans Memorial Trust came about in Worcester. This group was able to settle on a location for the memorial in Green Hill Park as well as the architectural company Harby, Rogers, and Catanzaro to design and build the memorial.

Sadly, fundraising efforts failed to meet the requirements for the memorial, but the Massachusetts Vietnam Veterans Memorial Committee was created to complete the fundraising. Through the MVVMC and four Massachusetts politicians, funding was secured through the state legislature to the tune of $1.4 million. Once funding was acquired, the project was able to begin and was finally delivered and dedicated on June 9, 2002.

The Vietnam War memorial is made up of multiple parts including the Place of Flags, the Place of Words, the Place of Names, and the War Dog Monument. The Place of Flags is a landscaped circular memorial that features the American Flag, the Massachusetts State flag, and the POW/MIA flag. It is the first section of the whole memorial that visitors see and during special ceremonies the flags of the five US military branches are also displayed. There is also a granite retaining wall that features an inscription that states "Vietnam – 1955-1975." The Place of Words is a section of the memorial that features actual text from letters written by fallen soldiers that were sent home from Vietnam. The words were etched into the granite of the monument. The Place of Names features the names of the all state service members who were killed in action or as a result of wounds suffered during the Vietnam War. Finally, the War Dogs Monument was added to the original memorial in 2011 to honor the war dogs who served alongside the service members in Vietnam.

The Names of those from the state of Massachusetts Who Made the Ultimate Sacrifice

AARON CHARLES EDWARD
ABOLTIN RICHARD D
ADAMS EDDIE MARTIN
ADAMS GEORGE FRANCIS
ADAMS PETER ROBERT
AGRI JOSEPH JOHN JR
AGRI SALVATORE JR
AHERN BRIAN PAUL
ALAMED WILLIAM ROBERT JR
ALBERT PETER
ALBERTINI JAMES CHRISTOPH
ALBRECHT JOSEPH ALFRED
ALDAM KEVIN GERRY
ALFANO RODNEY ARTHUR
ALLARD VAL GENE
ALLEN EVERETT ALBERT
ALLEN FRANCIS MONROE JR
ALLEN WAYNE CLOUSE
ALLERBY MILTON RICHARD JR
ALLISON ARTHUR RICHARD
ALMEIDA EDWARD JOSEPH
ALMEIDA RICHARD HENRY
ALMEIDA RUSSELL VIVEIROS
AMARAL MATTHEW PERRY III
ANDERSON CARL EDGAR
ANDERSON EDWARD
ANDERSON HARRY WILLIAM JR
ANDRADE EDWARD JAMES
ANDRESEN SCOTT FREDERICK
ANDREWS LAWRENCE THEODORE
ANGELL ALAN FRANCIS
ARAUJO RUDOLPH ERNEST
ARCAND DONALD LEONARD
ARCHER RICHARD CHARLES
ARENS FREDERICK V JR
ARGY EDWARD WILLIAM
ARKOETTE PETER ALLAN
ARRUDA RICHARD HATHAWAY
ARSENAULT RICHARD ROLAND

ASHTON DONALD MILLARD JR
ASHTON NORMAND JOSEPH JR
ATKINSON FREDERICK GEORGE
AUCOIN ROBERT JOSEPH
AUSTIN MICHAEL PAUL
AVERY ALLEN JONES
BABIN JACOB BENEDICT JR
BADCOCK ROBERT
BAIRD JAMES STEPHEN
BAKER KENNETH ALVIN
BAKER LINWOOD LEE
BALDWIN TERRY LYMAN
BARNES JOHN ANDREW III
BARRY GEORGE FRANCIS JR
BARRY JAMES MICHAEL
BARTLETT ARTHUR FRANCIS
BARTLETT CHARLES DENNIS
BASILIERE RALPH
BASTARACHE FIDELE JOSEPH
BATCHELDER WILLIAM KIMBAL
BAXTER BRUCE RAYMOND
BAZZINOTTI CHARLES A
BEALS LAWRENCE FREDERICK
BEAN CHRISTOPHER JOHN
BEAUBIEN WILLIAM ALEXIS
BEAUFORD WILLIS JR
BEAULIEU NORMAND LOUIS
BEAUREGARD KENNETH EDWARD
BEDROSIAN DAVID PETER J
BEECY GREGORY WILLIAM
BELANGER GEORGE HENRY
BELCHER ROBERT WINSLOW
BELLERIVE DAVID LESLIE
BELLINO PAUL GEORGE
BENGTSON FRANK WALTER
BENJAMIN GARY THOMAS

BENJAMIN KENNETH ROGER
BENNETT THOMAS EVANS
BENSON DENNIS GUY
BERESIK EUGENE PAUL
BERG RALPH RUSSELL
BERGERON ROBERT JAMES
BERGERON SIMEON JOSEPH A
BERGIN GERARD FRANCIS
BERNARD VINCENT
BERRIO JOHN ANTHONY
BERRISFORD RONALD E
BERRY ALAN WAYNE
BERTHIAUME PAUL DAVID
BERTULLI ALFRED LEON
BERUBE KENNETH ALLEN
BETTENCOURT DANIEL STEPHE
BETTENCOURT JOHN FRANCIS
BETTY CLAUDE CHARLES
BIFOLCHI CHARLES LAWRENCE
BIGELOW ROBERT FRANCIS
BINGHAM DAVID ANDREW
BISSAILLON FRANCIS HENRY
BLADES WILLIAM CEACON III
BLAIS ROBERT LAWRENCE
BLAKE DALE ADAMS
BLAKE WILLIAM H JR
BLAZONIS PETER VINCENT
BLOUGH DAVID ANTHONY
BOGACZ JAMES MITCHELL
BOIS RICHARD JOSEPH
BOLES WARREN WILLIAM
BOND FRANCIS ARTHUR
BOND RICHARD WILLIAM
BONNELL WILLIAM LAWRENCE
BONNETTE PAUL EUGENE
BORDEN LAWRENCE THOMAS
BOREY DAVID CHRISTOPHER
BORONSKI JOHN ARTHUR

BOROVICK RICHARD JOHN
BOSS ROBERT LEON
BOTT RUSSELL PETER
BOUCHARD MICHAEL PHILIP
BOUCHARD PETER JOSEPH
BOUCHARD WILLARD J JR
BOUCHET ROBERT LOUIS
BOURGEOIS WILFRID NARCISS
BOURQUE VALMORE WILLIAM
BOUSQUET ROBERT GEORGE
BOWMAN DAVID WINSLOW
BRADLEY GERALD GREGORY
BRADY MICHAEL JOHN
BRANDT RICHARD CARL
BRANN DANA E
BRAULT DENNIS JAMES
BREDBURY PETER MALCOLM
BRENNAN THOMAS JOHN
BRESNAHAN WILLIAM JOHN JR
BREWER ARTHUR LOGAN
BREWER GARDNER
BRIDGMAN CLEAVELAND FLOYD
BRINE CHRISTOPHER DAREING
BROOKS RICHARD WILLIAM
BROOKS ROBERT EVERETT
BROWN ARTHUR LEROY SR
BROWN KENNETH RAYMOND
BROWN ROGER DAVID
BROYER CLIFTON LEE
BRUGMAN PAUL FRANK
BRUNDRETTE RICHARD E JR
BRUSO RICHARD NORMAN
BUCHANAN WAVERIE HUGH
BUILAERT FRANCOIS JOSEPHI
BULGER JOHN DAVID
BUMPUS RONALD LEE
BURGESS JOHN B
BURKE JOHN MARTIN
BURKE PATRICK KEVIN

BURKE THOMAS JAMES
BURNHAM NEIL ROBERT
BURNS EARL KENNETH JR
BURNS MICHAEL CHRISTOPHER
BURRELL PHILIP EDWARD
BURT MICHAEL DAVID
BURWELL LANGDON GATES
BUSH THOMAS EDWARD
BUZZELL RICHARD HOWARD
BYRON MICHAEL JOSEPH
BYSZEK JAMES JACOB
CABRAL ANIBAL SYLVIA JR
CABRAL JAMES ANTHONY JR
CABRAL JOHN JOSEPH
CADORETTE MICHAEL JOHN
CAHILL KEVIN ARTHUR
CAHILL PAUL MATTHEW
CAHILL WILLIAM JOSEPH
CALHOUN JOHN CALDWELL
CALL DANA ROBERT
CALLAHAN DANIEL DAVID
CALLERY WILLIAM THOMAS
CAMERLENGO MICHAEL DENNIS
CAMPBELL GEORGE
CAMPBELL JOSEPH TIMOTHY
CAMPBELL WILLIAM H III
CANDEAS JOSEPH EDWARD
CANDIANO JOSEPH PAUL
CANN DOUGLAS ALLEN
CANNON WILLARD SPARKS III
CANOVA RICHARD JOHN
CAPUANO PAUL RICHARD
CARDINALI RICHARD WILLIAM
CARDONA RONALD WILLIAM
CARLSON GARY WILLIAM
CARON WAYNE MAURICE
CAROTA JOHN THOMAS
CARPENTER CLINTON R JR

115

Massachusetts

CARROLL JOSEPH FRANCIS
CARSON PAUL ROLAND
CARSON RICHARD JAMES
CARTER KENNETH ROBERT
CARTER PAUL C JR
CARTWRIGHT ROBERT
MICHAEL
CARVALHO GILBERT
CARVEN RUPERT SADLER III
CARVILLE JOHN JOSEPH
CASALE JAMES ERNEST
CASALETTO EDWIN JAMES
CASEY GEORGE WILLIAM
CASEY JAMES PATRICK
CASEY MICHAEL JAMES
CASEY THOMAS MICHAEL JR
CASPOLE RALPH WARREN
CASTELLANO SAMUEL
RODGER
CAVANAUGH EDWARD
JOSEPH
CAVICCHI JAMES HENRY JR
CERRONE JOSEPH CARMEN JR
CHACE GEORGE HENRY
CHADWICK FRANK W JR
CHAMPAGNE THOMAS
EUGENE
CHAPIN JOEL HENRY
CHARLAND ROGER OVIDE
CHASE CURTIS EDWARD
CHASE FREDDIE NICKLYS
CHASE ROBERT KENDRICK
CHASE WALTER WILLIAM
CHASSION PHILIP RONALD
CHAVES JOHN CLIFFORD
CHENEY RICHARD DANIEL
CHESLEY LEONARD GEORGE
JR
CHEVALIER HENRY ANTHONY
CHILD CHARLES CHRISTO-
PHER
CHISHOLM DAVID ANDREW
CHRISTIAN RUSSELL THOMAS
CHRISTIANSON PETER
BUGBEE
CIRIELLO BASIL LINCOLN
CLANCY WILLIAM EDWARD
CLONEY WILLIAM THOMAS III
CLOUTIER DAVID WILLIAM
COAKLEY WILLIAM FRANCIS
COCCHIARA JAMES STEPHEN
COCHRANE DEVERTON C
COGGESHALL WILLIAM AYER
COGILL PETER
COHEN GARY MARTIN
COHEN SHELDON ROBERT
COLE JOHN HENRY
COLEMAN ROBERT JOSEPH
COLLAMORE ALLAN PHILIP JR
COLLINA GEORGE WILLIAM
COLLINS BRIAN PATRICK
COLLINS THOMAS TIMOTHY
COLLOPY JOHN PATRICK
CONAXIS NICHOLAS S
CONCANNON JOHN FRANCIS
CONDON JAMES GREGORY III
CONNELLY EDWARD WALTER

JR
CONNOLLY RICHARD
GEORGE
CONNOR JAMES FRANCIS JR
CONRY DENNIS
CONSTANDE DONALD
CONSTANTINE MICHAEL
EUGEN
CONTARINO DONALD ALLEN
CONTESTABILE DANIEL J
CONWAY JOHN JAMES
COOK JOSEPH FRANCIS
COOK PETER ALLAN
COOKSON ROBERT MERLE
COOLEY WILLIAM
CORDEAU EDWARD RICHARD
CORMIER EDWARD JAMES
CORMIER EUGENE FRANCIS
CORRIVEAU GERARD
CORWIN FRANCIS HENRY JR
COSTA ROBERT JOSEPH
COTE ROBERT FRANCIS
COTTER RICHARD LANE
COUGHLIN JOHN PETER
COULOMBE FRANCIS JOSEPH
COUNIHAN MICHAEL BREN-
DAN
COUNTAWAY JOHN ALDEN JR
COUSINEAU HENRY CONRAD
COUTO JIMMIE MICHAEL
CRAVEN GARY ALAN
CREED BERNARD JAMES
CROCCO WALTER VINCENT
CROCE ROBERT JAMES
CRONK PAUL MARVIN JR
CROSBY ROBERT LEROY
CROWE DOUGLAS D
CROWE KEVIN ROBERT
CROWLEY JAMES ALLEN II
CROWTHER DONALD DAVID
CULVER PHILIP LEE
CUMMINGS CHARLES HENRY
CUMMINGS CHESTER AR-
THUR
CUMMINGS HAROLD WARREN
JR
CUMMINGS STEPHEN WIL-
LIAM
CUNNINGHAM ROBERT JAMES
CURRAN PAUL WILLIAM
CURRIER GERALD FRANCIS
CURRY JAMES JOSEPH
CURRY ROBERT LOUIS
CURTIN DONALD LEO
CURTIS FREDERICK N
CYR WILLIAM JOSEPH
CZERWONKA PAUL STEVEN
D AMATO PAUL JOHN JR
D AMICO FRANK ANTHONY
D ORSAY DOUGLAS HAROLD
DABREU DANIEL JOHN
DAIGLE JAMES CHARLES
DAIGNEAULT JOSEPH RICH-
ARD
DALEY PAUL MICHAEL
DALEY RICHARD JOHN
DALEY ROBERT F
DALEY RONALD PAUL
DALEY WALTER RALPH
DALEY WILLIAM MICHAEL
DALTON EDWARD JOSEPH JR
DALY EUGENE THOMAS JR
DANA ROGER JOSEPH
DARRIGAN RAYMOND MAU-
RICE
DASHNER GILFORD FRANK
DAVIS GREGORY CHALMERS
DAVIS RICHARD ROBERT
DAVIS RICHARD SHIRLEY JR
DAVIS STEPHEN WINFIELD
DAVIS WAYNE ROBERT
DAVIS WILLIAM WALTER JR
DE LORENZO FRED JOSEPH JR
DE LORENZO PHILIP T JR
DEANE MICHAEL LINDSEY
DEGNIS JAMES EDWARD
DEINLEIN LEONARD PETER
DELMONT JAMES LOVES
DELVERDE RONALD LEON
DEMARIS RICHARD ORIN
DEMERS RICHARD WILFRED
DEMUTH RICHARD LAW-
RENCE
DENTON RANDALL MORRIS
DEPROFIO MICHAEL ALLEN

DEROCHER FREDERICK
GEORGE
DEROSIER THOMAS ALBERT
DESCHENES THOMAS ALFRED
DESMARAIS DONALD ROGER
DESMOND JOSEPH FRANCIS
DESPER RICHARD LINCOLN
DESROCHERS ROBERT ALAN
DEVINE RICHARD DANIEL JR
DEVOE DAVID FRANCIS
DEYERMOND WARREN
CHARLES
DI BERARDINO PERRY
DI REDA ROBERT J
DI TULLIO FRANCO ANTONIO
DICK BRUCE DAVID
DICKEY THOMAS ROBERT
DOBBIN LOUIS DAVID II
DOIG DOUGLAS WILLIAM
DOLAN JAMES EDWIN
DOLAN THOMAS WILLIAM III
DOLIBER EDGAR SNOW
DONAHUE CHRISTOPHER C
DONAHUE JOHN JOSEPH
DONALDSON STEVEN ELLIS
DONATO PAUL NICHOLAS
DONDERO ROBERT ALFRED
DONOVAN PAMELA DORO-
THY
DONOVAN THOMAS STEPHEN
DORR GERALD BRIAN
DOTEN ROBERT ALAN
DOUGHTY JAMES ALDEN
DOWD LAWRENCE KENT
DOWDS ROBERT RAOUL
DOWNEY EDWARD JOSEPH JR
DOWNEY MICHAEL WAKE-
FIELD
DOYON PAUL FRANCIS
DRAKE DAVID LAWRENCE JR
DRAKE RICHARD KENNETH JR
DREW JOSEPH LAWRENCE
DRISCOLL FRANCIS MUR-
TAUGH
DRISCOLL PAUL RICHARD
DROWN DAVID ALAN
DUFAULT PAUL
DUFFETT EDWARD STEPHEN
DUFFY DANIEL BENJAMIN JR
DUFFY DONALD RAYMOND JR
DUFFY LAWRENCE RICHARD
DUHY HARVEY ALBERT JR
DULL EDWARD JAMES
DUMONT ROGER JOSEPH
DUNCAN JOHN DAVID
DUNN JOSEPH PATRICK
DUNN JOSEPH WESLEY
DUNN MICHAEL ROY
DUNNE PAUL HUBERT JR
DUNTON JAMES G
DUPERE EDWARD JOSEPH
DUPERE PAUL ANDREW
DUTTON BERNARD F JR
DWYER THOMAS RICHARD
EARLE JOHN STILES
EASTMAN ALLAN JOHN
EDGE PAUL JOSEPH II
EDMONDS JOSEPH
EGAN EDWARD THOMAS JR
EGAN STANLEY JOSEPH
ELA ALAN DAVID
ELLIOT ROBERT MALCOLM
ELLIOTT DONALD LYLE
ELLIS KENNETH WARREN
ELLISON RICHARD WRIGHT
ELLSWORTH NEIL ROBERT
EMERSON WILLIAM
EMERY STEPHEN BRADFORD
ENGLISH JAMES PATRICK
ENMAN DEVON MARDIC
ERBENTRAUT STEVEN
CHARLES
ERDELY RALPH GABRIEL
ERLANDSON DANIEL KEN-
NETH
EVANS GEORGE FREDRICK
EVANS JOHN DOUGLAS
FALARDEAU JOSEPH ERNEST
FALCO ANTONIO
FANNING MICHAEL FRANCIS
FARELLI LAWRENCE JOHN
FARRELL CHARLES DOUGLAS
FARRIS NORMAN CARL
FASSITT ERIC RICHARD

FAVUZZA LOUIS ANTHONY
FAY ROBERT JOSEPH
FEELEY EUGENE JOSEPH JR
FELL GEORGE FRANCIS JR
FERA JOHN ANTHONY
FERENCE EDWARD PAUL
FERRIS ROBERT CLARK
FERRO JAMES
FERRON FRANCIS RAYMOND
JR
FINNEY BOBBY LEE
FISHER ERIC ANDERS
FITTS RICHARD ALLAN
FITZGERALD JOHN FRANCIS
FITZGERALD JOSEPH EDWARD
FITZGERALD MARK JOSEPH
FITZGIBBON RICHARD B III
FITZGIBBON RICHARD B JR
FITZGIBBONS JOHN FRANCIS
FITZGIBBONS PAUL EDWARD
FITZPATRICK MICHAEL
THOMA
FITZPATRICK WALTER JOSEPH
FLEMING PAUL DENNIS
FLINT WILLIAM JOHN
FLOOD JOHN PATRICK JR
FLOOD THOMAS BERNARD
FLOOD WILLIAM JAMES JR
FLOYD PAUL EDWARD JR

FLUMERE KEITH MICHAEL
FLYNN GARY FRANCIS
FOLEY MARTIN FRANCIS
FOLEY ROBERT MICHAEL
FOLEY ROBERT PAUL
FOLEY ROBERT RAYMOND JR
FOLLETTE FREDERICK JOHN
FONTAINE JOHN ALBERT
FONTAINE NORMAND ED-
WARD
FOOTE PETER WELLESLEY
FORBES WALTER HENRY III
FORDI MICHAEL JOSEPH
FORGET RONALD EDMOND
FORTE RICHARD JOSEPH
FRANCIS DAVID ANTHONY
FRANK RICHARD WAGNER II
FRAWLEY WILLIAM DAVID
FRECHETTE FRANCIS GERALD
FREDA ARTHUR ANTHONY JR
FRIEL JOSEPH AUGUSTUS
FRINK PAUL JOSEPH
FRONGILLO JOHN RALPH
FROST DANA STANLEY
FULLERTON JOHN JOSEPH JR
FURTADO EDWARD JR
GAGNE DONALD
GAGNE LOUIS PHILLIP JR
GAGNE ROBERT OMER

GALE ALVIN RICHARD
GALLAGHER PHILIP S III
GALLANT FRANK JAMES
GAMBINO MICHAEL JAMES
GAMBLE PHILIP LYLE JR
GARDELLA WILLIAM KIRBY
GARIEPY ROBERT DAVID
GARRON LAWRENCE E JR
GARSIDE FREDERICK THOMAS
GASPAR ALFRED JOHN
GATES ROBERT SIDNEY
GAUDREAU CHARLES AR-
THUR
GAUGHAN ROGER CONRAD
GAUTREAU REGINALD JOSEPH
GAUVIN PETER JOSEPH
GEMBORYS JOHN CHESTER
GERO EDWARD W
GERRISH ALAN ROBERT
GERRY PETER JAMES
GHAIS TAHER FATHI
GHELLI THOMAS ALFRED
GIGNAC RAYMOND ALBERT
GILBERT HAROLD JEFFREY
GILBERT RICHARD JOSEPH
GILE JOSEPH THOMAS JR
GILLESPIE MARTIN L JR
GILLESPIE ROBERT JAMES JR
GIORDANO RALPH JOSEPH
GIROLIMON LOUIS MARIO
GIROUX RONALD
GIVEN MARTIN GEHRING
GLASSER JOHN MICHAEL
GLIDDEN RICHARD CUN-
NINGHA
GLINIEWICZ RICHARD
FRANCIS
GONNEVILLE ROBERT RO-
LAND
GONSALVES AUGUST JR
GOODWIN DANNY ERIC
GOODWIN PHILIP BENJAMIN
GORDON GLENN RAYMOND
GORMAN PAUL JAMES
GORMLEY PAUL LEO JR
GORRILL THOMAS ROY
GOTTWALD GEORGE JOSEPH
JR
GOULD WILLIAM C JR
GRADY LEO FRANCIS
GRANAHAN JOHN WILLIAM
GRANT CREIGHTON ROONEY
GRANT JOSEPH XAVIER
GRANT NORMAN WILLIAM JR
GRANVILLE RONALD LESTER
GRASSO PAUL VINCENT
GRAVES RICHARD CAMPBELL
GRAY RICHARD JOSEPH
GRAY ROBERT ROGER
GREEN ROBERT EUGENE
GREEN ROBERT PAUL
GREENE DONALD JOSEPH
GREENE KENNETH LAW-
RENCE
GREENLEAF JOSEPH GALES
GREGOIRE DAVID EDWARD
GRENHAM LAWRENCE AL-
PHONSE
GRESCH FREDERICK WILLIAM
GRIFFIN ROBERT
GRIFFIN THEDORE STEVEN
GRIFFIN THOMAS B JR
GRIGSBY MARK WELDON
GRITTE ROBERT JOSEPH
GROVER DANIEL LAWRENCE
GUERTIN DONALD ALAN
GUEST GARY RICHARD
GUIDA PAUL ANTHONY
GUILMETTE JOSEPH JR
GUIMOND PAUL DANIEL
GUNSET WILLIAM FRANCIS
GUZZETTI MICHAEL T JR
HAGERTY WILLIAM THOMAS
HAIN ROBERT JAMES
HALL DAVID COLIN
HALLETT ROBERT J
HAMEL WAYNE DOUGLAS
HAMILTON DAVID KENNETH
HAMLIN RALPH GERALD JR
HANLON GEORGE MARTIN
HANSCOM JOHN WILLIAM
HANSEN PETER MYKAL
HARDIMAN KEVIN BARRY
HARDY ARTHUR HANS

HARGETT JOHN JR
HARPER RICHARD K
HARRIMAN ALAN BATES
HARRINGTON FREDERICK E JR
HARRINGTON JOHN DANIEL
HARRIS ROBERT GEORGE
HARRISON ROBERT LOUIS
HARTLAGE JOHN PETER III
HARTLEY CHRISTOPHER
ROBER
HARTLEY ROBERT JOSEPH
HARTNETT MICHAEL GERALD
HASSEY PAUL ELIAS
HAUER ROBERT DOUGLAS
HAVEL MICHAEL DENNIS
HAYDEN JOHN JOSEPH JR
HAYES TRISTAN WHITNEY
HAZARD JAMES JOSEPH
HAZZARD FRANKLIN GEORGE
HEDGE ROBERT BLANCHARD
HELLARD RICHARD W JR
HENDERSON ROY JOHN
HENN JOHN ROBERT JR
HENNEBERRY JAMES CALVIN
HENNESSEY ARTHUR F JR
HENRY JAMES EDWARD
HENRY STEPHEN MICHAEL
HEPPLER JAMES HOWARD
HERRIN HENRY HOWARD JR
HICKEY JAMES PHILIP
HIGGINS DENNIS MICHAEL
HILL DAVID ALLEN
HILTZ JAMES FREDERICK
HINCKLEY CHARLES ALBON
HINCKLEY WILLIAM K JR
HINE GLEN DOUGLAS
HINES RALPH EARLE
HINGSTON WILLIAM E JR
HINKLEY STEPHEN
HIRTLE HAROLD HERMAN
HITCHCOCK WILLIAM F
HOBART GLENN EDWARD III
HODGE WILLIAM JOHN
HOLDEN DAVID CHARLES
HOLDEN WILLIAM DAVID
HOLMES DAVID HUGH
HOLSTER TIMOTHY
HOLT ROBERT ALAN
HORAN LEO JOSEPH
HOSNANDER CARL E
HOULIHAN JOHN RICHARD
HOWLETT NORMAN LOCKE JR
HUBBARD CORNELIUS
FRANCIS
HUBBARD GLEN DAVID
HUBICSAK FRANK CHARLES
HUBIS BRIAN ANDREW
HUBISZ JAMES FRANCIS
HUGHES KENNETH ROCK-
WELL
HUGHES PAUL JOSEPH
HUNT JOSEPH THOMAS
HUNTOON RICHARD WARREN
HURD ROGER MICHAEL
HURLEY WILLIAM PAUL JR
HURST RONALD CHARLES
HUSSEY GEORGE ELLERY
INGERSOLL DAVID PAUL
ISAACSON GERALD EDWARD
ITRI DOUGLAS JOHN
JABLONSKI JOHN ANDREW
JABLONSKI ZYGMUNT PAUL JR
JACK MICHAEL FRANCIS
JACKSON THOMAS FRANCIS
JACOBS JOHN PAUL
JAQUINS CHARLES EGBERT
JARRAS STEPHEN THEODORE
JARVIS EDWARD CARL
JASNOCHA ALFRED L JR
JASON BRUCE ELLSWORTH
JENCZYK FRANK PAUL JR
JENKS JAMES JOSEPH JR
JENNINGS GLENN ROBERT
JOHANSEN DONALD CHARLES
JOHNSON ALAN PAUL
JOHNSON BYRON STEVEN
JOHNSON DANIEL JOSEPH
JOHNSON EDWARD LEE
JOHNSON HUGH RICHARD JR
JOHNSON JAMES EDWARD JR
JOHNSON JOHN ROBERT
JOHNSON KENNETH RICHARD
JOHNSON KENNETH ROBERT
JOHNSON RICHARD CHARLES

JOHNSON RODNEY W
JOHNSON THEODORE FRED
JOHNSON WILLIS WAYNE
JOHNSTON BRUCE E III
JONES BENNIE FRANK
JONES BRUCE EDWIN
JONES CLIFFORD RAYMOND
JR
JONES ISAAC
JORDAN ALLAN HAROLD
JORDAN DUDLEY NORMAN
JOYCE JOHN GERARD
JOYCE JOHN H
JOYCE WALTER EDWARD JR
JOYCE WILLIAM EDWARD JR
JOYCE WILLIAM FRANCIS
JOYNER KENNETH RUSSELL
JUSTIN WILLIAM BARRY
KACSOCK WALTER JOSEPH JR
KADLEWICZ ZDZISLAW
BRUNO
KALEN JOHN JOSEPH
KAMP THOMAS KEITH
KANE TERRANCE FREDERICK
KANE THOMAS JOSEPH
KAPETANOPOULOS KOSMAS
PET
KASTER LEONARD LEE
KAZANOWSKI JOHN FRANCIS
KEARNS STEVEN JOHN
KEATING ALLEN FRANCIS
KEEFE MARTIN RUSSELL
KEEFE PAUL PATRICK
KEENAN JOHN SCOTT
KELLER JAMES MASON
KELLETT DANIEL MACAR-
THUR
KELLEY DANIEL MARTIN
KELLEY RICHARD JOSEPH
KELLEY RICHARD ROBERT
KELLY BRIAN RICHARD
KELLY DOUGLAS JOHN
KELLY PATRICK JOSEPH JR
KENNEDY BRUCE JAMES
KENNEY DAVID EDWARD
KENNEY JOHN JOSEPH
KENT DANIEL WILDER
KENT GREGORY PATRICK
KIERZEK STANLEY P
KILLILEA MARTIN FRANCIS
KING PAUL CHESTER JR
KINSMAN GERALD FRANCIS
KLEINBERG PETER SHELL
KMIT CHESTER JON
KNOWLTON PAUL DARYLL
KOLENDA PAUL MICHAEL
KOONTZ NOBE RAY JR
KOSTANSKI STEPHEN FRAN-
CIS
KOZACH JOHN ALBERT
KRAJESKI STEPHEN EDWARD
KRAWCZYK JAN
KRESESKIE FRANK THOMAS JR
KRISTOF PETER FRANK
KUHNS DAVID ALLEN
KULACZ DONALD EDWARD
KUPKA ANTHONY EDWARD
KUSTIGIAN MICHAEL JOHN
KUSY DAVID PAUL
KUZMA MARC JOHN
KYLE BARRY STUART
KYRICOS GEORGE ARTHUR
LA CHANCE CLIFFORD
DAMON
LA FLAMME ROBERT JAMES
LABONTE DONALD ARTHUR
LACUS GEORGE DONALD JR
LADEROUTE MICHAEL JOHN
LAIDLAW WILLIAM CLIVE
LAIDLER ERNEST HAMMOND
LAKE RONALD FRANCIS
LAMEIRAS RICHARD ARTHUR
LANDERS KEITH TERRELL
LANDRY PAUL JOSEPH
LANDRY PETER JOSEPH
LANE SIDNEY DANIEL JR
LANE STEPHEN LESLIE
LANGER ALAN KARL
LAPAN GEORGE FRANCIS
LAPOINTE RAYMOND RO-
LAND
LARRAGA ANGELO GENTRY
LARSON ROBERT MERCHANT
LASKOWSKI JOHN JOSEPH

LATANOWICH THOMAS
DANIEL
LATESSA ANDRE ROLAND
LAWSON JOHN DAVID
LAZAROVICH JOHN F JR
LE CLAIR WILLIAM GEORGE
LE GROW ARTHUR RUSSELL JR
LEAVER JOHN MURRAY JR
LEE EDWARD GILBERT
LEE PAUL RICHARD
LEE VINCENT BURKE
LEFEBVRE RUDOLPH H JR
LEGER GERALD ROGER
LEGERE EMILE JOSEPH
LEIGH LAWRENCE GRAHAM
JR
LEMAIRE DOUGLAS JAMES
LEMIEUX WALTER JOHN
LEONARD WILLIAM
LETENDRE GERALD ARTHUR
LETOURNEAU EDWARD R JR
LEVESQUE GEORGE ROBERT
LEVESQUE LEO WILLIAM
LEWIS THOMAS LAMAR
LIKELY RICHARD ALLEN
LIONETTA EDWARD ARTHUR
LITCHFIELD FRANK EDWARD
LITTLE WILLIAM HARRIS
LITWIN ROBERT RICHARD
LOANE ALLEN ROBERT
LOGAN JOSEPH PATRICK JR
LOHEED HUBERT BRADFORD
LONG JAMES THOMAS
LOONEY PAUL THOMAS
LOPEZ EDWARD
LORDITCH PATRICK MICHAEL
LOUGHLIN EDMUND MI-
CHAEL
LOVETT BERNARD JAMES JR
LOWE BARRY
LUDWIG JAMES MICHAEL
LUSCIER HOWARD HENRY
LYNAH TIMOTHY JOSEPH
LYNCH DANIEL FRANCIS JR
LYNCH KEVIN FRANCIS
LYON JOHN PAUL
MAC CANN HENRY ELMER
MAC DONALD JEROME JAMES
MAC KAY CALVIN RONALD
MAC MILLAN THOMAS
MACLEAN JOHN DONALD
KENNE
MACNEIL EDMUND LAMBERT
III
MADDEN JOHN MARTIN JR
MADIGAN JOHN EDWARD JR
MAGEE JOHN JOSEPH
MAGNUSSON JAMES A JR
MAGRASS JOEL MICHAEL
MAHLER JAMES WILLIAM
MAIATO JAMES COSTA JR
MAILLOUX JOHN JOSEPH
MAIN ROBERT JAMES
MALLOY JOHN JOSEPH
MANEY RALPH WARREN
MANFERDINI JOHN SEBAS-
TIAN
MANNION DENNIS JOHN
MANTOUVALES ANTHONY
RALPH
MANZARO DANIEL VICTOR
MARCIN PAUL JOHN
MARCOULIER LEO RENE
MARKS ANTONE PATRICK
MARRON BRUCE ALEN
MARSDEN ROBERT PAUL
MARTIN BRUCE EDWARD
MASSE RAYMOND GEORGE
MATERN ROBERT SCHRACK
MATRANGA ROBERT
MATTA BRUCE JOSEPH
MATTAROCCHIA JOHN F JR
MATTE ALAN LOUIS
MATTHEWS EARL MARTIN
MC AFEE DAVID ALFRED
MC ANDREW RICHARD T JR
MC ARTHUR JOHN DOUGLAS
MC BRIDE MORRIS RALPH
MC CAFFERTY MICHAEL
LESTE
MC CANN VINCENT OWEN JR
MC CARTHY EDWARD
CHARLES
MC CARTHY JOHN EDWARD

MC CARTHY JOHN HENRY
MC CARTHY ROBERT JOHN
MC CARTHY WILLIAM
FRANCIS
MC CLUSKEY ROBERT WIL-
LIAM
MC CORMACK WILLIAM
EDWARD
MC CORMICK THOMAS A JR
MC CUE JAMES EDWARD
MC DONALD GERALD FRAN-
CIS
MC DONOUGH ROBERT JAMES
MC EACHRON PAUL
MC EWING HARRY
MC GARRY JAMES BRIAN
MC GINN WALTER WILLIAM
MC GINNESS PAUL EDWARD
MC GOVERN KEVIN MICHAEL
MC INTOSH JOHN ARTHUR
MC INTYRE ARTHUR JAMES
MC KENNEY KENNETH
DEWEY
MC LAUGHLIN FREDERICK J
MC LAUGHLIN MARK MI-
CHAEL
MC LAUGHLIN WILLIAM F
MC MAHON FREDERICK
ALFRED
MC MAHON JAMES HAROLD
MC NABB JOHN JOSEPH
MC NALLY PAUL FRANCIS
MC NAMARA WILLIAM JAMES
MC NEIL JOSEPH DANIEL
MC NULTY JOSEPH DENNIS
MC NUTT CHARLES THOMAS
MC RAE WILLIAM JOSEPH
MCELENEY EDWARD RALPH
JR
MCLAUGHLIN ARTHUR VIN-
CENT JR
MCMAHON CHARLES JR
MCMURRAY PETER HINCH-
MAN
MEARS PETER JOSEPH JR
MEDEIROS MICHAEL JOHN
MEDEIROS WILLIAM CORREIA
MELLO EDWARD THOMAS JR
MELVIN JAMES LEONARD
MENDALL CARLTON JOSEPH
MENOWSKY GLENN ALFRED
MERCER ROBERT JOHN
MERRILL JOSEPH ADELBERT
MERRILL ROBERT FRANKLIN
MESSER JAMES ALLEN
METCALF GERALD ERNEST
MEUSE JOHN RICHARD
MILAN EDWARD WALTER
MILLER CARLETON PIERCE JR
MILLER CHARLES DAVID
MILLER CLARK ALAN
MILLER IRVIN GEORGE
MINAHAN DANIEL JOSEPH
MINEHAN MICHAEL PAUL
MIRANDA ROBERT
MITCHELL CYRIL JR
MITCHELL LAWRENCE
HOWARD
MOBILIA MICHAEL HOWARD
MOHRMANN DOUGLAS
ROBERT
MONAHAN MICHAEL JAMES
MONSKA BRUCE WILLIAM
MOONEY WALTER STEPHAN
MOORE CURTIS WAYNE
MOORE DOUGLAS FILLE-
BROWN
MOORE ROBERT EVERETT
MOORE THOMAS RICHARD JR
MOORE WAYNE PAUL
MOORES KENNETH FRED-
ERICK
MORAN JOHN FRANCIS
MORAN PAUL ROBERT
MOREAU EUGENE RAYMOND
MOREAU JOHN ALFRED
MORENO JOHN HERBERT
MOREY FRANK ERNEST JR
MORGAN WALTER WILLIAM
MORIN RICHARD GIRARD
MORRILL DENNIS LEROY
MORRILL EDWARD FRANCIS
MORRIS DAVID MICHAEL
MORRIS HAROLD HERBERT

117

Massachusetts

MORRIS ROBERT EDWARD JR
MORRIS THOMAS RICHARD
MORRISON WENDELL ALBERT
MORRISON WILLIAM JOHN
MOSKOS PETER
MOTTOLA VINCENT ANTO-
NIO
MOXLEY RICHARD STEPHEN
MULLEN WILLIAM FRANCIS
MULLIN WAYNE WILSON
MURACA PATRICK JOHN
MURACO FRANCIS JOHN
MURDOCK JOHN LEO
MURPHY ARTHUR PATRICK JR
MURPHY CHARLES JOHN JR
MURPHY DAVID WAYNE
MURPHY EDWARD JOSEPH JR
MURPHY MICHAEL
MURPHY RICHARD BRIAN
MURPHY TIMOTHY FRANCIS
JR
MURPHY WILLIAM CAMPBELL
MURRAY JAMES FRANCIS
MURRAY JOHN BUTLER
MURRAY THOMAS EDWARD
MURRAY WILLIAM DONALD
JR
MYERS JEFFERY PHILIP
NAPIERATA NORMAN JOSEPH
NASH PETER GARY
NASHAWATY RICHARD JOHN
NEALON JOHN MICHAEL
NEE PETER MARY
NELSON HAROLD BARNETT
NELSON ROBERT JOSEPH
NEWBERRY LEWIS JEWETT
NEWCOMB WAYNE PAUL
NEWMAN PAUL FRANCIS JR
NICKERSON THOMAS CAR-
ROLL
NIMIROSKI JOSEPH ELWIN
NOLAN PETER FRANCIS
NORMAN GORDON JOSEPH
NORTON RICHARD L
NOWICKI ROBERT PHILIP
NOYES RUSSELL WILLIS
NUTE RONALD WADE
O BRIEN ALAN JOSEPH
O BRIEN EDWARD STEPHEN
O BRIEN PHILLIP ANTHONY
O BRIEN WILLIAM JOSEPH
O CONNOR BRIAN RICHARD
O CONNOR FREDERICK J JR
O LEARY JAMES KEVIN
O LEARY RICHARD LAUGHLIN
O NEIL WALTER JAMES
O NEILL CHARLES LEO JR
O NEILL DENNIS MICHAEL
O NEILL GEORGE EDWARD
O NEILL THOMAS EDWARD JR
O REILLY FRANCIS JOSEPH
O REILLY JAMES C JR
O TOOLE JAMES EDWARD JR
O'BRIEN T CHRISTOPHER
FORD
ODIORNE GEORGE ALFRED
OFFLEY JAMES CLIFTON

OLINSKY WALTER STANLEY JR
OLSON RICHARD
OLSON RICHARD EMIL
ORDWAY WILLIAM DWIGHT
OSBORN THOM THURSTON
OUELLET DAVID GEORGE
OVERLOCK JOHN FRANCIS
OWENS DAVID LEE
PAGE ADDISON WILLIAM JR
PAGNANO ENRICO HENRY JR
PAINTER JOHN ROBERT JR
PALERMO GEORGE ROBERT
PALMA LUCO WILLIAM
PALMERI JAMES EDWARD
PALMIERI DAVID HAROLD
PARENT JEFFERY MARK
PARENT ROBERT WARREN
PARKER FREDERICK G JR
PARKER ROBERT KENNETH
PARMELEE BRUCE CARLTON
PARSONS CHARLES EDWARD
PASQUANTONIO JOHN
EMIDIO
PASSERELLO ANTHONY
JOSEPH
PATENAUDE HENRY EDWARD
PAULETTE JOSEPH RONALD
PEEL JOHN CHARLES
PEIXOTO GILBERT COROA
PELLETIER PAUL JOHN
PENNUCCI PETER JAMES
PENTA STEPHEN JOSEPH
PERKINS FREDERICK JOSEPH
PERRAULT ALAN JAMES
PERRY DANIEL
PETERSON BURTON W JR
PETERSON CARL ALFRED
PETERSON GERALD ROY
PETSCHKE ROBERT ELTON JR
PHILIPSON JOSEPH BION JR
PICANSO LEONARD JR
PIGNATO JOSEPH MICHAEL
PINA LUIZ JR
PINHEIRO JEFFREY ANTONE
PINSONNAULT RICHARD
NORMA
PINTO CAESAR AUGUSTUS
PISCITELLO SALVATORE JOHN
PITTS DAVID ALLEN
PIZZANO JAMES ROBERT
PLAZA BERNARD STANLEY
PLUNKETT ROBERT STEPHEN
POINTER RONALD JOSEPH
POIRIER PAUL EUGENE
POLENSKI EDMOND CHESTER
POPKIN STEVEN JAY
PORCELLA STEPHEN RICHARD
PORRAZZO LOUIS EDWARD
PORTER GARY THURSTON
PORTER KEVIN ANTHONY
POTTER RICHARD EDWARD
POWER RICHARD WILLIAM
POWERS FRANCIS EDWARD JR
PRATT JOHN LIONEL
PRESTON ROBERT EDWARD
PROTANO GUY JERRY JR
PROVENCAL ROLAND ANDRE
PRZELOMSKI PAUL ANTHONY

PUGH DAVID JAMES
PUTNEY EDWARD ALLEN
QUILL PAUL FRANCIS
QUINN JOHN PHILIP JR
QUINN MICHAEL PATRICK
QUINN RICHARD JAMES
QUINTAL JOHN VINCENT
RABAIOTTI ANDREW
CHARLES
RABIDEAU JOHN J
RABINOVITZ JACK
RAGO STEPHEN JOSEPH
RAINAUD JEFFREY WILLIAM
RAISIS LEONIDAS
RALEIGH LOUIS RAYMOND
RALSTON THOMAS MICHAEL
RAMSAY DAVID LEROY
RAMSEY JOHN LOUIS
RAND RICHARD PAUL
RANDO JOSEPH PAUL
RASENYUCK JAN IVAN
RATTEE CARL ALLAN
RAY WALTER DONALD
RAYMOND JOHN JAMES
REARDON DENNIS JOSEPH
REGAN WILLIAM KENNETH
REID PAUL FRANCIS
REIS TIAGO
RENAULD RALPH VICTOR JR
REZENDES PAUL ALLEN
RHOADES CLINTON MORELL
JR
RHODES ROBERT DAVID
RICE JAMES JOSEPH
RICH PETER BERNARD
RICHARD WILLIAM W
RICHARDS DONALD LAW-
RENCE
RICHEY NEAL OLIN
RINEHART RICHARD BEN-
NETT
RIVERA CARLOS JR
RIVEST MARK HENRY
ROBERTSON BRISTOL JR
ROBERTSON CHARLES WIL-
LIAM
ROBINSON CHARLES JOHN
ROBINSON EUGENE FRANCIS
ROBITAILLE PAUL EDWARD
RODERICK RONALD
RODERIGUES PAUL IRVING
RODRICK ROBERT LAWRENCE
RODRIGUES DANIEL EVER-
ETTE
RODRIGUES RICHARD
ROGERS EDWARD FRANCIS
ROGERS RICHARD LEO
ROHR JOHN WILLARD
RONGA ROBERT FRANK JR
ROSBECK RICHARD A
ROSE ALBERT JAMES
ROSE CARLOS JAMES
ROSENBUSCH CHARLES
RICHAR
ROSS ROBERT LESLIE
ROSSER EDWARD JOHN
ROULIER RUSSELL RENE
ROY DAVID PAUL

ROY PETER WILLIAM
RUMRILL PAUL WILLIAM
RUMSON SAMUEL JAMES JR
RUSSELL GORDON WARREN
RUST GARY ALFRED
RYAN ROBERT EDWARD JR
RYAN WILLARD R
SABATINELLI VINCENT F
SADBERRY SEYMOUR PATRICK
SAGERIAN BRUCE ELLIOTT
SAINT AMAND RICHARD
CARL
SAMARAS PETER NICHOLAS
SANCHEZ IGNACIO
SANSONE JAMES JOSEPH
SANTOS ALBERT WILLARD
SAS THEODORE FRANCIS
SAVAGEAU JOHN HENRY
SAVINO LAWRENCE NEIL
SCAHILL EDWARD JOHN
SCHENA ROBERT PETER
SCHMAUTZ FRANCIS PHILLIP
SCHOFIELD ALFRED VINCENT
SCHOLES WILLIAM HADLEY
SCHRAMM PETER FRYE
SCHULTZ GEORGE CLIFTON JR
SCOTT RICHARD ALLEN
SCULLY KENNETH WILLIAM
SEABORNE FREDERICK
VERNON
SEAMAN JOSEPH ANDREW
SEKLECKI THOMAS MARTIN
SERVENT HENRY JOSEPH JR
SESTITO ANTHONY JOHN
SEVENEY WILLIAM FRANCIS
SEVERINO WAYNE THOMAS
SHALLAH JOHN HERBERT
SHALLER RONALD WILLIAM
SHATTUCK HAROLD LEON JR
SHEA HAROLD JOSEPH
SHEEHAN CHARLES J III
SHEEHAN PAUL HENRY
SHERMAN RONALD EARL
SHINE DENNIS FRANCIS
SHIRAKA JOHN EDWARD
SHORTSLEEVES WILLIAM JOSE
SHOTWELL JAMES HUNTER
SHUFELT GEORGE JERRY
SHUMAN MICHAEL BERNARD
SICKEL JOHN AULDE III
SILVA ROBERT JOHN
SILVERBERG ARVID OSCAR JR
SIMEONE CRAIG MICHAEL
SIMMONS WAYNE CARL
SIMOES ANTHONY
SIMON DAVID LOWELL
SINTONI JOSEPH EUGENE
SIROIS LAWRENCE EVERETT
SKAPINSKY GEORGE JOSEPH
SKELTON RONALD ALBERT
SKINNER GORDON A II
SLACK RICHARD DON JR
SLEEPER DAVID FREDERICK
SLEIGH DUNCAN BALFOUR
SMALL ALFRED JOHN
SMART CEDRICK LOUVANE
SMIGLIANI DOMENIC
SMITH ALAN JOHN

SMITH EDWARD FRANCIS
SMITH WILLIAM MARK
SNYDER GUY FORD
SOARES MANUEL AGUIAR
SOKOLOWSKI FRANK MI-
CHAEL
SORRENTI JOHN ANTHONY
SOUSA LAURENCE NELSON
SOWINSKI ROBERT JOSEPH
SPEYER ALFRED WILLIAM
SPIERS STEPHEN ARTHUR
SPILLANE PAUL DONALD
SPIRES ROBERT EVERETT
ST CYR JAMES AUGUSTINE
ST JEAN BERNARD EDWARD
ST JOHN DAVID MICHAEL
ST LAWRENCE ALBERT
ALFRED
ST PIERRE MICHAEL LEONARD
STANISZEWSKI WLADYSLAW
STARK JAMES ALEXANDER
STEEL JOHN ALLEN
STEFANIK EDWARD PETER
STEINSIECK ROBERT T JR
STEPHENSON HOWARD
DAVID
STERITI STEPHEN JOSEPH
STEWART JAMES JOSEPH
STEWART PAUL JOHN
STEWART PAUL LEO
STOCHAJ PAUL JOHN
STONE EDWARD THOMAS JR
STOPYRA THOMAS JOHN
STOVALL GUS JR
STRAFELLO CHARLES
FRANKLI
STROUT ROGER HENRY
STROYMAN ARTHUR
SULLIVAN DAVID OWEN
SULLIVAN EDWARD MICHAEL
SULLIVAN FRANCIS C JR
SULLIVAN JAMES EDWARD
SULLIVAN JOHN JOSEPH
SULLIVAN JOHN MILLER
SULLIVAN LEO JOSEPH JR
SULLIVAN MARTIN JOSEPH
SULLIVAN MICHAEL XAVIER
SULLIVAN PAUL JOSEPH
SULLIVAN PETER MICHAEL
SULLIVAN STEPHEN THOMAS
SURETTE PAUL JOSEPH
SUTTON HUBERT DANIEL
SWAIN CRAIG FRANCIS
SWAN LEO EDWARD JR
SWEENEY THOMAS PAUL
SYLVIA WAYNE JOHN
TADEVICH EMIL JEROME
TAILLON JOHN PHILLIPS
TANASSO AMBROSE P JR
TAURISANO JAMES VINCENT
TAVARES BELMIRO JR
TAVARES CHARLES ALBERT
TAVARES MANUEL ANTO-
NIO D
TAYLOR PHILIP JOSEPH
TEDFORD ROBERT CHARLES
TERMINI JAMES MICHAEL
TERRY WILLIAM BICKERS

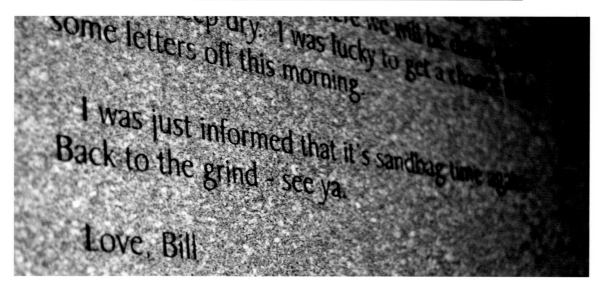

COMMONWEALTH OF MASSACHUSETTS

VIETNAM VETERANS MEMORIAL

PLACE OF NAMES

WOODLAND PATHS

PLACE OF WORDS

MEADOW

PLACE OF FLAGS

GREEN

TOZER ELDON WILLIAM
TRAINOR PAUL WILLIAM
TRIPP DONALD DELMORE
TSIROVASILES PETER
TUCKER EDWIN BYRON
TULLER ERIC LAWRENCE
TURBERT FRANCIS XAVIER
TURCOTTE PETER RUDOLPH
TURNER DANIEL ROBERT
TURNER DON ELDRIDGE
TURNER DONALD JOHN
TYNE JEFFREY GORDON
TYREE EARL EDWARD
UNDERWOOD GEORGE
WARREN
UPTON CARLETON WEBSTER
VALENTINE PERVIS B JR
VALLEE JOSEPH LEO
VANASSE PHILIP RICHARD
VANCELLETTE DAVID MI-
CHAEL
VASCONCELLOS RICHARD
JOHN
VAUGHAN MICHAEL PATRICK
VAUGHAN RONNIE GORDON
VELARDO ANTHONY GUY
VICALVI TIMOTHY LAW-
RENCE
VIEIRA JOSEPH
WADEN JOHN F
WAKEFIELD CARL DANIEL
WALKER DOUGLAS ALEX-
ANDER
WALKER ORIEN JUDSON JR
WALL STERLING AIDEN
WALLACE HARRY WILLIAM
WALLEN JOSEPH ROBERT
WALSH JAMES MICHAEL
WALSH RICHARD DAVID
WALSH ROBERT STEPHEN
WASHBURN FRED ZIMRI
WATERMAN CRAIG HOUSTON
WATERMAN MICHAEL J
WATTS RALPH O
WAX DAVID J
WEBB GARY JOSEPH
WEITZ MONEK
WELCH RICHARD WILLIAM
WENDLER RUSSELL WILLIAM
WEST RAYMOND JOHN
WHALEN MICHAEL CORNE-
LIUS
WHEELER JAMES KENNETH
WHITE ALBERT RONALD
WHITE JOHN CULLIN
WHITE JOHN EDWARD
WHITTAKER HAROLD
CHARLES
WIGHT RALPH CHESTER JR
WILK THOMAS JOHN
WILKINSON DONALD ALFRED
WILLARD ALAN WAYNE
WILLARD RALPH JOHN
WILLIAMS ALFRED LACY
WILLIAMS ARTHUR PAUL
WILLIAMS DAVID EDWARD
WILLIAMS RONALD
WILLIAMS WILBUR LEO JR
WILSON CALVIN RAY
WIRTH JOSEPH WILLIAM
WISSMAN RONALD EDWARD
WITANEK CHESTER LAWREN
JR
WOLFENDALE EDWARD
JAMES
WOLK BARRY LEE
WOODSON LAURENCE
OLIVER
WRIGHT ARTHUR EMERSON
III
WRIGHT WILLIE JOSEPH
YEOMANS CHARLES AGUSTUS
YOUNG DONALD RAYMOND
YOUNG DOUGLAS ALLEN
YOUNG GERALD FRANCIS
YUREWICZ STANLEY JOSEPH
ZEBERT JAMES DONALD
ZENGA RONALD PAUL
ZOZULA NICKOLAUS
CHARLES

TETREAULT ROBERT NAZAIRE
THALIN NEAL ROBERT
THERIAULT PAUL RAYMOND
THIBAULT KENNETH M
THIBEAULT JOHN LORNIE

THOMAS JOSEPH EUGENE
THOMPSON JERRY LENWOOD
THOMPSON JIMMY LEE
THOMPSON THEODORE A JR
THOMPSON WILLIAM F JR

THORPE FRANCIS JOSEPH
TIGHE JAMES EDWARD
TILLSON GARDNER JR
TOBEY MICHAEL JAMES
TODD ROBERT JACY

TOFFERI CHARLES EHNSTROM
TOLPA ROBERT RICHARD
TOLZMANN TED NORMAN
TOPHAM ROBERT WILLIAM JR
TOWNES ROBERT FRANCIS JR

119

Photos by Kayla Dusseau

Michigan Vietnam Veteran's Memorial Island Park

333 North Main Street Mount Pleasant, MI 48858

The Michigan Vietnam Memorial is located in Island Park in Mt. Pleasant, Michigan. Island Park is 50 acres of parkland situated behind the town's City Hall on the north side of the town. The memorial consists of a flag gallery, an information booth, a footbridge, and the 16 plaques that feature the names of the Michigan casualties and missing. The scene is made complete by a nearby riverbank and picnic shelters.

The Michigan Vietnam Memorial was a project taken under by the Vietnam Veterans' Mt. Pleasant chapter, VVA 438. Everything was built through volunteer labor done by veterans. It was dedicated originally in July 1990 as an official state memorial. It was rededicated in 1994 when a bronze statue called "War Cry" was added to the memorial. Both dedications were attended by a crowd of people including surviving veterans and families of the fallen and missing soldiers as well as local officials.

The memorial consists of a footbridge that connects the nearby Picken's ball fields to Island Park, as well as a flag gallery which were both built by the Mt. Pleasant chapter of Vietnam Veterans. The veterans also constructed the brick monuments and dedication plaques of the memorial which contain over 2,700 names.

The plaques hold the names of the Michigan men and one woman who died in captivity, were declared missing in action, or died in action. Each designation is denoted with either one star for missing in action or two stars for died in captivity.

The statue "War Cry" was cast in bronze and depicts a soldier cradling a wounded brother. It was created by sculptor Derek Rainey for the memorial's 1994 rededication. A Memorial day service is held every year.

The Names of those from the state of Michigan Who Made the Ultimate Sacrifice

ABRAMOSKI LEO BERT
ACHTERHOFF JAMES PATRICK
ACKERMAN LEONARD MICHAEL
ACKERMAN MAXIE EDWARD
ACTON GERALD RICHARD
ADAMS CLARENCE MATTUE
ADAMS DAVID LEE
ADAMS DAVID LEE
ADAMS DAVID VERNON
ADAMS LARRY
ADAMS PAUL VERNON
ADAMSON DONALD BRUCE
ADE DWIGHT I
AERTS DAVID LEE
AESCHLIMAN DAVID KEITH
AILI DAVID EMIL
AIREY GEORGE VERNON JR
AIRLIE WILLIAM CLARK
ALANDT CHARLES BYRON
ALBERTSON DONALD NORMAN
ALBERTSON ROBERT ALLEN
ALBERTSON RONALD DALE
ALDERSON MICHAEL EDWARD
ALDRED JAMES VINCENT
ALEXANDER WILLIAM LEE
ALFREDSON WILLIAM RICHARD
ALLARD RICHARD MICHAEL
ALLEN LARRIE CORNELIUS
ALLEN LYLE ERNEST JR
ALLEN RICHARD JAMES
ALLEN TERRANCE W
ALLEN WILLIAM JOHN
ALVAREZ BERNARDO RODRIGUE
AMES RONALD EDWARD
ANDERS JOHN ROYLE
ANDERSEN BARRY FRANK
ANDERSON JOHN KEITH
ANDERSON JOHN PERRY
ANDERSON RICHARD MERIDITH
ANDERSON ROBERT DALE
ANDERSON WARREN LEROY
ANDREWS DALE CHARLES
ANDRUS CARL JOSEPH
ANGEL TOMMIE RAY
ANGERMAN DONALD EDWARD
ANGLIM ADRIAN JAMES
ANKNEY SAMUEL FREDERICK

ANNIS THOMAS RICHARD
ANTHONY JOHN FREDERICK
ANTOL DAVID
APUTEN LESLIE GEORGE
ARBOGAST CARL FRANCIS JR
ARENAS REYNALDO
ARIZMENDEZ DANIEL MICHAEL
ARNOLD LOUIS GEORGE WASHI
ARNOLD ROBERT WILLIAM
ARVIN CARL ROBERT
ASBURY DONNIE DEWAYNE
ASHFORD DAVE EDWARD
ATKINS DOUGLAS PAUL
ATKINS MATTHEW DAVID III
ATKINSON ROBERT LOUIS JR
AUSTIN EDWARD PAUL
AUSTIN ELLIS ERNEST
AUSTIN GLENN FREDERIC
AUSTIN SCOTTY GENE
AUSTIN VICTOR LEROY
AVERY GERALD LAWRENCE
AVERY RALPH LEE
AXFORD JOSEPH WILLIAM
AYALA GEORGE HERMAN
AYLWORTH RANDAL RAY
BACHERT RICHARD CHARLES
BADER WILLIAM EDWARD
BAER RANDALL THOMAS
BAILEY DAVID ORIN
BAILEY JOHN J
BAIN BRUCE ARNOLD
BAIRD ALBERT FRANKLIN
BAKER BOBBY GENE
BAKER HARRY E JR
BAKER JERRY
BAKER PHILIP KENNETH
BAKER RUSTON LEE
BAKER THOMAS HARRY
BAKER WILLIAM EMANUEL
BAKKE LARRY NEIL
BALDONI LINDSAY DAVID
BALDWIN CHARLES LEROY
BALDWIN HENRY PHILIP
BALL HARRISON BRUCE
BALLINGER TIMOTHY J
BALOG LOUIS ROBERT
BANKOWSKI JOHN FRANCIS
BANKS JAMES C
BANKS RICHARD STEVEN
BANNISTER RUSSELL REID
BARBER RICHARD JOSEPH
BARCALOW RONALD RICH-

ARD
BARD MICHAEL
BARILI PETER LINO
BARKER GREG ALLEN
BARKLEY KIRK OWEN
BARLOW JEFFREY LAWRENCE
BARNABY DAVID W
BARNES GALE LYNN
BARNES GEORGE LEE
BARNES JAMES ALAN
BARNES JIMMY ONEAL
BARNES WILLIAM CAREL JR
BARRETT GEORGE PATRICK
BARRETT STANLEY FOSTER
BARRUS DAVID WILLIAM
BARTH BRUCE GEORGE
BARTON ALAN KEITH
BASS WILLIAM THOMAS JR
BAST PAUL G
BATOZYNSKI CHARLES HENRY
BATTIN DARRELL GENE
BAUCOM JAMES FREDERICK
BAUER ROBERT LOUIS
BAWAL ROBERT JOSEPH
BAXTER JOHN STANLEY
BEACH DEAN L
BEACH LEO ALBERT JR
BEALS MICHAEL ALLEN
BEARDSLEE TERRY HUGH
BEARDSLEY WILLIAM BURDON
BEATTY THOMAS WILLIAM
BEAUCHAMP ALBERT ALLEN
BEAVER JOHN DOUGLAS
BEAVER WESLEY
BEBO WAYNE RICHARD
BECANNEN BARRY J
BECK GLEN RAY
BECKER TOMMY JOE
BECKETT RONALD LEE
BECKMAN ROBERT CHARLES
BECKWITH HARRY MEDFOR III
BEEBE LARRY DWAYNE
BEHM DANIEL LOUIS
BELCHER FRANK EDWARD
BELFORD JOHN ARTHUR
BELINSKI JAMES GERALD
BELL DEXTER
BELL ELIAS JR
BELL JAMES EDWARD
BELL WAYNE MORRIS
BELLAIRE JOHN MICHAEL
BELLANT FRANK LEROY

BENEDETTI DENNIS EUGENE
BENNETT DONALD LEE
BENNETT KENNETH DEVON
BENNEY KENDAL LEE JR
BENOIT ROBERT CHARLES JR
BENSON ROBERT JOHN
BENT GORDON WILLIAM
BERG GARY RICHARD
BERG ROGER LEE
BERKFIELD THOMAS DUDLEY
BERKHOLZ DAVID DENNIS
BERNARD RODNEY ROYCE
BERNHEISEL DAVID ARNOLD
BERRY JAMES CRAIG
BERRY JAMES E
BESKE WILLIAM HENRY JR
BEST CAREY EDWIN
BETTS TERRY WADE
BICE DOUGLAS WYATT
BICKFORD THOMAS WAYNE
BIEHN MAURICE JOHN
BIES EDWARD ALAN
BIGELOW PAUL LEE
BIHLMEYER JAMES ROY
BINDER QUENTIN WAYNE
BINKLEY STUART MARSHALL
BINKOWSKI RONALD JOHN
BIRD CHARLES WESLEY
BIRDSALL THOMAS EDDY
BLAAUW JAMES EVART
BLACK RALPH ROLAND
BLACKMER WILLIAM EDWARD
BLACKMON JOHNNY
BLACKSMITH RONALD RAY
BLANDINO HOWARD
BLASKOWSKI RICHARD L
BLAUWKAMP ARLYN JAY
BLISSETT ROBERT ALLEN
BLOMFELT DANIEL JOHN
BLONDIN MICHAEL ANTHONY
BLOOMFIELD MICHAEL LEE
BLOSSEY RAYMOND ROBERT
BLOSSOM STEVEN CARL
BLUME GERARD JAMES JR
BODELL LARRY ALLEN
BODZICK WILLIAM JOSEPH
BOESKOOL ROBERT RAY
BOJARSKI GEORGE JOSEPH
BOLAK THEODORE NICHOLAS
BOLDING EDGAR LEE
BOLTER KENT ROBERT
BOLTZE BRUCE EDWARD

BONACCI LAWRENCE LOUIS
BONESTEEL DAVID LARRY
BONNER JOSEPH
BONNEY JOHN CLAIR
BONNICI ROBERT JOHN
BOOKER JIMMY
BORCHERS CARL WILHELM
BOROWSKI RAYMOND JOHN
BORR JEFFREY
BORROUSCH DEAN WALTER
BORTON ROBERT CURTIS JR
BOSENBARK SAMUEL GAROLD
BOSOWSKI MICHAEL ALAN
BOSS CHARLES FREDRIC
BOST MICHAEL JAMES
BOUCHARD ROGER HAROLD
BOUDREAU JOHN HENRY
BOUGHNER GARY WILLIAM
BOULWARE GEORGE WALTER
BOURDAGE NELSON JOSEPH
BOUSLEY DONALD GEORGE
BOVA EDWARD JAMES
BOVAN PAUL CLAYTON
BOVINETTE CHARLES E JR
BOWEN RALPH EDWARD
BOWENS JAMES TERRY
BOWER HOWARD JAMES JR
BOWERS BRUCE ELLIOT
BOWERS DANNY WARD
BOWMAN BRICKIE JR
BOWMAN DAVID FRANK
BOWMAN JAMES LAVERN
BOWMAN JOSEPH GEORGE JR
BOWMAN PAUL BARKLEY
BOWMAN TOMMY RAY
BOYD DON GAYNOR
BOYER THOMAS MICHAEL
BOYLESS JOSE JULIO
BRAATZ CURTISS EDWARD
BRADFORD THOMAS JOHNSON
BRADLEY MARTEE JR
BRADLEY ROBERT NEAL
BRADLEY ROBERT TIMOTHY
BRADSHAW DAVID ALLEN
BRADY JOSEPH JAMES
BRAID JOHN EDWARD
BRANCHEAU DANIEL ALLAN
BRANCHEAU FRANCIS EMIL II
BRAND THOMAS RICHARD
BRANDON TOMMIE
BRANDT FREDRICK KEITH
BRANHAM ROY LEE
BRANNON DAVID CRAIG
BRANT DAVE WILLIAM
BRDA JUSTIN PAUL
BREEDEN CLIFFORD LYNN JR
BREEDING EUGENE JR
BREWER JOHN WILLIS JR
BRIMMER DELBERT ELLERY
BRINKEY LARRY HOWARD
BRINKS KENNETH LEE
BRITTON MILTON DONALD
BROCK JAMES BARRETT
BROCKS EVERETT LEWIS
BROMLEY EDWARD LEWIS
BROOKS CHARLES ALLEN
BROOKS DAVID T
BROOKS LARRY EUGENE
BROSNAN RANDY DALE
BROTZ DANNY RAY
BROWN ALBERT LEE
BROWN CHARLES PATRICK
BROWN DAVID ALAN
BROWN DAVID PETER
BROWN JIMMIE DONOVAN
BROWN JOSEPH M
BROWN MAX EUGENE JR
BROWN ROBERT LEE
BROWN SYRES MATTSON
BROWN WARREN GENE
BROWN WILLIAM ARTHUR
BROWN WILLIAM LENNINGTON
BROWN-BEY LANCASTER
BRUECK RICHARD ALLEN

Michigan

BRUM PETER
BRUNING DAVID KENNETH
BRUSKE GARY LEE
BRYAN JAMES MICHAEL
BRYAN LARRY MICHAEL
BRYAN ROBERT LAMARR
BRYANT CHARLIE PAUL JR
BRYANT RONALD WILLIAM
BRYANT WILLIAM MAUD
BUCKLER TERRY WAYNE
BUCKLEY LOUIS JR
BUCZOLICH PAUL JOSEPH
BUGNI FLORIAN ANTHONY JR
BUKALA DANIEL SCOTT
BULIFANT ROGER DEAN
BUMP THOMAS EDWARD
BUMSTEAD DONALD ROYCE
BURCK WILFRIED
BURDICK DOUGLAS JOHN
BURGESS JOHN LAWRENCE
BURGESS SCOTT M
BURKELL GENE MICHAEL
BURKES JOSEPH
BURNETT DOUGLAS MCAR-
THUR
BURNETT WILLIAM ROBERT
BURNEY JAMES LARRY
BURNOR LEE ERVIN
BURNS HOWARD MICHAEL
BURRELL GEORGE HARRY
BURROUGHS TED WILLIAM JR
BUSCHLEITER WALTER
DENNIS
BUSH LEE RANDALL
BUSHARD WILLIAM DEAN
BUSHONG DONALD RICHARD
BUSICK LARRY RUSSELL
BUSSE DONALD GENE
BUSUTTIL JOSEPH
BUTGEREIT LARRY DUANE
BUTLER CHARLES LEWIS
BUTLER DENNIS LEE
BUTLER GERALD EUGENE
BUTLER GERALD THOMAS
BUTLER HARRY WILLIAM
BUTTERFIELD ROBERT A
BUURSMA DAVID
BUYNOSKI LAWRENCE J III
BYERS MELVIN JOHN
CABALA DUANE JACOB
CACCIA CARL HENRY
CADE BRUCE WAYMAN
CADY DOUGLAS MICHAEL
CAFFEY MICHAEL ALEXAN-
DER
CAHILL DANIEL FRANCIS
CAIN ROBERT DANIEL
CALDWELL HUGH PINSON JR
CALHOUN JOSEPH
CALLAHAN PATRICK RICH-
ARD
CAMERON JAMES LUTHER
CAMERON WILLIAM BURR
CAMPBELL DAVID GRAHAM
CAMPBELL DONALD A
CAMPBELL GEORGE LEE
CAMPBELL GEORGE SAMUEL
CAMPBELL STANLEY CLAUS
CANALES VICTOR JOEL
CANFIELD JESSE DEFOREST
CANN HORACE
CANTLON JOHN EDWARD JR
CAPANDA ROBERT JOHN
CAPLING ELWYN REX
CARCLAY JACK CRAIG
CARDINAL WAYNE MEDDIE

CARMONA JESSE JR
CARNES DONALD LLOYD
CARR ALVIN
CARR GEORGE LEE
CARR JAMES OTIS
CARRIER ALBERT JOSEPH III
CARROLL PATRICK HENRY
CARROLL THOMAS J
CARTER CLIFFORD RUSSELL
CARTER FRANKIE NATHANIEL
CARTER JAMES DEVRIN
CARTER TERRY ALFRED
CARVER RANDALL ALLEN
CASINO JOSEPH WALTER
CASSIDY MICHAEL PATRICK
CAST THOMAS EDWARD
CASTILLO DANIEL SANDUAL
CASTILLO JOHN JAMES
CASWELL EUGENE WILLIAM
CASWELL KENNETH LEE
CASWELL ROBERT LYNN
CAVAZOS RONALD THOMAS
CAVIS DAVID JUDE
CECIL ROBERT RANDALL
CHAFFIN THOMAS WILLIAM
CHALOU RONALD DAVID
CHAMBERLAIN ROY WARNER
JR
CHAMBERS RICHARD ALAN
CHAMBLISS ROGER RIDGELY
CHAPMAN CHARLES DANE
CHAPMAN RODNEY MAX
CHARBONEAU LE ROY HAR-
LAND
CHARLES NICHOLES WILLIAM
CHASE GARY LEE
CHASE RUSSELL DAVID
CHATMON NATHAN EUGENE
CHESTER HENRY J JR
CHILVERE ROBIN LEE
CHISHOLM JOSEPH CHARLES
CHMIEL ANDREW
CHRAN RICKEY LEE
CHRISTOFFER VERNON H JR
CHURCH RICKY WAYNE
CIECURA THOMAS PAUL
CLANCY DENNIS PATRICK
CLANCY TERRANCE BURTON
CLAPP GARY LYN
CLARK ANTHONY LAURENCE
CLARK JAMES NELSON
CLARK JAMES THOMAS
CLARK JAMES THOMAS
CLARK JERRY WAYNE
CLARK LUGENE JACKIE
CLARK RICHARD THOMAS
CLARK WALTER LEVON
CLARK WILLIAM HOWARD JR
CLARKE KEVIN MICHAEL
CLARKE WILLIAM VICTOR
CLAVIER DAVID MICHAEL
CLAWSON RODGER DEAN
CLAY CHRISTOPHER ELLIOTT
CLAY JAMES WILFORD
CLELAND THOMAS LEONARD
CLEVERLEY WILLIAM BERT
CLIFFORD GEORGE HENRY
CLIME RALPH JOHN
CLINE CURTIS ROY
CLINE ROBERT LOUIS
CLINE RODNEY BARRETTE
CLOUSE DUANE LEON
COATES KENNETH WILLIAM
COBEIL EARL GLENN
COBLEY WARREN W
COCHRANE JOHN FLOYD
CODY PETER GIRARD
COIN GREGORY CLEVELAND
COLATRUGLIO ROBERT F
COLBY BRIAN LYNN
COLE GORDON EUGENE
COLE JERRY RICHARD
COLE WILLIAM WINSTON
COLEMAN BONNIE LEE
COLLIER DONALD EARL
COLLIER GERALD JAMES
COLLINS SYLVESTER
COLLINS THOMAS RUSSELL JR
COLLIS GERALD ALAN
COLLYER DALE ELWYN
COLSTON EDWARD JEROME
COLWELL RONALD LEE
COMIS LARRY MELVIN
CONKLIN MICHAEL LEE

CONLEY DAVID LEE
CONLEY LARRY RAY
CONNORS DAVID THOMAS
CONNORS JACK LEE
CONNORS PATRICK JOSEPH
CONRAD ANDREW CHARLES
JR
CONSAVAGE RALPH EDWARD
CONWAY EDWARD JOHN
COOK CHARLES FRANCIS
COOK CURTIS KEITH JR
COOK DELFIN HILARIO
COOK DONALD MICHAEL
COOK JAMES JOHN
COOK JOEL LESLIE
COOK MILTON
COOK TIMOTHY ANDREW
COOLS JAMES HARVEY
COOPER GERALD ALLAN
COOPER MILES DENNIS
COOPER ROCKY LEE
COPELAND ARTHUR PERRY
COPELAND JERRY DON
COPELAND MELVIN
COPPO PATRICK BRIAN
CORNWELL THOMAS GLENN
COSSEY JOHN DWANE
COSTLEY LARRY LEE
COTES MICHAEL EUGENE
COTTEN OLLIE RAY
COTTER JOHN REDMOND
COTTRELL JOHN NELSON
COUGHLIN PATRICK CHARLES
COUSIN MOSES JAMES
COVEY JAMES HERBERT
COX FREDIE RAY
COX MICHAEL JOHN
COX MICHAEL LOU JR
COX WILLIAM JOSEPH
CRABBE ROBERT JOHN
CRAFT HAROLD GLEN
CRAFT JAMES ADOLPH
CRAGAR JAMES LEROY
CRAIG BRUCE KEITH
CRAIG WILLIAM THOMAS JR
CRAMER DONALD JAMES JR
CRANDALL RODNEY ALLEN
CRANDELL JAMES LEE
CRANE JOHN LA VERNE
CRANE PHILIP MATT II
CRAWFORD LAWRENCE
BERNARD
CREAGER RONALD LEE
CRIBELAR MICHAEL DEAN
CRIPE JACK LESTER
CRISMAN WILLIAM HAROLD
CRITES FRANKLIN THOMAS
CROMIE MICHAEL JOHN
CRONK RICHARD EDWARD
CROSSLEY EUGENE
CROSSMAN GREGORY JOHN
CROSSMAN WILLIAM HARRY
CROUCH NATHAN EUGENE
CROWLEY RALPH HEMAN
CRULL RAYMOND H
CRUTTS RALPH JOEL
CRUZ FRANK BRYAN
CUDLIKE CHARLES JOSEPH
CUDNIK EDMUND VICTOR
CUNNINGHAM DENNIS
ANTHONY
CUNNINGHAM JERRY MAX
CUNNINGHAM WALTER
WAYNE
CURL ROBERT GRAHAM
CURLEY ROOSEVELT C JR
CURRAN DANIEL JOSEPH
CURRETHERS JEFF
CURRY GLENN VERNARD
CURTIS BRUCE WAYNE
CURTIS TERRY MELVIN
CUSHMAN JOHN ROBERT
CUSSINS LOUIS WADE
CUTHBERT GEORGE RICHARD
CUTLER JAMES IRVING
CUTLER RALPH LOUIS
DAILEY DOUGLAS VINCENT
DAINS ROGER ALLAN
DALTON ROBERT LLOYD
DAMROW OLIVER PIERCE
DANIELS CHARLIES E
DANIELS MARK FRANCIS
DANKERT ROBERT SHELDON
DANKOWSKI JAMES HILLARY

DANNEELS ROBERT HAROLD
DARDEN WILLIAM HENRY
DARLING JOHN EDWARD JR
DARLING ROLLAND EUGENE
JR
DAUNIS JOSEPH
DAVENPORT JOHN CHARLES
DAVIDSON MICHAEL MURRAY
DAVIDSON THOMAS JAMES
DAVIDSON WILLIAM GALAN
DAVIES ALFRED JOHN JR
DAVIS ARNEL J JR
DAVIS BILLY SYLVESTER
DAVIS DALE L E
DAVIS DOUGLAS ONEILL
DAVIS ERNEST J JR
DAVIS FREDERIC HUTCHISON
DAVIS GERALD ARTHUR
DAVIS HAROLD MICHAEL JR
DAVIS JAMES LEONARD
DAVIS JAMES WILLIAM
DAVIS LEROY JR
DAVIS THOMAS ARTHUR
DAVIS WILLIE SONNY
DAVISON ROBERT GAYLE
DAWSON JOHN ROBERT
DAY EDWARD
DAY WESLEY DAVID
DE BOER LAWRENCE NEIL
DE DIE ROGER ALLEN
DE HOMMEL HANK JOHN
CONRA
DE NIKE STEVE SPENCER
DE TAMBLE THOMAS GLENN
DE VRIES KEITH ALLEN
DE WILDE PETER F JR
DE WULF PATRICK THOMAS
DE YOUNG ABE RICHARD
DEAN KENNETH BERNARD
DEAN WOODIE JUNIOR
DECKER DEWEY RUSSELL
DEEDS RICK DUANE
DEEL HAROLD BUTLER
DEERING GALE EDWARD
DEFER RICHARD HENRY
DEFER WILLIAM CHARLES
DEKKER DAVID ROSS
DELAPHIANO JOE B
DELLANGELO DAVID JOSEPH
DELLVON WILLIAM GRANT
DEMGEN ROBERT NICHOLAS
DEMOND DONALD ALLEN R
DEMOREST DAVID KEITH
DEMOROW ALAN GEORGE
DENNANY JAMES EUGENE
DERRICK ROBERT ALLEN
DESCO DENNIS A
DESORMEAUX HARRY HENRY
DEVINS RICHARD CHARLES
DEW PAUL ROBERT

DEWEY DANNY LEE
DI PIETRO ROBERT JOHN
DI RITA GENE
DIAZ BENITO JR
DICKEY ALAN EVERETT
DICKS MARVIN MERLE
DICKSON MARK LANE
DIDASKALOU GEORGE
ARTHUR
DIECKMAN JAMES HENRY
DIEKEMA ARNOLD RAYMOND
DIGGS WILLIAM FRANKLIN
DILLINDER RANDY EUGENE
DINGER JAMES ROBERT
DION THOMAS JAMES
DIX CRAIG MITCHELL
DIXON CARL DEAN
DLUGOKINSKI EDMUND
VALENT
DOBBS RONALD STEPHEN
DODGE MICHAEL JAMES
DOEZEMA FRANK JR
DOMKE PAUL LOUIS
DONALDSON HERBERT C JR
DONOHUE JOHN MARTIN
DORSE ROBERT EDWARD JR
DORSEY GEORGE HARRY JR
DOTSON MICHAEL ROBERT
DOTY CLAIR DUANE
DOUGLAS LARRY WAYNE
DOWNING DUANE AULDON
DOYLE HOWARD L
DOZIER JAMES EDWARD
DRABY LEROY JUNIOR
DRAKE MICHAEL JOSEPH
DRAKE MICHAEL LEON
DREW ROBERT DEARHART
DRINKARD DANNY GEORGE
DROB DAVID MICHAEL
DUDEK RICHARD ALAN
DUFFEY JERRY NORMAN
DUGGER DOUGLAS ALAN
DUKES ROY RAYMOND
DULYEA BARRY H
DUNAJ WILLIAM ANTHONY
DUNCAN ANDREW MCAR-
THUR
DUNDAS JERRY RICHARD
DUNN CHARLES CLIFFORD
DUNNEBACK MICHAEL
ARTHUR
DURAND DENNIS CHARLES
DURHAM JAMES CLAUDE JR
DURHAM JOHN MELVIN
DURTKA GERALD WILBERT
DUSSEAU ALBERT EUGENE
DUSSEAU JERRY JAMES
DUSSEAU RICHARD FRANK
DUSZYNSKI ANDREW JOSEPH
DUTY MELVIN DAROLD

IN HONOR & MEMORY
OF THOSE
WHO MADE THE
ULTIMATE SACRIFICE

This Memorial preserved by:
Forgotten Eagles
Chapter 3
&
VVA

DWORNIK VALENTINE MARION
DYE MELVIN CARNILLS
DYER JAY CEE
DYKEMA ROSS ALLEN
DYKES CLEVELAND E
EADIE GORDON PATTERSON
EADS RUSSELL WADE
EARLENBAUGH DANIEL LEE
EASTERN JOE BUTLER
EBEL WILLIAM MICHAEL
ECHOLS ALVIN
ECKERDT CHRISTIAN JOHN JR
ECKLES JAMES PATRICK
EDDY JOHN DAVID
EDGERLY JOHN WALLACE
EDINGER JAMES GARD
EDMONDS MONZIE DURREL
EDWARDS GEORGE RAY FAYFIE
EDWARDS KENNETH MILES
EDWARDS LEON GEORGE
EDWARDS ROBERT WAYNE
EDWARDS RODNEY CLINTON
EDWARDS ROGER WAYNE
EGLINSDOERFER LARRY JAMES
EGYED GERALD LEONARD
EHLERS LONNEY LEWIS
EHNIS KENNETH PAUL
EICHHORN MONTY JAY
EILERS ANTHONY MICHAEL
EKSTADT JOHN MILTON
ELAND JOHN FREDERICK
ELENBAAS JACK
ELLIOTT PHILLIP ALLEN
ELLISON NEVADA LARRY
ELLISON ROBERT LOOMIS
ELMORE GARY LEWIS
ELMY MICHAEL LEE
ELTING STEVEN VERNON
ELWART PAUL DEAN
ELZINGA LARRY LA VERN
EMEIGH MICHAEL GEORGE
EMERSON PHILIP BLAINE
EMERY ROBERT LEWIS
ENGEL GERALD WILLIAM
ENOS ROBERT RAYMOND JR
ERICKSON JOSEPH FRANK
ESCARENO ARMANDO LEO
ESCHBACH CHARLES LINWOOD
ESMAN DAVID HARM
ESTES WALTER O
ETTER PAUL QUAMMEN
EVANS DONALD PATRICK
EVANS GREGORY JAMES
EVANS THOMAS JOHN
EVERETT LUCIOUS LIONEL
EWING DAVID JAMES
EWING JERRY LEE
EX DAVID LEE
FACULAK GARY J
FAIRBOTHAM ROBERT LAWRENC
FALK DAVID JOHN
FANTE ROBERT GERALD

FARHAT ALAN JAMES
FARLEY JOHN HARLAND
FARMER WILLIAM NIAL
FARO JAMES ELLIS
FARRELL MICHAEL CHARLES
FARRO STANLEY DALE
FAUGHT DAVID LAWERANCE
FAULL CLIFFORD LEONARD
FAVOR JOHN ROBERT
FEENEY JAMES TERRANCE
FELKER GREGORY WAYNE
FENECH EMMANUEL SALVATORE
FENELEY FRANCIS JAMES
FERGUSON GARY SCOTT
FERGUSON MARION FRANKLIN
FERGUSON WALTER LEE
FERN JOHN CHARLES
FERNANDEZ EUGENIO ERASMO JR
FERRELL BILLY
FERZACCA MICHAEL
FETTERMAN GLENN LEROY
FEWLASS CALVIN JOE
FIELDER DONALD REED II
FIELDS JAMES THOMAS
FILLION WILLIAM HENRY
FISHER JAMES ELTON
FISK RICHARD OWEN
FITCH RONALD JAMES
FITZPATRICK PETER THOMAS
FLANIGAN ROBERT MORRIS
FLANNERY DAVID ELWOOD
FLEMING JAMES MARTIN
FLEMING JOHN J
FLETCHER CHARLES EUGENE
FLETCHER JAMES FERRELL
FLIPPEN HENRY COAKLEY
FOLDVARY JOHN JR
FORD GLENN JESSE III
FORD THOMAS VINCENT JR
FORMAN LEWIS MICHAEL
FORSBERG JAY EDWARD
FOSBURG RICK HAROLD
FOSTER BYRON JAMES
FOSTER DWIGHT DUNARD
FOWLER JAMES EDWARD
FOWLER JOHN KENNETH
FOX PHILLIP CARROL
FOXWORTH ROGER CHRISTOPHE
FRACKER DOUGLAS MONROE
FRALEY EUGENE THOMAS
FRANCIS JOHN PAUL
FRANCIS MICHAEL JAMES
FRANCIS WILLIAM JOSEPH
FRANCISCO DARRYL GRANT
FRANKLIN WILLIAM JOHNSON
FRANKLIN WILLIE
FRANKOWIAK ROBERT JOSEPH
FRANKS JOSEPH RONALD
FRASER THOMAS EDWIN
FRAZIER RICHARD JACOB
FRECHETTE TERRY ALLEN

FREDA NORMAN ALAN
FREDERICKSON PAUL LOWELL
FREDRICKSON ALAN DOUGLAS
FREEMAN ROBERT LEE
FREESTONE SPENCER SCOTT
FRIES DANIEL LESLIE
FRIESNER ROGER HUNTER
FRITZ JERALD DUANE
FRITZGERALD LARRY JOHN
FROEHLICH LAURENCE E
FRY JAMES HUGH
FUGETT HENRY J
FUITE RONALD JAMES
FULLER HERMON EUGENE JR
FULTON KENNETH LIGE
FUNK DALE LEE
FUSSEY GENE PAUL
FYAN RUSSELL RICKLAND
GAGNON PATRICK JOHN
GALLAGHER MICHAEL PATRICK
GALLIS STEVE SAMUEL JR
GALPIN RONALD DAVIS
GAMBOTTO LARRY LOUIS
GANDOLFO PHILIP NICK
GARDNER JAMES EDWARDS
GARRETT JONATHAN WAYNE
GARRICK JERRY ARTHUR
GARSIDE THOMAS EDWARD
GARTEN JAMES RAY
GASSER JAMES EDWARD
GAUSE CHARLIE
GAUTHIER DENNIS LEE
GAUTZ WAYNE JACOB
GAWEL JOHN LEONARD
GENITTI CHARLES THOMAS
GENTINNE THOMAS HENRY
GEORGE LOUIS AARON
GERALD BOZY
GEREAU RICHARD NORMAN
GERMANY FRANKLYN WALLACE
GERSTHEIMER HEINRICH
GERSTNER RONALD EDWARD
GERTEN RONALD EUGENE
GETTY DENNIS ALAN
GIBBS KITCHELL SNOW
GIBSON BRUCE STEWART
GIBSON KEITH EDWARD
GIBSON RAYMOND ALBERT
GIDDENS LA VON
GIDDINGS JOHN PATRICK
GILBERT RICHARD LEE
GILBREATH GRANT MADISON
GILES DENNIS LEE
GILIN GARY JOSEPH
GILLESPIE JAMES ROY
GINAL RICHARD
GIPNER GORDON PHILLIP
GIPSON CLARENCE FREDDIE
GIRARDOT FRANCIS MICHAEL
GIUSTA JOSEPH MICHAEL
GIVENS DAVID JERRY
GLADNEY HAROLD SAMUEL
GLASCO DAVID MAX
GLEASON DANIEL WARD

GLESENKAMP JOHN CARR
GLIME DONALD EUGENE
GLOWE STEPHEN
GNIEWEK KENNETH STANLEY
GOFF LARRY ALLEN
GOLEMBIEWSKI WALTER EDWAR
GOLOMBESKI WALTER BILLY
GOMEZ EVELIO ALFRED
GONZALES NICHOLAS VALERIA
GOOCH JERRY DALE
GOOD LARRY DEAN
GOODALL EARL WAYNE
GOODCHILD RAYMOND LEE
GOODMAN NORMAN ELAN
GOOSEN ROBERT HENRY
GORAL GERALD EUGENE
GORBE VAUN ARLEN
GORDON JAMES LEWIS
GORDON ROGERS STUART
GORDON WAYNE VICTOR
GOSCHKA LARRY HERMAN
GOSS HEZEKIAH JR
GOUGER WILLIAM DAVID JR
GOULD CARLYLE LEROY
GOULET RONALD DAVID
GRABOSKEY EDWARD ELLIOTT
GRACHTRUP JOHN NORBERT
GRAHAM DAVID DEL
GRAHAM GORDON J
GRANDAHL JACK WILLIAM
GRANT ANDREW CARL
GRAY ASA PARKER JR
GRAY LARRY GENE
GRAY THOMAS EDWARD
GREEN GEOFFREY EMMONS
GREEN KENNETH GERALD
GREEN LARRY EDWARD
GREEN LARRY VERNARD
GREEN RALPH LA VERNE
GREENE FREDERICK DAVID
GREENE JESSIE FLOYD JR
GREENHOUSE RONALD RAPHAEL
GREENWALD DENNIS
GREER FRANK MICHAEL
GREGORY ROBERT ARTHUR
GREGSON THOMAS ROBERT
GREILING DAVID SCOTT
GREKELA WILLIAM EINO
GRENNAY WILLIAM EFREN
GRESSEL JOHN VINCENT
GRIEVE MICHAEL A
GRIFFIN CHARLES FARRELL
GRIFFIN WILLIAM DONALD II
GRIGGS EDWARD LOUIS III
GROAT RICHARD JAMES
GROAT WAYNE DOUGLAS
GROF ROBERT LESTER
GRONAU DAVID JAMES
GRONOWSKI THEODORE JR
GROOM ALAN DAVIS
GROOVER JOHN WILLIAM O
GROSS ALAN HARRY
GROSS RICHARD ALBAN

GROSS ROBERT HENRY
GROTH WADE LAWRENCE
GROVE KENNETH ARNOLD
GROVER ROBERT JOHN
GROW GARY WARREN
GRUEZKE JAMES A
GRULKE BARRY RICHARD
GRYZEN GARY M
GUENTHER WILLIAM RICHARD
GUILMETTE DENNIS MICHAEL
GUINN ROBERT GEORGE
GULASH DAVID JOHN
GULICH DENNIS FRANSIC
GULLA DENNIS JAMES
GULLEY HOUSTON
GUTIERREZ HENRY L JR
GUTIERREZ OSCAR G
GUY GEORGE ANDREW
GUYTON MELVIN
GUZZO ALFREDO
GYDESEN GREGORY ALLEN
GYULVESZI THEODORE LOUIS
HAAG RICHARD HAROLD JR
HAGE MARK KELLOGG
HAGER THOMAS GARY
HAHN BRUCE EDWARD
HAINES JOHN LODA
HAKE WILBUR O
HALEY GARY ROBERT
HALL DAYLE RAYMOND
HALL DEAN ELLSWORTH
HALL DONALD ALLEN JR
HALL JEFFERY H
HALL WILLIAM GARDINER
HALLOWELL ALBERT GEORGE
HALSTEAD LEE MICHAEL
HAMES LAWRENCE EVERETT
HAMILTON JAMES RICHARD
HAMILTON PAUL JR
HAMILTON RICHARD LENARD
HAMMARSTROM ARTHUR F JR
HAMMOND CHARLES WELDON
HAMMOND DENNIS WAYNE
HAMMOND TERRY MICHAEL
HANEY BOBBY GENE
HANEY WILLIAM DAVID
HANEY WILLIAM THOMAS
HANNA ELGIE GEORGE
HANNING DONALD JERRY
HANSELMAN CHARLES LEON
HANSELMAN ROBERT ALAN
HARBISON SHERRON EVERETT
HARDY WILLIE CHARLES
HARKNESS THOMAS JONES JR
HARMON CHARLES
HARMON DENNIS LEROY
HARMS LOWELL EUGENE
HARPER GREGORY ALEXANDER
HARPER HAROLD OWEN
HARRINGTON JOHN CHARLES
HARRIS CARL ALLEN
HARRIS DOYLE LEE
HARRIS EDWARD LAWRENCE
HARRIS HAL
HARRIS PHILIP ANTHONY
HARRIS ROLAND LORENZO
HARRIS TERRENCE L
HARRIS WILLIAM LAWRENCE
HARRISON THOMAS NORMAN
HART RAYMOND LEONARD
HARTER JAMES WILSON
HARTMAN HOWARD JOHN
HARTNESS DONALD HARRY
HARTRY OTIS LAMONT
HARTSUFF LEO FRANCIS
HASFORD JOHN LAWRENCE JR
HASTINGS THOMAS WILLIAM
HATFIELD BOBBY RAY
HATFIELD MICHAEL JAMES
HATH JAMES STUART
HATTON RUSSELL ODELL
HAUER LESLIE JOHN
HAWKINS HAROLD FREDRICK
HAWLEY PETER SHELDON
HAWYER DONALD ROBERT
HAYDEN MICHAEL PYM
HAYES DALE LAMONT
HAYES NEIL BURGESS JR
HAYES NELSON LLOYD
HAYS KENNETH DOUGLAS

Michigan

HAZEN PAUL GORDON
HEAD DAVID FREEMAN
HEARNS ROGER CHARLES
HEARSCH JOHN PATRICK JR
HEATH NED ARTHUR
HECK NORMAN WALTER JR
HECKMAN CLARENCE ALVIN
HEDGER JAMES ROBERT
HEETHER JAMES JOSEPH II
HEISER JOHN LOUIS
HELIKER RUSSELL JAMES
HELKA GLENN OWEN
HENDERSHOTT THOMAS EDWARD
HENDERSON RICKY DONALD
HENDRICKSON GERALD RAY
HENLEY CHARLES RAY
HENNESSEY JAMES DALE
HENRICH BRUCE JAMES
HENRY BERNARD JAMES
HENRY THOMAS CARMEN
HENSLEY GARY LEE
HERING JAMES ALFRED
HERMANN GREGORY WILLIAM
HERNANDEZ RAMON SANCHEZ
HERNDON RICKY LYNN
HERNDON THOMAS HAYDEN
HERRADA GABRIEL
HERRANDO ROGER DANIEL
HERRERA JIMMIE ANDREW
HERRINGTON RAYMOND NEIL
HERRINGTON RICHARD LEE JR
HETRICK GERALD EVERETT
HEYER WALTER EARL JR
HICKS HIAWATHA
HIGBEE GARY LEE
HIGGINS JOHN FRANCIS
HIGGS DANNY TRENT
HIGHLAND BYRON GRANT
HILDERBRANT PHILLIP JAY
HILL PHILLIP ANDREW
HILL ROBERT HARDY JR
HILL ROBERT LA VERNE
HILL STERLING HAROLD
HILLIARD JAMES FRANCIS
HILYARD JAMES HAROLD
HIMEBAUGH LEE EDWARD
HINES RANDY VICTOR
HINKLE JOHN WARREN
HINTZ JAMES RAYMOND
HINTZ JAMES WESLEY
HJORTH WILLIAM HAROLD
HOADLEY LARRY FRANCIS C
HOAG EARL THOMAS
HOBBS CECIL R JR
HOBBS RONALD WAYNE
HODAL ROBERT JOHN
HODGE EDWARD L
HODGE MICHAEL LEONARD
HODGE RONALD ELLSWORTH
HOEKER JOSEPH ALAN
HOFF MICHAEL GORDON
HOFFMAN MELVIN ELMER
HOGLUND MICHAEL AUGUST
HOHN RODNEY ALLEN
HOLBROOK GARY WAYNE
HOLDEN LOWELL DEAN
HOLDERBAUM JOHN ARTHUR
HOLIDAY MICHAEL LEONARD
HOLLINGSWORTH RICHARD LEE
HOLLOWAY MICHAEL SCOTT
HOLMAN GERALD ALLAN
HOLMAN MARSHALL DANIEL

HOLMES JIM HENRY
HOLSTON ARVELL BERNARD
HOLSTON PAUL
HOLTREY DANIEL PERRY
HOLTZLANDER DOYLE EDWARD
HOLUPKO LON MICHAEL
HONEYCUTT DONALD EUGENE
HOOD RAYMOND
HOOGTERP STEPHEN JOSEPH
HOOPENGARNER BENJAMIN LEE JR
HOOVER GERALD DONALD
HOOVER ROGER JOSEPH
HOPKINS JACK MAYNARD
HOPSON JAMES HARVEY
HORSLEY LA MONTE VAN
HOSKINS HAROLD ORION
HOSKO GARY LYNN
HOSNEDLE ALAN ROGER
HOUGH MICHAEL PETER
HOWARD CHARLES VINCENT
HOWARD JAMES RAY
HOWE JOHN ALLAN
HOWER THOMAS ALLEN
HOWZE DAVID JR
HOYT NORMAN LEE ROY
HOYT VICTOR RONALD
HRISOULIS ROBERT
HUARD JAMES LINTON
HUBBARD MEREDITH GERALD
HUBBELL THOMAS SIMCOCK
HUCZEK GERALD ALBERT
HUEY HERMAN LEROY
HUGHES FERNANDO JAMES
HUGHES GREGORY JOHN
HUGHES ROBERT ALLEN
HULTQUIST EDWARD CHARLES
HUMPHREYS GILMER EARL
HUNTER BARRY ALAN
HUNTER BILLY RAY
HUNTER MICHAEL WOODROW
HURD SAMUEL EUGENE
HURLBUT THOMAS WILLARD
HURSTON HORATIO WILLIAM
HURT PAUL THOMAS III
HUSCHER JOHN RANDOLPH
HUSTED GARY ARZA
HUTCHINSON WALTER EUGENE
HUXTABLE RONALD LESTER
HYDER FLOYD ALLAN
HYETT KENNETH MONROE
ILLI DANIEL JOHN
INDRECC GREGORY THOMAS
ISAACS GARY NEAL
JACKEMEYER ROBERT RAYMOND
JACKSON DOUGLAS
JACKSON GARLAND DUANE
JACKSON KENNETH MCKINLEY
JACKSON WILLIAM EUGENE
JACKYMACK RUDOLPH
JACOBS DAVID PAUL
JACOBS RICHARD ALLEN
JACOBSON JOHN W SR
JACQUES ROBERT PAUL
JAKO JAMES LOUIS
JAKOVAC JOHN ANDREW
JAMES DONALD JR
JAMISON ROGER LEE
JAQUISH JAMES IVAN
JARRELL JOSEPH DANIEL
JARRELL WENDALL JOSEPH
JARRETT JAMES DALE
JARVIS JEREMY MICHAEL
JASKIEWICZ DANIEL JOSEPH
JASURA ROBERT WILLIAM
JAWOROWICZ LAWRENCE FRANK
JEFFRIES GERRIE GEORGE
JENKINS LARRY RUFUS
JENKINS ROBERT DOUGLAS
JENKS HARLEY JOHN
JENKS JOSEPH WILLIAM
JENKS ROBERT JAMES
JEROME STANLEY MILTON
JESSE CLIFFORD EARL
JEZIORSKI DENNIS ALFRED

JOHNSON BRUCE GARDNER
JOHNSON BRUCE MICHAEL
JOHNSON CARL IRVING
JOHNSON CHARLES TIMOTHY
JOHNSON CHRISTOPHER PAUL
JOHNSON CLARENCE EDWARD
JOHNSON DALE LLOYD
JOHNSON DANNIE LEWIS
JOHNSON DAVID KEITH
JOHNSON GERALD LEE
JOHNSON JOE D JR
JOHNSON LAWRENCE EUGENE
JOHNSON MICHAEL JAMES
JOHNSON PAUL EDWARD
JOHNSON PHILIP ALLEN
JOHNSON RICHARD ARNOLD
JOHNSON ROBERT BRUCE
JOHNSON WAYNE DAVID
JOHNSON WILLIAM HENRY
JOHNSON WILLIE JR
JONES BYRON NORRIS
JONES DANNY LEE
JONES HORATIO LEE
JONES JAMES
JONES JAMES DALE
JONES JON CARL
JONES LARRY ALLEN
JONES LEE FRANCIS
JONES MARVIN HAROLD JR
JONES PAUL DAVIS
JONES RICHARD ALFRED
JONES RICHARD STEPHEN
JONES ROBERT ERNEST
JONES TERRY AGUSTA
JONES TRACY LEE
JONES WILLIS GEORGE
JORDAN GRADY MERRIL
JOSE PAULL DAVID
JOZWIAK ROGER EDWARD
JUDD DAVID TERRENCE
JUNGA HAROLD JOSEPH
KAMINSKI KENNETH
KAMINSKI RICHARD DENNIS
KANAAR LOUIS KENNETH
KANGAS ARTHUR NELSON
KANSIK FREDERICK DANIEL
KARI JARMO ANTERO
KARR CHARLES LEE
KASNOW EDWARD
KAUFFMAN EARNEST LEE
KAY BRYAN THOMAS
KAYGA WILLIAM DUANE
KEATING RALPH AINSWORTH
KECK FRANK LESLIE
KEENE WALTER MARTIN
KEIL DUANE RICHARD
KEIRNS THOMAS LEE
KEITH DANIEL SCOTT
KELLER CHARLES HENRY II
KELLER PETER JOSEPH JR
KELLEY EDDIE RALPH
KELLEY JOHNNIE WOODROW
KELLEY THOMAS R
KELLY JOHN WILLIAM SIDNEY
KELLY MICHAEL DENNIS
KELLY MICHAEL JOHN
KELSEY CLIFFORD EARL
KELSEY J C
KELSEY RONALD KEITH
KEMBLE DONALD WILLIAM IV
KEMP JERALD WAYNE
KEMP THOMAS WILLIAM
KENNEDY DONALD LEE
KENNELL DANNY OWEN
KENT JESSE PHILLIP
KEPSEL ELMER FRED
KERR NORMAN THEODORE
KESKI KEITH
KEYES WILLIAM GEORGE
KIEL STEVEN TRACY
KILLIAN GARY MARTIN
KILLING RONALD JAMES
KIMZEY JOHN ALBERT
KING DONALD LEWIS
KING IVAN CLAUS
KING JAMES ROGERS
KING STEVEN ROSS
KINNEY CHARLES WILLIAM
KIPINA MARSHALL FREDERICK
KIRACOFE BURLEY DARREL
KIRBY JAMES EUGENE

KIRBY LEWIS ROY
KIRCHNER GARY ALLEN
KISH CARY MICHAEL
KITRILAKIS JOHN ANDREW
KIVEL ELMER MARVIN
KLCO JAMES EDWARD
KLETT JOHN EARLE
KLIGAR JOHN JOSEPH III
KLIMO JAMES ROBERT
KLINE GARY WAYNE
KLINE ROBERT JAMES
KLINGLER GARY LYNN
KLINSKI MICHAEL ROMAN
KLIPPEL DAVID JOHN
KLOS RONALD FRANK
KLUGG JOSEPH RUSSELL
KLUSENDORF HAROLD JOHN
KNAGGS JOHN CHRISTOPHER
KNARIAN DANIEL
KNIGHT MARTIN ROY
KNOBLOCH CRAIG GEOFFREY
KNOLL RAY EDWARD
KNOLL ROBERT EDWIN
KNOX IRVILLE J
KOCH RONALD LEE
KOEBKE JOHN LEE
KOENIG DAVID BRUCE
KOHLER PAUL JEROME
KOHLER TERRY
KOHUT ROGER SCOTT
KOIVUPALO ROBERT W JR
KOLAKOWSKI HENRY JR
KOLKA EDWARD LOUIS
KONING DOUGLAS LEE
KOOI JAMES WILLARD
KOONE JACK RUSSELL
KOOS NORMAN LAVERN
KOPP BARRY LORENZ
KORSON GERALD EDWARD
KOSAKOWSKI GERALD ANTHONY
KOSEBA DENNIS WILLIAM
KOSKI GENE RAYMOND
KOTEWA FLOYD WILLIAM JR
KOTKE LEO LEROY
KOTTYAN GEORGE EDWARD
KOVACEVICH THOMAS JAMES
KOVALCSIK RICHARD
KOWALSKI LEONARD J JR
KOWALSKI ROBERT ALLEN
KOWITZ DAVID RALPH
KOYL HARRY GLENN
KOZDRON CHESTER JOSEPH
KOZIOL JOHN THOMAS
KRAEMER MAURICE PETER JR
KRALOWSKI JAMES EDWARD
KRANSI RONALD TERRY
KREBS FRANK J
KREH GARY HAROLD
KRIDLER BERNIE CHARLE III
KROLIKOWSKI RICHARD
KRONBERG CHARLES AUGUST
KROPP GORDON GENE
KRUGER FREDERICK LYLE
KUCHEK RICHARD MICHAEL
KUCWAY ROBERT JOHN
KUCZYNSKI DAVID EDWARD
KUEBLER CLIFFORD A JR
KUICK STANLEY J
KULIK CASIMIR
KUNKLER HARRY GROVER III
KUNNA FREDERICK CARMEN
KUNST GENE ARTHUR
KUPIEC THOMAS MARK
KUSHMAUL ROBERT EDWARD
KUTCHEY LAWRENCE DAVID
KUZILLA DONALD G
LA BOHN GARY RUSSELL
LA COSSE JIMMY JOHN
LA COST REGNOLD JOSEPH
LA LONE JAMES CLIFTON
LA NORE DENNIS ARNOLD
LA PORTE BRUCE STEPHEN
LA ROCCA VINCENT MICHAEL
LA ROUCHE JAMES MICHAEL
LACKLAND LUTHER JAMES
LACLEAR JAMES PHILLIP
LADD ALBERT ALLEN
LAFFERTY THOMAS LEE
LAFLER JOHN JAMES
LAHR CLYDE DAVID
LAHTI JAMES WALTER
LAIRD JAMES FRANKLIN M

LAJKO ROBERT DENNIS
LAMAR WILLIAM ERNEST
LAMB ELWIN JAY
LAMEY LAVERN MICHAEL
LAMONT PETER ALAN
LAMS ALLEN JAMES
LANDER MARK ROBERT
LANDON VINCENT P
LANDWEHR DUANE HENRY JR
LANE ALAN
LANE ALBERT LEROY JR
LANE THOMAS ALLEN
LANE WARREN CLIFFORD
LANGE RICHARD ROSS
LANGLER STEPHEN DOUGLAS
LANIER CHARLIE LOUIS
LANING JOHN EDWARD
LANTEIGNE ARTHUR
LANTRY MERRILL LAGENE
LAPHAM ROBERT GRANTHAN
LARABEE BENJAMIN CARLTON
LARGE BRUCE EDWARD
LARGENT JOHN ALYN
LARRICK RICHARD ALLEN
LARSEN MICHAEL CONRAD
LAUDICINA JAMES RAY
LAUSE DALE MICHAEL
LAUTNER FRANCIS ANTHONY
LAVERTY STEVE L J
LAVIGNE STEVEN BRUCE
LAW BRENT ROBIN
LAWFIELD GLENN ROBERT
LAWRENCE RICHARD ALFRED
LAWSON WILLIAM EDWARD JR
LAZAR DANIEL STEPHEN
LEACH JAMES EDWARD
LEDFORD JEFFERY LEE
LEE GARY ELTON
LEE JOHN ALEX
LEHMANN PETER BODO
LEIGHTON THEODORE RICHARD
LEIJA MARIANO JR
LEITCH LARRY DUANE
LEONARD MARVIN MAURICE
LEPARD DONALD GEORGE
LEPTRONE FRANK
LESKY CHRISTOPHER ALLAN
LESS RANDALL PATRICK
LEWANDOWSKY STANLEY ROBER
LEWIS ALFRED JOHN
LEWIS RICHARD KENNETH
LEWIS WILLIAM DAVID
LICHOTA DENNIS
LIGHT WILLIAM MARVIN
LIMINGA FREDERICK HUGO
LINDEMANN JAMES WILLIAM
LINDSEY DENNIS PAUL
LINK DAVID JOHN
LINK GARY WILLIAM
LINN ROBERT LEWIS JR
LINNA STEVEN PAUL
LINVILLE MICHAEL THOMAS
LIPAROTO LEONARD JOSEPH
LIPSEY THOMAS WASHING III
LISKOW LARRY LEE
LITTLE RANDELL BLAKE
LIVINGSTON NORMAN JAMES
LIVINGSTONE DAVID MICHAEL
LOBBEZOO DENNIS LEE
LOBKER DANN JOSEPH
LOCKETT WILLIAM NORRIS
LOCKHART DOVER LEON
LOGAN BRADLEY JOHN
LOGAN JAMES DWIGHT
LOGAN JOSEPH LAWRENCE JR
LOISEL JAMES LEE
LOISELLE RICHARD J
LONG DOUGLAS EUGENE
LONG JAMES ROBERT
LOPEZ FRANK JR
LORD ROBERT RANDAL
LORD STEPHEN GEOFFREY
LOSO JAMES MICHAEL
LOUCKS HERBERT ALLEN
LOUGHREN MICHAEL EVAN
LOUX JAMES ARTHUR
LOWE ROBERT KINLOCH
LOWER LARRY LA MAR
LOZANO CARLOS FELIPE

MEND
LOZIER WILLIAM EARL
LUBAVS KONSTANTINS ADOLFS
LUCIER JOHN WILLIAM
LUGAR DENNIS WAYNE
LUKAS JEROME KRISTIN
LUKES THOMAS BURTON
LULOFS DENNIS JAY
LUNA ADOLFO
LUNDBERG WILLIAM RAYMOND
LUSTER LARRY
LUTZE JOHN EDWIN
LYDEN DENNIS M
LYNCH JAMES OLIVER
LYNCH JOHN CHARLES P
LYON CHRISTOPHER EDWIN
LYONS JAMES ANDREW
LYVERE RONALD LEE
MAC BETH KENNETH NEIL
MAC DONALD HAROLD LEE
MAC DONALD LARRY EDWARD
MACHUL JOHN FRANCIS
MACK DENNIS LEE
MADDOX JULIUS
MAES DANIEL JOHN
MAHARG EVERT RALPH
MAIDENS MICHAEL ROBERT
MAISANO JOSEPH ANTHONY
MAJOR LA MARRE ARTHUR
MAKAREWICZ DANIEL
MAKI FRANK RUDOLPH
MAKI GLEN ARVID
MALDONADO CARLOS O
MALEC DENNIS STANLEY
MALKUT STEFAN
MALLETT DOUGLAS MACKARTHE
MALLORY JERRY DOUGLAS
MANER HOWARD JOSEPH
MANIERE MICHAEL JOHN
MANK ROYAL CHARLES
MANNEROW PAUL DAVID
MANNING DENNIS CARROL
MARCH DONALD GEROLD
MARCHLEWICZ ARNOLD M
MARCOTTE DENNIS WILFRED
MARIA CHARLES ANTHONY
MARKEL RONALD JOE
MARKEY CHRISTOPHER HUGH
MAROSITES BRUCE LOUIS
MARSH LARRY LEE
MARSHALL BRIAN ALEXANDER
MARSHALL JOSEPH HENRY III
MARSHMAN MICHAEL JON
MARTIN BRUNO LEO
MARTIN ELMER
MARTIN HENRY CHARLEMONT
MARTIN JAMES LOUIS
MARTIN JERRY LEWIS
MARTIN KENNETH
MARTINE JAY BARKLOW JR
MARTINEZ REYNALDO
MARTINEZ TOMAS VASQUEZ

MARTZ MELVIN LOUIS
MARVIN ROBERT CLARENCE
MASSEY RALPH LAWRENCE
MASSUCCI MARTIN JOHN
MASTERSON EDMUND MACEO
MATARAZZO EVERETT ROBERT
MATCHETT JAMES STEVEN
MATHESON DOUGLAS ROY
MATHEWS CHARLES DONALD
MATTIE ANDREW MARION
MAWDSLEY DANNY JOSEPH
MAXAM JAMES ALAN
MAXSON CHARLES DANIEL
MAY MICHAEL FREDRICK
MAYBEE MICHAEL OWEN
MAYS THOMAS CURTIS
MC ARTHUR MELVIN LLOYD
MC CANN CECIL DARRELL
MC CARTHY BRIAN EDWARD
MC CLAIN JAMES HARRY
MC CLAIN WILLIE JAMES JR
MC CLURG CHARLES D
MC COMB TERRY RUSSELL
MC CORMICK JOHN VERN
MC CORVEY GERALD
MC COY RICKEY CLAUDE C
MC CREERY FLOYD SANFORD
MC CULLOUGH BENJAMIN LEE
MC CURRY ANDREAS
MC DAID JOHN MURL
MC DANIEL FRANKIE B
MC DONALD GEORGE E JR
MC GEE KENNETH WESLEY
MC GEE ROBERT LEE
MC GEE ROY DELL
MC GINNIS WILLIAM E II
MC GOVERN PATRICK EDWARD
MC GRATH JOHN AUGUST
MC HUGH FRED C JR
MC ILROY DOUGLAS STEVEN
MC ILROY PATRICK C
MC ILVOY JAMES LEE
MC INTIRE WALTER EDWIN JR
MC KEAGUE GREGORY DEAN
MC KEATHON DWIGHT PINZA
MC KEE KENNETH DALE
MC KELLAR DENNIS ALVIN
MC KENNA ROBERT CHARLES
MC KENZIE DAVID DAYLE
MC KINLEY PATRICK JAMES
MC KINNEY DWIGHT A JR
MC KINNIS CLARENCE EARL
MC LEAN DONALD KENT
MC LEAN JAMES MC MURRAY
MC LEAN TERRY R
MC LENNAN GARY ALFRED
MC MAHON THOMAS MARK JR
MC MANN ALVIN CHARLES JR
MC MANUS MICHAEL GEORGE
MC MEANS DENNIE
MC NABB DOUGLAS MEREDITH
MC NEES DWIGHT ALLEN
MC NEILL KENNETH REX

MC NEW BRIAN RICHARD
MC PHAIL MORRIS GENE
MC PHEE DONALD CAMERON
MC PHERSON MICHAEL LEE
MC QUADE WILLIAM VICTOR
MC QUEER MICHAEL PATRICK
MCCLOUD WILLIE JR
MCGEE CARL BARRY
MEAD LENUS EDWARD
MEADOWS MERL RUSSELL
MEAGHER ROBERT JOHN
MEASELL KENNETH WILLIAM
MEDARIS RICK EGGBURTUS
MEDLEY MICHAEL MILTON
MEDLIN RICKEY JOE
MEEHAN MICHAEL ALLEN
MEEK CHARLES EDWARD
MEEK THOMAS WESLY
MEGIVERON EMIL GEORGE
MEIRNDORF BERNARD JAMES
MELL FRANK RALPH JR
MELVIN JOSEPH ERNEST
MERINGA GARY PAUL
MERRILL HUGH WALLACE
MERRITT VERNON ALLEN
MERRYMAN DENNIS GARY
MERZ JAMES R JR
METZ GARY RAYMOND
METZLER PERRY
MEYER TERRY ROBERT
MEYER WILLIAM MICHAEL
MICHAEL PATRICK JOSEPH
MICHALKE RUSSELL ARTHUR
MICHALSKI JOHN MITCHELL
MICHELS DONALD MATHEW JR
MICULS JANIS
MICUNEK MICHAEL MARK
MIDDLETON KENNETH DALE
MIEDEMA MATTHEW GEORGE
MIEDZIELEC TIMOTHY R
MILES PATRICK CHARLES
MILKS RICHARD ALLEN
MILLAN ROBERT DENNIS JR
MILLER CHARLES IRVIN
MILLER CHARLES WAYNE
MILLER DAVID BRUCE
MILLER DAVID RAYMOND
MILLER DOUGLAS J
MILLER ERNEST LEE
MILLER EUGENE STUART
MILLER GEORGE ERNEST
MILLER JAMES EDWARD
MILLER JOHN ROBERT
MILLER KENNETH
MILLER KENNETH EDWARD
MILLER RALPH PETERSON III
MILLER RAYMOND P II
MILLER RICHARD DAVID
MILLER ROBERT J
MILLER RONALD JAY
MILLER WILLIAM LOUIS III
MILLIMAN DAIN W
MILLS RODNEY KENNETH
MILLS STEVEN BERNARD
MINCH ROGER CARL
MINUS RAYMOND BENJAMIN
MIS RONALD HENRY

MISZEWSKI DAVID MARTIN
MITCHELL LEROY GERALD
MITCHELL ROBERT E JR
MIZE WILLIAM DAVID
MODDERMAN PHILIP JOHN
MOE CHARLES MERLIN JR
MOEGGENBORG LENARD F
MOHAREMOFF MICHAEL GEORGE
MOILANEN DALE BURTON
MOLINE KEVIN EUGENE
MOLL ROGER RALPH
MOLNAR NICHOLAS MICHAEL
MONHOF AUGUST HAROLD
MONTROSS BURTON CHARLES
MOORE CHARLES BERNARD
MOORE DAVID ALLEN
MOORE DAVID CHARLES
MOORE ERNEST LAWRENCE
MOORE HAROLD
MOORE STANLEY LEROY
MOORE WILLIAM VINCENT
MOOREHEAD RICHARD L
MORAS ROBERT JOHN
MORE GARY KEITH
MOREY DARRELL H
MORGAN CALVIN CARL
MORGAN JAMES MARK
MORGAN LEONARD ANTHONY
MORLEY JAMES RICHARD
MORONEY ROBERT JEROME
MORRIS RAYMOND LESTER
MORRISON JACK ALLEN
MORRISON JAMES JOHN
MORROW JAMES FRANCIS
MORROW KENNETH PORTER
MORSE HOWARD EDWARD
MOSER KEITH MILTON II
MOSES DONALD SYLVESTER
MOSHER ALEX ROY
MOSIER CLIVE VERE
MOSLEY JOHN CHARLES
MOUNCE BARRY MITCHELL
MRAZIK JAMES PATRICK
MROSEWSKE ROY JAMES
MUELLER RALPH THOMAS
MUELLER ROBERT STEPHAN
MUELLER WOODROW JOHN
MULHOLLAND ARNOLD LEE ROY
MULLIN GERALD CARL
MUNN WILLIAM ARTHUR
MUNOZ JOSE
MUNSEY RALPH CHARLES
MUNSON EDWARD LOUIS
MURDOCK CARL THOMAS
MURPHY DONALD JOSEPH
MURPHY PATRICK MICHAEL
MURPHY PATRICK WILLIAM
MURPHY VINCENT FRANCIS
MURPHY WILLIAM JOSEPH
MURRAY MICHIEL DAVID
MURRELL JIMMY ROGER
MUSCYNSKI FRANK
MUSICH JOHN PAUL
MUSSIN ROBERT JAMES
MUTSCHLER JOHN LLOYD
MUTZ DENNIS HOWARD
MYERS GORDON E
MYERS JOHN MAURICE
MYERS LARRY DALE
MYERS MICHAEL LEE
NABOZNIAK MYRON RICHARD
NAGELKIRK DENNIS DALE
NAHAN JOHN BENEDICT III
NATALIE RONALD JOHN
NAUGHTON THOMAS DANIEL JR
NAUSS BRENT BRITTEN
NEESON BRUCE ROBERT
NEGRO DANIEL LEE
NEGUS JACK THOMAS
NEIL SAMUEL THOMAS JR
NELSON DANIEL ALAN
NELSON DAVID BRUCE
NELSON DAVID ELLSWORTH
NELSON DONALD ERWIN
NELSON DONALD LEE
NELSON FRED ARTHUR
NELSON JAMES RAYMOND
NELSON LARRY DOUGLAS
NELSON REX FRANKLIN JR

NELSON RICHARD ALLEN
NELSON WILLIAM HUMPHREY
NESBIT ROGER CHARLES
NEWHOUSE BERNARD JOSEPH
NEWSTEAD THOMAS EUGENE
NEWTON DONALD WILLIAM
NEWTON VERNON LEE
NICHOLL DALE ALLEN
NICHOLS ELI WAYNE
NICHOLS ERNEST JAMES JR
NICHOLS GARY BRUCE
NICKERSON GENE BERTAN
NICKLYN ROBERT JAMES
NICOL MICHAEL WILLIAM
NIEBOER DOUGLAS ALAN
NIEMI MARTIN ROY
NIEMI ROGER LYLE
NIEZGODA MICHAEL ALLEN
NIMOX BENNY FRANK
NIMPHIE MAX EDWARD JR
NITZ ROBERT FRANKLIN
NIXON LEN EVERETT
NOBLE RONALD GLEN
NOLDE WILLIAM BENEDICT
NOLFF DANIEL BENSON
NOLL DAVID ROGER
NOON JACK ALDEN
NORTH DONALD RICHARD JR
NORTHOUSE ROLLIE MELVIN
NORTON BENJAMIN PAUL
NORTON DEWIGHT EDWARDS
NOVAK BERNARD JOHN
NOVAK MICHAEL JOSEPH
NOVOTNY RICHARD DENNIS
NOWACZYNSKI NATALIE
NOWAK JOHN THOMAS
NOWAKOWSKI JOHN ALEXANDER
NOWRY RICHARD LOREN
NOZEWSKI ROBERT
NUBER RICHARD ANTHONY
O BYRN HERMAN JAMES
O CONNOR GARRETT TIMOTHY
O DONNELL DANIEL MARTIN
O DONNELL ROBERT WAYNE
O NEAL JERRY LEE
O NEAL RICHARD MARK
O NEIL VAUGHN THOMAS
OBNEY RONNIE LEE
ODENWELLER PETER EDWARD
OLEKSA CHRISTOPHER JAMES
OLENZUK KENNETH FRANCIS
OLENZUK PAUL GREGORY
OLINGER JAMES EDWARD JR
OLLIKAINEN ROBERT JOHN
OLNEY STEVEN IRA
OLOFSON PHILIP JOHN
OLSEN DONALD BRYAN
OLSON HENRY LOUIS
OMILIAN DENNIS ALLEN
ONKALO PHILIP GORDON
ORLANDO SAMUEL GIZZI
ORLOWSKI HEDWIG DIANE
ORR MICHAEL WALTER
ORRIS STEVE III
ORTWINE DENNIS ROYAL
OSIER ROBERT DALE JR
OSTERBERG BERNARD J JR
OTT ALAN ROBERT
OTTE RICHARD LEE
OUILLETTE THOMAS ROBERT
OUTMAN CHARLES EDGAR
OVERTON DANNY JR
OVERWEG GEORGE ALLEN
OVERWEG ROGER DALE
OVIST DAVID EMANUEL
OWENS JOHN WILLIAM
PACE GEORGE ALEXANDER
PACK SANFORD GENE
PAINTER HOWARD LEROY
PAKULA THOMAS VINCENT
PALMER BRUCE CAMERON
PALMER DAVID SCOTT
PARKER CHARLES JOHN
PARKER WILLIAM GENE
PARKIN HAROLD LESLIE
PARKS ALAN HUGH
PARKS JERRY EMMET
PARKS JOSEPH L
PARKS RAYMOND GEORGE
PARKS WILLIE ALBERT
PARMELEE JAMES EARL

Michigan

PARR LARRY DOUGLAS
PARRISH IVORY PERRY
PARTEE WARDLOW WESLEY
PARTIDA ANDREW
PATON RICHARD ALLEN
PATTERSON GEORGE WILLARD
PATTERSON JAMES BARNETT
PATTERSON MICHAEL RICHARD
PATTERSON WALTER MARCELLU
PAUL GARY MICHAEL
PAUL JAMES LEE
PAWLAK JAMES WILLIAM
PAYNE ROY CHARLES JR
PEARCE WILLIAM CALVIN IV
PEARSALL RICHARD MARK
PEASLEY GARY WAYNE
PECKHAM GEORGE ROBERT
PEEK JOHN FOREMAN
PELCH MICHAEL J D
PELLOSMA DAVID JOHN
PELTIER JAMES WARDEN
PENDELL DAVID ALLEN
PENKE RICHARD ALLEN
PENNELL WILBERT GENE
PEPIN JOHN FREDERICK
PERKETT DAVID LOUIS
PERMALOFF CHARLES WASSEL
PERRIGO STANLEY CHARLES
PERRY OTHA LEE
PERRY RANDALL LAWRENCE
PERRY ROBERT CONROY
PERRY WILLIAM EDWARD
PERSICKE ALLAN WAYNE
PERSON JAMES ALFRED
PERSONETTE MICHAEL DARWIN
PERSYN RONALD FRANK
PETELA THOMAS JOSEPH
PETERSEN PAUL JOSEPH
PETRE RONNIE JOSEPH
PETRIMOULX ROBERT GORDON
PETROLINE PAUL EDWARD
PHAIR JAMES W
PHELPS RONALD JOSEPH
PHENNEY GEORGE S
PHILLIPS JAMES CLIFFORD
PHILLIPS JAMES JR
PICCIANO TERRANCE ALAN
PIERCE ROBERT DUANE
PIERCY ROBERT CONOVER
PIERPONT WILLIAM MCGREGOR
PIETRZYK MARK HAROLD
PINEAU ROLAND ROBERT
PINGEL WAYNE EDWARD
PINKNEY ROBERTIS
PIOTROWSKI DANIEL JOSEPH
PIRKOLA PAUL HENRY
PITCOCK ELZIA RAY
PIZZUTI JOHN
PLANTE GARY WILLIAM
PLUM BILLIE NEAL
PLUMB CHARLES DONALD JR
PLUMM RICHARD DALE
POBLOCK BERNARD FRANCIS
POE JAMES WALKER
POE JOHN WAYNE
POET LAWRENCE
POHJOLA JEFFREY WILLIS
POLASEK JOSEPH JAMES JR
POLEGA GERRY ALBIN
POLISKI MICHAEL CHARLES
POLISKY THOMAS RICHARD

POLKINGHORNE ROBERT ELISH
POLLACK JOHN JOSEPH
POMEROY JACK WILLIAM
PONAK CORDELL JOSEPH
PONCE ANTONIO RAMON
PONTIUS MARK DURWOOD
POPP JAMES ARTHUR
PORCARO SALVATORE VINCENT
PORDEN LEE VICTOR JR
PORTA GERARD PAUL
PORTER RONALD HARRY
POSIUS ROBERT
POTAS ALEXANDER FRANK
POTTER NEIL WARREN
POTTER PAUL PRICE
POTTER WESLEY ROY
POTTS GEORGE HENRY
POWELL DAVID MICHAEL
POWELL RONALD LOUIS
POWERS JAMES WILLIAM
POWERS STEVEN JAMES
POXON ROBERT LESLIE
PRANGE THOMAS CHARLES
PRATT WALTER RAYMOND
PRECOUR RICHARD FRANK
PREUSS CARL JOHN
PRICE HUMPHREY JAMES
PRICE JACK LEON
PRICE RICHARD JOHN
PRICE THOMAS J
PRINCE DENNIS GLENN
PRINGLE DONALD IRVEN
PRINZ RANDALL BOYD
PROEHL PAUL ALLEN
PRUETT JAMES RANDALL
PRUIETT THOMAS PIERRE JR
PRZYBYLINSKI GERALD
PRZYBYLOWICZ WALTER JR
PURGIEL ROBERT CHARLES
PYNE ROGER DALE
PYNNONEN MICHAEL JONAS
QUAN KENNETH RAYMOND
QUANDT ROBERT FREDRICK
QUATTLEBAUM JOHN FRANKLIN
QUICK JOHN JAMES
QUINN ROBERT
RADABAUGH HAROLD W II
RADICS DONALD M
RADZIECKI MICHAEL ANTHONY
RAGLIN RONDA LEE
RAJALA STANLEY ROBERT
RAMIREZ HONORIO JR
RAMIREZ MARIO
RAMSBY JAMES EDWARD
RANDALL DONALD DAVID JR
RANDOLPH RICHARD DALE
RANSHAW DOUGLAS LE ROY
RASNICK SIDNEY MC ARTHUR
RATTA FELICE NICHOLS
RAUB FRANKLIN HARRISON
RAUBOLT THOMAS EDWARD
RAWLS ROBERT EDWARDS
RAY CHARLES
RAY DEWEY JUNIOR
RAY DEWEY VERN
RAY EDWARD GEAN
RAY JACKIE
RAY THOMAS FREDRICK JR
RAYMOND FRANK JR
REAMS TERRY D
REBITS JOHN RAYMOND
REED RONALD LEE
REEL J C
REESE ABRAHAM B
REGENHARDT ROBERT JOHN JR
REICH DONALD GEORGE
REICHARD GARRY LEE
REICHERT ROBERT D
REICHLE DWIGHT GERALD
REID DANIEL GEORGE
REITER CLYDE ALVIN
REMONDINI LEO ANGELO JR
RENFROE MATHEREW DENNIS
RENNER LYNN CARL
RESKA CRAIG THOMAS
RESPECKI DONALD GEORGE
REVOIR RICHARD RUSSELL
REYES WILLIAM
REYNOLDS CARL MITCHELL

REYNOLDS OSSIE
RHEAULT WILLIS CLIFFORD
RHODES STANLEY RUFUS
RICCI GERALD
RICE LARRY ALLEN
RICE ROBERT CHARLES
RICH JON WILLIAM
RICH MICHAEL ROBERT
RICHARDS DANIEL MARTIN
RICHARDS FRED EARL
RICHARDSON BENJAMIN
RICHARDSON FLOYD JR
RICHARDSON MARVIN KEITH
RICHMOND ROBERT STANLEY
RICHTER KARL WENDELL
RIDER EARL CONRAD JR
RIESBERG DANNY PAUL
RIGGS JOSEPH BURNITT
RIGGS THOMAS FREDERICK
RIGHTLER GORDON RAY
RIKER RICHARD JOHN
RILEY DAVID CLARK
RILEY THOMAS EUGENE
RIMSON MARTIN LUTHER
RINCK RICHARD JAMES
RINDY GREGORY ARNOLD
RITSEMA WARREN PETER
RIVERS CLARENCE
ROACH TERENCE RAYMOND JR
ROACH THOMAS JOSEPH JR
ROAT RODNEY ALLEN
ROBERTS ALAN RICHARD
ROBERTS JERRY LEE
ROBERTS RICHARD DEAN
ROBERTS WALTER JAMES
ROBERTSON MARK JOHN
ROBINSON JOSEPH EARL
ROBINSON LARRY MICHEAL
ROBINSON LEWIS MERRITT
ROBINSON RAYMOND CARL
ROBINSON THOMAS LEON
ROBISON GARY HERBERT
ROCHE JOHN DONALD
ROCK GERALD FRANCIS
ROCKEY MICHAEL CRAIG
RODDY DONALD BARRETT
RODRIGUEZ JOE
RODRIGUEZ ROMAN DURAN
ROE JEFFREY TERRY
ROE KENNETH ALLEN
ROERINK GARY DOYLE
ROGALLA GEORGE HENRY
ROGERS ROBERT DAVID
ROGERS ROY RUMSEY
ROGERS WILLIAM HENRY
ROLFE DARYL EDSON
RONDO RONALD LEE
ROOSSIEN ROBERT ALLEN
ROOT JAMES MICHAEL
ROOT ROGER DALE
ROOT RUSSELL LEE
ROOT VERN ERNEST
ROSALES BERNIE JR
ROSE ALBERT EUGENE
ROSE DAVID EARL
ROSEKRANS LESLIE DONALD
ROSENBERGER ROGER DALE
ROSS JLYNN JR
ROSS LARRY DAVID
ROTH JOHN HOWARD
ROTH RONALD ARTHUR BERT
ROUSSOS WILLIAM ROBERT
ROWLAND HARVEY LYN
ROWLEY DONALD ALBERT
RUDD DONALD LEE
RUDITYS EDWARD MICHAEL
RUEHLE DOUGLAS DUANE
RUFF GARY LYNN
RUGAR STEVEN DALE
RUGENSTEIN GREGORY P
RUITER JERRY LEE
RUIZ JOHN FRANCO
RULISON DANIEL GRANT
RUPLE HOMER ALFRED JR
RUSH GEORGE HENRY JR
RUSHA GARY EDWARD
RUSHLOW RICHARD LEONARD
RUSNELL DANIEL JOSEPH
RUSS JAMES ERWIN
RUSSELL DAVID PAUL
RUSZKIEWICZ PAUL FRANK
RYALS JIMMIE DALE

RYCKAERT ANTHONY LEE
RYDEN GARY ARDEAN
SACHEN WILLIAM GEORGE JR
SAGAN SYLVESTER STANLEY
SAIDE DAVID ALLIE
SALAZAR JOSE
SALES NATHAN RAY
SALINAS DAVID GREGORY
SALISBURY GARY EUGENE
SALMINEN PAUL JOHN
SALMOND RICHARD WILLIAM
SAMOLEJ GERALD
SAMSON JERRY ERNEST
SANBORN JACK RICHARD
SANDERS FRANCIS EUGENE
SANDERS RONALD WALTER
SANDERS TERRY LEE
SANDIFER RICHARD WELLS
SANFORD DAVID AMON
SANFORD GARY BERNERD
SANTELLAN TEODORO
SANTO PATRICK ANGELO
SARAH HUGH HENRY
SARNA ARNOLD PAUL
SAULS OLLIE LESLIE JR
SAWYER WILLIAM A
SCANLAN GEORGE JOSEPH
SCARBERRY DONALD YOUNG
SCARBOROUGH ELMER WAYNE
SCARMEAS JAMES SAM JR
SCHAFER DONALD FRED
SCHELL EDWARD EARL
SCHEMEL JERRY L
SCHIEVE PAUL EVERETT
SCHLICHTING VICTOR STEVEN
SCHMIDT DANIEL THOMAS
SCHMIDT DENIS GORDON
SCHMIDT DONALD FRANK
SCHMUDE JOHN ROBERT
SCHNAKE RICHARD MARTIN
SCHNEIDER TERRANCE H
SCHONFIELD JEFFREY ALAN
SCHOONMAKER LARRY
SCHOUWBURG GERRIT JOHN
SCHRANK KARL F
SCHRECONGOST FREDERIC LEE
SCHROCK PHILIP JOHN
SCHULTZ CHESTER JOSEPH
SCHULTZ DAVID CHARLES
SCHULTZ JOHN JOSEPH JR
SCHULTZ JOHN LA VERN
SCHUTZ PETER JOHN
SCHWARTZ TERRY E
SCOTT BARRY FRANK
SCOTT DAVID LEE
SCOTT HUGH DON
SCOTT JOHN MELVILLE JR
SCOTT MARTIN T II
SCOTT MARVIN
SCOTT RICHARD LEE
SCOTT STEVEN JOSEPH
SCOTT WILLIAM HENRY M
SCOVILL GARY ALAN
SCOVILLE WILLIAM WARD
SCOWDEN DUANE NEVADA
SCRIBNER GARY DAVID
SEABLOM EARL FRANCIS
SEABRIDGE RICHARD ROY
SEABROOKS ARTHUR
SEADORF MICHAEL J
SEBURG DONALD PAUL JR
SELKEY DONALD ANTHONY J
SELMAN CHARLES GEORGE
SERVANTEZ JOSEPH ANTHONY
SETTER JAMES ADRIAN
SETTLEMYRE JEFFERY COLIN
SEXTON ANDREW BOWMAN
SEXTON EDWARD CICERO
SEYMOUR GARY CARL
SHAFER ROGER DALE
SHAMBAUGH GREGORY RANDALL
SHAPLAND KENNETH WAYNE
SHARP BURTON IRVING
SHARP RAY LAVERN
SHARPE CHARLES DENNIS
SHARPE THOMAS EDWARD
SHAUVER TERRY DEAN
SHEAFFER DAVID LAWRENCE
SHEELER GREGORY WILLIAM

SHELLIE DARREL ANDRE
SHELTON CHARLES H
SHEPPERDSON GARY ROBERT
SHERMAN THOMAS ALAN
SHERMOS JOHN DANIEL
SHERROD WALTER JR
SHERWOOD JAMES ROBERT
SHERWOOD RICHARD GUY
SHETRON WILLIAM MACKS
SHIELDS MELVIN LEROY
SHIER RONALD JAMES
SHINELDECKER RAYMOND MACK
SHINN GARY JAMES
SHIRODA ROBERT LOUIS JR
SHOAPS KENNETH DUANE
SHORT J C LESLIE
SICKLES JOHN ANDREW
SIDOR MICHAEL EDWARD
SIETING STANLEY LAWTON
SIKORSKI LEO PETER
SIMISON TERRY CLEO
SIMMONS DAVID LEROY
SIMMONS KENNETH JEROME
SIMMONS ROY LEE
SIMMONS WILLIE JAMES
SIMMS LEON
SIMON TERENCE EDWARD
SIMONS ROBERT VINCENT
SIMPSON JOEL B
SIMRAU ROGER ALLEN
SINKLER MARVIN JOHN
SIPPERLEY LORNE JAY
SKALBA JOHN JOSEPH
SKINNER ERNEST MACK
SKINNER ROBERT CLARENCE
SKOVIAK RONALD FRANK
SKUTT DENNIS DWAYNE
SLACK LLOYD
SLANAKER ROBERT JAY
SLATER DONALD EUGENE
SLOAT BENNY DAVID
SMARSH JOSEPH II
SMITH ALTON
SMITH DAVID LEON
SMITH DELBERT RAY
SMITH DONALD ALLEN JR
SMITH EARL
SMITH ELLIOTT ROBERT
SMITH GARRY GREGORY
SMITH GARY KENT
SMITH GARY MARTIN
SMITH J T
SMITH JOHNNIE JR
SMITH KENNETH WILLIAM
SMITH LARRY CURTIS
SMITH LARRY ELLSWORTH
SMITH LARRY JAMES
SMITH MICHAEL LA VERN
SMITH PEDRO ANDRE
SMITH PRESTON LEE
SMITH RICHARD EDWARD
SMITH ROBERT LEE
SMITH RONALD C
SMITH SCOTT PHILIP
SMITH STEVEN ROBERT
SMITH THOMAS EMINGS
SMITH THOMAS PAUL
SMITH WARREN ALLEN
SMITH WILLIAM ARTHUR JR
SMITH WILLIE JAMES JR
SMITHERMAN FRANK DONALD
SMOLAREK KENNETH JAMES
SMYK FRANK BARTH
SNYDER CHARLES DAVID
SNYDER EARL SPENCER
SNYDER GEORGE EUGENE
SNYDER RICHARD ANDREWS
SNYDER ROCKY RAND
SOMERO KENNETH EDWIN
SOMERS RICHARD KEITH
SOMES RONALD REE
SOPER RICHARD ORRIN
SOROVETZ MICHAEL
SOSNOSKI RONALD FRANCIS
SOTZEN HAROLD JAMES
SOVA CONRAD ANDREW
SOVEY ELWOOD CHARLES JR
SOWA JAMES ANDREW
SPARKS RICKIE D
SPARRE LYN DWIGHT
SPATAFORE DONALD JAMES
SPAULDING DEAN FRANCIS JR

SPEAR FRED HAROLD
SPECK DENNIS JEROME
SPENCER ARLIE JR
SPENCER JAMES FREDERICK
SPENCER KENNETH JAMES
SPENCER PHILIP GLENN
SPENELLI DENNIS ARTHUR
SPENS WILLIAM EDWARD III
SPITZER HOWARD RAY
SPOONER EUGENE EDWARD
SPRADLIN EDDIE EUGENE
ST ONGE THOMAS GORDON
STACEY JAMES SHELTON
STAFFORD THOMAS STEPHEN
STALEY FREDDY KEITH
STAMPS GEORGE HARRELL
STANCROFF DENNIS CHARLES
STANDIFER ANTHONY
STANICH NADE MICHAEL
STANKO WALTER LEE
STANTON EDWARD RYLAND II
STAPLES JAMES ARTHUR
STAPLES THOMAS HAROLD
STARRY DOUGLAS C
STAUFF ERIC LOUIS WILLIAM
STAUFFER GORDON CHARLES
STEAD VERNON ROBERT
STEELE PATRICK MATTHEW
STEFFEN FREDERICK GEORGE
STEFFENS WALTER FREDERICK
STEFFLER CHARLES ERVIN
STEHLE HERBERT NEIL
STEPHENS DENNIS ARTHUR
STEPHENS JASPER JR
STERLING DAVID WALTER
STETTEN G LYLE
STEVENS DENNIS MICHAEL
STEVENS PHILIP PAUL
STEVENS ROBERT LOUIS JR
STEVENSON BOBBY DALE
STEVENSON RUFUS NEWTON J
STEWART CHARLES LEROY JR
STEWART JOHN WALLACE
STEWART RICK JAMES
STIGLITZ DENNIS LAWRENCE
STOAKLEY GORDON ALAN
STOCKDALE JAMES BYRON
STOCKHOLM DWIGHT ROSS
STODDARD KEITH ARTHUR
STOELT HAROLD EDWIN
STOGSDILL DELBERT RAY
STOKEN CHARLES ALBERT
STOKES ROY E
STOKKERMAN JON WILLIAM
STOLLEY WILLIAM R JR
STORY FRED DELL
STOTTS DONALD MAURICE
STOUT JERRY LEE
STOUT WILLIAM HENRY III
STOVER JAMES EDWARD
STRAHAN LARRY
STRANGE ROBERT ALLEN
STRECHA JAMES JR
STREMLER DAVID ALLEN
STRICKLAND THOMAS NEIL
STROVEN WILLIAM HARRY
STRUCEL JOSEPH JOHN
STUBBLEFIELD JAMES ED-
WARD
STUCKY RONALD
STUDARDS ROBERT LARRY
STUIFBERGEN GENE PAUL
STUPAR MITCHELL NICK
STURGILL MICHAEL JAMES
SUAREZ EUGENE RAVN
SULLIVAN CHARLES A JR
SULLIVAN ROBERT MICHAEL
SUMMERVILLE FREDERICK
BRU
SURBER MARK WAYNE
SUTTON JAMES THOMAS
SWEETLAND RONALD KEN-
NETH
SWOVELAND WILLIAM ALAN
SYLWANOWICZ CASIMIR
SYLWA
SYNOD MICHAEL JOHN
SZLAPA JOHN FRANK III
TAFFE THOMAS LEO
TAGLIONE ROBERT
TALLMAN ROGER LEE
TAMM RICHARD DAVID
TANGEMAN JAMES LEROY
TANGEN TERENCE RONALD

TANK CHARLES LOUIS
TANK PHILIP LEONARD
TANKERSLEY JAMES ESTILL
TANNER DOUGLAS HOWARD
TARASZKIEWICZ JOSEPH G
TATE BERNIE LEE
TATE ROBERT GERALD
TAYLOR BILLY JOE
TAYLOR CHARLES FRANKLIN
TAYLOR DARRELL DUANE
TAYLOR DAVID BERNARD
TAYLOR DONALD THOMAS
TAYLOR DONNIE CARL
TAYLOR JEROME MILTON
TAYLOR MARK ALLAN
TAYLOR RICHARD ALLEN
TAYLOR WILLIAM JOHN III
TAZELAAR JAMES ALLEN
TEAR GEORGE BERNARD
TEEPLE WAYNE WINSTON
TEETER GARY ALAN
TEEVENS RICHARD PAUL
TELLEFSEN TIMOTHY MARTIN
TELLING JACK EDWIN
TELLIS WILLIAM JAMES
TEMPLETON GARY DALE
TENNANT WILLIAM ALLAN
TERAN REFUGIO THOMAS
TERRY TOMMY J
TERWILLIGER DAVID WIL-
LIAM
TEWKSBURY JAMES LEE
THANE ROBERT LEE
THATCHER THOMAS MILTON
THELEN ROBERT JOSEPH
THIBAULT JAMES WILLIAM
THICK HOMER DANIEL
THIMM JOSEPH MICHAEL
THOMAS DAVID GEORGE
THOMAS GARY JOSEPH
THOMAS JOHNIE B
THOMAS JOSEPH MICHAEL
THOMAS MELVIN RAY
THOMPSON BARRY ALLAN
THOMPSON EVERETTE
ARTHUR
THOMPSON LEONARD LUKE
THOMPSON MICHAEL GUY
THOMPSON MICHAEL KELLY
THOMPSON NEIL STEWART
THOMPSON SAMUEL DWIGHT
THOMPSON STANLEY JAMES
THOMSON CARL ALLEN
THORESEN DONALD NELLIS
THORNELL RICHARD LLOYD
THORNTON BRIAN LEE
THORSON WALLACE R JR
TH-UOT HUBERT OWEN
THURNHAM JOHN BRENT
TIDWELL VOYD EUGENE
TIERNEY KENNETH PETER
TIGLAS THOMAS LEE
TILLEY JAMES A
TIMBERLAKE DWIGHT ELMER
TINKER GARY LYNN
TINSEY DAVID FREDERICK
TOBIE DAVID CARL
TOINS FRED
TOLER RICHARD GEORGE
TOMAKOSKI JAMES ROMAN
TOMLIN CARL DELBERT JR
TOMLINSON GERALD DOUG-
LAS
TOPOLINSKI DENNIS MI-
CHAEL
TOTTEN RANDY GENE
TOWNE TERRY ALLEN
TOWNER ALLEN RAY
TOWNLEY JAMES EDWARD SR
TOWNSEND FRANKLIN
ARTHUR
TOWNSEND ROBERT FRANK-
LIN
TRACY PATRICK
TRAPANI ANDREW
TRAVER CRAWFORD HENRY
TRAVIS WILLIAM HARRY
TRAYNOR STEPHEN MICHAEL
TREADWAY KENNETH EARL
TREADWAY WILLIAM MI-
CHAEL
TRESCOTT CHARLES ROBERT
TROMBLEY MICHAEL LAW-
RENCE

TROMP WILLIAM LESLIE
TROUGHTON PHILLIP NEIL
TROUTT LOUIE JAY JR
TROYAN MICHAEL JOSEPH JR
TRUELUCK GEORGE GUTHRIE
TUCCI ROBERT LEON
TUCKER GERALD ALEXANDER
TUCKER MICHAEL RAYMOND
TUNGATE NORMAN LEE
TURCOTTE DANIEL JOSEPH
TURNER JAMES PAUL
TURPIN RICHARD FLOYD III
TYLER GEORGE EDWARD
TYLER MARK DENNIS
U REN FRED THOMPSON
UEBLER ROY NICHOLAS JR
UHLIK FRANK ANTHONY JR
ULBRICH JOHN HAROLD
ULICNI JOHN
ULMAN EDWARD DELBERT
ULRICKSON PETER EDWARD
URBAN ALEXANDER JOHN JR
URQUHART GLENN ROSS JR
UTRIAINEN GARY ALBERT
VACZI ALEX E
VALENCICH PETER LYLE
VALERIUS MILLARD RUSSELL
VALLELONGA LARRY COSIMO
VAN BEUKERING MARK ALAN
VAN BEUKERING RONALD
DALE
VAN CLEAVE WILLIAM F
VAN DAM BRUCE ALLAN
VAN DE WARKER RICHARD L
VAN DONKELAAR GERALD
WAYNE
VAN DUSEN ROBERT EDWARD
VAN DYKE STEPHEN DENNIS
VAN GESSEL LARRY EUGENE
VAN GILDER ROBERT RAN-
DALL
VAN HAITSMA RANDALL
CRAIG
VAN HATTEM JAMES ROBERT
VAN HORN ALBERT JAMES
VAN LEEUWEN ROBERT JAMES
VAN OCHTEN TERRY JOSEPH
VAN RAEMDONCK RONALD
FRAN
VAN TONGEREN TIMOTHY
RAY
VAN WIEREN JACK ALAN
VANCE DE WITT STANLEY
VANCE WILLIAM CHARLES
VANCIL MICHAEL
VANDEN BOSCH DWAINE
WILME
VANDENBERG JOHN EDWARD
JR
VANDENBERG RONALD J
VANDER WEG PHILLIP JOHN
VANDERCOOK DAVID FRANK-
LIN
VANDERHAAG ALBERT JACOB
VAUGHN WENDELL GLEN
VEDRO EDMUND RONALD
VEIHL JOHN
VELLANCE RICHARD PAUL
VENTLINE LUKAS JOHN
VERRETT KENNETH EMERY
VESCELIUS MILTON JAMES JR
VIGH ALEXANDER JOSEPH
VILAS GERALD FRANK
VILLALOBOS JAIME
VILLAMOR ROMAN ROZEL JR
VILLARREAL ERNESTO
VINCENT DONALD W
VINCENT IVAN L JR
VINNEDGE JOHN ROBERT
VISSER JOHN JOSEPH
VLIEK RONALD CRAIG
VLISIDES GEORGE FREDERICK
VOGEL LAWRENCE SMITH
VOLLMAR CHRISTOPHER LEE
VON DER HOFF RALPH HENRY
VOORHEES RICHARD ALLEN
VORE RODNEY ALLEN
VORENKAMP DAVID R
VOS KENNETH RICHARD
VRUGGINK JOEL
WADDELL BOBBY LEE
WAGENER DAVID RAYMOND
WAGER RICHARD JAMES
WAGNER BRUCE ALLEN

WALERZAK WILLIAM THOM-
AS
WALKER ARTIE DELL
WALKER CLARK L
WALKER DALE ALLEN
WALKER FRANK MARK
WALKER KENNETH EARL
WALKER ROBERT LEE
WALKLEY ROBERT MARK
WALLACE EDDIE
WALLACE JAMES CLARENCE
WALLACE JOHN THOMAS
WALLACE MICHAEL JOHN
WALLACE RICHARD DEAN
WALSH ALLAN RAY
WALSH ROBERT PAUL
WALTON LOUIS EUGENE
WANGESHIK MELVIN UDEL-
SON
WARD ROGER ELGIN
WARFIELD DENNIS GORDON
WARGO DAVID ROY
WARNER ROBERT ALLEN
WARREN COLEMAN LITCH-
FIELD
WARREN JIMMIE SHERRILL
WASHINGTON CARL
WASZKIEWICZ DENNIS
LAVERN
WATSON ERNEST
WAWERSIK KENNETH WIL-
LIAM
WEATHERHEAD GARY
ROBERT
WEAVER GARY LEE
WEAVER GEORGE ANTHONY
WEAVER JERRY MICHAEL
WEBB GREGORY LYNN
WEDHORN DAVID EARL
WEEKS WALTER DARRYL
WEID RICHARD GEORGE JR
WEIHER ROBERT LESTER
WEIL LARRY STEVEN
WELCH ROBERT JOHN
WELCH ROBERT LEROY
WELLS JAMES ALLEN
WELLS JOHN CURTIS
WENDEL RICHARD LOUIS
WENDER TERRY ARTHUR
WENGER ROBERT LEE
WENRICK CLYDE ALLEN
WENTZEL WILLIAM CHARLES
WESSELS EDWARD JOHN
WEST CHARLES ROBERT
WEST JERALD DALE
WESTBROOK THEODORE
ELBA
WESTRATE ROBERT JAY
WETZEL JOHN THOMAS
WHEELER WILLIAM EUGENE
WHELPLEY RAYMOND LAW-
RENCE
WHINNERY DAVID VERNON
WHITBY THOMAS ALVIN
WHITCOME LARRY WILLIAM
WHITE ALGER LAWRENCE JR
WHITE BEDFORD FREDERICK
WHITE CHARLES BOYD
WHITE ROGER DUWAINE
WHITE TIMOTHY ALLEN
WHITE TOMMY LEE
WHITELAW GEORGE DAVID
WHITLOCK THOMAS DANIEL
WIAR JOSEPH CLERMAN JR
WIDMER RICHARD JAMES
WIDON KENNETH HARRY
WIELKOPOLAN DONALD
DAVID
WIGINTON LARRY MICHAEL
WILCOX RAYMOND LEE
WILDER BRUCE JEFFREY
WILDERSPIN DEAN ALLYN
WILDERSPIN VERNON
CHARLES
WILKERSON WILLIAM
MHOON
WILKIE DOUGLAS WILMER
WILKIE ROBERT GEORGE
WILKINS JAMES MARVIN
WILKINSON RICHARD
THOMAS
WILLIAMS BRUCE REGINALD
WILLIAMS CLIFFORD DAN
WILLIAMS DEWAYNE THOMAS

WILLIAMS EDWIN JEROME
WILLIAMS FRANK A
WILLIAMS FRANK EDWARD
WILLIAMS GERALD STUART
WILLIAMS JAMES RAYMOND
WILLIAMS PATRICK
WILLIAMS RICHARD WARREN
WILLIAMS ROY KENNETH JR
WILLIAMS TED
WILLIS ARCHIE VAUGHN
WILLIS JOSEPH F
WILLOUGHBY JULIAN B
WILSON DAVID REYNOLDS
WILSON DELVIN KEITH
WILSON EARL LEE
WILSON PHILLIP ALLEN
WILSON RICHARD EDWIN
WILSON ROBERT ALLAN
WILSON THOMAS EDWARD
WILSON WILLARD EUGENE
WILSON WILLIAM HENRY JR
WIMBERLY ARNOLD BRUCE
WINCKLER DONALD LEWIS
WINNER BRIAN CARL
WINNINGHAM RICHARD
DANIEL
WINSLOW LARRY A
WINTER CARL J
WINTERS ALLEN LANE
WIRT DENNIS ARTHUR
WIRT DENNIS HAROLD
WISE JAMES JOSEPH
WISHON DONALD RAY
WITTBRACHT MARK
CHARLES
WOFFORD WILLIAM PHILLIP
WOLFE PATRICK ROBERT JR
WOLFE RICHARD LAWRENCE
WOLOSONOWICH LARRY
JAMES
WOLOSZYK DONALD JOSEPH
WOLPERT LARRY MICHAEL
WOOD DONALD FRANK
WOOD JAMES ANTHONY
WOOD JOHN ALLEN
WOOD LAWRENCE JEFFREY
WOOD RODNEY GLEN
WOODS ROBERT EARL
WOODWORTH CLARK NEW-
ELL JR
WOOSTER ROGER EDSON
WORCESTER JOHN BOWERS
WORLEY THOMAS JAMES JR
WORTMAN DOUGLAS FRED-
ERICK
WOZNIAK FREDERICK JOSEPH
WRIGHT ARTHUR
WRIGHT GARRY
WRIGHT LESLIE EDWARD
WRIGHT RODERICK MICHAEL
WUNDER ROBERT LEE
WYDRA MILAN CHAUNCEY
WYMAN JERRY ALAN
YACKS RONALD G
YAGLE THOMAS NEIL
YAMASHITA RICK
YATES CRAIG EDWARD
YELLEY DANNY KEITH
YINGLING JOSEPH WALTER JR
YNTEMA GORDON DOUGLAS
YOKES FRANK JOSEPH
YOLKIEWICZ THOMAS JOSEPH
YOSHONIS GEORGE CHARLES
YOST RUSSELL CHARLES
YOUNG DONNIE WINFIELD
YOUNG JAMES BRUCE
YOUNGS JIMMIE WALKER
YOUTSEY RICHARD DUANE
ZALESNY HARRY FRANKLIN JR
ZAMIARA JOSEPH C
ZAYAS SAUL
ZELINSKI JOSEPH VINCENT
ZIMMERMAN JOHN RANDALL
ZLOTORZYNSKI GERALD
ZOMBERG GEORGE ALAN
ZOODSMA JACK ALLEN
ZSIGO ALEXANDER C JR
ZUPANCIC GEORGE PAUL
ZUREK MICHAEL ROBERT
ZYWICA GARY ROMAN

WE WERE YOUNG. WE HAVE DIED. REMEMBER US.

Photos by Sarah Soderland

Lakefront DMZ

12th St W, St Paul, MN 55155

The Minnesota Vietnam Veterans Memorial known as Lakefront DMZ was made to be less of a political statement and more of an insightful art installation. It was made with the intention of honoring the veterans and remembering their sacrifices and service while still offering local people a place to grieve for their lost loved ones.

In September 1987, a young lady named Teresa Vetter came up with the idea for Minnesota to have their own Vietnam Memorial to help the local communities heal. She reached out to organizations including the Veterans of America local Chapter 62 in Twin Cities, Minnesota. She met people who helped her dream materialize.

The project, however, didn't happen without its fair share of problems. Local companies didn't want to support a project that involved a controversial topic like the Vietnam War, so Vetter and her supporters turned to donations to fund the project. It wasn't until 1989 that legislature was passed to appropriate $300,000 and two-and-a-half acres of land for the Lakefront DMZ memorial.

To make the memorial special, a design competition was held and designers Nina Ackerberg, Jake Castillo, Rich Laffin, and Stanton Sears won. Money issues hit again in 1991 and it took a grandmother from Delano climbing a billboard in protest to collect the necessary funds to finish the project. The memorial was finally completed in 1992 and dedicated that same year.

The Lakefront DMZ features a granite wall with the names of the 1,120 lost Minnesota Vietnam veterans. It is also made up of a map of Indochina made out of red granite, a plaza of 68,000 granite stones (one stone for every Minnesotan that served). Dark green stones among those represent the deceased or missing. There is also a limestone house façade to represent the sense of coming home for the veterans who were fortunate. Memorial and Veterans day events are held at this site each year.

The Names of those from the state of Minnesota Who Made the Ultimate Sacrifice

AAMOLD DANIEL LAWRENCE
ABBOTT STEVEN GLENN
ACKERMAN EDWIN ARTHUR JR
ADES ARNOLD ALVIN
AGATHER FREDERIC GUSTAVE
ALGAARD HAROLD LOWELL
ALLDRIDGE GALE ARTHUR
ALLEN PAUL JAMES
ALLEN RONALD PAUL
ANDERS JOEL GARY
ANDERS ROBERT LEROY
ANDERSON BRUCE CARLYLE
ANDERSON DARRELL EUGENE
ANDERSON DAVID BRUCE
ANDERSON DENNIS KEITH
ANDERSON GORDON GUY
ANDERSON HERBERT ROY
ANDERSON JOHNNIE LEE
ANDERSON MARK ANTHONY
ANDERSON MARLYN RONALD
ANDERSON MELVIN WALLACE
ANDERSON ROBERT GARY
ANDERSON ROGER WILBUR JR
ANTONOVICH RICHARD ROBERT
APLAND RICHARD BRUCE
ARRIGONI RONALD LOUIS
ARVIDSON JAMES WARREN
ARVIDSON KENNETH ARVID
ASHBY JAMES WESLEY
ASP FRANK WALTER
AUGE DAVID CHARLES
AUWARTER EARL DEAN
BACKLUND JAMES VICTOR
BAER HERMAN JOHN
BAGAASON GERALD BENNETT
BAILEY JOHN EDWARD
BAKER DON CARTER
BAKER RAYMOND JOHN
BAKKE TONY LEE
BALDWIN SCOTT DOUGLAS
BANG JAMES CURTIS
BANGERT BYRON ALLEN
BANGERT ROGER CARL
BANKS JOHN LAWRENCE III
BARDUSON DAVID JULIEN
BARICKMAN LEON ROSS
BARNES GARY LESTER
BARNICK CHARLES EDWARD
BARON FRANCIS VINCENT
BARRETT ANDREW RYAN
BARTELS GARY LESLIE
BARUTH DAVID ARTHUR
BAST ALBERT FRANK JR
BAUER KENNETH LEROY
BAUER STEVEN ROBERT
BEAULIEU LEO VERNON
BEBUS CHARLES JAMES
BECKWITH WILLIAM HENRY
BEEBE JERRY RAY
BELLANGER JOHN GEORGE
BEMBENEK MARLIN EDWIN
BENGTSON ROBERT DAVID
BENOIT FRANCIS ARTHUR
BENSON MARTIN JOSEPH
BERG JULIAN WINSLOW
BERG THOMAS ALAN
BERINGER MICHAEL AUGUST
BERNIER ROGER JEROME
BESCH ROBERT DEAN
BEVAN JERRY EUGENE
BIEGERT RONALD LEE
BINA THOMAS MELVIN
BIXBY JACK DENTON
BJORKE ERLE LAWRENCE
BLACKBURN RICHARD VINCENT
BLANKSMA GERRIT LYNN
BODIN DANIEL ROGER
BOECK GARY RAYMOND
BOGGS CHARLES WILSON
BOLDT CHARLES DAVID
BOLTON WAYNE FRANKLIN
BOTTS DAVID MARTIN
BOWMAN DAVID LLOYD JR

BRAASCH GARY WILLIAM
BRADLEY THOMAS JAMES
BRANCH CHARLES ARTHUR
BRANDBORG JOHN RALPH
BREDE ROBERT WILLIAM
BRENDEN NORRIS LEE
BRENKE FREDRICK JOSEPH
BRICKMAN DEWAINE LAWRENCE
BRIESACHER MARVIN CARROLL
BRINDLEY THOMAS DREW
BRONCZYK LAWRENCE JOSEPH
BRONSON RICHARD TERRY
BROSE STEPHAN ROBERT
BROWN BRIAN CHARLES
BROWN DENNIS RICHARD
BROWN RICHARD GORDON
BROWN THOMAS RICHARD
BROWN WALTER STONEMAN
BUAN LEE BJARNE
BUCKINGHAM KEITH CHARLES
BUNNIS RICARD THOMAS
BURCH STEVEN RALPH
BURGESS RICHARD ALBERT
BURKE GARY LEE
BURNS ROBERT ALLEN
BURNS STEVEN CRAIG
BURNSIDE DONALD WAYNE
BURSAW CLARENCE HERBERT
CALLIVAS GUST
CAMPION EUGENE MICHAEL
CARDENAS MANUEL II
CARDINAL GARRYL DAVID
CAREY ROGER GARYLEE
CARLSON CARL LEONARD
CARLSON DONALD LE ROY
CARLSON PAUL VICTOR
CARLSON RICHARD ARNOLD
CARLSON VERNELL DWIGHT
CARLSON WILLIAM EUGENE
CARLTON DAVID JAMES
CARSON EDWIN EVERETT
CASPER RICHARD ALLEN
CASSERLY JOSEPH MICHAEL
CAWLEY PATRICK FRANCIS
CHAMPION WAYNE WILLIAM
CHAPMAN GARY MICHAEL
CHRISTENSEN BRUCE ARDEN
CHRISTENSEN JAN PAUL
CHRISTENSON WILLIAM LEE
CHRISTIANSEN THOMAS LEE

CHRISTOPHERSEN KEITH ALLE
CHRISTOPHERSON DAVID LYN
CLARK DOUGLAS MARK
CLARK NORMAN HARVEY
CLICKNER MICHAEL DUANE
CLITTY CHARLES GUST
CLOUD RONALD MYRON
COLEMAN RONALD JOHN
COLLINS GEORGE PORTEOUS
COLTON MICHAEL NORRIS
COMPTON MICHAEL JOSEPH
COOK WILLIAM RICHARD
COOREMAN RAYMOND ROBERT
CORLEY JERRY WAYNE
COUILLARD BRUCE ALVIN
COWDEN LESLIE LAWRENCE
COX RICHARD LEIGH
CRAN JAMES ALLEN
CULSHAW DONALD IGNATIUS
CUNNINGHAM RONALD CHARLES
CYSEWSKI GARY FRANCIS
DAHL WILLIAM JOHN
DAMSGARD CHARLES DE WAYNE
DANIELS THOMAS RAY
DANIELSON BENJAMIN FRANKL
DAU KENNETH RUSSELL
DAVIS RICHARD JOHN
DE BOER WILLIAM SYLVESTER
DE VINNEY ROBERT EUGENE
DEBNER DENNIS ERWIN
DECKER MELVIN JEROME
DEEDRICK CHARLES ORVIS JR
DEETZ BILL WAYNE
DEHN ARTHUR ANDREW
DENEEN EARL MERRILL
DENNY RICHARD EMERSON JR
DERBY EARL LEE
DES LAURIERS PHILIP GENE
DIERYCK JAMES LEO
DISRUD DAVID A
DONAHUE JOHN THOMAS
DORING LARRY ALLEN
DORMAN DARREL GENE
DOYLE LARRY R
DOYLE PATRICK LAWRENCE
DRESSEL KENNETH HAROLD
DRESSEN DOUGLAS STANLEY
DUFRESNE WILLARD J JR
DUMDEI CHARLES MARION

DUNCAN KURT WILLIAM
DUNKEL MICHAEL ROBERT
DUSBABEK JOHN ROBERT
DZIENGEL MICHAEL PETER
EASTMAN THOMAS DELL
EBBINGA HERMAN GERALD
EDMOND PAUL ROBERT
EDWARDS CHARLES HAROLD JR
EGGE ERIC CRAIG
EHLERS ROBERT FREDERICK
EKLUND MARK JAMES
ELLEDGE KEITH O NEIL
ELLIS JERRY NORMAN
ELWELL MICHAEL REID
ELZY JOHN CALVIN III
ENGEL TERENCE DEAN
ENGREN RUSSELL ALAN
ENZ HARVEY GORDON
ERBES JOHN HENRY
ERICKSON DAVID WAYNE
ERICKSON KENT DOUGLAS
ERICKSON RICHARD ANTON
ERICKSON THOMAS GUSTAV
ERTL RICHARD LOUIS
ESPARZA JOHN PAUL JR
ESSLER RONALD HENRY
EVANS THOMAS JAMES
EVENSON EDDIE LEE
EWALD RICHARD CLAYTON
EWALD WOODROW JOHNSEN JR
FAEHNRICH DAVID RAYMOND
FASNACHT DAVID ANTHONY
FAST ROGER THEODORE
FASTH KENNETH LEE
FEDDEMA CHARLES JOHN
FELLOWS ALLEN EUGENE
FENNEY DOUGLAS JAMES
FERGUSON DENNIS WAYNE
FINZEL JAMES WARREN
FIRKUS JAMES RONALD
FISH FRED KEITH
FIX MICHAEL DAVID
FLEMING MICHAEL JOHN
FOGARD RONALD DEAN
FORMAN WILLIAM STANNARD
FORSBERG DOUGLAS BRUCE
FOSS DUANE JOHN
FRANK NEAL RAY
FRANSEN RONALD CLIFFORD
FRASER RONALD MONTE
FREDERICKSON EARL WAR-

REN
FREDRICKSON GERALD GEORGE
FREEBERG RANDALL ROGER
FREESE ELMER EUGENE
FRITZ DAVID URBAN
FRITZ LAUREN DEAN
FRYE CLIFFORD KENNETH
FULLER EUGENE EDWARD
GAERTNER BYRL WILLIAM
GAGNE DALE FRANCIS
GAGNE JOSEPH JAMES
GAGNIER WILLIAM JOSEPH
GARDNER MICHAEL JOHN
GARNER BOYD GRAYSON
GARTNER ROBERT FREDRICK
GARVICK DERYL RAY
GEERDES DONNIE ADELBERT
GEHLING DONALD ANTON
GENS JONATHON LEE
GERDESMEIER JOHN ALOIS
GIBBS WILLIAM ARNOLD
GILBERT STANLEY DONALD
GILBERTSON ALAN DALE
GILBERTSON LARRY RUSSELL
GILLESPIE GEORGE ALLEN
GLIDDEN ROBERT WAYNE
GLYNN MICHAEL JOSEPH
GOAD ROBERT LEE
GOELZ STEVEN WILLIAM
GOLBERG LAWRENCE HERBERT
GOLDEN ROBERT WALTER
GOODMAN KENNETH VIRGIL
GOODNO KEVEN ZANE
GOSEN LAWRENCE DEAN
GOSSMAN KERRY RAY
GRABER GARY DAVID
GRAHAM DONALD TERRY
GRAHAM PATRICK JOHN
GRAHAM ROBERT LEE
GRANGER DALE GENE
GRAVES EDWARD STEPHEN
GREEN DONALD CARL
GREEN STEVEN LYNN
GRIFFITH WILLIAM WILLIS
GROTH DENNIS ARTHUR
GROVE EARL RUSSELL
GRUNDMAN ROBERT FRANCIS
GUCK RALPH STEPHEN
GUENTHER CARLYLE
GULSETH SHELDON LEE
GUNDERSON GERALD JAMES
GUNDERSON RICKIE NOR-

Minnesota

MAN
GUNDERSON THOMAS LA
VON
GUNHUS GORDON MARLO
GUSTAFSON JAMES ERNEST
HAAS HARRY JAMES
HAEFNER DAVID ALLEN
HAERLE JEFFREY WILLIAM
HAGEN JEROME ALFRED
HAIDER JAMES MICHAEL
HAIDER MICHAEL EDWARD
HAKES CLARENCE DEAN
HALBAUER DAVID MICHAEL
HALGRIMSON MARLOYE
KEITH
HALL WARREN STUART
HAMILTON MILBERT WALTER
HAMMER RICHARD JOSEPH
HAMMERSTROM RONALD
ROY
HAMSMITH ALLAN FREDRICK
HANDRAHAN EUGENE ALLEN
HANEY ROBERT ALAN
HANEY THOMAS WILLIAM
HANSEN BERNARD TIMOTHY
HANSEN PETER RAYMOND JR
HANSON ALAN MORRIS
HANSON DARRELL WAYNE
HANSON DENNIS GORDON
HANSON JAMES RICHARD
HANSON TERRANCE RAN-
DALL
HANSSEN WILLIAM DENNIS
HARBER STEPHEN JAMES
HARDER ERWIN JOHN
HARDT ROSS ERLE
HARPER TIMOTHY VAUGHN
HARRA LEE HAMILTON
HARRIGAN GREGORY MI-
CHAEL
HARRIS EDWARD LEWIS
HARRIS LAWRENCE HUBERT
HART MELVIN ELLSWORTH
HARTUNG CHARLES LEON-
ARD
HARWORTH ELROY EDWIN
HATCH LARRY G
HAUGEN EDWARD JOHN
HAUPERT WILLIAM JOHN
HAUSCHILDT CHARLES LEE
HAYASHIDA HERBERT REIJI
HAYES LYLE DENNIS
HEALY THOMAS MICHAEL
HEDIN GARY ORVILLE
HEDLUND PETER BURR
HEDSTROM HARVEY NOR-
MAN
HELLAND JERRY IRVEN
HELMKE DARREL BRUCE
HEMPHILL FREDRICK H
HENDRICKSON CURTIS LYNN
HENDRICKSON PATRICK
RAYMO
HENDRICKSON WESLEY PAUL
HENNEN PATRICK ERNEST
HENNESSY STEPHEN THOMAS
HERING WILLIAM ARTHUR
HERNANDEZ STEVEN
HERRON JOSEPH SAMUEL
HEUER MICHAEL WAYNE
HICKS JAMES RUSSELL
HIEBERT LYNN GREGORY
HILL HENRY JR

HILL JERRY JAMES
HILT RICHARD MICHAEL
HINDS STEPHEN JOHN
HINKLE NORMAN LEE
HINSCHBERGER LAWRENCE K
HOCHSTETTER JAMES JAY
HOFF DENNIS WAYNE
HOFFMAN ALLAN ROY
HOGLUND GARY WILLIAM
HOLAN ROBERT ANDREW JR
HOLCOMB ROBERT EARL
HOLLER ROGER EMIL
HOLTE MARK DELANE
HOLTE ROGER ALLEN
HOMSTAD MILO STEVEN
HONEK KENNETH JEROME
HOUSKER HAROLD DEAN
HUBERT STEVEN JAMES
HUBERTY DAVID JEROME
HUBERTY WILLIAM M
HUERD LAUREN DALE
HUGHES DAVID JAMES
HULWI WILLIAM GEORGE JR
HUOT RAYMOND CHARLES JR
HUSO WAYDE MURRAY
HUSTAD LARS PETER
ILSTRUP JOHN ALVIN JR
JACOBSEN DONALD JAMES
JAMES MICHAEL RAY
JAMROS RICHARD KENNETH
JANNETTA RODNEY ALAN
JANSEN MILES EDWARD
JANSKI RICHARD JOSEPH
JARVI RAYMOND LEE
JEDNEAK DANIEL JOHN
JENNIGES RONALD ARTHUR
JENSEN ARLIN ROGER
JENSEN JAMES ALLEN
JENSEN ROGER DALE
JENSEN TERANCE KAY
JENSON MICHAEL GREGORY
JERDET DENNIS CLARENCE
JETT RONALD GENE
JEWELL PHILIP LAWRENCE
JOHNS JEFFREY JAY
JOHNSON ALLEN ISAAC
JOHNSON BARTON WENDELL
JOHNSON BRUCE ERVIN
JOHNSON BRUCE MARK
JOHNSON CHARLES WALTER
JOHNSON DEAN RAYMOND
JOHNSON EUGENE CHARLES
JOHNSON FRANCIS DAVID
LEO
JOHNSON GERALD DEAN
JOHNSON GERALD JAMES
JOHNSON JAMES KENNETH
JOHNSON JERRY ALLEN
JOHNSON PAUL CONRAD
JOHNSON RALPH WILLIAM
JOHNSON RICHARD ALLEN
JOHNSON RUSSELL CARL
JOHNSON STEPHEN DUANE
JOHNSON TERRY JAY
JOHNSON WARREN DEAN
JOLY MITCHELL LEWIS
JONES LOYD ELLIS
JONES ROBERT EDWARD
JORGENSON JEROME DAVID
JUDSON HAMPDEN CUTTS JR
JUNTILLA HARRY WILLIAM
KAATZ BARNEY
KAISER FRANK MELVIN
KALIS GERALD LEONARD
KAPOUN TIMOTHY JOHN
KARAU RONALD DEAN
KARGER GREGORY SCOTT
KASHIEMER CARL FREDERICK
KASTER STEPHEN JOSEPH
KEARBY JEAN ARTHUR
KELLER DODD CLIFTON
KELM LARRY ROBERT
KELMAN WAYNE H
KELSEY MILTON GEORGE
KENT ERROL LYNN
KERN DAVID JOSEPH
KERVIN JOEL CHARLES
KETTNER ALAN ARTHUR
KIGER DENNIS DELMAR
KINDLE WILLIAM HENRY
KINGSTON THOMAS LLOYD
KINNEY LEE CHARLES
KITTLESON ROGER MICHAEL
KJOS TERENCE MICHAEL

KLANCKE CHARLES WILLIAM
KLEIN DANIEL FRANCIS
KLEIN DON ROBERT
KLEMMER SYDNEY WILLIAM
KLINKENBERG RICHARD
CARL
KLOEK LYLE ARCHIE
KLOMSTAD RONNIE GENE
KLUKAS BRADLEY WILFRED
KNAKE LLOYD E
KNOPIK THOMAS ALLISON
KNOTT JOHN CHARLES
KNOWLTON DON GLENN
KNUTSON LARRY LEE
KNUTSON RICHARD ARTHUR
KNUTSON VERNON G
KOCH JAMES ANTHONY
KOCHENDORFER MICHAEL J
KOEBERNICK ALLAN FRED
KOEHLER RONALD LEE
KOEPPEN ERIC R
KOKESH ANDREW FRANK
KOKOSH GEORGE GERALD
KOLINSKI THOMAS GEORGE
KOLLER MICHAEL JOSEPH
KOLLMEYER CARL
KOLSTAD THOMAS CARL
KONECNY JAMES FRANK
KOOB THOMAS JOHN
KOPP PATRICK DANIEL
KORDOSKY THOMAS JAMES
KORTESMAKI PATRICK LEO
KOSEL GENE MARLOW
KOSKI LARRY CHARLES
KOSKI RICHARD ARNE
KOSKOVICH MICHAEL L
KRAEMER FRED CHRIS
KRAGE BRUCE HERBERT
KRAUS KENNETH C
KRAUSE RUSSELL EMIL
KREBS LARRY EDWARD
KREBSBACH RONALD AL-
PHONSE
KRECH MELVIN THOMAS
KREKELBERG RAYMOND
JOSEPH
KRINKE STEPHEN MATHEWS
KRISPIN THOMAS ALBERT
KROSSEN RICHARD MARVIN
KRULL JAMES LEE
KRUMM RICHARD HENRY
KRUSE PAUL HARLAN
KUEFNER JOHN ALAN
KUEHN LLOYD MARTIN
KUFFEL CONRAD JOSEPH
KULAVIK RICHARD MARVIN
KUNKEL ROBERT HAROLD
KUNSHIER GARY LEE
KURTZ ALVIN EDWARD
KURZ JOHN PETER
KUSILEK LAWRENCE ROBERT
KYAR LARRY CLARENCE
LA PLANTE WILLIAM ROY III
LALLY MICHAEL JOHN
LAMUSGA MICHAEL ALAN
LANNING RONALD BARRY
LANNOYE NICHOLAS PIERRE
LANTZ GARY LEE
LARCHER ROGER WILLIAM

LARSON DUANE CLIFFORD
LARSON EDWARD DAVID
LARSON ROBERT DARREL
LARSON ROBERT JOHN
LARSON RONALD JOE
LARSON VERLE NORMAN
LATTMAN DONALD WAYNE
LAU JOEL THOMAS
LAVALLEE MICHAEL EUGENE
LAWRENZ DAROL EDWARD
LE DUC THOMAS GLEN
LE NOUE BRUCE VERNON
LE VASSEUR JEROME F JR
LEDIN DANIEL BING
LEE CLYDE MARVIN
LEHMANN DERLYN REYNOLD
LEINO VERNON LEROY
LEPPKE LYLE GORDON
LEWELLIN LAWRENCE FRANK
LEWER THOMAS CHARLES
LEWIS ROGER CHARLES
LEYDE THEODORE EDWARD
LIEBERMAN JAY LESLIE
LIEBL DONALD ALVIN
LILIENTHAL WILLIAM ED-
WARD
LILLIS RICHARD NED
LILLY JOSEPH DARRELL
LINDBERG BRIAN VICTOR
LINDGREN ROBERT WILLIAM
LINDSTROM RONNIE GEORGE
LIPETZKY DANIEL JOHN
LISLE TERRILL MICHAEL
LITZINGER DUANE EDWARD
LLOYD ALLEN RICHARD
LLOYD DANIEL WILLIAM
LOBSINGER JOHN FORMAN
LOEGERING DEAN CHARLES
LOEHLEIN ROBERT JOHN JR
LOFTUS CRAIG JOHN
LOHSE RICHARD LEE
LONGTIN MARK WARREN
LONGTINE JERRY ALLEN
LORIMER WILLIAM IV
LOSSING CLARENCE ERNEST
LOVELLETTE GARY VAUGHN
LUCAS WILLIAM HARVEY
LUECK DOUGLAS ROY
LUNDEQUAM DONALD JAMES
LUNDY GERALD VERNON
LURTH MELVILLE ALBERT JR
MAAHS MILO GEORGE
MAAS ROY FRANCIS
MACE BRADLEY THOMAS
MACKEDANZ LYLE EVERETT
MACKEY WILLIAM RUSSELL JR
MADLAND ROBERT LOUIS
MADSEN MARLOW ERLING
MAHOWALD MICHAEL ALLEN
MAKI DONALD DALE
MAKI ROGER LEE
MALLOY JAMES FRANCIS
MANDERFELD THOMAS
GEORGE
MANKA DOUGLAS THOMAS
MANNIE ROGER MICHAEL
MARIZ ROBERT J
MARKUS LARRY FRANK
MARTHALER ROBERT FRANK

MARTIN CARL RAYMOND
MARTIN TERRY LEE
MARTINSON DARRELL WAYNE
MARTINSON DELVIN CARL
MARTINSON LEROY CLAYTON
MATEL RONALD JAMES
MATHEIS RICHARD ALAN
MATHISON MICHAEL ALFRED
MAXIE CHARLES LEE
MAYER DWIGHT BENNIE
MC CARTY THOMAS HUBERT
MC CLELLAN MICHAEL JAMES
MC CLOSKEY JOHN JAMES
MC COLLUM WAYNE ADEL-
BERT
MC CUE GARY FRANCIS
MC DANIEL JOHN THOMAS
MC DONNELL JOEL WILLIAM
MC GARRY JEREMIAH D
MC GOVERN RICHARD DALE
MC INERNY ROGER JAMES JR
MC KEEVER MICHAEL ED-
WARD
MC KENZIE LARRY DEAN
MC KNIGHT JOSEPH PATRICK
MC LAUGHLIN DANIEL P JR
MCCLUSKEY PATRICK
CHARLES
MEAKINS CHARLES HENRY
MEES WAYNE EDWARD
MEGA JAMES FRANK
MEIER GARY MICHAEL
MELDAHL ALLEN ROBERT
MENSEN ANTHONY JOSEPH
MENSHEK STEPHEN ALBERT
MENZ CLYDE RONALD
MERCIER PATRICK TIMOTHY
MERRILL DOUGLAS CARROLL
METTLING CRAIG STEVEN
MEULEBROECK KENNETH
JOHN
MEYER JOHN FRANKLIN
MICHAEL LEO GENE
MICHEL DAVID GEORGE
MICHEL JOHN STEPHEN
MICKELSEN WILLIAM EMIL JR
MIDTHUN ROGER WAYNE
MILBRATH ROBERT KEITH
MILBRODT GERALD LEE
MILDE DAVID W
MILLER JAMES WALTER
MILLER STEVEN RICHARD
MISHUK RICHARD EDWARD
MOLTZAN WILLIAM JOHN
MONCRIEF WILLIAM GRADY
MONDYKE CHARLES AN-
THONY
MOORE ROLAND EDROY
MORNINGSTAR DUANE LEE
MOTT BARRY LEE
MUHICH CRAIG STANLEY
MULVANEY JAMES RAYMOND
JR
MUNDEN STEVEN DOUGLAS
MURPHY DANIEL JOSEPH
MURRAY PATRICK PETER
MUSIL CLINTON ALLEN SR
NABBEN ARTHUR S
NEHL JOSEPH ROBERT

NEHRING DENNIS DEAN
NELSON BRUCE JEFFREY
NELSON CHARLES CARTER
NELSON DANIEL IVAN
NELSON DAVID THOMAS
NELSON DENNIS LEROY
NELSON DENNIS ORVILLE
NELSON JAMES LESLIE
NELSON MAX ALVIN
NELSON MELVIN DOUGLAS
NELSON NOEL STEVEN
NELSON PAUL ALBERT
NELSON PAUL ARTHUR
NELSON ROGER THEODORE
NELSON RONALD ELWOOD
NELSON RUSSEL COURTNEY
NELSON STEPHAN JOSEPH
NELSON STEVEN HOWARD
NELSON THEODORE RUSSELL
NELSON THOMAS LEON
NESS PATRICK LAWRENCE
NEUDAHL MICHAEL LLOYD
NEWBROUGH GERALD RAY-
MOND
NICHOLS JAMES WILLIAM JR
NICKELSON MARTIN JOHN
NIEKEN LARRY LEE
NIEMI JAMES ARNE
NOBLE THOMAS GREGORY
NOGGLE STEPHEN M
NOLAN DAVID ALLEN
NORDQUIST JON HARRIS
NORMAN JAMES MICHAEL
NORMANDIN DUANE MI-
CHAEL
NORTHRUP MAURICE FRED-
RICK
NOTERMANN MICHAEL
WILLIAM
NOVAK RICHARD DANIEL
NYBLOM DUANE WILLARD
NYE WALLACE GREGORY
NYSTUL MICHAEL DEAN
O BRIEN TERRENCE PATRICK
O CONNOR DANIEL JEROME
O KEEFE RONALD THOMAS
O LAUGHLIN DANIEL THOM-
AS
O NEILL MICHAEL JAMES
O TOOLE GEORGE PATRICK JR
OAK GLEN EVERETT
OBENLAND ROLAND ROBERT
ODAFFER RICHARD DUANE
OESTRIECH JIM EDWARD
OGLESBY GERALD PHILLIP
OHM DAVID JAMES
OHMAN GARY ALAN
OKER DAVID PAUL
OLCOTT STEVEN JAMES
OLIN THOMAS M
OLIVE DONALD LEWIS
OLMSTEAD DALE FRANK
OLSON ALFRED RICHARD
OLSON BARRY A
OLSON BRUCE DENNIS
OLSON CHARLES EMMETTE
OLSON GENE JOHN
OLSON JEROME ANDREW
OLSON KENNETH LEE
OLSON RICHARD RALPH
OLSON ROBERT EUGENE
OLSON THOMAS PERCY
ONSLOW ROBERT CRANLEY
OOTHOUDT GARY ELDON
ORDNER JOHN ALBERT
ORLEMANN JOHN STEWART
OTT DOUGLAS RALPH
OTTO DALE LESTER
PAHL KENNETH ALLEN
PAINE PAUL WARREN
PALMER CLARENCE LEROY
PANNO RONALD WILLIAM
PANZER RONALD LEE
PAPESH DAVID C
PARKER DAVID LLOYD
PARSONS GREGORY ALLEN
PAULSEN LAWRENCE EDWARD
PAWLOWICZ DENNIS WAYNE
PEAK DAVID FRANCIS
PEDERSEN DENNIS IRWIN
PEDERSON ARTHUR CLIF-
FORD
PEDERSON KENNETH ALLEN
PEPPER JAMES THOMAS

PERKINS GEORGE PETER
PERSONS DANIEL BRUCE
PETERSON BRADLEY EUGENE
PETERSON DAVID MARTIN
PETERSON DELBERT RAY
PETERSON DONALD LEE
PETERSON DUANE ARVID
PETERSON GARY WAYNE
PETERSON JACK WALTER
PETERSON JEFFREY CHARLES
PETERSON LEROY EMANUEL
PETERSON MARLIN TRENT
PETERSON MICHAEL GERALD
PETERSON MICHAEL VIRGIL
PETERSON RENOLD WILLIAM
PETERSON STEPHEN EDWIN
PETERSON TERRILL GENE
PETERSON WALTER ARNOLD
JR
PFEFER ARTHUR THOMAS
PIERSON DENNIS LEROY
PIPER THOMAS LEIGH
PITTS DANA ALLEN
PLATH STEVEN DALE
POEPPING WILFRED NORBERT
POLZIN HENRY CLARENCE
POPP DAVID JOSEPH
POTTHOFF THOMAS ALBERT
POWER RICHARD DEAN
POWERS TRENT RICHARD
PRATHER CHRISTOPHER
DAVID
PREKKER GARY LEE
PRICE THOMAS JOHN
PROSZEK ANTON JR
PROUE JAMES THOMAS
PRUDEN ROBERT JOSEPH
PUARIEA JAMES FREDERICK
PURDUM RALPH SCOTT
QUILLIN WILLIAM THOMAS
QUINN MICHAEL EDWARD
QUINN THOMAS WAYNE
QUITMEYER TONY JOHN
RABEL LASZIO
RABEL VICTOR ART
RAIOLO JAMES JOHN
RAMAKER LAWRENCE FRED-
RICK
RANDALL ROBERT BRUCE
RANTHUM DALE HAROLD
RASSEL ROBERT HERMAN
RATHBUN GARY ALLEN
REDENIUS RONALD JAMES
REED SAMUEL LEE
REEVES RAYMOND STANLEY
JR
REHN GARY LEE
REIGSTAD DANNY RAY
REILLY JOHN FRANCIS
REILLY LAVERN GEORGE
REINHARDT BARRY THOMAS
REINKE JACK RAYMOND
REITMANN THOMAS EDWARD
REMER KEVIN RALPH
RICE THOMAS EVERETT
RICHARDSON PHILIP OWEN
RICK EUGENE MERLYN
RIDDLE LARRY LYNN
RIEDLBERGER GERALD FRANK
RIEHL HARLAN CYRUS
RILES DONALD EUGENE
RILEY THOMAS JAY
RINGHOFER CURTIS EDWARD
RINGOEN MARVIN LEE
RITTER ALLEN JEROME
RITZSCHKE DAVID AARON
ROBERG JAMES AUSTIN
ROBINSON TIMOTHY GEORGE
ROBINSON WARREN JAMES
ROMANKO MICHAEL JAMES
ROOD CRAIG ALLEN
ROSTAMO THOMAS DAVID
ROWELL LEE MILTON
ROWLAND ROGER LEE
RUBIN HERMAN FRANCIS
RUCKTAESCHEL GARY ARDEN
RUDLONG THELMER ARTHUR
RUSSEK JOHN JOSEPH
RUSSELL JERRY WILLIAM
RYAN THOMAS LAWRENCE
RYBERG CHARLES EDWARD
RYDER JOHN LESLIE
SAARELA WILLIAM GEORGE
SAATHOFF RAYMOND JOSEPH

SABA LESTER PAUL
SACK GERALD DUANE
SAHLBERG GREGORY IRVING
SALVESON SELMER ERNEST
SALZER GENE LEO
SAMPSON MICHAEL JOHN
SANDMANN RONALD LEE
SANGSTER GARY LAVERN
SARVELA MERREL GERALD
SATTER DONALD STEPHEN
SAVOREN WILLIAM MARTIN
SCHAAF JOHN RAYMOND
SCHAEFER ROGER BERNARD
SCHAEFER SYLVESTER ANT-
ONY
SCHAEFER THOMAS KOENIG
SCHANCK HENRY EDWARD
SCHELL RICHARD JOHN
SCHEUBLE MELVIN JOHN
SCHINDLER EUGENE DONALD
SCHLICHT JEROME JOSEPH
SCHMELING ERWIN ROSS
SCHMIDT DONALD HAROLD
SCHMITZ LOREN MICHAEL
SCHMITZ PHILLIP NICHOLAS
SCHMITZ WILLIAM DAVID
SCHNURRER REINHARD J JR
SCHOEBEN SCOTT DOUGLAS
SCHOOLMEESTERS JOSEPH A
SCHOUVILLER THOMAS JOHN
SCHRAMEL KENNETH MI-
CHAEL
SCHROEDER LYLE WILLIAM
SCHROM JOHN FRANCIS
SCHROM KENNETH R
SCHULTZ STEVEN OWEN
SCHUMACHER STEPHEN
LAWREN
SCHUMACHER WAYNE
THOMAS
SCHUMANN JOHN ROBERT
SCHUMMER DALE CLARENCE
SCHWAGEL KENNETH
FRANCIS
SCHWARZKOPF ALLAN
ALBERT
SEEMAN STEVEN CARL
SELENKA RUDOLPH CARL JR
SELLER JOSEPH JOHN
SELLNER CHARLES EDWARD
SERRANO THOMAS ROBERT
SETTERQUIST FRANCIS LESLI
SEVERSON JOHN EDGAR
SEYKORA WILLIAM JOSEPH
SHANNON THOMAS ERIC
SHANOR GERALD DELMAR
SHARLOW GUST J
SHARPE RICHARD ALLAN
SHEPERSKY VINCENT LLOYD
SHERECK JAMES JOHN
SHORTLEY DOUGLAS LYLE
SHURR ROBERT JAMES
SIKICH MICHAEL MATTHEW
SILVER JOHN CLYDE
SIRES ROBERT JOHN
SKAGGS WILLIAM FRANK
SKANSON LOUIS JAMES
SKARMAN ORVAL HARRY
SKOUBY RICHARD LOWELL
SKUZA ARVID BURDEEN
SLANDER RICKEY ALLAN
SLOAN MONTE THOMAS
SLOAN THOMAS NEWTON
SMITH DENNIS CAROL
SMITH HAROLD JOHN
SMITH JERROLD PATRICK
SMITH JOHN ALEXANDER
SMITH RICHARD WILLIAM
SMITHSON CRAIG DENNIS
SMOGER MICHAEL ARTHUR
SOLBERG KALE ARLAN
SOMA THOMAS EDWARD
SOMMERHAUSER JOSEPH
ALLYN
SONGLE CLAYTON ANDREW
SORENSEN ROBERT WILLIAM
SPARKS PETER ALLAN
SPENCE DONALD EDWIN
SPICER EUGENE DOUGLAS
SPICZKA ALOYSIUS F JR
SPIEROWSKI RUSSELL DEAN
SPILLMAN CHARLES OTTO
SPINLER DARRELL JOHN
SPINLER RAYMOND PAUL

SPONG ERNEST ALLAN
STAHL ROBERT HENRY
STARK HERBERT D
STEELEY MARK M
STEFANICH NICHOLAS C
STEFFES WILLIAM JOSEPH
STEIRO ROBERT EDWARD
STENBERG JOHN MARVIN
STEVENS RAYMOND JOHN
STEVENS WESLEY WARREN
STIYER DAVID ALAN
STOCKDALE MELVIN JAMES
STOEN MARCUS SHERWIN
STOLTZMAN GEORGE LEO
STOMMES KENNETH CLAR-
ENCE
STRADTMAN THOMAS LEE
STRANDBERG ERVIND CARL
STRAUSSER DARRY RICHARD
STRICKLIN THOMAS GRADY
STROSHANE MICHAEL ALLEN
STROUB STEVEN JOHN
STRUSS LARRY ANTHONY
STUART JOHN DESMOND JR
STURTZ DARWIN CLIFFORD
SULANDER DANIEL ARTHUR
SULLIVAN DOUGLAS J
SULLIVAN TIMOTHY EMMETT
SUNDQUIST DAVID HARRY
SUNDQUIST JOHN OLAF
SVOBODNY LAWRENCE
MARVIN
SWANSON DARREL THOMAS
SWANSON LAWRENCE HARRY
SWANSON LYNN CURTIS
SWANSON ROGER WESLEY
SWANSON WILLIAM EDWARD
SWART WALDON JEROME
SWEDEEN RICHARD ALLEN
SWEDENBURG ROBERT JOHN
SWEET RICHARD DONALD
TABOR DENNIS RICHARD
TAYLOR KERRY LAMONT
TAYLOR PHILLIP EDWARD
TEICH DAVID LEE
TENHOFF TRACY STEPHEN
TENNISON ALVIN GENE
TERHORST BERNARD REIN-
HOLD
TESCHENDORF RONNIE CARL
TESKE BERNARD ALBERT III
THEISEN JAMES ELMER
THOMAS BERNARD MONROE
THOMPSON LYLE JOHN
THOMPSON STANLEY WEN-
DELL
THOMPSON WESLEY ROBERT
THORSON ERNEST LEROY
THOTLAND JOHN ALFRED
THUET STEPHEN PAUL
TOENYAN FRANCIS HENRY
TOLLEFSON DWIGHT DUANE
TOMS DENNIS LEROY
TOSTENSON MICHAEL LEE
TRACY GERALD FRANCIS
TRACY JOHN WILLIAM
TRAMEL WALTER OTHO
TRISKO WALTER HENRY
TRONERUD STEPHEN LYLE
TRONNES ALVIN PHILLIP
TROXEL MARLON WADE
TROY PETER JOHN
TRUHLER BRUCE LEE
TSCHERTER VERNON S
TSCHUMPER ROBERT G
TURNER ARTHUR JOSEPH
TURNER JON ARNOLD
TUSKEY ROBERT WILLIAM SR
UECKER DAVID ARNOLD
UGELSTAD BRUCE ALLEN
UGLAND DAVID LEONARD
ULICSNI MICHAEL JOHN
UNGERECHT RICHARD
ALFRED
UTTERMARK JAMES FREDERIC
UUTELA DERRIS LEE
VAN VLEET JEFFERY HAROLD
VASQUEZ DEAN
VEDDER RICHARD JEROME
VERWERS ROGER LEE
VICK RICHARD DEWEY
VINCENT NORMAN WAYNE
VOGEL GARRITY
VOGELSANG JOHN KIM

VONASEK RICHARD JAMES
WACKERFUSS RICHARD
WILLIA
WAGNER JAMES EDWARD
WAGNER KENNETH JAMES
WAGNER WILLIAM OSCAR
WALDRON JAMES TAYLOR
WALDRON KARL MERRITT JR
WALDVOGEL ROBERT E
WALENSKY GORDON DAVID
WALLACE RONALD RAY
WALLIN DOUGLAS DEWEY
WALSH RICHARD AMBROSE III
WANKA CARL JEFFREY
WATSON LARRY WILLIAM
WATTERS PHILLIP DONALD
WAYRYNEN DALE EUGENE
WEAVER ROBERT DUANE
WEBB ALFRED JR
WEBER DAVID ALLAN
WEBER DAVID FRANK
WEBER JOHN KNUTE
WEBER WILLIAM JAMES
WEHRS DAVID WILBERT
WEISE RICHARD RAYMOND
WEISLER JAMES ROBERT
WEITZEL BILLY DEAN
WELCH LELAND DOUGLAS
WELIN DANIEL KENNETH
WELK LAWRENCE NORMAN
WELLMANN DENNIS WELDON
WENCL DAVID ALLAN
WENNES ROBERT ALLEN
WENZEL MARK ANDREW
WESTERBERG KENNETH GLEN
WESTMAN MYLES DALEN
WESTRA DIRK JON
WETTERGREN STEVEN
EDWARD
WHITE CHARLES CLINTON
WHITE CHARLES HENRY
WHITE JAMES DAVID
WHITE MICHAEL JAMES
WHITE RICHARD JOSEPH
WHITE RICHARD NEAL
WHITMAN RAYMOND LEE
WICKENBERG ERIK BERNARD
WICKLACE RANDALL JAMES
WIDEN JOHN GEORGE
WIEHR RICHARD DANIEL
WIELINSKI DONALD JAMES
WILBER WILLIAM FREDRICK
WILBRECHT KURT MICHAEL
WILDE RAYMOND CHARLES
WILLARD LLOYD LOREN
WILLEY PETER RAY
WILLIAMS DAVID LEIGH
WILLIAMS LARRY DALE
WILLIAMS RICHARD EARL
WILSON DANIEL KEITH
WILSON JOHN WALTER JR
WILSON LARRY EUGENE
WILSON MARVIN JAMES
WIMMERGREN EDMOND
DALE
WISE GORDON SCOTT
WITT KENNETH LEE
WITTKOP JOE ALLEN
WOEHNKER HARRISON E JR
WOLTER JAMES LESTER
WOOD ROBERT HAROLD
WORTHLEY KENNETH WAYNE
WRIGHT BRADFORD DWAIN
WRONSKI JOHN C
YANKOSKI ROBERT ALLEN
YATES JOHN CHARLES
YEAGER GREGORY LEE
YOUNG RANDALL LEE
YOUNK DAVID ALAN
ZAGER JOHN CARL
ZAITZ JACK MICHAEL
ZEMPEL RONALD LEE
ZIMMERMAN DEAN ROGER
ZUTTER DANIEL ROGER

131

Photos by Gordon Denman

Vietnam Veterans Memorial Park

Highway 90, Ocean Springs, MS

This large memorial features not only the names of fallen and missing veterans but also their pictures. It sits on over four acres of land that was granted to the Mississippi Vietnam Veterans Memorial Committee by the city of Ocean Springs and has since won landscaping awards.

According to the Mississippi Vietnam Veterans Memorial Committee, the Vietnam War Memorial was made possible only because a group of Vietnamese Americans were looking for a way to show their gratitude to the local service members who fought in the Vietnam War.

The Ocean Springs location was one of three locations that the Committee was considering in 1989. The size of the land offered by Ocean Springs convinced them to build on that spot. The proposal for the memorial was submitted to the State Senate and House Appropriations Committees and stated that the group was looking for $1.5 million from the state for the memorial.

In May 1995, ground was broken on the site as over a thousand people watched, but the actual construction didn't begin until December 1996. The memorial was dedicated in May 1997 and was meant to be a living legacy that would continuously help people heal and educate the younger generations about the Vietnam War and the men and women who made the ultimate sacrifice for their country.

The Mississippi Vietnam Veterans Memorial consists of a black granite monument that serves to honor the veterans. Two granite slabs face each other and feature the names and some of the pictures of the 668 Mississippians who were either killed in action, went missing in action, or were held as prisoners of war. These granite slabs are surrounded by concrete and limestone walls. The monument was coated in a protective material to make sure the laser-etched pictures remained in the granite over the course of time. A Huey has perched on a pedestal since the memorials dedication. Memorial day and Veterans day services are held annually.

The Names of those from the state of Mississippi Who Made the Ultimate Sacrifice

ADAMS AUGUSTUS
ADAMS DAVEY MARLIN
ADAMS LARRY EARL
ALLEN BILLIE ALVIN
ALLEN DANNY RAY
ALLEN JOE EBERT
ALLEN THOMAS BARRY
AMOS JAMES ALBERT
ANDERSON LEON JR
ANDREWS WILLIAM LARRY
ARD BOBBY JOE
ARMSTRONG TERRY LEE
AVANT JOE LYNN
BAILEY BOBBY LEE
BAKER WILLIE JAMES
BALL MICHAEL HENRY
BARFIELD JOHN R
BARRAS GREGORY INMAN
BARRITT WILLIAM EMMETT
BATCHELOR JOHN ELSEY JR
BATEMAN JAMES RONALD
BEACHAM EDWARD EARL
BEARD ASBERRY JR
BEASLEY EDGAR HUNTER
BEASLEY LUZON
BECKLEY GEORGE EDWARD
BELL ARTHUR FREDERICK
BENNETT ROBERT HORACE
BENNETT THOMAS WARING JR
BIZZELL RAYMOND ALBERT
BLACKWELL JOSEPH CARLTON
BLAIR IVY LOUIS
BLAKE EDWARD ALOYSIUS
BLALOCK JOHN HILTON
BLANDEN JAMES D
BLANKS CLARENCE
BLYTHE TERRY LEE
BOATWRIGHT GEORGE OLIVER
BOLAND MELVIN LYNN
BOLEN FREEMAN

BOLTON WILLIE EDWARD
BOOZER DON ALLEN
BOSTON JOHNNY B
BOUDREAUX JIMMY DALE
BOURRAGE I V
BOYD CHARLES
BRADY SHERMAN C
BRANTLEY WILLIAM OSLER JR
BRELAND CECIL DOUGLAS
BRELAND LEO MEYERS
BREWER JESSIE SEYMORE
BRITTON WILLIAM DAVID
BROCK PERCY GUY JR
BROWN CHARLES
BROWN JOE HENRY
BROWN JOHN WAYNE
BROWN JOHNNIE LEE
BRYANT MAURICE HERBERT
BRYANT NELTON RAYMOND
BUCHANAN JAMES ELSON
BURKETT ELIJAH WALLACE
BURKS JAMES CARL
BURNLEY EARL ROSEMOND JR
BURNSIDE DONALD RAY
BURRISS JOHNNY LEE
BURTON JAMES ALLEN
BURTON JOHNNY RAY
BUSH ELBERT WAYNE
BUTLER ALBERT CHARLES
BYRD ALTON DOYLE
BYRD DOUGLAS EVERETT
CANOY ERVIN PRESTON JR
CARLISLE BILLY PAT
CARPENTER KENNETH BRAXTON
CARTER JOE EDDIE
CARTER L C
CASEY LEO CARL JR
CATLING WILLIE B
CAUTHEN HENRY CLAY SR
CHALMERS DEMPSEY JR
CHAPMAN CLINTON
CHAPMAN DAVID THOMAS

CHAPMAN RONALD
CHERRY JAMES EDWARD
CHILDRESS CALVIN BUSTER
CLARK DENNIS EUGENE
CLARK JAMES GENIUS
CLARK ROBERT SAMUEL JR
CLEMONS JAMES NOEL
CLOPTON KENNETH RAY
COCHRAN CHARLIE LYNN
COCHRAN ROBERT FISHEL JR
COCKRELL JAMES WARREN JR
CODY HOWARD RUDOLPH
COFER EVERETTE EARL
COLE MOZIE LEE
COLEMAN ARTHUR JR
COLEMAN LOUIS WILSON JR
COLEMAN LYNN BAILEY
COLLIER GEORGE EDWARD
COODY GEORGE LA FAYETTE
COOK PETER BROWN JR
COOK WEYMAN TERRY
COTTON MOSES M
COWART JOHN WAYNE
CRAFT GRAYSON
CRAIN JOSEPH DEWEY JR
CREEL DAVID DE WITT
CROOM HUBERT
CROSBY ROBERT BARRY
CROTWELL BYRON HUGH
CROW DAVID REID IV
CROWDER HYLAN LYNN
CUMMINGS HAROLD VAN JR
CUMMINGS JAMES LONEL
CUNNINGHAM LARRY LA MONT
CURRY M L
CUTRER FRED CLAY JR
DANDRIDGE ALBERT
DANIEL OLLIE JAMES
DANIELS PAUL FREEMAN JR
DANNER JESSIE JAMES
DANTZLER WILL ROY
DAVIS ELLSWORTH I JR

DAVIS FRANK EDWARD
DAVIS JAMES WOODROW
DAVIS JERRY REED
DAVIS PORTER THAD
DAVIS RANDOLPH
DAVIS SAMUEL M
DAVIS WILLIAM JEWEL JR
DAVIS WILLIAM R
DE PRIEST DARRELL JAMES
DEDEAUX ALDON JAMES
DELANEY ALBERT LEE
DENLEY BILLY WAYNE
DEPREO WALLACE JOSEPH
DILWORTH ARTHUR WILLIAM
DILWORTH HENSLEY MC-FADDEN
DIXON JAMES C
DODDS LARRY FLOYD
DONALD HARMON ODELL JR
DOOLEY ROBERT ELLIS
DOUGLAS LESLIE FORREST JR
DOUGLAS PAUL MELVYN
DOZIER DEBROW
DRAKES CLARENCE EARL
DRANE WILBERT RAY
DUNCAN PHILLIP ALLEN
DUNNING DENNIS GYMAN
EARNEST WILLIE LEE
EASLEY DAVID ROY
EATON MARK HASKIN
EDMONSON BOBBY
EDWARDS ROBERT JAMES
ELLIOTT JERRY W
ELLIS SYLVESTER
ENTRICAN DANNY DAY
ERBY LESTER
EUBANKS CARL MARCUS
EVERETT ROCKFORD GREY
EXPOSE HENRY RAY
FARR JACK GRAHAM
FELSHER JOHN ALFRED
FELTON THOMAS MOODY
FERGUSON SAMUEL

FICKLIN ERIC
FICKLIN EXCELL
FINCH MICHAEL THOMAS
FINNEY CHARLES ELBERT
FISCHER ADAM
FLANAGAN TOM
FLEMING WILLIAM ELGIN JR
FOREMAN TAYLOR W JR
FORREST JIMMIE LEE
FORTENBERRY EDWARD EUGENE
FOSTER FRANK
FOX LEON VINCENT
FRANKLIN JEROLD
FRAZIER GARY LEE
FRYE ALFRED ALLEN
FULGHAM JOE HUGH
FULLER FLOYD EDWARD JR
FULLER JOEL
FULMER RONNIE DALE
GAINES CHARLES JERRY
GANCI SAMUEL JOSEPH
GANT EDDIE DEAN
GARDNER EDDIE AUGUSTUS
GARNER IRA L
GARRETT HOWARD
GARTH CLYDE JR
GIBSON WILLIAM ARTHUR
GILBERT TRUMAN JAMES W
GILES BARNEY MCKINLEY JR
GILLASPY THOMAS DAVID
GILLIARD EDWARD LEE
GILLISPIE LUCION JR
GILMORE EUGENE THOMAS
GOLDEN DONALD LEWIS
GOODWIN FORREST
GORDON ARTHUR MELVIN
GORE HAROLD DOUGLAS
GRAFTON JAMES CALVIN
GRAHAM BENNIE JOE
GRANTHAM ELY JR
GRAY JOHN TERRY
GRAYSON SAMUEL A III

Mississippi

GREEN ROBERT EARL
GREENE RAYMOND MILLER
GREER HARRY CHARLES
GREGORY DAVID CLARK
GRESHAM WILLIAM THOMAS JR
GRIFFIN ALLEN AVERY
GROOVER JAMES COMPTON
GRUBBS JERRY ROE
GUEST ROGER THOMAS
GULLEDGE ERNEST PEPPER JR
HADDOX GEORGE HENRY
HALFORD CALVIN DOUGLAS
HALL MARVIN LOUIS
HAMBURG MC ARTHUR
HAMBY JIMMY WAYNE
HAMMOND JULIAN DICKIE JR
HAMNER LEON
HAMPTON DELL GENE
HAMPTON FRED LEE
HAND LARRY EDWARD
HARDY THOMAS DOUGLAS
HARE JOHN HENRY
HARLEY ROBERT LEE
HARLOW DAVID HUGH
HARMON ARTHUR
HARPER BILLY FRANK
HARPER EDWARD BENJAMIN
HARPER JOHN CURTIS
HARPER WILLIE JR
HARRINGTON PAUL VINCENT
HARRIS CHARLES RICKEY
HARRIS RICHARD FLAMOND
HASSELL ULYSSES C
HATHORN CHARLES LEE JR
HAYES WILLIE JAMES
HAYNES FREDDIE NEIL
HELM HERSCHEL PITTMAN JR
HENDRIX KENNETH LEVON
HENLEY ROY LEE
HENRY ANDREW L
HESTER GUY WILSON JR
HICKMAN STEVEN MURDOCK
HILL CHARLIE III
HILL ELMER DEAN
HILL JAMES EDWARD
HILL JOHN KENNY
HILL ROBERT MORRIS
HINES LOUIS CLARK
HINTON JIMMIE DAVID
HODGES DWIGHT
HOFFMAN JAMES MICHAEL
HOLLIDAY CRIS
HOLLINGSWORTH JOSEPH K
HOLLOWAY JOHNNY RAY
HOLMES JAMES MICHAEL
HOLMES JOHN HARRIS
HOLT JAMES
HORNBURGER WILLIE ROGERS
HORST PHILLIP METZ
HOSEY TOMMY BRYAN
HOUSTON LATHAN
HOWARD CHESTER THEO JR
HOWARD GEORGE DOUGLAS
HOWELL ADRIAN EALON
HOWELL ROLAND HAYES
HOWELL WILLIAM ERAY
HUDSON BOBBY
HUDSON JAMES WILLIAM
HUDSON JOSEPH WILLIAM
HUMPHRES JIMMY DARREL
HUMPHREY JERRY DALE

HUNT EUGENE
HUNT SAMUEL L
HUNTER CHARLES LOUIS
HUNTER MELVIN TYRONE
HUX THOMAS MICHAEL
IDOM MAX RALPH
IRVIN THOMAS FRANKLIN
ISBELL JIMMIE RAY
IZARD PHILLIPS H JR
JACKS GLENN GATES
JACKSON CHARLES EDWARD
JACKSON ROBERT BUFORD
JAGGERS THOMAS MURL
JEFFRIES MACK SIMPSON
JENKINS JERRY WAYNE
JENNINGS JAMES DALE
JERMYN BOBBY RAY
JOHNSON DAVID E
JOHNSON JAMES CARL
JOHNSON JAMES JR
JOHNSON JOE LOUIS
JOHNSON JOHN FOSTER
JOHNSON PRINCE ARTHUR JR
JOHNSON VICTOR JR
JOHNSON WILLIAM NEWTON
JONES CHARLES RAY
JONES CLARENCE
JONES DAVIES LEE
JONES GEORGE EMERSON
JONES MOSES EDWARD
JONES WILLIE MORRIS
JORDAN JACK JOSEPH JR
KENEDY WILLIAM MICHAEL
KENNEBREW JOHN C
KENNEDY GLENN ALEXANDER
KENT LLOYD HENRY
KERNS DONALD RAY
KING JAMES ROY
KING ROBERT SHELTON JR
KUYKENDALL WILLIE CLYDE
LA GRONE WILLIAM NAPOLEON
LAMAR MELVIN STETTINIUS
LANGHAM HENRY JR
LANKFORD EVELYN FRANKLIN
LANNOM GARY KENNETH
LAWS ISAIAH JR
LAYTON CALVIN JEROME
LEACH JAMES WILLIS
LEE JOHNNIE GENE
LEHECKA JOHN ARTHUR
LICHTE JACK ROWLEY JR
LIDDELL BENJAMIN F III
LINT DARRELL CLIFFORD
LOFTON JERRY WAYNE
LOFTON JOE EDDIE
LUNA CARTER PURVIS
LUNSFORD HERBERT LAMAR
LURIE ROBERT MICHAEL
MACK JOSEPH DEAN
MAGEE HERMAN PAUL
MAJURE EUGENE JEHLEN
MALLETTE AVON NORRIS
MALONE HERBERT LEE
MANGUM WILLIAM THOMAS JR
MARLAR DONNIE JOE
MARTIN DAVID WAYNE
MARTIN DONAIL
MARTIN WILLIAM DAVIS
MATTHEWS DAVID EARL
MAUNEY GERALD CLINTON
MAXWELL JAMES EDWARD
MAY RICHARD EARL
MC ALLISTER ANGUS W JR
MC BRIDE EDWARD ERNEST
MC CARTHY EDWARD POLK III
MC CARTHY TIMOTHY CLAY
MC CARTY FREDERICK DONALD
MC CLAIN CECIL EVERETT
MC CLELLAN M L
MC CLOUD LAWRENCE BEVERLY
MC CRARY RONALD SMITH
MC CRAY MELVIN
MC DONALD LARRY JAMES
MC GEE CHARLES EDWARD
MC GEE WILLIAM ROYAL
MC GEHEE NOBLE DOUGLAS
MC GINNIS STEVEN LAVELLE

MC GUIRE JOHN EDDIE
MC INNIS HENRY DAVID
MC KENNEY PATRICK MICKAEL
MC KINNEY HOLLIS RAY JR
MC KINNON BOBBY RAY
MC LAURIN WILLIE JAMES
MC MILLAN EDDIE LEE JR
MC NAIR WILLIE CHARLES
MC NEIL WILLIE DAVIS
MC WHORTER JAMES DAVID
MCCOY LARRY
MELTON CHARLES EARL
MENEES RICHARD ALLEN
MERRIWEATHER T Z
METCALF CLAUDIE
MILLER ALVIN EDWARD
MILLER LAWRENCE SCOTT
MILTON CHARLES
MOONEY TOMMIE LEE
MOORE GEORGE WASHINGTON
MOORE HERMAN JR
MORRIS JAMES ALBERT JR
MORROW HERSHEL EUGENE
MUSE EDWARD GRADY
MYERS OLIVER WENDELL
MYERS R C
NARD JAMES PETER III
NEELY BILLY JOE
NELSON EARNEST GLYNN
NELSON HOWARD HAMILTON
NETTER DANNIEL JR
NEWCOMB LARRY FRANKLIN
NICHOLSON GERMAN LEE
NOBLES LAVELLE MILLARD
O DELL TIMOTHY LEE
O'BRIEN TERENCE DALE
ORNSBEY CHARLES RAY
OTT THOMAS DEVECMON II
OVERBEY MARION DEWAINE
PAGE LARRY LEE

PALMER JAMES HESTER
PALMER SAMMY RAY
PAPALE ARTHUR LAWRENCE
PARMLEY DONALD WAYNE
PEARSON FRANCIS LAURY
PEOPLES ALEXANDER A S
PERKINS CLYDE J
PERKINS WILLIAM DEWITT JR
PETERSON ROBERT VERNON
PETTIEGREW JAMES PAUL
PETTIT HUGH MICHAEL
PHILLIPS CLYDE RAYMOND
PHILLIPS HARRY V JR
PHILLIPS WILEY LAVELL
PINION DOCK JEFFERSON
POORE ROBERT EARL
POPE JAMES RUSSELL
PORTER THOMAS LAMAR
PORTIS ANTHONY JEROME
POWELL WALTER MERRILL
POWERS ROGER STEVEN
POWERS SPENCER BYRD JR
PRESLEY DONNIE DWIGHT
PUGLIESI DAVID JAMES
QUICK ROBERT GLYNN
QUINN ROBERT FRANK
RANDALL RICHARD DENNIS
RANDLE GEORGE JR
RAWSON JAMES HILTON
RAY DENNIS MICHAEL
RAY WILLIE JAMES
REED CHARLIE JR
REED SHELLIE JEAN
REESE DANIEL JR
RHODES JOSEPH LEE
RICHARDSON LOUIS DOUGLAS
ROADS DENNIS LEE
ROBERSON JIMMY DARRELL
ROBERTS JOE RAYMOND JR
ROBERTS MICHAEL LAND
ROBERTSON PAUL ALLEN

ROBINSON CHARLES WAYNE
ROBINSON HORRIS GENE
ROBINSON JOEQUIN
ROBINSON JOHNNY LEE
ROBINSON PAUL WILLIAM
ROECKL JOHN DANIEL
ROLAND HULAN DUANE
ROSS HARVEY TURNER JR
ROSS JIMMIE CALVIN
RUSH JAMES EDWARD JR
RUSHING EDWARD FRANKLIN
SALTERS LEE EARNEST
SANDERS JOHNNY CRAWFORD
SANDERSON JOHN DANIEL
SANDLIN RONALD LEE
SANFORD HOLLIS COLEMAN JR
SANTEE HENRY EDWARD
SARGENT STANTON GERALD
SAVELL MYLES CLAYTON
SCOTT ROBERT LEE
SEALS WALTER
SELDON JAMES LENVER
SHANNON JAMES HERVEY JR
SHARBER JOHN JR
SHARP DERRELL KEITH
SHAW JOHN DILLINGER
SHINGLER ROY DELL
SHORT ANDREW JONAH III
SHOWS JAMES JERRY
SHUMPERT CHARLES MC CLAME
SIGALAS GEORGE CURTIS
SIMMONS BENNIE LEE
SIMMONS WILLIAM PRESTWOOD
SISTRUNK CANOY LEWIS
SISTRUNK CREIGHTON WAYNE
SIZEMORE JAMES WILLIAM
SLAUGHTER FREDDIE L JR

SMITH AUDRON L
SMITH BENNY LEON
SMITH CLEO
SMITH DENNIS GERALD
SMITH GARY WILLIAM
SMITH GEORGE W JR
SMITH HUGH EDWIN
SMITH JAMES EDWARD
SMITH JAMES EDWARD
SMITH JOHNNIE EARL
SMITH KENNETH LAVELLE
SMITH LARRY HAYS
SMITH LEONARD DALE JR
SMITH RICHARD TERRY
SMITH ROBERT EUGENE
SMITH ROBERT WALTER
SMITH RONNY
SMITH SPENCER
SMITH WILLIAM EUGENE
SMITH WILLIE JAMES
SMITH WILLIE JAMES
SMITH WILLIS WILSON JR
SPANN LYNN
SPIERS FRANK
SPIKES A V
STEELE WILLIE LEE
STEVENSON JAMES RALPH
STEVERSON SIM SMEDDLEY
STINGLEY JAMES
STODARD WILLIAM TERRY
STOKES COLBEN BENJAMIN JR
STRIBLING VICTOR MICHAEL
STRINGER WILLIAM FRANK-
LIN

STUART LEE DAVIS JR
STUDWAY DEWITT JR
STUTTS THOMAS RICHARD
SULLIVAN DANIEL
SUMLIN THOMAS EARL
SUMMERVILLE WILLIE JR
SUMRALL ROGER DALE
SUTTON GARRETT GARLAND
JR
SWAIN MILTON TRUMAN
SWINFORD SYLVESTER JR
SZEKELY JOSEPH CHARLES
TALL WARREN LEE
TALLEY JERRY WAYNE
TANNEHILL CHARLES DE-
VEAUX
TARTT CARLOS LEROY
TAYLOR BERNELL
TAYLOR LEE CURTIS
TAYLOR WALTER JOSEPH JR
TERRELL LEMUEL EBB
TERRY CHESTER H JR
TERRY CORNELIUS
THARP TERRY EDWARD
THIGPEN WILLIE LEE
THOMAS ANTHONY
THOMAS HENRY EARL
THOMPSON TERRY NEIL
THORNELL LESTER JEFFERSON
THORNTON JOHN THOMAS
THORNTON JOSEPH RAY
TINGLE TOM KERMIT
TOWNSEND JONATHAN
TRANTHAM VAN VERNON III

TRAXLER TOMMY JR
TRIM JACK RILEY
TUCKER EUGENE
TUCKER JAMES TAYLOR
TUCKER THOMAS EDWIN
TUMMINIA GIOVANNINO
TURNER HAYZELL CALVIN
TURNER VAN SYLVESTER JR
TYLER WILLIE
TYNER JAMES DANIEL
ULMER DAVID JOSEPH
VANDERFORD GERHARD W C
VAUGHN CLIFTON FLOYD
WALDEN LARRY HUSTON
WALKER JAMES EDWARD
WALKER JERRY DOYLE
WALKER TROY LEE
WALKER WINSTON CHARLES
WALL GEORGE MICHAEL
WALLACE CHARLES FRANK-
LIN
WALLACE GEORGE F JR
WALLACE JIMMIE CARL
WALTERS CHARLES ALLEN
WALTERS THOMAS JERRY
WALTMAN RICHARD A
WALTON JOSEPH
WARBINGTON HOWARD
OTTO
WARD EARNEST SHELBY
WARD GEORGE HOWARD
WARD THOMAS LESLIE
WARE RICHARD ALEXANDRA
WASHINGTON CLARENCE

EDWAR
WASHINGTON EDWARD W JR
WASHINGTON SYLVESTER
WATT SAMMIE LEE
WEATHERSBY JAMES EARL
WEEMS RONALD CLIFTON
WELCH E J JR
WELLS JOHN ELMORE
WEST MELFORD WAYNE
WHEAT ROY MITCHELL
WHITE GARSON FRANKLIN
WHITE ROOSEVELT
WHITFIELD WILLIE JR
WICKLIFFE ROBERT LOGAN
WILDER FRANK
WILK CHARLES LEE
WILKINSON WILLIAM GIL-
BERT
WILKS JOHNNY LEE
WILLIAMS BOBBY RAY
WILLIAMS CARTER LEE JR
WILLIAMS HENRY BRAXTON
JR
WILLIAMS JAMES EARL
WILLIAMS JAMES ELLIS
WILLIAMS LEROY
WILLIAMS RICHARD OLIVER
WILLIAMS ROBERT LEE
WILLIAMS THOMAS JOSEPH
WILLIAMS WILLIAM CHARLES
WILLIAMS WILLIE LEE
WILLIAMSON EDWARD LARUE
WILLINGHAM CHARLES
WATSON

WILLIS HOWARD DANIEL
WILSON BILLIE JOE
WILSON BILLY JOE
WILSON CHARLES E JR
WILSON EUGENE
WILSON LOUIS HENRY
WILSON RICHARD LEWIS
WILSON WILLIAM REED JR
WINFIELD LUCIUS
WITT DANNY KEITH
WOOD JAMES ALBERT
WOODRUFF JAMES AMMONS
WOODS SAMUEL LEE
WOOLBRIGHT DONALD
EUGENE
WORTHEY OWEN WAYNE
WOZENCRAFT WARREN LYNN
WRIGHT BOOKER T
WRIGHT CHARLES HENRY
WRIGHT ROGER DALE
WYLEY NATHANIEL
YIELDING LARRY THOMAS
YOUNG CALVIN EDWARD
YOUNG EUGENE
YOUNG JOHN CURTIS
ZHE ARDEN DALE
ZOLLICOFFER FRANKLIN

Vietnam Veterans Memorial Fountain

47th St. & J.C. Nichols Pkwy, Kansas City, MO 64112

Kansas City, Missouri is home to the beautiful Vietnam Veterans Memorial Fountain. It was built along with a memorial wall to honor the 451 fallen veterans from the Kansas City area. It is located in the Vietnam Veterans Memorial Park near the Vietnam Veterans Memorial along Broadway between West 43rd St. and Vietnam Veterans Memorial Drive. The fountain is flanked by two pools representing the differing opinions on the Vietnam War.

The water for the Vietnam Veterans Memorial Fountain begins on the south end of the Vietnam Veterans Memorial Park and travels through the two pools before flowing the north end of the park.

The memorial fountain uses a series of overlapping pools that start at the top of a hill as smaller-sized pools pour into bigger ones as they travel down the hill before they empty into the two flanking pools. The increase in the pools' size is a representation of the country's increased presence and involvement in Vietnam. Stone columns of varying heights have water flowing from their tops back into the surrounding pools.

Besides the fountain, there is also a memorial wall that stands 10-feet high and 155-feet long. On the wall are the names of the 451 fallen or missing servicemen from Kansas City, Missouri (and the seven-county metropolitan area) who were in the Vietnam War. On either side of the wall are depictions of the Purple Heart medal and the Vietnam Service medal.

The Kansas City Vietnam Veterans Memorial Fund Inc. developed the Vietnam Veterans Memorial Fountain while the artist who created the concept for the fountain was a Vietnam veteran named David Baker. The fountain was dedicated in December of 1985. Memorial day and Veterans day events are held each year.

The Names of those from the state of Missouri Who Made the Ultimate Sacrifice

ABBENHAUS GERALD ROBERT
ABBOTT ROBERT ESTEN JR
ABERNATHY ROBERT LLOYD
ADAMS HARLAN FLOYD
ADAMS OLEY NEAL
ADAMS RICHARD LYLE
AGNEW JAMES WILLIAM
ALLEN CHARLES RICHARD
ALLEN ELVIN L
ALLEN JOHN LEE
ALLEN OTIS LEE
ALLEN RAYMOND EUGENE
ALLEN TERRY LEE ODIS
ALLEY MICHAEL MORRIS
ALY LESLIE MORGAN
AMANN MARK THOMAS
AMES JAMES DAVID
AMOS THOMAS HUGH
AMRHEIN HERBERT FRANK-LIN
AMSTUTZ WILLIAM JOSEPH JR
ANDERS JOHN WILLIAM
ANDERSON GEORGE ROGERS
ANDREOTTA GLENN URBAN
ANDRESEN TERRY LEE
ANSPACH ROBERT ALLEN
ARMOR LOYDE DEAN
ARNOLD STEVEN ERNEST
ASBURY DAVID CHARLES
ASHFORD BILL JR
ASPEY DARRELL WAYNE
ASQUITH WILLIAM ROBERT
ATKISON CHARLES LEON
AUSTERMANN RAYMOND A JR
AUSTIN TYRONE WAGNER
AYER HERLEY JR
AYERS JAREL WAYNE
BACKY THOMAS ALAN
BADGER THOMAS ALBERT
BAGLEY JACK LAWRENCE
BAKER GARY PAUL
BAKER MICHAEL RAY
BAKER RENNIE JOE JR
BAKER THOMAS MICHAEL
BAKER WILLIE CECIL
BALCOM JOEL ARNOLD
BALLAY JAMES VINCENT

BAMVAKAIS JOHN ROBERT JR
BANCROFT STEPHEN WAYNE
BARBER JOHNIE RAY
BARHAM LARRY GENE
BARKER LARRY LEE
BARNES JOHN HENRY
BARNETT BILLIE JOE JR
BARTELL LARRY MICHAEL
BARTLE RICHARD PAUL
BARTON JAMES PAUL
BASNETT JERRY DALE
BATEMAN RAYMOND
BATES ROBERT W
BATESEL DENNIS GORDON
BATTS PERCILL
BAUER JAMES PHILLIP
BAUM RORY MICHAEL
BAX BERNARD HERMAN
BAXTER BOBBIE RAY
BAXTER LARRY LEE
BEAN DONALD WAYNE
BECKMANN LOUIS MARTIN
BEELER GEORGE FREDRICK
BEERS CARL WILLIAM JR
BEESLEY GARY EVANS
BEHRENS PETER CLAUS
BEILE FRED
BELINGE RICHARD LEWIS
BELL DEAN ALLAN
BELL RONALD EUGENE
BELTZ JOHN DAVID
BEMBOOM HERBERT DONALD
BENNETT JAMES HARRELL JR
BENTON CARROLL JOE
BERHOWE MARVIN RICHARD
BERRY ELMER EUGENE
BETEBENNER DAVID LEE
BEZOLD STEVEN NEIL
BIGGS JIMMY DEAN
BIGLIENI CHARLES ROBERT
BIONDO MARTIN
BISHOP RONALD BURK
BLACKSTEN BILLY JOE
BLAIR TERRY LEE
BLAKE RICHARD THOMAS
BLANKENSHIP JAMES ORIS
BLANTON KENNETH GENE

BLASSIE MICHAEL JOSEPH
BLATTEL DAVID LEE
BLEVINS FRANK LEE
BLISS BENJAMIN CHARLES
BOARDMAN MICHAEL KEN-NETH
BOATMAN ELMER LEE
BOBO LEON NELSON
BOCKEWITZ CARL EDWARD
BOEVER DAVID RICHARD
BOGGS DONNIE REX
BOHON RONALD EUGENE
BONDERER THOMAS EDWARD
BONDS MICHAEL DAVID
BONNARENS FRANK OWEN
BONO BEN DOMINIC
BORAWSKI JAMES DAVID
BOSTON KENNETH DEAN
BOSWELL JOE ROSCOE
BOURNE LAWRENCE GILBERT
BOWDERN ROBERT JAMES JR
BOWMAN RONALD LEON
BOYD ROY BRADLEY
BOYER DENNIS MICHAEL
BOYER JAMES ROGER
BOYLES JERRY LEE
BOZIKIS RONALD HENRY
BRADEN TERRY LEE
BRADLEY SYLVAN KEITH
BRANDOM THOMAS M JR
BRANSON JERRY LEON
BRANSON RALPH ALTON JR
BRASHEARS RONALD LEE
BREEDING WAYNE PETER EARL
BRENNAN JAMES ALOYISUS
BRIGHT THOMAS JR
BROCK TERRANCE LEE
BROCKMAN VERNDEAN ARTHUR
BROOKS EDWARD ALLEN
BROOKS FRANKLIN EUGENE
BROOKS LARRY LEE
BROWN CLARENCE
BROWN DANIEL L
BROWN DIEROTHER
BROWN GALEN CHARLES

BROWN HARRY LEE
BROWN HARVE EDWARD
BROWN HARVEY LEE III
BROWN HOWARD EUGENE JR
BROWN JAMES RICHARD
BRUTON CARL LEON
BUCKNER ANTHONY EUGENE
BUDZINSKI LAWRENCE JOSEPH
BUELL CRAIG HAROLD
BUFFINGTON LARRY DANIEL
BULLERDICK GARY ALLEN
BUMILLER ROBERT OSCAR
BUNCH LARRY DALE
BUNTION CHARLES WAYNE
BURFORD JOHN SHELBY
BURKS VIRGIL JR
BURNETT CHARLES C JR
BURNETT CURTERS JOSEPH
BURNETT GARY RAY
BURNS JOHN ROBERT JR
BURROW LEONARD
BURRUANO SAMUEL VINCENT
BUSCH ELWIN HARRY
BUSH STEVEN CLARENCE
BYRD GARY DEAN
CABRINI JOHN RICHARD
CADY STEPHEN MICHAEL
CAFFERY HOWARD EUGENE
CAGLEY JAMES NELSON
CAHALL JAMES WARREN
CAIN GLENNIE WAYNE
CAIN JERRY MAURICE
CALDWELL FLOYD DEAN
CALLAHAN THOMAS FRANCIS
CALLIHAN LYNDAL RAY
CALMESE ALBERT
CAMPBELL JOHN ALLEN
CAMPBELL LARRY GENE
CAMPBELL ROBERT CRAW-FORD
CAMPBELL WILLIAM L JR
CANDRL BRUCE CHARLES
CAPLAN LAURENCE CURTIS
CARLYLE DONALD RICHARD
CARNETT DENNIE LYNN

CARNOSKE ROBERT THOMAS
CARRICO CHESTER CALVIN JR
CARROLL ROGER WILLIAM JR
CARSON LAWRENCE HOWARD
CARSON TYRONE BRUCE
CARTER JERRY RAY
CARTIER VICTOR JOHN
CARVER HAROLD LEROY
CASEBOLT HENRY CLAYTON
CASH JAMES RONALD
CASON GEORGE GILBERT JR
CASSMEYER VICTOR PAUL JR
CAWLEY RICHARD ERNEST
CHAMBERS UDELL
CHAMPION GERALD ALAN
CHANDLER CHARLES WIL-LIAM
CHANDLER JOE WAYNE
CHANNEL BILLY GENE
CHAPPELL JOHN MONROE
CHASTEEN ROGER WILSON
CHERRY ALLEN SHELDON
CHESHIRE GARY ALLEN
CHILD RONALD WILLIAM
CHILDERS ESTILL LEE
CHILDRESS GEORGE W
CHITTWOOD JAMES PHILLIP
CHITWOOD FRED ALLEN JR
CHORLINS RICHARD DAVID
CHRISTENSEN ROGER LEE
CHRISTOFFERSON SCOTT ANDR
CLAGGETT JOHN ALLEN
CLARK DALE LEE
CLARK LARRY GENE
CLARKSON JAY OWEN
CLASPILL LARRY VERNAL
CLAVERIE RICHARD LEE
CLAXTON RICHARD REX
CLAY CHARLES EDWARD
CLAYTON BENNIE CLIFFORD
CLEAVER DONALD GENE
CLEMMON EDWARD L
CLEVE REGINALD DAVID
CLIFTON ROBERT HARRISON
CLUBBS CHARLES EARL
COBB RONALD DAVID

Missouri

COFFMAN CLYDE LEE
COLBERT JOHN WAYNE
COLEMAN JAMES EDWARD
COLEMAN JAMES FRANCIS
COLEMAN LONALD RAY
COLLINS ARLIE RAY
COLLINS ARLIN DARRELL
COLLINS DONALD CLIFTON
COMBS CLIFFORD DALE
COMBS PHILLIP EUGENE
CONNER DONNIE RAY
COOK CHARLES HERMAN
COOK JAY ALAN
COOK ROBERT EDWARD
COOK SCOTT HOWARD
COOPER GARY ROBERT
COOPER RICHARD LEE
COPE CHARLES ALFRED
COPELAND NORMAN OTTIS
COPELAND ROBERT
CORDIA MICHAEL JAMES
CORP JERRY MARSH
CORTOR FRANCIS EDWIN JR
COUCH JACKY RAY
COULT GERRY DON
COWEN HAROLD EDWARD
COWSERT KENNETH WILLIAM
COX JAMES ALLEN
COX MARTIN
CRABTREE JAMES OTIS
CRAIG JAMES HERBERT
CRAIGHEAD TERRY DEAN
CRAMER DONALD MARTIN
CRAMER ROBERT MICHAEL
CRANE WILLIAM JOSEPH II
CRAWFORD JAMES J
CRAWFORD JOHN NELSON JR
CRAWFORD ROBERT DEAN
CRAWFORD WILLIAM
THOMAS
CROCKRAN JAMES
CROOK OREN LEE
CROOK THOMAS HIRAM
CROSBY LOUIS JOHN
CROSS GARY LEE
CROSS THOMAS JOHN
CROW LARRY EDWIN
CROW LINDSEY HOUSTON
CROWDER NEAL STEVEN
CULLERS RONALD KENNETH
CUNNINGHAM DONNIE LEE
CUNNINGHAM WELLS ELDON
CURETON RONNIE CHARLES
CURRIER GORDON LEROY JR
CURRY MARVIN ELLIS
CUTBIRTH RICHARD EUGENE
DAILEY BILLY JACKSON
DAINS PAUL LELAND
DALE JAMES MILTON
DALMAN LEONARD JAMES
DALTON MICHAEL FRANCIS
DALTON MICHAEL JOSEPH
DANIELS CURTIS RAY
DARNELL JIMMY ALLEN
DARROW DONNIE LEN
DAUGHERTY EVERETT LEROY
JR
DAVENPORT ROBERT DEAN
DAVIES STEPHEN GEORGE
DAVIS CARL RAYMOND
DAVIS CHARLES CECIL
DAVIS CLIFFORD MORRIS JR
DAVIS DON EDDIE
DAVIS GAIL LEE
DAVIS MICHAEL DE-WAYNE
DAVIS RAYMOND CARL

DAVIS STEVE
DAVIS WILBERT CLAUDE
DAY ROY JUNIOR
DE BRULER JAMES PAUL
DE WEESE RONALD GENE
DEAN DONALD CHESTER
DEAN TERRY LEE
DEATHRAGE DON LE ROY JR
DEEDS LELAND SAMUEL
DELASSUS CHARLES EDWARD
DENNIS JOHN ALLEN
DETERS DAVID STEPHEN
DIAN DON FAUROT
DICKENS FREDDIE DALE
DICKERSON OMER PAUL
DICKEY FORREST PITTMAN
DICKSON RONALD GEORGE
DICUS RICHARD LEE
DILLARD JOHN EDWARD
DINGUS MICHAEL JOE
DIRNBERGER LAWRENCE
ANDRE
DITTMER DAVID ALLEN
DITZFELD BOBBIE LEE
DOBBS RONALD GENE
DODSON WILLIAM NEAL JR
DOEBERT PHILLIP RAY
DOGGETT RONALD THOMAS
DOLLENS HAROLD RAY
DONALDSON JAMES ALLEN
DONNELLY JAMES WARREN JR
DOSSETT JOHN ADRIAN JR
DOUGAN CHARLES GARVIN
DOUGHERTY KENNETH
EUGENE
DOWNING JAMES LESLIE
DRAKE STEVEN COLE
DRAUT CHARLES BERNARD JR
DREW JAMES LEE
DROZ DONALD GLENN
DUFF JACK CECIL JR
DUFF ROBERT DARREL
DUNCAN GARY BERYL
DUNCAN WILLIAM ARTHUR
DUNN ROBERT TERRENCE
DYKES FRANK FAYETE
EADS JOHN PATRICK
EDMONDSON WILLIAM
ROTHROC
EDWARDS DOUGLAS GLYN
EIDSON RONALD LEE
ELLERMAN GARRY RONALD
ELLIOTT LARRY WILBERT
ELLIS EARL WAYNE
ELMORE LARRY EUGENE
EMRICK ERVIN JUNIOR
ENDICOTT MICHAEL LEE
EPPS CLINTON HURANSO
ERNST GARY JOSEPH
ERNST RALPH HERMAN
ESKEW CURTIS DEAN
ESSARY JAMES
EVANS RICHARD ALLEN JR
EWING KENNETH GENE
EZELL DONNIE D
FAHRENHORST THOMAS
KENNET
FAIRCLOTH JULIUS CLYDE
FANT LAWRENCE L
FARLEY WILLIAM DANIEL
FARRINGTON ROBERT DEAN
FARRIS MICHAEL J
FELDMANN BARRY EDWARD
FENNEWALD DANIEL FRANK
FERGUSON JAMES DONAHUE
FERGUSON LATNEY DEAN
FERGUSON RANDALL EUGENE
FERREN JERRY WAYNE
FILKINS RONALD MARION
FINKE STEPHEN PAUL
FINLEY CHARLES RICHARD
FINLEY DICKIE WAINE
FIREBAUGH ROBERT AN-
THONY
FISH GORDON ALIDEAN
FISHBACK WILLIAM EDWARD
FISHER DENNIS FAY
FITCH EARL FREDERICK
FLAHERTY WILLIAM F III
FLOYD BOGARD LAFAYETTE
FOLLETT ALLAN EUGENE
FORCK MICHAEL RICHARD
FORDYCE RAY
FOSTER THOMAS EUGENE

FOWLER DONALD LEON
FOWLER JAMES JEWEL
FRANCIS JAMES EDWARD
FRANCIS LINDELL
FRANKLIN FLOYD STANLEY
FRASCH ROBERT LOUIS
FRAZIER ALBERT WILLIAM
FRAZIER RONALD LEON
FREDWELL GARCLEE M
FREEMAN DAVID FRANKLIN
FREY DONALD
FRISK JOHNNY EARL
FRITTER JOHN WILLIAM
FULK BILLIE HOWARD JR
FUSSNER ALLEN GEORGE
GAFFNEY MC ARTHUR
GAITHER CURTIS
GALLINA ANTHONY JOSEPH
GALLOWAY DENIS WAYNE
GAMET RANDOLPH MERL
GARCIA JOSE OSCAR
GARNER GARY HAROLD
GARVEY DONALD JESS JR
GAYLORD DOUGLAS DRUE
GEARHEART MIKE DUANE
GENZLER AUGUST HENRY
GEORGE CHARLES MICHAEL
GIBBS CLIFFORD WARREN
GIBLER DONALD GENE
GIBSON JOHN A
GILMORE STANLEY RAY
GISH BENNY THOMAS
GIVEN FRANK ALLEN
GLADU ROBERT JOSEPH
GLEAR GARY LEE
GLENDENNING FRANK BARD
GLENN JACKIE D
GLOVER DAVID CYRIL

GLOVER JAMES LEAR JR
GODLEY RALEIGH LEE
GOEDDE ROBERT JOSEPH
GOEGLEIN JOHN WINFRED
GOETZ JAMES GORMON
GOFORTH JACKIE LEE
GOINS MICHAEL RAY
GOLDEN BARRY LEIGH
GONZALEZ DOUGLAS DAVID
GOODALE THOMAS LEE
GOODEN GERALD LYNN
GOODWIN WILLIAM FRANK-
LIN
GORRELL DAVID EUGENE
GOSS JACK EUGENE II
GOSSAGE DOUGLAS EUGENE
GOSZEWSKI THOMAS WALTER
GOWER WILLIAM RAY
GRAGNANI THOMAS J
GRAHAM SEBERN EMLIS JR
GRANT BILL WAYNE
GRAVES JERRY LEE
GRAY HAROLD LEROY
GRAY ROY VIRGIL
GRAYS DEMETRIUS JEROME
GRAYSON JERELL LEE
GREB JAMES J
GREEN ALLEN RUSSELL
GREENE LLOYD ROLLAND
GREGORY PHILIP LEE
GREGORY ROBERT RAYMOND
GREIFE DALE EDWARD
GRENIER JOSEPH KENT
GRIFFIN DONALD ORTHEL
GRIFFIN RODNEY LYNN
GRIFFITH JOHN GARY
GRIGSBY JOE WALTER
GRIMES MICHAEL

GRISSOM GARY L
GRIZZLE CHARLES WENDLE
GROSSHART ROBERT STEVEN
GROVE STEVEN EUGENE
GRUTSCH JOHN WILBUR JR
GUDISWITZ EUGENE RICH-
ARD
GUITTAR DONALD HARVEY
GUNN ALBERT LEONARD
GUTIERREZ CHRISTOPHER
HACKLEMAN LARRY L
HADDEN ROBERT BRUCE
HADLEY GARY PATRICK
HAGEN JAMES JOSEPH
HAILEY JERRY LEE
HAILEY JOSEPH CARLTON
HALBERT EDWARD JOSEPH
HALL CHAUNCEY IKE
HALL GEORGE THOMAS
HAMILTON BERT ABNER JR
HAMILTON DOUGLAS BLAKE
HAMILTON MICHAEL EUGENE
HANCOCK JERRY EDWARD
HANEY WILLIAM THOMAS
HANLIN GARY LEON
HANSEN JOHN MARK
HAPPEL JERRY LEE
HARDIE ANTHONY ROY
HARDY JOHN CHARLES
HARLAMERT MICHEAL RAY
HARPER GEORGE DALE
HARPER JIMMY CHESTER
HARPER MARVIN
HARPER WILLIAM MICHAEL
HARRIS HAROLD RAY
HARRIS JAMES BRADDOCK
HARRIS LEE RUSSELL
HARRIS MICHAEL STEVENS

HARRIS ROBERT JOHN
HARRIS STEPHEN WARREN
HART GREEN LEE
HARTMAN DONALD OWEN
HARTWICK FLOYD WAYNE JR
HASENBECK PAUL ALFRED
HASTINGS ANDREW LALONE
HATFIELD JIMMY DALE
HAWKS RONNIE LEE
HAYES GARRY LEE
HAYES HAROLD UTAH
HAYMES RICHARD SCOTT
HAYNES SIDNEY JR
HAYNIE ROBERT RAY
HAZLEY MELVIN
HEAD NOBLE THOMAS
HEADRICK WILLIAM DAVID
HEARD HOWARD CHARLES
HECK RICHARD MICHAEL
HEIBEL DANIEL JOSEPH
HELSEL RODNEY GLENN
HEMMEL CLARENCE JOSEPH
HEMMITT TERRY EUGENE
HENDERSON RUFUS Q
HENDRIX ELWOOD RANDALL
HENN NORVILLE MARTIN JR
HENSON CLARK LEE JR
HENSON ROGER LEE
HERRICK BENNETT JAMES
HERRICK DENNIS HALDANE
HERRING BILLY DALE
HESS FREDERICK WILLIAM JR
HEUGEL FREDDY PAUL
HICKMAN ARTHUR EDWARD
HICKS JEFFREY LYNN
HICKS LARRY DAVID
HIGGERSON TOMMY DOYLE
HIGGINS HERSHEL
HIGHFILL REX WHEELER
HILL ALAN JR
HILL CHARLES DALE
HILL EUGENE DONALD
HILL GARY
HINDERKS GREGG CLIFTON
HIRNI TROY EDWARD II
HIRTLER ERNEST LLOYD
HIX KEITH EUGENE
HOBBS CHARLES MICHAEL
HODGES WILLIAM JESSE SR
HOFFEDITZ DONALD RAY
HOFFMAN DAVID PAUL
HOGAN BILLY JACK JR
HOLDER FREDRICK LEE
HOLLINGSWORTH GARY LYNN
HOLLOWAY LYLE D
HOLMES JAMES ROBERT
HOLMES MICHAEL DOUGLAS
HOLMES ROBERT HAROLD
HOLSWORTH JAMES MICHAEL
HOLT CLARENCE RAY
HOLT CLAY JR
HOLTMAN JOHN THOMAS
HOLTSCHNEIDER GEORGE ALEX
HOLTZHOUSER RONALD LEE
HOLZER RICHARD EUGENE JR
HONNOLD STEPHEN JEFFRY
HOOD DERALD JOE
HOOD WILLIAM WILLIS
HOPPERS MICHAEL EUGENE
HORN DOUGLAS LEE
HORST ROBERT LOUIS
HORTON CHARLES BRENT
HOUSE OSCAR LEE
HOUSER CARL RAY
HOUSH RICHARD HENRY
HOUSTON JOHN ROBERT
HOUTZ JOSEPH MERLE
HOWARD MARK THOMAS
HOWELL LARRY L
HUBBARD ROBERT STEPHEN PO
HUBBARD ROGER LEROY
HUDGINS CARL WILLIAM JR
HUDSON DALE FRANCIS
HUDSON GARY DUANE
HUDSON GARY LEE
HUDSON PHILIP LONNIE
HUDSON THOMAS GORDON
HUDSON THOMAS HAROLD
HUFF TERRY KENNETH
HUGHES JERRY DANIEL
HUGHES JESSE RAY JR

HUGHES THOMAS EDWARD
HULL JAMES ALBERT JR
HULTS GARY DEAN
HUMPHREY GALEN FRANCIS
HUMPHREYS CHALMERS CLAUDE
HUNSLEY DENNIS ROGER
HUNT PHILIP WADE
HUNTER HAROLD HENRY
HUNTER JOHN LOUIS
HUNTER MICHAEL RAY
HUPP RICHARD LEWIS JR
HURLEY TIMOTHY LAWRENCE
HUTTER ROBERT NELSON JR
HUTTON WILLIAM JAMES
HYDE WAYNE
IGERT JOHN WILLIAM
IHRIG GARRY LYNN
IJAMS DENNIS EARL
IMPERIALE RONALD JOSEPH
INGRAM WARREN G
INLOW CHARLES EDWARD
IRELAN KENNETH RAY
IRELAND ROBERT NEWELL
IRVIN STEPHEN LEE
ISBELL JAMES RUSSELL
ISGRIG DENNIS EDWARD
IVES PHILLIP THOMAS
JACKSON BOBBY GENE
JACKSON CECIL JR
JACKSON DALE RAYMOND
JACKSON DEARING MICHAEL
JACKSON HUGH MAR
JACKSON LAMOND JOSEPH
JACKSON LEWIS JAMES
JACKSON RICHARD ALBERT
JACKSON ROBERT JR
JAMES GERALD LYNN
JAMES LEE CHRISTOPHER JR
JAMES ROBERT LEE
JASMINE CHARLES
JENNER RICHARD LEE
JENNINGS LARRY JO
JENRY ROBERT EUGENE
JILES JAMES JR
JINES ROBERT ALLAN
JOANIS KENNETH JOSEPH
JOHNSON EDWARD DEWEY
JOHNSON EMMET LEE
JOHNSON FRANK EDWARD
JOHNSON GREGORY BERT
JOHNSON LOUIS
JOHNSON ROGER LEE JR
JOHNSON WILLIAM JAMES
JOHNSON XAVIER
JOHNSTON GEORGE ELDON
JOLLY EUGENE
JONES CURRAN M
JONES DOUGLAS WAYNE
JONES ISAAC
JONES JOHN ORA JR
JONES JOHNNY EUGENE
JONES KENNETH LOREN
JONES MICHAEL GILBERT
JONES ROBERT LEWIS
JONES WAYNE IRA
JORDAN JAMES SAMUEL
JORDAN LAWRENCE WICKS
JORENS EVERETT RALPH JR
JOSLEN PHILLIP DALE
JULIAN JAMES JULIUS JR
JUNE WILLIAM ALBERT
KARNES LESLIE LEROY
KARR DAVID RAY
KASTER ROBERT LEE
KEENEY GERALD ROBERT
KEESLING GERALD EDWARD
KEEVEN LOUIS FERDINAND
KEITH CLYDE LEE
KEITH RICHARD HENRY
KELLEY OWEN C
KELLY WILLIAM PATRICK
KEMPKER PATRICK BENJAMIN
KENNEDY GENE RANDOLF
KENNEY OTIS
KEROHER GAYLAND EUGENE
KESHNER KEO JOE
KESTER THOMAS DUFAUX
KETHE HENRY JAMES
KIGAR LARRY EUGENE
KIGER GEORGE ALAN
KILGORE LARRY WYATT
KILLABREW ROBERT LEROY

KIMBLEY ROBERT GLENN
KIMBRELL LOUIS CLEVELAND
KINDLE WILLIAM DOYLE
KINDRED LAWRENCE JOSEPH
KING DAVID MICHAEL
KING GARRY EUGENE
KING GUY RICHARD
KING REGINALD DAVID
KINKEAD MAURICE HARRISON
KINKEADE RONALD JAY
KIRKLAND CHARLES S JR
KIRKSEY WILLIE JAMES
KITCHEN RUSSELL HAROLD JR
KLAGES ROBERT JOHN
KLEIN JAMES MORTON
KLIPFEL JOE PAUL
KNIGHT DAVID MARSHALL
KNOLL ANTHONY
KNOX LARRY WAYNE
KOENIG DAREN LEE
KOGER SIDNEY KEITH
KOLZ JOHN JORDAN
KRAFT JERRY BERNARD
KRAM HAROLD ANDREW JR
KRATZBERG JIMMIE LYNN
KRAUS JEAN MASON
KREBS STANLEY GENE
KRELL ROYAL TINDORF
KREUTZ KENNETH JOSEPH
KUHNS WILLIAM JOSEPH
LAKEY GEORGE LEO
LAND LARRY PAUL
LAND RICHARD LEON
LANDERS BLAINE WILSON
LANE FAMOUS LEE
LANG JAMES FRANKLIN
LANGSTON MICHAEL GARY
LAWHON CHARLES R
LAWRENCE JAMES LARRY
LAWS BILLY WAYNE
LAWS DELMER LEE
LAWSON GERRY WAYNE
LAY MELVIN W
LAYNE HOWARD WILSON JR
LAYTON PATRICK ARTHUR
LE CLAIR TIMOTHY KIM
LEDBETTER JAMES RILEY
LEDERLE MICHAEL ALAN
LEDFORD VIRGIL MADISON
LEE WALTER CLARENCE
LEECH ROBERT VOYD
LEEK THOMAS JR
LENLEY JESSE LEE
LEON PEDRO JR
LESTER JIMMY DON
LEUTHOLD DONALD FREDERICK
LEWIS FRANK FREDERICK
LEWIS LEONARD LEROY
LEWIS RONALD EUGENE
LIESE TIMOTHY FRANCIS
LINGLE ROBERT DEAN
LINTNER DARRYL CHARLES
LIVINGSTON JOHN JOSEPH
LIVINGSTON LESLIE E III
LLOYD MARTIN ROGER
LOCKHART WILLIAM LON
LOGAN DONALD GORDON
LOGAN DOUGLAS ALFRED
LOGAN RONALD CHARLES
LOGWOOD CLARENCE
LOLLAR BYRON CLIFTON
LOLLAR THOMAS ARTHUR
LOMAS JOHNIE
LONG DONALD EUGENE
LONG HARRY LEROY
LONG RAY STEPHEN
LONG THOMAS ARNOLD
LONGO DENNIS MICHAEL
LONGSTON HENRY RALPH
LOOS THOMAS WALTER
LOWE AARON HARVEY
LOWREY CHUBBY DEAN
LOYD HAROLD IVAN
LOYD MELVIN
LUBBERS THOMAS LAMBERT
LUCAS WILBUR RAY
LUEBBERS RALPH JOSEPH JR
LUEBKERT BERNARD MICHAEL
LUKITSCH FRANK JOSEPH JR
LUNN EUGENE AUSTIN
MADDOX ROBERT BRUCE

MAGEL JAMES EDWARD
MAHAN DOUGLAS FRANK
MAHURIN ELMER WAIN
MAIZE WILSON JUNIOR
MALIN LOUIS NATHANIL
MALLINCKRODT ARTHUR T JR
MANLEY RICHARD JOSEPH
MANN EDWARD HAROLD
MANN NATHAN JAMES
MANNING PATRICK PEARSE
MAPLE ARCHIE JAMES JR
MARACZI ANTHONY
MARCO JERRY ROY
MARCUM GILBERT GEORGE
MARIK CHARLES WELDON
MARLIN ELLIS SANFORD
MARRION JIMMIE CHARLES
MARSHALL JOHN GRADY
MARSHALL LARRY HUNTER
MARTENS STANLEY WAYNE
MARTINEZ RICHARD PAUL
MASDEN STEPHEN KNIGHT
MATTHEI PETER KARL
MATTINGLY TIMMY G
MATTSON RICHARD DEAN
MAUNE FRANCIS EDWARD
MAXWELL DENNIS RAY
MAY GARY WAYNE
MAY RAYMOND ALLEN
MAYBERRY DONALD RICHARD
MAYBERRY MICHAEL JOSEPH
MAYES HARRY LEROY
MAYS CARL SHERRELL
MC ADOO MICHAEL DOUGLAS
MC BRIDE DONALD WAYNE
MC CALLISTER ROBERT LYNN
MC CARTNEY DARRYL EUGENE
MC CLAFFERTY JAMES EDWARD
MC CLATCHEY JEWEL EDWARD
MC CLELLAN EDWARD EUGENE
MC CLUSKEY JOHN DAVID
MC COIN KENNETH DALE
MC COWN ROBERT DEWAYNE
MC COY JAMES WILLIAM
MC COY PETER JOSEPH
MC CRAY PLEASANT JR
MC CREA LAWRENCE
MC CULLOUGH MARVIN L JR
MC DANIEL JOHNNIE LEE
MC DANIEL MICHAEL EUGENE
MC DANIEL ROY DEAN
MC DERMOTT LEWIS E
MC FALL ROBERT DALE
MC GUIRE MICHAEL JOSEPH
MC GUIRK CHARLES ANTHONY
MC KEE JULIAN ALLAN
MC KEEVER LEROY
MC KINNEY EUGENE PHILLIP
MC LEARY ORVAL WADE
MC MAHON THOMAS W JR
MC MILLIN GARY DON
MC MULLIN CHARLES ERNEST
MC SWINE JOHN HENRY
MCNUTT WINDOL WILSON
MEAD SAMMY LOUIS JR
MEADOR BILLY JAY
MEADOWS MILLARD FRANKLIN
MEIER CARL FREDRIC
MENLEY EARNEST DALE
MENNINGER ROBERT PATRICK
MENSCH CHARLES R
MERKLE ELLIOTT LYNN
MERTELL JAMES RICHARD JR
METZ JAMES HARDIN
MEYER GREGORY LEO
MEYERKORD HAROLD DALE
MICHAEL WILLIAM ARTHUR
MIDDLETON RICHARD WAYNE
MILLER BERTMANN EARL
MILLER BURKE HOLBROOK
MILLER CARL DEAN
MILLER CLINTON EUGENE
MILLER EUGENE LEWIS
MILLER LARRY FLOYD

MILLER LARRY LEE
MILLER MICHAEL WESLEY
MILLER PAUL JAMES
MILLER WILLIAM HOWARD
MILLER WILLIAM JULIUS
MILLS LAWRENCE STEVEN
MILLS TERRY WAYNE
MITCHELL DANIEL LEE
MOLLER GLENN LOREN JR
MONIA TERRY ROBERT
MONTGOMERY JACKIE GENE
MONTREY REAVIS A JR
MOODY JIMMY DALE
MOORE CHARLES JAMES
MOORE CHARLES THOMAS
MOORE JESSE LOUIS
MOORE JOSEPH LEE
MOORE LOUIS CHARLES
MOORE PAUL VINCENT
MOORE ROBERT WAYNE
MOORE TERRY LEE
MORAN WALTER C B
MORGAN DAVIS JUNIOR
MORLEY CHARLES FRANK
MORRIS DOYLE ANTHONY
MORRIS JOHN HENRY JR
MORRIS ROBERT JOHN JR
MORRIS TOMMY GENE
MORRISON JOSEPH WALTER
MORSE RICHARD DEAN
MORSE STEVEN PAUL
MORT DANIEL LEON
MOSLEY GLEN HERBERT JR
MOSLEY WALLACE JEROME
MOULDER LARRY THOMAS
MOYERS RICHARD MICHAEL
MULLEN CLIFFORD TRUMAN
MULLER EDWARD JERRY
MURFF HERBERT STERLING
MURPHY JOHN WILLIAM III
MURRAY BERNARD PHILLIP
MURRAY JOSEPH VAUGHN
MYERS GEORGE ARTHUR
MYERS TONY HOWARD
NACY JON O
NAILE THOMAS GLEN
NELMS PATRICK IRVIN
NELSON BRUCE ANTHONY
NELSON LARRY THOMAS
NEU WILLIAM ALLEN
NIEMEYER LOUIS ANDREW JR
NIXON DONALD LEE
NIXON JEROME
NOELLSCH ROBERT DONALD
NOONAN JOHN MICHAEL
NORE KENNETH HAROLD
NORFLEET BRIAN ROSS
NORTH DALE EUGENE
NORTHROP RONALD ROBERT
NOTO ROBERT JOSEPH
NOWACK THOMAS MICHAEL
NULL ARTHUR ELLIOTT JR
O BRIEN JOHN MICHAEL
O HARA JAMES LOYD
O NEAL LEROY
O NEAL MARSHAL JUNIOR
O NEILL TIMOTHY MICHAEL
O TOOLE MICHAEL JOSEPH
OAKS STEVEN BOYD
OFFIELD REX KAYE
OFFUTT GARY PHELPS
OJILE MICHAEL RAYMOND
OLIVER CHARLES
OLSON STEVEN RICKY
ORTON MATTHEW THEODORE
OVERKAMP NORBERT ALVIN JR
OWEN CLYDE CHILTON
PAIGE ROBERT EDWARD
PARENT JOSEPH W
PARKER HERMAN JR
PARKER KENNETH WAYNE
PARKER MARVIN RUSSELL
PARNELL PETER PAUL JR
PARSONS JAMES LLOYD
PARSONS WARREN CECIL JR
PASCHAL LESLIE CALVIN JR
PATRICK JERRY
PATTEN JEARL RAY
PATTERSON CHARLES EDWARD
PATTON ROGER WAYNE
PAULE PHILLIP ARTHUR

Missouri

PAYNE WENDLE L
PEAK EARL ARCHER
PECK ROBERT WILLIAM
PEEL STEPHEN BLAKE
PEMBERTON GENE THOMAS
PENNEL LAWRENCE PAUL

PENNINGTON THOMAS JACK
PERRY JACK ARMOND
PERRYMAN WILLIAM JOSEPH
PETRECHKO EDMUND A JR
PETTY ERNEST DE FOREST
PETTY JERRY LEON
PHELPS LARRY LEE
PHILIPS BURTON KEENEY JR
PHILLIPS CHARLES EDWARD
PHILLIPS JAMES RILEY
PHILLIPS RICHARD GREGORY
PHILLIPS SHEPHEN HIETT
PIERCE DANNY RALPH
PIERCE DARREL GENE
PILCHER WILLIAM GEORGE
PIPKIN THOMAS DEWEY JR
PLASSMEYER BERNARD
HERBER
PLATTER GEORGE RICHARD
PLEASANT EDDIE LEE
PLILER LARRY DEAN
PLUMMER JOHN DAVID
PODMANICZKY CHRISTO-
PHER

POESCHL JOHN EDWIN
POFF BILL DEAN
POLLARD GERALD RAY JR
POMEROY DAVID KEITH
PONDER JOHN DAVID
POPE CHARLES ALFRED JR
POPPLETON CHARLES AR-
THUR
POSCOVER GARY STUART
POSEY DALE L
POSS GARY STEVEN
POTTS LEONARD LEE
POWELL MICHAEL ALLAN
POWELL RAYMOND LEE
POWELL ROBERT ALLEN
POWERS CHARLES RAY
POWERS HARRY LEE
PRAGMAN DONALD EUGENE
PRATHER GARY W
PRENTICE ALAN NEIL
PRESSON JAMES DAVID
PRESSON WILLIAM PAUL JR
PRESTON MACK LEE JR
PREVEDEL CHARLES FRANCIS

PRICE DERRILL LE ROY JR
PRICE GARRY OWEN
PRICE JOHN CHAD
PRIEST FRANKIE LEON
PRINCE RONALD PERSHING
PROFFER GEORGE FLOYD
PROSE WILLIAM THOMAS
PRUITT GEORGE ALAN
PRYOR LARRY ROY
PUCKETT DENNIS RAY
PUMPHREY DONALD LEE
PURCELL RICHARD MICHAEL
QUICK RALPH RICHARD JR
RABER JOE EDWARD
RAGLAND DAYTON WILLIAM
RANDOLPH CORTEZ ALLEN
RANKIN ANDREW BRYAN
RATHBUN CRAIG
RAULSTON RILEY DAVID
RAWLINGS JEROME
RAY JOHN EDWARD
RAYFIELD GREGORY RUSSELL
RAYSKI LARRY ALLAN HENRY
REECE STACEY DANA

REED EARL DONALD
REESE DELBERT LEON
REEVES LONNIE MICHAEL
REGISTER DORSIE EUGENE
REICHARDT STEVEN JOHN
REIFF MICHAEL DEAN
REIFSCHNEIDER ELMER J JR
REILLY DONALD JOSEPH
REINBOTT HAROLD W JR
REITER DEAN WESLEY
REITHER PHILIP HENRY JR
REITZ KEITH HAROLD
REMBOLDT RONALD PAUL
RENFRO RICHARD ALVIN
RENNE MYRON KEITH
REYNOLDS EDWARD LEE
RHODE EDWARD ANTHONY
RICE IRA ALBERT
RICE JAMES BURNEL JR
RICH CHARLES RAY
RICHARDS LON DAVIS
RICHARDSON JAMES EVERT
RICHARDSON MARVIN
NELSON
RICHARDSON RAYMOND LEE
RICHARDSON RICHARD ELVIN
RICHARDSON ROBERT DANIEL
RICHARDSON ROBIN WILLIAM
RICHARDSON RONALD
DOUGLAS
RICKMAN DWIGHT GRAY
RIDEN FRANK LEE
RIDINGS LESTER LEON
RIEDE RONALD EDGAR
RILEY CHARLES FRANKLIN
RILEY LESTER JR
RILEY MELVIN JOSEPH JR
RIPPEE JON ALAN
ROBERSON SAMUEL LOUIS
ROBINSON ALAN JOSEPH
ROBINSON EDWARD
ROBINSON O DELL
RODGERS ROBERT LOUIS
ROEPKE PHILLIP WRAY
ROGERS DAVID CLYDE
ROGERS LESTER A
ROSENBACH ROBERT PAGE
ROSS ALAN
ROSS DALE RAY
ROTHER ROBERT DAWSON JR
ROYAL FRANCIS PATRICK
RUDDELL ALAN JAMES
RUFF GILBERT OLIVER JR
RUHL ROBERT JACK JR
RUHLMANN HEINRICH
RUPERT LEO FRANKLIN
RUSH CLIFFORD JAMES
RUSSELL CLARENCE DEAN
RUSSELL RANDALL KERWIN
RUTHERFORD RICHARD
EUGENE
SAILOR EDDIE
SAMPLE MICHAEL RAY
SANAZARO ERNEST JR
SANDERS ARTHUR EDWIN
SANDERS JERRY JOSEPH
SANDERS MARVIN HOWARD
SARAKAS RICHARD THOMAS
SATHOFF DALE ERVIN
SAVOY M J
SCHEMEL GARY LEROY
SCHENE TERRANCE RICHARD
SCHERRER LAWRENCE
FRANCIS
SCHEULEN GARY JEROME J
SCHLOEMER CARL WAYNE
SCHLOTTMAN ALVERN
WARREN
SCHMICH JOSEPH JR
SCHMIDT FREDERICK
CHARLES
SCHMIDT HERBERT ELLIS
SCHMIDT JOHN GEORGE
SCHMIDT JOSEPH VINCENT
SCHMITZ CRAIG ALAN
SCHRADER PETER ANTHONY
SCHRAND ROBERT LEE
SCHWARTZ CHARLES GLEN-
NON
SCOTT DANIEL R
SCOTT DAVID LEE
SCOTT EUGENE C
SCOTT LLOYD M JR
SCOWDEN CURTIS DEAN

SCRUGGS DAVID L
SEABOURNE BENNY ELLIS
SEAMAN HAROLD LA VERN
SEATON CHARLES EVERETT
SEAWEL WARREN PAUL
SEUELL JOHN WAYNE
SEWELL MONTY RAE
SHANNON BILLY EUGENE
SHANNON GEORGE DAVID
SHARP DANIEL FRANKLIN
SHEA THOMAS WELCH
SHEETS ORVILLE ALLAN
SHEGOG WILLIE LEE
SHELTON JAMES EDWARD
SHEPPARD RONALD EUGENE
SHERMAN WILLIAM WARREN
SHOCKLEY BOBBY JOE
SHORT BILLY DALE
SHORTT WALTER RUBEN
SHUH FREDERICK JOHN
SIEGLER BOBBY TRUMAN
SILVEY HAROLD RAY
SIMMONS DONALD LEE
SIMMONS HAROLD JOSEPH
SIMMS JAMES WILLIAM
SIMPSON BOBBY GENE
SIPES JAMES LELAND
SIRON JAMES LLOYD
SISLER GEORGE KENTON
SKALLY THOMAS MICHAEL
SKINNER COURTNEY A
SKINNER LARRY RICKFORD
SLAGLE DAVID RODDY
SLANKARD WAYNE ALBERT
SLAUGHTER PHILLIP EDWARD
SLAYTON RONALD DENNIS
SLY JOHNNIE RAE
SMITH CARY CARSON
SMITH CHARLES FRANKLIN
SMITH DAVID WILLIAM
SMITH DONALD C
SMITH EDDIE LOUIS
SMITH EDWARD JR
SMITH GARY EDWARD
SMITH GEORGE CRAIG
SMITH JAMES ANDREW
SMITH JAMES WESLEY
SMITH KENNETH RAYMOND
SMITH LARRY DEAN
SMITH ROBERT WILLIAM
SMITH SCOTT GARY
SMITH THEODORE
SMITH THOMAS MONTGOM-
ERY
SMITH WALKER JR
SMITH WILLIAM DOUGLAS
SMITH WILLIAM WALTER
SMYTHE JAMES EDWARD
SNODGRASS NORMAN ED-
WARD
SNYDER JOHN MARSHALL JR
SNYDER MICHAEL BRYANT
SOOTER GARY ERCIL
SOUTHWICK HAROLD KEN-
NETH
SPENCER DANNY RAY
SPIERS RANDOLPH
SPINDLER JOHN GATES
SPINNICCHIA JOSEPH FRANK
SRADER CHARLES WESLEY JR
STACEY GARY ROSS
STACKHOUSE JOHN E
STAFFORD LEE ROY
STAMP GEORGE RILEY
STANTON RICHARD EUGENE
STARKS WARNER
STEMMONS BIRCH UDELL
STEPHENS ARTHUR ALLYN
STEPHENS BING FOREST
STEPHENS LARRY ALAN
STEPHENSON KENNETH RAY
STEPP CHARLES HAROLD
STEPP WILLIAM D
STEVENS EDWARD HOWARD
STEWART GEORGE EDWARD
STEWART JERRY DEAN
STEWART STEPHEN JAMES
STILWELL ROY MILES
STOCKBAUER CHARLES
THOMAS
STOCKLIN GARY DENNIS
STOCKWELL PAUL MARION
STODDARD RUSSELL MERRILL
STOKES JAMES DOYLE

STONE CHARLES H
STONEKING HERBERT RALPH
STONER WILLIAM DENNIS
STORIE WILLARD GENE
STRAIT LAFFEY FRANKLIN
STRAKER GARY ENNIS
STREET ROBERT ANDREW
STUMP EDWARD EARL
SUBER RANDOLPH BOTH-
WELL
SUGGS JOHN FENTON JR
SUHR ALFRED HENRY
SULLIVAN MELVIN
SUMMERS PHILLIP PAUL
SUMMERS ROBERT RAN-
DOLPH
SUTHERLAND CHARLES
EDWARD
SUTHERLIN WILLIAM REGI-
NAL
SWAFFORD ROBERT WAYNE
SWAIM CHARLES MICHAEL
SWANGUARIM LAWRENCE
ALFRE
SWANN HOWARD ERNEST
TANNEHILL RAY EDWIN
TATE JACKIE LEE
TATE LEE BERNARD
TATUM RICHARD LEE
TAYLOR DANNY GENE
TAYLOR JAMES OTIS
TAYLOR JOHN VERNON JR
TAYLOR ROBERT LEE JR
TEMPLETON CLARENCE
WAYNE
TENNILL LARRY EARL
TESTORFF THOMAS EDWARD
THARP EARL WATSON JR
THOMAN FLOYD NICKOLAS
THOMAS DAVID EUGENE
THOMAS TOBY ARTHUR
THOMPKINS MICHAEL LAROY
THOMPKINS RONALD WIN-
STON
THOMPSON DENNIS HUGH
THOMPSON JAMES EDWARD
THOMPSON JAMES MICHAEL
THOMPSON NATHANIEL
THOMPSON ODIS
THOMPSON SAMMY LEE
THOMURE LARRY LEE
THREET HOWARD ANDREW
THURMAN CURTIS FRANK
TIBBETT CALVIN B
TILLEMAN PAUL ROBERT
TILLERY RONALD DEAN
TIMS FREDERICK HOWARD

TINDLE DANIEL WAYNE
TINSLEY JAMES E
TITSWORTH CARREL JEAN
TODD VERNON BERNARD
TOLLIVER THOMAS JAMES
TOMEK GLEN DALE
TOOLOOSE DALE LEROY
TOOMEY SAMUEL KAMU III
TOUSLEY GEORGE HENRY III
TRAMMELL RODGER LEON
TRAVERS WALLACE OLDHAM
JR
TREVISANO ANTHONY
TRINKLER DICKIE DAVIS
TROTTER SHELBY MILES
TRUITT JERRY BOB
TSCHAMBERS JOSEPH L
TULLIS JAMES CLEVELAND
TURLEY MORVAN DARRELL
TURNER MERLE DEANE
TURNER MICHAEL DENTIS
TURNHAM CLAY SAMUEL
TYE MICHAEL JAMES
TYRON WILLIAM DAVID
UNDERWOOD HARRY WIL-
LIAM
UNDERWOOD ROBERT
STEPHEN
UNZICKER GREGORY DEAN
USSERY CARL RICHARD
UTLEY DAVID WAYNE
UTLEY MICHAEL LEWIS
VALENTINE LLOYD EARL
VAN HORN DONALD THOMAS
VAN SKIKE MONTE EUGENE
VANBUSKIRK HAROLD
DENNIS
VANDIVER HARRY MELBORN
JR
VARVELL DAVID LEE
VEALE RALPH DEAN
VELAZQUEZ FRANK
VENABLE BILLY RAY
VERSTRAETE MICHAEL JAMES
VINCENT JOHN LEROY
VOGLER GREGORY RAYMOND
VOLLMER HERMAN JOHN
VORIS RUSSEL EARL
VOSS WILLIAM ARTHUR
WAITE CAROLD REX
WAKE RUSSELL DEAN
WALDEN TRAVIS GARY
WALKER CURTIS
WALKER LAVALLE
WALKER RANDALL EDWARD
WALKER ROSS JEROME
WALLACE JIMMIE LEWIS JR

WALLACE JOHN CLAYTON
WALLER HAROLD DEAN
WALLING ROGER PAUL
WALLS JERRY FRANKLIN
WALTERS ROBERT DANIEL
WALTERS WILLIAM HAROLD
WARD ALBERT
WARD DAVID EUGENE
WARD IVORY JR
WARD LEROY
WARREN RONALD JOHN
WATERFIELD RICHARD F
WATKINS GLENN ALLEN
WATSON CURTIS LEE JR
WATSON KENNETH LAW-
RENCE
WATSON SAMMIE LEE JR
WEAVER JOHN FORREST
WEBB JAMES WILLIAM JR
WEBB LEONARD JR
WEBB WALLIS WAYNE
WEEKS WALKER NORWOOD
WEESE RONNIE GENE
WEHDE GERALD ALBERT
WEICHE LAWRENCE MICHAEL
WEIDNER DAVID EDWARD
WELCH JOHN HAROLD
WELLS JOHN CHARLES
WELTY CARROLL LEON
WENGER JEFF LYNN
WERLEY ROBERT WAYNE
WESTBROOKS ALLISON A JR
WESTFALL ROBERT LOUIS JR
WESTFALL RUBIN WILBERT JR
WESTLAKE CLAIR LLOYD JR
WHEELER CHARLES EDWARD
WHEELER RAYMOND LEE
WHICKER DENNIS RAY
WHITAKER DONALD EUGENE
WHITE CORDIS RAY
WHITE DONALD EUGENE
WHITE JOHN MICHAEL
WHITE ROBERT LEE
WHITE RONALD GENE SR
WHITE TOMMY RYAN
WHITLEY FREDDIE LEE
WHITTED JOE RAY
WIDEMAN ELVIN JOSEPH
WIESE ROBERT JAMES
WIGGINS JERRY LEE
WILCOX CHARLES KIRBY
WILKINS ROBERT JOHN
WILLIAMS BILLIE JOE
WILLIAMS CURTIS LEE
WILLIAMS GEORGE ANTHONY
WILLIAMS JOHN DILLARD
WILLIAMS JOSEPH JEROME

WILLIAMS LARRY JOE
WILLIAMS LEMUEL TAYLOR
WILLIAMS LESTER LEE
WILLIAMS WILLARD LOYD
WILLIS PAUL MITCHELL
WILLIS WILLIE CLIFTON
WILSON CHARLES JACKSON
WILSON DAVID OLSEN
WILSON JAMES CLAIBORNE
WILSON JERRY LEE
WILSON MICHAEL RICHARD
WILSON MICHAEL ROY
WILSON ROBERT HENERSON
JR
WILSON WILLIAM D
WILSON WILLIAM RALPH
WINFREY JAMES ARTHUR
WINKEL WILLIAM DANIEL
WINKLER BOBBY JOE
WINTER ROY ALAN
WINTERS RONALD PAUL
WISEMAN JOHN SAMUEL
WITHERS STEVEN RICHARD
WOLFE THOMAS HUBERT
WOOD ARTHUR WINDELL
WOOD FREDERICK DERL
WOOD JAMES LEWIS
WOOD JAMES SCHENLER
WOOD LOREN EDWIN JR
WOOD PATRICK HARDY
WOOD STEPHEN DUANE
WOODEN DAVID WAYNE
WOODS DAVID ALEXANDER
WOODS GREGORY
WOODSON JAMES LIONEL JR
WOOLF DWIGHT D
WOOTEN BOBBIE GENE
WORKMAN DONALD RENAY
WORRELL DAVID ALLEN
WRAY STEVEN CHARLES
WRIGHT JOHN WESLEY
WRIGHT JOHN WILLIAM
WRIGHT LANNY GAYLE
WRIGHT TOMMY DEE
WYATT BILLY JOE
WYATT JAMES EDWARD
YOUNG DENNIS LEE
YOUNG JIMMY RAY
YOUNG JOHN DELBERT
YOUNGS JACK M
ZIY GERALD WAYNE
ZUMALT TERRY LESTER
ZUNIGA MARTIN HARRY

KANSAS CITY
VIETNAM VETERANS MEMORIAL FOUNTAIN

WATER, LIKE TIME, HAS THE POWER TO CLEANSE AND HEAL. THIS MEMORIAL FOUNTAIN STANDS AS A SYMBOL OF THAT HEALING, FROM THE DEVASTATING DIVISION CAUSED BY THE VIETNAM WAR. THE FOUNTAIN'S POOLS REPRESENT THE COUNTRY'S GROWING INVOLVEMENT IN THE WAR, CULMINATING IN TWO POOLS SYMBOLIC OF THE DIVIDED OPINIONS AT THE TIME.

AMERICANS TOOK DISTINCT AND DIFFERING STANDS ON THE WAR, AND CAUGHT IN THE MIDDLE WERE THE THOUSANDS OF MEN AND WOMEN FROM THE KANSAS CITY AREA WHO SERVED IN VIETNAM, HUNDREDS OF WHOM WERE KILLED OR ARE MISSING AND UNACCOUNTED FOR. THIS MEMORIAL IS TO HONOR THEM, AND BRING US ALL TOGETHER IN TRIBUTE TO THEIR DEDICATION AND BRAVERY.

THE PARK IS FOR ALL OF KANSAS CITY TO ENJOY, AND TO REMEMBER. FOR ONLY BY REMEMBERING CAN WE ASSURE THAT IT NEVER HAPPENS AGAIN.

Photos by Terry J. Cyr

Vietnam Veterans Memorial

200 Blaine St, Missoula, MT 59801

The Vietnam veterans' memorial is located in Rose Park in Missoula, Montana. The memorial was created to honor the men and women who served in Vietnam. The memorial is accessible to visitors regularly, but there are special events at the memorial commemorating days of remembrance including Armed Forces Day, Veterans Day, and Memorial Day.

In 1988, the Vietnam veterans' memorial was added to the Memorial Rose Garden. It was created by Deborah Coperhaven after a long battle to raise the money necessary to fund the memorial from the ground up. The final design was selected by the committee after they heard the back story from a veteran about how he felt like an angel protected him during his time in Vietnam.

The veterans looking to build the memorial also faced a lot of opposition due to the polarizing nature of the Vietnam War. The effort finally paid off and the memorial was dedicated officially on Veterans Day in 1988.

The Vietnam War memorial in Missoula features a sculpture of an angel carrying the soul of a fallen serviceman to the afterlife. The statue, known as the Warrior Angel, is made of bronze and is supposed to symbolize the angel that a Vietnam veteran believes saved him on multiple occasions while he was on his tour in Vietnam.

To honor the 316 of 36,000 local veterans who served and passed in the Southeast Asian conflict, their names were engraved on a memorial plaque. The wall is made of seven plaques that are also made of bronze. The memorial also features a marble plaque that features an inscription written by Tom Crosser that reads, "Together we flew to battle, Hearts and blades pounding, Sharing the fear and pain... Parted.. Yet together.... We faced the unknown You Death, I the future.."

The Names of those from the state of Montana Who Made the Ultimate Sacrifice

ALLINSON DAVID JAY
ANDERSON DAVID ANTHONY
ANDERSON DAVID GEORGE
ANDERSON GEORGE ROLAN
ANDERSON JACK HERBERT
ANDERSON MITCHELL LESTER
APPELHANS RICHARD DUANE
ARCHER SANFORD KIM
ASHALL ALAN FREDERICK
BABICH RONALD GREGORY
BACKEBERG BRUCE BURTON
BAKER JACK MARVIN
BARBER DAVID LYNN
BARNUM WAYNE ALAN
BARTON JIM ALBERT
BEARY DANIEL WARREN
BECKER HARRY MATHIAS
BENSON JOSEPH HENNING
BERCIER KENNETH SAND-FORD
BEST RICHARD JAMES JR
BIBERDORF DENNIS FLOYD
BIRKLAND WILEY COLE
BOUCHARD MICHAEL LORA
BOYER ALAN LEE
BOYER CHARLES GOODHUE
BRISTOL CLARENCE FRANK
BROWN ROBERT RAYMOND
BURKHARDT LARRY JAMES
BURNS JAMES LYNN
BYFORD GARY D
BYINGTON STEVEN L
CADWELL ANTHONY BLAKE
CARROLL ROBERT HUGH
CASEY DENNIS LEE
CATON LEONARD ROGER
CAWLEY ROBERT WILLIAM
CECH LEROY CHARLES
CHAVEZ PAUL EDWARD
CHILDERS JOHN KENNETH
CHOPPER FRANKLIN DELANO
CHRISTENSEN WILLIAM MURRE
CLARK RICHARD DEWYATT
COLE RAYMOND ALLEN
COTTET DUANE LEE
COURVILLE ROGER MARVIN
CRASE LIONEL RUSSELL
CURL FRANKLIN NEWTON
CYR WILLIAM LOUIS
DARCY JAMES LEO

DEITCHLER RUSSELL FLOYD
DELLWO THOMAS ALBERT
DEMPSEY JACK ISHUM
DERENBURGER RONALD HAL
DERHEIM KENNETH LEE
DOAN LESTER ALLAN
DOANE MICHAEL LEO
DORRIS DAVID WALTER
DUDLEY CHARLES GLENDON
DUNBAR DOYLE DANIEL
DUNN DAVID HAMILTON
ECKSTEIN RODGER DEAN
EDDEN GEORGE EDWARD
EHNES RICHARD LEE
ELL ALLEN CHARLES
ELMORE KENNETH GLENN
ELSHIRE TERRY MICHAEL
EWING RONALD ARTHUR
FASCHING LEROY JAMES
FERGUSON RONALD BRUCE
FISH GLENN CHARLES
FISHER WILLIAM JOHN
FLANAGAN RUSSELL DAVID
FLEMING PATRICK JAY
FRANK EDWARD ROY SR
FRAZIER RICHARD BERYL
FRIED DOUGLAS LAWRENCE
FUHRMAN JAMES FRANCIS
GALLAGHER RAYMOND LEROY
GARCIA FRANK JR
GARRITY WILLIAM JOHN JR
GIFFORD GREGORY ALLEN
GORE JAMES RAYMOND
GRASSER ARTHUR
GRAYSON REID ERNEST JR
GREEN ROBERT WILLIAM
GREENO GERALD THOMAS JR
GREINER GARY JAMES
GRIFFIN GARY O NEAL
GROSE THOMAS NEIL
HAAKENSEN DAVID ARNOLD
HABETS GREGORY LEE
HAGL EDWARD JOSEPH
HAN CHARLES WILLIAM
HARNESS ANTHONY GENE
HARRIS LEWIS CRAIG
HAVRANEK MICHAEL WIL-LIAM
HEALY LOUIS GLENN
HELSLEY GREGORY PHILLIP

HENDERSON GREG NEAL
HENDERSON HAL KENT
HENDRICKSON MICHAEL FRANC
HENSLEY MARK ALAN
HEVERN RUSSELL JAMES E
HINKLE MARK GORDON
HINTHER GARY ROGER
HOERNER RAYMOND DALE
HOLTON ROBERT EDWIN
HOOD ROGER WILLIAM
HOOPER BILLY DEAN
HOWARD WALTER JOHN JR
HUNT JAMES D
INCASHOLA JEAN BAPTISTE
ISAACS DEAN ROBERT
JANHUNEN DANIEL JOHN
JODREY WILLIAM MICHAEL
JOHNSON CALVIN LEE
JOHNSON LESTER WESLEY JR
JOHNSON LYLE ALBERT
JOHNSON RICHARD MICHAEL
JOHNSON WILLIAM MICHAEL
JORDET RONALD GEORGE
JUEL DARRYL RICHARD
KEGLEY JOE DAVID
KELSO TODD DUANE
KENDALL GEORGE PERCY JR
KERN DOUGLAS DUANE
KILWINE RICHARD JAMES
KLEIV MANFORD LLOYD
KLEMENCIC JOSEPH GORDON
KNUDSON KENNETH MAX
KOJETIN ROGER JOHN
KRISKOVICH RAYMOND GEORGE
KUCERA RICHARD RALPH
KULTGEN ALAN JOSEPH
LANGAUNET BRUCE MAGNUS
LAWRENCE DELMAR LEON
LEHUTA DONALD ALEXAN-DER
LETCHWORTH EDWARD NORMAN
LIGHT GLEE ROY
LILLIE RICHARD ARTHUR
LITTLE WILLIAM GREGORY
LOCHER WALTER NORVEL
LOCKWOOD HAROLD SPEN-CER
MAGEE PATRICK JOSEPH

MARGRAVE DANIEL W II
MARTIN RONALD STEVEN
MATTHEISEN JOHN CHARLES
MATTOCKS GEORGE ELI
MAYES RICHARD LE OTIS
MC CARVEL STEPHEN LEWIS
MC DOUGALL HIMA DUNCAN JR
MC NALLY EUGENE FLOYD
MEHLS LELAND MC GEE
MELNICK STEVEN BERNARD
MEYER LEWIS DONALD JR
MILLER BERNHARDT WIL-LIAM
MILLER GEORGE DANIEL
MILNE RONALD JAMES
MOE RONALD JOHN
MOOER GARY OWEN
MURPHY JON MICHAEL
MURPHY STEVEN PATRICK
MURREY TRACY HENRY
NAASZ EMIL JOHN
NAASZ LARRY DUANE
NATHE MICHAEL LEO
NEIBAUER ALEXANDER DUANE
NEISS TERRY DUANE
NELSON LOUIS HOWARD
NELSON RAY LA GRANDE
NELSON STEPHEN CARL
NICHOLS PHILLIP ARTHUR
NORDAHL LEE EDWARD
O NEILL DANIEL JOHN
OSTERLOTH JAMES ALLAN
OVIATT STEPHEN STANFORD
PADILLA MICHAEL DAVID
PARKER JOSEPH E JR
PERRY RANDOLPH ALLEN JR
PETERS GEORGE EDWARD JR
PETERSON DUANE KENNETH
PETRIE JOHN JAMES
PICARD MICHAEL W
PICKETT RICHARD DALE
PICKLE JIMMY DEE
PIRKER VICTOR JOHN
PISENO RAYMOND RICHARD JR
POGREBA DEAN ANDREW
POKERJIM JOSEPH LOUIS
POLK PRESTON WAYNE
POMEROY ALEXANDER P

RAMBO ARTHUR JOHN
REECE WESTON HENRY
RICHARDSON ROGER PAUL
ROBBINS WILLIAM JAY
ROBERTSON MARVIN KENT
ROBERTSON RAYMOND L JR
ROBINSON TIMOTHY CHARLES
ROGERS JACK
ROLLINS DALE FRANKLIN
ROWLAND ZACK OSCAR
RUMMEL JAMES DOUGLAS
SALYER STANLEY WILLIAM
SAMPSON LESLIE VERNE
SANDERS STEVEN ROY
SATTERTHWAITE RICHARD D
SCHMIDT EDMUND JOSEPH
SCHNOBRICH ANTON JOHN
SCHULTZ DANNY CARL
SCHWARZ LARRY EDWARD
SEIDEL DONALD WILLIAM
SLIFKA JOSEPH JOHN JR
SLUSHER STEVEN
SMITH GARY MICHAEL
SMITH LARRY MAX
SMITH MILTON WARREN
SNIDER CHARLES CALVIN JR
SNYDER JERRY WAYNE
SONSTENG DENNIS WAYNE
SRB ERVIN RYNOLT JR
STEMBRIDGE WAYLAND DAN
STENGEM PETER MICHAEL
STEPAN JACOB FRANCIS
STEPHENS JAMES WILLIAMS
STOCKBURGER ARTHUR LEE
STREET DOUGLAS GERALD
STUBE RICHARD HURRELL
STYER MICHAEL EDWARD
SULLIVAN DAVID PATRICK
SUMMERS JON RAY
SWENSGARD WILLIAM ELLING
TALLON DOUGLAS WAYNE
TANNER RONALD RUSSELL
TAYLOR WILLIAM EUGENE
TEETH AUSTIN
THATCHER GARY DAVID
THOMAS ROBERT JOSEPH
THOMAS ROY STEPHEN
TILLOTSON ROBERT VIRTUS
UHREN BERNARD JEFFERY
ULSTAD DENNIS ELMER
UNDERWOOD JAMES EDWARD
URBAN JOHN ROBERT
UTTER KEITH EDWARD
VALLANCE DAVID CLARK
VANDENACRE HOWARD DANIEL
VOLK BARCLAY LEONARD
WALSH TRUMAN J
WANDLER LOUIS JOHN
WEAR DENNIS WILLIAM
WEBSTER CHRISTOPHER C
WEIGAND PAUL GARY
WELCH ROBERT EDWARD
WEST KENNETH PETER
WESTERVELT JOHNNIE BOWEN
WESTFALL RICHARD EARL
WHETHAM VERNON E
WIEST JOHN ROBIN
WILLETT ROBERT VINCENT JR
WILLIAMS RALPH LEROY
WOLFE DONALD FINDLING
WOOD ALVY EUGENE
WOODS ALBERT CLARENCE JR
YARGER JOHN ROBERT
ZAHN FLORIAN J
ZERBST GILBERT LEROY
ZIEBARTH DENNIS LEROY
ZINDA FRANCIS JOHN

Photos by James Palmer

Korea-Vietnam Peace Memorial

6005 Underwood Ave, Omaha, NE 68132

The Korea-Vietnam Peace Memorial is located in Omaha, Nebraska and was created at the end of the Vietnam War in the mid-1970s. Due to the controversial political times in the US, the memorial took on one of promoting peace and honoring the soldiers who fought instead of memorializing the actual war. It's located within Memorial Park south of the WWII Memorial in the Omaha park.

The Peace Memorial was finished and dedicated in 1976 and placed in Omaha's Memorial Park. The final monument stands at around five feet high and nine feet wide. It's made of bronze as well as brick. The memorial was created by John Snider and Keith James, but Korean War veteran William Ramsey organized the entire project including the fundraising for the actual monument.

Thanks to the polarizing opinions on the Vietnam War, the memorial committee decided to air on the side of caution and create a memorial that focused more so on the servicemen and their efforts to preserve peace in a foreign nation, instead of using the memorial to commemorate a war that so many people were against.

The Korea-Vietnam Peace Memorial needed $15,000 to cover the cost of the sculpture. The first $1,000 came from a contribution from the Omaha Post 1 American Legion. Private donations began to roll in from civilians and families of Vietnam veterans; actor John Wayne even contributed to the memorial's fund. In 1998, $25,000 was paid to restore the monument. However, despite those efforts, a 2014 study showed that the sculpture was showing significant signs of instability and wear. The memorial committee can either restore the statue to a stable condition or replace it completely with more durable materials.

This memorial is considered to be iconic due to its position in the park at the top of a hill. The featured statue depicts a US soldier holding a Vietnamese child who is looking trustingly at the soldier. The soldier's helmet and weapon are placed to the side while the magazine for his weapon is missing completely from the statue. By neutralizing those polarizing elements, the memorial takes on a more peaceful meaning and shows the soldier as more of a humanitarian and less of a combat soldier. Veterans day and Memorial day services are held here annually.

The Names of those from the state of Nebraska
Who Made the Ultimate Sacrifice

ADKINS RONALD EUGENE
ADLER HENRY
ADOLF LARRY EUGENE
ALLEN JERRY L
ALTSCHAFFL STEPHEN ALLEN
ANDERSEN BUEL EDWARD
ANDERSON DENNIS WILLIAM
ANDERSON JOHN LOUIS
ANDERSON LEE DAVID
ANDERSON WARREN CHARLES
ANTHONY RAYMOND F JR
AUMAN ERVIN LEWIS
BAADE CLIFFORD KEITH
BACKHAUS STEVEN EUGENE
BAHNSEN KENT EUGENE
BAHR DENNIS KEITH
BAILEY ALLEN CHARLES
BAILEY BYRLE BENNETT
BALES RONALD EUGENE
BALLANTINE RICHARD REED
BALLINGER WILLIAM JOSEPH
BARNES HAROLD DUANE
BARNETT CARL EUGENE
BARNEY TERENCE EDWARD
BARNHILL ROBERT EUGENE
BARRON FLORENTINO CIPRIAN
BARTZ ROGER CHARLES
BAUMANN LANNY ROSS
BAZAR PAUL THOMAS
BEAMS JAMES WOODSON
BECKER MICHAEL PAUL
BEESON ROBERT BRUCE
BENZE PATRICK HENRY
BERLETT THEODORE JAMES
BERMINGHAM JAMES CHARLES
BERNAL VINCENT
BERNEY TERRY LYNN
BIBER GERALD MACK
BIERMA LYNN SEATON
BIGLEY RICHARD RAY
BISCAMP MARVIN LYNN
BISCHOFF JOHN WILLIAM
BOOZE DELMAR GEORGE

BOSILJEVAC MICHAEL JOSEPH
BOYLE ROBERT RAY
BRAGG PAUL JOSEPH
BRANSTROM DAVID JOSEPH RI
BREDENKAMP DAVID JOE
BRENNAN HERBERT OWEN
BRENNING RICHARD DAVID
BRING JOHN DALE
BRUHN JAMES WILLIAM
BRUNCKHORST ROBERT L JR
BUCKLES DONALD RAY
BULL KENNETH R
BUSSELMAN DUANE LORENZ
CALDWELL LARRY GAIL
CAMPBELL JAMES ROBERT
CARPENTER DONALD EUGENE
CARR DANIEL LEE
CHANDLER JEROME DEE
CHURCH RALPH LEE
CLARK CHARLES CHAPMAN
CLARK KENDALL HANSON
CLEMENTS MILO DEAN
COEN ROGER LEE
COKER RONALD LEROY
COLE MURIL STEVEN
CONDON ROBERT EUGENE
CONFER MICHAEL STEELE
CORDOVA ROBERT JAMES
COVEY LAWRENCE LAVERN
COWLES GARY TWYMAN
COZAD JERRY LEE
CRAYNE KENNETH EUGENE
CRUMLEY ELDON GENE
CULBERTSON GARY MORTEN
CUNNINGHAM RICHARD IRA
DAGLEY GARY GENE
DANGBERG ROBERT LEE
DAVIS JAMES DEAN
DAVIS JOHN CLINTON
DAVIS JOHN EDWARD
DE FORD DALE DARREL
DEWOLF DALE LEE
DOAK STANLEY WAYNE
DOEDEN NICOLAUS AUGUST

DOOLITTLE JON HILIARE
DRAPER CLIFFORD ARVIN
DUGAN EDWARD MICHAEL
DUNN GARY WAYNE
EATON ROBERT LEROY
EDMOND COIL JR
EISENHOUR DWIGHT DAVID
ENGEL HARVEY LEROY
ENGEL RODNEY LOUIS
ENQUIST ARTHUR JOHN
ESTRADA RICHARD ALLEN
FARLEY MICHAEL MARION
FARRELL TIMOTHY CHARLES
FIELDER PAUL WESLEY
FISHER CARL NELSON JR
FLANAGAN DAVID DALE
FLORANG LARRY DEAN
FLOURNOY JAMES KAISER
FOLEY JAMES WILLIAMS
FONTAINE LARRY LEE
FORD OMAR RAY
FORK NORMAN KERMIT
FOUS JAMES WILLIAM
FOWLER THOMAS LEE
FRUHLING DALE ERVIN
FRYC DAVID CHARLES
FUSS ROBERT EDWARD
GAETH JOHN CEPHAS
GAGE JOHN THOMAS
GARAMILLO ELDON
GARCIA JERRY FRANK
GATHMAN GORDON KAYE
GEHRKE DARRELL DEAN
GERRY RONALD LEE
GEVARA RAY JR
GIITTINGER RICHARD FREDER
GILLHAM JAN ROYCE
GILLHAM RICHARD GERALD
GOC PAUL STEPHEN JR
GOLDEN JACK DUANE
GRAHAM HARLAN LEE
GREEN NORMAN DUANE
GRELLA DONALD CARROLL
GRIFFIN GERALD CHARLES
GRONBORG MARTIN WAYNE

JR
GRUEBER RANDALL ROMAN
GUBBELS STANLEY DONALD
HAAKENSON ROBERT W JR
HAEGELE WOLFGANG ALBRECHT
HAGOOD JOHN ROBERT
HALL RICHARD LE ROY
HAMILTON GERALD LOUIS
HANCOCK WILLIAM HOWARD II
HANSEN ROBERT GREG
HARGENS DAVID ALLEN
HARIG DEAN ALLEN
HARRIS JOHN HENRY JR
HARVEY LAWRENCE DANIEL
HATFIELD GARY CLARK
HAYES JOHN COOK
HEESACKER VICTOR ROMAN
HEINZ JOHN DIETRICH
HEMPEL THOMAS EUGENE
HENK JAMES LYNN
HILEY THOMAS CHARLES
HILFIKER HERBERT ALLEN
HOBSON JOHN KING
HOLLAND JOHNNY ROBERT
HOLTZ LARRY WILLIAM
HOLTZ PAUL AUGUST
HORNELAS ISMAEL FERNANDO
HOVENDEN DARREL LEROY
HOYT ERVIN JAMES
HUDSON DANNY CHARLES
HULTQUIST LEONARD ASHBY
HUNDT ROGER LEE
HUNTER HENRY DAVID
HURT WILLIAM C
ILER KENNETH MARVIN
JACKSON EDDIE LEE JR
JAMES DANIEL RAYMOND
JESSEN ROBERT DUANE
JOHNSON FLOYD DEAN
JOHNSON GARY LEE
JOHNSON JOHN ERNEST
JOHNSON JOHN PAUL
JOHNSON KENNETH LEE

JOHNSON LANE CARSTON
JONES THEODORE R JR
KAHLER HAROLD
KAMINSKI RAYMOND DONALD
KAVULAK JOHN HENRY
KEITH MIGUEL
KELLER KENNETH LAVERN
KELLEY HARVEY PAUL
KEMPKES ROBERT LOUIS
KIER LARRY GENE
KILDARE WILLIAM JAMES
KIMM CLARENCE ALFRED
KINGMAN DAN CHRISTIE JR
KINKAID FRANK W JR
KINSMAN ALLEN EUGENE
KLABUNDE ARTHUR JOHN JR
KLABUNDE JOHN PAUL
KLINGNER MICHAEL LEE
KNIPPEL LARRY DON
KOCANDA JERRY JOSEPH III
KOCH DALE ROY
KONWINSKI RONALD EUGENE
KORINEK JOHN CHARLES
KOT MYRON
KOTRC JAMES CARL
KOTROUS EUDELL LEO
KROUS KENNETH WAYNE
KUDLACEK EDWIN ALLEN
KUHLMAN MELVIN ERNEST
KURZ DENNIS LEE
LACKAS MONTY GILBERT
LAIRD JAMES ALAN
LAMBOOY JOHN PATRICK
LAMERE ANTHONY JOHN
LANE ROGER LEROY
LANGAN LARRY MILTON
LANGE DEAN RICHARD
LANGSTON MELVIN DOYLE
LARSON DAVID NEIL
LARSON DAVID WAYNE
LEACH DEAN KENT
LEICHLEITER THOMAS ALLEN
LEIGHTON EARL LA ROY
LEMASTER LARRY D
LENTZ JERRY FRANCIS

Nebraska

LEWIS DONALD GENE
LILES LARRY JOE
LIMBACH HENRY LEE
LINDELL LARRY ALBERT
LOECKER MARLOW MARTIN
LOOBY LAWRENCE CLARENCE
LUEDKE WILLIAM
MADDOX PHILIP NEIL
MAGERS PAUL GERALD
MALONE WALLACE JAMES
MARCHAND WAYNE ELLSWORTH
MAREZ FREDERICK
MARRS CARL ROBERT
MARSH ALAN RICHARD
MARTIN MARVIN HENRY
MARTIN MICHAEL TERRY
MATSON WILLMER ARDEN
MAXWELL SAMUEL CHAPMAN
MC ADAMS GEROLD JEROME
MC ALLISTER CAMERON TRENT
MC FADDEN PAUL RAY
MC KNIGHT THOMAS EDWIN
MC LEESE KENNETH RICHARD
MC QUAY ROGER DILLON
MCCURDY JOHN A
MEIROSE DAVID ALLEN

MEISINGER JEROLD WERNER	MURPHY ROBERT DENNIS	NOWAK ROBERT VIRGIL	PEETZKE RONALD EUGENE
MICKNA JOHN RONALD	NACHTIGALL DAVID JOSEPH	OGDEN HOWARD JR	PERRIN RICHARD THOMAS
MOHR ROY JOHN	NAPIER LEE ALLAN	OHM ERIC GEORGE	PETERS CHARLES HENRY
MOORBERG MONTE LARUE	NEEDHAM RUSSELL DEAN	OHNESORGE THOMAS HERMAN	PETTY ERNEST FLOYD
MOORE DANIEL EUGENE JR	NELSON BILLY DEAN	MAN	PHILSON WILLARD ARLIN
MOORE JAMES ELDON	NELSON ROBERT WARREN	OONK LESTER EUGENE	PINA LOUIE PETE
MOORE RONNIE GENE	NEUBAUER FRED ALLEN	ORR MERLIN GEORGE	PINEGAR WILLIAM DENNIS
MORENO DENNIS RALPH	NEWMAN LARRY JEROME	OSTRANDER MORRIS EDWARD	PLAHN JACK CHARLES
MORRISON JAMES ANTON	NEWMAN MICHAEL CARL	WARD	POESE NIGEL FREDERICK
MOSER JAMES MYRON	NIEBUR EDWARD LEROY	OTTE KENNETH MICHAEL	POGGEMEYER JAMES ROBERT
MUELLER STEVEN WAYNE	NOVAK CLARENCE JOSEPH	PARSON DOYLE HALL	POLT ERWIN ANDREW
MURPHY JOHN PATRICK	NOVAK LARRY DEAN	PEARSON MICKEY DON	POSPISIL MARVIN LEROY

LIAM
THIEM WILLIAM RAYMOND
THOMPSON JOHN CLYDE JR
THOMPSON ROBERT CHARLES
TIPPERY TERRY LEE
TWEHOUS GENE LEANDER
UTTS WILLIAM WARNER
VAN ANDEL CLAUDE RICH-
ARD
WAGNER ROBERT ALFRED
WAITE DONALD STEVEN
WALKER ELBERT BERTON
WALKER MICHAEL ALLEN
WALTERS DONOVAN KEITH
WALTERS GERALD LEROY
WARD CARL RAY
WARNICK LEONARD CHARLES
WELDING CLIFFORD KAY
WEMHOFF MICHAEL LYNN
WIDICK MAURICE GENE
WIESE THOMAS ARTHUR
WIESER LYNN JAY
WIESNETH ROBERT PAUL
WIGTON PHILIP GREGORY
WILKEN BRYAN LEE
WILKERSON STEVEN DOUG-
LAS
WILKINSON HARLAND LYLE
WILLIAMS ROBERT FLOYD
WILSON GALEN LLOYD
WILSON MICHAEL JOSEPH
WINCHELL DOUGLAS JAMES
JR
WITT MARK STEVEN
WOJTKIEWICZ RONALD
JOSEPH
WOLF JACK MORSE
WOLFE MATHEW
WRIGHT DELBERT PAT
YOUNG ROBERT LEE
ZABROWSKI LOUIS
ZICH LARRY ALFRED
ZICHEK RICHARD LANSING
ZIEHE GERALD DEAN
ZUEHLSDORF JOHN WILLIAM

RADIL RONALD LUDWIG	SANDSTEDT DANIEL JOSEPH	SHRADER HAROLD WILLIAM	STAFFORD RONALD DEAN
RAUBACH WILLIAM PIERCE	SAWICKI RICHARD P	SHUEY GLENN COLIN	STARK LARRY ALLEN
REISER STEVE RONALD	SAYER TERRY LYNN	SKAVARIL THOMAS JOSEPH	STARK WILLIE ERNEST
RIEKEN LARRY RIEK	SCHEURICH THOMAS EDWIN	SLATER JAMES ALLEN	STEEL KENNETH LEE
ROARK WILLIAM MARSHALL	SCHMIDT GARY RUSSELL	SMILEY STANLEY KUTZ	STEWART FRANCIS ERNEST
ROBERTS WILLIAM	SCHMIDT KENNETH WAYNE	SMITH LARRY EUGENE	STOEHR DAVID LOREN
ROBERTSON JOHN CRAIG	SCHMIDT RICHARD LEROY	SMITH MICHAEL FRANCIS	STOLINSKI JAMES FRANCIS
ROBINSON LARRY WARREN	SCHNEIDER ROGER LLOYD	SMITH PAUL RICHARD	STONER LARRY LEE
ROOTH CHARLES WILLIAM	SCHRODER JACK WAYNE	SMITH THOMAS LEROY	STRAUS ALLEN ARTHUR
ROSS MILTON ALAN	SCHULTZ JAMES RONALD	SMOCK DARYL EUGENE	STRUBE STEVEN DREW
RUHTER MICHAEL ALLEN	SCHWARTZ ALLAN EDWARD	SOBOLIK KARL DAVID	STRUEBING DEWEY IRVIN
RUSTINE DOUGLAS CECIL	SCOTT IRA EDWARD	SOLOMON WILFRED L SR	STUBBE WILLIAM LEROY
SALYARDS PATRICK JOHN	SCOTT MICHAEL MONROE	SOTO THOMAS GABRIEL	SUTTON TERRY JAMES
SAMUELSON RONALD EARL	SEADORE LARRY LEWIS	SPENCER FRANK III	SWAIM JAMES LEE
SANDERS MACK ROYAL	SHELDON LEROY ELLSWORTH	SPERLING WESLEY WILLIAM	TAYLOR LESTER KEITH JR
SANDERS ROBERT NEIL	SHELTON CRAIG STEPHEN	SPRICK DOYLE ROBERT	TEGTMEIER LA VERN WIL-

Photos by Sharon Green

Nevada State Vietnam Memorial

1111 E William St, Carson City, NV 89701

Constructed of local Nevada stones, the Vietnam War Memorial is located in the northeast corner of Mills Park and was created to remember and honor the servicemen and women who made the ultimate sacrifice for their country.

The Nevada State Vietnam Memorial was originally dedicated in 1992 and simply consisted of a flag pole paired with a memorial plaque in Mills Park. The plaque honored Nevada's Vietnam War prisoners of war as well as the missing in action. It was created through the efforts of three separate chapters of the Vietnam Veterans of America. The Carson Area Chapter 388 was responsible for financing and sponsoring the memorial, while Chapter 545 (made up of incarcerated veterans) quarried the native Nevada sandstone for the monument at the Nevada State Prison, and Chapter 719 (made up of veterans housed at the Northern Nevada Correctional center) who was responsible for the actual labor and construction of the memorial. This unusual team was put together by prison warden and former US Marine Terry Hubert when he heard that the original humble memorial was looking to be expanded upon.

The three VVA chapters worked hard to complete the memorial which was rededicated on Veterans Day of 2002. The park received an addition of a sandstone bench (also constructed by Chapter 545) in 2004 in order to honor the servicewomen who lost their lives in Vietnam and was further refurbished in 2012 by Chapter 388.

While it started with that flag pole and plaque (which are the center of the current memorial), the memorial now consists of five sculpted sandstone boulders that are each marked with individual bronze plaques. Each plaque features the names of Nevada servicemen who either passed or went missing during the Vietnam War. The memorial also consists of the previously mentioned sandstone bench that honors the servicewomen of the Vietnam War.

148

The Names of those from the state of Nevada Who Made the Ultimate Sacrifice

ACKERMAN DANIEL LEVERNE
ALECK JOHN IRA
ALLEN ROBERT CHRISTIAN
ANDREWS ROBERT WARREN JR
BARBER BARRY MORRIS
BARGER LARRY EARL
BELL JIM GLENN JR
BERGER LORAN LEON
BIDART DAVID LOUIS
BLEA MICHAEL DELANO
BODAMER MICHAEL ANTHONY
BOGUE JEFFREY LYNN
BOWER JOSEPH EDWARD
BRAEUTIGAN MICHAEL L
BROWN LARRY DONALD
BUIS DALE RICHARD
BURGESS STANLEY WAYNE
BURKE LARRY ERWIN
BURT GLEN GEORGE
BYLINOWSKI MICHAEL DAVID
CAMPBELL DONALD R
CARRUTHERS EDWARD ANTHONY
CARTER TIMOTHY GENE
CARTWRIGHT PATRICK G
CATES ROBERT MATHEW JR
CHRISTIAN ROBERT M JR
CLARK JAMES WOODFORD
CLARK KERRY EDWARD
COLLINS DAVID LEROY
COSSINS JACK EDWARD
CRANE RODNEY LANE
CUE CARL JAMES
DARRAH MICHAEL LEE
DAVIES ARTHUR JR
DAVIS THOMAS JOSEPH
DEDMAN RONALD EUGENE
DRAKEN OTTO JAMES
DU BOIS GREG ALAN
EVANS RAY FRANCIS
FARNOW JERE DOUGLAS
FOX CHARLES BURTON JR
FRANSEN ALBERT MERK JR
GARCIA JOSEPH ANDREW
GARDNER LAWRENCE LEE
GAREY ROBERT EARLE
GERRY JERRY LAWSON
GETZ PAUL ROBERT
GOMEZ JOSEPH JAMES
GRANT MELVIN LEE
GREEN ROBERT JAMES
GRIFFITH RICHARD OWEN
HAGEMEIER THOMAS VANCE
HAMMOND FRANK DALE
HASTINGS MICHAEL KENNETH
HEATER PAUL LEO
HENRY ROBERT JOHN
HESS JACK WAYNE
HIGGINSON LARRY CRAIG
HILL BILLY DAVID
HODGE CHARLES LYNN
HOGAN WALTER DE WAYNE
HOLLY GEORGE JOSEPH III
HOPKINS WALLACE W JR
HORTON DANIEL EUGENE
HOWE STEVEN TIMOTHY
HUTCHINS DALE EUGENE
HYDE MICHAEL LEWIS
JACKSON LLOYD WILNER
JAMISON DAVID
JENSEN RICK V
JOHNSON STERLING PRICE
JOHNSON WILLARD VERNON
JUDD GARY DEAN
KIRK ROBERT LEE
KNITTLE HAROLD JOSEPH
KRAMER JAMES LEE
LAKEY LARRY LEE
LARSEN STEPHEN EARL
LE BEAU ANDREW ERNEST JR
LINVILLE HAROLD LEE
LONDON DENNIS W

LUNZMANN LOWELL EUGENE
MAC KAY WILLIAM MICHAEL
MAC KINNEY PHILIP V
MAYBERRY RONALD JAMES
MAYHALL ALONZO EARL
MC ANDREW JAMES DELMAS
MC GEE STEVEN WESLEY
MC KNIGHT JAMES BRUCE
MEGINN MICHAEL MERIDITH
MINETTO ROBERT NEIL
MOLINO EDDIE JR
MONAHAN WILLIAM S III
MONTGOMERY GEORGE WESLEY
MORGAN ROBERT LEROY JR
MORRISON HOWARD GLENN
MUNCEY JAY ALLAN
MURRAY THOMAS EDGAR
NEWMAN CLYDE EDWARD
ORTEGA JESUS F JR
PARKER LARRY
PARNELLE SAMUEL W III
PARSONS MICHAEL DUANE
PERKINS STEPHEN JOHN
PERRY RICHARD CLARK
PIERINI JOHN ROBERT
POE CHARLES ALTON
QUILICI PETER JR
REESE DENNIS DEAN
REID DANIEL FRANCIS
RHODEHAMEL JOHN RAY II
RICE MAXIE ROSS
RODRIGUES RONALD
ROGERS DAVID ROBERT
ROGNE WILLIAM ROBERT
RUCKER CARLOS WILSON
SAM WILFRED GERALD
SANDERS ROBERT BRUCE
SAUNDERS KEITH FRANK
SCHWORER RONALD PAUL
SHEA MICHAEL FRANCIS
SHOCK JACK DEAN
SKIVINGTON WILLIAM E JR
SMITH JAMES HERBERT JR
SMOTHERS DANNY LEE
STEELE ROGER ALLEN
STIERWALT LADDIE C
STONE LARRY GEORGE
STONE RAYMOND EDWARD JR
STUDDARD DANNY GERALD
TAYLOR KEITH DEGERO
THARP ALEXANDER
THEMMEN MICHAEL JAMES
THOMAS MORRIS E
THOMPSON DALE EARL
THOMPSON GREGORY MALCOLM
TRASK LEWIS ARTHUR
VANRENSELAAR LARRY JACK
WALKER RICHARD HOWARD
WALLACE CHARLES JAMES
WARD DAVID JAMES
WARREN RICHARD MICHAEL
WESSEL MICHAEL DANIEL
WEST DANIEL FLOYD
WHITLOCK THEODORE JAY
WHITTEMORE FREDERICK H
WILKINS TERRY KENNETH
WILLIS JAMES RONALD
WORLEY ROBERT LEE
WRIGHT WALTER CLARENCE
YELLAND RICHARD MAX

1959
DALE R. BUIS

1963
JAMES D. McANDREW

1965
JOSEPH E. BOWER · THOMAS E. MURRAY · JERRY L. GERRY

1966
ALEXANDER THARP · TERRY K. WILKINS · PETER QUILICI, JR.
MICHAEL L. HYDE · PAUL L. HEATER · HOWARD G. MORRISON
LARRY L. LAKEY · ROBERT E. GAREY

1967
MICHAEL A. BODAMER · MICHAEL L. BRAEUTIGAN
LARRY G. STONE · GREGORY M. THOMPSON · LARRY E. BURKE
GLENN G. BURT · MICHAEL D. BYLINOWSKI · DONALD R. CAMPBELL

1967
EDWARD A. CARRUTHERS · MICHAEL F. SHEA · RICHARD M. YELLAND
LAWRENCE L. GARDNER · JOHN R. RHODEHAMEL, II · RICHARD C. PERRY
KERRY E. CLARK · RONALD J. MAYBERRY · JACK E. COSSINS
LADDIE C. STIERWALT · CARL J. CUE · WALTER C. WRIGHT
ARTHUR DAVIES, JR. · ROBERT J. GREEN · CLYDE E. NEWMAN
DANIEL E. HORTON · LEWIS A. TRASK · WALLACE W. HOPKINS, JR.
WALTER D. HOGAN · STEVEN W. McGEE · RAYMOND E. STONE, JR.
JAMES L. KRAMER · MICHAEL D. WESSEL · GEORGE W. MONTGOMERY
ROBERT J. HENRY · DAVID JAMISON

1968
ROBERT M. CATES, JR. · GREG A. DU BOIS · CHARLES B. FOX, JR.
JOSEPH A. GARCIA · DAVID L. COLLINS · THEODORE J. WHITLOCK
MICHAEL K. HASTINGS · WILLARD V. JOHNSON
FREDERICK H. WHITTEMORE · JACK W. HESS · LARRY C. HIGGINSON
DANIEL F. WEST · JERE D. FARNOW · DANIEL L. ACKERMAN

1968
GEORGE J. HOLLY, III · JAMES R. WILLIS · ROBERT L. WORLEY
RICHARD O. GRIFFITH · KEITH D. TAYLOR · DAVID L. BIDART
DAVID J. WARD · MORRIS E. THOMAS · JAMES H. SMITH, JR.
STERLING P. JOHNSON · ROBERT B. SANDERS · ROBERT N. MINETTO
CHARLES J. WALLACE · HAROLD J. KNITTLE · DAVID R. ROGERS
DANIEL F. REID · MAXIE R. RICE · ANDREW E. LE BEAU, JR.
DENNIS D. REESE · SAMUEL W. PARNELLE, III · LOWELL E. LUNZMANN
DANNY L. SMOTHERS · JOHN R. PIERINI · ALONZO E. MAYHALL
JIM G. BELL, JR. · ROBERT C. ALLEN · LARRY E. BARGER
MICHAEL M. MEGINN · BARRY M. BARBER · RONALD P. SCHWORER
JESUS F. ORTEGA, JR. · DAVID B. HILL ** · LARRY J. VANRENSELAAR *

1969
PHILIP V. MAC KINNEY ** · JAMES B. McKNIGHT · MELVIN L. GRANT ·
WILLIAM M. MAC KAY · ALBERT M. FRANSEN, JR.

*Date Declared M.I.A.

1970
FRANK D. HAMMOND · RAY F. EVANS
EDDIE MOLINO, JR. · ROBERT L. KIRK · JAMES O. DRAKEN
LLOYD W. JACKSON · DALE E. HUTCHINS · HAROLD L. LINVILLE

1971
PATRICK G. CARTWRIGHT · DANNY G. STUDDARD · MICHAEL L. DARRAH
RICHARD M. WARREN · JOSEPH J. GOMEZ

1974
WILLIAM E. SKIVINGTON, JR.

1975
MICHAEL D. PARSONS · DENNIS W. LONDON

Photos by Alexander Azzi

Vietnam Veterans Monument

110 Daniel Webster Hwy, Boscawen, NH 03303

Since its 1997 opening, the New Hampshire State Veterans Cemetery is nestled nicely in a grove of trees in Boscawen, New Hampshire. Within that cemetery, thousands of veterans and their partners rest, but within the confines there is also a somber place known as the Grove of Memorials. That place hosts war memorials including the state's Vietnam War Memorial. This monument was built to honor the New Hampshire service members who lost their lives in the Vietnam War. Visitors to the cemetery can take a self-guided foot tour of the land and stop by the Vietnam War Memorial as well as the other monuments that were erected along the cemetery's Memorial Walkway which is conveniently located near the entrance of the cemetery.

New Hampshire's beautifully constructed Vietnam War Memorial was designed by Manchester, New Hampshire native Ann Goulet. Dedicated May 22nd, 2004 it was designed to look like an open book with the names of the 227 fallen service members engraved across the two large pages. Across the top of the two pages read the words "We Give to You Our Final Gift;" a message that is meant to emphasize the sacrifice made by those who passed during, or as a result of, the Vietnam War. There is also an inscription across the bottom of the pages that reads "In memory of those New Hampshire Vietnam Veterans who gave their lives or are still missing 1959 – 1975." The stone monument is flanked by two pieces of black granite; one depicting the outline of New Hampshire, the other featuring an outline of Vietnam.

Goulet designed the memorial as a way to bring the Vietnam War and that whole time period full circle. She described that passed era as one that tore America apart due to the polarizing opinions on the war. For that reason, New Hampshire veterans, like most veterans across the country, didn't receive the recognition they rightly deserved. Goulet hoped the Vietnam War Memorial would remedy that in a small way. More recently VVA Chapter 992 dedicated a memorial bench. Veterans day and Memorial day services are held each year.

150

The Names of those from the state of New Hampshire Who Made the Ultimate Sacrifice

ABBOTT TERRY MICHAEL
ACKERMAN DAVID ALAN
AHERN ROBERT PAUL
ALBERT DANIEL JOHN
ALLOWAY CLYDE DOUGLAS
ARCHBOLD JOHN CHRISTOPHER
BADOLATI FRANK NEIL
BALL GARY WAYNE
BARNETT GLENDON ROMAN
BARTLETT JAMES B
BEAUDOIN GAETAN JEAN GUY
BEAUPRE GILBERT THOMAS
BENNETT MICHAEL E
BLANCHETTE GUY ANDRE
BLOOM LAWRENCE CLIFFORD
BOUCHARD RICHARD GEORGE
BRADY MICHAEL ERVAN
BROOKS RICHARD ALBERT
BROOME CECIL ANGUS JR
BROWN BRUCE WADLEIGH
BROWN JAMES WARREN
BROWN MARK LARRY
BROWN WARREN RICHARD
BUNKER DAVID ELVIN
BURNETT SHELDON JOHN
CABANA JOHN BISHOP JR
CAHILL GEORGE EUGENE
CAMIRE PAUL JOSEPH
CASS FRANK LEE
CASTELOT ROBERT SHEEHAN
CATE WILLIAM EARL
CHAMBERLIN HOWARD ARTHUR
CLOUGH ARTHUR EDWARD
CORMIER RONALD RAYMOND
CORONIS MARTIN JAMES
CUMMINGS RALPH RONALD
CUTTING WILLIAM STANLEY
DALE GEORGE LOUIS
DALEY RAYMOND COYLE
DAVIS RONALD CHARLES
DAVIS WILLIAM THOMAS
DELANO DARWIN JAMES
DEMERS ARTHUR EMILE JR
DEMERS RICHARD ARTHUR
DEROSIER RICHARD TERRANCE
DESCOTEAUX MAURICE CLAUDE
DESMARAIS GEORGE PHILIP
DICKEY CHARLES JOSEPH
DIONNE ROBERT PAUL
DOUCET LEON NORMAND
DOUILLETTE WILLIAM R JR
DUBIA LAWRENCE NORMOND
DURLING JOSEPH A III
DYER ORRIN LEONARD JR
EMRO ROBERT BENNETT
FALCONE JOHN PAUL JR
FEASTER WILLIAM NEWCOMER
FECTEAU RALPH BARNARD JR
FINAN ROBERT EDWARD
FINN JAMES NORMAN
FLANAGAN GEORGE FRANCIS
FLETCHER PETER
FORD MARSHALL H
FRASER WILLIAM GEORGE
FRATUS EDWARD FRANCIS
FROST GERALD JAMES
GALBREATH ROBERT GENE
GAMELIN ERNEST ULRIC JR
GANLEY RICHARD OWEN
GARDNER DAVID ERNEST
GARDNER ROBERT LOUIS
GAUDET THOMAS WILFRED
GEISTER MICHAEL LEWIS
GELINAS JOSEPH ARMAND ROG
GENEST RICHARD EDGAR
GIROUARD YVON ELDMOND
GODBOUT RICHARD GERALD
GODFREY BARRY WILLIAM

GREELEY VERNE MILTON
GUILD ELIOT FRANKLIN
HAINES ROBERT FREDERICK
HALGREN RICHARD LEE
HALL KENNETH ROBERT
HALL KENNETH WALTER
HARRIMAN EUGENE HOWARD
HARVELL RICHARD KENEFICK
HEBERT YVON ANDRE
HELMICH GERALD ROBERT
HERSEY CARROLL FRANKLIN
HILDRETH DAVID WAYNE
HILLSGROVE BARRY MALCOLM
HOGAN JOHN LAWRENCE
HOWARD RALPH ARTHUR
HURD JAY ALLEN
INDYK FRANK ALAN
JEWETT STEPHEN DYER
JOHNSTON RICHARD BRUCE
JORDAN KENNETH BRADLEY
JOY WILLIAM ARTHUR
JOY WILLIAM CLYDE
JOYCE GEORGE EDWARD
KAISER HOWARD WALKER
KELLER RONALD NORMAN
KELLEY VERNE CARL
KENISON BENJAMIN ALBERT
KILTON STANLEY ROY JR
KILUK EDWARD GEORGE JR
KNIGHT ROBERT LOUIS JR
KOWALCZYK CZESLAW
KREITZER DAVID A
LA FAVE RUSSELL THOMAS
LABONTE ROLAND CHARLES
LAROCHE ERNEST ALBERT
LAWRENCE JOHN WINSLOW JR
LE HOULLIER PAUL RAYMOND
LEAHY DANIEL MICHAEL
LEE JOHN F
LEIGHTON GREGORY A
LETENDRE RICHARD EDWARD
LOOMIS WILLIAM NICHOLAS
LORDEN DENNIS FRANKLIN
LOZEAU NORMAN GERARD
MALENFANT WILLIAM ARTHUR
MANN ROBERT BERNARD
MARCOTTE ANDRE EDWARD
MARGARITIS SOTORIOS MILTO
MARSHALL DENNIS HARDIE
MARSHALL LAWRENCE JAY
MARTEL NORMAND RICHARD
MARTIN STEVEN WAYNE
MC ALLISTER ROGER J JR
MC GLONE MICHAEL THOMAS
MC GUIRE JOHN WINCHESTER
MC KEON JAMES PATRICK
MICHAEL THOMAS
MILLER WILLIAM MICHAEL
MILOT LARRY JOSEPH
MORRISON PETER WHITCOMB
MORRISSEY THOMAS J JR
MOURTGIS ARTHUR C JR
MROCZYNSKI RAYMOND CHARLE
MULLEAVEY QUINTEN EMILE
MURZIN WALTER ALECK
MUZZEY CHARLES EDMOND
NADEAU PAUL ERNEST
NADEAU THOMAS DENNIS
NUTE LEONARD KING
O NEIL WILLIAM WAYNE
O NEILL THOMAS PHILIP
OLSON RONALD LEON
PAGE ALBERT LINWOOD JR
PARADIS RAYMOND LOUIS
PAUL ERNEST GEORGE
PEARSON WILLIAM ROY
PELLETIER RICHARD WILLIAM
PERREAULT DAVID B
PHILBRICK STEVEN JAY
PILLSBURY JERRY DEAN

WE GIVE TO YOU OUR FINAL GIFT

IN MEMORY OF THOSE NEW HAMPSHIRE VIETNAM VETERANS WHO GAVE THEIR LIVES OR ARE STILL MISSING 1959 - 1975

VETERANS HELPING VETERANS

PLOURDE ROBERT JAMES
PORTER KARL DENNIS
PORTER RICHARD CHARLES
PRATT PHILIP AVERY
PROVENCHER WAYNE THOMAS
RAYMOND LAWRENCE ROBERT
RAYMOND RICHARD PAUL
RAYNO JOSEPH ANDREW
RIBICH MICHAEL P
RIORDAN GEORGE WILLIAM
RIVARD RICHARD NORMAN
ROACH RONALD D
ROBERGE EDMUND EDWARD
ROBICHAUD ROGER EDWARD
ROBILLARD WILFRED ROLAND
ROBINSON JOSEPH ROBERT
ROENTSCH ROBERT QUENTIN
RUNNELLS EVERETT PORTER
RUSS ALFRED BAYARD
SALTMARSH THOMAS JOHN
SANDFORD BRADLEY ELLIOTT
SANTY STEVEN CRAIG
SANVILLE ERNEST EUGENE
SAUNDERS MICHAEL JORN
SAWYER JAMES EVERETT JR
SAWYER ROBERT WILLIAM
SCHUNEMANN JAMES EDWARD
SCIBILIA ROBERT PETER

SENECHEK JOHN
SEYBOLD GERALD CALVIN
SHAREK FRANK JOSEPH JR
SHAW ROBERT ERNEST
SMITH GARY ROY
SMITH MURRAY LAWRENCE
SOUTHER DOUGLAS S JR
STANLEY RAYMOND ERNEST
STEER JOHN CLIFTON
STEVENS HAROLD KENNETH JR
STICKNEY PHILLIP JOSEPH
STOVER DOUGLAS EARL
SUCCI MICHAEL LAWRENCE
SULLIVAN ALLAN FRANCIS
SULLIVAN ROBERT JOSEPH
SULLIVAN TERRENCE COLIN
SWEENEY MICHAEL BERNARD
SWEET EUGENE FREDERICK JR
SYLVESTRE ARMAND ALVIN
TAGGART WINSTON ADAMS
TESSIER LUCIEN CHARLES
THERIAULT SAMUEL SILVER
TITCOMB ROBERT PAUL
TOWLE GARY CHESTER
TOWNSEND JOHN A JR
TRUDEAU RAYMOND L
VATISTAS DENNIS NICK
VEZEAU THOMAS JOSEPH
VIEL ALFRED
VILLIARD JOSEPH GEORGE
WAINIO ALEXANDER GEORGE

WALKER JOHN RAYMOND
WELLER DAVID HOWARD
WHIPPLE GARY NORMAN
WHITE ROBERT FREDERICK
WHITTAKER TERRY JAMES
WHITTICOM JONATHAN CHARLE
WILKOWSKY WILLIAM JR
WILLARD THOMAS ALAN
WILLEY ALDEN BERTRAM
WRYE BLAIR CHARLTON

DEDICATED TO
ALL VETERANS
PAST AND PRESENT

ARMY
AIR FORCE
COAST GUARD
MARINES
NAVY

Photos by Matt Ymbras

New Jersey Vietnam Veterans Memorial

1 Memorial Ln, Holmdel, NJ 07733

The New Jersey Vietnam Veteran's Memorial is located in Holmdel, New Jersey at the PNC Bank Arts Center. It is not only a memorial, it's also a museum that attracts visitors from all walks of life. The museum holds exhibits coinciding with the memorial, including timelines of events that happened in both the US and Vietnam during the years of the Vietnam War and an exhibit of letters and photos sent home by veterans during the conflict.

The memorial, like other monuments, was created as a way for the state to honor the local men and women who served bravely in the war. The memorial, which is outside of the museum, is accessible 24 hours a day, seven days a week with no fee.

The Vietnam War memorial was complete on May 7, 1995, and was dedicated around the same time. It's conveniently located on the grounds of the PNC Bank Arts Center (formerly known as the Garden State Arts Center). It was designed by a refugee from South Vietnam named Hien Nguyen.

The construction was made possible thanks to a collaborative effort of individual donors, businesses, civic organizations, multiple New Jersey counties, and veterans' groups. The cost of the whole project was estimated to cost around $5 million and was made possible through donations of materials and labor by generous private contractors. To offset the cost of the memorial, people were also able to apply their donation to either the cost of paving stones for the walkway or to the cost of one of the black granite wall panels.

The memorial is expansive (it takes up 200 feet in diameter) and consists of different components. There is a retired 1964 Huey helicopter complete with two mannequins inside as well as functioning lights and speakers that play helicopter sounds. Surrounding the area is a granite wall made up of 366 panels in order to symbolize each day of the year. The wall features the names of local service members who passed on each individual day. In the middle of the open-air pavilion are statues honoring the veterans including the nurses of the Vietnam War. The sculpture made by Thomas J Warren consists of a wounded soldier reaching out for another soldier while a nurse tends to his wounds. This memorial hosts thousands of visitors per year including class trips, and a great turnout for the Memorial and Veterans day services.

The Names of those from the state of New Jersey Who Made the Ultimate Sacrifice

ABARA JOSE GENE
ABBATEMARCO JOHN BENJAMIN
ABENE CHARLES FREDERICK
ABERNATHY DANIEL OWEN
ABRAMOFF ARTHUR JOHN
ABRAMS LEWIS HERBERT
ACKERMAN DAVID FOLEY
ACKERMAN THOMAS ALAN
ADAMS PHILLIP CURTIS
ADDICE FRANK PAUL
ADRIAN JOSEPH DANIEL
ALAMO GABRIEL RALPH
ALBANESE ROBERT
ALBERT DAVID
ALBERTS FRANCIS JOHN
ALBIETZ RAYMOND PETER
ALEXANDER CALVIN EUGENE
ALEXANDER ELEANOR GRACE
ALFONSO JOHN
ALI ARFIEN CLIFFORD
ALLAWAY DONALD
ALLEN GARY JOHN
ALLEN ROBERT SAMUEL
ALLEN ROY RANSOM
ALVES MOSES LOPES
AMBROSE LOUIS ALLEN
AMEJKA JOSEPH EDWARD
AMENDOLA JAMES JOSEPH
ANASIEWICZ RICHARD JOSEPH
ANDERSON MARCUS PETER
ANDERSON STEVE
ANDREASEN ROBERT WAYNE
ANDRISANO FRANK JR
ANDUJAR CHARLES MANUEL
APONTE EDWIN
APPLEGATE ROSS
ARMENTO FRANKLIN CHARLES
ARNTZ WILLARD LEE
ARRIBI DONALD
ASHFORD HOWARD HERRELL
ASHNAULT RAYMOND JOHN
ASMUTH ROBERT LABUDDE JR
ASSELTA CHARLES CARL
ATKINS DAVID BRUCE
ATKINS JOHN
ATKINSON FRANKLIN G JR
ATTARIAN ALAN
AYRES WILLIAM FRANCIS
BABBITT WALTER LEE JR
BACH COLIN JAMES
BACHMAN ROGER JOSEPH
BACKES BRUCE RICHARD
BACON PAUL DAVID
BACON ROBERT FRANKLIN
BADAVAS THOMAS EDWARD
BADER ARTHUR EDWARD JR
BAGLEY DENNIS
BAKER GEORGE ARTHUR
BAKER JON ALLEN
BALDWIN MICHAEL RICHARD
BALL ROBERT LEE
BALMER ROBERT OLIVER
BAMBRICK RICHARD GEORGE
BARANOSKI JOHN FRANK
BARDET RAYMOND FREDERIC
BARKER JEDH COLBY
BARKER JEFFREY LAWRENCE
BARNES ALFRED
BARNES LAWRENCE MERRIDITH
BARNHART JACK ADRIAN
BARRIOS BERNARD
BARSCH JOHN PAUL
BARTON JAMES JOHN
BASHAW DAVID
BASS SEYMOUR R
BASTIAN MICHAEL FRANCIS
BATES GLEN DOUGLAS
BATES RICHARD STANLEY
BATES RONALD JOSEPH
BATTEL ANTHONY BRIAN
BATTISTA FRANCIS DUANE
BAUER ALFRED

BAUMANN LUDWIG GEORGE
BAUMANN OTTO WILLIAM JR
BAUSCH DAVID ALAN
BAXTER DENNIS WARREN
BEATTIE DAVID ROWLAND
BEAUMONT HERBERT MICHAEL
BEDROCK ALAN
BEERES GEORGE KEVIN
BEKIEMPIS THOMAS CHESTER
BEKSI WILLIAM JOSEPH
BELCHER FRED ARTHUR
BELICOSE RICHARD J
BELL LARRY GENE
BELL LEON EARL
BELL WILLIAM BRENT
BENN PHILIP CRAIG
BENN WILLIAM PAUL
BENNETT ROBERT ELWOOD III
BERENWICK WILLIAM MICHAEL
BERG GEORGE PHILLIP
BERG JOHN STEPHEN
BERKERY MICHAEL WAYNE
BESCHEN JAMES
BETANCOURT-MOJICA CARLOS
BETZ ROBERT JOSEPH
BEZEGA MICHAEL STEPHEN
BIBBS WARREN LARRY
BIDDLE JOSEPH LENORD
BIERLEIN PATRICK M R
BIESANTZ HOWARD STANLEY
BIESIADA RICHARD EDWARD
BIGHAM CHARLES FREDERICK
BILENSKI JOHN CHARLES
BILLERO MICHAEL JAMES JR
BILLINGHAM FREDERICK A JR
BINGENHEIMER JAMES
BINGER GERALD A
BIRCH THOMAS H
BIRD JOHN THOMAS
BLEVINS THOMAS LEE JR
BOCHE GARY ALLEN
BOLAND JAMES ROBERT
BOLTZ RICHARD LEONARD
BOND RONALD LESLIE
BONINE THOMAS MARVIN
BONNER FREDERICK NEIL
BORDEN TIMOTHY ZANE
BOROSS LASZLO JR
BORREGO ANTHONY J
BOSKO MICHAEL JOHN JR
BOWEN LARRY HANSEN
BOWMAN HARRY THOMAS II
BOYCE JAMES FRANKLIN
BOYD DAVID LEROY
BOYD SAMUEL JR
BOYDEN THOMAS ROBERT
BOYE HENRY JOSEPH JR
BRADMAN JOHN FREDRIC
BRADY EDWARD FRANCIS III
BRADY ROBERT JAMES
BRAGUE EDWIN STEVEN JR
BRANCH WILLIAM ANDERSON
BRANIN MICHAEL FRANCIS JR
BRANNON HARRY G
BRASWELL JAMES HILLIARD
BRAUNER HENRY PAUL
BRAYBROOKE CHRISTOPHER
BREEN GERALD JAMES
BRELLENTHIN MICHAEL JOHN
BREMS PATRICK JOHN
BRENNAN JOHN PATRICK
BRENNER RICHARD IRVING
BRERETON RAYMOND JAMES
BRETSCHNEIDER HANS KARL
BREWSTER GLENN RICHARD
BRICE WILLIAM FRANCIS JR
BRIDGES LESTER
BRINCKMANN ROBERT EDWIN
BRITTEN ROGER GEORGE
BROCKMANN ROBERT JAMES
BROOKS DAVID WILLIAM

BROPHY DENNIS JAMES
BROWER DONALD HARRY
BROWN CHARLES PAUL
BROWN GERALD FRANCIS
BROWN JEFFREY JOSEPH
BROWN RICHARD JAMES
BROWN ROGER CLINTON
BROWN TYRONE
BRUCH DONALD WILLIAM JR
BRUNN RICHARD CONRAD
BRUNNOW RICHARD ALBERT
BRYANT DAVID THEODORE
BRYANT SOLOMON HERBERT
BRYDUN BOHDAN PETER
BUCIOR ANDREW ZBIGNIEW
BUCK FRANK HENRY
BUCK PAUL JOHN
BUCZYNSKI GREGORY THOMAS
BUKOWSKI RONALD
BULLWINKEL ALDEN JOHN
BULMER ROBERT ARTHUR
BURD GEORGE JAMES
BURGANS RICHARD
BURKE WILLIAM GREGORY
BURNHAM JOSEPH FRANCIS
BURNS CHARLES STUART III
BURR STEWART SAMUEL
BURROUGHS ROBERT JAMES
BURSIS JOSEPH THOMAS JR
BURTON WILLIAM RUSSELL JR
BUSCH JOHN EDWARD
BUSCH THOMAS LEOPOLD III
BUSHEY FRANK HARRY
BUTLER GREGORY WILLIAM
BUTTENBAUM GARY RICHARD
BYRNE JEFFREY R
BYRNE JOHN PATRICK
CALLAHAN MICHAEL PATRICK
CALLAN GEORGE ALLAN
CAMA DENNIS ROCCO
CAMPBELL PATRICK FRANCIS
CAMPBELL RANDALL M III
CAMPEAU FRANCIS
CANCELLIERE FRANK ANTHONY
CANNITO DENNIS JOHN
CAPORALE MICHAEL JOSEPH
CAPPARELLI GEORGE GUY
CARDEN CHARLIE ALFRED
CARLOUGH GEORGE GERALD
CARLSON RICHARD LEE
CARLTON RANDALL MARK
CARNEY GEORGE AUSTIN
CAROVILLANO ROBERT
CARTER ROBERT NEL
CARTWRIGHT JOHN STANBOROU

CASEY ROBERT MICHAEL
CASTALDI JAMES
CATLING ROBERT PHILIP
CEMELLI SALVATORE PETER
CERRATO NICHOLAS FRANK
CERVERA MICHAEL BERNARD
CHAMBERS HILLMAN GLEN
CHAPMAN SHERMAN JR
CHARD SALUM EDWARD JR
CHEEKS JOHN HERBERT
CHEESEMAN ALAN B
CHRISTENSON WILLIAM B
CHRISTIANSEN BERNHARD M
CHRISTIE DONALD
CHRISTMAS PAUL
CHRUPCALA WALTER JOHN
CHURCH ROBERT EDWARD
CHWAN MICHAEL DANIEL
CIALLELLA JOHN WILLIAM
CICHON WALTER ALAN
CLARK BARRY EDWIN
CLARK BRADLEY ELLIS
CLARKE CLIFFORD LESLI III
CLARKEN THOMAS HENRY III
CLAYTON BRIAN DOUGLAS
CLAYTON GEORGE DONALD
CLEMENTS WAYNE DOUGLAS
COFFARO ANTHONY CHARLES
COFRANCESCO LOUIS J JR
COHEN SIDNEY
COINER CHARLES FREDRICK
COLANTUONO WAYNE ALBERT
COLASURDO JOSEPH PETER
COLEMAN GEORGE WILLIAM
COLES ALEXANDER JR
COLES GEORGE EUGENE JR
COLES VINCENT SAMSON
COLEY BRUCE EDWARD
COLL DENNIS JOSEPH
COLL JOHN THOMAS JR
COLLINS THEOTHIS
COLON-SANTOS RAFAEL
COMLY WILLIAM ALVIN
CONE REGINALD LOUIS
CONLAN BRIAN DALY JR
CONNER GERALD WILLIAM
CONNOR PETER SPENCER
CONRAD HARRY FLOYD
CONSTANTINO CLIFFORD JOHN
COPPEDGE LAWRENCE
CORBIN ANDREW PHILLIP
CORBIN DONALD LEE
CORCORAN KEVIN
CORCORAN RICHARD FRANCIS
CORNISH RUSSELL HUBARD
CORREA ANGEL MERESI
COSGROVE CHESTER

COSTA MARIO
COVINGTON LAWRENCE CORNEL
COWEN CHRISTOPHER
COYLE GARRY
COYLE HUGH
COYLE JAMES MICHAEL
COYLE JOHN
CRAIG EDWARD JOSEPH
CRANE DENNIS
CRAWFORD CHARLES J JR
CRESSMAN PETER RICHARD
CREWS JOHN DIVINE JR
CRIKELAIR JOHN FRANCIS
CRITCHFIELD WILLIAM ROBER
CRITELLI ALFRED JOSEPH
CROUTER ROBERT
CROWELL ROGER BRIAN
CRUDEN DONALD JOSEPH
CUCCINELLI ROBERT ALVANDR
CUMMINGS DANIEL TERRY
CURRY GEORGE DEVER
CURTIN JOHN III
CUSTODE RALPH
CYGON STANLEY JOHN
CYRAN RICHARD EDWARD
CZARNOTA CHRISTOPHER ZENO
D ADAMO ALBERT L JR
D ADAMO JOHN JR
DA PONTE ANTHONY
DA SILVA HELDER ARTHUR C
DABONKA JOHN ANTHONY
DALEY DANIEL WILLIAM
DALEY GERALD CHARLES
DALEY MICHAEL JAMES
DALTON JAMES ALBERT
DALTON JOHN
DALY TIMOTHY
DANBERRY CHARLES LABAW
DANCHETZ LESTER
DANDO THOMAS J
DANDURAND JAY THOMAS
DANIELS CHRISTOPHER MICHA
DANIELS EDWARD STEVEN
DANIELS JOSHUA MARICE
DANNA JOSEPH JOHN JR
DANOWSKI THOMAS GEORGE
DAVERN MATTHEW JOHN
DAVIS CHARLES HENRY
DAVIS HARLAND M JR
DAVIS RICHARD WAYNE
DAVIS ROBERT CHARLES
DAVIS WILLIAM SHELDON III
DAWSON STEVEN JAMES
DE ANGELIS ADAMO ERMINO
DE CROSTA JOSEPH FRANCIS

New Jersey

DE GARMO GORDON EARL
DE HAAS PETER
DE JESSA JOSEPH CARMINE
DE JESUS-COLON JOSE CELSO
DE LANGE FREDERIC R
DE LORENZO RONALD
DE LUCA GEORGE ABRAHAM
DE MAGNIN MICHAEL ANDRE R
DE MERCURIO ROCCO JERRY
DE MORE KENNETH EDWARD JR
DE RIGGI ANTHONY
DE ROSE GERALD LOUIS
DE VORE CRAIG JESSE
DEAL WILLIAM LEANDER
DEGE RAYMOND WILLIAM III
DEGENAARS BRADLEY RICHARD
DEITMAN EDWARD
DEL GUIDICE GREGORY
DELASANDRO DENNIS FRANCIS
DELIKAT EDWARD JOHN JR
DELL ARENA RICHARD M
DEMATTIO MARIO FRANK
DEMBOSKI STANLEY T
DEMSEY WALTER EDWARD JR
DEPAUL MICHAEL JOSEPH S
DERBYSHIRE JAMES WILBERT
DESIMONE ALFRED
DEVLIN THOMAS ROGER
DI ANTONIO MARTIN M JR
DI CAVALLUCCI VICTOR
DI NAPOLI JOHN JR
DIANI FRANCO
DICESARE ANTHONY JR
DICKERSON DOUGLAS R JR
DIDURYK MYRON
DILLARD HAROLD JEROME
DILLON FRANCIS THOMAS
DINAN DAVID THOMAS III
DIXON CECIL F
DOBBINS ISAIAH ANTHONY
DODSON ROBERT GERALD
DOHERTY JOHN WILLIAM
DOLAN JIMMY MICHAEL
DONATIELLO JERRY RICHARD
DONOVAN JOHN DENNIS
DORAN JAMES DONALD
DORIO JOHN WILLIAM ALLEN
DORN PHILIP KENNETH
DOTY WESLEY GEORGE
DOUGHERTY ROBERT JOSEPH
DOUGHTY ROBERT THOMAS
DOWD JOHN ALOYSIUS
DOWLING WILLIE JR
DRAKE DONALD WILLIAM
DREWES RICHARD CHARLES
DROZDZ STANISLAW JOSEPH
DUDASH JOHN FRANCIS
DUFFY JOHN
DUGAN JOHN FRANCIS
DUTCHES WILLIAM GEORGE
DWYER MATTHEW MURICE JR
EALEY DOUGLAS
EBERHARDT WILLIAM HENRY
EDLEY GEORGE STEVEN
EGAN JAMES THOMAS JR
EGGENBERGER WILLIAM GARY
EHRLICH DENNIS MICHAEL

EISTER WILLIAM
ELFENBEIN ERNIE JON
ELICHKO DEAN JOSEPH
ELLIOTT ROBERT WILLIAM
ELLIS CHARLES PAUL
ELLIS HARRY JOSEPH III
ELLISON JASPER JR
ELMAN DAVID HERBERT
ENGEDAL JOHN
ENGLE RUSSEL WARREN
EPIFANIO NEAL DAVID
EPPINGER GEORGE
ESPOSITO JAMES MICHAEL
ESPY JOHNNIE BEE
ESTERGREN JAMES HOWARD
ETTZ MICHAEL CHARLES
EUCKER FRANKLIN CHARLES
EVANS HAYDN
EVANS SAMUEL JAMES
EVERETT NORMAN ROY
EVERT BARRY EDWARD
FAITH WALTER DANIEL
FALATO JOSEPH ANTHONY
FALLON MICHAEL JAMES
FANNING EDWARD CHARLES
FARAWELL GEORGE THOMAS
FARMER MICHAEL LEE
FEATHERSTONE RICHARD ALLI
FEDOR ANDREW
FEENEY JOSEPH MICHAEL
FEISTNER STEPHEN ELY
FELDER JESSE CLARANCE
FELVER GALE HERBERT
FERGUSON KEVIN LEE
FERRELLI ROBERT THOMAS
FERRUGGIA RICHARD GEORGE
FESKEN WILLIAM
FETT DENNIS JAMES
FIDUCIOSO STEPHANO JAMES
FIELD LEON ROY
FINNERTY FRANCIS M JR
FIRTH THOMAS ELWOOD
FISCHER ROBERT PHILIP
FLACK REGINALD
FLAHERTY PAUL JAMES
FOLEY JOHN JOSEPH III
FOLGER JOHN VINCENT
FONT MANUEL LOUIS
FORD DOUGLAS OAKLEY
FORD RICHARD EDWARD
FORE ALEXANDER
FORMICA GARY PETER
FORSMAN JAMES ESKEL
FOSTER CARL
FOSTER STEVEN JOEL
FOULKS CHARLES JR
FOXWORTH ARTHUR
FRAMBES JOHN MALHON
FRANCIS JOSEPH WILLIAM JR
FRANCIS THOMAS EARL
FRANCISCO JAMES LEONARD
FRANCISCO WILLIAM JR
FRANKE WILLIAM THOMAS
FRECH THOMAS WILLIAM
FREED DAVID BRUCE
FREITAG DIETER KUNO
FRITZ RAYMOND WILBERT JR
FROLICH LESLIE JAMES
FRYAR BRUCE CARLTON
FUNICELLI ERNEST D JR
GA NUN PAUL HUNTINGTON
GABRIEL VINCENT JAMES JR
GABURO GEORGE W
GADDA ANTHONY JOSEPH JR
GAINES JAMES JR
GALLAGHER JOHN MICHAEL
GANDIL ROBERT PATRICK
GARRIDO ROBERT JACOB
GARRISON RUSSELL G
GASKO ROBERT JOHN JR
GASPARD CLAUDE JOSEPH JR
GATTI DENNIS JOSEPH
GEBHART DONALD WILLIS
GEIB ALLEN
GENOVESE CARMINE VINCENT
GEOGHEGAN GERALD DALY
GERWATOWSKI JOSEPH
GIACOBBE ANGELO
GIEGEL JAMES LLOYD
GILCH JAMES XAVIER
GILCHRIST RICKY DEAN

GILES FRANKLIN N JR
GILLIES ROBERT KNELL
GILRAY ROBERT BRUCE JR
GIORDANO DANIEL J III
GIRTANNER JULES T
GLANVILLE JOHN TURNER JR
GLAWSON GEORGE HOWARD JR
GLEASON DENNIS STEPHEN
GLOVER MANZIE JR
GLYNN AARON GEORGE
GOBLE NORMAN ROBERT
GOCZAL FREDERICK
GODFREY CHARLES FAIR-CHILD
GODFREY WILLIAM ROLLAND
GOFF FLOYD HAROLD
GOINES ROBERT
GOLDBERG HOWARD STANLEY
GOLDBERG JOSEPH A
GOLDSBORO STEVEN MICHAEL
GOLEMBSKI PAUL JOSEPH
GOLON WAYNE LEONARD
GONDER KENNETH WALTER
GONZALEZ FRANCISCO HERNAN
GOODING WILLIAM PHILIP
GORDON JAMES THOMAS
GOTTHARDT ROBERT WILLIAM
GRABOWSKI JAN JOSEPH JR
GRAF ALBERT STEPHEN
GRAHAM BARRY FRANCIS
GRAHAM DAVID TIESON JR
GRANGER WILLIE EARL
GRANT WAYNE AUGUSTUS

GRASSIA JOSEPH JR
GRAU ANTONIO AMBROSIO
GRAY EDWARD JAMES
GRAY ROBERT LEE
GREEN DOUGLAS BARTON III
GREEN LEO FRANK JR
GREEN OTIS
GREENE JOHN PRESTON
GREENE KENNETH JOHN
GREENE KEVIN LESLIE
GREENE LLOYD VINCENT
GREENSPAN RICHARD
GREGORY DAVID
GRIM MALCOLM JONATHAN
GRIMSTAD SIGARD RICHARD
GRINER JOHN ARTHUR
GRISARD JOHN ROBERT
GRIX THOMAS E
GROHMAN JOHN JOSEPH
GROSS VICTOR MAHLON
GROVE LOUIS CANCIAN
GROVER THOMAS ROY
GRUCA PETER ALAN
GUARINO SALVATORE
GUENTHER THOMAS ANDREW
GUERRA DARIO DAVID
GUNDAKER FRANK JOSEPH
GUNSTER DAVID JAMES
GURDCILANI BORIS WALTER
HAARWALDT ERWIN JOHN
HAAS LEON FREDERICK
HABER CHARLES HARRY JR
HADDOCK EDWARD
HADLEY STEPHEN JAMES
HAGELSTEIN JAMES DAVID
HAHN JEFFREY CHARLES
HAINES JOHN CHARLES JR

HALEY JOHN MATTHEW JR
HALLADAY JOHN ANTHONY
HALPIN MICHAEL PATRICK
HALSEY Macdonald Brooke
HALVORSEN DONALD KELCEY
HAMACHER WILLIAM BERNARD
HANCOCK JOHN ALBERT
HANDERHAN PAUL WAYNE
HANLON JAMES PAUL
HANNIGAN UDO
HANSON ROBERT
HARBIENKO ANDREW
HARDIN WILLIAM RICHARD
HARGROVE DALE VERNON
HARRIS LANTIE LAWRENCE JR
HARRIS WALTER
HARRISON HERMAN CLYDE JR
HARTMAN RICHARD DANNER
HASHAGEN WILLIAM LOUIS
HAUSER ROBERT CHARLES
HAVER DALE HARRY
HAYES FRANCIS JOSEPH JR
HAYES JEREMIAH MICHAEL JR
HAYES JOSEPH FRED
HAYES MICHAEL JOHN JR
HAYNES BARTON EDWARD
HAYWARD ARNOLD COURTNEY
HAYWARD PHILLIP BRUCE
HEALEY ROBERT CHARLES JR
HEATHCOTE CLIFFORD S JR
HECK RONALD DAVID
HEFFRON JAMES BROOKS
HEGGAN DONALD ERNEST
HEIL BRUCE HUPPERT
HEIN ANTHONY
HEINZE KELLY KARL

HENASEY HAROLD
HENDRICKS EUGENE WILLIAM
HENRY JOHN PATRICK
HERMANSON GARRY W
HERNANDEZ-PENA AUDELIZ
HERROLD NED RAYMOND
HERRON ROCKWELL SELDEN
HESS PHILLIP HOWARD
HESSON JOSEPH LEONARD III
HETZEL NORMAN RALPH
HEYMACH HAROLD FRANK
HEZEL KARL D
HICKS JOSEPH LONNIE
HICKS LEROY
HIGDON DAVID DARRIN JR
HIGH LARRY JAMES
HILDEBRANDT DANIEL ALAN
HILL EUGENE JOHN JR
HILL JOHN CHARLES
HILL TYRONE
HIMMELREICH HARRY EDWARD
HINSON ALVIN CRAWFORD
HIPPIE BRADFORD JOHN F
HOAR JOHN MICHAEL
HOCKNELL HENRY ROBERT JR
HOFFMANN ROBERT JAMES
HOFFMANN THOMAS MARTIN
HOLDEN THOMAS JAMES
HOLJES FREDERICK Y
HOLLAND CHARLES JAMES
HOLLAND WILLIAM L JR
HOLLOWAY JAMES OWENS JR
HOLMES HAROLD
HOOPER VINS RONALD
HOPKINS MARION MARSHALL
HORNBY THOMAS FRANK
HORNER WALTER DENNIS
HORVATH ANDREW
HOSKING CHARLES ERNEST JR
HOVANEC DONALD FRANCIS
HUBBARD WILLIAM HOBSON
HUBBS DONALD RICHARD
HUGHES ROBERT
HUGHES TONY HOWARD
HUMES FRANK WILLIAM
HUMPHREY KEVIN RICHARD
HUNTER DONALD LEE
HUSTER ROBERT RICHARD
HUTCHINSON GEORGE ROBERT
HUTTON WALTER WESLEY
HUYLER WILLIAM D JR

HYMAN LINWOOD EARL
IANDOLI DONALD
IANNUZZI CHARLES EARNEST
IASELLO DENNIS ANTHONY
IHNAT MICHAEL JOHN
IKE THOMAS ROBERT
INFERRERA LOUIS JOSEPH
IVAN ANDREW JR
IVES RICHARD V
JACKSON HARRY JOHN JR
JACKSON JOSEPH LOUIS
JACKSON KEITH MICHAEL
JACKSON ROBERT LEONARD
JACKSON WILLIAM
JACOBS DEL RAY
JACOBS JEROME EDWARD
JACOBS VINCENT LAWRENCE
JACOBSON JON CHRISTOPHER
JACOBUS WILLIAM THOMAS
JAMES DUTLEY
JAMES HENRY
JAMES JESSE JR
JAMES RODNEY ALVIN
JANOWITZ ROBERT LAWRENCE
JARMOLINSKI CHESTER JR
JENKINS CLIFFORD JR
JENKINS LANCE NORMAN
JENNINGS MICHAEL
JENSEN GARY EDWARD
JERVIS JOHN LEROY III
JOHNSON ANTHONY KENT
JOHNSON CHARLES EDWARD
JOHNSON DONALD LEE
JOHNSON EDWARD BRUCE
JOHNSON GUY FREDERICK
JOHNSON HAVART EARL
JOHNSON HOWARD WARNER JR
JOHNSON JAMES ALLEN
JOHNSON LESTER JR
JOHNSON RALPH EDWARD
JOHNSON RICHARD
JOHNSON ROBERT IRVIN
JOHNSON SYLVESTER
JOHNSON WILLIAM THOMAS
JONES CLIFFORD JR
JONES GARLAND
JONES JOHN LEWIS
JONES ROY MORGAN JR
JONES STEPHEN CRAWFORD
JONES THOMAS HUBERT
JORDAN ALLAN H
JORDAN ARTHUR
JORDAN KENT DOUGLAS

JOYNES FRANK DENNIS JR
JUDGE CHARLES MARK JR
JURANIC FRANCIS JOSEPH JR
JURSZA WILLIAM JR
KALIVAS JOHN ANGELO
KAMINSKI EDWARD J
KANE RICHARD RAYMOND
KAPELUCK JOHN MICHAEL
KAUS WLADISLAW
KEARNS BRENDAN JOHN
KEARNS JOSEPH THOMAS JR
KEELER WILLIAM GILBERT
KEEN ARTHUR
KEENAN ROBERT JAMES
KEIN ROBERT JOSEPH
KELLER FRANCIS JOSEPH
KELLER JOSEPH JOHN JR
KELLER LEONARD
KELLY CHARLES WESLEY
KENNARD JAMES HORACE
KENNEDY JAMES EDWARD
KENNEY EDWARD
KERBL FRANK RONALD
KERI ROBERT CHARLES
KERNAHAN GREGORY P JR
KIATKIN NIKOLAI
KIERNAN JOSEPH M JR
KILROY MICHAEL WINSTON
KIMBALL WILLIAM B JR
KINSLER FREDERICK C JR
KIRKBY JAMES KENT
KIRSCHNER STEPHEN BENJAMI
KISIELEWSKI JOHN WILLIAM
KISSAM EDWARD KNELL JR
KLANIECKI EDWARD MATTHEW
KLECZ STANLEY STEPHEN
KLEIN DENNIS W
KLEIN JOSEPH
KLINE DENNIS
KLINGAMAN BRUCE DAVID
KLOSSEK GERALD
KNAPP HERMAN LUDWIG
KNAUS JOHN RICHARD
KNOSKY RONALD WAYNE
KNUCKEY THOMAS WILLIAM
KOCH KENNETH JOHN
KOCHER LAWRENCE HENRY
KOHLMYER FRANK JOSEPH
KOMMENDANT AADO
KONYU WILLIAM MICHAEL
KOOB JOHN PETER
KOONCE JEFFREY WAYNE
KOPCINSKI STANLEY JOHN

KORONA ALBERT III
KOVACH PETER FRANK
KOWAL BOHDAN
KOWALESKI GREGORY STANLEY
KOZAK DAVID MICHAEL
KRAMER LEON JOSEPH
KRAVCHAK MICHAEL STEVEN
KROSKE HAROLD WILLIAM JR
KRUEGER JOHN KENNETH
KRUGER ROBERT HENRY JR
KRUPINSKI FREDERICK JOSEP
KUBISKY EDWARD
KUGELMANN ROBERT CHARLES
KUKOWSKI THOMAS
KULACZKOWSKI LESZEK A
KULBATSKI FRANCIS KENNETH
KURTZ CHARLES JOHN
KUSPIEL KENNETH EDWARD
KUTKOWSKI GREGORY MITCHEL
KYLE THOMAS ROBERT JR
LA DUKE JOHN HENRY
LA FASO JOSEPH STEPHEN
LAIRD JERRY PROCTOR
LAMANNA JOHN MICHAEL
LAMBERSON CARL EDWARD JR
LAMON WILLIAM CHARLES JR
LANCE ALFRED FRANK
LAND SYLVESTER
LANG CHARLES VANDERBILT
LANG JAMES L
LANGFORD ALVIN HUGH
LANNING HAROLD JAY
LANZONE MARCHELLA RAYMOND
LATOURETTE PAUL E
LAUER JOSEPH EDWARD
LAVELLE JOHN JOSEPH
LAW EUGENE
LAWLESS THOMAS ALOYSIUS
LAWLOR PATRICK EUGENE
LAWRENCE BRUCE EDWARD
LAWRENCE MICHAEL JAMES
LAWSON ALBERT C
LAWSON BIRDEN JEROME
LAYTON ROBERT ALLEN JR
LAZARO ROBERT JAMES
LE BLANC FRED JOSEPH
LE DONNE LAWRENCE JOSEPH
LEARY JOHN DENNIS
LEBRON LUIS ANGEL

LEE BOBBY EUGENE
LEEDS CLYDE A
LEGETTE O'NEAL
LEHEW DONALD LEE
LENZSCH ROLF FRED
LEONARDIS STEPHEN WILLIAM
LETA DONALD
LEVERING EDWIN HARRY
LEWIS BENJAMIN F JR
LEWIS RICHARD GARY
LIGHT JERRY CLIFTON
LILLEY THOMAS EDWARD
LINDABERRY JOHN LANCE
LINDSLEY DONALD PETER
LITTLE WILLIAM F III
LITTLEHALES ROY CHARLES
LIVELY WARREN II
LOATMAN RODNEY ELLIS
LOFGREN JAMES ESKEL
LOLLIS CHARLES W
LONG GEORGE FRANCIS
LONG RICHARD LYTLE
LOPEZ JOSE LIUS
LOPINTO FRANK THOMAS
LOVE FREDERICK EUGENE
LOWDEN THOMAS ALLEN
LUBONSKI LAWRENCE FRANK
LUMPKIN GARY
LUNAPIENA NATHAN CHARLES
LUPU JOHN WILLIAM
LUTTGENS JAMES
LUTZ WERNER ERHARD
LYONS FRANK ELLIS
LYONS GEORGE MICHAEL
LYONS WILLIAM MICHAEL III
LYTAL JAMES FRANCIS
MAC MANUS COLIN DAVID
MAC VEAN STEPHEN SHERWOOD
MACARELL MICHAEL JOSEPH
MACK FRANCIS WILLIAM
MACZULSKI WACLAW JOZEF
MADDEN FRANCIS BERNARD JR
MADDEN PAUL BERNARD
MAGLIARO CHARLES LOUIS
MAGNUSON ERIC C JR
MAGUIRE WILLIAM A JR
MAHER EDWARD MICHAEL JR
MAHURTER LAWRENCE WILLIAM
MAIURO JOSEPH
MAJESKI MICHAEL THOMAS

155

New Jersey

MALLON THOMAS JOHN
MANGANELLO ANTHONY JR
MANIAS ROBERT JAMES
MANNERS VAN DYKE WILLIAM
MANTHEY BARRY ARTHUR
MANUEL ROLAND WILL
MARASON JOHN EDGAR
MARCANTONI ROBERT JOHN
MARCHUT THOMAS ANDREW
MARKOVICH DOUGLAS JOSEPH
MARSH FREDERICK CURTIS
MARSHALL DONALD RICHARD
MARSHALL WILLIE JUNIOR
MARTER EZRA BUDD
MARTIN DENNIS PHILIP
MARTIN JOHN BERNARD II
MARTIN JOHN C
MARTINEAU MICHAEL WILLIAM
MARTORELLA GARY MARIO
MASCARI PHILLIP LOUIS
MASON ALPHONZA
MASON BENJAMIN H JR
MASTEN JAMES ARTHUR
MATHEWS CLAUDE WESLEY
MATHEWS HAROLD JOSEPH JR
MATHEWS JAMES MICHAEL
MATUSCSAK GEORGE EDWARD
MATYAS ANDREW
MAURO VINCENT CARMEN JR
MAUTERER OSCAR
MAXEY EASON JASPER
MAYER FRANCIS JOHN JR
MAYER HOWARD HERCHER
MAYERCIK RONALD MICHAEL
MAYSEY LARRY WAYNE
MAZZILLO PETER JR
MC BRIDE CLAUDE WILLIAM
MC BRIDE PATRICK EUGENE
MC CALL GERALD ANTHONY
MC CALLUM PETER JOHN JR
MC CANTS JOSEPH JR
MC CARTHY JOHN JOSEPH
MC CAULEY DENNIS JAMES
MC CLAIN RICHARD LARRY
MC CLELLAND GEORGE
MC CLOSKEY ROBERT ALLEN
MC COLLUM JAMES PATRICK
MC CONNELL JAMES T III
MC DERMOTT THOMAS ANTHONY
MC DONALD JAMES
MC DONOUGH JOHN RICHARD
MC DOWELL DONALD FRANCIS
MC FADDEN GREGORY WALTER
MC FADYEN BRUCE SEARIGHT
MC GUIRE FRANCIS MICHAEL
MC INTYRE GREGORY
MC KAY GILMAN WILLIAM
MC KENNAN CLIFFORD ABDUL
MC KIM WILLIAM RITCHIE
MC LAUGHLIN THOMAS MICHAE
MC MANUS ROBERT FRANCIS
MC WILLIAMS GEORGE

LINWOO
MEAD PETER FRANCIS
MEADE THOMAS ALLERTON
MEARA WILLIAM DANIELS JR
MEEKER EDWARD HOWARD JR
MEESTER EVERETT JACOB
MEISTER WILLIAM ALFRED
MELADY RICHARD RAPHAEL
MELENDEZ RAFAEL
MELNICK STEWART ARTHUR
MELNYK MIKOLAW
MENA SAMUEL
MENDEZ JOHN WILLIAM
MENTER JEROME
MERLINO CARL STEVEN
MERSCHROD LAWRENCE RICHAR
METTINGER ALBERT ELWOOD
MICHEL ROBERT ALFRED
MICHELS LESTER GEORGE
MICKENS CARL LAWRENCE
MIDUSKI FRANCIS CHARLES
MIKA VICTOR GEORGE
MIKULA EMERY GEORGE
MILAN GEORGE LEONARD
MILES WELDON JOHN
MILEY BRUCE MICHAEL
MILEY EUGENE
MILLAN RICHARD
MILLAR PETER EDMUND
MILLER DONALD
MILLER EDWARD KENNETH
MILLER GEORGE WILLIAM
MILLER ROBERT THEODORE
MILLER STANLEY JOSEPH JR
MILLER THOMAS CRAIG
MILLER WALTER PETER JR
MILLS KARL WILLIAM
MINES JAMES JR
MINNOCK JOSEPH PATRICK
MIONE ANTHONY V
MIRRER ROBERT HENRY
MOBUS JOSEPH PATRICK
MOKUAU KENNETH WILLIAM JR
MOLNAR ALBERT RUSSELL
MOLNAR FRANKIE ZOLY
MONAHON ROBERT EDWARD
MONGILARDI PETER JR
MONGILLO PAUL JOHN
MONROE VINCENT DUNCAN
MOON THEODORE EDWARD
MOORE HERBERT HUBERT
MOORE JEROME
MOORE LEON DAVID
MOORE MANUEL
MOORE ROBERT JOSEPH
MOORE ROBERT JR
MOORER BOBBY
MOORHOUSE WILLIAM CURTIS
MOORMAN FRANK DAVID

MORAN VINCENT
MORGAN BRUCE
MORGAN DONALD THOMAS
MORGAN JERRY JR
MORGAN JOHN D
MORGAN RAINER K
MORGAN ROBERT WEST
MORLEY JOHN JOSEPH JR
MORRIGGI JOSEPH
MORRIS DANIEL EUGENE
MORRIS JAMES ROBERT
MORRIS ROBERT JOHN
MORROW EDWARD CY
MORSE CHARLES FRANCIS JR
MORVAY JON RICHARD
MOSELEY WILLIAM FRANCIS
MOSES DONALD HARVEY
MOUNT JOHN EDWARD
MOURITZEN DONALD ANDREW
MOYE FLOYD
MULLINS WILLIAM F JR
MURNER PETER PATRICK JR
MURPHY ROBERT L
MURPHY TIMOTHY JOHN
MURRAY STEPHEN BRIAN
MUSER LOUIS CHARLES II
MUSSELMAN JOSEPH HENRY
MYERS DANIEL JOHN
MYERS JAMES EDWARD
MYERS THOMAS WAYNE
NASH DAVID ROBERTSON
NASH JAMES ROBERT
NAWROSKY MICHAEL ROBERT
NEAL JOHN HALL JR
NEGRON VICTOR MANUEL
NELSON PAUL VINCENT
NEMCHIK JOHN JOSEPH JR
NEMETH MICHAEL
NESTOR FRANK RODNEY
NEWMAN THOMAS MCKNETT
NEWTON BARRIE MYRON
NICHOLS DANIEL CLEMENT
NICHOLS RICHARD ALLEN
NICHOLS WILLIAM WARD JR
NICKENS CECIL BERNARD
NICKERSON RONALD WILLIAM
NIEDERMEIER ARTHUR ALAN
NOFFORD CLARENCE
NORDMAN ERIC REINHARD
NOVEMBRE CARMINE
NUDENBERG DAVID ALAN
NUGENT JAMES PATRICK
NUGENT RICHARD FRANCIS
O CALLAGHAN MAURICE JOSEP
O CONNELL EUGENE GEORGE
O NEAL MELVIN JR
O SHAUGHNESSY JAMES JOHN
O SHAUGHNESSY JOHN FRANCI

O SHEA WILLIAM II
OBERMEIER GEORGE RICHARD
OCHS TIMOTHY CARL
OGBURN FRANK JR
OHLINGER JAMES
OLESON JOSEPH JR
OLIPHANT JOSEPH B JR
ONEILL DOUGLAS LEE
ORTA RAUL
ORTIZ ROBERT WILSON
ORTIZ-BURGOS JOSE ALBERTO
ORTIZ-CORREDOR LUIS
OSSMEN JOHN DONALD
OSTENFELD OTTO JOHN
OUTWATER ALBERT ALEYA JR
OVAITT RICHARD ARTHUR
OWEN DAVID B
PAARZ GARY FREDRICK
PACKER JOSEPH EVERETT JR
PALL JOHN JOSEPH
PALMA GERARD VINCENT
PAPE JOHN DAVID
PAQUIN PAUL EVERETTE JR
PAREDES ISMAEL JUSON
PARMERTER MICHAEL JAMES
PARTON CARL
PATTERSON JAMES GORDON
PATTERSON JEFFERY SCOTT
PATTERSON RICHARD
PAUL EDWARD JOSEPH
PAUL FRED JOHN
PAVLOCAK MICHAEL PETER JR
PAWLOWSKI EDWARD WESLEY
PEARCE HENRY ELLWOOD II
PERRELLI KEITH FRANCIS
PERRONE JAMES PAUL JR
PERRY LOUIS EDWARD
PETERSON JOHN B JR
PETRACCO ROBERT
PETRICK FRANK EDWARD
PETRONE LOUIS GENE JR
PFEFFERLE WARREN WALTER
PFEUFER MICHAEL ANTHONY
PFROMMER STANLEY DENNIS
PHILHOWER CHARLES ALBERT
PHIPPS LANNY WILLIAM
PIANO RALPH ERNEST JR
PIASCIK MICHAEL
PICARELLI JOSEPH HENRY
PIERCE IRVING CLARENCE JR
PIERSANTI ANTHONY J JR
PIERSON ROBERT EMMETT
PINE FREDERICK ANDREW
PINNELL ROBERT MERRITT JR
PINO ALFRED
PIPPENBACH JOSEPH
PISCIOTTA WAYNE CARLYLE
PIZZI CHARLES DANIEL
PLEASANT WILLIAM ANDREW
PLOTTS RICHARD

POLLARD WILLIAM ISAAC
POLLIN GEORGE JOHN
POLLOCK SEVENTY J
POLONKO JOSEPH JOHN JR
PONTO AUGUSTUS J III
PONTY STEPHEN CHESTER JR
POOR GEORGE ALBERT JR
PORTER RONALD WILLIAM
POST VERNON JR
POTTER ALBERT RAYMOND
POTTER PAUL D
POWELL ELMER FRANKLIN
PREMOCK DENNIS
PRESLEY AVEY
PREZIOSI JOHN PATRICK
PRICE FRANK APPERSON III
PRICE JOHN WILLIAM
PRISET JOHN FREDRICK
PRIZGINTAS ANTANAS ARVIDA
PROCOPIO PETER LOUIS
PRYOR WILLIAM JACKIE
PTAK THOMAS JOHN
PUGGI JOSEPH DAVID
PYPNIOWSKI LARRY
QUARLES WAYNE ROBERT
QUATRONE FERDINAND JOSEPH
QUEEN WALTER LOUIS
RAAB JAMES DONALD
RAITT ALBERT HAROLD
RAM CORNELIUS HERBERT
RAMIREZ NELSON R
RAMOS ANGEL LUIS
RAMSAY CHARLES JAMES
RAMSEY STEVEN GEORGE
RAND DWIGHT FRANCIS
RANGES ROBERT HENRY JR
RANKIN JOHN ROBERT
RASMUSSEN PETER TERENCE
REAMER DONALD PAGE
REBELO JOAQUIM VAZ
RECK JOHN
REDDICK WILLIAM CARL
REED ROBERT BRUCE
REED STANLEY MAJURE
REEVES MICHAEL DAVIS
REEVES WAYNE PAUL
REGO ARTHUR
REID JOHN MICHAEL
REILLY JOHN NORMAN JR
REILLY JOSEPH JOHN
RENCEVICZ CHESTER MICHAEL
RENZ RAYMOND ALLAN
RETZLAFF ARTHUR CLIFTON
REYNOLDS ROBERT GEORGE
RHOADES FRANCIS STEVEN
RHODES WILLIE JOE
RICHARDSON CHARLES HENRY
RICKS JAMES LUTHER

RIDGE WILLIAM FRANCIS
RIGGINS SIM HENRY JR
RILEY RICHARD STEPHEN JR
RILK HARLAN CARL
RINGWALL RONALD WALTER
RIOS NOEL LUIS
RISOLDI VINCENT F
RIVERA EUCLIDES
ROBBINS RONALD
ROBERTS GARY LEE
ROBERTSON DAVID WILLIAM
ROBINSON JAMES P
ROBINSON MICHAEL JAMES
ROBINSON MITCHELL
ROBINSON ROBERT JAMES
ROCKY ROBERT EDWARD
RODGERS JOHN JOSEPH
ROELL MICHAEL CONRAD
ROGERS DOUGLAS MANUEL
ROGERS THOMAS SAMUEL
ROMAINE THOMAS GILBERT
ROMAN EULALIO ARTURO
ROMAN-AGUILAR CARMELO
ROMANO MICHAEL JR
ROMERO RICARDO IBRAHIN
ROSE JOHN CHARLES
ROSS ARTHUR JAMES JR
ROSS MYRON RUDOLPH II
ROSS ROGER ALAN
ROSSELL FRANCIS L JR
ROUGHGARDEN RICHARD J
ROWE ARTHUR MORTIMER
ROWE SALVATORE ALFRED
RUBY STEPHEN CHARLES
RUNYON BARRY LEE
RUSCH STEPHEN ARTHUR
RUSSELL PETER FRANSSON
RUSSELL WAYNE HOWARD
RUSSO WILLIAM
RUTH DENNIS
RUTTER LYNNE HARLAN
RUTTER THOMAS CLAYTON
RUVOLIS EDWARD JOSEPH
RUZILA PETER JR
RYAN TERRENCE PATRICK
RYAN WILLIAM CORNELIUS JR
SACHARANSKI FRANK ERIC
SALEMI VINCENT RALPH
SALERNO ANTHONY JOHN
SALUGA STEPHEN JOHN III
SANCHEZ VIDAL JR
SANTIAGO LUIS SANTIAGO
SANTIAGO-CRUZ RAFAEL
SANTORI JOSEPH
SANTORO RONALD PETER
SARGENT GORDON LEROY JR
SATTERFIELD WILLIAM
HURLE
SAUNDERS DONALD BARON
SAVOTH TERRY LEE
SAWRAN RICHARD ARTHUR
SCATUORCHIO DOMINIC N JR
SCAVUZZO PETER GARY
SCHAEFFER GUY LAWRENCE
SCHARIBONE DAVID JOHN
SCHAUBLE KENNETH WILLIAM
SCHAVELIN HUGH ERNEST
SCHELLER JEFFREY LYNN
SCHERDIN ROBERT FRANCIS
SCHIESS THOMAS CHARLES
SCHLINGER JAMES IRWIN
SCHMALZ CARL FREDRICK JR
SCHMID JAY JULIUS
SCHMIDT DENNIS RICHARD
SCHMIDT WALTER JAMES
SCHMUTZ ANTHONY MICHAEL
SCHNABOLK HOWARD JON
SCHODERER ERIC JOHN
SCHOELIER TJEERD
SCHOENBERG RICHARD C
SCHOETTNER GEORGE CRAIG
SCHOPMANN RAYMOND
FRANK
SCHORNDORF KENNETH
FRANCI
SCHULTZ CHARLES JOSEPH
SCHULZ WILLIAM ARTHUR
SCHUSTER FRANK
SCHWEYHER JOHN WILLIAM
JR
SCIVOLINO ANTHONY
CHRISTO

SCOTT DONALD BLUE
SCOTT MIKE JOHN
SCOTT PETER W
SCOTT WILLIAM ALEXANDER
SCUITIER JAMES
SECOR GILBERT ARTHUR
SEEL WALTER PHILLIP JR
SEIBERT RICHARD J
SELF EUGENE LAWRENCE
SELIG RONALD JOHN
SELLERS WILLIAM CLESSON
SELLITTO MICHAEL JOSEPH
SERVEN PAUL ELLIOTT
SERVICE JOHN ANDREW
SEVELL ROBERT LEE
SEVENSKI ALFRED
SEXTON LEONARD EARL
SHARP BRUCE DAVID
SHAUGER HARRISON BENJAMIN
SHAW JOHN JAMES
SHELLEM ROBERT PATRICK
SHELLEY GREGORY ALLEN
SHEPPARD LONNIE JR
SHEPPARD ROBERT LEE JR
SHEPPARD ROBERT PORTER
SHIELDS DAVID
SHOOK BOYD LEROY
SHORTT WILLIAM
SICKLER CHARLES STEVEN
SIEGWARTH DONALD EDWIN
SIERCHIO ALFONSO DONATO
SIMCHOCK THOMAS PETER
SIMMONS WILLIAM
SIMON JOSEPH LOUIS JOHN
SIMONE DENIS LAVERN
SIMPSON WALTER STEPHEN
SINCAVAGE RICHARD
SINIBALDI MICHAEL WILLIAM
SIPE ROBERT ERNEST
SISCO ARTHUR CLARENCE JR
SKILES JAMES ARTHUR
SKINNER DONALD ALVAH
SKODMIN ANTHONY
SLATER JERALD ALBERT
SLATTERY ROBERT JOHN
SLOMIANY KAZIMIERZ
HENRYK
SMITH ALFRED JAMES
SMITH DENNIS
SMITH ERNEST WILLIAM
SMITH FORTUNE
SMITH GEOFFREY STEPHEN
SMITH JOHN DAVID
SMITH JOSEPH JOHN
SMITH ROBERT BARRY
SMITH TERRANCE EDWARD
SMITH THOMAS F JR
SMOYER WILLIAM STANLEY
SMYRYCHYNSKI GEORGE
MICHA
SNAITH THOMAS RANKIN
SNODGRASS GEORGE EDWARD
SNYDER THOMAS WAYNE
SOLARI STEVEN
SOLOMON ROBERT GEORGE
SOROKA DOUGLAS MARTIN
SOSNOWSKI JAMES FRANCIS

SPENCE ROGER JAMES
SPIKES STANLEY
SPILMAN DYKE AUGUSTUS
SPINA FRED CONCETTO
SPRINGSTEADAH DONALD
KENN
SPRUILL OVELL
STALEVICZ GREGORY HENRY
STANLEY CHARLES GERALD
STEFANIAK STEPHEN ROBERT
STEFFEN CARL ROBERT
STEFKO WILLIAM CHARLES
STEIDLER JOHNSON AUGUSTUS
STEPHAN RICHARD EDWARD
STEPHANAC MARK JOHN
STEPHENS GEORGE JOSEPH
STEPHENSON WILLIAM JAMES
STERLING CHARLES WESLEY
STERNIN EDWARD MARVIN
STEVENSON CLEMENT OLIN
STEWART RICHARD CORTLANDT
STOUT CLIFFORD RUSSELL
STRANGEWAY JAMES J JR
STRUPP DAVID ALAN
STYBEL CONRAD ANTHONY
SUYDAM JAMES LAWRENCE
SWANGIN MICHAEL DEWITT
SWAYKOS WILLIAM ERNEST
SWAYZE JOSEPH J
SWENSON SWANTE AUGUST
SYDOR DENNIS WILLIAM
SYKES JONATHAN EDWARD
SYLVIA JERRY
SZAWALUK NICKOLAS
SZCZEPANCZYK GEORGE V
SZYMANSKI JOHN STEPHEN
TALMADGE THOMAS ROBERT
TAMAGNINI JOSEPH EDWARD
TANGARIE JOSEPH THOMAS
TASSEY MALCOLM FAIRCHILD
TATE SCIP
TAYLOR ANTHONY
TAYLOR CHARLIE WILLIAM
TAYLOR JAMES R
TAYLOR LOUIS ANTHONY
TENCZA ANTHONY JOHN
TERRY FREDERICK G JR
TERRY JAMES WILLIE
TETKOSKI LEON ANTHONY
THIBAULT JEFFERY ALLEN
THOMAS ALTON JR
THOMAS DANIEL WAYNE
THOMPSON CALVIN EUGENE
THOMPSON DANIEL FRANCIS
THOMPSON DENNIS WAYNE
THOMPSON GERALD RONALD
THOMPSON JAMES EDWARD
THOMPSON OTIS FRANKLIN
THORN JOSEPH MEREL
TIEFENTHALER JOSEPH
THOMA
TIEMAN WILLIAM EDWARD
TIPTON JOHN EDWARD
TOLBERT DELANCY DU BARRY
TOMENY JOHN HAROLD
TOOKE JOHN KARL
TORRES ANTHONY WILFRED

TORSIELLO WAYNE LOUIS
TOZOUR MARVIN GEORGE
TREMBLAY RICHARD
TRIVISONNO ROBERT
TRUEX GLENN ELLSWORTH
TUFTS ROBERT BRUCE
TULLY WALTER BUSILL JR
TULP GUYLER NEIL
TUNICK FRANKLIN MICHAEL
TURNBULL ROBERT CHESTER
TURNER PRESTON HARRY JR
VALLECILLO EDGAR HENRY
VALLONE RICHARD JOSEPH
VALT RALPH WESLEY
VAN BARRIGER RONALD
ERNES
VAN HOUTEN NELSON OMAR
VAN HOUTEN THOMAS
EDWARD
VAN VLIET HOWARD ELMER
VAN WINKLE HAROLD J JR
VANDERHOFF GEORGE A JR
VANDERHOOF ALLEN WALTER
VANDERSKI NORMAN JAMES
VARNER CHARLES ALFRED
VARS JONATHAN R
VAUSE JAMES EDWARD
VENABLE WESTOVEL
VENNIK ROBERT NICHOLAS
VICHOSKY WALTER JOSEPH JR
VIEHMANN GEORGE JOHN JR
VIGGIANO ROBERT EDWARD
VIRBICKAS ANTHONY A
VIRGILIO LAWRENCE JOSEPH
VOGEL EDWARD BARRY
VOHRINGER WILLIAM
THOMAS
VOLPONE DANTE
VON BISCHOFFSHAUSEN
ROBERT A
WADE ROBERT JOHN
WAGNER HARRY EDWARD
WALD GUNTHER HERBERT
WALKER GERARD JOSEPH
WALKER IRVIN
WALKER LAWRENCE PERCELL
WALTERS MICHAEL ARTHUR
WALTERS RONALD C
WARNER STEPHEN HENRY
WARNETT RONALD LEONARD
WASHINGTON ALBERT B JR
WATSON GREGORY ALTON
WATSON MARVIN LEROI
WATTERS CHARLES JOSEPH
WEBB JAMES ARTHUR
WEBER WILLIAM PAUL
WEDLAKE BRIAN FRANCIS
WEEDO VINCENT JAMES JR
WEISS WALTER
WELSH THOMAS H
WEMPLE EARL SCOTT
WEST DONALD FREDERICK
WEST EDWARD TYRONE
WEST JOHN HAYDEN
WETZEL CHARLES ROBERT
WHEELER FREDERICK
GEORGE
WHELAN JOSEPH VINCENT
WHITAKER FREDDIE

WHITE HERBERT FRANKLIN
WHITE LOWELL FRANKLIN
WHITE RONALD LEE
WHITE WESLEY WILLIAM
WHITE WILLIAM GEORGE
WHITING JUSTIN RICE IV
WICKLINE DONALD LEE JR
WICKWARD WILLIAM J
WIDDIS JAMES WESLEY JR
WIENCKOSKI DAVID RAYMOND
WIGGINS ALFRED FRANCIS JR
WILKINS BOBBY RAY
WILKINS RANDOLPH RECARDO
WILLIAMS ALLEN
WILLIAMS C W RICHARD
WILLIAMS FRED THOMAS
WILLIAMS GLEN RAYMOND
WILLIAMS LEROY JR
WILLIAMS LESTER JR
WILLIAMS ROBERT ALTON
WILLIAMS WILLIAM LYNN
WILSON ELROY
WILSON SYLVESTER WILLIAM
WINTERS CHRISTOPHER
MICHA
WINTERS JOHN
WITT JAMES
WOHLRAB BRUCE
WOLFE JOHN THOMAS
WOLFF RICHARD GLEN
WOOD RICHARD ALAN
WOODARD PAUL LEROY
WOODROW ROBERT A
WOODS JAMES BERNARD JR
WOODS ROBERT WALTER
WOODSON ARNOLD
WORSHINSKI ROBERT MATTHEW
WORTHINGTON ROBERT
WARD
WRIGHT FREDERICK W III
WRIGHT LEROY NORRIS
WROBLESKI WALTER FRANCIS
WYATT RONALD
WYNDER EDWARD ORLANDO
WYNN GERARD MICHAEL
YAWORSKY MICHAEL
YOHN THOMAS LEEONAS
YOHNNSON GEORGE SALVATORE
YORK ROBERT LEO
YOUNG STEPHEN WALTER
YOUNG WILLIAM VINCENT
ZALEWSKI WILLIAM JOHN
ZAMORSKI GLENN JOHN
ZAPOROZEC JULIUS
ZIBURA MICHAEL EDWARD JR
ZICCHINO DARRON FREDERICK
ZIMMERMAN ALAN HARRY
ZUBAR WLADMIR WILLIAM
ZUKOV STEPHEN ANDREW
ZYCK FRED JOSEPH

Photos by Ryan Swanzey

Vietnam Veterans Memorial State Park

34 Country Club Rd, Angel Fire, NM 87710

The Vietnam Veterans Memorial Park is the only state park in the country that is completely dedicated to the Vietnam veterans. The on-site chapel and visitor's center offers educational resources as well as space for people to reflect on the sacrifice of all veterans in order to help people learn and heal through remembrance.

This unique memorial came about due to the efforts of Victor and Jeanne Westphall. The two lost their son, Marine First Lieutenant David Westphall in May 1968 when his unit was attacked in Vietnam. They used David's insurance policy to start the construction of the Peace and Brotherhood Chapel overlooking the Moreno Valley. This is the same chapel that stands in the Memorial Park today and is open to the public 24 hours a day.

Since its dedication in 1971, the Vietnam Veterans Memorial Park was supported by the David Westphall Veterans Foundation. In 2005, the park became an official state park, but it's still operated by the David Westphall Veterans Foundation in partnership with the Vietnam Veterans Memorial State Park Friends Group. Due to this partnership, the park is accessible without a fee to the public.

The park has multiple features including the aforementioned chapel. There is also a visitor center that boasts a media room where an 86-minute documentary on Vietnam is shown. There is also a research library with photos of service members who died in action or who are still considered missing.

On the site, there is also a Huey helicopter that was used in Vietnam. It was damaged during a March 1967 rescue mission and returned to the US for repairs. It ended up in the care of the New Mexico National Guard who donated it to the Memorial Park. In addition to these features, the park also houses a Doug Scott statue named "Dear Mom and Dad" as well as a model of the Vietnam Women's Memorial that was dedicated in 1993 in the Washington Mall in Washington D.C. Visitors will also have access to the cemetery in the park, a water garden, and a memorial walkway. Memorial and Veterans day services are held here each year.

The Names of those from the state of New Mexico Who Made the Ultimate Sacrifice

ABEYTA JERRY DELBERT
ADAMS GEORGE DAYTON
ADAMS JOHN K
ADAMS MICHAEL THOMAS
ADLER WOODROW DENNIS
AGUIRRE CARLOS CRUZ
ALEXANDER DONALD RAY
ALEXANDER GEORGE W JR
ALLEN RICHARD LEE
ANAYA GEORGE MICHAEL
ANDLER MARION BRYAN
ANTONIO JOHNNIE JR
ARAUJO ABELARDO
ARCHULETA JOSEPH
ARELLANO ANTHONY WILLIAM
ARMIJO FRANK CHARLES
ARTHUR JOHNNY
ARVISO HERBERT
ATOLE FLOYD SAMUEL
AUSTIN ROLLIN RANDOLPH
AUTEN FRANK LEROY
BACA GABRIEL
BACA ISIDRO
BACA JOHNNY LAWRENCE JR

BAKER MICHAEL O BRIEN
BALDONADO SECUNDINO
BARBOUR JOHN RAMAGE
BARELA IGNACIO
BARNEY LUTHER
BAZAN ISIDRO SIGFREDO
BECKETT JOHN WESLEY
BEGAYE EDDIE CHARLES
BEGAYE FELIX DOHALTAHE
BELL GEORGE BENJAMIN
BENAVIDEZ BENJAMIN JOHN
BERGFELDT DAVID EDWARD
BILBREY EDMOND DAVID
BLACK RODNEY JOE
BLOOMFIELD NORMAN HUBERT
BOYER MONTY DOYAL
BRANCH FREDDIE ISIDORE
BROWN KENNETH RAY
BRYANT JERRY HAROLD
BULLOCH JAMES GRADY
BUNYEA WALTER CLIFFORD JR
BUSTAMANTE PAUL
CABE JOHNNY DWAIN
CABRERA ANDY ANASTACIO

CABRERA EDWARD A
CAMPOS LARRY PAUL
CARNAHAN STEPHEN MICHAEL
CARPENTER DAVID CLYDE
CARRILLO MELVIN
CASERIO CHARLES DOMINIC
CASS ANTHONY MAC
CHAMBERLAIN ROBERT F
CHARLIE PETER
CHAVEZ DANIEL JOSEPH
CHAVEZ DAVID CRUZ
CHAVEZ FREDDIE PAUL
CHAVEZ GLEN ALEX
CHINO GERALD GREGORY
CIPRIANI ALAN BRADLEY
CISNEROS CHARLES CASTULO
CLOUGH KENNETH RICHARD
COCA ANDREW
CONANT GREGORY C
CORDOVA CHRIS B
CORFIELD STAN LEROY
CRAWFORD DAVID WESLEY
CRESPIN ARTHUR
CROCKETT WILLIAM JAMES

CROW ENNIS EUGENE
CRUCE LEONARD ERWIN
CRUZ SAM
CUMMINS JOHN RUDOLPH JR
CURLEY ALBERT ALLEN
DALE CHESTER DONALD
DAVID ROBERT
DAVIES EDWARD EARL
DAVIS EDWARD DANIEL
DAVIS RICARDO GONZALEZ
DE SHURLEY GEORGE ROBERT
DEFOOR FREDDIE CARVIAL
DELORA PEDRO ASCENCION
DEMARCO BILLY JOE
DEMPSEY WARREN LEIGH
DENNEY DONALD GENE
DENNIS DOUGLASS J
DERDA JAMES MICHAEL
DEVINE CAMERON JOSEPH
DIREEN KEVEN THOMAS
DOW ROBERT MELVIN
DOZIER JOBIE CLAYTON
DUGAN BEN GOOLMAN
DURAN RICHARD LOSOYA
DURAN STEVE GONZALES

DYER HARRY GORDON
EARL MICHAEL RANDALL
EDWARDS DANIEL WINSLOW JR
EGGERT SAM
ESQUEDA ANTONIO ALVARADO
ETSITTY VAN
FAIRCLOTH RICHARD DWAYNE
FANNING MARTIN VINCENT
FERNANDEZ DANIEL
FLETCHER LON M
FLORES CHARLE CORDOVA
FLORES JERRY
FOLEY CHARLES DANIEL
FORGETTE DUANE GARTH
FOSTER GEORGE ARTHUR III
FOSTER JAMES LESTER
FOWNER JACOB HENRY
FRAGUA GEORGE LEONARD
FRINK JOHN WESLEY
GADZIALA GARY LEE
GALBREATH TERRELL ROBERT
GARCIA ANDRES
GARCIA DAVID JOSE
GARCIA EDDIE LEONARD
GARCIA FRANCISCO M JR
GARCIA ISIDRO
GARCIA JOE CECILIO
GARCIA LOUIS MAGIN
GARCIA LUPERTO
GARCIA RAMON
GARLEY FRANK ELOY
GARRAPY DAVID EARL
GASS CHARLES LEE
GHAHATE LUTHER ANDERSON
GIBSON ROY ALLEN
GONZALES JOSE BERNARDINO
GOODING LLOYD LEE
GRAHAM ROBERT LEE
GRIEGO JESUS
GRIEGO JOHN FRANK RAY
GRIEGO RICHARD EDWARD
GRIFFITH THURSTON A JR
GRIJALVA DAVID CENTENO
GRUBB EARL GILBERT
GURULE RICHARD ALBERT
GUTIERREZ JUAN FEDERICO
GUZMAN REYNALDO
HAGER HAROLD EUGENE
HAGMAN RICHARD HAROLD
HAMILTON JOHN SMITH
HAMILTON RICHARD ELMER
HANAWALD LEN MARTIN
HARRISON RONALD EDWARD
HARVEY OCTAVIANO MARTINEZ
HAYES THOMAS
HEISTER RICHARD EUGENE
HERN WILLIAM BURCH
HERNANDEZ SALOME
HERRERA FREDERICK DANIEL
HERRERA JOSE BENJAMIN
HERRERA NARCISO FRANCIS
HODGKINS GUY MERRILL
HOHSTADT JIMMY ROSS
HOLLAND RUSSELL JAMES
HOLLEY LARRY DOUGLAS
HORN RONALD DAVID
HOWLAND LEROY LARKIN
HUBBARD GERALD MONROE
HURTA JOSEPH DANIEL
ISLER REID ALLEN
JACKSON FREDERICK G JR
JACQUES JOSEPH ARTHUR
JAMES BILLIE
JENNINGS STEPHEN KENNETH
JOHNSON ARTHUR HARRY
JOHNSON LARRY DEAN
JOHNSON ZANE EVERETT
JONES MICHAEL THOMAS
JONES WILLIAM COY
JORY EDWARD LEWIS JR

New Mexico

LOVATO MICHAEL LEON
LOVATO RUDOLPH DANIEL
LUJAN ENRIQUE
MADRID FRANK DODGE
MADRID FRANK JESSE LEE
MADRID GABRIEL HERNAN-
DEZ
MAGBY LLOYD BURNEY
MALINS DAVID REAY
MALL RONALD AVERY
MARCHBANKS R B JR
MARKLAND GERALD DAVID
MARLING BILLIE JAYE
MARQUEZ JULIAN ERNEST
MARTIN EMERSON
MARTIN GUY WAYNE
MARTINEZ ALEX EZEQUIEL
MARTINEZ BILLY RICHARD
MARTINEZ BOBBY JOE
MARTINEZ DANIEL TIOFILIO
MARTINEZ EDDIE ANTHO-
NY JR
MARTINEZ JIM DANIEL
MARTINEZ JUAN HENRY
MARTINEZ MANUEL
MARTINEZ WILLIE DANIEN
MASCARENAS ALCADIO
NORBER
MATHER HARRY MICHAEL
MATHIS JIMMY CLIFTON
MAXWELL CALVIN WALTER
MAY CLOVIS LEE
MC CLURE CHRISTABOL TOBY
MC CRAW RONALD GENE
MC FARLANE RICHARD DEAN
MC KEAN GUY EDWIN JR
MC PHERSON STANLEY W
MCINTIRE SCOTT WINSTON
MEADOWS JOHN WILLIAM
MECHEM JESSE
MEHLHAFF RICHARD WAYNE
MEUTE HOWARD MICHAEL
MIKE STEVEN
MIRANDA PETER KALANI
MONTANEZ MIGUEL F

MONTOYA EUSEBIO
MONTOYA JOE HERMAN
MONTOYA JOE NED
MONTOYA JOSE ALBINO
MONTOYA ROBERT GONZA-
LES
MONTOYA VICTOR H JR
MOORE JAMES MICHAEL
MORALES GILBERT
MORALES SAMUEL
MORENO ANDRES JR
MORENO HILARIO
MORRIS MICHAEL JOHN
MORRISSEY ROBERT DAVID
MUNIZ DANIEL HAROLD
MUNSON ALVIN JAMES
MUSKETT WAYNE
NABOURS JIMMIE FLOYD
NAVA FRANCIS XAVIER
NEELD BOBBY GENE
NOSEFF RONNIE LEE
NUNEZ GEORGE HENRY
NUNN SAMUEL JOHN
OLDHAM JOHN SANDERS
OROSCO STEPHEN
ORTEGA RAMON FELIX JR
PACHECO ANDREW JOSE
PACHECO JAIME
PADILLA PEDRO
PATCH DONALD CHARLES
PEARSON KURT BYRON
PEINA ERNEST DELBERT
PENA JOHN
PEREA JUANITO
PETERS JOHN THEODORE
PHELPS LARRY DELTON
PHILLIPS GREGORY LEE
PIERCE ROBERT LIVINGSTON
PLATERO RAYMOND
PLATO JIMMIE LEON
PLATT RUSSELL LOWELL
PORTER FRANK SOLIS
PUCKETT HARRY LEE
PYLE CHRIS MONROE
QUEVEDO ANGEL ALARID

QUINTANA SANTIAGO V E
RAMIREZ SAMUEL MEDINA
RANSDELL CURTIS H
REA BILLY MC CALL
RIBERA ANTONIO
RICE HOWARD JACOB
RICKELS JOHN A
RISNER JOHN MILTON
ROBERTS FREDDIE JOE
ROBERTS JERRY MARCO
ROBERTS VIRGIL JESSIE
ROBERTSON KENNETH LEE
RODGERS LUIA
ROGERS JOHN DAVID
ROMAN VICTOR MUNOZ
ROMERO CHARLES ANTHONY
ROMERO SAMMY CHACON
ROMERO TIMOTEO FRED
ROMERO TRINE JR
ROWE SHARBER MAYFIELD
RUBIO RUBEN
SAENZ HECTOR MARIO
SAENZ RICHARD
SAIZ FRED ROMAN
SALAZAR CRES PADILLA
SALAZAR MEL ERNEST JR
SALAZAR PATRICK
SANCHEZ CAMILO JAMES
SANCHEZ CHARLES ANTHO-
NY
SANCHEZ CRESENCIO PAUL
SANCHEZ JOSE L
SANCHEZ JUAN DIEGO
SANCHEZ UVALDO
SANDERS JAMES GARLAND
SANDERS JULIUS MITCHELL
SANDFER WILLIE J JR
SANDOVAL PHILLIP JAMES
SAWYERS ROGER THURSTON
SEGURA MANUEL TIODORO
SENA BENNY
SERNA RAYMOND
SERRANO FILEMON
SHAW JOE CARL
SILLIMAN JACK LLOYD

SIMBOLA JOSE SCOTTY
SIMONS GERALD SHIELDS
SIMPKIN WALLACE FRED-
ERICK
SIMPSON MAX COLEMAN
SISK ROBERT ALAN
SISNEROS ARTURO SYLVESTER
SISNEROS ROMAN
SKEET PATRICK
SMITH DANNY LE MOYNE
SMITH JEFFERY NOLAN
SMITH JOL NEBANE
SMITH LLOYD EDGAR
SMITH THOMAS FRANKLIN
SPURGEON ROY STEPHEN
STAGER KENNETH L
STAKE KENDALL ALBERT
STANLEY DON SCOTT
STOLL DAVID LOUIS
SUMMERS DONALD L
TAFOYA FLORENTINO JR
TAFOYA FRANK
TAFOYA GEORGE ELOY
TAFOYA JOHN OLIVIO
TAFOYA MARK ALVAN
TARANGO MAGDALENO
TAYLOR BOBBY ALLEN
TEETER KENNETH WARREN
TEJADA HENRY LEROY
TENORIO RAFAEL GABRIEL
TENORIO SAM
TETER RANDALL KEITH
THARP HAROLD ALLEN JR
THOMAS WILLIAM MICHAEL
THOMPSON JERRY ELMER
THORNTON LEO KEITH
THORNTON STEPHEN H
TICE WAYNE ARTAMUS
TOLEDO THOMAS AMBROSE
TORREZ MANUEL ANTONIO
TOSA ANTONIO TONY
TRAINOR TERRY LEO
TRUJILLO GABRIEL
TRUJILLO GREGORIO JR
TRUJILLO JOSEPH FELIX

KAUFMAN DONACIANO
FRANCIS
KEFFALOS CHRIS ALBERT
KELLER GEORGE RICHARD
KEMP JOE MAC
KLEIN JEROME DON
KOZAI KENNETH BRUCE
KOHEI
KRUG MICHAEL JOE
LANE MITCHELL SIM
LARGO CALVIN DAVID
LE COMPTE JOHN AULT
LEE WILLIE B
LEISURE JACKIE GLEN
LEONARD KENT ALAN
LERMAN CONRAD
LEYBA RAMON
LOPEZ ROBERT CHARLES
LOTT CHARLES ALLISON

160

TRUJILLO PAUL
TRUJILLO ROBERT STEVEN
VALDEZ FRANK
VALDEZ LEROY FRANK
VALDEZ PHIL ISADORE
VANN GARY STEVEN
VELASQUEZ JULIAN VICTOR
VIGIL LAURENCIO
WAIDE DONALD GILES
WALKER BURTON KIMBALL
WALTER ALBERT MARION
WALTON WILLIAM LEROY

WEBBER BRIAN LEE
WEST BENNIE LEE
WESTPHALL VICTOR D III
WHITLOW RONALD DAVID
WILLARD JAMES MONROE
WILLIAMS DENNIS ALAN
WILLIAMS THOMAS RALPH
WILSON JOHN STANTON
WILSON JUAN JAY
WILSON LAVON STEPHEN
WINKLES HARVIE PERRY III
WISEMAN BAIN WENDELL JR

WOLFE PAUL EDWARD
YAZZIE DAN
YAZZIE JONES LEE
YAZZIE RAYMOND
YOUNG STEPHEN ANDREW
ZAMORA CARLOS JR
ZAMORA JUAN MANUEL ALBA

"Doc's Journey to Vietnam"

Photos by Ronny and Matt Ymbras

Vietnam Veterans Memorial

2610 NY-35, Katonah, NY 10536

The Westchester Vietnam Veteran's Memorial is located in Lasdon Park in Westchester County, New York. It was set up in a way to offer visitors a serene and natural setting for reflection and remembrance. Besides the four monuments that make up the Vietnam memorial, there is also a museum dedicated to honoring the Westchester County men and women who served their country during the Southeast Asia conflict.

Lasdon Park came about after Westchester Count purchased the Lasdon Estate after the death of William Lasdon in 1986. It was purchased in order to preserve the area's history instead of allowing investors to build it up with impersonal condominiums.

It was developed as parkland with different components and, in 2001, Lasdon's daughter Nan Laitman made a sizeable donation in order to make a garden within the park. The Lasdon Garden was considered to be the perfect location for the Vietnam memorial and contains multiple monuments to local service members. The county itself is responsible for establishing the memorial and its upkeep.

The memorial starts with a pathway called the Trail of Honor which is dedicated to the local veterans of all wars. On the trail are individual busts representing the various wars including one for the Vietnam War. There is a hill on the grounds with a clearing where the main portion of the Vietnam memorial can be found. Here visitors find a large sculpture made by Julia Cohen. It depicts a nurse running toward a soldier carrying a wounded brother. There is also an obelisk featuring the names of the fallen 217 service members from Westchester County along with a granite stone honoring the eight nurses who were lost during the war. There is also a walkway consisting of 5,900 pavers, each symbolizes ten US soldiers who were killed during the war.

The Names of those from the state of New York Who Made the Ultimate Sacrifice

ABBATE ROSARIO RUSSEL
ABBOTT ROBERT WILLIAM
ABDULLAH GHALIB AHMED
ABREU-BATISTA MIGUEL A JR
ABRUZESE ANTHONY JOSEPH
ABRUZESE ROBERT ALEXANDER
ABRUZZESA MICHAEL JOHN JR
ACEVEDO HECTOR SANTOS
ACEVEDO ROBERTO
ACKWOOD WALTER JAMES
ADAMO RICHARD CHARLES
ADAMS CHARLES WESLEY
ADAMS GEORGE HARTWELL
ADAMS PHILIP FRANCIS
ADAMS THOMAS EDWARD
ADAMS WILLIAM JR
ADAMS WOODROW WILLIAM
ADDISON RICHARD EDWARD JR
ADIUTORI RICHARD
AGAR ANTHONY PHILIP
AGARD ROWLAND NATHANIEL
AGAZZI DAVID MICHAEL
AGUGLIARO MATTHEW JOHN
AGUILAR DOMINGO IGNACIO
AHLMEYER HEINZ JR
AHRENS RUSSELL GEORGE
AIKEN LARRY DELARNARD
AIKEN LEROY BENJAMIN
AIKEN WILLIAM LESLIE

AKERLEY DENNIS
AKINS CHARLES JAMES
ALAIMO HOWARD JAMES
ALBANESE JOHN ERNEST JR
ALBANO PAUL EDWARD
ALBERICI MICHAEL
ALBERT SERGIO EDITH
ALDAG WILLIAM ARTHUR
ALDERMAN JAMES MURIEL
ALDRICH ROBERT HENRY
ALEXANDER WOODROW
ALFEROFF IVAN
ALFONSO RONALD JOSEPH
ALICEA DAVID
ALICEA ISRAEL
ALICEA MANUEL JR
ALICEA ROBERT
ALIVENTO FRANCIS DOMINICK
ALLEE RICHARD KENNETH
ALLEN BRUCE JOSEPH
ALLEN CHANNING JR
ALLEN DEAN BROOKS
ALLEN EUGENE
ALLEN GERALD WILLIAM
ALLEN LEONARD PETER
ALLEN TERRY ERNEST
ALLING JOHN STEPHEN JR
ALLISON DARRELL GENE
ALLSOPP STEPHEN ALLISON
ALSEVER MICHAEL HADWIN
ALSTON FRANKLIN JR
ALSTON WILLIE EDWARD

ALTERWISHER ARTHUR CARL
ALVAREZ-DELGADO LUIS F
AMATO EDWARD MATHEW
AMATO MICHAEL JOHN
AMBROSE GREGORY FRANCIS
AMBROSIO FRANK CARL
AMENDOLA WILLET RANKIN
AMES ALEXANDER AUDREY
AMES THOMAS ROBERT
ANABLE HAROLD JAMES
ANDERS HERMAN E JR
ANDERSEN ANDREW CARL
ANDERSON BOYD WELLINGTON
ANDERSON DAVID PAUL
ANDERSON DOUGLAS RAY
ANDERSON ERIC ARNOLD
ANDERSON HENRY JR
ANDERSON JOHN AUSTIN
ANDERSON LARRY
ANDERSON MICHAEL PATRICK
ANDERSON PETER NEWELL
ANDERSON RICHARD ANDREW
ANDERSON ROBERT LEE
ANDERSON RONALD DAVID
ANDERSON STEVEN RICHARD
ANDINO NELSON
ANDREWS ALAN WAYNE
ANDREWS JAMES EDWARD
ANDREWS WALTER EUGENE JR

ANDRUS FLOYD EDWARD III
ANDRYSIAK FRANCIS HOWARD
ANGELIDES JAMES JOSEPH
ANGRISANI CHARLES JOSEPH
ANTWINE RONALD MICHAEL
APRILLIANO ANJELO JOSEPH
ARAUJO ROBERT JOSEPH
ARGENZIO NESTOR LORENZO
ARIMENTO JOSEPH A
ARMATO SALVATORE JOSEPH
ARMENIO ROBERT WILLIAM
ARMLIN LOREN AARON
ARMSTEAD JAMES DOUGLAS JR
ARMSTRONG HAROLD KINGSLEY
ARMSTRONG ROBERT GEORGE
ARMWOOD JESSE JAMES
ARNETT FRANCIS IENATIUS
ARNIOTIS DIMITRIOS G
ARNOTT DAVID BRUCE
ARROYO RAMON JAIME
ARTHUR LAWRENCE KENNETH
ARZUAGA JOAQUIN
ASEP MICHAEL
ASHER ALAN
ASIP EDWARD VINCENT
ASPINALL WILLIAM ALBERT
ATKINSON JOHN F JR
ATWELL ROBERT WAGNER
AUGUSTINAS WALTER PETER
AULETTI PETER PAUL
AUTORINO JOSEPH G JR
AVILES ALFREDO EDWARDO
AVILES ANIBAL FELIPE JR
AVILES PETER
AVOLESE PAUL ANDREW
AYRES ALBERT BOYD
BABEY DAVID PAUL
BACHMANN LYNN JR
BACKUS KENNETH FRANK
BACO JOHN
BACON WILLIAM IVOR TENNEY
BAGSHAW WILLIAM MICHAEL
BAILEY DENNIS MICHAEL
BAILEY DONALD GERALD
BAILEY FRED MC KINLEY
BAILEY JOSEPH THOMAS
BAILEY KENNETH NORMAN JR
BAILEY RAE ARVID
BAIN THOMAS ARTHUR
BAINES TOMMIE
BAIZ LEE THOMAS
BAJIN ENVER
BAKER CHARLES ALFRED
BAKER GARY BRUCE
BAKER LARRY JAMES
BAKER MICHAEL DEAN
BAKER PAUL JOSEPH
BALAMOTI MICHAEL DIMITRI
BALAZY GEORGE STEPHEN
BALDERA BARTOLOME ALFONSO
BALDWIN CLARENCE JAY
BALDWIN GERALD LEE
BALDWIN PETER NELSON
BALDWIN WILLIAM MCKINLEY
BALES RICHARD LEE
BANNA WILLIAM THOMAS JR
BANSAVAGE JOHN GEORGE
BARANOWSKI BISHOP SKIP
BARASH LOUIS ABBEY
BARBARIA LOUIS JOSEPH
BARBER CHRISTOPHER JAMES
BARBERA PETER
BARBERY ROBERT NELSON
BARBIERE CHARLES LOUIS
BARBOUR JAMES WESLEY
BARCA JOHN JR
BARCKLOW LAWRENCE ANTHONY

BARETTI ALAN GEORGE
BARIGLIO RICHARD LOUIS
BARILLO JOSEPH WILLIAM
BARISIC LAWRENCE WILLIAM
BARLOW CLARK EUGENE
BARNARD RICHARD GEORGE
BARNES GARY ALAN
BARNES JAMES WILLIAM JR
BARNES MARK ALBERT
BARNES RICHARD FRANK
BARNES ROBERT SEWELL
BARNEY ALEXANDER LORENZO
BARNHILL JAMES EUGENE
BARON FRANTZ MARIO
BARONE COSMO LEONARD
BARRETT STANLEY HOWARD
BARRETT STEPHEN CLARK
BARRIMOND ERROL MICHAEL
BARRY CRAIG NICHOLAS
BARRY EDWARD FRANCIS
BARRY JOHN FRANKLIN
BARRY ROBERT OWEN
BARRY THOMAS R
BARSLOW KENNETH WILLIAM
BARTOCCI JOHN EUGENE
BARTOLF NOEL MICHAEL
BARTON ROBERT W JR
BARZAN JOHN JOSEPH
BASSO MICHELE
BATEMAN NEIL ELLIS
BATES JAMES JOHN
BATES ROBERT MICHAEL
BATISTA-RODRIGUEZ JORGE L
BATTERSON JOHN PEDDIE JR
BAUER CHARLES JAMES JR
BAUER GREGORY CHARLES
BAUER ROBERT ERNEST
BAUER WILLIAM HENRY
BAURLE MATTHEW JOHN
BAUSCH BARRY RALPH
BAXTER PETER WALTER
BAYES THOMAS JOSEPH
BEACH MYRON STANLEY JR
BEAGLE HOWARD EUGENE
BEALE ROBERT BOUGHTON
BEANE HAROLD GEORGE JR
BEATTIE ERICK WALTER
BEATTY DONALD EDWARD
BECHARD JOHN COWIE
BECHTOLD FRANCIS SCOTT
BECK EDGAR PETER JR
BECK JOHN ROBERT
BECKER WILLIAM JOHN
BECKMAN ROBERT CARL
BECKWITH RICHARD EARL
BEDIENT ROSS EDWARD
BEDNAREK JONATHAN BRUCE
BEDWELL WAYNE JOSEPH
BEEBE RICHARD WILLIAM
BEECHING EARL PETER
BEEK ERWIN
BEGLINGER THOMAS EDWIN
BELASCO CHARLES THEODORE
BELL LEONARD JONATHEN
BELLACH LOUIS WILLIAM JR
BELLAMY ROLAND ROBERT
BELLEMARE ANDRE REMI
BELLINGER RONALD LEE
BELLWOOD RICHARD ROY
BELT MARVIN MARK
BELTRAN FRANK JOSEPH
BENBOW EVANS JR
BENNETT CLIFFORD RAYMOND
BENNETT JOHN JAY
BENNETT RICHARD JAY
BENVENUTO THEODORE F JR
BENZ ROBERT JOSEPH
BERDY MICHAEL EDWARD
BERG DALE RUSS
BERGER DONALD JOSEPH
BERGER ROBERT FRANCIS
BERGIN THOMAS JAMES
BERLANGA RAFAEL ANGEL

New York

BERMINGHAM DANIEL JOSEPH
BERMUDEZ JOSE DAVID JR
BERNARD JOHN EDWIN
BERNARD-ROBLES ANTONIO RA
BERNHARDT WAYNE WILLIAM
BERNREUTHER WALTER JOHN
BERNSTEIN ALAN MARTIN
BERNSTEIN JACK
BERNSTEIN LESLIE PAUL
BERNTSEN ROBERT
BERRIOS MICHAEL
BERTHEL JOHN JOSEPH
BEST CHARLES HYMAN
BEST OLIVER ADRIAN JR
BEST THOMAS EMANUEL
BESTMANN CHARLES EDWARD
BETANCOURT JAMES
BETHEA HENRY
BETHEA RAYMOND LEWIS
BEVERLY FRANCIS M
BEY NELSON
BIALKOWSKI JOHN JOSEPH
BICKEL ROBERT JOHN
BIELICKI GREGORY CHESTER
BIENKOWSKI WALTER JOSEPH
BIGELOW LAWRENCE CARROLL
BIGELOW RALPH WILLIAM
BIGTREE JAMES VICTOR
BILLINGS JAMES ARTHUR JR
BILOTTA RICHARD GALE
BINK JAMES CLEVELAND JR
BIRENBAUM BERNARD
BISONETT LAWRENCE EDWARD
BITEL BEN STANLEY
BITTENBENDER DAVID FRITZ
BIVENS HERNDON ARRINGTON
BIVETTO CHARLES FRED
BIXBY THOMAS EUGENE
BLACKMON JAMES ARTHUR
BLAIR ALAN LEE
BLANCHARD WILLIAM GEORGE
BLANDING HENRY ARTHUR
BLODGETT DAVID WILMER
BLOEMHARD ANTON D
BLUE JONATHAN JR
BOARDMAN DAVIS JAMES
BOATWRIGHT TOMMY LEE
BOBO JOHN PAUL
BOBOWSKI JAN EDWARD JR
BOCHNEWETCH SHERMAN II
BOHNWAGNER PETER PAUL
BOISE RICHARD HOWARD
BOJANEK ROBERT ARTHUR
BOKINA ROBERT JOHN
BOLSON JAMES JOSEPH
BOMBERRY GREGORY LEE
BONAPART PAUL
BONEY BERNARD
BONNETT EUGENE EDWARD
BONVENTRE THOMAS S
BOONE ROBERT EDWARD
BOORAS PETER WILLIAM
BOOTHE RONALD CHARLES
BORCZYNSKI FREDERICK EARL
BORDEN CHARLES E
BORLAND DENNIS ALLEN

BORNHEIMER RICHARD IRVING
BOROWICZ KENNETH
BORS JOSEPH CHESTER
BORTLE JONATHAN R
BORYSZEWSKI STEPHEN J
BOSCH JOHN ARTHUR
BOSSMAN PETER ROBERT
BOSSONG FRANK W
BOSTOCK JAMES EDWARD
BOSTON RONALD
BOSWELL DAVID HENRY
BOULE THOMAS MICHAEL
BOURDEAU VINCENT CARMEN
BOURNE RICHARD E
BOUTTRY CHARLES EDWARD
BOWDLER GARY LEE
BOWEN DONALD
BOWEN HOWARD LEWIS
BOWEN RAYMOND LEWIS JR
BOWERS STEVEN WAYNE
BOWMAN ARTHUR BAILEY JR
BOYD DAVID STEWART
BOYD STEPHEN LESLIE
BOYER CHARLES DENNIS
BOYLE FRANCIS MICHAEL
BOZIER WILLIE JR
BRABANT WILLIAM ANDREW
BRACEY LESTER
BRADLEY JAMES JEROME
BRADLEY RICHARD BURTON
BRADLEY WILLIAM MARTIN
BRADY EDWARD MARK
BRADY JOHN PATRICK JR
BRADY MICHAEL EDWIN
BRADY THOMAS RICHARD
BRAITHWAITE ARNIM N
BRANCATO PETER JOSEPH JR
BRANCH JAMES
BRANDES KENNETH NEIL
BRANES EDUARDO PAUL
BRANT DONALD GENE
BRASILE TERRENCE CARMINE
BRASWELL JAMES PORTER JR
BRATHWAITE ROGER CLAYTON
BRAVIN JOSEPH SIMON
BRAY BERNARD
BRAYE LLOYD HERBERT
BREAULT RODERICK WAYNE
BRENNAN JAMES JOHN
BRENNAN TIMOTHY FRANCIS
BRENNAN WILLIAM ROBERT
BRESE ANTHONY ARTHUR
BREUER DONALD CHARLES
BREWER EDWARD JOSEPH
BREZINSKI CHARLES ANTHONY

BRIALES MIGUEL EUGENIO
BRICKHOUSE EMANUEL KRIS
BRIDENBAKER PATRICK G R
BRIDGE WILLIAM DAVID
BRIGGS DAVID IVAN
BRITO ALFONSO ANTONIO
BRITTON BERNARD BRUCE
BRITTON THOMAS WESLEY JR
BRODIE RAYMOND HERBERT JR
BROEKHUIZEN ALLEN PAUL
BROGDON MARGIE
BROOKS ANDRE MAURICE
BROOKS NICHOLAS GEORGE
BROOKS WALTER HARM JR
BROPHY JAMES JOHN
BROPHY MARTIN EARL
BROULLON ANTHONY JOSEPH
BROW CHRISTOPHER
BROWN ANDREW THOMAS
BROWN DANIEL MARTIN
BROWN DONALD CALVIN
BROWN EDWARD FREDERICK JR
BROWN EDWARD LEE
BROWN GEORGE ALLEN
BROWN GEORGE WASHINGTON
BROWN JAMES ARTHUR
BROWN JAMES AZALOU JR
BROWN JOEL ANDREW
BROWN JOHN THOMAS
BROWN KARL EUGENE
BROWN LAWRENCE GEORGE
BROWN RAYMOND LEE
BROWN RICHARD
BROWN RICHARD CHARLES
BROWN ROBERT IRWIN
BROWN ROBERT JOSEPH JR
BROWN ROBERT LEE
BROWN ROGER
BROWN RONALD LEWIS
BROWN STANLEY ALTON
BROWN STEVEN MERLE
BROWN THOMAS
BROWN WILLIAM ANTHONY
BROWN WILLIAM JOSEPH
BROWNE EARL FREDERICK
BROWNE GORDON FRANCIS
BROWNFELD PHILIP
BRUCE JEFFREY RICHARD
BRUCE ROBERT
BRUCHER ANDREW CARL
BRUCK DONALD WILLIAM
BRUCKNER HOWARD RUSSELL
BRULE GORDON JOSEPH JR
BRUNELLE JOSEPH EMILE
BRUNN CHRIS FREDRICK
BRUNO EDWARD

BRUNSON JACK WALTER
BRUSH RICHARD BERNARD
BRUST GLENN ROY
BRUSTMAN DOUGLAS JOHN
BRYAN LIONEL JOHN JR
BRYANT PELLUM JR
BRYSON ROBERT EUGENE
BUCHANAN BENJAMIN JOHN J
BUCHNER JAMES IRVING
BUCKLEY MICHAEL FRANK
BUCKLEY ROBERT WALTER
BUCKLEY THOMAS EDWARD
BUDKA RICHARD WALTER
BUKOVINSKY ANDREW THOMAS
BUKOWSKI DAVID FREDERICK
BULL ROBERT GEORGE II
BULLOCK DAN
BULMAN WILLIAM CHARLES
BUNDY WAYNE PHILIP
BUNK FRANCIS XAVIER
BUONAIUTO JAMES JOSEPH
BUONO MATTHEW JOSEPH
BURCH HENRY
BURCHELL EDGAR BROWER III
BURDICK BRIAN HARRY
BURDICK DANIEL JOSEPH
BURDICK HOWARD EARL
BURGESS ALEX LEROY
BURGESS DAVID ROY
BURGOS JUAN R
BURKART CHARLES WILLIAM JR
BURKE JAMES FRANCIS JR
BURKE JOHN JOSEPH
BURKE ROBERT ALLEN
BURKE THOMAS CHARLES
BURKE WALTER FRANCIS
BURKE WALTER LAVERTE
BURKETT EDWARD DALE
BURKHARDT WILLIAM JAMES
BURNETT JAMES SANDFORD JR
BURNEY ELMO JR
BURNS ERNEST DOOM
BURNS FREDERICK JOHN
BURNS JAMES EDWARD
BURNS JAMES PHILLIP
BURNS JAMES T
BURNS JOHN JAMES JR
BURNS ROCKY AUGUST
BURRELL ROBERT GEORGE
BURRELL ROBERT LANSING
BURRIS BERNES EDWARD
BURRIS FREDERICK
BURROUGHS JUDGE JR
BURT WILLIAM ROBERT JR
BURTON CHRISTOPHER

LEONAR
BUSCEMI ANTHONY PETER
BUSCHMANN JOHN RICHARD
BUSHEY PETER B
BUSHEY WILLIAM TIMOTHY
BUSKEY ORRIE JULIUS
BUTCHER REUBEN
BUTLER JAMES CLIFFORD JR
BUTLER MERLE FLOYD II
BUTLER PETER MARK
BUTLER ROBERT EDWARD
BUTLER STEVEN ANDREW
BUTLER TERRENCE EDWIN
BUTLER THOMAS J JR
BUTTON HOWARD EARL
BUTTS ROY JOHN
BYNOE MIGUEL ANTONIO
BYRNES RALPH WILLIAM
BYRNES ROBERT JOHN
CABALLERO HENRY JOHN
CACCIOLA DOMENICO
CACCIUTTOLO MICHAEL
CACERES ADALBERTO
CACERES ADALBERTO
CACIOPPO JOHN RICHARD
CADENHEAD THEODORE L
CADILLE FREDERICK FRANK
CADY BRIAN THOMAS
CAFIERO LESTER VINCENT JR
CAIN ROBERT EMMET
CAINES FREDERICK ALFRED
CALAMIA JACK
CALDWELL EDWARD CLARK III
CALHOUN STEVEN BRIAN
CALLAGHAN DENNIS PATRICK
CALLANDER CECIL EUCLED
CALLIHAN BLAINE EDWARD
CALLWOOD GLADSTON
CAMBRELEN JAIME
CAMERLENGO JOSEPH VINCENT
CAMERON DARRELL ADEN
CAMERON JOHN IRWIN
CAMMARATA SALVATORE
CAMPANIELLO ANTHONY VICTO
CAMPBELL THOMAS FRANCIS
CAMPESTRE ALBERT JOHN
CANADY ROY BILLY
CANAMARE GEORGE JOSEPH
CANAN HAROLD JEFFREY
CANNAN DENNIS CHARLES
CANNATA GEORGE ANTHONY JR
CANNIZZARO VINCENT JUNIOR
CANNON SHAWN GLEN
CAPERS LEE MARVIN

CAPOZZI ANTHONY LOUIS	CARMODY JAN ARTHUR
CAPRIGLIONE ANTHONY	CARMODY ROBERT J
CAPRIO MICHAEL JOSEPH	CARMODY TIMOTHY LEE
CAPUANO FRANK PHILIP	CARNEY WALTER JOHN
CAPUTO JAMES WILLIAM	CARPENTER FRED WILLIAM
CARABALLO HECTOR LUIS	CARPENTER GEORGE WHITNEY
CARABBA RICHARD ALOYSIUS	CARPENTER WALTER ANDREW
CARACCILO ANTHONY JOSEPH JR	CARR LEN E
CARAPEZZA RICHARD ALLAN	CARR LON GALE
CARAVELLO VINCENT JAMES	CARR MICHAEL PETER
CARBAJAL-AZMITIA RENE	CARRATURO FREDERICK JAMES
CARBONE RICHARD	CARROLL KEVIN JAMES
CARDINAL DAVID CHARLES	CARROLL MICHAEL
CARDONA GABRIEL JR	CARRUTH DAVE SCOTT JR
CARDOT JOHN ANDREW	CARSON BRADLEY JAMES
CAREW FARRELL RICHARD	CARTER DUANE ELWOOD
CAREY BRUCE LEO	CARTONIA CARMEN PAUL
CAREY WILLIAM JAMES	CARTWRIGHT THOMAS CLARK
CARLIN JAMES COOK	CARVAJAL FRANCISCO TERONI
CARLIN STEPHEN BERNARD	CARVALLO CESAR EDUARDO
CARLISI IGNATIUS	CASE DAVID DUANE
CARLISLE THOMAS G II	CASE ORSON HOWARD
CARLO GILBERT	CASE THOMAS JOSEPH
CARLONI JAMES FRANCIS	
CARLSON FREDERICK JOSEPH	
CARLSON WAYNE LOUIS	
CARLUCCI ANTHONY JACK	

CASEY FRANCIS JOSEPH	COCKERL JAMES CALVIN	COWDRICK HORACE W JR
CASEY PAUL WILLIAM	CODRINGTON STEPHAN	COX EUGENE THOMAS
CASEY THOMAS JEROME JR	COFFEY EDWARD VINCENT	COX FRANCIS PATRICK
CASIANO JUAN	COFFIN DONALD A	COX FRANK WILLIAM JR
CASILLA-VAZQUEZ MANUEL JR	COFFINO THOMAS PAUL	COX JAMES ALAN
CASILLO CARMINE	COHEN GERALD	COYE ROGER HERBERT
CASS WILLIAM DAVID	COLASUONNO VINCENT	COYLE GARY JOSEPH
CASSANO RICHARD ANTHONY	COLBERT DOUGLAS ROBERT	CRAGIN ROBERT STUART JR
CASSIDY DONALD THOMAS	COLBURN RICHARD EUGENE	CRAIG BENJAMIN JR
CASSIDY RAYMOND SENTER	COLE PATRICK LERVILLE	CRAIG PHILIP CHARLES
CASTAGNA JOSEPH PHILIP	COLE RICHARD MILTON JR	CRAMER PARKER DRESSER
CASTELLANOS JUAN CARLOS	COLEMAN CLARENCE BERNARD	CRANDALL RONALD JAY
CASTILLO-LIMA BENJAMIN	COLEMAN LEO ALFRED	CRANDALL WAYNE STEPHENS
CATALANO GEORGE FRANCIS	COLEMAN ROBERT LAURENCE	CRANE DAVID CHAUNCEY
CATANZARITI RONNIE S	COLEMAN WILBERT EVANE	CRAWFORD ANDY PAUL
CATHER TERRENCE JAY	COLES HOWARD FRANKLIN JR	CRAWFORD CURTIS EUGENE
CATHERMAN ROBERT RAY	COLES LEONARD ASHWORTH	CRAWFORD DOUGLAS JAY
CAUGHMAN MC KINLEY JR	COLICCHIO PETER	CRAWFORD JOHN CALVIN
CAVALARATOS GEORGE ANASTA	COLLAZO CARLOS MANUEL	CRAWFORD RICHARD WAYNE
CAVALLI ANTHONY FRANK	COLLER WILFORD PAUL	CREAMER FRANCIS P
CAVANAGH ARTHUR	COLLIER JAMES WILLIAM	CREED BARTON SHELDON
CAYEY EDWARD CECIL JR	COLLINS ARNOLD	CREGON KEVIN FRANCIS
CAYLOR RANDY LEE	COLLINS MICHAEL	CRESPO JOSE
CELESTE RAYMOND	COLLINS NATHANIEL	CRIBBS MARTIN JOSEPH
CELMER LAWRENCE JOSEPH	COLLINS THOMAS EDWARD	CRISCI LARRY ANTHONY
CENTENO HERMINIO GENOVA	COLLINS WALTER MONROE	CRISE PERRY ROCCO
CERIO JOSEPH ANTHONY	COLON ALBERTO	CRISSELL EARL LEON JR
CERRANO LOUIS FRANCIS JR	COLON HARRY JOSEPH	CROCKER DAVID ROCKWELL JR
CERVELLINO CARMINE ANTHON	COLON LUIS ANGEL	CROCKER DENTON WINSLOW JR
CESTARE JOSEPH ANGELO	COLONNA PHILIP GEORGE	CROSBY BRUCE ALLEN JR
CHADEE NYROON	COLON-PEREZ ABRAHAM LINCO	CROSLEY CHARLES RAYMOND
CHAMBERLAIN MICHAEL JOHN	COLORIO JOSEPH	CROSSLEY ORMAN LEE JR
CHAMBERS RICHARD THOMAS	COLOTTI JOSEPH LEONARD	CROSSON GERALD JOSEPH JR
CHAMBERS-QUIROZ GEORGE	COLSON BRUCE NORMAN	CROWLEY JOHN EDWARD
CHAMBLEY THEODORE ROOSEVE	COLTER KENNY LAWRENCE	CRUDO RICHARD FRANK
CHANDLER CHARLES	COLVIN GENE FRANCIS	CRUMM WILLIAM JOSEPH
CHAPMAN WARREN NELSON	COLWELL WILLIAM KEVIN	CRUSIE WILLIAM MICHAEL JR
CHASE OLIVER C JR	COMSTOCK ARTHUR EDWIN JR	CRUZ JOSE MANUEL
CHAVEZ GARY ANTHONY	CONKLIN BERNARD	CRUZ LUIS
CHAVIS HENRY LEWIS	CONKLIN JOSEPH PETER	CRUZ LUIS ANTONIO
CHENEY WILLIAM CHARLES	CONKLIN LARRY JAMES	CRUZ LUIS PHILIP
CHERRICK JAMES WESTON	CONKLIN RONALD RAYMOND	CUBERO HECTOR
CHIARELLO VINCENT AUGUSTU	CONLEY MICHAEL FRANCIS	CUCCIA DOMINICK LAWRENCE
CHIASERA AUGUST JR	CONLIN PETER EDWARD	CUDWORTH ALBERT WAYNE
CHIMERI LOUIS	CONNELL VAUGHN DAVID	CULHANE GERALD AUGUSTINE
CHIPCHASE PHILIP GRANT	CONNERS LEE ALEXANDER	CULLEN MARK JAMES
CHISHOLM ALEXANDER	CONNOLLY MICHAEL DENNIS	CULLEN THOMAS JOSEPH
CHISHOLM HOWARD	CONNOLLY TERRENCE CHARLES	CUMMINGS JAMES EDWARD
CHISOLM RONALD	CONRAD MARTIN JAMES	CUMMINGS KENNETH THOMAS
CHOMYK WILLIAM	CONROY MICHEAL EUGENE	CUMMINGS ROBERT GEORGE JR
CHRISTEN RONALD ARTHUR	CONROY PATRICK J	CUNNINGHAM CLARENCE BENO
CHRISTIANO JOSEPH	CONROY PAUL AMES JR	CUNNINGHAM EDWARD
CHRISTIE LARRY EDWARD	CONTREROS ALBERT D JR	CUNNINGHAM PRINCE CHARLES
CHUBBUCK MICHAEL FRANCIS	CONWAY LEROY	CUNNION MICHAEL ALFRED
CIBOROWSKI THOMAS PAUL	COOK DONALD GILBERT	CURLESS EUGENE JEROME JR
CICIO ROBERT DANIEL	COOK JOHN PATRICK	CURRY FRANCIS MICHAEL
CIFELLI DOMINIC JOSEPH	COOK RANDALL VINCENT	CURRY WILBUR JR
CIRILLO PHILIP M	COOK RAYMOND LEE	CURTIS DAVID ALLEN
CLARE BERNARD ERROYL	COOK ROGER JOHN	CUSHMAN HAROLD EDWARD
CLARK BRADLEY RUSSELL	COOK THOMAS STANLEY	CUSICK MICHAEL PETER
CLARK JAMES EARL	COON DAVID WILLIAM	CUSUMANO ANTHONY MICHAEL
CLARK JAMES MCKENZIE	COON JOHN LEMOINE	CUTRI MICHAEL JOHN
CLARK JERRY FURMAN	COONEY JAMES	CZAJAK DANIEL JOSEPH JR
CLARK MELVIN E	COONS HENRY ALBERT	D AGOSTINO JOHN
CLARK THOMAS OSBORN	COONS ROBERT WAYNE	D AGOSTINO NORMAN THOMAS
CLARKE IRVIN JR	COOPER JAMES RALPH	D AMBRA JOSEPH NICK
CLARKE RITSON LEWIS Y	COOPER JOSEPH HENRY JR	D AMICO ROBERT JOSEPH
CLARKIN FRANCIS THOMAS	CORBETT LINWOOD CALVIN	D ANGELICO JOSEPH MICHAEL
CLAVIO PETER ANTHONY JR	CORBETT MARK CHARLES	D ANGELO RAYMOND ANTHONY
CLAYTON ALFRED PATRICK	CORCORAN EDWARD WALTER	D EUSTACHIO THOMAS GERARD
CLEARWATER NORMAN WILBUR	CORDOVA OSCAR	DACEY BERTRAND JAHN
CLEELAND DAVID	CORE JAMES ALBERT	DAIELLO VINCENT THOMAS
CLEMENS ROGER O	CORLEW ROY KENNETH	DAIGLE LOUIS VINNIE
CLERKIN JOSEPH	CORLEY JOHN THOMAS JR	DAILEY GERALD LEE
CLOKES ROBERT	CORNELL STEVEN THOMAS	DALEY WALTON GARLAND
CLORE LEE WILLIAM	CORNISH LARRY IRVING	DALKE BURTON WARD
CLOSSON JAMES STANLEY	CORNWELL JOHN BRUCE	DALY JAMES JOSEPH
CLUNE BRIAN J	CORREA LUIS FELIPE	DALY RICHARD VINCENT
CLUTE MICHAEL ALLEN	CORREA MICHAEL STEVEN	DAMMER WILHELM KARL
COACHMAN JAMES LEE JR	CORSON RICHARD P	DANCE ALAN
COATES PAUL JAMES	COSTANZA KENNETH DAVID	DANIEL ANDREW JAMES
COATES ROBERT EDMUND	COTTERILL MICHAEL J	
COBB JAMES PAUL	COTTO MODESTO JR	
	COULON JOHN GERARD JR	
	COUTURE JOHN VICTOR	
	COVELLA JOSEPH FRANCIS	
	COVENY DAVID PAUL	
	COVEY ELWOOD D JR	
	COVINGTON CLAUDE HENRY	
	COWAN JOHN R	

165

New York

DANIELS CHARLES E
DANIELS DAVID RICHARD
DANIELS REX MARTIN
DANILUCK PETER JOSEPH
DARLING ALBERT JAMES III
DARLING LLOYD THOMAS
DARNELL GEORGE RICHARD
DART WALTER JOSEPH JR
DASARO AGOSTINO WILLIAM
DASH SAMUEL LEE JR
DASHNAW JOHN MYRON
DATENA VINCENT ANTHONY
DAUBER JERRY MARTIN
DAVENPORT ALBERT ASHLEY
DAVENPORT RICHARD
DAVIES KEITH RICHARD
DAVILA ALFREDO
DAVILA FRANK
DAVIS AARON JR
DAVIS ARTHUR RAYMOND
DAVIS ERNEST RAY

DAVIS EVERARD AARON
DAVIS JAMES
DAVIS JAMES GREGORY
DAVIS JOHN HENRY
DAVIS JOHN HENRY
DAVIS MICHAEL EDWARD
DAVIS RICHARD JR
DAVIS STANLEY ROY
DAVIS WILLIAM LOUIS
DAY OSCAR ALFRED
DE ANGELIS DOMINIC A
DE ANGELIS DOMINIC JOHN
DE BONO ANTHONY JAY
DE DOMINIC ROBERT MARIO
DE FELICE LAWRENCE JOSEPH
DE FRANCO JAMES CLINTON
DE GENNARO JOSEPH
DE HIMER MARTIN JAMES
DE LA HOZ CARLOS A M
DE LOOZE JERALD FREDERICK
DE LUCA THOMAS STEVEN JR
DE MARCHI FRANK JR
DE MARCO MICHAEL GREGORY
DE MARIA FRANK F JR
DE MARINIS THOMAS JOSEPH
DE MARSICO MICHAEL JAMES
DE MEOLA RAYMOND WARREN
DE MICHAEL DENNIS JOHN
DE MICHELLE CRAIG NORMAN
DE NISCO THOMAS JOSEPH
DE RUBEIS FERNANDO
DE RUE DAVID JOHN
DE SOCIO DANIEL JOSEPH
DE VAULT MARVIN ANDREW
DE VEGA DUANE ALFRED
DE WALD JOHN FRANCIS
DE WITT LAWRENCE
DEAN THOMAS JOLLEY

DEARDEN ALLEN KENNETH
DEBLASIO RAYMOND VINCE JR
DECKER BERTON
DECKER DAVID FRANKLIN
DECKER MICHAEL THOMAS
DEDEK JOHN FRANCIS
DEDRICK DWIGHT A
DEE KENNETH SAMUEL
DEGEN ROBERT
DEITCH DAVID
DEL TERZO COLOMBO PHIL
DELANO PETER FRANK
DELAPLAINE DONALD LYNN
DELGADO REINALDO
DELL THOMAS CARL
DEMETRIS VASILIOS
DEMUNDA GERALD ANTHONY
DENARDIS CLAUDE CHARLES
DENGLER JOHN LEO
DENHOFF ALAN BRIAN
DENISOWSKI STANLEY GEORGE
DENNISON JAMES RICHARD
DENSLOW GEORGE ROBERT
DESMORE LAWRENCE
DESO BERTRAM ANTHONY
DETOMASO CHARLES PHILIP
DEUSEBIO FRANK CESARE
DEVERS PAUL ANTHONY
DEVINE RICHARD WILLIAM
DEVINE THOMAS EDWARD
DEVLIN JOSEPH WILLIAM
DEVOS WILLIAM MARINUS
DEWANE RICHARD ALLEN
DEWEY LARRY RICHARD
DEWISPELAERE REXFORD JOHN
DEWYEA RONALD RICHARD
DEYNEKA CARL

DI CAPRIO PAUL JOSEPH
DI DOMIZIO JOHN
DI FATE RALPH DOUGLAS
DI FINIZIO LOUIS CARL
DI GREGORIO JOSEPH
DI GUARDIA ALEXANDER NICH
DI LANDRO JOSEPH JOHN
DI NIRO WILLIAM WARREN
DI PACE RALPH JOSEPH
DI ROBERTO ROBERT
DI SAPIO DONALD ANTHONY
DI TOMMASO ROBERT JOSEPH
DIAKOW ROBERT
DIAMOND STEPHEN WHITMAN
DIAZ ALEJANDRO
DIAZ FRANCISCO
DIAZ GILBERT
DIAZ PEDRO
DIAZ RAFAEL ANGEL
DIBBLE MORRIS FREDERICK
DICE CARL RICHARD
DICKERSON STANLEY HEMAN
DICKINSON JOHN ALBERT
DIDAMO RALPH ANTHONY JR
DIKER GEORGE JR
DILALLO JOHN LAWRENCE
DILGER HERBERT HUGH
DILIBERTO KIM MICHAEL
DILLETT LENO RENALDO
DILLON RAYMOND LAWRENCE
DIMOULAS CHRISTY TED
DINGWALL JOHN FRANCIS
DINUNZIO CARL LAWRENCE JR
DISCHHAUSER DIETER HERBER
DISTEFANO FERDINANDO
DITORO WILLIAM FENTON

DIXON DAVID ALLEN
DIXON JOHN ALANSON
DOANE STEPHEN HELDEN
DOBASH JOHN ERNEST
DOCKERY ROOSEVELT GEORGE
DOCTOR GARY DEAN
DODD JOSEPH JAMES
DODGE JEFFREY BRUNS
DODSON LEONARD
DOHERTY MARTIN STEPHEN
DOILEY ARTHUR LEROY JR
DOLVIN JAMES RICHARD
DOMIANO PETER PAUL
DOMINKOWITZ MICHAEL JOHN
DONLON MICHAEL PATRICK
DONNELL ROBERT A II
DONNELLY ALAN CHARLES
DONOHOE CHARLES VINCENT
DONOHUE WILLIAM EDMOND
DONOVAN ARTHUR EDMUND
DONOVAN JOSEPH MICHAEL
DONOVAN MICHAEL G III
DONOVAN PETER MICHAEL
DONOVAN ROBERT JOHN
DONOVAN WILLIAM JOSEPH
DOODY ALBERT CHARLES
DORAN WILLIAM JOSEPH
DORIA RICHARD ALBERT
DORNER ROBERT ANDREW
DORSEY WILLIAM BANFIELD
DORSEY WILLIAM TIMOTHY
DOSTER HENRY JAMES
DOTY VAUGHN ORMON
DOUGHER THOMAS EDWARD
DOUGLAS CLARK ROBERT
DOUGLAS ROBERT EDWARD JR

NURSES OF VIETNAM

CAROL DRAZBA
ELIZABETH ANN JONES
DIANE ORLOWSKI
ELANOR GRACE ALEXANDER
PAMELA DONOVAN
ANNIE RUTH GRAHAM
SHARON LANE
MARY KLINKER

DOWD THOMAS BROWN
DOWELL GILBERT
DOWNEY GERALD JOSEPH
DOWNEY JOHN FRANCIS JR
DOWNIN RAYMOND CHARLES
DOYLE MICHAEL WALTER
DRAGONE JAMES VINCENT
DRAGOTI JAMES ROBERT
DRAKE JOHN PETER
DRAPER WILLIAM MICHAEL
DRAY DONALD BARRY
DREW THOMAS FRANCIS
DREYER THEODORE HENRY
DRUM THOMAS
DRUZINSKI KARL WALTER
DRY MELVIN SPENCE
DUELK JOSEPH DAVID JR
DUFFNER WILLIAM FRANK
DUFFY DANIEL WALTER
DUFFY KEITH WILLIAM
DUGAN JOHN FREDERICK
DUGAN KEVIN HOWARD
DUGAN KEVIN JOHN
DUGNESS PETER
DUMAS DAVID DONALD
DUMOND DAVID EDWARD
DUNCAN LEON TIMOTHY
DUNHAM RICHARD FRANCIS
DUNKENBERGER DAVID
GEORGE
DUNLOP JOHNSTON
DUNN GERALD
DUNN MICHAEL JOHN
DUNNE GERARD JOSEPH
DUPONT RALPH PETER JR
DUPRE CHARLES VAUGHN
DUPRE LARRY DAVID
DUPREE WILBERT SHELBY JR
DUPREY ARTHUR RAYMOND
DURANT RICHARD HENRY
DURHAM WILLARD DUANE JR
DURR BRIAN FRANCIS
DURRETT THADDEUS
DURYEA ARNOLD MAX
DUSART KENNETH WALTER
DUTCHER JIMIE DALE
DUTTON CHARLES MATHEW
DWYER ROBERT MARTIN
DYCZKOWSKI ROBERT RAY-
MOND
DYER MARTIN BARRY JR
DYKE KENNETH
DYKE STANTON RICHARD
DYMERSKI ALFRED JOHN
DZIARCAK WILLIAM WALTER
DZIENCILOWSKI JAMES
EARLEY CLARENCE ANDREW
EASLEY TIMOTHY
EASTMAN JESSE GEORGE
EATON CLIFFORD LYMAN
ECHEVARRIA JOSE ANIBAL JR
ECHEVARRIA RAYMOND

LOUIS
EDDY RICHARD NELSON
EDELMAN IRWIN LEON
EDWARDS EUGENE
EDWARDS JOHN JAY
EDWARDS RANDOLPH A
EDWARDS RONALD CHARLES
EDWARDS THOMAS WILLIAM
EGAN DONALD JASON JR
EGAN FRANCIS XAVIER
EGLIN CHARLES WILLIAM III
EHRMENTRAUT JOHN E JR
EISERT HAROLD BERNARD JR
EISNER JAMES WILLIAM
EKWELL THOMAS JANES
ELAM WALTER ALAN
ELBERT GEORGE STEVEN
ELDER EUGENE
ELDRIDGE THOMAS FARRELL
ELFLEIN MICHAEL FREDRICK
ELIA GARY LAWRENCE
ELIOT BRUCE JR
ELLIOTT RICHARD
ELLIS FRANK JOSEPH G JR
ELLIS JOHN MICHAEL
ELLISON GREG BENSON
ELLSWORTH ELMER EDWARD
ELMANDORF ARTHUR DEWEY
EMERLING JOHN PATRICK
EMERY CHARLES HENRY JR
ENGELHARDT ALBERT ALOIS
ENGLAND GARY LLOYD
ENGLE RICHARD EUGENE
ENGSTROM BRUCE EINAR
ENNERS RAYMOND JAMES
ENZINNA JOHN JOSEPH
EPPS JAMES
EPPS PATRICK BEVERLY
ERENSTOFT DAVID KARL
ERIKSEN ALF EDWARD
ERSKINE ALBERT
ESNAULT JEAN CLAUDE T E
ESPOSITO FRANK CARL
ESPOSITO WILLIAM
ESPOSITO WILLIAM JR
EVANGELISTA FRANK PAUL
EVANS ERIC WILLIAM
EVANS GEORGE AUGUSTA
EVANS JERRY BRIAN
EVANS JERRY DEWAIN
EVANS JOHN HARPER JR
EVANS PAUL MICHAEL
EWING ARTHUR RICHARD
FACCIO ROBERT DANIEL
FAKO JOHN STEPHEN
FALDERMEYER HAROLD JOHN
FALK THOMAS EDWARD
FALLON THOMAS J JR
FANELLA LAWRENCE AN-
DREW
FANNING JOSEPH PETER
FARINARO GUIDO

FARR DAVID LE ROY
FARRAR ERROLD RUFUS
FARRELL ALBERT JAMES JR
FARRELL GERALD MARTIN
FARRELL KENNETH JAMES
FARRELL WILLIAM PETER
FASSEL GARY CARL
FAUCETT GARY LEE
FAULK PAUL
FAUSER RUSSELL JAY JR
FAVROTH CHARLES
FEBO-BETANCOURT IVAN
ROBE
FEBUS OCTAVIO
FEDASCH PETER
FEGAN RONALD JAMES
FELDEN ANTHONY WAYNE
FELICIANO NOEL JESUS
FELLINGER WILLIAM G JR
FELLOWS DAVID THOMAS
FELSHAW JOHN ARTHUR
FELTER ROBERT CHARLES
FENNESSEY DAVID LEE
FERGUSON RICHARD HAROLD
FERGUSON WALTER JR
FERNANDEZ GARY DENNIS
FERNANDEZ JORGE L
FERNANDEZ MANUEL FOR-
TUNAT
FERNANDEZ RENE
FERNHOFF CURTISS
FERO RONALD MILLER
FERRARA MICHAEL JOHN
FERRAZZANO JOHN RAY-
MOND
FERRIL JOHN HENRY II
FETNER HAROLD EVERETT
FETTER KENNETH LLOYD
FETTUCCIA FRANK
FICARA JOSEPH
FIDIAM AARON GREGORY JR
FIELD GARY EDGAR
FIELDS JAMES EDWARD
FIELDS MICHAEL DAVID
FIELLER RICHARD BURDICK
FIFFE JOHN CHARLES
FIGUEREDO CARLOS
FIGUEROA ADAN
FILIBERTI RUSSELL LOUIS
FILIPPELLI ALFRED ANDREW
FINGER DAVID HAROLD
FINGER SANFORD IRA
FINK HUBERT JOSEPH
FINK WILLIAM MICHAEL
FINKEL CHARLES
FINLAY EDWARD ARTHUR
FINNEGAN DENNIS WILLIAM
FINNEGAN JOHN JOSEPH
FINNEGAN ROBERT MICHAEL
FINNEY JAMES JR
FINSTERWALDER RICHARD
KEI

FINTER GEORGE AIKMAN
FISCHER GEORGE WARREN JR
FISCHER LOUIS HAROLD
FISHENDEN ARTHUR ERIC
FISHER ARTHUR
FISHER FRANK CLARK
FISHER RICHARD JAMES
FISHER RONALD EZELL
FISK BARRY KEVIN
FITZGERALD PATRICK VIN-
CEN
FITZGERALD ROBERT MI-
CHAEL
FITZGIBBON THOMAS
GEORGE
FIUME JAMES ROCCO
FLAHERTY KEVIN MICHAEL
FLAHIVE WILLIAM JOSEPH JR
FLANIGAN THOMAS F II
FLAVIN PATRICK JAMES
FLEMING MORRIS LAFOND
FLIEGER GERARD JOHN
FLINT RAYMOND LLOYD
FLORES-JIMENEZ ANGEL
RAMO
FLORIO FRANK
FLORIO ROLAND LOUIE
FLOWERS DANIEL THOMAS
FLYNN DANIEL JOSEPH
FLYNN FREDERICK HAROLD
FLYNN WILLIAM PATRICK
FODARO THOMAS ANTHONY
FODEN JOHN JOSEPH
FOLEY BRENDAN PATRICK
FOLEY BRIAN ROBERT
FONDA PETER FRANCIS
FOOTE FERNANDO VICENTE
FORAN PATRICK JOSEPH
FORBES KEVIN LYNN
FORBUSH ROBERT WALDRON
JR
FORCE RODGER DENNIS
FOREMAN JOHN WILLIAM
FOREST DONALD STEVEN
FORKL ROBERT WAYNE
FORTE RICHARD MICHAEL
FOSTER CLEVELAND
FOSTER DANIEL JOHN
FOSTER EVERETT EDWARD
FOSTER GEORGE
FOSTER JAMES CLAIR
FOSTER ROBERT EUGENE
FOTI PAUL JOHN
FOURNIER NELSON EDWARD
FOX DAVID NELSON
FOX GERALD LAWRENCE
FOX LARRY ROSS
FOX THOMAS JOSEPH
FRANCO CHARLES STEPHEN
FRANKLIN KEITH KOY
FRANKS IAN JACK
FRANTZ MAXWELL STOWELL

FRASCA RICHARD PATRICK
FRASIER DENNIS WILLARD
FRATELLENICO FRANK
ROCCO
FRAZIER TIMOTHY JOSEPH JR
FREDERICK PETER JOSEPH
FREEMAN FURNACE JR
FREEMAN LESTER
FREEMAN RONALD WILLIAM
FREEMAN WILLARD
FRENCH ALBERT LEROY
FRESE STEVEN ROBERT
FREYNE BERNARD ANTHONY
FRINK RICHARD W
FRISBIE JARED ARTHUR
FRISBIE WALTER CHARLES
FRITZ NICHOLAS HEFER
FRY GEORGE HAROLD
FRYE JOHN R
FRYER EDWARD ALBERT JR
FUENTES FRANCISCO
FUERST GEORGE JOSEPH
FULKERSON ROGER ALAN
FULLAM JOHN JOSEPH JR
FULLER JOHN F
FULLER MICHAEL ALLAN
FUNK JOSEPH JOHN
FUNK ROBERT NELSON
FURLONG EDWARD FRANCIS
JR
FUSCO PAUL RICHARD
GABRYS STEPHEN MICHAEL
GAGLIARDI GREGORY
GAGLIARDO FRANK ANDREW
GAISER LEWIS BERNARD
GAJAN ALTON LOUIS
GALANTE RONALD ALFRED
GALBRAITH MICHAEL JOSEPH
GALEA MICHAEL
GALENO ANTHONY MICHAEL
GALINDEZ MANUEL ANTO-
NIO
GALL ROBERT JOSEPH
GALLAGHER GERALD THOM-
AS
GALLAGHER LARRY HERBERT
GALLAGHER MICHAEL
JOSEPH
GALLAGHER PATRICK
GALLAGHER PATRICK JOSEPH
GALLERY RICHARD MULROY
GALLOWAY EMMITT
GALUTZ JAMES ANTHONY
GAMBINO JOSEPH JR
GAMBOA DAVID HERCLIFF JR
GANNON EUGENE RICHARD
GANNON GERALD WILLIAM
GANTT GRADY JR
GANZY ALLAN ALPHONSA
GAPINSKI ROBERT VICTOR
GARCIA BENJAMIN
GARCIA EDWARD
GARCIA EDWARD MARC
GARCIA JUAN RAFAEL
GARCIA LEANDRO
GARCIA LUDIN
GARCIA ROBERT IRA
GARDELIS NICHOLAS LEWIS
GARDNER ALEN LOUIS
GARDNER WILHIMON
GARI STEPHEN LOUIS
GARIEPY CRAIG BARRY
GARITY CHARLES JOSEPH JR
GARLAND RONALD EDWARD
GARLO MICHAEL
GARRAHAN ERNEST EDWARD
GARRETT ALLEN MORGAN
GARRISON WILLIAM LAW-
RENCE
GARRITY ANDREW JAMES
GARY CYE
GASCON GARY LYNN
GASTON JUAN
GATES ALBERT HENRY JR
GATES RICHARD PALMER
GATTI DENNIS ALBERT
GATTO DANIEL ARTHUR
GAY CHARLES ELBERT
GAZARD WILLIAM
GEARING WILLIAM CARL JR
GEARY JOHN MICHAEL
GEARY JOHN WESLEY
GEARY ROBERT FRANCIS JR
GEBBIE RONALD JACKSON

New York

GEDDES KERRY RICHARD
FOST
GEIGER WALTER THOMAS
GELB ALAN STUART
GELUSO SALVATORE ANTHO-
NY
GENAU CLARENCE HAROLD JR
GENCHI BERNARDINO
FRANCIS
GENERAL LESLIE NEIL
GENTILE HAROLD FRANCIS
GENTILE JAMES RAYMOND
GEOGHEGAN JOHN LANCE
GEOGHEGAN PETER DANIEL
GEORGE HEZEKIAH
GEORGE KENDALL EMANUEL
GEORGE-PIZARRO ARTHUR
GERMAIN JAMES THOMAS
GERSHNOW STEVEN ANDREW
GERSPACH PETER JOSEPH III
GERSTEL HOWARD MARTIN
GERSTENLAUER PETER F
GHERARDINI SERGIO JOHN
GIACONE JOHN ALBERT JR
GIACOPPO STEVEN JR
GIAMMARINO VINCENT
FRANK
GIANNELLI ALAN ROBERT
GIARDINA STEFANO
GIBBONS BRIAN FRANCIS
GIBBONS JOHN MICHAEL
GIBBS CHARLES EDWARD
GIBBS HAROLD DOUGLAS
GIBBS MATHEW
GIBLIN WALTER EDMUND JR
GIDEON RICHARD EDWARD
GIERAK GEORGE GREGORY JR
GIFFORD ROBERT ALLEN
GIGLIO PHILIP
GILBERT GLENN RAYMOND
GILBERT LARRY
GILBERT PATRICK JOSEPH
GILL WILLIAM DANIEL
GILL-BEY MEHMET ALI
GILLEN ROBERT C
GILLETTE WILLIAM JAMES
GILLEY RICHARD ALAN
GILLIES JAMES FRANCIS
GILLINGHAM RICHARD KIRK
GILROY WILLIAM THOMAS
GIOIA JOHN ALBERT
GIORDANO ANDREW MIX
GIORDANO JOSEPH CHARLES
GIRETTI ANTHONY ALFRED
GIULIANI RICHARD ANTHO-
NY
GLADDING DANIEL GEORGE
GLADNEY JOHN WILLIE
GLASSEY JOHN GIRARD
GLAZER HERBERT JAY
GLEASON ARTHUR A
GLENN EDWARD FRANCIS JR
GLESSING FERDINAND W JR
GLIDDEN ARTHUR
GLIM RALPH THEODORE
GLINNEN EDDIE DANIEL
GLOVER DOUGLAS J
GLOVER LARRY
GLOWACKI DANIEL NORBERT
GLYNN JOHN JOSEPH JR
GLYNN PETER JOHN
GLYNNE MICHAEL THOMAS
GODDEAU THOMAS ARTHUR
GODFREY JOHN JR

GODLEY LOUIS HENRY
GOETZER JOSEPH JAMES JR
GOGGIN JOHN PHILIP
GOLDA EDWARD
GOLDSMITH DAVID PETER
GOLDSTEIN STEVEN VICTOR
GOLWITZER RONALD AN-
THONY
GOMEZ JESUS EPHRAIM JR
GOMEZ-MESA LUIS G
GONZALES CARLOS LUIS
GONZALEZ CONRAD NICH-
OLAS
GONZALEZ RAMON HER-
NANDEZ
GONZALEZ ROBERT
GOODHEART WILLIAM
GOODIN DANIEL EUGENE
GOODIN SYDNEY UEL
GOODMAN BRUCE TED
GOODRICH EDWIN RILEY JR
GOODRICH THOMAS WEN-
DELL

GOODSELL OWEN DAVID
GORDIAN LARRY BERNARD
GORDILS LOUIS ALFREDO
GORDON JOHN PATRICK JR
GORE GREGORY
GORMLEY JAMES
GORTON GARY BRUCE
GOSS BERNARD JOSEPH
GOSS JEFFREY KENNETH
GOTT HERBERT D III
GOULD CARLTON EDGAR
GOULD FRANK ALTON
GOULD WILLIAM IRVING
GRABOW OTTO CHARLES
GRACE DENNIS FREDERICK
GRAEBNER SIEGFRIED LOUIS
GRAHAM ARMAND ROY
GRAHAM BURDETTE DELROY
GRAHAM FLOYD JR
GRAHAM JAMES
GRAHAM JOHNNIE JR
GRAHAM RICHARD FRANCIS
GRAHAM SAMUEL HENRY II

GRANATO FRANK
GRANDE JOSEPH JOHN JR
GRANDE ROBERT JOSEPH
GRANIELA JOSE ANTONIO JR
GRANOFF ROBERT HOWARD
GRANT KELLUM WARREN
GRASSI ERNEST JOSEPH
GRASSO PAUL DAVID
GRAUERT HANS HERBERT
GRAVER RAYMOND CHARLES
JR
GRAVES DONALD LAVERNE
GRAVES TERRENCE COL-
LINSON
GRAY HAROLD EDWIN JR
GRAY JAMES ANTHONY
GRAY RALPH
GRAY THOMAS ALAN
GRAY WILLIAM EARL
GRAY WILLIAM RUSSELL JR
GRAYSON JOE EDWARD
GRAZIANO ANDREW ALBERT
GRAZIER RUSSELL ALLAN

GRAZIOSI FRANCIS GEORGE
GREELEY TIMOTHY MARTIN
GREEN CARL FRED
GREEN CLAUDE HEWITT
GREEN GEORGE RICHARD JR
GREEN JOHN JR
GREEN LARRY
GREEN MOSES
GREEN RICHARD HERSHEL
GREEN RICHARD JR
GREENE CHARLES RICHARD
GREENE JOSEPH
GREENE RICHARD HAYWARD
GREENIDGE MICHAEL ROD-
NEY
GREENWOOD PAUL JOHN
GREGORY THOMAS ESTON
GREILING JOHN FREDRICK
GREINER DONALD HENRY
GRENZEBACH EARL WILFRE JR
GRESENS JOHN CARL
GRIER LAIFELT
GRIER RONALD EUGENE

168

GRIFASI JAMES ANTHONY
GRIFFIN BRADFORD THOMAS
GRIFFIN HALLIA LEON JR
GRIFFIN LEANDER
GRIFFIN RUDOLPH WILLIAM
GRIFFITH JOHN HOWARD
GRIMES RANDOLPH CLINTON
GROH CHARLES DIETER
GROMPONE JAMES JOHN
GROSS MARK IRWIN
GROUF JACK STEVEN
GRUBB PETER ARTHUR
GRUHN ROBERT AYERS
GRUNBERG RICHARD HENRY
GRZEGOREK JAMES ANDREW
GUASP GARY ARNALDO
GUCWA JOSEPH JOHN
GUENETTE PETER MATHEW
GUICHAUD FRANK JOHN
GULBRANDSEN ROBERT
EIVEND
GULEY DAVID ANTHONY
GULLIVER JOHN JOSEPH
GUNN GEORGE BRUCE
GUNNING LEO BRENT
GUSTAFSON RANDALL JOHN
GUTLOFF EDWARDO LEOP-
OLD
GUTLOFF PETER EMMANUEL
GYORE ALLAN RONALD
HACK RONALD GORDON
HAIGHT STEPHEN HAROLD
HAIRSTON CLIFTON ODELL
HALEN JAMES PAUL
HALL CLARENCE JAY
HALL JAMES HENRY
HALL WILLIE LEE JR
HALLOCK DOUGLAS PAUL
HALLOWS DANIEL JOHN
HALPIN DAVID PAUL
HAMILTON ANDREW LEROY
HAMILTON CHARLES GARY
HAMILTON JOHN DAVID JR
HAMILTON ROBERT THEO-
DORE
HAMM ADOLPH BRINKMAN
JR
HAMMER WILLIAM JOHN
HAMMERSCHLAG WALTER
LUDWI
HAMMOND GERALD ED-
MUND JR
HAMMOND RICHARD MARK
HAMMOND TIMOTHY
ROWLEY
HAMPTON OTIS JAMES
HANDLEY CRAIG WILLIAM
HANLEY THOMAS JOSEPH
HANNIGAN TIMOTHY
CHARLES
HANNIGAN WILLIAM FRAN-
CIS
HANNO MARTIN LARRY
HANSHAW EDWARD PAUL
HARDIMAN LA FRANCIS
HARGRAVE TRACY WALLACE
HARLEY MICHAEL NATHAN
HARNED RICHARD DOUGLAS
HARPER RICHARD WALKER
HARRELL LOVETT LEE
HARRIGAN LAWRENCE
COLBURN
HARRINGTON JAMES A JR
HARRIS GREGORY JOHN
HARRIS JOHN OLIVER
HARRIS ROY GREEN JR
HARRISON CHARLES E JR
HARRISON CHARLES FRANCIS
HARRISON DONALD
HARRISON JOHNNY
HARRISON ROBERT ALAN
HARRISON ROBERT HEERMAN
HARSTER RAYMOND JAMES
HART JOSEPH BRENDAN
HARTLEY JOHN THOMAS
HARTMAN GARY RICHARD
HARTWELL ROBERT ALLEN
HARTZ JOSEPH EDWARD
HASELBAUER JOHN IRVINE
HASENFLUG JAMES MICHAEL
HASSELL NORMAN WINSTON
HASTINGS CARLETON PHILIP
HATCHER DAVID LEE
HATCHETT KYLE HENRY

HATTER LARRY RICHARD
HAUPT WILLIAM HENRY III
HAUSS JAMES ROBERT
HAVENS KENNETH GAGE
HAVERS LARRY RONALD
HAVILAND ROY ELBERT
HAWCO RICHARD JOSEPH
HAWKINS HENRY B JR
HAWKINS NORMAN LEVERN
HAWTHORNE RICHARD
WILLIAM
HAY JAMES STEWART
HAYES BRUCE ROBERT
HAYES WILLIAM JOHN
HAYES WILLIAM THOMAS
HAYNES JIMMY LAWRENCE
HEALY JOSEPH
HEBERT ROBERT W
HEERDT RANDY LEIGH
HEFNER JAMES JOSEPH
HEIKA JOHN ALLEN
HEINZ DENNIS RALPH
HEISER DUANE KENNETH
HEISSENBUTTEL PETER
HERMA
HEITNER DENNIS EDWARD
HELSTROM KENNETH JAMES
HELT HARRY PHILIP JR
HEMER JOSEF
HEMMINGS SEAFORD NA-
THANIE
HEMPEL CHARLES ROBERT JR
HENAGHAN WILLIAM FRED-
ERIC
HENDERSON FREDERICK
HOWAR
HENDERSON GARY LLOYD
HENDERSON HENRY F III
HENDERSON ROBERT CAU-
FIELD
HENDRIX ROBERT EDWARD
HENNING DOUGLAS ALLEN
HENRICH RICHARD FRED-
ERICK
HENRIQUEZ JOSEPH STEPHEN
HENRY DANIEL BENEDICT
HENRY EUGENE EARL
HENRY WILLIAM JAMES
HENS JOHN MICHAEL
HENSON FRANK THEODORE
HENSON THOMAS GEORGE
HENTSCHEL ROBERT EDWARD
HERBERT REGINALD MILMEN
HERING MARK RICHARD
HERLIHY JOHN HENRY JR
HERNANDEZ FELIX
HERNANDEZ JULIO JR
HERNANDEZ MANUEL
HERNANDEZ PEDRO JR
HERNANDEZ-CARRION
GILBERT
HERNANDEZ-DIAZ MIGUEL
ANGEL
HEROD ARTHUR WEINMANN
HEROLD RICHARD WALTER
HERSHAN WILLIAM
HESSE GEORGE BERNARD
HEUSEL ALBERT FRANCIS JR
HICKEY JAMES WILLIAM
HICKMAN VINCENT JOSEPH
HIGGINS DENNIS EDWARD
HIGGINS EDWARD C III
HIGGINS JOHN LAWRENCE
HILBURGER MICHAEL J
HILERIO-PADILLA LUIS A N
HILL DALE ARTHUR
HILL ERNEST JAMES
HILL HERMAN LINWOOD
HILL HUGH GILBERT JR
HILL JAMES WALLACE JR
HILL ORVILLE EDWARD
HILL RICHARD ALFRED
HILL VICTOR C
HILLARD WILLIAM JAMES II
HILLMAN RONALD JOSEPH
HILLS RICKY J
HILTS DAMION RANDOLPH
HINES JOHN CHARLES
HINES WILLIAM JOSEPH
HINSON DON
HINTON RICHARD
HIRST KENNETH LEWIS JR
HITRO BERNARD GEORGE JR
HIVELY ROBERT LYNN

HOAG PAUL RICHARD JR
HOARE THOMAS JOSEPH JR
HOCK RONALD FRANCIS
HOCKRIDGE JAMES ALAN
HODGE CLAUDE ARTHUR
HODGES ERNEST MAEHUE
HOFFMAN IRWIN LEWIS
HOFFMAN KENNETH JAMES
HOGAN EDWARD JOSEPH
HOGAN JOHN BERNARD
HOGENBOOM DENNIS NOR-
MAN
HOLL GEORGE WILLIAM
HOLLEDER DONALD WALTER
HOLLENBACH DONALD
WALTER
HOLMES DAVID
HOLT EDWARD EUGENE
HOMEYER FREDERICK
HOMINICK HOWARD HUGH
HOMMEL DANIEL JOHN
HOOTS RICHARD MAXWELL
HOOVER GORDON WOOD
HOPKINS JAMES FREDRICK
HOPKINS PAUL ROBERT
HOPKINS THOMAS
HOPPOUGH DENNIS KARL
HOPSON FREDERICK WAYNE
HORAN JOHN WILLIAM
HORDERN DAVID JAMES
HORN DONALD FRANCIS
HORNE WAYNE MORRIS
HORNYAK JOHN JOSEPH
HORTON JOHN MARTIN JR
HORTON JOHN RICHARD
HOSTEN CLIFFORD ARTHUR
HOTALING DENNIS MICHAEL
HOTTELL JOHN A III
HOUGHTALING FLOYD W III
HOURIGAN WILLIAM JOSEPH
HOUSE JOHN ALEXANDER II
HOUSE RICHARD ALLAN
HOVEY VERNON FLETCHER III
HOWARD ELI PAGE JR
HOWARD TAYLOR BROOKS JR
HOWCOTT HENRY GRANT
HOWDEN ROBERT WILSON
HOWE CHARLES LEE
HOWE FRANK ROBERT
HOWE LEROY CHARLES
HOYT LARRY LEONARD
HOYT LAWRENCE WILLIAM
HUBBARD THOMAS
HUBSCHMITT ELBERT R JR
HUCKS WALTER HERMAN
HUDSON GEORGE ALEX JR
HUDSON JOHN BARDEN
HUDSON LESSAINT PETER
HUEFFNER RICHARD ALAN
HUESTIS JOHN EDWARD
HUESTIS ROGER EDWARD
HUEY DONALD RAYMOND
HUFFMAN RONALD PETER
HUFSCHMID ROBERT GEORGE
HUGGS HAROLD SYLVESTER
HUGHES EDWARD JOHN JR
HUGHES GRAHAM
HUGHES JOHN JAMES
HULL ARNOLD MELVIN
HULSE GEORGE EDWARD III
HUNT DANIEL THOMAS
HUNTER MARVIN LYNN
HURLIHE RICHARD RAY-
MOND
HUTCHINS FRANK JOHN
HUWYLER JOSEF S
HYLER NELSON MICHAEL
HYMAN WILLIAM ALTON
HYNES ROBERT JOHN
HYSON RAYMOND LEE
IGLESIAS JULIO A III
IMERESE JAMES DAVID
IMPELITHERE ALAN JOHN
INGLESTON STARET JOHN
INGRAM ARTIE
INSLEE RAYMOND STEPHEN
INTIHAR JOHN THOMAS
IOANNI LORENZO JOSEPH
IORIO LEWIS PATRICK
IOZZIA SALVATORE
IRIZARRY JOSE ANGEL
IRVING STANLEY NIXON
IRWIN ROBERT HARRY
IVEY GLEN SIMMANG

JACARUSO FRANK
JACKOWIAK HENRY PATRICK
JACKOWSKI DENNIS EUGENE
JACKSON ABRAHAM
JACKSON ALFRED
JACKSON EDWARD HENRY JR
JACKSON EDWARD JR
JACKSON JOHNNIE
JACKSON LAMONT
JACKSON LEON JEROME
JACKSON NOBLE
JACKSON RAYMOND COLUM-
BUS
JACKSON SANFORD LEVON JR
JACKSON STANLEY ALLEN
JACKSON TERRENCE TURNER
JACKSON THOMAS PETER JR
JACOBS CHRISTOPHER
JACOBS EDWARD DANIEL JR
JACOBS JAMES
JACONETTI KENNETH RICH-
ARD
JACQUES DONALD
JAKEL CRAIG JAMES
JAMES JOHN HENRY JR
JAMES LEE ALLEN JR
JAMES MARC STEVEN
JAMES SAMUEL JR
JAMES THELBERT ALLYSON
JAMES THOMAS
JAMIESON GARY LEE
JAMILSKI MARIAN
JAMRO ROBERT JAMES
JAMROCK STANLEY M
JANKOWSKI RICHARD JOHN
JANOSKA JOHN JAY JR
JANOWSKY CARL EMIL JR
JANSEN ARTHUR RUSSELL
JANTZEN LEONARD FRED-
ERICK
JARVIS DAVID LEONARD
JEDNAT ERIC JOHN
JENKINS BARRY DOUGLAS
JENKINS ISADORE
JENNINGS JOHN MICHAEL
JENNINGS LAWRENCE MAR-
TIN
JERNIGAN CHARLIE MIZZELLE
JERRO WILLIAM GEORGE
JESSIE MARSHALL
JESSMAN JAMES HENRY
JESZECK JOSEPH COBDEN
JIMENEZ ANASTACIO
JIRSA PETER JOSEPH
JOHANNSEN GUSTAV ALFRED
JOHANSON WAYNE
JOHNS MICKY JAMES
JOHNSEN WILLIAM ARTHUR
JOHNSON ALAN HOWARD
JOHNSON BERNARD DEREK
JOHNSON CHARLES WILLIAM
JOHNSON COLLIE JR
JOHNSON DAVID ARNOLD
JOHNSON DONEL RAY
JOHNSON EDWARD A JR
JOHNSON GERALD
JOHNSON HENRY LOUIS JR
JOHNSON HERBERT BURTON
JOHNSON HERBERT NICH-
OLAS
JOHNSON JACK
JOHNSON KENNETH
JOHNSON KENNETH
JOHNSON LARRY WAYNE
JOHNSON LAWRENCE EV-
ERETT
JOHNSON LEROY
JOHNSON MATTHEW JR
JOHNSON RICHARD HERMAN
JOHNSON ROBERT EDWARD
JOHNSON ROBERT LEE
JOHNSON ROBERT LEONARD
JR
JOHNSON THEODORE W
JOHNSON WEBSTER BEREAL
JOHNSON WILLIAM LOVETT
JOHNSON ZEBULON MURPHY
JR
JONES CHARLES JR
JONES HAROLD DANA
JONES JAMES BRUCE
JONES JERRY
JONES JOHN MONROE JR
JONES LESTER

JONES MARVIN CARL
JONES ORVILLE NELSON
JONES ROGER LARRY
JONES RONALD RUSSELL
JONES RONNIE LEE
JONES THOMAS GEORGE
JONES THOMAS PAUL
JONES WALTER CHAPMAN III
JONES WILLIAM BARTON
JORDAN HENRY CRAWLEY
JORDAN PATRICK MICHAEL
JORDAN ROBERT PATRICK
JORDAN TERENCE PATRICK
JOSEPH AUSTIN RAYMOND
JOSEPHS NOEL FITZROY
JOURDANAIS THOMAS F JR
JOUVERT VICTOR MODESTO
JOY CHESTER JOSEPH
JOY RICHARD DENNIS
JOYCE DANIEL THOMAS
JOYCE WALTER ALOYSIUS
JOZEFOWSKI THOMAS JOSEPH
JUDD DONALD R
JUDGE WILLIAM CHARLES JR
JUERS ROY JAMES
JULES GEORGE HENRY
JUSTICE WILLIAM PAUL
JUSTINIANO VICTOR A JR
KADETZ GARY STEVEN
KALER RICHARD DAVID
KANE BRUCE EDWARD
KANE COLEMAN JOHN JR
KANE DENNIS JAMES
KANE WILLIAM GERARD JR
KANGRO LAURI
KARAMAN FRED
KARDASH KENNETH MI-
CHAEL
KARINS JOSEPH JOHN JR
KARN WAYNE DOUGLAS
KAROPCZYC STEPHEN ED-
WARD
KASAI THOMAS TARO
KASSATKIN PAUL
KASTENDIECK WILLIAM
PETER
KATAVOLOS ROBERT
KATZ ELKER GURTH
KAUFMAN HAROLD JAMES
KAUFMAN JAY ALLEN
KAULBACK PETER JON
KAUS HARRY LEONARD JR
KAZEKEVICIOUS JOSEPH
HENR
KAZMIERCZAK ROBERT
JOSEPH
KEARNEY DONALD BRIAN
KEARSE JULIUS JOEY
KEATING DANIEL JAMES JR
KEAVENEY THOMAS ROBERT
KEDENBURG JOHN JAMES
KEEBLE EDWIN AUGUSTUS JR
KEELER WILLIAM CHARLES
KEELER WILLIAM HOWARD
KEELS MARLOWE EUGENE
KEENAN DENNIS JOSEPH
KEEPNEWS JOHN ARTHUR
KEHOE DOUGLAS BERNARD
KEHOE MICHAEL JOSEPH
KEITH KENNETH ARCHIBALD
KEITT CHARLES JOSEPH
KELLER LEROY HENRY
KELLER NORMAN LAWRENCE
KELLEY BERNARD JAMES
KELLOGG GREGORY JAMES
KELLY FATHIES JR
KELLY GLENN ERROLL
KELLY JAMES KEVIN
KELLY MICHAEL JOSEPH JR
KELLY ROBERT FRANCIS
KELLY ROGER EDWARD
KELLY WILLARD DOUGLAS
KELLY WILLIAM MARTIN
KEMELMACHER ROBERT
KEMP FREDDIE
KEMP FREDERICK DONALD
KENDRICK JAMES MICHAEL
KENEALLY CORNELIUS PAUL
KENNY JOHN HENRY
KENT KENNETH ROSS
KEOGH MARTIN JEROME
KERN DANIEL OLMSTEAD
KERNAN MICHAEL ROBERT
KERNDL BRUCE EDGAR

New York

KERR WESLEY SHEPPARD
KESSLER JULIUS ALLEN III
KESTER RICHARD LEE
KEYES ARNELL
KIBBEY RICHARD ABBOTT
KIDD PETER ALAN
KIDD PHILLIP MERIDITH
KIEFHABER ANDREW JOHN
KIENER KENNETH RICHARD
KILKENNY FRANK JOSEPH
KIMMEL ROBERT CHARLES
KINDEL JAMES CARL
KING JON MARC
KING JOSEPH CEPHUS JR
KING LESTER
KING THOMAS GEORGE
KINGSTON GEORGE HENRY JR
KIRCHER ALFRED GEORGE
KIRKLAND EUGENE H
KISCH ROBERT ANTHONY
KISSINGER RONALD CLAYTON
KITSON JOHN FRANCIS
KLEIN HENRY IRVING JR
KLEINHANS LAWRENCE CHARLE
KLENERT WILLIAM BLUE
KLESTINEC ALBERT F JR
KLETINGER HANS
KLINGEN JOHN EDWARD
KLINGMAN RONALD ARTHUR
KLOS DANIEL EDMUND JR
KLOSTER THOMAS HENRY
KLOTZ MICHAEL PETER
KMETYK JONATHAN PETER
KNAPP DAVID JOHN
KNAPP FREDRIC WOODROW
KNAUS RICHARD A
KNECHT ADAM DYCKMAN
KNEELAND PAUL JAMES
KNICKERBOCKER IRWIN LEE
KNIGHT ALBERT S III
KNIGHT BRYAN THEOTIS
KNIGHT JAMES ROY
KNIGHT KEVIN PETER
KNOBLOCK JOSEPH M JR
KNOPF JOHN FRANCIS
KNOTT KEITH ROBERT
KNUTSEN DONALD PAUL
KOBERLEIN CHARLES ERNEST
KOCH KENNETH EDWIN
KOCH LAWRENCE GEORGE
KOEHLER DAVID JAMES
KOENIG ROY ROBERT
KOHLAND RICHARD GLEN
KOHLMEIR GEORGE JOHN III
KOHLRUSCH WILLIAM FREDERI
KOMAROWSKI PETER MARK
KONEVAL ARTHUR PAUL
KOOMAN GARY ROGER
KOPACSKA JOHN CARL
KOPIK EDWARD STANLEY
KORDASIEWICZ HARRY JAY
KOZMA CARL NOEL
KRAFT MICHAEL ALBERT
KRANZ WILLIAM FRANCIS JR
KRASHES HAROLD DAVID
KRASNOFF ARNOLD ROSS
KRAUSS RONALD IRWIN
KRAUSS WALTER JOSEPH JR
KRAXNER FRANK IMRE
KREC FRANK
KRESSE WOLFGANG EDWARD
KREUSCHER DONALD ED-

WARD
KRIEGER FRANK ANTHONY
KRISCHE JOHN DANIEL
KROBETZKY RAYMOND GEORGE
KROL JOHN LEWIS
KRONTHALER PAUL JOHN
KROUSLIS JOHN DAVID
KRUKOWSKI EDWARD EU-GENE
KRUKOWSKI EDWARD STE-PHEN
KRUMM ROBERT CHARLES
KRYSZAK THEODORE EUGENE
KUBINCIAK ROBERT JOSEPH
KUCICH JOHN ANDREW
KUHNE WILLIAM
KUIPER JOHN FREDERICK
KULPA RICHARD WALTER
KUPCHINSKAS PAUL NORMAN
KUPKOWSKI JOHN WALTER
KURTOWICZ JAMES DAVID
KYSER DOUGLAS MASON
KYZER RAYMOND BERT
L HUILLIER JOSEPH ANDRE
LA BARBERA RICHARD F
LA BARR EDWARD LYNN
LA BOUNTY GENE ALFRED
LA FLAIR RICHARD LEON
LA FLEUR JAMES GEORGE
LA FOUNTAIN ROBERT ALAN
LA GRANGE LANCE
LA GRAY ERNEST JAMES
LA LAND GEORGE EUGENE
LA POLLA JOHN ANTHONY
LA PORTE DAHL JAMES
LA ROCCO ANTHONY
LA ROSE JOSEPH RHUBEN
LABIANCA MICHAEL
LABOMBARD CLIFFORD GEORGE
LABOY JAIME
LABOY NEFTALE JOHN
LACKNER MICHAEL ALEX-ANDER
LACOSTE MICHAEL THOMAS
LADOUCEUR LANNY GUY
LAFFERTY JOHN ARTHUR
LAGERWALL HARRY ROY
LAGODZINSKI ROGER THOMAS
LAGUER JOSE ENRIQUE
LAIRD RICHARD FRANCIS
LAIRD ROBERT MURRAY
LAJEUNESSE CLEMENT FOSTER
LALLAVE ALFRED
LAMBERT DENNIS MICHAEL
LAMBERT GARY RAMOND
LAMBY CHARLES MICHAEL
LAMITIE TYRONE FRANCIS
LAMPMAN KENNETH WAYNE
LANCASTER RICHARD P JR
LANDERS DONALD FRANCIS
LANDI GEORGE FRANCIS
LANDOR JOHN JOSEPH
LANDRINGHAM ROBERT GEORGE
LANE DENNIS WILLIAM
LANE JAMES THOMAS
LANE LOUIS MICHAEL
LANGHORN GARFIELD M
LAPARDO ANTHONY N
LAPES DONALD ARTHUR
LAPORT LEONARD OSCAR
LARACUENTE ERNESTO LUIS
LARKIN JOHN PATRICK
LARKIN WILLIAM RONALD
LAROCK REXFORD ADELBERT
LARSON BRUCE STANLEY
LARUE DONALD EDWARD JR
LASHER ERNEST REGINALD JR
LASSEN DAVID HENRY
LATHAN GEORGE
LATOUR CARL JOSEPH
LAUREANO-LOPEZ ISMAEL
LAVERY GREGG EUGENE
LAWENDOWSKI JOHN JACOB
LAWRENCE SEWALL KENT
LAWTON MICHAEL EUGENE
LAWYER ALFRED LEWIS
LAYMON MICHAEL DIGNON
LEAHY RICHARD JAMES
LEAHY ROBERT MICHAEL

LEAKE JOHNNY H
LEBITZ STEPHEN JR
LECASTRE KENNETH JOHN
LEDERMAN MELVIN
LEE JAMES HOWARD
LEE MARZEL RAY
LEESER LEONARD CHARLES
LEHMAN PETER ALLEN
LEIBA LAWRENCE E
LEIGH THOMAS ANTHONY JR
LEMCKE DAVID EARL
LENHARD HOWARD THOMAS
LENNON JERRY
LEO THEODORE THOMAS
LEON WILLIAM
LEONE JOHN FRANK
LEONOR LEONARDO CAP-ISTRAN
LERCH JOHN CHRISTIAN JR
LESANDO NICHOLAS PETER JR
LESLIE PHILLIP WILLARD
LESS REUBEN ANTHONY
LESURE ERNEST ESTELL
LESZCZYNSKI WITOLD JOHN
LEVATO FRANK
LEVINE ROBERT
LEVINSON JAY BARRY
LEVULIS JOHN JOSEPH
LEVY BRUCE
LEVY NORMAN STANLEY
LEVY WALTER NEVILLE
LEWIS ERIC OAKLEY
LEWIS FRANKLIN CHARLES
LEWIS GARY
LEWIS MICHAEL LOUIS JR
LEWIS MOSES JOHN
LEWIS PAUL
LIA NICHOLAS ANTHONY
LIFRIERI PAUL J
LIGAMMARI NICHOLAS PAUL
LIGHTBOURNE RICHARD GREGO
LILLA JOHN THOMAS
LINK JOHN JOSEPH
LIPTON JOSEPH PRICE
LISBOA RAFAEL
LISCUM RONALD FRANCIS
LISIEWSKI FREDRICK ALLEN
LISZCZ ROBERT STANLEY
LITTLE PAUL FREDERICK
LIVERMORE KEITH WARREN F
LIVINGSTON JOHN DEWEY
LIVINGSTON PETER B
LJUNG CARL LOUIS
LO GRASSO RALPH ANTHONY
LO MAURO ROBERT BRUCE
LOBACK THOMAS JOHN
LOCKHART JOHN THOMAS
LODGE ROBERT ALFRED
LOESCHNER THEODORE R JR
LOFARO MARCELLO JAMES
LOGSDON CLIFFORD DOUG-LAS
LONERGAN HAROLD SHER-MAN
LONEY ASHTON NATHANIEL
LONG JOSEPH LEROY
LONG PATRICK JEROME
LONGABARDI MICHAEL JOSEPH
LOONEY EDWARD MICHAEL
LOPEZ HECTOR
LOPEZ RICARDO
LOPEZ-GARCIA GEOVEL
LOSCUITO NED NATALE
LOVAN PETER JOHN
LOVE DANIEL HALEY
LOVETT PETER LOUIS
LOW GEORGE
LOWE THOMAS MICHAEL
LOYA PAUL NELSON
LOZADA CARLOS JAMES
LUCAS GLENN ALLEN
LUCAS JAMES FRANCIS
LUCCHESI GIANCARLO
LUCISANO ROCCO ROSARIO
LUCIW HENRY
LUCKENBACH RICHARD M JR
LUDECKER ROBERT
LUDWIG CHARLES
LUDWIG LARRY GEORGE
LUMM CHARLES LAVERN
LUNA ANGEL RAFAEL
LUND RALPH JAMES

LUND WILLARD SPENCER
LUPIEN DAVID G
LUPO JOSEPH CHARLES
LUPOLI ALBERT FRANCIS JR
LUTHER ROBERT BENJAMIN
LUTZ HANS PETER
LYNCH BERNARD
LYNCH CARL DONALD
LYNCH CHARLES AARON
LYNCH JAMES JOSEPH III
LYNCH MICHAEL
LYNCH MICHAEL JOSEPH
LYNCH PETER
LYONS JOHN JOSEPH
LYONS THOMAS JOSEPH
LYSAGHT ROBERT JOHN
MAC LEAN WESTON DAVID
MAC LEOD ROBIN DOUGLAS
MAC MILLAN GORDON ALAN
MAC NEIL DOUGLAS GERALD
MACCIO DONALD JOSEPH
MACEDONIO CARMINE ANGELO
MACEY EARL FRANCIS JR
MACFARLANE WILLIAM JOSEPH
MACK WILLIAM JAMES
MACKAY PAUL ALFRED
MACKEY DONALD ANDREW
MACKO CHARLES
MACOMB ORRIE E JR
MADEL ROBERT THOMAS
MADISON FRANK ANTHONY
MADISON HENRY JR
MAGISTRO ANTHONY PHILIP
MAGRI GIUSEPPE
MAGUIRE DANIEL JOHN
MAGUIRE KEVIN JAMES
MAHER PAUL IVAN
MAHON RICHARD MICHAEL
MAIN RICHARD HAROLD
MAINARDY GEORGE WILLIAM
MAIORANA RONALD VIN-CENT
MAJER CHARLES ANTHONY
MAJOR ALLAN STANLEY
MAJOR KENNETH CARROLL JR
MAKOWSKI WILLIAM JOHN
MALABE JULIO
MALAVE-RIOS ABELARDO
MALDONADO ABRAEL
MALDONADO JORGE JOSEPH
MALDONADO JOSE
MALEWICZ EDWARD A JR
MALEY CHARLES THOMAS
MALIN MICHAEL LEE
MALLOY THOMAS VINCENT
MALMANIS ULDIS JACK
MALONE LEO FREDRICK
MALONE RICHARD LEE
MALONEY DANIEL JOSEPH
MALONEY JAMES PATRICK JR
MALONEY JOHN EDWARD JR
MALONEY THOMAS ROBERT
MALOY RICHARD FRANCIS
MAMMOLITTI JOSEPH A
MANCA RONALD JOHN
MANCINI RICHARD MICHAEL
MANCINO SALVATORE JOHN
MANCUSO ANTHONY JOHN
MANCUSO JOSEPH ANTHONY
MANCUSO SALVATORE
MANDARINO JOSEPH
MANDRACCHIA PAUL SCOTT
MANGUAL JOSE MANUEL
MANINO SALVATORE PATRICK
MANN RUSSELL CLAIR
MANNERS RALPH WILLIAM
MANNINA MICHAEL CAR-MELO
MANNING BRUCE K
MANNING ROBERT THOMAS
MANNO MICHAEL RALPH
MANNS EDWARD EMIL
MANSFIELD JOHN MON-TAGUE
MANSFIELD WILLIAM GRAN-VIL
MANZI JOHN PETER
MARABLE WALTER A JR
MARASCO JOSEPH ALLEN
MARASCO PHILIP
MARASCO WILLIAM FRED-ERICK

MARCANO CARLOS ALBERTO
MARCANO WINSTON EL-VETTE
MARCANTONIO JOSEPH FRANK
MARCHESE THOMAS VIN-CENT
MARCIANO LOUIS VINCENT
MARCIN WILLIAM KEITH
MARCONI FRANK JOSEPH
MAREK PAUL STEVEN
MARFURT RICHARD AGUST JR
MARGRO JAMES ANTHONY
MARINO ARIEL
MARKS GEORGE ALFRED JR
MARKS RICHARD EDWARD
MARKUM ROBERT BAILEY
MARKUNAS THOMAS WIL-LIAM
MARLOW JAMES EDWARD
MARQUARDT WAYNE JOHN
MARRONE JOSEPH VIETO
MARSDEN TYRONE CECIL
MARSH DAVID JOSEPH
MARSH JOHN ROBERT
MARSH RICHARD CHARLES
MARSH ROBERT ALLEN
MARSHALL EDDIE LESTER
MARSHALL JACKIE EVERETT
MARSHALL JAMES ALFRED
MARTIN ALBERT
MARTIN CHARLES ROBERT
MARTIN JOHN FRANCIS
MARTIN LAWRENCE
MARTIN MICHAEL JOSEPH
MARTIN ROBERT WILLIE
MARTINEZ ANTHONY VIN-CENT
MARTINEZ FRANK
MARTINEZ ISRAEL JR
MARTINEZ JOSEPH RAYMOND
MARTINEZ RICARDO RAUL
MARTINEZ-MERCADO EDWIN J
MARTURANO JOSEPH A JR
MASHLYKIN KENNETH HENRY
MASIN MERRILL HOWARD
MASINSKI JOHN GEORGE
MASON HAROLD JR
MASON JOHNNIE
MASON ROBERT SCOTT JR
MASOTTI JAMES JOSEPH
MASSA LUIS ALBERTO
MASSARI RICHARD D
MATARAZZO PETER DAVID
MATARAZZO STEVEN
MATEJOV JOSEPH ANDREW
MATHIS JAMES RUFUS
MATHIS WILLIAM LEE
MATIAS-SANTANA FEDERICO
MATTEI-SANTIAGO DANIEL
MATTERA GERALD
MATTESON THOMAS WIL-LIAM
MATTHEW HARRY ERIC
MATTHEWS ALFRED RUSSELL
MATTHEWS GEORGE RUSSELL
MATTRACION PHILIP REGI-NAL
MATULONIS JOHN
MAVROUDIS ANTONIO MICHAEL
MAY ROBERT WALTER
MAY THOMAS ANDREW
MAYER NORMAN ROBERT
MAYER ROBERT P
MAYERS RALPH EMERSON III
MAYNARD GREGORY VAL-ENTINE
MAYNARD JOHN
MAZAL ROGER JAMES
MAZYCK RAYMOND JOHN JR
MAZZONE JOSEPH MARK
MC ALISTER JAMES DAVID
MC ARDLE KEVIN JOSEPH
MC BRIDE ALBERT
MC BRIDE KENNETH GERARD
MC BROOM WILLIAM STAN-LEY
MC CABE HUGH ROBERT
MC CABE JOHN FRANCIS
MC CABE MICHAEL RICHARD
MC CAFFREY CHARLES PATRIC
MC CAFFREY GERALD WIL-

LIAM
MC CAFFREY JAMES WILLIAM
MC CAGG CARLTON F JR
MC CARRON WILLIAM P JR
MC CARTHY BRIAN FRANCIS
MC CARTHY CARL RICHARD
JR
MC CARTHY JAMES JOSEPH
MC CARTHY JOHN NILES
MC CARTHY PETER ROVERT
MC CARTHY ROBERT ALAN
MC CARTHY TIMOTHY JOHN
MC CARTHY WALTER R JR
MC CARTY GLENN MURRAY
MC CARTY WILLIAM JOSEPH
MC CASKEY ROBERT WALTER
MC CLAIN ROY HOWARD
MC CLEAN JOHN HOWARD
MC CLENNAHAN CHARLES
HENR
MC CLURG JAMES WALTER
MC CONNELL GERARD
ROBERT
MC CONNELL JERRY
MC CONNYHEAD JAMES JR
MC CORMACK HUGH JOHN
MC CORMACK JAMES JOSEPH
MC CORMICK PATRICK JOHN
MC COY JAMES GLENDALE
MC CRACKEN RONALD
MC CULLOUGH SYLVESTER
MC CUNE EDWARD JAMES
MC CUTCHEN MARL W JR
MC DANIEL MURAL
MC DAVITT GEORGE FRANCIS
MC DERMOTT PATRICK
THOMAS
MC DONALD GERARD MORRIS
MC DONALD HAROLD JOHN
MC DONALD HENRY
MC DONALD MICHAEL
WILLIAM
MC DONALD RONALD IRVIN
MC DONNELL MARTIN
GERARD
MC DONOUGH JAMES MI-
CHAEL
MC DOWELL ROBERT J JR
MC EACHIN JOHN JR
MC ELROY JOHN LEE
MC ELYNN THOMAS JOSEPH
MC FARLAND TOMMIE LOUIS
MC GARRY JOHN THOMAS
MC GEE HENRY HERBERT
MC GINNIS MICHAEL JOSEPH
MC GOEY JAMES FRANCIS
MC GOLDRICK MICHAEL
JOSEP
MC GONIGAL JOHN P JR
MC GOVERN KEVIN BERNARD
MC GOVERN MICHAEL JOHN
JR
MC GRADE GERARD
MC GRATH DANIEL WILLIAM
MC GRATH EDWARD ALBERT
MC GRATH JAMES PATRICK
MC GRATH PAUL MARTIN
MC GRAW THOMAS EDWARD
MC GREGOR RICHARD
MC GUIRE PATRICK JOHN
MC HALE JOHN BUNCE
MC INERNEY RICHARD NASH
MC INTOSH JAMES CRABB
MC KELLIP ROBERT JR
MC KENNA NELSON WILLIAM
MC KINNEY HUGH RUFUS
MC KINNEY RICHARD HENRY
MC KNIGHT WILLIAM JR
MC LOUD DOUGLAS LYLE
MC LOUGHLIN MILES JOSEPH
MC MAHON LAWRENCE
VINCENT
MC MANUS FRANK JOSEPH
MC MANUS JOHN
MC MULLAN EDWARD MI-
CHAEL
MC NALLY ARTHUR GERALD
MC NAMARA JOHN FRANCIS
MC NEILLY JAMES H
MC NULTY WILLIAM FRANCIS
MC PARLANE MICHAEL
JOSEPH
MC QUEEN FREDDIE
MC RAE ADELBERT EARL

MC ZEAL MARTIN ALLEN
MCDONALD JOSEPH WILLIAM
MCGOVERN MICHAEL DON-
ALD
MCGUIRE MICHAEL KELLY
MCINTYRE JAMES ANTHONY
MCLEAN RODNEY W
MCLEOD ARTHUR EDWARD
MEAD NORMAN ARTHUR
MEADE DANIEL
MEAGHER CHRISTOPHER W
MEARNS ARTHUR STEWART
MEARS JOSEPH HARRY
MEDER PAUL OSWALD
MEDINA CARLOS JUAN
MEDINA ISRAEL
MEDINA ORLANDO
MEEHAN JAMES MICHAEL
MEEHAN RAYMOND PATRICK
MEEKER ROBERT IRWIN
MEERHOLZ CHARLES J JR
MEGLIO ROBERT FRANK
MEIN MICHAEL HAMMOND
MELAHN PETER T
MELECA FRANK
MELISH ARNOLD EDWARD
MELNICK JOEL
MENCHISE MICHAEL J JR
MENDEZ ANGEL
MENDEZ DAVID
MENDEZ ISMAEL JR
MENDEZ-QUINTANA EDWARD
MENZIES ALEXANDER JOHN
NE
MEOLA ANTHONY PAUL
MERCADO DIEGO
MERCADO GEORGE
MERCADO-COLLADO LUIS
ROLA
MERCADO-SANTOS WILFRE-
DO
MERCHANT CARL LEE
MERCURIO JOHN A
MERINO LOUIS PHILLIP
MERKLE EDWARD DANIEL
MESSENGER KENNETH
EDWARD
METCALF LARRY DUANE
METOTT GERALD PATRICK
METTY RAYMOND JAMES
METZ ALAN RUSSELL
MEYER BURT JOSEPH JR
MEYER LOWELL WAYNE
MEZZATESTA FREDERICK
MICHAEL TIMOTHY SHAWN
MICHAEL VINFORD FRANK-
LYN
MICHALSKI JOHN STEVEN
MICHEL DANIEL VICTOR
MICKA RUDOLF
MICKO MICHAEL ALBERT
MIDDLETON RONALD
MIDDLETON WAYNE LEE
MIDGYETTE JOSEPH HAR-
GROVE
MIGLIORE LUCIEN JOSEPH
MILAN-ANAVITARTE LUIS ENR
MILES LARRY ALLEN
MILES RAYMOND GENE
MILES RICHARD ROBERT
MILEY REUBEN JR
MILICH JAMES HENRY
MILK ALLAN ARLYN
MILLE WALTER
MILLER ALLEN ROBERT
MILLER CECIL VERNON
MILLER CHARLES
MILLER DONALD JOHN
MILLER GREGORY L
MILLER HERMAN EUGENE
MILLER JEFFERY ALLEN
MILLER PETER THOMAS
MILLER RICHARD ARTHUR
MILLER ROBERT ARTHUR
MILLER ROBERT HERBERT
MILLETT STEVEN LAWRENCE
MILLINER ROBERT
MILLS DAVID MICHAEL
MILLS DENZELL RAY
MILLS DONALD
MILOS JOSEPH LOUIS
MINARD EARL CHESTER
MINER GEORGE LOUIS
MINO ROBERT E

MINOGUE THOMAS FRANCIS
MINOTTI ANTHONY JOHN
MINUTOLI JOHN ROBERT
MIRANDA OSWALDO LUIS
MISKIMMON JONATHAN JR
MITCHELL ALBERT COOK
MITCHELL CHARLES LEROY JR
MITCHELL DANA WESSON
MITCHELL HENRY ALBERT
MITCHELL MICHAEL JOHN
MITCHELL PETER
MITCHELL PHILIP DANIEL
MITCHELL RICHARD A
MITCHELL THOMAS WILLIAM
MOCKER WILLIAM FRANCIS
MODEN RICHARD SHELDON
MOELLER VINCENT GERALD
ST
MOHAMED MACK PAUL
MOHAMMED NAZIR
MOHR VICTOR ALLEN
MOINESTER ROBERT WILLIAM
MOLANO CHARLES EDWARD
MOLESE DENNIS PATRICK
MOLLICA BENJAMIN GEORGE
MOLLICONE DONALD ALLAN
MONAHAN WILLIAM BRIAN
MONEY WILLIAM WALLACE
MONG WILBUR LEROY
MONGELLI ALEXANDER A
MONIN FRANCIS GEORGE
MONISH RONALD ANTHONY
MONK THOMAS
MONKELBAAN TIMOTHY
JAMES
MONKS JOHN
MONTAGUE DENNIS EDWARD
MONTAGUE WILLIAM JOSEPH
MONTANO WILLIAM AN-
DREW
MONTERO IGNACIO
MONTONE KENNETH MI-
CHAEL
MONTOYA ANTHONY JOHN JR
MONTZ ROGER ELLIS
MOODY FRANCIS
MOODY STEPHEN TRUE
MOORE CARTER LEE
MOORE CHARLES EDWARD JR
MOORE CHARLES L
MOORE DENNIS FRANCIS
MOORE GILLIAM
MOORE JAMES CHARLES JR
MOORE JAMES RODNEY
MOORE JOHN BIGELOW
MOORE JOHN JOSEPH
MOORE ROBERT VICTOR
MOORE RONALD KELVIN
MOORE THOMAS JON
MORALES VICTOR DAVID
MORAN DAVID ALFRED
MORAN TERRENCE
MORANO THOMAS LAW-
RENCE
MORGAN DENNIS EDWARD
MORGAN MELVIN DAVID JR
MORGAN MICHAEL ROY
MORGAN WILLIAM S II
MORIARTY THOMAS WILLIAM
MORINA ANTHONY JOSEPH
MORK THOMAS LEE
MORKA PETER JOSEPH
MORRIS GARY KEVE
MORRIS WILLIAM
MORRISSEY RICHARD
THOMAS
MORROW MICHAEL JOSEPH
MORROW SAMUEL THOMAS
MORROW WILLIAM WALLACE
MORSE JAMES EARL
MORSE RICHARD LUCIAN
MORSE WILLIAM JOSEPH
MOSBACH MICHAEL P
MOSBY JERRY
MOSES JAMES JR
MOSHER HARRY VAN ARNAM
MOSHIER CHESTER JOHN JR
MOSLEY RICHARD JOHN
MOSS CHARLES LEWIS JR
MOSS WILLIAM VANCE
MOSSMAN HARRY SEEBER
MOTA PEDRO JUAN TOMAS
MOTT JOHN JAMES
MOTT JOSEPH ANTHONY

MOTT TERRY WARD
MOTTO THOMAS NICHOLAS
MOWBRAY DOUGLAS RON-
ALD
MOYER CHARLES ALBERT
MRAVAK THOMAS A
MUENCH JOSEPH EARL
MULARZ JOHN BRUCE
MULDOVAN WILLIAM JEF-
FREY
MULLENS ROBERT JOSEPH JR
MULLER JAMES VAN NESS
MULLERVY MICHAEL
MULROONEY GEORGE
MULVEY LAWRENCE PATRICK
MUNIZ CARLOS NOBERTO
MUNIZ-GARCIA LUIS ERNES-
TO
MUNOZ LUIS R
MUNTZ GIRAUD DOMENICO
MURPHY CORNELIUS F JR
MURPHY DENNIS GERARD
MURPHY EDWARD THEO-
DORE
MURPHY JAMES HOWARD
MURPHY JOHN ROBERT
MURPHY JOSEPH THOMAS JR
MURPHY PATRICK JAMES
MURPHY ROBERT DENNIS
MURPHY TERENCE MERE-
DITH
MURPHY WALTER MICHAEL
MURPHY WILLIAM
MURRAY DENNIS BRIAN
MURRAY MARVIN WINSTON
MURRAY MERRITT LEWIS
MURRAY ROBERT CHARLES
MURRAY STEVEN
MURRAY THOMAS J
MURRAY WAYNE PAUL
MURRIN THOMAS JR
MUSCO VINCENT JAMES
MUSICK FRANK FREDERICK
MUSSENDEN GEORGE AD-
OLFO
MUSTO RICHARD FRANK
MUSZYNSKI MICHAEL JOHN
MYERS THOMAS WAYNE
MYERS WAYNE CHESTER
MYSKYWEIZ RICHARD JOHN
NADEAU HAROLD BRADLEY
NAMER MARTIN YALE
NASCHEK MARVIN JOEL
NASS WINFORD ALLEN
NATALE NICHOLAS ANTHONY
NATALE PATRICK HENRY
NAVARRO CARMELO
NEAL CARY
NEAL JOSEPH E R
NEDERLK MICHAEL ALEX-
ANDER
NEIDLINGER JAMES JOSEPH
NELSON CURTIS LEE
NELSON DAVID CHARLES
NELSON DENNIS WAYNE
NELSON RICHARD WILLIAM
NELSON WAYNE HERMAN
NESBITT CALVIN IAN
NESKE JOSEPH EDWIN
NETO JOSE TORRES
NEUBAUER JOHN FRANK
NEUBIA WILLIAM JAMES
NEUSS WILLIAM HENRY
NEVIDOMSKY FRANK THE-
ODORE
NEWELL MICHAEL THOMAS
NEWKIRK THOMAS CLIFTON
NEWMAN DANIEL JAMES JR
NEWSOME WILLIAM LESTER
NEWTON KENNETH PURCELL
NICHOLS MAX E
NICKLAS GILBERT MICHAEL
NICOLINI RICHARD DOMENIC
NIDDS DANIEL RUSSELL
NIEDERMEIER THOMAS
DAVID
NIELSEN CHARLES JOSEPH
NIELSEN ROBERT
NIEVES DAVID
NIGHTENGALE TIMOTHY
JAMES
NILSEN ERIC BJARNE
NISKI LEONARD EDWARD
NIX HENRY LEWIS

NOEL MAURICE THOMAS
NOGIEWICH WILLIAM PETER
NOLAN MICHAEL FRANCIS JR
NOLDIN RICHARD JOHN
NOLTE WILLIAM HARRY
NOONAN THOMAS PATRICK
JR
NORRIS ALIN EMILE
NORTON GEORGE HAROLD
NORTON THOMAS
NORTON THOMAS FRANCIS
NOVEMBER DWIGHT MYLES
NOVOTNY JOHN RAYMOND
NOWLIN FLETCHER JACOB JR
NUEBEL WILLIAM GEORGE JR
NULTON JAMES EDWARD II
NUNZIATO ANIELLO CARLO
NURZYNSKI JOSEPH ANTHO-
NY
NUTLY DANIEL THOMAS
NYE DANIEL EUGENE
O BRIEN KEVIN
O BRIEN PATRICK RORY
O BRIEN THEODORE
O CONNELL DANIEL GERARD
O CONNELL KEVIN GERALD
O CONNER RICHARD EDWARD
O CONNOR EDMUND AN-
THONY
O CONNOR MICHAEL PATRICK
O DONNELL JOHN MICHAEL
O GRADY JOHN FRANCIS
O GRADY MARTIN EDWARD
O HALLORAN WILLIAM BRIAN
O KEEFE MICHAEL ANDREW
JR
O KEEFE PATRICK FRANCIS
O LEARY PAUL FRANCIS
O NEIL TERRENCE EDWARD
O NEILL ANTHONY JOSEPH
O NEILL JAMES RAYMOND
O SHEA STEPHEN JOHN
O SULLIVAN CHRISTOPHER JO
O TOOLE LAWRENCE P II
OAKDEN TERRY LEE
OAKLEY JAMES RONALD
OBERLE STEWART WILLIAM
OBEY DONALD ALTON
OCASIO FELIX
OCHAB ROBERT
ODDO ANTHONY PHILIP
ODIERNO JOHN WILLIAM
ODIOT EDGAR WILFREDO
OHANESIAN VICTOR
OHLER FREDERICK RICHARD
OJEDA NESTOR
OKEEFE RICHARD WILLIAM
OLGYAY ROY CHRISTOPHER
OLIVER BERNARD GEORGE JR
OLIVER KENNETH EARLSTON
OLIVO RAFAEL
OLSEN GEORGE CHARLES
OLSEN GEORGE THOMAS
OLSEN JOHN ANDREW
OLSEN JOHN LOUIS
OLSON TIMOTHY ARTHUR
OLT JOHN PAUL HARRIS
OQUENDO FRUTO JAMES
ORBINO DENNIS MICHAEL
ORLANDI JOSEPH ELISIO JR
ORLANDO ANTHONY MARIO
ORLANDO GASPER
ORNELAS JACK MICHAEL
OROSZ ANDREW JOHN
ORPHANOS THEOPHILOS
ORR ROBERT THOMAS
ORTEGA ANIBAL JR
ORTEGA WILLIAM
ORTEGA WILLIAM JR
ORTIZ DOMINGO
ORTIZ EUGENIO
ORTIZ-RAMIREZ JUAN
ORTIZ-RIVERA ANIBAL JR
ORTIZ-RODRIGUEZ JAIME
ORTON KENNETH WILLIAM JR
OSBORNE LAWRENCE ELSTON
OSTERHOUDT CLIFFORD ROY
OTT JOSEPH STANLEY
OTTMAN TODD WHITNEY
OUELLETTE LEWIS CHARLES
OVERTON JEROME
OWCZARCZAK MELVIN
JOSEPH
OWEN TIMOTHY SAMUEL

New York

OWENS JAMES JOSEPH
OWENS PERCIE EDWARD
OWENS VERNELL
OWENS WALTER ALBERT
OXFORD HARRY EDWARD JR
OYOLA HECTOR DAVID
OZGER ISLAM
OZIMEK RONALD ROBERT
PABST EUGENE MATTHEW
PACETTA COSMO FRANCIS
PACIO GEORGE HENRY
PACKARD ROBERT FRANK
PADDOCK DAVID ALLEN
PADUCHOWSKI PAUL RICHARD
PAGAN MIGUEL
PAGAN-LOZADA WILFREDO
PAGE JOHN GEORGE
PAGE WILLIE LEE
PAGLIARONI ALAN PAUL
PAIGE DOUGLAS ALAN
PALAZZOLA STEPHEN FRANK
PALCIC ERNEST PATRICK
PALENSCAR ALEXANDER J III
PALENSCAR ROBERT JOSEPH
PALLADINO THOMAS ARTHUR
PALMER WALTER
PALMIERI ERNEST
PALUMBO ANTHONY PAUL
PANICCIA RONALD JAMES
PANNELL TYRONE SIDNEY
PAPA FRANK J
PAPE JOHN CHARLES
PAPKE THEODORE ARTHUR
PAPPAS ELEFTHERIOS PANTEL
PAQUIN HOWARD ROBERT
PARADA-BARRERA ANTONIO
PARAMATTO PAUL ANGELO
PARASILITI NICHOLAS
PARDO THOMAS ANTHONY
PARKER GEORGE JOSEPH JR
PARKER JAMES EDWARD
PARKER JAMES LEONARD JR
PARKER MARK EDWARD
PARKER MICHAEL
PARKER RICHARD EUGENE
PARKER STEPHEN VANCE
PARKER WILLIAM HILL
PARMELEE JEFFREY MATHEW
PARSONS DON BROWN JR
PARSONS GARY LEE
PARTINGTON WILLIAM JAY
PARTLOW KENNETH
PASCASCIO RODNEY GUSTAVUS
PASCO ALLEN
PASKINS WAYMAN E
PASTORE ROBERT JOSEPH
PASTRANA VICTOR RAPHAEL
PATENAUDE HAROLD MICHAEL
PATRIZIO CHARLES JOSEPH
PATRONE JOHN THOMAS
PATTERSON RICHARD ALAN
PATTERSON STANLEY
PATTON GEORGE
PAULINO CARL ARTHUR
PAULLEY LARRY
PAULOS FRANK WOLGO
PAULSEN GERARD FRANCIS
PAULSEN NORMAN MACLEOD
PAVAN KENNETH ALAN
PAWLICK HENRY JOHN JR

PAWLOWSKI THEODORE J JR
PAYNE JOHN ALLEN
PEAKE JOE LOUIS
PEARSON DAVID ALLEN
PECK JEFFREY LLOYD
PEDA ROBERT CHARLES
PEDERSEN RUSSELL ALFORD
PEDROSA CARLOS ALBERTO
PELLEGRINO BERNARD MICHAE
PELLEGRINO MICHAEL PHILIP
PELLEW DAVID SEELEY
PELLIZZARI LOUIS JOSEPH
PENFOLD PETER ALLAN
PENNA JOHN ANTHONY
PENNEY DONALD THOMAS
PENSYL DONALD NEIL
PERALEZ LOUIS FABIAN
PEREA EDWARD
PEREZ DANIEL TORRES
PEREZ DAVID
PEREZ JEFFREY
PEREZ LOUIS ANTONIO
PEREZ WILFRED
PEREZ WILFRED M
PEREZ-RIVERA MILTON
PERKINS BOBBY JAMES
PERKINS CHARLES HAROLD
PERRETTA JOHN ROCCO
PERRICHON DONALD HAROLD
PERRINE ELTON LAWRENCE
PERRY FRANK MICHAEL JR
PERRY GRAFTON LAWRENCE
PERRY KARL FREDERICK
PERSHING RICHARD WARREN
PESCE PAUL JOHN
PETERKIN THOMAS DOUGLAS
PETEROY BRUCE EDWARD
PETERS LAWRENCE DAVID
PETERSEN CARL ROBERT
PETERSON ALBERT EUGENE
PETERSON CARL JERROLD
PETERSON ROY KEITH
PETERSON THOMAS PAYNE
PETRAGLIA ANGELO ANDREW
PETRAMALO THOMAS
PETRASHUNE MICHAEL JAMES
PETRILLO JOHN JAMES
PETSOS PHILLIP CHRIS
PETTEYS CORNEL
PEZZULO JOHN FRANCIS
PFEIFER RONALD EDWIN
PFEIFFER JOHN CLIFFORD
PHELPS DAVID CLAYTON
PHELPS DAVID HARLOW
PHELPS WALTER WILLIAM
PHELPS WILLIAM
PHILLIPS DAVID JEFFERY
PHILLIPS WILLIAM RONALD
PIAMBINO JOSEPH ROBERT
PIAZZA ROBERT GARY
PICARAZZI JAMES VINCENT
PICCOLELLA CHARLES VICTOR
PICKARD DENNIS LEE
PICKEL GEORGE WILLIAM
PICKENS JOHNNIE JR
PICKETT STEPHEN WILLIAM
PIERCE RONALD SHAFER
PIERRE CARRIER
PIETRAS FRANK MARTIN
PIGNATARO JULIUS PHILIP
PIKE PETER XAVIER
PINDER JOHN JOSEPH
PINN ARNOLD
PINO ANTHONY CARLOS
PINTA RICHARD THOMAS
PIPPINS GUS
PIREZ-BERGES CARLOS
PITCHES JAMES SUTHERLAND
PITT ALBERT
PITTMAN JAMES SHERWIN
PIXLEY RICHARD GORDON
PIZARRO VIC MANUEL
PIZZUTO LOUIS EDWARD
PLATTNER ERNEST MELVIN
PLAZA JUAN JOSE
PLOTKIN MARTIN LOUIS
PLOTKIN STEPHEN LEWIS
PLUMADORE KENNETH LEO
PLUMEY RAYMOND
POGGI MICHAEL LOUIS

POHL FLOYD WILLIAM
POLCHOW WILLIAM ALFRED
POLDINO THOMAS
POLIZZI SALVATORE FRANK
POLLASTRO DOMINICK
POLNIAK ROBERT JOSEPH
PONTUCK HOWARD SAMUEL
PORTER WILLIAM ROY
PORTERFIELD DAVID EDWARD
POST THOMAS FRANK
POSTIGLIONE JOSEPH JOHN
POTTER JAMES FRANK
POWELL ALFRED LEE
POWELL TONY GORDON
POWELL WILLIAM
POWERS MARTIN ROBERT
POWERS ROBERT LAWRENCE
PRAST MARTIN THOMAS
PRCHAL CHARLES ROBERT
PREISS ROBERT FRANCIS JR
PRENTICE EDWIN PAUL
PRESCOTT STEVEN JAMES
PRETE ROBERT NICHOLAS
PRETTER THOMAS
PREZIOSI JAMES LAWRENCE
PRICE BARRY FRANCIS
PRICE WILLIE CAPAHAS
PRIEST DONALD JAMES
PRINGLE JAMES EDWARD
PRIOR ANTHONY GEORGE
PRIVITAR RICHARD JOSEPH
PROCIDA RICHARD NICHOLAS
PROSCIA RICHARD MICHAEL
PROSSER IRVIN WILLIS JR
PROTHERO WILLIAM HENRY
PRUDEN RENE THOMAS
PRUNER JOHN MARK
PRUNKA ALEXANDER E JR
PTASNICK WALTER JAMES
PUDERBAUGH CHARLES KAY
PUGH ROGER LESLIE
PUGLIESE FRANK
PULASKI PETER JR
PULLEN THOMAS RICHARD
PUMA WAYNE PAUL
PUMILLO MICHAEL
PURDY RANDALL BREWARD
PURELIS JOSEPH KENNETH
PURVIS BERNARD GEORGE
PURYEAR JOSEPH A
PYLE HOWARD MACDONALD JR
QUAMO GEORGE
QUARLES FLOYD ELMER
QUEALY MICHAEL JOSEPH
QUICK ADRIAN ALLEN JR
QUICK ISHAM IKE
QUILLEN JOHN EDWARD JR
QUILLEN LLOYD DANIEL
QUINN DANIEL
QUINN JOHN MICHZEL
QUINN RICHARD FLOYD
QUINN RONALD GENE
QUINN WILLIAM DANIEL III
QUINONES DAVID
QUINONES EDWARD
QUINONES JOSE LUIS
QUINONES JUAN MANUEL
QUINONES-RODRIQUEZ LUIS A
RACHON CHARLES JOSEPH
RAETZ ROBERT WILLIS
RAGUSA FRANK RICHARD
RAHILLY ANDREW STEPHEN
RAINES WARREN HENRY
RAKENTINE KENNETH CARL
RAMIREZ ALBERTO ANTONIO
RAMIREZ NELSON
RAMOS ROBERTO
RAND MICHAEL
RANDALL JAMES LAVERNE
RANDAZZO EDWARD D
RANDOLPH JAMES TIMOTHY
RANDOLPH VAN LA SALLE JR
RANELLUCCI RAYMOND ANTHON
RANSOM ROBERT CRAWFORD JR
RAPPAPORT HAROLD KENNETH
RAPPLEYEA TUNIS E JR
RAPTIS ANGELO CESARE JR

RARRICK JOHN EDWARD
RAUSCH ROBERT ERNEST
RAVER CHARLES DAVID
RAYMOND CARL ROGER
RAYMOND PAUL DARWIN
RAZ FRANK VINCENT
RAZZANO ROBERT THOMAS
REARDON RICHARD JOHN
REDMOND WILLARD THOMAS
REED ANTHONY ERICH
REED BRUCE EDWARD
REED CHARLES MICHAEL
REED DAVID NEAL
REED LOUIS JOSEPH
REEDER BRENT ALEXANDER
REESE GOMER DAVID III
REESE WILLIAM PHILIP
REGAN MARTIN JOSEPH
REGGIO GERARD MICHAEL
REGINALD ROBERT JAMEISON
REHDER ROBERT EDWARD
REICHELT JAMES LOUIS
REICHERT JOSEPH R
REICHERT WILLIAM FRANCIS
REICHLIN JOSEPH ALBERT JR
REID JOSEPH H
REIKMANIS VIESTURS
REILLY JAMES RICHMOND
REILLY JOHN THOMAS
REILLY MARTIN DANIEL
REILLY ROBERT JUDE
REILLY WILLIAM RAYMOND
REINER CHARLES EDWARD
REINHARDT ARTHUR WELKER
REITER BRUCE MARTIN
REITER LESLEY STEVEN
REITHMANN TIMOTHY CHARLES
REITZ MICHAEL ROBERT
RENTAS JOSE CARMELO JR
REPETA HENRY JAMES
RERA ROBERT
REVAK ANTHONY NEAL
REYES ANGEL LUIS JR
REYES HUMBERTO
REYNOLDS DAVID RICHARD
REYNOLDS GEORGE F JR
REYNOLDS RICHARD PETER JR
REYNOLDS THOMAS YORK
RHODES CLIFFORD M JR
RHODES JOHN JOSEPH
RIAL RICHARD FRED
RIALE RICHARD WILLIAM
RIBEIRO JOSEPH FRANCIS
RIBITSCH ERIC
RICCIARDO RONALD FRANCIS
RICCIONE STEVEN BLAINE
RICE FRANCIS DAVID
RICE HERBERT CHARLES
RICE ROBERT
RICETTI CHRISTOPHER JOHN
RICHARDS DANIEL PAUL
RICHARDS DONALD JUAN
RICHARDSON BRUCE
RICHARDSON EUGENE
RICHARDSON JOHNNIE BRYANT
RICKMERS ROLF ERNST
RIEGEL ARTHUR WILLIAM JR
RIES WILLIAM STUART
RIGGS STEVEN JAMES
RIJOS TONY
RILEY JOHN PATRICK
RILEY PAUL WILLIAM
RILEY THOMAS JOHN
RINGHOLM JOHN AZEL
RINKER FRANCIS M
RIOS JOSE TOMAS
RIPEL JOHN KENNETH
RISING ALBERT CHARLES
RITCHIE DAMON LIGOURI
RITTLINGER DONALD ANDREW
RITZ DAVID GERALD
RIVERA CARLOS MANUEL
RIVERA DAVID
RIVERA EMILIO
RIVERA FERNANDO A JR
RIVERA JAMES
RIVERA JESUS
RIVERA JOHN ASDRUBAL
RIVERA JUAN
RIVERA MIGUEL ANGEL
RIVERA MIGUEL ANGEL

RIVERA RAUL
RIVERA SANTOS JR
RIVERA-BALAGUER RAFAEL L
RIVERA-CRUZ CRISTOBAL
RIVERA-DELVALLE MANUEL A
RIVERA-GALARZA BENIGNO
RIVERA-GARCIA WILLIAM
RIVERA-REYES JOSE ALBERTO
RIVERA-TRINIDAD NESTOR JU
RIZZI RALPH JOSEPH
RIZZO JOHN MICHAEL JR
ROACH SYLVESTER
ROBBINS ARNOLD LEE
ROBENA CHARLES EDWARD
ROBERTS CYRUS SWAN IV
ROBERTS HARLEY RICHARD
ROBERTS JOHN HENRY
ROBERTS JULIUS JR
ROBERTSON LEONARD
ROBILOTTO GEORGE FRANCIS
ROBINSON DONALD FREDERICK
ROBINSON GEORGE RAY
ROBINSON JOHN
ROBINSON LEROY
ROBINSON STANLEY A JR
ROBINSON TERRY ALAN
ROBLES CECILIO JR
ROBLES-MIRANDA JOSE ANTON
ROCCO RICHARD MICHAEL
ROCHE JON PATRICK
ROCHE MATTHEW PETER JR
ROCHEZ ESTEBAN VALERIANO
ROCKEFELLER RONALD EDWARD
ROCKENSTYRE RICHARD
ROCZEN ALEXANDER ANTHONY
RODNEY CARLISLE ANTHONY
RODRIGUEZ ALBERT EDUARDO
RODRIGUEZ CARLOS MARIO
RODRIGUEZ COLON RICARDO
RODRIGUEZ DAVID
RODRIGUEZ EDWARD
RODRIGUEZ ISRAEL
RODRIGUEZ JACK CHARLES
RODRIGUEZ JOSE ESTABAN
RODRIGUEZ NICK NATHANIEL
RODRIGUEZ RALPH O
RODRIGUEZ RAMON SAUL
RODRIGUEZ RAYMOND
RODRIQUEZ JOAQUIN
ROE JOHN PHELEN
ROECKL CHARLES
ROEDERER JOHN STEPHEN
ROEMER DONALD PETER
ROESCH HEINZ KURT
ROESE ALAN JOHN
ROGERS PHILLIP
ROGERS WILLIAM JAMES IV
ROHRING KEVIN MICHAEL
ROHTVALI ARVI
ROMAN JEREMIAS
ROMAN MARK JOSEPH
ROMANO AUGUST
ROMANO GERALD MICHAEL
ROMANO MICHAEL STEPHEN
ROMEO DUANE CLARK
ROMERO-DE-JESUS BENJAMIN
ROMESSER RICHARD JAMES
RON GRIJALBA HUMBERTO
ROPETER LESTER EARL
RORABACK KENNETH M
ROSA JOHN MICHAEL
ROSA JUAN ANTONIO
ROSADO ERNEST SR
ROSARIO AGUSTIN
ROSE ANDREW CLAYTON
ROSE LAWRENCE OLIVER
ROSEBRUGH FRANCIS PAUL
ROSEDIETCHER HOWARD
ROSENBERG KENNETH
ROSETO JOHN
ROSOLIE WALTER WILLIAM
ROSS DAVID SETH
ROSS GEORGE BACON JR
ROSS KENNETH
ROSS KEVIN HENRY
ROSSANO RICHARD JOSEPH
ROSSI RUDOLPH
ROSSI VINCENT LOUIS SR

ROSSINI RONALD STEPHEN
ROSSOTTO VINCENT JOSEPH
ROST JAMES FRANCIS JR
ROTGER GUSTAVO JR
ROTONNELLI JOHN
ROUNTREE HARVEY F JR
ROUSE GORDON ARTHUR JR
ROUSSELL CLARENCE A JR
ROWCROFT MICHAEL J.
ROY GERALD RAYMOND
ROZELL EDWARD ARNOLD
ROZO JAMES MILAN
RUANE MICHAEL PATRICK
RUBADO CHARLES FRANCIS
RUBIN ROY GARLAND
RUBINS JOHN CHARLES
RUCH FRANCIS WILLIAM II
RUDERSON ANDERSON
LINWOOD
RUDOLPH WALTER WILLIAM
RUDON JOSE ANTONIO
RUDY PAUL CHARLES
RUGGERO VICTOR JOSEPH JR
RUGGIERO ROBERT JOHN
RUIZ ANGEL O
RUIZ CARLOS HERIBERTO
RUIZ JOSE
RUIZ JOSE JR
RUIZ THOMAS
RUMINSKI PHILIP EDWARD JR
RUMMEL DONALD EUGENE
RUMSEY JAY DEE
RUNDLE JAMES JR
RUPERT JOHN MICHAEL
RUSCITO JOHN ANDREW
RUSS PAUL EDWARD
RUSSELL BRIAN PATRICK
RUSSELL GARY LEE
RUSSELL PETER JOHN
RUSSO DENNIS JOSEPH
RUSSO JOSEPH CHARLES
RUSSO MICHAEL CANDIDO
RUSSO MICHAEL L III
RUSSO MICHAEL PHILLIP
RUSSO RONALD SALVATORE
RUST JAMES HENRY
RUTIGLIANO ANTHONY
RUTTAN JAMES EARL
RYAN DONALD JAMES
RYAN JOHN ALOYSIUS JR
RYAN JOHN THOMAS
RYAN LAWRENCE BRENDAN
RYAN THOMAS KEVIN
RYBAK FRANCIS PAUL
SAAVEDRA LUIS FORERO
SACCO JAMES DOMINICK
SADLER RONALD FRANCIS
SAGE ROBERT DAVID
SALANITRO GARY CHARLES
SALAS ORLANDO ALBERTO
SALAZAR ERNESTO VICTOR

SALMIERI JOHN DOMINICK
SALTER JAMES WILLIAM
SALTMARSH JAMES JOHN
SALTZ ERIC DONN
SALVANI RONALD LANDON
SALVO JOSEPH MICHAEL
SAMORAY RICHARD MARTIN
SAMPLES STEPHEN HENRY
SAMUELS JAMES
SAMUELSON ROBERT L
SANABIA OSCAR ENRIQUE
SANCEVERINO GARY AN-
THONY
SANCHEZ ANGEL MANUEL
SANCHEZ CESAR ERNESTO
SANCHEZ DAVID
SANCHEZ JOSE RAMON
SAND JAMES EDWARD
SANDERS THOMAS
SANDERS THOMAS ANDREW
SANDERSON JACK JOHN-
STONE
SANDMAN MITCHELL HAR-
VEY
SANDSTROM HUGH THOMAS
SANFILIPPO FRANK
SANGER STEPHEN CARROLL
SANSEVERINO ANTHONY
SANSONE DOMINICK
SANTANA ANTHONY JOHN
SANTANA JOSE JR
SANTANA JOSE MANUEL
SANTANGELO SAMUEL JOHN
SANTANIELLO VINCENT
BENOR
SANTIAGO ALAN ANGEL
SANTIAGO ALEXANDER P JR
SANTIAGO FELIPE OBED
SANTIAGO HUMBERTO RUIZ
JR
SANTIAGO JOSE JUAN
SANTIAGO ROBINSON
SANTIAGO-COLON HECTOR
SANTIAGO-LUGO JOSE C JR
SANTIAGO-VAZQUEZ BER-
NARDINO
SANTINELLO RALPH MICHAEL
SANTOROSKI MICHAEL PAUL
SANZONE ROBERT BENJAMIN
SANZOVERINO WILLIAM
EUGEN
SAPORITO MICHAEL CHARLES
SAPORITO RONALD
SARDINA FRANK
SAUNDERS BRUCE
SAWTELLE PAUL COBURN
SAXBY JAMES FRANCIS
SAYER JOHN STEPHEN
SCALA RICHARD MICHAEL
SCAMARONI LUIS GUILLERMO
SCAVELLA ALLAN NAPOLEON

SCAVELLA JESSE ELLISON JR
SCHAEFER ALAN FRANCIS
SCHAMPIER ROBERT BRUCE
SCHAPANICK CHESTER
SCHARLACH STEVEN ED-
WARD
SCHEIB RALPH EUGENE
SCHELL ROBERT CHARLES JR
SCHERER CHRISTOPHER J
SCHERLAG ROBERT
SCHETTIG ROBERT SCOTT
SCHIAVONE RALPH
SCHIFRIN RAYMOND RICH-
ARD
SCHILLER JOSEPH FREDERICK
SCHLECHT JOHN III
SCHMID ROBERT ANTHONY
SCHMIDT DARYL JAY
SCHMIDT MARK VEDDER
SCHMIDT RICHARD CARL
SCHMIDT RICHARD MARTIN
SCHMIDT ROBERT GUSTAVE
SCHMIDT WALTER ROY JR
SCHNEIDER GERARD JOSEPH
SCHNEIDER KENNETH
EUGENE
SCHNEIDER KENNETH
EUGENE
SCHOCK HAROLD HENRY
SCHOENER ROGER HARRY
SCHOFER KARL ANDREW
SCHRAMM WILLIAM GEORGE
SCHULER GARY FREDERICK
SCHULER ROBERT HARRY JR
SCHULTE HENRY GERARD
SCHULTZ ROBERT CHARLES
SCHULZ JAMES WILLIAM
SCHUMACHER JEFFREY DAVID
SCHUSTER JOSEPH JOHN
SCHWARTZ ABRAHAM
SCHWARTZ CALVIN ELLIOT
SCHWENDY RANDALL JAMES
SCIBELLI THOMAS ANTHONY
SCICUTELLA JOSEPH
SCOGNAMILIO PATRICK JOHN
SCOLNICK DAVID
SCOTELLARO MICHAEL
BERTRA
SCOTT DENNIS LEE
SCOTT DUANE CARL
SCOTT GARY ARNOLD
SCOTT GARY JAMES
SCOTT HAROLD
SCOTT JAMES GUINAN
SCOTT THOMAS LASANDA
SCOTT WILLIAM
SCOTT WILLIAM GRAVELLE JR
SCRITCHFIELD DAVID ALLEN
SCRUTON KENNETH CHARLES
SCULLY EDWARD ANTHONY
SEABROOK ROY MICHAEL

SEARIGHT JAMES ARNOLD
SEBAST WILLIAM MICHAEL
SEBASTIAN LOUIS JOSEPH
SEDA FERNANDO JR
SEDA PABLO ISREAL
SEDDIG WALTER S
SEEFELDT CHARLES L JR
SEEL JOHN CHARLES
SEELEY WILLIAM ARTHUR
SEGAL JEFFREY BERNARD
SEGARRA LUIS ERNESTO JR
SEIDENSTICKER JAMES
SEILER WILLIAM JOSEPH
SELANIKIO LEONARD
SEMIDEY HECTOR LUIS
SEMINARA CHARLES BEN-
JAMIN
SEMMLER DAVID ALBERT
SENESE CHRISTOPHER LEIGH
SENOR JOHN JOSEPH
SERAVALLI JOHN ANTHONY
SERIO ROBERT FRANK
SERRANO JOHN REYITO
SERRANO MARCO ANTONIO
JR
SERRANO RENE
SERRANO-ECHEVARRIA RAUL
SERWINOWSKI RICHARD EARL
SESSA MICHAEL JR
SETTER RICHARD ALLEN
SETTIMI RONALD MARK
SEUFERT ROBERT JOHANN
SEVERSON ROBERT DARYL
SEYMOUR JAMES THOMAS
SHAFFER JOHN ANDREW
SHAFFER WILLIAM PAUL
SHANKS JAMES LEE
SHATTUCK BERNARD MERLE
SHATTUCK RONALD LAW-
RENCE
SHAVEL FREDRICK STANLEY
SHAW RICHARD EARL
SHEA GARY JOHN
SHEARES JOHNNIE N JR
SHEEHAN ALLEN PAUL
SHEEHY DONALD JAMES
SHELTON TIMOTHY JOHN
SHEPHERD RICHARD DEWITT
B
SHEPHERD THOMAS CHRIST
SR
SHEPPARD GLENN ANDRE
SHERADIN ROBERT DONALD
SHEREDOS ROLAND MIKE
SHERIDAN EUGENE RAY-
MOND
SHERIDAN PHILIP FRANCIS
SHERIDAN ROBERT ROY
SHERLOCK STEPHEN ANDREW
SHERMAN VICTOR P JR
SHERRILL HERBERT

SHINE ANTHONY CAMERON
SHINE JONATHAN CAMERON
SHIPMAN JAMES ROBERT
SHOEMAKER DONALD ELTON
SHORTEN TIMOTHY JOHN
SHORTS WILLIAM VINCENT
SHOUP WILLIAM KING
SHRAMKO MICHAEL ANGELO
SHROPSHIRE RONALD LEE
SHUBBUCK ROLLAND BER-
NARD
SHUMBRIS EUGENE PAUL
SHUMWAY GEOFFREY RAY-
MOND
SHUTT CARL ALVIN JR
SIBILLY JOHN RICHARD
SICKLER HARRY JOSEPH
SICKLES ROBERT PAUL
SIDELKO GEORGE
SIEGWALT MARLIN LYNN
SIGMAN CHRISTOPHER
SCOTT
SIGNA ANTHONY ROBERT
SIKORSKI SIGMOND MICHAEL
SILBERT LEO VINCENT
SILON JOSEPH ARTHUR JR
SILOS FRANKLIN ROSADO
SILVEIRA LEONEL MENDON-
CA
SILVERNAIL DOUGLAS
HAROLD
SILVERSTEIN GERALD LEON
SIMANCAS LUIS JOSE
SIMIELE DONATO JOSEPH
SIMMONS BRADLEY JOSEPH
SIMMONS ELLIOTT JR
SIMMONS JAMES BENJAMIN
SIMMONS ROBERT EUGENE
SIMON PAUL RICHARD
SIMON VICTOR
SIMONDS HAROLD RILEY
SIMONE JOSEPH RALPH
SIMPKINS TIMOTHY HAYES
SIMS HARRY
SINCLAIR GARY PHILIP
SINCLAIR JOHN JAMES
SINCLAIR ROBERT HENRY JR
SINGER MORTON HAROLD
SINGLETON CLIFFORD
RICHAR
SIPE ROBERT VINCENT
SIPOS WILLIAM GEORGE
SIPPEL WILLIAM JAMES
SIRIANNI DANIEL EDWARD
SISARIO FELIX ANTHONY
SISLEY WILLIAM EDWARD
SISSON RONALD PAUL
SITEK THOMAS WALTER
SITO RICHARD ANTHONY SR
SIVATTA MARC ANTHONY
SKEBECK EDWARD JOHN JR

New York

SKOCH EUGENE RICHARD
SKOLITS WAYNE E
SKOMSKI JAMES MARK
SKUMURSKI DAVID LEONARD
SLATTERY JAMES DENNIS
SLAVEN RICHARD E
SLAVIN RICHARD NEAL
SLAYTON CHARLES DEWANN
SLINGERLAND GERALD HOWARD
SLINGERLAND HAROLD J JR
SMALL DONALD BRUCE
SMALLIDGE JEFFERY RONALD
SMASO JACK
SMEAL ROBERT
SMITH ALLEN LLOYD
SMITH ANTHONY ROOSEVELT
SMITH BARRY JAMES
SMITH BRUCE MARTIN
SMITH CARL ARTHUR
SMITH DANIEL J
SMITH DOUGLAS MARK
SMITH EDGAR ARMSTRONG
SMITH EDMOND EUGENE III
SMITH EDWARD BRUCE
SMITH EDWARD DEWILTON JR
SMITH GEORGE EUGENE
SMITH GEORGE JULIUS JR
SMITH GREGG ALLISON
SMITH JAMES LEE
SMITH JAMES ROBERT
SMITH JAMES WILLIAM JR
SMITH JOHN THOMAS
SMITH JOSEPH BERNARD JR
SMITH JOSEPH EARNEST
SMITH JOSEPH FRANK
SMITH JOSEPH JOHN
SMITH LARRY EARL
SMITH LEWIS BENJAMIN
SMITH MICHAEL
SMITH MICHAEL FRANCIS
SMITH MICHAEL THOMAS
SMITH REGINALD EDWARD
SMITH RICHARD ALBERT
SMITH RICHARD FLOYD
SMITH RICHARD ROBERT
SMITH RICHARD VROMAN
SMITH ROBERT CHARLES
SMITH ROBERT CHARLES
SMITH ROBERT JAMES
SMITH ROBERT JEREMIAH
SMITH SIDNEY COURTNEY M
SMITH STEPHEN SCOTT
SMITH WAYNE MICHAEL
SMITH WILLIAM HENRY JR
SMITH WILLIAM PAUL
SMOLAREK EDWIN JOSEPH JR
SNELL RALEIGH JOHN JR
SNOW JOHN FRANCIS
SNOW MILTON JR
SNOW RONALD M
SNYDER CHARLES JOHN
SNYDER DALE MARVIN
SNYDER HAROLD JR
SNYDER ROY HARRISON
SOBEL IRWIN ROSS
SOCHACKI NICHOLES
SODAITIS GEORGE FRANK
SOKAL IRWIN NORMAN
SOLER RAFAEL
SOLIS FELIX
SOLOMON SIDNEY MORTON
SOLTAN LAWRENCE WILLIAM
SOMMA RYUZO

SOMMERER ROBERT JOHN
SONNER EUGENE VINCENT
SOPER JOHN CAMDEN
SORANNO VINCENT MICHAEL
SORCI MARK TIMOTHY
SORRELL SHERMAN AMOS
SORRENTINO GERALD DAVID
SOSA ARISTIDES
SOSA MARCOS JR
SOSINSKI JOSEPH
SOTO CHARLES
SOTO EFRAIN SR
SOTO FELIX F
SOTO ISMAEL
SOTO JOHNNY
SOTO-CONCEPCION JOSE
SOUTHWORTH RONALD HUBERT
SOWLE NED ALEXANDER
SPADARO THOMAS
SPARK MICHAEL MELVIN
SPAULDING JACK DOUGLAS
SPEAR MICHAEL SHELDON
SPENCE ALEX C JR
SPENCE GEORGE ANTHONY
SPICER DONALD FAYE
SPIRES JOHN ALBERT
SPISTO JUSTIN RICHARD
SPITZFADEN ALFRED LOUIS
SQUIRES SIDNEY CHESTER
SROKA JOHN MICHAEL JR
SROKA RICHARD MARION
STACHOWSKI ARTHUR THOMAS
STACY WALTER ROBERT
STAHL JOHN JOSEPH
STAHLECKER GARY ROBERT
STANKIEWICZ KENNETH DAVID
STANNARD DARYL KENNETH
STARBUCK ROBERT FRENCH
STARKES JOHN MILTON JR
STARKEY RICHARD WILLIAM
STARR EDWARD IRWIN
STASIO RICHARD PETER
STATELMAN EDWARD CHARLES
STEC ROBERT MICHAEL
STEED GERALD
STEELE RAYMOND THOMAS
STEEN ANTHONY MICHAEL
STEFFANS MARSHALL GEORGE
STEFFEK EDWARD STEPHEN
STEGELAND JOHN JOSEPH III
STEIER WILLIAM EDWARD
STEIGER WILLIAM FREDRICK
STEIN ANDREW PAUL JR
STEINER JOSEPH R III
STEINKIRCHNER JAMES LEWIS
STEMPER PHILIP JON
STEPHENS MICHAEL JEFF
STERN LARRY
STERN ROBERT ALAN
STETTER RONALD THOMAS
STEVENS RICHARD DURAND
STEWART LAWRENCE
STEWART ROY KENNETH
STIEVE WILLIAM JOHN
STIGER GREG PATTERSON
STIMSON JOHN THOMPSON JR
STIRPE JOHN
STIRRUP WILLIAM DAVID
STOCKFEDER MARTIN
STODDARD MARCUS WILLIAM
STOKES FRANK EDWARD
STOKES RONALD T
STOLZ JAMES EDWARD JR
STONE DANIEL MELVIN
STONE DEE WAYNE JR
STONE LESTER RAY JR
STONEHOUSE ALFRED LEE
STORELLI JOHN
STORZ RONALD EDWARD
STOVER DAVID DONALD
STOW JOHN LEWIS
STRACK LAWRENCE
STRASSNER CORNELIUS WILLI
STRATE JOHN DELBERT
STRIDIRON GEORGE THOMAS
STRIPPOLI JOSEPH PAT JR
STROBEL WILLIAM ERIC
STRONG ANDREW CARNEGI III
STRONG JACK MERRIEL

STROUSE GARY LEE
STRYKER ROBERT FRANCIS
STUART MILES BOYD
STUCKEY HENRY JAMES
STUDIER RICHARD ERWIN
STYMUS GARY LEE
SULLIVAN DANIEL JOHN
SULLIVAN JAMES MICHAEL
SUMMERS CHARLES G H
SUPINO LOUIS VINCENT
SUPPLE JOHN PHILIP
SURETTE WILLIAM WARREN JR
SURLES LOREN CLEVELAND
SUSI ANDREW PAUL
SUSI RAYMOND PETER
SUSSMEIER JAMES JOSEPH
SUTHERLAND REGINALD J
SUTTLEHAN LAURENCE CHRIST
SWAGLER CRAIG EVERETT
SWAN WAYNE ROBERT
SWANE BRIAN EDWARD
SWANKER NELSON CHRISTAN
SWANSON RAYMOND WILLIAM
SWANSTROM DOUGLAS GAYLORD
SWARTZ WILLIAM JOSEPH
SWED ROY FRANCIS
SWEDA JOSEPH R
SWEENEY BRUCE ROBERT J N
SWEET JAMES NEWTON
SWEET JERRY ALAN
SWIDONOVICH NICHOLAS JOHN
SWIECZKOWSKI MICHAEL JOHN
SWISHER LARRY RAYMOND
SYGNATUR JOSEPH JOHN
SZOR HENRY
TAISLER JOSEPH ANDREW
TAMBURRI JOHN RICHARD JR
TAMILIO THOMAS
TANNEY JOHN MICHAEL
TANZOLA CARL JOSEPH JR
TAPPE KENNETH WILLIAM JR
TARANTO DAVID WILLIAM
TARANTO ROBERT JOSEPH
TARBELL CLIFFORD LAWRENCE
TARBELL WILLIAM M
TART CLIFTON LEE
TATARSKI LESLIE MILES
TATE ALEXANDER JR
TATE CHARLES THOMAS JR
TAVAREZ JOSE RAFAEL
TAWIL AARON
TAYLOR DWIGHT JOSEPH
TAYLOR ERIC WYCKOFF
TAYLOR HARRY EDWARD
TAYLOR JACK EDWIN
TAYLOR LOUIS GAINES
TAYLOR PHILIP CHARLES
TAYLOR RAYMOND RALPH JR
TAYLOR ROBERT
TAYLOR VINCENT ANDREW
TEASLEY ROBERT
TELFER ROBERT RAY
TEMPLE MALONE BENNETT
TEREJKO BENJAMIN JOHN JR
TERRILL PHILIP BRADFORD
TERRY RONALD TERRANCE
TESTA DONALD ANTHONY
TESTA RICHARD
TETTE JOHN BERNARD
THACKER JAMES
THIBODEAU WALLATE FRED
THIBOU ALLAN COURTNEY
THIELGES CHARLES THEODORE
THOMAS DANIEL
THOMAS DANIEL PATRICK JR
THOMAS HARRY JR
THOMAS JAMES WELDON
THOMAS JOSEPH HAROLD
THOMAS LEONARD ALAN
THOMAS NATHANIEL
THOMAS NORMAN ARNOLD
THOMAS WYATT STEPHEN
THOMPSON DONALD EARL
THOMPSON FRANCIS JAMES
THOMPSON HARRY NATHANIEL

THOMPSON JOHN BRYAN
THOMPSON JOHN L JR
THOMPSON LAWRENCE CURTIS
THOMPSON ROBERT VINCENT
THOMPSON WILLIAM BERNARD
THOMPSON WILLIAM MATT
THORNHILL WILLIAM JOSEPH
THORNLOW GARY WILLIAM
THORNTON CURTIS FRANCIS
THORNTON WILLIAM D JR
THORP JOHN WILLIAM
THORPE DAVID ALBERT
THORSTEINSON VERNON JOSEP
THREET PIERRE ANATOLE
THRUSH OLIN RICHARD
THURSTON WESLEY GEORGE
TICE PAUL DOUGLAS
TIERNO JAMES
TIMIAN FRANK EDWARD
TIMMS ALFRED
TIMPA JOSEPH JR
TIMS ANDRE BARRY
TINES FRANZ
TINGLEY PHILIP ALLISON JR
TINNEY DONALD WARREN JR
TINSLEY FRANKLIN DENIS
TIRADO DANIEL
TIRICO RICHARD LOUIS
TISSIER RICHARD HENRY
TITUS KARL WILLIAM
TIZZIO PASQUALE JOSEPH
TODARELLO FRANCIS VINCENT
TODD FRANKLIN GODFREY
TODD WILLIAM ANTHONY
TODI JOHN ANTHONY
TOGNERI DANIEL ERNEST
TOKARSKI STANLEY RICHARD
TOMASEK MICHAEL JOSEPH
TOMASOVIC STANLEY ROBERT
TOMASZEWSKI STANLEY JR
TOMASZEWSKI THOMAS DAVID
TOOLE TERRY EDWARD
TOONKEL BENJAMIN RICHARD
TOPORCER ANDREW JAMES JR
TORELLO CARL HARVEY
TORI THOMAS JOSEPH
TORO JOSE MIGUEL
TORPIE WILLIAM JAMES
TORRANCE FREDDY LEE
TORRE PASQUALE
TORRES FERNANDO LUIS JR
TORRES JESUS M
TORRES JOSE ENRIQUE
TORRES ROBERTO
TORRES SANTIAGO JR
TORRES SANTIAGO JR
TORRES VICTOR LUIS
TORRES VINCENT
TORREY RAYMOND DELANO JR
TORTORICI FRANK
TOTTEN KENNETH ROMAINE JR
TOWE EDWARD SCOTT
TOWNES LEROY
TOWNLEY CYRIL HARRIS
TOWNSEND BURDETTE D JR
TOWNSEND GARY RAY
TOWNSLEY THOMAS EDWARD
TRANI FREDERICK EUGENE JR
TRAVIESO JOSE ANTONIO
TRAVIS JON PAUL
TREMBLAY ALAIN JOSEPH
TREMBLAY PATRICK JOSEPH
TRIER KENNETH ROBERT
TRIMM ARCHIE EDWARD
TRINCHITELLA FRANCIS A
TRIPODO BENEDICT JOHN
TROJAHN DARRELL CARL
TROTTER RICHARD BARRY
TROVATO ROSS ANGELO
TRUGLIO ROBERT
TRYON GARY PAUL
TUBBY ROBERT WILLIAM
TUCKER DAVID BRUCE
TUMINO JOHN JOSEPH
TURIANO BENJAMIN ROBERT

TURNBULL JUSTIN GLASTON
TURNER BRENDAN XAVIER
TURNER CHARLES WONDREWS
TURNER JOHN RICHARD
TURNER WILLIAM RICHARD JR
TURNER WILLIE GEORGE
TURSO DONALD ARTHUR
TURZILLI STEPHEN EDWARD
TUTHILL CHARLES PRESTON
TUTTLE HERBERT LEROY JR
TWOREK GERALD JOHN
TYLER LESTER
TYMESON RAYMOND W JR
TYRCZ WALTER FREDRICK JR
TYRRELL WALTER RIPLEY
UGARTE CARLOS
UGINO JOHN JOSEPH
UHL THOMAS FRANCIS
ULRICH GEORGE HENRY
UNDERDOWN GEORGE MICHAEL
UNDERWOOD PAUL GERARD
UPRIGHT EDWIN FRANCIS
URBANCZYK JOSEPH MICHAEL
URQUHART THOMAS
URRUTIA ANTHONY JOHN
VAD HENRY JOSEPH
VALENTE ANTHONY NICHOLAS
VALENTIN MARTINIANO JR
VALENTIN MIGUEL ANGEL JR
VALENTIN RAFAEL
VALENTINO ANTHONY ROBERT
VALERIO THOMAS
VALESKO JOSEPH JR
VALLE GUILLERMO
VALLE HECTOR
VALLEN DONALD WILLIAM JR
VALLIERE STEPHEN CHARLES
VALLONE FRANK
VAN ALST HARRY L JR
VAN ANTWERP WILLIAM M JR
VAN CLIEF LARRY
VAN COOK DONALD F JR
VAN DER SCHANS DONALD EDW
VAN DEUSEN PHILLIP ANDREN
VAN DUSEN JOHN PHILLIP
VAN DUYNHOVEN PATRICK FRA
VAN GELDER WILLIAM H JR
VAN REYPEN ROBERT JULIUS
VAN TASSEL WILLIAM D JR
VAN TASSELL JAMES WARREN
VANDERBROOK GARY LAURENCE
VANDERHEID MARK EDWARD
VARGAS JULIO CESAR
VARNEY KENNETH ARTHUR
VASQUEZ ANGEL RUDY
VASQUEZ CARMELO
VASQUEZ EDDIE
VAUGHAN DONALD FRANKLIN
VAZQUEZ FELIX JR
VAZQUEZ JOSE ANGEL
VEGA FRANCISCO
VEGA-DIAZ HECTOR MANUEL
VELEZ BERT
VELEZ LUIS FELIPE
VELEZ PAUL
VELEZ VICTORIANO
VELILLA WILLIAM
VENNARD JOHN JOSEPH
VER PAULT KEVIN EDWARD
VERA PEDRO ANGEL
VERCOUTEREN EDWARD ARNOLD
VERGALLITO JOHN ANTHONY
VERGARA-ARBIL AUGUSTINE
VERRY FREDERICK ALFRED
VETRANO GERALD MICHAEL
VICICH CHARLES EDWARD
VICKERY GARRY FRANCIS
VICKERY GARRY FRANCIS
VICKS EDWARD JAMES JR
VICTORIA FREDERICK PEARCE
VILLANUEVA HILARIO PIZARR
VINSCOTSKI ROBERT VALENTI

174

VIRGONA JOHN ANGELO
VIRUET JOSE GALENO
VISCONTI FRANCIS EDWARD
VITACCO MICHAEL
VITRO VITO
VOHS WILLIAM FRANCIS
VOJIR JAMES PAUL
VOLLMAR GEORGE THOMAS
VOLLMER DAVID STEPHEN
VONDERCHEK WALTER EARL
VOORHEIS HAROLD ROBERT
VULTAGGIO ANTHONY
WADE MELVIN ALEXANDER
WAGNER WILLIAM WALTER
WAGSTAFF JAMES DONALD
WAINZ ROBERT MICHAEL
WAIT BERNARD JOSEPH
WAKLEE DUANE A
WAKULICH GREGORY PAUL
WALKER JOHN FREDERICK
WALKER JOHN JOSEPH
WALKER RICHARD DUANE
WALKOWSKI PAUL DOUGLAS
WALLACE BRIAN FRANCIS
WALLACE ROBERT CHARLES
WALLENBECK FRANK C
WALLNER FRANZ XAVIER
WALLS ALBERT CALVIN JR
WALSH JEROME WILLIAM
WALSH JOHN MICHAEL
WALSH MICHAEL PATRICK
WALSH THOMAS ROY
WALTERS BRUCE ELLIOT
WALTERS JAMES REESE
WALTON RICHARD FREEMAN
WALTON ROGER EDWIN
WANZEL CHARLES JOSEPH III
WARD FORREST EDWARD
WARD JOHN FRANCIS
WARD JOHN LAWRENCE
WARD LUTHER BURNETT JR

WARD RICHARD HENRY
WARREN JOHN EARL JR
WARREN RICHARD JOHN
WARREN STEPHEN EDWARD
WARSHAWSKY JOEL BARRY
WASHINGTON ALBERT
ALLEYNE
WASHINGTON COLEY LOUIS
JR
WASHINGTON ERICK LEE
WASHINGTON GLENN
WASHINGTON ROBERT
WASHINGTON WILLIE JAMES
WATERS LEON ELDRED JR
WATINGTON RALPH H JR
WATKINS GREGORY H
WATKINS MICHAEL
WATSON ARTHUR
WATSON JAMES THOMAS
WATSON JOHN ELMO
WATTS RICHARD ALLEN
WAUCHOPE DOUGLAS
WAUGH JOHN LOUIS
WAXMAN SAUL
WAY THOMAS URBAN
WEARING MARION BERNARD
WEAVER JERALD BRUCE
WEBBER FREDERICK CARL
WEEKS DAVID L
WEEMS RICHARD QUENTIN
WEIDNER RICHARD JOHN
WEIGLE THOMAS HERMAN
WEINPER ARTHUR J
WEISMAN ALAN N
WELCH RICHARD DENNIS
WELCH STEPHEN MARTIN
WELENOFSKY ERICK RU-
DOLPH
WELKER THOMAS EDWARD
WELLS ROBERT JAMES JR
WELLS WILLIAM

WELSCH GERALD
WEMETTE SCOTT FRANCIS
WENZEL CARL RICHARD
WERNER GREGORY EDMUND
WERNIG RANDY RICHARD
WESIGHAN LESTER ARTHUR
WESTERN RICHARD ALAN
WESTPHAL JAMES FRANCIS
WETZEL WALTER JOSEPH
WHALEN RICHARD D
WHEATON JAMES
WHEELER JAMES CHRISTO-
PHER
WHEELER JOSEPH KEITH
WHITAKER THOMAS EARL
WHITE GREGORY LEE
WHITE JAMES LEE
WHITE JOSEPH
WHITE JOSEPH RUMMEL JR
WHITE STEPHEN ROBERT
WHITE WHITNEY LEE
WHITE WILLIAM SAMPSON
WHITEHEAD ESAU JR
WHITEHILL DAVID HUGH
WHITELEY WAYNE ANTHONY
WHITFIELD CHARLES F JR
WHITFORD LYNN CECIL
WHITLEY ROBERT LEE
WHITMAN THOMAS MICHAEL
WHITTIER JAMES BENJAMIN
WICK GERALD PAUL
WICK MICHAEL RAYMOND
WICKHAM RALPH ARTHUR
WICKS WILLIAM ARTHUR
WIDENER JAMES EDWARD
WIDGER GEORGE JAMES
WIDMANN RAYMOND
WIDOMSKI DANIEL ALBIN
WIEDEMAN ROBERT ARTHUR
WIER MICHAEL BRODERICK
WIESNEIFSKI PETER ROBERT

WILCOX CHARLES CHESTER
WILCOX GARY LEE
WILCOX WILLIAM EIDMAN JR
WILDER AVERY
WILDERS WILLIAM JAMES
WILENSKI STANLEY JR
WILEY GILBERT
WILHELM RICHARD THOMAS
WILKENS JOHN HERMAN
WILKIE CHARLES DAVID
WILKINSON JACK WILLIAM
WILKINSON JAMES JOSEPH JR
WILLETT LOUIS EDWARD
WILLIAMS AMOS LEVERN
WILLIAMS FREDERICK
THOMAS
WILLIAMS GEORGE JOSEPH
WILLIAMS HOWARD
WILLIAMS JAMES THOMAS JR
WILLIAMS JOSEPH JEREMIAH
WILLIAMS NATHAN C
WILLIAMS NEIL STEPHEN
WILLIAMS REGINALD JR
WILLIAMS ROBERT JOHN
WILLIAMS SAMUEL WILLIE
WILLIAMS VAN
WILLIAMS WALLACE
WILLIAMS WILBERT JR
WILLIAMS WILLEY EDGAR JR
WILLIAMS WILLIE ROGERS
WILLIAMSON JOEL STEPHEN
WILLIAMSON WILLIAM
CURTIS
WILSON DAVID WALTER
WILSON GERALD ANTHONY
WILSON HERBERT JR
WILSON NATHANIEL
WILSON PETER JOE
WILSON REGINALD EUGENE
WILSON ROBERT CHARLES
WILSON ROBERT GRANT

WILSON ROBERT LEE
WILSON WILLIAM BERNARD
WILTSIE JOSEPH CARL
WINDHAM JAMES EUGENE JR
WINGENFELD ROBERT JOHN
WINKLER GARY JOHN
WINSLOW JOHN KEMPE
WINSTON CHARLES C III
WINTERHALTER HUGH
FRANCIS
WINTERS MICHAEL JOHN
WINTERS ROBERT J
WINTERS WILLIAM JOHN
WISHER HERBERT JAMES
WISNIER GARY
WISSIG EDWARD SIMON
WISTRAND ROBERT CARL
WITHERSPOON THOMAS JR
WITKOP MICHAEL ERWIN
WITTMAN GORDON RICHARD
WITTMAN WILLIAM
WITZEL ROBERT CHARLES
WOGAN WILLIAM MICHAEL
WOHLMAN STANLEY ROCKY
WOLFENDEN HARRY
WOLFRIES LOWELL ASTLEY
WOLPE JACK
WOOD EDWARD CHARLES
WOOD MELVIN
WOOD RAYMOND CHARLES
WOOD ROBERT VICTOR
WOODHOUSE ROBERT F JR
WOODS EDWARD JOHN
WOODWORTH MARC ALAN
WOYNARSKI RICHARD MI-
CHAEL
WOZNIAK ROBERT ANDREW
WRATTEN GARY PATTERSON
WRAY WILLIAM CLAYTON
WRAZEN GERALD
WRIGHT MARTIN WILLARD
WRIGHT O NEAL
WRIGHT ROBERT
WRIGHT WILLIAM CLAY JR
WROBEL ROBERT JOSEPH
WUERTENBERGER CHARLES
EDG
WUEST LOUIS ARTHUR JR
WULFFERT JOHN LAWRENCE
WUNDERLICH HENRY
WURTENBERG JOHN RICH-
ARD
WYNNE JOHN ROBERT
WYNNE THOMAS EDWARD
WYSOCKI WOJCIECH
WYSZOMIRSKI JOHN DONALD
YANCY JOSEPH STANLEY
YASKANICH WILLIAM ROBERT
YATES DONALD FRANCIS
YATES ROBERT SR
YATTEAU RICHARD FRANKLIN
YONTZ STEPHEN LEO
YORK HENRY
YOUNG ANDREW KUNG
YOUNG CHARLES LUTHER
YOUNG JOSEPH ROBERT
YOUNG WELDON HORACE
YOUNGKRANS ALLAN T JR
ZAPOLSKI LAWRENCE ED-
WARD
ZAWTOCKI JOSEPH STANLEY
JR
ZELDES MARK HILLARY
ZEMANICK WILLIAM JOSEPH
ZENKEWICH GEORGE WALTER
ZERILLI ROBERT JOSEPH
ZEWERT EDWARD JOSEPH JR
ZIELINSKI JOHN PETER
ZIMMER JAMES LEON
ZIMMER JERRY ALLEN
ZIMMERMAN TERRY
ZIMPFER FRED CHARLES
ZIMULIS JOHN JAUTRIS
ZISSU ANDREW GILBERT
ZORNOW ROBERT LAWRENCE
ZUPAN JOHN
ZWERLEIN ROBERT LOUIS
ZYDEL RONALD WALTER

Photos by Jessica Crawford

Vietnam Veterans Memorial

1 E Edenton St, Raleigh, NC 27601

The Raleigh, North Carolina Vietnam Veterans Memorial is entitled "After the Firefight" and was created to honor the hundreds of thousands of North Carolina service members who served during the Vietnam War. This memorial was the first one to be authorized on the Capitol State grounds since the World War II time period. It's located near the "Presidents North Carolina Gave the Nation" monument where Edenton Street and North Wilmington Street meet.

The memorial was designed by Abbe Godwin from Guilford County and was officially dedicated in May of 1987 during the North Carolina Vietnam Veterans Homecoming Salute. The project took four years and Godwin took her time studying the history of the war and collecting artifacts from the period. To make sure she understood exactly what she wanted to portray in her design, she also spent a lot of time talking to local Vietnam veterans. While Godwin designed it, Johnpaul Harris sculpted the monument. The North Carolina Vietnam Veterans Inc. sponsored the Vietnam Veterans Memorial while the necessary funds were collected through private donations and contributions.

The featured monument designed by Godwin and sculpted by Harris depicts two combat soldiers waiting for medical help as they carry a third wounded soldier between them. They are in full combat gear and carrying their weapons which were sculpted using the guidance of actual combat equipment loaned to the artist by cooperating Vietnam veterans.

The monument is made of bronze situated on a stone base. It is an impressive six-foot three-inch tall and is complete with dedication plaques near the monument and on a nearby bench. The plaque near the monument holds a dedication to the North Carolina Vietnam veterans while the plaque on a nearby bench boasts a dedication specifically for General Alfred M. Gray of the United States Marine Corps. Along with annual Veterans day and Memorial day event there is also a Christmas Eve candle lighting service each year.

The Names of those from the state of North Carolina Who Made the Ultimate Sacrifice

ADAMS DWANE LONNIE
ADAMS JESSE LEWIS
ADAMS JOSEPH BOYCE
ADAMS WILLIAM OTHELLO JR
ADDIS BILLY WAYNE
ALEXANDER BARRY KENNETH
ALLAIRE JOHN KEVIN
ALLISON JOHN ROBERT
ANDERSON ARCHIE
ANDERSON WALTER H
ANTON TERRY LYNN
ARDIS JOHN COLEMAN
ARNOLD JAMES
ARNOLD LARRY FRANKLIN
AYERS JAMES WESTLEY
BACOT DOUGLAS MONROE
BAIR DONALD RAY
BAKER CLARENCE EUGENE
BAKER JESSE RUTLEDGE
BALDWIN WILLIAM ROBERT
BANKS MICHAEL FRANCIS
BANKS VINCENT NORVELL
BARBARE JAMES MICHAEL
BARBEE FRANK LEROY
BARNES ALLAN GEORGE
BARNES JOE WILSON
BARNETT TONEY ANTHONY
BARNETTE ROY GRANT
BARNHART JOHN LOUIS
BARRS SHELTON FERRELL
BARTON MICHAEL GEORGE
BELL NEWTON THOMAS JR
BELLAMY SIMMIE JR
BELTON JAMES
BELTON THEODORE
BERGESS FREDERICK WILSON
BEST NEAL IRA
BETHEA JIMMY CARLTON
BETHEA LUTHER JR
BEVERLY WILLARD FRANKLIN
BICKLEY WILSON CHARLES
BINGLEY JOHN LEE JR
BISCHOFF JOHN MALCOLM
BISHOP JAMES MATTHEW
BLACK DAVID FORREST
BLACKBURN HUGH FRANK
BLACKWELL ROBERT LAW-
RENCE
BLACKWELL ROY JAMES JR
BLANDIN RAYMOND WEL-
LINGTO
BLANDING AARON
BLANTON BURTON ALEX-
ANDER
BLAS FRANK
BOAN JIMMY E
BODISON JAMES CALVIN
BOLES HARRY LEE
BOUTON WILLIAM INNES JR
BOWEN MARVIN WHITNER
BOWEN MATTHEW ANDER-
SON
BOWERS GROVER COLEMAN
JR
BOWMAN CURTIS
BOWMAN FRANK
BOWMAN MELVIN
BOYCE EUGENE RUSSELL
BOYD SAM HENRY
BOYD WAYNE
BOYTER GEDDES CHARLES JR
BRADLEY JAMES
BRADLEY KENNETH RAY
BRAGG CLAUDE EDWARD
BRANHAM JOHNNY THOMAS
BRANNON WALTER LEE
BRANTLEY LEROY
BRASINGTON JACK WILLIAMS
BRATTON ROY DONALD
BREWER THOMAS COLEMAN
JR
BRICKLE DONALD LEVER
BRIDGES RICHARD HAMIL-
TON
BRIGMAN JOHNNIE LEE
BROCK THOMAS DEAN

BROCKMAN RICHARD
BELTON
BROCKWELL LEYBURN W JR
BROOKS DAVID LEE
BROOKS JAMES ROY JR
BROWDER PAUL ROGER
BROWN CHRIS JR
BROWN HARRY WILLIS

BROWN IRVIN
BROWN JAMES HENRY JR
BROWN JAMES LEE
BROWN SAMUEL JUNIOUS
BROWN WILSON BOYD
BROWNING ROBERT EUGENE
BRUNSON DAVID LEROY
BRUSTER WILLIAM EARL

BRYANT GARY RAY
BRYANT JAMES ROY
BRYANT JERRY
BURDETT CLARENCE HENRY
BURDETTE HILBURN M JR
BURGESS TITUS LEVEN
BURNS LUTHER
BURNS WALTER

BURROUGHS ULYSSES G
BURTON HENRY LEE
BURTON WILLIAM JR
BUTLER JOHNNIE ELMER
BUTLER LINNELL
BUTTON DONALD B
BYERS JERRY WALTER
BYRD RALPH EUGENE

North Carolina

CAIL GLENN ALFRED
CAIN ROBERT JR
CALDWELL WILLIAM JAMES
CALLAHAN CLIFTON EUGENE
CAMP JAMES STEVEN
CAMPBELL BILLY WAYNE
CAMPBELL RICHARD MICHAEL
CANTRELL JAMES WESLEY
CAPE JERRY
CARNELL ARCHIE DENNIS
CARPENTER JAMES ALVIN
CARPENTER ROGER LEE
CARTER FRED JOSHUA
CARTER JOHN LEWIS
CARTER WILLIAM THOMAS
CASSEL KENNETH WAYNE
CATHEY J B
CATO HERBERT HUGO III
CAULDER DURWOOD
CAUSEY WILLIAM HARVEY
CHANDLER THOMAS LEROY
CHAPLIN LAWRENCE
CHASTAIN DONNIE RAY
CHASTAIN GERALD EDWARD
CHASTAIN WOODROW BEAL
CHAVIS RONNIE
CHESEBROUGH JOHN LANE
CHESTNUT LELAND MC LANE
CHILDS BOBBY RAY
CHISOLM ALEXANDER
CHISOLM ARTHUR LEE
CHRISTMAS MICHAEL LYNN
CHRISTOPHER SAMUEL JR
CLACK CECIL JAMES
CLARDY JAMES DONALD
CLARK ALLEN HOWELL
CLARK CHARLES
CLARY JOHN WILLARD
CLAYTON BENNY DEAN
COATS WILLIAM G
COAXUM THEODORE
COBB GEORGE LEE
COE HEWETT FRANK EVASTUS
COKER BILLY LEE
COLEMAN OLAN DAN
COLEMAN WYMAN BYRD
COLLINS CLAUDE LAVERNE
COLLINS JULIUS JR
CONE LLOYD ALFORD
COOK JAMES EDWARD
COOK JOHN I
COOKE EDDIE BOYD JR
COOLER SIDNEY HOMER
COOPER CALVIN EMANUEL
COOPER FAY KENNY
COSOM LEVERN
COSTNER JOHNNY PHILLIP
COX JOSEPH WILLIAM
COX ROGER DALE
CRAIG ROBERT MITCHELL
CRAPSE WAYNE FRANKLIN
CREAMER ALBERT EUGENE
CRIBB FLOYD ALLEN

CROFT JIMMY O NEAL
CROSBY JACKIE LAWRENCE
CROTTS DONALD COLEMAN
CULP JOHN PAUL
CULVEY KENNETH LEROY
CURETON JOHNNY JACOB JR
CURRIE JAMES JR
DANIEL JOHNNIE LINCOLN
DANIELS DANNY EARL
DARNELL DANA CORNELL
DASH DAVID WINTHROP
DAVENPORT DAVID D JR
DAVES DONALD CHARLES
DAVIS CHARLES EDWARD JR
DAVIS CORNER MACK
DAVIS JOHN K
DAVIS LEWIS ANTHONY
DAVIS RANDY MAYO
DAVIS ROLAND K
DAVIS RONNIE LEE
DAVIS WILLIE JAMES
DAVIS WILSON
DAVIS WOODROW JR
DAWKINS CALVIN DONALD
DAWKINS CLARENCE JR
DEARING PHILIP RAY
DENNIS DELMAR CLAUDE
DERRICK ALVIN JOSEPH
DERRICK BRUNSON A SR
DIETZ WALLACE JAMES
DILL JAMES ARTHUR
DINGLE EARL
DIXON CORDIE LEE
DIXON FRAZIER THOMAS
DOLLARD JAMES
DOOLEY MARVIN LOUIS
DORSEY HARRY JAMES
DOUSE JAMES LOUIS
DOVER JOHNNY LEWIS JR
DOZIER WILLIE CLAY
DRIGGERS ARTHUR M JR
DRIGGERS JERRY TRUMAN
DRUMMOND AUSTIN LEON
DUCKER RONALD DWIGHT
DUDLEY BRUCE WESLEY JR
DUKE THOMAS WAYNE
DUKES GEORGE BENNIE
DUNMORE ONEAL
DUSING CHARLES GALE
DYCHES CHARLES HENRY
EDDLEMAN ROYCE EDSEL
EDMONDSON HAROLD T JR
EDWARDS AUSTIN IVAN
EDWARDS EDWIN RAY
EDWARDS FREDFOR
EDWARDS HARRY JEROME
EDWARDS JAMES MERTON
EDWARDS WILLIAM EDGAR
ELDRIDGE ROBERT BURCH
ELLIOTT THAROLD WASHINGTO
ELLIS JAMES MARION
ELLIS RANDALL SHELLEY
ELLIS RAYMOND DEAN
ELLIS WILLIAM JR
ELMORE ROBERT LOVIS
ESTELLA ANTHONY JOHN
ETHERIDGE COLIE JR
EVANS ALONZA
EVANS ERNEST
EVANS HENRY FRANKLIN
FAISON EVERSON BENJAMIN
FANNING RICHARD HENRY
FEDERLINE AUDLEY M JR
FELTY JAMES LEE
FERGUSON BENNY HAROLD
FERGUSON LEROY
FERRELL CHARLES REGINALD
FIELDS RONALD ELWOOD
FINCH FORDHAM E JR
FLAHERTY STEVE
FLANDERS LEON D
FLEMING LARRY JR
FORD CHARLES EVANS
FORE WILLIAM C
FOSTER EARL WILLIAM JR
FOSTER JAMES BYRD JR
FOSTER SAMUEL EDWARD
FOSTER WILLIE FRANK
FOXX RICHARD L
FRANKS ANTHONY L
FREDERICK ARTHUR DONALD
FREEMAN JOSEPH LLOYD JR
FRIERSON KENNETH

FURSE WARREN RANDOLPH
FYALL VERNON ROBERT
GAILLIARD HERMAN BERNARD
GAMBRELL FRANKLIN DOUGLAS
GARNER JOHN HENRY
GARNER WILLIAM DONALD
GARRETT ALONZO
GARRETT JACKY LEROY
GARRETT LLOYD EARL
GAY JAMES NATHANIEL
GEDDINGS JOHN HUGHIE
GELL JACK EARL
GENES LUTHER ALLEN
GENTRY ROBERT LEVETT
GEORGE EMMITT ROOSEVELT
GIBBS ISREAL
GIBBS WILLIAM JR
GIBSON DAVID
GIBSON JAMES DONALD
GIBSON SILAS EUGENE
GILMORE JAMES ROBSON JR
GILSTRAP DWIGHT ONEAL
GINN DAVID LANDRELL
GIRARD CHARLES JACK
GLEATON HOMER LAYNE
GLENN DONALD NELSON
GOLDSMITH DONALD LAWRENCE
GOLSON ANTHONY
GOOD ROBERT SAYE JR
GOODSON JOSEPH ALAN
GOODWIN RONALD NELIONAL
GORE HORACE ROSCOE
GOUDE CHARLES MELVIN
GOURDINE LARRY RONALD
GRAHAM EARL CLAYTON
GRAHAM ROBERT LEE
GRANT PHILLIP
GRANT THOMAS
GRAVES JEHOVAH
GRAVES MICHAEL LEROY
GREEN DAVID HARRY JR
GREEN DAVID NATHANIEL JR
GREEN EDWARD JR
GREENE WILBUR
GRIFFIN FRANCIS LEKIRKLAS
GRIGGS HARLEY FRANKLIN
GROOMS JIMMIE LEE
GRUBER CLEMENT BRADLEY
GUNNELLS WILLIAM ASHLEY
HACKETT HARLEY B III
HAINES ADHERENE LOUIS
HALL BOYCE LEE
HALL JIMMY WILLIAM
HAM TERRELL THOMAS
HAMBY PAUL CHARLES JR
HAMES BOBBY JOE
HAMILTON WINSTON CLINTON
HAMMOND LELAND EMANUEL
HANNA KENNETH
HANNON RICHARD LAMAR
HARDEE JOSEPH EDWARD
HARDIN CURTIS LEVENCE
HARRILL RONALD WILLIAM
HARRIS ABRAHAM
HARRIS CARL E
HART JOSEPH FELDER
HARVIN JIMMIE LEE
HATCHER KENNETH MARVIN
HAWKINS WILLIAM HENRY
HAYNES JOHN WALLACE
HAZLE BRUCE EDWIN
HEARON TERRY SHAVON
HELMS JERRY DONALD
HENDERSON MONTIE H JR
HENDERSON ROBERT LEE
HENNEGHAN ROBERT LEE
HENRY WILLIAM JAMES
HENSLEY GAREY LEE
HENSON LARRY KEITH
HERRING ALFRED JIMMY JR
HERRING BRADY WILLIAM
HIGBEE CHARLES E JR
HIGHT WILLIAM REID
HILL RALPH OWEN
HILL VERTIS JAMES JR
HILTON STEPHEN RANDOLPH
HINES JOHN THOMAS
HINGLETON EUGENE JR

HINO MICHAEL LYNN
HINSON JOHN ROBERT
HINSON RONALD DOUGLAS
HODGES HARKLES LEROY
HOFFMAN SHULER ADOLF
HOLDER CARL LEWIS
HOLLIFIELD VERNON GILBERT
HOLLINGSWORTH JERRY G JR
HOLMES JOHN HENRY
HOLMES THOMAS EUGENE
HOLMON ALPHONZO JR
HOOKER JOHN ARTHUR
HOOKS DAYTON JOSEPH
HOPKINS MYLON RAY
HORLBACK FRANCIS D
HOUGH MATTHEW
HOWARD DAVID LAFATE
HOWARD JULIUS JAKE JR
HOWE JAMES DONNIE
HOWELL ROBERT LEE
HOWELL SAMMIE
HOWLE ERNEST CLARENCE
HOWZE CHARLES CROCKETT
HUCKS LLOYD JUNIOR
HUGGINS EUGENE
HUGGINS GORDON SAMUEL
HUMPHRIES HAZEL H III
HUMPHRIES RONALD EDWARD
HUNTER LEROY
HUTCHERSON GARY CAROL
HYNDS WALLACE GOURLEY JR
JACKSON BENJAMIN FRANKLIN
JACKSON JAMES CLEVELAND
JACKSON NATHANIEL E L
JACKSON SYLVESTER JR
JACOBS ERNEST LINWOOD JR
JAMES EDDIE LOUIS JR
JEFFERS JOHN LARRY
JEFFERSON LEROY
JENKINS FRED HARVEY
JENKINS KENNETH BRUCE
JENKINS REGINALD ROCKEFEL
JENKINS WILLIAM
JENNINGS ROBERT LEE
JETER CURTIS LEE
JOE WILLIE LEE
JOHNSON ALBERT JR
JOHNSON CHARLES FRENCH JR
JOHNSON CHARLES JR
JOHNSON CLEVELAND OSBORNE
JOHNSON FRANKIE B JR
JOHNSON GEORGE
JOHNSON JACOB
JOHNSON LEMUEL
JOHNSON LEROY
JOHNSON NATHANIEL LERVERN
JOHNSON RALPH HENRY
JOHNSON WILLIE
JOHNSTONE JAMES M
JONES CHARLES A
JONES CLIFTON RANDAL
JONES ELIZABETH ANN
JONES HORACE
JONES JERALD LOUIS
JONES WILLIAM JUNIOR
JORDAN LARRY LEON
JORDON ARTHUR L
JOSEPH JAMES
KAPP RICHARD WORRELL JR
KEEFE DOUGLAS O NEIL
KEENE GERALD BRICE
KEITH WILLIE LEE
KELLEY DEWEY WILLIAM
KELLY BENJAMIN EDWARD JR
KEMP JIMMY
KENNEDY MATTHEW DEVANUGHAN
KEY ANDERSON HAROLD
KILBURN WILLIAM HUNTER
KING ROBERT CARL
KING ROBERT EARL
KING ROBERT LOUIS
KING THOMAS PICKETT BYRD
KING WYLIE CLARENCE
KINSEY HARVEY JUNIOR
KITCHENS JOEL RHYNE
KNIGHT RICHARD

KNUPP WAYNE WOOD
KOHN ALAN SPENCE
KOTT STEPHEN JAY
LAKASZUS HELMUT GUSTAV
LAMA IVARS
LAMKIN FREDDIE LEE
LANCE JOHN HENRY
LANE MICHAEL S
LANKFORD HENRY DEAN
LATIMER CLARENCE ALBERT
LAWLESS WILLIAM RALPH JR
LAWSON LAMAR ALVIS
LAWSON LARRY EDWARD
LAWTON BILLY JAMES
LAYTON WEBB HERMAN JR
LAZARUS SIDNEY GILBERT JR
LAZICKI JOSEPH CHARLES
LEAPHART HAROLD PAUL
LEDFORD DON KENNETH
LEE BILLY
LEE MELVIN
LESAINE JIMMY WILSON
LEWIS JIMMIE
LEWIS LEE
LEWIS WALTER WAYNE
LILES EPHRIAM RUTLEDGE II
LIPSCOMB ROY LOLLIS
LISBON JOHNNY
LIVINGSTON ERSKIN DAN
LLEWELLYN JOHNNY WILLIAM
LOCKLAIR ALLISON WAYNE
LONG RAYMOND ERVIN
LONG THOMAS IRA
LOPEZ LUIS BELT RAN
LOUALLEN JACK NEECE
LUCAS JOHN WILLIE
LUCAS PATRICK DONOVAN
LUSTER MILTON BERNARD
LYLES BELTON JR
LYLES CHARLIE JR
LYONS JAMES JR
MAC MICHAEL CHARLES EDWAR
MACK DANIEL JAMES
MACK HAROLD JR
MADDEN LEON SHIRLEY
MADDEN WILLIE ERSKINE
MAGAHA DANNY ROY
MAGBEE G W
MAHY HAROLD EUGENE
MAKIN JAMES BRIAN LAWRENC
MAKIN WOODROW JR
MALICHI BOBBY SPENCER
MANGUM ROBIN
MARTIN CLYDE THOMAS
MARTIN HOYLE
MASON KENNETH ALLEN
MASSEY JOHN WILLIAM JR
MATTHEWS EARL JR
MATTHEWS ROY GIBSON
MATTISON BENJAMIN FRANKLI
MAULDIN MELVIN CALVIN JR
MAULDIN THOMAS JASPER
MAYES JOSEPH
MC CLARY SAMUEL DONALD
MC COY ELEC
MC COY JAMES RAYMOND
MC CRARY DOUGLAS MAC ARTH
MC CULLOUGH BILLY RAY
MC CULLOUGH JOHN EARNEST
MC DANIELS BILLY CLAYTON
MC DONALD JERRY SYLVESTER
MC DOWELL CHARLES ELVIN
MC DOWELL HAROLD GUINN
MC DOWELL SAMUEL T JR
MC FADDEN CARL JR
MC FADDEN HARRY BERNARD
MC GAHA HAROLD F
MC GEE JOSEPH O NEIL
MC KAY JOHN ROLAND JR
MC KIE JACOB
MC KINNEY THOMAS ALAN
MC LENDON MICHAEL RYAN
MC MAHAN CHARLES LARRY
MC MAHAN THOMAS EDWARD JR
MC MAKIN WALLACE THOMSON

MC MILLIN ROBERT ALLEN
MC PHERSON ALFRED
MC RAE JIMMIE LEFON
MCKITTRICK JAMES CLIFFORD
MCMURRAY FRED HOWELL JR
MEADOWS CARROL FAYNE
MEETZE DENNIS RAY
MEGGS MARION LEE
MELVIN JAMES EDWIN JR
MERCK ROGER EUGENE
MESSER JACK WILLIAM
MICKLE MOSES
MIDDLETON STEVEN ALFRED
MILES GALEN SPINKS
MILES HAROLD GENE
MILES JOE JUNIOR
MILLER JAMES
MILLER JERRY LEE
MILLER PAUL
MILTON CHARLES RUDOLPH JR
MIMS GEORGE IVISON JR
MINCEY JAMES MARSHALL
MINOR CARROL WILLIAM
MITCHELL DAVID LEE
MITCHELL DONALD THOMAS
MITCHELL ISAIAH JR
MITCHELL MACK LEE
MITCHELL WILLIE JAMES JR
MONTAGUE JESSE WILLARD JR
MOODY ADGER EUGENE
MOODY HERBERT WAYNE
MOORE JACK DONALD JR
MOORE JIMMY LEE
MOORE LEONARD LEE
MOORE NORMAN JAMES
MOORE THOMAS ANTHONY
MOORE WILLIAM LEWIS
MOORER CLARENCE LARRY
MORENO RAMON
MORGAN CHARLES LEWIS
MORGAN CLYDE EDWARD
MORROW MERLE BRUCE
MORSE REGINALD GEORGE
MOSER SAMUEL RALPH
MOULTRIE JOE DAVIS
MULLINAX JAMES CARLTON JR
MUMFORD JIMMY EARL
MUNGER JOHN ROBERT
MURRAY CAESAR
MURSCH JOHN WILLIAM
NASH ANTHONY PRESTON
NEDD HEYWARD WINDELL
NELSON DICKIE OWENS
NELSON EUGENE
NELSON SANFORD LEWIS
NETTLES VICTOR LEE JR
NICKENS JAMES
NORRIS CHARLES BENJAMIN
NORRIS GRADY LEE
NORRIS RONNIE EUGENE
NORWOOD THOMAS LEE JR
NORWOOD WILLIAM ARNOLD
NUNNERY CLARENCE E JR
O CONNOR THOMAS DUCKETT
O CONNOR WILLIAM EUGENE
OLIVER CLIFTON
OLIVER RANDY DEWITT
OSBORNE SAMUEL WILLIAM JR
OWEN RAY WILLIAM
OWENS HENRY LAWRENCE
OWENS JAMES DOUGLAS
OWENS JAMES EUGENE
OWENS MARTIN LEE
OWENS ROBERT LEE
OXNER MARION LUTHER
PACE GARY LYNN
PACE RONALD EUGENE
PAGE PHILLIP ALLEN
PAINTER CURTIS WAYNE
PAIRIS ARNOLD
PARKER ALBERT EDEN
PARKER ARTHUR M III
PARKER EDGAR EUGENE
PARKER ROBERT
PARKER WESLEY
PARKS CHARLES H JR
PATE RONALD DALE
PATTERSON PATRICK CHASIE
PAULEY WASHINGTON

PAYNE WILLIE JAMES
PEAGLER WAYNE DONALD
PEARSON JESSE JAMES
PEEPLES BILLY HAMMOND
PELZER BENJAMIN F II
PENDARVIS ROBERT
PENN FRANKLIN HAMILTON
PERKINS WILLIE JAMES
PERRY ROBERT DALE
PESSIER STEVEN LEROY
PETERSON MATTHEW
PETTIT STANLEY RUSSELL
PETTY WILLIAM CLARK
PHIFER FREDDIE JOE
PHILLIPS CHARLES W JR
PHILLIPS JERRY ALFRED
PHILLIPS LAWRENCE
PHILLIPS WESLEY LEE
PHILLIPS WYLIE ORIA
PHOENIX ALONZA WILLIE
PIGATT HARMON JULIOUS
PLATT ROBERT LENWOOD JR
POE JIMMIE CLYDE
POLSTON ERNEST ELIJAH
POOLE WILLIAM DAVID JR
POORE THOMAS WYATT
POPE DONALD BURRIS
PORTER LARRY JAMES
POSTON JAMES
POTTER LARRY EMERSON
POTTS JOHNNIE WYLIE
POUND THEUS JOSEPH
PRESTON JOSEPH JR
PRICE MAXIE LANE
PRICE ROBERT HUGH
PRIEST JOHN HENRY JR
PROCTOR WAYNE SHELTON
PRUITT DAVID MONROE
PRUITT OSIER LAWRENCE
PUCKETT TROY MURL
PUGH GERALD RALPH
PUSSER THOMAS WILSON
QUICK GEORGE DEWEY JR
QUIGLEY TERRY LYNN
RABON JOSEPH LEVERN
RAILEY GEORGE EDMUND
RAINS FORREST DE VERE JR
RAMBERT FRANKLIN
RAMEY JORDAN EUGENE
RAWLS CHARLES GLENN
READY JOHN III
REAVES JAMES LOUIS
REAVES JOHN SHEPARD JR
REEDER MELVIN
RENWICK HAROLD MCGILL JR
REVELL WILLIAM JAMES III
RHINEHART CLYDE A
RHODES FERRIS ANSEL JR
RHODES KENNETH
RICE THOMAS JR
RICH JOSEPH WALTER
RICHARDSON JIMMIE JENKINS
RICHARDSON THEODORE
RIKARD CHARLES DAVID
RILEY EDDIE LEE
RIVERS NATHAN
ROBERSON WILBURN
ROBERTS GARY KENNETH
ROBERTS KENNETH EUGENE
ROBERTS LLOYD VERNON
ROBERTSON BOBBY LEE
ROBINSON CALVIN
ROBINSON CHARLES HENRY
ROBINSON FREDDIE LEE
ROBINSON JIMMIE LEE
ROBINSON JOHN WILLIAM JR
ROBINSON LUCIEN
ROBINSON MELVIN
RODGERS TILLMAN DAVID JR
ROWELL FRANKLIN DELANO
RUFF WILLIE JEROME
RUTLAND WARREN LESTER
RUZICKA JOSEP L JR
SALE HAROLD REEVES JR
SALLEY JAMES JR
SAMUELS ISAIAH
SANDERSON BOBBY
SANFORD ARNOLD
SANFORD JAMES WALTER
SARTOR LEONDA
SAWYER JOHNNIE PAUL
SAYLOR WILLIAM JR
SCHAFFER BILLY JOE
SCHELLIN JAMES WILLARD F

SCHOPER GREGORY CARLYLLE
SCOTT LAWRENCE EDWARD
SCOTT RANDOLPH
SCOTT RANDOLPH CLINTON
SCOTT VERNON ELBERT
SEASE WILLIAM DANIEL
SEGAR CHARLES JR
SELLERS ROBERT
SHAW WILLIAM FREDERICK JR
SHEPHERD THOMAS GLENN
SHUMPERT JOE THOMAS
SIMMERS GAROLD RAY
SIMMONS EDDIE LEE
SIMMONS FAY CLYDE III
SIMMONS ISIAH
SIMMONS RICHARD
SIMMONS ROBERT LOUIS
SINGLETARY HILBERT M JR
SINGLETON RAYMOND
SIZEMORE DONALD RAY
SKIPPER MICHAEL RAY
SLOAN VERNAR
SMALLS BENJAMIN ALONZA
SMALLS JOSEPH
SMARR ALBERT WARD JR
SMITH CARY JOSEPH
SMITH DAVID GERALD
SMITH DAVID WAYNE
SMITH DAVID WESLEY
SMITH HAROLD MCRAE
SMITH JOHN CLIFFORD III
SMITH JOHN LEWIS
SMITH R J
SMITH RICHARD LEE
SMITH TERRY LEE
SMITH WILBUR EUGENE JR
SMITH WILLIAM HARRY
SMITH WILLIAM THOMAS
SMOAK JAMES THURSTON JR
SMYLY DUNCAN PADGETT
SNEED SAMMIE RAY JR
SNOWDON RICHARD ATWOOD
SOSSAMON EDWARD DE CAMP
SOWELL HARRY LEE JR
SPARKMAN ISAAC
SPEIGHTS ROOSEVELT
SPENCER JAMES ALBERT JR
SPIVEY WILLIE DALPHUS
SPRINGS ANDREW
SPROUSE RONALD EDWARD
STARKS JAMES EDWARD
STATON DAVID WALDEN
STEGALL LINDELL RAY
STENHOUSE J LYNN JR
STEVENSON BOBBY GENE
STEWART DAN ROGERS
STEWART ROBERT HENRY
STILL FLOYD LAVERNE
STILL FRED HAROLD
STILL JERRY MELTON
STONE DAVID RONALD
STONE JAMES EDWARD
STRICKLAND JOSEPH ODELL
STROUD DENNIS CARROLL

STUKES ISAIAH TRUMAN
STUMP HAROLD OLIVER
SULLIVAN GERALD DEWAYNE
SULLIVAN JOSEPH HARRY
SWEATT CLYDE STANLEY
SWEENEY MICHAEL MURPHY
SWOFFORD DANNY RAY
TALLENT HERSHALL
TAPSCOTT KENNITH WALKER
TATE WALTER REAVES JR
TATE WILLIE JAMES
TAYLOR CECIL FRANKLIN
TAYLOR GEORGE THOMAS JR
TAYLOR PRESTON JR
TAYLOR TED JAMES
TAYLOR THOMAS MARCELLUS
TAYLOR TOMMY LEE
TEAGUE JOHN WALTER
TEER WILLIAM EDWARD
THOMAS DOUGLAS JR
THOMAS JACK JR
THOMAS JAMES MYER
THOMAS RUDOLPH CALVIN
THOMPSON ALBERT C
THOMPSON CARL
THOMPSON CARROLL U
THOMPSON CHARLES LEE
THOMPSON DANNY STEWART
THOMPSON THERMALL
THOMPSON WILLIAM FRANK
TIMMONS BOBBY DANIEL
TODD CARL EDWARD
TOUCHBERRY MILES D JR
TOWNSEND CHESTER DAVIS
TRAPP BOBBY RAY
TRIPLETT JOHNNY RAY
TRUESDALE CHARLES KENNETH
TRUSTY MICHAEL JEFFERSON
TUCKER DAVID
TUCKER JOE NATHAN
TURNER ARTHUR JR
TURNER LINDSAY CLINTON
TURNER MARCUS SHARPE JR
TURNER ROBERT LAWRENCE
TUTEN MICHAEL HAMILTON
TYLER EARTHELL
TYLER JESSIE JAMES
TYNER JAMES PHILEMON
VILLANUEVA LARRY
VILLEPONTEAUX JAMES H JR
WAITERS LAWRENCE
WALDEN ROBERT DAVID
WALKER EDWARD
WALKER EVANS S
WALKER JOHN HENRY
WALKER LUTHER JR
WALKER THOMAS EDWARD
WALL JAMES ALLEN
WALLINGTON GEORGE HEYWARD
WALTERS EUGENE
WALTERS RICHARD EDWIN
WALTERS RICHARD FLOYD
WARD BEN CALHOUN
WARD HERSCHEL RUDOLPH
WARNER ARTHUR LEE

WASHINGTON GEORGE RAYMOND
WASILOW JOHN STEPHEN
WATERS FRANKLIN DELANO
WATSON CHARLES BRYANT JR
WATSON LARRY ELLIOTT
WAY CLARENCE L
WEATHERFORD ROY JULIAN JR
WELDON LIBERT JAMES JR
WELLONS HUGH WILLIAM
WELSH STEPHEN JACKSON
WESTPOINT THOMAS LEE
WHITAKER JERRY
WHITE BEN
WHITE ISAIAH
WHITE LEON
WHITE MELVIN ELIJAH
WHITE NATHAN JR
WHITE WILLIE
WHITT BROADUS ALFRED
WIGFALL HERBERT JR
WILKS JAMES LEE
WILLARD HUGH GREY
WILLIAMS BEN
WILLIAMS BILLY
WILLIAMS CALVIN
WILLIAMS CAROL EDWARD
WILLIAMS CLARK LEE VERN
WILLIAMS DANIEL III
WILLIAMS DOYLE
WILLIAMS EDDIE KENNETH
WILLIAMS FRANK NORMAN
WILLIAMS HARRIS LEE
WILLIAMS JAMES EDGAR
WILLIAMS JAMES U III
WILLIAMS JERRY HIOTT
WILLIAMS JOHNNY EDWARD
WILLIAMS LEONARD
WILLIAMS RONALD ANTHONY
WILLIAMS WILLIE AMOS
WILLIAMSON BENJAMIN JEFFE
WILLIAMSON EDWARD LOUIS
WILSON LAWRENCE W
WILSON WILLIAM EARL
WINE HARRISON JR
WITHERSPOON MARION
WOOD JAMES EDWIN
WOOD WILLIAM ESLEY JR
WOODS ADVERT JR
WOODS WILLIAM STEPHEN
WORTHY JERRY DEAN
WRIGHT CLIFTON
WRIGHT CURTIS RONALD
WRIGHT JAMES
WRIGHT JOE DAVID
WUESTENBERG LEWIS CURTIS
YATES CHARLES RUSSELL
YELDELL DAVID
YOUNG CLARENCE C
ZEIGLER GLEN ALLEN

Photos by Robert Knutson

All Veterans Memorial

600 E Boulevard Ave, Bismarck, ND 58505

Located on the state capitol grounds in Bismarck, North Dakota, the All Veterans Memorial is a war memorial honoring all North Dakota residents who served in the first century of North Dakota's existence. This includes all wars from the Spanish-American War to the current Global War on Terrorism.

The All Veterans Memorial was originally dedicated in June of 1989, the state's centennial, and was dedicated to all North Dakotan citizens who served in some capacity in the state's first 100 years of existence. It was commissioned by the All Veterans Centennial Memorial Association.

It features a cube building structure sitting on stone pillars. It is domed but features a hole in the center which is positioned over a globe. That globe sits on its own pedestal that features a plaque describing the significance of the globe and the hole in the dome. That significance is that on Veteran's Day (November 11), the sun will hit the globe directly at 11 a.m. and show off the physical characteristics of the model, the main one being a raised image of North Dakota.

The memorial also features 49 bronze tablets that hold the engraved named of the 4,050 people who died in the various wars. Since the memorial's creation, the names of recently fallen service members who were lost in the Persian Gulf War and the War on Terror were added to the tablets. These panels are arranged in a circle around the centered globe. The All Veterans Memorial is located on the southeast of the State Capitol and is situated protectively amidst a copse of trees. Its cubed and domed structure are said to be a representation of unity, stability, and purity.

The Names of those from the state of North Dakota Who Made the Ultimate Sacrifice

ALBERTS ROGER DUANE
ALDERSON THOMAS EARL
AMUNDSON DALE HARLAN
BACKER WILLIAM PAUL
BARGMANN GILBERT RAY
BARTOLINA ERNEST E JR
BEAN KENYON ELROY
BEIER ELROY EUGENE
BELCHER GLENN ARTHUR
BERDAHL DAVID DONALD
BERG HAROLD EDWARD JR
BERGER CARL STEPHEN JR
BEYER THOMAS JOHN
BINSTOCK PETER JR
BOEHM RICHARD JOHN
BOND RONALD DALE
BOOTS CURTIS EUGENE
BORGMAN RICHARD LEE
BRENNO WESLEY CRAIG
BRINKMEYER JOHN WILLIAM
BRISS MARVIN CLARENCE
BROTHEN ROBERT ALVIN
BUJALSKI DAVID ALLAN
BURINGRUD RICHARD ALLEN
CARTER LESLIE LOUIS
CHARNETZKI PAUL FREDERICK
COONS CHESTER LEROY
COPELAND RALPH A
CORCORAN DAVID JAMES
COULTHART GERALD FRANK
COX LEON DAVID
CRARY JOSEPH WILLIAM
CUSHMAN CLIFTON EMMET
DAVIS CHRISTOPHER WILMER
DE PRIEST DAVID REED
DECKER GERALD ANTHONY
DOWLING FRANCIS ELLSWORTH
ECKES WILLIAM CARL
EICHELBERGER STEPHEN JOHN
EINARSON LOWELL GREEMER
ELLENSON JEROME WILLIAM
ELLINGSON JOEL ARDEN
ELSBERND DAVID DUANE
EMINETH NORMAN ANTHONY
ESCALLIER STEVE LOUIS
ESSER LAWRENCE ROBERT
EVANS WARD CECIL
EVENSON MICHAEL ARTHUR
FINLEY VALARIAN LAWRENCE
FISCHER JOSEPH DENNIS
FLECK WILBERT CLEMENS
FOREMAN ROGER EARL
FOWLER JAMES ALAN
FREIDT JAMES CHRISTIAN
FROEHLICH NORBERT LOUIS
FROST RAPHEAL JOHN
FULLMER ROBERT MICHAEL
GABLE ARLAN DEAN
GAFFANEY RICHARD JAMES JR
GEIGER FRANCIS EDWARD
GIETZEN GENE THOMAS
GILL RICHARD DENZIL JR
GOODIRON RONALD CHRISTY
GREANY VIRGIL RAYMOND
GREENLEY JON ALFRED
GROTH STEPHEN JAMES
GUNDERSON GUNDER PETER
HAEGELE DAVID PETER
HAGEN LOREN DOUGLAS
HANSEN RANDY LEE
HANSEY MITCHEL CAREY
HANSON DANNY LEROY
HANSON STEVEN REED
HARDMEYER LOWELL
GEORGE HERDEBU DAN LEON
HERTEL RODGER RAY
HILLYER LOUIS
HIMLER ROBERT JOHN
HIMMERICK MICHAEL DUANE
HINZPETER ALAN ROLF
HONCHAROFF GENE EDWARD

HOVLAND RICHARD DALE
IVERSON GERALD ALLEN
JACOBSON LARRY BRUCE
JANSONIUS FRED WALTER
JOHANNES LYLE MAYNARD
JOHNER KENNETH LEO
JOHNSON DAVID FRANCIS
JOHNSON FRED LEROY
JOHNSON MYRON BLAINE
JOYCE JOHN MORRIS
JUST GERHARDT
KELLER WENDELL RICHARD
KELLY DOUGLAS MILTON
KELLY JOHN EDWARD JR
KENT RONALD LEROY
KESSEL MICHAEL HENRY
KIRKEBY DAVID LYNN
KLEIN GARRY DEAN
KLEIN GERALD DEAN
KLINE DAVID BRUCE
KLOSE DOUGLAS CLEMENS
KNIPPELBERG IRVIN DALE
KNUDTSON ROGER DOUGLAS
KOPSENG JAMES CLAIRE
KOROM ALLAN JAMES
KRAFT ROBERT LEO
KRAMER RAYMOND EUGENE
KRISTJANSON WILLIAM DAVIS
KRUEGER GREGORY KEITH
KUHN DAVID JAMES
KULLAND BYRON KENT
KUSS FLORIAN HENRY
LABER MERLIN JAMES
LAPP HERBERT
LARSON DAVID ALLEN
LARSON GERALD LEE
LATRAILLE DAVID JOHN
LAVOY EUGENE LOUIS JR
LEETUN DAREL DEAN
LEMBKE MELVIN DENNIS
LEVANG CLEO LARRY
LEVINGS JAMES M
LINDSAY GARY WAYNE
LOCHTHOWE LEON LEROY
LOKKEN GARY DAN
LUNDE GREGORY HOWARD
LUNDIN JOHN CHARLES
MAIER GLENN ERVIN
MANGIN MARK DANIEL
MANSON DENNIS JAMES
MARTHE RANDOLPH LEE
MC CABE PATRICK JOSEPH
MC COWAN RALPH CHARLES
MC NEILL RONALD PATRICK
MEDUNA DENNIS LEE
MEYHOFF MICHAEL FREDERICK
MILLER PIUS LEO
MUELLER WESLEY ERWIN
MYERS GARY FREDRICK
NADEAU ERIC DARYL
NARUM THOMAS LEROY
NESSET DAVID JON
NEUENSCHWANDER DAN LEE
OLSON DELBERT AUSTIN
OLSON LARRY REX
OLSON RICHARD JAMES
ORSUND RICHARD WAYNE
OTTMAR STANLEY JOHN
PAULSON MERLYN LEROY
PIERCE DOUGLAS JACK
PIPER CHARLES HERMAN JR
POLING LARRY STERL
PORTER ALLEN WAYNE
POTTER WILLIAM TOD
RAAUM JOHN VILNIS
RENNER JOHN MICHAEL
ROBBINS JON PIUS
SCHMITZ ROBERT EUGENE
SCHOSSOW DENNIS ROBERT
SELBY DE WAYNE MICHAEL
SENNE THOMAS ALFRED
SIKORSKI LARRY JOSEPH
SIME ROBERT JOY
SOBY DONALD JEAN

SPITZER THOMAS EDMUND
STEEN MARTIN WILLIAM
STOLTENOW RONALD GILBERT
STOREY ROBERT LEE
SVEEN BRENT WILLIAM
SVIR ROGER LEE
SWANSON ROBERT EDWIN
TAGUE JOHN ROBERT
TINGLEY JOHN CHARLES
TONGEN GEORGE ELWOOD
TRISKE RICHARD FRANK
VALKER GEORGE ERNEST III
VAN DE VENTER BLYTHE NOEL

VIDLER MURRAY DEAN
VOLK RICHARD ANTHONY
VOLLMER DONALD GENE
WAGNER ROY CARL
WALDERA DAVID ARLEN
WALTER WARD GUNARD
WARBIS LARRY LYLE
WEBER WILLIS WILLIAM
WELKER THOMAS A
WENAAS GORDON JAMES
WENDT CHARLES DONALD
WERMAN EDWARD ALEC
WILLIAMS NORMAN PAUL
WOLD BRUCE LLOYD
WOLF MICHAEL FERDINAND

WOLOS PAUL HARVEY
WOODS CARL JULIUS
WOSICK DENNIS STANLEY
ZIETLOW LAURENCE CRIS
ZUBKE DELAND DWIGHT

Photos by Noelle Bye

Ohio Veterans Memorial Park

8005 Cleveland Massillon Road, Clinton, OH 44216

The Ohio Veterans' Memorial Park is located in Clinton, Ohio and is dedicated to the local heroes from each branch of the military who have served in both times of peace as well as war. It was created to help everyone remember that there are men and women who sacrifice themselves for the freedom of their country. The park is accessible seven days a week, 24 hours a day and is lit at night for the convenience and safety of the visitors.

While visitors can come and go as they please, there are special events held at the memorial on days of remembrance. Veterans can also pay to have a piece of the memorial customized for themselves (families of fallen veterans can do this for their lost loved ones, too). Besides the Vietnam memorial, the park also holds other war memorials.

The park itself was funded by donations from individuals, businesses, civic groups, houses of worship, and other organizations who were looking to thank the local veterans. Fundraisers were set up for the initial start of the memorial and continued through the construction of the monument. The Vietnam memorial was dedicated in 2009 following the dream of Ohio resident Dan Delarosa who was looking for a way to honor his brother who died during his tour in Vietnam.

The main portion of the Vietnam memorial is the memorial wall. It is a 125-foot long black granite wall that features the names of the over 3,000 service members who lost their lives in the war. There is also an inscription on the wall that simply says "Lest We Forget." Recently more casualties surfaced and will soon be added to the wall besides that the memorial is constantly evolving and has other components including a mounted helicopter and a military truck. There are plans to add a tank to the memorial, too. There is also a serene reflection pond dedicated to the prisoners of war and missing in action service members. This scene is made complete by a free standing field cross.

182

The Names of those from the state of Ohio Who Made the Ultimate Sacrifice

ACHOE LEEVERNE RICHARD
ACKERMAN JAMES CARROLL JR
ACREE ROGER LEE
ADAMS RICHARD LEE
ADAMS ROGER DEAN
ADAMS STANLEY LEE
ADAMS STEVEN JACK
ADAMS WAYNE ROGER
ADAMS WILLIAM J
ADAMS WILLIAM RICHARD
ADKINS CHARLES LEROY
ADKINS HENRY DALE
ADKINS JAMES DALE
ADKINS JOHN
ADKINS LLOYD MARVIN JR
ADKINS MICHAEL DUANE
ADKINS NORMAN DALE
AGEE JOHN CHARLES
AILES WILLIAM EUGENE
AKINS RONALD PAUL
ALDRICH DAVID ALAN
ALEXANDER DALLAS C JR
ALEXANDER JAMES BLAIR JR
ALEXANDER JASPER MARION
ALFERINK JERRY LAVERN
ALGIRE ROGER DEAN
ALLAN DONALD EUGENE JR
ALLEN CHARLES DELMAR JR
ALLEN DALE CHARLES
ALLEN JAMES WARREN JR
ALLEN JESSE WAYEN
ALLEN KENNETH JEFFREY
ALLEN LUECO JR
ALLEN SAMUEL R
ALLEN THOMAS
AMADOR RAYNALD JIMENZ
AMATO DENNIS FLOYD
AMBROSIO JOSEPH GEORGE
AMHEISER DAVID JAMES
AMMON WILLIAM RESOR
ANDERS RICHARD ALAN
ANDERSON CHARLES E
ANDERSON EARL ERNEST
ANDERSON GEORGE DONALD
ANDERSON JAMES THEODORE
ANDERSON MILLARD RAY
ANDERSON NORMAN RALPH
ANDERSON STEVEN RAY
ANDERSON WILLIAM EDWARD
ANDERSON WILLIAM JOHN JR
ANDREWS DAVID LYNN
ANDREWS JOHN MICHEAL
ANDREWS MICHAEL ALLEN
ANDREWS WILLIS NORWOOD
ANGE CARMELLO JR
ANTEAU KARL THOMAS
ANTHONY WARD LEROY
APPLE GLENN WILSON
APPLEGATE DONALD LEE
ARMENTROUT STANLEY WILLIA
ARMSTRONG MICHAEL DAVID
ARMSTRONG RAYMOND
ARNETT RAY JR
ARNOLD GEORGE DALE
ARNOLD ROBERT JOSEPH
ARQUILLO JOHN DOMINICK
ARTHUR JAMES RAYMOND
ASBECK GREGORY JOSEPH
ASH RONALD KEITH
ASHER DENNY LYNN
ASHER FRANK LOUIS
ASHER SAMUEL QUENTON
ASKAM ROBERT EUGENE
ASTON JAY STEVEN
ATER ROBERT ALLAN
ATWOOD RICHARD
AU KEITH WARREN
AUER EDUARD ADOLPH
AURADY MICHAEL VICTOR
AUSMUS ROBERT ARTHUR
AUSTIN TOM LEW
AVERY LEWIS EUGENE

AVERY MARVIN DOUGLAS
AYDLETT JAMES QUINEL
AZARA CHARLES F JR
BABYAK ANDREW JOHN JR
BACH MICHAEL ROBERT
BACHMAN PAUL JOHN
BACHUS JOSEPH RICHARD
BACORN KEITH RAY
BADGLEY DALE ERNEST
BADON JOHN WAYNE
BAGEN RONALD S
BAGO JOHN STEVEN
BAILEY DERWIN MICHAEL
BAILEY MICHAEL A
BAILY PHILLIP RAY
BAIRD JACKIE RANDLE
BAKER BOBBY RUSSELL
BAKER DONALD ALLEN
BAKER EDGAR JR
BAKER ELWOOD CHARLES
BAKER JAMES MICHAEL
BAKER LA BROSSIE LUCIEN
BAKER PAUL WILLIAM
BAKER RICHARD ALLAN
BAKER SAMUEL THEODORE
BAKER WALLACE EDWIN
BALDAUF RAYMOND JOSEPH
BALISTERI CODY ALLEN
BALL CHARLES HOMER
BALL DWIGHT HERBERT
BALL EDWARD MEARL
BALL ROBERT
BALL THOMAS ELROY JR
BALLARD PAUL ALLEN
BANAR MARVIN DALE
BANKS DINGUS JR
BANKS RICHARD ANTHONY
BANKS ROBERT ALLEN
BARB MANVILLE LAWRENCE
BARBARINO DAVID ANGELO
BARBER FLOYD EDWARD
BARCUS TERRY LEE
BARDEN HOWARD LEROY
BARGA SAMMY A
BARGAHEISER LAWRENCE GILB
BARGAR RICHARD M
BARGY MORRIS LEE
BARKLEY LAWRENCE WILLIAM
BARNES LAURIE EUGENE
BARNETT EUGENE MELVIN
BARNETT MEREDITH LEE
BARNHART EARL EDWARD JR
BARNITZ DOUGLAS WANNER
BARRITT WILLIAM STEPHEN
BARRY ROBERT JAMES
BARSCHOW WILLIAM MARCUS
BARTEK DONALD EUGENE
BARTHELMAS WILLIAM J JR
BARTLETT DAVID ALLAN
BARTLEY JOHN PETER
BARTLEY WALTER CARL JR
BARTON LARRY DEAN
BARUZZI MARCO JOSEPH
BASALLA DONALD ALBERT
BASINGER RICHARD LOUIS
BASS HARRY WAYNE JR
BATOR WILLIAM HENRY
BATT MICHAEL LERO
BATTLES CHARLES EDWARD
BAUER DARYL CHARLES
BAUER TIMOTHY PAUL
BAUGHN PHILIP WAYNE
BAUMAN RICHARD LEE
BAUMBERGER RICHARD L JR
BAUMER JAMES CHARLES
BAYLOR HAROLD BOOKER T
BEAGLE FRANCIS WAYNE
BEAL GEORGE WILLIE JR
BEALE GEORGE EUGENE
BEAM JACK EVAN
BEAN RICHARD RAYMOND
BEARD JEFFREY LEE

BEASLEY ROY CLAUDE
BEATTY JAMES RUSSELL
BEAUCHAMP RAYMOND FREDERI
BEAVER ROBERT LYNN
BEBOUT STEPHEN EUGENE
BECHTEL HERBERT STEPHEN
BECK EDWARD EUGENE JR
BECKER JOHN JOSEPH JR
BECKSTED RONALD JAMES
BEEDY GEORGE
BEESLER CHARLES WILLIAM
BEGAN JOHN LESTER
BEHM CHARLES JOEL JR
BEHNFELDT ROGER ERNEST
BELCHER ROLAND
BELCHER TOMMY JOE
BELEY JERRY
BELL DAVID THOMAS
BELL JOE EDGAR
BENCH CLIFFORD EUGENE
BENNETT DARL D
BENNETT DONALD CASPER
BENNETT JOHN WILLIE
BENNETT ROBERT DAVID
BENNETT VICTOR RAYMOND JR
BENSON JOSEPH
BENTFELD JOHN JOSEPH
BERGENSTEIN DENNIS PAUL
BERGER JOHN EDWARD
BERKHEIMER DENVER JOSEPH
BERNHART CARL HANS
BERRY TIMOTHY DALE
BERRY WILLIAM ANTHONY
BERTSCH KENNETH RAY
BERZINEC WILLIAM EDWARD
BEST RONALD LEE
BETHUNE ROBERT EDWIN
BEVIER MELVIN EDWARD
BEVILACQUA RENATO MARTIN
BEWLEY THOMAS EUGENE
BEZEAU RICK WILLIAM
BIBEY DWAIN LEE
BIEHL JAMES ALBERT
BILLINGS TERRENCE ROY
BILLINGSLEA DAMON EARL
BINGAMON DAVID LEE
BINGMAN PETER RUBEN
BIONDILLO JOHN CARL
BISHOP RONNIE HAROLD
BIXEL KENNETH BRUCE
BLACK DE WAYNE RODNEY
BLACK PAUL JR
BLACKWELL FREDERIC DELANO
BLACKWELL THOMAS MICHAEL
BLAIR RONNIE
BLAIR WILLIAM EARL
BLAKE JAMES WILLIAM
BLAKESLEE THOMAS WAYNE
BLANKENSHIP DONALD RAY
BLANKENSHIP EDGAR WILLIAM
BLANKENSHIP OVIE EARCIL
BLANTON BILL EDWARD
BLANTON CALVIN JR
BLANTON RUSSELL LEE
BLANTON WALTER CLAY
BLASKIS JAMES LAWRENCE
BLEVINS JAMES EVERETT
BLOOMFIELD WILLIAM DAVID
BLOSSER ROBERT KEITH
BLOUGH ROBERT DEAN
BLY ROBERT TILDON
BOBANICH JOSEPH A JR
BOBKOVICH STEPHEN JOSEPH
BOBO WILLIAM CHARLES
BOCKBRADER JERRY ALLAN
BOCOOK RONALD EDWARD
BODDIE JAMES EDWARD
BODNAR GEORGE JOSEPH
BODNAR JOSEPH A

BOEHM ALLEN THOMAS
BOGARD JACK CROSBY
BOGGESS RALPH M III
BOGGS IRA C JR
BOHANNON RONALD
BOLAN EDWARD WILLIAM
BOLAND DENNIS MICHAEL
BOLEY RONALD MARTIN
BOLIN DANNY ARNOLD
BOLING CHARLES L
BOLMAN DENNIS LOUIS
BOND GEORGE ALAN
BOND THEODORE CHARLES
BONIFANT SAMUEL HAROLD
BONKO DONALD RAYMOND
BONNELL GEORGE HARRISON III
BONNER IKE OTHEL
BONNETT SHERL KENT
BORDER WILLIAM EDWARD
BORDNER WILLIAM HAROLD
BORGER ROBERT LEE
BOROMISSZA CSABA FERENC
BORON DAVID JOSEPH
BOSWORTH RICHARD LEE
BOTTS ROBERT EUGENE
BOTTS THOMAS H
BOULWARE SHERMAN JAMES
BOWEN GROVER
BOWERS JEROME EARL
BOWERS LARRY HUTTON
BOWERSMITH CHARLES GEORGE
BOWERSOCK STEVEN EDWARD
BOWLING GARY DEAN
BOWMAN JAMES ERNEST

BOWMAN LARRY EARL
BOWMAN MICHAEL LEE
BOWMAN ROBERT EUGENE
BOWMAN RUDY GORDON
BOWMAN WILLIAM THOMAS
BOYD HURLEY MILLARD
BOYER CHARLES THOMAS
BOYER JOHN WILLIAM
BOZARTH TERANCE M
BRACKETT EVERETT LEE
BRAGG FRED GARLAND JR
BRAGGS ROOSEVELT JUNIOR
BRAINERD FLEMING B III
BRAM RICHARD CRAIG
BRANDON DARWIN OTHEL
BRANDON PHILLIP MICHAEL
BRANNON PAT GERARD
BRANSON DAVID RUSSELL
BRASSFIELD ANDREW THOMAS
BRAUGHTON CLYDE B JR
BRAZIK RICHARD
BREWER DON RAYLAN
BREWER JAMES DALE
BREWSTER CARL WARDEN
BRIGHT PAUL GLEN
BRISCOE CHARLES
BRISTER JAMES STANLEY
BRITTON STEVEN MICHAEL
BRITTON WILLIAM NED
BROBST JAMES ROBERT
BROCK DILLARD
BROCK JAMES PATRICK
BROCKMEIER THOMAS MICHAEL
BROERMAN BARRY BERNARD
BROGAN ROBERT HENRY

Ohio

BROOKS ALFRED IVRY
BROOKS CHARLES EDWARD
BROOKS DAVID LEE JR
BROOKS JERRY EDWARD
BROOM ERNEST ODELL
BROUGHTON ROBERT BALLARD
BROUGHTON WILLIAM ERNEST
BROUMAS ANDRE GEORGE
BROUSE PAUL ANDREW
BROWER RALPH WAYNE
BROWN ARTHUR DANIEL
BROWN CHARLES WILLIAM JR
BROWN DAVID ALLEN
BROWN EDDIE STEPHEN
BROWN FRANK LESTER
BROWN FRED EDWARD
BROWN GARY LEE
BROWN GREGORY LYNN
BROWN JACKIE RAY
BROWN JAMES MICHEAL
BROWN MERLE DEWAYNE
BROWN RICHARD ALLEN
BROWN RICHARD ALLEN
BROWN RICHARD SAMUEL
BROWN ROBERT DENNIS
BROWN RONALD DOUGLAS
BROWN WILLIAM EDWARD
BROWN WILLIAM ERNEST
BROWN WILLIAM FRANKLIN
BROYLES RICHARD ALAN
BRUCE RICHARD PETER
BRUCE RONALD DWIGHT
BRUCK THOMAS FREDERICK
BRUCKER LESLIE L JR
BRUCKNER DONALD RICHARD
BRUNDAGE MICHAEL LESTER
BRUNNER MICHAEL CARL
BRUNO VITO VINCENT
BRYAN CLIFFORD EDWARD
BRYAN DAVID ANDREW
BRYAN JERRY WAYNE
BRYAN LAWRENCE GEORGE
BRYANT DAVID EDWIN JR
BRYANT DONALD RAY
BRYANT JAMES ROBERT
BUCHANAN ROGER ALLAN
BUCK ARTHUR CHARLES
BUCK ROBERT RONALD
BUCKLEW DENNIS
BUDBILL GERALD JACOB
BURGESS JOHN PETER
BURGOON WILLIAM PAUL
BURKE JAMES EDWARD
BURKE JAMES ROBERT
BURKES BRUCE WAYNE
BURLILE THOMAS EDWARD

BURNETTE GARY WAYNE
BURNS JOHN FRANCIS
BURROUGHS ROBERT NELSON
BURROUGHS WALTER L
BURTON CECIL WAYNE
BURTON JOHNNY EDWARD
BUSBY RONALD DEAN
BUSBY RONALD DEAN
BUSCH JON THOMAS
BUSSEY MARVIN WILLIAM
BUTCHER DAVID AUSTIN
BUTSKO ALBERT MICHAEL
BYERS CLAYTON HENRY JR
BYOUS MARCUS RANDOLPH
BYRNE JAMES PATRICK
BYRUM DONALD EDWARD
CAHALANE MICHAEL JOSEPH
CAHILL CARL THOMAS
CALDWELL GARY LESLIE
CALDWELL WILLIAM MILES
CALENTINE RONALD LEE
CALEY MICHAEL SHANE
CALKINS DAVID EARL
CALLAHAN CLYDE
CALLAN ROBERT THOMAS
CALLAWAY MC ARTHUR
CALLOWAY LARRY JAMES
CALVITTI DAVID
CAMERON ROBERT CHARLES
CAMPBELL DAVID DANA
CAMPBELL DAVID JAMES
CAMPBELL DONALD ALLEN
CAMPBELL EARNEST EUGENE
CANAS ROBERTO LUIS
CANIFF JOHN R
CANNON ROBERT BYREL
CANTU ESIQUIO AIRNALDO
CARDER DENZIL MASON JR
CAREY DAVID LEE
CARLTON LAVALLE ERNEST
CARMAN JAMES CONRAD
CARNEY ROBERT ARTHUR
CARPENTER CHARLES EDWARD
CARPENTER CLIFFORD LEE
CARPENTER HOWARD B
CARPENTER KENNETH HAROLD
CARPENTER NICHOLAS MALLORY
CARPENTER SAMUEL DAVID
CARPENTER TERRY WAYNE
CARPENTER WILLIAM JR
CARR JAMES ALLEN
CARRION JOSE ANTONIO
CARROLL JAMES NATHAN III
CARSON RUSSELL BERTON
CARSTENS THOMAS HENRY
CARTER DAVID EDWARD
CARTER EDWARD EUGENE
CARTER ERNEST MACK
CARTER GREGORY
CARTER HUBERT CLAYTON
CARTER RALPH DWAIN
CARTER ROBERT LESTER
CARTER RONALD JAMES
CARTER ROY LYNN
CARTER WILLIAM EDWIN
CARTWRIGHT MICHEAL GLENN
CARTWRIGHT RICHARD

CORTEZ
CARUSO DAVID RAYMOND
CASAREZ RAUL
CASEY MAURICE ALOYSIUS
CASHMAN CORNELIUS JAMES
CASSIDY MICHAEL OLIVER
CASTLE LARRY FLOYD
CASTLE VIRGIL LEE
CASTO CLARENCE LEROY
CATLIN THOMAS DAVID
CAUDILL DONNIE WAYNE
CAUDILL ELMON C II
CAUDILL JAMES
CAUDILL ORVILLE
CAVINEE RONALD C
CEGIELSKI RICHARD JOSEPH
CHADWICK JAMES EDWARD
CHAFFEE VAN
CHAHOC DAVID KEITH
CHAMBERS DONALD EDWARD
CHAMBERS JAMES LARRY
CHAMBLISS JIMMY LEE
CHANDLER CHARLES PERRY
CHANDLER RONALD EUGENE
CHANEY STEPHEN JOHN
CHAPMAN PETER HAYDEN II
CHAPMAN THOMAS TODD
CHARLES EARL EUGENE
CHARLES RONALD
CHARLESWORTH JAMES W JR
CHASE MICHAEL LYN
CHASSER RAYMOND MICHAEL
CHAVEZ FILIBERTO
CHEADLE HAROLD LEE JR
CHEARNLEY JOSEPH MICHAEL
CHEMIS CHARLES ROBERT
CHILCOTE BRYAN MICHAEL
CHILDRESS MARTIN DEAN
CHILTON RICHARD KENNETH
CHINN JAMES RUSSELL
CHIVERS JAMES LEE
CHMEL DENIS MICHAEL
CHOPPA RICHARD ANTHONY
CHRISTIAN DANIEL KIETH
CHRISTIAN VERNON WEBB JR
CHRISTY JAMES ARTHUR
CHRISTY RICHARD NEIL II
CHURCH JIMMY KERMIT
CHURCH REX FILLMORE
CLACK HOWARD LEE
CLARK BARRY ROBERT
CLARK BRIAN DALE
CLARK DONALD ROBERT
CLARK JAMES LEE
CLARK JAMES ROGER
CLARK LORINZER PAUL JR
CLARK RALPH E
CLARK ROGER DALE
CLARK THOMAS EDWARD
CLARK THOMAS ELMER
CLARKE CORNELL RICHARD R
CLAYPOOL GEORGE ROBERT
CLEM EDWARD
CLEMENTZ RICHARD JOSEPH
CLEMSON GERALD RICHARD
CLIMER DAVID LEROY
CLINE RONALD GREER
CLINGERMAN JOSEPH ALLAN
CLINTON LARRY ELZA VAN
COATES RONALD PERRY
COATS JAMES PRESTON

COBB WILLIS
COE RONALD RAY
COFFMAN ROGER LEROY
COLE EMERSON PAUL
COLE MARVIN EUGENE
COLE NATHAN JOHN JR
COLEMAN MC ARTHUR
COLEMAN RICHARD FREEMAN
COLEMAN WILLIAM FRANK
COLLENE CHARLES EDWARD
COLLINS DAVID JIM
COLLINS GARY EDWARD
COLLINS RALPH RAYMOND JR
COLN RAY EUGENE
COLOPY STEPHEN LYNN
COMBS DENNIS ALAN
COMBS JAMES STEPHEN
COMBS JOHN ASHER
COMBS LEE ROY
COMBS THOMAS EUGENE
COMBS TYRONE
COMPA JOSEPH JAMES JR
CONAWAY LAWRENCE YERGES
CONDIT WILLIAM HOWARD JR
CONGER JOHN EDWARD JR
CONKEL THOMAS EUGENE
CONKLIN THOMAS ARTHUR
CONLEY EUGENE OGDEN
CONLEY MONROE JASON
CONNOR PETER MICHAEL
CONRAD JOHN WILLIAM
CONRAD PAUL LEWIN
CONRADY MICHAEL JOSEPH
CONWAY JAMES THADDEUS
COOK CHARLES ROBERT
COOK EARL LLOYD
COOLEY SHELBY EMERSON
COON CALVIN KERMIT
COONROD ARNOLD LEE
COOPER TERRY LEE
COPLEY BRUCE
COPLIN SCOTT RONDAL
CORE DERRICK
CORFMAN DARYL RAYMOND
CORLETT GERALD ERNEST
CORLEY THOMAS EUGENE
CORNELL DONALD FREDERICK
CORRELLO SCOTT DENNIS
CORSI BOBBY GLYNN
CORSINO EDDIE NELSON
CORWIN EDWIN HUGH
COSGROVE COURTNEY JAMES
COTTERMAN HARRY ANDREW
COTTRELL SIDNEY ALLEN
COTTRELL TIMOTHY JAMES
COTTRILL GEORGE W JR
COUK KARL HENRY
COURTNEY MICHAEL JOSEPH
COWLEY THOMAS REGINALD
COX RAYMOND PRATER
COX TIMOTHY ROBERT
COYNE WILLIAM FRANCIS
CRAMER JEFFREY THOMPSON
CRAMPTON GARY LEE
CRAVENS DANNY CAROL
CRAVENS THOMAS LLOYD

CRAWFORD CHARLES HUGH
CRAWFORD CHARLES MARION
CRAWFORD STEPHEN EARL
CRIST KENNETH LEE
CRIST STEPHEN EDWARD
CRITCHFIELD REECE A JR
CRITES RICHARD LEE
CROSS EDWARD JOHN
CROSS JAMES EMORY
CROSSLEY JOSEPH
CROTHERS DANNY KAY
CROUSE JOHN RAYMOND
CRUM CURTIS RAY
CRUMBAKER LARRY HOMER
CRUZ JESUS ROSAS
CRYSEL KENNETH LEE
CRYTZER RALPH WOODWARD JR
CUE WILLIAM CHARLES
CULOTTA ANTHONY THOMAS
CULP RICHARD THOMAS
CULP THOMAS DALE
CULWELL JAMES RONALD
CUNNINGHAM CHARLES ROBERT
CUNNINGHAM STEPHEN EARL
CURLEE JOHNNIE M
CURRAN MICHAEL PATRICK
CURRAN PHILIP ROBERT
CURTIS CHARLES CLINTON JR
CURTIS ROBERT JOHN
DADANTE LEONARD JOHN
DADISMAN GORDON ALAN
DAGGER CARL RICHARD
DAHILL DOUGLAS EDWARD
DAILEY RONALD CHARLES
DAISHER DAVID CHARLES
DALEY TERRENCE JOSEPH
DALTON GORDON THOMAS
DALTON JAMES WENDALL
DALTON ROBERT ALAN
DANIEL FREDDIE LAWS
DANIELS JOSEPH RAY JR
DANIELS RICHARD GALE
DANIELS WILLIAM MARCUS
DANISON JOSEPH WILLIAM
DARNELL GEORGE W JR
DASEN GERALD RANDAL III
DAUCH GARY LEONARD
DAUGHERTY THOMAS LEE ROY
DAVENPORT JAMES MICHAEL
DAVID CHARLES EDWARD JR
DAVIDSON DENNIS ROBERT
DAVIDSON RONALD LEE
DAVIS ABRON EARL
DAVIS ALBERT J
DAVIS CHARLES R JR
DAVIS DUANE MICHAEL
DAVIS ELWOOD WILLIAM JR
DAVIS GARY RAY
DAVIS GENE THOMAS
DAVIS JOHN CLAYTON JR
DAVIS KENNETH
DAVIS PAUL PATRICK
DAVIS PHILIP GEORGE
DAVIS ROBERT DENNIS
DAVIS ROBERT HENRY
DAVIS ROBERT WILSON

DAVIS SYLVESTER
DAVIS TERRY LEE
DAVISON NORMAN RAY
DAWSON FRANK WILLIAM
DAY CHARLES KEITH
DE BOARD ROBERT DARRELL
DE CAMP MICHAEL DAVID
DE FOSSE THOMAS GLENN
DE FRANGE MARK JOHN
DE JARNETT GEORGE WESLEY
DE LONG WILLARD JR
DE VILLE FRANCIS XAVIER
DEAN KENNETH LEE JR
DEAN LARRY LAMARR
DEAN WILLIAM EDWARD
DEARING LARRY GENE
DEBO WILLIAM LOUIS
DECKER JOSEPH NICHOLAS
DECKER WILLIAM BERNARD
DECKER WILLIAM THOMAS JR
DEETER DAVID KIM
DEETER MICHAEL ALAN
DEIKE ROBERT JAMES
DELANEY JAMES PATRICK
DELANEY JOHN PATRICK III
DELGADO-MARIN ARTURO
DEMALINE PAUL ALLEN
DENIG JOSEPH HENRY
DENKINS FRED JR
DENNEY TERRY LEE
DENNIS MARK V
DENNIS WILLIE ROSS
DENNULL EDWARD MICHAEL
DETRICK GARY GENE
DEVERS DAVID RONALD SR
DEVINE DAVID EUGENE
DEWEESE BILLY CLARENCE
DEWINE ROBERT BRUCE
DICE ROBERT FLOYD
DICKEY DOUGLAS EUGENE
DICKEY WILLIAM RONALD
DICKMAN DAVID MICHAEL
DIECKMANN JOHN E
DIEHL PATRICK REGAN
DILLEY DANA ALLEN
DIMAGARD WILLIAM
CHARLES
DIMMERLING ROME EDWARD
DISCEPOLO ANTHONY
ALBERT
DISHMAN JERRY
DOAK TOMMY ALLEN
DOBBINS GARY LEE
DODSON JOHN LARRY
DOMBROSKI DARRYL TOD
DOODY DOUGLAS WILLIAM
DOOLITTLE GARY WAYNE
DOPP RICHARD ERNEST
DORAN TIMOTHY PATRICK
DOSECK RICHARD ALLEN
DOUGLAS DELBERT
DOUGLAS THOMAS EVAN
DOUGLAS WILLIAM LOWELL
JR
DOWNEY CLAY EDWARD
DOWNEY EDWARD FRANCIS
JR
DOYLE DAVID LEE
DRAHER CLIFFORD EARL
DRAKE CARL WILSON
DRAKE ROGER KENNETH
DREHER RICHARD E
DREWICZ ROBERT CHARLES
DROWN LARRY GENE
DU PONT JAMES CAMIL
DUBACH GARY LYNN
DUCHNOWSKI JOHN PAUL
DUDLEY HARVEY JR
DUFFY JAMES PATRICK JR
DUGGER ALFRED
DULEBOHN DENNIS LEE
DULEN RENDLE
DULIK THOMAS WILLIAM
DUMAS WILLIAM RICHARD
DUNCAN WALTER EARL JR
DUNN CARL EDWARD
DURHAM GEORGE RAY
DURHAM JOHN ALBERT
DURHAM WILLIAM JAMES
DUTKIEWICZ ROBERT JOHN
DYBVIG NED TURNER
DYCE DONALD MYRON
DYCKS RONALD KING
DYE DAVID ALAN

DYE EDWARD PHILLIP
DYE JAMES HERBERT
DYE TIMOTHY ELDEN
DYER ALLEN JOHN
EAKINS MELVIN WARREN
EALY CARL
EARICK JAMES ALLEN
EARLY JAMES MICHAEL
EARLY WILLIAM DAIL
EATON DAVID LEE
EATON JERRY ARNOLD
EBEL WILLIAM EARNEST
EBRIGHT WILLIAM RAYMOND
O
ECHOLS DAVID ALLEN
ECHOLS TIMOTHY DAVID
ECKERFELD MICHAEL DAVID
ECKLE STEPHEN JOHN
ECTOR JERRY
EDER ROBERT OTTO
EDMONDS ARTHUR LEE JR
EDRIS RICHARD JOHN
EDWARDS CHARLES KENNETH
EDWARDS JOHN THOMAS
EDWARDS RICHARD LYON
EDWARDS THOMAS RAY
EGGER WALTER JACOB
EGGERS CHARLES RONALD
EGGERT RUSSELL WILLIAM
EGGLESTON RODNEY LEE
EICHELER GARY ERNEST
EICKHOLT ROBERT LEO
EIDUKAITIS GEDIMINAS JUST
EISENBRAUN DAVID LAW-
RENCE
ELCHERT JAMES MELVIN
ELDRIDGE WETZEL LONNIE
ELLERBROCK MARVIN
CHARLES
ELLIS JESSE LEONARD
ELLIS ROBERT LEE
ELLIS RUSSELL HAROLD
ELLIS WALTER EUGENE
ELLWOOD EUGENE LEE
ELSTON ROY DAVID JR
ELSWICK ROBERT WAYNE
EMERY OWEN RAY
EMMERT CHARLES WILLIAM
ENCZI RAYMOND MICHAEL
ENDERLE CLYDE WILSON
ENDICOTT DANNY G
ENDRESS WILLIAM JAMES
ENGELHARD ERICH CARL
ENGLERT JAMES RAYMOND
ENIX JACK GENE
ENSELL JOHN ROBERT
ENSIGN WALTER LYMAN JR
EPPLEY GERALD VERNON
ERBLAND NORMAN JOSEPH

ERDOS DENNIS KEITH
ERFORD DENNIS CHARLES
ERFORD JOHN LEWIS JR
ERHART MICHAEL DAVID
ERVIN BAXTER FRENCH
ERVIN GARY LEE
ESPENSHIED JOHN LEE
ESTERLY LAWRENCE ALAN
ESTES KENNETH
EUTSLER JOHN WESLEY
EVANS CLYDE SAMPSON
EVANS JAMES WILLIAM
EVANS JOE FRANKLIN
EVANS LARRY EDGAR
EVANS RONALD D
EVANS RONALD LEE
EVANS WILLARD JAMES
EVILSIZOR RALPH RAYMOND
EWALT DONALD THOMAS
FAILS EDWARD LEE JR
FALK GARY DAVID
FARLEY GARY LEE
FARLOW CRAIG LEE
FARLOW GARY ALLAN
FATICA ROBERT JOSEPH
FAUL KENNETH WAYNE
FAULKNER LARRY ALLEN
FAULKNER MICHAEL LEE
FAVERTY ALVIS RAY JR
FAVOURITE RONALD LEE
FAZZINO JAMES DOUGLAS
FEARN GUY VICTOR
FEATHERSTON FIELDING W III
FECK DANIEL EDWARD
FEEMSTER COLINNA
FEKETE JAMES CHARLES
FELICIANO GILBERT
FELL DAVID GLEASON
FELT RICHARD WAYNE
FELTZ KEITH A
FENKO STEVE BRIAN
FERGUSON DAVID CHARLES
FERGUSON LYNN MICHAEL
FERGUSON RALPH
FERRO JOSEPH
FERRY DANIEL SAMUEL
FETHEROLF JOHN LAWRENCE
FETTY CLARENCE EDWARD
FIELDS LARRY EDWARD
FIELDS RONALD CLARK
FILIPPI JOHN CHARLES
FILLIATOR RICHARD AN-
THONY
FINCHER JULIAN A JR
FINKEL WILLIAM ARTHUR
FINNICUM JOHN OTIS
FIRST MICHAEL BRUCE
FISCHBACH ALLAN RUSSELL
FISCHER GREGORY WILLIAM

FISCHIO JOHN ANTHONY
FISER DIETER JAMES
FISH GEORGE WILLIAM JR
FISHER DANNY JAY
FISHER DARRELD EDWARD
FISHER DAVID FRANCOIS
FISHER DAVID HERBERT
FISHER MARSHALL WAYNE
FISHER RICHARD OTIS
FISHER RONALD JAY
FISHER THOMAS GAYLON
FITCH PHILIP
FITZPATRICK THOMAS M
FLAGELLA JAMES POTITO
FLANIGAN JOHN DAVID
FLECK GARY LEE
FLEMING DENNIS K
FLEMING KENNETH CLAIR JR
FLEMING RAYMOND E JR
FLESHER RUSSELL RAY
FLESHMAN RANDY ALLEN
FLETCHER ROBERT WENDELL
FLONNOY FRANK WARREN JR
FLOOD CHARLES DALE
FLORA LARRY VINSON
FLORES RAMON JR
FLORES ROBERT JR
FLOREZ FRANK OCHOA JR
FLOYD ROBERT EUGENE
FLYNN RAYMOND JOSEPH JR
FORD KENNETH ALLISON JR
FOREE JOSEPH HERMAN
FORRISTAL RUSSELL PATRICK
FORRY JEFFREY SCOTT
FOSNAUGH CAREY ALLEN
FOSTER ALFONZA
FOSTER DOUGLAS EUGENE
FOSTER WILLIAM EARL JR
FOUST KENNETH EDWARD
FOUTZ KENNETH LEE
FOWBLE ROBERT L JR
FOWLER DAVID ALLEN
FOX HOWARD THOMAS
FRALEY DAVID FORREST
FRANCE WILLIAM RICHARD
FRANCIS CARRIS MICHAEL
FRANKENSTEIN JACKIE
FRANKHAUSER CHRIS WAL-
TER
FRANKLIN CHARLES EDWARD
FRANKS DAVITT JOHN
FRANZER NICHOLAS LEAN-
DER
FRASHER JOSEPH EDWARD
FRASURE HURSHEL
FRAVEL DAVID WARD
FRAZEE GEORGE HOWARD JR
FREDERICK CHARLES EM-
METT

FREDERICK DAVID ADDISON
FREDERICK LAMAR DONALD
FREDERICK WILLIAM V
FREELAND CHARLES JEFFERY
FREEMAN JEFFREY ALEXAN-
DER
FREEMAN JOSEPH WARREN JR
FRENCH DAVID LEE
FREPPON JOHN DENNIS
FREUND WILLIAM CARL
FRILEY ARTHUR TIMOTHY
FRILLING JEROME RAYMOND
FROMANT KENNETH B
FRYE DANNY LEE
FUCHS GREGORY GERALD
FUCHS RICHARD ELLSWORTH
FUGATE GARLAND G
FULLER CHARLES MC DON-
ALD
FULLER GARY LEROY
FULLER LARRY LEE
FULLER ROBERT JOSEPH
FULLUM DARRYL BLAKE
FUNK EMMONS EDWARD JR
FUQUA ROBERT LEE JR
FUSILE MARK LOUIS
FUSON JACK ANTHONY
FUSS THEODORE FRANK
FUTO JOHN ANTHONY
GAARDER DAVID EIDNES
GABEL GARY LEE
GABLE RONALD HOWARD
GAGE ROBERT HUGH
GALBRAITH RUSSELL DALE
GALLANT ROY DALE
GALLAUGHER DARRYL ALAN
GALLAWAY WILLIAM DENNIS
GALLOWAY GEORGE K JR
GALLUP ROBERT DARYL
GALVAN RICARDO
GAMBLE DAVID LESLIE
GANT HERMAN EUGENE
GANTZ KARL RAY
GARCIA RAYMOND CHARLES
GARDNER ROBERT LINLEY
GARRARD JESSIE JEROME
GARRETT ROBERT LEE
GARRISON DANIEL LEE
GARRISON RONALD MILLARD
GARVEN CHARLES DANIEL
GARVEN WAYNE ERIC
GARZA ANTONIO
GASE JAMES FLORIAN
GASKINS LARRY LEE
GASSMAN GERALD LYNN
GATES THOMAS LAWRENCE
GATWOOD MICHAEL OWEN
GAUCH DAVID ALAN
GAWEL WALTER L

185

Ohio

GAY DAVID AUSTIN
GEARHEART RALPH ALLAN
GEIB JEFFERY LYNN
GEIGER WALDEMAR JOHN
GELDIN JEFFREY LEE
GENTKOWSKI JOHN STEVEN
GEORGE LEO ALLEN
GEORGE WILLIAM MICHAEL
GERBER LEROY LYNN
GERHARDT KEITH EDWARD
GEUY JAMES ROBERT
GIBSON ALTON GEORGE
GIBSON BURRELL
GIBSON EDDIE CARL
GIBSON JAMES VAL
GIBSON JOHN EDDIE
GIBSON RONALD WARREN
GIFFORD DAVID RANDOLPH
GILDOW THOMAS JOSEPH
GILGER WILLIAM GRANT III
GILL DAVID ROYDEN
GILL JOSEPH GEORGE
GILLESPIE ALBERT ALOYSIUS
GILLIAM THOMAS EDWARD
GILLUM RICKEY LEAROY
GIRON MARK JOSEPH
GLANTON ALBERT THOMAS
GLEISINGER STANLEY RICHAR
GLENN WILLIAM STUART
GLOVER CALVIN CHARLES
GLOWACKI EUGENE THOMAS
GODFREY MC VINCENT JR
GODFREY PARKER KEITH
GOEBEL MARVIN E
GOINS GORDON L
GOINS HUGH CRAIG

GOLD ROBERT JOSEPH
GOLDEN KENNETH EDWARD
GOLLAHON GENE RAYMOND
GOLLING CHARLES RONALD
GONZALEZ AMALIO
GONZALEZ ANDRES AVALOS
GOODSPEED WILLIAM
HUNTER
GOODWIN JOHN MASON
GOONAN PAUL EDWARD JR
GOPP THOMAS ALAN
GORDON JAMES EDWARD
GORNEY JERRY EDWARD
GORSLENE TERRY EUGENE
GORTON THOMAS FREDERICK
GORVET WILLIAM ANTHONY
GOSSARD DAVID EUGENE
GOUDY GARY ROY
GRABER SCOTT THOMAS
GRACE EARLEY CARTER
GRAFF THOMAS GEORGE
GRAHAM KENNETH ERROL
GRANT THOMAS RICHARD
GRAY CLARENCE HENRY
GRAY JAMES H
GRAY RAYMOND ANTHONY
GRAY ROBERT VERNON
GRAY RONALD K
GREAVU BILLY JOEL
GREEN MELVIN RICKEY
GREEN PHELBON MICHAEL
GREEN WILLIE F
GREENDYKE GERALD BRUCE
GREENE CHARLES WILLIAM
GREENE JOE WILLIAM
GREENE PAUL HARRISON
GREENLEE DUANE THEO-
DORE
GREENLEE JAMES EDWARD
GREGG JOSEPH GENE
GREGORY HENRY
GREGOVICH PAUL MICHAEL
GRIER RICHARD EUGENE
GRIESER PHILIP LEE
GRIFFIN BRUCE FRANKLIN
GRIFFIN RONALD LEWIS
GRIFFIS ROBERT DALE
GRIFFITH MICHAEL LYNN
GRIFFY VIRGIL D

GRONSKY DALE ANDRE
GROOMS CHARLES EARL JR
GROOMS ROBERT LEE
GROOMS WILLIAM DAVID
GROSCOST ROBERT MILLARD
GROSS GARY WAYNE
GROSS OLLIE JAMES
GROTH WILLIAM GEORGE
GROVER GENE DELANO
GROVES LOWELL ROGER
GROW GARY LEE
GRUBER MARTIN STEVE JR
GRUBER MARTIN STEVE JR
GRUBER MILTON DARRELL
GRUNDER ROLAND JOHN
GUENTHER JOSEPH ELLIS
GUERIN ROBERT LOUIS
GUERRA BERT III
GUIST JOHN JOSEPH
GUMBERT ROBERT WILLIAM
JR
GUNDOLF STEVEN DEAN
GUNTHER ROBERT LOUIS
GUTIERREZ LOUIS SAM
GUYER WILLIAM HARRIS
HABEN MERLE WILLIAM
HABERMAN DAVID
HACKER RONALD VENTION
HADDIX DOUGLAS BOYD
HAGER JACK LEONARD
HAGERTY PATRICK MICHAEL
HAGLAGE ANDREW MARTIN
HAHN RAYMOND GEORGE JR
HAIFLEY MICHAEL FIRESTONE
HAINES PAUL ALLEN
HAINES WILLIAM ALLEN JR
HAINLEY WILLIAM ROBERT
HALE JOHN JR
HALL ALBERT
HALL DONALD DALE
HALL JAMES ALBERT
HALL ROBERT KENNETH
HALL SAMUEL CHRISTIAN
HALLAS JOSEPH MICHAEL
HALMAN JOHN HENRY JR
HAMBLETON BARRY NEWTON
HAMBLIN RICHARD ALAN
HAMILTON DAVID ALLEN
HAMILTON JEFFREY GILES

HAMILTON MARCUS JAMES
HAMILTON ROBERT RICHARD
HAMILTON RONALD JOA-
QUINE
HAMILTON RONALD LLOYD
HAMILTON RUSSELL LEE
HAMMACK ORLA DANIEL
HAMMOND LAWRENCE CLAIR
HAMNER MICHAEL KEITH
HAMPSHIRE ROBERT CLOYCE
HANCOCK EDWARD DEAN
HAND WILLIAM HARRY
HANDEL LIBERO CHARLES
HANDY TODD ARTHUR
HANEY KEITH EUGENE
HANIOTES STEVEN MICHAEL
HANLAN ALLEN DEWEY
HANN CHARLES EDWARD
HANNA DAVID RUSSELL
HANNAH BYRON MARK
HANNAH SAMUEL JAMES
HANNAMAN ROBERT ALLEN
HANSEN CARL VALENTINE
HANSLEY TIMOTHY WHAR-
TON
HANSON ROBERT TAFT JR
HARBIN CARL ROSS JR
HARDIN DENNIS IVAN
HARDY JOSEPH EDGAR
HARGREAVES JOHN JAMES
HARLOW JOHN BRAYTON JR
HARMON PAUL OLIVER JR
HAROULAKIS ANDRE
HARP DONALD EUGENE
HARRIS DAVID STANLEY
HARRIS EDWARD JAMES JR
HARRIS ERVIN ELLIS
HARRIS JAMES FRANKLIN
HARRIS JAMES WALDEN JR
HARRIS PAUL WINIFORD
HARRIS REUBEN BEAUMONT
HARRIS RICKEY ELTON
HARRIS SAMUEL GARY
HARRISON PAUL RAYMOND
HARRISON RONALD EUGENE
HARROW DAVID NELSON
HARSHMAN STEPHEN WARD
HARTKEMEYER JOHN RAY-
MOND

HARTLEY DAVID WILLIAM
HARTLEY ROBERT CARL
HARTPENCE DENNIS RAY
HASKINS HARRY DONALD
HATFIELD JACK
HATFIELD JOHN FREDERICK
HATTON ROBERT WILLIS
HAUCK JAMES MICHAEL
HAUF JERRY WAYNE
HAUSERMAN LEONARD
STEPHEN
HAUSMAN HENRY RICHARD
JR
HAWKINS ALBERT WILLIAM
HAWKINS ARTHUR LEE JR
HAWKINS JOHN LEWIS JR
HAWKINS TERRY LEE
HAWLEY JACK ALLEN
HAWLEY ROGER LEE
HAY GERALD WAYNE
HAYES DANNY CARLTON
HAYMAN ARCHIE ANDREW
HAYNES CHARLES FRANCIS
HAYNES WILLIAM THOMAS
HAYS WAYNE ALLEN
HAYWARD GEORGE ERNEST
HAZZARD LOUIS TIMOTHY
HEALY RICHARD JOHN
HEALY THOMAS EDWARD
HEDGPATH WILLIAM THOM-
AS
HEIDRICH GREGG WILLIAM
HEIFNER EDWARD WILLIAM
HEIFNER KENNETH RICHARD
HEIGHTLAND ORVILLE W JR
HEIL LOUIS GEORGE
HEIM JAMES PHILLIP
HEINTZ NED RICHARD
HEISER EDWARD MICHAEL
HEISER TERRY RICHARD
HELBER LAWRENCE NEAL
HELFENSTINE SAMUEL JAMES
HELGESON ELLIS EUGENE JR
HELSEL PAUL ELROY JR
HELTON DONALD LEONARD
HELTON DWAYNE
HELTSLEY JOSEPH JUSTIN
HELTSLEY PAUL R III
HENDERSON JOHN LESLIE

HENDERSON WILLIAM ROY
HENDLE DAVID WALLACE
HENDRICKS CLARENCE O III
HENDRICKS HARRY LANN
HENDRIX KENNETH WAYNE
HENNINGER KENTON EL-
WOOD
HENSLEY JAY THOMAS
HENTHORN ALVA DEAN
HERALD CLABE JR
HERBERT LARRY EUGENE
HERINGHAUSEN ARTHUR J JR
HERKINS THOMAS FRANKLIN
HERRON DENNIS
HESKETT JERRY WAYNE
HESLER STEVEN EUGENE
HESS DALE JOSEPH
HESSELSCHWARDT HERMAN
ALO
HEWIT RUSSEL HOWARD JR
HEWITT BURL DENTON
HEWITT ROBERT EUGENE
HEYDINGER LAUREN JOSEPH
HEYDORN CHARLES GEORGE
HICKMAN DALLAS EDWIN
HICKMAN ROBERT EUGENE
HICKMAN WANDLE LEWIS
HIDO RICHARD LEE
HIGGINS JEROME
HIGHLAND FOREST G JR
HILDENBRAND CHARLES
JOHN
HILL GERALD WILLIAM
HILL JERRY L
HILL LAMONT DOUGLAS
HILL PAUL JEWELL
HILL ROBERT ALLEN
HILL ROBERT ALLEN
HILLIARD ROBERT RICHARD
HILTON DAVID LYNN
HILTON EUGENE JR
HINDERMAN ANDREW JACOB
HINDS KENNETH WILLIAM
HINERMAN WILLIAM RUSSELL
HINES JOHN WAYNE
HINKEL DANIEL KENNETH
HINKLE DOUGLAS LEE
HINKLE JACK LEE
HINKLE TERRY LEE
HINSON HERBERT STEPHEN
HINTERLONG LEO EDWARD
HISEY JOHN EDWIN
HISEY TYRONE WADE
HITER VIRGIL LEMAR
HITES ALLEN LYNN
HIVELY DANIEL RICHARD
HOBBS GLEN THOMAS
HOCHSTETLER TERRY LYNN
HOCK RICHARD JAMES
HODGE DENNIS RAY
HODGE JAMES EDWARD
HODOROWSKI RAYMOND
HOEWELER JAMES EDWARD
HOFFERT DAVID EDGAR
HOFFMAN CARL DEAN
HOFFMAN DONALD ROBERT
HOGARTH RICHARD DOUG-
LAS
HOGE FRANK LEE
HOGSTON ROGER LEE
HOHMAN DANIEL JOHN
HOLBROOK BENNIE H
HOLBROOK WILLIAM R
HOLEMAN RAY WALTER
HOLLAND FARIS E
HOLLER CARL WAYNE
HOLLEY PAUL RICHARD
HOLMES JAMES CECIL
HOLSINGER GARY OLSON
HOLYCROSS RICHARD LAKE
HOOD BUDD EDWARD
HOOD CHARLES ALAN
HOOD DALE ROBERT
HOOD JERRY WAYNE
HOOK MARK LOREN
HOOVER JAMES
HOOVER THOMAS EUGENE
HOPE SAMUEL JR
HOPKINS RICHARD LEE
HORN DAVID MICHAEL
HORN EDWARD ANDREW JR
HORNER ALBERT LEROY
HORVATH WAYNE STANLEY
HOSKEN JOHN CHARLES

HOSKINS ALVIN
HOSKINS DANNY
HOUSEHOLDER RICHARD
WAYNE
HOUSER DAVID ROBERT
HOUSTON RICHARD PAUL
HOUX LESTER JR
HOWARD CLAUDE
HOWARD DAVID RAY
HOWARD DAVID RAY
HOWELL EDWARD MICHAEL
HOWERTON JERRY RUSSELL
HOXWORTH WALTER BRUCE
HRINKO WILLIAM JOHN
HUBBARD GEORGE ALLEN
HUDAK ANDREW MICHAEL
HUDAK FRANK PAUL
HUDDLESON RODNEY LEROY
HUDDY DANNY JOE
HUDSON GEORGE HOWARD
HUEBNER TERENCE ARTHUR
HUEY GUY WINFRED
HUFF FRANK CALVIN
HUFF JAMES HENRY
HUFF RAY GENE
HUFFER ALBERT EUGENE
HUFFMAN GERALD
HUFFMAN GLEN MICHAEL
HUFFMAN SAMUEL LEWIS
HUGHES GORDON KAY
HUGHES KENNETH RICHARD
HUGHES LEWIS EUGENE II
HUGHES MARVIN THOMAS
HUGHES ROBERT LAURENS
HUGHES WILLIAM JOSEPH
HUGHEY LLOYD RAY
HULL EDISON DENNIS
HULL GERALD EDWARD
HULL RICKY LEE
HUME JOSEPH SYLVESTER JR
HUME KENNETH EDWARD
HUMMEL HARRY LYNNE
HUNLEY JAMES WILLIAM
HUNT RALPH EDWARD JR
HUNTER JOHN CLARK
HURD JOHN LAWRENCE
HURLEBAUS LESLIE VERNARD
HURLEY JERRY LEE
HUSTON CHARLES GREGORY
HUSTON HARRY D JR
HUTCHINS CHARLES E JR
HUTCHINSON RANDOLPH
SCOTT
HUTCHINSON RICHARD JR
HUTSON GEORGE GLENN
HUWEL MICHAEL FRANCIS
HUZICKO CHARLES JAMES
HYATT JERALD MICHAEL
IDING GREGORY THOMAS
IDLE THOMAS GEORGE
INBODEN ROGER LEE
INGELS CHARLES WILLIAM
INGLES DANNY LEE
INGRAM ISRAEL LONZO
INLOW RICKY GENE
INMAN WILLIAM IVAN
INSANA SALVATORE CAR-
MELO
INTIHAR JOSEF PAUL
IRVIN OPHREY AUSTIN
IRVIN RICHARD LOWELL
IRVING LEE
ISAACS JOHN PAUL
ISAACS WAYNE LEE
ISER KENNETH EUGENE
IVEY SHERMAN LEE
JACKSON CHARLES WILLIAM
JACKSON DAVID LEE JR
JACKSON DAVID RUSSELL
JACKSON DONALD ALLEN
JACKSON JERRY LEE
JACKSON JOHN HERSTON
JACKSON JOHN WILLIAM
JACKSON LAWRENCE DAVID
JACKSON RONALD
JACOB PHILLIP
JACOBS RALPH WAYNE
JACOBS RICKIE JEROME
JAMISON TED RAY
JANEDA STEVEN MICHAEL
JAQUA MICHAEL DOUGLAS
JARRETT MICHAEL DONALD
JASSO JOHN
JATEFF WILLIAM ALBERT

JATICH GARY LEE
JELINEK ALLEN LEO
JELKS CARLOS DENNIS
JENKINS DONALD RAY
JENKINS KENNETH CLIFFORD
JENKINS MORRIS E
JENKINS RANDALL LEE
JENKINS TERRY LEE
JENKS GARY LEE
JEWELL JAMES CLARENCE JR
JILEK LOUIS HENRY
JINDRA ROBERT JAMES
JOECKEN RICHARD KENNETH
JOHN ROLAND RALPH
JOHNSON ALFRED LEWIS
JOHNSON DENNIS GEORGE
JOHNSON EVERETT EUGENE
JR
JOHNSON GUS WINSLOW JR
JOHNSON HARRY WILBUR
JOHNSON JOHN MARTIN
JOHNSON JOHNNY L
JOHNSON LARRY DEAN
JOHNSON LARRY RICHARD
JOHNSON RAY ELLSWORTH
JOHNSON RAYMOND JUNIOR
JOHNSON RONALD GENE
JOHNSON SAMUEL ARLON
JOHNSON STANLEY
JOHNSON THOMAS ALAN
JOHNSON THOMAS EUGENE
JOHNSON THOMAS WILLENE
JOHNSTON DAVID ALLEN
JOHNSTON DAVID NEAL
JOHNSTON DENNIS NEIL
JOHNSTON EVARISTO PACKE-
CO
JONES ARTHUR ELLIOTT
JONES BRUCE R
JONES CHARLES THOMAS
JONES DAVIS ALLEN
JONES GREGORY THOMAS
JONES HOWARD WILLIAM
JONES JOHN HOWARD
JONES MONTE RICHARD
JONES NORMAN JR
JONES ROBERT ARTHUR
JONES RONALD TRENT
JONES RONALD WEAVER
JONOZZO THOMAS CHARLES
JORDAN JAMES ELDON JR
JORDAN LARRY CHRISTOPHER
JORGENSEN DAVID WAYNE
JOY EDGAR DALE
JUDD MICHAEL BARRY
JUDKINS TERRY WILLIAM
JULIAN MICHAEL HENRY
JULIAN PERCY
JUSTICE DONALD LEE
JUSTICE EDWARD JAMES
JUSTICE WILLIAM ALLEN
KALETTA BARRY PAUL
KANDEL JAMES EDWARD
KANE JOSEPH LEON
KANE LARRY WAYNE
KANTER EDWARD LEE
KAPP PAUL LASZLO
KARNEHM STEVEN DALE
KATTERHENRY LEROY W JR
KATTERHENRY TERRY FISHER
KEA ANDREW MILLARD
KEAN BILLIE ORR
KEARSLEY RONALD CHARLES
KEATON EVERETT DENNIS
KEATON JOHN LAWRENCE
KEEFER DAVID CHARLES
KEEFER DAVID CHARLES
KEEFER KENNETH RAY
KEENE DANIEL ARTHUR
KEERAN WILLARD DAVID
KEETLE JEFFREY CHARLES
KEGG DONNIE STANLEY
KEISTER DAVID EARL
KEISTER JOHN LOY
KELLER DAVID RICHARD
KELLEY GEORGE ROBERT
KELLY MICHAEL JOHN
KELLY WILLIE J
KEMER ROBERT PATRICK
KEMP EDWARD
KEMPEL MICHAEL RICHARD
KEMPER JOHN RICHARD
KEMPF DOUGLAS SCOTT
KEMPLE GILBERT VERNON JR

KENNEDY JOHN FRANKLIN
KENNEY HARRY JOHN
KERLIN WILLIS EUGENE JR
KERNER RONALD BRIAN
KERNS JOHN EDWARD
KERNS ROGER RAY
KERR ERNEST CLANEY JR
KERR GAYLORD GERALD
KESTER JAMES JOSEPH
KESTERSON DAVID MICHAEL
KETCH MICHAEL HAYWARD
KETTERING ROBERT PAUL
KEYER DENNIS LEE
KIJOWSKI ROBERT GEORGE
KILBANE TERENCE JOSEPH
KILBANE TERRENCE PATRICK
KILLENS RICHARD
KILLIAN DAVID EDWARD
KIMSEY DONALD WAYNE
KINDSVATTER WARREN EARL
KING DANNY RAYMOND
KING JAMES EDWARD
KING LEE RAY
KING LEROY ALAN
KING MICHAEL LEE
KING RICHARD LEE
KING ROBERT LARRY
KING ROBERT LEE
KINGERY PAUL JAY
KINGSLEY THOMAS EDWARD
KINIYALOCTS CHARLES M
KINNARD DANIEL LEE
KINNARD DENNIS RAY
KIRKPATRICK ELDON JOHN JR
KITCHEN ORVILLE EUGENE JR
KLANN MARTIN DOUGLAS
KLEMM DONALD MARTIN
KLEVENOWSKI ROBERT
MICHAE
KLINK JAMES MARION
KLINZING THOMAS LEE
KLUG HERBERT WHEELER
KLUMP JOHN THEODORE
KLUTE JERRY CRAIG
KLYNE JAMES ARNOLD
KNAPIC BERNARD RICHARD
KNAPP RICHARD CHARLES
KNAUS WILLIAM CAMPBELL
KNEECE CHARLES LEROY
KNICKERBOCKER RICHARD J
KNIGHT RICK LEE
KNOCH DENNIS RICHARD
KNOTT DOUGLAS HUGH
KNOX WILLIAM EDWARD
KOCAK JOHN ANTHONY
KOCHENSPARGER JOHN
EDWARD
KOHR PAUL THEODORE
KOLAROV MICHAEL CAREY
KOLTER BRUCE
KOLY ROBERT JAMES
KONOFF KENNETH GLEN
KOON ALBERT LEWIS
KOON CHARLES MARION
KOONS DALE FRANCIS
KORECKI EUGENE M
KORNS ROBERT ORAL
KOS JOHN JAMES
KOSCHAL GREGORY ANDREW
KOSS FREDERICK M
KOTORA JOHN LEWIS
KRAFT LARRY WILLIAM
KRAMER DONALD EUGENE
KRAMER ROBERT DEAN
KRANSHAN TIMOTHY MI-
CHAEL
KRICK DONALD WILLIAM JR
KRIEGEL PAUL HENRY
KROTZER LYNN ROBBIN
KRUEGER RANDALL LEE
KRUKEMYER KENNETH
WARREN
KUDRO TERRENCE JOSEPH
KUHN CHARLES EDWARD JR
KUNEY JERRY DEAN
KUTSCHBACH STEPHEN RAY
LA CHAPELLE GARY GEORGE
LA LONDE HARRY FRANK JR
LA POINTE JOSEPH GUY JR
LA TELLE RONALD LON
LA TORRE EDGARDO RAFAEL
LABBE ROBERT BERG
LACEY DAVID MICHAEL
LACKEY BILLY JAY

LAIRD ERVIN LEONARD
LAKIN RICHARD DENMAN
LAMAR WILLIE JAMES
LAMBERT CECIL WAYNE
LAMBERT JAMES CALEB JR
LAMMERS WILLIAM JOSEPH
LAMP ARNOLD WILLIAM JR
LAMPLEY LEON PARNELL
LANDRUM THOMAS WILLIAM
LANE SHARON ANN
LANGLOIS JAMES THOMAS
LANTER KENNETH WAYNE
LANTER RAYMOND EDWARD
LANTZ CHRISTOPHER JOSEPH
LAPLANTE NOEL CHARLES
LAPPIN DENNY RAY
LARKIN THOMAS JOHN II
LASKAY DONALD THOMAS
LASURE DANNY LEWIS
LATTIN JOHN H JR
LAUBACHER ROBERT FRANCIS
LAUER CHARLES ARTHUR
LAVELLE TERRENCE MICHAEL
LAVEROCK PAUL STUART
LAWSON AMOS DAVID
LAWSON ROGER DALE
LAY ROGER MINETT
LAYAOU ERNEST E JR
LAYFIELD DONALD EDWARD
LAYPORTE OSCAR ROBERT
LE BEAU DAVID ALLEN
LE GRAND WILLIAM FRANCIS
LEA ROBERT EDWARD
LEACH GARY PAUL
LEACH RICHARD STEPHEN
LEACH STEVEN LAWRENCE
LEAK JERRY DAY
LEASE RICHARD FRANKLYN
LEASURE DELBERT LOUIS
LEASURE JOHN EDWARD
LECHAK FREDERICK JAMES
LEE ALBERT EUGENE
LEE GENE FRANCIS
LEE GUY EUGENE
LEE LARRY EUGENE
LEE PHILLIP LEWIS
LEES PAUL ERIC
LEFEVER DOUGLAS PAUL
LEFLER CLIFFORD JOHN T
LEHMAN JIMMY FRANCIS
LEMMON RICHARD KEITH
LENGYEL DAVID GEORGE
LENIO DALE JAMES
LENNER JACK RONALD
LEVIER DAVID JAMES
LEWIS BARRY WAYNE
LEWIS CHARLES ALBERT JR
LEWIS DAVID
LEWIS DAVID HARRY
LEWIS DELBERT O
LEWIS JAMES FREDERICK
LEWIS ROGER DALE
LIBBEE LARRY LEE
LICATE DAVID LOUIS
LIEURANCE DAVEY ALAN
LIGHTFOOT JAMES EDWARD
LINCOLN GARY GENE
LIND RALPH RICHARD JR
LIND THOMAS REINO
LINDE RICHARD VICTOR
LINDSEY ELMER R JR
LINNEN BENEDICT J III
LINVILLE DENNIS WAYNE
LITTLE CECIL EUGENE
LITTLETON RICHARD WIL-
LIAM
LIVESAY RALPH HOWARD
LIVINGSTON BRUCE BERNARD
LIVINGSTON WILLIAM
MICHAE
LLOYD KENNETH EDWIN
LLOYD RANDALL LYNN
LOAN THOMAS LEE
LOCKER JAMES DOUGLAS
LOCKHART HARLAN NA-
THANIEL
LOFTON JOSEPH ALAN
LOFTUS RAYMOND SHARP III
LOGAN JACK WILLIAM JR
LOGAN RICHARD MATTHEW
LOGUE ROBERT DONALD
LOITZ MICHAEL NELSON
LOMAX RICHARD EUGENE
LOMBARDO RICHARD MYRON

Ohio

LONG DONALD RUSSELL
LONG EDWARD EUGENE
LONG JAMES ALLEN
LONG RICHARD PAUL
LONG ROBERT DAVIS
LONG ROBERT LESTER
LONG ROBERT WESLEY
LONG SHELBY MARCENE
LOONEY JAMES WESLEY
LOONEY PHILLIP R
LOPEZ DONACIANO GUTI-
ERREZ
LORD BARRY DAVID
LORENCE JOHN EDWARD
LOTHMAN JAMES EDWARD
LOUDIN DALE RUSSELL
LOVE GARY LEE
LOVE JAMES THOMAS
LOVE JOHN WAYNE
LOVEDAHL CHARLES ROBERT
LOVELACE KENNETH
LOVELL EDWARD API
LOVETT GLENN ALAN
LOWDER JARVIS CRAWFORD
LOWE ROBERT ERNEST
LOWMAN JONATHAN FAYE
LOZANO EDWARD ROBERT
LUC CHESTER ANTHONY
LUCA PATRICK CHARLES
LUCAS ALAN FRANK
LUCAS WILLIAM ROBERT
LUCKETT JAMES SAWYER II
LUDWIG BYRON NELSON
LUECKE ROBERT WAYNE
LUKER RUSSELL BURR
LULLA ROBERT ALLEN
LUNDELL JACKIE LINN
LUNSFORD JAMES WILLIAM JR
LUNSFORD PAUL R
LUPAS GERALD ALLEN
LUSTER ROBERT LEE
LUTE JAMES ROBERT
LUTZ GENE MILTON
LUTZ LARRY EUGENE
LUTZ ROBERT EDWARD JR
LYKINS DANIEL CLYDE
LYNN JACK DALE
LYONS GARY DEAN
LYONS ROGER GENE
MABEE DOUGLAS CRAIGLOW
MACK ALVIN ANTHONY JR
MACK ROBERT LEWIS
MADDEN ERNEST GARY
MADDEN JOHN PAUL
MADDEN RICHARD JR
MADDOX PAUL RAY
MADER RICHARD MICHAEL
MAHAN DAVID ALLAN
MAHER MARTIN JOSEPH
MAHONE WILLIAM BENJAMIN
MAHONEY ERNEST
MAHONEY RALPH GEORGE
MAIN WILLIAM GENE JR
MAKSIN MIKE A
MALCOLM WILLIAM EDWARD
JR
MALECKE JAMES ALLEN
MALEWSKI DENNIS W
MALICEK DONALD JOSEPH
MALINOWSKI EDWARD
MAMIE RICHARD NORMAN
MANDLER JAMES THOMAS
MANGAN MICHAEL L
MANGINO THOMAS ANGELO
MANLEY RONALD LEE
MANNING RONALD JAMES
MANTON BRUCE ARTHUR
MARCUM HAROLD LEE
MARCUM WALTER
MARHEFKA DUANE JOSEPH
MARLEY KENNETH CHARLES
MARMIE ROBERT THEODORE

MAROSCHER ALBERT GEORGE
MARSH LARRY GLENN
MARSH RICHARD ALBERT
MARSHALL DENNIS CRAIG
MARSHALL DOC HENRY
MARSHALL MARK DUANE
MARSHALL RONNIE SHINYA
MARSHAND KENNETH LLOYD
MARTELL GARY WILLIAM
MARTICH THOMAS MARK
MARTIN ALAN DAVID
MARTIN DAVIE JOE
MARTIN EDWARD DEAN
MARTIN LARRY CHARLES
MARTIN LEONARD RAY
MARTIN NAPOLEON
MARTIN PATRICK ROBERT
MARTIN RICHARD LEE
MARTIN TONY LEE
MARTIN WILEY LOUIS
MASSA DAVID LYNN
MASTEN ARMAND DOMINIC
MASTERS WILLIAM RICHARD
MASTERSON JOHN PATRICK
MASTROIANNI THOMAS
FRANCI
MATHEWSON ROGER MI-
CHAEL
MATHIS ARNOLD
MATTERN CHARLES DUANE
MATTHES PETER RICHARD
MATTHEW CECIL LEROY JR
MATTHEWS WILLIAM L JR
MATTINGLY JOHN EUGENE
MATUSKA JOHN JAMES
MAURER JAMES ROBERT
MAXWELL WILLIAM EARL
MAYBERRY SQUIRE N JR
MAYNARD GREGORY JOHN
MAYS JAMES EDWARD
MAZE DAVID LEE
MAZITIS VICTOR ALLEN JR
MC AFEE CARY FRANCIS
MC CABE ROBERT WARREN JR
MC CAFFERTY CORNELIUS
A JR
MC CAMMON GLENN EUGENE
MC CANDLESS MICHAEL
DAVID
MC CANN EDWARD DEAN
MC CARROLL OREN B
MC CARTY EDWARD WESLEY
MC CAULEY DALE MARTIN
MC CLELLAND GEORGE
DENNIS
MC CLUNG WAYNE OLAND
MC COMAS HOBART WILSON
JR
MC CONAUGHEAD HARVEY
R JR
MC CONNAUGHEY DAVID
LYNN
MC CONNELL JOHN STEVEN
MC CONNELL ROBERT
MUELLER
MC CORD HAROLD RAYMOND
JR
MC CORD JOHN RICHARD
MC CORKLE STEPHEN ALAN
MC CORMICK MICHAEL P
MC CORVEY ROBERT KEN-
NETH
MC COY BOBBY LEE
MC COY JOHN WILLIAM
MC DANIEL GEORGE WILLIAM
MC DONALD THOMAS MI-
CHAEL
MC ELROY THEODORE R JR
MC FARLAND RICHARD
WESLEY
MC FARLAND TERRENCE W
MC FARLIN CHARLES RICH-
ARD
MC GARVEY CHARLES ED-
WARD
MC GHEE DENNIS OLIVER
MC GILL ROBERT WARREN
MC GLEW JOHN JOSEPH
MC GOVERN MICHAEL LEWIS
MC GRATH DANIEL EDWARD
MC GRAW DONALD ORIN
MC GREW WILLIAM WALLACE
III
MC GUIRE MITCHELL

MC INTOSH CHARLES GLENN
MC INTURF SAMUEL DUANE
MC INTYRE DAVID ALLEN
MC INTYRE RAYMOND NEAL
MC JUNKIN RONALD LEE
MC KEE CHARLIE MEARL
MC KENZIE WENDELL HOW-
ARD
MC KIDDY GARY LEE
MC KILLOP WILLIAM DION
MC KINNEY FORREST ADRIAN
MC LEAN WILLIAM EDWARD
MC LELLAN STUART MURRAY
MC LEMORE JOSEPH
MC MAHAN DANIEL JACKSON
MC MAHON WILLIAM LAW-
RENCE
MC MILLION CHARLES
EUGENE
MC NAUGHTON MICHAEL
DEAN
MC PHERSON DAVID LEE
MC PIKE JAMES EDWIN
MC SWAIN HARVEY JOSEPH
MC VEY MICHAEL LEE
MCCARTNEY ANDREW C
MCCOY ALBERT JR
MCELFRESH ALLEN KEITH
MCKEE LARRY WILLIAM
MCKIETHAN DONALD
FRANCIS
MCKINNEY CLEMIE
MCLEAN TIMOTHY L
MEADOWS CHAD DAVID
MEADOWS CHARLES THOMAS
MEANS DANA EDWARD
MEDLEY CLARENCE
MEHL RICHARD EARL
MELOTT CHARLES EDWARD
MENART JAMES JOSEPH
MENDEZ THEODORE SR
MENDOZA DAVID LOUIS
MENGES GEORGE BRUCE
MERRIAM DAVID HENRY
MERRIMAN MICHAEL GENE
MERRIMAN THOMAS BRUCE
MERSCHMAN JOHN WILLIAM
MESENBURG TERRANCE R
MESSER BOBBY GENE
MEYER KENNETH ANTHONY
MEYER ROBERT JEROME
MEYER TERRY LYNN
MICHALK ROBERT BRUCE
MICK FRED GEORGE
MICKLE JOHN RICHARD
MIDDLETON CLAYTON
MIDDLETON DONALD AR-
THUR
MIDDLETON TEDDY EUGENE
MIKA STEPHEN ADAM
MIKITIS MICHEAL ALLEN
MIKOLAJCZYK DENNIS LEE
MILANO JOSEPH JOHN
MILBURN MICHAEL DRENNEN
MILES THOMAS EDWARD
MILLARD LARRY DAVID
MILLER BENTON LEWIS
MILLER BURT EVERETT
MILLER CHARLES EUGENE
MILLER DANA LEE
MILLER DARRELL EDWIN
MILLER FOSTER BISHOP
MILLER FRANK HAROLD JR
MILLER FREDRICK WAYNE
MILLER HAROLD DWIGHT
MILLER HEBER JOSEPH
MILLER JAMES GARRETT
MILLER JAMES GREGORY
MILLER JAMES HOWARD
MILLER JAMES IRVIN
MILLER JAMES RUSSELL
MILLER MILLS CRAFT
MILLER ROBERT JACOB JR
MILLER RONALD DARRELL
MILLER TERRY BROWN
MILLER VICTOR RAYMOND
MILLER WILLIAM HARVEY
MILLER WILLIAM LEE
MILLINGER GLEN ALLAN
MILLS THOMAS WAYNE
MINGLE ROBERT LOUIS
MINK BOYD CARL
MINNIEAR HAROLD NORMAN
MINNIX LEROY FRANKLIN

MINOR MICHAEL JAMES
MIRICH JOHN
MIRICK STEVE JR
MITCHELL CLARENCE E JR
MITCHELL DAVID ARTHUR
MITCHELL JOHN ALBERT
MITCHELL JOHN LOUIS
MITCHELL PAUL HOLLAND JR
MITCHELL THOMAS ALLAN
MIZNER DARRELL CONDIE
MOBLEY DANIEL M
MOLZON ERNEST ALVIN
MONAT DONALD HENRY JR
MONROE FRANCIS MARION
MONTGOMERY JOHN THOM-
AS
MOON THOMAS HENRY
MOONEY CLARENCE ALLEN
MOONEY FRED
MOONEY PATRICK THOMAS
MOORE CHARLES THOMAS JR
MOORE JAMES CECIL
MOORE PHILLIP ALEXANDER
MOORE RAYMOND GREGORY
MOORE RICHARD ALLEN
MOORE RONALD JAMES
MOORE WAVERY
MOORHEAD MICHAEL
EUGENE
MOREE BARRY RUSSELL
MORELAND JOHN LEE
MORGAN RICHARD
MORGAN RONALD CURTIS
MORGAN THEODORE JR
MORGAN THOMAS RAYMOND
MORRIS GARY WILLIAM
MORRIS NEIL JAY
MORRIS ROBERT L
MORRISON GENE FRANCIS
MORROW TERRY PATRICK
MORTON DAVID EUGENE
MORTUS PATRICK CLINTON
MOSES WALTER LEWIS JR
MOSHER ALDEN GRAY JR
MOSS THOMAS JOHN JR
MOSSFORD GREGORY FRED-
RICK
MOWERY CARL FRANCIS
MOWREY GLENN WILLIAM
MUCHA HOWARD ALLEN
MUELLER STEPHEN MICHAEL
MUGAVIN MARTIN M
MULDROW ROBERT LEE
MULLINS JAMES RAY
MULLINS RICHARD ALLEN
MUNOZ RUDOLPH PINA
MURPHY JOHN LYLE
MURPHY ROBERT EDWARD JR
MURPHY THOMAS RALPH
MURRAY ARTHUR JOSEPH JR
MURRAY BRIAN THOMAS
MURRAY JAMES EDWARD
MYERS ALBERT C
MYERS PAUL JUNIOR
MYERS PAUL RICHARD
MYLANT STEVE VICTOR
MYRICK WILLIE J
NAGY JOSEPH RALPH
NAJMOLA JOHN HENRY
NALL CARL DAVID
NALLEY CHARLES THOMAS
NAPIER DARREL GENE
NAPIER ZACK WILLIAM
NAPOLI DANIEL LUKE
NAU JAMES CHRISTIAN
NEAL CHARLES MARION JR
NEAL ROBERT EUGENE
NEAVES CLAYTON WILLARD
NECE HERBERT JAMES
NEFF DAVID RUSSELL
NEILL JOHN MAUTZ
NEMETH JOSEPH JR
NERAD WALTER JOSEPH JR
NERVIE KENNETH JOHN
NESS MICHAEL LESTER
NESSELROTTE JAMES MI-
CHAEL
NESTICH FRANK JOSEPH
NEUTZLING WILLIAM P
NEW ZACHARY PHILLIP
NEW ZACHARY PHILLIP
NEWCOMER JAMES HENRY
NEWELL CALVIN EUGENE
NEWELL EDWIN GRANT

NEWLAND MICHAEL DWAINE
NEWLIN MELVIN EARL
NEWMAN FRANK ALLEN
NEWMAN MAURICE GLENN JR
NEWPORT SCOTT HERBERT
NEWTON RICHARD ERIC
NICEWANDER OSCAR FRANK-
LIN
NICHOLAS DEAN EDWARD
NICHOLS JERRY RUSSELL
NICKELS DARIS WAYNE
NIEHAUS JAMES EDWARD
NIGH FREDRICK ELLIS
NISSENBAUM MICHAEL DAVID
NOEL MICHAEL DAVID
NORRIS WILLIAM THEODORE
NOTTAGE MICHAEL LEWIS
NOVAK THOMAS EUGENE
NULPH WILLIAM LEE JR
NUTTER FREDERICK LEROY
NYSTROM BRUCE AUGUST
O CONNOR DAVID LEE
O DONNELL BERNARD JACK
O DONNELL JOHN PATRICK
O KEEFE GARY MAURICE
O LAUGHLIN JAMES FRANCIS
O NAIL ROBERT PAUL
O NEAL DENNIS RAY
O NEAL ROY DAVID
O REILLY ANTHONY PAUL
OATES ROBERT JAMES
OBENOUR RONALD MICHAEL
OBERDIER LYN DOUGLAS
OEN MICHAEL LYNN
OGRINC RONALD ROY
OLIVER CARL W
OLIVER EDDIE VAN JR
OLIVER KENNETH ROY
OLIVER WALTER B
ONEY DANIEL LUTHER
OQUENDO JOHN ONOFRE JR
ORELL QUINLAN ROBERTS
ORLANDO RICHARD DUANE
ORT STEVEN MICHAEL
ORTIZ JOSE HECTOR
ORTIZ ZENEIDO JR
ORWIG DAVID THOMAS III
OSBORN JERRY WAYNE
OSINSKI RONALD ANTHONY
OSTAPCHUK WALTER MI-
CHAEL
OSTRAKOVIC GUSTAV
OWENS WILBERT
PACIOREK ROBERT EDWARD
PAINTER MARVIN REED
PAINTER WAYNE ALLEN
PAIRAN WALTER ALLAN
PALEY NORMAN FRED
PALISKIS EUGENE MICHAEL
PALLAYE LOUIS DALE
PALM DALE ARDEN
PALMORE DONALD STEVEN
PALUMBO NICKOLAS R JR
PANKUCH BRUCE ALAN
PAONESSA MICHAEL DOM-
INIC
PAPA WILLIAM JAMES
PARKER DONALD LEE
PARKER ERIC
PARKER LEON VICTOR
PARKER LESTER EUGENE
PARKER ROGER LOUIS
PARKER SAMUEL LEE
PARKER WILLIAM E III
PARKER WINSTON GLEN
PARKS JAMES KERMIT
PARKS RAYMOND FRANCIS
PARKS SYDNEY
PARSLEY RONALD LEE
PARSONS PAUL GENE
PARSONS RONALD NEAL
PARSONS TERRY LEE
PASTVA MICHAEL JAMES
PATRICK DONNIE LEE
PATRICK MARINER
PATTERSON JAMES W JR
PATTERSON KEITH ALLEN
PATTERSON RICHARD STUART
PATTON JAMES ALAN
PAUL CRAIG ALLAN
PAUL JOE CALVIN
PAVLAKOVICH NICHOLAS
ALLE
PAWLAK RICHARD VICTOR

PAYNE HUBERT JACKSON
PAYNE NORMAN
PEACE JOHN DARLINGTON III
PEACOCK THOMAS EDWARD F
PEARCE DALE ALLEN
PEARCE WAYNE WILLIAM
PEARL RICHARD MAX
PEARL RONALD LEE
PECHAITIS MATTHEW JOHN
PENDERGRAFT RAY DANIEL
PENLAND FRED DANIEL
PENNEY CHARLES OTIS
PENNY JAMES MELVEN
PEPPER LARRY JAMES
PEPPER WILLIAM FRANKLIN
PERETIATKO JERALD PAUL
PEREZ HILARIO OCHOA
PERKINS DONALD ROBERT JR
PERKINS GARY WILLIAM
PERKINS MICHAEL DAVID
PERKO TERRY JOHN
PERRY CHARLES LEON
PERRY EDWARD LEE
PERRY JOHN EVERETTE
PERRY KENNETH LEE
PETAL JOHN DARRYL
PETERS BILLY LEE
PETERS CARL HARMAN JR
PETERS DANIEL ALLEN
PETERS TOMMY RALPH
PETERSON JESSE EARL
PETERSON MARK ALLAN
PETERSON MICHAEL HAR-
RELD
PETRIC JOHN ANTHONY
PETTAWAY LARRY CHARLES
PETTY ROY ANDREW JR
PHILLIPS DEAN ANTHONY
PHILLIPS DENNIS L
PHILLIPS MARSHALL W JR
PHILLIPS ROBERT PAUL
PHILLIPS ROGER LEE
PHILLIPS WILLIAM GRIGABY
PHIPPS ROY LESTER
PHLEGER ROBERT CRAIN
PIATT CHARLES WILLIAM
PICELLE FRANK JOHN JR
PICKWORTH JERRY LEE
PIERCE DAVID WAYNE
PIERCE HERBERT LEE JR
PIERCE HOMER EARL JR
PIERRE NORMAN WALLACE
PIERSON GROVER CECIL JR
PIERSON LEROY
PIES JOHN DAVID FREDRICK
PIETRASZAK DAVID ALOYSIUS
PIETRZAK JOSEPH RAY
PIETSCH ROBERT EDWARD
PIGOTT JAMES HAROLD
PINTOLA JAMES MICHAEL
PIPHER CARL DALE
PIRRUCCELLO JOSEPH S JR
PITSENBARGER WILLIAM
HART
PITTMAN JACK
PITTS FRED EARL
PIZZINO EUGENE II
PIZZINO THOMAS CARMEN
PLANTS THOMAS LEE
PLAVCAN KENNETH MICHAEL
PLEIMAN JAMES EDWARD
PLUMMER RALPH WILLIAM III
POHL RICHARD SHARON
POLAND RONALD LEE
POLING JACKIE RAY
POLING KENNETH
POLITO GENE ALBERT
POLLEY GARY PAT
POLLEY RICHARD ALAN
POLLEY ROGER DALE
POLLOCK LAWRENCE ED-
WARD
POLSTER HARMON
POOLE HARTWIG RALPH
POOLE RONALD DEAN
POPPAW MICHAEL ROBERT
POREA ROBERT GEORGE
PORTER JACK EUGENE
PORTER LAWRENCE EUGENE
PORTER LAWRENCE WILLIAM
PORTER RICHARD LEE
PORTER ROY LYNN
POSS TRAVIS O NEAL

POTTER RAYMOND GEORGE
POTTKOTTER JAMES VIN-
CENT
POWELL JOHN PARKER
POWELL RICHARD LEE
POWERS KENNETH
POZMANN ALEXANDER JR
PRATER ROY DEWITT
PRATHER WILLIAM HARLEY
PREDOVIC WILLIAM MARK
PRESSLER CHARLES EDWARD
PRESTON THOMAS RAY
PRETNAR ALLEN JOHN
PRETNAR ALLEN JOHN
PRICE BILLY RAY JR
PRICE DAVID J
PRICE DENNIS ALTON
PRICE ELBERT FORD JR
PRICE KENNETH RANDAL
PRICE WILLIAM EDWARD
PRIDEMORE DALLAS REESE
PRIEBE JAMES EDWARD
PRIESER ROBERT SHERMAN
PRIEST DONALD WAYNE JR
PRINCE DANNY DEAN
PRITCHARD DONALD RAY
PROCTOR DANIEL VAUGHAN
PROKOP FRANK JOSEPH
PROMMERSBERGER JAMES
EDWI
PROSE CHARLES WILLIAM JR
PROTAIN DAVID ALAN
PRUDHOMME JOHN DOUG-
LAS
PRUITT WILLIAM HENRY JR
PRYOR JEROME
PTACEK TIMOTHY RICHARD
PUCCI DANIEL LOUIS
PUCHALSKI WALTER MARTIN
PUDULS JURIS
PULTZ ROBERT LEWIS
PURDON GERALD WAYNE
PUSKARCIK RONALD JOSEPH
PUTMAN THOMAS ANDREW
PYLE NICHOLAS IRVIN
PYLES HARLEY BOYD
QUIMBY DANIEL LEE
QUINN ROBERT JOSEPH
QUINN ROGER ALLAN
QUINT ANTHONY PETER
RADCLIFF ROBERT PAUL JR
RADER ALAN REED
RADU STEVEN NICHOLAS
RAINS CHRISTOPHER LEE
RALICH RONALD
RALSTON THOMAS JOSEPH
RAMEY SONEY

RAMEY SONEY
RAMON DENNIS MICHAEL
RAMOS FRANK JR
RAMSEY RANDOLPH RAY-
MOND
RAMSEY SAMUEL VIRGIL JR
RANC WILLIAM EDWARD
RANDOLPH CLIFFORD L
RANKE ALLEN JAMES
RAPPOLD ALBERT JOSEPH JR
RATHBUN ROBERT FRANK
RATLIFF FREDERICK R JR
RATLIFF JOHNNY
RAUPACH KIM
RAUSCHENBERG DOUGLAS
EDWA
RAVENCRAFT JAMES ALVIN
RAY TIMOTHY
REAM PAUL EUGENE
REED CLYDE JR
REED JAMES WILLIAM
REED RALPH EUGENE
REED ROBERT THOMAS
REED WILLIAM CLEMON
REES JOSEPH MAURICE
REES RICHARD MORGAN
REESE PAUL HENRY
REEVES WILLIAM DOUGLAS JR
REID DARRELL LEE
REISING DALE
REITER WILLIAM FRANCIS
RENO LAWRENCE GERALD
RESPRESS THOMAS
REYNOLDS GEORGE THOMAS
REYNOLDS HARVEY MICHAEL
REYNOLDS LESLIE JR
REYNOLDS ROBERT CLAR-
ENCE
RHOADES DAVID
RHODES GRANT A
RHODES JOSEPH JOHN
RICE FINLEY AUSTIN
RICE MICHAEL PHILLIP
RICE ROBERT THOMAS JR
RICE WALTER GARLAND JR
RICHARD JOHN WAYNE
RICHARD PHILIP EUGENE
RICHARDS DENNIS R
RICHARDSON FARRIS LEE
RICKARD RONALD LEE
RICKARDS CLARENCE HOW-
ARD
RICKEY LAWRENCE DAVID
RIECK JOHN JAMES JR
RIELLY DAVID
RIEMER DAVID WALTER
RIFFLE STANLEY

RIGDON WILLIAM FRANCIS
RIGGINS EDDIE
RIGHTER ROBERT LE ROY JR
RILEY CURTIS RAY
RILEY VERNON RAY
RINEHART TIMOTHY HOW-
ARD
RINGEL JAMES ROBERT
RIPLEY LARRY DEAN
RIPLEY WILLIAM L
RIPLIE GEORGE HENRY
RITCH ERNEST EUGENE
RITCHEY GARY WAYNE
RITCHEY LUTHER EDMOND JR
RITTICHIER JACK COLUMBUS
RITZLER RICHARD PAUL
RIZZO JAMES PATRICK
ROACH JOHN HAROLD
ROACH RICHARD FRANKLIN
ROARK EDWARD LEE
ROBAR STEPHEN FRANK
ROBBINS RICHARD JOSEPH
ROBERTS ARTHUR JAMES JR
ROBERTS ERVIN BRADLEY
ROBERTS MICHAEL EDWARD
ROBERTS THOMAS WARREN
ROBERTS WALLACE
ROBERTS WILLIAM JOHN
ROBERTSON ALVIN WARNER
ROBERTSON WILLIAM LEE
ROBEY RICHARD NEAL
ROBINSON JAMES LLOYD
ROBINSON LUTHER
ROBINSON RANDALL
CHARLES
ROBINSON ROY RAY
RODZEN BERNARD JAMES
ROGERS ARCHIE DEE
ROGERS JERRY LEE
ROGERS KENNETH FAULKNER
ROGERS LARRY LEE
ROHLER SIDNEY EARL
ROLF GERALD R
RONALD THOMAS ALAN
ROOT CLYDE DEAN
ROSATO JOSEPH FRANK
ROSE JERRY GENE
ROSE JOSEPH SHEPHERD JR
ROSE RAYMOND ALFRED
ROSE THOMAS ELDEN
ROSEBERRY MICHELE MC
CORD
ROSHON ROBERT BROWN
ROSS GREGORY MARK
ROSS JAMES ARTHUR
ROSS RONALD CARL
ROSS THOMAS ARTHUR

ROSS THOMAS MICHAEL
ROSS WILLIAM KEITH
ROSSER GARY EDWARD
ROUSE WILLIAM CLARENCE
ROWLAND WAYNE HULEN
ROWLEY THEODORE TEXAS
ROY DANIEL THOMAS
ROYDES KRAG BARRY
RUCKER RICHARD LEE
RUEHLE MEDARD A J
RUGGLES LARRY DEAN
RUGH FRED PLYMOUTH
RUNZO RICHARD FRANCIS
RUOFF ROGER DALE
RUPCIC RAYMOND ELLS-
WORTH
RUPP JEFFREY DAVID
RUSEK RONALD LEE
RUSH JAMES LEROY
RUSH JAMES THEODORE
RUSHTON WAYNE STERLING
RUSS JAMES LEE JR
RUSSELL BOBBY
RUSSELL JAMES LOWELL
RUSSELL JOHN JOSEPH
RUSSELL WILLIAM JOHN JR
RUSSIN DONALD JOHN JR
RUTLEDGE JAMES BENSON
RUTTER JOSEPH DELMAR JR
RYAN FREDERICK LEE
RYAN JOSEPH ROBERT JR
RYAN SAMUEL FRANKLIN
SABEC DAVID LOUIS
SABLOTNY RICHARD ALAN
SACCOMEN EDMOND RAY
SADICK RICHARD JOHN
SAMPSON JOSEPH C JR
SANCHEZ ANGEL LUIS
SANCHEZ HECTOR LOUIS
SANDER THOMAS WOODROW
SANDERS DONALD ROBERT JR
SANDERS GEORGE AUSTIN
SANDERS RICHARD WAYNE
SANDERS WILLIAM JACOB
SANEDA JOHN
SANFORD HENRY CHARLES JR
SANFORD JAMES RUSSELL JR
SANTORA RAYMOND PAUL
SAPP BENNY JAMES
SARGENT GARY LEE
SARJEANT DWIGHT CUTLER
SAROSSY STEVE SANDOR
SAS ROBERT LOUIS
SAUER WALTER JR
SAUNDERS CLYDE WILLIAM
SAVICK JOSEPH JAMES JR
SAVIEO RICHARD HUGH

189

Ohio

SAYERS THOMAS RALPH
SAYRE LESLIE BERKLEY
SCARBROUGH DAVID CLIFTON
SCHAAF RONALD JOSEPH
SCHAICH DONALD BRUCE
SCHALK THOMAS MICHAEL SR
SCHARF RONALD JAMES
SCHAUB TERRY LEE
SCHEELER VICTOR RAY
SCHEELY ROBERT JAMES
SCHEUER BOBBY DALE
SCHIELE CRAIG BRIAN
SCHLUEB STEVEN MICHAEL
SCHMALTZ DOUGLAS RALPH
SCHMEES WILLIAM F JR
SCHMIDT JOSEPH
SCHMIDT MICHAEL
SCHMITZ RICHARD TRAVIS
SCHNEBEL ROBERT FRED
SCHNEE DONALD LAWRENCE
SCHNEEMAN CLIFFORD W JR
SCHNEGG CHARLES GLENN
SCHNEIDER DAVID ALAN
SCHNEIDER DAVID FRANCIS
SCHNEIDER GARY GENE
SCHNEIDER THOMAS JAMES
SCHOBER JACK ERVIN
SCHOENHOFF ROBERT JOHN
SCHOEPPNER LEONARD JOHN
SCHOOLCRAFT CHARLES EARL
SCHRECKENGOST FRED THOMAS
SCHRINER JUNIOR LEE
SCHROEDER GLENN MICHAEL
SCHULZ ALLAN HENRY
SCHWAB THOMAS PAUL
SCHWAN DANIEL GEORGE
SCISLO ROBERT TED
SCOTT BILLY JOE
SCOTT EDWARD EARL JR
SCOTT JAMES LEE
SCOTT MICHAEL
SCOTT TERRENCE DUANE
SCOTT WARREN TAYLOR
SCULLEN THOMAS JAMES
SCULLY JOHN MICHAEL
SCURLOCK LEE D JR
SEALL GEORGE MELVIN
SEBASTIAN JOSEPH WILLIAM
SEELEY DOUGLAS MILTON
SEERY JOHN JOSEPH
SEGER VERNON JOSEPH
SEIFERT THOMAS LEONARD
SEIFERTH STEPHEN ERIC
SEKNE SYLVESTER VICTOR
SELLERS LLOYD ANDREW III
SEMERARO DAVID ALEXANDER
SEMPLE WILLIAM EUGENE
SENS PHILIP MARION
SERENA JAMES DAVID
SEXTON DAVID MASON
SEXTON JOHN JUNIOR
SGAMBATI PAUL ANTHONY
SHAEFFER CHRISTOPHER L
SHAFER DONALD MAILY II
SHAFER JAMES DUDLEY
SHAFER LESLIE HOMER
SHAFER THOMAS JAMES
SHAFFER VICTOR THOMAS
SHAFFER WILLIAM EMERSON
SHAGOVAC PETER WILLIAM JR
SHANER STEPHEN PAUL
SHANK RALPH
SHANOWER TIMOTHY EDWARD
SHARP CURTIS HENRY JR
SHARP RICHARD DENNIS
SHARP STEPHEN LAMONT

SHARROCK EDWARD ALVA
SHARTZ FRANK JR
SHAUM RAY THOMAS JR
SHAVER RONALD LEE
SHAW GARY FRANCIS
SHAW JEFFREY MICHAEL
SHAW LARRY LEE
SHAW STANLEY SERGEANT
SHEEN WILLIAM EDWARD
SHEHL MICHAEL JOHN
SHELINE ALLEN CURTIS
SHELLITO WALTER CHARLES
SHELTON EDWARD ARNOLD
SHELTON MATTHEWS
SHEPHERD CLIFFORD
SHEPHERD GORDON
SHEPHERD LARRY EUGENE
SHEPHERD ROGER EUGENE
SHEPPARD JOHNNIE ARNOLD
SHERMAN ANDREW MARCO
SHERMAN PETER WOODBURY
SHILLING DEAN RICHARD
SHINGLEDECKER ARMON D
SHOEMAKER RAYMOND A II
SHORT RANDALL CHARLES
SHORT WILLIAM MICHAEL
SHOUP ROY NEAL
SHOVER BRUCE CHARLES
SHOVER WILLIAM
SHRACK ROBERT VENARD JR
SHROYER PERRY VERNANDO
SHULTZ JERRY LEE
SHUMINSKI STANLEY JOHN
SHY GARY NOLAN
SIBERT DARRELL WAYNE
SICILIANO JOSEPH A JR
SIEBENALLER ROBERT CHARLE
SIEGLER WILLIE JAMES
SIGMON HAROLD WAYNE
SIGWORTH RICHARD JACOB
SIKORSKI ELMER GERALD
SILLS FRANK RICHARD
SIMKO ANDREW MICHAEL
SIMMONS MICHAEL LEE
SIMONS LEROY EUGENE
SIMPSON DOUGLAS EDWARD
SIMPSON MICHAEL
SIMS ERWIN BRUCE
SINCHAK ANDREW RICHARD JR
SINES TIMOTHY DAVID
SINGER MICHAEL ERNEST
SINGLETON DANIEL EVERETT
SININGER TEDDY RAY
SINNOCK JOHN ROBERT
SITTNER RONALD NICHOLIS
SIZEMORE THOMAS JEFFERSON
SKAGGS FLOYD PETER
SKAGGS RAYMOND GENE
SKALA DAVID FRANCIS
SKINNER OWEN GEORGE
SKOVRAN WILLIAM MICHAEL
SKYLES NYLES BERNARD
SMALL SAM JARRELL JR
SMIDDY KYLE
SMITH ALAN RAY
SMITH ALLEN DEWAYNE
SMITH ANDREW RICHARD JR
SMITH ARTHUR WAYNE
SMITH CHARLES ROBERT
SMITH DAVID ARLIE
SMITH DAVID RONALD
SMITH DAVID ROSCOE
SMITH DENNIS
SMITH DENNIS ARTHUR
SMITH ELIJAH HENRY
SMITH FRANK NORMAN
SMITH FRANKLIN WAYNE
SMITH GARY LEE
SMITH GEORGE ARTHUR
SMITH HERSHEL CLIFFORD
SMITH JAMES ALVIN
SMITH JAMES DAVID
SMITH JAMES DELVIN
SMITH JOHN LEWIS
SMITH JOHN WILLIAM
SMITH JOHNNIE CECIL JR
SMITH KENNETH SHELDON JR
SMITH MARSHALL ROY
SMITH MICHAEL EDWARD
SMITH NELSON LEE
SMITH PATRICK EDWARD JR

SMITH PHILLIP JOE
SMITH RICHARD EUGENE
SMITH ROBERT GEORGE
SMITH ROBERT LEE
SMITH ROBERT LEE
SMITH ROGER LEE
SMITH STEPHEN JAY W
SMITH STEVEN DEAN
SMITH STEVEN EUGENE
SMITH THOMAS DAVID
SMITH THOMAS EUGENE
SMOKE BRUCE ALLEN
SNAKOVSKY LOUIS ALLAN JR
SNELSON JOHN WILLIAM
SNIDER MARVIN DALE
SNITCH JOHN HERBERT
SNYDER DUANE HAROLD
SNYDER GARY FOSTER
SNYDER JAMES DALE
SNYDER LAWRENCE DAVID
SNYDER ROBERT DUANE
SNYDER THOMAS LYNN
SOBCZAK JOSEPH S II
SOLLARS FRANKLIN ELLWOOD
SOMBATI ROBERT STEPHEN
SONNEBERGER RICHARD G
SONNICHSEN EDWIN CHARLES
SONNKALB CHARLES DAVID j
SOULE JOSEPH PAUL
SOUTHARD JERRY LEE
SOWERS RANDAL GENE
SPAK GEORGE STEPHEN JR
SPANGLER JOHN FLANAGAN
SPANGLER LARRY KIETH
SPANGLER MICHAEL ROBERT
SPARKS RICHARD L
SPARKS STEVEN LEE
SPARKS WILLIAM DOUGLAS
SPARROW CARL WILLIAM
SPAULDING LARRY EUGENE
SPEAKMAN RICHARD PAUL JR
SPEAKS PAUL EDWARD
SPEAR HOWARD JOSEPH
SPEARS RONDALL PRESTON
SPECK GEORGE EDGAR
SPEIDEL LOUIS JOHN
SPEIR DALE LLOYD
SPENCE RONALD LEE
SPENCER HARRY HERBERT
SPENCER JAMES HERBERT
SPICER MICHAEL BRUCE
SPILKER JAMES DENNIS
SPILLER CLIFTON
SPITLER JERRY ROBERT
SPITLER NELSON EVERETT
SPRADLIN JERRY DEAN
SPRING BRUCE WAYNE
SPRING TIMOTHY LANZER
SPRINGFIELD WILLIAM VAL
SPROWL JAMES EDWARD
STAATS GERALD MARTIN
STAHL JOHN WELFRED
STAINER WILLIAM EDWARD
STAIR GLENN ROBERT
STAMPER RICHARD G JR
STANEART RONALD KEITH
STANKO ROBERT GEORGE
STANLEY CHARLES IRVIN
STANLEY JACKIE G
STANLEY THEODUS MORRIS
STANTON EDGAR DOUGLAS J
STANTON RONALD
STAPLES ALTON LEON III
STAPLETON CLIFFORD
STAPLETON LAWRENCE GEORGE
STARK LAWRENCE J
STARKEY BLAIR WILLIAM
STARKEY DANIEL LEE
STARKS GEORGE LARRY
STARR RONALD DEAN
STEBNER ROBERT LYLE JR
STECKER RICHARD E
STEFANIC RUDOLPH MICHAEL
STEFFEN ALAN RALPH
STEIN ALAN ALBERT
STEIN DONALD VEARL
STEMEN FREDERICK MILTON
STEPANOV ROBERT DUANE
STEPHENS JAMES
STEPHENSON RICHARD C
STEPP DONALD EUGENE

STEPP EUGENE HENRY
STEVENS FORESTAL ALONZO
STEVENS RODNEY FRANKLIN
STEVENSON LARRY
STEWARD ANDREW RICHARD JR
STEWART GARY LEE
STEWART HENRY MATT JR
STEWART JAMES HERBERT JR
STEWART JIMMY GOETHEL
STICKEL GARY STEPHEN
STICKLE TIMOTHY DAVID
STIGER HAROLD EUGENE
STILES JAMES LEO
STILL RICHARD LOUIS
STIRNKORB CYRIL EDWARD
STOCK DALE LOUIS
STOCKLIN CURTIS ROBERT
STOFFER BENJAMIN F II
STOLL GEORGE GERALD
STOLLAR LARRY DAVID
STOLZ ROBERT LARRY
STONE JAMES EMMETT
STONE MELVIN LOUIS JR
STONE PAUL AARON
STONE THOMAS DAVID
STONEBURNER JOHN FREDRICK
STOPPELWERTH DAVID HENRY
STOTSBERY RICHARD PAUL
STRAHM PAUL DOUGLAS
STRAHM ROBERT EUGENE
STRAIT BENNIE HOWARD
STRASSHOFER STEVE OTTO
STRAUSBAUGH HOWARD ALBERT
STRAWBRIDGE JOSEPH EDWARD
STRAYER LAWRENCE EDWARD
STRAYER PATRICK JOSEPH
STRAZZANTI ALAN PETER
STRIPLING JOHN DAVID III
STRIZZI PHILLIP ARTHUR
STROBL JOHN GREGORY
STROHMAIER JOHN RICHARD
STROISCH LOYD EDWARD
STRONG HAROLD E JR
STRUCHEN THOMAS MICHAEL
STULL LARRY WARREN
SUBLER GERALD FRANCIS
SUHAR WALTER
SUKARA MICHAEL THEODORE
SULLINGER JAMES EDWARD
SULLIVAN ROBERT JOSEPH
SULSER DAVID WESLEY
SUNDAY JAMES MICHAEL
SUSTERSIC LOUIS ROBERT
SVANOE KENNARD ERROL
SWAFFORD KENNETH WAYNE
SWANEY LARRY DEAN
SWANSON JAMES CLIFFORD SR
SWARTZ CHARLES DELANO
SWEESY JOHN EARL
SWEET DAVID ARTHUR
SWEINSBERGER THOMAS EDWAR
SWIGER BERNARD LEROY
SWIHART DAVID EUGENE
SWINFORD RONALD DEAN
SZAHLENDER JULIUS NICHOLA
SZYMANSKI FRANK ADAM IV
TACKETT GARY DOUGLAS
TAKACS THEODORE NELSON J
TALLENTIRE GARY LEE
TALLION JOHN MICHAEL
TALLMAN DONALD CHARLES
TAMER RICHARD EDWARD
TANNER RAY EUGENE
TANNER ROGER LEE
TARPLEY NORMAN WESLEY
TASKER JAMES BRUCE
TATUM JOSEPH STEPHEN
TAYLOR EARL EUGENE
TAYLOR EDMUND BATTELLE JR
TAYLOR EDWARD EUGENE
TAYLOR ERNEST RAY JR
TAYLOR FREDERICK WAYNE
TAYLOR GARY LEE
TAYLOR GARY LYNN

TAYLOR GRANT CARL
TAYLOR HAROLD
TAYLOR HENRY LUSCIOUS
TAYLOR JAMES TIMOTHY
TAYLOR KENNA CLYDE
TAYLOR RICHARD LEE
TAYLOR RONALD LEE
TAYLOR THEODORE JR
TAYLOR WILLIAM KERRY
TECCO MICHAEL JAMES
TEDESCO LEONARD VITO
TERRELL JOHN WESLY
TERRY ARLIE
TERWILLIGER VIRGIL BYRON
TEUTSCH DAVID CHARLES
THAYER JOHN MERL
THEDFORD LUTHER JAMES
THEIS FREDDIE EDWARD
THEIS LAWRENCE WILLIAM
THEOBALD DAVID EDWARD
THIEL JOHN EDWARD
THOMAS ALGERNON PAUL
THOMAS ALLEN
THOMAS CHARLES EDWARD JR
THOMAS CLYDE EUGENE
THOMAS DALE DANIEL
THOMAS DAVID ROY
THOMAS EARL
THOMAS FREDDIE LEE
THOMAS GEORGE JR
THOMAS GLENN WILLIAM
THOMAS GREEN
THOMAS JERRY GALE
THOMAS JERRY LYNN
THOMAS NORMAN EUGENE
THOMAS RICHARD GEORGE
THOMAS ROBERT VIRGIL
THOMAS ROBERT WAYNE
THOMAS TOM MICHAEL
THOMPSON DALLAS EUGENE
THOMPSON DOUGLAS
THOMPSON GERALD RICHARD
THOMPSON HOWARD MICHAEL
THOMPSON JAMES ESCOL
THOMPSON JENNINGS MILROY
THOMPSON JERRALD RICH
THOMPSON JIM ALLEN
THOMPSON KENNETH DAVID
THOMPSON RANDALL ALAN
THOMPSON ROBERT RAYMOND
THOMPSON TIMOTHY JOSEPH
THOMPSON TOMMY RAY
THORNE KEVIN GARNER
THORNTON KENNETH CHARLES
THORNTON KENNETH EUGENE
THREET TROY TONY
THUM RICHARD COBB
THURMOND JAMES
TIMMONS JAMES MICHAEL
TIMMONS MICHAEL VINCENT
TIMMS TERRY LYNN
TISCHLER THOMAS JOSEPH
TITUS FIRMAN ANDREW
TITUS JAMES ELROY
TITUS TERRENCE RICHARD
TODD CHARLES MICHAEL
TOLLEY EDWARD ROBERT
TOMALKA VINCENT MILO
TOMPKINS ERNEST GALE
TOMSIC THOMAS T
TONGRET THOMAS EDWARD
TONON JAMES ANTHONY
TONTI MARK EDWARD
TOOPS FRANCIS IVAN
TOTH ANDREW JOSEPH JR
TOTH DAVID MC BRIDE
TOTH JOHN PAUL
TOWNSLEY STEVEN DOUGLAS
TOWSLEE EDWARD LAWRENCE
TRAMMELL HARRY MICHAEL
TRENT ALAN ROBERT
TRIPLETT RALPH MORGAN
TRITTSCHUH GERALD F
TROTTA FRANCIS JEFFREY
TROXELL DONALD RICHARD
TROXELL ROGER LEE

TROYER JOHN MICHAEL	VAUGHAN HOWARD JAMES	WASSENICH STEPHEN GEORGE	WHITE HERMAN JR

Photos by Don Henke & Melissa Hall

Vietnam Memorial

NW 3rd St, Lawton, OK 73507

Like most Vietnam War memorials, the Vietnam Veterans Memorial of Lawton, Oklahoma was created to honor the service members who bravely served their country during the controversial Vietnam War. The soldiers who weren't lucky enough to come home especially evoke a somber and reflective mood for visitors who come to the memorial to pay their respect to the local Vietnam veterans.

The Lawton, Oklahoma Vietnam Veterans Memorial was completely created and funded through public support and through the efforts of the local Chapter 751 of the Vietnam Veterans of America. The veterans and civilians of Lawton combined their efforts as well as donations to purchase the necessary supplies for the Vietnam Veterans Memorial including bricks and stone benches. Certain veterans' names and business names adorn the memorial as a thank you for larger monetary contributions. It brings about a sense of pride in the town due to the grassroots nature of the project. The town of Lawton as a whole helped bring the Vietnam Veterans Memorial to life. The memorial was officially dedicated in November 2001.

The Lawton, Oklahoma Vietnam Veterans Memorial stands at 12 feet high and is constructed of beautiful black granite. It is shaped in the outline of South Vietnam and features the names of the regions in the Asian country engraved in the stone.

On the base of the memorial, the six seals of the branches of the US military forces are proudly featured along with the words "Duty – Honor – Country" and "Welcome Home Vietnam Veterans." Surrounding the large memorial are flagpoles which fly the American Flag, the POW/MIA flag, the Oklahoma state flag, and Vietnam Veterans of America flag. Stone benches encircle the memorial in order to give visitors a place to sit and reflect on the sacrifices of the local veterans.

The Names of those from the state of Oklahoma Who Made the Ultimate Sacrifice

ABMEYER KENNETH RONALD
ACKERSON VENCEN
ADAIR THURMAN
ADAMS JOHN TERRY
ADAMS NORMAN EDWARD
ADDINGTON ROYCE LEE
ADKINS CARL EDWARD
AKINS DONALD WAYNE
ALEXANDER MICKEY ROY
ALLEN GARY
ALLEN ROBERT EUGENE
ALLEN RONALD STEWART III
ALLEN THOMAS RAY
ALLISON GEORGE BRIAN
AMSPACHER ROBERT ALAN
ANDERSON KENNETH RAY
ANDERSON ROY L
ANTLE MICHAEL LOUIS
ARANDA EUGENE LEONARD
ARMSTRONG BILLY CARL
ARMSTRONG DEAN EDWARD
ARMSTRONG EDWIN LAW-
RENCE
ARMSTRONG JAMES LEONARD
ARMSTRONG SHERMAN
FELTON
ARNOLD KENNETH HAROLD
ARTHUR GREGORY KENNETH
ARTMAN GARY RAY
ASHBY JEDD EDWARD
ASHER HAROLD E
ASHFORD JAMES ANTHONY
AUSBERN JOHN RAYMOND
AVERA JOHN ADAMS
AZLIN LUKE JUNIOR
BAILEY JAMES EDWIN
BAKER EDWARD GLEN
BAKER ELWOOD
BAKER FRANKIE GUY
BAKER GARRY WAYNE
BAKER THOMAS HUGH

BALLARD CARL HERSHEL
BALLEW CHESTER LLOYD
BARBEE THOMAS JOSEPH
BARBER BOB
BARE WILLIAM ORLAN
BARGER IVAN LLOYD JR
BARKER LARRY DALE
BARNARD THOMAS WALTER
BARNES HERBERT SPENCER
BARNETT GARY KEITH
BARNETT PAUL WAYNE
BARNUM GARY LANE
BARTLING TERRY NOBLE
BAXTER TERRY DON
BAYNE JAMES TERENCE
BEAM ERNEST EUGENE
BEAR DONALD EARL
BEAVER MAX RUSSELL
BECK LARRY MONROE
BECKHAM JERRY LEE
BEELER CLIFFORD DOIL
BEGLEY JACK PERRY JR
BELL EDGAR DEWAYNE
BELT CECIL DELBERT JR
BENIEN JOHN DAVID
BENNETT ANTHONY HER-
CULES
BENNETT DWIGHT LLOYD
BENNETT JERRY CLAUD
BENNETT LARRY DARRALL
BERNARD DONALD LEE
BERRY KENNETH BERYL
BESS BENNY DALE
BETTIS JAMES WILLIAM
BIRKS JAMES P
BLACK JOHN ENOCH
BLACK WALTER CURTIS JR
BLACKBURN EDMOND SMITH
JR
BLACKFOX ROBERT LEE
BLAIR WILLIAM WEBB JR

BLAKELY MELFORD KEITH
BLANKENSHIP JAMES THOM-
AS
BLANTON CLARENCE FINLEY
BLEVINS RICHARD LEWIS
BOATMAN LARRY NEAL
BOGGS JIMMIE WAYNE
BOGLE DENNIS DEAN
BOGUSKI PAUL ARTHUR
BOHANNON JOHN CALVIN
BOLDING BENJAMIN FOREST
BOLIN FORREST LEE
BOLLMAN ROBERT VINSON
BOLTE WAYNE LOUIS
BONDI CHARLES NICK
BOOKOUT CHARLES FRANK-
LIN
BOON MURLIN EUGENE
BOSTON GROVER WESLEY
BOWERS BRADLEY D
BOWMAN HAROLD E
BOYD SAMUEL LEE JR
BOYETT PAUL DEWAYNE
BOYLES DONALD RAY
BRADFORD EDWARD LEWIS
BRADLEY ALFRED LEE
BRADSHAW DAVID ALFORD
BRANNON GARY MICHAEL
BRASIER CHARLES DAVID
BRIGHTMAN DONALD LA-
VOYCE
BRISENO JOHNNY CHARLES
BROAD WILLIAM RAY
BROADHEAD LARRY IVAN
BROCK LARRY DEE
BROCK WILLIAM TONY
BROWN GORDON RICHARD
BROWN LARRY LEE
BROWN REX LEE
BROWN THAL ANTHONY
BRUNER DAVID

BRUNS ROBERT HARRIS
BRYANT CHRISTOPHER
BRYANT LARRY KENNETH
BUCHANAN GARY WAYNE
BUCK JAMES MARION
BUCKMASTER MICHAEL GENE
BUFORD ALPHA LEE
BULLARD CURTIS HERMAN
BURGESS DONALD RAY
BURKETT GARY LEE
BURKETT JOSEPH WILLIAM
BURLESON JOHN ALLAN
BURNES ROBERT WAYNE
BURNETT EDWARD DENZEL
BURNS GERALD RAY
BURNSED RANDELL HEATHE
BURTON THOMAS LEE
BUSH THOMAS BURKE
BUTLER ELMO LARRY
BUTTS GEORGE LESSIE
BYNUM NEIL STANLEY
BYRD ELMER DON
BYRNS GERALD WINSTON JR
CALDWELL LARRY EUGENE
CALLISON JIMMY RAY
CALVIN GLENN HENRY
CAMPBELL DWIGHT STANLEY
CAMPBELL EDGAR ALLEN
CAMPBELL JIMMY LEE
CAMPBELL THOMAS EUGENE
CANADY DEE OKEY NELSON
CANFIELD LEON
CARDWELL JOHNNIE WAYNE
CARNEY JOSHUA ELI
CARPENTER RAMEY LEO
CARTER CLYDE ELMER JR
CARTER ERNEST LEE
CARTER MERLE KEITH
CARVER BILLY KAY
CASEY MICHAEL DALE
CASTILLO GEORGE RALPH

CASTLEBERRY BILLIE MAC
CAUTHRON R G
CAZARES JAMES STEVEN
CECIL ALAN BRUCE
CHALAKEE RUDY YORK
CHAMBERS BILLY CLAYTON
CHAMBERS JACKIE DEAN
CHAMBERS JERRY LEE
CHAMBERS LORANZEY PAUL
CHANDLER QUINNEN T JR
CHAPMAN DAVID LEE
CHAPMAN GARRY RAYMOND
CHILDS FORREST CLIFFORD
CHITWOOD JERRY MICHAEL
CHRISS GARY DOYLE
CHRISTIANSEN ROBERT
DOUGL
CHRISTIE EDWARD EUGENE
CHRISTY GILMORE WILSON
CHURCHILL WENDELL
EUGENE
CLAY RAYMOND
CLAYTON GARY EVERET
CLEMENTS RICHARD BART
CLEMMER DERRELL W
CLOSE FLOYD EUGENE
CLOWER HUGH JR
COAST ALBERT FRANK
COBB MILFORD EUDENE
COE BENNY BOB
COLE WILLIE JR
COLEMAN ROBERT LEWIS
COLLINS ALBERT EUGENE
COMBS VIRGIL CARLYLE
CONSTIEN JOHN RICHARD W
COOK LARRY DEAN
COOPER TOMMY DALE
COPELAND JERRY DON
CORBO AL DOUGLAS
CORNETT CHARLES RANDELL
COSTELLO STEPHEN RAN-
DALL
COUCH GAYLORD MARTIN
COURTNEY RONNIE
COVEY JERRY K
COWAN DARRELL WAYNE
COWAN ROBERT LE RHEA III
COX LEWIS EARL
COX STANLEY GILBERT
CRADDOCK RANDALL JAMES
CRAWFORD MICHAEL ALAN
CRAWFORD WILLIAM LLOYD
CRAYTHORNE ROBERT EARL
CRISP WILLIAM HENRY
CROCKETT STANLEY GENE
CROSBY HERBERT CHARLES
CROSS SAMMY JOE
CROSSLIN GAILEN CHEEK
CRUSE STANLEY JOE
CULLUM DENNIS OWEN
CUMISKEY JAMES LEE
CUTTER WILLIAM SCOTT
DACY JAMES WESLEY
DAILY SAM WEBSTER
DALTON JAMES GILBERT
DANIELS CHARLES WESLEY JR
DART DANNY JOE
DASHER GARY JOHN
DAVENPORT JOHN SANFORD
DAVEY LOREN KEITH
DAVIE THOMAS EARL JR
DAVIS BUREN RAY
DAVIS JERRY WILLIAM
DAVIS JOHN LOUIS
DAVIS RAY ELBERT
DAVIS RAY GENE
DAVIS ROBERT WENDELL
DAVIS WALTER EMERSON
DAVOULT GAYLON DARYL
DAY CLINTON LEE
DAY DENNIS IRVIN
DAY JOLLY J
DEATON CARL WOODROW
DEATON JOHN CLAUD
DEER TERRY LOUIS
DEERE CHARLES KENNETH

Oklahoma

DEERINWATER BRUCE EDWARD
DEEVERS DONALD JAMES
DEMINGS DAVID EUGENE
DEMPSEY JACK TAYLOR
DERRICK RANDY WAYNE
DEWBERRY JERRY DON
DIEHL WILLIAM CALVIN JR
DIRICKSON MARION LEE
DOBRY STEVEN LOUIS
DODD RICHARD EUGENE
DOIRON WILFRED ALCIDE
DOKE JAMES ALLEN
DOMINE MANUEL DE LEON
DONOHUE STEPHEN SCOTT
DOWLING JESSE WILLARD
DOWNING WILLIAM KELLY
DRESHER HARRY EVERETT JR
DRINNON BEDFORD LEE
DRYDEN MICHAEL THEODORE
DUKE DOUGLAS OVYLE
DUMAS LONNIE EUGENE
DUNCAN GALVIN LEE
DUNCAN JAMES PAUL
DUNCAN ROGER EVANS
DUNN DONALD LOUIS
DURANT FORBIS PIPKIN JR
DURHAM DWIGHT MONTGOMERY
EAST JAMES BOYD JR
EATMON JAMES LARKIN
EATON NORMAN DALE
EDDY JOHN ARTHUR
EDMONDS WILLIAM ORVILLE
EDWARDS CHARLES DAVID
EGGER JOHN CULBERTSON JR
EISENBERGER GEORGE JOE BU
ELKIN JAMES FREEMAN
ELKINS ROGER LYNN
ELLIOTT BILLY RONALD
ELMORE CLAUDE EUGENE
ELSTON JACKIE LINDELL
EMERSON TOM
EMMETT GARY WILLIAM
EOFF WILLIAM BRADFORD JR
ERVIN CHARLES DWAYNE
EULITT LEONARD ELZY
EVANS CLIVE LEROY
EVANS DANNY LEO
EVANS JOHN TROY
EVERETT TONY
EZELL BURLEY DEAN
FAIRES ROBERT DON
FARBRO MILLARD WADE
FARRIS DENNIS CLAUDE
FARRIS GARY BRUCE
FAUGHT FRANK EDWIN
FERGUSON WILLIE C JR
FERRELL CHARLES ELTON
FIELDS ELMER EUGENE
FIELDS JAMES BENJAMIN
FIELDS JOHN CURTIS
FINERTY MICHAEL ROY
FOILES FRANCIS IVAN
FOLEY WILLIAM LOYD
FORD RICHARD WILLIAM
FORMAN CLARENCE GENE
FORNEY DENNIS RAY
FORRESTER JORDEN DUWAYNE
FOWLER JAMES HARRELL

FRANKLIN MARVIN LYLE JR
FRAZIER FLOYD WENDELL JR
FRAZIER FRED RAYMOND JR
FRAZIER GENE ALLEN
FRAZIER JOHN DUDLEY
FREEMAN JOHN OLIVER
FREEMAN REX BRADFORD
FROST ROBERT DEAN
FRY NASH
FRYER CHARLES WIGGER
FULLER MICHAEL BRUCE
GAINER JOHN ROBERT
GAINES PHILIP FALCONA
GAINES WILLIAM FRANKLIN
GALINDO EDWIN GENE
GAMBLE DEXTER NUNTON JR
GARDNER DANIEL ELI
GARDNER MARION LORA
GARRETT ALLEN MATTHEW
GAULEY JAMES PAUL
GAY HERBERT LYMAN
GEORGE S W
GIBBONS CLAUDE ROBERT
GIBSON AUSTIN DALE
GIBSON OTHA DOUGLAS
GILMORE WILLIAM ALLEN
GILMORE WILLIAM F JR
GIST TOMMY EMERSON
GLEGHORN JERRY WAYNE
GLOVER EDWARD LEE
GODDARD ROBERT GORDON
GOING WALLACE
GOINS BILLY LEE
GOINS GERALD
GOOD JOHN DUDLEY
GOODELL JIMMY LEON
GORDON DRANNON RAY
GOUCHER EDWARD LOUIS
GOUDEAU JEFF JR
GOUGH HURSHELL HARRY
GRAMMAR WILLIAM MICHAEL
GREEN JAMES ARVIL
GREEN JIMMIE RAY
GREEN WESLEY
GREER EDMOND JUNIOR
GREGORY BOB LEROY
GRESHAM WILBERT JAMES
GRITTS WILLIAM ARCHIE
GRUNDY ANTHONY WARREN
GUFFEY JAMES DALE
GUILD LEROY J JR
GUINN ALLAN
GUSTAFSON DONALD LEE
GUTHRIE DENNIS HAROLD
GUTHRIE EDWARD F
HALEY JACK WAYNE
HALEY TOMMY WAYNE
HALL BILLIE ALLEN
HALL DONALD JOE
HALL GARY NEAL
HALL ROY RAY
HAMILTON FLOYD WAYNE
HAMILTON JAMES LEON
HAMILTON LARRY EDWARD
HAMMER ROBERT WAYNE
HAMMONS HERBERT DON
HAMPTON DAVID CONRAD
HARBERT RONALD VINCENT
HARGER CHARLES F JR
HARGROVE TEDDY EARL
HARJO KENNETH DEWAYNE
HARRISON BUFFARD CLIFTON
HARTNESS ROGER DALE
HARVEY LARRY DREW
HARVEY ROBERT LEON
HAVLICK JOHN CHARLES
HAWKINS JOHN LEE JR
HAYS THOMAS EARL
HEIDEBRECHT DALE ROGER
HEIDERICH DANIAL GUY
HELVEY JOE DEAN
HENDERSON MARION F
HENDERSON ROBERT KNAPP J
HENDERSON TIMOTHY
HENDON WILLIAM ATTLEE
HENSHAW LARRY ROY
HENSLEE JAMES EUGENE
HERSCHBACH PAUL DELL
HEWITT WILLIAM FRANK
HICKMAN JOSHUA
HICKS GLEN RAY
HICKS PAUL EVERETT
HIGDON JIMMY RAY

HIGH RONALD CLYDE
HILDEBRAND ALFRED DEAN
HILDERBRAND RONALD LEE
HILL DOUGLAS WAYNE
HILL FOSTER EUGENE
HILL JERRY WILLIAM
HILTERBRAN DANNY LEE
HINCKLE JAMES NELSON
HISAW TEDDY LEE JR
HODGES RAYMOND LEON JR
HOFFMAN JOHN PAUL
HOLBROOK JAMES NEWTON
HOLDEN ELMER LARRY
HOLDER JAMES EDWARD
HOLDER SAMUEL LOYD
HOLDERBY VERLIN DON
HOLDING DARRELL EUGENE
HOLLAND GARY R
HOLLEY ROBERT GORDON
HOLLOWAY JOHN MARSHALL
HOOD JOHN EDWARD
HOPE MICHAEL CLINT
HOPKINS ALVIN JR
HORINEK BRIAN ANTHONY
HORSMAN GEORGE LESLIE II
HOSKINS GARY LEE
HOWARD A W JR
HOWARD MICHAEL DAVID
HUBBARD CHARLES AUSTIN
HUBBARD DAVID LEE JR
HUBBARD THEODORE JR
HUDGENS EDWARD MONROE
HUDSON ROBERT BENJAMIN
HUDSPETH JAMES L
HUMPHREYS LARRY DON
HUMPHRIES WAYNE WARREN
HUNTER MICHAEL J
HUTSON RONALD WAYNE
HUTTON CHARLES PHILLIP
HYATT MICHAEL DALE
HYDE JIMMY DON
HYSLOP LELAND WAYNE
INGRAM JOHN LEE
IRSCH WAYNE CHARLES
ISAACS ROYAL GEORGE JR
ISAACS SAMMY FLOYD
ISHMAEL JOHNNIE LEROY
JAMES RICKY LYNN
JANTO PAUL CHALMERS
JENKINS EUGENE RAY
JENNINGS EARL WAYNE
JENT BILLY GENE
JESSE WILLIAM CLIFTON
JOHNSON CLIFFORD CURTIS
JOHNSON DARYL LINN
JOHNSON DENNIS VAN
JOHNSON JAMES EARL
JOHNSON JAMES EARL
JOHNSON JAMES JR
JOHNSON JERRY DEAN
JOHNSON JOHN KIRBY
JOHNSON LEE GRANT
JOHNSON MICHAEL LEE
JOHNSON PAUL WILLIAM
JOHNSON ROBERT LEE JR
JOHNSON RONNIE LLOYD
JOHNSON STEVEN EDWARD
JOHNSON TURNER COLEMAN
JOHNSON VERNON JOE
JOHNSTON STEVEN BRYCE
JONES ARLAND JASPER
JONES BILLY CHARLES
JONES GARY CLAUD
JONES GARY HOWARD
JONES HAROLD LESLIE
JONES HOWARD LEE
JONES JERRY ROBERT
JONES RICHARD WARREN
JONES ROBERT LEE JR
JONES SAMMY JR
JORDAN JOE RITCHARD
JUNGER WALTER JOSEPH JR
KALSU JAMES ROBERT
KEAHEY CARL JOHN III
KECK GARTH WAYNE JR
KECK RUSSELL FORREST
KEELER LARRY DEAN
KELLEY JOE C
KELSEY D J
KENDALL ALBIN LEE
KENNEDY MARCUS TRUMAN
KESTER FRED DUANE
KIDD JOHNNY LEE
KIELY BILLY RAY

KINDRED RONNY KAY
KING FRANCIS J R
KING LARRY DOUGLAS
KING VERLON DONALD JR
KINGSBURY DAVE ROYCE
KINKLE BOBBY GENE
KINNEY JOHN WADE
KIPP RAYMOND SIDNEY
KISER JERRY ALLEN
KNIFFIN ARNOLD DEAN
KNIGHT LARRY COLEMAN
KNIGHT MICHAEL PERRY
KOEHN ARLIN WAYNE
KOUPE GREGORY LANCE
KRIEG RONALD JAMES
KRIEGER ELDON EUGENE
KRUMREI DONALD ALAN
KUMMELL ROBERT MICHAEL
KUSCH WILLIAM HOWARD
LA FEVRE DARREL EUGENE
LAKEY HOWARD WALLACE
LANDKAMER MICHAEL GEORGE
LANDRUM JAMES ALFORD
LANGSTON JOHN ALAN
LANNOM WADE ANDREW JR
LANSDEN THOMAS JACK
LASITER LAWRENCE RAY
LASKEY JOHN BENNIE
LATIMER RICHARD ELI JR
LAUDERDALE ARTHUR LEON
LAUINGER JOSEPH MARK
LAWRENCE CLYDE WESLEY JR
LAWRENCE JOHN ROBERT
LAWSON TOMMY ROSS
LE CLAIR PRENTICE DALE
LEATHERS CLIFFORD W JR
LEDFORD ALVIE JUNIOR JR
LEE ROGER GAIL
LEEMHUIS DONALD J
LEMLEY JIMMY DAVIS
LEONARD JAMES MICHAEL
LEOPARD JACK DAVID
LESTER EARL ROY JR
LEWIS BENNY JOE
LEWIS DONNIE GORDON
LEWIS JERRY D
LINAM MAXIE DEAN
LINDBERG JOHN DAVID
LING WILLIAM CLIVE
LISENBY MAX
LITTLE SUN THOMAS LEE
LLOYD DONALD LEE
LOCKE JAMES LEE
LOEFFLER NORMAN F JR
LOVE DON WAYNE
LOVE VERNON GLEN
LOWERY FREDDIE LEON
LOYD LONNIE DOUGLAS JR
LUCAS PAUL DAVID
LUMAN RONNIE DEAN
LUNA ABEL
LYNCH BRUCE ANTHONY
LYNN ROY EUGENE
MABRY RALPH EDGAR JR
MACKEY TALTON LEE
MADDEN RORY ANTONIO
MANDEVILLE ROSS EDWARD
MANESS MARTIN ROWLAND
MANNERS DAVID PAUL
MAREADY TERRY KAY
MARLOW JOHN P
MARR NOEL DON
MARTIE ERNEST RAYMOND
MARTIN DONALD ARTHUR
MARTINO STEPHEN LEE
MASON DENNIS RAY
MASON GEORGE ARDEN
MASSIE LARRY GLEN
MATLOCK NELSON ALLEN
MAYER ALEXANDER LEO
MC BRIDE BEN K
MC BRIDE JAMES LARRY
MC BURNETT LARRY TURNER
MC CLURE JAMES M
MC DANIEL GILBERT ELLIS
MC DONALD ALBERT JR
MC DONELL TERRY KEITH
MC DONOUGH GEORGE WATSON
MC DOWELL EARL WAYNE
MC ELHANNON JAMES PHILLIP
MC ELREATH RANDALL LEE

MC FARLAND ARTHUR RAY
MC GEE DARWIN DALE
MC GEE GEORGE WILLIAM
MC INTIRE DON RAY
MC KEE WESLEY RAYMOND
MC KINZIE THOMAS LEON
MC MILLAN JAMES M JR
MC VAY THOMAS MARTIN
MC VEY JOHN WESLEY
MC WHORTER JERRY MONROE
MC WILLIAMS RICHARD EUGEN
MCNEILL MICHAEL SIDNEY
MEIXNER EDWIN GEORGE
MENDENHALL MICHAEL JOSEPH
MERCER GARY LYNN
MERRIMAN JOHN DAVID
MESSER FERRELL EDWARD
MICHALIK WILLY ROBERT
MIDKIFF GARY BRUCE
MIKULECKY DONALD HENRY
MILLER ALLEN PERDA
MILLER CARL ROBERT
MILLER CHARLES DANIEL
MILLER JERRY LAVON
MILLER PHILLIP
MITCHELL ARTHUR G
MITCHELL JAY ANDERSON
MOCK DONALD RAY
MOFFETT JAMES DELTON
MOORE CLARENCE EARL
MOORE FRANKLIN EDWARD
MOORE JAMES CURTIS
MOORE TERRY ENGLEBERT
MORAN LONZO JOSEPH JR
MORGAN JACKIE RAY
MORGAN LARRY GENE
MORRIS EDWARD
MORRIS MICHAEL EUGENE
MORRISON JACKIE LEE
MORRISON RANDY STANTON
MOSBURG HENRY LEE
MOSES LESLIE DON
MOSS RONALD GENE
MOUNT CHARLEY LE MEAR
MOYER ROBERT W
MOYERS MURL ALVIN
MULLINS JIMMY MERYL
MULLINS WILLIAM DONALD
MUNCY ROBERT WILLIAM
MURPHY JESSE ALLEN
MYERS JAMES HOWARD
NABORS J C
NABORS PAUL HOWARD
NAYAR WALTER HODGKINGSON
NEAL ARTHUR DARNELL
NEAL BARNEY KING JR
NEELY DONALD LEE
NEISLAR DAVID PHILLIP
NEWFIELD JIMMY CHARLES
NIEVES JORGE LUIS
NIMAN ROBERT O NEAL
NOAH JOSH CAIN
NOAH MARVIN TIDWELL
NOE MARVIN LEWIS
ODOM HENRY DUANE
OGLE LEWIS MILTON
OGLESBY RONALD CHARLES
OKEMAH JOHN
OLIVER CHARLES EDWARD
OLMSTEAD STANLEY EDWARD
OSBORN CHARLES EDWARD
OSBORN ROBERT JAMES
OWENS FRED MONROE
PACK ROBERT VAN
PADBERG LARRY GENE
PADGETT SAMUEL JOSEPH
PAHCHEKA ROBERT CARLOS
PAPPIN JOHN PATRICK
PARKER DANNY LYNN
PARKER JOHNNY RAY
PARKER RICHARD HOWARD
PARKS A L
PARSONS GERALD LOYD
PARSONS LIONEL EUGENE
PARSONS ROY BROWN
PATTERSON JERRY LEROY
PATTERSON ROBERT WAYNE
PATTON RONALD WADE
PATTON THOMAS JAMES
PAYNE ELDON RAY
PEARL RAYMOND JR

PECK STEVEN RUSSELL
PENDERGRASS JAMES WILLIAM
PERDUE RICHARD W
PERRYMAN DALLIS
PESEWONIT RUSSELL EUGENE
PETERS DAVID ARTHUR
PETERS JOSH
PHILLIPS OSCAR C JR
PHILLIPS RANDALL SCOTT
PICKETT HOMER LEE
PICKETT MORRISON LOUIS
PIERCE JERRY DEAN
PIERCE OSCAR WAYNE
PITTS DERWIN BROOKE
PITTS RILEY LEROY
PLATO ROBERT DEAN
POINTER WALTER LEON
POOLAW PASCAL CLEATUS SR
POOLE FRANKLIN WILLIE
PORTER BOBBY L
PORTER ROBERT LEE
POSPISIL ALFRED FRANK
POWELL DAVID BRUCE JR
POWELL PETER EARL
POWELL SAMUEL HERBERT
POYNOR DANIEL ROBERTS
PRATT GUY LEON JR
PRAY VERN LEE
PREWITT LARRY GENE
PRICE HUBERT JR
PRICHARD JOHN LEE
PRINCE EUGENE JR
PULLEY JAMES EDWARD
PULLIAM EDGAR RUSSELL JR
PULSE DOYLE GEAN
RAGSDALE DONALD RAY O
RANGEL FLORENTINO
RANKIN KENNETH DEAN
RANKINS DONALD LEE
RANSBOTTOM FREDERICK JOEL
RANSTEAD JAMES TERRY
RAPP WILLIAM HENRY JR
RATCLIFF LENOX LEE
RAY JOHN MACK
RAYMOND ROBERT KENNETH
REED JOHN ARTHUR
REED WILLIAM ELBERT
REEDER JAMES EDWARD
REESE CHESTER ROY JR
REEVS JOHN CURTIS
REID CARL J
REMBERT HARVEY LEE
REYMAN LAWRENCE FRANCIS
REYNOLDS DAVID MACK
REYNOLDS EARNEST LANE
REYNOLDS ELDON LEE
REYNOLDS WILLIAM DONALD
RHAMY RAYMOND DALE
RHODES WAYNE A
RICHARDS WAYNE
RIDDLE WALTER RAY
RIDGE JESSE LEE
RIEGER RODNEY L
RILEY JOE ED JR
RILEY LARRY LLOYD
RITCHIE HERMAN HIRAM
ROBBINS LAWRENCE STEPHEN
ROBERTS ALBERT C
ROBERTS CHARLES DWAINE
ROBERTSON DON MARK
ROBINSON GEORGE BERNARD
ROBINSON THOMAS DALE
ROBIRDS PATRICK DALE
RODGERS BILLY GENE
ROLEN SAMUEL FLOYD
ROLLINS BOBBY JOHN
ROMERO WALTER DAVID
ROSE JAMES ELDON
ROUSE CLARENCE LEON
ROUSE FREDERICK EUGENE
ROWLETT HAL JONES
ROY BILLY DUANE
ROYAL JERRY CHARLES
RUHLAND KLAUS DIETER R
RUSSELL KENNETH TRUMAN
RUTLEDGE JAMES ROBERT
RYCROFT LARRY WAYNE
SALES HARLIS CALVIN
SANDEFUR TOMMY GERALD
SANDERS FREDERICK WRIGHT
SANDERS JIMMY DOYLE
SANDERS KENNETH EUGENE

SANDERS PHILLIP DUANE
SANDERS RONALD LLOYD
SAUNDERS BASIL LEE
SAUNDERS WILLIAM MICHAEL
SAVAGE WILLIAM ROSS
SAWNEY JACKIE LEE
SCALF JAMES RAY
SCHOUWEILER DAVID LEE
SCOTT DONALD EUGENE
SCOTT MARTIN RONALD
SCOTT PERRY JAY
SCROGUM JIMMIE CHARLES
SEALS CLIFFORD
SEHESTED RONALD ALLEN
SEIGLE WILLIAM ARTHUR
SHACKELFORD IVAN J JR
SHACKELFORD RICKY LEE
SHAFER FRANCIS LOE JR
SHANNON PATRICK LEE
SHAW CLARENCE LEE
SHAW ROLAND JUNIOR
SHAW RONNIE DEAN
SHAWNEE CLARK VERNON
SHEARER DON TYRONE
SHELTON JAMES HARVEY
SHERRILL JOHN OTIS
SHERROD EDWARD HERBERT
SHIELDS JAMES CURTIS
SHIELDS RONALD WAYNE
SHIPLEY DREW DOUGLAS
SHIRLEY DALE EDWARD
SHOOPMAN KENNETH DOYLE
SHORES DANNY JEAN
SHORT LARRY RAY
SHUYLER JAMES EARNEST
SIDES CHARLES KENNETH
SIKKINK ROY DEAN
SILLAWAY CHARLES EUGENE
SIMMONS RONALD WAYNE
SINGER NORMAN PAUL
SINGLETON J D
SITTON TROY NELSON
SKINNER KENNETH W III
SLOAN JOHNNIE LEE
SLOAN MICHAEL LEE
SLOAT DONALD PAUL
SMART ARVEL RAY
SMITH ALBERT JOSEPH
SMITH ALLAN LESLIE
SMITH ALLEN JAY
SMITH CHARLES LESLIE
SMITH DONALD BOYD
SMITH ERVIN DALE
SMITH FREDERICK PHILLIP
SMITH HOWARD HORTON
SMITH JACK MILTON
SMITH JAMES HOWELL
SMITH JAMES ROBERT
SMITH JAMES RONALD
SMITH MYRON FRANCIS
SMITH RICKEY DOVIE
SMITH SAMMY RAY
SMITH TOMMY DAVE
SMITH WILLIAM GENE
SMOOT RAYMOND EUGENE
SNEED CARL MICHAEL
SNODGRASS JACK LEE
SNOOK JAMES ARTHUR
SOCKEY RONALD
SODOWSKY MELVIN DEWAYNE
SONAGGERA FREDDIE LEON
SPARKS STEPHEN DUANE SR
STANBERRY JERRY WAYNE
STANDRIDGE HARLEY ROY
STARRETT JOHN DELBERT
STAYTON COY G
STEDMAN LEE ALLEN
STEELE DAVID MARK
STEELE WALTER EDWIN
STEPHENS CURTIS ADRON
STEPHENSON DAVID RICHARD
STEWARD STEVE LEE
STINSON GEORGE WILLIAM JR
STIVERS GEORGE EDWARD
STIZZA JOHN BONAT
STOCKDALE JOHN ROBERT
STOW LILBURN RAY
STOWERS AUBREY EUGENE JR
STREET LENARD JR
STROME JOHN CLARENCE
SULLENS GEORGE BUSTER JR

SUMTER FORREST DARRYL
SUTTON ARTHUR LAVERN
SUTTON LOWHMAN SOLON
SWEETEN R C EARL
TABER JERRY DEAN
TAGMAN JOHNNY RAY
TARKINGTON RICHARD JR
TARVER LLOYD ROBERT
TAYLOR CLYDE DAVID
TAYLOR ERNEST RAY
TAYLOR LARRY
TEFFS JAMES RICHARD
TEFFT GEORGE EDWARD
TERRY DELTON EUGENE
THOMAS BILLY LEE
THOMAS CHARLES ELBERT
THOMAS JOE MINOR
THOMAS MICHAEL HOWARD
THOMPSON FRANK ALBERT
THOMPSON GEORGE RAY
THOMPSON NATHANIEL ANTHON
THOMPSON RONALD EUGENE
THOMPSON RUDY MICHEL
THOMPSON TURNER L JR
THOMPSON VENEY EWELL
THORNTON TERRY ALLEN
THURMAN RAYMOND DALE
TIFFANY JOHN MICHAEL
TIFFT DANNY WILLIAM
TIPTON MARTINIS GENE
TOLBERT CLARENCE ORFIELD
TOMLINSON JAMES RICHARD
TRACY JOHN WAYNE
TREAT FLOYD GENE
TREEN HARLIN PERRY
TRENT JIMMIE EDWARD
TRIGALET ROBERT ERNEST
TRIZZA SAM RICHARD JR
TROYER RODNEY PHILLIP
TRUESDELL JOHN LEROY
TUCKER JAMES HALE
TUELL ROBERT LEE III
TULL MARTIN NELSON
TURNER EUGENE
TYLER EDWARD
TYNER JAMES ANTHONY
UMDENSTOCK MICHAEL LANE
UPCHURCH JAMES GLENN
VAUGHAN THOMAS CECIL
VENABLE ELTON RAY
VETTER ERNEST JR
VINCENT KEITH DAN
WABLE SAMUEL LEE
WADE BOYD LEE
WADE THOMAS JOE
WAGES JAMES LEWIS
WAGNER FARL DUANE
WALKER JACKIE DALE
WALLACE BARTLEY ALLEN
WANN DONALD LYNN
WARD CARL RAY JR
WARD LEONARD DANIEL
WARD RONALD JACK
WARD WILLIAM LEE
WARE ORTEN LEE
WARNICK MICHAEL GENE
WASHINGTON DONALD LEROY
WATKINS ANTHONY RONALD
WATSON BILLY JOE
WATSON FRANK PETER
WATSON RONALD LEE
WEBB BILL ALAN
WEBSTER HENRY WAYNE
WELCH GARY MAX
WELLMAN CECIL ALBERT
WELLS ELROY FREDERICK
WEST DANNY GENE
WEST JIMMY DON
WHEELER CARL EUGENE
WHEELER CLINTON LEE
WHILES FRED LAMAR JR
WHITE ALLEN THOMAS
WHITE GARY SIDNEY
WHITED JAMES LAFAYETTE
WHITFIELD RICHARD K
WHITLOCK JIMMIE DALE
WIEGERT LARRY ROBERT
WILBANKS DONALD MELVEN
WILBURN JOHN EDWARD
WILCOXSON ROBERT FRANKLIN

WILLEFORD FRANKLIN PATRIC
WILLIAMS BOBBY RAY
WILLIAMS CLAUDE NATHANIEL
WLLIAMS DANNY
WILLIAMS JAMES
WILLIAMS JAMES EDWARD JR
WILLIAMS JERRY LEONARD
WILLIAMS LARRY GLEN
WILLIAMS LAVESTER LEE
WILLIAMS MACK WILBERT
WILLIAMS MARK EVERETT
WILLIAMS NOEL DEAN
WILSON BILLY LEO
WILSON ISAIAH HERMAN
WILSON JAMES CLAIR
WILSON JOHN CHARLES
WILSON RAY GENE
WILSON RICHARD LEE
WILSON ROGER GLENN
WILSON WAYNE MICHAEL
WILSON WILLIAM DEAN
WINER ROY LEE

WINGET HAROLD WILLIAM
WINTERS STEVEN ANDREW
WOLFE HULSA D
WOOD ROSS W JR
WOODS FLOYD WILLIAM
WOODWORTH SAMUEL ALEXANDER
WOOLSEY JACK LEE
WRIGHT ROBERT CARROL
WRIGHT ROSCOE JR
YANCEY CRAIG MARTIN
YEAROUT DONALD EUGENE
YORK DANIEL WEBSTER
YOUNG RONALD HERMAN
YOUNG RONALD WAYNE
YOUNG WILLIAM GLENN
ZORNES HAMP EDWIN

Photos by Corey Terill

Vietnam Veterans of Oregon Memorial

4000 SW Canyon Rd, Portland, OR 97225

Formally named The Garden of Solace, this Vietnam War memorial was dedicated in 1987 in Portland, Oregon. The outdoor memorial features the names of hundreds of Oregon-based service members and personnel who are now passed or are still listed as missing in action.

This beautiful memorial features a spiral 1,200-foot walking path (that is wheelchair accessible) that is set in a depressed bowl shape. Several statues inscribed with names of some of the fallen or missing service members are along the path before the actual monument which features all of the names of the Vietnam veterans deceased or missing. This black granite wall is separated into three sections which each represents segment time period of the Vietnam War. On the wall, there are also descriptions of local Oregon events that occurred during the same time period. This was done to show what was going on in Oregon during war time.

Besides the memorial itself, there are beautifully landscaped lawns and hedges as well as peach trees which are meant to stand for sacredness and life. There are also water features that represent hope, life, and purity. The Oregon Vietnam Veterans Memorial was created by the architectural company Walker and Macy based out of Portland, Oregon. But the majority of the construction, along with the labor and the materials, were actually supplemented by volunteer donations.
The memorial is maintained by the Oregon Vietnam Veterans Memorial Fund, led by President Jerry Pero, who meets at the memorial around Memorial Day weekend. These volunteers help keep the landscape well-manicured and clean the monument so it's in prime condition.

The idea for the Oregon Vietnam Veterans Memorial was conceived after a handful of Vietnam veterans along with relatives of a fallen soldier visited the dedication of the Vietnam Veterans Memorial in Washington, D.C. in 1982. They were so moved by the memorial that they were inspired to create their own in Oregon. Reports show that the veterans were able to raise close to $1 million in order to convert the amphitheater in the Hoyt Arboretum to the memorial. The memorial is accessible during park hours which begin at 5 a.m. and end at 10 p.m.

196

The Names of those from the state of Oregon
Who Made the Ultimate Sacrifice

ABRAMSON ANDREW JOHN
ACHISON TIMOTHY EUGENE
ADAY RICHARD DONALD
AKIN RICHARD ARTHUR JR
ALLEN MELVIN ARLEIGH
ALLEN TERRY LEE
ALTUS ROBERT WAYNE
AMES GARY DENNIS
ANDERSON DALE ARTHUR
ANDERSON FRANKLIN VANCE
ANDERSON RONALD STANLEY
ANDERSON WALTER GILMORE
ANDREWS WILLIAM RICHARD
APPLEBURY MELVIN LYNN
ARMSTRONG LEVI LESTER
ARNESON KEITH SAM
ARNOLD RODNEY KEITH
AUSTIN CARL BENJAMIN
BADLEY JAMES LINSDAY
BALDON RUDY LEE
BALL LESLIE ARNOLD
BAMFORD GEORGE ARTHUR
BARBER RONALD LEE
BARKER ELVIS GORDON
BARKER JAMES HAROLD
BARNETT ALAN LYNN
BARNETT IRIA DANIEL
BARNETT MELVIN DONALD
BARRETT DAVID MORRIS
BARROW MICHAEL EDWARD
BARRY KENNETH DONALD
BARTELL MICHAEL RICHARD
BARTHOLOMEW ROGER JAY
BATEMAN JAMES TERENCE
BAUER WILLIAM LYLE
BAUGHMAN WESLEY GENE
BAUMAN CHARLES W
BEAM EARNEST LEE
BEAR CHARLES MARTIN
BECKWITH WILLIAM ARNOLD
BEDDOE PAUL MELVIN JR
BEGLAU DAVID BERNARD
BELLAMY PAUL ROBERT
BENSON STANLEY J
BENTLEY DUANE RUSSELL
BENZEL RICHARD DALE
BERGE JAMES MAYNARD
BERGERON DOUGLAS HUGH
BERN HAROLD STANLEY
BETZ SAMUEL
BEUTLER RONALD EUGENE
BIGELOW RONNIE O
BLACKMAN DAVID RAWSON III

BLACKWELDER KIT
BLAIR ANTHONY BURDETTE
BLANCHARD DAVID MELVIN
BLANTON MICHAEL MERLE
BLOCHER RUSSELL GLEN
BLOCK WILLIAM JOHN
BOLDEN ROLLIE LEE
BOSSOM JOHN AUSTIN
BOWDISH KEVEN WAYNE
BOWMAN BARRY ALAN
BOWMAN JOHN OTTO
BOYCE JOHN MERTON
BOYER MICHAEL DAVID
BRADLEY GLEN WAYNE
BRAMSEN DAVID EUGENE
BRANDON DAVID BRUCE JR
BRANNFORS ERIC ARTHUR
BRANNON JOHN LESLIE
BRATTAIN THOMAS LAIRD
BREHM TOMMY JOE
BRETT ROBERT ARTHUR JR
BRIGHT RICHARD HALL
BRILES JAMES WALTER
BRITTON GARY WILLIAM
BROCKMAN JOHN NELSON
BROOKS CLARENCE HERBERT
BROOKS JACKIE RAY
BROWN DENNIS EARL
BROWN JOHN PATRICK
BROWN JOSEPH GORDON
BROWN LESTER EUGENE
BROWN MARVIN H
BROWN RAYMOND EARL
BROWN TIMOTHY JOHN
BRUBAKER DONALD DEAN
BRUCHER JOHN MARTIN
BURGARD PAUL EDWARD
BURKE WILLIAM ERVIN III
BURKHART RONALD WAYNE
BURNHAM MASON IRWIN
BURRELL CHARLES FRANKLIN
BUSHNELL BRIAN LEE
BUSWELL ROBERT DALE
BUTLER KENNETH DORAN
BUTTON MONTY DUWAYNE
BUXTON DELOS RICHARD
BYSTEDT DAVID JOHN
CADWALLADER PATRICK A
CALDWELL JAMES BRUCE
CALKINS VIRGIL ALLEN JR
CALLOWAY RONALD DUANE
CAMPBELL FRANCIS DUNCAN
CANTU FELIPE JR
CAREY JOHN DOUGLAS

CARLSON JAMES BLAIN
CARPER JOHN WILLIAM JR
CARTER BRUCE LANDON
CARTER GERALD LYNN
CARTWRIGHT JAMES WARREN
CARTWRIGHT RONALD JOSEPH
CASEY DANNY CURTIS
CAVANAGH MICHAEL HOWARD
CAVANAUGH RICHARD FRED
CHAMBERS CHRISTOPHER LEE
CHANDLER LARRY TOM
CHATBURN THOMAS W III
CHLOUPEK MELVIN EUGENE
CLARK DAVID LEROY
CLARK DAVID LLOYD
CLARKE ROBERT WAYNE
CLEMENTS JAMES WALTER
CLIFTON KENNETH CHARLES
CLIFTON LAYNE FARELL
COATES DONALD LEROY
COATS CHARLES THOMAS
COCHRAN PAUL JEFFEREY
COCHRAN SCOTT EDWARD
CODY FRANCIS WARREN
COFFMAN DOUGLAS DAVIS
COLE JERRY JEROME
COLLINS FRANCIS LEO
COLLINS MICHAEL STEPHEN
COLLINS ROBERT KNAPP
COLQUHOUN TED D
CONDIT DOUGLAS CRAIG
COOK DOUGLAS ALEX
COOK LEWIS COLLIN
COOK MELVIN BRUCE
COOPER DANIEL DEAN
COOPER DAVID LAWRENCE
COUCH MICHAEL ALFRED
COULTER ROBERT LLOYD
COX JAMES WILLIAM
CRABTREE MICHAEL ANDREW
CRAIGMYLE FLOYD JOSEPH
CRAM ROY ROBERT
CRAMER HENRY LEE
CRANE WILLIAM RANDALL
CRARY MORRELL JOSEPH
CRAUN GALE EUGENE
CRAWFORD TERRY LEE
CRAWN RONALD MARCEL
CROCKER DAVID STEPHEN
CUMMINGS DAVID NEWTON
CURRY ROY JERRY

DAFFRON JIMMY SHERMAN
DAFLER DEAN BLAIN
DAHL LARRY GILBERT
DALTON DONALD EVERETT
DARNALL DON EDWARD
DAVIS KELLY RAY
DAVIS WILLIAM WESLEY
DE BUTTS DANIEL FRENCHY
DE SULLY MAX FRANCIS JR
DEATON CHARLES THOMAS
DEDMORE GERALD GLEN JR
DELANO JIMMY LYNN
DEXTER BENNIE LEE
DICKSON JIM LEE
DIETZ LEWIS RAY
DIKEMAN LARRY ERNEST
DILLON JACK HOWARD
DIMICK HARLEY DANIEL
DIMMITT FRANK ROBERT
DIXON DAVID LLOYD
DOBY HERB
DOOLEY RICHARD LEE
DOOLITTLE RONALD LEE
DORMAN GEORGE STANTON
DOWLING CLIFFORD FRANKLIN
DOWNS EDWIN ALFAY
DREW KENNETH LEE
DUNCAN EDWARD FRANCIS
DUNCAN LLOYD ALVAN
DUPONT KENNETH FRANCIS
DYRESON DONALD LEE
EAGLESON ROBERT WILLIAM
EBBS RALPH ELDON
ECHANIS JOSEPH YGNACIO
ECKLEY WAYNE ALVIN
EDGE DENNIS EUGENE
ELFORD GARY EUGENE
ELLEFSON DAVID JOHN
ELLIOTT ARTHUR FLOYD
ELLIS ALDWIN ARDEAN JR
ELLIS STEVEN JOHN
ELSTEN WILLIAM JAMES
ELZINGA RICHARD GENE
EMERY DONALD CARLTON
ERWIN ARTHUR ALBERT
EVANS CURTIS NEIL
EVANS DAVID LYNN
EVANS NORMAN FRANCIS
EVENHUS GERALD WALLACE
FARMER THOMAS HOYT
FAULKNER ARNOLD JOE
FELLOWS ROBERT DAWYNE
FERGUSON WAYNE ARDELL

FILLMAN WALTER CHARLES SR
FINCH TERRY DEAN
FINDLAY ROBERT BRUCE
FISHER DONALD ELLIS
FISHLEIGH ROBERT JUNIOR
FLEMING JOHN FREDERICK
FLETCHER BRUCE JAMES
FLIEGER HAROLD NORMAN
FLOREN JIMMY ERIK
FLORES FIDENCIO JR
FOSTER JAMES WILBUR
FOSTER LARRY RAY
FOUTZ MICHAEL GEORGE
FRAY EARL RICHARD
FREEMAN WALTER DAVID
FRIEND GARY RALPH
FULLER ROBERT GENE
FULTON JOHNNY LEE
GABRIEL JOEL LYNN
GARDNER JAMES LEE
GARNER LYDES JAY JR
GASSNER LARRY MICHAEL
GATLIFF LARRY ALLEN
GAUCHE DAVID PETER
GEORGE EDWARD LEE
GESSEL GREG LOW
GHEER JAMES MELVIN
GIBSON THOMAS MICHAEL
GIFFORD LARRY HOWARD
GILLASPIE LARRY ALAN
GILLERAN DENNIS MICHAEL
GILLILAND GEORGE ERNEST
GLECKLER STEPHEN DONALD
GOEDEN GENE WILLIAM
GOLZ JAMES ROLAND
GONSALVES GEORGE GREART
GOODWIN JACK LEROY
GORHAM MARC CHARLES
GOUGH STANFORD MORRIS
GRATEN FREDERICK DUNHAM
GRAY CARLTON COE
GRAY RICHARD PAUL
GRAY TERRY ADAM
GREEF LAWRENCE DAY
GREELEY MICHAEL JEFFERSON
GREEN CARL JOHN JR
GREEN CRAIG STEVEN
GREENFIELD GUY EMERY
GROGAN KEVIN DOUGLAS
GULLETT KENNETH RAY
GULLIXSON RICHARD OWEN
GUMP TERRY LEWIS
HADDOCK ARTHUR
HAMES HENRY MC NEAL T JR
HAMILL WRIGHT BARTWYN
HANDLEY TERENCE ARNOLD
HANDY EDWARD LA MONT
HANN GAROLD ARTHUR
HANSEN JOHN CURRIE
HANSON EDWARD MAYNARD
HARDEN DONALD LEWIS
HARDENBROOK ROBERT HOMER
HARR GERRY ARTHUR
HARRIS DENNIS DAY
HARSCH RODNEY CECIL
HART DANIEL LESTER
HASLINGER PAUL MICHAEL
HASSENGER ARDEN KEITH
HASSLER HARVEY JOE
HATCHER CHARLES LAVERNE
HAUSHERR CHARLES RAYMOND
HAYES DWIGHT
HEBERT MELVIN DERWARD JR
HECKMAN LAWRENCE EUGENE
HEINTZ HERBERT CHARLES
HEINTZ RONALD ARTHUR
HENDERSON ROGER LEO
HENJYOJI GRANT HIROAKI
HESS JAMES DONALD
HILL JOE LAWRENCE

Oregon

HILL RONALD JAMES
HILLS EARL H
HIUKKA GERALD ALLEN
HOAGE GERALD CURTIS
HOCK STEPHEN LOUIS
HOFF MICHAEL GEORGE
HOLCOMB JOHN NOBLE
HOLLADAY GEORGE ALFRED
HOLMAN RICHARD JEROME
HOLMES JOHN LEE
HOPPER GERALD LEE
HORNBACK RICHARD JERRY
HORNER MICHAEL MERVIN
HOSKINSON ROBERT EUGENE
HOYEZ JAMES KENNETH
HUDELSON JAMES E
HUFFMAN GERALD DON
HUGHES THOMAS STEVEN
HURSE KENNETH CHARLES
IANNETTA LARRY ALBERT
IMLAH JACK SELWYN
IRELAN DANIEL ALBERT
IRWIN WILLIAM EDWARD
IYNDELLIN EDWARD ALLEN
JANIGIAN RICHARD ALLEN
JENSEN NORMAN A
JENSEN WILLIAM NORMAN JR
JEREMIAH RANDALL CRAIG
JOHNSON CHARLES EUGENE
JOHNSON CLIFFORD THOMAS
JOHNSON DAVID ALLEN
JOHNSON EDWARD HARVEY
JOHNSON GARY RAY
JOHNSON JAY DEAN
JOHNSON RONALD HAROLD
JOHNSON WILLIAM MELVIN
JOHNSTON EDWARD CHARLES
JONES CECIL LEE
JONES KENNETH HAROLD
JONES MARLIN MARK JR
JONES THOMAS JAKE
JUDY DAVID LEROY
KAMPH MICHAEL CLYDE
KARR GEORGE GEOFFREY
KEKEL JERRY EDWARD
KELLEY PATRICK GENE
KELLY JOE DUSTIN
KEMPKE SANTFORD BERNARD
KESTLER GARY LYLE
KILGORE DANNY RAY
KING GEORGE LOUIS JR
KINKADE WILLIAM LOUIS
KINYON RODNEY EDWIN
KLAWITTER WILLIAM RICH-
ARD
KLEIN GLEN CHARLES
KLEINSMITH ROBERT LLOYD
KLINDT DAN THOMAS
KLINEFELTER GAYLORD
NATHA
KNIGHT LARRY DALE
KOHO WILLIAM HARMON
KOLB CALVIN WILLIAM
KORSMYER GARY ROBERT
KREBS KENNETH MARTIN
KROLL ERNEST NICK
KURTTI STEPHEN WILLIAM
LA FLEMME DELBERT
CHARLES
LA GRAND WILLIAM JOHN
LADD GARY MELVIN
LAMB GARY GRANT
LANE AUSTIN CLIFFORD
LANGMAN WILLIAM JAMES
LANGSTON MARK MITCHELL
LARSON LAWRENCE DONALD
LARSON WILLIAM FRANCIS
LAWSON MICHAEL LESTER
LAYTON JON WALTER III

LE CLERC PERRY ANDRE
LEAMEN ROBERT EDWARD
LEE JOHN RAYMOND
LENTZ DAVID BURNETT
LESSEG JAMES ALFRED
LEWIS ARTHUR EUGENE
LEWIS DAVID MARION
LIDDYCOAT WILLIAM ROW-
LAND
LINDLAND DONALD FRED-
RICK
LINDSEY LARRY ALAN
LINVILLE ROBERT EDISON
LITTLE PETER CLARK
LLOYD DANIEL EDWARD
LOCHNER VERNE ELDON
LOCKE WILLIAM EDWARD
LOEW DAVID WILLIAM
LONGANECKER RONALD LEE
LOOMIS RICKIE ALLAN
LOUVRING CARL FREDRICK
LOVE CLYDE CURTIS
LOVEGREN DAVID EUGENE
LOVETT DONALD WALTER
LOVETT RONNIE RAY
LOWERY JAMES ALLEN
LOWRY TYRRELL GORDON
LUKER RICKIE
MAC DONALD JOHN ALAN
MACHAU JOHNIE BOYD
MALLON RICHARD JOSEPH
MAMBRETTI DANIEL IRVIN
MANELA RANDALL PAUL
MANGAT FREDRICK CHARLES
MARCUM LEONARD GERALD
MARTIN CHARLES LEROY
MARTIN JAMES EMMETT
MARTINI GARY WAYNE
MATHES EDWARD ARTHUR
MATHEWS FRANK JAMES
MATHEWS GROVER C JR
MATSON HAROLD EUGENE
MAYNE STEPHEN WOOD-
THORPE
MC CLAFLIN ROBERT F
MC CLELLAN BRUCE MAYO
MC CLELLAN PAUL TRUMAN
JR
MC CLURE DWAYNE CHARLES
MC DONALD JERRY DUANE
MC GINN EDWARD CHARLES
MC MULLEN LYMAN ALLISTER
MC NINCH PHILIP AARON
MC STRAVICK RICHARD P JR
MC WHORTER JAMES ELMER
MCKIBBAN MICHAEL JAMES
MEADE DAVID ERNEST

MEEKER TIMOTHY JAMES
MEIER TERRANCE LEO
MENZIES CLIFFORD LEROY JR
METCALF EDWARD WALTER
MICHAELIS JOHN DAVID
MICKELSON DENNIS ERWIN
MILBERGER RUSSELL DALE
MILLER DONALD GENE
MILLER RICHARD THOMAS
MINER STEVEN JAY
MITCHELL TOMMIE LEE
MOGCK DARYL MILTON
MOLLENHOUR ROBERT
CARROL
MONTAG LEE EDWARD
MOORE CHARLES LARRY
MOORE LONNIE DEAN
MOORE PAUL MARTIN
MOORE RONALD STANLEY
MORELAND TERRY LEE
MORELOCK REX DEWEY JR
MORGAN DAVID ALLEN
MORRIS ROBERT J
MORROW JOSEPH EDWARD JR
MOUNTS BOBBIE JOE
MUIR WILLIAM GUY
MUNDHENKE DOUGLAS O
MURPHY JOHN WILLIAM
MURPHY PATRICK EDWARD
MURPHY WAYNE STEPHEN
MUSGROVE JOHN DAVID
MUTH JAMES RAY
NEAL RONALD FORREST
NEASHAM ROBERT DEAN
NEHER ROBERT WILLIAM
NELSEN DEAN RICHARD
NELSON BRADLEY ALBERT
NELSON GRADY RAY
NEWPORT GARY LEE
NEWTON WARREN EMERY
NEWTON WILLIAM WALLACE
NOPP ROBERT GRAHAM
O BRIEN JOHN HENRY
O BRIEN MICHAEL STEVEN
OBERG WILLIAM ARTHUR
OKAMOTO ROGER THOMAS
OLIVER MICHAEL DEE
OLSON GROVER KONARD
ONCHI CURTIS
ORFIELD DAVID CHARLES
ORTIS JOSEPH RAYE
OSBORNE AMOS ROY
OSBORNE DONALD RAY
OTT LARRY FREEMAN
OWNBEY TIMOTHY ROBERT
PAGE JIM CAREY
PARKER JAMES DALTON

PARTSAFAS TERRYL GLENN
PATRICK REESE MICHAEL
PATTERSON BRUCE MERLE
PATTERSON DONALD LEE
PEARSON RODNEY SHAYNE
PEARSON WILLIAM DELBERT
PECK DARRELL VERNON
PERKINS ALLEN DEAN
PERKINS DALE ALLEN
PERKINS KEITH CHARLES
PETERS RODNEY WALTER
PETERS STEVEN LLOYD
PETERSON ROBERT LEE
PHARES KENNETH DUANE
PHILLIPS ROBERT JAMES
PHILLIPS WALTER MACK
PIERCE ALLEN LINN
PIERCE MERRICK ROBERT
PITNER MONTE GALE
PLAEP ALFRED EDGAR JR
POCHEL GERALD DEVER
POPP DAVID FRED
PORTERFIELD CHARLES
WILBU
POST KARL WALTER
PRATHER RONALD ROBERT JR
PRENTICE GARY GALE
PRYOR ROBERT EDWIN
PYLE JESSE ANDREW
RADER REX EARL
RASMUSSEN JON SIDNEY
RAVA HENRY TONY
REBER KENNETH NEAL
REED TED QUINTON JR
REINECKE WAYNE CONRAD
RHOADES EUGENE BRUCE
RHODES GARY ARTHUR
RICHTER DALE RAY
RICKER WILLIAM ERNEST
RIDENOUR WILLIAM ALBERT
RIPPY TERRY ALLEN
RITZAU AUGUST KARL
ROBERTS CHARLES LEROY
ROBERTS DAVID WILLIAM
ROBERTS HAROLD JAMES JR
ROESNER DANA HYLAND
ROGERS DEAN FRANCIS
ROGERS SCOTT CAMERON
ROHDE LLOYD HANS
ROLAND LARUS WAYNE
ROMANSKI JAMES HENRY
ROSE ROBERT FRANCIS
ROSS DAVID LYLE
ROWDEN JAMES HERBERT
ROWDEN JOHN WAYNE
RUSSELL DAVID GORDON
SALISBURY JAMES RUSSELL

SANDEFUR BILLIE E
SCHAFER DONALD RAYMOND
SCHAUERMANN ARTHUR
GARRY
SCHNEIDER DENNIS PATRICK
SCHON JOHN EDWARD
SCHRIVER THOMAS CLYDE
SCHROCK VERNON EARL
SCHULTZ JACK ELSWORTH
SCHUMACHER ROBERT JAMES
SCHWAB RICHARD MICHAEL
SECRIST FRED JASON
SEELIG GERD FRANZ
SHADBURNE BROOKE MCKAY
SHARP TED LEROY
SHARPE ROBERT ERNEST
SHEER PAUL ARTHUR
SHELTON LARRY DEAN
SHEPHERD PETER MERRILL
SHERRY THOMAS
SHIPLEY ROGER WILLIAM
SHIPP KEITH LEROY
SHIVELY DENNIS CARL
SHORACK THEODORE JAMES
JR
SHRIVER ROBERT S JR
SIEBEN THOMAS RICHARD
SILLS DAREL LEE
SILVER EDWARD DEAN
SIMMONS GLENN HAROLD
SIMONSEN RICHARD HAROLD
SLANE RONALD ALLEN
SMITH CHARLES ERNEST
SMITH DAVID LEE
SMITH EARL FREDERICK
SMITH EVERETT HAROLD JR
SMITH FERROL SHANE
SMITH GARY WAYNE
SMITH HALLIE WILLIAM
SMITH MARVIN R
SMITH PHILIP EDWIN JR
SMITH STEVEN ADRIAN
SNOW CHARLES HARRY
SODERSTROM WILLIAM E
SOLLERS FRANCIS CRAIG
SORENSEN DALE EDWARD
SPEARE WALTER RICHARD III
SPEARMAN GORDON KEITH
JR
SPENCER DANIEL EUGENE JR
SPERB WILLIAM LYLE
SPOHN KENNETH RAYMOND
SPRINKLE VERNON PATRICK
SPROUL ROBERT LEE
STAMAN TERRY LA VERN
STAMM ERNEST ALBERT
STANLEY ROBERT WILLIAM

STAPLES GREGORY JOE
STEARNS JERRY SHELDON
STEELE DANIEL SCOTT
STEELE ROBERT CHARLES
STENBERG JERRY OSCAR
STEPHENSON WILLIAM WILLAR
STEVENS MICHAEL DAVID
STINNETT RICHARD DOIL
STOCKTON DON EUGENE JR
STORM EDWARD REYNOLD
STRATTON MILO HERSEY
STRAUSER JOHN CHARLES
STRAWN JOHN THOMAS
STROHM TIMOTHY LAWRENCE
STUART EDWARD HAROLD
STUART JAMES HENRY
STYLES DAVID IRA
SUDBOROUGH MICHAEL G
SUMERLIN TERRY LEE
SUMICH FRANKLIN JOHN
SUNDEEN TERRY ALLAN
SUTTON LAWRENCE EDWIN
SWIFT DERALD DEAN
TABABOO DANIEL JOHN JR
TASHNER WALTER A

TATE LYLE SCOTT
TAYLOR JACK CLINTON
TAYLOR ROBERT WAYNE
TAYLOR STEVEN EARL
TEBAULT BENJAMIN LEE
TEETOR JOHN HAROLD
TEMPLE KIRK IRWIN
TERRELL GORDON LEE
THOMAS KENNETH LEON
THOMPSON JIMMIE MALCOLM
THOMPSON LELAND HERBERT
THOMPSON ROBERT BRUCE
THOMPSON ROBERT NOEL
THOMPSON TERRY LEE
TIMMERMAN PETER STEVEN
TOLBERT DALE WILLIAM
TORRES DAVID
TOVEY DONALD LEE
TRASTER RICHARD EUGENE
TROUT MICHAEL RICHARD
TROXEL EDWIN NEWTON
TRUSSELL LARRY HUGH
TUBB JAMES CALVIN JR
TURNER MICHAEL GLENN
TWOREK JOHN RENFIELD

UDELL EDGAR JOHN
ULLBERG VICTOR VANCE
ULM DOUGLAS RAYMOND
VAN AVERY RONALD FRANCIS
VAN POLL HUBERT CLARENCE
VANCE THEODORE ROOSEVELT
VAUGHN HERBERT LEE
VEST ROBERT LEE
VIESTENZ KREG ARTHUR
VILLANUEVA FELIPE
WABSCHALL ARCHIE CARL III
WADE GERALD ROBERT
WAGNER CLIFTON FRED
WALDORF ARTHUR LOUIS
WALKER LLOYD FRANCIS
WALKER THOMAS TAYLOR
WALLACE JOHN EDWARD
WALSH THOMAS CHARLES
WARE JOHN ALAN
WATROUS ROBERT ROLAND
WATSON GEORGE WILLIAM
WAYT SCOTT WILLIAM
WEBB FREDERIC PEERS
WEBB MICHAEL RAY
WEBB MICHAEL WILLIAM
WEEKS MICHAEL DOUGLAS

WELLS ALLEN GLAINE
WELLS JERRY DAN
WEST GARFIELD JR
WHEELER JOHN CLARK
WHEELER LARRY JAY
WHISNAN JAMES CARL
WHITAKER JOSEPH LEON JR
WHITE LOREN DOUGLAS
WHITMORE WILLIAM LEE
WHITNEY DICK EDWARD
WHITTON EDWARD JAMES
WIBBENS JOHN EDWARD
WILKINS MICHAEL LEE
WILLIAMS FRED JOSEPH JR
WILLIAMS RALPH GENE
WILLIAMS ROBERT A JR
WILLIAMS STEVEN JAMES
WILLIAMS TERRY ALLEN
WILLIAMS THOMAS HENRY
WILLOUGHBY JIMMY STEWART
WILSON BRIAN LYLE
WILSON JOHN LANNING JR
WILSON LAWRENCE HUMES JR
WILSON MICHAEL LANCE
WILSON WILLIAM LARRY

WINFREY AUTHRAN WAYNE
WINTER GARY JAMES
WIRTH GORDON LEE JR
WISDOM SELWIN DEROY
WISEMAN LANE WAYNE
WOLFE JOSEPH GEORGE
WOOD CHARLES VICTOR
WOOD DARRELL GEORGE JR
WOOD JOE IRVIN
WOOD STRATHER FRANKLIN
WOODEN VICTOR ROBERT
WOODROFFE TERRY SCOTT
WOODS GERALD ERNEST
WOODS JAMES ROBERT
WOODS PATRICK LEONARD
WRIGHT GEORGE NATHAN
WRIGHT JAMES ALFRED
YABES MAXIMO
YOUNG GARY NORMAN
YOUNG JOHN EDWARD
YOUNGMAN PAUL ARNOLD
ZIMMERLE GORDON LEE

Pennsylvania

Photos by Scott Warren

Philadelphia Vietnam Veterans Memorial

Spruce St, Philadelphia, PA 19147

The Philadelphia Vietnam Veterans Memorial came about to honor the local fallen heroes in the same fashion that all of the fallen soldiers were recognized through the national memorial in Washington DC. Veterans in Philadelphia wanted to create their own monument to their lost brothers and sisters in order to give them the thanks that they didn't properly receive when the Vietnam War finally came to an end.

Dedicated originally on October 26, 1987, the Philadelphia Vietnam Veterans Memorial had its fair share of ups and downs. The local Vietnam veterans began their crusade for their memorial in the spring of 1984. They were able to get a city ordinance in order to establish the Philadelphia Vietnam Veterans Memorial Fund and picked a site for the memorial before launching a design competition. In November of 1985, a committee picked a design submitted by a young architect named Perry M. Morgan who worked for Sullivan Associates of Philadelphia. The vets then had to undertake a large fundraising effort. They threw gold tournaments, picnics, and similar fundraising events in order to get donations for the creation of the memorial. They successfully raised $1.2 million and the memorial was finally able to get underway. The project was finally completed in October of 1987 when it was dedicated. The memorial started out with just over 600 names of fallen local service members on it, but over the years other names were added bringing the total up to 646.

The memorial is managed by the Philadelphia Vietnam Veterans Memorial Fund. Over the years, the memorial has suffered weather damage, normal wear and tear, and sadly vandalism. Funds were raised to fix the damages done by the vandalism, but it sadly continued after the 2007 restoration. Measures continue to be taken to up the security of the site and keep the memorial in top shape.

The Philadelphia Vietnam Veterans Memorial consists of 10 granite panels that are etched with scenes from the battlefield in a timeline sequence while the south wall holds the names of the local service members who lost their lives overseas. The memorial consists of four levels that are complete with granite ledges with the names of local veterans, too. The memorial plays host to remembrance events throughout the year including Memorial and Veterans day.

IT IS OUR DUTY TO REMEMBER.....

The Names of those from the state of Pennsylvania Who Made the Ultimate Sacrifice

AARON THOMAS MILTON JR
ABBOTT DENIS EUGENE
ABBOTT RAYMOND LAW-
RENCE
ABEY GEORGE WAYNE
ABRAHAM JAMES JOSEPH
ACALOTTO ROBERT JOSEPH
ADAM BARRY L
ADAMOLI ROWLAND JOSEPH
ADAMS DENNIS MICHAEL
ADAMS LEE SCOTT
ADAMS ROBERT LEE JR
ADAMS ROBERT LELAND
ADDIS FRANCIS RAY
ADDISON JOHN EDWARD
AFFLERBACH MARK
AHERN RAYMOND JOSEPH JR
AHLUM WILLIAM JOHN
AIGELDINGER ELDRIDGE
CHAR
AIKEY TIMOTHY WAYNE
ALAIMO JOHN CHARLES
ALBERTSON BERNARD
GEORGE
ALBRECHT GEORGE HENRY
ALBRIGHT WALTER LEROY
ALESHIRE KENNETH EDWARD
ALEXANDER DAVID LEE
ALEXANDER RICHARD CARL
ALEXANDER ROBERT SAMUEL
ALINCIC RONALD ELI
ALLARD PAUL EDWARD
ALLEN ANTHONY
ALLEN JAMES JOSEPH
ALLEN JOHN BAXTER
ALLEN RICHARD C
ALLEN ROY
ALLEN WILLIAM EUGENE
ALLESSIE JOSEPH
ALLUM DANIEL E
ALMASY ROBERT
ALMONEY JOHN STANLEY
ALTHOFF RODNEY EUGENE
ALTHOUSE EARL IRVIN
ALWINE RAY ERNEST
AMANTEA SAMUEL DONALD
AMBROGI ALLEN ROBERT
AMOS JOE
ANDERSON GEORGE JOHN
ANDERSON JAMES
ANDERSON JAMES ALBERT
ANDERSON JAMES GERALD
ANDERSON JOHN HARRISON J
ANDERSON WILLIAM EDGAR J
ANDERSON WILLIAM LEE
ANDERTON SAMUEL LEE
ANDRE DOUGLAS VERNON
ANDREW DENNIS RICHARD

ANDREWS DENNIS DEE
ANELI JOHN ROBERT
ANELLO BRUCE FRANCIS
ANGERT PAUL EDWARD
ANGSTADT RALPH HAROLD
ANTONACE JOHN JR
ANTONELLI JOSEPH PAUL
ANTONELLY CHARLES JOSEPH
AREY WILLIAM NOVAK
ARNOVITZ RICHARD MI-
CHAEL
ASH RONNIE EDWARDS
ASHERMAN ALDON MACKAY
ASHLOCK CARLOS
ASHMAN JOHN FREDERICK
ATHERTON FRANK WILLIAM
ATKINSON HOWARD
ATKUCUNAS EDWARD
AUEN DAVID OLIVER
AUFIERE ARMAND JAMES
AULT DANIEL LEE
AUMILLER AARON BUCKLEY
AYERS DENNIS MICHAEL
AYERS GEORGE BERNARD
BAAL CARL THOMAS
BABINSACK JOHN DAVID
BABULA ROBERT LEO
BACH LAWRENCE EDWARD
BACHMAN ALBERT CARL JR
BAER GLENN CHARLES
BAGENSTOSE TOM JAY
BAGGS WILLIAM F JR
BAGSHAW JAMES MALCOLM
BAHL ROBERT FRANCIS JR
BAILEY JAMES ALBERT
BAILEY THOMAS EARL
BAKER BERNARD GERALD
BAKER DAVID
BAKER DAVID RICHARD
BAKER DENNIS RALPH
BAKER DONALD
BAKER JACK AMOS
BAKER JERALD LAVERN
BAKER JOSEPH WILLIAM
BAKER JOSEPH WRIGHT
BAKER RONALD BOYSEN
BAKEWELL RONALD CHARLES
BALDAUF FREDERICK WIL-
LIAM
BALDINO FRANCIS
BALITCHIK MICHAEL JOSEPH
BALKIT DONALD
BALLOU CHARLES DAVISON
BALMER WAYNE ASHLEY
BALUKONIS RICHARD
CHARLES
BALZARINI DAVID RAYMOND
BANCROFT PHILIP SEAN

BANKS ROBERT LEE
BARBARINO ANTHONY
ADAMS
BARBER DAVID LEON
BARGER GEORGE HAYES
BARGER KENNETH ALLEN
BARKER RAY MILTON
BARKLEY EARL DUANE
BARKLEY KENNETH RAY
BARNER LARRY KENNETH
BARNES CHARLES RONALD
BARNES DAVID THOMAS
BARNES DONALD JOSEPH
BARNES MICHAEL ALLEN
BARNES WALTER EDWARD
BARNETT JOHN DANIEL JR
BARNETT SAMUEL HOYT
BARNHART ROGER ALAN
BARONOWSKI MICHAEL
ALEXAN
BAROTT WILLIAM CHAUNCEY
BARR ELMER EDWARD
BARR ROBERT CHARLES
BARR WILLIAM JAMES
BARRETT DREW JAMES III
BARRETT JAMES ALLEN
BARRICK HAROLD EUGENE
BARRON JOHN ELDREW
BARTASCH WALTER
BARTHOL JEFFREY CLAYTON
BARTHOLOMEW DAVID
RUSSELL
BARTHOLOMEW HARRY
ROBERT
BARTHOLOMEW WILLIAM
H JR
BARTOCK DAVID
BASEHORE HAROLD EDWARD
BATTISTA ANTHONY JOSEPH
BAUER KAROL RAYMOND
BAUGH FRED OTIS JR
BAUN DAVID ELROY
BEADLE HARRY JOSEPH JR
BEAM ROGER LEROY
BEAN DAVID ELTON
BEANNER ROBERT RANDOLPH
BEARDSLEY JEFFREY RAN-
DOLP
BEASLEY MICHAEL LAW-
RENCE
BEATTY FREDERICK LEE
BEATTY JERRY ALLEN
BEAUMONT WARREN MARTIN
BECK CARL GARY
BECK JAMES ROBERT
BECK JOSEPH ROBERT JR
BECK RICHARD JAMES
BECK TERRY LEE

BECKER THOMAS ALEXAN-
DER
BEDNAR STEPHEN ANDREW
BEEBE LARRY CHARLES
BEECH HARRY DAVID JR
BEERS EDWARD NELSON
BEHAN WILLIAM GERALD
BELANCIN GEORGE JOHN
BELARSKI RONALD DALE
BELL CHARLES ARTHUR
BELL RICHARD WILLIAM
BELL WILLIAM
BELLETTI ANTHONY JOHN
BELSAR KENNETH RAY
BEM WALTER PAUL
BENEDIK NORMAN FLORIAN
BENIGNI ALFREDO
BENNETT DONALD CHARLES
JR
BENNETT JOSEPH RICHARD
BENNING WILLIAM DONAVAN
BENSE JOHN FREDERICK JR
BENSON ARNOLD JR
BERG GERALD LEROY
BERGER ELDIN GEORGE JR
BERKEBILE JACK
BERNESKI LAWRENCE AU-
GUSTI
BERNSTEIN JOEL
BERRY FLOYD JOSEPH JR
BEVICH GEORGE MICHAEL JR
BEYRAND JOHN MICHAEL
BEZENSKI STEVEN MICHAEL
BIANCONI NICHOLAS
CHARLES
BIDWELL BARRY ALAN
BIELEK RUDOLPH JOHN JR
BIERLINE THOMAS RALPH
BIGLEY CHRISTOPHER JOHN
BILBO WILLIAM JOHN JR
BILKO TIMOTHY JAMES
BILLINGS DAVID VERN
BINGHAM MICHAEL FRANCIS
BIRCHAK FRANCIS JOSEPH
BIRD THOMAS ARNOLD JR
BIRELEY KENNETH PAUL
BISCHOFF EDWARD ALLEN
BISH LEONARD THOMAS
BITTING JACK
BIXLER MARTIN EDWARD
BLACK ROBERT JACOB
BLACK ROBERT JAMES JR
BLACKMON WILLIAM B JR
BLACKWELL KENNETH
BLADEK JOHN EMERY
BLANCHETT STEPHEN PAUL
BLANCO CHARLES JOSEPH
BLANK FRANK HUFFORD

BLESSING LYNN
BLEVINS LURAL LEE III
BLEWITT WILLIAM A JR
BLICKENSTAFF JOSEPH W JR
BLISS THOMAS ROBERT
BLOOM DARL RUSSELL
BLOOM RONALD NORMAN
BLOSCHICHAK JOHN ROD-
MAN
BLOTZER EDWARD JOSEPH
BOBULA JEFFREY LOUIS
BODISH JAMES ROBERT
BOLGER LAWRENCE JOSEPH
BOLICH KENNETH CHARLES
BOLLINGER NEAL GEORGE
BONDROWSKI DARREL AN-
THONY
BONGARTZ CHARLES JOSEPH
BONNEY ALAN WAYNE
BOOKS JAY KARL
BOORMAN JAMES EDWARD
BOOTH JOHN ROBERT
BOOTH TERRY LYLE
BOOTH WILLIAM DOUGLAS
BOOTHE BAY BENTON
BORICK JOSEPH JAMES
BORRELL CLIFFORD GLENN
BOSWELL BRADLEY LLOYD
BOTTESCH JOHN RICHARD
BOUCHER ROBERT CHARLES
BOWDREN JAMES IGNATIUS JR
BOWER IRVIN LESTER JR
BOWERS DONALD IRA
BOWERS JAMES DALE
BOWERS JEROME EDWARD JR
BOWERS WILLIAM JAMES
BOWERSOX CHARLES W JR
BOWMAN JOHN DAVID III
BOWMAN JOSEPH MICHAEL
BOWMAN LESTER ELWOOD JR
BOWMAN ROBERT WAYNE
BOWMAN WILLIS SHEPHERD
JR
BOYANOWSKI JOHN GORDON
BOYD RICHARD KLEMM JR
BOYER DAVID EUGENE
BOYER JAMES IRVIN
BOYER LARRY EUGENE
BOYER STEVEN HESS
BOYLE JAMES ROBERT
BOYLE WILLIAM
BRACKEN ALAN LEE
BRADFORD LEONARD ED-
WARD
BRADLEY JOHN ALLAN
BRADY DANIEL WILLIAM
BRANIGAN LAWRENCE AN-
THONY
BRANTNER WAYNE EUGENE
BRAXTON JOHN ALAN
BRAYBOY BRYANT JR
BRAZEN HAROLD JOSEPH
BREECE WILLIAM WARREN JR
BREEDEN WILLIAM RAYMOND
BREEN DAVID THOMAS
BREIGHTMYER WILLIAM
DENNI
BRENKER ECKHARD GER-
HARD
BRENNAN CHARLES EUGENE
BRENNAN GARY O
BRENNAN JAMES FRANCIS JR
BRENNAN JOHN FRANCIS
BRENNEMAN RICHARD
EUGENE
BRENT LARRY THOMAS
BRESKI JOSEPH JR
BRESLIN PATRICK JOHN
BREWER THOMAS RICHARD
BRICE ROBERT KENNETH
BRIDGE JOSEPH LEONARD
BRIDGEFORD WILLIAM
MICHAE
BRIGGS RONALD DANIEL
BRILLO ALBERT JR
BRINZO ANDREW JOSEPH III

Pennsylvania

BRISKIN ROGER STEVEN
BRISUDA STEPHEN CHARLES
BRIZZOLI LOUIS EMIDIO
BROCHETTI FRANK THOMAS
BROCKMAN PHILLIP LLOYD
BROMLEY ALBERT LEROY
BROMS EDWARD JAMES JR
BRONAKOSKI JAMES DENNIS
BROOKINS FREDDIE
BROOKINS ZACKRIE JR
BROOKS JAMES FRANCIS JR
BROOKS JOHN RICHARD
BROOKS JOHN WESLEY
BROOKS RICHARD W III
BROPHY PATRICK JOSEPH
BROSIUS DONALD EDWARD
BROUGHT DALE EDWARD
BROWN ALFRED LEE
BROWN DAVID LYNN
BROWN EMMETT RUBEN
BROWN FRANK MONROE JR
BROWN JAMES BRENT
BROWN JAMES FREDERICK
BROWN KENNETH WILLIAM
BROWN LAWRENCE JAMES
BROWN MARTIN
BROWN MICHAEL PAUL
BROWN RALPH WAYNE
BROWN ROBERT GUY
BROWN ROBERT LESLIE
BROWN ROGER ALLEN
BROWN ROGER THOMAS
BROWN SHERRILL VANCE
BROWN THEODORE JR
BROWN THOMAS MICHAEL
BROWN WALTER
BROWN WILLIAM JOSEPH
BROWN WILLIAM LEROY
BRUBAKER JOSEPH HAROLD
JR
BRUDER JAMES ROBERT
BRUGGEMAN DAVID CHARLES
BRULTE ROBERT FRANCIS JR
BRUNN WILLIAM EDWARD
BRUZNACK NICHOLAS
EDWARD
BRYAN HECTOR WARREN
BRYANT ROBERT ELMER
BUCCILLE RICHARD GARY
BUCHANAN ROY OTIS
BUCHER HARRY LUTHER
BUCKA WALTER HERBERT JR
BUCKLEY CHARLES JOSEPH
BUCKLEY WILLIAM ROBERT
BUCKWALTER JAY Q III
BUGAR JOSEPH EDWARD JR
BUGMAN DAVID CHARLES
BULLOCK DENNIS JOHN
BULLOCK GLEN F
BUNCH FRANCIS JOSEPH
BURD HARMON CHARLES
BURGDORFER STEPHEN
WALTER
BURIAN DENNIS WAYNE
BURINDA JOSEPH FRANK JR
BURKART CHARLES KENTON
JR
BURKE EARL FREDERICK
BURKE MARSHALL JR

BURKE ROY JEFFREY
BURKETT HAROLD ELMER
BURKETT SCOTT MC CLEL-
LAND
BURKHARDT THOMAS ALAN
BURNETTE GARY RAY
BURNITE BARRY TYSON
BURNS BERNARD JOHN JR
BURRIS DONALD DEANE JR
BURSE TYRONE GREGORY
BURTON DONALD RUSSELL
BURTON SAMUEL NURRELL
BURTON THEODORE HUGHES
BUSH FRANK KENNETH
BUSH PAUL WILLIAM
BUSINDA CHARLES ARTHUR
BUTLER ALLEN LEROY
BUTLER DOYLE LEROY JR
BUTLER WILLIAM GRANT JR
BUTTS GARY RICHARD
BUTZ CLAIR BERNARD
BUTZ ROBERT ALLEN
BUZA FREDERICK ANDREW
BYERLY JAY MARTIN
BYHAM DAN RAE
BYRNE CONAL JOSEPH JR
BYRNE JOSEPH LEON JR
CABBAGESTALK EUGENE
CABOT ANTHONY JOHN JR
CAFFARELLI CHARLES JOSEPH
CAFRELLI ALFRED BENNETT
CALLAHAN DAVID PATRICK
CALLAHAN MICHAEL JOHN
CALLAHAN RAYMOND W JR
CALLEN JAMES GRANT
CAMAROTE MANFRED
FRANCIS
CAMINO JOHN EDWARD
CAMPBELL DONALD BRUCE
CAMPBELL PERCY LEROY
CAMPBELL ROBERT JOSEPH
CAMPBELL RONALD JACOB
CAMPBELL WILLIAM ROGER
CAMPOS MICHAEL WILLIAM
CANCILLA NICHOLAS
CANDY JOHN ELTON
CANNON KEVIN GEORGE
CANNON ROBERT EARL
CANNON WILLIAM EUGENE
CAPITANI DANIEL CARL
CAPPELLO DANIEL PETER
CAPUTO MICHAEL ANTHO-
NY SR
CARA ROBERT JOSEPH
CARDIFF THOMAS N JR
CARDWELL TYREE
CAREY JAMES EDWARD
CAREY JOHN LEROY
CAREY JOHN PATRICK JOSEPH
CARN ROBERT MARION JR
CARNELL PATRICK J
CARNEY JOHN CHARLES
CARNEY THOMAS EARL
CAROTHERS CECIL WAYNE
CARPENTER WILLIAM H JR
CARR DENNIS ROBERT
CARR GERALD REID
CARR ROBERT HARDY
CARR ROBERT HOWARD JR
CARROLL DAVID
CARROLL FERGUS JOSEPH
CARROLL JAMES RICHARD
CARSON CARL LEE
CARTER GLENN
CARTER RICHARD ALBERT
CARTER WALLACE SPERGON
CARTNEY PATRICK CYRIL
CASEY DAVID WARRINGTON
CASHLEY JOHN EDWARD
CASHMAN HAROLD EDWARD
JR
CASP MICHAEL ALLEN
CASSEL RONALD ROY
CASSIDY JOSEPH J JR
CASSIN FRANK ANDREW JR
CAUCCI STEVEN RICHARD
CAUTHEN CALDWELL M JR
CAVAROCCHI JOSEPH
CAWLEY JAMES PATRICK
CERENE AMBROSE JOSEPH
CESTARIC JOSEPH ANTHONY
CHALLENER ROBERT JOSEPH
CHAMAJ ANDREW PETER
CHAMBERLAIN LESLIE ALLEN

CHAMBERS ROBERT JOHN
CHANDLER WILLIAM GARY
CHAPMAN RONALD JAMES
CHAPMAN WILLIAM JR
CHARLES FRANCIS
CHARLEY MICHAEL JOHN
CHARTERS GEORGE W JR
CHASE RAYMOND HOWARD JR
CHATBURN JOSEPH THOMAS
CHATMAN NATHANIEL
CHAVOUS ALVIN RICHARD-
SON
CHEEK KENNETH NORRIS
CHILDRESS ROBERT MORRIS
CHISKO JOSEPH JOHN
CHITTESTER NORMAN PHILIP
CHRIN JOHN STEPHEN
CHRISTMAN RONALD S H
CHRISTY ALBERT KRISUNAS
CHUBB RICHARD CHARLES
CHULCHATSCHINOW
WALERIJA
CHUTIS JOHN VINCENT
CICCHIANI WALTER ANTHO-
NY
CIESIELKA MICHAEL J JR
CIMORELLI JOHN JOSEPH JR
CLAMPFFER ROBERT LEE
CLAPPER GEAN PRESTON
CLARK REUBEN L JR
CLARK TERRY LEE
CLARK THOMAS EDWARD
CLARK THOMAS LEE JR
CLARKE EDWARD ALLEN
CLAY JAMES HENRY
CLAYBORNE MILTON GAY
CLAYCOMB CLARENCE JAMES
CLEAVER FRANCIS CRAIG
CLEVELAND DAVID LUHVER
CLEVER LOUIS JOHN
CLICKNER LEE FULTON
CLIFFORD MICHAEL JAMES
CLINGER GUY WESLEY JR
CLINGER WILLIAM C III
CLOUGH TONY
CLYDESDALE CHARLES
FREDRI
COATES STERLING KITCH-
ENER
COBARRUBIO LOUIS ANTO-
NIO
COBB ALBERT JR
COBB BRUCE ALAN
COFFEY EDWARD AUBREY
COFFIELD JOHN DAVID
COHEN CHARLES MITCHELL
COHEN LOUIS GEORGE
COLBERT JAMES HAMILTON
COLE JAMES WILLIAM
COLE JOHN MATTHEW
COLE WILLIAM NOEL
COLEMAN JOEL DANIEL
COLEMAN PETER MICHAEL
COLL WILLIAM PATRICK
COLLETTO ALBERT V JR
COLLINS EDWARD W III
COLLINS JAMES FREW
COLLINS JAMES GILBERT
COLLINS JOHN JAMES
COLLINS MICHAEL TIMOTHY
COLTMAN WILLIAM CLARE
COMBER DAVID WAYNE
COMBS DAVID JOHN
COMFORT RAY THOMAS
CONCANNON JAMES P JR
CONLEY GERALD DONALD
CONLEY RONALD CLARENCE
CONLEY WILLIAM THOMAS
CONLIN RICHARD JOSEPH
CONLON JOHN FRANCIS III
CONNELL MICHAEL JOSEPH
CONNOR FRANCIS JOSEPH
CONTI ANTHONY NOAH
CONTI ROBERT FREW
CONVERY JOSEPH FRANCIS JR
CONWAY JOSEPH QUINTON
COOK LESLIE
COOK ROBERT EMERY
COOK THOMAS RAY JR
COON JESSE JAMES
COONON DANIEL JAMES
COONS RICHARD WILLIAM
COOPER ALEXANDER
COOPER DAVID H II

COPE STANLEY SMITH JR
CORCORAN EDWARD JOSEPH
COREY WILLIAM GEORGE
CORL FRANKLIN MATTHEW
JR
CORLE JOHN THOMAS
CORNMAN CHARLES NOR-
MAN
CORNWELL LEON LAWRENCE
JR
CORRELL JOSEPH CLAIR
COSGRIFF PAUL LEONARD
COUCH JAMES ROBERT
COUCH LESLIE CRAIG
COVINGTON HOPSON
COVINGTON RORY ARN
COWART DAVID LAWRENCE
COX DAVID LEE JR
COX EDWARD JAN
COX ROBERT IVAN
COYLE GERALD A
COYLE GERARD
COYLE RICHARD DENNIS
CRAFT TOMMY LEWIS
CRAFTON JAMES J

CRAIG DAVID III
CRAIGE AMOS MARK
CRAMER HARRY GRIFFITH
CRAMER JAMES WALLACE
CRANE DONALD ELLIS
CREEP GUY BARE JR
CRESCENZ MICHAEL JOSEPH
CREW JAMES ALAN
CRISWELL JAMES JOSEPH
CRISWELL RICHARD K III
CROCKER RICHARD ANTHO-
NY
CRON JODY ALLEN
CRONE GARY LEE
CRONRATH STEVEN MARK
CROPPER RAY D
CROSS JOSEPH ALEXANDER
CROWDER RAYMOND D JR
CRYSTER JAMES PERRY III
CULLEN RICHARD IVORY
CUMMINGS RICHARD MI-
CHAEL
CUMMINGS THOMAS FRAN-
CIS
CUNEEN MICHAEL RAY

CUNNANE DENNIS THOMAS
CUNNINGHAM BRUCE EDWARD
CURRY HOVEY RICE
CURRY RICHARD JOHN
CURRY WENDELL PAUL
CURTISS EDWIN HARRY
CUSHMAN KENNETH GEORGE
CZZOWITZ THOMAS EUGENE
D AMICO PHILIP ANTHONY JR
DAHR JOHN WESLEY
DALOLA JOHN FRANCIS III
DALTON THEODORE HUBERT
DALY JOSEPH FRANCIS
DAMBECK ROBERT CARL
DANIELLES DEIGHTON ALONZO
DANIELS FRANCIS ANTHONY M
DANIELS GEORGE W JR
DANIELS JOHN FRANCIS JR
DANIELS LEROY BUDDY
DANIELSON CHARLES F JR
DANOWSKI DAVID LEE
DARLING ROBERT HARRY
DASCOMBE RONALD EUGENE
DASILVA RAYMOND M JR
DAUBERT EDWARD LYNN
DAUGHERTY JAMES ALEXANDER
DAUT CHARLES WILLIAM
DAVID GARY CHARLES
DAVID JEFFREY JAY
DAVIDHEISER STANLEY JR
DAVIDSON WILLIAM JAMES JR
DAVIS CHARLES AUGUSTUS
DAVIS GERALD EDWARD
DAVIS JOHN LARRY
DAVIS JOSEPH EDWARD JR
DAVIS MARTIN JOSEPH
DAVIS RICHARD LARRY
DAVIS ROBERT ARNOLD
DAVIS THEODORE H
DAVISON WILLIAM A JR

DAWSON DANIEL MILLARD
DAY ARTHUR MICHAEL
DAY DENNIS PATRICK
DAY EDWARD
DAY WENDELL LEWIS
DAYHOFF RALPH PAUL
DE BOARD BLAINE A JR
DE BOW EDWARD CARL
DE HART SOLOMON WILLIAM
DE MARCO PATRICK THOMAS
DE MILIO LAWRENCE
DE VOE ROBERT LEE
DE WALT VICTOR MONROE
DEAN CHARLES ROBERT JR
DEAN THOMAS JOSEPH III
DEBOLD REGIS PETER
DECKER DAVID JOHN
DEENY MICHAEL FRANCIS III
DEETER JACK EARL
DEFAZIO PHILLIP FRANK
DEGEROLAMO ANTHONY JR
DEISHER LAWRENCE JAMES
DELL KENNETH JOHN
DELLAPINA CHRISTOPHER L
DELLINGER CHARLES HILTON
DELOZIER DAVID VINCENT
DEMKO LEONARD RICHARD
DENEEN JOHN FRANKLIN JR
DENNIS PAUL LESLIE
DENNIS WILLIAM ROY
DERRICO JACK EDWARD
DETRICK DONALD GLEN
DETRIXHE JAMES B W
DETWILER LAWRENCE R JR
DEVINE FRANCIS STANLEY JR
DEVOR KENNETH LEE
DEWAR JAMES CRAIG
DEWEY JAMES ELLIOTT
DEWITT DAVID EUGENE
DEXTER VAUGHN LEROY
DI BARTOLOMEO RONALD J
DI SANTI RAYMOND JAMES
DI STEFANO RONALD MICHAEL

DIAMOND CHARLES EDWARD
DIAS RALPH ELLIS
DICK TIMOTHY MORGAN
DICK WILLIAM EDWARD JR
DICKIE GUY DOUGLAS
DICKSON EDWARD ANDREW
DIEHL DANA EDWARD
DIEHL ROBERT ERNEST
DIEMLER RICHARD LEE
DIETZ DIETER WALTER
DIFFENDERFER TERRY EUGENE
DILE STEVEN ORLANDO
DILLER JAY THOMAS
DILLMAN WAYNE THOMAS
DILLON DENNIS EARL
DILMORE JOHN HARRY
DINEEN THOMAS GERARD JR
DISSINGER GARY FRANK
DIXON JOHN T
DIXON STEPHEN DOUGLAS
DOBSON CAREY LEE
DODDY VICTOR LOUIS
DODSON WESLEY ELLSWORTH
DOERING ROBERT
DOERRMAN CHARLES ELLSWORT
DOLBY MELVIN LESTER
DOLOUGHTY JAMES CORNELIUS
DOMAN HAROLD ARTHUR
DOMER GLENN WILSON
DOMIAN EDWARD THOMAS JR
DONAGHY EDGAR STOMS
DONAHUE ROBERT WILLIAM JR
DONALD HOWARD ARTHUR
DONALDSON DONALD ROBERT
DONALDSON LAWRENCE GERARD
DONAVAN TIMOTHY CHARLES
DONICS WILLIAM CALDWELL
DONNELLY JOHN JOSEPH III
DONOHUE FRANCIS DAVID
DONOVAN DENNIS GEORGE
DORCHAK GEORGE ROBERT
DORN MICHAEL LEWIS
DORSCH RICHARD STEPHEN
DOUGHERTY THEODORE ALOYIS
DOUGHTIE RONALD EDWARD
DOUGLAS FRANK FREDERICK
DOWD THOMAS JOSEPH
DOWNEY CHARLES ROBERT JR
DOWNEY EARL GARLAND
DOYLE JOSEPH CLARENCE
DOYLE MICHAEL WILLIAM
DOYLE RAYMOND E JR
DRAGOSAVAC DAVID GEORGE
DRAK ROBERT
DRAKE DONALD JOSEPH
DRAKE GLENN FRANKLIN
DRAVIS JAMES STEVENS JR
DRAZBA CAROL ANN ELIZABET
DRENNEN NILS ARDEN
DRESSLER EMMETT L
DRIGGERS VESTIE TIMOTHY
DRIVERE RICHARD JOSEPH
DRIZA STANLEY WILLIAM
DUBBS RAYMOND ARTHUR
DUCKETT THOMAS ALFRED
DUDDY CHARLES STEVEN
DUDLEY JOHN MITCHELL
DUFFORD PAUL EDWARD
DUGAN PATRICK JAMES
DUGAN THOMAS WAYNE
DUGGER JAMES DOWEL JR
DUMOND ROLAND DENNIS
DUNGEE RUDOLPH FRANCIS
DUNKLE JAMES ROBERT
DUNLAP FRANCIS EDWARD JR
DUNLAP JOHN TURNER III
DUNLAP WILBUR TURBY
DUPELL ROBERT JOSEPH JR
DUPLESSIE ALEXANDER WILLI
DUPLESSIS RICHARD JAMES
DURANT WILLIE
DURLIN JOHN STEWART
DURRWACHTER HERMAN K JR
DURST LARRY BLAINE

DWYER PATRICK PETER
DYER JOSEPH FRANCIS JR
DYKE CHARLES EARL
DYSON CHARLES E JR
EAKER DENNIS KEITH
EAKIN HOWARD MAXWELL JR
EANS LAWRENCE GEORGE
EARNESTY JOHN WILLIAM
EAST VERNON WAYNE
EASTER DENNY RAY
EASTON DAVID STEARNS
EATON GEORGE ELWOOD
EBALD MICHAEL LEO
EBERT CHARLES DANDRIDGE
EBY EDWARD LEE
ECCARD HARRY LEE
ECKENROAD RONNIE LEE
ECKENRODE DAVID JOHN
ECKER ROBERT RAYMOND
ECKERT HAROLD LEE JR
ECKERT RONALD LEE
ECKHART LEON DELBERT
EDSALL JAMES
EDWARDS BERNARD W JR
EDWARDS JOSEPH
EDWARDS TED LAVERN
EFAW ROBERT THOMAS
EGGLESTON HARRY H
EHRHART MELVIN GRAYSON
EICHELBERGER BARRY LEE
EICHER MERLE CLAYTON JR
EISAMAN DALE LEON
EISENHART GUY LEE
EISENHOWER WILLIAM JACK
ELLIOTT LEROY
ELLIS ADOLPHUS
ELLIS RAYMOND
ELMORE HUGH WILLIAM
ELTRINGHAM WILLIAM DAVID
ENGEL MEIR
ENGELMEIER JAMES FRANCIS
ENGLISH GLENN HARRY JR
ENGLISH RUBEN
ENGROFF RICHARD CHARLES
ENOS BLAINE WILBERT JR
EPSTINE LARRY DAVID
ERKES WILLIAM JAMES JR
ESBENSEN CHARLES JOSEPH
ESHLEMAN DENNIS CHARLES
ESTOCIN MICHAEL JOHN
ETCHBERGER RICHARD LOY
EUBANK CHARLES HORTON
EVANCHO RICHARD
EVANS DONALD JERRY
EVANS GORDON EDWARD
EVANS JOHN R
EVANS JOSEPH GEORGE JR
EVANS WALTER C
EVERETT JAY LEROY
EXUM EDMUND GARDNER JR
FAHEY WILLIAM PAUL
FAIR RONALD
FALK RICHARD WILLIAM
FALLER JOEL EDWARD
FALLON PATRICK MARTIN
FALLSTICH JAMES ROLAND
FAMILIARE ANTHONY JOHN
FARNSWORTH JOHN JOSEPH JR
FARNSWORTH NEVIN OAKLEY JR
FARRINGTON HERBERT L III
FAWCETT DONALD JAMES
FAWKS ERNEST EUGENE
FAY RICHARD EUGENE
FEDOR TERRENCE EUGENE
FEEHERY RICHARD JOSEPH
FEEMAN JAMES OSCAR
FEESER JOHN RAYMOND
FEGELY TERRY GRANT
FEIT CHRISTIAN FRANZ III
FELD RAYMOND GENE
FELL CARL EUGENE
FELTON WALTER
FENNELL ROBERT HARRY
FENNER STANLEY STEWART
FENUSH THOMAS PAUL
FERGUSON ROBERT FRANCIS
FERRARO DAVID ALLEN
FERRELL MARK JR
FEY GLENN THOMAS
FIELDING WAYNE JAMES
FIGUEROA MICHAEL ANGEL

FIKE DANIEL EUGENE
FINDLAY WILLIAM THOMAS
FINDLEY ROBERT GAYLORD
FINE NORMAN ELLSWORTH JR
FINK RICHARD ELWOOD
FINN ALBERT MAURICE
FINNEGAN DAVID GARTH
FISCHER DONALD ERNEST
FISCHER JOHN RICHARD
FISCHER THEODORE LAUER
FISHER DALE CHARLES
FISHER DONALD GARTH
FISHER DUAINE KARL
FISHER EDWIN FREDERICK
FISHER JAMES ROY
FISHER THOMAS WILLIAM
FLADRY LE ROY EDWARD
FLAGIELLO RICHARD JAMES
FLAHIVE THOMAS FRANCIS
FLANDERS DANNY GEORGE
FLANNERY JAMES KENNETH
FLIZANES VAUGHN PAUL
FLOOD JOHN JOSEPH JR
FLORES DAVID
FLOWERS FLOYD TYRONE
FLUHARTY DONOVAN RUSSEL
FLYTE FORREST JAY
FOLEY ROBERT JOHN JOSEPH
FOLKS EDWARD LEROY
FONTANA ADAM ANTHONY
FONZI DONALD OLIVER
FORD HAROLD ANDREW
FORD VICTOR JAMES
FORRESTER CARL JAMES
FORSYTHE DALE RICHARD
FOSTER TONY CURTIS
FOUGHT PAUL EARL JR
FOULK PAUL FREDERICK
FOWLER LARRY LYNN
FOX GARY WAYNE
FOX JOHN WILLIAM
FOY JERRY
FRAKER ROGER LEE
FRANK TIMOTHY GEORGE
FRANKLIN PHILIP GILBERT
FRANKLIN ROBERT ORME
FRANKS WILLIAM J
FRANTZ CURTIS RUSSELL
FRANTZ LARRY EDWARD
FRANTZ WILLIAM DAVID
FRAZIER BARRY LYNN
FRAZIER KEITH EUGENE
FREELAND GEORGE EDWARD
FREEMAN EARL MARVIN JR
FREY WILLIAM AUSTIN
FREY WILLIAM JOSEPH
FRIDAY LORRENCE TEALOA
FRIEDMANN GARY WAYNE
FRITSCH ANDREW JOSEPH III
FRITZ MARTIN CHARLES
FRY JOSEPH PATTON
FRY RICHARD LEROY
FRYE LOUIS ARTHUR
FUELLHART ROBERT HOWARD
FUHRMAN JAMES MICHAEL
FUHRMAN ROBERT MICHAEL
FULLERTON FRED SAMUEL JR
FULMER NICHOLAS JOSEPH
FUNELLI RICHARD ARTHUR JR
FUNK BRUCE ELLIOTT
FYOCK TERRY LOUIS
GABLE ROBERT LEE
GAINES BERNARD LAVERNE
GAISER JAMES ALFRED
GALATA JOHN MICHAEL
GALATI JAMES FRANCIS
GALBRAITH RAYMOND CLARENC
GALKA VINCENT EDWARD
GALKOWSKI JAMES LEONARD
GALLAGHER DANIEL PATRICK
GALLAGHER JOHN JOSEPH
GALLAGHER JOSEPH THOMAS
GALLAGHER RICHARD
GALLMAN SAMUEL III
GAMBER ROBERT ALLEN
GAMBLE DAVID JOHN
GAMBLE RONALD RICHARD
GAMBLE WILLIAM H
GARCIA ANTONIO
GARDNER ALAN DAVID
GARDNER LARRY WAYNE
GARIS GARY WILLIAM
GARIS LE ROY DELANO M

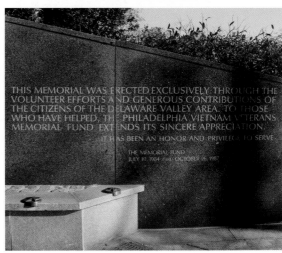

Pennsylvania

GARLAND DENNIS
GARLAND MARCELLUS JR
GARMS PETER HENRY
GARNETT REUBEN LOUIS JR
GARRETT MAURICE EDWIN JR
GARRIS GEORGE ROBERT
GARRITY EUGENE JOSEPH JR
GARRITY JAMES JEFFREY
GARSTKIEWICZ WALTER J JR
GARTLAND THOMAS
GARVEY VINCENT FRANCIS
GASTON STANLEY STEPHEN
GATES MONTE LEROY
GAUGHAN AUSTIN MICHAEL
GAULT ALAN ROBERT
GAY CURTIS TAYLOR
GAY GARY PAUL
GEARHART DONALD LEE
GEARY WILLIAM STANLEY
GEBHARD ROY ALLEN
GEHRIG JAMES MONROE JR
GEIGER GARY GEORGE
GEIGER LAWRENCE
GENDEBIEN WILLIAM RAY-

MOND
GENNOCRO ANTHONY
ANGELO
GEORGE GARY RICHARD
GEORGE MICHAEL THEO-
DORE
GEORGE STEPHEN FREDERICK
GERMEK JOHN ALAN JR
GERNERT EDWARD HARRY
GERTSCH JOHN GARY
GERTZEN FRED
GERWIG RONALD EUGENE
GEYER JAY FRANKLIN
GIACOBELLO FRANK A JR
GIAMBRONE RICHARD LAW-
RENC
GIANNANGELI ANTHONY
ROBER
GIANNINI MICHAEL ANTHO-
NY
GIBBLE ALVIN RALPH
GIBEL RAYMOND
GIBILTERRA CHARLES J JR
GIBSON CLIFFORD MICHAEL
GIBSON WILLIAM TERRY
GIEBELL FLOYD STEPHEN
GIGLIOTTI MICHAEL
GILBERT DAVID ARTHUR
GILBERT LAWRENCE JOSEPH
GILBERT PAUL ARTHUR
GILKEY TIMOTHY WILLIAM
GILLEN FREDRICK PAUL
GILLESPIE WILBERT LEE
GILLIAM RICHARD
GILLINGHAM WILLIAM DAVID
GILLIS HAROLD THOMAS
GILPIN JAY BARRY
GILPIN TERRY LEE
GINGERY JOHN BERNARD
GINTER EDWARD JOHN
GIORGIANNI JON JOSEPH
GIPE J RANDALL
GISE DONALD RICHARD
GISH HENRY GERALD
GLASGOW ROBERT LEE
GLATFELTER LARRY EUGENE
GLEASON LARRY FRED
GLEIXNER WILLIAM ALLEN
GLENN CHARLES JOSEPH III

GLENN GEORGE EARL JR
GLICKMAN DONALD ERIC
GLORIOSO JOHN ANTHONY
GLOVER RALPH LEWIS
GLOVER ROSCOE JR
GLOWIAK FRANK ANTHONY
GOBER CLARENCE JR
GOEDEKER WILLIAM HOW-
ARD
GOELZ FRANK GERALD
GOFF DALTON TRURO
GOLASZEWSKI WALTER
GOLEBIEWSKI RONALD
FRANK
GONANO JAMES MARTIN
GONTERO ROBERT CLYDE
GOOD PAUL EUGENE
GOOD RAY LYNN
GOODMAN DONNIELL
GOODMAN ROBERT JAMES
GORDON CLEVELAND WIL-
LIAM
GORMAN HENRY WILLIAM
GOSHORN WALTER L
GOSSELIN JAMES EDWARD
GOTTIER ROBERT CARL
GOUGH LINWOOD
GOZDAN MICHAEL STEPHEN
GRADEL JOSEPH
GRAESER CALVIN KYRLE JR
GRAHAM BARRY LEE
GRAHAM JAMES EVERETT JR
GRAHAM JAMES SCOTT
GRAHAM RICHARD SCOTT
GRASSER HAROLD PHILLIP
GRASSI CLEMENT JOHN
GRASSI LAWRENCE GARY
GRAY GARY GERALD
GRAY GEORGE ALBERT
GRAY GEORGE CHRISTIAN
GRAY HAROLD PAUL
GREELEY DENNIS ANTHONY
GREEN VERNON ANDREW
GREENE BRUCE BRIANT
GREENE BRUCE GREGORY
GREER LAWRENCE FREDER-
ICK
GREER WAYNE STEVENSON
GREGA GEORGE WILLIAM

GREGORY CHARLES LESTER
GREGORY FRANK EDWIN
GRETH ROBERT EUGENE
GRICE LARRY JAMES
GRIFFIN ALLAN GEORGE
GRIFFIN SAMMIE
GRIFFIS MICHAEL DANIEL
GRIFFITH LARRY DONALD
GRIMENSTEIN JOHN PAUL JR
GRINE PAUL RAY
GRINER THOMAS EUGENE
GRISAFI JOSEPH
GRIST WILLIAM ANTHONY
GROFF RONALD HOWARD
GROSICK PAUL DAVID
GROVE STANLEY COLVILLE
GRUBB WILMER NEWLIN
GRUGAN JOSEPH PATRICK
GRUMLING RONALD GARY
GUARDINO STEPHEN AN-
THONY
GUEST DOUGLAS WILLARD JR
GUEST EDWARD ROBERT
GUEST JAMES WALKER
GUILLERMIN LOUIS FULDA
GULA PAUL RICHARD
GUSEMAN WILLIAM E III
GUTEKUNST JOHN THOMAS
GUTHRIDGE JOHN HOWARD
GUY LEONARD ALLEN
GUY THOMAS EDWARD
HAAS RAY IRA
HAAS THOMAS VENCENT
HABBLETT EDWARD F JR
HABECKER GERALD LLOYD
HACKETT DAVID SPENCER
HACKETT WILLIAM CLAYTON
HAEFNER DAVID RAYMOND
HAGARA LESLIE PAUL
HAGINS GREY LYNN
HAHN LEON HENRY JR
HAINES ROBERT LEE
HAIRSTON MELVIN LEE
HALE RALPH DAVID II
HALL ALFRED FLOYD
HALL DAVID CHARLES
HALL LINDY ROLAND
HALL WILLIAM JR
HALLER LEROY CLAYTON

HALLOW DONALD WILLIAM
HAM GEOFFREY LAWRENCE
HAMILTON FOSTER
HAMILTON GEORGE BARKER
HAMILTON GEORGE W JR
HAMILTON JOSEPH THOMAS
HAMILTON WALTER WADE
HAMILTON WAYNE DAVID
HAMMEL RALPH LEWIS
HAMMOND KEITH TAIT
HAMMOND RUSSELL EARL
HAMPTON ROBERT POST JR
HANKINS THOMAS FRED
HANKS JOSEPH HENRY III
HANLEY RICHARD WILLIS
HANLON MARTIN JOSEPH
HANNA ROBERT
HANNIGAN THOMAS M JR
HANNINGS WILLIAM ELWOOD
HANNON PATRICK KEITH
HANS NEIL RONALD
HARAH FRANK A
HARBAUGH ROY ARBEN JR
HARDEN LARRY OWEN
HARGRAVE JOHN KING
HARGROVE LANE KORNEGAY
HARING WALTER WALTON
HARKANSON JAMES PHILLIP
HARMAN ROBERT HENRY
HARMON DENNIS GUY
HARNED GARY ALAN
HARNER RICHARD EDWARD
JR
HARPER THOMAS O JR
HARRINGTON KYLE TURNER
HARRIS BENJAMIN HARRY
HARRIS CHARLES LOUIS
HARRIS EDWARD LOUIS
HARRIS HARRY KENDALL JR
HARRIS MATTHEW N JR
HARRIS RONALD LEE
HART LARRY EUGENE
HART WILLIAM DARRYL
HARTGEN WILLIAM CLAYTON
HARTMAN THOMAS JOSEPH
HARTSOE DAVID EARL
HARTZEL GERALD LESTER
HARTZELL DONALD F JR
HARVEY PAUL EUGENE

HASHIN JOSEPH MICHAEL JR
HAWK CHARLES EDWARD
HAWK RAY GLENWOOD
HAWKINS ANTHONY
HAWKINS CHARLES E JR
HAWKINS WAYNE R
HAWLEY RICHARD A JR
HAWORTH WILLIAM HARRY
HAWRYSHKO DAVID WILLIAM
HAYLETT LARRY CLARENCE
HAYNES ALBERT RANDELL
HAYNES CLIFFORD EARL JR
HAYNES ROBERT EMMETT
HAYWARD LEWIS MORRISON
HEAGY JAMES WILLIAM
HEAPS CHARLES WILLIS
HEARD JAMES CORNEL
HEASLEY EDWARD FRANCIS
HEATH JAMES ROBERT
HEATH RUSSELL M
HECTOR GEORGE ANTHONY JR
HEDGLIN MILES BRADLEY
HEEMAN GARY LEE
HEFFERNAN THOMAS FRANCIS
HEFFNER KEITH DARRELL
HEILMAN WILLIAM EMORY
HEINE CHARLES EDWARD
HEISER DAVID FRANKLIN
HELD KEITH ARTHUR
HELLER EMERSON E
HELRING CARL JOHN
HEMKE DAVID LEE
HENDERSON CARL
HENDERSON CHARLES EDWARD
HENDERSON ROBERT MICHAEL
HENISS FRANK AMOS
HENNESSY DANIEL A
HENRY LEONARD IRA
HENRY SCOTT ORVILLE
HENRY TERRY LYNN
HENSEL DAVID WILLIAM
HENSINGER ARTHUR JAMES
HEPLER FRANK MONROE
HERBSTRITT RONALD LEE
HERMAN ROBERT CURTIS
HERMANN LOUIS VINCENT JR
HERR JAY DONALD
HERRIN LARRY FRANKLIN
HERRON PATRICK
HERSHEY CHARLES THOMAS
HERSHMAN KENNETH
HERTRICK DONALD JOSEPH
HERTZ ALLEN CONRAD
HESS KENNETH CHARLES
HESS LARRY LEROY
HESS PAUL J JR
HESSOM ROBERT CHARLES
HESTER WILLIAM WALTER
HETRICK RAYMOND HARRY
HETZER JOSEPH EDWARD JR
HEUSTON GERALD WILFORD
HEWLETT DENNIS HENRY
HEWLETT JAMES JOSEPH
HEYDT RICHARD BROWN
HIBBARD GARY EDWARD
HICE ROBERT KENNETH
HICKEY EDWARD ROBERT
HICKS JAMES ROBERT
HICKS WILLIAM DONALD
HIGBEE ROBERT DANIEL
HIGGINS MATTHEW
HIGGINS MERLE ROBERT
HIGHBERGER FRED DEAN JR
HILDENBRAND LESLIE IAN
HILL DAVID WILLIAM
HILL GERALD RAY
HILL HOWARD SCOTT
HILL JOHN RICHARD
HILL ROBERT OREN JR
HILL RONALD LEE
HILTE FRANK ELTON
HIMES BERNARD MALCOLM
HIMES EARL W
HIMES LLOYD ALLEN
HINCEWICZ EDWARD JOSEPH
HINKLE TERRY RICHARD
HINNANT BENJAMIN LOWELL
HIPPLE DUANE ALLEN
HIPPO DENNIS WILLIAM
HIRST ROBERT LEE

HITE FRANKLIN DANIEL
HITESHUE FRANK RICHARD
HIXSON CARL EDWARD
HOCH LARRY DEAN
HOCKENBERRY EDWIN CLOYD
HOCKENBERRY JOSEPH LESTER
HOFFMAN DENNIS EUGENE
HOFFMAN JOHN DANIEL
HOFFMAN LYNN ARTHUR
HOGENMILLER FRANK
HOHMAN LARRY LOUISE
HOJNOWSKI STANLEY BERNARD
HOLDEN JOHN PARKER II
HOLDER KENNETH LEROY
HOLLAND CLAYTON MONROE JR
HOLLAND DAVID HERMAN
HOLLAND FRANK RODNEY
HOLLAND GARY DAVID
HOLLAR COURTNEY PRICE JR
HOLLENBACH KENNETH RALPH
HOLLENBACH MERLIN CHARLES
HOLLER ROGER GUY
HOLLIDAY JAMES WILLIS
HOLLISTER HOMER WARREN
HOLLOWAY EDWIN NEWLIN III
HOLOKA JOHN CHARLES
HOLTZMAN EDWARD MONROE
HOLZ JOHN FREDERICK
HOLZAPFEL NORBERT PAUL
HOMER ERNEST CRAIG
HOMER WARD ELLIOTT
HOMMEL DAVID ELSON
HOMSCHEK ROBERT WILLIAM
HONAN JOSEPH PAUL
HOOK WILLIAM FOSTER JR
HOPES GLENN CHALFANT
HOPKINS DAVID LEE
HOPKINS IRVIN JAMES
HOPPER BARRY VORRATH
HOPPER RICHARD WHAN
HORCHAR ANDREW ANTHONY JR
HORNBAKER KENNETH EUGENE
HORNE AUSTIN ALBERT
HORNER LARRY MARK
HORTON DONALD MULLALY
HORVATH CHARLES WILLIAM
HORVATH WILLIAM FRANCIS
HORWITZ STANLEY LOUIS
HOSKINS GEORGE JR
HOUCK ALLEN PAUL
HOUCK LAWRENCE EMANUEL II
HOUSER CLYDE RICHARD JR
HOUSTON WILLIAM JOSEPH
HOVANCIK ANDREW M JR
HOWARD HORACE
HOWARD LAWRENCE PAIGE JR
HOWARD ROBERT CLARENCE
HOYER MICHAEL GERARD
HRUTKAY MICHAEL STEPHEN
HUBBARD MERLE GRIFFIN
HUBER RANDY S
HUBLER EDWARD LINCOLN JR
HUDSON JOSEPH ROBERT JR
HUFF LOUIS HOWARD II
HUFF WILLIAM EDWARD
HUFFMAN RICHARD ALLEN
HUGHES CHARLES GEORGE
HUGHES FREDRICK JOSEPH
HUGHES JOHN EDWARD D JR
HUGHES JULIUS BRADLEY
HUMM RONALD JOSEPH
HUNSICKER JAMES EDWARD
HUNSINGER CHARLES EDWARD
HUNT JOSEPH FRANCIS
HUNTER DONALD EUGENE
HUNTER PAUL CARLING
HUNTZINGER GEORGE RAYMOND
HUPP JAMES EARL
HUTCHINSON PAUL LEE
HYMERS CHARLES SUTHERLAND

HYNEK CARL FREDERICK III
IANIERI RICHARD JOSEPH
IGNASIAK DAVID JOSEPH
IGOE WILLIAM JOHN
ILLER RONALD
ISAAC ROCCO RENELL
ISABELLE LEMUEL
ISHMAN JOHN EDWARD
ISHMAN MICHAEL RAYMOND
ISLEY CHARLES LESTER III
ISOM DENNIS ROSS PAUL
IVES EDWIN ORRIN
IVORY KENNETH JOHN
JACKSON ALVIN EDWARD
JACKSON JOHN GLEN
JACKSON JOSEPH EUGENE
JACKSON LEHRON JR
JACKSON MALAKIA JR
JACKSON PAUL HOWARD JR
JACKSON RICHARD BERNARD
JACOBS DONALD WAYNE
JAMISON ERNEST CECIL
JAMISON FRANK
JANDERSHOVITZ PAUL WILLIA
JANKOWSKI WALTER JOSEPH
JARZENSKI JAMES HENRY
JAUER VICTOR CARL
JEDRZEJEWSKI HARRY FRANCI
JEFFERIS GARY DWIGHT
JEFFERSON RANDOLPH THOMAS
JEMISON JOHN L
JENERSON RALPH MATTHEW
JENKINS CHARLES OWEN JR
JENKINS JULIUS EDGAR
JENNINGS THOMAS EDWARD
JENSEN ROBERT ARTHUR
JEREMICZ GREGORY
JERMANY EMMETT JR
JESTER THOMAS STORY
JOHNSON ANTHONY
JOHNSON ARTHUR ANTHONY
JOHNSON ASA THOMAS
JOHNSON CHARLES WILLIAM
JOHNSON CLIFFORD ALVIN
JOHNSON DWIGHT DAWSON
JOHNSON GEORGE HARRY
JOHNSON GEORGE RUSSELL
JOHNSON GERALD VICTOR
JOHNSON HOWARD WESTLEY JR
JOHNSON JAMES JR
JOHNSON JOE THOMAS
JOHNSON JOSEPH JR
JOHNSON NORMAN WALLACE
JOHNSON OSCAR GIBSON JR
JOHNSON THEODORE
JOHNSTON CLEMENT B JR
JOHNSTON GARY FRANCIS
JOHNSTON RANDY LEE
JOHNSTON RICHARD WILLIAM
JOHNSTON ROBERT PATRICK
JOHNSTON RONALD LEE
JOHNSTON THOMAS PATRICK
JOLIET DAVID LOUIS
JOLLEY JOHN WILLIAM JR
JONES CARL ALVIN
JONES CHARLES RICHARD
JONES DANIEL JOHN JR
JONES DAVID LAWRENCE
JONES GREGORY
JONES JIMMIE JACK JR
JONES LEONARD ALEXANDER
JONES MICHAEL LEON
JONES PERRY KINARD
JONES REESE ALVIN
JONES RICHARD JOSEPH JR
JONES ROBERT NELSON
JONES ROOSEVELT
JONES WAYMON LOUIS JR
JONES WILLIAM
JONES WILLIE LEE JR
JORDAN JOHN EDWARD
JORDAN REGINALD ARCHIE
JOYCE VAN JOHN
JULIUS WILLIAM F III
JURICH WILLIAM AGNER
KACHLINE JAMES LEE
KACHMAN EDWARD MICHAEL
KAELIN CHARLES WRAY
KAHLER CHARLES EDWARD
KAIRAITIS FRANCIS

KAISER DENNIS DALE
KALEIKINI THEODORE K JR
KANE CHARLES FRANKLIN JR
KANE FRANCIS XAVIER
KANONCZYK RICHARD WALTER
KAPLAFKA MICHAEL JOHN
KAPLAN ALBERT
KAPP JOHN FRANCIS
KAPSHA RICHARD RUDOLPH
KAPUSTA EDWARD JOHN
KARDOS JOSEPH FRANCIS
KARPIAK MICHAEL JR
KARPY JOSEPH RUBEN
KARRAS JAMES MICHAEL
KASPAUL ALFRED AUGUST
KASTLER CURTIS CHARLES
KAUFFMAN RICHARD JOHN
KAUSE WILLIAM RAYMOND
KEARNEY DAVID GEORGE
KEE DANIEL PETER III
KEENAN DONALD WAYNE
KEENER LARRY LEE
KEESLER STEPHEN JOSEPH
KEGLOVITS EDWARD JOSEPH
KEGLOVITS RONALD EDWARD
KEIPER JOHN CHARLES
KEISLING DERVIN JOHN
KELLER RAYMOND E JR
KELLER RICHARD ALDEN
KELLER ROBERT CRITCHLEY
KELLEY LOUIS JAMES
KELLEY RONALD JAMES
KELLY GERALD JOHN JR
KELLY GREGORY RICHARD
KELLY JAMES ANTHONY JR
KELLY JAMES PATRICK
KELLY LAWRENCE LEE
KELLY LEO JOHN III
KELSO JAMES MICHAEL
KEMERER THOMAS BLAIR
KEMMERER DONALD RICHARD
KENDALL NEIL SCOTT
KENDLE RANDY TRUMAN
KENDRA GEORGE JOHN
KENNEDY ARLAN JOSEPH JR
KENNEDY THOMAS JOSEPH JR
KEOGH THOMAS PATRICK
KEPHART RUSSELL EDWARD
KERCHNER ROBERT BARD
KERCSMAR ROBERT CALVIN
KERL MICHAEL JAMES
KERR RICHARD ALLEN
KESSEL ROBERT LESTER
KIBBE GERRITH LOWELL
KIBLER ALFRED JAMES
KIDD VICTOR ELDEN
KIEFEL ERNST PHILIP JR
KIEHNE JAMES WESLEY
KIEZKOWSKI EDWARD THOMAS
KILDERRY MICHAEL JOSEPH
KILLION THOMAS JOSEPH JR
KILPATRICK DONALD ROBERT
KIMBROUGH GOLSBY JR
KIMMEL GORDON LEE
KINARD LARRY VERGESS
KINARD LESTER STEPHEN
KINDER WILLIAM ARTHUR
KING RONALD RICHARD
KINGHORN STEPHEN JOHN
KINIRY ANDREW JOHN
KINNEAR LAWRENCE FRANK
KINSKY RONALD CHARLES
KIRBY STEVEN
KIRCHNER JOHN HENRY JR
KIRK HERBERT ARTHUR
KIRSCH WARREN MICHAEL
KISCADEN MICHAEL EDWARD
KISER ROBERT THOMAS
KISNER THOMAS RUSSELL
KISSELL BERNARD FRANCIS JR
KISSINGER HAROLD JAMES
KISTLER BERNARD FRANCIS
KITE HARRY TURNER JR
KITTLE FREDRICK MARTIN
KIZER CARL SANFORD
KLAIBER FRANCIS EARL
KLARIC TERRANCE EDWIN
KLEIN GEORGE PAUL
KLEIN SZOLTON SIGMOND
KLINE DAVID SAMUEL
KLINE JAMES JOSEPH

KLINE KENNETH GORDON
KLINE ROBERT EARL
KLINGENSMITH CLYDE WALTER
KLINGENSMITH THEODORE R
KLORAN THOMAS WALTER
KLOTZ CRAIG GORDON
KNEPP GLENN DONALD JR
KNIGHT CLAUDE ARTHUR
KNIGHT RONALD EUGENE
KOCH DENNIS EARL
KOEHLER ROBERT THOMAS
KOHR WILBUR LINWOOD
KOLENC RONALD JAMES
KOLLER HAROLD JUNIOR
KOMAN LAWRENCE RYLAND
KOONS MICHAEL BOMBERGER
KOOSER KENNETH BRIAN
KOPRIVNIKAR JAMES JOSEPH
KORNICK FERDINAND J JR
KORPICS ANTHONY FRANCIS
KOSLOSKY WALTER NORMAN
KOSTICK PAUL FRANCIS
KOTIK ROBERT JOHN
KOVACS FRANCIS STEVEN
KOWALSKI ROBERT JOSEPH
KRALL ROBERT WILSON
KRAUS ROBERT LEE
KRAYNAK JAMES CLYDE
KREIDLER JOHN ROBERT
KREISHER SIDNEY GEORGE
KRESHO STACY
KRETZCHMAR PETER
KRILL BRIAN STUART
KROH CARL FRED
KROLIKOWSKI JAMES JOSEPH
KRUPA FREDERICK
KRUPINSKI RAYMOND JOHN
KUBELUS ANTHONY GEORGE JR
KUBICA THOMAS MICHAEL
KUCAS STEPHEN THOMAS
KUCHTA JOHN VINCENT
KUHN ANDREW ALAN
KUHNS DAVID PAUL
KUHNS VICTOR EUGENE
KUKTELIONIS JOHN LEON
KUKURUDA ANDREW JOSEPH
KULICKE FREDERICK W III
KULIKOWSKI EDWARD JOSEPH
KULL JOSEPH JOHN JR
KULWICKI RICHARD STANLEY
KUNER ROBERT MARTIN JR
KUNF JOHN THOMAS JR
KUNTZ RICHARD LORRAINE
KUPREVICH WILLIAM ALAN
KURTIK RAYMOND PAUL
KUSICK JOSEPH GEORGE
KUSTABORDER RONALD LEE
KUSTABORDER THOMAS WILLIA
KUTZER FREDERICK ROBERT
KUZAK TERRENCE MICHAEL
KUZER DENNIS
KUZMANKO ROBERT JOSEPH
KWORTNIK JOHN CHARLES
LA CHANCE ARTHUR ELVIN
LA PISH ROY ROBERT
LABANISH GEORGE MICHAEL
LACEY RICHARD JOSEPH
LACEY WILLIAM GIRARD
LACHER MARTIN JAMES
LAIR ELLIS EDWARD
LAMELZA MARIO
LAMON FRANCIS WILLIAM JR
LAMPLEY JAMES JR
LANDIS CLAUDE BRUCE II
LANE ALLEN G
LANE ROBERT CARL
LANGER FREDERICK PETER
LANGERIO MICHAEL LUKE
LANGLEY JERRY RAY
LANINGER LEON LAVERNE
LANNON JOSEPH JR
LANSKI JOSEPH WALTER
LARGE GEORGE WAYNE
LARIMORE JOHN RICHARD
LASHINSKY STEPHEN M JR
LASSITER KENNEY EARL
LATSCH DAVID RUDOLPH
LATSHAW HARRY KENNETH
LAUCHMAN MICHAEL ALAN

Pennsylvania

LAUER GENE ALLEN
LAUER MICHAEL DENNIS
LAUGERMAN LLOYD CHARLES
LAUGHLIN THOMAS JOHN
LAUGHLIN THOMAS WILLIAM
LAWS DWIGHT WILLIE
LE FEVER DAVID BRAUCH
LEAR RICHARD DAVID
LEBRUN LAWRENCE P
LECRONE PAUL ALBERT
LEDERER ANTHONY JOSEPH
LEE DAVID
LEE NATE FRANCIS
LEED CARL ROBERT
LEHMAN DENNIS RAY
LEIGHTON GARY WILLIARD
LEJEUNE THOMAS MILTON
LENARTOWICZ CHARLES
LEONARD KENNETH EDWARD
LEONBERG ROBERT CHARLES
LEOPOLD FREDERICK ERIC
LESH TERRY LEE
LESNIK WILLIAM ELGIE
LESSIG DANIEL KEPNER
LEVAN ALVIN LEE
LEWIS GARY FRANKLIN
LEWIS JOHN WESLEY JR
LEWIS STANLEY
LEWIS WAYNE EUGENE JR
LEWIS WILLIAM RUSSELL JR
LHOTA ROBERT ALLAN
LIELMANIS ATIS KARLIS
LIGHT JOSEPH MARION
LIGHTCAP JOSEPH MICHAEL
LIGHTMAN SAMUEL
LINCH LEE FRANCIS
LINDECAMP HOWARD S JR
LINDERMAN MARK THOMAS
LINDSAY PHILIP TRIESTE
LININGER GARY LEE
LINK GEORGE ARTHUR
LINK RAYMOND PATRICK
LIPP THOMAS STEVEN
LIPTOCK MICHAEL
LISHCHYNSKY GEORGE
LISTORTI JOSEPH ANTHONY
LITTLEJOHN GREGORY LAWREN
LITTS JAMES GARRIS
LITZ TERRY RICHARD
LODISE JOSEPH FRANCIS JR
LOERLEIN RONALD JOSEPH
LOGUE JOHN EDWARD JR
LONG CHARLES ELBERT
LONG EARL WAYNE
LONG JAMES JOSEPH
LONG JOHN HENRY SOTHORON
LONG PAUL MICHAEL
LONGFELLOW RONALD ANTHONY
LOPOCHONSKY JOHN HENRY JR
LORD ERIC ANTHONY
LORDI LOUIS ROBERT
LORENZINI DENNIS JOSEPH
LOUNSBURY WILLIAM DAVID
LOUTHAN DAVID CARL
LOVE BURGESS ALLEN

LOVSNES NEAL WALLACE JR
LOWAS JOHN
LOWERANITIS JOHN LEON
LOWNES CHARLES DAVID
LOWRY RONALD RAYMOND
LUCAS DONALD RAY
LUCAS MOSES B
LUDWIG FRANCIS JOSEPH
LUDWIG GALEN GEORGE
LUECKING PAUL J
LUFF EDWARD RICHARD JR
LUKASIK BERNARD FRANCIS
LUND WILLIAM EDWARD
LUTHER NELSON CHARLES
LUTZ WILLIAM
LYBRAND CARL FREDERICK
LYDIC DAVID ALLEN
LYDON RALPH JOSEPH JR
LYLE ALAN DAVID
LYNCH FREDERICK GEORGE JR
LYNCH GERALD JAMES
LYNCH JAMES MARTIN
LYNCH JOHN EDWARD
LYNCH RICHARD THOMAS
LYNN DOYLE WILMER
LYNN RICHARD ROBERT
LYONS JAMES E
LYONS JOHN MICHAEL
LYONS THOMAS KEVIN
MAC DONALD CHARLES JOSEPH
MAC NAMARA EDWIN JOSEPH
MACHRISTIE ANDREW
MACK ALLEN GLENN
MADONNA DOMINICK JOSEPH
MAGERR WILLIAM LEO III
MAGGS ROBERT HOWARD
MAGRIE DENNIS LOUIS
MAGRUDER DARRELL ZANE
MAGUIRE CALVIN GENE
MAGUIRE GERALD JOSEPH
MAGUIRE JACK IVAN
MAHER HAROLD WILLIAM
MAHONEY MICHAEL THOMAS
MALASPINA RICHARD THOMAS
MALLOY THOMAS WILLIAM
MALMQUIST PIERCE
MALONE CHARLES KENNETH
MALONEY CHARLES DEO
MALSEED FRANK JAMES JR
MANCILL DONALD BRYAN
MANGIOLARDO MICHAEL ANTHO
MANGOLD LEO JOSEPH
MANGUS ARLIE ROBERT
MANN THOMAS CHARLES
MANSON JAMES EDWARD JR
MANSOR THOMAS NICKOLAS
MARATTA CRAIG
MARCAVAGE ROBERT JOHN
MARCHAND THOMAS MICHAEL
MARGLE THOMAS JOSEPH
MARINO NICHOLAS III
MARISKANISH CHARLES EDWAR
MARK RICHARD STRODE
MARK THOMAS RICHARD
MARKEL JAMES CALVIN JR
MARKEY JAMES PAUL JR
MARKOSKI GERALD MICHAEL
MARLATT ROY WAYNE
MARPO JOHN ERNEST
MARROQUIN ELADIO R JR
MARSH EDWARD K
MARSH JOHN A
MARSHALL HAROLD B
MARSHALL RICHARD WILLIAM
MARSHALL THOMAS ROBERT
MARTELL TERRY JACK
MARTIN EUGENE JOSEPH
MARTIN GEORGE ROLAND
MARTIN JOHN JR
MARTIN JOHN MURRAY
MARTIN LAWRENCE SAMUEL
MARTIN LEONARD JR
MARTIN MICHAEL PETER JR
MARTIN ROBERT PHILLIPS JR
MARTIN RONALD ANDREW
MARTIN RONALD ROBERT

MARTINEZ ADOLFO
MARTZ WILLIAM HENRY JR
MASON THEODORE RAYMOND
MASON WILLIAM PAUL
MASTER WILLIAM STANLEY
MASTRAMICO PHILIP
MASTROMATTEO FRANK JAMES
MATARAZZI JOHN JOSEPH JR
MATTHEWS KERMIT LESLIE
MATTHEWS NATHANIAL CARL
MATTHEWS ROBERT WILLIAM
MATTY THOMAS RICHARD
MATYKIEWICZ DAVID BENJAMI
MATYLEWICZ LEO JOHN
MAURONE WILLIAM GREGORY
MAUTHE WILLIAM HAYES
MAY CRAIG NOLAN
MAY REED MC KINLEY JR
MAYER OSCAR CLEMENT III
MAYES JAMES WILLIAM
MAYNARD BRUCE CALVIN
MAYNARD RICHARD LEE JR
MAYS PICARDO RAMONLZY
MC ALEER JAMES K III
MC ATEER THOMAS JOSEPH
MC BRIDE EARL PAUL
MC CABE JAMES LOUIS JR
MC CAFFREY JAMES J JR
MC CAHAN MARLIN E
MC CAHAN WALTER LEE
MC CAMMON DONALD WILLIS
MC CANN JACK WILLIAM
MC CANN OWEN FRED
MC CARRICK WILLIAM W
MC CARTHY HOWARD C JR
MC CLARY BENJAMIN FRANKLI
MC CLELLAN BRENT A
MC CLELLAND JAMES RICHARD
MC CLELLAND RONALD EDWARD
MC CLINTIC GEORGE PATRICK
MC CLINTOCK GERALD
MC CLUSKEY KENNETH JAMES
MC CONAHY THOMAS ARTHUR
MC CORMICK JOHN W JR
MC CORMICK RICHARD H JR
MC CORMICK ROBERT PATRICK
MC CREIGHT JOSEPH THOMAS
MC CUEN WILLIAM DAVID JR
MC CULLOUGH ALFRED
MC CULLOUGH JOHN JAMES
MC DANIEL PATRICK ELSWOOD
MC DONALD HENRY III
MC DONALD SAMUEL LEE
MC DONALD WILLIAM EARL JR
MC DONNELL WILLIAM HERBERT
MC DONOUGH JAMES ROBERT
MC DOWELL LAURENCE THOMAS
MC ELHANEY RODGER DENNIS
MC ELROY JOHN JAMES
MC ENTEE THOMAS
MC EWEN JAMES ARTHUR
MC FARLAND RICHARD SCOTT
MC FARLAND WILLIAM LEROY
MC GARVEY RAYMOND LEE
MC GARVEY WILLIAM BERNARD
MC GILL JAMES BARRY
MC GILL MICHAEL GREGORY
MC GINNIS HARRY F JR
MC GINNIS LEONARD DAVID
MC GINNIS MICHAEL BRIAN
MC GINTY LAWRENCE MICHAEL
MC GRATH CHARLES FRANCIS
MC GRAW LARRY JOE
MC GUIGON WILLIAM EDWARD

MC GUIRE DENNIS FRANCIS
MC HENRY PAUL VINCENT
MC HUGH JOHN J
MC HUGH TIMOTHY DAVID
MC KERNS THOMAS PATRICK
MC KINNEY GERALD LEE
MC KNIGHT PAUL DAVID
MC LAUGHLIN JOHN ROBERT
MC LAUGHLIN PETER FRANCIS
MC LAUGHLIN WILLIAM LEE
MC LAY JOHN JACOB JR
MC MAHON RAYMOND PAUL
MC MASTER GLENN LEON
MC MINN RICHARD LEE
MC MONEGAL JOHN JOSEPH JR
MC MORRIS JIMMIE LUE
MC MULLEN GENE SMEDLEY
MC NAMARA WILLIAM JAMES
MC NEISH RICHARD LEE
MC NELIS PATRICK ROBERT
MC NELLIS ANTHONY FRANCIS
MC NICHOLS RICHARD FRANCI
MC PHERSON ROBERT ALAN
MC RAE LAWRENCE JOSEPH
MC SWEENY RICHARD JOSEPH
MC TIER KENNETH CHARLES
MC VAY RICHARD WAYNE
MC WILLIAMS ROBERT H JR
MCCANN FRANCIS JOSEPH JR
MCCORMICK WILLIAM C JR
MCCRACKEN JAMES MUIR
MCDANIEL EDGAR
MCDONALD MARTIN TERRANCE
MCGONIGLE CHARLES D
MCILVAIN EDWARD M III
MCKELVEY WILLIAM R
MCWHINNEY HARRY DEWITT JR
MEADOWS CALVIN JR
MEADOWS LEE DAVID
MEALY DAVID HOWARD
MEBS FRANK MARTIN
MECHLING DANIEL GARY
MECKLEY RONALD EUGENE
MEEHAN ROBERT EUGENE
MEISBURGER JOSEPH STEVEN
MEISS ROBERT WARREN JR
MEISTER DAVID WILLIAM
MELLINGER CARL B JR
MELLO ERHARD JAMES
MELNICK PETE
MELNYK JOSEPH JAMES JR
MELOY PAUL HOOVER
MELTON RODNEY WAYNE
MENTZER GERALD LERVERNE
MERRILL CHARLES LE ROY JR
MERRILL WILLIAM FRANKLIN
MERRIMAN JOHN SARGENT
MERRITT EDWARD JEROME
MERSCHEL LAWRENCE JAMES
MESSINO DAMIEN JOHN JAMES
METZGER ANTHONY JOHN JR
METZLER JOHN EDWARD
MEYER FRANKLIN DELANO
MEYER JOSEPH JOHN JR
MEYERS CHARLES CARTER
MICHAELS JOHN JAY
MICHELLI ANGELO FRANCIS
MICHELS JOHN JAY
MICHINOK FRANCIS MICHAEL
MICOLA DANA J
MIECZKOWSKI JOSEPH
MIHORDIN DONALD STEPHEN
MIKOSZ WALTER JOSEPH JR
MIKULA CARL STEPHEN
MILEK MARTIN HEINRICH
MILES PHILLIP ALBERT JR
MILEY FREDERICK JAMES JR
MILIKA GEORGE AUREL
MILLER ARLEN JAY
MILLER CHARLES EMIL
MILLER DANIEL AUGUST
MILLER DENNIS J
MILLER EDWARD CHARLES
MILLER FREDERICK C III
MILLER GARY LEONARD
MILLER GEORGE LIVINGSTON
MILLER HARRY J

MILLER JAMES EDWARD
MILLER JEFFREY HAROLD
MILLER JESSE J
MILLER JOHN JEROME JR
MILLER JOHN RUSSELL
MILLER JOSEPH ANTHONY JR
MILLER MARLIN MCCLELLAND
MILLER PHILIP CHANTRY JR
MILLER PHILLIP DANIEL
MILLER ROBIN BREWER
MILLER RONALD JOHN
MILLER STEPHEN PETER
MILLER TED ROGER
MILLER WYATT JR
MILLIKAN JOHN RUSSELL
MILLISON DENNIS KEITH
MILLISON EDWARD JAMES III
MINDYAS EDWARD ANDREW
MINER MICHAEL DAVID
MINICK STEPHEN MICHAEL
MINKUS DENNIS JAMES
MINNICH RICHARD WILLIS JR
MINOR ROBERT PATRICK
MITCHELL BYRON JOSEPH
MITCHELL DAVID GEORGE
MITCHELL PAUL JOSEPH
MITCHELL RALPH BURTON
MITCHELL THOMAS VICTOR
MITCHELL WILLIAM A II
MITCHELL WILLIAM BROOKS
MITZEL LONNY LEROY
MOHN RICHARD SAMUEL
MOHR RICHARD ALLEN
MOLETTIERE BARRY ALAN
MOLETTIERE JOSEPH ANTHONY
MOLL WAYNE TYRONE
MONAGHAN JOSEPH THOMAS
MONAHAN DANIEL FRANCIS
MONISMITH WAYNE EUGENE
MONTALVO MANUEL GUALVERTO
MONTEITH ROBERT F II
MONTGOMERY CLARENCE WILLI
MONTROSS CHARLES PAUL
MOORE ABRAHAM LINCOLN
MOORE DENNIS WESLEY
MOORE EARL THOMAS JR
MOORE FRANK HARRIS
MOORE HERBERT WILLIAM JR
MOORE JAMES BUCKSON
MOORE JAMES HARRISON
MOORE MICHAEL KEITH
MOORE PETER CHARLES
MOORE RANDALL WHIT
MOORE ROBERT CLAYTON
MOORE ROBERT IRVIN
MOORE ROBERT THOMAS
MOORE WILLIAM JUNE
MORALES JULIO VICTOR
MORAN BERNARD JOSEPH JR
MORAN BRUCE JAMES
MORAN EDGAR C II
MORAN JOHN WILLIAM
MOREIRA RALPH ANGELO JR
MORGAN GEORGE
MORGAN JOHN PATRICK JR
MORGAN WILLIAM DAVID
MORIN JAMES THOMAS
MORNINGSTAR ROBERT LEE
MORRIS CARL MICHAEL
MORRIS JEFFREY LYNN
MORRIS ROBERT DEAN
MORRIS ROBERT JOSEPH
MORRIS WALTER F
MORRIS WILLIAM HENRY JR
MORRISON CHRISTIAN HERMAN
MORRISON JAMES ALBERT
MORRISSEY JAMES JOSEPH
MORROW BOYD ELLIS
MORROW HAROLD EUGENE
MORSE EUGENE JOSEPH
MOSCRIP ARTHUR DAVID JR
MOSER DAVID LLOYD GEORGE
MOSER HARRY JULIUS IV
MOSER TERRY LEE
MOSHER MAURICE WILLIAM
MOSSMAN JOE RUSSELL
MOSSO ROBERT BRUCE
MOSTOWSKI THEODORE

MOWERY ROBERT ALLEN
MOYER BARRY LEE
MOYER CECIL GERALD JR
MOYER DENNIS LEE
MOYER DOUGLAS ISAAC
MOYER LAWRENCE RICHARD
MOYER MERRHAGE MICHAEL
MOYER WARREN JR
MOYLAN DAVID JOHN
MOYLE WESLEY ALLEN
MUDD LEROY BERNARD
MULLEN FREDERICK WILLIAM
MULLEN GILBERT GREGORY
MULLINEAUX BARRY THOMAS
MULLINS EDWARD PATRICK
MUMMERT GEORGE LEON-
ARD
MURPHY DENNIS JAMES
MURPHY DONALD LEROY
MURPHY JAMES JOHN
MURPHY JOHN JAMES
MURPHY JOHN PAUL
MURPHY RALPH OLIVER III
MURPHY RONALD JAMES
MURPHY WILLIAM
MURRAY WILLIAM JOSEPH JR
MURTAUGH BARRY WAYNE
MUSCARA CARMEN
MUSSER RICHARD LAVERNE
MUTTER ALVIN GEORGE
MYERS DANIEL LEROY
MYERS DAVID GEPHART
MYERS DAVID ROSS
MYERS GEORGE NERVIN
MYERS RICHARD VAUGHN
NADANY FRANK JOSEPH JR
NADOLSKI ROBERT
NAFE TIMOTHY MARK
NAHER STEPHEN CHARLES
NAHODIL DONALD A JR
NAPIER ROBERT WAYNE
NATOLI JOSEPH R
NAUGLE RUSSELL WAYNE
NEAL REUBEN JAMES
NEFF LARRY LEE
NEFF PHILLIP ERNEST
NEIDRICK JACK LEE
NEILL TERRY JOSEPH PATRIC
NEIMAN GARY PRESTON
NELSON CLYDE KEITH
NELSON JOHN CHARLES
NELSON RICHARD CRAW-
FORD
NEMCHICK MICHAEL JOSEPH
NESMITH LEROY
NESS LESTER MILTON
NESTERAK NORMAN LOUIS
NESTLERODE GEORGE HER-
BERT
NETHERLAND ROGER MOR-
TON
NEUMAN RONALD M
NEUMYER OWEN FRANCIS
NEVEL ROBERT JOSEPH
NEWBY FREDERICK ALBERT JR

NEWELL CRAIG ALLEN
NEWELL RONALD EUGENE
NEWMAN CHARLES DAVID
NEY DAVID CHARLES
NICHOLAS DAVID LAMPREY
NICHOLS BRUCE JOSEPH
NICHOLS COLIN KEITH
NICHOLSON GEORGE JAMES
NICHOLSON JAMES CLIFFORD
NICKERSON LEWIS RAYMOND
NICKLOW ROBERT JAMES
NICKOL ROBERT ALLEN
NIEMANN DAVID LEE
NIERER JOHN EDWARD
NIGRO ANTHONY JOSEPH
NITKA JOSEPH STANLEY
NIXON ROBERT JOHN
NOBLES NORMAN JAMES
NODEN TIMOTHY JOSEPH
NOEL JOSEPH PAUL
NOLAN CHARLES ALBERT JR
NOLAND KENNETH EUGENE
NOLDER CHARLES JAMES
NOLT CALVIN EUGENE
NORMAN GARY LESLIE
NORRIS ROBERT NORMAN
NORTON WARD III
NOSTADT FRANK JOHN JR
NOTICH ANTHONY MICHAEL
NOVAK WALTER MARK
NUHFER WILLIAM DANIEL
NULL RICKY LEE
NUSCHKE EDGAR ERWIN
NUTTER GREGORY LEROY
NYE HAROLD CURTIS
NYE JERRY WARREN
O BRIEN JOHN LAWRENCE
O BRIEN WILLIAM JOSEPH
O DONNELL JOHN THOMAS
O FARRELL JOHN MICHAEL
O FARRELL WILLIAM PATRICK
O NEIL ROBERT ANDREW
O TOOLE PETER JOSEPH
OAKES JACK WAYNE
OATES SAMUEL ARTHUR JR
OBERLE CHARLES G
OBMAN JOSEPH HOWARD
OCHS VALENTINE AMBROSE
OCKEY BRUCE GORDON
ODONNELL SAMUEL JR
OESTERREICH TILO RUDOLF
OGDEN WILLIAM STEPHEN
OGRIZEK JOHN ANTHONY
OLDS JOHN HENRY
OLEARNICK THOMAS
OLESNANIK JOHN FRANCIS
OLIVER CHARLES OTIS
OLIVER FRANK GEORGE II
OLMEDA EDWIN JOSEPH
OLSON ROBERT CHARLES JR
OLSON WILLIAM CRAIG
OLSZEWSKI JOHN MICHAEL
OLZAK RAYMOND DENNIS
ONDERKO JOHN PATRICK
OPLINGER DAVID PAUL

ORENDORFF ERNEST E JR
ORISON LOUIS JAMES
ORLEY WALTER FRANK
ORSINO JOHN GEORGE
ORTALS DAVID JOHN
ORTIZ ELIEZER
ORTIZ LOUIS THOMAS
OSCELUS JOHN ALBERT
OSTIFIN JOSEPH LYNN
OSWALD DONALD ROY
OTT WAYNE HARRY
OTTEY CARLTON MATTES
OWENS GARY LEE
OXLEY JAMES KEITH
PAINTER ROBERT GLEN
PALANDRO RAYMOND JOSEPH
PALM ALLEN NEIL
PALMA FRANCIS MICHAEL
PALMER GARY JAMES
PALMER JAMES LAMONT
PALMER LARRY DALE
PALUSCIO JOHN JOSEPH
PANAK JOHN JR
PANELLA NICK JR
PANNABECKER DAVID ERIC
PANNO DONALD DAVID
PANTALL JAMES ROBERT
PAOLANTONIO BENNIE JOE
PAPE ROBERT PAUL
PARADA EDWARD JOHN
PARK JOSEPH CONARD JR
PARK RICHARD LEWIS
PARKER DAVID ALBERT
PARKER FRANK C III
PARKER VERNON HOWARD JR
PARKHILL FRANCIS EDWIN JR
PAROBEK SILAS WILLIAM
PAROLA JAY WAYNE
PAROPACIC JOHN PAUL
PASEKOFF ROBERT EDWARD
PASS JOHN III
PASSANANTE WILLIAM JAMES
PASTORINO MICHAEL AN-
THONY
PATRICCA ANTHONY
PASQUALE
PATRILLO ALBERT JOHN
PATTERSON EARL ALLEN
PATTERSON ROBERT DE-
WAYNE
PATTON DAVID ALAN
PATTON FRANCIS G
PATTON KENNETH JAMES
PAUL JAY
PAULL CHESTER DONALD
PEACE CHARLES LAMONT
PEAGLER LEROY W
PEARCE EDWIN JACK
PEARSON GEORGE B III
PEDEN CLARK EDMUND
PELLEGRINO JOSEPH D
PELULLO LEONARD SALVA-
TORE
PELUSO PAUL RENATO JR
PENN RAYMOND BISHOP JR

PENNETTI FRANCIS
PENTLAND JAMES DOUGLAS
PEOPLES JAMES DALE
PERECKO PAUL JOHN
PERINOTTO ERNEST DAVID
PERRINS ROBERT RICHARD
PERRY TIMOTHY EUGENE
PERSELY RICKY EDWARD
PERUSO LAWRENCE DAVID
PETERS CHARLES EDMUND
PETERS KENNETH WALTER
PETERS RONALD JAY
PETERSON ALBERT ALLEN
PETERSON DILLARD ERIC
PETRARCA JOSEPH A
PETREY JAMES JIM
PETRILLA JOHN JOSEPH JR
PETTEYS JAMES BIRCH
PETTIFORD JAMES LLOYD
PETTUS KENNETH
PFORDT CHARLES C JR
PFOUTZ MYRON MCCLEL-
LAND
PHILLIPS ANDREW MARK
PHILLIPS DANIEL RAYMOND
PHILLIPS EDISON RICHARD
PHILLIPS NATHANIEL JAMES
PHILLIS DONALD R JR
PIATKOWSKI ROBERT J
PICONI PIETRO
PIERCE DONALD JAMES JR
PIERCE LARRY WENDELL
PIERCE WALTER MELVIN
PIERSOL JOHN LAURENCE JR
PIFER ROGER LEE
PILSON THOMAS VICTOR
PIOTROWICZ DAVID
PISCAR VINCENT JR
PISKULA RICHARD
PITTS JOSEPH WADE JR
PLANK JAMES DUANE
PLATT ROBERT LLOYD
PLESAKOV LUCIANO PAUL
PLESH RAYMOND NICHOLAS
PLISKA MICHAEL DENNIS
PLUMB JACK CLARE
POCKEY JAMES JODY
PODLESNIK WAYNE A
POLEFKA JOHN ARN
POLEY DAVID ALLAN
POLICASTRO MARK EDWARD
POLUSNEY JAMES FRANCIS
PONIKTERA STANLEY F JR
PONTIERE JOHN RANDALL
POPOWITZ GREGORY FRAN-
CIS
PORT HYRUM BARRY
PORT WILLIAM DAVID
PORTER JAMES FRANK
PORTER RAYMOND JAMES
PORTER TIMOTHY MICHAEL
PORTER WILLIAM ROBERT JR
POST DANIEL GIBSON
POST DOUGLAS ARTHUR
POTEMPA LOUIS WILLIAM

POTTER WILLIAM JOSEPH JR
POTTS JERRY
POVEY JOHN THOMAS
POWELL RICHARD EDWIN
POWERS JAMES WILLARD JR
POWERS WILLIAM MAXWELL
PRAZINKO ROBERT JAMES
PREAUX THOMAS ALFRED
PREISENDEFER HAROLD ALAN
PRENTICE DAVID GRAY
PRESLIPSKI MICHAEL JR
PRICE ARNOLD W
PRICE BARRY CARLTON
PRICE FREDERICK
PRICE RODNEY ALLEN
PRIVIECH ROBERT MICHAEL
PROBST DELMAR WAYNE
PROM WILLIAM RAYMOND
PROUDFOOT LEWIS H III
PUGH STEPHEN BRIAN
PURCELL HOWARD PHILIP
PURCELL MICHAEL JOSEPH
PURVIS ALFRED ALEXANDER
PYE SAFFORD SMITH
PYSHER GERALD JOHN
PYSZ ALEX DENNIS
QUICK ROBERT LEE
QUIGLEY JOHN MARTIN
QUINN JAMES ANTHONY
QUINN JOHN ARNOLD
QUINN RAYMOND FRANCIS
RABINOVITZ BARRY IVAN
RACKOW ANDREW CHARLES
RADECKI PHILIP HENRY
RADGOWSKI CHESTER J JR
RADZELOVAGE JAMES MI-
CHAEL
RAFFERTY BERNARD JOSEPH
RAFFERTY EDWARD JOHN
RAGLAND FRED MICHAEL
RAHN DONALD KEITH
RAIFORD CHARLES LEROY JR
RAILING CHARLES DAVID
RAKER RONALD LEE
RALLS RAYMOND BERNARD
RALPH DAVID EDWARD
RALSTON JAMES VINCENT
RAMEY VERNON LEMAR
RAMPULLA TERRY JAMES
RANALLO CHARLES EDWARD
JR
RANSOM RODNEY LEE
RATHMELL HENRY PORTER
RATZEL WESLEY DALLAS
RAUBER WILLIAM
RAY FREDERICK FRANKLIN JR
REAGLE JOHN LOUIS
REALE JOHN BATTISTE III
REAM ERIC ALLAN
REAM GARY LEE
REDDINGTON JAMES THOMAS
REDMON STANLEY EUGENE
REDMOND CARTER
REECE PETER EDWARD
REED ALBERT MARSHALL
REED GARY DEWAYNE
REED GEORGE JOSEPH JR
REED JOHN BRUCE
REED LARRY BRUCE
REED MICHAEL CHARLES
REED PAUL EDWARD
REED SCOTT DOUGLAS
REEDER EDWARD JAMES
REEFER CHARLES LENARD
REES DONALD BRUCE
REES WILLIAM ALLEN
REESE RAYMOND RICHARD
REESE WILLIAM RICHARD JR
REEVES M RAYMOND
REHM TERRY MICHAEL
REICH THOMAS ALAN
REICHERT LAWRENCE JOHN
REID JAMES MURRY
REIGLE AARON HENRY
REILLY EDWARD DANIEL JR
REILLY EDWARD WILLIAM
REILLY JAMES JOSEPH JR
REINHARDT JAMES MICHAEL
REINHART PETER SIMMONS
REITER GERALD ANDREW
REMPER GERALD NEAL
RESINGER DENNIS MICHAEL
RESNICK ROBERT ALBERT
REYES ANGEL

Pennsylvania

REYNOLDS ARTHUR JR
REYNOLDS JACK EDWARD
REYNOLDS JOHN EUGENE
REYNOLDS ROBERT MICHEAL
RHEN DENNIS HENRY
RHOADS THOMAS VERNON
RICARD FRED LAYTON JR
RICE CALVIN CHARLES JR
RICE JACK WALTER
RICE MICHAEL PAUL
RICHARDS CHARLES H JR
RICHARDS JOHNNY LEE
RICHARDSON HERMAN JR
RICHARDSON JAMES
RICHARDSON JEFFERY ALLEN
RICKERSON STEVEN ALLEN
RICKERT GLENN DALE
RICKMAN WILLIAM JOEL
RIDDLE OLIVER JOHN
RIEGEL JOHN FRANKLIN
RIEGER CHARLES A III
RIFFEY TRACY HARLEY
RIGGLE JOSEPH DALE
RIGHTMYER JACK LEE
RILEY DENNIS LEROY
RILEY HOWARD GEORGE
RILEY NATHANIEL JULIUS JR
RILEY NEIL EDWARD
RIMEL MELVIN LEWIS
RINGLER ROBERT LEWIS JR
RIPKA HERBERT A
RISNER WAYNE ERIC
RITCHEY CLAIR F JR
RITSICK EDWARD
RITTER RICHARD FRANK
RITZ MARSHALL LEROY
RIVERS HARRY EUGENE
RIVERS SANDY MITCHEL
RIZOR DAVID LEE
ROBERTS CHARLES G
ROBERTS JAMES ALLEN
ROBERTS RICHARD DANIEL
ROBERTSON ROBERT GLENN
ROBINSON MARTIN ROBERT
ROBINSON REMBRANDT
CECIL
ROBINSON WILLARD MI-
CHAEL
ROBISON WILLIAM RANDALL
ROBLE JOSEPH EDWARD
ROCCO WILLIAM FRANK
ROCHOWICZ WAYNE CARL
ROCKOWER HENRY NEIL
RODGERS GREGORY WAYNE
RODGERS LARRY MORGAN
RODKEY WILLIAM EUGENE
RODMAN DAVID B
RODRIGUEZ LOUIS
RODRIQUEZ SAMUEL
ROEDER CHARLES THOMAS
ROGERS KENNETH LEE
ROKASKI MARK CHARLES
ROLDAN WILLIAM JUNIOR
ROLLER BENJAMIN C JR
ROMANCHUK MICHAEL
GEORGE
ROMANELLI LOUIS VINCENT
ROMANKO DANIEL ROBERT
ROMIG EDWARD LEON

ROMIG LEROY HENRY
RONAN PATRICK JOSEPH
RONCA ROBERT FRANCIS
ROOT CLARENCE ROBERT
ROPCHOCK THEODORE
MATTATW
ROSEN MAX EMMANUEL
ROSENBERRY FRED BRYAN
ROSENWALD ROBERT JOHN
ROSS REID REX JR
ROTH FRANK THEODORE
ROTHHAAR BRUCE LEE
ROUBA EDWARD S
ROUSCHER JOHN MARTIN
ROUSE JOHN WILLIAM
ROUSH RONNIE RAY
ROVINSKY RICHARD MI-
CHAEL
ROVITO GILBERT ALLAN
ROWE CHESTER EARL JR
ROWE ERNEST LEROY
ROWLES ALLEN DUANE
ROWLETT GARY PAUL
ROY PATRICK ROBERT
ROYER RICHARD HOWARD
ROYER ROBERT HENRY
ROYSTER DOUGLAS
ROZELLE DAVID THOMAS
RUBBO KENNETH WILLIAM
RUCH ROBERT STEPHEN
RUDD JAMES EARL
RUDINEC JOHN JOSEPH
RUDZIAK ERIK NILES
RUMBAUGH ELWOOD EU-
GENE
RUMMEL FRANCIS CLAIR
RUNDLE CARY FRANK
RUNEY LAWRENCE F
RUNK GARY WESLEY
RUPINSKI BERNARD FRANCIS
RUSNAK GEORGE BERNARD
RUSNAK ROBERT JOSEPH
RUSS JOSEPH BLAIS
RUSSO AUGUSTINE DANIEL
RUSSO THOMAS PETER
RUTBERG FRANKLIN STEVEN
RUTTLE ROBERT PRESTON JR
RYAN ROBERT ANTHONY
RYDER CARL EDWARD
RYKACZEWSKI STANLEY K
RYKOSKEY EDWARD JAY
RYLEE JAMES SIDNEY
RYNEARSON KARL FRANCIS
RYNKIEWICZ RICHARD
ROBERT
SABO LESLIE HALASZ JR
SADOWSKY LLOYD J
SAIN JEROME ROBERT
SALDANA RICHARD E
SALERNO RALPH DENNIS
SALISBERRY LARRY GORDON
SALLEY WALTER JUNIOR
SALONISH EDWARD GEORGE
SAMPSON GERALD HILBERT
SANDBERG CHARLES H
SANDERS GERARD JUDE
SANDERS ROBERT JAMES
SANDNES LARRY GORDON
SANDS KENNETH EARL
SANFORD JOHN FRANCIS
SANTIAGO ANGELO CAR-
MELO
SANTILLI RAYMO
SANTONE JOSEPH ANTHONY
SAS LOUIS
SATCHELL RONALD EDWARD
SATTERFIELD ROBERT W
SATTERWHITE RUFFIN J JR
SAUBLE MARTIN G JR
SAUBLE THOMAS EUGENE
SAUERS GERALD
SAUKAITIS JOSEPH STEPHEN
SAUNDERS JOHN LLOYD
SAVAGE DANIEL
SAWICKI ANTHONY PETER
SAYLOR SCOTT EDWARD
SCADUTO RICHARD LEE
SCAIFE KENNETH DOYLE
SCALISE THOMAS RANDAL
SCHAAF LEE RICHARD
SCHAEFFER PAUL HENRY
SCHAFFER BLAINE CLARENCE
SCHALL CHARLES NELSON JR
SCHATZMAN ROBERT JAMES

SCHECKLER PAUL
SCHEETZ JOHN ELLWOOD
SCHEIB LAWRENCE ELWOOD
JR
SCHELL RANDY STEPHEN
SCHELL WILLIAM LEROY
SCHERER JAMES MICHAEL
SCHERMANN HERMAN
WILLIAM
SCHEU GUNTER WILFRIED
SCHIFFHAUER JOHN CHARLES
SCHIMPF JOSEPH FRANCIS
SCHLEE HARRY LEE
SCHMERBECK DAVID JOHN
SCHMIDT RICHARD HERMAN
SCHOENIG EDMOND DAVID
SCHOFF LEO RICHARD
SCHOLL CLIFFORD PAUL JR
SCHRAMM CHRISTOPHER
JOSEP
SCHRECKENGOST HAROLD
LEE
SCHREFFLER CLEON LARRY
SCHROEDER TIMOTHY
RICHARD
SCHROEFFEL THOMAS AN-
THONY
SCHUBERT WILLIS JUNIOR
SCHULER HAROLD RICHARD
SCHULTZ ALAN ROBERT
SCHULTZ GEORGE JOSEPH
SCHULTZ SHELDON D
SCHULTZ THOMAS RUSSELL
SCHUSSLER WILLIAM JAMES
SCHUSTER JOSEPH WILLIAM
SCHWANGER FREDERICK JAY
SCHWARTZ JOHN GUSTAVE
SCHWARTZ SAMUEL BRUCE
SCHWARTZ WAYNE GILMORE
SCHWEIGHOFER REED JAY
SCHWESINGER RAYMOND
PAUL
SCIARRETTI VINTURE
SCOTT BILLY EDWARD
SCOTT DAIN VANDERLIN
SCOTT EDWARD DRAKE
SCOTT GREGORY JOHN
SCOTT JAMES BERNARD
SCOTT LEROY HARRY
SCOTT ROBERT L
SCOTT WILLIAM
SCUNGIO VINCENT ANTHO-
NY
SEAMAN DONALD JOSEPH
SEASHOLTZ RONALD J
SEAWRIGHT WALTER LEE
SECREST EDWARD WILLIAM
SEDGWICK RICHARD BRUCE
SEE EDWARD EUGENE
SEEDES HARRY BATON III
SEELY RICHARD CLAIR
SEGICH MICHAEL PAUL
SEIBERLING KARLHEINZ S
SEIBERT WILLIAM ROBERT
SELGRADE STEPHEN FRANK
SELKREGG EDWARD M III
SELL CLIFFORD LLOYD
SELTZER JACKIE RALPH
SEMENTELLI DOMINIC M JR
SENG RICHARD MICHAEL
SENTMAN DONALD WARREN
SESSIONS WILLIAM ROBERT
SETH CHARLES WILLIAM
SETZENFAND CHARLES
FREDER
SEXTON RICHARD JARRETT II
SEYMOUR LEO EARL
SHADE GEORGE EVERETT
SHAFFER CHARLES
SHAFFER WALLACE CLAIR JR
SHAKLEY GERALD WAYNE
SHANE WALLACE WILLIAM
SHANKS THOMAS FRANK
SHARPNACK MATTHEW F
SHATTUCK FRED WILLIAM JR
SHEFFEY RONALD DAVID
SHEIBLEY CLARENCE DAVID
SHEMORY KENNETH CHARLES
SHENK LESLIE FRED
SHERIDAN MICHAEL FRANK-
LIN
SHERLOCK ROBERT EUGENE
SHEROKE JOHN RICHARD JR
SHERWOOD ROBERT JAMES JR

SHIELDS ROBERT EARL
SHIELDS WILLIAM JOHN
SHIKO RAYMOND JOSEPH
SHILLER ALBERT
SHIMEK SAMUEL DALE
SHIPE THOMAS ALLEBACH
SHIPLEY WALTER W JR
SHOBER TIMOTHY ALLEN
SHOEMAKER JOHN STOUDT
SHOEMAKER ROBERT LEE
SHOGAN PAUL FRANCIS
SHOLL ROBERT LEE
SHOOP JACK HENRY JR
SHORT PAUL THEODORE JR
SHOVLIN FRANK JOSEPH
SHOWALTER JAMES EDWARD
SHOWERS JOHN ELLSWORTH
JR
SHRADER JAMES GAYLORD
SHUBERT DARNAY
SHUBIAK JOSEPH EDWARD
SHUEMAKER MICHAEL
THOMAS
SHULTZ DALE EDWARD
SHULTZ WILLIAM HARRY
SIBSON SCOTT MEYER
SICKLES JAMES ARTHUR
SIEGRIST WILBUR JERRY
SIEMON DAVID ALAN
SIENGO RONALD JAMES
SIETZ RICHARD MARTIN
SIEVERS FRANCIS EUGENE JR
SIGAFOOS WALTER HARRI III
SIGEL LEWIS WILLIAM
SIGG JOHN CHARLES
SIKON ROBERT ARCHIBALD
SILFEE JAMES EVERETT
SIMMONS CLARENCE JIMMIE
SIMMONS HARRY JENNINGS JR
SIMMONS RICHARD STANLEY
SIMONS DAVID RICHARD
SIMONS RAY OTIS JR
SIMPKINS ROBERT LEE JR
SINCHAK WILLIAM ANDREW
SINGER DONALD MAURICE
SINGER SAMUEL ARNOLD
SINGLER DELBERT LEO JR
SINGLETARY NEELY JAMES
SINKEWICZ JOSEPH MICHAEL
SIPPEY WAYNE KEITH
SIVITS CHARLES E
SKOCICH FRANK ALBERT
SKONIECKI LEONARD F JR
SKUNDA EDMUND
SLACK CHARLES LEROY JR
SLAGLE LARRY RAY
SLESH JOHN DANIEL JR
SLICHTER DONALD JAMES
SLOAN BOBBY LOUIS
SLOCUM QUENTON EDWARD
JR
SLUSSEAR ALEXANDER
MARTIN
SMALL TERRY SIDNEY
SMELTZER CHARLES E III
SMILEY FRANCIS EDWARD
SMITH ALBERT EDWIN
SMITH BARRY LEE
SMITH BOOKER JR
SMITH CHARLES EDWARD
SMITH CHARLES FRANK
SMITH CLAUDE ALLEN
SMITH DAVID FRANCIS
SMITH DAVID LELAND
SMITH FREDERICK JOSEPH
SMITH GEORGE W III
SMITH HARLEY ALBERT JR
SMITH HAROLD
SMITH HARRY CHARLES
SMITH HENRY EDWARD
SMITH JAMES WALTER
SMITH LEO BRIAN
SMITH LEWIS PHILIP II
SMITH MARK JR
SMITH OLEN WAINWRIGHT
SMITH PAUL RICHARD JR
SMITH RALPH EDWARD
SMITH RALPH WENTZ
SMITH RAYMOND JULIUS
SMITH RICHARD
SMITH ROBERT HAROLD
SMITH ROBERT JR
SMITH ROBERT NORMAN
SMITH RONALD CARLTON

SMITH STANLEY RICHARD
SMITH WALTER THOMAS
SMITH WILLIAM FRANKLIN JR
SMITH WILLIAM ROBERT
SMOYER JOSEPH RONALD
SNAVELY ROBERT AMMON
SNELL HERBERT DONALD
SNELL MARC EDWARD
SNOCK JAMES EDWARD
SNYDER CHARLES OWEN
SNYDER GERALD ALLISON
SNYDER JAMES RALPH
SNYDER LAWRENCE JAMES
SNYDER STEPHEN FRANCIS
SOKALSKY STEPHEN W JR
SOLIS ANTONIO ABEL
SOLLENBERGER DENNIS
MILTO
SOLOMON MILTON
SOPKO ROBERT MICHAEL
SOTAK TIBOR
SOUZON JEAN PIERRE
SOVIZAL ROBERT JAMES
SOWELL RONALD
SPADARO VICTOR ANTHONY
SPAHN DENNIS M
SPANGLER STANLEY E JR
SPEAR EDWARD BRUCE
SPENCE JOHN ANDREW III
SPENCE RICHARD BRUCE
SPENCER GLENN EUGENE
SPIELMAN JOHN MARK
SPILLERS WILLIAM ROBERT
SPINA ELMER FRANK
SPRENKLE DENNIS ALLEN
SPRINGFIELD ALFRED C JR
SPRINKLE THOMAS THOMA
SPROULE WILLIAM C JR
SPROUT RICHARD MICHAEL
SRAL LEONARD WALTER
STAFF JOHN STANLEY
STAFFORD FREDERICK
STAHL DONALD EUGENE
STAHL GEORGE HENRY JR
STAHL ROGER WILLIAM
STAIR WILBUR THOMAS
STALNECKER WILLIAM JOHN
STAMATO VINCENT JAMES JR
STANCELL JAMES JR
STANCHEK EDWARD MILTON
STANCIL REGINALD ALFONSO
STANCIU KENNETH ALLAN
STANKEVICH EDWARD JOHN
STANLEY DAVID CARL
STANLEY MICHAEL JOHN
STANTON JAMES
STASKO PAUL JR
STATES DAVID PERSHING
STAUFFER HERBERT
HOLLINGE
STAUFFER ROBERT EARL
STAYER HARRY SHERMAN
STAYROOK DONALD GLENN
STEARNS ALLAN JULIUS
STEELE THOMAS DONALD
STEIGHNER JAMES THOMAS
STEIGLEMAN DERWOOD D JR
STEIN LEON CHARLES
STELL JAMES ARTHUR
STELLMACH STANLEY R JR
STEPHEN PHILIPPE BRUCE
STEPHENS GARY BENNETT
STEPHENS JAMES CALVIN
STEPHENSON LYNN LADELLE
STEPHENSON RONALD DEE
STEPSIE RONALD STEVENS
STERN GARY WAYNE
STEVENS DAVID JR
STEWART JAMES WESLEY
STEWART JOHN JOSEPH
STEWART LESLIE JAMES
STIBBINS MILO BENETT
STILES DONALD LAVEREN
STILLEY RONALD JOSEPH
STINE JOSEPH MILLARD
STITELY CARL MICHAEL
STOCKMAN GENE WALLACE
STOCKMAN JOHN FRANK
STOCKTON CLIFFORD
GEOFFRE
STOJINSKI JOSEPH JOHN JR
STONE JOSEPH CHARLES
STONEMETZ GERALD DUANE
STOUDT GORDON EDWARD

208

STOUDT JOSEPH GEORGE
STOVALL WILBERT
STRALEY JOHN LEROY
STRATHMANN THOMAS WILLIAM
STRATTON THOMAS ALLAN
STRAUB JOHN EDWIN
STRAUB TERRY GORDON
STRAUSS HOWARD DAVID
STROHLEIN MADISON ALEXAND
STRONG HENRY HOOKER JR
STROTHERS THOMAS F JR
STUDER FLOYD
STUHL ALOYSIUS JOHN
STYS STANLEY ALBERT
SUDLESKY THOMAS FRANCIS
SUGDEN WILLIAM JAMES
SUGHRUE PATRICK J
SULLIVAN HUGH JOHN JR
SULLIVAN JEREMIAH JOSEPH
SULLIVAN JOHN BERNARD III
SULLIVAN PIERRE LEROY
SULLIVAN WILLIS M JR
SURGALSKI JOHN ANTHONY
SUSMARSKI KENNETH JOHN
SUTTON MATTHEW EARL JR
SUVARA FRANK CARL JR
SWANHART RUSSELL JAMES
SWARTZ GARY LEE
SWARTZ JAMES ALBERT JR
SWEENEY JOSEPH EDWARD
SWEENEY THOMAS JAMES
SWEGER RICHARD HAUSE
SWENSON PEDRO ARNADO
SWIFT JAMES THEALBEART JR
SWIGART ROBERT WILLIAMS
SWINDELL WILBUR EUGENE
SYKES KENNETH BERNARD
SYNKOWSKI VALENTINE JOHN
SZEYLLER EDWARD PHILIP
SZOSZOREK GERALD JAMES
SZYSZPUTOWSKI GERALD ADAM
TAGLIEBER LEONARD JOSEPH
TALIAFERRO NAPOLEON ENOCH
TALIANA JOHN BARRY
TALLMAN RICHARD JOSEPH
TARANTOWICZ JOHN EDWARD
TAYLOR DONALD RICHARD
TAYLOR KARL GORMAN
TAYLOR ORIS CAMILLUS
TAYLOR RICHARD HENRY
TAYLOR TERRY LEE
TEASLEY HENRY EZRA
TEMPLE THOMAS RICHARD
TERLA LOTHAR GUSTAV T
THIROWAY PATRICK JAMES JR
THOMAN TYRONE GARY
THOMAS AARON LEON
THOMAS DONALD LEROY
THOMAS GREGORY JOSEPH
THOMAS HENRY BENNY
THOMAS JOHN HENRY JR
THOMAS JOHN JOSEPH
THOMAS JOHN WILLIAM
THOMAS LEE DANIEL
THOMAS MICHAEL CLAIR
THOMAS MILTON HUMPHERY JR
THOMAS RAYMOND BRUCE
THOMAS WILLIAM ARTHUR JR
THOMPSON BILLY ALBERT
THOMPSON CHARLES MICHAEL
THOMPSON DAVID MATHEW
THORNTON JAMES VINCENT
THREATS GEORGE EDWARD
THURSTON DANIEL TUCKER
TICE EDWARD JOSEPH III
TICE FRED ROST
TIFFANY CLARENCE JAMES
TIGUE PAUL EDWARD JR
TILLERY JERRY THOMAS
TINKER NORMAN LEE
TINKO GEORGE DONALD
TIPPING HENRY ALBERT
TOAL ALONZO R
TOMASCHEK ARTHUR
TOMASKO DAMIAN THOMAS
TOMKO JOSEPH ANDREW
TOMON F RONALD

TONER LOUIS JOSEPH
TOOMEY JOSEPH PATRICK
TOOMEY JOSEPH PATRICK
TORCIVIA ANTHONY RICHARD
TORRENCE JAMES EDWARD
TORRES ROBERT
TORZOK JOSEPH
TOTH DONALD BONNEY
TOWNER JOHN GARTH
TOWNSEND GEORGE HARRY
TOWNSON ARTHUR CLARENCE
TRBOVICH DAVID JOHN
TREDINNICK CHARLES NICHOL
TREIBLEY KENNETH EUGENE
TRIEVEL CLYDE EDWARD JR
TRINKALA DAVID ALLEN
TRITT JAMES FRANCIS
TRUANCE FRANCIS PATTON
TRYPUS FRANK DONALD
TUBBS EDWIN FRANKLIN
TURCHI LOUIS
TURNER EDWARD PHILLIP JR
TWEEDY STUART KING
TWIGG MICHAEL WILLIAM
ULRICH RAY LEONARD
UMBENHAUER DALE E
UNRUH JAMES HOWARD
UPRIGHT BRIAN DALE
URBANSKI RONALD MICHAEL
URMANN JOSEPH HERMAN
URQUHART PAUL DEAN
USILTON JOHN CLANNAHAN
VALENTE GLENN CURTIS
VALENTINE JAMES RUSSELL
VALENTINE LEWIS RUSSELL
VALERIO DAVID N
VALINT JULIUS JOSEPH JR
VALKOS FRANCIS J
VALUNAS MICHAEL
VAN ARTSDALEN CLIFFORD DA
VAN BLARCOM RICHARD WILLI
VAN DYKE DEANE S JR
VAN KEUREN DEPUY RAYMOND
VANCOSKY MICHAEL ANTHONY
VANDERVORT WILLIAM F JR
VANLEW KENNETH LESLIE
VANNOY DAN PAGE
VANOVER EDWARD CHARLES
VASEY WILLIAM CHARLES
VAUGHAN DONALD SYLVESTER
VAUGHN JOHN MYRON
VAUGHN ROBERT LEE JR
VENDITTI NICHOLAS LOUIS
VERBILLA DAVID
VERDINEK GEORGE THOMAS
VERLIHAY FRANK T JR
VERNER SCOTT MITCHELL
VERNON PAUL LAWRENCE
VICKERS CHARLES GRIFFIN
VINAS GARY LIONEL
VITALE MICHAEL NICHOLAS
VITANZA CHARLES JOSEPH
VOGEL HAROLD RAYMOND
VOGEL TIMOTHY PETERSON
VOKISH JERALD ANTHONY
VOLTZ ROBERT CARL
VONTOR THOMAS JOSEPH
VRANKOVICH NICHOLAS SAMUE
VROMAN MERLIN HOWARD
VUGA STEPHEN MICHAEL
WADDLE WILLIAM SAMUEL
WADE LARRY
WAGNER BRUCE DAVID
WAGNER WILLIAM JOHN JR
WAHL PHILIP RAYMOND
WAHLEN GERALD JOHN
WALKER CARL LYNN
WALKER JAMES RICHARD
WALKER ROBERT LAMONT JR
WALKER SAMUEL FRANKLIN JR
WALKER WAYNE HOWARD
WALKO DANIEL STEVEN
WALL ALBERT CHARLES JR
WALL ROBERT ALBERT

WALLER JAMES
WALLICK RICHARD ALLEN
WALLS CARL WILLIAM
WALLS RONALD RAY
WALLS WILLIAM HENRY
WALSH CHARLES SUMNER
WALSH JEFFREY MICHAEL
WALSH WAYNE EMERICK JR
WALTER CLIFTON MARTIN
WALTER CLYDE ELMER JR
WALTERS SHERMAN LEE
WALTERS WILLIAM
WALTZ JAMES ROBERT
WALTZ LARRY THOMAS
WANNER CARL JOSEPH
WANTO JOHN PAUL
WARD PATRICK EDWARD
WARDELL HORACE LEE
WARGO DENNIS MICHAEL
WARGO VINCENT JOSEPH JR
WARK WILLIAM EDWARD III
WARNER MICHAEL PATRICK
WARNER THOMAS CRAIG
WARREN DAVID BRANIARD
WARREN ERVIN
WARREN LARRY
WARREN ROBERT MARION
WARREN THOMAS ALLEN
WARSING CHARLES GRAFFIOUS
WASHBURN DAVID ANDREW
WASHINGTON NATHANIEL
WASHINGTON SHERMAN THOMAS
WASHINGTON WILLIE J JR
WATKINS MAHLON HUGH
WATSON DENNIS ALLEN
WATSON JAMES ARTHUR
WATSON TYRONE CALVIN
WATTS THEARTIS JR
WEATHERBY JOHN GEOFFREY
WEAVER DALE LARRY
WEAVER FRANKLIN FLOYD
WEAVER GEORGE ROBERT JR
WEAVER HENRY LUE
WEAVER RONALD LEE
WEBER JEROME PAUL
WEBER JOSEPH ALAN
WEBSTER JAY DENNIS JR
WECKER HARRY HERR
WEEST JAMES JOSEPH
WEIAND RAYMOND D
WEIDLE ROBERT JAMES
WEIGNER DAVID RALPH
WEINTRAUB NEIL WILLIAM
WEIR PHILIP GRANT
WEISS FRANK ENZER
WEISS THOMAS JOSEPH
WEISS WILLIAM CONRAD JR
WEITKAMP EDGAR WILKEN JR
WEIXEL DANIEL JOSEPH
WELCH DONALD WALTER
WELESKI MARTIN W III
WELKER ABRAM JOSEPH
WELLER TERRY LEE
WELLINGS EDWARD ALFRED
WELLS ROBERT JAMES
WELLS ROGER ORRIE
WELSH EARL RAYMOND JR
WELSH JAMES RAYMOND
WELSH JOHN O NEIL JR
WELSH LEWIS NEAL
WELTZ HERBERT F JR
WENGER DAVID ALLEN
WENSEL MILFORD HOMER
WENTZ DONALD RAY
WENTZEL RALPH MICHAEL
WENZEL ROBERT LEE
WENZLER JOSEPH R
WERTMAN MICHAEL LEE
WERTS GREGORY IRA
WESCOTT FREDERICK DEVILLA
WESCOTT RICHARD LEE
WESCOTT ROBERT HYATT JR
WEST JAMES EDWARD
WESTFALL ROBERT LEE
WESTRA LEROY JAMES
WEYANDT IRVIN GRANT
WHALEN EDWARD EUGENE
WHARTON THOMAS MICHAEL
WHEELER LOUIS GERARD
WHEELER MORRIS EUGENE

WHEELER RALPH D III
WHIPPLE GARY EUGENE
WHITE ARNOLD SYLVANUS
WHITE CRAIG PRESTON
WHITE DANFORTH ELLITHORPE
WHITE EUGENE
WHITE GENERAL
WHITE JAMES BROADUS
WHITE JOHN WILLIE
WHITE MICHAEL MATTHEW
WHITE ROBERT RICHARD
WHITE WILLIAM IVAN
WHITEHEAD WILLIAM C JR
WHITEHOUSE RICHARD JAMES
WHITEMAN WILLIAM EARL II
WHITMORE GREGORY BRIAN
WHITTEKER RICHARD LEE
WHITTLE ALBERT ALLAN
WICKEL KENNETH WILLIAM
WIDENER MICHAEL EDWARD
WIEGAND DEAN MICHAEL
WIELER JAMES LAWRENCE
WILBUR WILLIAM JR
WILDMAN MILES GREGORY
WILHELM JOHN LESTER
WILHELMI HENRY JOSEPH JR
WILKINS WILLIAM GEORGE
WILL WILLIAM ANTHONY
WILLIAMS DUANE GREGORY
WILLIAMS FRANKLIN BRUCE
WILLIAMS JOHN KIRBY
WILLIAMS JOSEPH THOMAS
WILLIAMS KERRY LEE
WILLIAMS LONNIE CLIFFORD
WILLIAMS RICHARD C
WILLIAMS RICHARD HARRY
WILLIAMS RONALD AARON
WILLIAMS STEPHEN
WILLIAMS THOMAS ALBERT
WILLIAMS WOODROW
WILLIAMSON ERVIN HOWARD
WILLINGHAM NATHANIEL
WILLIS JOHN HENRY
WILLOUGHBY EARL CHARLES
WILLOW ROBERT GLENN
WILLS ROBERT JOHN
WILSON DAVID RALPH
WILSON HARRY CONARD II
WILSON JOHN HENRY
WILSON JOHN WILLIAM JR
WILSON LLOYD CALVERIA
WILSON ROBERT ALLYN
WINDFELDER JOHN EDWARD
WINDSHEIMER RICHARD LEE
WINGERT JAMES ALBERT
WINK MELVIN RALPH
WINKELVOSS THOMAS JOHN
WINOWITCH THEODORE ALAN
WINTER EDWIN THOMAS
WINTERS WILLIAM FREDRICK
WISE JAMES DAVID
WISE JAMES LEROY JR
WISSLER RICHARD LAVERN JR
WITKO DANIEL ANDREW
WITKOWSKI DENNIS EDWARD
WITMER KENNETH EUGENE
WITMER OMAR DAVID JR
WITTS JOHN JOSEPH JR
WITYCYAK GLEN ROBERT
WODARCZYK MATT JOHN
WOEHLCKE BERNARD RICHARD
WOJTYNA ROBERT ANTHONY
WOLF PAUL DEIHL
WOLFE FRANK JESSE
WOLFKEIL WAYNE BENJAMIN
WONN JAMES CHARLES
WOOD DONALD CHARLES
WOODARD WAYNE HOWARD
WOODLAND WAYNE KARL
WOODS GERALD
WOODS GREGORY WAYNE
WOODSON GEORGE WILSON JR
WOOLHEATER JOHN STEVEN
WORDEN ROBERT LEE
WORKMAN DAVID FRANK
WORKMAN JAMES HERBERT
WORLEY WILLIAM PAUL
WORMAN CHESTER EUGENE

JR
WORMAN KENNETH GLEN
WORRELL PAUL LAURANCE
WORTHINGTON ROBERT LEE
WRIGHT LESTER ALLEN
WRIGHT MICHAEL DALE
WRIGHT PAUL THOMAS
WRIGHT ROBERT RICHARD JR
WRIGHT TYRONE MELVIN
WUNSCH MICHAEL CHARLES
WYANT ALFRED LEROY
WYCINSKY GEORGE JR
WYLES DONALD CLAIR
WYLIE GLENN ROBERT
YADOCK DANIEL JOSEPH
YANTIS KENNETH RICHARD
YAPSUGA EDWARD F JR
YARBINITZ BERNARD FRANCIS
YARD BENJAMIN EDWIN JR
YARTYMYK MICHAEL HARRY
YASENOSKY ANDREW RICHARD
YATSKO JOSEPH PAUL JR
YEAST JOHN
YECKLEY CYRIL THOMAS
YEINGST PETER JOEL
YEUTTER DANIEL JOHN
YINGER WAYNE LEROY JR
YINGLING HARRY PATRICK
YODER LARRY EUGENE
YONIKA THADDEUS M JR
YONKIE PAUL E
YORK LARRY LEE
YOUNG FREDERICK ANTHONY
YOUNG GEORGE ALBERT
YOUNG ROBERT MILTON
YOUNG ROBERT WILLIAM
YOUNG RONALD EDWARD
YOUNG WILLIAM LLOYD JR
YUHAS RONALD PETER
ZACKOWSKI EDWARD FRANCIS
ZAREMBA THOMAS HENRY
ZAVACKI FRANCIS
ZAVISLAN BARRY ALAN
ZBOYOVSKI JAMES ROBERT
ZEIGLER ROGER DAVID
ZELENICK JOHN MALCOM
ZELTNER WILLIAM J III
ZERFASS JEROME VINCENT
ZERGGEN FRANCIS ALBERT
ZERR KENT MARTIN
ZIEGLER JOHN PAUL
ZIEGLER LAWRENCE GORIC
ZIEGLER STEVEN WILLIAM
ZIERDEN ROLAND STEVEN
ZOOG CHARLES LOUIS
ZOOK HAROLD JACOB

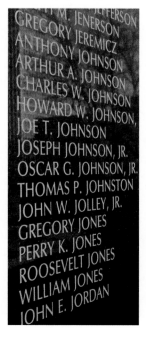

Photos by Ronny Ymbras

Vietnam War Memorial

301 S County Trail, Exeter, RI 02822

The Vietnam War Memorial located in Exeter, Rhode Island is situated within the Rhode Island State Veterans Memorial Cemetery on South County Trail. The memorial was created to honor the Rhode Island citizens who lost their lives during or as a result of the Vietnam War. The monument consists of multiple stone tablets as well as engraved rocks. The Vietnam War Memorial is one of many war memorials housed within the cemetery.

Less than a decade after the Veterans Memorial Cemetery opened in Exeter in 1974, Vietnam veterans pushed for one of the first memorials to be commissioned. In July of 1982, the Vietnam Veterans Memorial Monument and Grove was dedicated. It is said that the idea came about through Rhode Island State Representative Irving Levin. This original Vietnam War memorial features a single granite rock with an inscription honoring the service members who were missing in action. The 10 names of the Rhode Island missing are on the back of the rock. The rock proclaims that the surrounding grove is dedicated to the missing.

After some fundraising, the Vietnam Era Veterans Association in Rhode Island was able to create an Honor Roll memorial featuring the granite slabs with the engraved names of those lost in the war. The Vietnam Memorial Honor Roll was dedicated in 1986 by the then-governor Edward DiPrete. Bonner Monument Company was responsible for creating the memorial.

The memorial is understated and features 11 granite tablets in a "V" formation. Ten of the tablets feature the names of 224 deceased service members while the middle tablet features the dedication. Unlike other memorials, these tablets are made of Westerly blue-white granite instead of the polished black granite. The Rhode Island state flag and the POW/MIA flag fly above the memorial.

The Names of those from the state of Rhode Island Who Made the Ultimate Sacrifice

ADAMS CARROLL EDWARD JR
ALLEN ROBERT JOHN
ALSUP STEPHEN JOHN
ANDRADE JOHN DUTRA
ANDREOZZI VICTOR PATRICK
ANDREWS WILBERT ISOM
ANTER ALBERT GABRIEL
APPOLONIA JOHN JOSEPH
ARGENTI ROBERT LEE
ATTERIDGE LEON JOSEPH JR
AUGUSTINE FRANK FRANCIS
BAGLINI THOMAS EDWARD
BANEK LAWRENCE BENJAMIN
BEAUREGARD RICHARD MAURIC
BEDROSSIAN GEORGE J
BELVER DAVID EUGENE
BENEDETTI VINCENT MARIO
BENNETT FRANK EVERETT
BLAIR GERALD ALLAN
BLAKE RONALD EARL
BLANCHARD JAMES A
BOIS RENE ARMAND
BORGES JOSEPH WILLIAM
BOSCO FRANK JOSEPH
BOSSE LAURIER GERARD
BOURNE GEORGE LEANDER
BOYER BRUCE E
BRAGA JOHN PAUL JR
BRISSETTE RONALD JOSEPH
BRULE RICHARD CHARLES
BULPITT JOHN A
BURKE DENNIS EDWARD
BURLINGHAM ROBERT GENE
BURTON STEPHEN E
CABRAL PAUL ANTHONY
CALLAHAN CHARLES L III
CAMPBELL JOSEPH
CAPPELLI CHARLES EDWARD
CARPENTIER LUCIEN GERARD
CARROLL RAYMOND FRANK
CARTER FREDERICK THOMAS
CARUOLO RICHARD ANTHONY
CAVANAUGH JAMES VINCENT
CAVANAUGH WILLIAM THOMAS
CAZANAS-DIAZ EDWARDO ENRI
CHESEBROUGH FREDERIC READ
CLARK BRUCE ALAN
COCHRANE BLANCHARD WARD
COTTENIER ROBERT WILLIAM
COUTU RENE RAYMOND
COX HAROLD ANTHONY
CURTIS ALAN DENNIS
DALTON MICHAEL MORAN
DAVIS WILLIAM RUSSELL
DE CUBELLIS CARMEN JR
DE PALMA THOMAS CARMINE

DE RISO LESTER MICHAEL
DEAN JAMES WILLIAM
DECOTA WALTER JOSEPH
DION LAURENT NORBERT
DONOVAN ROBERT MARTIN
DRUMMOND PAUL ROBERT
DUCHARME RICHARD EDWARD
DUNSMORE LEO PAUL
DUPERE JOSEPH RENE
DURAND PAUL LIONEL
DYE DANIEL ROBERT
DYER RICHARD
EATON CURTIS ABBOT
EDWARDS JAMES HERBERT
EVANS DAVID PAUL
EXUM NEIL HARRIS
FARLEY PATRICK MICHAEL
FEGATELLI PETER FRANK
FRISK ROBERT JOHN
FULLER RONALD FRANCIS
GALLAGHER ROBERT PATRICK
GARDNER FRANK MAYNARD
GILL THOMAS PATRICK III
GLOVER JOHN
GOFF KENNETH B JR
GREENE RICHARD EDWARD JR
GUADAGNO GUY PAUL
HANLEY KEVIN CARROLL
HASLAM ALBERT WILLIAM
HEADLEY JOHN BRYANT
HICKS GENE DANIEL
HILL FRANK ALLEN III
HOLT CRAIG BARKER
HORNSTEIN EDMUND HENRY
HOULE ROBERT KENNETH
HULME JOHN WILLIAM
JACKSON ROBERT ANDREW
JALBERT DAVID MICHAEL
JAMES EDWARD ARTHUR
JEROME PAUL ANDREW JR
JOHNSON MAX ARDEN
JOURDENAIS GEORGE HENRY
JURCAK RICHARD ALAN
KANACZET JOHN FRANCIS JR
KAPAS PETER JR
KELLEY JOHN PATRICK
KNOWLTON GEORGE FRANK
KOSTER JOHN KNOWLES
KRAMER JOSEPH P
KRAWCZYK EDWARD CHESTER
LA PIERRE EDWARD ARTHUR
LA SCOLA VALENTINO J JR
LA VALLEE ROBERT C JR
LABRECQUE PAUL E JR
LAMBERT HENRY RAYMOND
LANCASTER ROBERT WEST
LANCTOT RICHARD LOUIS
LAUZON ROBERT WILLIAM
LAVOIE GERALD HENRY
LEBRUN ROBERT NORMAND

LEWIS JAMES EARL
LILLEY DAVID WILLIAM
LITTLEJOHN MC GEARY
LOPES LAWRENCE RENALDO
MAC NAUGHT ROBERT WILLIAM
MACCHIONI ALPHONSE JOSEPH
MACIMINIO ANTONIO PAUL
MACK GEORGE JACOB
MALONEY JOHN FRANCIS JR
MANCHESTER JAY HARRISON
MANSFIELD BRUCE ELWIN
MARSHALL JOSEPH LOUIS
MARTIN ROBERT WILLIAM
MC CUTCHEON ALLAN BRUCE
MC KENZIE RICHARD DOUGLAS
MC LAUGHLIN JOHN BERNARD
MC QUADE ARTHUR F JR
MELLOR FREDRIC MOORE
MELLOR MARK ELDREDGE
MENCONI WILLIAM LEE
MICHALOPOULOS RAYMOND WIL
MILLER ROBERT THOMAS
MILLIARD DENNIS EVANS
MISIASZEK JOSEPH PETER
MOORE WILLIAM HOWARD III
MORETTI ANTONIO LOUIS
MORGERA DOMENICO JR
MOSHER ROBERT LLOYD
MULLEN LEO ROBERT
MYLLYMAKI CARL W III
NELSON DANIEL RAYMOND
NIGRELLI THOMAS LYNWOOD
NIHILL RUSSELL EDWIN
NOEL JOSEPH DONAT
NORBERG WILLIAM GUNTHER
O BRIEN ROBERT EDWARD
O NEILL JOHN JOSEPH JR
OUILLETTE DONALD HENRY
PACHECO FRANK MANUEL
PALMER KENNETH OSCAR
PANZARELLA JAMES FRANCIS
PARMENTIER GERALD VICTOR
PENDER ORLAND JAMES JR
PERRY ERNEST MANUEL JR
PHILLIPS HENRY RICHARDSON
PIERCE RONALD GERARD
PIGEON JOSEPH THOMAS JR
PIGOTT CHARLES WILLIAM
PILTON GAVIN WILLIAM
PLANTE NORMAND AURELE
POTTER ALFRED N
PRITCHARD WALTER LEO JR
RAY JAMES MICHAEL
REILLY JOHN CHARLES
RENAUD ROBERT WILFRED

RIVET PAUL ROBERT
ROUNSEVILLE JOSEPH WILFRE
SAULNIER JEREMIAH JOHN
SCHANCK WILLIAM G JR
SCURFIELD DANNY VINSON
SHAW EDWARD BRENDAN
SILBA ANTHONY
SILVA WILLIAM GREGORY
SILVIA CLIFFORD WILLIAM
SISSON DONALD HENRY
SIVO ANTHONY JOHN
SMITH ANDREW DAVID III
SMITH JAMES EDWARD
SMITH THOMAS JOEL
STANLEY WILLIAM CHARLES
STEPHENSON GARY LUCKY
STERLING ROBERT JAMES
SYINTSAKOS PETER CHARLES
SYLVIA MICHAEL ALAN
SZYDLO THOMAS JOSEPH
TAYLOR ROBERT THOMAS
THIBEAULT FRANCIS JOHN
TRIPP ALFRED LEONARD

UPTON STEPHEN LOUIS
VAILLANCOURT EDWARD JOHN
VANDEVENDER JOSEPH TOMMY
VARDNER JOHN JOSEPH
WALLING LEWIS METCALFE JR
WALTON LEWIS CLARK
WANG ANDREW JACOB
WARNER JAMES LEO
WEBSTER DENNIS WADE
WEIR DAVID ANTHONY
WILKINSON RONALD JAMES
YAGHOOBIAN CHARLES JR
ZARBO MICHAEL

Photos by Keta Forrest

Vietnam War Memorial

700 Hampton St, Columbia, SC 29201

At the corner of Gadsden Street and Hampton Street in Columbia, South Carolina stands Memorial Park. The area is meant to be a space to honor all of the local soldiers from all of the US wars. In the park, lies the Vietnam War Memorial as well as other memorials for WWI, WWII, the Holocaust, Pearl Harbor, the Korean War, and the USS Columbia. The Vietnam War Memorial offers visitors a chance to reflect on the sacrifices made by the men and women who fought in the controversial conflict all those years ago.

The Memorial Park was dedicated in November of 1986 as was the actual South Carolina Vietnam War Memorial. According to records, this specific memorial is the biggest monument of its kind outside of Washington D.C. The Vietnam War Memorial is made of a tall granite column in the center with two granite walls standing on either side of the column.

The Vietnam War Memorial at Memorial Park consists of the granite column and freestanding walls. On the two granite walls are the inscribed names of the 980 South Carolinians who lost their life in Vietnam or who were deemed missing in action while on their tour of duty. All of the names are grouped by county and listed by the person's date of death or by the date they were declared missing in action. The MIA servicemen are designated with an asterisk before their name. The back of the granite walls boast Biblical scripture inscriptions.

The granite column features specific scenes that are reminiscent of the conflict that have been etched into the stone. A map of South Vietnam and the names of the Corps areas are also set into the actual plaza of the memorial. The Vietnam War Memorial is complete by stone benches around its perimeter as well as planted flags and a manicured surrounding landscape.

The Names of those from the state of South Carolina Who Made the Ultimate Sacrifice

ADAMS DWANE LONNIE
ADAMS JESSE LEWIS
ADAMS JOSEPH BOYCE
ADAMS WILLIAM OTHELLO JR
ADDIS BILLY WAYNE
ALEXANDER BARRY KENNETH
ALLAIRE JOHN KEVIN
ALLISON JOHN ROBERT
ANDERSON ARCHIE
ANDERSON WALTER H
ANTON TERRY LYNN
ARDIS JOHN COLEMAN
ARNOLD JAMES
ARNOLD LARRY FRANKLIN
AYERS JAMES WESTLEY
BACOT DOUGLAS MONROE
BAIR DONALD RAY
BAKER CLARENCE EUGENE
BAKER JESSE RUTLEDGE
BALDWIN WILLIAM ROBERT
BANKS MICHAEL FRANCIS
BANKS VINCENT NORVELL
BARBARE JAMES MICHAEL
BARBEE FRANK LEROY
BARNES ALLAN GEORGE
BARNES JOE WILSON
BARNETT TONEY ANTHONY
BARNETTE ROY GRANT
BARNHART JOHN LOUIS
BARRS SHELTON FERRELL
BARTON MICHAEL GEORGE
BELL NEWTON THOMAS JR
BELLAMY SIMMIE JR
BELTON JAMES
BELTON THEODORE
BERGESS FREDERICK WILSON
BEST NEAL IRA
BETHEA JIMMY CARLTON
BETHEA LUTHER JR
BEVERLY WILLARD FRANKLIN
BICKLEY WILSON CHARLES
BINGLEY JOHN LEE JR
BISCHOFF JOHN MALCOLM
BISHOP JAMES MATTHEW
BLACK DAVID FORREST
BLACKBURN HUGH FRANK
BLACKWELL ROBERT LAW-
RENCE

BLACKWELL ROY JAMES JR
BLANDIN RAYMOND WEL-
LINGTO
BLANDING AARON
BLANTON BURTON ALEX-
ANDER
BLAS FRANK
BOAN JIMMY E
BODISON JAMES CALVIN
BOLES HARRY LEE
BOUTON WILLIAM INNES JR
BOWEN MARVIN WHITNER
BOWEN MATTHEW ANDER-
SON
BOWERS GROVER COLEMAN
JR
BOWMAN CURTIS
BOWMAN FRANK
BOWMAN MELVIN
BOYCE EUGENE RUSSELL
BOYD SAM HENRY
BOYD WAYNE
BOYTER GEDDES CHARLES JR
BRADLEY JAMES
BRADLEY KENNETH RAY
BRAGG CLAUDE EDWARD
BRANHAM JOHNNY THOMAS
BRANNON WALTER LEE
BRANTLEY LEROY
BRASINGTON JACK WILLIAMS
BRATTON ROY DONALD
BREWER THOMAS COLEMAN
JR
BRICKLE DONALD LEVER
BRIDGES RICHARD HAMIL-
TON
BRIGMAN JOHNNIE LEE
BROCK THOMAS DEAN
BROCKMAN RICHARD
BELTON
BROCKWELL LEYBURN W JR
BROOKS DAVID LEE
BROOKS JAMES ROY JR
BROWDER PAUL ROGER
BROWN CHRIS JR
BROWN HARRY WILLIS
BROWN IRVIN
BROWN JAMES HENRY JR

BROWN JAMES LEE
BROWN SAMUEL JUNIOUS
BROWN WILSON BOYD
BROWNING ROBERT EUGENE
BRUNSON DAVID LEROY
BRUSTER WILLIAM EARL
BRYANT GARY RAY
BRYANT JAMES ROY
BRYANT JERRY
BURDETT CLARENCE HENRY
BURDETTE HILBURN M JR
BURGESS TITUS LEVEN
BURNS LUTHER
BURNS WALTER
BURROUGHS ULYSSES G
BURTON HENRY LEE
BURTON WILLIAM JR
BUTLER JOHNNIE ELMER
BUTLER LINNELL
BUTTON DONALD B
BYERS JERRY WALTER
BYRD RALPH EUGENE
CAIL GLENN ALFRED
CAIN ROBERT JR
CALDWELL WILLIAM JAMES
CALLAHAN CLIFTON EUGENE
CAMP JAMES STEVEN
CAMPBELL BILLY WAYNE
CAMPBELL RICHARD MI-
CHAEL
CANTRELL JAMES WESLEY
CAPE JERRY
CARNELL ARCHIE DENNIS
CARPENTER JAMES ALVIN
CARPENTER ROGER LEE
CARTER FRED JOSHUA
CARTER JOHN LEWIS
CARTER WILLIAM THOMAS
CASSEL KENNETH WAYNE
CATHEY J B
CATO HERBERT HUGO III
CAULDER DURWOOD
CAUSEY WILLIAM HARVEY
CHANDLER THOMAS LEROY
CHAPLIN LAWRENCE
CHASTAIN DONNIE RAY
CHASTAIN GERALD EDWARD
CHASTAIN WOODROW BEAL

CHAVIS RONNIE
CHESEBROUGH JOHN LANE
CHESTNUT LELAND MC LANE
CHILDS BOBBY RAY
CHISOLM ALEXANDER
CHISOLM ARTHUR LEE
CHRISTMAS MICHAEL LYNN
CHRISTOPHER SAMUEL JR
CLACK CECIL JAMES
CLARDY JAMES DONALD
CLARK ALLEN HOWELL
CLARK CHARLES
CLARY JOHN WILLARD
CLAYTON BENNY DEAN
COATS WILLIAM G
COAXUM THEODORE
COBB GEORGE LEE
COE HEWETT FRANK EVAS-
TUS
COKER BILLY LEE
COLEMAN OLAN DAN
COLEMAN WYMAN BYRD
COLLINS CLAUDE LAVERNE
COLLINS JULIUS JR
CONE LLOYD ALFORD
COOK JAMES EDWARD
COOK JOHN I
COOKE EDDIE BOYD JR
COOLER SIDNEY HOMER
COOPER CALVIN EMANUEL
COOPER FAY KENNY
COSOM LEVERN
COSTNER JOHNNY PHILLIP
COX JOSEPH WILLIAM
COX ROGER DALE
CRAIG ROBERT MITCHELL
CRAPSE WAYNE FRANKLIN
CREAMER ALBERT EUGENE
CRIBB FLOYD ALLEN
CROFT JIMMY O NEAL
CROSBY JACKIE LAWRENCE
CROTTS DONALD COLEMAN
CULP JOHN PAUL
CULVEY KENNETH LEROY
CURETON JOHNNY JACOB JR
CURRIE JAMES JR
DANIEL JOHNNIE LINCOLN
DANIELS DANNY EARL

DARNELL DANA CORNELL
DASH DAVID WINTHROP
DAVENPORT DAVID D JR
DAVES DONALD CHARLES
DAVIS CHARLES EDWARD JR
DAVIS CORNER MACK
DAVIS JOHN K
DAVIS LEWIS ANTHONY
DAVIS RANDY MAYO
DAVIS ROLAND K
DAVIS RONNIE LEE
DAVIS WILLIE JAMES
DAVIS WILSON
DAVIS WOODROW JR
DAWKINS CALVIN DONALD
DAWKINS CLARENCE JR
DEARING PHILIP RAY
DENNIS DELMAR CLAUDE
DERRICK ALVIN JOSEPH
DERRICK BRUNSON A SR
DIETZ WALLACE JAMES
DILL JAMES ARTHUR
DINGLE EARL
DIXON CORDIE LEE
DIXON FRAZIER THOMAS
DOLLARD JAMES
DOOLEY MARVIN LOUIS
DORSEY HARRY JAMES
DOUSE JAMES LOUIS
DOVER JOHNNY LEWIS JR
DOZIER WILLIE CLAY
DRIGGERS ARTHUR M JR
DRIGGERS JERRY TRUMAN
DRUMMOND AUSTIN LEON
DUCKER RONALD DWIGHT
DUDLEY BRUCE WESLEY JR
DUKE THOMAS WAYNE
DUKES GEORGE BENNIE
DUNMORE ONEAL
DUSING CHARLES GALE
DYCHES CHARLES HENRY
EDDLEMAN ROYCE EDSEL
EDMONDSON HAROLD T JR
EDWARDS AUSTIN IVAN
EDWARDS EDWIN RAY
EDWARDS FREDFOR
EDWARDS HARRY JEROME
EDWARDS JAMES MERTON
EDWARDS WILLIAM EDGAR
ELDRIDGE ROBERT BURCH
ELLIOTT THAROLD WASH-
INGTO
ELLIS JAMES MARION
ELLIS RANDALL SHELLEY
ELLIS RAYMOND DEAN
ELLIS WILLIAM JR
ELMORE ROBERT LOVIS
ESTELLA ANTHONY JOHN
ETHERIDGE COLIE JR
EVANS ALONZA
EVANS ERNEST
EVANS HENRY FRANKLIN
FAISON EVERSON BENJAMIN
FANNING RICHARD HENRY
FEDERLINE AUDLEY M JR
FELTY JAMES LEE
FERGUSON BENNY HAROLD
FERGUSON LEROY
FERRELL CHARLES REGINALD
FIELDS RONALD ELWOOD
FINCH FORDHAM E JR
FLAHERTY STEVE
FLANDERS LEON D
FLEMING LARRY JR
FORD CHARLES EVANS
FORE WILLIAM C
FOSTER EARL WILLIAM JR
FOSTER JAMES BYRD JR
FOSTER SAMUEL EDWARD
FOSTER WILLIE FRANK
FOXX RICHARD L
FRANKS ANTHONY L
FREDERICK ARTHUR DONALD
FREEMAN JOSEPH LLOYD JR
FRIERSON KENNETH
FURSE WARREN RANDOLPH

South Carolina

FYALL VERNON ROBERT
GAILLIARD HERMAN BERNARD
GAMBRELL FRANKLIN DOUGLAS
GARNER JOHN HENRY
GARNER WILLIAM DONALD
GARRETT ALONZO
GARRETT JACKY LEROY
GARRETT LLOYD EARL
GAY JAMES NATHANIEL
GEDDINGS JOHN HUGHIE
GELL JACK EARL
GENES LUTHER ALLEN
GENTRY ROBERT LEVETT
GEORGE EMMITT ROOSEVELT
GIBBS ISREAL
GIBBS WILLIAM JR

GIBSON DAVID
GIBSON JAMES DONALD
GIBSON SILAS EUGENE
GILMORE JAMES ROBSON JR
GILSTRAP DWIGHT ONEAL
GINN DAVID LANDRELL
GIRARD CHARLES JACK
GLEATON HOMER LAYNE
GLENN DONALD NELSON
GOLDSMITH DONALD LAWRENCE
GOLSON ANTHONY
GOOD ROBERT SAYE JR
GOODSON JOSEPH ALAN
GOODWIN RONALD NELIONAL
GORE HORACE ROSCOE
GOUDE CHARLES MELVIN
GOURDINE LARRY RONALD
GRAHAM EARL CLAYTON
GRAHAM ROBERT LEE
GRANT PHILLIP
GRANT THOMAS
GRAVES JEHOVAH
GRAVES MICHAEL LEROY
GREEN DAVID HARRY JR
GREEN DAVID NATHANIEL JR
GREEN EDWARD JR
GREENE WILBUR
GRIFFIN FRANCIS LEKIRKLAS
GRIGGS HARLEY FRANKLIN
GROOMS JIMMIE LEE
GRUBER CLEMENT BRADLEY
GUNNELLS WILLIAM ASHLEY
HACKETT HARLEY B III
HAINES ADHERENE LOUIS
HALL BOYCE LEE
HALL JIMMY WILLIAM
HAM TERRELL THOMAS
HAMBY PAUL CHARLES JR
HAMES BOBBY JOE
HAMILTON WINSTON CLINTON
HAMMOND LELAND EMANUEL
HANNA KENNETH
HANNON RICHARD LAMAR

HARDEE JOSEPH EDWARD
HARDIN CURTIS LEVENCE
HARRILL RONALD WILLIAM
HARRIS ABRAHAM
HARRIS CARL E
HART JOSEPH FELDER
HARVIN JIMMIE LEE
HATCHER KENNETH MARVIN
HAWKINS WILLIAM HENRY
HAYNES JOHN WALLACE
HAZLE BRUCE EDWIN
HEARON TERRY SHAVON
HELMS JERRY DONALD
HENDERSON MONTIE H JR
HENDERSON ROBERT LEE
HENNEGHAN ROBERT LEE
HENRY WILLIAM JAMES
HENSLEY GAREY LEE
HENSON LARRY KEITH
HERRING ALFRED JIMMY JR
HERRING BRADY WILLIAM
HIGBEE CHARLES E JR
HIGHT WILLIAM REID
HILL RALPH OWEN
HILL VERTIS JAMES JR
HILTON STEPHEN RANDOLPH
HINES JOHN THOMAS
HINGLETON EUGENE JR
HINO MICHAEL LYNN
HINSON JOHN ROBERT
HINSON RONALD DOUGLAS
HODGES HARKLES LEROY
HOFFMAN SHULER ADOLF
HOLDER CARL LEWIS
HOLLIFIELD VERNON GILBERT
HOLLINGSWORTH JERRY G JR
HOLMES JOHN HENRY
HOLMES THOMAS EUGENE
HOLMON ALPHONZO JR
HOOKER JOHN ARTHUR
HOOKS DAYTON JOSEPH
HOPKINS MYLON RAY
HORLBACK FRANCIS D
HOUGH MATTHEW
HOWARD DAVID LAFATE
HOWARD JULIUS JAKE JR

HOWE JAMES DONNIE
HOWELL ROBERT LEE
HOWELL SAMMIE
HOWLE ERNEST CLARENCE
HOWZE CHARLES CROCKETT
HUCKS LLOYD JUNIOR
HUGGINS EUGENE
HUGGINS GORDON SAMUEL
HUMPHRIES HAZEL H III
HUMPHRIES RONALD EDWARD
HUNTER LEROY
HUTCHERSON GARY CAROL
HYNDS WALLACE GOURLEY JR
JACKSON BENJAMIN FRANKLIN
JACKSON JAMES CLEVELAND
JACKSON NATHANIEL E L
JACKSON SYLVESTER JR
JACOBS ERNEST LINWOOD JR
JAMES EDDIE LOUIS JR
JEFFERS JOHN LARRY
JEFFERSON LEROY
JENKINS FRED HARVEY
JENKINS KENNETH BRUCE
JENKINS REGINALD ROCKEFEL
JENKINS WILLIAM
JENNINGS ROBERT LEE
JETER CURTIS LEE
JOE WILLIE LEE
JOHNSON ALBERT JR
JOHNSON CHARLES FRENCH JR
JOHNSON CHARLES JR
JOHNSON CLEVELAND OSBORNE
JOHNSON FRANKIE B JR
JOHNSON GEORGE
JOHNSON JACOB
JOHNSON LEMUEL
JOHNSON LEROY
JOHNSON NATHANIEL LERVERN
JOHNSON RALPH HENRY
JOHNSON WILLIE
JOHNSTONE JAMES M

JONES CHARLES A
JONES CLIFTON RANDAL
JONES ELIZABETH ANN
JONES HORACE
JONES JERALD LOUIS
JONES WILLIAM JUNIOR
JORDAN LARRY LEON
JORDON ARTHUR L
JOSEPH JAMES
KAPP RICHARD WORRELL JR
KEEFE DOUGLAS O NEIL
KEENE GERALD BRICE
KEITH WILLIE LEE
KELLEY DEWEY WILLIAM
KELLY BENJAMIN EDWARD JR
KEMP JIMMY
KENNEDY MATTHEW DEVANUGHAN
KEY ANDERSON HAROLD
KILBURN WILLIAM HUNTER
KING ROBERT CARL
KING ROBERT EARL
KING ROBERT LOUIS
KING THOMAS PICKETT BYRD
KING WYLIE CLARENCE
KINSEY HARVEY JUNIOR
KITCHENS JOEL RHYNE
KNIGHT RICHARD
KNUPP WAYNE WOOD
KOHN ALAN SPENCE
KOTT STEPHEN JAY
LAKASZUS HELMUT GUSTAV
LAMA IVARS
LAMKIN FREDDIE LEE
LANCE JOHN HENRY
LANE MICHAEL S
LANKFORD HENRY DEAN
LATIMER CLARENCE ALBERT
LAWLESS WILLIAM RALPH JR
LAWSON LAMAR ALVIS
LAWSON LARRY EDWARD
LAWTON BILLY JAMES
LAYTON WEBB HERMAN JR
LAZARUS SIDNEY GILBERT JR
LAZICKI JOSEPH CHARLES
LEAPHART HAROLD PAUL
LEDFORD DON KENNETH

LEE BILLY
LEE MELVIN
LESAINE JIMMY WILSON
LEWIS JIMMIE
LEWIS LEE
LEWIS WALTER WAYNE
LILES EPHRIAM RUTLEDGE II
LIPSCOMB ROY LOLLIS
LISBON JOHNNY
LIVINGSTON ERSKIN DAN
LLEWELLYN JOHNNY WILLIAM
LOCKLAIR ALLISON WAYNE
LONG RAYMOND ERVIN
LONG THOMAS IRA
LOPEZ LUIS BELT RAN
LOUALLEN JACK NEECE
LUCAS JOHN WILLIE
LUCAS PATRICK DONOVAN
LUSTER MILTON BERNARD
LYLES BELTON JR
LYLES CHARLIE JR
LYONS JAMES JR
MAC MICHAEL CHARLES EDWAR
MACK DANIEL JAMES
MACK HAROLD JR
MADDEN LEON SHIRLEY
MADDEN WILLIE ERSKINE
MAGAHA DANNY ROY
MAGBEE G W
MAHY HAROLD EUGENE
MAKIN JAMES BRIAN LAWRENC
MAKIN WOODROW JR
MALICHI BOBBY SPENCER
MANGUM ROBIN
MARTIN CLYDE THOMAS
MARTIN HOYLE
MASON KENNETH ALLEN
MASSEY JOHN WILLIAM JR
MATTHEWS EARL JR
MATTHEWS ROY GIBSON
MATTISON BENJAMIN FRANKLI
MAULDIN MELVIN CALVIN JR
MAULDIN THOMAS JASPER
MAYES JOSEPH
MC CLARY SAMUEL DONALD
MC COY ELEC
MC COY JAMES RAYMOND
MC CRARY DOUGLAS MAC ARTH
MC CULLOUGH BILLY RAY
MC CULLOUGH JOHN EARNEST
MC DANIELS BILLY CLAYTON
MC DONALD JERRY SYLVESTER
MC DOWELL CHARLES ELVIN
MC DOWELL HAROLD GUINN
MC DOWELL SAMUEL T JR
MC FADDEN CARL JR
MC FADDEN HARRY BERNARD
MC GAHA HAROLD F
MC GEE JOSEPH O NEIL
MC KAY JOHN ROLAND JR
MC KIE JACOB
MC KINNEY THOMAS ALAN
MC LENDON MICHAEL RYAN
MC MAHAN CHARLES LARRY
MC MAHAN THOMAS EDWARD JR
MC MAKIN WALLACE THOMSON
MC MILLIN ROBERT ALLEN
MC PHERSON ALFRED
MC RAE JIMMIE LEFON
MCKITTRICK JAMES CLIFFORD
MCMURRAY FRED HOWELL JR
MEADOWS CARROL FAYNE
MEETZE DENNIS RAY
MEGGS MARION LEE
MELVIN JAMES EDWIN JR
MERCK ROGER EUGENE
MESSER JACK WILLIAM
MICKLE MOSES
MIDDLETON STEVEN ALFRED
MILES GALEN SPINKS
MILES HAROLD GENE
MILES JOE JUNIOR
MILLER JAMES
MILLER JERRY LEE

MILLER PAUL
MILTON CHARLES RUDOLPH JR
MIMS GEORGE IVISON JR
MINCEY JAMES MARSHALL
MINOR CARROL WILLIAM
MITCHELL DAVID LEE
MITCHELL DONALD THOMAS
MITCHELL ISAIAH JR
MITCHELL MACK LEE
MITCHELL WILLIE JAMES JR
MONTAGUE JESSE WILLARD JR
MOODY ADGER EUGENE
MOODY HERBERT WAYNE
MOORE JACK DONALD JR
MOORE JIMMY LEE
MOORE LEONARD LEE
MOORE NORMAN JAMES
MOORE THOMAS ANTHONY
MOORE WILLIAM LEWIS
MOORER CLARENCE LARRY
MORENO RAMON
MORGAN CHARLES LEWIS
MORGAN CLYDE EDWARD
MORROW MERLE BRUCE
MORSE REGINALD GEORGE
MOSER SAMUEL RALPH
MOULTRIE JOE DAVIS
MULLINAX JAMES CARLTON JR
MUMFORD JIMMY EARL
MUNGER JOHN ROBERT
MURRAY CAESAR
MURSCH JOHN WILLIAM
NASH ANTHONY PRESTON
NEDD HEYWARD WINDELL
NELSON DICKIE OWENS
NELSON EUGENE
NELSON SANFORD LEWIS
NETTLES VICTOR LEE JR
NICKENS JAMES
NORRIS CHARLES BENJAMIN
NORRIS GRADY LEE
NORRIS RONNIE EUGENE
NORWOOD THOMAS LEE JR
NORWOOD WILLIAM ARNOLD
NUNNERY CLARENCE E JR
O CONNOR THOMAS DUCKETT
O CONNOR WILLIAM EUGENE
OLIVER CLIFTON
OLIVER RANDY DEWITT
OSBORNE SAMUEL WILLIAM JR
OWEN RAY WILLIAM
OWENS HENRY LAWRENCE
OWENS JAMES DOUGLAS
OWENS JAMES EUGENE
OWENS MARTIN LEE
OWENS ROBERT LEE
OXNER MARION LUTHER
PACE GARY LYNN
PACE RONALD EUGENE
PAGE PHILLIP ALLEN
PAINTER CURTIS WAYNE
PAIRIS ARNOLD
PARKER ALBERT EDEN
PARKER ARTHUR M III
PARKER EDGAR EUGENE
PARKER ROBERT
PARKER WESLEY
PARKS CHARLES H JR
PATE RONALD DALE
PATTERSON PATRICK CHASIE
PAULEY WASHINGTON
PAYNE WILLIE JAMES
PEAGLER WAYNE DONALD
PEARSON JESSE JAMES
PEEPLES BILLY HAMMOND
PELZER BENJAMIN F II
PENDARVIS ROBERT
PENN FRANKLIN HAMILTON
PERKINS WILLIE JAMES
PERRY ROBERT DALE
PESSIER STEVEN LEROY
PETERSON MATTHEW
PETTIT STANLEY RUSSELL
PETTY WILLIAM CLARK
PHIFER FREDDIE JOE
PHILLIPS CHARLES W JR
PHILLIPS JERRY ALFRED
PHILLIPS LAWRENCE
PHILLIPS WESLEY LEE
PHILLIPS WYLIE ORIA

PHOENIX ALONZA WILLIE
PIGATT HARMON JULIOUS
PLATT ROBERT LENWOOD JR
POE JIMMIE CLYDE
POLSTON ERNEST ELIJAH
POOLE WILLIAM DAVID JR
POORE THOMAS WYATT
POPE DONALD BURRIS
POSTON JAMES
POTTER LARRY EMERSON
POTTS JOHNNIE WYLIE
POUND THEUS JOSEPH
PRESTON JOSEPH JR
PRICE MAXIE LANE
PRICE ROBERT HUGH
PRIEST JOHN HENRY JR
PROCTOR WAYNE SHELTON
PRUITT DAVID MONROE
PRUITT OSIER LAWRENCE
PUCKETT TROY MURL
PUGH GERALD RALPH
PUSSER THOMAS WILSON
QUICK GEORGE DEWEY JR
QUIGLEY TERRY LYNN
RABON JOSEPH LEVERN
RAILEY GEORGE EDMUND
RAINS FORREST DE VERE JR
RAMBERT FRANKLIN
RAMEY JORDAN EUGENE
RAWLS CHARLES GLENN
READY JOHN III
REAVES JAMES LOUIS
REAVES JOHN SHEPARD JR
REEDER MELVIN
RENWICK HAROLD MCGILL JR
REVELL WILLIAM JAMES III
RHINEHART CLYDE A
RHODES FERRIS ANSEL JR
RHODES KENNETH
RICE THOMAS JR
RICH JOSEPH WALTER
RICHARDSON JIMMIE JENKINS
RICHARDSON THEODORE
RIKARD CHARLES DAVID
RILEY EDDIE LEE
RIVERS NATHAN
ROBERSON WILBURN
ROBERTS GARY KENNETH
ROBERTS KENNETH EUGENE
ROBERTS LLOYD VERNON
ROBERTSON BOBBY LEE
ROBINSON CALVIN
ROBINSON CHARLES HENRY
ROBINSON FREDDIE LEE
ROBINSON JIMMIE LEE
ROBINSON JOHN WILLIAM JR
ROBINSON LUCIEN
ROBINSON MELVIN
RODGERS TILLMAN DAVID JR
ROWELL FRANKLIN DELANO
RUFF WILLIE JEROME
RUTLAND WARREN LESTER
RUZICKA JOSEP L JR
SALE HAROLD REEVES JR
SALLEY JAMES JR
SAMUELS ISAIAH
SANDERSON BOBBY
SANFORD ARNOLD
SANFORD JAMES WALTER
SARTOR LEONDA
SAWYER JOHNNIE PAUL
SAYLOR WILLIAM JR
SCHAFFER BILLY JOE
SCHELLIN JAMES WILLARD F
SCHOPER GREGORY CARLYLLE
SCOTT LAWRENCE EDWARD
SCOTT RANDOLPH
SCOTT RANDOLPH CLINTON
SCOTT VERNON ELBERT
SEASE WILLIAM DANIEL
SEGAR CHARLES JR
SELLERS ROBERT
SHAW WILLIAM FREDERICK JR
SHEPHERD THOMAS GLENN
SHUMPERT JOE THOMAS
SIMMERS GAROLD RAY
SIMMONS EDDIE LEE
SIMMONS FAY CLYDE III
SIMMONS ISIAH
SIMMONS RICHARD
SIMMONS ROBERT LOUIS
SINGLETARY HILBERT M JR

SINGLETON RAYMOND
SIZEMORE DONALD RAY
SKIPPER MICHAEL RAY
SLOAN VERNAR
SMALLS BENJAMIN ALONZA
SMALLS JOSEPH
SMARR ALBERT WARD JR
SMITH CARY JOSEPH
SMITH DAVID GERALD
SMITH DAVID WAYNE
SMITH DAVID WESLEY
SMITH HAROLD MCRAE
SMITH JOHN CLIFFORD III
SMITH JOHN LEWIS
SMITH R J
SMITH RICHARD LEE
SMITH TERRY LEE
SMITH WILBUR EUGENE JR
SMITH WILLIAM HARRY
SMITH WILLIAM THOMAS
SMOAK JAMES THURSTON JR
SMYLY DUNCAN PADGETT
SNEED SAMMIE RAY JR
SNOWDON RICHARD ATWOOD
SOSSAMON EDWARD DE CAMP
SOWELL HARRY LEE JR
SPARKMAN ISAAC
SPEIGHTS ROOSEVELT
SPENCER JAMES ALBERT JR
SPIVEY WILLIE DALPHUS
SPRINGS ANDREW
SPROUSE RONALD EDWARD
STARKS JAMES EDWARD
STATON DAVID WALDEN
STEGALL LINDELL RAY
STENHOUSE J LYNN JR
STEVENSON BOBBY GENE
STEWART DAN ROGERS
STEWART ROBERT HENRY
STILL FLOYD LAVERNE
STILL FRED HAROLD
STILL JERRY MELTON
STONE DAVID RONALD
STONE JAMES EDWARD
STRICKLAND JOSEPH ODELL
STROUD DENNIS CARROLL
STUKES ISAIAH TRUMAN
STUMP HAROLD OLIVER
SULLIVAN GERALD DEWAYNE
SULLIVAN JOSEPH HARRY
SWEATT CLYDE STANLEY
SWEENEY MICHAEL MURPHY
SWOFFORD DANNY RAY
TALLENT HERSHALL
TAPSCOTT KENNITH WALKER
TATE WALTER REAVES JR
TATE WILLIE JAMES
TAYLOR CECIL FRANKLIN
TAYLOR GEORGE THOMAS JR
TAYLOR PRESTON JR
TAYLOR TED JAMES
TAYLOR THOMAS MARCELLUS
TAYLOR TOMMY LEE
TEAGUE JOHN WALTER
TEER WILLIAM EDWARD
THOMAS DOUGLAS JR
THOMAS JACK JR
THOMAS JAMES MYER
THOMAS RUDOLPH CALVIN
THOMPSON ALBERT C
THOMPSON CARL
THOMPSON CARROLL U
THOMPSON CHARLES LEE
THOMPSON DANNY STEWART
THOMPSON THERMALL
THOMPSON WILLIAM FRANK
TIMMONS BOBBY DANIEL
TODD CARL EDWARD
TOUCHBERRY MILES D JR
TOWNSEND CHESTER DAVIS
TRAPP BOBBY RAY
TRIPLETT JOHNNY RAY
TRUESDALE CHARLES KENNETH
TRUSTY MICHAEL JEFFERSON
TUCKER DAVID
TUCKER JOE NATHAN
TURNER ARTHUR JR
TURNER LINDSAY CLINTON
TURNER MARCUS SHARPE JR
TURNER ROBERT LAWRENCE
TUTEN MICHAEL HAMILTON

TYLER EARTHELL
TYLER JESSIE JAMES
TYNER JAMES PHILEMON
VILLANUEVA LARRY
VILLEPONTEAUX JAMES H JR
WAITERS LAWRENCE
WALDEN ROBERT DAVID
WALKER EDWARD
WALKER EVANS S
WALKER JOHN HENRY
WALKER LUTHER JR
WALKER THOMAS EDWARD
WALL JAMES ALLEN
WALLINGTON GEORGE HEYWARD
WALTERS EUGENE
WALTERS RICHARD EDWIN
WALTERS RICHARD FLOYD
WARD BEN CALHOUN
WARD HERSCHEL RUDOLPH
WARNER ARTHUR LEE
WASHINGTON GEORGE RAYMOND
WASILOW JOHN STEPHEN
WATERS FRANKLIN DELANO
WATSON CHARLES BRYANT JR
WATSON LARRY ELLIOTT
WAY CLARENCE L
WEATHERFORD ROY JULIAN JR
WELDON LIBERT JAMES JR
WELLONS HUGH WILLIAM
WELSH STEPHEN JACKSON
WESTPOINT THOMAS LEE
WHITAKER JERRY
WHITE BEN
WHITE ISAIAH
WHITE LEON
WHITE MELVIN ELIJAH
WHITE NATHAN JR
WHITE WILLIE
WHITT BROADUS ALFRED
WIGFALL HERBERT JR
WILKS JAMES LEE
WILLARD HUGH GREY
WILLIAMS BEN
WILLIAMS BILLY
WILLIAMS CALVIN
WILLIAMS CAROL EDWARD
WILLIAMS CLARK LEE VERN
WILLIAMS DANIEL III
WILLIAMS DOYLE
WILLIAMS EDDIE KENNETH
WILLIAMS FRANK NORMAN
WILLIAMS HARRIS LEE
WILLIAMS JAMES EDGAR
WILLIAMS JAMES U III
WILLIAMS JERRY HIOTT
WILLIAMS JOHNNY EDWARD
WILLIAMS LEONARD
WILLIAMS RONALD ANTHONY
WILLIAMS WILLIE AMOS
WILLIAMSON BENJAMIN JEFFE
WILLIAMSON EDWARD LOUIS
WILSON LAWRENCE W
WILSON WILLIAM EARL
WINE HARRISON JR
WITHERSPOON MARION
WOOD JAMES EDWIN
WOOD WILLIAM ESLEY JR
WOODS ADVERT JR
WOODS WILLIAM STEPHEN
WORTHY JERRY DEAN
WRIGHT CLIFTON
WRIGHT CURTIS RONALD
WRIGHT JAMES
WRIGHT JOE DAVID
WUESTENBERG LEWIS CURTIS
YATES CHARLES RUSSELL
YELDELL DAVID
YOUNG CLARENCE C
ZEIGLER GLEN ALLEN

215

Photos by Derald Gross

Vietnam War Memorial

Capitol Lake, East Broadway Ave. Pierre, South Dakota 57501

The South Dakota Vietnam War Memorial is located in Pierre, South Dakota and was dedicated in 2006. It's meant to be a tribute to all those who served in Vietnam. The memorial consists of bronze, full-size soldiers in combat gear similar to that worn by actual military members in Vietnam. There is also a fighting eagle bronze statue to go with the soldiers as well as a memorial wall.

The South Dakota Vietnam War Memorial uses one soldier to stand for all of the local military members who served in some capacity in the Vietnam War. It was created as a way to elevate those from feeling less than honored when they arrived home, to hero status as they originally deserved.

The sculptures were commissioned by South Dakota governor Mike Rounds who partnered with the South Dakota War Memorial. Artists Lee Leuning and Sherri Treeby created the statues with the help of a few Vietnam Veterans who aided in the overall design of the sculpture. Leuning and Treeby were also the artists responsible for the state's WWII and Korean War memorials. The artists were picked to execute the Vietnam War Memorial based on their merits and their past work on the other war memorials. The monument was installed and dedicated in September 2006 with the help of local volunteers and the South Dakota Bureau of Administration.

The sculpture itself is a life-size soldier wearing full combat gear like that worn in the jungles of Vietnam. It features fine details including the ammunition rounds slung across his shoulders and the dog tags and cross pendant around his neck. The soldier is also shown holding the dog tags of a fallen brother in a partially outstretched hand. The full-size soldier stands near the previously erected memorial wall which features the names of the 210 local fallen Vietnam veterans. To go along with these, there is also a fighting eagle bronze sculpture with its talons ready to strike.

The Names of those from the state of South Dakota Who Made the Ultimate Sacrifice

AADLAND GERALD L
ADAM RAYMOND ALVIN
ADAMS PHILIP J
ALDERN DONALD DEANE
ANDERSEN CURTIS LEE
ANDERSON DANIEL LEONE
APPERSON GERALD FRANK-LIN
ASCHENBRENNER DENNIS DALE
AUKLAND LEO CURTIS
BALTEZORE THEODORE ELLIS
BARBEE LARRY HULAN
BARTON JIMMIE WOODROW
BAUER LEO ALLEN
BECKERS JOHN PAUL
BERRIGAN BRENDON JAY
BETTELYOUN PERCY JR
BIBBY JOHN FRANCIS
BIEVER WILLIAM DENNIS
BOLHOUSE DEAN FRANKLIN
BOYER STEPHEN GREGORY
BRECH RICHARD LEE
BRIDGES ROBERT JAN JR
BROWNOTTER LAWRENCE DEAN
BUSSE DANIEL DEAN
CALLIES MARLIN JOSEPH
CALLIES TOMMY LEON
CAMERON ROGER SLETTEN
CASE EDWIN HARRY
CHRISTENSEN ALLEN DUANE
CHRISTENSEN ALVIN PETER
COATES HARRY JAY JR
CUMMINS RICHARD LEROY
CUNNINGHAM LOUIS JAMES
CUTSHALL DAVID WARREN
DEAL GARETH JOHN
DEXTER RONALD CLIFFORD
DOUGHERTY KIRBY JON
DRAIN HOWARD ELMER
EIDSMOE NORMAN EDWARD
EVANS PAUL OLYNN
FANTLE SAMUEL III
FENENGA TERRY HOWARD
FERDIG RUSSELL NORMAN
FJERSTAD DAVID ORSON
FLOWERS EDGAR ALLEN

FLYING HORSE CONRAD LEE
FORTIN ROBERT GENE
FRAHMAN LAWRENCE JOHN
FRENG STANLEY JON
FRIED VERN JACOB
FUCHS JAMES LEE
GAPP ALVIE WAYNE
GATTON DAVID RAY
GEARHART MICHAEL EUGENE
GEHLER RONALD CHARLES
GLAESMAN RICHARD E
GREEN RODNEY R
GRUENWALD MICHAEL JEROME
HAIDER JAMES FRANCIS
HALLSTROM CHARLES MAURY
HARMON JAMES CRAIG
HARRIS VERN ALLEN
HARTMAN DARRELL ELMER
HATLE THEODORE MAGNUS
HEIMES RICHARD THEODORE
HEINZMAN PETER GEORGE
HEVLE DAVID EUGENE
HILL WILLIAM LEO
HOLM DENNIS LEE
HORNER MARK ROLAND
HUNTER ARLEN JOHN
HURNEY JOSEPH EMANUAL
HUTCHISON JOHN WILLIAM
HYSELL HOWARD ROBERT
JACKSON MICHAEL MERE-DITH
JACOBS ROBERT MILTON
JAMERSON KENNETH ROBERT
JANSSEN DOUGLAS DUANE
JEALOUS-OF-HIM FRANK W
JOHNSON RONNIE WAYNE
JOHNSON ROY ALBERT
JORGENSEN SAMUEL JOSEPH
KAHLER LE LUND MORRIS
KAUPP CURTIS JAMES
KAYSER RUSSELL WILLIAM
KENYON DALE DEAN
KIDD DONNY RAMON
KIMMEL EUGENE WILLIAM
KIRCHGESLER DANIEL JAMES
KJELLERSON MYRON DALE
KLOOS RICHARD NICHOLAS

KNUTSON DENNIS CLARK
KORB DONALD DUANE
KRAGE LANNY RAY
KUSTER STEVEN MARK
KVERNES ROGER WENDELL
LABAHN DARWIN LYN
LANE CHARLES JR
LANGENFELD CHARLES THOMAS
LARSEN GARY ALVIN
LARSON FRED DUANE
LARSON MARVIN DEAN
LARSON RICHARD ANDREW
LAWVER DENNIS D
LAYTON DONALD DEAN
LEBEAU LOREN DALLAS
LEWIS ROBERT RUSSELL
LIEN JAMES LAWRENCE
LIPPMAN GORDON JOSEPH
MARKS MICHAEL DAVID
MATTHEWS EDGAR DONALD
MATTHEWS GILBERT LEWIS JR
MAYNARD RICHARD RAY
MC DOWELL JOHN CLARK
MC ILRAVY RONALD DEAN
MC INNIS DALE RICHARD
MC PHERSON DENNIS CRAIG
MEANS MICHAEL EDWARD
MEYER VERLYN GWEN
MICHELS MICHAEL RONALD
MILBRANDT CHARLES J
MILLER RICHARD LEE
MILLETTE HARLENE EUGENE
MILLS ARTHUR LEE
MILTON RICHARD DWAYNE
MOREHOUSE DAVID LLOYD
MOSER LEROY
MUHM ANTON LEONARD
NELSON ROGER MILLER
NOELDNER DANIEL MORRIS
NOLDNER RONALD LEE
NORGAARD LARRY WAYNE
NOTEBOOM IVAN
OLESEN RONALD ANDREW
OLLILA DONALD WARREN
OMMEN PETER RICHARD
OPSAHL JAMES DEAN
OVERACKER EARL JOHN

PASCH WILLIAM ERNEST
PETERSEN ROGER ALLAN
PETERSON JOHN ARTHUR
PLATE JAMES RICHARD
PORTER ROGER LEE
RADA TERRY GENE
RED HAWK JESSE MILTON
REIL RONALD LE ROY
REKER ROBERT VINCENT
RENELT WALTER A
RENNOLET RICHARD FRED-RICK
RENVILLE ARDEN KEITH
ROACH ORLANDO SILAS
ROBESON EVART EUGENE
RUSSELL ROY DEAN
SANDVE DONALD RAYMOND
SASSE PATRICK T
SCHNAIDT RONALD RUSSELL
SIMONSON DONALD WAYNE
SIP RAYMOND LEE
SOYLAND DAVID PECOR
SPIDER ALVIN RICHARD
SPRAGUE STANLEY GEORGE
STEINEKE JAMES LEE
STERN LONNIE LEE
STEWART WENDELL WARREN
STOCKWELL DENNIS BER-NARD
STURDEVANT WILLIAM DEAN
SUNDET GARY LEE
SURMA STEVEN JOHN
SWAYZE GERALD CLIFFORD

SYROVATKA ARNOLD DEAN
TAYLOR RONALD JAMES
THORMODSGARD ARVID PALMER
THORNE JOSEF LLOYD
TRANT STEVEN ALLEN
TWO CROW BLAIR WILLIAM
TWOEAGLE GABRIEL LAW-RENCE
UHLS WILLIS GRANT
VAN REGENMORTER RONALD RA
VENENGA DARRELL DEAN
VOIGT THEODORE HAROLD
VOLK DENNIS RICHARD
VOSS CURTIS MALROY
WATSON LESTER ARTHUR
WHITE MAUSE JOSEPH LEWIS
WHITES ROBERT JOSEPH
WHYTE RICHARD ALAN
WILLIAMS CURTIS LELAND
WILLIAMS RANDALL LEE
WINTERTON LARRY DEAN
WRIGHT JAMES PAIGE
YELLOW ELK CARLOS NICH-OLA
ZEIGLER THOMAS LEE
ZIMPRICH DENIS JAMES

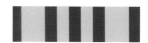

Photos by Janet Smith

Tennessee Vietnam Veterans Memorial

301 6th Ave N, Nashville, TN 37243

Located in Nashville, Tennessee's Legislative Plaza, the Vietnam Memorial Wall and the Vietnam Veterans Memorial were created to honor those who served and lost their lives during the Vietnam War. It is situated in the Vietnam Veterans Park, south of the War Memorial Building on Sixth Avenue. The park also plays host to the sculpture honoring the fallen Vietnam Veterans.

Following the dedication of the national Vietnam veterans' memorial in Washington D.C. in 1982, many states decided to make their own memorials to honor their local fallen service members. Tennessee took on their own effort in 1983 by forming the Tennessee Vietnam Veterans Leadership Program which began collecting money to honor the veterans. The effort proved to be successful when on Veteran's Day of 1985 the Vietnam Veterans Park was dedicated in the War Memorial Plaza. That day, a plaque was issued to honor the 49,000 local service members, especially the 1,289 men and women who passed as a result of the Vietnam War.

The memorial didn't come about without hiccups. Tennessee's governor Lamar Alexander and the legislature were happy to donate the space for the monument, but the necessary funds ($250,000) were posing to be a problem as people were shying away from donating to a standing monument instead of a continuing operation. After all was said and done, the Park's official dedication came in 1986.

The Tennessee Vietnam Memorial Wall is made of black granite and features the names, branches, and ranks of the fallen. There is also a bronze monument featuring three soldiers representing the honored combat veterans. The sculptures were made by sculptor Alan LeQuire who was chosen by a committee who examined his work along with other local sculptors. The three soldiers were made to represent varied ethnic backgrounds to cover all of those lost in the war. George Paine, a platoon leader in Vietnam wanted to battle negative feeling about the war by focusing on the sacrifice of Tennessee's civilians too.

The Names of those from the state of Tennesee Who Made the Ultimate Sacrifice

ABBOTT CARROLL DAVID
ABBOTT JAMES MICHAEL
ABBOTT WALLACE ADRION
ADAMS DWIGHT LEE
ADAMS FRANK HOUSTON
ADAMS JAMES RICHARD
ADCOCK BILLY ANTHONY
AKE HOMER LEE JR
AKINS JAMES FRANKLIN
ALDERIDGE JAMES CURTIS
ALEXANDER CHARLES PHILLIP
ALEXANDER ELTON HARROLD
ALEXANDER J H
ALEXANDER ROBERT LEE
ALLEN ADRIAN LAURENCE
ALLEN DAN S III
ALLEN DAN STEVEN
ALLEN WILLIAM CORDELL JR
ALLEY DON EARL
ALSTON BILLY CLYDE
AMICK RICHARD MICHAEL
ANDRE HOWARD VINCENT JR
ANDREWS JERRY LYNN
ANGUS WILLIE JAMES
ANTHONY GERALD DOUGLAS
ARCHER DAN WILLIE JR
ARCHER JESSE HAROLD
ARMES REXIE LEO
ARMS WILLIE DEWITT
ARMSTRONG BRUCE ELLIS
ARMSTRONG EVERETT
ARMSTRONG JAMES HAROLD
ARMSTRONG ROBERT DALE
ARWOOD LARRY RANDALL

ASHBURN RONALD WAYNE
ASHER SAMUEL EARL
AVERY JOHN PAUL
AYERS HAROLD GENE
AYRES CHARLES HASKELL
BABSON MARK ALBERT JR
BACON NILON KAY
BAILEY JAMES RAY
BAILEY WILLIAM EUGENE
BAKER GERALD D
BAKER MICHEAL ROGER
BAKER WILLY SCOTT
BALDRIDGE JOHN ROBERT JR
BALITSARIS JOHN BOMAR
BANNISTER RICHARD WAYNE
BARBEE JERRY PAUL
BARBEE JOHN WESLEY
BARBER BOBBY JOE
BARNARD LARRY WAYNE
BARNARD LEON EDWARD
BARNES CLARENCE EDWARD
BARNES JACKSON DILLON
BARNES MITCHELL ODELL
BARNETT DAVID WILLIAM
BARNETT JEFF THOMAS SR
BARRETT THOMAS A
BASHAM HAROLD LAWTON
BASS JAMES HENRY JR
BATCHELOR CHARLES EDWARD
BEARD CHARLES C
BEARFIELD CLUSTER LEE
BEAUREGARD SILVESTER
BEAVERSON HAROLD A JR
BECKMAN KENNETH BRYANT

BEDWELL SAMMIE LEE
BEERS JACK BLAINE
BELL DAVID LYNN
BELL HOMER B JR
BELL JERRY LEE
BELL PAUL M
BENNETT BILLY JOE
BENNETT PRENTICE J
BENTLEY JAMES E JR
BENTON JAMES AUSTIN
BERNARD THOMAS D
BIFFLE WILLIAM CALVIN
BIGGS JAMES EDWARD
BILLS KENNETH DALE
BISHOP WILLIAM BUEL II
BLACK WILLIAM RAY
BLACKBURN WILLIAM ALFRED
BLAKELY BRUCE WILLIAM
BLAKLEY EDWIN JR
BLAKLEY JAMES AUBREY
BLALACK JIMMY DALE
BLANTON JAMES LEE JR
BLAYLOCK BERYL STANLEY
BLEDSOE MILARD LUTHER E
BLEVINS DANNY EUGENE
BLUNKALL EARL JEROME
BOLDING LANNY ROSS
BOLTON DAN ARTHUR JR
BOLTON JESSE JAMES
BONNER CHARLES LARRY
BORDEN JAMES ARTHUR
BOSHEERS JAMES LARRY
BOWEN CLIFTON LEE
BOWERS JAMES DAVID

BOWMAN JAMES EDGAR
BOYD JOHN JOSEPH JR
BOYD ROBERT RAY
BRADFORD JOHN LESLIE
BRADFORD KIRBY WAYNE
BRADFORD TERRILL EDWARD
BRADLEY JOSEPH KEITH
BRADLEY KENNETH EUGENE
BRADLEY LARRY GRANT
BRADLEY RICHARD ALLEN
BRADSHAW CONLEY ARLEN
BRADY DAVID HARVEY
BRADY JAMES ALFRED SR
BRAMWELL RAYMOND SANDERS
BRANAM LARRY ANTHONY
BRANAM RONNIE FRANKLIN
BREMMER DWIGHT AMOS
BREWER JAMES LEWIS
BREWER SAMMY L
BRIDGES JERRY GLEN
BRIDGES WILLIAM EARL JR
BRINKLEY ROBERT THOMAS
BRINKMAN JAMES MARTIN III
BRITT JAMES JAY
BROCK JOHN HARRY
BROOKS HESSIE ALLEN
BROOKS ROY MAURICE
BROOME WADE LAMAR
BROWDER JEROME ALBERT JR
BROWN DALE FRAZIER
BROWN DAVID HAROLD
BROWN ERNEST LYKURGUS JR
BROWN JAMES AUSTON
BROWN KARL ANTHONY

BROWN MARSHALL EDWARD
BROWN NED RAYBURN
BROWNLOW ERNEST R III
BROYLES ALVIN KLASON JR
BROYLES FREDERICK PHILLIP
BRUBAKER HAROLD RAYMOND
BRYANT JOHNNY LEON
BRYANT WALTER TARVER
BUFORD LARRY GRAHAM
BUFORD LEROY
BULLARD STEPHEN EUGENE
BURDETTE LARRY WAYNE
BURKHEART GEORGE WILLIAM
BURKS HARMON WAYNE
BURNS BENNY CHARLES
BURNS CHARLES CALVIN
BURNS HOWELL WAYNE
BURNS JOHN ROBERT
BURNS RONDAL LEE
BURTON JACK EDWIN
BURTON JOHN THOMAS
BUSH JAMES EDWARD
BUTLER ROBERT EARL
BYRD JAMES EDWARD JR
BYRD RALPH
CABE PAUL PHILIP
CALL JIMMY OWEN
CAMPBELL JERRY ALBERT
CAMPBELL ROBERT WAYNE
CAMPBELL THOMETT DARTHAN
CANNING RICHARD BRUCE
CANNON RALPH

Tennesee

CANTRELL LESLIE HOWARD
CARLTON DANNY E
CAROTHERS RICHARD LEE
CARR DANNIE ARTHUR
CARR HAROLD EDWARD
CARROLL DWIGHT WAYNE
CARTER GARY MICHAEL
CARTER RONALD LEE
CASEY DONALD FRANCIS
CASEY JOHN MICHAEL
CASH JERRY MICHAEL
CASTEEL JAMES DENNIS
CATES GARY RAY
CHAFFIN WILLIAM T III
CHAMBERLAIN CARL EUGENE
CHAMBERS JAMES WAYNE
CHAMBERS THOMAS BEEB
CHAMBLEE DICKEY
CHANDLER ROBERT LARRY
CHANDLER THOMAS J JR
CHENEY DAVID PAUL JR
CHERRY JERRY LEE
CHESNUTT JEFFERSON CLIFFO
CHESTNUT JOSEPH LYONS
CHILDRESS BENJAMIN V JR
CHILDRESS IVY GALE
CHILDRESS ROBERT JR
CHILDS VANDIVER L
CHITWOOD HAROLD LYNN
CLARDY JASPER DUEL
CLARK BOBBY JOE
CLARK CHARLES EDWARD
CLARK GEORGE EDWARD JR
CLARK JIMMY RAY
CLARK JOHN JOSEPH
CLARK LORENZO
CLARK RAYMOND GORDON
CLAY AMBROSE WILKIE J JR
CLEAVE LONNIE LEONARD
CLIBURN HALQUA DALE
CLIFTON RANDY LEE
COALSTON ECHOL W JR
CODDINGTON JAMES PATRICK
COFFEY JESSE J
COFFEY WILLIAM LOUIS
COGDILL RANDY RALPH
COGGINS WILLIAM RAY
COLEMAN WILLIE JR
COLLIER JUNIUS COLUMBUS
COLLIER LAWRENCE HENRY
COLLIER WILLIAM FLOYD
COLLINS JAMES WILFORD
CONNER JEROME
CONNER KENNETH LEE
CONNER MICHAEL RAY
CONWAY JAMES BENNETT
COOK DONALD RICHARD
COOK GARRY KENDELL
COONE GEORGE W JR
COOPER WILLIE GENE
COOPERWOOD JACK J III
COPELAND EUGENE
CORBIN RUSSELL BIGBEE JR
CORNETT JAMES MITCHELL
COTHRAN ROBERT CARLIES
COTTEN LARRY WILLIAM
COUCH GEORGE MICHAEL
COULTER DONALD CLAY
COURTNEY JIMMY DARRELL
COVINGTON HOBART EARL
COX JOSEPH LEE
COX STERLING EDWARD
CRABTREE DAVID ALLEN
CRABTREE GEORGE RONALD

CRABTREE HARVEY C JR
CRABTREE STEPHEN C
CRAIN CARROLL OWEN JR
CRAVEN FLOYD CARL
CRAWFORD GALE VERNON
CRAWFORD JAMES DAVID
CRAWFORD JAMES EDWARD
CREEKMORE JESSE CARL
CREIGHTON PAUL BAREN
CRENSHAW OLLIE EDWARD JR
CREWS PHILIP MARVIN
CROWE CHARLES DOUGLAS
CRUTCHFIELD TERRY WAYNE
CULVER DARYL CHESTER
CUMMINGS DAVE JR
CUNNINGHAM JAMES ANDREW
CUNNINGHAM JAMES EARL
CUNNINGHAM JOHN EDD JR
CUNNINGHAM LEONARD DWIGHT
CURRIE ANDREW
CURTIS GARY ALLEN
CURTIS HAROLD GENE
CURTIS LARRY GENE
CUTSHAW WILLIAM
DAIL WILLIE FRED JR
DALLAS RICHARD HOWARD
DAMEWOOD DONNIE LEE
DANIEL JAMES LEE
DARLING THOMAS DUANE
DARNELL MICHAEL EDWARD
DAUGHERTY RONNIE JOE
DAUGHERTY WILLIAM DON
DAVIDSON GERALD WAYNE
DAVIDSON VERBIN EUGENE
DAVIS AUBREY GUY
DAVIS J C
DAVIS JAMES THOMAS
DAVIS JAMES THOMAS
DAVIS JOHN EDWARD
DAVIS JOHN PAUL
DAVIS JOHN TRAVIS
DAVIS KENNETH JOE
DAVIS LESTER RAY
DAVIS LUTHER EUGENE
DAVIS RICHARD HAROLD
DAWSON MICHAEL DAVID
DAY BILLY BROWN
DAY MICHAEL ROBERT
DE LONG RONALD LAWRENCE
DEAN SIMON JR
DELANEY DONNEY
DELONG JOE LYNN
DENSON JERRY EDWARD
DERRY DAVID WAYNE
DIAL JAMES WILLIAM
DICKEY JAMES MARCELL
DICKINSON THOMAS MORTON
DICKSON KENNETH ODELL
DILLARD TERRY LEE
DINGUS JOHN WILLIAM JR
DISON EDWARD DEAN
DOBBS GERALD THOMAS
DOCKERY STEVE JULIUS
DODSON BILLY
DODSON SEAN PAUL
DORRIS CURTIS EUGENE
DOSS LARRY DONNELL
DOTSON DONALD LUTHER
DOUGHERTY LON JR
DOUGLAS JAMES THOMAS
DOWDY WILLIAM
DUFFER ERIC THOMAS
DUMAS SAMUEL ALEXANDER
DUNCAN ROBERT LEE
DUNLAP RICHARD LANCE
DUNNAVANT JAMES M JR
DUTY CHARLES HOWARD
DYE RALPH VICTOR JR
DYER WILLFORD LEON
DYKES WILLIAM FRANK
EARLS LARRY DON
EDMONDS JERRY BAXTER JR
EDNEY DAVID LEE
EDWARDS CHARLES LEE
EDWARDS GARY LEE
EDWARDS GILBERT
ELAM JOHN JR
ELLIS BILLY JOE
ELROD WAYMON CLAY
EMORY THERMON HENRY JR
ENGLE PHILLIP HENRY

ENSLEY RONALD JOHN JR
ERVIN JERRY GLENN
ERWIN DANNY MAX
ERWIN YOUEAL DEAN
ESSARY GEORGE ARTHUR
ESTES DOUGLAS DALE
EUBANKS RAYMOND CARL JR
EVANS WILLIAM LARRY
FARLEY JAMES CABELL
FARMER JOSEPH LYLE
FARRIOR BILLY RANDY
FELDHAUS JOHN ANTHONY
FERGUSON JERRY ROGER
FERGUSON MICHAEL LYNN
FERGUSON RICHARD LEE
FERRELL JOHN WESLEY
FERRELL WILLIAM ALFORD
FIELDER CALVIN
FIELDS JERRY L
FILLERS DONALD JAY
FINK PHILIP RUSH
FINLEY NICK ALLISON
FISHER HARRY
FITZGERALD MANFRED
WILLY
FLANIGAN WESLEY ELMER
FLEMING WILLIE JAMES
FLETCHER DONALD FRANK
FLETCHER HERMAN RAY
FLETCHER JERRY
FLOYD CHARLES GRADY
FLOYD LONNIE ALLEN
FLURRY JAMES DURWARD
FORD CHARLES JESSE
FORD FREDDIE DARREL
FORD GEORGE B
FORD WILLIAM WALLACE
FORKUM GARRY MICHAEL
FORT MELVIN FRANK
FOSTER DOYLE

FOSTER LARRY EDWARD
FOSTER PAUL LEONARD
FOWLER WILL LEE DENNIS
FOX GARY REGAN
FRANCE MACK LEMUL JR
FRANKLIN EUGENE DELANO
FRANKLIN WILLIAM E JR
FULGHUM JOE RAYMOND JR
FULLAM WAYNE EUGENE
FULTON CHARLES EDWARD
FUQUA JOHN EDWARD
GALBRAITH HUGH CAMPBELL
GAMBLE BOBBY GENE
GARANT ROBERT OLIVER
GARCIA AUGUSTO JOSE
GARDNER JAMES ALTON
GARRIGUS HARRY
GARRISON GEORGE ALBERT
GARTH JESSIE JAMES
GAY WAYNE OLIVER
GENTRY LENNIS CLYDE
GEORGE JOHN WILLIAM
GERVAIS DONALD PETER
GIBBS JACK RONALD
GIBBS MICHAEL GERALD
GIBSON ERNEST
GIBSON LONNIE LOWELL
GILBERT CARL EDWARD
GILBERT JAMES CAROLL
GILL WILLIAM RALPH JR
GILREATH LUTHER VERMONT
GLANDON GARY ALVEN
GLASS BILLY WAYNE
GLOSSUP LOWELL THOMAS
GLOVER ARTHUR WAYNE
GOBBLE CHARLES HARTSELL
GOETSCH THOMAS AUGUST
GOINS WILLIAM HARRISON
GOLD FRED EDWARD
GOODMAN CHARLES OBA-

DATH
GOODMAN THOMAS HILL JR
GOSSETT HERSHEL LEE
WALTO
GRADY JAMES WILLIAM
GRANTHAM ROY EUGENE
GRAVES GEORGE W III
GRAY FREDDY LYNN
GRAY JAMES T
GRAY JESSE ALEXANDER
GRAY JIMMIE DELL
GREEN DALLAS EUGENE
GREEN WILBUR LEON
GREENE FREDDY
GREENWELL WILLIAM
LEONARD
GREGG DANIEL LEE
GRIFFIN JAMES LLOYD
GRIFFIN JIMMY RICHARD
GRIFFIN ROBERT EUGENE
GRIFFITH JOE EDD
GRIFFITH TONY LEE
GRINDSTAFF THOMAS
JACKSON
GRISHAM CHARLES COLE
GRISSOM HAROLD GLENN
GRUBB DONALD LEE
GUINN FREDDIE RAY
GUINN JOSEPH WADE
GUNTER ALVIN FLYNN
GUTHRIE ROBERT ELDRIDGE
HAAK WILLIAM LEWIS
HAGGARD WILLIAM ELMER
HALL ACIE DAVID
HALL FRANK JR
HALL GARY VAN
HALL JOHN LOUIS
HALL PRESTON LEE
HALL WILLIAM GARY
HALL WORLEY WAYNE

HAMILTON ROBERT E LEE
HAMLETT BYRON DWAYNE
HAMM EDDIE DEAN
HAMMOCK JERRY WENDELL
HAMPTON WALTER JAMES
HANKINS ALBERT RAY
HANKINS THOMAS MAURICE
HANN DAVID LEE
HARDEN DANIEL DAVID
HARDISON CHARLES HER-
BERT
HARDISON ROBERT SMITH
HARGROVE WILLIAM EUGENE
HARLAN JOHN ALVIN
HARPER STEVEN FRANCIS
HARPER WILLIAM CLYDE
HARR MICHAEL EDWIN
HARRINGTON CHARLES J
HARRIS EARL WAYNE
HARRIS JOHN CARLOS
HARRIS LARRY RAY
HARRIS LESLIE EARL JR
HARRIS WILLIAM LEE
HARRIS WILLIAM THOMAS
HARRISON BILLY GERALD
HARRISON BILLY JOE
HARRISON RICHARD EARL
HARRISON WALTER CORNEL
JR
HART WILLIAM EDMOND LEE
HARTLEY WILLIAM LEE
HARTMAN BENJAMIN C JR
HASSLER FREDRICK ANDREW
HAWKINS GORDON ABNER
HAYES JESSE BOYD
HAYES LYNN CAROL
HAYES RAY ALLEN
HAYNES BOBBY GENE
HAYNES RON JACKSON
HAYWORTH DENNIS TRUMAN
HEADLEY CHARLES PAUL
HEADRICK VERNON LEON
HEATHERLY DARRELL W
HEATHERLY GEORGE GLEN
HEDDEN HAROLD C JR
HEISKELL LUCIUS LAMAR
HEISSE EDWIN
HELTON JAMES CARLOS
HENDERSON BRUCE DALE
HENDERSON DONNELL
HENRY EDWARD EARL
HENRY GEORGE EDWARD
HENRY JIMMY LYNN
HENRY NELSON PAGE
HENSLEY A G
HERRIN CLIFFORD RAY
HESSON JIMMY DALE
HESSON LARRY DEAN
HIBBARD RONALD EDWARD
HICKEY JERRY THOMAS
HICKEY JOHNNY
HICKS ELVIS GORMEN
HICKS JAMES LARRY
HICKS MANUEL ARTLAN JR
HICKS PAUL J
HICKS RANDOLPH TRUMAN
HIEMER JERRY ALLAN
HILL CLEABERN WILLIAM JR
HILL LONNIE O NEAL
HILL RAYMOND LEE
HILL TOMMY EDWARD
HILLMAN COLEMAN GEE
HIMES EDWARD LOUIS
HINNANT KENNETH LEE
HITCHCOCK TILLMON
POWELL
HIXSON RANDALL LEE
HODGE KENNETH RAY
HODGES HARRY GAINES
HODGES LARRY LEON
HODGES ROBERT GENTRY
HOLCOMB DOYLE
HOLEYFIELD ROBERT ERIE
HOLLAND CHARLES RALPH
HOLLOWELL DALE MITCHELL
HOLMES DAVID WILLIAM
HOLMES EDWARD WAYNE
HOLT DENNIS EDWARD
HOLT HERSCHEL CYLE
HOPPER JOSEPH CLIFFORD
HOPPER WILLIAM CARL
HORNBUCKLE ALTON LEE
HORNER HERBERT DAVID
HORTON DONNIE EDWARD

HOSKINS GOMER DAVIS JR
HOUSE DOUGLAS ARTHUR
HOUSLEY CHARLES LARRY
HOVER JOHN MICHAEL
HOWARD DONNELL
HOWARD ERNEST
HOWARD JAMES VAN
HOWELL JOHN WILLIAM
HOWLAND HOWARD P JR
HUBBARD JAMES RAY
HUBBS DANNY EUGENE
HUDDLESTON ROBERT
JOSEPH
HUFFINE MELVIN THOMAS
HUFFMAN WALTER LEE
HUGHES CHESTER STACY
HUGHEY CHESTER LYN
HUGHLETT JOHN ALBERT
HUMPHREY LAWRENCE
JAMES
HUNT BOBBY EARNEST
HUNT ISAAC E JR
HUNT ROBERT EARL
HUNTER JAMES D
HUSKEY ESTEL
HUTCHERSON JIMMIE CLAY
HUTCHERSON RONALD
DAVID
HUTCHINGS ALVIN DALE
HYLMON JAMES EDWARD
INGLE NATHAN LAMAR
IVORY MICHAEL THOMAS
JACKSON ARNOLD BRYAN
JACKSON CHARLES SID
JACOBS AUBREY EUGENE JR
JACOBS DANNIE DICK
JACOBS GEORGE EDGAR
JAMES RAY DON
JAMES SAMUEL LARRY
JARRELL ROGER DALE
JAYNES JAMES OAKLEY
JENKINS LEN MCKINLEY
JENKINS ROBERT DONALD
JERSTAD LESLIE ARTHUR
JOHNSON ALEX LEE
JOHNSON DALE ALONZO
JOHNSON EDWARD
JOHNSON EVERETT WILSON
JR
JOHNSON HOWARD LEON
JOHNSON WILLIAM LEROY
JOHNSTON RICHARD KEITH
JONES BENNIE RAY
JONES CECIL BEN JR
JONES DANNY LEE
JONES DOUGLAS LEE
JONES EMANUEL JR
JONES JIMMY LEWIS
JONES JOHN FREDERICK
JONES JOHN HUBERT
JONES JOHN IVORY JR
JONES JOSEPH WESTER III
JONES LENNIS GODDARD JR
JONES MICHAEL ALLEN
JONES RAY MORGAN KEITH
JONES SAM RAYMOND
JORDAN RAYMOND ROBERT
JOWERS RAY RAMSEY
KALLAHER CHARLES THOMAS
KEASLING ELMER LEO
KEELEY FREDDIE JOE
KEEN JASPER LEE
KEETON TOMMIE
KEITH JAMES KELLY III
KELLEY FREDDIE RAY
KELLEY GLENN HOWELL
KELLEY JAMES DANIEL
KELLY CARL EUGENE JR
KELLY EDDIE JR
KEMMERLING JOE THOMAS
KENNEDY MILTON REGINALD
KENNEDY THOMAS MARTIN
KENNEDY WILLIAM HENRY
KENT WILLIAM WAYNE
KILGORE GARY BREWSTER
KIMSEY WILLIAM ARTHUR JR
KING DANNY EUGENE
KING GERALD EUGENE
KING HAROLD B
KING LONNIE RALPH
KING ROBERT EARL
KIRKES KENNETH LEE
KIRKLAND VIRGIL JR
KNIGHT JAMES WILLIAM

Tennesee

KNIGHT RONALD HAROLD
KNOX EDDIE L
KOLWYCK JOHN A
KOONCE MICHAEL EARL
KOPPEL REDLICK SIMS
LAMB FLOYD WATSEL JR
LAMMEY LLOYD GENE
LANCASTER JERRY DAVID
LAND LARRY ADRIAN
LANDRY JOHN PATRICK
LANE NORMAN EDWARD JR
LANE ROBERT HARRISON JR
LANKFORD BILLY EUGENE
LANNOM RICHARD CLIVE
LARSON ANDREW MARTIN
LATHAM THOMAS EUGENE
LAWSON DALLAS
LAWSON THOMAS ANDREW
LAY JOHN EARL
LAYNE DILLARD RAY
LAYNE THOMAS ELSWORTH
JR
LEA FRED STANLEY
LEAMON LARRY DEWAYNE
LEDFORD JAMES HARVEY
LEE BENJAMIN IV
LEE HAROLD EUGENE
LEE JOHNNY ANDREW
LEFTWICH WILLIAM GROOM
JR

LEMONS BOBBY JOE
LEONARD CHARLIE MURPHY
LEWIS AL RICKEY
LEWIS BOBBY DWIGHT
LEWIS THOMAS
LIGHTFORD WILLIE JUNIOR
LIKENS ARTHUR EMMITT
LINDLEY RONNIE DEAN
LINSON ROBERT WYLIE
LITTLEJOHN TROY A
LIVELY MARVIN EUGENE
LIVERMORE ROSS WHITTIER
LOCKHART KENNETH EU-
GENE
LOCKRIDGE JAMES T
LODEN LARRY DAVID
LOFTON RAYFON
LOGAN JIMMY MORRIS
LONDON WILLIAM THOMAS
LONG BILL BROOKS
LONG FREDDIE LERON
LONG JAMES DAVID
LONG WAYNE THOMAS
LOONEY ROBERT
LOVE HUGH ALLEN
LOVE LARRY DALE
LOVELACE ROBERT ALAN
LOVELL JERRY MICHAEL
LOWE EARNEST LON
LUMLEY DONALD RICHARD
LUNSFORD JAMES ROBERT
LUNTSFORD JACK EDWARD
LUSCINSKI JAMES TIMOTHY
LYLES OSCAR BURL JR
LYNN JOHNNY RALPH
LYONS LARRY JEROME
MADDEN DONALD EUGENE
MADISON WILLIAM CURTIS
MAJORS DANIEL WILLIAM
MALONE CLIFTON
MALONE WILLIAM FRANKLIN
MANIS EDDIE
MANNS ROY NANCE
MAPLES PAUL EDWARD
MARCUM JIMMY ALLEN
MARINE DAVID HARLON
MARKLAND JAMES HARRY
MARLIN LEONARD THOMAS
MARLIN ROBERT DOUGLAS

MARLIN WILLIAM LUNN JR
MARLOW DONALD RAY
MARPLE REECE LESLIE
MARTIN DANNY GALE
MARTIN GEORGE WILLIS
MARTIN IRVIN EUGENE
MARTIN JAMES HENRY
MARTIN JOHN DAVID
MARTIN JOSEPH VENSON
MARTIN KENNETH WAYNE
MARTIN ROBERT THOMAS JR
MASON ROBERT
MATHIS DONALD ROBERT
MATTHEWS THOMAS W JR
MATTHEWS WILLIS ALANZO
MATTOCK JOHN LEE
MAYS JAMES JR
MC ARTHUR ROBERT LAMAR
MC BROOM EDDIE O'DON-
ALD JR
MC CALL CLAIBORNE PARKS
MC CARRELL JOHN EDWARD
MC CARTER JERRY
MC CARTER ROBERT LEON-
ARD
MC CARTER THOMAS LUTHER
MC CORKLE BENNIE EUGENE
MC CORMICK DONNIE RAY
MC CORMICK RONNIE LEON
MC COULLOUGH BEN JR
MC COY CARL THOMAS JR
MC COY JAMES LARRY
MC CRARY JACK
MC CULLOUGH JERRY WEN-
DELL
MC DANIEL ANDREW LEROY
MC DERMOTT BERNARD A III
MC ELHANEY LEE ROY
MC ELROY THOMAS LEE
MC EWEN THOMAS C JR
MC FALLS JERRY ARNOLD
MC GARRITY JAMES ERLEY JR
MC GARRY THOMAS STEWART
MC GHEE BILLY WALKER
MC JUNKIN ROBERT TAYLOR
MC KEE JAMES EVERETT
MC KINNEY CHARLES MI-
CHAEL
MC LEMORE JAMES ROBERT

MC LOUGHLIN ROBERT A JR
MC MURTRY RALPH DAVID
MC NEAL RICHARD
MC NISH JAMES RONALD
MC PHAIL WILLIAM THOMAS
MC REE JOHN HENRY
MC REYNOLDS GEORGE
WAYNE
MCDONALD WILLIAM E
MEADE JOSEPH LYNN
MEDLEY HOMER LANDUS
MEEK DONALD HOWARD
MEISTER BERNARD EDWIN
MELTON RONALD DAVID
MEREDITH HUBERT ARTHUR
MERONEY RAPHNELL J
MERRELL LOWELL HOWARD
MERRELL ROBERT DELL
MERRIMAN CHARLES ED-
WARD
MERRIWEATHER GENE
OPERIE
METCALF TOM ANDREW
MEZZLES TOMMY
MIDCALF THOMAS EDWARD
MILENDER JOHN EMERSON
MILES STEPHEN LEE
MILHORN LARRY DAYTON
MILLER BILLY LEE
MILLER CARL SOCRATES JR
MILLER GLENN STANFORD
MILLER RICHARD DENNIS
MILLER WAYNE TERRY
MINTON PHILLIP EDWARD
MITCHELL ALEX LOUIS
MITCHELL CLARENCE
MITCHELL FRED EVANS III
MITCHELL GLENN EDWARD
MITCHELL ROCHESTER
MIZELLE JOHN MARSHALL
MOLLEY CHESTER ANDREW
MONEYMAKER WAYNE
MONGER OTHA LEE
MOONEYHAM BILLY FRANK-
LIN
MOORE BILLY DALE
MOORE FULTON BEVERLY III
MOORE HERBERT LEE JR
MOORE JAMES MICHIAL

MOORE PERCY
MOORE WILLIAM HYRAM
MOOREHEAD JOE HOWARD
MORELOCK WILLARD FRANK-
LIN
MORGAN HUBERT HARROL JR
MORGAN JAMES HENRY
MORGAN JAMES RAYMOND
MORGAN JOHN HENRY
MORGAN MICHAEL LYNN
MORROW WILLIAM DANNY
MORTON JAMES EDWARD JR
MOSES CLIFTON
MOTT WILLIAM LARRY
MULLEN ELVIS EARL
MULLINS LARRY EUGENE
MURDOCK LARRY
MURPHEY DOUGLAS WAYNE
MURPHY B L JR
MURRAY GARY
NAPPER CHARLES CRAWFORD
NAVE BILLY JOE
NAYLOR GEORGE EDWARD
NEAL CHARLES OTTIS
NEAL EDWARD LEON
NEAL ROY WILLIAM
NEAL WILBERT HOYT JR
NEILL LARRY WAYNE
NELMS DANIEL EARNEST
NELSON EARL
NELSON ROBERT DWAYNE
NESBITT JOSEPH
NEWMAN JAMES CLIFFORD JR
NEWMAN LARRY EDWARD
NICASTRO CHARLES EDWARD
NICHOLS CHARLES EDWARD
NICHOLSON GEORGE P
NIPPER DONALD EDWARD
NOE JERRY LYNN
NORMAN ARTHUR EUGENE
NORRIS CALVIN ANDREW
NORVELL JEFFREY WOODROW
NOVEL CHARLES EDWARD
NOWLIN CHARLES DOUGLAS
NUNLEY WALTER WILLIAM JR
NUNNERY TRAVIS EDWARD
O DONNELL GEORGE MAU-
RICE
O DONNELL JAMES PATRICK

222

O KIEFF WILLIAM BRANDON
O NEAL JAMES ELTON
O NEAL NELSON MONTAGUE
O SHELL DON MANUEL
OAKES ARNOLD GLEN
OLIVER BOBBY GLENN
OLIVER GARY LEE
OLIVER MICHAEL PIERCE
ORR JIMMIE RAY
OSBORNE GEORGE D
OWEN CHARLES THOMAS
OWENS HOWARD
OWENS LARRY THOMAS
PACE JAMES RALPH
PALK BOBBY LEE
PALMER JAMES KENNETH
PARHAM JOHN WILLIAM
PARKER ALVIN G
PARKER ANDREW DAYTON JR
PARROTT OSCAR ROBERT
PATE JOHN H JR
PATTEN CARL EUGENE
PATTERSON BILLY J
PATTERSON JOHNNIE HUGH
PATTERSON MARK
PATTERSON RICHARD LEE
PATTON JOHN HENRY
PAUL BRINSON IRA
PAYNE JEPPIE JOSEPH
PAYNE MONTE LYNN
PEASE HOMER LEFTERAGE
PEAY JAMES EDMUND
PEEPLES HARRY FRANK
EDWAR
PEGRAM RICHARD EPPS JR
PEMBERTON WILLIAM LARRY
PERKINS GARY ELDON
PERRY R T
PERRY RANDALL EARL
PERRY RONALD DWIGHT
PETTITT JAMES ALLEN
PETTY HOWARD PALMA
PETTY JOHN ROBERT JR
PHELPS RAY WILLIARD JR
PHILLIPS MARVIN FOSTER
PILKINGTON CHARLES H JR
PIPKIN ERNEST GERALD
PITCHFORD L C
PITT WILLIAM LYNN
PLESS WILLIAM HUDSOL
PLUNKETT GERALD W
PODY JOHN CHRISTOPHER III
POE ROBERT EDWIN
POOLE WILLIAM CLAUDE
POOLE WILLIAM GUY III
POPE WALTER GLENN
POTTER JAMES RALPH
POWERS DONALD HOWARD
PRESLEY RONNIE CALVIN
PRICE ROBERT GLEN
PRIDDY WILLIAM F
PRINCE JOHN R
PROCTOR ERVIN
PROFITT HARVEY JUNIOR
PUGH MICHAEL LAVERNE
PUTNAM RONALD VIRGIL
QUALLS ARTHUR GERALD
QUILLEN EARL THOMAS
QUILLEN ROGER DELL
QUINN BOBBY JOE
RABEY ROGER WILLIAM
RAINES CHARLES RANSOME
RAINWATER JAMES RONALD
RANDOLPH WILLIAM JR
RANDOLPH WILLIAM LEWIS
RATLIFF FRANKLIN DELANO
RATLIFF JERRY SCOTT
RAULSTON CHARLES ALLEN
RAWLS JERRY DOUGLAS
RAY DAVID ROBERT
RAY WILLIAM LEE
REAMS WILLIAM BLAIR JR
REASONS JAMES ALTON JR
REED CHARLES OSCAR
REED JACKIE KENNETH
REED JAMES EDDIE
REED JOE ALLEN
REED ROGER DALE
REED WILBERT
REEDER RONNIE ELLIS
RENSHAW ANDERSON NEELY
III
REYES DOUGLAS COOPER
REYNOLDS JOHN DAVID

RHODES HU BLAKEMORE
RHODES JOHN DAVID III
RICH ROY WAYNE
RICHARDS JAMES PAUL
RICHARDSON DAREK NICH-
OLAU
RICHARDSON EUGENE P JR
RICHARDSON FRED LEWIS
RICHARDSON RAYMOND
WILKIE
RICHARDSON WILLARD D JR
RIDDLE MICHAEL DEAN
RIDLEY GLENN THOMAS
RIMMER JEARL EDWARD
RIPPETOE RAE KELLAND
RIVERS JETTIE JR
ROBBINS LONNIE JUNIOR
ROBERTS CHARLES CAMILLE
ROBERTS DANNY RAY
ROBERTS HERBERT JR
ROBERTS JOHN LEONARD
ROBERTS WILLIAM JACKSON
ROBERTSON ALLEN HARVEY
ROBINSON DALLAS DEAN
ROBINSON DONALD RICHARD
ROBINSON JAMES EDWARD
ROEBUCK ROBERT LENNON
ROGERS ALLEN TEBBS JR
ROGERS JOHN AVERY
ROGERS ROBERT LEE
ROGERS THOMAS FRANKLIN
JR
ROGERS WILLIE JR
ROLLINS EDWIN CHARLES
ROLLINS GLENN HASKELL
ROLLINS HOBERT TRUMAN
ROSS CARLTON
ROSS FRANK MILAN JR
ROSS WILLIAM GRAY
ROSS WILLIAM SIDNEY JR
ROULETT JAMES HOUSTON
ROUSSEAU DUANE MICHAEL
ROWLETT GARY STEVEN
ROYSTER JOSEPH EDWARD
RUCKER MACEY LEE
RUGGLES JOHN RICHARD III
RUSH CHARLES GLYNN JR
RUSH MARVIN GENE
RUSHTON BRIAN WAYNE
RUSSELL FRED CALVIN
RUSSELL JAMES A III
RUSSELL SAMUEL
RUTLEDGE JAMES ROBERT JR
RYLAND WILLIAM PATRICK
SADLER CARL J
SALLER DONALD VINCENT
SASSER GEORGE FREDERICK
SAVAGE VARIS JR
SCALLIONS CARL WAYNE
SCHILLER MARTIN SULLY JR
SCOTT JAMES ELLISON
SCOTT WILLIAM HENRY
SEEBER FLETCHER JR
SEGINE RONALD EUGENE
SEIBER DAVID ANDREW
SELF JAMES EDWARD
SENSING RONALD LESLIE
SESLER JOHN JOSEPH
SEXTON JOHN DAVID
SEXTON LARRY LEE
SHAFFER ROBERT LEE
SHARPE DENNIS CLAUDE
SHASTEEN KENNETH PARKER
SHAW WADE THOMAS
SHEFFIELD JOHN NOBLE
SHELL MARVIN
SHELTON BOBBY JAMES
SHELTON FRANK TIMOTHY
SHELTON WILLIAM ARTHUR
SHELTON WILLIE
SHENEP KARL EDMOND
SHERLIN FREDDIE MICHAEL
SHERRILL JAMES J
SHERROD DONALD ANCKER
SHETTERS JOHN HENRY
SHINAULT JOHN MICHAEL
SHIPLEY THOMAS FREDERICK
SHIRLEY DONALD LEE
SHOCKLEY THURMAN B JR
SHOEMAKER ROBERT DALE
SHORT JOSEPH WILLIAM
SHRUM KENNETH EDWARD
SHULL SANDY LEE
SILBERBERGER PAUL JOHN

SIMPSON JOHN HARRISON
SIMS FREDERICK AUGUSTAS
SINGLETON WALTER KEITH
SITZ EDWARD R
SKYLES GORDON RAY
SLOAN HAROLD MARTIN
SLUDER DONALD TED
SMALL WILLIAM DALE
SMALLING CHARLES LEE
SMELSER ROGER MYERS
SMITH ANTHONY
SMITH BOYD WAYNE
SMITH CHARLES EDWARD JR
SMITH CHARLES WALLACE
SMITH CLARENCE ELVIN
SMITH DENNIS WAYNE
SMITH DON
SMITH GARY D
SMITH GARY RAY
SMITH GERRAL AUBREY
SMITH JACKIE GLENN
SMITH JERRY LYNN
SMITH RICKY EDWARD
SMITH ROBERT L
SMITH RONALD GORDON
SMITH TERRY LEE
SMITH WALTER DANIEL
SMITH WALTER LEWIS
SMITH WILLIAM MARTIN JR
SMITH WINSTON OSBORNE
SMITH YANCEY JR
SNEED WARD GRAY
SNYDER THOMAS DEAN
SOUTH JOHN HERSHEL
SOUTHERLAND ROY EDWARD
SPAIN HUGH FRED
SPAKES ESTEL DENNY
SPANGLER CARL C
SPEARS DAVID PAUL
SPEARS JERRY WAYNE
SPRINGER CHARLES A
SPRINGS RALPH RONALD JR
SPRINKLES WILFORD LESLIE
STACKS RAYMOND CLARK
STAFFORD ROBERT BERYL
STALLINGS JAMES D
STANCIL KENNETH LEON
STANDLEY THOMAS GARY
STANEK ROBERT LEE
STASHONSKY JOHN RAY
STEAGALL EDSEL WAYNE
STEED WILLIAM OWEN
STEPHENS THOMAS ALLEN
STEPHENSON DONALD RAY
STEPHENSON KEITH POWELL
STEPP JOEL RICHARD
STEPPEE LARRY ELMER
STEVENSON BILLY EDWARD
STEWART DAVID GLENN
STEWART HARRY RAY
STIGALL CHARLES BENNETT
STILLS HAROLD CLIFFORD
STOKES JAMES DOUGLAS
STONE BEN WADE
STORY EDDIE B
STOUT SAM EUGENE
STOVALL ALVIN RAMSEY JR
STRAUB MARK ALAN
STRAUSSER PAUL JOSEPH
STRINGFELLOW JOHN D JR
STRINGFIELD CHARLES DEAN
STRONG JAMES LARRY
STROYE FERDINAND
SULLIVAN LAWRENCE MI-
CHAEL
SULLIVAN R D
SULLIVAN STANLEY HOUSTON
SUTTON JACK LENN
SWANSON WILLIAM HENRY
SWATSELL DONNIE JAY
SWINT CHARLES JUNIOR
SWONER ERNEST WILLIAM
TABOR MOSES CLARK
TALBERT CLAUDE JR
TANKSLEY CLIFTON
TANNER RAYMOND
TARTE JAMES LAFON
TATE CHARLES EDWARD
TATE JAMES E
TATUM LAWRENCE BYRON
TAYLOR CLIFFORD MC
ARTHUR
TAYLOR DENNIS WAYNE
TAYLOR GEORGE DAVID

TAYLOR GEORGE DENNIS
TAYLOR HOMER JR
TAYLOR JAMES GLENN
TAYLOR JOE KENNETH
TAYLOR JOHN HENRY
TAYLOR LOUIS ROBERT
TAYLOR RAY
TAYLOR TOMMY LEE
TAYLOR TYRONE
TAYLOR WALTER MINOR
TAYLOR WENDELL GENE
TEMPLETON DONALD LEE
TERRY HOYLE JR
THOMAS JACKSON
THOMAS ROBERT LEE
THOMASON JAMES CALVIN
THOMPSON OLIVER NATHAN
THOMPSON RUSSELL LEE
THORNBURG SCOTT WILLIAM
TILLMAN CECIL WAYNE
TIPSY HAYWOOD WADE JR
TOLLETT ELIJAH GOAR JR
TOLLEY MICHAEL
TORBETT STEPHEN JUSTENE
TRAVIS JAMES DAVID JR
TREADWAY THOMAS
CHARLES
TREECE JAMES ALLEN
TREVATHAN ROBERT LEWIS
TRIER ROBERT DOUGLAS
TRISDALE ROBERT LEE
TROGLEN JACKIE WAYNE
TRUELOVE THOMAS WILLIAM
TUCK HUBERT JR
TUCKER CHARLES GILBERT
TUCKER THOMAS CECIL
TUCKER TOMAS C
TURNER CHARLES
TURPIN GORDON JAMES JR
TUTTLE JAMES WALTER
TUTTLE ROBERT ERVIN
UDELL MARK FOSTER
UNDERHILL BENJAMIN S
UNDERWOOD JACKIE SHIRL
UZZELL FRANK NELSON
VADEN WOODROW WILSON
VANDERGRIFF RODGER ALAN
VARNER THOMAS ALLISON JR
VAUGHN DENNIS WAYNE
VAUGHN KELLY PATRICK
VAUGHN WILLIAM OREL
VAUGHT JOHNNIE L JR
VICKERY MICHAEL CLARENCE
VINSON BOBBY C
VOILES GASPER ALLEN
VOLNER JOHN DELANE
WADE DOUGLAS BUREM
WADE RICHARD EDWARD
WALDEN DANIEL EDGAR
WALKER JACKIE CARROLL
WALKER JOHN HENRY JR
WALKER MANLEY GLEN
WALKER VERNON LEWIS
WALLACE ULYSSES
WALSH GORDON O DELL
WALTON EUGENE
WANAMAKER JOHNNY
WAYNE
WARD CLYDE ANDERSON
WARD GEORGE ROBERT JR
WARD JOHN DOUGLAS JR
WARD RANDY NEAL
WARFIELD PHILLIP RAY
WARRICK CLARENCE RAY
WASHAM DENNY LEE
WASHINGTON LARRY ED-
WARD
WATKINS CHARLES EMANUEL
WATKINS JOHN WILLIAM
WATSON CLARENCE EDWARD
WATSON JAMES CHARLES
WATSON JAMES FRANKLIN
WATSON JAMES HAROLD
WATSON JOE NATHAN
WATSON WILLIAM B
WATTS BRADLEY KEITH
WAUGH RONNEL LOUIE
WEAVER GARRY LYNN
WEAVER JERRY LEE
WEBB ALFONSO AUGUSTAS
WEBB JACKIE JOE
WEBB LARRY DALE
WEBER GREGORY JOHN
WEBSTER REGERNAILD

WELLS HARRY LEON
WELSHAN JOHN THOMAS
WEST CARL LYNN
WEST JOHN EDWARD JR
WESTBROOK JAMES BAR-
RINGTO
WHEELER GORDON LEE
WHITAKER KELLY EUGENE
WHITE ALLEN EUGENE
WHITE CARROLL WAYNE
WHITE DONNIE RAYMOND
WHITEAKER JOHNNY LA-
VERNE
WHITEHEAD LARRY GENE
WHITING J C JR
WHITMAN DAVID STEWART
WHITTHORNE PAUL LUCIUS
JR
WHITTINGTON PAUL TIM-
OTHY
WILBURN WILLIAM LEVY JR
WILES ALVIN EUGENE
WILKES JOHN GRADY
WILKINSON JOHN TERRELL
WILKS JAMES ALAN
WILLIAMS CLAUDE ARTHUR
WILLIAMS DAVID CLARK
WILLIAMS DAVID RICHARD
WILLIAMS DORSEY BURWIN
WILLIAMS EDDIE JONES JR
WILLIAMS JAMES PRITCHARD
WILLIAMS LABON RAPHAEL
WILLIAMS MAXIE R JR
WILLIAMS REUBEN CHARLES
WILLIAMS ROBERT D JR
WILLIAMS ROBERT EUGENE
WILLIAMS TERRY EUGENE
WILLIAMS TERRY WAYNE
WILLIAMS WAYNE CHARLES
WILLIAMS WILLIAM JACK
WILLIAMSON CHARLIE C JR
WILLS ROY SHANNON
WILSON CHESTER WAYNE
WILSON HAROLD WENDELL
WILSON JOHN WESLEY
WILSON JOHN WILLIAM
WILSON LEWIS BRACY
WILSON ROY HASKEL
WILSON TOMMY ROBERT
WISE JAMES EDWARD
WISEMAN MALCOLM RICH-
ARD
WITT EARNEST LYNN
WOLFE JIMMY RAY
WOOD ARTHUR MURRY
WOOD LARRY T
WOOD WILLIAM COMMO-
DORE JR
WOODARD JACKIE LAVANDA
WOODARD JOHNIE KENNETH
WOODARD JOSEPH WILBERT
JR
WOODARD MICHAEL DAVID
WOODS JOHN WILLIE JR
WOODS LAWRENCE
WOODS RANDLE TOM
WOODSON BILLY BARNELL
WOOLIVER CHARLES WIL-
LIAM
WORKMAN LANCE DAVIS
WORLEY GARRY LEE
WORLEY JAMES RONALD
WORTH TIMOTHY LANE
WRIGHT ALBERT N JR
WRIGHT BILLY LEE
WRIGHT JAMES FRANK
WRIGHT JOHNIE J JR
WRIGHT MELVIN ROY
WRIGHT RAYMOND EARL
WRIGHT ROBERT LEE
WRIGHT RONALD LEE
YARBROUGH GEORGE ALLEN
YEARY RANDALL DOUGLAS
YEWELL BOBBY JOE
YODER JAMES STRONG
YOUNG DANNY STEPHEN
YOUNG JAMES EDWARD
YOUNG THOMAS DUDLEY
YOUNG WILFORD AVON
YOUNT WILLIAM HENRY JR

223

Photos by Sheldon Reynolds

Vietnam War Memorial

West Pioneer Parkway, Arlington, Texas

One of the more recent additions to America's many Vietnam War memorials, the Vietnam War Memorial in Arlington, Texas honors the sacrifices by both the US forces and the allied South Vietnamese forces. This is a unique aspect of this memorial as most monuments for the Vietnam War deal solely with the American veterans of the conflict. The project was completed in 2015 and was dedicated in October of the same year.

The Vietnam War Memorial features a sculpture consisting of two soldiers standing at the ready side by side in combat. One American and one South Vietnamese. The memorial was placed in the 103-acre Veterans Park in Arlington, Texas where it is part of a continuous memorial project for veterans under the direction of the Heroes of South Vietnam Memorial Foundation. This monument joins the existing bronze soldier who solely represented Arlington veterans as a whole.

The nonprofit group Heroes of South Vietnam Memorial Foundation commissioned the Vietnam War Memorial that took four years to see completion. The idea came about as leaders of the local Vietnamese community sought to honor the service of both the US soldiers and the South Vietnamese forces. To see their dream come true, the group raised a necessary $500,000 in contributions and donations. The design was approved by the City Council who stated that the monument would fit well with the existing memorial and the city's ongoing plans for the memorial space.
In 2015, the memorial foundation donated the monument to Arlington and it was placed in Veterans Park where it's positioned on a raised platform. Mark Austin Byrd, a Marine, executed the sculpture. Byrd was a helicopter pilot during the Vietnam War.

The Names of those from the state of Texas Who Made the Ultimate Sacrifice

AALUND JAMES DOWNING
AARON MICHAEL PETER
ABLE DAVID FLOYD
ABRAHAM PAUL HAROLD
ACOSTA JESSE RODRIQUEZ
ACTON TOM PERRY
ACUFF EDDIE DUANE
ADAME ARTHUR PINA
ADAMES SANTIAGO D JR
ADAMS BERT MORRIS III
ADAMS CARL TURNER
ADAMS GEORGE GAYRAL
ADAMS HUGHIE DARELL
ADAMS KENNEY MILTON JR
ADAMS MICHAEL DRUE
ADAMS RONNIE LEE
ADAMS ROSCOE DAVID
ADAMS TED WANE
ADAMS TERRANCE DEAN
AGUILAR ADOLFO
AGUILAR ARMANDO
AGUILAR ARNOLD
AGUILAR NICK ALFRED JR
AGUILAR PEDRO RAMIREZ
AGUILAR REIMUNDO
AGUILAR ROBERT
AGUILLON JOSE JESUS
AGUIRRE ARTHUR CECILIO
AGUIRRE FIDEL JOE
AHLSTROM ROBERT ERNEST
AILSTOCK JIM LAMARR JR
AKIN JOHN VINCENT
ALANIZ AMADO JR
ALANIZ BENITO V
ALANIZ LUIS ANGEL
ALANIZ PAUL GILBERT JR
ALANIZ RAYMOND
ALBA JESSIE CHARLES
ALBAREZ SEFERINO JR
ALBRECHT ADOLPH WILLIAM
ALDERSON TERRY HOWARD
ALDRICH LAWRENCE LEE
ALEWINE LEMUEL LENOEL
ALEXANDER DEWEY LEE
ALEXANDER SAMMIE ED-

WARD
ALEXANDER STAMATIOS G JR
ALFORD GEORGE ALLEN JR
ALFORD TERRY LANIER
ALLBRIGHT RONALD HAR-
RISON
ALLEN ANDREW AUGUSTUS
III
ALLEN CHARLES DAVID JR
ALLEN DANIEL WEBSTER JR
ALLEN DONALD RAY
ALLEN EARNEST JR
ALLEN EDDIE HUGH
ALLEN JAMES LOUIS
ALLEN JOHN DOSS
ALLEN JOHN WILLIS
ALLEN KEITH WESLEY
ALLEN TERRY DE LA MESA JR
ALLISON JAMES SAMUEL
ALLISON SAM STEPHEN
ALLMOND BARRY KENNETH
ALMANZA JUAN
ALMANZAR IGNACIO JR
ALMENDARIZ SAMUEL
ALONZO JULIAN
ALONZO MANUEL BUSTOS
ALTAMIRA LUIS ANTONIO
ALVARADO RAMIRO
ALVARADO RAUL JR
ALVAREZ FRANCISCO
ALVAREZ GUADALUPE MASIAS
ALVAREZ IGNACIO JR
ALVAREZ JULIAN MARTINEZ
ALVAREZ ROBERT
ALVERSON ROBERT WARREN
JR
AMADOR ERNEST BALDO-
NADO
AMADOR SEVERIANO
ANDERSON JOHNNY MAC
ANDERSON RICHARD ALLEN
ANDERSON VERNON RAY JR
ANDREWS WILLIAM LARRY
ANGERMILLER JAMES ALLEN
ANGERSTEIN MICHAEL

EDWARD
ANGUIANO RUBEN
ANGUIANO TONY
ANTU JUAN
ANZALDUA ALBERTO TORRES
ANZALDUA OSCAR
ARANDA JUAN FRANCISCO
ARENAS MANUEL V JR
ARMBRUSTER ANTHONY
CLARK
ARMSTEAD ROCKY D
ARMSTRONG JOHN WILLIAM
ARMSTRONG WILLIAM
PRESTON
ARNOLD ROBERT MILTON JR
ARNOLD WILLIAM HENRY
ARNWINE EDWARD RAY
ARONCE JOSEPH CHARLES
ARRANTS MICHAEL LORRELL
ARREDONDO JESSIE
ARROYO JOSE FRANCISCO
ARTHINGTON MARVIN S
ASHER ROBERT FRANKLIN
ASHFORD HENRY LEWIS
ASHMORE LAURENCE RAY
ASHTON CURTIS MORRIS
ASHTON JAMES ODELL
ASTON BLAKE EDWARD
ASTON JAMES MICHAEL
ATHANASIOU RONALD S
ATTERBERRY EDWIN LEE
AUDILET FRANKLIN DELANO
AUSTIN RILEY CLAYTON
AUSTIN WILLIAM OLEN
AUTEN NORMAN DWANE
AUTREY JAMES HAROLD
AVALOS ALBERTO ANGEL
AVERITTE WILLIAM CLAYTON
AVILA JUAN JR
AVILA RAFIEL
AWALT JIMMY ARDELL
AYALA EDUARDO
AYALA TONY JOHN
AYRES JAMES HENRY
AYRES JESSE STEPHEN

AZORE DAVID
BACIK VLADIMIR HENRY
BAER WILLIAM CLAY
BAILEY ELLIS MILLER
BAILEY ROY DEE
BAIRD MICHAEL HARRY
BAKER ALLEN JAMES
BAKER ARTHUR DALE
BAKER BILLY RAY
BAKER JOHN HOUSTON
BAKER KENNETH EARL JR
BAKER ROBERT BENTON JR
BAKER WAYNE ROLAND
BALDWIN CLIFTON ADAIR
BALDWIN NELLO JR
BALDWIN ROBERT EARL
BALL ARTHUR WYMAN
BALL DAVID MARTIN
BALLARD GERALD ROY
BALSLEY ROBERT F JR
BANDA MACARIO S
BANKS JAMES R
BANKS RAY CARROL
BANKSTON RONALD NEIL
BARBER GEORGE L III
BARBER HENRY EDWARD JR
BARBOLLA RICHARD AN-
THONY
BARCHAK JOHNNIE F JR
BARGER LEE MELLINGTON
BARGER PHILLIP DENNIS
BARLOW ROSS OWEN
BARNES CHARLES PETER
BARNES SHELDON ORA
BARNES STEPHEN WESTLEY
BARNETT CHARLES EDWARD
BARNETT JIMMY DALTON
BARNETT ROBERT RUSSELL
BARNS LAWRENCE RAY
BARNWELL JACKIE WAYNE
BARR WILMA J
BARRERA GILBERTO
BARRERA TOMAS ANTONIO
BARRETT JOHN DANIEL
BARRON ROBERT BRUCE

BARTLETT ARTHUR WAYNE SR
BARTLEY HOWARD LYNN
BARTON JAMES RAYBON
BARTON JAMES WESLEY
BARTRAM FORREST LA
WAYNE
BASDEN JERRY DON
BASS GEORGE CLINGER
BASS JOE HARRELL
BATEMAN WILLARD THUR-
MAN
BATSON MICHAEL OLAN
BATSON ROBERT FILMORE
BATTLE JOSEPH CHRISS
BAUMGARDNER THOMAS
EDI JR
BAXTER ROGER BRUCE
BEALS STEPHEN CARL
BEAN JAMES FRANCIS
BEAN JIMMY DALE
BEARD DONALD WAYNE
BEASLEY EDWARD RUSSELL
BEATY ARTHUR LEE
BEAVERS ROBERT ALLEN
BECERRA JAVIER
BECERRA RUDY MORALES
BECK JERRY DON
BECKER JAMES CHRISTOF
BECKER WALTER WARD
BEDFORD CHARLES
BEETS RONNIE D
BEGGS TERRY KENT
BEJARANO ADOLFO MAR-
TINEZ
BELL ALBERT LEE
BELL DOYLE LYNN
BELL EDWARD JAMES
BELL HARDY LEE
BELL HARRISON
BELL HOLLY GENE
BELL LESLIE RAVEN
BELL LEWIS DOUGLAS
BELL MICHAEL DEAN
BELL ROBERT GRAHAM
BELT ROBERT ERIC

Texas

BELTRAM AUGUSTINE JR
BELTRAN ANASTACIO HERNAND
BENAVIDEZ TRINO BALTAZAR
BENGE LARRY WAYNE
BENITEZ JUAN
BENNETT CHARLES EDWIN
BENNETT GEORGE WILLY JR
BENNETT MARVIN DALE
BENSON GUYE RAYMOND
BERAN FRANK HENRY III
BERGEN JAMES THOMAS III
BERGER DIXIE CARL
BERGQUIST ERIC EMANUEL
BERRIER DANNY CLARENCE
BERMEA VICTOR D
BERNAL ENRIQUE MUNOZ
BERNARDY THOMAS G
BERRIER DANNY CLARENCE
BERRY CHARLES RAY
BERRY DAVID LOYALL
BERRY JACK ALBERT
BERRY TOMMY LOYD
BERRY VANCE ALYN
BEST HUGH VICTOR
BEVELS LEONARD LEROY
BEYER EDWARD HUGO
BICE QUINTON MORGAN
BIEDIGER LARRY WILLIAM
BIFFLE JOE LESLIE JR
BILLS RUFUS WILSON
BIRD JACKIE DEAN JR
BIRD MICHAEL DE VERNE
BISHOP DANNY RAY
BISHOP TED JASON
BLACK VICTOR LEE
BLACKBURN JERRY EDWARD
BLACKMAN THOMAS LEE
BLACKMON EDWARD GEE
BLACKMON KENNETH LEON
BLAIR DONALD RAY
BLAIR PATRICK LYNN
BLAIR THOMAS ARTHUR
BLAKE JACK PATRICK
BLAKE L C
BLAKELEY ROY JAMES
BLALOCK GHERALD EDWARD
BLALOCK WALTER ROGERS
BLANCO HERIBERTO
BLETSCH WILLIAM PETE
BLISSETT JIMMIE RAY
BLOMSTROM WAYNE ALDEN
BLOODSWORTH LARRY WILL
BLUE RONALD MICHAEL
BLUNT PAUL BOREN JR
BLUNT SAMUEL
BOAZ DONALD JOE
BOEGLI STEVEN WARREN
BOEHM LARRY JOE
BOGGS DAVID LEONARD
BOLDEN CARL EUGENE
BOLDEN DANIEL HYMAN
BOLTON ANDERSON DON
BOLTON BILLY CARROLL
BONE LOSSIE FRANKLIN
BONE ROBERT LYNN
BONETTI FREDDIE ALLEN
BONEY WILLIAM
BONIFAZI GERARD REX
BONIN BOBBY JOE
BONNER JOHN SIDNEY JR
BOOKER ALBERT N
BOOTH JOHN BINGHAM
BOOTH ROY ROBERT
BOROSKI ANTON WALTER
BORREGO LUIS CARLOS JR
BORREGO RUIZ FRANCISCO J
BOSWELL RICHARD WELDON JR
BOSWORTH RAYMOND PAUL JR
BOTELLO JUAN JOSE
BOTHWELL WILLIAM DAVID
BOURGEOIS LEROY ANTHONY

BOVIO RICHARD STEPHEN
BOW MICHAEL WAYNE
BOWEN BILLY ROY
BOWENS TOMMY
BOYD JAMES P
BOYD RANDALL JAMES
BOYD THOMAS MASSIE III
BOYKIN KENNETH LEE
BOYLE RICHARD EUGENE
BRADBERRY DUDLEY FRANCIS
BRADDOCK STEPHEN LEE
BRADFORD JOHN TRAVIS
BRADLEY FRANKLIN S JR
BRADLEY LARRY ALAN
BRADSHAW FAYBERT RAY
BRADSHAW ROBERT S III
BRADY JOSEPH CLINTON
BRAGER ROY LEE
BRAMLET WILLIE JOE
BRANNING WILBUR RALPH JR
BRANNOM MORRIS II
BRANNON PHILLIP ARTHUR
BRASHER JIMMY MAC
BRASWELL BOBBY JOE
BRATCHER CLIFFORD SHERAN
BRAY ARNOLD REX
BRAZZEAL TED OLAND
BREUER LOUIS KARL IV
BREWER RICHARD DENNIS
BREWER THOMAS NEAL
BRIDGES CECIL BENNY
BRIGGS ERNEST FRANK JR
BRISCOE LOUIE DON JR
BRISTER BILLY MAC
BROACH EARL DAVID
BROCK HARRY GILES
BROOKS CHARLES EDWARD
BROOKS JAMES EDWARD
BROOKS MAURICE
BROOKS WILLIAM LESLIE
BROSSMAN EDGAR JAMES
BROUMLEY TERRY HUGH
BROWN BILLY EDWARD
BROWN BILLY RAY
BROWN BRIAN DALE
BROWN CLARENCE ARTHUR
BROWN CLINTON RAY
BROWN CURTIS CHARLES
BROWN DEWEY HEARRELL JR
BROWN DONALD GENE
BROWN DONALD GEORGE
BROWN EARNEST WAYNE
BROWN FRED JR
BROWN GERALD AUSTIN
BROWN JAMES DAVID
BROWN JAMES DOUGLAS
BROWN JAMES GARLAND
BROWN JAMES GREGORY
BROWN JAMES RONALD
BROWN JAMES WILLIAM
BROWN JIMMY RAY
BROWN JOHN CHARLES
BROWN JULIUS LAVERN
BROWN LARRY PAUL
BROWN OWEN DAVIS JR
BROWN ROBERT ALLON
BROWN ROBERT RAY
BROWN ROSS ANDREW
BROWN WALTER OTHO JR
BROWNE FRANK HAROLD II
BROWNE ROBERT GODWIN
BROWNING FRANK LEON
BROWNLEE ROBERT LEON
BRUCE SAMMY BRYAN
BRUMLEY JOHNNY EDWARD
BRUMLEY MERRELL EUGENE JR
BRYAN CHARLES WILLIAM
BRYANT ALVIA GRADY
BUCHANAN HERMAN DALE
BUCKELEW EARNEST JACK
BUCKLEY MAC CURTIS
BUCKNER ROBERT OLEN JR
BUENDIA JUAN VILLEGAS
BUENTELLO LEONEL
BULLARD CHARLES DORIAN
BULLARD HOWARD
BULLER RENE ALDO
BULLOCK HERSHEL JOE SR
BUNKER WILLIAM REUBEN III
BUNTE WILLIE EARL
BURCHFIELD JIMMY FRED
BURCHWELL ASHLAND FREDERI

BURCIAGA ALBERT
BURGE BEN CARLOS
BURGESS JOHN HARLIE JR
BURGESS RUSSELL DAVID
BURK JIMMY REA
BURK TERRY PAUL
BURKE JOSEPH SCOTT
BURKS LEROY JR
BURLESON MICHAEL FINNIE
BURNETTE MICHAEL ROBERT
BURNS DEWEY RAY JR
BURNS JOHN D JR
BURNS MICHAEL PAUL
BURNS MICHAEL ALLEN
BURROUGH JESSE CLARENCE
BURTON JAMES BILLY
BURTON ROBERT THOMAS
BUSBY RICHARD CURTIS JR
BUSENLEHNER RICHARD THOMA
BUSH JOSEPH KERR JR
BUTCHER DAVIS CARROLL
BUTLER ALBERT JR
BUTLER HENRY
BUTLER ROBERT D
BUZZARD LLOYD LYNN
BYARS EARNEST RAY
BYARS JERRY DAN
BYFORD LARRY STEPHEN
BYLER STEPHEN HAWLEY
BYNUM ALANSON GARLAND
BYRD ARTHUR MALCOLM
CABALLERO DAVID JOE
CABALLERO GILBERTO JR
CABALLERO JOSE LUIS
CABARUBIO JAMES
CADELL ERNEST WOODY JR
CAIN JAMES CALWINN
CAIN JAMES DOUGLAS
CAIN JIMMY RAY
CAIN WILLIAM MICHAEL
CALABRIA DAVID MICHAEL
CALDERON CESARIO
CALDWELL MERLIN FRANCIS
CALFEE JAMES HENRY
CALHOUN EDWIN GERALD
CALHOUN WILLIAM STEVE
CALLAWAY MICHAEL ROGERS
CALLISON DONALD JOSEPH
CAMARGO JUAN HIPOLITO
CAMARILLO FELIPE DURAN
CAMARILLO FERNANDO JR
CAMERON GERALD WAYNE
CAMERON VIRGIL KING
CAMP JACK
CAMPBELL CLYDE WILLIAM
CAMPBELL IVAN J
CAMPBELL RONALD STEVEN
CAMPBELL WILLIAM EDWARD
CAMPOS JOSE BALLENTINE
CAMPOS LUIS BARRON
CAMPOS LUIS HECTOR
CAMPOS RICARDO
CANADA SAM JR
CANALES REFUGIO
CANDELARIA RAUL
CANDLER DONALD PRIESTER

CANNON JOHN WAYNE
CANO JOSE RAMON
CANTER RONALD M
CANTRELL KEITH NOLAND
CANTU ADAM
CANTU ERNESTO SOLIZ
CANTU FLORENTINO JR
CANTU REFUGIO JOSE
CANTWELL KENNETH JAMES
CARAWAY EARNEST WESLEY
CARAWAY THOMAS GLENN
CARBAJAL RUBEN JOSE
CARDENAS PAUL H JR
CARDENAS RUDY
CARDOSA CRECENCIO
CARMAN ROBERT LEON
CARR DONALD GENE
CARR GEORGE DARE
CARRANZA HORACIO
CARRILLO ARNOLDO LEONEL
CARRILLO JUAN
CARRIZALES DIONISIO G
CARROLL MICHAEL DAVID
CARROLL WESLEY WOMBLE III
CARSON ALAN DALE
CARSON JOHN HARVEY
CARTER ARDON WILLIAM
CARTER FRED DOUGLAS JR
CARTER GARY DON
CARTER JAMES DOUGLAS JR
CARTER STEVE DWAYNE
CARTLEDGE ALBERT J III
CARTWRIGHT BILLIE JACK
CARTWRIGHT JAMES HOWARD
CARUTHERS THOMAS HOWARD
CASSELL KEVIN RAY
CASTANON ALFREDO
CASTILLO ANTONIO GONZALES
CASTILLO APOLINAR JR
CASTILLO ERASMO CAMARGO
CASTILLO JOSE JAIME
CASTILLO RICHARD
CASTLEBERRY JAMES ANDREW
CASTON JAMES CALVIN
CASTRO ERNESTO F JR
CASWELL RAYMOND M
CAUBLE ARTURO ALVARADO
CAUDILLO PEDRO JAIME
CAVAZOS DANIEL GUTIERREZ
CAVAZOS MARTIN
CAVER JOHN WAYNE
CAVIN DOUGLAS JAMES
CAYCE JOHN DAVID
CERNA NARCISO REZA JR
CERVANTEZ JUAN JOSE
CEVALLOS ROBERT G
CHADWICK KENNETH RAY
CHAFFIN CLARENCE RAY
CHAMBERLAIN HENRY
CHAMBERS DAVID WAYNE
CHAMBERS JAMES THOMAS
CHAMBERS JOHN LUTHER
CHAMBERS LESTER EUGENE

CHAMBERS TOMMIE ALLEN
CHAMPION CHARLES DON
CHAMPION JAMES ALBERT
CHANNON BUDDY EUGENE
CHAPA LORENZO JR
CHAPPELL DANIEL L
CHATELAIN RAY AUGUSTINE
CHAVEZ JOHN MANUAL
CHAVEZ RAFAEL ALBERT
CHAVEZ RUBEN GARZA
CHEEK JAMES WILLIAM
CHENAULT ROBERT GLEN
CHENAULT THOMAS DUDLEY
CHERRY JAMES L
CHILDRESS BILLY W
CHIMINELLO THOMAS JAMES
CHITWOOD ROY DUKE
CHRISTIAN PETER KARL J
CHRISTIE ZANE
CHURAN RONALD BRUCE
CHURCHILL THOMAS HENRY
CHUTER JOHN DAVIS
CISNEROS ROY
CLAIBORNE DELL ROSS
CLARK EARL GLENN
CLARK GARY VAUGHN
CLARK HOWE KING JR
CLARK JOHN CALVIN II
CLARK TOMMIE JOE
CLARKE JAMES PHILLIP
CLAY EUGENE LUNSFORD
CLAY HERMAN ALLEN JR
CLAY MELVIN EUGENE
CLEAVELAND MELVIN RAY
CLEMENT GREGORY C
CLEVELAND WALTER K
CLIFCORN JAMES RICHARD
CLINARD CHARLES WAYNE
CLINE JOSEPH OLIVER III
COBB CHARLES MICHAEL
COBB HUBBARD DON
COBLE CLYDE WAYNE
COCHRAN ISOM CARTER JR
COCHRAN PATRICK SHELDON
COCHRAN ROBERT EDMUND
COCKRELL WILBERT R
COFFELT BOBBY J
COFFEY BILLY RAY
COGGINS TERRY LEE
COKER TROY ORION
COLE CLAUDE JR
COLE ROBERT EARL
COLEMAN BOBBY WALTER
COLEMAN JAMES LARRY
COLEY BILLY JOHN
COLGAN DANIEL PAUL
COLLEPS CHARLES LEE
COLLETT PAUL RAYMOND JR
COLLIER CHARLEY HOLTON
COLLIER TONY WAYNE
COLLINS HARRIS LESTER
COLLINS MARSHALL BARB
COLLINS NOBLE JR
COLLINS WILLIAM ELICE JR
COLUNGA GEORGE
COMBEST JERRY WAYNE
COMBS LOWELL THOMAS

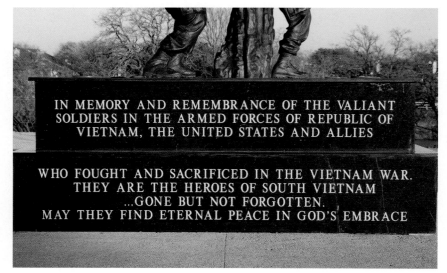

IN MEMORY AND REMEMBRANCE OF THE VALIANT
SOLDIERS IN THE ARMED FORCES OF REPUBLIC OF
VIETNAM, THE UNITED STATES AND ALLIES

WHO FOUGHT AND SACRIFICED IN THE VIETNAM WAR.
THEY ARE THE HEROES OF SOUTH VIETNAM
...GONE BUT NOT FORGOTTEN.
MAY THEY FIND ETERNAL PEACE IN GOD'S EMBRACE

COMEAUX JOSEPH BERNILLE
CONCHOLA BENITO
CONDON RUSSELL WILLIAM
CONN JAMES DOUGLAS
CONN RONALD RAY
CONNACHER RONNIE ED-
WARD
CONNEL DAVID ARNOLD
CONNER EDWIN RAY
CONNER STEPHEN GRANT
CONNEVEY LAYNE HALE
CONNOLLY VINCENT JOHN
CONOLLY SIDNEY MCLEAN JR
CONTRERAS JUAN LEONARDO
CONTRERAS PABLO GUERECA
CONTRERAS VALERIANO
DAVID
CONVERSE PHILIP HOWELL
COOK KENNETH LYNN
COOK LESTER CHARLES
COOK NATHANIEL
COOKS MELVIN EUGENE
COOLEY HARVEY LYNN
COOLEY MONTE RAY
COON MICHAEL RAY
COOPER GARY WAYNE SR
COOPER GEORGE GRADY
COOPER JAMES WILLIAM
COOPER OTIS JR
COOPER ROBERT WESLEY
COPE CHARLES RICKY
COPELAND JOE MIKEL
COPP BARRY ALAN
CORDERO JULIAN GARZA
CORDOVA JOHN BARELAS
CORES THOMAS RICHARD II
CORKILL ROBERT ARNOLD
CORMIER FRANCIS JOSEPH
CORMIER WILLIS
CORNELIUS SAMUEL BLACK-
MAR
CORONA JOEL
CORONADO ROBERT
CORRALES RICHARD MEN-
DOZA
CORTEZ ALBERTO GUTIERREZ
CORTEZ JOSE G
CORTEZ JUAN ESQUIVEL
CORZINE BOBBY WAYNE
COTTEN ROBERT BRYAN
COTTIN LELAND RICHARD
COTTON CHARLES MICHAEL
COTTON THOMAS III
COURSON CHARLES TRUITT
COURTNEY ALLEN WESLEY JR
COVARRUBIAS JUAN ALONSO
COWLEY BENNYE WARREN
COWLEY JEFFRY EDWARD
COX CLAUDIE LEE
COX GARY LEE
COX OMMIE TRUMAN JR
COX ROY ALLEN
COY BEN
COY DWIGHT CLIFFORD JR
CRAFT HARLAN MERDEAN
CRAFT JAMES
CRAIG WAIN PERRY
CRARY DAVID WAYNE
CRAVEN DONALD RAY
CRAVEN JAMES EVERETT III
CREECY LARRY RAYMOND
CREED EDWARD GAFFNEY
CREEK HENRY LEE
CREEK THOMAS ELBERT
CRENSHAW JAMES LEE
CRIBBS JAMES WESLEY
CRIM CHARLES RAY
CRISP JIMMY WAYNE
CROCKER EVANS BLANE JR
CROCKETT CHARLES D JR
CROCKETT TRAVIS RICHARD
CROLEY JAMES ROBERT
CRONE CARL RICHARD
CRONKRITE CHARLES LIGON
CROSBY ROBERT MICHAEL
CROSSLAND RICHARD GUINN
CROSSLEY MICHAEL LEE
CROUCH JIMMY LEELAND
CROWE HAROLD MICHAEL
CRUM ROBERT H JR
CRUMP CHARLES ALVIN
CRUZ RICHARD PEREZ
CRUZ VIRGIL GALAN
CRYDER ROBERT D

CUELLAR JULIAN CASTILLO
CULP KARL HOWARD
CULVER ARCHIE GLENN
CULVER DICK DAVIS
CULVER WILLIAM RONALD
CULWELL JIMMY LEE
CUMBRY JOHN EDWARD
CUNNINGHAM DENNIS LANE
CURIEL SAM TRINIDAD
CURRIE ANTHONY EUGENE
CURRY DOUGLAS RAY
CURTIS HERBERT RAY
CUSHING DAVID ROY
CUTBIRTH KENDELL DWAYNE
D AGRELLA MICHAEL LOUIS
DABBS ALAN COURTNEY
DABNEY RICHARD EARL JR
DACUS FREDDIE LOUIS
DAILEY HAROLD CARL II
DAILEY JERRY MICHAEL
DALGLIESH MARK ANTHONY
DANIEL STEPHEN ARTHUR
DANIELS HAROLD DWAINE
DANIELS WILLIAM ALVIN JR
DANSBY CHARLES M
DARNELL HAROLD DALE
DARRETT TYRONE
DAUGHERTY ROBERT LEE
DAVENPORT ROGER DALE
DAVILA GUILLERMO JR
DAVILA JOSE
DAVILA TEODORO
DAVIS BENNY EARL
DAVIS CLIFFORD GORDON
DAVIS EDGAR SYLVESTER
DAVIS EDWARD LEE
DAVIS EDWARD LEE
DAVIS ELTON JR
DAVIS JAMES ROBERT
DAVIS JAMES WELDON
DAVIS JOHN SEVIER
DAVIS JOHN WESLEY
DAVIS JON ERIC
DAVIS LEROY JR
DAVIS MARVIN HOMER
DAVIS RICHARD BOUCHE JR
DAVIS ROBERT CHARLES
DAVIS ROBERT FORD
DAVIS ROBERT ORLIFF
DAVIS STEPHAN ANDREW
DAVIS STEVE CLAYTON
DAVIS WILLIE EDWARD
DAVISON DENNIS ALLEN
DE BAULT JOE ROBERT
DE FOOR VICTOR LEE
DE LA CERDA ANTONIO H JR
DE LA CRUZ FERNANDO
DE LA PAZ HILARIO JR
DE LA ROSA GUMESINDO
DE LA ROSA JESUS JR
DE LEON GUILLERMO JR
DE LEON JESUS HERNANDEZ
DE LEON MARIO ONTIVERO
DE LEON RAFAEL JR
DE LOACH LLOYD DWAIN
DE LOS RIOS PABLO G P JR
DE VOE MICHAEL EUGENE
DEATHERAGE DENNIS RAY
DEBUSK RAY B JR
DEERE DONALD THORPE
DEES CURTIS CLEVELAND
DEGNER HAROLD PAUL
DEHART DONNIE RAY
DELANO HENRY HARRISON
DELCAMBRE TERRY LEE
DELEON MARIO P
DELGADO CARLOS MARTINEZ
DELGADO CHRISTOPHER
GEORG
DELGADO FRANCISCO H
DELGADO GILBERT TREVINIO
DELUNA MANUEL JR
DEMERSON JOE EDDIE
DEMORY RAYMOND FRANK
DENNIS CHARLES
DENNIS DAN MICHAEL
DENNIS LARRY WAYNE
DENNIS PAUL JONES
DENT BILLY RAY
DENTON BOBBY LEE
DENTON MANUEL REYES
DENTON NORRIS JAMES
DENTON ROBERT ANTHONY
DESSELLE THOMAS WILLIAM

DEWLEN MICHAEL LEE
DEWVEALL JERALD GLENN
DEXTER RONALD JAMES
DIAZ JOSE RENTERIA
DIMOCK JAMES ALBERT JR
DISHEROON BILLY WAYNE
DIXON CHARLES O'NEIL JR
DIXON WILLIAM ALFRED JR
DOBBS ROBERT ARTHUR
DOBROSKI JOHN LEE
DODD EDDIE LEROY
DODD LAWRENCE ADDINSON
DODSON FREDDY DEAN
DOMINGUEZ CARLOS
DOMINGUEZ JOE REINI
DOMINGUEZ ROBERTO
DONALDSON DARRELL
WAYNE
DOOSE GARY LEE
DORMAN MICHAEL RODNEY
DORNAK LEONARD EDWARD
DORRIES CARL WAYNE
DOTSON WILLIAM THOMAS
III
DOYLE REX WAYNE
DOZIER JERALD LEON
DREWRY NOLAN FRANKLIN
DRISCOLL VICTOR MICHAEL
DROBENA MICHAEL JAMES
DROIGK MARTIN WAYNE
DRYNAN ARTHUR W
DUARTE JOHN
DUCKETT LARRY THOMAS
DUCOTE LONNIE JOSEPH JR
DUDLEY LAWRENCE WESLEY
JR
DUGGAN WILLIAM YOUNG
DUKE GEORGE G
DULAK RAYMOND ROBERT JR
DUNAWAY ROBERT LEON JR
DUNBAR ALLEN SEVARN
DUNCAN BOYCE LOWRANCE
JR
DUNES ALBERTO JR
DUNLAP JOHN WALTER
DUNLAP RAYMOND EARL
DUNN JOE DANIEL
DUNN ROBERT WAYNE
DUNN TEDDY REX
DUPRE NORMAN LEE
DUPREE DOUGLAS
DURAN IGNACIO
DURAN SALVADOR GUTIER-
REZ
DURHAM OLIVER EARL
DUSBABEK GLENN HENRY
DWORACZYK WALLACE
STANLEY
DWYER ALFRED THOMAS
DWYER LAWRENCE LEE JR
DYER IRBY III
DYKES MONTE DALE
EADEN WILLIAM HENRY
EAGLIN JOHN HENRY
EARLEY WILEY B
EARLL DAVID JOHN
EARNEST JAMES DALE
EASLEY DENNIS BOYD
EASLEY LEONARD EUGENE
EATON BOBBY LYNN
EDDY THOMAS EARL
EDGEMON JAMES EDWARD
EDINGTON PAUL RICHARD
EDMUND EDWARD JOSEPH
EDWARDS ANTHONY JOHN
EDWARDS KINNETH GLENN
EDWARDS ROY WILLIAM
EGAN WILLIAM PATRICK
EHLERS DOUGLAS GARY
EIBER ROBERT ALLAN
ELDER ALLEN THOMAS JR
ELDER HOWARD LEE
ELIZONDO DAVID
ELLEDGE DON THOMAS
ELLEDGE WAYNE CLARENCE
ELLERD CARL JOSEPH
ELLIOTT JULIUS R
ELLIOTT THOMAS MCCLURE
ELLISON JESSE ROGER
ELLISON WAYNE EDWIN
ELLISON WILBERT ALLEN
ELMORE WILLIAM H JR
ELSENBURG WILLIE EDWARD
EMERT TOMMIE D

EMERTON WILLIS WAYNE
EMMONS JOHN WARREN JR
ENGLAND TONIE LEE JR
ENGLISH ERNEST ERVIN
ENMON DAVID JERRELL
EPHRAIM EDDIE LEE
EPPS JOE HERBERT
EPPS TITUS LEE
ERVIN JERRY LYNN
ERWIN RICHARD EUGENE
ESCAMILLA JOSE
ESCANDON JOE ALEXANDER
ESCOBEDO JULIAN JR
ESCOBEDO ROBERTO
ESPARZA FELIX JR
ESPARZA ISRAEL
ESPARZA NICHOLAS JR
ESPINOSA MIKE
ESPINOZA MARTIN
ESPINOZA VICTORIANO JR
ESQUIERDO JOHNNY RAY-
MOND
ESSARY MARTIN WILLIAM JR
ESTEIN DALTON MAIN
ESTES EDWARD STANLEY
ESTRADA CARLOS ALBERT JR
ESTRADA ESTEBAN PENA
ESTRADA JUAN VARGAS
EUDALY F M
EVANS ALFRED KINDELL
EVANS CLARENCE LOVICE
EVANS EDWARD LOUIS
EVANS HENRY ELMER
EVANS LLOYD WILLIAM JR
EVANS ROBERT DAVID
EVERETT EVERETT WHITE
EVERETT GARY WAYNE
EYRING KENNETH ROBERT
FABACHER SAZIN DALE
FAGGETT CHARLES EARL
FAHEY JOSEPH MICHAEL JR
FAIN GARY LEE
FANCHER JIMMIE ALVIN
FANNING HUGH MICHAEL
FANNING THOMAS F JR
FAREWELL ROGER WILLIAM
FARLEY ROBERT JERRY SR
FARMER JAMES DALE
FARRIS DALE WAYNE
FARROW JAMES EUGENE
FAULKNER TROY DAVID
FEDRO JAMES RAY SR
FEE EDWARD FRANCIS JR
FELLERS ROGER WAYNE
FENDLEY JOEL DAVID
FERGUSON DEWEY LINDON
FERGUSON JAMES P
FERGUSON THOMAS ALTON
FERGUSON THOMAS WAYNE
FERNANDEZ MARGARITO JR
FERNANDEZ REYNALDO
SALINE
FERNANDEZ SANTANA S JR
FERNANDEZ WILLIAM MAT-
THEW
FIELDEN WAYNE SAMUEL
FIELDS SAMUEL JR
FIESZEL CLIFFORD WAYNE
FIGUEROA ALBERT MARTINEZ
FIKE RONALD EDWARD
FINDLEY ROBERT DENNIS
FISHER JAMES LOUIS
FISHER ROYAL CLIFTON JR
FITTS CHARLES MILTON
FITZSIMMONS LARRY LEE
FLAGG ALTON ONEIL
FLEITMAN GLENN RAY
FLEMING SIDNEY WADE
FLETCHER KENNETH JACK
FLORES ANTONIO JR
FLORES ARTHUR MERINO
FLORES DANIEL PORRAS
FLORES FLORENTINO
FLORES GUADALUPE
FLORES JIMMY
FLORES JOSE MARIA
FLORES JUAN JR
FLORES RAMON AGUILAR
FLORES RAUL
FLORES RAYES CISNEROS
FLORES ROBERTO C
FLORES VICTOR JR
FLOURNOY MAURICE W
FLOYD JAMES MILTON

FLOYD LARVON
FLOYD MELVIN FRANKLIN
FLYNN MICHAEL FRANK
FOERSTER RAYMOND CARL
FOMBY JIMMY LEE
FORBES MICHAEL
FORD BOB W
FORRESTER RONALD WAYNE
FORSBACH RONALD CARL
FORSYTHE DAVID ALLEN
FORSYTHE THOMAS LYNN
FORTENBERRY JAMES RICH-
ARD
FORTNER GARY DUANE
FOSTER HADLEY
FOSTER JAMES EARNEST
FOSTER JEAN CLIFTON
FOSTER MARVIN LEE
FOSTER MARVIN RAY
FOSTER ROBERT L
FOX EDWARD HAROLD
FRACIONE FREDERICK R
FRALICKS LARRY DOUGLAS
FRANCE RICHARD WAYNE
FRAUSTO NOLBERTO JR
FRAZE JERRY WAYNE
FRAZIER EDWARD LEE
FRAZIER JOHNNIE LEE
FREASIER THOMAS HALL
FREE JOHNNY WAYNE
FREEMAN JAMES PAUL
FREEMAN OLLIE CURTIS
FREEMAN STEVEN FORREST
FREESTONE DAVID EDWARD
FRENCH DOUGLAS ROBERT
FRIEL JOHN CHARLES
FRIERSON EDDIE TYREE
FRITTS FREDERIC WILLIAM
FRITZ ALVIN RAY
FRITZ GERALD W
FROST BOBBY GENE
FROST WOODY JOE
FUENTES ROBERT MARTINEZ
FULKERSIN ROBERT DOUGLAS
FULLER JAMES RAY
FULLER WILLIAM OTIS
FULLERTON RONNIE JOE
FURPHY KENT PALMER
FYFFE THOMAS CLEO
GABLE CHESTER LEWIS
GABRIEL CHARLES DAVID
GABRIEL HERBERT JAMES
GACHES CHARLES WILLIAM
GAGE JAMES ROBERT
GAGE NORMAN GLENN
GAJDOSIK ERNEST WAYNE
GALBREATH BOBBY FRANK
GALBRETH EMME R II
GALINDO GUADALUPE JR
GALLARDO ARMANDO
GALLARDO ERNESTO R
GALLEGOS OSCAR CONANDO
GALLOW RYAN JUDE
GARCIA ABEL D JR
GARCIA ALEJANDRO JR
GARCIA ANDRES
GARCIA ANTONIO
GARCIA ANTONIO VARGAS
GARCIA ARTURO
GARCIA CARLOS HILL
GARCIA CRECENCIO CASAREZ
GARCIA DAVID ADAME
GARCIA DAVID Z
GARCIA EDELMIRO LEONEL
SR
GARCIA EDUARDO JR
GARCIA EMILIO GAMBOA
GARCIA FRANCISCO
GARCIA GEORGE ARRIAGA
GARCIA GILBERTO
GARCIA HENRY R
GARCIA HERIBERTO ARNAL-
DO
GARCIA JESSE EULOJIO
GARCIA JOE ROBERT
GARCIA JOSE
GARCIA JOSE GILBERTO
GARCIA JOSE JR
GARCIA JUAN
GARCIA MIGUEL JR
GARCIA OSCAR
GARCIA PEDRO GALLARDO
GARCIA PEDRO INCARNA-
CION

Texas

GARCIA RAUL JR
GARCIA RICARDO MARTINEZ
GARCIA RICHARD
GARCIA RICHARD
GARCIA ROBERT
GARCIA SALVADOR BORREGO
GARD DANNY D
GARDNER PHILLIP D
GARGUS ROY PHILLIP
GARNER JOHNNIE LINTON
GARNER RICKEY DEAN
GARRETT EUGENE JR
GARRETT FRANK DAVID
GARRETT LEROY E
GARRETT NORMAN RAY
GARRETT THOMAS STEVEN
GARZA ANTONIO GUERRA JR
GARZA CARL EDWARD
GARZA DAVID
GARZA ELIAZAR EFRIEN
GARZA FRANCISCO
GARZA GENARO
GARZA HENRY ALLEN
GARZA JOSE JR
GARZA JOSE JR
GARZA JOSE SALUSTINO
GARZA MARCELLO C JR
GARZA MARGARITO
GARZA PABLO BENITEZ
GARZA RAMON
GARZA VICENTE
GATLIN IVAN WEBSTER
GAUNA DANIEL JR
GAYTAN EDWARD RAY
GEISERT CHARLES PRICE
GENTRY DAVID ANTHONY
GEORGE JAMES EDWARD JR
GEORGE MICHAEL DEAN
GETTER WAYDELL
GEURIN STEPHEN BURL
GIBBS JERRY DON
GIBBS RAYMOND ANDREW
GIBBS SAMUEL DAVID
GIBNER GEORGE THOMAS
GIBSON JOHNNY ALLISON
GIBSON RICHARD HOPKINS
GIBSON ROY LEE
GIDEON WILBURN CHASTAIN
GIESECKE JERRY DON
GILBERT DON RAY
GILBERT FRANKLIN DELANO R
GILBERT PAUL FARIS
GILFORD JAMES ARNOLD
GILLESPIE JOE
GILLESPIE LLOYD DEAN
GILLEY CHARLES TEDDY
GILLINGS JOHNNY RAY
GILLOCK RAYMOND RICHARD
GILMORE ANTHONY
GILMORE HARLON JOE
GILMORE KENNETH DEE
GILSTRAP DANNY LYNN
GIPSON F G JR
GIPSON GAYLON G
GIPSON R W
GLADDEN MICHAEL JAY
GLASSCOCK CARL LEE
GLASSCOCK DAVID LEWIS
GLAZE JERRY WAYNE
GOBER CHARLES WESLEY JR
GOBER RONALD BERNARD
GOBLE PATRICK MICHAEL
GOEN THOMAS SANFORD
GOGGAN HERBERT GARY
GOLDHAGEN BOBBY GENE
GOLEMON FLOYD EDWARD JR
GOMEZ ATANACIO JR
GOMEZ BASILIO
GOMEZ JOSE MANUEL
GOMEZ OSCAR JOE
GOMEZ RAYMUNDO

GOMEZ VALENTINE BERMEA JR
GOMEZ XAVIER
GONSALEZ MARIO
GONZALES AGAPITO JR
GONZALES ARTURO PEREZ
GONZALES DOMINGO RAUL
GONZALES EDWARDO JOSE
GONZALES ELIGIO RICE JR
GONZALES FELIX G JR
GONZALES FRANK CAVOSOS
GONZALES JOSE LUNA JR
GONZALES LUIS GARCIA JR
GONZALES MANUEL
GONZALES MANUEL MARTINEZ
GONZALES OSCAR CRUZ
GONZALES PAUL GUTTERREZ
GONZALES PEDRO CHAVARRIA
GONZALES ROY JR
GONZALES SANTIAGO RODRIGU
GONZALES TOMAS
GONZALEZ ALFREDO
GONZALEZ AMADOR L
GONZALEZ BENITO REYNA
GONZALEZ CARLOS SAAVEDRA
GONZALEZ GUADALUPE
GONZALEZ HECTOR MANUEL
GONZALEZ JESUS
GONZALEZ JESUS ARMANDO
GONZALEZ JOSE JESUS
GONZALEZ JUAN ANTONIO
GONZALEZ JUAN JOSE
GONZALEZ PABLO RENE
GONZALEZ PEDRO ACEVEDO
GONZALEZ RAMIRO MEDINA
GONZALEZ RODOLFO GUADALUP
GONZALEZ RODOLFO MARCIANO
GONZALEZ VINCENTE RAMIREZ
GOOCH CALVIN LIONEL
GOODMAN CHARLES EDWARD
GOODMAN JAMES A
GOODRICH JOHN MATTHEW
GOODWIN CHARLES BERNARD
GOODWIN CHARLES RAY
GOOLSBY WILLIAM RAY JR
GORDON GARY GENE
GORDY JESSE ARNEL
GORE THOMAS
GRACE LARRY
GRAHAM ALAN RAY
GRAHAM EARNEST WILMER
GRANT BENJAMIN DAVIS
GRANT ROBERT LEE
GRAVES STANLEY EDWIN JR
GRAY JAMES KENNETH
GRAY MICHAEL DOUGLAS
GRAY RICHARD ARLINGTON
GRAY RUZELL
GREATHOUSE JULIUS JR
GREEN AUTRY
GREEN CLYDE RAY
GREEN FRANK CLIFFORD JR
GREEN JIMMY LEON
GREEN MELVIN
GREEN MILFRED RAY
GREEN ROBERT BAILEY
GREEN ROBERT EARL
GREENE BILLY RAY
GREENLEE STEVEN JOSEPH
GREENMAN DREW MARLIN
GREENSAGE ROY LEE
GREENWOOD DALE EDWARD
GREENWOOD JAMES WILLIAM
GREER MONROE
GREER WADE ANTHONY
GREGG JOHNNY GLEN
GRIER JIMMY LEE
GRIFFIN CEPHUS JR
GRIFFIN DOUGLAS HOLTZ
GRIFFIN JAMES T JR
GRIFFIN KENNETH WAYNE
GRIFFIN ROBERT ALLEN
GRIFFIN WALTER JOE LOUIS
GRIFFIS WILLIAM A III
GRIFFITH ROBERT ELWIN

GRIGGS BRENT IKE
GRIMES GARY DEMPSY
GRIMES GARY LYNN
GRISBY DON LEE
GRISSETT EDWIN RUSSELL JR
GRIZZLE WENDELL RAY
GROSSE CHRISTOPHER A JR
GUAJARDO HILARIO H
GUENTZEL LARRY RAY
GUERRA GEORGE JR
GUERRA ROBERT ROCHA
GUERRERO ANDREW CASTRO
GUERRERO JESSE
GUERRERO RICHARD JR
GUERRERO WILEY
GUILLEN GILBERTO LUIS JR
GULLETT GORDON ELDON
GUNN ALAN WENDELL
GUNN CHARLEY EDWARD
GUNN DANIEL MCNEIL
GUSMAN FRED GRABIEL JR
GUTHRIE FRANK LYNN
GUTIERREZ ALBERT R JR
GUTIERREZ ARTURO B
GUTIERREZ ERISTEO JR
GUTIERREZ FERNANDO
GUTIERREZ GEORGE JR
GUTIERREZ JOSE ANTONIO
GUTIERREZ RAUL CAMPOS
GUTIERREZ RAUL GRIMALDO
HACK BENNY GLEN
HACKER THOMAS EWALD
HACKWORTH DWIGHT LEE
HADNOT RICHARD LEE
HADNOTT GARY ANDERSON
HAILEY ODDIE C
HAINES GLENN BRANSON JR
HALE HENRY MAURICE STAFFO
HALE LANNY EARL
HALE ROBERT LAWRENCE
HALE TERRELL WILLIAM
HALE TERRY ALLEN
HALE WILLIAM THOMAS
HALEY CLIFFORD EUGENE
HALEY GARLAND GENE
HALFORD MICHAEL DEAN
HALIBURTON NATHANIEL JR
HALL BRUCE
HALL CHARLES WILLIAM JR
HALL JACKIE BURL
HALL JOHN STANLEY
HALL MICHAEL JIM
HALLAM DURWOOD MICHAEL
HALT ARDON
HAMIL LOUIS WILLIAM
HAMILTON AMBERS ANDREW
HAMILTON AUGUST FRANKLIN
HAMILTON ROBERT DAVID
HAMILTON ROBERT LEE JR
HAMLIN WILLIAM LLOYD
HAMMOND EARL NEWSOM
HAMMONDS ROY LEE
HAMMONS JAMES LUTHER
HANCOCK JOHN DAVID
HAND FRANK EDWARD III
HANEY PERRY EUGENE
HANSARD JAMES BURL
HANSON LOWELL RAY
HANVEY ROBBY DAVID
HARDIE CHARLES DAVID
HARDIN JAMES RICHARD
HARDWICK ROCKNE LAMAR
HARDWICK WILLIAM HIXSON
HARDY ABRAHAM LINCOLN
HARDY DAVIS EDWIN
HARGIS DANNY WAYNE
HARGRAVE KENNETH LEE
HARGROVE JAMES WELDON
HARGROVE WILLIAM S
HARKINS ROBERT CHARLES
HARLAN RICHARD ELLIOTT
HARLAND WAYNE LYNN
HARLOW CLARENCE LOUIS
HARNESS JAMES WILLIAM
HARPER ROBERT EDWARD
HARRELL J D
HARRELL JAMES RANDOLPH
HARRIS BILLY DEAN
HARRIS BOBBY GLENN
HARRIS EDDIE CLAYTON
HARRIS ELTON ODIS

HARRIS MAX GILBERT
HARRIS NED HENRY
HARRIS WILLIAM THOMAS
HARRISON JAMES ROY
HARRISON WILLIAM MILAM
HARROTT RICHARD LEONARD
HARSON EDWARD EARL JR
HART BENNY EUGENE
HARTMAN HENRY WILBURN
HARTMAN TIMOTHY JAMES
HARTNESS AARON
HARTNESS GREGG
HARWOOD JAMES ARTHUR
HASDORFF DENTON JOSEPH
HATFIELD LARRY DEAN
HATFIELD WOODWARD S JR
HATTON FRANKLIN DELANO R
HATTON WILTON NEIL
HAUGER KEVIN JEFFREY
HAVEL DONALD JAMES
HAVEMANN JAMES EDWARD
HAVERKAMP AUSTIN WILLIAM
HAWKINS EDGAR LEE
HAWKINS PHILIP III
HAWKINS STARLING G
HAWKINS WILLIE GEORGE JR
HAWLEY DONALD REY
HAWLEY KENNETH RAY
HAWTHORN JOHN EDMON
HAWTHORNE MARVIN DALE
HAYES DONALD RAY
HAYNES MARTIS LEON
HAYNES ROBERT MARION JR
HAYS ROBERT BRADFORD
HAYTON BRENT ALLAN
HEATH RICHARD FARLEY
HEAVIN WILLIS RAY
HEBERT RODGER DALE
HEBRON CHARLES EDWARD
HEFLIN JOHN DARRACOTT
HEINEN EDGAR PAUL
HELEMS KENNETH EUGENE
HELM DAVID EARL
HENDERSON DAVID B JR
HENDERSON ISAAC LEE
HENDERSON WILLIE
HENDRICKS STERLING CRAIG
HENDRY DAVID EUGENE
HENLY CARL O NEAL
HENRY DAVID FRANKLIN
HENRY DAVID PAUL
HENSON GEORGE RICHARD
HERD RONALD WARD
HERNANDEZ ANTONIO
HERNANDEZ ANTONIO BENAVID
HERNANDEZ ARTHUR
HERNANDEZ HERIBERTO SEGOV
HERNANDEZ JOHN ALBERT
HERNANDEZ JOSE G
HERNANDEZ JOSE JR
HERNANDEZ MANUEL JR
HERNANDEZ MARCOS
HERNANDEZ RAMON N
HERNANDEZ RAYMOND RODRIGU
HERNANDEZ RICHARD
HERNANDEZ ROBERT
HERNANDEZ ROBERT REYES
HERNANDEZ ROLANDO
HERNANDEZ RUDOLPH VILLALP
HERNANDEZ WILLIAM ANTHONY
HERNDON ARTHUR ROBERT
HERNDON HARRY H
HERRERA AURELIO GARZA
HERRERA BEN LOPEZ
HERRERA FELIPE
HERRERA FRANCISCO
HERRERA JESSE EMIL
HERRERA MOISES ROMERO JR
HERRERA RAMIRO JR
HERRINGTON CARWAIN L
HERRINGTON CHARLES EDWARD
HERRON LEE ROY
HERST WILLIAM DONALD JR
HESS ZAN
HESSLER JAMES JOHN
HIBBLER RICHARD WAYNE

HICKLEN GRAHAM RAY
HICKS LEVIL
HIDALGO JAMES YBARRA
HIGGINBOTHAM JOHN BILL
HIGGINBOTHAM MICHAEL JOE
HILBRICH BARRY WAYNE
HILL IRVIN HUGH
HILL JAMES LEROY
HILL RAYFORD JEROME
HILL RICHARD DALE
HILL THOMAS
HILLIARD JOSEPH ROLLINS
HILLIARD WILLIAM EARL
HILLIN DOUGLAS WAYNE
HILLMAN RONALD ARWED
HINCH JAMES GARY
HINES NAMON JR
HINES RONALD DICKERSON
HINOJOSA FERNANDO AMYO
HINOJOSA JOSE ANTONIO
HINOJOSA MARCOS
HINSLEY EARNEST RICHARD
HINSON THOMAS ALLEN
HINZ DAVID LEE
HOBBS GARY LYNN
HOBBS KIMMEY DEAN
HOCHMUTH BRUNO ARTHUR
HOCKADAY JIMMY LEON
HODGES HOMER LEE JR
HODGES LEE ROY
HODGES RUFUS WELDON
HODGES WESLEY EUGENE
HODGSON CECIL J
HOELSCHER JOHN MICHAEL
HOFF SAMMIE DON
HOFFMAN CHARLES EDWIN
HOFFMAN RONNIE JOE
HOGGATT JOSEPH LEE
HOLDBROOKS THOMAS BERNARD
HOLDER HENRY EMIL JR
HOLDER LEONARD DONALD
HOLGUIN ISMAEL
HOLGUIN JOSE JR
HOLLAND ROBERT LOW
HOLLAND WILLIAM DELBERT
HOLLANDSWORTH EDDIE DEAN
HOLLE JOHN WILLIAM
HOLLEMAN JOE EARL
HOLLEY GLYNN BYRON
HOLLEY TILDEN STEWART
HOLLINGSWORTH DON RAY
HOLLIS JOHN EDWARD
HOLLIS THEODORE ROBERT
HOLLOWAY RICHARD ALTON
HOLLY CHARLIE DELNO
HOLMAN BOBBY FOSTER
HOLT BILLY JOE
HOLT RONALD WALTER
HOMSLEY IVAN D
HONEY RICHARD LANCE
HONEYCUTT BENJAMIN ALLEN
HONEYCUTT JAMES DON
HOOD CHARLES PERRY JR
HOOD RUFUS
HOOK ROBERT W
HOPKINS GARY WAYNE
HOPKINS ROBERT LOUIS
HORACE ALBERT C JR
HORN EMMETT HARVEY
HORTON HARRY WADE JR
HOSEA MICHAEL LEE
HOSEY SANDRA
HOTCHKISS KENNETH EUGENE
HOTCHKISS LEROY CASE III
HOUSE JOHN K
HOUSE JOHN LEE
HOUSTON THOMAS EUGENE
HOWARD BRUCE LEE
HOWARD JAMES GEORGE JR
HOWARD RAY JR
HOWARD WALTER LEE
HOWE OLAN JOSEPH
HOWELL A T
HOWELL DWIGHT SANFORD
HOWELL PERCY WRAY
HOWELL RANDALL DUMON
HOWELL ROBERT MALICHI JR
HOWISON CALVIN DANIEL
HOWISON GRAHAM HENRY

HUBBARD NATHANIEL
HUBBARD W D
HUBER WILLIAM FREDRICK JR
HUCKABEE JAMES EDWARD
HUDDLESTON LYNN RAGLE
HUDSON JERRY DOUGLAS
HUDSON KENNETH WAYNE
HUDSON RAYMOND HOYT
HUGGINS JAMES FREDERICK
HUGHES BEN ALLEN JR
HUGHES BILLY EUGENE II
HUGHES JAMES OLIVER
HUGHES JERRY LYNN
HUGHES JOHN RAYMOND III
HUGHES JOHN S JR
HUGHES MICHAEL NORMAN
HUGHES RICHARD RAMSEL
HUGHES ROBERT DOUGLAS
HULL JAMES LARRY
HULSEY JAMES AUBREY
HUMMEL JOHN FLOYD
HUMPHREY CHARLES EVERETT
HUMPHREY ROBERT LOY
HUMPHREY VICTOR JAMES
HUNT RICHARD
HUNTER GERALD N
HUNTER HERBERT PERRY
HURST JAKE EDWARD
HURST JOHN CLARK
HUSTON JOE STEPHEN
HUTCHINGS STEVEN WYLIE
HUTH PHILIP NICHOLAS II
HYDEN DEE AARON
HYLAND PAUL EDWARD
IBANEZ ALFONSO
IBARRA MIKE GOMEZ
IBARRA RODRIGO FUENTES
IBROM ADRIAN JOSEPH
INSALL BILLY GLEN
IRBY DONALD REECE
ISAACS MILO CLINTON
ISBELL DAVID GENE
ISOM THEODORE
ITUARTE ROBERTO
IZARD B C
JABLONSKY EDMOND A JR
JACKSON ALPHA RAY
JACKSON AMIL JR
JACKSON DAVID LEON
JACKSON EDWARD MERL
JACKSON JAMES DONALD
JACKSON JAMES HERMAN
JACKSON JOHNNIE BRUCE
JACKSON LAWRENCE
JACKSON MAXIE JR
JACKSON TOBY LEE
JACKSON WILLIAM BRAXTON
JACOBS BOBBY JOE
JAFFE BERNARD
JAGELER CHARLES DAVID
JAIME ANTONIO BARRERA
JAMES BOBBY JOE
JAMES EMMETT DWAINE
JAMES GARY LYNN
JAMES KENNETH EARL
JAMES RUFUS LA DELL
JANAK JOE JOHN
JARRELL KENITH LEWIS
JARRELL RANDALL DAVID
JASSO MARTIN
JENKINS BERT MC CREE
JENKINS DON
JENKINS ROLAND HAYES
JENKINS VERNELL
JENNINGS TIMOTHY PAUL
JENNINGS WILLIAM A III
JERGENSON RICKEY LAYNE
JERNIGAN RICHARD LEE
JESKO STEPHEN EDWARD
JIMENEZ ANTONIO
JIMENEZ JOSEPH ARTHUR
JIMENEZ JUAN MACIAS
JIMENEZ THOMAS ORTEGA JR
JINKS WILLIAM DONALD
JOBE BOBBY W
JOHNSON ADRIAN JOSEPH JR
JOHNSON AUGUST DAVID
JOHNSON BEN ODELL
JOHNSON BENJAMIN F III
JOHNSON CAL DUAIN
JOHNSON CARL THOMAS
JOHNSON CHARLES RAY
JOHNSON CLAUDE L

JOHNSON DENNY LEE
JOHNSON DONALD RAY
JOHNSON DONALD RAY
JOHNSON DONALD VERN
JOHNSON DUANE AARON
JOHNSON GARY LYNN
JOHNSON GARY MORGAN
JOHNSON HERBERT LAWRENCE
JOHNSON JAMES DOYLE
JOHNSON JAMES GRADY
JOHNSON JAMES LOUIS
JOHNSON JOE ALAN
JOHNSON JOHNNY VENT
JOHNSON LARRY TRAVIS
JOHNSON PHILLIP DALE
JOHNSON ROBERT ALAN
JOHNSON ROBERT DENNISON
JOHNSON ROBERT HENRY
JOHNSON ROY L
JOHNSON TAYLOR DOUGLAS
JOHNSON W C
JOHNSON WILLIAM
JOHNSTON CHARLES KENNETH
JOHNSTON GARY CLARENCE
JOLIVETTE MATHEW L
JONES ALAN PETER
JONES CHARLES CLIFTON
JONES FREDDIE LEE
JONES GARY ALLEN
JONES GARY WILLIAM
JONES GEORGE ALLEN JR
JONES HARVEY WAYNE
JONES JIMMIE DOUGLAS
JONES JOHN ROBERT
JONES JOHNNY CARL
JONES JOHNNY CARROLL
JONES KENNETH RAY
JONES LEMEN EARL
JONES LOUIS FARR
JONES SAM
JONES SANDERFIERD ALLEN
JONES SEABORN DAN
JONES VICTOR WAYNE
JONES WILLIAM ARTHUR
JONES WILLIAM EUGENE
JORDAN CHESTER GALE
JORDAN ROBERT CLAYTON
JORDAN TEDDY ROOSEVELT
JORDAN THOMAS LEE
JOY RAYMOND STANLEY JR
JUAREZ JOE MANUEL
JUAREZ OSCAR REINA
JUMPER STEPHEN FRANKLIN
JURADO ELIAS CASTRO JR
JURADO FRED V
JURADO RAMON
JURECKO DANIEL EDWARD
JURECKO DANIEL EDWARD
JUREK DALMER DOLAN
JUSTUS ROGER GALE
KAASE FLOYD WAYNE
KAHLA VICTOR DAVISON JR
KAISER LARRY KURT
KALKA CHARLES CLINTON
KAMENICKY GEORGE WAYNE
KAPLON PHILLIP FELIX JR
KASPRZYK GERALD BENEDICT
KEAL CLEVELAND JR
KECK CARL RANDOLPH
KEEFE RICHARD CARLYSLE
KEEL DAVID LATTIMORE
KEELER HARPER BROWN
KEESEE ARTHUR EARL
KEITH LEE ALBERT
KELLEY DANIEL THOMAS
KELLEY PAUL GLEN
KELLUM NORMAN WADE
KELLY DONALD GLENN
KELLY JOHN FRANKLIN
KEMPNER MARION LEE
KENDRICKS DOY RAY
KENNEDY JUDD WAYNE
KENNEDY LEE DONNIE
KENNON DONALD NEAL
KENT ROBERT DUANE
KERNS ARTHUR WILLIAM
KERR J L JR
KERSEY ARDEN ELLSWORTH JR
KESTLER JESSE LYNN
KEY ANTHONY WAYNE
KEYES SCOTTY LEE

KIEFER STUART OTIS
KIKER DOUGLAS HUGH
KILGORE CHARLES HOWARD
KIMBALL WILLIAM ROBERT
KIMLING MILES WAYNE
KINCER ALFRED LEMUEL III
KINDLEBERGER HAROLD PAUL
KING BOBBY
KING BRUCE THOMAS
KING CARSON MILO
KING CHARLES LEWIS
KING JACK LLOYD
KING LESLIE GENE
KING THOMAS KEITH
KINMAN TERRY DEWAYNE
KINSEY MICHAEL CHRISTOPHE
KIRBY GERALD
KIRKLAND LARRY JAMES
KISSLING BENJAMIN KAON
KITTRELL LARRY DON
KNADLER ROBERT STANLEY
KNAPP RICHARD
KNETSAR GEORGE ARTHUR
KNIGHT ROY ABNER JR
KNIGHT TERRY VASCAL
KNIGHT THOMAS WILFORD
KNOBLES JAMES LEONARD
KNOBLOCK GLEN LESTER
KNOWLES CHARLES MILFORD
KOHANKE LANCE JACK
KOHLER JOEL R
KOLIBA HERBERT
KOLLENBERG CHARLES LOUIS
KOONCE TERRY TRELOAR
KOROLZYK RALPH STANLEY
KOSCHKE MICHAEL EDWARD
KOZIK RAYMOND JIM
KOZUCH JOHN CLARENCE
KRUSSOW DONALD JOHN
KULHANEK ARNOLD JOHN
KUNZ ANTHONY EDMOND
LA BONTE GARY LEE
LA COMBE ROBERT LEE
LABOUNTY CHARLES RICHARD
LADELL JOE EARL
LAFIELD WILLIAM TRUMAN JR
LAIRD JAMES EUGENE
LAMAS RAUL RUBEN
LAMBERT JERRY WILLIAM
LAMBERT WALTER DENNIS
LANCASTER EDDIE LYNN
LANDRY ROBERT ANTHONY
LANDRY ROBERT ANTHONY
LANE GLEN OLIVER
LANE JAMES EVERETT
LANGFORD JAMES MINTER
LANGHAM HOLLAND IRWIN
LANGLEY JODY MAC
LANKFORD ROBERT MITCHELL
LANKFORD WALTER MERL JR
LARGENT LOEL FLOYD
LARREMORE PAUL WILLIAM
LASATER LUTHER MCKIND III
LASHER RICHARD ALLEN
LATHAM DANNY RICHARD
LAUREL DESIDERIO C JR
LAURENCE WILLIAM H JR
LAW ROBERT DAVID
LAWRENCE BERT OTTO
LAWRENCE EARL DAWSON
LAWRENCE JOHNNIE LEE
LAWSON JOHNNIE CARL JR
LAXSON RICHARD DOUGLAS
LAYNE VICTOR LEE
LAYNE VICTOR LEE
LAZICKI RONALD WAYNE
LE MASTER MICHAEL EUGENE
LEAL ARMANDO GARZA JR
LEAL FERNANDO
LEAL GUADALUPE MENDOZA
LEARY PAUL EDWARD JR
LEATHERBURY DAVID WARREN
LECHUGA MARTIN
LEDESMA ENCARNACION
LEE CALVIN RAY
LEE CHARLES EDWIN
LEE HENRY C
LEE HOMER HARDY

LEE HOWARD W
LEE JERRY DWAIN
LEE MARION LEONARD JR
LEE MILTON ARTHUR
LEE NATHANAEL
LEE RONALD PAUL
LEE WILLIAM ALLEN
LEHMAN DAVID JOHAN III
LEIJA LOUIE ZAPATA
LEMMONS BRIT P
LENZ THOMAS WAYNE
LEOS LEONARDO
LEOS NARCISO JR
LERMA GUADALUPE
LETBETTER BOBBY WELDON
LEVINGSTON JAMES ARTHUR
LEVIS CHARLES ALLEN
LEWIS CHARLES EDWARD
LEWIS CONVERSE RISING III
LEWIS DARREL GENE
LEWIS EARL LEROY
LEWIS FREDERICK HARRY
LEWIS JAMES WIMBERLEY
LEWIS LESLIE A
LEWIS MICHAEL LEE
LEWIS STEPHEN HERMAN
LEWIS TEDD MCCLUNE
LEWIS WILLIE GEORGE JR
LEYVA RICHARD
LICON FRANCISCO
LIGHTFOOT BELVIN
LILE JOE CHARLES II
LIMBRICK ALLEN ISSAC
LIMERICK BOBBY FRANK
LIMON ANDRES
LIMONES JESUS MARIO
LINDLEY BOBBY PAT
LINDSEY JACK WAYNE
LINDSEY WILLIAM ROYAL
LINEBERGER HAROLD BENTON
LINK ROGER MARK
LIRA ALFRED GEORGE
LISERIO JOE FRANK
LITTERIO ROBERT DALE
LITTLE DANNY LEONARD
LITTLE GARLAND PAUL
LITTLE NORMAN EARL
LOCKET ROBERT JR
LOCKHART FLOYD BARNEY JR
LOCKHART HARRY JAMES
LOFTIN TEDDY CARL
LOFTIS JOEL CONRAD
LOFTON BOOKER T JR
LOFTON GLEN DORSE
LOFTON RONALD HARRY
LOGAN FRANCIS MARRION III
LONG CARL EDWIN
LONG WILLIAM LOUIS
LONGORIA JOE GILBERT
LOPEZ ALFREDO JR
LOPEZ ANTONIO JR
LOPEZ ARTURO JR
LOPEZ AUGUSTINE JR
LOPEZ JOSE
LOPEZ MANUEL
LOPEZ PAULINO GUTIERREZ
LOPEZ PETE
LOPEZ RICHARD
LOPEZ RICHARD
LOPEZ RUDY
LOUIS ROBERT YOUGETE JR
LOVATO JOE JR
LOVE CLARENCE LEE
LOVE JOE L JR
LOWE DONALD W
LOWE WALTER BEDFORD JR
LOYD MICHAEL GLENN
LOZANO JOSE REYMUNDO
LOZANO MATTHEW T JR
LOZANO PAUL RODRIGUEZ
LOZANO RICHARD BILL
LULL HOWARD BURDETTE JR
LUMMUS FRANKLIN JACK
LUNA ANGEL
LUNA ARMANDO CERVERA
LUNA DONALD ALFRED
LUNA FORTUNATO JR
LUNA FRANCISCO
LUNA JULIAN
LUSK SAMMY RAY
LUTE HARRY GENE
LYMAN ALAN RICHARD
LYNCH SAMUEL ROY

LYNN ADRIAN LOUIS JR
LYONS JOE L
MAAS CLARENCE F III
MABERRY CALVIN DWIGHT
MABLE WASHINGTON CARVER
MABRY DONALD HENRY
MACHEN BILLY WAYNE
MACIAS JOE
MACIAS ROBERTO JAVIER
MACIAS TRISTAL
MACK DOUGLAS DULANE
MACOMBER CLIFFORD F JR
MADDOX MARCUS WAYNE
MADDUX DAVID ALLEN
MADISON CYRIL HYMAN
MAESE JORGE V
MAGALLAN NOE
MAGNON MYRON WILLIAM
MAHAN DARREL ULDRIC
MAHANA VANNY CHRIS
MAHONEY TIMOTHY KEITH
MAJORS JAMES RAY
MAKINTAYA ALEJANDRO
MALDONADO BALTAZAR A
MALDONADO PATRICIO JR
MALONE ROBERT EARL
MALONE WILLIAM WALTER
MALONE WILLIE EDWARD JR
MANGOLD CARL JOSEPH
MANGUM SAM HENRY
MANNING CHARLES EDWARD
MANNING JERRY
MANNING JOHN EDWARD
MANSFIELD CLAYTON JOHN JR
MANSIR PAUL WAYNE
MANSKE CHARLES JEROME
MANTOOTH JIMMIE HUGH
MANZANARES WILLIAM JR
MAPLES FRANCIS LEROY
MARECK RAYMOND DONALD
MARESH JAMES ANTHONY
MARIN FRANCISCO SANDOVAL
MARKER MICHAEL WAYNE
MARKOS GEORGE
MARLOWE DANIEL PAUL
MARQUEZ EDUARDO JR
MARROQUIN TOMAS JR
MARRS RONALD WAYNE
MARSH RONALD ALTON
MARSH WILLIAM CLIFTON
MARSON RICKEY JOE
MARTIN AUBREY GRADY
MARTIN DONNIE RICHARD
MARTIN DOUGLAS KENT
MARTIN HARRY WILLIAM
MARTIN JOHN D
MARTIN LARRY WAYNE
MARTIN LONNIE GENE
MARTIN RALPH
MARTIN RONALD LYNN
MARTIN SAMMY ARTHUR
MARTINEZ ANGEL
MARTINEZ ARMANDO DANIEL
MARTINEZ ENRIQUE
MARTINEZ EUGENE OSCAR
MARTINEZ EVARISTO III
MARTINEZ FLORENTINO JR
MARTINEZ GILIVALDO A JR
MARTINEZ GUADALUPE
MARTINEZ ISIDRO
MARTINEZ JAKE
MARTINEZ JESUS
MARTINEZ JOHN
MARTINEZ JOHN ANDREW
MARTINEZ JORGE
MARTINEZ JUAN JOSE
MARTINEZ LOUIS ALVARADO
MARTINEZ MANUEL GODINE
MARTINEZ MARGARITO
MARTINEZ PEDRO
MARTINEZ RAFAEL
MARTINEZ ROBERT R
MARTINEZ SIXTO R JR
MARTINEZ SYLVESTER C
MARXMILLER GARY EDWARD
MASEDA ROBERT
MASON DAVID LEE
MATELSKI LEONARD JAMES
MATHEWS CLYDE JR
MATHIS FOY MANION

Texas

MATIAS WENCESLAO ROSAS JR
MATLOCK WILLIAM TRAVIS
MATOCHA DONALD JOHN
MATTESON GLENN
MATTHEWS JAMES ERICH
MATULA VALENTIN GEORGE
MAUGHAN GEORGE LEE SR
MAURER JERRY EUGENE
MAURICE ROBERT CHARLES
MAXWELL EVERETT LEE
MAY ALFRED BYRON
MAY CHESTER HOWARD
MAY JOEL AUSBIN JR
MAY LARRY ALLAN
MAYER THOMAS J
MAYER WALTER CHRISTIAN
MAYFIELD JIMMY GENE
MAYO PIKE POWERS
MAYS RAYMOND
MC ADAMS THOMAS ARTHUR
MC ANINCH MICHAEL ALAN
MC BETH RONALD GENE
MC CAIN JOHNNY WAYNE
MC CARTER JIMMY CARL
MC CARTNEY JOSEPH BYRON
MC CARTNEY KEN ALLEN
MC CASKILL FREDRIC CECIL
MC CLAIN WILLIAM DAVID
MC CLAIN WILOFARD A II
MC CLATCHY JEFFERY JR
MC CLATCHY PERCY W
MC CLELLAND AUBREY DAVID
MC CORD BURTON KYLE
MC CORKLE DOUGLAS P JR
MC CRONE JAMES ROLAND
MC CULLOUGH GARRY MICHAEL
MC CUMBER RAYMOND
MC DAVIS CALVIN LEE
MC DERMOTT JOHN FREDERICK
MC DONALD BILLY WALLACE
MC DONALD CHARLIE RAY
MC DONALD JERRY CECIL
MC DONALD LONZO O
MC DONNELL JOHN TERENCE
MC ELHANON MICHAEL OWEN
MC ELHANON WARREN SHELBY
MC ELROY RONALD LENEAR
MC FARLAND CHARLES HENRY
MC FERON ERNEST
MC GEE PAT WELDON
MC GHEE GEORGE WILLIAM
MC INTYRE ROBERT LEWIS
MC KAY HOMER EUGENE
MC KENZIE JERALD THOMAS
MC KIBBEN LARRY SIMS
MC KIM EDWARD ALTON
MC KINLEY PAUL BLOUNT
MC KINNEY CECIL CURTIS
MC KINNEY RONALD GENE
MC LEMORE DAVID EUGENE
MC LENDON KENNETH HAYES
MC LENNAN ROY DEWAYNE
MC NAIR WILLIAM TERREL
MC NEIL HAROLD LOYD
MC NELIS FRANK CHARLES JR
MC SHAN DOYLE ALLEN
MC SWAIN BAYNES BALLEW JR
MC VEA ROBERT MINOR
MC VEA WILLIE DEE
MC WRIGHT DALE STEPHEN
MC WRIGHT EDWARD ARTHUR
MCCARTHY LOYD VAN JR
MCCARTY GLENN WELDON
MCCARTY JAMES LON
MCCUNE RAY EUGENE
MCDANIEL SAMUEL WAYMON II
MCDONELL R D
MEADOR FRANCIS ELMORE
MEADOR KENNETH BRUCE
MEADOR PHILLIP WAYNE
MEANS JOHN A
MEARES CECIL A
MEDINA ALFREDO JR
MEDINA ARTHUR
MEDINA CARLOS
MEDINA RAYMOND
MEDLEY JOHN R
MEDLEY TOMMY RAY
MEDLIN JOHN WILLIAM
MEDRANO JOSE JR
MEEKS DUSTAN WILLIAM
MEERDINK GEORGE JR
MEIER CARL LOUIS
MEJIA JESUS
MELENDEZ HUMBERTO C E
MELONSON JOSEPH DUDLEY JR
MELTON EDGAR ROBERT
MELTON GEORGE CECIL
MELTON MICHAEL DENNIS
MENDEZ ERINEO MENDEZ
MENDEZ JULIAN
MENDEZ ROBERTO
MENDIAS MARIO JUAN
MENDIOLA RICARDO
MENN ARTHUR JOHN
MENNINGER GEORGE EDWARD
MENTON ALBERT DAVID
MEREDITH CLYDE PEYTON
MERRIMAN REGINALD WALTER
MESHIGAUD ANDREW HARRY
METCALF JIMMY ALLEN
MICAN ALLEN STANLEY
MICHEL WILLIAM LEIGH
MICHULKA FRANK GEORGE
MIDDLETON HOMER RAY
MILBURN ALBERT
MILES MORRIS CALVIN JR
MILLER CURTIS DANIEL
MILLER DAVID MICHAEL
MILLER DOYCE GENE
MILLER EDDIE LEE
MILLER FRED ANTHONY
MILLER HERMAN A II
MILLER LARRY T
MILLER ROBERT WARREN JR
MILLER RONALD ALAN
MILLIGAN GENE CHARLES
MILLS JAMES DALE
MILLS LEONARD MARK
MILLS ROBERT PERRY JR
MILSTEAD ANTONIO
MILTON GARY ANDREW
MILUS EDWARD LEE
MIMS FELTON LEE
MINDACH WILLIAM R
MINKS RAYMOND CRANSTON
MINOR ARMANDO ALVEREZ
MINOR JOHN MICHAEL
MINTER WILBUR LOVING JR
MINTON DON WAYNE
MIRAMONTEZ ENRIQUE
MIRANDA FILIBERTO GUILLER
MITCHAM CHARLES EMMETT
MITCHELL GEORGE GROVER
MITCHELL JOHN EDWIN
MITCHELL MALCOLM EVERETT
MITCHELL MICHAEL DENNIS
MITCHELL MICHAEL LANG
MITCHELL RONALD EARL JR
MITCHELL TORRANCE JR
MITCHELTREE ROBERT G JR
MIXSON JOSEPH GARY
MODISETTE THOMAS GLENN
MOHN LAURANCE RICHARD JR
MOLINA GILBERTO MENDEZ
MOLPUS JAMES DAVIS
MONEACHI DAVID KEITH
MONSEBAIS LUPE
MONTALVO SIGIFREDO JR
MONTANA ROSENDO
MONTANIO ANDREW
MONTEMAYOR FRANK DE LEON
MONTEMAYOR JOSE SANCHEZ
MONTEZ ANASTACIO
MONTEZ JESSE BARRERA
MONTEZ JOE
MONTOYA DAVID
MONTOYA LUIS ALBERTO
MOON ROBERT WAYNE
MOORE ALBERT JR
MOORE AMON FRANKLIN JR
MOORE DEAN
MOORE ELDON WAYNE
MOORE JAMES D
MOORE JOHN TERRY JR
MOORE LAWRENCE HAMILTON
MOORE SCOTT FERRIS JR
MOORE THOMAS DEWEY JR
MOORE WALTER LEE
MOORE WALTER LEE JR
MORA ROBERT CHARLES
MORADO ANTONIO
MORADO DOMINGO FLORES
MORALES ANTONIO JR
MORALES BENITO
MORALES FELIPE
MORAN JOE MICHAEL
MOREIDA MANUEL JESUS
MORENO ANGEL JOSE
MORENO DAVID J
MORENO FRANCISCO HERNANDE
MORENO JESUS JR
MORENO JOE
MORENO JOSE LOUIS
MORENO RICHARD LAWRENCE
MORGAN AUBRA ERLE JR
MORGAN LOWELL EDWARD
MORIN SILBANO
MORRIS CLENZELL
MORRIS DONALD DURWOOD
MORRIS ELROY
MORRIS HOOVER
MORRIS JESSE DON JR
MORRIS JOHN LEE
MORRIS JULIUS WILLIAM JR
MORTON BILLY WAYNE
MORTON JERRY WAYNE
MORTON WILLIAM ACE BILLY
MORTON WILLIAM HOWARD
MOSER CECIL JOE
MOSHER HARVEY MILFORD JR
MOSLEY A D
MOSLEY IRVIN WILLIAM
MOSSNER DAVID CAMPBELL
MOTLEY PAUL WILLIAM
MOUDRY CHARLES RAY
MOUTON WILLIAM WAYNE
MOWER JOHN WAYNE
MOYA HERMANDO SANCHEZ
MOYA JOE
MOYA RAMON JR
MUETING MICHAEL JOSEPH
MULLEN WALTER STEPHEN
MULLER WALTER JR
MULLINAX HOMER LAMAR
MULLINS JAMES MICHAEL
MUNDT HENRY GERALD II
MUNOZ DOMINGO
MUNOZ ERNEST CEDILLO
MUNOZ GUILLERMO
MUNOZ JOHNNY
MUNOZ JUAN
MUNOZ PEDRO
MURPHY ALFRED WALKER
MURPHY FRANK MONROE
MURPHY MICHAEL THOMAS
MURPHY ROBERT D JR
MURPHY WILLIAM ELLIOTT
MURRAY THOMAS E
MUSE MICHAEL DENNIS
MUSICK THOMAS WAYNE
MUSSELMAN STEPHEN OWEN
MYATT JOHN CARNUL
MYERS DONALD WAYNE
MYERS JIMMY LEE
NAJAR ALFRED SATURN JR
NAJAR MIGUEL FERNANDO
NANCE RICHARD ALAN
NARVAEZ PAUL REYES
NARVARTE PETER E JR
NATION JIMMY LEE
NAVARRO ARMANDO SANCHEZ
NAVARRO DANIEL LEON
NEAL BURNETT JR
NEAL WILLIAM RICHARD
NEARY JOHN RUNYON II
NEASBITT LARRY DOUGLAS
NEEL ROBERT RAY
NEELY JAMES ELGIN
NELSON FRANKLIN KING
NELSON HARLAN CLAUDE
NELSON MARSHALL DAYLE
NELSON ROY LANE
NELSON SYLVESTER JR
NEVELS JOHN ALTON
NEW CITY THEODORE C JR
NEW MORRIS DWAINE
NEWCOMB CLIFTON CURTIS
NEWMAN ALLEN TRUMAN
NEWMAN ERMAN MILFORD JR
NEWMAN GARY KEN
NEWMAN ROBERT NELSON
NEWTON CHARLES VERNON
NICHOLAS JOHN ALVIE
NICHOLAS PAUL RUSSELL JR
NICHOLS THOMAS EDWARD
NICKERSON CURTIS CARL
NICKLEBERRY CLIFFORD
NICOLAISEN JAMES ELLSWORT
NIEDECKEN RAYMOND ALVIS
NIEDECKEN WILLIAM CLINTON

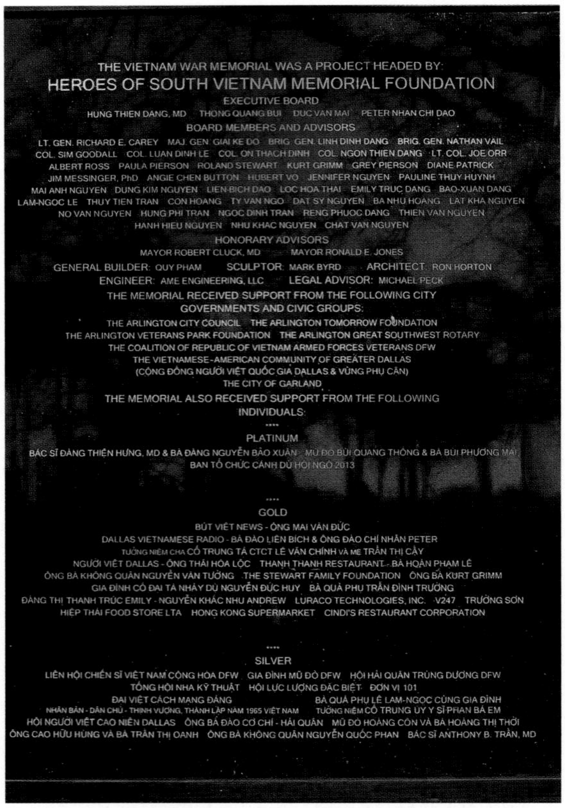

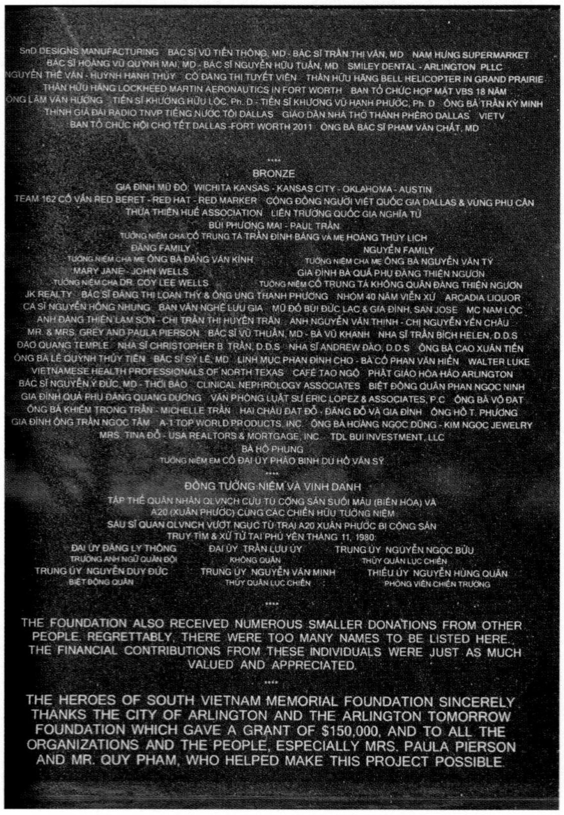

NIELSEN ROLAND ALBERT
NINO AMELIO
NIX ROBERT MICHAEL
NOBLES AUBREY ELDON
NOLAND JERRY LYNN
NOLLEY LEE ROY
NOONAN MICHAEL DENNIS
NORMAN MARION HENRY
NORTH BENNIE LEE
NORTHCUTT DANNY RAY
NORTHCUTT WILLIAM
BUCKELE
NORTON GERALD WAYNE
NORWOOD HUGH
NORWOOD JOE WILLIAM
NOVISKI BERNIS J
NUCKLES C GREGORY
NUNEZ JESSE MANUEL
NUNEZ SANTOS SILVAS
O BRIEN MICHAEL MACKIE
O HARA JOHN PATRICK
O NEAL DOYD DEWAYNE
O TOOLE GERALD ARNOLD
OAKES BRUCE DONALD
OAKS ROBERT LARRY
OATES EUGENE C III
OATMAN LEO CLARK
OCHOA ALFREDO JR
OCHOA LUPE P
OCHOA ROBERT
ODSTRCIL WILLIAM JOSEPH
OGDEN EDWARD PAUL
OJEDA JOE B
OLDFIELD JAMES STANLEY JR
OLDHAM ROBERT LEE
OLIVER ROBERT LYNN
OLIVO RAMIRO FROLIAN
OLSOWSKI GARY NEAL
OLSSON JON MICHAEL
OLVERA ADAN MONSIVAIS
OMAN RICHARD ARLEN
OREM ALLEN WAYNE
ORNELAS BENJAMIN
OROZCO ANTONIO
OROZCO REYNALDO NAVA
ORSINI JOHN JOSEPH JR
ORTEGA HILBERTO
ORTEGA JOE LUCIO JR
ORTIZ ALBERTO JR
ORTIZ JOSE ANGEL
ORTIZ JOSE ESPIRIDION
ORTIZ LAWRENCE JR
ORTIZ MARCELO JR
ORTIZ MELECIO
ORTIZ PEDRO MORENO
OSBON TOMMY EUGENE
OSBORN JACK WILLIAM
OSBORNE ELVIS WELDON JR
OTTMERS RAYMOND MAX JR
OVALLE PORFIRIO M JR
OVERSHINE GEORGE EDWARD
OVERSTREET DAVID DE-
WAYNE
OVERSTREET WILLIAM
DANIEL
OVERSTREET WILLIE JR
OVIEDO HIGINIO OVALLE
OVNAND CHESTER MELVIN
OWEN SAMUEL TAYLOR
OWEN THURMAN WAYNE
PACE JAMES TAYLOR
PACE RONALD GENE
PACHECO DONALD GON-
ZALES
PACK JUNIOR B
PACK WILLARD ORVAL
PADGETT DALLAS LANDON
PADIER WILTON JR
PADILLA DAVID ESEQUIEL
PADILLA FIDEL
PADILLA GILBERTO
PADILLA MICHAEL RAYMOND
PAGAN GARY DON
PAGE EDGAR DE WITT
PAGE JOHN WILLIE
PAGE M C
PALACIO GILBERT GONZALES
PALACIOS OSCAR H JR
PALMER JESSE JAMES
PALMO JIMMIE CHARLES
PALMORE ROBERT DUANE
PALOS ERASMO
PANNELL PHILLIP RANDALL
PANTOJA CIPRIANO J JR

PARADISE CHARLES ALONZO
PAREDEZ AUGUSTIN CHAVEZ
PARHAM JAMES WAYNE
PARHAM LOUIE SNYDER
PARK AUBREY G
PARKER CARL SYLVESTER
PARKER DAVID ALLEN
PARKS JOE
PARNELL BILLY RAY
PARR MICHAEL GRAMBLING
PARRISH FRANK COLLINS
PATE MILTON DALE
PATINO ROBERTO LERMA
PATRICK CALVIN RAY
PATRICK DEREK WILKERSON
PATRICK J V
PATTERSON DAVID Q
PATTERSON GORDON LEE
PATTERSON OSCAR BERNARD
PATTON JERRY DON
PAUL HAMILTON JR
PAVLICEK JAMES EMIL JR
PEARCE JERRY DOYLE
PEEPLES HARDY WINSTON
PEGROSS LEROY
PELAJIO ARTURO
PELTON GLENN EUGENE
PENA DANIEL JR
PENA JOE JR
PENA MANUEL JUAN
PENLAND RAY LEE JR
PENNELL ARVIN DOUGLAS
PENNINGTON EDWARD LEE
PEOPLES DAVID DOUGLAS JR
PEPPLE CARL FRANKLIN JR
PERCY DONALD LEE
PERDUE GEORGE EDWARD
PEREA ERNESTO SALVADOR
PEREA ROBERTO
PEREZ ADOLFO MORENO
PEREZ ALBERTO L
PEREZ ANTHONY
PEREZ ASCENSION ROSALES
PEREZ BENITO
PEREZ DANIEL FLORES JR
PEREZ ESPIRIDION
PEREZ FREDERICO
PEREZ HOMERO
PEREZ ISRAEL
PEREZ JAMES SANDERS
PEREZ JESUS RAMON
PEREZ JOSE MANUEL
PEREZ RAUL BAUTISTA
PEREZ RAYMOND
PEREZ ROBERTO
PEREZ RODOLFO
PERKINS WALLACE SAM
PERRY BILLY EARL
PERRY CARROLL WAYNE
PERRY ELMER JOSEPH JR
PERRY JACKIE RAY
PERRY R C JR
PERRY RODDIE LEE
PERRY WILLARD ALTON JR
PERSON DAVID EUGENE
PESEK THOMAS JOHN
PETEET CHARLES LEONARD
PETERS BERYL GENE
PETERSON EDGAR LEWIS JR
PETERSON KENNETH AUBREY
PETTWAY PATRICK HENRY II
PETTY ROY LYNN
PFEUFFER MICHAEL LAW-
RENCE
PFEUFFER RONALD HOWARD
PHEARS RONALD GENE
PHELPS HERMAN ROY
PHILIBERT BRIAN HARDMAN
PHILLIPS DENNIS MICHAEL
PHILLIPS DONNELL
PHILLIPS JERRY
PHILLIPS JOHNNY WENDELL
PHILLIPS NORRIS ARTHUR
PHILLIPS ROGER LEE
PHILLIPS ROY EDWIN
PHILPOTT HAROLD DEAN
PICKARD ALFRED
PIERCE JOE JR
PIERCE MICHAEL ABEL
PIERCE WILLIAM WESLEY
PIKE NIXON DEWAYNE
PILKENTON CLARENCE
WESLEY
PILOTTE JOSEPH MARION

PILSNER JOHNNY MACK
PINKARD ROBERT LEE
PINSON CLOYDE CYRIS JR
PIPER SIDNEY JR
PITTS ROBERT ARDELL
PLAMBECK PAUL WANDLING
JR
PLASTER BILLY JOE JR
PLATA JOHNNY MORRIS
PLATT BILLY WAYNE
PLATT WAYNE B
PLEMMONS ROBERT
COLQUITT
PLUMLEE JAMES LEO JR
PLUMMER HERBERT JR
POCS LESLIE MARTIN
POE CLIFFORD EARL JR
POHL EHRHARD HANS
KONRAD
POLENDO RAYNALDO
POLING JOHN EARL
POLK CHARLES QUINTEN
POLK ROBERT LOUIS
POMEROY DEANE ALVA
PONCE BENITO ANDRADE
PONGRATZ RONALD EUGENE
POOL THOMAS JOHN
POOLE OTHA LENSEY
POOLE RONALD FELTON
POORE LEONARD BURTON
POPE ROBERT DALE
POST DANIEL ZACHARY JR
POSTON RAYMOND ROGER
POSTON WILLIAM THOMAS
POWELL GEORGE EDWARD
POWELL JOHN DEE JR
POWELL LARRY KEITH
POWELL ROBERT CLYDE
POWELL THOMAS STOKES
POWELL WAYLEN LEE
POWELL WILLIAM ELMO
POWERS EDWARD DOYLE
POWERS WILLIAM JAMES
PRADO GUADALUPE JR
PREDMORE DAVID MARTIN
PRICE CHARLES ALLEN
PRICE ELVIN
PRICE JHUE FRANK
PRITCHARD VICTOR HEENAN
PROCTOR GEORGE RICHARD
PRYOR DONALD RAY
PRYOR MELVIN SR
PUENTES MANUEL RAMERIZ
PULLEN ROBERT DALE
PUMPHREY JAMES J L
PURIFOY RAY WARREN
PYLE CHARLES RICHARD
PYLE LARRY GENE
QUALLS TED WAYNE
QUEEN CARY PAUL
QUEEN CECIL WAYNE
QUINN JOHN FRANCIS
QUINN MICHAEL COURTNEY
QUINTANILLA FRANCISCO JR
QUIROZ ALEXANDER
RACKLEY INZAR WILLIAM JR
RADFORD GARY MONROE
RAGLAND ROBERT EUGENE
RAGSDALE ROBERT LOUIS

RAINER CURTIS HALL
RALPH THOMAS HENRY JR
RAMIREZ DIEGO JR
RAMIREZ FLORENCIO JR
RAMIREZ JOSE HERIBERTO
RAMIREZ JUAN JOSE
RAMIREZ RAMIRO RIOS
RAMON ANDRES LOPEZ
RAMON AURELIO R JR
RAMON EUGENE DOMIN-
GUEZ
RAMOS EDWARD
RAMOS FELIX RICO JR
RAMOS FIDEL JR
RAMOS JESUS LOPEZ JR
RAMOS JOSE JR
RAMOS JOSE PABLO
RAMOS JUAN MANUEL
RAMOS LEONARDO JR
RAMSEY BILL EDWARD JR
RAMSEY CHARLES MARLIN
RAMSEY DON MICHAEL
RAMSEY HENRY CHARLES
RAMSOWER IRVING BURNS II
RANDALL DELBERT BRYAN
RANDALL GARLAND JERONE
RANDOLPH SETH EARL
RATCLIFF ROY
RATH ROBERT EMIL
RAVENNA HARRY M III
RAWLINSON TERRELL LEE
RAY ROLAND WOOLDRIEDGE
RAY RONALD EARL
RAY RUFUS
REAGAN NORMAN REX
RECK DAVID LYNN
RECTOR ROY JACK
REDFORD JAMES ROBERT
REED DELMA LEE
REED TERRY MICHAEL
REED WILLIAM VAL
REEDER PHILIP DALLAM
REESE RUBEN DWIGHT
REESE WILLIAM ROBERT
REEVES ALVIS OREN
REEVES DENNIS LEE
REEVES HAROLD RAY
REEVES HAROLD RAY
REEVES LARRY RAY
REEVES SAMUEL DAVID JR
RELF WILLIAM CHARLES
REMMERS KENNETH LEE
REMMLER MILTON WILLIAM
JR
RENDON THOMAS
RESENDEZ AUGUSTINE
REYES ALFREDO VICENTE
REYES ANTONIO
REYES JOSE ANGEL
REYES MOISES A JR
REYES PETER C
REYNA JOE JR
REYNA JUAN MANUEL
REYNA SAMUEL
REYNER DAVID ELLIOT
REYNOLDS LARRY ALLEN
REYNOLDS OLIVER EUGENE JR
REYNOLDS RONALD BURNS
RHODES TIMOTHY V

RICE JAMES R
RICE PATRICK L
RICE VIRGIL RAY
RICHARD CURTIS
RICHARD DONALD WAYNE
RICHARDS JOHNNY FRANK-
LIN
RICHARDS PAUL ALLEN
RICHARDSON DONALD LOYE
RICHARDSON HARRY TRACY
JR
RICHARDSON LARRY EUGENE
RICHARDSON ROBERT EARL
RICHARDSON ROBERT
WAYNNE
RICHARDSON WILLIAM F
RICKELS DAVID LEE
RICKS LARRY EUGENE
RICKS RONALD GLENN
RIDDLE CHARLES LLOYD
RIDDLE ROBERT THOMAS
RIDGE FELIX DENNIS
RIGGINS JAMES PATRICK
RIGGINS ROBERT LUCIAN JR
RINARD KEVIN ALONZO
RIOS FIDENCIO GARZA JR
RIOS IGNACIO ELENO
RIOS ROBERTO PENA
RIOS SALVADOR DE LOS S
RIOS TEOFILO CARMONE
RISINGER JERRY LEROY
RITCH HAROLD JUNIOR
RIVAS ARTURO BROWN
RIVAS JOSE LUIS
RIVERA ALFREDO
RIVERA ARNOLD JAVIER
RIVERA JOE LEWIS
RIVERA RUBEN
RIVERS MICHAEL ROSS
ROACH JOHNNY FRANKLIN
ROBALIN ALBERT SIMON JR
ROBBINS HENRY EARL
ROBBINS JAMES WALTER
ROBBINS RUSSELL LINDSEY
ROBERSON JIMMY DON
ROBERSON JOHN WILL
ROBERSON ROBERT SIDNEY JR
ROBERSON WILLIAM THOMAS
ROBERTS BILLY DALE
ROBERTS BILLY JACK
ROBERTS BOBBY LEE
ROBERTS CHARLES ALAN
ROBERTS GERALD RAY
ROBERTS HOWARD TAYLOR
ROBERTS JOHN ALLEN
ROBERTS JOHN CLYDE
ROBERTS THURSTON CRAIG
ROBERTS WALTER EUGENE
ROBERTSON TOMMY WAYNE
ROBINETTE DANNY LEON
ROBINSON FRANK EUGENE
ROBINSON GORDON LEE
ROBINSON LEONARD JR
ROBINSON MARVIN RAY
ROBINSON ROBERT DOUGLAS
ROBINSON WALTER R JR
ROBLEDO EFRAIN JULIO
ROBLEDO JESUS JR
ROBY CHARLES DONALD

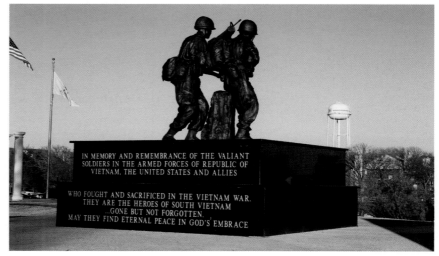

Texas

ROCHA RAYMOND GONZALEZ
ROCHA ROBERTO JR
ROCHA RUBEN LOPEZ
ROCHA RUDOLFO LEONARD JR
RODEN JOHN JOSEPH WILLIAM
RODGERS JERRY PAUL
RODGERS JOHNNY MICHAEL
RODGERS LARRY JOE
RODGERS MARTIN LEROY
RODRIGUES JOE G JR
RODRIGUEZ ARTURO
RODRIGUEZ CALIXTRO S
RODRIGUEZ CASIMIRO JR
RODRIGUEZ DOMINGO JR
RODRIGUEZ ELIAS RANGEL
RODRIGUEZ FRANCISCO JR
RODRIGUEZ JESSE NICKLUS
RODRIGUEZ JOE
RODRIGUEZ JOEL
RODRIGUEZ JUAN ARMANDO
RODRIGUEZ JULIAN ROBLES
RODRIGUEZ MANUEL JR
RODRIGUEZ MARGARITO JR
RODRIGUEZ MATIAS T JR
RODRIGUEZ PEDRO JUAN
RODRIGUEZ REYNALDO SALAIS
RODRIGUEZ ROBERT
RODRIGUEZ SAMMY PINA
RODRIGUEZ VICENTE QUINTAN
RODRIQUEZ FELIX
RODRIQUEZ LEONDIS ENRIQUE
RODRIQUEZ PEDRO S JR
RODRIQUEZ TOBY S JR
ROE DONALD JAY
ROE JERRY LEE
ROE JOHN ELMER
ROGERS DANIEL GORDON
ROGERS JIMMIE DALE
ROGERS ORVILLE CURTIS JR
ROGERS WILLIAM LOUIS
ROLAND JOHN PAUL
ROLF TOMMIE ALLEN
ROLLER CHARLES BENNETT JR
ROLLINS WILLIAM PAUL
ROMERO HECTOR MANUEL
ROMERO MANUEL VILLAREAL
ROMERO RONALD M
RONJE JOE LUIS
ROPER ROBERT CLARANCE
ROSAS JOSE ANTONIO
ROSE DONALD RAY
ROSE LUTHER LEE
ROSEMOND JOHN L
ROSEN PATRICK DEAN
ROSS GENE AUTRY
ROSS MICHAEL ROBERT
ROSS THOMAS EDWARD
ROTHEL LARRY WAYNE
ROUSH WILLIAM WAKEFIELD
ROWELL DAVID LOU
ROWELL KEITH WILLIAM
ROWLETT JAMES WESLEY
ROY DAVID PAUL
ROY HENRY JOHN JR
ROYSTON ROY LEE
RUBIO JUAN AMADOR
RUCKER EMMETT JR
RUCKER JOHN ONEAL
RUCKER KENNETH RAY
RUDD ROBERT CHARLES
RUELAS MATEO
RUIZ ANTONIO ELIZONDO
RUIZ FELIX ALVARDO
RUIZ MARTIN JR
RUIZ RICHARD PETER
RUSH CLAUDE BENJAMIN JR
RUSHING GEORGE WILLIAM

RUSHING WILLIAM LENDELL
RUSSAW PRESTON IVORY
RUSSELL JAMES LEROY
RUSSELL JOHN ERNEST
RUSSELL PETER LOWELL
RUSSELL RICHARD LEE
RUSSELL ROBERT THOMAS II
RUVALCABA-LOPEZ MIGUEL AN
RYAN LIONEL ALVAREZ
RYAN ROBERT DALE
RYLANDER ROBERT J
RYZA WAYNE DAVID
SACKETT ERIC
SADLER HOWARD JR
SAENZ ALFREDO JOSE
SAENZ RODOLFO ANDRES
SAFFELL RONALD CORY
SAINT CLAIR LEONARD RAY
SALAS FELIX JUAN
SALAZAR ALFREDO
SALAZAR ARTURO
SALAZAR ELIAS JR
SALAZAR JOSE LUIS
SALAZAR RENE JAVIER
SALAZAR ROBERTO
SALAZAR RUDY JESSIE
SALDANA FERMIN JR
SALDIVAR JOSE ANGEL
SALINAS JAIME ARTURO
SALINAS JOSE CONTRERAS
SALINAS MERCEDES PEREZ
SALINAS RAMIRO LOPEZ
SALINAS ROBERT LONGORIA
SALINAS ROY RODRIGUEZ
SAMANIEGO JOE HENRY
SAMANIEGO ROBERTO
SAMARIPA JESSE
SAMFORD JESSE LEROY
SANCHEZ ALBERTO VASQUEZ
SANCHEZ EDWARD CHARLES
SANCHEZ ERNESTO JR
SANCHEZ JAVIER ARTURO E
SANCHEZ JOSE GUADALUPE JR
SANCHEZ JOSEPH SEBASTIAN
SANCHEZ JUAN OSCAR
SANCHEZ MACARIO JR
SANCHEZ MICHAEL
SANCHEZ PABLO DEMEO
SANCHEZ PEDRO JR
SANCHEZ REYNALDO AYALA
SANCHEZ ROBERT HUERTA
SANCHEZ WILBERTO CABRERA
SANDERLIN WILLIAM DALE
SANDERS ALAN EARL
SANDERS CLYDE DOUGLAS
SANDERS LARRY TRUMAN
SANDOVAL EVARISTO
SANDOVALL ANTONIO RAMOS
SANTIAGO TIMOTEO MUNOZ JR
SANTOS JOSE CARLOS
SANTOS RENE ANTHONY
SAPP FREDDY LEE
SAPP JEFFERY TRUETT
SAPP STANLEY L
SATTERFIELD HOWARD EUGENE
SATTERWHITE DWIGHT KING
SAUCEDA JOE A
SAUCEDO ROGELIO
SAURINI JAMES PAUL
SCHAEFER JOHN STEVE
SCHAFERNOCKER MICHAEL E
SCHAUTTEET LOUIS L JR
SCHERER JAMES LEE
SCHILLING GEORGE DON
SCHMIDT DALE W JR
SCHMITTOU EUREKA LAVERN
SCHOLZ KLAUS DIETER
SCHONBERG DENNIS WAYNE
SCHOPPAUL ROBERT EARNEST
SCHOPPE FRANKLIN DALE
SCHOPPE SHERWIN CRESCENT
SCHROEDER JOE LAWRENCE
SCHROLLER LEO JOE JR
SCHULTE NORMAN DOUGLAS
SCHULTZ DAVID PAUL
SCHUMACHER LARRY DEAN
SCOGGINS ROYCE GLENN
SCOTT DANNY RAY

SCOTT MICHAEL FREDRICK
SCOTT PATRICK RAY
SCOTT PAUL
SCROGGINS CARREY EUGENE
SCROGGS JERRELL DAVID
SCRUGGS JAMES ARTHUR
SEAGRAVES MELVIN DOUGLAS
SEBRING RONALD WAYNE
SECREST JACK MCCOMBS JR
SEE MICHAEL DUANE
SEGOVIA RAUL LERMA
SEIGLER STEVEN LEE
SEILHEIMER HOWARD ALLEN
SELLS TERRY STEPHEN
SEPULVEDA JESUS GARCIA
SERENIL RICARDO
SERNA PHILIP JOSEPH
SEWELL RAYFORD NEAL
SEYMOE JOSEPH PHILLIP
SHADWICK ALVIN LEE
SHARP MICHAEL ANTHONY
SHARP PRESTON DOUGLAS
SHARP STEPHEN CARL
SHARP THOMAS BOYD
SHARP VALDEZ
SHAUGHNESSY JOHN F JR
SHAWN RAYMOND BEN
SHEA MICHAEL JOHN
SHELTON DARWIN HUGH
SHELTON JAMES DALLAS
SHELTON SHERRELL VANCE
SHEPARD RONALD WAYNE
SHIMEK ALBERT LAWRENCE
SHIVER CHARLES JR
SIDES HAROLD ERWIN
SIEGERT WILLIAM FRY
SILER MANLEY EUGENE JR
SILVA JOE REYES
SILVA MANUEL
SILVA RITO
SILVEE HERMAN WILLIAM
SIMMONS ARTHUR D
SIMMONS JAKE A
SIMMONS NOLAN LESTER
SIMMONS RICHARD CHARLES
SIMMONS TOM WILLIS JR
SIMMONS TRAVIS A JR
SIMON THOMAS JAMES
SIMONS GARVIS KEITH
SIMPSON ADAM ERNEST JR
SIMPSON ELMORE ROBERT
SIMPSON LOYDE HAROLD
SIMPSON MELVIN RICHARD
SIMPSON MORRIS ALFRED
SIMS JOHN CHARLES JR
SIMS KIRK WAYNE
SIMS WILLIAM A
SINEGAL LARRY JAMES
SINER WALLACE KINGSLEY
SINGER KENNETH EDWIN
SINGLETON ELWIN EARL
SINGLETON THOMAS ARNOLD
SISSON BENNIE JOE
SKELTON PAUL DARRELL II
SKILES THEODORE VAN
SLADE JAMES L JR
SLAY RONNIE GLYNN
SLAYMAKER LARRY STEPHEN
SLEDGE DOUGLAS ROY
SLOAN LARRY EUGENE
SLOUGH RUSSELL EUGENE
SLUSSER HARLAN RAY
SMART ROBERT HALL
SMELSER ROGER WAYNE
SMITH AARON LEE
SMITH ALBERT CHARLES
SMITH ALFRED JOHN
SMITH CLIFTON THOMAS
SMITH CURTIS ORAN JR
SMITH DEANE FRANKLYN JR
SMITH DONALD CLAYTON
SMITH DONALD RICHARD
SMITH DONALD WAYNE
SMITH DONNIE PAUL
SMITH DOUGLAS WAYNE
SMITH EUGENE
SMITH GENERAL DEWAYNE
SMITH JAMES WARREN
SMITH JERRY WALTON
SMITH JERRY WAYNE
SMITH JOHNNY LEE
SMITH JOSEPH EWING

SMITH L C JR
SMITH LONNIE LEO
SMITH MICKEL MELVIN
SMITH PHILIP THOMAS
SMITH RALPH MACK
SMITH RAYBURN LESTER III
SMITH ROY MILTON
SMITH THOMAS KING
SMITH VARDE WESTON III
SMITH WARREN PARKER JR
SMITH WILLIAM EDWARD
SMITH WILLIE FRANKLIN
SMITHEE RONALD GAIL
SMOOTS NORMAN CARTER
SNELL ROBERT MICHAEL
SNITKO JOE ANTHONY
SNOWDEN BEN DAVID
SOLIS EXTRUMBERTO
SOLIS OSCAR ABREGO
SOLIZ GEORGE
SOLIZ JULIAN
SOLIZ ROLANDO LUIS
SOLOMON FLOYD DEAN
SOLOMON MICHAEL VERNON
SOMBELON ALBERT EDWARD
SONNIER ALBERT WILBER
SONY THOMAS ANTHONY
SOSA GEORGE RAMIRO
SOSA JORGE
SOSA SECUNDINO GARCIA JR
SOSA VICTORIANO PEREZ JR
SOWDER BERNARD ALLEN
SOWELL COTIES R
SOZA REYNALDO
SPANGLER GEORGE OWEN
SPANGLER MAX RAY
SPANGLER RICHARD ALLEN
SPARKS CHARLES PIERCE
SPARKS JAMES HENRY
SPARKS PAUL ALLAN
SPARKS THOMAS JAMES
SPEARMON J B
SPEARS JOHNNY CLARENCE
SPEER ROBERT FRITZ
SPENCER FLOYD BROWN JR
SPENCER ROBERT DALE
SPIEKER GARY LYNN
SPIER HARRY DIWAIN
SPILLER LEROY III
SPINKS ALLEN ROBERT
SPIVEY ELMER LYNN
SPRATLEY GLENN EUGENE
SPRING HOMER DOYLE
SPRINGER LOUIS DANIEL
SPRINGER ROBERT L
STAFFORD FRED PATRICK
STAHLSTROM ALLAN EMILE
STALEY JOHN ARTHUR
STALL WILLIAM ROBB
STANDEFER JAMES GLENN
STANLEY BOBBY DWAYNE
STANSBURY THOMAS RODGERS
STANUSH THOMAS JOSEPH
STARK ALFRED
STARNES MILBURN HINES
STARNS DAN CLIFTON JR
STARR ALLEN EUGENE
STAVINOHA ROBERT JAMES
STEARNS HARREL EARL
STEED JERRY LYNE
STEEL RICHARD EDWARD
STEGALL DOUGLAS WAYNE
STEINBACH THOMAS RAYMOND
STEINDAM RUSSELL ALBERT
STEINER LAWRENCE TERRELL
STEINFELD HOWARD MARSHALL
STELTER NYMAN WILLIAM JR
STEPHEN VIRGIL LYNN
STEPHENS ANDREW LEWIS
STEPHENS BEN WESLEY
STEPHENS BENNIE VORICE JR
STEPHENS CLYDE WAYNE
STEPHENS WILLIAM F JR
STEPTOE RAYMOND
STEVENSON CHARLES ROBERT
STEVENSON GREG DOUGLAS
STEWART MORGAN EUGENE
STIEFERMAN CURTIS EDWARD
STILES THOMAS NELSON
STILLEY JOHN WAYNE
STOCKER DANIEL LEO

STOKER HUELYN BERNARD
STOLL GEORGE LUDWIG
STONE BYRON CLARK
STONER CLARENCE MOODY JR
STOUT KEVIN ARLEY
STOVALL JAMES TUCKER
STRAIN JAMES PAUL
STRAIT DAVID LEON
STRATTON CHARLES WAYNE
STRATTON SIDNEY TAYLOR
STRAUSS KLAUS JOSEF
STRICKLAND GAIL LYNN
STRIDE JAMES DANIEL JR
STRINGER OTTIS EDWARD
STRONG DAVID ALLEN
STROTHER CLAUD PAUL
STROUD ROGER LEE
STROUD SANDERS KEY II
STROUD STEVEN ARNOLD
STUCKEY BENNY DAVIS
SUAREZ VALENTINE BERRONES
SUBLETT JOHN KENNETH JR
SULLINGER WILTON JAMES JR
SULLIVAN FARRELL JUNIOR
SUMMERS FRANKLIN DALLAS
SUMMERS RONALD LEE
SUMMERS WILLIAM ELVIN
SUNIGA RUBEN BOSQUEZ
SWANCY JAMES ANDREW
SWANSON BOBBY GENE JR
SWEAT DONALD JEANE
SWEET LARRY EUGENE
SWINDELL BOBBY DALE
SWINFORD FRANK LEVI III
SWINNEA THOMAS HENRY
SWONKE EDWARD ANTONE JR
SYMANK TOMMIE LEE
TABOADA FRANK OLIVARES
TAFF GEORGE THOMAS JR
TAJCHMAN ADOLPH WILLIAM
TALKINGTON DENNIS LEE
TAMAYO JOEL
TAMEYOZA NOE
TAMEZ NOE
TANNER DONALD JAY
TANNER RAYMOND MARSHALL
TARIN EDWARD JAMES
TARIN ELISEO ESPINOZA
TARKENTON JAMES C III
TATE ANTHONY GARY
TATE DANIEL HARRISON
TAYLOR DAVID EARL
TAYLOR ERNEST VERNON
TAYLOR JESSE ALLEN
TAYLOR MICHAEL GEORGE
TAYLOR RODNEY ALAN
TAYLOR SELVWYN RISHER
TEAGUE CHARLES E
TEAGUE MICHAEL AUTREY
TELLEZ DANIEL
TELLO JOAQUIN RODRIGUEZ
TEMPLE LAMAR HAYES
TEMPLIN ERWIN BENARD JR
TERRAZAS NICHOLAS E
TERRELL LOUIS WAYNE
TERRONEZ DOMINGO MENDOZA
TERRY ALLEN LEE
TERRY CONDON HUNTER
TERRY ROBERT ISAAC III
TERRY RONNIE LEE
THACKREY WADE E JR
THARP JERRY DONALD
THOMAS ALLEN WALKER
THOMAS CLYDE
THOMAS DAVID JOHN
THOMAS DOUGLAS MCARTHUR
THOMAS EARL WILLIAM JR
THOMAS ELMER WAYNE
THOMAS GERALD LYNN
THOMAS JAMES EDWARD
THOMAS JERRY LEE
THOMAS JONATHON E JR
THOMAS JULIUS
THOMAS L V JR
THOMAS LEWIS MCCOY
THOMAS MATTHEW ALONZO JR
THOMAS THEODORE DAVE JR
THOMAS WAYNE LEWIS

THOMPSON CHARLIE EARL
THOMPSON DON CARTHAL JR
THOMPSON DONALD BRUCE
THOMPSON JOHN MICHAEL
THOMPSON JOSEPH DAVID
THOMPSON OTHA THEANDER
THOMPSON PETER GARLAND
THOMPSON ROGER ALLEN
THOMPSON ROY EUGENE
THOMPSON VICTOR HUGO III
THOMPSON WAYLAND KENT
THOMPSON WILLIAM JAMES
THOMPSON WILLIE RAY
THORN CLIFTON CARDELL
THORNTON CHARLES ED-
WARD
THRASHER LARRY GLEN
THREADGILL DAVID ELLIS
THURMAN LARRY PRESTON
TICE JIMMIE RAY
TIDWELL DONNY GAY
TIDWELL EARL CARL E JR
TIENDA DANIEL
TIJERINA ALBERT JR
TIJERINA ARTHUR CASTILLO
TIJERINA HOMERO ELIUD
TIJERINA JOSE BENIGNO
TILGHMAN JIMMIE MACK
TILL RALPH GARY
TINAJERO JOSE ANTONIO
TINNEY JOHNNY MACK
TINNIN EUGENE SANFORD
TINSLEY RONALD ETHRIDGE
TIPTON FREDDIE LEON
TISCHLER HOMER ERICK
TITUS TOUSSAINT LEO
TOBIAS BILLY LEE
TODD FREDRICK WELTON
TOLIVER WILLIAM LEE
TOLLESON LYNDOL EARL
TOM GEORGE WILLIAM
TOMAS DAVID RAY
TOMCHESSON TEDDY JAMES
TONEY WILLIE LEE
TORREROS JOSE
TORRES EZEQUIEL JR
TORRES GILBERT GARCIA
TORRES IGNACIO JR
TORRES JOE D
TORRES JOSE
TORRES JUAN
TORRES LOUIS FERNDEZ
TORRES RAMON HERNANDEZ
TORRES REYNALDO LERMA JR
TORRES REYNALDO SANDO-
VAL
TOWNSEND CHARLES DWYNE
TOWNSEND FRANCIS WAYNE
TRAVNICEK EDWIN RAY
TREJO JOSE MANUEL
TREJO MIGUEL
TREVINO CARLOS V
TREVINO ESTEBAN ANGEL JR
TREVINO FAUSTINO
TREVINO GREGORIO JR
TREVINO MANUEL VAILLIDO
TREVINO RODOLFO
TREVINO SAVAS ESCAMILLA
TRIANA SALVADOR PUGA
TRICKEY JOE H JR
TRIGGS FOSTER F
TRIGGS WAYMON LEON
TRIMBLE TOMMY LEE
TRISTAN ALBERT FLORES
TRITICO MICHAEL JOSEPH
TROLLINGER JIMMY MICHAEL
TRUMBLE DARRELL LYNN
TUBBS GLENN ERNEST
TUCH JIMMIE
TUCKER BOBBY DAN
TUCKER JAMES EDWARD JR
TUCKER OTTO DALE
TUCKER WESLEY GRIFFIN
TURK JOSEPH MICHAEL
TURN HENRY LON
TURNER CLARENCE S III
TURNER JOHNNY CHARLES
TURNER OTIS
TURNER TONY RAY
TUTTLE ERVIN LEE
TUTTLE NELSON PAYNE
TWEEDLE KEVIN EDWARD
TWINN LOUIS BELL SR
TYLER ERNEST KENNETH

TYLER LARRY JEROME
TYLER MITCHEL RAY
TYRONE WILLIE DONALD
TYSON LARRY PRESTON
TYSZKIEWICZ ARTHUR
KASIMI
UBERMAN RODNEY RAY
ULLOA HUGO HECTOR
UNDERWOOD ANDREW
FILLEBRO
URBANOVSKY ROBERT
EUGENE
URDIALEZ RUBEN
URSERY MICHAEL TERRY
VADEN WILLIAM KENNETH JR
VALANDINGHAM EVERETT
JOSE
VALDEZ GREGORIO JR
VALDEZ RODOLFO
VALE TONY
VALENCIA AMADO ACOSTA
VALENZUELA JUAN
VALLE ELOY RUBEN
VALTR JAMES ROBERT
VALUSEK DENNIS WAYNE
VAN CLEAVE WALTER SHELBY
VAN ORDEN EDWIN WARD JR
VAN SANT JOHN WILLARD
VAN ZANDT THOMAS MILTON
VANCE JAMES SIDNEY
VANDEVENDER JERRY WAYNE
VANDEWALLE GREGORY
JEROME
VANLANDINGHAM RONALD
LEE
VANN RONALD BRYSON
VANZANDT RAY LOUIS
VAQUERA ALBERT ALVARADO
VARA PAUL MARTINEZ
VARGAS JORGE
VASQUEZ ALBERTO RIOS JR
VASQUEZ ENRIQUE
VASQUEZ EPHRAIM
VASQUEZ JESUS ROBERTO
VASQUEZ RODOLFO ARTURO
VASSAUR FRANKIE CARL
VAUGHAN CARVER JOE
VAUGHN DELBERT LEE JR
VAUGHN GENE
VAUGHT MICHAEL EUGENE
VAUGHT WILBURN FRED
VEARA JOHN VINCENT
VEGA ANGEL
VELA VITALIO JR
VERA ABELARDO
VERASTIQUE JOHNNY RALPH
VERGARA ELISEO
VICKREY CLARKE KEMBLE
VIEREGGE WALTER III
VILANO EVARISTO
VILLA FELIBERTO
VILLA RAUL
VILLAFRANCO RODOLFO
VILLALOBOS ELISEO MO-
RALES
VILLANUEVA ALFREDO
JULIAN
VILLARREAL JOHNNY
VILLARREAL ROLANDO G
VILLASANA FERNANDO
VILLASENOR GONZALO H
VINCENT THOMAS DEAN
VINEYARD GEORGE MICKEL
VINSON BOBBY GENE
VIVIAN JOHN HALL
VOIGT ARNO JOSEPH
VOLLMAR SAMMIE J
VOSS RALPH
VOTAW MARVIN LYNN
VRBA JAMES MATHEW J
VRBA RAYMOND JOE JR
WACKER JOSEPH HENRY
WADDELL JAMES DARRELL
WADLEY JACK DALE
WADSWORTH DEAN AMICK
WAID BILLY GENE
WAITES SHERMAN RAY
WALKER HOLLIS ALLEN
WALKER J C JR
WALKER RALPH BAMFORD II
WALKER ROBERT DONALD
WALKER ROBERT HARVEY
WALKER ROY
WALKER TOMMY DALE

WALL ARLON DANIEL JR
WALL GEORGE ELTON
WALL JAMES ARTHUR
WALL JERRY MACK
WALLACE JAMES EARL
WALLACE JERALD D
WALLACE KEM L
WALLEY TERRY CLINTON
WALTERS ROBERT JAMES
WARD CHARLES DWIGHT
WARD NEAL CLINTON
WARDLOW JAMES DILLON
WARMSLEY HAROLD JAMES
WARR JAMES MILTON
WARREN TOMMY RAY
WARREN WILLIE CRAIG
WASH NATHAN JR
WASHBURN LARRY EUGENE
WASHINGTON DONALD RAY
WASHINGTON JAMES B
WASHINGTON JAMES BELL
WATERS GLENDON LEE
WATKINS GARY WINSTON
WATKINS JAMES LEE
WATSON BILLY FRANK
WATSON CARNELL EARL
WATSON DAVID WARREN
WATSON GARY EUGENE
WATSON RONALD LEONARD
WATT ROBERT LEE
WATT WILLIAM ROY
WATTS AFTON M
WATTS WILLIAM E
WATTS WILLIAM SCOTT
WAUGH MARION EDWARD
WEATHERBY JACK WILLIS
WEATHERFORD JERRY GLENN
WEATHERFORD JOHN MI-
CHAEL
WEATHERS BOBBY LYNN
WEAVER ALLEN PRICE
WEAVER CHARLES EDWARD
WEAVER GREG
WEAVER WILLIAM CARRELL
WEBB JAMES EDWARD
WEBB WILLIAM MATTHEW
WEBBER FLOYD DEAN
WEBSTER JAMES ROBERT JR
WEBSTER MICHAEL WARREN
WEEKS GEORGE DALE
WEIK MICHAEL JOSEPH
WELBORN JOE THOMAS
WELCH CLYDE RAY
WELCH NORMAN GENE
WELDY GEORGE W JR
WELLMAN KENYON GARY
WELLS DAVID CLAUD
WELLS JUDSON ARTHUR JR
WELLS ROBERT OLIVER
WELLS RUSSELL LEE JR
WENDT ROBERT WAYNE
WESLEY ROBERT EARL
WESOLICK HAROLD JAMES JR
WESSINGER LARRY ALLEN
WEST DAVID RICHARD
WEST GEORGE A
WEST ROBERT LEWIS
WESTBROOK DONALD ELLIOT
WESTER ALBERT DWAYNE
WESTERFIELD FRANK BROWN
WESTON JAMES EDWARD
WHEAT WENDELL RAY
WHEELER CONRAD JACK
WHEELER KENNITH WAYNE
WHEELER OSCAR LEE
WHELESS JIMMY RAY

WHETSEL JACK ALLEN JR
WHIRLOW ROGER DALE
WHITE BARNEY JOE
WHITE CHARLES THERON
WHITE JERRY MORGAN
WHITE LARRY JOE
WHITE LENWOOD JR
WHITE RAYMOND AUSTIN III
WHITE TOMMIE VAUGHN
WHITEHOUSE GREGORY
KENT
WHITLOCK ALAN D
WHITLOCK HALLEY DON
WHITTEN TOMMIE JOE
WHITTLESEY ROY LEE
WICKERSHAM HARRY W JR
WIDNER DANNY LEE
WIEBURG WILLIAM WARREN
WILBANKS JAMES HARDY
WILBANKS TIMOTHY MARCIA
WILBURN WOODROW
HOOVER
WILDER RONALD FREDERICK
WILEY FRANK DAVID
WILHELM MACK HOUSTON
WILKERSON LARRY WAYNE
WILKERSON LAWRENCE
WILKEY EMMITT JAMES JR
WILKINS CALVIN WAYNE
WILKINSON BILLIE WELDON
WILKINSON CLYDE DAVID
WILLEFORD JIMMY WAYNE
WILLIAMS AUGUSTUS LOUIS
WILLIAMS BOBBY LEE
WILLIAMS BOBBY RAY
WILLIAMS DAVID
WILLIAMS DAVID EDWARD
WILLIAMS EDDIE EARL
WILLIAMS FREDDY THOMAS
WILLIAMS GAYLE EDWARD
WILLIAMS GERALD DAN
WILLIAMS HERBERT
WILLIAMS HILLARD EVANS
WILLIAMS HOWARD C JR
WILLIAMS JAMES WESLEY
WILLIAMS JOE BUCK
WILLIAMS JOHNNY BEE
WILLIAMS LARRY LEE
WILLIAMS LAWRENCE C JR
WILLIAMS MARSHALL WAYNE
WILLIAMS NORMAN COLUM-
BUS
WILLIAMS RAY LEE
WILLIAMS RAYFIELD
WILLIAMS RICHARD JR
WILLIAMS ROY CHARLES
WILLIAMS RUFUS TIMOTHY
WILLIAMS SAMUEL HARRY
WILLIAMS VANCE GEORGE
WILLIAMSON BENTON
CLAUDE
WILLIAMSON JOHNNY
GORDON
WILLIAMSON THOMAS
DARRELL
WILLIS HINEY
WILLIS M L
WILLMAN GARY LYNN
WILLS ROBSON WARD
WILLSON LOYD MEREDITH
WILSHER JOSEPH MICHAEL
WILSON ALFRED MAC
WILSON BILLY WAYNE
WILSON CHARLES EDWIN
WILSON GAIL FRANCIS
WILSON HARRY TRUMAN

WILSON KERRY FRANK
WILSON MICKEY LOUIS
WILSON PHILLIP MARK
WILSON ROBERT EUGENE
WILSON RODNEY JOSEPH
WILSON WILMER DWAYNE
WILT JOHN WILLIAM JR
WIMBERLY BENNY EARL
WINDHAM MELVIN GEORGE
WINFIELD GEORGE EDWARD
WINFREY JOHNNIE PAUL
WININGHAM JERRY LYNN
WINTER PETER LOUIS
WINTERS DANIEL EARL
WINTERS JOHN LANE
WINTERS TINEY W
WISDOM JESSE ALLAN
WITCHET FRED DOUGLAS
WITHROW MICHAEL DENNIS
WITT CHARLES DON
WITT MORRIS BOWDOIN
WITZKOSKI BILLY JOE
WOOD CALVIN KNIGHT JR
WOOD JAMES WILBURN
WOOD JOHNNY MACK
WOOD LEROY
WOOD LESTER LEE
WOOD ROBERT TINSLEY
WOOD RONALD WILLIAM
WOODCOCK MICHAEL KEITH
WOODMANSEE RONNY LOUIS
WOODRUM JOHN JAMES
WOODWARD RICHARD
HENRY
WOODY VERNON WAYNE
WOOLF ALTON KENNETH JR
WOOLFOLK JIMMY LEE
WOOLLARD RUSSELL DAN
WOOLLEY JAMES NED
WOOTEN PHILIP MILTON
WORTH RICHARD A
WORTH ROBERT EARL
WORTHAM MURRAY LAMAR
WORTHINGTON EDWARD
LLEWEL
WORTMANN FREDERICK
EDWARD
WRANOSKY ROBERT WAYNE
WRIGHT ARTHUR P
WRIGHT JACK LEE
WRIGHT KENNETH HAROLD
WRIGHT LINCOLN ROY
WRIGHT MICHAEL VINCENT
WRIGHT RICHARD JOHN
WRIGHT ROBERT JOSEPH
WYATT EDWARD WILLIAM
WYATT EVERETT ALBERT JR
YANEZ JESUS JR
YARBROUGH BILLY EDWARD
YARBROUGH WILLIAM P JR
YATES CHARLES MICHAEL
YATES ROBERT CLYDE
YBARRA MARIO
YBARRA RICARDO
YBARRA SAMUEL GARCIA
YEATTS JOHN MARSHALL
YORK EMMETT LEE JR
YOUNG GARY EUGENE
YOUNG GERALD LEE
YOUNG JAMES EDWARD
YOUNG JAMES MICHAEL
YOUNG JOHNNY

233

Photos by Aaron Shaw

Utah Vietnam Veterans Memorial

350 State St, Salt Lake City, UT 84111

The Utah Vietnam Veterans Memorial is located in Salt Lake City, Utah on the land of the grounds of the State Capitol. It was created to honor the soldiers who served in Vietnam from Utah and features the 388 names of the people who died or went missing during the conflict. There is also a full-size sculptor of a soldier and a curved memorial wall that make up the whole memorial.

The Utah Vietnam Veterans Memorial was officially dedicated in October 1989. All in all, the memorial cost more than $300,000. The State of Utah paid $116,000 of the bill while the remaining amount was paid through donations. The Vietnam Veterans Memorial was executed by Neil Hadlock from the Wasatch Bronzeworks of Lehi who cast the statue, sculptor Clyde Ross Morgan who made the bronze soldier, and Mark Davenport who created the granite circular memorial wall. Mark H. Bott Monument Company of Ogden was responsible for the granite inscription work along with Dave Bott.

The Utah Vietnam Veterans Memorial features a bronze statue of a soldier that stands around 8-feet high. The soldier is supposed to be returning home from war holding his fallen friend's rifle. The memorial also consists of a gray granite curved wall. That wall has black granite panels on which the names of the fallen or missing 388 men and women from Utah are inscribed. Other black granite panels feature a depiction of the Purple Heart medal and an outline of Vietnam, Laos, and Cambodia. The gray granite used in the memorial was imported from Georgia while the black granite used for panels was imported from southern India (as was the black granite used for the national memorial).

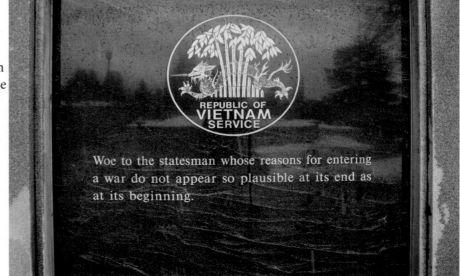

The Names of those from the state of Utah Who Made the Ultimate Sacrifice

ACKERMAN BILL R
ADAMS RAYMOND SPENCER
AITKEN DEAN L
ALDOUS LILO ELMER
ALLGOOD RONALD KEVIN
ANDERSON JAMES K
ANDERSON ROBERT RALPH
ANDRESEN HAAKON WILLY
ANDRUS DANIEL FRANCIS
ANGELL VAUGHN MARVIN
ARCHULETA JESUS MAGIN
ARCHULETA RODOLFO JOSE
ARVESETH BRENT LARSEN
ATWOOD DOUGLAS EDWARD
BACCA RONALD VICTOR
BAILEY SCOTT JAY
BARBURY JOHN
BEAN LARRY DAVID
BELANDER DONN WHITNEY
BELEW GREG BLAINE
BENALLIE DAVID HOWARD
BENNETT RICHARD BOYCE
BENTSON RUSSEL DEAN
BERGERA DEE
BIGHAM THEODORE LEWIS
BINGHAM ORAN LOTHIER
BLAIN JAMES ALLEN
BOARDMAN CURTIS
BODELL KENNETH A
BOHN DAVID J
BOLINDER ARNOLD LEE
BOND LAWRENCE FREDRICK
BONZO JOHN CLIFTON
BOURNE JOHN NOLAN
BOWCUTT SCOTT CANNON
BOWLER SHELDON DON
BROADHEAD DAVID J
BROCK ROBERT LEE
BROWN KENNETH HYRUM

BROWN LARRY LYNN
BROWN MICHAEL DEAN
BROWN MICHAEL GEORGE
BROWN NEIL SHIPP
BROWN THOMAS TAD
BRUDERER STEVEN LEE
BUGGER CURTIS BURKE
BURR GEORGE WALLACE
BUTCHER DEWEY FRANK
CALLISTER ARTHUR ALLEN
CAMPBELL REED EARL
CARAS FRANKLIN ANGEL
CARLQUIST BRIAN FIZTGERAL
CARPENTER BILL DUAYNE
CASEY EDDY RAY
CEDERSTROM DAVID ORIN
CERVANTEZ LUIS GODINEZ
CHESNUT GERRY GEORGE
CHIPMAN RALPH JIM
CHRISTENSEN DALE ELLING
CHRISTENSEN DICK HOOTEN
CHRISTENSEN JOHN MICHAEL
CHRISTENSEN QUENTON LEE
CHRISTENSEN WARREN LEE
CLAYTON MICHAEL MAR-
SHALL
COBBLEY EARL WILLIAM JR
COLBERT GEORGE H
COLES KYLE J
CONLEY ROBERT ALAN
CONNELL STEVEN MARK
CONNOR CHARLES RICHARD
CORNETT ROGER LARRY
COUCH STEVEN WILLIAM
COWDELL MELVIN THOMAS
COX GERALD WAYNE JR
CRAFT ROBERT LEE
CRANDALL BRET FLETCHER
CROW RAYMOND JACK JR

CROWLEY LELAND STEPHEN
JR
CROWLEY ROBERT EDWARD
CUCH WILBERT WAYNE
DALTON BILL NORMAN
DAVIS RAYMOND ALEXANDER
DAVIS WENDLE CLYDE
DAY JERROLD BERNELL
DAY TERRY BUCKLES
DE WAAL HOWARD JACOB
DELGADO ROBERT MON-
TOLVE
DELMARK FRANCIS JOHN
DUNC
DITTMER LEWIS ALLEN
DOCKSTADER RANDELL L
DRAPER MARION LEON
DUDLEY DONALD KIETH
DUFFIN REY L
DUNYON DAVID PHERRAL
DYRDAHL RAYMOND ERNEST
ECKMAN KENNETH WAYNE
ECONOMOUS GEORGE J JR
EICHBAUER EARL KENT
ELLISON JOHN COOLEY
EVERTS JACK CHARLES
FACER RICHARD MICHAEL
FEINAUER WAYNE OWEN
FERGUSON AARON FLOYD
FIELDING CRAIG PYPER
FITZGERALD HOWARD KIM
FONGER LYNDSEY FRANK
FOSTER GARY NEIL
FOSTER GLENN WILCOX JR
FOWLKE EARNEST WESLEY
FOX DEAN FRANKLIN
FRYE STARR FREDERICK
GAILEY ALLEN DALE JR
GARDINER ROY WILLIAM

GARNER LARRY D
GAY ALVIN LEON
GLINES ALLEN BRUCE
GONZALES TOM JR
GOODMAN RUSSELL CLEM-
ENSEN
GOSS JEFFERY ALAN
GREEFF REMI HENDRICUS
GREEN LAURENCE BURTON
GULLEY WILLIAM JEFFERY
HALES RAYMON DRAPER
HALL GARY DODDS
HANSON JACK DELE JR
HARDMAN DEAN WILLIAM
HARDY KEITH L
HARKER JACK ALBERT JR
HARRIS RUSSELL LEE
HAYNIE GALEN EARL
HEIN RICHARD AUGUST
HENDERSON WILLIAM GLADE
HILL RICHARD ALLEN
HINER FRANKLIN JOE
HIPPACH MICHAEL HARVEY
HOLFELTZ KYLE CLINTON
HOOD DON RICHARD
HORSPOOL ROBERT KENT
HOWES DOUGLAS GREGORY
HUBLER GEORGE LAWRENCE
HUGHES MICHAEL DONALD
HUNTER LYNN ELMO
HUNTINGTON BRUCE
HYDE DANNY EVERETT
INGMAN BRUCE EDWIN
JACKMAN ROGER DAHL
JACOBS PHILLIP HAROLD
JACQUEZ JOSEPH EDWARD
JARVIS DEE RANDALL
JEFFS CLIVE GARTH
JENKINS RAY G

JENNE ROBERT EARL
JENNINGS JIM FRANK
JENSEN JAMES CHRISTIAN
JENSEN JAMES PAUL
JENSEN REED GEORGE
JOHNSON LARRY DU WAYNE
JOHNSON RONALD LEE
JOHNSON VERNE DE WITT III
JOHNSON WILLIAM WAYNE
JOHNSTON RICHARD CRAIG
JONES KENNETH LEE
JONES RONALD
JORGENSEN EMORY LEE
KANGAS CLIFFORD F D
KANOSH KENNARD KING
KANOSH WILBERT DWAYNE
KELLER BRUCE M
KENNEDY CHARLES FLOYD
KEOWN BLAIR LOGAN
KERKHOFF RICHARD LEE
KILLIAN MARVIN CLYDE
KIMBER TERREL OLIN
KNOPPERT ANDRE LOUIS
KREK PHILIP JAMES JR
KRUSI PETER HERMAN
LA JEUNESSE DAVID LYNDALL
LACEY FRANK JAY
LAFON VAL LYNDON
LAMKIN STUART BASSETT
LARSEN JIMMY LEE
LAWRENCE MICHAEL D
LAWS JERRY DEE
LAYNE BOB RAYMOND
LINDLEY MARVIN LEROY
LOW JAMES BERNARD
LUCKI ALBIN EARL
LUKE STEVE RALPH
LUND STEVEN DANNY
MACE JAMES DOYLE

Utah

MADDY KENNETH LYNN
MAGYAROSI JOHN JOSEPH
MAIR ALLAN LEON
MARTIN JAMES EDWARD
MARTIN JERRY WAYNE
MARTINEZ JOHN JAMES
MARTINEZ JOHNNY SALAS
MASCHER BRENT THOMAS
MATHIAS RANDY LEE

MATHIAS STEVEN FRANKLIN
MAURIN CHARLES DENNIS
MAXWELL KEN SWAIN
MAYS THOMAS MONROE
MC ARTHUR BRENT HAL
MC BROON JIMMY
MC COY DENNIS RAY
MC FARLANE JOHN WILLIAM
MC GREGOR DONALD VERNON
MC NEIL ALLEN W
MEINERS PAUL ALBERT
MERMEJO JOSEPH MICHAEL
MILLER JERRY LEE
MILLER WILLIAM LEE
MOON RAYMOND ROSS
MOORE LYLE THOMAS
MORGAN WAYNE D
MORTENSEN GENE AL
MOWER GARY RUEL
MUNSON CHRIS DELANO
MYERS PAUL DAVID
NANCE SHIRL BRAD
NEWBOLD BOYD KEITH
NEWBRAND JERRY LESTER
NEWELL GREGG A
NICHOLES HAROLD JAMES
NICK OTIS LEE
NIELSEN TERRY LEE
NOBLE RICHARD EDWIN
OAKEY JOHN RUSSELL
ORR BYRON DAVIS
ORTON ELLIS JR
OSNESS VICTOR WILLIAM
OVESON JAMES RAYMOND
PALMER LYLE CLINT
PARKER EUAN JOHN ERNEST
PARKER PAUL ELMER
PARKER RICHARD DENNIS
PECK JOE RUSSELL
PENNINGTON JOHN CHARLES
PERCIVAL ALTON D
PERRY STEVEN J
PETERSEN HARRY THOMAS
PIERCE TED
PINKERTON MICHAEL DAVID
PINSONAULT FRED JOHN
POLLOCK GARY JOE

POOL GARY GLEN
POPE MORELL JOE
POWELL LYNN KESLER
PRASZYNSKI STEPHEN JAY
PRICE DARREL L
PRICE TERRY HUNTER
PROCIV RICHARD MICHAEL
RASMUSSEN DAVID NILSSON
REEVE DAVID LEO
REID HAROLD ERICH
RENSHAW ROBERT FRANCIS
REX ROBERT ALAN
RHEAD JIM MARBLE
RICHARDSON ROY LEE
RICHTSTEIG DAVID JOHN
RIGGS NIEL BURNS
ROBB MARION C
ROBBINS LARRY OLIVER
ROBERTS RONNY DEAN
ROBINSON VAL CLARK
ROGAN JAMES PAUL
ROGERSON GARRY EARL
ROLAND THOMAS MILTON
ROSE DANNY LEE
RYAN GERALD SCOTT
SALTER SCOTT BRUCE
SAUNDERS RANDALL LEROY
SAWAYA ROBERT MITCHELL
SCHIELE JAMES FRANCIS
SCHOFIELD THOMAS HARVEY
SEARLE JOSEPH KENT
SHANNON ROBERT CONRAD
SHEPHERD BLAINE JOSEPH
SIMMONS SERGE BENSON
SIMPSON BLAIR H
SKEWES ROBERT JOSEPH
SMITH GENE ALBERT
SMITH GERALD ALLEN
SMITH JERRY DEAN
SMITH KENT ANDREW
SMITH PATRICK JACKSON
SNOW CRAWFORD
SNOW KELLYNN VAL
SNYDER FREDERICK DON
SNYDER ROBERT WILLIAM
SOMMER DOUGLAS JOHN
SORENSEN KENNETH JAY
SORENSEN RICHARD LEE

SOWER DONALD MICHAEL
SPENSKO LOUIS PAUL
STEINER MARK STEPHEN
STEPHENSEN MARK LANE
STEVENSON JESSE BRENT
STEWART JOSHUA B
STITT GREGORY CARLYLE
STOKES ROBERT LEE
STUEWE CHRISTOPHER S
SUNDQUIST JACK DONALD
SUTPHEN JACK B
TAFOYA FRANK LEROY
TAFOYA VICTOR ARNALDO
TALBOT THOMAS PAUL
TAYLOR KENT CHILDS
TELFORD JOHN WILLIAM
THORNTON JOHN BRUCE
THORNTON LARRY LEE
THORNTON RODNEY GARDNER
THORPE GARY WILFORD
TIPPETS LENNY MAURICE
TRUJILLO RICHARD TOBY
TUELLER JAMES ALBERT
UDINK ALLEN DAWAN
UFFORD ROBERT LYNN
VAN STAVEREN THEODORE D
VAN VLEET ROBERT CLAY
VANDYKE RICHARD HAVEN
VASQUEZ MICHAEL
VASQUEZ TONY MARIA
VIGIL HENRY ORLANDO
VINCENT MARK DEE
VOGRINEC JOSEPH GARY
VOLZ ROGER WAYNE
WALKER HENSON FRANK
WALKER KURTESS HOWARD
WALKER LESLIE ELROY
WALLACE MICHAEL WALTER
WALSTER THOMAS GAVIN
WARD BRENT F
WARD RANDOLPH BUCK
WARD TERRY J
WATKINS SAMUEL EUGENE
WEBSTER FRANCIS MARION
WEEKS MICHAEL DALE
WELCH BLAINE ALFRED
WELCH GREGORY JOHN

WETZLER ROBIN KIRMEYER
WHITNEY WILLIAM ARTHUR
WIDDISON IMLAY SCOTT
WIECHERT ROBERT CHARLES
WILKEY BLAIR CECIL
WILKS GREGARY ALAN
WILLIAMS DOUGLAS CANDIT
WILLIAMS GREGORY J
WILLIAMSON DON CLAUDE
WILSON EDWIN EUGENE
WINKLE DAVID RYAN
WOLTERMAN GERARD THE-
ODORE
WOOD DON CHARLES
WOOD RICHARD STEVEN
WOODS ROBERT FRANCIS
WORTHEN ROBERT KENT
YARBROUGH DAN BURGESS
YARDLEY RODNEY BIRDELL
YARRINGTON DONALD P JR
YATES LEWIS RICKEY
YEATES MICHAEL HOWARD
YONGUE WILLIAM RAYMOND
YOUNG BYRANT HENRY JR
YOUNG LOGAN DALE
YOUNG ROBERT FRANCIS

237

Photos by Tim Peters

Sharon Welcome Center & Vietnam Veterans Memorial

Vietnam Veterans Memorial Hwy, Sharon, VT 05065

The Sharon rest area on Highway I-89 plays host to the Vermont Vietnam Veterans Memorial. The memorial is the main attraction of the Sharon North Welcome Center and honors the sacrifice as well as the service of Vermont's brave men and women in all of the branches of the military who were deployed to Vietnam. The memorial is open to visitors continuously, but it holds special events including an all-night vigil annually on Memorial Day.

This memorial is believed to be the first state-sanctioned Vietnam Veterans Memorial in the country and was officially dedicated on October 30, 1982. The location of the memorial was selected for a variety of symbolic reasons. For one reason, it is said to resemble a highway in the highlands of Vietnam. It was also selected because it is the closest spot to mile marker 138 which represents the 138 Vermont service members lost in the war.

While the memorial was successfully constructed, the memorial was not without its problems. In the 1990s, the rest stop was almost closed by uninformed politicians who didn't seem to know that the Vermont Vietnam Veterans Memorial was located at that location. In order to save the memorial, supporters and veterans wrote letters to officials as a way to lobby for the preservation of the memorial. Not only was it saved, but in the year 2000, it was turned into a welcome center focused on veteran remembrance.

The Vermont Vietnam Veterans Memorial has grown over the years and hosts a great Memorial day services. A welcome banner featuring photos of Vermont soldiers during Vietnam hangs over a column honoring the Vietnam veterans. These photos allow visitors to be reminded of the fact that there are faces behind the 7,236 names that are etched into the column. There is also a display case filled with other photos as well as a timeline of events regarding what was going on during the Vietnam War in the US, Vietnam, and Vermont. Panels on the memorial also show how the war escalated from 1957 through 1968 and finally through the evacuation of refugees in 1975.

The Names of those from the state of Vermont Who Made the Ultimate Sacrifice

ARCHER ALLEN H
ASHLINE PAUL STUART
BACON CLIFTON LEROY
BADGER BRUCE LYLE
BAILEY KENNETH DEAN
BAKER BRUCE ALLAN
BAKER STANLEY MARTIN
BASSIGNANI WILLIAM JOHN
BEGNOCHE REGINALD PETER
BIRMINGHAM EDWARD ARNOLD
BLAKE JOHN CHARLES
BOVE HARMON JOSEPH JR
BRILEYA DAVID ALAN
BURNHAM ROGER CLARK
BYRNE WAYNE EUGENE
CALLAHAN DAVID FRANCIS JR
CARR FREEMAN ABRAHAM
CARR STEPHEN DOUGLAS
CLARK ROGER WILLIAM
COLVIN GERALD SELAH
CURRIER PHILIP BUCHANAN
DARTT PAUL RICHARD
DE LANEY RICHARD LAWRENCE
DEUSO CARROLL JOSEPH
DICKINSON EUGENE HAROLD
DONER PATRICK RALPH
DOOLEY JAMES EDWARD
FIELDER ROBERT FLETCHER

FISH JOSEPH KENNETH
FITZGERALD WILLIAM CHARLE
FRAPPIEA FRED C H JR
FRIZZELL MARSHALL RAY
GABORIAULT SANFORD RENE
GALLAGHER GEORGE FRANCIS
GERMAIN PHILIP MICHAEL
GILLESPIE DANIEL CRAIG
GOMEZ GELASIO NICANOR JR
GREEN RICHARD ALBERT
GREENE ALLEN JOSEPH
GREENE PHILIP F
HALL GARY ALBERT
HASELTON JOHN HERBERT
HENRY EDWARD DOUGLAS
HILL GARY PAUL
HOLLAND JOHN HENRY
HOWARD HARVEY RICKERT
HUMPHREY HARVEY EDWARD
INKEL GUY MARCEL
IVANOV WILBUR WILLIAM
JOHNSON DEAN HERBERT
JORDAN ROGER FRANCIS
LA COURSE DAVID ANTHONY
LACROIX PAUL DOUGLAS
LAFAYETTE JOHN WAYNE
LAMOTHE GEORGE ANDREW
LAROCQUE LESLIE HOWARD

LAVIGNE STEWART JAMES
LESAGE ARMAND PAUL
LESTAGE WILLIAM FRED
LEVESQUE STEVEN DOUGLAS
LONG PERLEY MILFORD JR
MAXHAM RALPH ARDEN JR
MC KENZIE EDWARD AUSTIN
MORIN DONALD WILLIAM
MORRISON BRUCE AUSTIN
MORSE DURWARD GLENNIE
NOWICKI JOHN PAUL
O CONNOR EDWIN THOMAS JR
ORR BRIAN STANLEY
ORVIS DOUGLAS GORDON
PAQUETTE BRUCE ARMAND
PETTIS STEVEN GENE
RANDALL WAYNE MICHAEL
RITCHIE BERNARD FREDRICK
RIVERS NELSON KEITH
ROCK ALLEN CLARENCE
RUSSELL JOHN MALCOLM JR
RUTHERFORD LEROY
SARGENT ROLLIN CHESLEY JR
SHELDON EDWARD CLARENCE
SHOMPANY ERNEST VON
SILVA JOSEPH ANTHONY
SMITH HAROLD ROGER
ST GEORGE FRED DAVID

STONE ROBERT DOUGLAS
SWEENEY JOHN EDWARD
THOMPSON FRANCIS LLOYD
TREADWELL EUGENE DURWOOD
TUTTLE ALFRED JOSEPH
ULMER JAN ALAN
VANDEVENTER JOHN WILTON
WALBRIDGE WILLIAM ROBERT
WALKER GEORGE EDWARD JR
WASHINGTON LOUIS WEL-

DON
WEIGHTMAN KENNETH G JR
WESTON WENDELL ALLEN
WILBUR DENNIS
WILKINS ALLAN FRANCIS
WITHAM JAMES GEORGE
WOOD BARRY RUSSELL

Photos by Joseph McCormick

Virginia War Memorial

621 S Belvidere St, Richmond, VA 23220

The Virginia War Memorial is the state's official monument that was created in order to remember and honor the local men and women who served during the Vietnam War (as well as other wars). It provides education through its Education Center in order to teach the public about the sacrifices made through war and Virginia's contribution to the country's military. The Education Center uses a number of resources including artifacts, documentaries, and exhibitions to teach visitors about Virginia's service members. The memorial holds monthly programs as well as special ceremonies to keep the memory of the fallen alive. It is open to the public, but visitors are not permitted on the property between midnight and 5 am.

The Virginia War Memorial was funded through both private and public funds. The Paul and Phyllis Galanti Education Center consists of an 18,000 square foot space that holds the memorial as well as the education facilities. As a whole, the space is made up of classrooms, conference rooms, an amphitheater, a research library, and exhibit space. There is also a Memorial Store on the location. The Center was finished and dedicated in September of 2010.

There are different memorials on site to honor the veterans of the different US wars including the Vietnam War. The Shrine of Memory, the statue known as "Memory," and the Flag Court honor the Vietnam veterans. The Shrine of Memory is a memorial wall that features the names of Virginian service members who were killed in action in World War II, Korea, Vietnam, and the Persian Gulf. The names of the Vietnam veterans were added in 1981. Each name is engraved into the walls which are made of glass and stone. To make finding the name of a loved one more convenient, the names are arranged by county and city and are then alphabetically. The statue "Memory" is located at the end of the Shrine of Memory.

It was designed by Leo Friedlander and physically sculpted by sculptors Joseph Campo and William Kapp. It is made of white marble and is an impressive 23 feet tall. It was made to represent the mixture of pride and sorrow felt by the local citizens for their lost service members. Finally, Flag Court is located outside of the Shrine of Memory. It flies seven flags including the Virginia War Memorial flag, the Marine Corps flag, the US Army flag, the Navy flag, the Air Force flag, the Coast Guard flag, and the flag of the Merchant Marines. In the center of everything, the POW/MIA flag flies to honor prisoners of war and the missing in action. The US flag and Virginia state flag also fly here.

The Names of those from the state of Virginia Who Made the Ultimate Sacrifice

ABBOTT DAVID FRANCIS
ADAMS DONALD BEN JR
ADKINS DONALD WAYNE
AKERS E G JR
AKERS EDWARD DARRELL
ALDERMAN WILFORD HARLESS
ALLEN CHARLES ERVIN
ALLEN HERVEY HARRIS
ALLEY DONALD RAY
ALLEY LONNIE DOUGLAS
ALLEY WILLIE WARNIE
ALPHIN TALMADGE HORTON JR
ALSUP TERRY DANE
AMBURGEY ALFRED JUNE
AMMONS WALTER NORRIS
ANGE RONALD EDWARD
ANGELL MARSHALL JOSEPH
ANGLE PETER JASON
ANGUS CLARENCE RAY
ARANN RICHARD MAXWELL
ASH EDWARD GARLAND JR
ASHBY DONALD ROBERTS SR
ASHLEY UPTON FINLEY
ASHLEY WILLIE FRANK
ATKINSON ROGER DALE
AUSTIN JAMES FRANKLIN
AUSTIN VIRIL LEE
BABB KENNETH ALVIN
BACH JOHN JOSEPH III
BAHR RICHARD DUNCAN
BAILEY GEORGE EDWARD
BAILEY JESSE THOMAS JR
BALDWIN SANDERS RAY
BALFOUR DENNIS R
BALL ALBERT THOMAS
BALL JAMES EDWARD III
BALZER MICHAEL ARLIN
BANDY RAYMOND DOUGLAS
BANKS IRVIN SYLVESTER
BANKS STERLING CLARK
BARBOUR JAMES C JR
BARNES MERRILL
BARNETTE FRED EDWARD

BARONE SANDRO NICHOLAS
BARR JUNIOR WAYNE
BARRINGTON ALVIS T JR
BARTLEY DON LAVERNE
BARTON LANCE BRYAN
BARTONE JOHN PATRICK
BASNIGHT RALPH WOOD
BASS JOHN DABNEY
BASS RONALD WAYNE
BATTEN JAMES EARNEST
BAYLISS PAUL M
BAZEMORE WILLIAM HODGES
BEADNELL WILLIAM LEE
BEALE MILLS III
BEARD LEON
BEAUCHAMP ERNEST MICHAEL
BECKMAN DOUGLAS MARTIN
BECKNER JAMES MALCOLM
BELL WILLIAM JR
BENOIT PAUL BRIAN
BERGMAN CLIFTON BALLANTYN
BERNHARDT ROBERT EDWARD
BESSOR BRUCE CARLTON
BETTIS JOHN CALVIN
BIGGERSTAFF HENRY CHARLES
BITNER DANNY LEE
BLACKBURN HARRY LEE JR
BLACKWELL MILTON
BLAIR CHARLES EDWARD
BLAIR THOMAS GEORGE JR
BLAKEY HOWELL FRANK
BLANDING JOHN WESLEY
BLANKENSHIP GODFRED
BLANKS TONY PAGE
BLATZ RUSSELL KEITH
BLEVINS HUGH BRADLEY JR
BLEVINS RONALD WAYNE
BLODGETT DOUGLAS RANDOLPH
BLOW JAMES LYNELL JR
BLY PERCY EUGENE JR

BOBBITT WILLIAM E G
BOFFMAN ALAN BRENT
BOHRER LEROY PRESTON
BONNET CHARLES CHRISTOPHER
BOOKER JOSEPH OTIS
BOONE JOHN THOMAS
BOONE WILLIAM EDWARD
BOOTH LAWRENCE RANDOLPH
BORDEAUX JAMES PRESTON JR
BOVE ROGER GERHARD
BOWDEN CARLYLE MILLER III
BOWERS JAMES E
BOWLES DWIGHT POLLARD
BOWLES FRANK JOSEPH
BOWLIN PAUL MADISON
BOWLING JOHN RAYMOND SR
BOWMAN JOSEPH B
BOWMAN ROGER DALE
BOYD WALTER
BOYKIN ROBERT THOMAS
BRADNER JACK RAY
BRADSBY KERNELL PERSONE
BRAGG DONNIE JAY
BRANDON JAMES MILES JR
BRANHAM HARRY WALTER
BRANTLEY LESTER J
BRECKENRIDGE FRANKLIN U
BREWER ELZA MORTON JR
BRICKEY BARRY WAYNE
BRICKHOUSE WILLIE T JR
BRIGGS EVERETTE WAYNE
BRIM JOHN LARUE
BRODA RICHARD DALE
BROEGELER HERMAN C III
BROOKS JAMES LLOYD
BROOKS LONNIE ALLAN
BROOKS MONTE D
BROOKS TERRY HUDGINS
BROUGHMAN RALPH WAYNE
BROWN CHARLES EDWARD JR
BROWN CHARLES WILLIAM JR
BROWN DAVID CARLTON

BROWN EDGAR CLARENCE
BROWN ELMER WILLIAM
BROWN EUGENE ONEIL
BROWN GEORGE ARTHUR
BROWN GEORGE LAWRENCE
BROWN HERMAN
BROWN IRAN COURTLAND
BROWN JOHN ALPHONZO
BROWN JOSEPH WHELTON III
BROWN ROBERT MACK
BROWN RONALD LEE
BROWN WILLIE LEE
BROWNE RAY BURMASTER
BROWNING PERRY NATHAN
BRUBAKER NORMAN CURTIS
BRUMFIELD STEPHEN MICHAEL
BRYAN AUBREY ALLEN
BRYANT DAVID BANKS
BRYANT ROGER SMITH
BRYANT THOMAS MELVIN
BRYANT TINSLEY
BUCHANAN CHARLES C
BUCHANAN CHARLES DON
BUCHANAN JOSEPH WILLIAM
BUCHANAN ROBERT DAVID
BUCKLEY VICTOR PATRICK
BULLINGTON FREDERICK CURT
BULLOCK RICHARD WILLIAM
BUNTING BERTRAM ARNOLD
BURCH DAVID FELIX
BURCHETT LONNIE MORRIS
BURD DOUGLAS GLENN
BURKHALTER RALPH JR
BURKHEAD JERRY CLARK
BURKS GARY ALLEN
BURNETT WILLIAM A
BURNLEY DILLARD REED
BURTON FRED DOUGLAS
BURTON LUTHER WILLIAM
BUTLER CHARLES KING
BUTT HERBERT HAMBLY JR
BUTT RICHARD LEIGH
BYRD GEORGE ELLIS

BYRNE PAUL RANDOLPHE
CAIN LEWIS RODNEY
CALLIS JAMES HAROLD
CAMPBELL FRANK WILLIS JR
CAMPBELL KEITH ALLEN
CANNADAY MICHAEL D
CANTRELL GERALD WAYNE
CARDWELL ERNEST DANIEL
CAREY DANIEL LESTER
CAREY FRANKLIN LEE
CARKIN HARVEY MC KEE
CARMICHAEL ROBERT EDWARD
CARPER LORING WILLIAM JR
CARR JAMES WILLIAM
CARTER LEONARD ALEXANDER
CARTER LINWOOD CHARLES JR
CARTER RICHARD THOMAS
CARTER ROBERT JEROME
CARTWRIGHT RALPH WINDALL
CASTELDA ANDREW THOMAS
CASTLE HAL CUSHMAN JR
CASTLE RUSSELL LEONARD
CATES NORMAN LOUIS
CATON GERALD LEWIS
CAVIN STEVEN IKE
CHANEY ARTHUR FLETCHER
CHANEY DAVID LEE
CHAPMAN JOHN RICHARD
CHAPPELL KENNETH LEE
CHARLES MICHAEL LANE
CHARTIER RAYMOND ALLEN
CHATTEN MARYLAND
CHEIVES CALVIN L
CHERRY WILLIAM LOUIS
CHESSER ROBERT RICHARD
CHITTUM RONALD HENRY
CLARK DANNY TAYLOR
CLARK DENNIS MC COY
CLARK DONALD E JR
CLARKE GEORGE WILLIAM JR
CLARKE KENNETH GEORGE JR

Virginia

CLARKE LEE TILSON
CLARKE WILLIAM MOSBY JR
CLAUD PERNELL RUSSELL
CLAYTON GEORGE MILTON JR
CLEM THOMAS SAMUEL
CLEMENT JAMES WILFRED
COATES FLOYD BURNETT
COBB PAUL FREDERICK
COBURN WILLIAM H
COCHRAN GARY DUANE
COGHILL MILO BRUCE
COLE FRED VINCENT
COLE RANDALL EARL
COLGAN GEORGE BURTON III
COLLIER CHARLES MICHAEL
COLLINS EUGENE
COLLINS LARRY ELBERT
COLLINS ROSS WILLARD JR
COLVIN PAUL SILVEY
COMPTON FRANK RAY
CONLEY ALEX BOYD
CONNOR GLENN MARSHALL
CONSOLVO JOHN WADSWORT JR
CONWAY RAYMOND LESTER
COOKE CALVIN EDWARD
COOKE CHARLES THOMAS
COOKE DOUGLAS RUDOLPH
COOKE ERNEST FRISSELL JR
COOKE ROBERT MORRIS
COOLEY DAVID LEO
COOLEY DICKEY LARUE
COOPER ANDREW JONES
COOPER JAMES WILLIAM
CORBETT THOMAS LOUIS
CORBITT WALLACE THOMAS

CORDLE CHARLES LINWOOD
CORNS BOBBY LARRY
COSBY DAVID FRANKLIN
COX ELBERT ELISAH JR
COX GEORGE JOSE
COX HENRY THOMAS
COX JAMES MICHAEL
COX SHERBERT LEON
CRABTREE ROBERT ANDREW
CRAGHEAD THOMAS JAMES JR
CRAIG JOHN PHILIP
CRAIG WAYNE SHELBY
CREECH ROBERT JR
CRESSEL TERRY WALKER
CRICKENBERGER RICHARD WAY
CRIGGER HENRY GLEAVES
CRITZER RONAL EDWARD
CROCKETT FREDDIE ISIASH
CROSS FRANK WARREN
CROSS FREDERICK WILLARD
CROUSE EDGAR FRANKLIN JR
CROWDER HERBERT HAROLD
CROWDER JOSEPH BERKLEY
CUMMINGS JAMES E JR
CUNNINGHAM WILLIAM NEAL
CUPP ROBERT WILLIAM
CURRY ALVIN CHRISTOPH
CURRY DICKIE CARSON
CURRY LARRY EDWARD
CURTIS HENRY THOMAS II
CUTCHINS ROLAND A
DAMERON ROBERT WOODROW
DANCE LAWRENCE RUSSELL
DANIEL CHARLES LINCOLN
DANIELS EDWARD EARL
DAUGHERTY JAMES OLIVER JR
DAVIDSON JOHN CLARK
DAVIDSON JOHN WAYNE
DAVIES JOSEPH EDWIN
DAVIS CHARLES WILLIAM
DAVIS CLIFTON HENRY
DAVIS GEORGE NATHAN
DAVIS HOLBERT EUGENE
DAVIS HUGH MOZELL
DAVIS WARREN K
DE LUCA RAYMOND PAUL
DE PRIEST DAVID WAYNE
DE WITT SPOTSWOOD
DEAN ROBERT WILLIAM
DEEL STONEY LEE

DEHART JACKIE CLYDE
DELPH JERRY
DENTON GUY THOMAS
DEVERALL GEORGE NOBLE
DICKENS JAMES AARON
DICKERSON JAMES CAROL
DICKEY JAMES WHEELER
DIGGS MICHAEL RONELL
DILLARD JERRY ALLEN
DILLENSEGER BERNARD GUY J
DILLMAN ROGER L
DISHMAN DOUGLAS EDWARD
DIXON JOHN HENERY
DIXON ROGER ALLEN
DODGE EDWARD RAY
DOERING LLOYD DOUGLAS
DONAHUE MORGAN JEFFERSON
DONLAN RICHARD MICHAEL
DONNELL PETER FRANCIS
DOSS ROBERT WILLIAM
DOTSON JEFFERSON SCOTT
DOUGLAS DWIGHT SAMUEL
DOVE JACK PARIS SR
DOWDY RUFUS JOHN
DRAPER WILLIAM LLOYD
DRIVER DALLAS ALAN
DUNHAM BRUCE JOEL
DUNN JAMES HARLOW III
DUNN MORRIS GORDON
DURHAM THOMAS WYATT
EADS WALTER TASMAN
EARNHARDT CLIFFORD JERRY
EASLEY SAMUEL HARRISON II
EAST LEON NELSON
EDENTON HIRAM EURIAS JR
EDMUNDS CALVIN
EDMUNDS ROBERT CLIFTON JR
EGGLESTON DAVID LEROY
EIGHMIE RONALD WILLIAM
ELLEN WADE LYNN
ELLER JOHN ARTHUR
ELLINGER VICTOR LEE
ELLINGTON HERBERT L
ELLIOTT CHARLES HENRY JR
ELLIOTT GEORGE L III
ELLIOTT NORMAN JR
ELLIOTT VANDERBILT JR
ELLIS OTIS RANDOLPH JR
ELSWICK JAMES TIPTON JR
EMBREY DAVID NORMAN
EMMANS WILLIAM ROBERT
EMORY CHARLES ROBERT

EPPERSON THOMAS EDWARD
EPPS CECIL WAYNE
EPPS RICHARD MAYNARD
ESKRIDGE WARREN REED
ESTRADA DAVID
EVANS BILLY KENNEDY JR
EVANS LONNIE BERNARD
EVANS THOMAS J JR
FAIDLEY JOHN CHARLES
FALWELL DONALD WAYNE
FARMER JAMES BRYON II
FARRIS BLAKE WILEY JR
FARROW DAVID ASHBY
FARROW FRANKIE LEE
FAUGHT WILLIAM AVENER JR
FAULKNER CHARLES LONG
FEAGAN MICHAEL JOHN
FEE DONALD FRED
FELTON GARLAND PARIS
FENTRESS LEON AUBREY
FERRELL HUGH JAMES
FIELD MICHAEL FINLAY
FIELDS LLOYD JR
FIELDS WILLIE JR
FINCH MELVIN WAYNE
FINCHAM WILLIAM EDWARD
FINK CHARLES KENNETH
FINLEY GUY MARVIN
FINNEY STEPHEN
FIRTH ALLEN EDWARD
FITZGERALD WOODROW MELVIN
FLEMING THOMAS RYAN
FLETCHER LAWRENCE EUGENE
FLOYD ROGER LEE
FOLEY DOUGLAS LEE
FOLLAND MICHAEL FLEMING
FORAME PETER CHARLES
FORBES HARRY BURKLEY
FORD CHARLES LEWIS
FORT JEROME
FOWLER ROY GILLMAN
FOX RICHARD W B JR
FOX THOMAS AMISS
FRANCIS LARRY EDWARD
FRANKLIN DOUGLAS M
FRANKS ERNEST RICHARD
FREEMAN GARY
FREUDENTHAL RICHARD HOLT
FRYE HERBERT ALANSON
FULCHER DOUGLAS EDWARD
FULCHER JOHN HENRY

FURROW SHERMAN ALVIN JR
GAITHER THOMAS MARK
GALLOWAY ARTHUR LEE JR
GARBER CHARLES WILLIAM JR
GARLAND CONLEY R
GARMAN THOMAS ALFRED
GARNER RONALD LEE
GARRETT HENRY WAYNE
GARRETT ROBERT GILMER
GARRETT ROGER LEE
GATEWOOD CLARENCE MELVIN
GAY ALBERT LUMMIS JR
GEE MAC ARTHUR G
GERALD GEORGE ADEN
GIBBONS DARRELL LEE
GIBSON CARL REED
GIBSON CORNELL HARRISON
GIBSON MICHAEL THURSTON
GIDDINGS ALBERT HUGH
GILBERT JAMES SILAS
GILL DONALD WADE JR
GILLIAM LARRY EDWARD
GILLS LEWIS CLARENCE
GILMER CHARLIE MACK
GILMER GILES WILSON
GIMBERT MARTIN JOSEPH
GINN JAMES MICHAEL
GIRARD CHRISTIAN GEORGE
GIVENS ROY NATHANIEL
GLENN HAYLEN FOSTER
GOFF HENRY LARRY
GOFF ROGER EARL
GOFORTH CHARLES DAVID
GOODING ORANGE
GOODLETT JOHN FLETCHER
GOODMAN EDWARD LEE
GOODMAN RAYMOND LEE
GORHAM WALTER PRESTON
GOSS WARREN JUDGE
GOSSELIN ROBERT JOSEPH
GOULDIN THOMAS MILTON
GRAHAM BRUCE ELLIOT
GRAHAM WILLIAM RICHARD
GRAJEWSKI JERRY FRANCIS
GRAMMER HENRY BRIAN
GRAVLEY JAMES THOMAS
GRAY ARTHUR POWELL IV
GRAY DOUGLAS TAYLOR III
GRAY RICHARD TENNEY
GRAY SAMUEL
GRAYSON WELBY HERBERT III
GREEN MARTIN L JR
GREENWOOD ROBERT ROY JR

GREGORY PAUL ANTHONY
GREGORY SHERMAN WILLIAM
GRIFFITH WILLIE ROGER
GRIMES JOHN LEONARD
GRINNELL THOMAS D III
GROSHONG ALLEN EBERLY
GRUBB STEVE FREEMAN
GRUBBS ROGER WAYNE
GUNTER CALVIN DOUGLAS
GWALTNEY GERALD WAYNE
HACKLEY LAWRENCE EARL
HAIRSTON JOHNNY MICHAEL
HALE CHARLES CHAPLIN JR
HALL BLUCHER RAY
HALL ROBERT EDWARD
HALLENBECK TED B
HAMBLETT ROBERT BRYANT
HAMILTON CHARLES HENRY
HAMILTON KYLE STEVENS
HAMMERBECK EDWARD COX
HARDIN JOHN EDGAR
HARDY FRANK EARLE
HAREFORD BASIL LEE
HARLEY LEE DUFFORD
HARPER TERRY L
HARRELL JOHN REHILL
HARRELL RAYMOND DALE
HARRIS CHARLES EDWARD
HARRIS JACK JR
HARRIS JACKIE LOUIS
HARRIS LARRY CORNELIOUS
HARRISON DANIEL WALLACE
HARRISON DOUGLAS LEE
HARRISON WILLIAM H III
HARTMAN VERNON LYNN JR
HARVELL MC KINLEY H JR
HARVEY CLEVELAND RAY
HARVEY EUGENE DAVID
HASSLER CHARLES EDWARD
JR
HATCHER LARRY DAVIS
HAWKER TOMMY MELVIN
HAWKINS FELIX BOYD
HAWKINS JOHNNY LEE
HAWTHORNE JAMES LYN-
WOOD
HAYDEN HAROLD RICHARD
HAYDEN TROY RAY
HAYES THOMAS JAY IV
HAYES WILLARD FAYETTE
HEBERT FREDERICK CONRAD
HEIZER JERRY HOGUE
HELBERT ROY LEE
HELM DAVID FRANKLIN
HEMP STUART FRANKLIN
HENDERSON WILLIAM WAYNE
HENSEL ERNEST VICTOR JR
HENSLEY SHELBY GLEASON
HERIOT THEODORE S JR
HERMAN CLYDE RUSSEL
HERNDON DONALD LEE
HERRING BOYD LEMUEL
HERRING THOMAS LEANDRA
HERRING WILLIAM EARL
HERSHBERGER DAVID
HARPER
HICKS DAN SELLMAN
HIGDON RALPH TAYLOR
HIGGS RALPH EDWARD
HILL JOHN EDWIN
HINES JOHN LESTER
HINSDALE GERALD CLINTON
HINTON LUTHER ANDERSON
HITTINGER FRANCIS R JR
HOAGLAND JEFFREY KAY
HODSON CHARLES HERBERT
JR
HOFFLER RICHARD WILLIAM
HOGGARD JAMES LUTHER
HOGGE DOUGLAS WARREN

HOLDAWAY GUY
HOLLAND CLARENCE MI-
CHAEL
HOLLAND KERMIT W JR
HOLLAND ROBERT VERNON
HOLLAR HOWARD ESLIE
HOLLIFIELD ROGER DALE
HOLMES WILLIAM DAVID
HOLTZMAN RONALD LEE
HONSE GEORGE EMILE
HOOD JAMES GARY
HOOPER WILLIE JR
HOPKINS MICHAEL EDWARD
HOPKINS MICHAEL WAYNE
HOPSON WILLIAM DOUGLAS
HOSKINS ROBERT LEE JR
HOUSTON ELWOOD LAYTON
HOWARD GREGORY MAR-
SHALL
HOWARD HARLEY MICHAEL
HOWE THOMAS JOHN
HOWELL PHILIP
HOY ROBERT ELVIN
HUBBARD SAMUEL BURNELL
HUDNALL WILLIAM LEON
HUDSON WILLIE JUNIOR
HUFF PAUL LLOYD
HUGGETT RICHARD THOMAS
HUGHES JAMES EDWARD
HUMPHREY ROY DARRELL
HUNDLEY THEODORE
LANGSTON
HUNT WILLIAM SPRAGGINS
HUNTER JAMES DOHERTY
HUNTER WILLIAM KENNETH
HURLEY JAMES K
HURT VASSAR WILLIAM III
HUTTON JOHN KENDRICK JR
INGRAM JIMMY ANDERSON
IRVING NATHANIEL
ISLER CHARLES C JR
JACKSON DONALD EUGENE
JACKSON PAUL VERNON III
JACKSON WITHERS THEO-
DORE
JAMES CLAYTON WADE
JAMES GENERAL FIRD JR
JAMES HARRY LEE JR
JAMES WILLIE LEE
JARRELL JOHN WAYNE
JARRETT STEVEN ANDREW
JEFFERSON BILLY RANDOLPH
JEFFERSON LOUIS ALLEN
JENKINS CHARLES ARTHUR
JENKINS CLAUDE THOMAS
JENKINS JERRY MALONE
JENKINS JOHN ALLEN III
JENKINS LEWIS FRANKLIN
JENKINS MICHAEL LEE
JENKINS VINCENT EARL
JENKINS WILLIAM CLIFTON
JESTER RICHARD GARVER
JETER BENTON ARTHUR
JINKINS GEORGE W III
JOHNSON ALVIN SAMUEL
JOHNSON ANTHONY LEE
JOHNSON ARTHUR LOUIS
JOHNSON BERNARD LEVERN
II
JOHNSON BOBBY RAY
JOHNSON CALVERT JAMES
JOHNSON CALVIN
JOHNSON CHARLES EDWARD
JOHNSON CHARLES EUGENE
JOHNSON DALLAS LEMON
JOHNSON DAVID CURTIS
JOHNSON GUFFEY SCOTT
JOHNSON HARVEY III
JOHNSON HENRY ALSTON
JOHNSON JAMES ROBERT

JOHNSON JERRY HAMPTON
JOHNSON JESSIE LEE
JOHNSON JOHN ALVIN
JOHNSON JOHN VICTOR JR
JOHNSON KEITH GEOFFREY
JOHNSON LORENZO RAY-
NARD
JOHNSON NATHAN JR
JOHNSON RICHARD SHERWIN
JOHNSON SAMUEL JR
JOHNSON THOMAS JOSEPH
JOHNSTON DONALD REGI-
NALD
JONES CARROLL HENRY
JONES GRANDVILLE R JR
JONES IVAN WAYNE
JONES LAVOYN AUGUSTUS
JONES MARYUS NAPOLEON
JONES ORVIN CLARENCE JR
JONES PAUL ELDEN
JONES RAYMOND PARKER
JONES RICHARD JUNIOR
JONES ROBERT HENRY
OSBORN
JONES ROGER KENNETH
JONES RONALD WAYNE
JONES RUDOLPH
JONES WILBERT JASPER JR
JONES WILLIAM OLIVER
JORDAN WAYNE LAMONT
JOURNELL ROBERT MASON III
JUDY HERMAN LEROY JR
KARAS PAUL RICHARD
KARDOS JAMES MARION
KEATON DANNY GARTH
KEBERLINE MICHAEL JOHN
KEEGAN RICHARD MICHAEL
KEEN ALBERT MASON JR
KEESEE JOSEPH TIMOTHY
KELLAM GEORGE LEE
KELLAS ROBERT LOUIS
KELLEY FRED ALLAN
KELLY JAMES EDWARD
KEMP JOHN ALDA
KENDRICK RICHARD SMITH
KENNEDY ALTON RAY
KENNEDY JOHN WILLIAM
KESTERSON CHARLES ROBERT
KETNER HAROLD K JR
KIDD LEONARD WHITLEY
KIDD MELTON LAVONE
KIDWELL ROGER GENE
KIDWELL WAYNE MINOR
KILBY RAYMOND MORGAN
KILDUFF MICHAEL JOHN
KING HAROLD WAYNE
KING JAMES EDWARD
KING JAMES ISRAEL
KING LYELL FRANCIS
KING ROBERT LEWIS
KINTON JOHN LEONARD
KIZZIE LEON EDWARD
KNACK RICHARD CARL
KNIGHT MICHAEL KAY
KNISELY RANDALL CLAI-
BORNE
KNUTSON ROBERT BRUCE
KOEHLER WILLIAM EDWIN
KOREL EMERY LOUIS
KOSKO WALTER
KOSTER ANTHONY ALBERT
KRAMER STEPHEN ARTHUR
KRIMONT NICHOLAS
KUYKENDALL RICHARD
WAYNE
LACKS CORNELIUS CLAYTON
LAIRD ROBERT LEE
LAMB WILLIAM LLOYD
LAMBDIN DANIEL ALVEY
LAMKIN LEWIS DEAN

LANCASTER HERMAN JR
LANDMAN THOMAS PAUL
LANGFORD LEWIS NELSON
LARSON GARY WAYNE
LASSITER HERMAN EARL
LASSITER RICHARD LEON
LATTIMORE CHARLES JR
LAWHORNE DONNIE JACK-
SON
LAWRENCE JOHN FRANKLIN
LAWSON BOYCE EUGENE
LAWSON CHAMP JACKSON JR
LAWSON JAMES GARFIELD
LAWSON RAYMOND CHRIS-
TOPHE
LAYNE DAVID DANIEL
LEDFORD HENRY ALVERSON
LEE ALFRED
LEE DANIEL L
LEE FRED VINCENT
LEE LEONARD MURRAY
LEE WILLIAM ROBERT
LEICHLITER VYRL EUGENE JR
LENCHNER DAVID ALLEN
LEONARD JOHN CHARLES
LERNER DAVID ATWOOD
LESTER GRADY RUDOLPH JR
LESTER JAMES LEROY JR
LEVENDIS WILLIAM MC-
NAMARA
LEWIS ARTHUR
LEWIS FLETCHER LEON
LEWIS JAMES HAROLD
LEWIS ROBERT LEE
LEWIS ROBERT RAYMOND
LIGHT EVERETT EARL
LINGLE JOSEPH M JR
LIPSCOMB DAVID LEE
LOCKHART GEORGE BARRY
LOGAN CHARLIE MATTHEW
LOGAN JOHN TYLER
LONG NORMAN LACY JR
LONG PHILLIP MICHAEL
LOONEY DOUGLAS OSCAR
LOTRIDGE GERALD STEPHEN
LOUNDERMON RALPH E
LOVELL LEWIS RANDOLPH JR
LOWE RONALD SIDNEY
LOWE ROY DALLAS JR
LOWERY CLEM SPENCER JR
LOWERY WILLIAM LEE
LUCAS ROBERT EUGENE
LUDWIG MICHAEL EUGENE
LUNDY RICHARD COLEN
LUNSFORD GLEN THOMAS
LYBERGER ARDEN RUSSELL
LYNCH PINK MILTON JR
MABE CARL MARION
MABE ROGER DALE
MAC LEOD SIDNEY B JR
MADISON RICHARD CARL
MAHONE HAYWOOD JR
MAIN CHARLES REID
MALONE JIMMY MCDONALD
MALONE PHILIP NEWMAN
MANN GLENN DILL
MANNS CHARLES EDWARD
MANUEL FRANCIS EVERETT
MANUEL JESSE STEPHEN JR
MARCHBANK LARRY OVER-
TON
MARR GEORGE RICHARD JR
MARSHALL ROBERT EDWARD
MARSHALL ROLAND TRENT
MARTIN ALPHONSO S
MARTIN BUDDY RAY
MARTIN ERIE A JR
MARTIN FLOYD NEWTON
MARTIN GARY ALAN
MARTIN HARRY PEMBERTON

MARTIN LINWOOD DWIGHT
MARTIN PAUL RIVERS
MARTIN STEVEN WAYNE
MARTIN WILLIAM REYNOLDS
MARTINEZ DONALD LYN-
WOOD
MASHBURN TSCHANN SCOTT
MASON ALFRED LEE
MASON LARRY MAURICE
MASSENGILL LARRY DALE
MATT JOSEPH WALTER
MATTOX DENNIS MAYON
MATTOX DWAINE ELBYRNE
MATTSON ROBERT KENT
MAYES ROBERT GRESHAM
MAYO JAMES RUSSELL
MAYS AUBREY REID
MC BRIDE WILLIS LEONARD
MC CANN DONALD WAYNE
MC CANTS LELAND S III
MC CARRON MICHAEL
JOSEPH
MC CAULEY WAYLAND F JR
MC CLARY GORDON STUART
MC CLUNG JIMMY HARRISON
MC DANIEL ROBERT THOMAS
MC DONALD ROBERT F II
MC DORMAN DARL KENNETH
MC ELHANNON KEVIN C JR
MC FARLAND WILLIAM
JOSEPH
MC GEE CURTIS J
MC GOVERN CHARLES
VENTON
MC ILWEE JAMES R
MC INTOSH RICHARD ROBERT
MC KIBBIN HUGH R JR
MC KINNEY DAVID LEE
MC LARNON THOMAS THE-
ODORE
MC LAUGHLIN FRANCIS
MC NULTY CHARLES RICH-
ARD
MC RAY WAYNE DABNEY
MC WILLIAMS WILLIAM G III
MCGINNES CHARLES DENNIS
MCPHERSON EVERETT ALVIN
MEADOR DANIEL R
MEALER FERRELL EUGENE JR
MEARS RALPH JUDSON JR
MEEK WILLIAM CHESTER
MEEKINS RAYMOND C
MEIGGS RICHARD RAY
MELLAR FRANCIS JOHN JR
MESSER THOMAS HUBERT
MICKLES CHARLES EDWARD
MIDGETT DEWEY ALLEN
MILBOURNE RALPH WILLIAM
MILES JAMES EDWARD
MILLER CURTIS O NEIL
MILLER DELANEY ERNEST JR
MILLER DONALD WAYNE
MILLER GARY LEE
MILLER JOHN EDWARD
MILLER LLOYD ASHTON
MILLER NORMAN NORWOOD
MILLER WILLIAM N
MILLNER CARLTON BRAN-
DARD
MILLNER CHARLES HEWETT
MILLS GRAHAM LONNIE
MILLS WARD WARREN JR
MINOR CALVIN M
MITCHELL RODGER HYLER
MITCHEM CHARLES CLIF-
FORD
MOFFETT JERRY LEE
MOLES THOMAS HARRY
MONAGHAN JOHN JOSEPH JR
MONETTE NEAL E

Virginia

NEWMAN STANLEY VICTOR
NEWSOM BENJAMIN BYRD
NICELY NELSON TALMADGE
NINOW WILLIAM CHARLES
NIXON JOHN ARLEIGH
NORRIS KENNETH EARL
NORRIS TRUMAN DENNIS
O CALLAGHAN BRIAN JOSEPH
O KEEFE MICHAEL JOSEPH
O KUSKY HENRY JOSEPH JR
OBERSON FRANCIS SHERMAN
OBIE CLARENCE WILLIS III
OLIVER THOMAS TUCKER
OLZER JAMES OSCAR JR
OQUINN ARVIN LEE
OSCAR CECIL THOMAS
OSTEEN THOMAS LELAND JR
OSTERHOUS JOHN GARDNER
OVERBAY CLARENCE M JR
OWEN LARRY JAMES
OWEN ROBERT DUVAL
OWENS LARRINGTON
OXENDINE RODNEY GLENN
PAGE GILBERT WAYNE
PAGE HENRY LINDSAY III
PAIER HELMUT WALTER
PAINTER DAVID OLIVER
PALM TERRY ALAN
PALMER JAMES EDWARD
PANGLE WILLIAM MEDFORD
PARDEE SCOTT KENTON
PARISH CHARLES CARROLL
PARKER ARNOLD RAY
PARKER EARL EPHRAIM
PARKER LONNIE THOMAS
PARRISH SAMUEL JOSEPH
PASCOE ROBERT EDWARD
PATTERSON FRED HENRY
PATTON CURTIS RAY
PAYNE DARNELL MILTON
PEELE LLOYD WILLIAM JR
PENNINGTON PHILIP EUGENE
PEPPER ANTHONY JOHN
PEPPIN DAVID DAWSON JR
PERDUE RICHARD WAYNE
PERDUE ROBERT DECKER
PERKINS CECIL CARRINGT JR
PERRY GEORGE EVERETT III
PETTIGREW FRED LAFAY III
PHILLIPS JAMES EDWARD
PHILLIPS LEROY JACKSON
PHIPPS HERBERT CHARLES
PHIPPS NORMAN IRA
PICKETT RODNEY DOUGLAS

MONROE CARLTON LEE
MOORE DAVID NED
MOORE DONALD R JR
MOORE JOHN MARSHALL JR
MOORE LARRY RICHARD
MOORE WILLIAM CLARENCE
MOORMAN MICHAEL AUBREY
MORAN CHARLES KENNETH JR
MORBITZER CHRISTOPHER GEO
MORGAN ARTHUR EUGENE
MORRIS ALVIN GARVIE
MORRIS BILLY RAY
MORRIS RONALD EDWARDS
MORRIS RONALD LEWIS
MORRISON JOHN FRANKLIN JR
MOSS ROBERT EUGENE
MULHAUSER HARVEY
MULL GERALD CRAWFORD
MULLINS DANIEL LEE
MUNDY GEORGE LINWOOD JR
MURCHISON JAMES EMANUEL
MURPHEY DOUGLAS WOODFIN JR
MURRAY CHARLES EDWARD
MURRAY MICHAEL GARY
MYERS JOHN SAMUELS
MYNES THOMAS WILMER
NADEAU ROLAND HAROLD
NANCE ELMER MASON
NAPIER DAVID LAWRENCE
NEAL HARVEY RAY
NEBLETT LYNELL
NEISZ CHARLES WARREN JR
NELSON JULIUS DALE

PIERCE MORRIS WOODWARD JR
PIERCE TERRY PAUL
PINTER WILLIAM JAMES
PIPER DONALD CHANDLER
PIPES JAMES LEE JR
PITSENBARGER DENNIS STOVE
PLEDGER DONALD ALLEN
PLUM CARROLL STEVEN
POINDEXTER MOSES LEON
POOL CHARLES WINFRED JR
PORTER JOSEPH SAMUEL JR
POTTER WILLIAM STEVEN
POWELL DONALD KEITH
POWELL JOSEPH LEWIS JR
POWELL STEVEN REED
POWERS MONROE ALAN
PRESTON LUTHER ELMER
PRICE DAVID EDGAR JR
PRICE JOHNNY PAUL
PROCHASKA WILLARD FLOYD
PROSTELL RICHARD LOUIS
PRUETT WILLIAM DAVID
PRUHS ROBERT L
PRUITT JAMES THOMAS
PULLIAM ROBERT LEE
QUINN STEPHEN WAYNE
RADER JAMES DOIL
RAMSEY CALVIN WETZEL
RANDOLPH HOWARD EDWARD
RANKIN EDWARD GARRY
RANSON JOHN NORMENT
RASH DONALD RAY
RASH MELVIN DOUGLAS
RASNIC OLEN WESLEY
RAY GUY EDWARD JR
RAY KERMIT ANTHONY
RECTOR WILLIAM THOMAS JR
REED GARY ROBERT
REED PAUL MARTIN
REEDY WILLIAM BOYD
REESE DANIEL CORTEZ
REGAN THOMAS FRANCIS
REID AUBREY ARCHIE JR
REID RALPH HENRY
REMEIKAS JOSEPH JOHN JR
REYNOLDS GARY LEE
RHODES THOMAS HENRY
RICHARDS THOMAS JOSEPH JR
RICHARDSON HAROLD OWEN
RICHARDSON HARRY F JR
RICHARDSON NORWOOD

ROLAND
RICKETTS JAMES ELLSWORTH JR
RIDDICK DANIEL ALEXANDRIA
RIDDLE JOHN ROBERT
RIEK JEFFRY RANDAL
RILEY RICHARD WINFRED
RIPORTELLA FRANK J JR
RISINGER PAUL WILLIAM
ROARK JAMES DAVID
ROBERTS JAMES RICHARD
ROBERTSON DONALD REED
ROBERTSON MARSHALL EUGENE
ROBERTSON RONALD EDWARD
ROBINSON ALFRED WILLIAM
ROBINSON BRUCE ELTON
ROBINSON DONALD RAYFORD
ROBINSON JOSEPH LUTHER
ROGERS HOWARD LEONARD JR
ROHLINGER JOSEPH EARL
ROMANO NICHOLAS ANTHONY
ROSE HARRY QUINCY
ROSE ONSBY RAY
ROSS WALTER JR
ROUSE PHILLIP LEONARD
ROYAL COUNCIL LEE
ROYE GEORGE EDWARD
RUCKER JOHN WILLIAM
RUDD JAMES WALLACE
RUDOLPH RONALD CLEMENCE
RUGGLES ROBERT HOYT
RUMBLE JON MAC GILLIVRAY
RUNION MARION GILMER
RUSH JOSEPH BRADLEY
RUSH THOMAS CLYDE JR
RUSSELL HENRY EUGENE
SAINT CLAIR ELISHA REEVES
SAMANS WALTER A JR
SAMUELS DONALD RAY
SARGENT RUPPERT LEON
SAUNDERS BRUCE ALLAN
SAVAGE JAMES WADDELL
SAWYER KENNETH ROBERT
SAWYER MICHAEL KENNETH
SAYERS PAUL FREDERICK
SCANLAN WARREN LEE JR
SCARBOROUGH EDMUND BAGWEL

SCHLIE KENNETH MARTIN
SCHLIEBEN KLAUS DIETER
SCHNELL JOSEPH RICHARD
SCOTT NATHANIEL
SCOTT RONALD
SCOTT VINCENT CALVIN JR
SCURLOCK ALLEN GORDON
SEAMANS OTTO ANDREAE JR
SEAY TONY ELWOOD
SEEKFORD JOSEPH LEVI
SEVIGNY GEORGE WOLFGANG
SEXTON CARL HOWARD JR
SEYMORE PAUL JESSIE
SHANNON KENNETH ARTHUR
SHARTZER JOSEPH CLARENCE
SHAW CHARLES E
SHELDON CHARLES MILLS
SHELTON CLYDE DOUGLAS JR
SHELTON LESLIE LEWIS
SHERRELL MELVIN LEON
SHIELDS MARTIN DEAN
SHOOK KENNETH WILSON
SHREWSBURY PAUL WAYNE
SIDDALL JIMMIE
SIEGRIST WILLIAM LEROY
SIGHOLTZ ROBERT H JR
SIGNETT JAMES GUERDON
SILVER WILLIAM F JR
SIMMERS GEORGE WILLIAM JR
SIMON DONALD ROBERT
SIMON ROBERT LEE JR
SINGLETON JAMES PERRY
SINK OTIS BEVERLEY
SIROCCO WILLIAM DAVID JR
SLAUGHTER HARVEY NEWTON
SLAUGHTER WILLIAM SHELLEY
SLEMP FREDERICK ALBERT
SMITH ALTON
SMITH ARTHUR BURMAN
SMITH CHARLES PORTER JR
SMITH DAVID WAYNE
SMITH DONALD LEE
SMITH GENE DARRELL
SMITH JOHNNY JEROME
SMITH JOSEPH PRESTON
SMITH MARVIN
SMITH MARVIN BONNEY JR
SMITH ROBERT LEE JR
SMITH RODNEY HOWE
SMITH THURMAN HORACE
SMITH WALTER LEE

SMITH WILLIAM PROSPER JR
SMITH WINFRED LEE
SNEAD DOUGLAS LEE
SNEAD LEONARD HARRISON JR
SOMERS GREGORY S
SOUTHER JAMES ALLEN
SPARKS RONALD DAVID
SPEARMAN DAVID GLENN
SPENCE JOSEPH C JR
SPENCER BUFORD RONALD
SPENCER EDWARD ODELL
SPENCER KENNETH DARRELL
SPENGLER HENRY MERSHON III
SPINNER ALFRED WILLIAM
SPIRES JOHN MILTON
SPITLER FORREST F S
STAINBACK MACK DONALD JR
STALLARD DON GENE
STALLARD GILES WARREN
STANCIL GREGORY HALE
STANLEY DENNIS RALPH
STANLEY THOMAS LEE
STAPLES LOUIS FRANKLIN
STARKES ROBERT B JR
STARKEY LLOYD MARTAIN
STEPHENS HARRY EDWARD
STEVENS GERALD
STEWARD JERRY WAINE
STEWART JOHN FRANCIS
STEWART MICHAEL EDWARD
STEWART WILLIAM LOUIS JR
STICHER JOHN THOMAS
STIGLICH MICHAEL LEON
STINNETT JIMMY CLIFORD
STINSON RODNEY LEROY
STONE GORDON ELLIOTT
STONE LEWIS LYNN
STONE RODNEY HAROLD
STONEMAN DONALD LOUIS
STRANGE RICHARD LEE
STRAUSS ROBERT STEPHAN
STRICKLER DAVID FRANCIS
STRONG CALVIN MORRIS
STULLER JOHN CHARLES
STURGILL HAROLD J JR
SULLIVAN MICHAEL NELSON
SUMMERLIN JOHN WRIGHT
SUMMERS HARRY LEE
SUMNER BUFORD ELLIS
SUPINGER CLAUDE CARROLL
SUTHARD CHARLES LEE JR
SUTTLE FREDERICK N JR

SUTTON GEORGE STANLEY
SUTTON WILLIAM JOSEPH
SWANN JAMES CECIL
SWEAT LORAN EDGAR JR
SWEET DONN LAFAYETTE
SYBERT ROSCOE
TAFT PHILIP JEFFREY
TALLEY LARRY JAMES
TARPLEY WILLIAM JUNIOR
TATE ALENN MERRITT
TATE BRADLEY HAYNES
TATEM HAROLD PAUL
TAYLOR FRED
TAYLOR HOWARD FRANKLIN
TAYLOR JOSEPH GORDON
TAYLOR STANLEY EDWARD
TENNANT BYRON LEE
THARPE SAMUEL CHARLES
THOMAS ARTHUR ISIAH
THOMAS EDGAR DURPHY
THOMAS GEORGE DOLBRYN
THOMAS MICHAEL OLIVER
THOMAS WILSON DECOSTA
THOMPSON CARL WAYNE
THOMPSON DAVID BENTON
THOMPSON GROVER WILLIS
THOMPSON JOHN KIRKLAND
THOMPSON JOSEPH WAYNE
THOMPSON RAYMOND MASSIE
THOMPSON THOMAS MICHAEL
THORNHILL JOHN R III
THORNTON JAMES HOLMES
THORNTON MATTHEW WINSTON
THRUSTON ROBERT READE III
TISDALE DONALD WAYNE
TOLLEY CALVIN COOLIDGE JR
TOLLEY LEE G
TOLLIVER SAMUEL STANLEY
TOOMBS ALVIN CARNALL JR
TOUART FOSTER JEWELL G JR
TOWARD RONALD JOSEPH
TULLY ROBERT EDWARD
TURLEY CHARLES VAN
TURNER JAMES EDDY
TURNER LLOYD KENNETH
TYLER ALLEN
TYLER JOHN DEVON
TYNES REGINALD BERNARD
TYNES WILBERT A JR
TYREE WILLIAM EVERETT
TYSON STUART HANCKEL

ULREY KESTER
UMHOLTZ DARRELL RAYMOND
UNDERWOOD HARRY EDWARD
UNRUE ROBERT DANIEL
URBANI ROGER STANLEY
URICK JOHN WILLIAM
VALENTINE JERON FRANKLIN
VALENTINE WILLIAM MARTIN
VAN DEUSEN FREDERICK FREN
VANN CARL REGINALD
VANNATTER LLOYD ALLEN
VASSEY GEORGE CREIGHTON
VAUGHAN COUNCIL DELANO
VAUGHAN MILTON LEE
VELVET WALTER C JR
VENCILL EDDIE WAYNE
VERSACE HUMBERT ROQUE
VICTORY WILLIAM THOMAS
VIDALES ALEXANDER
VINES CLEVELAND
VRABEL JOHN ROBIN
WADE ROBERT MILTON
WADE THOMAS LEE
WALKER CLOVIS BERNARD
WALKER JOE FRANKLIN
WALLACE DOUGLAS DELANO
WALLACE GEORGE DAVID
WALLACE RICHARD COURTNEY
WALLER CASEY OWEN
WALTON WILLIAM HENRY JR
WAPINSKI DANIEL KEITH
WARCZAK DAVID JAMES
WARD JAMES CLINTON
WARD RONALD WAYNE
WARD RUDOLPH NATHINAL
WARE ROBERT DENNIS
WARNER RICHARD THOMAS
WARREN THOMAS WAYNE
WASHINGTON RALPH LEVON
WASHINGTON ROBERT ALLAN
WASHINGTON THOMAS MELVIN
WATERMAN THOMAS L
WATERS WILLIAM WALTER JR
WATKINS CHARLES HENRY JR
WATKINS FRANKLIN ROOSEVEL
WATKINS LAWRENCE D JR
WATSON DELMIS CLAYTON
WEBB FRANK WRIGHT

WEBB THEODORE WOOD
WEBB WILLIAM WINTON
WEBSTER DAVID
WEISS RICHARD EARL
WELLS KENNETH RAY
WELLS THOMAS RALPH
WESSELLS WILLIAM DAVID
WEST JAMES EDWARD JR
WESTFALL BRONSON LEE
WESTON OSCAR BRANCH JR
WHEATON ALLEN THOMAS
WHEELHOUSE CLIFTON P JR
WHIKEHART MARK ANDREW
WHITBECK ROBERT EARL
WHITE CARROLL EUGENE
WHITE DANIEL WESLEY
WHITE DAVID LEE
WHITE EDDY EUGENE
WHITE ROBERT ALEXANDER
WHITE ROBERT WESLEY
WHITEHEAD MORRIS ALFRED
WHITESELL DENHAM A JR
WHITLOCK HARRY OWENS JR
WHITLOW LEROY ALLEN
WHITMIRE WARREN TAYLOR JR
WHITMORE GARLAND D II
WHITTED CURTIS
WHITTEN THOMAS WILLIAM
WIGGINS HIRAM ELI MARVIN
WIGGINS OVID KEITH
WILBORN CHARLEY ANDREW
WILDE ERSKINE BUFORD
WILK JOSEPH ANTHONY JR
WILKINS ALTON III
WILLARD LEON DAVID
WILLEY BURR MCBRIDE
WILLIAMS CHRISTOPHER
WILLIAMS DAVID MICHAEL
WILLIAMS DONALD D JR
WILLIAMS EDWARD J
WILLIAMS GEORGE HARDY JR
WILLIAMS HAROLD ALLEN
WILLIAMS JAMES
WILLIAMS LEONARD TAYLOR
WILLIAMS RAYMOND LEROY
WILLIAMS ROGER DALE
WILLIAMS WALTER JOSEPH JR
WILLIAMS WALTER JR
WILLIAMS WAYNE RICHARDSON
WILLIAMSON DAVID THOMAS
WILLIAMSON RICHARD WESLEY

WILLIS KENNETH MAX
WILSON DAVID WAYNE
WILSON GEORGE L
WILSON JONATHAN TRAXLER
WILSON JOSEPH G III
WILSON REXFORD EARLE
WILSON ROBERT LEE JR
WILSON ROGER EUGENE
WILSON STEVEN WAYNE
WIMMER ROY DEAN
WINKLER JOHN ANTHONY
WIRT LARRY FRANCIS
WITCHER ALBERT
WITCHER SAMUEL EARNEST
WITHERSPOON LEON DUGAN
WOLFE HIRAM MICHAEL IV
WOMBLE WILLIAM THOMAS JR
WOOD DANIEL LEWIS
WOOD JAMES C
WOODFIN DONALD PIERCE
WOODS STERLING SADLER
WOODWARD DOUGLAS MORRIS
WOODWARD RICHARD RANDOLPH
WOODY THURMAN JR
WOOLDRIDGE LAWRENCE O
WORLEY ROBERT KEITH
WORRELL ROBERT EARL
WRIGHT CLAYTON DAVIS
WRIGHT GARLAND FLENDOLPH
WRIGHT GARY WAYNE
WRIGHT GROVER C JR
WRIGHT JAMES AMBLER
WRIGHT JOHNNY WAYNE
WRIGHT RUSSELL L III
WYATT MARVIN LEON
WYATT RICHARD COLEMAN
WYNDER CLEVELAND GYE
YATES DAVID EARL
YATES RICHARD WOODROW
YOST HARRY JAMES
YOUNG CARL L
YOUNG GORDON PRESTON
YOUNG HAROLD EARL
YOUNG JAMES RAY

Photos by Chris Dunnaway

Washington State Vietnam Veterans Memorial

215 Sid Snyder Ave SW, Olympia, WA 98501

The Washington State Vietnam Veterans Memorial, located in Olympia, was made to honor the 1,116 Washington residents who were either killed or went missing while serving in the Vietnam War. This memorial came about when surviving veterans voiced their displeasure with an earlier war memorial. They lobbied for a better memorial in order to honor their fallen brothers and sisters. The memorial site attracts visitors from all walks of life who leave trinkets to honor the sacrifice made by the Washington-based soldiers.

The original Vietnam War memorial, dedicated in 1982, consisted of a granite case that housed a scroll that listed the names of fallen or missing Washington veterans. Surviving veterans didn't think that the memorial served the lost service members properly so this second installment of a Vietnam War monument was dedicated in 1987. A necessary $178,000 was raised by way of private and corporate donations in order to build the memorial. It was created out of green granite by architect Kris Snider who worked with EDAW Inc. The semicircular memorial was placed on a grassy knoll on the Capitol Campus in Olympia.

The Washington State Vietnam Veterans Memorial is a semicircular wall that partially encloses a 45-foot base situated on a rolling hill. Due to its hilly shape the memorial ranges from one foot in height to seven feet at its tallest. This rise and fall are said to represent the good and bad times in the nation before the outbreak of the war. The wall is made up of 16 green granite pieces. The granite was inscribed with the names of over 1,000 fallen or missing Washington service members in chronological order; a cross next to a name marks those who are still missing. There are additional inscriptions containing the memorial's official name and a dedication to the soldiers.

246

The Names of those from the state of Washington Who Made the Ultimate Sacrifice

AARDE JAMES RAYMOND
AASEN DAVID KIM
ACRE LAWRENCE DALE
ADAIR WILLIAM MICHAEL
ADAMS MICHAEL EUGENE
AITON GERALD DAVID
AKEHURST HOWARD DAVID
ALAKULPPI VESA JUHANI
ALBANESE LEWIS
ALBANESE LUIGI FRANK
ALEXANDER TERRY LEE
ALFRED GERALD OAK JR
ALFSTAD KENNETH ORVILLE
ALM RICHARD ANDREW
ALURA RUDOLFO RESTA
ANDERSON CHARLES C JR
ANDERSON CHARLES EUGENE
ANDERSON CHARLES LEON
ANDERSON CHARLES T JR
ANDERSON DELOSS WILLIAM JR
ANDERSON DOYLE TRAVIS
ANDERSON LARRY EDWARD
ANDERSON LYNN DENNIS
ANDERSON RICHARD WILBUR
ANDERSON WAYNE MARSHALL
ANDREASSI CIRO JOHN
ARCHER DANNY LEE
ARENS DAVID LE ROY
ARMITAGE ROBERT LAYMON
ARMSTRONG PEDER WALTER
ARNDT ROBERT DARRELL
ARNEY RANDALL NAVE
ATCHLEY KEITH NOEL
ATOR RICHARD DENNIS
AUSTIN WILLIAM KENNETH
AVERY KENNETH VARSALL
AXSOM HOBART JR
AYERS DARRELL EUGENE
BAKER DAVID WALLACE
BAKER DUANE SCOTT
BAKER SAMUEL THOMAS
BALCOM RALPH CAROL
BALL THOMAS LESLIE SNIDER

BANGS LAWRENCE GENE
BARBER ROBERT FRANKLIN
BARKER JERRY EDWIN
BARKER STEPHEN PETER
BARNES DANNY CLEON
BARNHART BEVERLY LEE
BARNHILL GLEN ROBERT
BARTLETT LARRY PAUL
BARTON WILL PAGE II
BAYSINGER DONALD FREEMAN
BEARGEON DAN WILLIAM
BEATTY DEWEY LLOYD
BELDING CARL FRANK JR
BENEDETT DANIEL ANDREW
BENGEN ARTHUR BURTON
BENKERT PAUL ANTHONY
BERECH LAWRENCE PAUL
BERG BRUCE ALLAN
BERGERSON JOHN FRANCIS
BERNARD GUY NORTH
BETHEA TROY
BETTS DAVID PAUL
BIARUM DONNIE
BILSIE EDWARD ORVILLE
BINDER FREDRICK MARLTON
BINNS GEORGE MICHAEL
BLACK MARK STEPHEN
BLACKFORD JOHN MELVIN
BLAINE JAMES GRAHAM
BLANKENSHIP LEROY IRVIN
BLOOMER DONALD HUGH
BOLSTER DAN ARTHUR
BOOTH GARY PRESTON
BORNSTEIN ANTON THOMAS
BOSSIO GALILEO FRED
BOWDEN JAMES EDWARD
BOWEN GROVER CLEVELAND
BOWEN STANLEY E
BOYKIN PRENTIS BARNEY JR
BOYLE MICHAEL PETER
BOZZELLO FRANK MARIO
BRADLEY DAVID MICHAEL
BRADLEY DENNIS DALE
BRADLEY ROBERT RICHARD

BRADSHAW PAUL LESLIE
BRADY JAMES PATRICK
BRADY JOHN JAMES
BRAGG ROGER DALE
BRAIS JIMMY GENE
BRAND WILLIAM EDWARD
BRANDT KEITH ALLAN
BRENTON PRENTICE FAY
BRIGGS GORDON MICHAEL
BRIGGS JAMES RAYMOND
BRISTOL GUY RAY
BRIX JOHN ELMER
BROCK EDWARD LEROY
BROOKS GUY FRANKLIN
BROWER PATRICK EARL
BROWN ROBERT JAY
BROWN WAYNE GORDON II
BROWN WILLIAM FLOYD
BROWNING DENNIS JAMES
BROWNING JOHN C
BROZ GEORGE MICHAEL
BRUCE DAVID RAYMOND
BRUCE ROBERT GRAHAM
BRUTSCHER RONALD WAYNE
BRYANT JAMES WESLEY
BRYANT SIDNEY LEE JR
BUCHANAN JOSEPH MICHAEL
BURCHETT TIMOTHY GORDON
BURGESON VERNON WALTER
BURKE HOWARD D
BURNETTE ARCHIE JR
BURNEY NILES
BURNS DARRELL EDWARD
BURNS WENDELL MELVIN
BURTNESS ALAN CLARENCE
BUSBY STEPHEN LEE
BUTLER LARRY WAYNE
BUTLER LIONEL SR
BUTLER RUSSEL E
CACERES EDGARDO
CADY GARY ROBERT
CADY MICHAEL MORRIS
CAIN RODGER KENNETH
CALDWELL TIMOTHY BRUCE

CAMPBELL DONALD DUANE
CAMPEN GARY LYNN
CANNON STEVEN LEE
CARLSON GARY LEE
CARMICHAEL GERALD LANE
CARPENTER RAYMOND EARL
CARTER ALAN GLEN
CARTER STANLEY ALAN
CAVALLIN LESTER MELVIN
CAVAZOS REYNALDO ROY
CAYWOOD GARY STEVEN
CHAMBERS JOHNNY A
CHARETTE MARK OWEN
CHARVET PAUL CLAUDE
CHENEY DANIEL BERNARD
CHILDRESS WILBUR HERBERT
CHRISTENSON DANIEL BRIAN
CHRISTIAN BRUCE CALVIN
CHURCH STEVEN ANTHONY
CLARK JOHN JAMES
CLARK JOSEPH THAXTER
CLARK PHILIP SPRATT JR
CLARK RICHARD CHAMP
CLAUSEN HUGH CONRAD
CLEARWATERS CHRISTOPHER L
CLEMENTS ROBERT STEVEN
CLEVELAND JAMES ARTHUR
CLEVELAND LANCE JOSEPH
CLOWE ROBERT EARL
COEN WILLARD GILSON
COFFROTH ALFRED PATRICK L
COKER HORTON SISLER JR
COKLEY GARY WAYNE
COLE DENNIS WAYNE
COLE PHELON HERMAN
COLFACK LLOYD ARTHUR
COLITO JAMES MARTIN JR
COLLINS JAMES BRUCE
COLOMBO GARY LEWIS
COMBS JACKIE RANDALL
COMBS PAUL REX
COOK DELMAR FREDRICK
COOK WILLIAM HAROLD

COPE ROBERT JOE
CORDINER DUANE GORDON
COTE DONALD RICHARD
COTTINGHAM DUANE ROGER
COURCHANE DALE LOUIS
COWAN HARLEY RICHARD
CRANDALL GREGORY STEPHEN
CRANE ROBERT BRENDEN
CREVELING ZED CONNOR
CROSBY RICHARD ALEXANDER
CROSIER STEVEN SEBASTIAN
CROUT KENNETH MILES
CRUZ FRANK
CUNEO STEVE CLYDE
CURTIS GREGORY PAUL
CURTIS RONALD GAY
CYR RANSOM CRAIG
DAILEY DAVID LEON
DAILY DAVID CHRISTOPHER
DALRYMPLE WILLIAM RAY
DAMITIO MARTIN LEO
DAMSCHEN RICHARD A JR
DANIELS GEORGE HOWARD II
DAVENPORT DONALD GENE
DAVIES DAVID MARSHALL
DAVIS RICHARD GLEN
DAVISON GUY ALLEN
DAWSON LAWRENCE MICHAEL
DAZEY WILLIAM LESLIE JR
DE GRAF DICK
DE VERE MONTE RAOUL
DE WATER PATRICK LEE
DEGEN ROBERT PAUL
DENHOFF WILLIAM MICHAEL
DENNISON TERRY ARDEN
DENNY JERRY DAVID
DEPEW VERNON EUGENE
DERKSEMA WILLIAM ARTHUR
DESKINS RONALD DEAN
DEVIK DAVID RALF
DEVORE WILLIAM ROBERT
DIBBERT BERNARD WAYNE

Washington

DIEU GARY ALLEN
DILLARD DONALD GARY
DOLEN JIMMIE ALAN
DONNELLAN DANIEL PAUL
DOUGLAS DONALD DAVID
DOWD WILLIAM DAVID
DOWLING ROBERT MOFFETT
DOYLE PATRICK MICHAEL
DRAKE CLANCY GEORGE
DUBB DEWAIN V
DUFFY MICHAEL BERNARD
DUGGAN GARY LEE
DUNBAR ROBERT SIDNEY
DURHAM VAN LESLIE
DURO IGNACIO ESCOBAR
DUTRO RICHARD THOMAS
DYE HENRY ALBERT JR
DYER BRUCE HERBERT
DYVIG ARTHUR HARRIS JR
EAST MELVIN DOUGLAS
EDDY GARRETT EDWARD
EISENBEISZ ROBERT ARTHUR
EKLUND PAUL HERBERT
ELBERT JOE A
ELLENWOOD STEPHEN A JR
ELLING ROGER WILLIAM
ELLIS JAMES FRANCIS
ELLIS MELVIN RUPERT
ELLMAN JOSEPH RAYMOND
EMCH JAMES KENNETH
EMERSON JAMES WAYNE
ENGEBRETSON GARY LYNN
ENGEBRETSON LARRY
DOUGLAS
ENGELHART LESLIE EUGENE
ENGLISH STEVE CRAIG
ENGMAN DARWIN HAROLD
ENRICO ENRIQUE THOMAS
ERICKSON ALAN CLIFFORD
ERICKSON WILLIAM L JR
EVANS CLIFFRED MELVIN
EVELAND MARK W
EVERETT MARK ROSS
EVERETT STANLEY OLIVER
FABIAN WILLIAM HILRIC
FALCK CARL LEONARD JR
FANNING THOMAS GARRET
FARREN MARK
FAULCONER DAVID ROSS
FEIRO RICHARD DALE
FELTON MELVIN JAMES
FENN DANIEL RICHARD
FERGUSON DOUGLAS DAVID
FERGUSON JAMES ALLEN
FERNAN WILLIAM
FERRELL WALTER LARRY
FESER JEFFERY EVAN
FIFE JAMES HERBERT JR
FINCHER LARRY LEONARD
FISHER JAMES TED
FITZGERALD RONALD EU-
GENE
FLIEGER HARRY GREGG
FLYNN DANIEL LEOPOLD
FORCUM KEVIN PAUL
FORESTER RICHARD THOMAS
FORS GARY HENRY
FOSTER DOUGLAS GENE
FOSTER THOMAS RICHARD
FOWLER VICTOR ORIN JR
FOWLER VIRGIL JAMES
FOX CRAIG JAMES

FRANCAVILLA JOHN FRANCIS
FRANCISCO SAN DEWAYNE
FRANCK RALPH HENRY JR
FRANK HAROLD LEROY
FRANK RODNEY GALE
FRANKLIN AMOS LEE
FRANKLIN LAWRENCE ANDRE
FREDRICKSEN ALLAN MAR-
CUS
FRENIER FREDERICK IRVING
FRINK STEVEN ARTHUR
FRITTS LOUIE GENE
FRODSHAM EDWARD THOM-
AS DA
FROST MICHAEL DENNIS
FULLAWAY LAWRENCE LEE
FULTON RONALD JOE
FUNK LESLIE HAROLD JR
FURSTENWERTH ROBERT
EDWAR
GALBRAITH MARVIN EARL
GARBER EDWIN SIDNEY
GARCIA ENRIQUE LORENZO J
GARDELLA MARK JEFFREY
GARDNER STEPHEN MARK
GASSELING JAMES LEE
GAYNOR KURTIS LANE
GENTRY BILL W
GERTH PETER HUDSON
GEYER RICHARD WEBB
GILLARD GARY LEE
GILLMER STEVEN MARTIN
GINDER RICHARD SAMUEL
GLASFORD MICHAEL RICH-
ARD
GLORE STEPHEN LESLIE
GODDARD RONALD WILLIAM
GODSEY JAMES MARVIN
GOODSELL WILLIAM JOSEPH
GORDON GLENN ALLYN
GORDON GUY LEE
GRAFFE PAUL LEROY
GRAHAM BRADFORD MARK
GRAHAM LARRY ELLSWORTH
GRAHAM WENDELL JOHN
M JR
GRANDSTAFF BRUCE ALAN
GRAVES LEONARD OLSEN
GRAY CHRISTOPHER JAMES
GREEN CHARLES DEE
GREENHALGH TERRY LYNN
GREENWALD RONALD ALBERT
GREWELL LARRY IRWIN
GRIMSHAW DANNY LEE
GROVES WILLIAM E
GRUBB KENNETH WILLIAM
GRUNSTAD STANLEY LLOYD
GRYDER MICHAEL STEVEN
GUILMET DANIEL JOHN
GUSTAFSON BRUCE GORDON
HAKE DAVID TERRANCE
HALL HARLEY HUBERT
HALL KIMBER LYNN
HALL TIMOTHY JOHN
HAMILTON CHARLES ODEAN
HAMLIN WILLIAM ROBERT
HAMMOND JACK MICHAEL
HANCOCK JESSE LEROY
HANLEY JOHN JOSEPH
HANLEY LARRY JAMES
HANNA ROCKY WADE
HANNEMAN MICHAEL IRVIN
HANSEN GERALD STEVEN
HANSEN STEPHEN MICHAEL
HARALSON WILLIAM SCOTT
HARBIN MONTY LEWIS
HARBISON ROY F
HARDEN JAMES ARNO
HARDING THOMAS FORD
HARDY LARRY JOSEPH
HARKE LARRY ARNOLD
HARMON DANIEL WILLIAM
HARNER DAVID IRA
HARRIS HARVEY C
HARRIS JESSE LEE
HARRIS WILLIAM LEE
HART DONALD WAYNE
HART GREG EUGENE
HART RONALD DAVID
HARTMAN MARVIN LEO
HARTY THOMAS JOHN
HARVEY GARY WAYNE
HAUG EARL WARREN
HAWK MICHAEL ALLEN

HAWKINS DONALD DALE
HAWLEY ORIL WILLIAM
HAYDEN ROBERT ALLEN
HAYES RONALD MORRIS
HEATH RUSSELL JAMES
HEBERT DAVID NELSON
HEDBLUM DAVID ARTHUR
HEEN LARRY MICHAEL
HEIDER DONALD GEORGE
HEIMAN JOSEPH EDWARD
HEINDSELMAN MICHAEL
JAMES
HELLWIG STEVEN LOUIS
HEMNES ROBERT BERNARD
HEMPHILL CRAIG MANSFIELD
HENCE WILLIAM WASHING-
TON
HENDERSON EARL
HENRICKSON KEITH RICH-
ARD
HENRY WALTER MAURICE
HENSHAW PATRICK LEE
HENSLEY WAYNE GEORGE
HERNDON W COLE
HEROLD ERIC GARY
HESKETT BRUCE WILLIAM
HIBLER RUSSELL CRANSTON
HICKOX ROBERT DAVIS
HICKS JIMMY ISHMAEL
HIGGINS DENNIS MICHAEL
HIGGS DAVID A
HIGHTOWER GEORGE MAL-
COLM
HIGMAN JOHN EVERETT
HILL GORDON CLARK
HILL MARVIN CHARLES
HINES LESLIE EDMUND III
HINSON BERT HOWARD
HOBAN MICHAEL NOEL
HOBERT WILLIAM JOSEPH
HOCK LELDON EDWIN JR
HODGES KENNETH ALLEN
HODSON VICTOR M
HOEL RONALD EDWIN
HOGGATT JOHN ANDREW
HOGUE JOHN MICHAEL
HOLES JASON AIREAL
HOLIEN RICHARD PAUL
HOLKE DONALD STEVEN
HOLLAND MELVIN ARNOLD
HOLLIS JAMES FAY
HOLM ALAN HANS
HOLMAN ADAM JR
HOLMES NORMAN WARD
HOLT DANIEL JAMES
HOLT GARY RICHARD
HOLZ GARY LEROY
HOLZHEIMER DENNIS ALLEN
HOMBEL RAY EARL
HONEA STANLEY RAY
HOOVER ALVIN RUTHER-
FORD III
HOPKINS EDWARD ARTHUR
HOPKINS RONALD FRANK
HORN CHARLES HENRY
HORTON BARRY DEVERE
HOSKINS ROBERT EDWARD
HOSTIKKA RICHARD AUGUST
HOUCK STEPHEN CHARLES
HOWARD DOUGLAS ALLEN
HOWLEY WESLEY CHARLES JR
HULBERT JEFFREY LEE
HUMBERT JEAN PIERRE
HUTTULA CARL RICHARD
HYATT GEORGE JACKSON
ILSLEY ROBERT PATRICK
INAY CHRISTOPHER HENRY
INGRAM WILLIAM CARLYLE
JACKSON DARRELL ASA
JACKSON ROBERT ALAN
JACOBS EDWARD JAMES JR
JACOBSON KENNETH JAMES
JAMERSON LARRY ALLEN
JAMES MORRIS KEITH
JANES WILLIAM CAREY
JARVIS DANNY WAYNE
JARVIS PAUL ROGER
JARVIS RICHARD MARK
JENS TERRY ROY JR
JENSEN DANA BRUCE
JENSEN GEORGE WILLIAM
JENSEN JOHN ACE
JESKE JAMES ROBERT
JOHANSEN JAMES ARTHUR

JOHNSEN LARRY VERNON
JOHNSON BRADLEY JAMES
JOHNSON FRANKIE RAY
JOHNSON GUY DAVID
JOHNSON HENRY DAVID
JOHNSON MELVIN EDWARD
JOHNSON RONALD JAMES
JOHNSON ROY MARVIN
JOHNSON WILFRED C JR
JONES DOUGLAS ROBERT
JONES JAMES RANDLE JR
JONES MICHAEL CLAY
JORGENSEN ROLF WALLACE
JOY ROBERT HOLBROOK
KADOW PATRICK DENNIS
KARLSTROM SIGFRID R
KAUFFMAN KEITH WALTER
KEARNEY ROBERT CURT
KELBY WESLEY ROBERT
KELLER GARY DALE
KELLOGG PETER PATRICK W
KEO DANIEL WILLIAM
KERR ROBERT GEORGE
KERR STANLEY JESSE
KESSINGER KENNETH MAR-
TIN
KEYS MICHAEL HENRY
KIELLEY BYRON ALICK
KIMBALL PIERCE MALLORY
KIMES LOUIS D
KING ROBERT D ORR
KING RONALD DEAN
KINGHAMMER STEVE WIL-
LIAM
KINNE ALLEN GENE
KITTLESON RANDY GENE
KLEIN STEPHEN LOUIS
KNIGHTON PAUL GORDON
KNOLLMEYER MARK ALAN
KNOWLES DAVID DU WAYNE
KOENIG EDWIN LEE
KOHN WAYNE EDWARD
KOITZSCH RONALD NORMAN
KRAABEL JOHN SPAULDING
KRAFT DONALD RAY
KRAUSE KENNETH J
KREGER PAUL DENIS
KROGH RICHARD OTIS
KROSHUS LEONARD JOSEPH
KRUSE DALE LYNN
KUJAWA DONALD LEE
KULM GERALD ALBERT
LAAN JACOB CLARK
LAFAYETTE JERRY OWEN
LAFROMBOISE MICHAEL S
LAIPPLE JOHN ELDEN
LAMBERT DALE LEE
LANCASTER DAVID CLYDE
LANDERS BILLIE DWAINE
LANDIS BRUCE RANDOLPH JR
LANE JOHN TIMOTHY
LANG DEAN LAVERNE
LANG TIMOTHY MICHAEL
LANGSJOEN RICHARD CLAY-
TON
LANGWORTHY JAMES SCOTT
LARKIN SAMUEL JAMES
LARSEN CHRIS JOHN III
LARSEN FREDRICK ELLIS
LARSEN JOE PAUL
LARSON PAUL NOBLE
LARSON ROGNER ANDRE
LAURITSEN DAVID WAYNE
LAWSON DONNY RAY
LAWSON JESSE HUGH
LEACH ANTHONY MICHAEL
LEACH WALTER DARYLE
LEISY ROBERT RONALD
LENZ LEE NEWLUN
LEROY HOWARD LLOYD JR
LESTER RODERICK BARNUM
LETTERMAN LAWRENCE
ALLEN
LEUNING VERNON LEE
LEWIS EARL LLOYD
LEYERLE BILLY BOB
LICKEY MICHAEL LEWIS
LIKKEL DUANE ALLEN
LINDBERG DALE RAYMOND
LINDERMAN MICHAEL EDWIN
LINDSAY MICHAEL CLAUDE
LINDSTROM PATRICK EUGENE
LINNELL DENNIS RICHARD
LIVINGSTON RICHARD ALLEN

LOCHRIDGE ROBERT ERIC
LOCK MOON WAI
LODHOLM NORMAN ELLIOTT
LOGAN GORDON WESLEY JR
LOGAN JACOB DRUMMOND
LOMEN RALPH TERENCE
LONG JAMES ARTHUR
LONO LUTHER ALBERT
LOOBEY MERLE E
LOPEMAN STEPHEN RAY
LOPEZ ROBERT
LOVE KERRY BRENT
LOWE DONALD EVERETT
LOWERY ALVIN LEROY
LUND HARRY D
LUNN TOMMIE RICHARD
LYNN JOHN THOMAS
MAC ARTHUR DALE ALAN
MAC CALLUM STEPHEN
MORLEY
MAC LURG DAVID WEBSTER
MACE DAVID LESLIE
MACK GARY LEIGH
MALARZ RENE LEE
MALATESTA LARRY JOE
MALONE LAWRENCE MI-
CHAEL
MANNERY RICHARD CHARLES
MARCO ROBERT DONALD
MARINSIC ALLEN HENRY
MARKWITH GERALD WIL-
LIAM
MARTIN MERLE JAMES
MARTINEZ RICHARD EARL
MARVIN GREGORY ALLEN
MASON CHARLES BUCKLEY
MASON TERRY DEAN
MASTERSON MICHAEL JOHN
MATHEWS PATRICK T
MATTER MARK ALLEN
MATTERN RICKY PALMER
MATTSON TIMOTHY GEORGE
MAXIM THIERRY TIMOTHY G
MAYNARD ROBERT DEE
MC AFERTY ROBERT EUGENE
MC CALL WILLIAM ARTHUR J
MC CANDLIS OWEN TED
MC CARTHY JAMES IRVIN JR
MC CLANAHAN LARRY BYRON
MC CLINTOCK JAMES RICH-
ARD
MC CLINTOCK TED ERNEST
MC CONNEL GERALD WAYNE
JR
MC CURDY ROBERT LOWELL
MC DANIEL ARCHIE HUGH JR
MC DONALD EMMETT RAY-
MOND
MC GARVEY PATRICK GEORGE
MC GLOTHLEN JERRY WAYNE
MC HUGO DONALD LYLE
MC LAUGHLIN MICHAEL PAUL
MC LEOD PATRICK ALAN
MC MAHON TIMOTHY JAMES
MC MURRY RODERICK DANE
MC NEAL MICHAEL E
MC NEIL DONALD K
MC RAE DAVID LE ROY
MCMICAN M D
MCQUADE JAMES RUSSELL
MEHEGAN RICHARD HAROLD
MEIDINGER DARYL GENE
MELDAHL CHARLES HOWARD
MENO GEORGE SABLAN
MERRILL LARRY ARTHUR
MESOYEDZ HOWARD AN-
THONY
MESSERSMITH DEAN HAROLD
METTERT ROBERT ARNOLD
MEYER RONALD RAY
MICHEL WILLIAM FREDERICK
MILLER FREDERIC WILLIAM
MILLER JAN DEVOE
MILLER ROBERT LEE
MILLER THOMAS PAUL
MINA JUAN VELASCO JR
MINKLER STEVEN JEFFERY
MINOR DANIEL JAMES
MIOTKE STEVEN MICHAEL
MIRAMONTEZ LEONARD
MISNER KENNETH GENE
MOCK MAURICE KARL
MOE LESTER JAMES
MOEN JOSEPH ALLEN

MONTGOMERY ROBERT ALLEN
MOORE THOMAS MICHAEL
MORISETTE CLEMENT JOSEPH
MORIWAKI KAZUTO
MORRIS ARCHIE W
MORRIS PATRICK BENNETT
MORROW BRIAN JOHN
MORTON CHARLES TIENEREY
MOSELEY DAVID WESLEY
MOSER GREGORY PHILLIP
MOSS LARRY ALLEN
MOULTINE CHARLES RAY
MUKAI BRYAN THOMAS
MULFORD ALAN CRAIG
MULLAN JOHN TURNER
MURDEN STEPHEN BROOKS
MURDOCK MICHAEL GEORGE
MURPHY JOHN FRANCIS
MURPHY WILLIAM PATRICK
MURR CLYDE EDWARD
MURRAY LESLIE EUGENE
MYERS ROBERT LESLIE
NAKKERUD ARNOLD OLAF
NALLY ROBERT GERALD
NANSEL JAMES DAVID
NEAL DENNIS WADE
NEAS STEPHEN EDWARD
NEISESS JAMES A
NELSON DAVID LINDFORD
NELSON DONALD CARL
NELSON LEWIS CHARLES
NELSON RICHARD DEAN
NETH DANNY A
NEVILLE STEVEN WAYNE
NEWBERN MICHAEL R
NEWBY KENNETH LEROY JR
NICCOLI GREGORY JEROME
NICHOLS RICHARD ALLEN
NICKERSON PHILIP EUGENE
NILES JEFFERY CHARLES
NIXON RAY
NOBLE DENNIS RAY
NOBLE GARY PAUL
NOKES JOHN DARRELL
NORRIS CHARLES STEVEN
NUSSBAUMER JOHN JOSEPH
NYMAN LAWRENCE FREDERICK
O BRIEN JOHN JOSEPH
O LEARY TIMOTHY MONROE
ODEGARD DELL COLEMAN
ODELL JOHN MICHAEL
ODEN ROYAL PRESTON
OKERLUND THOMAS RICHARD
OLLOM ROBERT LEE
OLMSTEAD ROBIN LEE
OLMSTED GERALD RAY
OLSEN KEITH
OSBORNE RODNEY DEE
OTTO ROGER DEAN
OWEN STEPHEN BOYD
OWENS JOY LEONARD
OZUNA JUAN SANCHEZ
PADDOCK GARY CLIFFTON
PADDOCK JOHN EVERETT
PADILLA GEORGE ISAAC
PAGE RICHARD LEE
PARKER RONALD WAYNE
PARKHURST GREIG ROBERT
PARRANTO LAWRENCE W JR
PARRIS DOUGLAS HAROLD
PARRISH BILLY JOE
PARROTT BRIAN GREGORY
PASCHALL RONALD PAGE
PATTERSON GARY LEE
PAULSON MARVIN JR
PAYNTER THOMAS BERNHARD
PEARSON ANTHONY JOSEPH
PEARSON ROBERT HARVEY
PEARSON RONALD RUSSELL
PELTON WILLIAM FRANK
PENDER DONALD L
PERDUE JOHN HARRY
PERRY ROBERT KENT
PETCHNICK CHARLES RUSSELL
PETERS ELLIOTT LEE
PETERSEN HARRY ALLEN
PETERSEN ROBERT BRUCE
PETERSON CHARLES C
PETERSON THOMAS LAW-

RENCE
PETERSON WARREN GARY
PETERSON WILLIAM J
PETTERSEN WAYNE ADOLPHUS
PETTIGREW JOHN FLOYD
PHILLIPS GLENN ROSS JR
PHIPPS LEONARD MORRIS
PICK DONALD WILLIAM
PIERCE CLIFTON PALMER
PIERCE JERRY LEE JR
PIPER WALTER JR
PIPKIN DENNIS NEWMAN
PITTS ROBERT PATRICK
PLAYFORD RONALD EDGAR
POFF DANIEL LOYD
POLLARD JAMES FREDERICK
POLLARD WAYNE RICHARD
POMERINKE RICHARD ALLEN
POMPELLA PATRICK OWEN
POTTER ROBERT ALLAN
POWELL CARROLL WAYNE
POWERS JOHN ROGER
PRATT RICHARD EMMETT
PRENTICE KENNETH MORTON
PRICE DAVID STANLEY
PRICE LARRY LEE
PRINCE STEPHEN ROBERT
PRUETT DONOVAN JESS
PUISHIS DALE SCOTT
PURSEL THOMAS RONALD
QUIGLEY RONALD LEEROY
QUINN JAMES JOSEPH III
RABER PAUL J
RABER RALPH DONALD
RAMOS FORREST LEE
RAMOS RAINER SYLVESTER
RANDLES JOHN PETERS
RASH LYNLEY LEE
RAUEN JOHN VERNON
RAY DARRELL THOMAS
RAY MICHAEL GEORGE
RAY RONALD EDWIN
RAYBURN EDWARD LEE
REAUME WADE RUSSELL
REED JIMMIE LYNN
REESE DAVID PHILLIP
REGALADO RICARDO WAYNE
REID KENNETH WAYNE
REID LEON
REYNA THOMAS O
REYNOLDS JOSEPH LEE
RHODES FRANK MOSS
RICE JOHN CLIMATH
RICHARDSON STEPHEN GOULD
RIDER JAMES AUSTIN JR
RIEBLI JOSEPH ROBERT
RIORDAN JOHN MICHAEL
RISHER CLARENCE IRWIN
ROBERSON LEONARD WADE
ROBERTSON JOHN LEIGHTON
ROCK DON LESLIE
RODRIGUEZ GEORGE
ROGERS RANDAL LEE
ROLSTAD THEODORE S
ROPER JOHN MORTON
ROSE MARK RICHARD
ROSS MORRIS JEROME
ROTH FRED STEWART
ROUNDTREE RICHARD RALPH
ROUSKA DENNIS LEON
ROWLAND THOMAS PATRICK
RUD KENNETH HANS
RUFF WILLIAM HERMAN
RUIZ HECTOR LOPEZ
RUSSELL RONALD JAMES
SABINE JOHN SHAW IV
SALAZAR RICHARD FRANK
SANDERS DAUNT BRUNELL
SANDVIG DAVID JAMES
SANFORD ROBERT RAY
SARGENT KENNETH EUGENE
SARGENT STEVAN ROY
SAUX ROGER DOUGLAS
SAVARE HOWARD LEROY
SCHELL DUANE CHARLES
SCHELVAN DAVID ERIC
SCHIMANSKI KENNETH ALFRED
SCHMIDT JAMES DREW
SCHNEIDER SCOTT EDWARD
SCHOOK GEORGE WASHINGTON

SCHOOLER STEVEN THOMAS
SCHULTZ DAVID JOEL
SCHULTZ LOWELL EUGENE
SCHULZ RONALD KENNETH
SCHWEIKL JEFFREY ALLAN
SCHWINTZ BOBBITT
SCOBY RICHARD WILLIAM
SCOTT GREGORY EDWARD
SCOTT WILLIAM BLAKE
SCRANTON ALLEN FRANK JR
SEBERS FREDERICK THEODORE
SECOR WILLIAM DALE
SEDIES RICHARD SAMUEL
SELDEN FRANK WILLIAM
SEXTON TROY LAVERNE
SHANDS MICHAEL ANTHONY
SHATTUCK RONNY DEAN
SHEETS WINFRIED ALBERT
SHERBURN HUGH LESLIE
SHERMAN JOHN HAROLD
SHIELDS MARVIN GLEN
SHREWSBERRY ROGER LYNN
SHRINER THOMAS JOHN
SHULTS ROY EARL JR
SHURTLEFF BRUCE WARREN
SIGURDSON JOHNNY ALLEN
SILLS KENNETH HOWARD
SILVESAN DENNIS RAY
SIMES ROBERT GARLAND JR
SIMKINS GARY BEDE
SIMONSON LARRY ARNOLD
SLUSSER CHARLES RODNEY
SLYE GEORGE DALE
SMILEY EDWARD ROWE JR
SMITH ALBERT HEUGH
SMITH DAVID WALTER
SMITH FRANK LEE
SMITH GUS JR
SMITH HARRY ERNEST
SMITH JOHN MICHAEL
SMITH MITCHELL BRUCE
SMITH PHILLIP ROBERT
SMITH RICHARD DEANE
SMITH STEVEN MARTY
SMITH TERRY HUGH
SMITH WILLIAM THOMAS
SNOW EARL PATRICK
SONNER KALEY ALFRED
SOULE WILLIAM FRED
SOUTHWICK JOHN PAUL
SPAFFORD JOHN WAYNE
SPARKMAN LEONARD PETER
SPARKS JOHN W
SPARKS ROGER HOWARD
SPEARS MILTON EARL
SPENCER RICHARD CHARLES
SPIKER PATRICK JR
SPINELLI DOMENICK ANTHONY
SPINK WARREN LEE
SPRINGSTEEN DENNIS EUGENE
ST LAURENT LANCE WILFRED
STALLINGS ROBERT ELVIS
STARK GERRY LYLE
STARKEL MAX PAUL
STARKEY KURT L
STARR BENNY ARNOLD
STAUNTON JOEL PAUL
STEADMAN STERLING DWIGHT
STEARNS LLOYD PALMER
STEINBRUNNER DONALD THOMA
STEPHENS TOMMY LEE
STEWART JAMES BARTIE
STEWART JOHN LEONARD
STICKEL BRUCE JACOB
STITH PETER LEWIS
STOHLMEYER CHARLES JOSEPH
STRAIT DOUGLAS FRANK
STRENGTH NORMAN HOWARD
STRONG RICHARD WILLIAM J
STROOMER RONALD LEE
STUBBS WILLIAM W W
SULLIVAN JOHN ANTHONY
SUTHERLAND SCOTT EUGENE
SWALLEY ROBERT EUGENE
SWAN JERALD DAVID
SWANTAK DENNIS RAY
SWEENEY CLARENCE JOSEPH

SWINSON LONNIE MELROE
TANGUAY ALAN MICHAEL
TATE JOHN CULLEN
TAUALA TAGIPO VAOGA
TAYLOR GEORGE MICHAEL
TAYLOR GLENN DEAN
TAYLOR THOMAS EUGENE
TAYLOR WILLIAM A
TEAGUE THOMAS NICKELL
TEJANO RICARDO ROBERT
THEODORE J ATHAN
THIBAULT RICHARD GARY
THODE LAWRENCE GREGORY
THOMAS JOHN CHARLES
THOMPSON DENNIS MICHAEL
THOMPSON EVERETT BARL
THOMPSON GREGORY CARL
THOMPSON JAMES PATRICK
THOMPSON JOHN FRANKLIN
THOMPSON RICHARD LEE
THOMPSON ROBERT MICHAEL
THOMSEN GAIL WARD
THULIN DONALD FREDRICK
TILL JOHN JEREMIAH
TJERNBERG ROGER BLAKE
TORRES MANUEL VEGA
TOSCHI RICHARD WILLIAM
TRAW JIM SILAS
TREMBLEY J FORREST GEORGE
TRIMBLE DENNIS ARTHUR
TRIMBLE LARRY ALLEN
TRIPLETT GORDON MARSHAL
TROTTER DOUGLAS EARL
TRUSLEY JASPER H JR
TUCKER DARRELL LEE
TURNBOUGH CHARLES DANNIE
TURNER ALFRED LEE
TURNER DONALD EUGENE
TURNER RANDY VAN
TYSON CLIFFORD EARL
UNDERWOOD PERRY LUKE
USHER TERRY MAXWELL
VAN ALLEN CHARLES CLIFFOR
VAN HORN EDWARD LINDLEY
VANHULLE ANTHONY F II
VARNEY ROBIN LEE
VAUGHAN ROBERT LESTER
VERCRUYSSE GREGORY PAUL
VERNOR JAMES EDWIN
VIGIL ANTHONY
VINGE RICHARD LONNIE
VISE GEORGE FRANCIS
VISINTIN KENNETH HENRY
VISKER THOMAS PETER
VIVETTE LEON LOUIS
VOSS RAYMOND ALLAN
WAALEN JOHN HOWARD
WADE ARCHIE NORMAN
WAGENAAR DANIEL LEONARD
WAGNER EDWARD JOHN JR
WAGNER WILLIAM PETER III
WAINWRIGHT MICHAEL JAMES
WALDOWSKI JAMES RICHARD
WALKER JOHN DAVID
WALKER RICHARD HAROLD
WALKER RUSSELL BERNDT
WALL WILLIAM PENN III
WALLING HARRY ALLEN
WALMSLEY WILLIAM MORRIS
WALTERMAN LARRY JOHN
WARE CLIFFORD OTIS
WARNER DAVID HOWARD
WARREN GALEN EUGENE
WATSON RICHARD DALE
WAUGH GRANT REED
WEAMER ALLEN RAY
WEBB ROBERT JAMES
WEBER WILTSE LEE
WEDRICK LONNIE MARK
WEED JAMES ALLAN
WEED RODNEY RICHARD
WEGNER DENNIS RAY
WEIDERMAN CLAUDE FREDRICK
WEIGHTMAN GREG EUGENE
WEISSER ROBERT LEE
WEITZ HENRY KENNARD
WELLS EDWARD WILLIAM
WELSFORD JOHN AUGUST JR
WESSLER DANIEL GUY
WEST CHARLES EDWARD

WEST NOEL THOMAS
WEST STANLEY EUGENE
WEST WILLIAM RICHARD
WHITBY JOE ALAN
WHITE JACK LEE
WHITE MELVIN RICHARD
WHITE MILES EUGENE
WHITE RICHARD EDWARD
WHITLOCK PATRICK A
WHITNEY ROBERT ARNOLD
WHITTINGTON MERREL P
WHYTE CHARLES JAMES
WIDMER KIM WILLIAM
WIGHT CHARLES EDWIN
WILFORD PERCY LEE
WILHELM DAVID KENNETH
WILKENING ARNOLD GRIMM III
WILKINS GARY LEE
WILKINS RICHARD EDWARD
WILLIAMS KENNETH R JR
WILLIAMSON DONALD LEE
WILLIAMSON JAMES D
WILSON ALBERT RICHARD
WILSON DARRELL WAYNE
WILSON NORMAN RAYMOND
WILSON WALTER GENE
WIMMER JAMES ALLEN
WINGERT DOUGLAS GENE
WINKLES GEORGE WILLIAM J
WISE DONALD A
WISE ELWIN CLAUDE
WISE ROBERT EVANS
WOLF JOHN ROBY
WOLFE RONALD GALE
WOOD STUART JOHN
WOOD THOMAS EUGENE
WORKMAN TIMOTHY E
WORTHINGTON RICHARD C JR
WRAY ELMER ODELL
WRIGHT DARREL ZANE
WRIGHT DARRYL WHITNEY
WRIGHT MICHAEL LEE
WRIGHT ROBERT FRANK
WRIGHT ROBERT NORMAN
WULFF ROBERT WAYNE
YATEMAN DALE ARNOLD
YOUNG STEPHEN ROGERS
ZACHER LYLE DAVID
ZAMBANO QUENTIN DENNIS
ZELLER DOUGLAS LEE
ZEYEN WILLIAM RAYMOND
ZIMMER WALTER JOHN
ZORNES VERNON GLEN
ZYPH JAMES LOUIS

Photos by Garvey Strong

Vietnam Veterans Memorial

186 Greasy Ridge Rd, Princeton, WV 24739

The Vietnam Veterans Memorial in Princeton, West Virginia is located near the Tourist Information Center on Greasy Ridge Road and honors the veterans from the surrounding eight counties in West Virginia and neighboring Virginia. The memorial is accessible 24 hours a day to all visitors who wish to pay their respects to the veterans, both fallen and missing, of the Vietnam War.

The Vietnam Veterans Memorial was dedicated on Veteran's Day in November 1996. This was possible after volunteers from the Blue-Gray Chapter (Chapter 628) raised enough funds to design, build, and erect the memorial which is made of eight different granite plaques. While the Blue-Gray Chapter is the known developer of the Vietnam Veterans Memorial in Princeton, the actual creator of the physical monument is unclear.

The Vietnam War memorial is a circular structure split in two to allow visitors to enter the memorial. Once inside the granite ring, a fountain made of the eight black granite slabs creates an octagon shape that features the bubbling water of the fountain in its center. The water flows into a pool under the granite slabs. Stone benches were placed within the granite ring in order to give visitors a place to sit and reflect on the memorial and its significance.

Each of the eight slabs represents one of the four counties in West Virginia (Mercer, McDowell, Monroe, and Summers County) and the four represented counties in Virginia (Bland, Giles, Tazewell, and Wythe). Each granite slab displays the engraved names of the fallen and/or missing Vietnam War veterans from each location. In addition, there are 10 flag-poles flying the state flags, the POW/MIA flag, the five armed services flags, and two veteran association flags.

The Names of those from the state of West Virginia Who Made the Ultimate Sacrifice

ABRAHAM ARLEY GEORGE
ACORD EWELL EDGEL
ADAMS DARRIUS WAYNE
ADAMS EDWARD CODY
ADAMS JAMES CLARENCE
ADKINS BOBBY RAY
ADKINS KENNETH DALE
ADKINS MARVIN JARRELL
AGAR ROBERT LEE
ALBRIGHT JOHN SCOTT II
ALBRIGHT TERRY ROBERT
ALKIRE THEODORE A JR
ALTIZER ALBERT HAROLD
AMICK FREDDY L
AMOS WILLIAM LEE
ANDERS CHARLIE
ANDERSON CHARLES E JR
ANGLIN ROBERT LEE
ANKROM EVERETT LEE
ANTOLINI JAMES VINCENT
ARBOGAST RANDALL
ARMENTROUT RAYMOND LEE
ARMES BOBBY WAYNE
ARTHUR RICHARD THORN-
TON
ASHLEY FRANKLIN D II
ASTON LYLE GLENN
AUSTIN JOSEPH CLAIR
AUXIER JERRY EDWARD
AYERS JOHNNIE MARVIN
BALCOE CHARLES WALTER
DEW
BALL JOHN ROBERT
BALL ROSCOE WILLET JR
BANNISTER HOWARD WIL-
LIAM
BARKER KENNETH MONROE
BARRETT CHARLES ARTHUR II
BAYS PAUL EUGENE
BEAVERS FRANK ARVIS
BELCHER TED
BENNETT DOUGLAS ALVIN
BENNETT THOMAS WILLIAM
BERRY JAMES GRAYSON
BERRY RONALD LEE
BESS CHARLES RAY
BETHEL JAMES WALTER
BIAS CLIFFORD
BIGGS EARL ROGER
BINION CURTIS ESTILL
BISE ROGER ALLEN
BLACKWELL WILLIAM ALLEN
BLAKE TIMOTHY MORGAN
BLANKENSHIP CLAYTON
MITCH
BLANKENSHIP DENCIL RAY
BLANKENSHIP JACKIE LEE
BLEIGH ALFRED HARLEN JR
BOARD STEPHEN DOUGLAS
BOBLETT MACK CLIFFORD
BOGGS CHARLES EDWARD
BOGGS ROBERT SIDNEY
BOLEN JACKIE EVERRETT JR
BORSAY PETER SAMUEL
BOSLEY JAMES GILBERT
BOWARD KENNETH WILLIAM
BOWEN JOHN JR
BOWYER REX ALLAN
BOYER WAYNE DOUGLAS
BRADY JOSEPH MARTIN
BRAGG RAYMOND DALE
BREEDLOVE RODNEY ALLEN
BROOKS JAMES HARRISON JR
BROOKS THOMAS JOSEPH
BROWN EARL FREDERICK
BROWN EDWARD WALLACE JR
BROWN HARON LEE II
BROWN RONALD LEE
BROWN WENDELL LEE
BROWNING GEORGE EDWARD
BRUMBAUGH JOHN LOUIS JR
BRUNO ROGER LEE
BRYANT RICHARD WAYNE
BUNNER LESTER EARL
BURDETTE CLIFFORD GERALD
BURDETTE JAMES RONALD

BURDETTE ROBERT LEE
BURGESS GARRY LEE
BUTCHER LARRY R
BYUS ROGER LEE
CABELL DARRELL LEE
CAIN PORTER RAY
CAMPBELL WILLIAM HENRY
CANTERBURY MARVIN
DEWAYNE
CARDEN ALBERT PARKER
CARPER EDDIE DEAN
CARR WILLIAM LEE JR
CARROLL KENNETH AUTRY
CARSON DAVID RICKEY
CARWITHEN ALBERT MOR-
GAN
CASHDOLLAR GLENN FRAN-
CIS
CASTLE ROGER ALLEN
CAVINS SAMUEL MC ARTHUR
CHACALOS GEORGE MANUAL
CHAPMAN BILLY GENE
CHAPMAN HENRY LEE

CHARNOPLOSKY JOHN
ANDREW
CHATTIN AL JUNIOUS
CHESS LLOYD ALLEN
CHILDERS ROGER DALE
CHRISTIAN TED HOWARD
CHRISTIAN THOMAS BARRY
CLARK ROBERT ARTHUR
CLARK ROY EDWARD
COBB EARL RUSSELL
COFFMAN FREDDIE LEE
COLE ROGER DALE
COLEMAN THOMAS KEITH
COLLINS CLAYTON
COLLINS ROBERT ORVILLE
COLLINS RODNEY RAY
COMPTON LORN DAVID
CONN FRANKLIN L
CONNER DAVID LELAND
CONNOR JAMES KENNETH
COOK DAVID SAMUEL
COOL MARK DOUGLAS
COOPER MICHAEL LINN

COOPER ROGER EDWARD
CORDER JAMES RUSSELL
CORK CLIFFORD MARKWOOD
COX EARNEST LEE
COX JAMES BLAINE
CRABTREE VARISE HELTON JR
CRAFT CLAYTON ANDREW
CRAFT JOSEPH RODNEY
CRAFT WILLIAM EDWARD
CRAIG ROGER GENE
CREWS JOHN W JR
CRIGGER RELL JR
CRONIN WILLIAM BERNARD
CROSE JAMES CHARLES
CROSIER CARL ROGER
CROTHERS HOWARD ROBERT
CROW JESSIE FRANKLIN
CUMMINGS LONZO SILAS
CURD RICHARD LOWELL
CURRENCE EVERETT AUSTIN
CURRY KEITH ROYAL WILSON
CUSTER GEORGE PAUL
DAILEY LARRY EUGENE

DAILEY PAUL MARION
DALTON MAJOR ROY JR
DANEHART EDWIN RUSSELL
DAVEY GLEN VINCENT
DAVIS JOHN MICHAEL
DAVIS WILLIAM THOMAS
DAWSON DANNY LEE
DAWSON EUGENE
DAWSON HAROLD CARL JR
DEAN JAMES HOWARD
DEAN RONALD PHILIP
DEAN THOMAS WILLIAM
DELY WILLIAM
DENNEY WILLIAM HERMAN
JR
DICKENS JACKIE LEE
DILLOW JERRY WAYNE
DODD DANNY JOE
DOLIN DANNY JOSEPH
DOTY JAMES MARSHALL
DOTY LOYAL BARON
DROWN SAMUEL ROBERT
DUNCAN JAMES EDWARD

West Virginia

DUNFORD DAVID WILLIAM
DUNITHAN THOMAS LAW-
RENCE
DUNLAP DARRELL EDWARD
DUPLAGA JOHN STANLEY
DUPONT ERNEST THOMAS
DURBIN ROBERT VERNON
DURST JOHN BERNARD
DUTY EDWARD
DYE RONALD HARVEY
EAST FRANKLIN
EDDY JERRY WAYNE
EDMONDS LAWRENCE NA-
THANIE
EDWARDS DANIEL LYNN
ELLIS CLARENCE EDWARD
ELLIS RANDALL LEE
ELLYSON ARCHIE MERLIN
EMERSON ERVIN JUNIOR
EMMERT JAMES RICHARD
EMRICK STEVEN ERIC
ENNIS JAMES LESTER
EPLIN JAMES LEONARD
ESTEP EARL B JR
ESTERS FREDDIE
EUBANKS GEORGE F
EVANS ROGER DALE
EYSTER GEORGE SENSENY JR
FENSTERMACHER RONALD
LEE
FERREBEE RUSSELL EDWIN
FISHER CARROLL DEAN
FITZWATER JOHN CURTIS
FLANAGAN WARREN JUNIOR
FLECK ROBERT LEE
FORD BILLY KEITH
FORD KENNETH RAYMOND
FORD WALLACE ADDISON
FOSTER JULIUS CARTWRIGHT
FOSTER SHELBY GENE
FRANZINGER KURT WALTER
JR
FREEMAN GLENN WAYNE
FRIEL LUSTER CLARK
FROE GEORGE WASHINGTON
FULLERTON GLEN LEE
FUNK JOE ALBERT II
GAINER GARY LEE
GARRETT MICHAEL STEVEN
GENTRY TERRANCE NEIL
GIBSON HERMAN DANNY
GILL KENNETH LEE
GLOVER RAYMOND EDWARD
GOGGIN PAUL STEPHEN
GOLDSMITH CARL EDWARD
JR
GOOD BILLY DUANE
GOODMAN ROBERT ONNIE JR
GOODSON CARL BRADFORD
GORBEY JACK EUGENE
GORE EVERETT JR
GOULD WARREN LEE
GRAHAM HAROLD EDWARD
GREENE DANNY MARVIN
GREENE LLOYD EARLE JR
GREENFIELD KENNETH
EUGENE
GREEVER HAROLD LEE

GRELLA PATRICK MARTIN
GRIFFITH RICHARD WAYNE
GRIFFITH ROGER DALE
GROVES DAVID LIVINGSTONE
GRUBB GARY HOWARD
GUMM ROBERT HUGH JR
GUNTHER CLARENCE M JR
HACKNEY DONNIE LEE
HAGEY CLARENCE E
HAINES CRAIG WARD
HAIRSTON CHARLES MCKIN-
LEY
HALL CLYDE
HALPENNY JERRY LEE
HALSTEAD BENNY RAY
HAMRICK BENJAMIN NEAL
HAMRICK EDWARD JOSEPH
HAMRICK KENNETH JAMES
HANCOCK WILLIAM EDGAR
HANNA MARVIN JIM
HANNAH CHARLES MITCHELL
HARPER DENNIS JR
HARRISON RANDOLPH
MONROE
HARRISON RICHARD DARRELL
HARSANYI JIMMY ROGER
HARVEY THOMAS PRESTON
HARVILLE LAWRENCE
HATCHER JAMES LEWIS
HATFIELD DRUEY LEE
HATISON JEFFREY STEPHEN
HAUGHT GARY LEE
HAUGHT HOWARD THAD-
DEUS JR
HAVERLAND MARK JOSEPH JR
HAYES DANNY MARTIN
HAYNES GARRY DWIGHT
HAYNES MICHAEL WAYNE
HAYNES RICHARD WAYNE
HAYS GALE JACKSON
HEASTER ROY DWIGHT
HEATER LARRY STEVEN
HEETER JAMES RALPH
HELMICK ALAN DALE
HENDRICK LARRY EMERSON
HENRY GEORGE WARD JR
HENSLEY RONNIE LEE
HESS THOMAS G
HESSON DANNY ROBERT
HESSON RONALD EUGENE
HICKMAN JAMES RUSH
HICKMAN ROBERT DAVE
HICKS ARCHIE EVERETT
HIGHLEY RAYMOND HOWARD
HILL WILLIAM OMER
HIVELY GUY RICHARD
HOBACK DOUGLAS EDWARD

HOFFMAN CHARLES DAVID
HOLCOMB JAMES LEE
HOLMES BILLY RAY
HOLT MERRIL
HOOVER MELVIN SYLVESTER
HOPKINS DAVID MICHAEL
HORN MICHAEL LEE
HORTON JAMES HARRISON
HOSKINS ROBERT SULLIVAN
HOSTUTTLER HERMON R
HOWELL DANNY RAY
HOWELL HAL KENT
HUDDLE CHARLES EDWIN JR
HUFFMAN EDDIE GRAY
HUFFMAN ISAAC PAUL
HUGHART ROMEY EARL JR
HUNDLEY JAMES FREEMAN
HUNT DAVID RAY
HUNT ROBERT WILLIAM
HURD CHARLES EVERETTE
HUTCHINSON KENNETH P JR
HUTCHISON ROBERT LEE
ICE WESLEY GENE
INGRAM LAFE
JACKSON HOWARD WADE
JACKSON KENNETH EDWARD
JACKSON LARRY ALLEN
JACKSON LAWRENCE ED-
WARD
JACKSON LAWRENCE HENRY
JAMES WASHINGTON L
JAMES WILLIAM CALVIN
JESSEE SAMUEL RICHARD
JOHNSON EVERETTE R
JOHNSON GEORGE FRANKLIN
JOHNSON JOHNNY MALCOLM
JOHNSON THOMAS ALAN
JOHNSON WILBERT HERSHELL
JONES BENJAMIN ALLEN
JONES BERNARD FRANCIS
JONES DELMER R
JONES HOWARD WILLIAM
JOSH WHYLEY E
JOY WILLIAM CHARLES
KARICKHOFF WILLIS ARNOLD
KAUFFER WILLIAM THOMAS
KEATON DAVID ROGER
KEENER JAMES LEE
KEITH MASON ALAN
KENNEDY LARRY SCOTT
KERNS FRED MICHAEL
KERR EVERETT OSCAR
KESLING RAYMON DALE
KIDD WAYNE HUFFMAN
KINCAID PAUL EDDIE
KING FLOYD D SR
KING JAY WILLIAM

KING LEWIS MILTON JR
KING ROBERT LEON
KINNEY DAVID WASHINGTON
KISER DAVID BUTLER
KITTLE CECIL WILBERT JR
KLEIN JACK WEBB SR
KLUG JOSEPH RONALD
KLUG PAUL FRANCIS
KNAPP MARTIN C
KNIGHT WALTER GRANT
KNISELY ROBERT LEE JR
KOERNER FRANK MICHAEL
KOVAC DAVID ALLAN
KOVAL ROBERT GARY
LANE GLENN MCARTHUR
LANHAM DONALD GENE
LANTZ CHARLES WESLEY
LARGENT WILLIAM ALAN
LAUCK HARRY ELMER
LAUZON LAWRENCE JOHN
LAW JAMES NEWTON
LAWRENCE WILLIAM AUBREY
LAZEAR ROBERT LEROY
LEACH EARL GENE
LEACH LARRY KEITH
LEACH RAY ALLEN
LEGG ROGER DALE
LEMONS ROBERT LEE
LESTER EDWARD
LESTER WILLIAM WAYNE
LIKENS BOBBY DALE
LILLY CARROLL BAXTER
LILLY ROBERT C
LIPSCOMB THOMAS DELANO
LOCKE GEORGE W JR
LOCKETT JAMES EDWARD
LOCKHART FREDDIE LEWIS
LOCKHART ROBERT LEE
LONG BRIAN LEWIS
LONG RONALD JAMES
LOUGH ROBERT MELVIN JR
LOWTHER LARRY JOSEPH
LUCAS LARRY FRANCIS
LYNCH STEPHEN MICHAEL
MAHONEY JOHN MORRISON
MARCUM ERNEST DELBERT
MARPLE TERRANCE DUANE
MARSHALL DANNY G
MARTIN DARRELL G
MARTIN LARRY RAYMOND
MAYHEW ROBERT OLAN
MAYNARD LESTER EUGENE
MAYNARD RALPH
MAYO GEORGE OTHEL
MC CARTHY DAVID PAUL
MC CARTY DOUGLAS WAYNE
MC CLANAHAN CLEATUS

WAYNE
MC CLUNG JOHN AMBROSE
MC CLUNG RONALD OLIN
MC CORD DAVID PAUL
MC CORMICK RONALD LEE
MC CROBIE GEORGE EDWARD
MC DANIEL ROGER PAUL
MC DONALD DENNIS EL-
WOOD
MC DONALD JAMES MAT-
THEW
MC DONALD WILLIAM
FREDERI
MC GHEE RICHARD DALE
MC IE JOHNNY ELLIS
MC INTOSH JOHN RANDOLPH
MC KINNEY BERNARD B JR
MC KINNEY JOSEPH STANLEY
MC LAUGHLIN RUSSELL
FRANK
MC MELLON ARTHUR NELLO
MC MICKEN HARRY CARLYLE
MC MILLON JACKIE
MC NEAR TERRY LEE
MC NICOL GARY DOUGLAS
MC PHERSON WILLIAM
RICHAR
MCCARTNEY HARRY CURTIS
MCCLANAHAN TERRY LEE
MEADOWS LESTER LEE JR
MENENDEZ LEO JR
MESSENGER JAMES EDWARD
MIKELS JAMES HERBERT JR
MILAM ARLIE BROOKS
MILAM DALE E
MILLER CHARLES WILLARD JR
MILLER CHRISTOPHER A
MILLER JAMES CALVIN
MILLER JAMES RAY
MILLER JERRY RAY
MILLER RANDALL BRUCE
MILLER TERRY LYNN
MILLS FAIRLEY WAIN
MILLS JOHNNY RAY
MIRACLE DANIEL L
MITCHELL CHARLIE HOWARD
MITCHELL DANNY JOE
MOLES LEWIS DAYTON
MOLLETT CHESTER AUBREY
MOLLOHAN STEVEN P
MOORE ALAN RANDAL
MOORE LEWIS WAYNE
MOORE STEPHEN ALAN
MOORE TEDDY RAY
MOORE THOMAS WAYNE JR
MOORE WILLIAM JAMES
MOOREHEAD RONALD JOHN

MORELAND THOMAS LEE
MORGAN GARY WAYNE
MORGAN RODNEY EUGENE
MORRIS CHARLES H JR
MORRIS CHARLES RODNEY
MORRIS WAYMAN DEWEY
MOSS GARY REX
MOSSER CHARLES DENVER
MOSSGROVE ROBERT BOYD
MOYERS RICHARD LEE
MUIR JOSEPH EUGENE
MULLINS STEPHEN RALPH
MUNSEY CARL L
MURPHY HERBERT BURGESS
MYERS GRAT GELEANE
NEAL RONALD KEITH
NEEL FRANKLIN WYLIE
NELSON DANA EDWARD
NELSON HOMER DOUGLAS
NEWHOUSE GARLAND
ANDREW
NICHOLAS DAVID LYLE
NICHOLS DARRELL EUGENE
NICHOLS PHILIP LARRY
NORRIS CHARLES RAYMOND
NORRIS THOMAS ANDREW
NORTON MICHAEL ROBERT
NOSS JAMES THEODORE
NOWELL CHARLES KEITH JR
NULL WILLIAM EUGENE
O BOYLE SHIRLEY WAYNE
O BRIEN DWIGHT PRESTON
O BRIEN FRANK ANTHONY III
O SHAUGHNESSY PATRICK J
OHLER HERBERT
OLENICK JOHN DAVID
OLSON CHARLES ROBERT
PARKULO DANNY RICHARD
PARSLEY EDWARD MILTON
PATRICK TEX DELANO
PATTON BARRY MICHAEL
PAULEY MARSHALL IRVIN
PEMBERTON JAMES ALEX-
ANDER
PENNINGTON RONALD KEITH
PERITO JOSEPH
PERRY GERALD LESLIE
PERRY GORDON DEAN
PERRY LARRY BRUCE
PERSINGER ROBERT MOR-
RISON
PESIMER DANIEL
PETERS LYNN WAITMAN
PHELIX STEPHEN RAY
PHELPS RANDALL CARL
PHIFER CLYDE EDWARD JR
PHILLIPS EARL GENE
PICKETT JOHN PRICE
PIERCE ANDREW STARRETT JR
PIERPOINT DONALD EVERETT
PILSON WALLACE EDWARD
PLUMMER CHARLES DEAN
POFF ELBERT DARRELL
PORTER ARCHIE ANDREW
PORTER FRANKLIN DELANO
POWELL DAVID LEE
POWELL GEORGE RALPH JR
POWELL WAVEL WAYNE
PRATER LAWRENCE BUFORD
PRINGLE JOE HAROLD
RACEY KENZEL MEREDITH
RAGER DANA LEE
RAMEY ROY LINDSEY
RANSBOTTOM MICHAEL LEE
RANSOM ROY CARLAS
RANSON RODNEY KENT
RAPP BILLY WAYNE
RATLIFF BOBBIE JOE
RATLIFF DALLAS
RAY LANDON CLAIR
REDD CHARLES EDWARD
REDMAN SYLVESTER WILLIAM
REED LESLEY WAYNE
REGER WILLIAM LEWIS
REXROAD LOEL FRANKLIN
REXRODE JACK LEE
REYNOLDS LOUIS JAMES
RHINEHART JOSEPH LEE
RICHMOND JAMES ROSS
RICHMOND LAWRENCE
DOUGLAS
RICHMOND WILLIE BUREL
RIDER SAMUEL DEWEY JR
RIFFE CHARLES DAVID

RILEY JAMES LEWIS
ROBERTS JOHN EDWARD JR
ROBERTSON CHARLES ED-
WARD
ROBINSON LOYD EUGENE
ROBISON DAVID LEE
ROGERS WILLIAM PAUL JR
ROLLINS LEWIS CHARLES
ROLLINS WILLIAM BLAINE JR
ROMINE ROGER DALE
ROSE GERALD BRUCE
ROSS DON LEWIS
ROSS DONALD EDWARD
ROSS ELMER TIM
ROSS SAMUEL
ROSS WILLIAM ROBERT JR
ROWSEY RONALD DUANE
SABLE BROOKS EDWARD
SACKETT DAVID LEE
SALERNO PAUL LOUIS
SALISBURY ROBERT JAMES
SALZARULO RAYMOND PAUL
JR
SAMPLES HERBERT CLEVE-
LAND
SANDERS DARRELL W
SANDS OKEY LEE
SANFORD JACKIE WILLARD
SARGENT JAMES RAY
SAUNDERS GEORGE THOMAS
JR
SAUNDERS LOWELL RAY
SAWYER JAMES HOWARD
SAWYERS CHARLES DOUGLAS
SAYLOR WAYMOND ANDREW
SCARBERRY LARRY DALE
SCARBOROUGH JACK WADE
JR
SCATES CHARLIE KENNETH
SCHAFER JOSEPH RICHARD
SCHNABLY DONALD FRANCIS
SCHUBERT GARY EDWARD
SCRAGG BRUCE HASSELL
SEE OTTO WILLIAM
SELL JOSEPH WAYNE JR
SHAFFER EDDIE LOU
SHAFFER RANDALL DALE
SHAMBLIN KENNETH WAYNE
SHAMBLIN THEODORE
SHANNON GARRY MONZEL
SHELLEMAN KENNETH HYDE
SHELTON ARTHUR DAVID
SHERMAN REX MARCEL
SHIFLETT DAVID HENRY
SHINGLETON THEODORE JR
SHUMATE NILE DEAN
SHUPE HERBERT CARSON

SIMMONS JAMES ROBERT
SIMMONS ROBERT LEE
SIMONS EDWARD JUNIOR
SIMPSON GERRY GLEN
SIMPSON ROGER LEE
SINE HARRY RICHARD JR
SINNETT ALBERT MERREL
SISLER WILLIAM DOUGLAS
SLATE DONALD ANTHONY
SLAVENSKY JOSEPH JR
SMITH BILLY JAKE
SMITH HOMER LEROY
SMITH JAMES LEROY
SMITH JOHN DARRELL
SMITH JOHN GERDES
SMITH JOHNNY WILLIAM
SMITH LAWRENCE LEON
SMITH ROBERT LEE
SMITH WILBUR ALLEN
SMITH WILLIAM EUGENE
SNIDER HUGHIE FRANKLIN
SOMERS GENE WILLIAM JR
SPENCER DEAN CALVIN III
SPENCER KENNETH CLINTON
SPROUSE LEE ROY DAVID
SPURLOCK LON ARNOLD II
SQUIRES ROY BENJAMIN
STAMPER DAVID HIRAM
STARCHER EDDIE DEAN
STATES WILLIAM CODAR
STATON RODNEY DALE
STEPHENS MICHAEL EUGENE
STEPHENSON ROBERT CLAY-
TON
STEPP JOHN PAUL
STEVENSON WILLIAM MAYO
STEWARD FLOYD LESTER
STEWART ARNOLD LEE
STEWART BYRON DUNCAN
STEWART EDWARD LARRY
STILLWAGONER RALPH
LLOYD
STOTLER RAY W
STOUT EUGENE EDWARD
STOVER SHELBY DEAN
STOWERS JOE D
STRAFACE JEFFREY DENNIS
STRAHIN ARTHUR RONALD
STRICKLAND JOHN LEE
STUART MARVIN BLAIR JR
STUTLER DANIEL WAYNE
STYERS REID TYRONE
SUMMERFIELD SAMUEL REED
SUMMERS EUGENE C
SWIGER HARRY RAY
SWIGER RICHARD JACKSON
TALLMAN DANIEL FERREL

TARASUK VICTOR
TATE ROBERT ARNOLD JR
TAWNEY GARY WAYNE
TAYLOR EMORY LE ROY
TAYLOR GORDON LEE
TAYLOR GROVER R
TAYLOR JAMES LAWRENCE
TAYLOR ROBERT ELWOOD
TAYLOR ROBERT THOMAS
TAYLOR RUSSELL ALLEN
TEMPLETON DAVID LEE
TENNANT JOHN RANDY
TERRY THOMAS L
THOMAS BILLY DEAN
THOMAS JAMES CARL
THOMAS JOHN DERRAL
THOMAS MICHAEL DALE
THOMAS WILLIAM ARCHABLE
THOMPSON GEORGE WINTON
THOMPSON JAMES
THOMPSON ROBERT DEWEY
THONEN JAMES LEO
TICHNELL KENNETH EUGENE
TOMBLIN TROY FRANKLIN
TOWNSEND DELMAS SHER-
WOOD
TUCKER DANNY EUGENE
TUCKER GEORGE LESLIE JR
UMSTOT SAMUEL GILMORE JR
UNDERWOOD WATSON JR
URBAN PAUL RICHARD JR
VALENTINE JOHN WESLEY
VAN METER JAKE HAROLD JR
VANOVER CHARLES LINDON
SR
VICKERS BILLY JOE
VILONE PHILIP JR
VINCENT RICHARD ELS-
WORTH
WADSWORTH CHARLES
DAVID
WALKER ROY LEACY
WALLACE HOBART MCKIN-
LE JR
WALLACE WENDELL LEVERN
WALLS THURMAN TRACY
WALTERS WILLIAM PORTER
WARE CECIL Y
WARE OLIVER ARGYLE
WAYMIRE JACKIE L
WEARS JAMES CRAIG
WEBB DONALD
WEEKLEY GARY WAYNE
WEISSMAN VICTOR BARRY
WELLS MICHAEL ALONZO
WHARTON WAYNE ALLEN
WHEELER BOBBY LEE

WHEELER WILLIAM TIMOTHY
WHITE CHARLES FRANKLIN
WHITE COLEY PHILLIP
WHITE DANNY CARL
WHITE GARY RICHARD
WHITE TERRY ROGER
WHITLATCH WILLIAM CARL
JR
WICKHAM DAVID WALLACE II
WILLIAMS JOHN RAY
WILLIAMS RONAL LOYD
WILLIAMS TROY BYRON
WILLIAMSON CHARLES
ALTON
WILLINGHAM WILLIAM EARL
WILMOTH LEWIS DIXON
WILSON ARTHUR JR
WILSON MICHAEL JACK
WILSON WILLIAM WAYNE
WINCHELL CHESTER A JR
WINES THOMAS LOWELL
WITHROW PAUL RICHARD
WOLFE ALFRED MELVIN
WOLFE DAVID NORMAN
WOOLRIDGE THORNTON
LEWIS
WOOTEN JOHN WESLEY
WOOTON GARY LEE
WORKMAN GLENN ROGER
WORKMAN JAMES EDWARD
WORLEY STEPHEN MICHAEL
WRIGHT CHARLES HERMAN
YATES CHARLES LEONARD
YOHO KERMIT HAROLD
YOUNG JACK BERNARD
YOUNG ROGER DUANE
ZELASKI LEONARD JOE JR

Photos provided by The Highground

Wisconsin Vietnam Veterans Memorial

7031 Ridge Rd, Neillsville, WI 54456

The Wisconsin Vietnam Veterans Memorial Project, Inc., also known as The Highground, is located near Neillsville, Wisconsin. A 155-acre manned veterans memorial park that pays tribute to the dead, and honors the survivors, their service, and their sacrifices. The park is open 24/7/365. Besides the Vietnam memorials, the park also contains tributes for WWI through the present, Wisconsin Persian Gulf War. The Highground continues to be a focus of healing for all who come, regardless of the name of the battle which left the scars.

Specific to Vietnam veterans, there are several tributes including: "Fragments"; the National Native American Tribute; Fountain of Tears includes a Vietnam GI; a Vietnam Nurse; the study of the "Three Soldiers"; and the Effigy Dove Mound. A future tribute is the Military Working Dog Tribute based on Vietnam dogs/handlers.

The expansive memorial was started through the efforts of Wisconsin Vietnam veteran, Tom Miller. As Tom held his dying buddy on the battlefield, Miller knew that it was his duty to make sure his friend and other service members didn't die in vain. In 1983, he teamed up with the Wisconsin delegation of Vietnam Veterans of America to make his dream a reality. With their help, Miller was able to turn The Highground into a memorial park to honor the men and women in the US armed forces with a mission of healing and education.

There are several monuments dedicated to the Vietnam veterans at The Highground. "Fragments," was created by sculpture Robert Kanyusik, dedicated in 1988 and is the first veteran's tribute in the US to include a woman in the statuary. She bears the weight of the names of all the Wisconsin service personnel who gave their lives in Vietnam. The 1244 names are etched on the bronze bundles which are interspersed with wind-chimes.

Many visitors completely miss the Earthen Dove Effigy Mound in the valley below Fragments, designed by David Giffey. The Dove honors Prisoners of War (referred to as POWs) and those who remain Missing in Action (referred to as MIAs). A monument made to honor the women who served in Vietnam was designed by Roger Brodin and is one of the first memorials to commemorate the service of the many women who risked their lives for their nation.

The National Native American Vietnam Veterans Memorial was created by Harry Whitehorse, dedicated in 1995, as the first National Memorial to come to The Highground. The study of the "Three Soldiers", Artist Fred Harp, is the Learning Center's signature tribute. It stands proudly in the Gallery dedicated in 2011.

The Names of those from the state of Wisconsin
Who Made the Ultimate Sacrifice

ABBOTT PAUL DENNIS
ABRAMS JOHN LEON
ACKERMAN ROGER CARL HENRY
ADAMS CLARENCE CLIFTON
ADAMS LEE CHESTER
ADAMSKI DENNIS JAMES
AGARD TIMOTHY CHARLES
AGIUS VINCENT JAMES
AHRENS JAMES JOHN
ALBRIGHT TERRY LEE
ALLARD MICHAEL JOHN
ALLEN DONALD WILLIAM JR
ALLEN MERLIN RAYE
ALLMERS ROBERT ROGER
AMMERMAN ROSCOE
ANDERSON DALE EDWARD
ANDERSON ERLING ALTON
ANDERSON GERALD ROBERT
ANDERSON JACK WILLIAM JR
ANDERSON RICHARD LEE
ANDERSON THOMAS LESLIE
ANDERSON WILLIAM JOSEPH
ANDERSON WILLIAM JR
ANDRUS WILLIAM GEORGE
ANTOINE DENNIS LLOYD
ARENS TIMOTHY GEORGE
ARMITAGE THOMAS LEON
ARNOLD ROBERT DWAIN
ARNOLD WILLIAM TAMM
ARRIES JAMES MICHAEL
ARTEAGA JOHN J
ASHBURN JERRY ALLEN
ASHER JAMES LOUIS
ATKINSON ROGER CARL
BABCOCK DENNIS LEE
BABEL DWIGHT FABIAN
BABICH NIKOLA
BACH LYMAN CONRAD
BAHRKE RUSSELL LEROY JR
BAIR ROBERT VOLNIE
BAITINGER DAVID JAMES
BAKKEN WILLIAM DONALD
BALDWIN ORVAL ARTHUR

BALDWIN ROBERT LANOUE
BALTHAZOR RICHARD JOHN
BANASZYNSKI RICHARD MICHA
BANKS DAVID LENOX
BANNACH GERALD JOSEPH
BANOVEZ MICHAEL JOSEPH JR
BARANCZYK ALBIN ANTON
BARNES DAVID GREGORY
BARR ROBERT HOWARD
BARTELME MICHAEL PAUL
BARTHOLOMEW RICHARD D JR
BARTKOWSKI GREGORY JOSEPH
BAUER LEONARD WILLIAM
BAUMGART ROBERT LEE
BEARWALD ORLAND ORRIN
BEATY JEFFREY LANDIS
BECK ROBERT JAMES
BECK ROBERT MILTON
BECK TERRENCE DANIEL
BECKER JOHN PAUL
BECKER THOMAS LEWIS
BEDFORD WILLIE
BEGOTKA JOSEPH LLOYD
BEHLKE GERALD DENNIS
BEHNKE RICHARD CARL
BEHRENS WILLIAM CHARLES
BEHRENT MARK SYLVESTER
BEILFUSS EDWARD ALAN JR
BEITLICH JOHN WILLARD
BELLILE WILLIAM MARVIN
BELONGER DENNIS MICHAEL
BENDER LARRY WARREN
BENDORF DAVID GLEN
BENEDICT ROBERT JOHN
BENICEK JAMES MILTON
BENISHEK FREDERICK LEE
BENNETT DAN MICHAEL
BENSON GERALD ALLEN
BENWAY JAMES DWIGHT
BERANEK CHARLES SYLVESTER

BERANEK DEAN MITCHELL
BERGER RAYMOND REX
BERKHOLTZ LARRY WAYNE
BERTSCHINGER DENNIS LEE
BERWEGER ALLAN FREDERICK
BEST PATRICK WALLACE
BEYER WILLIAM ARTHUR
BIEHL LESTER OSCAR JR
BIEROWSKI REINER WALTER
BILLIPP NORMAN KARL
BILMER KRIS
BINDER PAUL LAROY
BJERKE GLEN ALLEN
BLACK NOLAN EUGENE
BLACKMAN THOMAS JOSEPH
BLAESE RONALD PAUL
BLAHA THOMAS JOHN
BLANK ROBERT GERDES
BLAVAT JAMES NORBERT
BLEXRUDE GORDON HARRY
BLOHM RONALD ROY
BLUMER KRIS
BOEHLER JAMES LEONARD
BOEING RONALD FRANK
BOETTGER TERRI MARTIN
BOHMER ROBERT JAMES
BOHRMAN MICHAEL DENNIS
BONNEAU DEAN LOUIS
BOOTH WALTER CLAY
BORZYCH DAVID RUSSELL
BOWERS RICHARD LEE
BOZINSKI JOHN MICHAEL
BRADLE JAMES DENNIS
BRANDES THOMAS GLENN
BRANNON JAMES EDWARD
BRANTMEIER BERNARD GEORGE
BRATZ WAYNE ALLEN
BRAUN MICHAEL WILLIAM
BREDESEN DAVID JOHN
BREFCZYNSKI EDWARD JOSEPH
BRENWALL KENNETH WAYNE

BREUER ANTHONY JOSEPH
BRINES GERALD RAYMOND
BRIXEN GARY MAURICE
BRODHAGEN FREDERICK HAROL
BRONKEMA JOHN MITTCHEL
BROOME THOMAS EDWARD
BROWN DENNIS LEE
BROWN EDWIN FAY
BRUNKE RICHARD JOSEPH
BRUNNER GARY EDWARD
BRUNNER MICHAEL JAMES
BRUX GARY H
BULIN JERRALD JOSEPH
BULKLEY DAVID JUSTUS
BUNGARTZ FREDERICK WILLIA
BURBACH RICHARD
BURBEY EUGENE LEROY
BURGERT ROBERT
BURLINGAME WYNNE LEONARD
BURNS THOMAS RAYMOND
BURR DANIEL LEE
BUSCHKE JOHN ALLEN
BUSH ROBERT IRA
BUSS ROGER LEE
BUTLER LAWRENCE JOSEPH
CALTON DENNIS ARNOLD
CALVERLEY ANTHONY GEORGE
CAMPBELL GIOVANNI HENRY
CAPELLE GERALD CARL
CAPEZIO FRANCIS JOHN
CARLSON PETER JOHN
CARR GEORGE JOSEPH
CARROLL PATRICK JOHN
CARSTENS THOMAS JAMES
CARTER RICHARD KENNETH
CASPER FREDERICK RAYMOND
CASPERSEN ROBERT P II
CERRA RICHARD RALPH
CHAMBERLIN DENNIS DEAN

CHAPMAN GEORGE ANTHONY
CHAPMAN JOHN THOMAS
CHATOS WALTER ALEX JR
CHICANTEK ANDREW JAMES
CHITKO BENJAMIN ALBIN
CHITWOOD WAYNE CECIEL
CHMIEL MARK ANTHONY
CHRIST DONALD ALFRED
CHRISTJOHN PAUL EMERSON
CHURCHILL RAYMOND JOHN
CLARK PHILLIP HENRY
CLEEREMAN DAVID FRANK
CLOUTIER ROBERT LOUIS
CONNELLY PATRICK ALLEN
COREY JAMES ALLEN
CORK RAYMOND LEE JR
COTTER KENNETH JAMES
COTTRELL THOMAS LEWIS
CRABB BRUCE WAYNE
CRAMER DAVID ARTHUR
CRANE DEAN DENNIS
CRAWLEY ROBERT LEO
CRESS TOM JOSEPH
CROOK THOMAS HARRY
CROWLEY CARL LESLIE
CUNNINGHAM KENT ALAN
CUNNINGHAM WILLIAM LEDFOR
CURTIS THOMAS MICHAEL
CWIKLA LEROY WALTER
D AGOSTINO JOHN R JR
DAANE DOUGLAS JACK
DAHL JAMES STEPHEN
DAHL TIMOTHY ALLEN
DAHLMAN GEORGE CLARENCE
DAHM RALPH ALBERT
DAILEY GEORGE FREDERICK
DALBERG DEAN LAVERNE
DAMM THOMAS WILLIAM
DANIELSON LEE ROGER
DANN DAVID BRIAN
DAVIDSON CHARLES LEON
DAVIDSON MICHAEL JOHN
DAVIES ROBERT JOHN
DAVIS RICHARD LLOYD
DAWSON CLYDE DUANE
DAZEY THOMAS FRANCIS JR
DE BOCK JOHN ALBERT
DE CORA ELLIOTT LEO
DE GALLEY JEROME ANTHONY
DE GRAY JERRY FREDERICK
DE LAAT DAVID WILLIAM
DE LANGE JACK PETER
DE LONG JERALD STEVEN
DE WINDT CHARLES ROSS
DEAN ALBERT
DEAN GLENN FREDRICK
DECKER ROBERT HUGH
DEFENBAUGH FRANKLIN D
DEL CAMP ADRIAN LEROY
DEMATA BRUNO WALTER
DEMOE RAYMOND ROGER
DENTON ARTHUR GERALD
DERBY PAUL DAVID
DEUEL CHARLES FRANK
DEVNEY JAMES ROBERT
DEXTER RICHARD AUGUSTINE
DIBB STEPHEN KEITH
DICKENSON LLEWELLYN PAUL
DIEDRICH JAMES NICHOLAS
DIEDRICH ROBERT JAMES
DINGELDEIN DONALD GLEN
DOBISH JAMES THOMAS
DOBOSZ DAVID GEORGE
DOBRENZ LAWRENCE CARL
DOBRINSKA THOMAS EARL
DOLL JEROME NORMAN
DOLLAR EUGENE DOYCE
DONAHOE DAVID JOHN
DONSTAD JAMES MARVIN
DOPP GARY RUSSELL
DOWLING JEAN PIERRE

Wisconsin

DOWNING DONALD WILLIAM
DRAEGER WALTER FRANK JR
DRAPP ROBERT GEORGE
DREA TERRANCE LEE
DROUGHT DAVID LEE
DU LONG FRANKLIN ROOS-
EVEL
DUELLMAN HENRY RALPH
DUMKE ALLEN WILLIAM
DUNBAR ROY WILLIAM JR
DUNIFER DELFERD BENJAMIN
DUTCHER LEONARD EARL
DWYER DALE DON
DZIEDZIC MARK ROBERT
EBERHARDT PHILLIP JOHN
EDELSTEIN ROY L
EIDEN EDWARD VALENTINE
JR
EITEL DENNIS
ELLIS ROGER ALLEN
ENGEL ALLEN NORBERT
ENSSLIN OTTO ROBERT
ERDMAN DALE ARTHUR
ERICKSON MARVIN LE ROY
ERICKSON ROBERT DALE
ERTEL LOREN LESLIE
ESSMANN ROBERT CHARLES
EVANS WILLIAM ANTHONY
EVELAND JOSEPH NORMAN
FABER THOMAS WALTER
FAHRNI DALE ALLEN
FARVOUR WILLIAM HAROLD
FEDDER FRED ANDERSON
FEDER LLOYD ARTHUR
FEIERABEND PETER MAT-
THEW
FELCH ARLEIGH FRANCIS
FELLENZ CHARLES RICHARD
FERDIG RICHARD CHARLES
FERGUSON TED SCOTT
FICKLER EDWIN JAMES
FIEDLER GARY JAMES
FIEDLER JOHN JUNIOR
FINA RICHARD CARL
FISCHER JAMES ROBERT
FISCHER RICHARD WILLIAM
FITZGERALD DAVID BART-
LETT
FITZPATRICK JOHN DOUGAL
FLYNN ROGER JOHN
FOLEY THOMAS HAROLD
FOLKERS LA VOUGHN HER-
MAN
FOLZ GARY LEE
FORAN JOSEPH PAUL
FORTNEY KENDALL THOMAS
FRAZIER PAUL REID
FREDENBERG RALPH
FREUND TERRENCE JAY
FRICKE EUGENE MARSHALL
FRISCHMANN DAVID JOSEPH
FUCHS WILLIAM JR
FULLER EUGENE OTTO
GAHAGAN JAMES MILAN
GALLAGHER DANIEL F
GALLAGHER DONALD LOUIS
GANNON JOHN PATRICK
GARSKI KENNETH JAMES
GAUTHIER GERALD ALAN
GEARY HARRY EUGENE
GEE PAUL STUART

GEHRKE GARY BERNARD
GEIGER LARRY FREDERICK
GEIS RANDALL HAROLD
GEISE DELL CONLEY
GERCZ FRANCIS GARY JR
GERG THOMAS ARTHUR
GERLACH PAUL EDGAR
GEROU JAMES ALLAN
GIEJC ALEXANDER
GIESE MICHAEL EVERETT
GIESEN WALLACE LEE
GILBERTSON RICKY MARK
GILBERTSON TERRY ALAN
GILLETT JERRY CECIL
GLUECKSTEIN WILLIAM
ROBER
GMACK JOHN ROBERT
GODFREY WILLARD ANSEL
GOLDEN RONALD DUANE
GOLDSMITH ROGER DWIGHT
GOMEZ FRANK
GONZALEZ WILFREDO LOUIS
GOODNESS KENNETH GLEN
GORGES RICHARD JOHN
GORSUCH WILLIAM DALE
GOSLIN GREGG MICHAEL
GRADECKI GLENN RICHARD
GRASSL KENNETH JOSEPH
GRAY HARVEY DUNCAN
GIBNEY ALLEN RICHARD
GRAY JOHN PATRICK
GREBBY ROBERT WILLIAM
GREEN DENNIS JOSEPH
GREENWOOD BRUCE JOHN
GREETAN ROGER WILLIAM
GREGORASH LON PAUL
GREGORIUS MICHAEL JON
GREIGER DONALD LEONARD
GREINKE NEIL NORMAN
GREISEN THOMAS ANDREW
GRENIER RONALD LOUIS
GRESHAMER LEON G
GRESKOWIAK ROBERT
GRIMES CARL WAYNE
GROFF DENNIS ALLEN
GROSS JOHN ALBERT
GROTZKE ALLEN FREDERICK
GRUBER MICHAEL ALFRED
GRUDZINSKI WILLIAM
THOMAS
GRUNEWALD JEROME E
GRUSCZYNSKI EDWARD ROY
GUDEN THOMAS CHARLES
GUDLESKE GUSTAVE FRANK-
LIN
GUELIG PAUL JOSEPH
GUEX BRUCE JOHN
GUILETTE LEONALD GEORGE
GUKICH MICHAEL MARTIN
GUNDERSON JAMES JOHN
GUNDERSON MELVIN WIL-
LARD
GUSTAFSON DENNIS RUSSEL
GUTKE RONALD LAWRENCE
HAAKENSON KENNETH
WAYNE
HAAS KENNETH DANIEL
HAAS MAURICE JOHN
HAAS RUSSELL CARL
HACKETT JAMES FRANCIS JR
HAGEN JAMES ROBERT
HAGEN RONALD JAMES
HAGUE GERALD CHARLES
HAJMAN PETER OSCAR
HALFMAN BLAKE HENRY
HALLBERG CARL RAYMOND
HALVERSON ALVIN LEONARD
HALVERSON GARY JOSEPH
HAMBLETON HARRY B III
HAMLET JAMES LEWIS
HANSBROUGH LYLE CLEVE-
LAND
HANSEN STANLEY RAYMOND
HANSON WILLIAM HENRY
HARFF WILLIAM HENRY JR
HARRIES JIMMY RIED
HARRIS JACK HAROLD
HARTEAU JAMES PETER
HARTER DENNIS MICHAEL
HARTMAN BRUCE BRADLEY
HARTZHEIM JOHN FRANCIS
HARVEY MICHAEL ANTHONY
HASKO RONALD JON
HASZ ROBERT LEE

HAUKENESS GLENN SHELDON
JR
HAUPT RONALD JOHN
HAUSCHULTZ JERRY LEE
HAUSWIRTH GERALD RICH-
ARD
HAWLEY KENNETH BRUCE
HAYES PATRICK JOHN
HAYES WAYNE MICHAEL
HEIDEN CARL WILLIAM
HEIDER ANDREW L
HEIDER WILLIAM STEPHEN
HEINECKE RONALD MATHIAS
HEINRICH GREGORY ALLEN
HEINZ DONALD E
HEINZ PAUL WALTER
HEISER ROBERT ALLEN
HELLENBRAND DAVID PETER
HELMICK DUANE A
HENKE KENNETH LEE
HENKE VERNON LEE
HENNING ARTHUR ROBERT
HENTZ RICHARD JAY
HERFEL LAURENCE JOHN
HERING ROBERT HENRY
HERMSEN MICHAEL EDWARD
HERNANDEZ JOSEPH J JR
HERRERA FRANK VINCENT
HESSING JAMES WILLIAM
HEUER JERRY WAYNE
HEWITT RALEIGH L II
HEYNE RAYMOND THOMAS
HICKEY JOHN JOSEPH
HIERLMEIER DONALD ALVIN
HILDEBRANDT JAMES

GEORGE
HILL CHARLES HERMAN
HILL WILLIAM RAY
HIPKE HARRY ALLAN
HODGE MICHAEL ALLARD
HOFFMAN LEROY DAVID
HOGLE RICHARD LEE
HOLTZ ALFRED JOSEPH JR
HONDEL WILLIAM JAMES
HORN ALEC HENRY
HORTON RUBEN LEE
HOULE DANNY WILLIAM
HOWARD GENE JAY
HOWIE LLOYD GEORGE
HREN TIMOTHY LOUIS
HUDIS JAMES BRIAN
HUGHES CHARLES FREDRICK
HULBERT JOHN ROY
HURKMANS WILHELM S JR
HUSS ROY ARTHUR
IMRIE JOHN CHARLES
ISAACSON GARY ALLEN
JACKSON RICHARD THOMAS
JACKSON TODD R
JACOBS VERNON DUANE
JACOBSON HARVEY GEORGE
JACOBSON MARK NELS
JAECK RICHARD ELMER
JAHNKE RONALD EDWARD
JAJTNER RAYMOND CHARLES
JANKA WILLIAM ROBERT
JANKE CHARLES JULIUS
JANKE KEITH BRIAN
JARDINE ROBERT AXEL JR
JAVA DANIEL M

JELICH JOHN ANTHONY
JENKINS CLAYTON DEAN
JENSEN ALAN THEODORE
JENSEN LARRY SCOTT
JENSEN RICHARD
JOHNSON CHARLES ALLEN
JOHNSON CHARLES LEO
JOHNSON DONALD PETER
JOHNSON GARY ALAN
JOHNSON LELAND CRAIG
JOHNSON RANDOLPH LEROY
JOHNSON RONALD PETER
JOHNSON ROSS ARNOLD JR
JOHNSON TIMOTHY HOLTON
JOHNSTONE KENT LEROY
JOOSTEN ROBERT WALTER
JORDAN JEFFREY ROBERT
JUDKINS LARRY DUANE
JUNK RICHARD HENRY
JURGELLA JOSEPH PETER
KAKUK ALLEN JOHN
KALHAGEN PHILIP ALFRED
KANAMAN KENNETH HARVEY
KARPENSKE DALE RODNEY
KASTEN DANIEL MARK
KEARNS JAMES THOMAS
KEEFE DENNIS WRIGHT
KELLENBENZ BARRY CHARLES
KELLER KENNETH LEE
KELLEY MICHAEL JAMES
KELLY ERIC STEVEN
KELPINE RANDALL WAYNE
KENNEDY TIMOTHY JOEL
KESSELHON JAMES EDWARD
KESSINGER JOHN MC FAR-

256

LAND

KETTERER JAMES ALAN
KIELPIKOWSKI RONALD LEE
KIES DAVID F
KIHL PATRICK JAMES
KIMPEL PHILIP JOHN
KINK DAVID ROBERT
KIRCHMAYER ANDREW GREGORY
KIRCHNER JOHN WARD
KIRKHAM DONALD ALAN
KISSINGER NORMAN CHARLES
KITZKE RONALD FREDERICK
KLAVES JEFFREY JOHN
KLEMP THOMAS JOHN
KLEPPIN KENNETH THOMAS
KLEVER MARK EDWARD
KLIMPKE DENNIS LEE
KLOC JOHN THOMAS
KMETZ DAVID WILLIAM
KNORR JOHN ROY
KNOUSE DAVID WALTER
KNOX BRUCE NEAL
KNOX JAMES RICHARD
KNUTSON EARL WILLIAM JR
KOCH DARRYL JAY
KOEHN BRIAN ROBERT
KOHLBECK TERRENCE EUGENE
KOHLBECK VICTOR JOSEPH
KOKALIS NICK
KOPKE ROGER JOSEPH
KOSSOWSKI DAVID STANLEY
KOSTKA ROGER JOSEPH
KOSTROSKI MARVIN DAVID

KOTNIK WILLIAM MAX
KRAMER DOUGLAS LEE
KREBS JOHN THOMAS JR
KRECKEL JOHN WILLIAM
KRESIC JOSEPH JR
KREUZIGER ROBERT ALAN
KRITZ EUGENE RICHARD
KROMREY DENNIS JOHN
KROPIDLOWSKI GERALD
KRUEGER CHARLES WILLIAM
KRUEGER DEAN WILBUR
KRUEGER DUNCAN FREDERICK
KRUEGER WAYNE DALE
KRUG RAYMOND HENRY JR
KRUMBINE LEO FREDERICK
KRZMARCIK JOHN EDWARD
KUBE JOSEPH BERNARD
KUBLEY ROY ROBERT
KUCZEWSKI WILLIAM ROBERT
KUEHL PAUL DAVID
KUHNLY GERALD LOREN
KURTH JAMES PETER
KURZ SIDNEY ALLEN
LA DUKE REX ALFRED
LA HAYE JAMES DAVID
LACAEYSE LARRY GENE
LAHNER THOMAS ALLAN
LALAN LARRY RALPH
LAMARR WALTER LOREN
LANG WILLIAM OTTO
LANGE KARL FERDINAND
LANGENFELD CHRISTIAN ALAN
LANGER MICHAEL WALTER

LARSEN THOMAS CHARLES
LARSON JAMES EDWARD
LARSON RANDOLPH LOUIS
LAST DONALD ROY
LAUX MICHAEL DEANE
LAWSON RODNEY JOHN
LAYTON STEVEN JAMES
LE BOSQUET CHARLES R
LEAHY JAMES ALEXANDER
LECHNIR PETER GERALD
LEDDEN TERRANCE EDWARD
LEDEBUR MICHAEL T
LEDEGAR RUSSELL OLE
LEDERHAUS DONALD HERMAN
LEDIN JAMES LARS
LEET DAVID LEVERETT
LEFEBER WAYNE ROBERT
LEICHT ROMAN HENRY
LEINDECKER LARRY JAMES
LEIS JOHN EUGENE
LEISING BRUCE CHARLES
LENZ JAMES WARREN
LEON MARIO ROBERT
LEONARD RICHARD JAMES
LEPAK DONALD CHESTER
LEWIS RAYMOND ROY
LEX MICHAEL EDWARD
LIBERSKY WILLIAM BERTRAM
LIEBHABER KENNETH GEORGE
LIEBNITZ JAMES TERRY
LIND JAMES JEROME
LINSKI THEODORE PAUL
LINTON LEE ROY EDWARD

LITKE JEROME WALTER
LODUHA GARY
LONGMIRE KENT WILLIAM
LOPEZ MANUEL TORRES
LORENZ TERRY WAYNE
LOY JAMES RICHARD
LUBENO JEROME DEANE
LUCIANI LAWRENCE ANTHONY
LUDVIGSEN LEO JOHN JR
LUKE JOHN ALBERT
LUND TERRY BRUCE
LUPE EUGENE KENNETH
LUTHER CLAYTON J
LUTZ ROBERT STEVEN
LUTZKE MICHAEL JON
LYNN JAMES E
LYONS PATRICK NICHOLAS
MADDUX DAVID THORNTON
MADISON THOMAS VERNON
MAGER VINCENT LEO
MAHNER LIN ALBERT
MAHONEY RONALD J
MALUEG MICHAEL PAUL
MANN ROBERT JAMES
MANSKE DENNIS RUDOLPH
MANSKE PAUL EDWARD
MANTHEI JAMES WALTER
MANZ TERRY LEE
MARKEVITCH ANTHONY G JR
MARSHALL JOHN KEITH
MARTIN DENNIS ROBERT
MARTIN TERRY LYNN
MARTIN VERNAL GLEN
MARTIN WILLIAM PAUL
MARTINEZ WILLIAM JOSEPH
MASON DENNIS RAE
MATSON HOWARD V JR
MATUSEK JOEL ALOIS
MATUSH THOMAS ERWIN
MATYAS RICHARD EDWARD
MAZURSKY BERNARD RICHARD
MC CALVY JAMES A
MC CANN WILLIAM GEORGE
MC CARTNEY ROBERT ALLEN
MC CLURE PATRICK RYAN
MC CONNELL JAMES PAUL
MC CORMICK DENNIS LEE
MC GEE DARRELL EUGENE
MC GILVARY DANIEL J JR
MC GREW LLOYD ARTHUR
MC GURTY TIMOTHY ARTHUR
MC KEE RICHARD CHARLES JR
MC LEISH CHARLES EDWARD
MC MASTER ROBERT PAUL
MEAD JEFFERY EVANS
MEE RANDALL ALAN
MEENAN THOMAS JAMES
MEHNE RICHARD ALLEN
MEIDAM THOMAS LAWRENCE
MEINECKE WILLIAM FREDERIC
MEINEN BERNARD PHILLIP JR
MELLON MICHAEL OWEN
MELTON TODD MICHAEL
MENGEL KENNETH RAYMOND
MERCIER JOHN CHARLES
MERRILL DAVID B
MERRIWEATHER NATHANIEL
MESICH MICHAEL STEPHEN
MEYER JAMES FREDRICK JR
MEYER LEO ROLAND
MEYERS VICTOR BERT
MEYSEMBOURG DANIEL LLOYD
MEZERA TERRY FRANCIS
MICHALSKI STEVEN
MICHAUD RONALD ALBERT
MIETUS JOHN ANDREW
MIHALOVICH JOHN MICHAEL
MILADIN RAYMOND EDWARD
MILANOWSKI JOHN EDWARD
MILLARD LOREN RAY
MILLER ANDREW J
MILLER KEITH NORMAN
MILLING LARRY DEAN
MIRACLE GARY RAYMOND
MITCHELL JAMES STEPHEN
MITCHELL THOMAS C
MOE HAROLD JOHN
MOLKENTINE RANDY WARREN

MONSON PHILLIP DEAN
MOODY ANDREW LESLIE JR
MOORE KEVIN WALKER
MORACK GENE CHARLES
MORAN MICHAEL PETER
MORGAN LARRY HAROLD
MORIARTY JAMES MICHAEL
MORNEAU JEROME DALE
MORRIS JOHN D
MORRISON EDWARD ARNOLD
MOSHER LARRY CLARENCE
MOUSEL WAYNE CHARLES
MUELLENBACH ROBERT JOSEPH
MUELLER JOSEPH BERNARD
MUELLER MARCO FRANCISCO
MUELLER RANDY ROY
MUELLER ROBERT GILBERT
MUELLER TOM RICHARD
MULCAHY MICHAEL LEE
MULDER RUSSELL WESLEY
MULVEY FRANCIS TRAINOR
MURPHY THOMAS JOSEPH
MURPHY TIMOTHY JAMES
MURPHY WILLIAM HENRY III
MURRAY GEORGE THOMAS
MYERS CHESTER ARTHUR JR
NAGEL GORDON LAVERN
NATZKE NICHOLAS LEE
NAUERTZ RANDALL KEITH
NELSON ORLAN MARVIN
NELSON VERNON LEE
NESOVANOVIC JOHN LAURENCE
NETTESHEIM BRUCE PAUL
NETZOW EARL JEFFREY
NEUBAUER JAMES R
NEUMANN MARK WILLARD
NEUMEIER TERRY JAMES
NEWMAN GREGORY EUGENE
NICHOLS JERRY ALLEN
NICHOLSON LARRY JAMES
NICKERSON GILBERT RONALD
NICOLAI RUSSELL CHESTER
NIELSEN MICHAEL CHARLES
NILES RONALD ROBERT
NOEL DONALD WILLIAM
NORMAN TIMOTHY JOHN
NORTH JOHN ALEX
NOTH WAYNE LOUIS
NOVAKOVIC GEORGE D
NOVOBIELSKI DUANE ANDREW
NOWAK LEONARD MICHAEL
NOWAKOWSKI GLENN EDWARD
O HARE RICHARD JAMES
OERTEL LARRY HUGH
OESTREICHER PAUL ANTHONY
OLDFIELD CARL EVERETT
OLMSTED JEROME EDWIN
OLSEN GREGORY JON
OLSON CRAIG SEIMON
OLSON DUANE VIRGIL
OLSON GARY WAYNE
OLSON RODNEY JAMES
ORLIKOWSKI DANIEL JOHN
ORLOWSKI CHARLES FRANCIS
OSUSKI ROBERT EUGENE
OTT STEVEN ROBERT
OTTE THOMAS WILLIAM
OVERBECK PHILIP MOREY
PACHE HARLAN T
PAGE JAMES HENRY
PAMANET PAUL JOSEPH
PAMONICUTT MARTIN JAMES
PAPENFUS ALLEN DARYL
PASKOWICZ DONALD
PATRICK DANIEL GARRY
PATTERSON WILLIAM ANTHONY
PAULICH PATRICK JAMES
PAULSON JOHN PAUL JR
PAVLACKY LOUIS A JR
PAWELKE RICHARD CHARLES
PAYNE TERRY JOHN
PEARSON BRADLEY ALAN
PEAT GARY LAVERNE
PEDERSON MARVIN CLIFFORD
PEDERSON ROGER ALLEN
PEREZ RICHARD ELOY
PERLEWITZ BRIAN SCOTT
PERLEWITZ STEVEN OWEN

257

Wisconsin

PERZ TERRY LEE
PETERLICH JOSEPH JAMES
PETERS DAVID E
PETERS LEE RAYMOND
PETERSON CARL ELVING
PETERSON DARWIN STUART
PETERSON DENNIS NEWELL
PETERSON JOHN ALFRED
PETERSON LOWELL TODD
PETERSON TIMM CONRAD
PFISTER DAN LEON
PHILLIPS LEON MILTON
PICKART RONALD ERNEST
PIERSON WILLIAM C III
PIERSON WILLIAM EDWIN
PINNEKER JERALD LEE
PITZER RICHARD LYLE
PLEASANT MURPHY JR
PLECITY JAMES DONN
PLIER EUGENE JOHN
PLINER RICHARD DUANE
POCHRON RONALD EDWARD
PODEBRADSKY ANTHONY
JOHN
POFF JOHN ROBERT
POKEY FRANK MICHEAL JR
POLAK PETER PAUL
POLASKI LEON CRAIG
PONATH KURT FRANCIS
PRICE DAVID MERRILL
PRICE JAMES ALLAN
PROPSON BERNARD AMBROSE
PROPSON MARVIN NORBERT
PROTHERO MICHAEL EUGENE
PROVEAUX RICHARD BLAINE
PRZYBELSKI THOMAS F
PULASKI ROBERT ALLEN
PUNDSACK TERRY LYNN
QUAST WILLY VASCILLE
QUESADA JESUS
QUINLAN DAVID PATRICK
QUIRK JEFFERY MICHAEL
RADES ROBERT RAYMOND
RADLEY LELAND EUGENE
RADONSKI KENNETH WAYNE
RADTKE ERIC RUDOLPH
RAIH ROGER WILLIAM
RANCE STEVEN PAUL
RANK DENNIS ROBERT
RATAJCZAK ROBERT EDGAR
RAUBER DALE EUGENE
RAUSCH JOHN ALEX
RAWLING BRUCE H
RAY RONALD JOHN
REAMER JAMES CHARLES
REED DENNIS WAYNE
REED JON EDWARD
REILLY RONALD HENRY
REINKE ROBERT HARVEY
REINKE RONALD RICHARD
REUTER NEIL GEORGE
RHOADES LOUIS GEORGE
RICE JAMES ROY
RICHARDSON DALE WAYNE
RICHARDSON EDMOND
WILLIAM
RICHTER DONALD JOSEPH
RICKERT ROGER ALLEN
RICKLI RODNEY HOWARD
RIEDERER CARL JOSEPH
RILEY JAMES G

RINGLE JAMES MYRON
RIOS SEVERIANO
RISCH JAMES MICHAEL
RISCHE KARL BALTHASAR JR
RITSCHARD ROGER LEE
RIXMANN EDWARD HAROLD
ROBERTS KENNETH DAVID
ROBERTS THOMAS JOHN
ROBINSON GERALD ARDEN
ROBINSON LANCE ALLEN
ROBINSON RAYMOND DOUG-
LAS
ROBSON TIMOTHY FRANCIS
ROCHA JOSE MARIE
ROE JOHN MARSHALL
ROE ROYCE EVERT
ROEGLIN WILLIAM JOHN
ROEHL ELWOOD JOHN
ROESLER RICHARD ALFRED
ROEST DOUGLAS RAY
ROGALSKE PAUL FRANK
ROHAN WILLIAM JAMES
ROMANSHEK JOHN CHARLES
ROSENOW ROBERT JAMES
ROSENOW THOMAS ARTHUR
ROSS RONALD ALAN
ROSSOW GERALD JOHN
ROUM STEVEN JEROME
ROUSE JEROME MICHAEL
ROYSTON ALAN MICHAEL
RUDOLF MARK PHILLIP
RUENGER CARL DENNIS
RUETH JOHN LEONARD
RUMINSKI ROBERT PAUL
RUNGE FRANKLIN JAMES
RUOHO JOHN RONALD
RUSH JACK RAYMOND
RUTOWSKI DENNIS DAVID
RYAN EDWARD KENNETH
RYDLEWICZ JOHN MICHAEL
SADLER JOHN WELDON
SAGEN THOMAS ALVIN
SALAMONE JAMES ALBERT
SAMZ FRANCIS MARK
SANCHEZ ROBERTO
SANDEL RONALD S
SANDERSON JOHNNIE D
SARTOR JOHN VICTOR
SCHACHTNER JAMES ALOY-
SIUS
SCHAEFER DAVID ROY
SCHAEFER KENNETH LEE
SCHAEFER ROY ANTHONY JR
SCHAEFFER GEROLD
SCHEPP DALE ALLEN
SCHETTL DAVID LEROY
SCHIESL GERALD RAYMOND
SCHLEY ROBERT JAMES
SCHLIESMAN JERROLD
JOSEPH
SCHLIEWE FLOYD ABNER
SCHLUTTER WILLIAM DAVID
SCHMID JOHN STEPHEN
SCHMID RONALD KENNETH
SCHMIDT DAVID JEROME
SCHMIDT DENNIS ROBERT
SCHMIDT LARRY ROMAN
SCHMIDT LAWRENCE ED-
WARD
SCHMIDT PETER ALDEN
SCHMITT FREDERICK
SCHMOLL JAMES KENNETH
SCHNEIDER HARRY WARREN
SCHNEIDER WILLIAM JOSEPH
SCHOLD RAY ARTHUR
SCHROEDER GARY LEE
SCHROEDER ROBERT EMIL JR
SCHUELLER JAMES PATRICK
SCHUETT JEROME ALAN
SCHUETTE DAVID FRANCIS
SCHUH ARNOLD RAYMOND
SCHUH DAVID MICHAEL
SCHULTZ DAVID CHARLES
SCHULTZ GERALD WAYNE
SCHUMACHER MICHAEL
WAYNE
SCHURRER JOHN RODNEY
SCHUSTER DANIEL CARL
SCHWARTZ RUSSELL ALBERT
SCHWEFEL DALE WAYNE
SCHWENDLER RICHARD
WILLIA
SCOTT BRUCE RICHARD
SCRIVER JAMES MICHAEL

SEARS MICHAEL
SEEKAMP ROBERT LEE ROY
SEIBERT JOSEPH DEAN
SENGSTOCK GARY DAVID
SENZ DENNIS LEON
SEVERSON DONALD JON
SEVERSON THOMAS EUGENE
SHAW THOMAS
SHAW THOMAS FRANCIS

SHELLUM JOHN CHARLES
SHEPARD JAMES MERRILL JR
SHEPHERD RONALD DEAN
SHEW DENNIS WAYNE
SHORT BARRY JAN
SIEGEL DENNIS LEE
SIEMANOWSKI DAVID ALBERT
SIJAN LANCE PETER
SIKORSKI DANIEL

SIMON RICHARD CHARLES
SINGER ALAN EDWARD
SINGERHOUSE ROBERT ALLEN
SKAAR WILBUR ARNOLD
SMEESTER DANIEL RAYMOND
SMITH DONALD LAVERN
SMITH FRED D
SMITH GARY WENDELL
SMITH JACK RUSSELL

258

SMITH JAMES LEONARD
SMITH KENNETH EUGENE
SMITH LARRY ALAN
SMITH LLOYD STEVEN
SMITH LYNN HUDSON
SMITH STEVEN JAMES
SMITH THOMAS HERBERT
SMITH TIMOTHY JOHN
SMITH WILLIAM THOMAS
SMITHWICK DAVID GEORGE
SONNENBERG ARDEN GENE
SOULIER DUWAYNE
SPENCER GENE B
SPOEHR WINFIELD AUGUST JR
SPRINGER JAMES ROBERT
SPURLEY JAMES VIRGIL JR
STANGEL LAWRENCE NOR-
BART
STECKBAUER CURTIS JOHN
STECKER DENNIS EUGENE
STECKER JOHN CHARLES
STEDL WILLIAM JOHN
STEGER JAMES ALVIN
STEIN PHILIP CLARENCE
STEIN RICHARD WILLIAM
STEINBERG GEORGE CHARLES
STEINER TERRY MICHAEL
STELPFLUG MERLIN CLAR-
ENCE
STEPHENS ARTHUR CHARLE
JR
STICH VERNON GENE
STICKS STEVEN MICHAEL
STINDL RICHARD WILLIAM
STOBER HERBERT J
STODDART GREGORY WIL-
LIAM
STOEBERL JAMES ARNOLD
STOFLET GORDON WAYNE
STOFLET MICHAEL HOWARD
STOLTZ DONALD ROBERT
STOPPLEWORTH DENNIS M
STOZEK GERALD STANLEY
STRACHOTA JOHN GREGORY
STRECKERT RONALD JOHN

STRITTMATER KENNETH
LEROY
STROUD WILLIAM HAROLD
STROUSE HOWARD DALE
STURGAL THOMAS JOHN
STYER FREDERIC CLARENCE
SUCHOMEL FREDERICK V
SUCHON CLARENCE MYRON
SUDBRINK DONALD ALBERT
SUKOWATEY STANLEY JOSEPH
SULLIVAN JOHN FRANCIS
SULLIVAN MIKAL JAMES
SUND TERRENCE LEE
SUTTON LARRY IVAN
SWAGER GENE STANLEY
SWATEK STEVEN PAUL
SWIGGUM LARRY WILLIAM
SZYMANSKI ROBERT THOMAS
TASCHEK KARL JOSEPH JR
TAUSCHEK LEONARD JOHN
TAYLOR ANDREW JAMES
TECHMEIR LARRY LESTER
TEGELMAN DALE FRANCIS
TERESINSKI JOSEPH ALVIN
TESSMER DAVID LEE
TEWS ERNEST WILLIAM
THEISEN WILLIAM ANTHONY
THELEN LE ROY EDMUND
THEYERL CLAYTON JOSEPH
THIEX RONALD CHARLES
THIRY SCOTT LOUIS
THOMA CHARLES JOHN
THOMAS LARRY EDWARD
THOMAS WAYNE ROY
THOMAS WILLIAM PHILIP
THOMPSON LEONARD DEAN
THOMSON STUART HAROLD
THRESHER KENNETH EUGENE
TIMM DAVID WILLIAM
TIMMONS RICHARD RUSSELL
TOBER PAUL HENRY
TOMCZAK THOMAS JAMES
TOMCZYK VICTOR DAVID
TORGERSON BARRENT OTTO
TOUSEY GEARWIN PHILLIP

TRAASETH LARRY DUANE
TREBATOSKI THOMAS HARRY
TREIBLE THOMAS CHARLES
TREMAINE CURTIS LLEWEL-
LYN
TREWEEK CHARLES JOHN
TRUDEAU ALBERT RAYMOND
TRUNKHAHN PEKKA
TUCKER BYRON CLAIR
TUINSTRA DENNIS
TURNER MICHAEL BRUCE
TYCZ JAMES NEIL
ULRICH JAYSON FRED
UNSINN MICHAEL JOSEPH
URBAS STANLEY FRANK
UTEGAARD THOMAS HAROLD
UTHEMANN ROBERT ERICK
VAIL THOMAS EARL
VALDEZ FRANCIS PEDRO
VALENTA RUDOLPH GLENN
VAN DEN HEUVEL DAVID F
VAN KEUREN ALLEN L
VAN LONE MURRAY WAYNE
SR
VAN MATER WILLIAM WAYNE
VANBENDEGOM JAMES LEE
VANDEHEI JOSEPH ROBERT
VANDER HEYDEN CHARLES G
VANDERBOOM PAUL WARD JR
VANDERSTERREN CORNELIS A
VANDERWEG PETER MICHAEL
VANEREM DAN ALLEN
VENET GLEN ORVILLE
VERHAEGHE MICHAEL J
VESER EDWARD
VIELBAUM JAMES MICHAEL
VIS JOHN PAUL
VIZER GERALD ARLENE
VOGEL DUANE ARLEN
VOGELI RALPH LEE
VOLLMER VALENTINE BER-
NARD
VOLTNER DONALD CHARLES
WAGIE DENNIS RICHARD
WAGNER MICHAEL JAMES

WAGNER PHILIP ROBERT
WAGNER RANDALL WILLIAM
WAHLER PAUL WINTON JR
WALDORF GARY ALAN
WALKER THOMAS DAVID
WALSH BLAINE HENRY
WALSH ROBERT THOMAS
WALZ GARY THOMAS
WALZ NICHOLAS GEORGE
WANGERIN LEON ARTHUR
WARD ALLAN CURT
WARD JAMES CALVIN
WARDEN RICHARD JOHN
WASCHICK WALTER RAY-
MOND
WASHKUHN JAMES WILLIAMS
WASSON STEVEN EDWARD
WATSON ALBERT CLAY JR
WATSON DAVID JAMES
WATSON DONALD DAVID
WEAVER HAYDEN EDWARD
WEBB JOHNNY LEE
WEBBER JAMES THOMAS
WEBER KARL EDWIN
WEEDEN ROBERT LEE
WEHRS DENNIS DUANE
WEIHER DOUGLAS RICHARD
WEINBERG DENNIS EDWARD
WEISS THOMAS RAY
WELSH LARRY MICHAEL
WESTLIE DANIEL LEE
WESTPHAL GARY LEE
WESTPHAL SCOTT BRIAN
WESTPHAL STEPHEN JOHN
WHITE MICHAEL LAWRENCE
WHITEHEAD RICKEY JACK-
SON
WIDDER DAVID JOHN WICK
WIECKOWICZ ANTHONY
JOSEPH
WIEGEL LLOYD GEORGE
WILCOX DAVID JOHN
WILKE ROBERT FREDERICK
WILL GORDON WALDEMAR
WILLEMS JOHN GUSTAVE

WILLIAMS TERRY JOE
WILLIAMSON LARRY ALLEN
WILSON MICHAEL JAY
WIMMER SCOTT THOMAS
WISCH ROBERT JAMES
WIXSON RALPH MICHAEL
WOJAHN ARTHUR EDWARD
WOJTKIEWICZ JEREMY
ROBERT
WOLF ROBERT CLARENCE
WOOD AARON LEE
WOOD DENNIS MELVIN
WOODS CURTIS STEVEN
WOODS THEODORE R JR
WOOLCOTT RANDALL ALAN
WOOLRIDGE LARRY ROGER
WOZNIAK JAMES KENNETH
WOZNICKI DAVID JAMES
WRIGHT VINCENT
WUCINSKI RICHARD THOMAS
WUSTERBARTH CLINTON
CARL
YOUNG GLEN HARRY
YOUNG JOHN E
YOUNG MICHAEL EDWARD
ZAHN WILLIAM FRANCIS JR
ZAJAC THADDEUS
ZEHNER THOMAS HOWARD
ZEICHERT HENRY JAMES
ZEIMET JAMES GEORGE
ZESKE ROBERT EDWARD
ZINK ROBERT GEORGE
ZOBEL STEVEN LYNN
ZWIRCHITZ DENNIS JAMES
ZYDZIK FRANK JR
ZYWICKE DAVID LEE

259

Photos by Rick Moser

Wyoming Vietnam Veterans Memorial

2519 26th Street Cody, WY 82414

The Vietnam War Memorial located in Cody, Wyoming honors the service members of the Vietnam War who resided in Wyoming. It's located in Wyoming Veterans Memorial Park. The memorial is meant to be a smaller version of the National Vietnam War Memorial in Washington D.C. but only features the 138 names of the Wyoming service members who passed in combat.

The Vietnam War Memorial was placed by the citizens of Wyoming in November of 1986 on Veteran's Day. The concept for the memorial came about in 1986 when the formation of the Veterans Memorial Park began. Vietnam veterans created the memorial to honor the fallen with Korean War veterans following suit to create a Korean War Memorial. From there, the idea for the memorial park grew.

Vietnam veteran Lt. Col. Buck Wilkerson is the committee chairman of the Wyoming Veterans Memorial Park project. He came up with the idea and the design of the Vietnam War Memorial in Cody. The city happily donated the land for the project and the Wyoming Legislature continues to appropriate funds to complete further monuments in the memorial park.

Wilkerson served in Vietnam after serving for two years in the Korean War. The last 14 years of his tenure before retirement were spent as a Green Beret. His service and the losses he witnessed moved him to dedicate his time to honoring the sacrifices of the servicemen and women who didn't make it home from their tours of duty, no matter the conflict.

The Vietnam War Memorial, like the national memorial, features a black granite wall with the names of the fallen engraved proudly. There is a ledge where visitors regularly leave tokens of their gratitude. The monument is surrounded by a semi-circle of benches in order to offer visitors a place to sit and reflect. There is also an engraved dedication centered on the ledge of the monument.

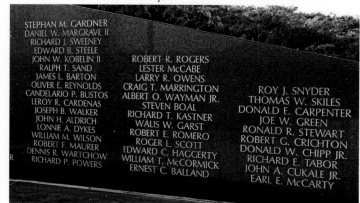

The Names of those from the state of Wyoming Who Made the Ultimate Sacrifice

ALDRICH JOHN HERRICK
ANDERSON ELTON GENE
ANDO CURTIS TADASHI
BALLAND ERNEST CLAUDE
BARNES JAMES FREDERICK
BARNES ROBERT EUGENE
BARTON JAMES LEE
BARTZ GARY EUGENE
BECK MICHAEL RAY
BENSON RAYMOND EDDIE
BLACKNER CRAIG SLADE
BOAL STEVEN
BRAUN EDWARD RAYMOND
BROWN KENNETH LLOYD
BROWN RICHARD STEVEN
BUSH GILBERT BYRON
BUSTOS CANDELARIO PATRICK
BYERS JERRY DUANE
CARDENAS LEROY ROBERT
CAZIN RICHARD PAUL
CHIPP DONALD WARREN JR
CLAYBURN MERRELL J
COEN HARRY BOB
COOLEY ORVILLE DALE
CRESSEY DENNIS CLARKE
CRICHTON ROBERT GARY
CUKALE JOHN ANTHONY JR
DARLING FRANK MAXWELL
DELLOS SAMUEL LEE
DYKES LONNIE ALLEN
ENDICOTT RICHARD LEROY
ESSLINGER WILLIAM BERTUS
EVANS BENNETT EDWARD
EVERT LAWRENCE GERALD
FALER ALLEN LEE
FARRIS DENNIS BARRY
FETZER TERRY LEE
FORD DONALD LEE
FOX GARY DUANE
FUQUA GARY JAMES
GARST WALIS WARREN
GIBSON HAROLD LEE
GLASSPOOLE RANDALL JOHN
GRAVES WILLIAM BOYD
GREEN JOE WORTH
GROVE ROBERT WOODROW
GUTHRIE ROBERT FRED
HAGGERTY EDWARD CHARLES
HANDY WALTER ELMER
HANSEN BARRY ANDRE
HARRISON GEORGE ROBERT
HART JOSEPH LESLIE
HOLLAND CARLTON JAKE
JENSEN BRUCE ALLAN
JOHNSON DALE WILLIAM
KASTNER RICHARD THOMAS
KING DENNIS DWAIN
KING KENNETH WALTER
KOBELIN JOHN WILLIAM II
KROGMAN ALVA RAY
LAIRD DANIEL REX
LANDES VICTOR REID
LANTOS LESLIE JOHN

LARSON TERRANCE HENRY
LAUCK ELMER DALE
LAWTON EDWARD LESTER
LUCAS DAVID GUY
LUJAN JOSE LEOPOLDO
MARRINGTON CRAIG THOMAS
MARTIN RICHARD M
MAUL HENRY EUGENE
MAURER ROBERT FRANKLIN
MAY LEONARD DON
MC ATEE WILLIAM JOSEPH
MC CABE LESTER
MC CORMICK WILLIAM T
MC NALLY EDWARD
MCCARTY EARL EDWARD
MILTNOVICH EMIL MAX
MONTANEZ PEDRO RODRIGUEZ
MOORE NORMAN LEE
MORGANFLASH ROBERT LEE
MOSS WELDON DALE
NIX VERNON WALTER III
OWENS LARRY RAY
PADILLA JOSEPH ANTHONY
PANTIER JAMES EDWARD
PATINO PABLO
PATRICK DOUGLAS TYRONE
POWERS RICHARD PAUL
REBERG CHARLES WAYNE
ROBINSON PHILIP OWEN
ROGERS DOUGLAS EUGENE
ROGERS ROBERT RICHARD
ROMERO ROBERT EUGENE
ROY CHARLES SULLIVAN
SAUNDERS TIMOTHY JUDD
SCHROEDER DONALD BENJAMIN
SCOTT ROGER LEE
SELDERS WILLIAM DEAN
SHUCK ROBERT LE ROY
SKILES THOMAS WILLIAM
SLAGOWSKI BENJAMIN EUGENE
SLOCUM STEPHEN ELLIS
SMITH DENNIS WAYNE
SNYDER ROY JASPER
STARK STEPHEN WILLIAM
STEELE EDWARD BERNARD
STEWART RONALD RICHARD
STUMPP ALMA JACK
SWEENEY RICHARD JOHN
TABOR RICHARD EUGENE
TAYLOR ERNEST EDWARD
TORREZ LAWRENCE DANIEL
WALKER JOSEPH BENSON
WARNOCK LARRY LYMAN
WARTCHOW DENNIS RUSSELL
WASHUT WALTER JUNIOR
WAYMAN ALBERT ORLANDO JR
WILSON WILLIAM MICHEAL

Photos by Matt Ymbras

The Traveling Wall

Visit americanveteranstravelingtribute.com for appearence schedule

The American Veterans Traveling Tribute, known as the AVTT, is a veteran owned small business that started with their well-known and well-respected Traveling Vietnam Wall. The organization's mission is to pay respect and to honor the fallen American heroes on a local scale throughout the country. This is especially important for all of the people who want to honor their fallen relatives or friends but can't physically make it to the actual national memorial in Washington DC.

The American Veterans Traveling Tribute wall is an 80 percent scale replica of the Vietnam Memorial Wall that is located in Washington DC. The construction of the wall was completed in 1998 and began traveling to different towns across the US in that same year. While there are a number of other traveling walls, the AVTT wall is the largest traveling replica wall to date.

The idea for the wall came from the veterans who own and operate the AVTT. The veterans are committed to their project and aim to continually pay tribute to their fallen brothers and sisters. The wall is available for leasing through the AVTT. When a town or organization leases the traveling wall, they will get a five-day event in order to give their local community time to enjoy the beauty of the memorial and to properly honor the fallen soldiers. The memorial arrives on a Wednesday, usually in the afternoon, and it completely setup by Thursday morning. Once the memorial is set up, the community is invited to view it and hold any ceremonies they'd like. The traveling wall will stay up until Sunday afternoon when it will be taken down.

The AVTT's memorial has multiple components known as the Cost of Freedom memorials and exhibits. The tribute wall is the focal point of the overall memorial. The wall is a replica of the national Vietnam War memorial. As stated, it's an 80 percent scale replica that measures 370 feet across and a stately 8-feet high at its highest point. The wall contains all of the names of the fallen that appear on the wall in Washington DC. The wall also features informational panels that have specific facts about the war. There is also a Vietnam timeline on multiple panels of the tribute wall. In addition, there is a flag display that is flown over the wall and two wall books to help people locate names of the fallen.

262

WILLIAM R PHILLIPS · DANNY E TUCKER · JOSEPH M ROMERO · WILLIAM M SAUNDERS · EDWARD A SHARROCK · THOMAS Y REYNOLDS · KENNETH M BARKER · DAVID J BARDUSON · DOUGLAS M BACOT · JOSEPH G ARTAVIA · GERALD G CHINO · WALTER L GAWEL · VAL E BARTON · DANNY P BOUCHEZ · GEORGE CAMPBELL · CHARLES E DAVENPORT II · SAMUEL E FOSTER · RICHARD A FRIEND · JOHN J ENZINNA · FRANKIE L FARROW · DARRYL B FULLUM · BERNARD L GAINES · DONALD L GEARHART · FRANCIS J GAULOCHER · WILLIAM R BREEDEN · TROY R HAYDEN · STEPHEN L HALSTEAD · JACK W HALEY · WILLIAM G HALL · DARWIN D GORDON · PAUL L HUTCHINSON · WILLIAM C JENKINS · ALBERT F HEUSEL JR · ROBERT L JACKSON · LOUIS J MAHER JR · ROBERT J LORENZO · GEORGE R MARTIN · GERALD R LEGER · RONALD R LAKE · EUGENE P MCKINNEY · MICHAL A MERKEL · KENNETH L NEAL · ERIC B NILSEN · GEORGE J NICHOLSON · FRANK M O'BRIEN III · CHARLES W PORTERFIELD · ARTHUR O PRENDERGAST · GEORGE F PROFFER · ERWIN A POLT · ROBERT D REICHERT · ALBERT J PETERS · JOHN B REALE III · LAWRENCE R RAYMOND · CARL R RASORI · RUDOLFO L ROCHA JR · DANNY S THOMPSON · TERRANCE E SMITH · DENNIS C SHIVELY · RONALD F SADLER · FRANCIS A BARNES · JOHN B WEILL · TERRY R WHITE · JEFFERY H VAN VLEET · DAVID L WILKERSON · JOHN M CASEY · LEONARD E BORCHARD JR · MCARTHUR CALLAWAY · PATRICK M BERWERT · RICHARD E BOYD · BRUCE A COBB · EUGENE J CURLESS JR · LLOYD A COLFACK · RONNIE J CHARLES · GLENN F CASHDOLLAR · DAVID E DEWITT · DAVID R DE PRIEST · TOM DAVIS JR · LOUIS M GARCIA · LARRY M GRONEWOLD · RICHARD E HEIL · TRACY L HARGRAVE · JOHN E HEFTY · JOHN D HARPER JR · BENNETT J HERRICK · ROY L JOHNSON · CHARLES A HINCKLEY · WALLACE B JOHNSON III · JOHN L HIGGINS · WILLIAM J SHORTSLEEVES · MONTE C KINASZ · MICHAEL J KELLEY · DOWARD L JONES JR · EDWARD J JUREK II · DONALD R KINTON · ALFRED F LANCE · BALFOUR O LYTTON JR · JOHN M MIHALOVICH · FRANK J MALSEED JR · FRANKLIN E MOORE · PATRICK C MCILROY · JOHN H MCCARTHY · DONALD L MCHUGO · JAMES E MOORE · JAMES M MULLINS · MICHAEL L PFEUFFER · WILLIAM L RAY · DAVID T ORWIG III · ROBERT D ROBINSON · WADE R REAUME · RICHARD W HOPPER · JOHN O SHERRILL · JOSEPH P TANGARIE · RONALD SOCKEY · DENNIS K SHOWERS · JIMMY J TESSADRI · FLOYD N THOMAN · HERBERT F WELTZ JR · BROADUS A WHITT · ROBERT J WIEDEMANN · LONNIE D ALLEY · FRANKIE E ALLGOOD + JIM L AILSTOCK JR · GARY R WIGINTON · LARRY E WORKMAN · JOEL G ANDERS · GENE T BAILEY · ROSS APPLEGATE · CRAIG P AVERILL · EDWIN L ARMSTRONG · DAVID P STARK · JERALD L BAKER · BENJAMIN J BELARDE · MICHAEL J BANDY · LARRY E BARGER · JIMMY D BARNETT · ROBERT J PIATKOWSKI · RONALD W BURKHART · RICHARD L BELINGE · JOSEPH M BROWN · STEVEN M BEZENSKI · WILLIAM H COOK · MICHAEL W CAMPOS · BARTON W CAREY · ROBERT L BUTLER · JAMES H CAMPBELL JR · ROBERT E CUNDIFF · JACK E DERRICO · ROBERT E DAVIS · RICHARD EVANCHO + LARRY E ELMORE · EARL W FRYE · WAYNE E FIELDING · DONALD C FOUST · MICHAEL A FAY · JOHNNY E GANTT · WILLIAM W DI NIRO · JAMES T GORSICH · JOHN P GIDDINGS · DENNIS S GLEASON · ALVIN R GIBBLE · STEVEN H GERLACH · PAUL D GRASSO · JOHNNIE D HARRIS · BARRY L GRAHAM · GERALD J GUNDERSON · LAWRENCE SKLODOSKI · CHARLES L HATCHER · JOHN R HORTON · WILLIE JACKSON · CHARLES L HOWE · GLEN D HUBBARD · ALAN S KOHN · KENNETH KAMINSKI · PHILIP J KREK JR · ERNEST C KERR JR + WAYNE D KRUEGER · RICHARD E LOMAX + LEO J MATYLEWICZ · LEONARD LONG · ROGER M LINK · MARVIN L MAYO · JAMES M MOSER · ROBERT W MORRIS · ROGER L OLSON · CHARLES R MOOMEY · MAURICE MOORE · CARL MCFADDEN JR · GLENN W MOWREY · JAMES A OSTERLOTH · ROBERT B MOSSO · CECIL R MILLSPAUGH · GARY M REEDY · LAWRENCE P PENNEL · ERNESTO S PEREA · ROBERT D POPE · PHILIP C BENN · WAYNE E RISNER · HAROLD J SIMMONS · RALPH SIBLEY · ROBERT J RYLANDER · RAYMOND F SCHOPMANN · LARRY E GREEN + HARVEY J TOMPKINS · JOE W SMITH · JAMES F STOLINSKI · HOYLE TERRY JR · JAMES L BADLEY + HARVEY G ADAIR · HERBERT F WHITE · JOHNNY W WANAMAKER · SYLVESTER WRIGHT JR · DENNIS E CHESTER · ROBERT J FRISK · WILLIAM E CARROLL · JOHN H BARNES · THOMAS E BIXBY · MICHAEL L DOANE · DONALD C EMERY · MARVIN E GALBRAITH · STEVEN J GAFTUNIK · JOHNNY C CALHOUN · FRANKLIN N GILES JR · JAMES C HAUGH · CHARLES K HENSON · RONALD J HASKO · LOUIS W HAMIL · RALPH J HORNADAY · WILLIAM C JUDGE JR · ROBERT E LUCAS · RICHARD LOPEZ · LAWRENCE R LOPES · RICHARD W ORSUND · PAUL J MOODY · ROGER W OVERSTREET · CARTER L MOORE · BERNIE J MOSLEY · WILLIE B RICHMOND · WAYNE W PEARCE · FRANK SCHUSTER · RICHARD L SCHMIDT · MICHAEL C RUSSO · FREDRIC B DAVIS · IRWIN R SOBEL · JOSEPH E SINTONI · CARL A SHUTT JR · MAC W SPEAKS · EXTRUMBERTO SOLIS · JACK D SUNDQUIST · THOMAS L TAFFE · LARRY D WEBB · RICHARD L WHITTEKER + JAMES D WHITE · PETER N BALDWIN · JAMES L BOWMAN · PEDER W ARMSTRONG · ALAN L BOYER + CLIFFORD L WILLIAMS · RAYMOND C DOWNIN · WILLIAM C DIMAGARD · GEORGE P DESMARAIS · LEE M LAMBERT · GEORGE R BROWN + ROY D ELSTON JR · DENNIS L GRAHAM · CHRISTOPHER A GROSSE JR · GERALD F GILBERT · RAYMOND R GILLOCK · CHARLES C HUSTON + RAY J HAAS · JAMES A HARRINGTON JR · RICHARD A HEWETT · GREGORY P KENT · EMERSON P COLE · ROBERT A LA FOUNTAIN · PAUL D KUEHL · LARRY C KYAR · FRANKLIN D LACEY · HENRY E MACCANN + JIMMIE L MCRAE · PAUL E MAYER · DOUGLAS T LOUDENBACK · FREDERICK D MCCARTY · LAWRENCE W O'MEARA · BERNARD M RICHARDSON · JOSEPH C PARK JR · VERNON D SANDVIG · DAVID L ROSS · ELMORE R SIMPSON · VINCENT B SANTANIELLO · JOSEPH F SCHLUCK · ROBERT S SHELTON · JACK M WOLF · JOHN T SUMMERS III · ARTHUR L TUCKER · FRANK T WATTS · HOWARD D WEEKS · MICHAEL W WALLACE · WALTER G WILSON · THOMAS C CARLISLE II · MICHAEL M BARR · FRANCIS A ANDERSON · HANS W BRUNNER · HOWARD M GIFFORD · MICHAEL N GANDY · CORNELL H GIBSON · THOMAS C HENRY · CARL L CARSON ·

46E   23 MAR 1968 - 29 MAR 1968

The Names of those from U.S. Territories
Who Made the Ultimate Sacrifice
American Samoa, Canal Zone, Guam, Puerto Rico, the US Virgin Islands and Elsewhere

ACEVEDO-RECHANI RAFAEL
ACOSTA GERMAN PORTACIO
ADENIR RESTITUTO POBLETE
AGNES MANOLO BRIONES
AGUON JOSE QUINATA
ALCOCER-MARTINEZ HECTOR M
ALDRICH JOHN HERRICK
ALGARIN-RIVERA RAFAEL ANG
ALVARADO-RIVERA JERONIMO
ALVAREZ-BUZO ELIAS
ALVAREZ-TAPIA JOSE LUIS
ANDERSON ELTON GENE
ANDO CURTIS TADASHI
ANDRADE ELISEO A JR
APEROCHO REGALADO M D
ARANDA-SANTOS EDUARDO
ASANOMA FRANCISCO M
AUBAIN ROY ANTONIO
AVILES-AVILES JUAN PASCUA
AYALA-MERCADO JUAN
BAGASOL ALEJANDRO BIRRI
BALLAND ERNEST CLAUDE
BARBOSA-VILLAFANE ANTONIO
BARNES JAMES FREDERICK
BARNES LEROY
BARNES ROBERT EUGENE
BARTON JAMES LEE
BARTZ GARY EUGENE
BAUZA-PEREZ JUAN
BECK MICHAEL RAY
BENAVENTE DAVID GUERRERO
BENCHER ALVIN KENNETH
BENITEZ-RIVERA JOSE EMILI
BENSON RAYMOND EDDIE
BERMEJO RICHARD ISMAEL
BERMUDEZ-PACHECO ENRIQUE
BERMUDEZ-QUINONES LUDIN
BERNARD RAMON
BERRY LOUIS EDWARD
BEVERHOUDT CLARENCE VEREN
BIAGINI MARK FREDERICK
BIRCO JOSE GOTERA
BLACKNER CRAIG SLADE
BLAIR DONALD D
BLAS ANTHONY MARTIN M
BLAZ JAMES LUJAN
BOAL STEVEN
BOLDUC DANIEL ALPHONSE
BONILLA-VIERA FELIPE
BORJA JUAN SANTOS
BRAUN EDWARD RAYMOND
BRENES NERY JACINTO
BRENES-ESCOBAR JOSE
BRIGGS FRANK HOWARD
BRODEUR DAVID LEE
BROWN KENNETH LLOYD
BROWN PETER H
BROWN RICHARD STEVEN
BROWN THOMAS EDWARD
BRUYERE PETER NORBERT
BRYAN PATRICK
BURGOS-CRUZADO ANGEL LUIS
BURTON HARRY PAYNE
BUSH GILBERT BYRON
BUSTOS CANDELARIO PATRICK
BUTT GARY
BYERS JERRY DUANE
CABRERA JOAQUIN PALACIOS
CABRERA-RODRIGUEZ CANDIDO
CABRERA-RODRIGUEZ MARCELI
CAMACHO DAVID BITANGA
CAMACHO RODRIGUEZ PEDRO J

CAMPBELL MICHAEL FRANCES
CAMPBELL RANDALL KENNETH
CANLAS SEBASTIAN PIADOCHE
CARABALLO-GARCIA MEGDELIO
CARDENAS LEROY ROBERT
CARMONA-MEDINA RAFAEL CEC
CARON BERNARD JOHN
CARRASQUILLO-DENTON ALBERTO
CARTER DONALD SUMINGUIT
CASAS BONNIE PATALINGHUNG
CASTILLO MANOLITO WISCO
CASTRO JOSE ANTONIO
CASTRO JUAN PASCUAL R
CASTRO-CARRASQUILLO MIGUE
CASTRO-MORALES RAMON
CAZIN RICHARD PAUL
CEPEDA JUAN DUENAS
CHAPARRO-VILLANUEVA GERMA
CHIPP DONALD WARREN JR
CINTRON-MENDEZ WILFREDO
CLAYBURN MERRELL J
COEN HARRY BOB
COLLINS LARRY RICHARD
COLLINS MARK PAINE
COLON-RIVERA JOSE RAMON
CONCEPCION-CHAPMAN JIMMY
COOLEY ORVILLE DALE
CORBIERE AUSTIN MORRIS
CORBIN NORMAND ALFRED
CORREA-MORALES FRANCISCO
CORTES-CASTILLO JUAN
CRABBE FRANK EDWARD
CRESSEY DENNIS CLARKE
CRICHTON ROBERT GARY
CRUZ EDWARD CRUZ
CRUZ ENRIQUE SALAS
CRUZ JOSEPH AGUIGUI
CRUZ JOSEPH WILLIAM
CRUZ PEDRO AFLAGUE
CRUZ-CRUZ RAFAEL
CRUZ-VAZQUEZ ANGEL MANUEL
CUASITO RONALD PEREZ
CUKALE JOHN ANTHONY JR
DACANAY FRANCISCO DE LA C
DAMIAN ALLAN JAMES
DARLING FRANK MAXWELL
DAVIES DONALD PAUL
DAYAO ROLANDO CUEVAS
DE JESUS-CARRERAS EFRAIN
DE JESUS-MUNOZ ALEJANDRO
DE JESUS-SANCHEZ ANIBAL
DE LEON HERMAN BORJA
DE LISA WILLIAM JOSEPH
DEARBORN PATRICK JOHN
DELA-CRUZ FREDERICO V
DELLOS SAMUEL LEE
DEVOE DOUGLAS WAYNE
DEXTRAZE RICHARD PAUL
DIANA-DIAZ JOSE RAMON
DIAZ EDWARD REYES
DIAZ-COLLAZO MIGUEL ANGEL
DIAZ-ROMAN CARMELO
DOMINGUEZ-CORTES ELIEZER
DONA BIENVENIDO GENIZA
DOYLE ALBERT BARCINAS
DRIVER JOHN CECIL
DUENAS JOSE BAMBA
DYER JEFFREY STEPHEN

DYKES LONNIE ALLEN
EAY RUDY EDEJER
ENCARNACION-BETENCOURT JESUS
ENDICOTT RICHARD LEROY
ESCANO JUANITO MAIQUEZ
ESPINOSA VICENTE T
ESSLINGER WILLIAM BERTUS
ESTEVES FERNANDO BARCINAS
ESTRADA ADOLFO MEDARDO
ESTRADA-COSTAS HERMAN
EUSTAQUIO JOSEPH MARTIN
EVANS BENNETT EDWARD
EVERT LAWRENCE GERALD
FAKIN ZLATKO M
FALER ALLEN LEE
FARRIS DENNIS BARRY
FAVOR ROBERT FRANCIS
FELIX-TORRES JUAN RAMON
FERNANDEZ MAXIMO PAULITE
FETZER TERRY LEE
FIGUEROA CABALLERO FERNAN
FIGUEROA JOSE JUAN
FIGUEROA-MELENDEZ EFRAIN
FINNEY HAROLD JAMES JR
FLORES BENNY SAN NICOLAS
FLORES DAVID CRUZ
FONTANEZ-VELEZ JOSE LUIS
FORD DONALD LEE
FORDHAM BENJAMIN STEPHEN
FOX GARY DUANE
FRAGOSA-GARCIA ANGEL LUIS
FRANCIS JAMES AUGUSTUS
FRANCIS JOHN FREDRIC
FRIGAULT JOSEPH O
FUNES DAVID JOHN
FUQUA GARY JAMES
GABANA ROBERTO LAY
GALINDO EVERARDO JR
GAN LEONARDO MEDINA
GARCIA-DIAZ JUAN ENRIQUE
GARCIA-MALDONADO JOSE I
GARCIA-SOTO JERONIMO
GARST WALIS WARREN
GAUTHIER GERARD LOUIS JOS
GIBSON HAROLD LEE
GLASSPOOLE RANDALL JOHN
GLEI ROGER LEE
GOMEZ-BADILLO DAVID
GOMEZ-RIVERA JUAN
GONZALES-MADERA ANGEL L
GONZALEZ-DROZ EDUARDO
GONZALEZ-MALDONADA MANUEL
GONZALEZ-MARTINEZ ANGEL L
GONZALEZ-MORALES ROBERTO
GONZALEZ-PEREZ ARAMIS
GONZALEZ-RIVERA MIGUEL A
GONZALEZ-VELEZ JOEL HUMBE
GOODMAN BARRY JASON
GORTON DAVID ATOIGUE
GRANADO-AVILES ALFREDO D
GRAVES WILLIAM BOYD
GREEN ARTHUR HAYWOOD
GREEN JOE WORTH
GROTTKE EDWIN REYNOLDS JR
GROVE ROBERT WOODROW
GUERRERO PEDRO ROSARIO
GUERRERO VINCENT FEJA
GUTHRIE ROBERT FRED
GUTIERREZ-OLIVERAS ELVING P
GUTIERREZ-VELAZQUEZ

JOSE D
GUZMAN-LUGO EDUARDO
GUZMAN-PAGAN JORGE LUIS
GUZMAN-RIOS ANTONIO
HAGGERTY EDWARD CHARLES
HALIBURTON MICHAEL R
HANDY WALTER ELMER
HANSEN BARRY ANDRE
HARRISON GEORGE ROBERT
HART JOSEPH LESLIE
HATTON RANDOLPH EDWARD
HAWES WAYNE LINDSAY
HEISEL RODNEY G
HENLEY AUBREY RUDOLPH
HERNANDEZ-FELICIANO DAMAS
HERNANDEZ-VELEZ ALBERTO
HERRERA JOSE BABAUTA
HOLDITCH ROBERT WILSON
HOLLAND CARLTON JAKE
HURTAULT CUTBURT
IBANEZ ARISTOTELES DEL R
INFANZON-COLON RAMIRO
IRIZARRY-ACEVEDO DANIEL
IRIZARRY-HERNANDEZ ANGEL
IRIZARRY-PEREZ JAIME
ISALES-BENITEZ JORGE LUIS
JARA-VERANO ALBERTO I
JAVINES JOSE JALOCON
JEFFRIES GABRIEL AUGUS JR
JENSEN BRUCE ALLAN
JIMENEZ LUIS RAFAEL
JIMENEZ-GONZALEZ ISABELO
JMAEFF GEORGE VICTOR
JOBEY ANDREW JOHN
JOHNSON DALE WILLIAM
JORDAN-MOLERO ADRIEN MANU
JURADO AMBROSIOS SANTIAGO
KASTNER RICHARD THOMAS
KELLAR HARRY DAVID CHARLE
KENNEDY BRUCE THOMAS
KENNY ROBERT WAYNE
KING DENNIS DWAIN
KING KENNETH WALTER
KINTARO JOHN JULLIANO
KOBELIN JOHN WILLIAM II
KROGMAN ALVA RAY
KROISENBACHER ADOLF J
KUILAN WENCESLEO
KUILAN-OLIVERAS RAMON
LAIRD DANIEL REX
LAMOURT-TOSADO PEDRO LUIS
LANDES VICTOR REID
LANG ANDREW ALPHONSO
LANTOS LESLIE JOHN
LARSON TERRANCE HENRY
LAUCK ELMER DALE
LAWSON DARRYL DEAN SMITH
LAWTON EDWARD LESTER
LEBRON-DOMENECH OMAR
LEBRON-LOPEZ ISMAEL
LEBRON-MALDONADO ROGELIO
LECOMPTE RUSSELL MARTINEZ
LEON FELIX JR
LEONARD LISTON RAPHEAL
LEON-DE JESUS EFRAIN
LESHEN LEE MYRL
LEVI LANE FATUTOA
LICIAGA-CONCEPCION LUIS A
LOKENI FAGATOELE
LOPEZ RAMON
LOPEZ-AGOSTO FELIX MANUEL
LOPEZ-COLON JUAN ANTO-

NIO
LOPEZ-DEL TORO SAUL
LOTT JAMES EDWARD
LOW KEVIN DOUGLAS
LUCAS DAVID GUY
LUGO-MOJICA HECTOR
LUJAN JOSE LEOPOLDO
LUKEY GEOFFREY JOHN
MAC GLASHAN JOHN WILLIAM
MALDONADO-AGUILAR BENJAMI
MALDONADO-TORRES LIONEL
MANNING DAVID KARL
MARCANO-DIAZ GAMALIEL
MARIANO JESUS ROSA
MARIER MAURICE JOHN
MARIN-RAMOS HECTOR RAMON
MARQUEZ-LOPEZ LUIS MANUEL
MARQUEZ-QUINONES RAIMUNDO
MARRERO-BAEZ FLOR
MARRERO-RIOS JOSE ANTONIO
MARRINGTON CRAIG THOMAS
MARTIN ALAN CRAIG
MARTIN RICHARD M
MARTINEZ-FELICIANO JOSE L
MARTINEZ-QUILES JUAN A JR
MARTINEZ-SANTIAGO RAFAEL
MARTINOVSKY MILOSLAV JOSE
MARTIR-TORRES JULIO IGNAC
MASSO-PEREZ JULIO
MATOS-CORREA JOSE ANTONIO
MAUL HENRY EUGENE
MAURER ROBERT FRANKLIN
MAY LEONARD DON
MAYMI-MARTINEZ PEDRO ANTO
MC ATEE WILLIAM JOSEPH
MC AULIFFE ALBERT JOSEPH
MC CABE LESTER
MC CORMICK WILLIAM T
MC INTOSH IAN
MC NALLY EDWARD
MC SORLEY ROB GEORGE
MCCARTY EARL EDWARD
MEDINA-GONZALEZ RUPERTO
MEDINA-RIVERA ANGEL M
MEDINA-TORRES VINCENTE
MELENDEZ CRISTOBAL
MELENDEZ-GONZALEZ JOSE D
MENDIOLA ROBERT L G
MENDY STAN
MENENDEZ-OCASIO ISMAEL
MENO JESUS QUINENE
MERCADO-GUTIERREZ RUBEN D
MESA TOMAS REYES
MILTNOVICH EMIL MAX
MIRANDA-CUEVAS LUIS ANTON
MIRANDA-ORTIZ JOSE LUIS
MIRANDA-PEREZ NOE
MOLINA-ROSARIO OCTAVIO
MOLYNEAUX JOHN LOUIS JR
MONETTE REGEN ALBERT
MONTANEZ PEDRO RODRIGUEZ
MONTES ANTHONY JOHN
MONTES JOSE L
MOORE NORMAN LEE
MORA LUIS GUILLERMO
MORALES-GONZALEZ JULIO ER
MORALES-LUCAS LESLIE ISMA
MORALES-MERCADO JUAN

BAUT
MOREHAM VINCENT PINAU-
LA
MOREU-LEON MARIO
MORGANFLASH ROBERT LEE
MOSS WELDON DALE
MULVANEY MICHAEL TER-
ENCE
NAME
NARVAEZ-MARRERO ANDRES
LU
NAZARIO JUAN JOSE
NEDEDOG EMILIO NINAISEN
NEGRON-RODRIGUEZ JOSE
NEGRON-RODRIGUEZ MI-
GUEL A
NERIS-APONTE JOAQUIN
NIEVES-COLON MARCELINO
JR
NIX VERNON WALTER III
OQUENDO-GUTIERREZ
RAMON
ORTIZ-COLON ULISES
ORTIZ-LOPEZ ALFONSO
ORTIZ-NEGRON JOSE JUAN
ORTIZ-ORTIZ CEFERINO ADRI
ORTIZ-PEREZ LUIS ANTONIO
ORTIZ-RIVERA JUAN
ORTIZ-RODRIGUEZ ANGEL
OSTOLAZO-MALDONADO
ALFREDO
OVERMAN-RODRIGUEZ
JOSE R
OWENS LARRY RAY
OYOLA-RABAGO ANIBAL
PADAYHAG AL SUMINGUIT
PADILLA JOSEPH ANTHONY
PAGADUAN GUILLERMO
BAUTIS
PAGAN-CARTAGENA JOSE
RAMO
PAGAN-PAGAN AMALIO
PAGAN-RODRIGUEZ EVAN-
GELISTA
PANGELINAN GREGORIO L
PANGELINAN PEDRO CABRE-
RA
PANTIER JAMES EDWARD
PARRILLA-CALDERON JAIME
PATINO PABLO
PATRICK DOUGLAS TYRONE
PELLOT-RODRIGUEZ RAMON

AL
PENA JOSE MANUEL
PENA-CLASS RAUL
PEREDA HENRY PANGELINAN
PEREIRA SOCORRO
PEREZ VICENTE DUENAS
PEREZ JOHN ANTHONY
PEREZ-CRUZ LUIS ANTONIO
PEREZ-RIVERA MANUEL
ANTON
PEREZ-VERGARA ALBERTO
PETERSON FRANCOIS
ACHILLE
PINTO-PINTO SIGFREDO
PISACRETA ROGER MELVIN
POWERS RICHARD PAUL
PUMAREJO-COLON WILFRE-
DO
PUMPELLY WALTER LEE
QUENGA JOHNNY CRUZ
QUIDACHAY JESUS AQUIN-
INGO
RAMOS ARMANDO
RAMOS-JIMENEZ RAUL
RAYMO WINSTON GLEN-
WOOD
REBERG CHARLES WAYNE
REEVES JOHN HOWARD
REID ROGER GLEN
REIMILLER THOMAS EVANS
REYES TOMAS GARCIA
RIOS-ROSARIO TEODORITO
RIOS-VELAZQUEZ LEONAR-
DO JR
RIPPEL EUGENE RAYMOND
RIVERA THOMAS ANTONIO
RIVERA THOMAS SALAS
RIVERA-BARRETO JOSE FERMI
RIVERA-COLON HECTOR
RIVERA-FERNANDEZ SAMUEL
RIVERA-LOPEZ JAIME ALBERT
RIVERA-MARTES CONFESOR
RIVERA-MELENDEZ JESUS D
RIVERA-VELAZQUEZ ANGEL A
ROBERTSON ELLIS ANDRE
ROBINSON EUGENE MAJOR
ROBINSON PHILIP OWEN
ROBLES JOAQUIN
ROBSON WILLIAM REID
RODRIGUEZ LUCAS HERRERA
RODRIGUEZ-ACEVEDO JOSE
RODRIGUEZ-ESTREMERA

ANGEL
RODRIGUEZ-LEBRON SAN-
TIAGO
ROGERS DOUGLAS EUGENE
ROGERS ROBERT RICHARD
ROMERO ROBERT EUGENE
ROMERO-OYOLA HERIBERTO
ROSADO-RODRIGUEZ EU-
GENIO
ROSARIO-CRUZ MIGUEL JR
ROSARIO-SALABERRIOS ELMO
ROSA-SEIN ROSARIO
ROSAS-SANZ SAMUEL
ROSA-URBINA VICENTE
ROY CHARLES SULLIVAN
RUBIO EURIPIDES JR
RUIZ JOSE MANUEL
RUIZ-BERNARD GUILLERMO A
RUIZ-DEL PILAR RAFAEL ANG
RUIZ-PEREZ ROBERTO
RYBICKI FRANK ANTHONY JR
SABLAN ANTONIO QUICHO-
CHO
SABLAN IGNACIO ESPINOSA
SABLAN JOHN TENERIO
SABLAN THOMAS QUICHO-
CHO
SAEZ-RAMIREZ ANGEL PERFIR
SALAZAR FIDEL GARCIA
SAN NICOLAS RUFO SANTOS
SAN NICOLAS VICTOR P
SANCHEZ CARLOS J
SANCHEZ GEORGE SANTIAGO
SANCHEZ-BERRIOS CARMELO
SANCHEZ-ORTIZ DIONISIO
SANCHEZ-ROHENA HECTOR
M
SANCHEZ-SALIVA RAFAEL
SANTIAGO-APONTE NELSON
SANTIAGO-MALDONADO
JUAN A
SANTORO ROBERT JOHN
SANTOS ENRIQUE ROSARIO
SANTOS ERNEST PABLO
SANTOS JAMES EDWARD
ANDER
SANTOS RAFAEL SALAS
SANTOS-IZAGAS DIOSDADO
SANTOS-PINEDO PEDRO
SANTOS-TRUJILLO DANIEL
SANTOS-VEGA MARCELINO
SAULER CHARLIE F

SAUNDERS TIMOTHY JUDD
SAUVE DANIEL LOUIS PAUL
SCARANO CHARLES PATRICK
SCHOTT RICHARD SIMPSON
SCHROEDER DONALD BEN-
JAMIN
SCOTT ROBERT JAMES
SCOTT ROGER LEE
SELDERS WILLIAM DEAN
SEMENIUK LARRY STEPHEN
SERRANO-RIVERA JULIO
SHARPE EDWARD GERALD
SHELL JOHN ROBERT
SHERIN JOHN C III
SHUCK ROBERT LE ROY
SIMPSON ROBERT LEWIS
SKILES THOMAS WILLIAM
SLAGOWSKI BENJAMIN
EUGENE
SLOCUM STEPHEN ELLIS
SMITH DENNIS WAYNE
SMITH LLEWELLYN ANTONIO
SNYDER ROY JASPER
SOMERS FRANK J
SOSA-HIRALDO CARMELO
SOSNIAK TADEUSZ
SOTO-GARCIA GILBERTO
SOTO-RODRIGUEZ ANGEL
MIGU
STALINSKI STEFAN ZBIGNIEW
STARK STEPHEN WILLIAM
STEEL ROBERT JAMES
STEELE EDWARD BERNARD
STEWART RONALD RICHARD
STUMPP ALMA JACK
SULATYCKI HENRYK TADUESZ
SUTHONS MELVIN HAROLD
SWEENEY RICHARD JOHN
TABOR RICHARD EUGENE
TAITAGUE JOHNNY SALAS
TARDIO RONALD ENRIQUE
TAYLOR ERNEST EDWARD
TEO FIATELE TAULAGO
THIESFELDT-COLLAZO WIL-
LIAM J
TORRE FRANCIS SAN NICOLAS
TORRES PRISHARDO JOSE T
TORRES-ACEVEDO JUVENCIO
TORRES-LOPEZ RIGOBERTO
TORRES-RIVERA RAFAEL
TORRES-RODRIGUEZ JULIO A
TORREZ LAWRENCE DANIEL

TOSADO-HERNANDEZ VIC-
TOR M
TRUJILLO-TRUJILLO ABRA-
HAM
TUAZON SIMEON ANDRADE
JR
VALPAIS-MORALES RAFAEL A
VARGAS-VARGAS ISRAEL
VAZQUEZ JUAN FRANCISCO
VAZQUEZ-BERRIOS RUBEN
ANT
VAZQUEZ-GONZALEZ PEDRO
VEGA-LOPEZ CARLOS
VEGA-MAYSONET RAFAEL
VELAZQUEZ-FELICIANO
ROD JR
VELAZQUEZ-LOPEZ VICTOR R
VELAZQUEZ-ORTIZ CARLOS A
VELEZ-RIVERA LUIS ALFONSO
VELEZ-RODRIGUEZ ELLIOTT
VELEZ-VILLAMIL JUAN
MANUE
VERA-DURAN MIGUEL DE
JESU
VIADO REYNALDO ROCILLO
VIGO-NEGRIN LUIS
VINLUAN DOMINGO BAL-
ANSAY
VIOLETT JAMES EDWARD
WALKER JOSEPH BENSON
WALLACE MERVIN EDEN
WARNOCK LARRY LYMAN
WARREN BAXTER
WARTCHOW DENNIS RUSSELL
WASHUT WALTER JUNIOR
WAYMAN ALBERT ORLANDO
JR
WELSH RUTHERFORD J
WHEATLEY JOHN ALBION
WHITE GORDON GLENN
WILLIAMS ALAN EDWIN
WILLIAMS THOMAS MURRAY
WILLIS BENJAMIN GALU
WILSON WILLIAM MICHEAL
YOKOI RALPHAEL SGAMBEL-
LUR
ZAYAS-CASTRO REINALDO

The men and women who served in Vietnam fought for freedom in a place where liberty was in danger. They put their lives in danger to help a people in a land far away from their own. Many sacrificed their lives in the name of duty, honor, and country. All were patriots who lit the world with their fidelity and courage.
-Ronald Reagan 11.11.1984

Acknowledgements

We would like to extend a big thank you to the following people for without their help, information and guidance this book would not exist.

Bill Tunnell Executive Director of the USS Alabama Battleship Memorial Park.

Tim Benetendi Chairman of Anchorage Veterans Memorial Renovation.

Billy Culin VVA Chapter 1043 US Army Vietnam Sun City, AZ

Tom Kleck, Quartermaster, VFW post 9095 Little Rock, AR

Jim Doody founder of Western Slope Vietnam War Memorial Fruita CO

Frank Potter VFW post 8762 West Sacramento California, United States Navy Aviation Vietnam 66-70

Jean Risley President Connecticut Vietnam Veterans Memorial Inc.

Paul Davis President VVA State Council, DE

Ben Humphries President VVA State Council - FL US Army Corps of Engineers, US Military Academy

Dotty Etris Executive Director Roswell Convention & Visitors Bureau GA

Gene Naesias Public Affairs National Cemetery of the Pacific - HI

David J Hollingshead Vice Commander VFW Post 755 Springfield, IL 363rd AF Security Forces Prince Sultan AF Base Saudi Arabia

Sherri Holt, Developement and Special Events Coordinator Community Foundation of Northwest Indiana

Dan Gannon Iowa Commissioner. of Veterans Affairs IA

Chuck Ford VVA chapter 344 Junction City, KS

Kristie Klemens VFW Post 4075 Frankfort, KY

Jay Walsh Commander Metairie American Legion Post 175 - LA

Jim Neville Director Cole Transportation Museum, Bangor, ME

Dana Hendrickson, Director of Outreach, Dept of Veterans Affairs, MD

Karen Greenwood, Veteran's Agent City of Worcester, MA

Jim Uphouse 1/327th Inf 101 Abn Div Vietnam 67-68 - MI

Chuck Miller VVA chapter 320 - St Paul MNC
Larry Lucas, Chairman Names and Pictures Committee, Mississippi Vietnam Veterans Memorial

Pete Slusarczyk Vice Commander VFW Post 9997 - Kansas City, MO USAF 363rd Security Force Saudi Arabia

Jack Reneau Trustee VFW Post 209 - Missoula, MT

Dan Raabe VFW Post 2503 Omaha, NE US Army Vietnam 69-70

Ron Guzman 3rd Marine Division Korea, American Legion Post 4 Carson City, NV

New Hampshire, Ray Goulet VVA State Council President

Sarah Taggert Curator New Jersey Vietnam Veterans Memorial
New Mexico, Gail McCutchen staff AngelfireVietnamMemorialVVM.org

New York, Dan Griffen Executive Director Chapter 49 VVA Pleasentville, NY

Frank Stancil, American Legion state, Adjutant NC -United States Navy 59-63 USS Fisk ADR 842

Pat Mischel President Chapter 150 VVA Bismarck, ND 2/27th Wolfhounds, 25th Infantry Division Vietnam

Barbara Freeman, Ohio Veterans Memorial Park

Nate Washington, State Council President, OK 7/15 1st Field ForceVietnam 68-69

Terry Williamson, President PVVM, Philadelphia, PA

Rhode Island, Jonathan Rascoe, Assistant Administrator Exeter, Rhode Island Veterans Memorial Cemetery

Tao Rivas Quartermaster post 641 VFW, Columbia SC, past State Commander, 1952 Korea 25th Inf Div. 1964, 66,68, 72,73. Vietnam 196th Light Infantry Brigade

Terry Meyer VVA State Council President, SD - 377th Transportation Squadron 7th Air Force Vietnam

Dennis Howland, UT - Third Marine Aviation Wing Vietnam 1966-67

Perry Melvin, State Council President VVA 1st memorial in US - VT

John Hatfield Executor Director Virginia War Memorial Richmond, VA

Washington, Ed Chapin Quartermaster VFW post 318, United States Army 1970-91 - WA

West Virginia, David Simmons state VVA Council President and Chapter 985 , 23rd Infantry Division Vietnam 1969-70

Kay, The Highground Memorial, Neilsville Wi

Jim Hoobler, Curator , Tennessee State Museum, Nashville,Tn

Dick Kemp, Adjutant Post 1 American Legion Portland, OR. US Army 1966-69 32nd Signal Battalion Germany

Bill Buntyn Commander VFW Post 2673 Cody, WY, USMC Lebanon 1958

Jim Runzheimer, TX - US Army Special Forces Detachment Europe

Bob Seal, US Navy USS Enterprise 1967-91 - ID

Don Henke, Lawton Oklahoma VVA Chapter 751

Would you like to be part of our next book?

RU Airborne's next publication will be a photobook chronologically documenting the war in Vietnam. If you would like to be part of the book you can submit your photos for consideration to the book.

Requirements: They must be your original photos and you must give an accurate date, location, names and description of all photos submitted.

Deadline is Sept 01, 2016

Send your photos to:

RU Airborne Inc
57 Sleight Plass Rd
Poughkeepsie, NY
12603

All those whose photos are chosen for the book will receive a free copy of the publication upon completion.

Include your full return address to if you would like us to return your photos to you.

For more books and other great products visit our website:

ruairborne.com

Find all sorts of products from all US Military branches at:

buymilitaryproducts.com

267